Principles and Applications of Molecular Diagnostics

PRINCIPLES and APPLICATIONS of MOLECULAR DIAGNOSTICS

EDITED BY

NADER RIFAI, PhD
Professor of Pathology
Harvard Medical School
Louis Joseph Gay-Lussac Chair in Laboratory Medicine
Director of Clinical Chemistry
Boston Children's Hospital
Boston, MA, United States

ANDREA RITA HORVATH, MD, PhD
Professor, Department of Clinical Chemistry & Endocrinology
New South Wales Health Pathology
School of Medical Sciences
University of New South Wales
Sydney, Australia

CARL T. WITTWER, MD, PhD
Professor of Pathology
University of Utah School of Medicine
Medical Director, Immunologic Flow Cytometry
ARUP Laboratories
Salt Lake City, UT, United States

ASSOCIATE EDITOR

JASON Y. PARK, MD, PhD
Associate Professor
Joint Appointment, Pathology and the Eugene McDermott Center for Human Growth and Development
University of Texas Southwestern Medical Center
Director, Advanced Diagnostics Laboratory
Department of Pathology
Children's Medical Center
Dallas, Texas, United States

Elsevier
Radarweg 29, PO Box 211, 1000 AE Amsterdam, Netherlands
The Boulevard, Langford Lane, Kidlington, Oxford OX5 1GB, United Kingdom
50 Hampshire Street, 5th Floor, Cambridge, MA 02139, United States

Copyright © 2018 Elsevier Inc. All rights reserved.

No part of this publication may be reproduced or transmitted in any form or by any means, electronic or mechanical, including photocopying, recording, or any information storage and retrieval system, without permission in writing from the publisher. Details on how to seek permission, further information about the Publisher's permissions policies and our arrangements with organizations such as the Copyright Clearance Center and the Copyright Licensing Agency, can be found at our website: www.elsevier.com/permissions.

This book and the individual contributions contained in it are protected under copyright by the Publisher (other than as may be noted herein).

Notices

Knowledge and best practice in this field are constantly changing. As new research and experience broaden our understanding, changes in research methods, professional practices, or medical treatment may become necessary.

Practitioners and researchers must always rely on their own experience and knowledge in evaluating and using any information, methods, compounds, or experiments described herein. In using such information or methods they should be mindful of their own safety and the safety of others, including parties for whom they have a professional responsibility.

To the fullest extent of the law, neither the Publisher nor the authors, contributors, or editors, assume any liability for any injury and/or damage to persons or property as a matter of products liability, negligence or otherwise, or from any use or operation of any methods, products, instructions, or ideas contained in the material herein.

Library of Congress Cataloging-in-Publication Data
A catalog record for this book is available from the Library of Congress

British Library Cataloguing-in-Publication Data
A catalogue record for this book is available from the British Library

ISBN: 978-0-12-816061-9

For information on all Elsevier publications visit our website at
https://www.elsevier.com/books-and-journals

  Working together to grow libraries in developing countries

www.elsevier.com • www.bookaid.org

Publisher: Susan Dennis
Acquisitions Editor: Kathryn Morrissey
Editorial Project Manager: Carly Demetre
Production Project Manager: Paul Prasad Chandramohan
Cover Designer: Miles Hitchen

Typeset by TNQ Technologies

CONTENTS

Contributors, vii
Preface, ix

1. **Principles of Molecular Biology,** 1
   John Greg Howe
2. **Genomes and Variants,** 17
   Carl T. Wittwer and Jason Y. Park
3. **Nucleic Acid Isolation,** 35
   Stephanie A. Thatcher
4. **Nucleic Acid Techniques,** 47
   Carl T. Wittwer and G. Mike Makrigiorgos
5. **Molecular Microbiology,** 87
   Frederick S. Nolte
6. **Genetics,** 125
   Cindy L. Vnencak-Jones and D. Hunter Best
7. **Solid Tumor Genomics,** 191
   Elaine R. Mardis
8. **Genetic Aspects of Hematopoietic Malignancies,** 201
   Todd W. Kelley and Jay L. Patel
9. **Circulating Tumor Cells and Circulating Tumor DNA,** 235
   Evi Lianidou and Dave Hoon
10. **Circulating Nucleic Acids for Prenatal Diagnostics,** 283
    Rossa W.K. Chiu and Y.M. Dennis Lo
11. **Pharmacogenetics,** 295
    Gwendolyn A. McMillin, Mia Wadelius, and Victoria M. Pratt
12. **Identity Testing,** 329
    Victor W. Weedn, Katherine B. Gettings, and Daniele S. Podini
13. **Amino Acids, Peptides, and Proteins,** 345
    Dennis J. Dietzen
14. **Proteomics,** 381
    Andrew N. Hoofnagle and Cory Bystrom

Index, 403

CONTRIBUTORS

D. Hunter Best, PhD
Associate Professor of Pathology
University of Utah School of Medicine
Medical Director, Molecular Genetics and Genomics
ARUP Laboratories
Salt Lake City, Utah

Cory Bystrom, BS, MS, PhD
Vice President, Research and Development
Cleveland HeartLab
Cleveland, Ohio

Rossa W.K. Chiu, MBBS, PhD, FHKAM, FRCPA
Choh-Ming Li Professor of Chemical Pathology
Department of Chemical Pathology
The Chinese University of Hong Kong
Shatin, New Territories
Hong Kong, China

Dennis J. Dietzen, PhD
Professor of Pediatrics
Washington University School of Medicine
Medical Director, Core Laboratory and Metabolic Genetics Laboratory
St. Louis Children's Hospital
St. Louis, Missouri

Katherine B. Gettings, PhD
Research Biologist
Applied Genetics Group
National Institute of Standards and Technology
Gaithersburg, Maryland

Andrew N. Hoofnagle, MD, PhD
Professor
Head, Division of Clinical Chemistry
Department of Laboratory Medicine
University of Washington
Seattle, Washington

Dave Hoon, MSc, PhD
Professor of Translational Molecular Medicine
Division of Molecular Oncology
John Wayne Cancer Institute
Providence Health Systems
Santa Monica, California

John Greg Howe, PhD
Associate Professor of Laboratory Medicine
Yale University School of Medicine
New Haven, Connecticut

Todd W. Kelley, MD, MS
Associate Professor of Pathology
University of Utah
Medical Director, Molecular Hematopathology
ARUP Laboratories
Salt Lake City, Utah

Evi Lianidou, PhD
Professor of Analytical Chemistry—Clinical Chemistry
National and Kapodistrian University of Athens
Athens, Greece

Y.M. Dennis Lo, DM, DPhil
Li Ka Shing Professor of Medicine
Department of Chemical Pathology
The Chinese University of Hong Kong
Shatin, New Territories
Hong Kong, China

G. Mike Makrigiorgos, PhD
Professor and Director of Medical Physics and Biophysics
Radiation Oncology, Dana Farber Cancer Institute
Harvard Medical School
Boston, Massachusetts

Elaine R. Mardis, PhD
Professor of Pediatrics
Ohio State University College of Medicine
Co-Director, Institute for Genomic Medicine, Research Institute
Nationwide Children's Hospital
Columbus, OH, United States

Gwendolyn A. McMillin, PhD
Professor of Pathology
University of Utah
Medical Director, Toxicology and Pharmacogenomics
ARUP Laboratories
Salt Lake City, Utah

Frederick S. Nolte, PhD, D(ABMM), F(AAM)
Professor and Vice-Chair for Laboratory Medicine
Department of Pathology and Laboratory Medicine
Director, Clinical Laboratories
Medical University of South Carolina
Charleston, South Carolina

Jason Y. Park, MD, PhD
Associate Professor
Joint Appointment, Pathology and the
  Eugene McDermott Center for Human Growth and Development
University of Texas Southwestern Medical Center
Director, Advanced Diagnostics Laboratory
Department of Pathology
Children's Medical Center Dallas
Dallas, Texas

Jay L. Patel, MD
Assistant Professor of Pathology
University of Utah School of Medicine
Salt Lake City, Utah

Daniele S. Podini, PhD
Associate Professor of Forensic Sciences
George Washington University
Washington, D.C

Victoria M. Pratt, PhD
Associate Professor of Medical and Molecular Genetics
Director, Pharmacogenomics Laboratory
Indiana University School of Medicine
Indianapolis, Indiana

Stephanie A. Thatcher, MS
Director of Systems Integration
BioFire Diagnostics
Salt Lake City, Utah

Cindy L. Vnencak-Jones, PhD
Professor of Pathology, Microbiology and Immunology
Vanderbilt University School of Medicine
Medical Director, Molecular Diagnostics Laboratory
Vanderbilt University Medical Center
Nashville, Tennessee

Mia Wadelius, MD, PhD
Associate Professor of Medical Sciences
Clinical Pharmacology
Uppsala University
Uppsala, Sweden

Victor W. Weedn, MD, JD
Professor and Chair of Forensic Sciences
George Washington University
Washington, D.C

Carl T. Wittwer, MD, PhD
Professor of Pathology
University of Utah School of Medicine
Medical Director, Immunologic Flow Cytometry
ARUP Laboratories
Salt Lake City, Utah

PREFACE

Biology is complex. Over 3 billion (3,000,000,000) nucleic acid base pairs are present in two copies within each human cell. Their sequence and redundancy determine the expression and regulation of proteins that determine human traits.

Molecular diagnostics usually refers only to nucleic acids, but that is unfair. Indeed, most laboratory analytes are molecules, except some electrolytes and heavy metal ions. In particular, proteins execute the orders of nucleic acids and deserve at least as much notoriety. Proteogenomics is an emerging field that combines proteomic, genomic, and transcriptomic data to better define the molecular signatures of disease.

Both nucleic acids and proteins are macromolecules made up of 4 (nucleic acid) or 20 (protein) components whose sequence determines function. Both are typically modified beyond their basic structure. The cytosines in DNA are variably methylated, and proteins are often glucosylated and may be reversibly phosphorylated. The combination of nucleic acids and proteins in the right environment is a highly regulated, yet adaptable, self-generating system that we call life.

We start with an introductory chapter on molecular biology, laying the groundwork for both nucleic acids and proteins. This is followed by a detailed look at the human genome, whose sequencing is perhaps the defining achievement of the early 21st century. Chapter 3 focuses on nucleic acid isolation as a preamble to analysis. This chapter is unique in that very few reviews on sample preparation are available from the chemistry perspective. The chapter on nucleic acid techniques provides a detailed look at methods ranging from basic PCR to massively parallel sequencing. We then present various applications of nucleic acid analysis, including infectious disease, genetics, and cancer. First presented is molecular microbiology: analysis of DNA or RNA from pathogenic organisms. Microbiology has been revolutionized by nucleic acid tests replacing phenotypic and culture methods. Then, genetics is covered in detail, including autosomal recessive, dominant, and X-linked diseases, in addition to mitochondrial disease and inherited cancer predisposition. Chapters on solid tumor genomics, genetic aspects of hematopoietic malignancies, and circulating tumor cells and circulating tumor DNA follow. Specialty topics of circulating nucleic acids for prenatal diagnostics, pharmacogenetics, and identity testing complete our coverage of nucleic acids.

As mentioned earlier, if nucleic acids are the brain of the organism, proteins are the muscles. No story of macromolecules is complete without proteins. We end our volume with an introductory chapter on amino acids, peptides, and proteins, and a final chapter on proteomics, assessing the spectrum of proteins present by conventional gel methods and powerful mass spectroscopy tools.

We thank the authors and our publisher who have made this compilation possible. The chapters presented here were first published in the 6th edition of the Tietz Textbook of Clinical Chemistry and Molecular Diagnostics in 2017. However, 4 of the 14 chapters included here were only published electronically at that time. We are now pleased to present all 14 chapters in one printed volume and hope you will enjoy consuming them as much as we enjoyed assembling them.

Principles of Molecular Biology

John Greg Howe

ABSTRACT

Background

Molecular diagnostics and its parent field, molecular pathology, examine the origins of disease at the molecular level, primarily by studying nucleic acids. Deoxyribonucleic acid (DNA), which contains the blueprint for constructing a living organism, is the centerpiece for research and clinical analysis. Molecular pathology is an outgrowth of the enormous amount of successful research in the field of molecular biology that has discovered over the last seven decades the basic biological and chemical processes of how a living cell functions. The success of molecular biology, as noted by the large number of Nobel prizes awarded for its discoveries, is now used for clinical diagnosis and the development and use of therapeutics.

Content

The following chapters are devoted to describing this field and the specific applications currently being used to characterize and help treat patients with a variety of ailments, including hereditary genetic diseases, cancer neoplasms, and infectious diseases. In this chapter the fundamentals of molecular biology are reviewed, followed by a focus on genomes and their variants in Chapter 2. In Chapters 3 and 4 techniques for isolating and analyzing nucleic acids are discussed. The clinically important subdivisions of molecular diagnostics are then reviewed and include microbiology in Chapter 5, genetics in Chapter 6, solid tumors in Chapter 7, and hematopoietic malignancies in Chapter 8. Chapters 9 and 10 are devoted to the molecular diagnostic analysis of circulating tumor cells and circulating nucleic acids. Finally, pharmacogenetics and identity assessment are the focus of Chapters 11 and 12.

HISTORICAL DEVELOPMENTS IN GENETICS AND MOLECULAR BIOLOGY

Molecular diagnostics would not be possible without the many significant pioneering efforts in genetics and molecular biology. Earlier observations in genetics began with the discovery of the inheritance of biological traits made by Gregor Mendel in 1866 and the observation in 1910 that genes were associated with chromosomes by Thomas Morgan. The initial findings that contributed to determining that DNA was the transmittable genetic material were performed by Griffith in 1928 and Avery, McLeod, and McCarty in 1944.[1,2] The definitive studies, published by Hershey and Chase in 1952, demonstrated that radiolabeled phosphate incorporated into the DNA of a bacteriophage was found in newly synthesized DNA containing bacteriophage instead of radiolabeled sulfur in protein, which showed that DNA and not protein was the genetic material.[3]

Deciphering the structure of DNA required several crucial findings. These included the observation by Erwin Chargaff that the quantity of adenine is generally equal to the quantity of thymine, and the quantity of guanine is similar to the amount of cytosine[4] and the pivotal x-ray crystallography results produced by Rosalind Franklin and Maurice Wilkins.[5,6]

Molecular biology has historically traced its beginnings to the first description of the structure of DNA by James Watson and Francis Crick in 1953.[7,8] The description of the DNA structure initiated the dramatic increase in the knowledge of the biology and chemistry of our genetic machinery. The impact of the Watson and Crick discovery was so significant that it is considered one of the most important scientific discoveries of the 20th century.[9]

One reason the work of Watson and Crick had such a dramatic impact on scientific discovery was that they not only described the structure of DNA, but hypothesized about many of its properties, which took decades to confirm experimentally.[7,8,10] One of those properties was the replication of DNA, which was shown to be semiconservative by Meselson and Stahl[11] in 1958. At the same time, DNA polymerase, which replicates the DNA, was discovered by Arthur Kornberg.[12] Deciphering the genetic code was vital for understanding the information stored in DNA, and cracking the code in 1965 required many scientists, most prominently Marshall Nirenberg.[13] Additional studies described the transcription and translation processes and uncovered several startling findings. One finding was the isolation of reverse transcriptase, an enzyme that synthesizes DNA from ribonucleic acid (RNA), which demonstrates that genetic information can be transferred in part in a bidirectional manner.[14,15] Another finding showed that the eukaryotic gene structure was composed of alternating non–protein-encoding introns and protein-encoding exons.[16,17] Along with the discovery of the basic biology of genes and their expression, many important techniques were invented. For example, the isolation of restriction enzymes[18] and DNA ligase allowed for the construction of recombinant DNA,[19] which could be transferred from one organism to another, leading to the cloning of DNA[20] and the emergence of genetic engineering. The Southern blot method, which identified specific electrophoretically separated pieces of DNA, participated in many discoveries and was one of the first molecular diagnostics methods to be used to test for genetic diseases.[21] DNA sequencing technologies were invented[22,23] and further advances in these technologies led to the first large biological science research undertaking, the Human Genome Project. Along with DNA sequencing, further technical discoveries, including the polymerase chain reaction in 1986[24] and microarray technology in 1995,[25] became methodologic foundations for molecular diagnostics.

MOLECULAR BIOLOGY ESSENTIALS

Whether it is a bacterium, virus, or eukaryotic cell, the genetic material located in these organisms dictates their form and function. For the most part the genetic material is DNA, which is composed of two strands of a sugar-phosphate backbone that are bound together by hydrogen bonds between two purines and two pyrimidines attached to the sugar molecule, deoxyribose, in a double helix (Figs. 1.1 and 1.2). DNA in human cells is wrapped around histone proteins and packaged into nucleosome units, which are compacted further to form chromosomes (Fig. 1.3). There are 23 pairs of chromosomes, two of which are the sex chromosomes, X and Y. Each chromosome is a single length of DNA with a stretch of short repeats at the ends called telomeres and additional repeats in the centromere region. In humans, there are two sets of 23 chromosomes that are a mixture of DNA from the mother's egg and father's sperm. Each egg and sperm is therefore a single or haploid set of 23 chromosomes and the combination of the two creates a diploid set of human DNA, allowing each individual to possess two different sequences, genes, and alleles on each chromosome, one from each parent. Each child has a unique combination of alleles because of homologous recombination between homologous chromosomes during meiosis in the development of gametes (egg and sperm cells). This creates genetic diversity within the human population. If a child has a random DNA sequence change or mutation, the child's genotype is different from that inherited from either of the parents (de novo variant). If the child's genotype leads to visible disease, the child has acquired a different phenotype from the parents.

Human cells have a limited lifespan and die through a process called apoptosis. Therefore most cells replace themselves as they progress naturally through their cell cycle. As a cell moves through phases of the cell cycle, its DNA doubles during the synthesis phase when the double-stranded DNA molecule separates. Each strand of DNA is used as a template to make a complementary strand by DNA polymerase in a process called DNA replication. Eventually during the cell cycle, two cells are created from one during the final mitotic phase.

DNA is composed of genes that code for proteins and RNA. For DNA to convert its store of vital information into functional RNA and protein, the DNA strands need to separate so that RNA polymerase can bind to the start region of the gene. With the help of transcription factors that bind upstream to promoters, the RNA polymerase produces single strands of RNA that are further processed to remove the introns and retain the protein-encoding exons. The mature, processed RNA molecule, the messenger RNA (mRNA), migrates to the cytoplasm, where it is used in the production of protein.

To start the process of protein synthesis or translation, the mRNA is bound by various protein factors and a ribosome, which contains ribosomal RNA (rRNA) and protein. The mRNA-bound ribosome begins to produce a polypeptide chain by binding a methionine-bound transfer RNA (tRNA) to the mRNA's initiating AUG codon or triplet code. The conversion of the nucleic acid triplet code to a polypeptide is accomplished by the tRNA, which contains a

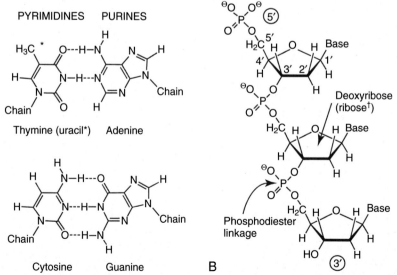

FIGURE 1.1 A, Purine and pyrimidine bases and the formation of complementary base pairs. *Dashed lines* indicate the formation of hydrogen bonds. (*In RNA, thymine is replaced by uracil, which differs from thymine only in its lack of the methyl group.) **B,** A single-stranded DNA chain. Repeating nucleotide units are linked by phosphodiester bonds that join the 5′ carbon of one sugar to the 3′ carbon of the next. Each nucleotide monomer consists of a sugar moiety, a phosphate residue, and a base. (†In RNA, the sugar is ribose, which adds a 2′-hydroxyl to deoxyribose.)

CHAPTER 1 Principles of Molecular Biology

nucleic acid triplet code (anticodon) in its RNA sequence that is specific for an amino acid bound to one end of the tRNA molecule. After synthesis, the protein migrates to its functional location and eventually is removed and degraded.

NUCLEIC ACID STRUCTURE AND FUNCTION

DNA is a rather simple molecule with a limited number of components compared to those of proteins. DNA is composed of a deoxyribose sugar, phosphate group, and four nitrogen-containing bases. Deoxyribose is a pentose sugar containing five carbon atoms that are numbered from $1'$ to $5'$, starting with the carbon that will be attached to the base in DNA and progressing around the ring until the last carbon that is not part of the ring structure. The bases consist of the purines, adenine and guanine and the pyrimidines, cytosine and thymine; an additional base, uracil, replaces thymine in RNA. A basic building block is the nucleotide, which consists of a deoxyribose sugar with an attached base at the $1'$ carbon and a phosphate group at the $5'$ carbon. The triphosphate nucleotide is the building block for making newly synthesized DNA. Newly synthesized DNA forms a polynucleotide chain that connects the individual nucleotides through the $5'$ and $3'$ carbons of each deoxyribose sugar via phosphodiester bonds.

Structure of Deoxyribonucleic Acid

DNA is double stranded, and the two strands bind to one another through hydrogen bonds between the bases on each strand. Hydrogen bonding is augmented by hydrophobic attraction (stacking) between bases on adjacent rungs of the DNA ladder. Both hydrogen bonds and base stacking are not covalent, but are weak bonds that can be broken and reestablished. This important property is exploited by many of the methods that are used in molecular diagnostics. The composition of DNA is equal quantities of guanine and cytosine and equal quantities of adenine and thymine, because, in general, guanine binds to cytosine and adenine binds to thymine.[4,7] There are two hydrogen bonds between adenine (A) and thymine (T) and three hydrogen bonds between cytosine (C) and guanine (G), and because of this difference in the number of hydrogen bonds, separating a guanine-cytosine (G-C) pair takes more energy than an adenine-thymine (A-T) pair (see Fig. 1.1).

Each of the two DNA strands is formed by a phosphate sugar backbone that starts at the $5'$ phosphate and ends at a $3'$ hydroxyl group with the complementary bases binding to one another between the two phosphate sugar backbones. Each strand is therefore a polar opposite of the other (see Fig. 1.2). When the two strands are bound to one another they progress in opposite $5'$ to $3'$ directions in an antiparallel configuration. By convention, the DNA sequence is denoted in a $5'$ to $3'$ direction. As discussed later, both the replication of new DNA and the transcription of DNA to RNA progress in the $5'$ to $3'$ direction. In addition, the conversion of RNA to protein, a process called translation, proceeds from the $5'$ end of the RNA to the $3'$ end. The combination of the base pairing and the directionality of the two DNA strands allows for the deciphering of the DNA sequence of one strand of DNA when the other complementary strand sequence is known.

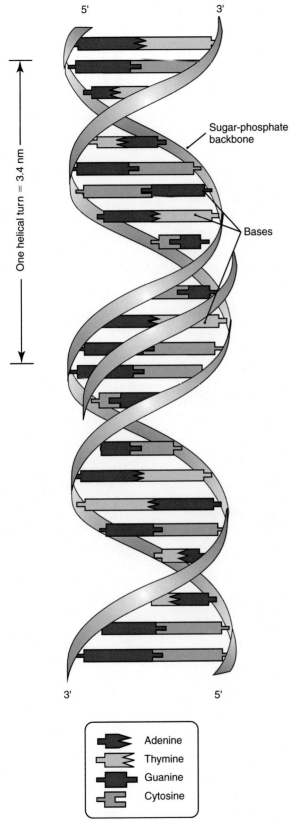

FIGURE 1.2 The DNA double helix, with sugar-phosphate backbone and pairing of the bases in the core-forming planar structures. (From Jorde LB, Carey JC, Bamshad MJ, editors: *Medical genetics.* 4th ed. Philadelphia: Mosby; 2010.)

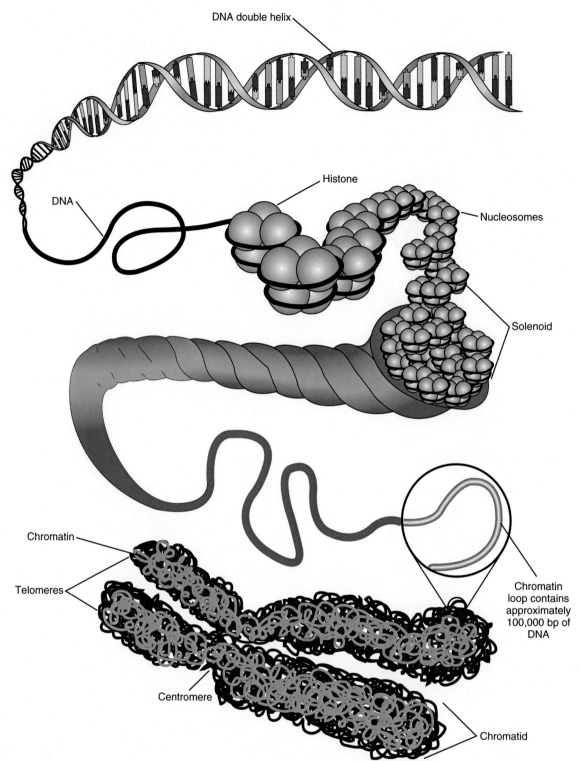

FIGURE 1.3 Structural organization of human chromosomal DNA. Double-stranded DNA is wound around the octamer core of histone proteins to form nucleosomes, which are further compacted into a helical structure called a solenoid. Nuclear DNA in conjunction with its associated structural proteins is known as chromatin. Chromatin in its most compact state forms chromosomes. The primary constriction of a chromosome is the centromere, and the chromosome's ends are the telomeres. (From Jorde LB, Carey JC, Bamshad MJ, editors. *Medical genetics*. 4th ed. Philadelphia: Mosby; 2010.)

Types of Deoxyribonucleic Acid

Double-stranded DNA in living cells is generally found as the right-handed B-DNA helical structure, which has specific dimensions. Each turn of the helix is 3.4 nm long and consists of 10 bases. The DNA sugar-phosphate backbone is on the outside of the helix, and the bases of each strand are inside bound to their complement on the other strand by hydrogen bonds. Other conformational structures of DNA occur, mostly associated with DNA sequences that are repeated. These non-B DNA forms include a left-handed Z-form,

A-motif, tetraplex G-quadruplex, i-motif, hairpin, cruciform, and triplex and are abundant in the human genome because a large percentage of the genome contains various repeats. Non-B DNA is associated with many biological processes, including transcriptional control. However, these structures also can create genetic instability, which can lead to various diseases such as neurologic disorders.[26]

Molecular Composition of Ribonucleic Acid

The composition of RNA is similar to that of DNA because it contains four nucleotides linked together by a phosphodiester bond, but with several important differences. RNA consists of a ribose sugar with a hydroxyl group at the 2′ carbon instead of the hydrogen atom in DNA. The bases attached to the ribose sugar are adenine, cytosine, and guanine, but not thymine because RNA uses another pyrimidine—uracil—as a substitute for thymine.

Structure of Ribonucleic Acid

One significant difference between DNA and RNA is that RNA does not normally exist as two strands bound to one another, although a single strand can bind internally to itself creating functionally important secondary structures. Although in the past several decades the complexity and number of different RNAs has greatly expanded, the majority of cellular RNA is composed of a rather small number of RNA types. These include mRNA, rRNA, and tRNA.

Ribonucleic Acids Associated With Protein Production

mRNA is the most diverse group of the three major types of RNAs, but constitutes only a small percentage of the total RNA. mRNAs are transcribed from DNA that codes for proteins and therefore are used as the template for the translation of proteins. In the case of prokaryotes the mRNA is colinear with the protein that is translated; however, in eukaryotes the mRNA begins as a precursor RNA called premessenger or heterogeneous nuclear RNA (hnRNA) that includes untranslated intron and translated exon regions. After transcription the hnRNA is spliced into mature mRNA lacking the introns. The mature mRNA contains only exons and can be further modified by the addition of a 7-methylguanosine cap at the 5′ end, which protects the mRNA from degradation, and a polyadenosine (polyA) sequence at the 3′ end. In eukaryotes the production and processing of the hnRNA to mRNA takes place in the nucleus, and the final form of the mRNA is then transported to the cytoplasm to be translated.

rRNA is associated with ribosomes, which are the primary structures that produce protein through the biological process of translation. rRNA, unlike mRNA, does not code for proteins. The ribosome is composed of two structures, the 50S and 30S subunits found in prokaryotes and the 60S and 40S subunits found in eukaryotes. The "S" stands for Svedberg units and is determined by the centrifugal sedimentation rate. The Svedberg unit measures the mass, density, and shape of an object. The ribosome is a mixture of RNA and protein. In eukaryotes there are four major rRNAs: the 18S rRNAs found in the 40S subunit and the 28S, 5.8S, and 5S rRNAs found in the 60S subunit. In prokaryotes, the 50S subunit contains the 23S and 5S rRNAs and the 30S subunit contains the 16S rRNA. Synthesis of eukaryotic rRNA occurs as a large 45S precursor RNA that is enzymatically cleaved to form all the rRNAs except the 5S RNA, which is transcribed separately. Ribosomal RNAs have secondary and tertiary structures that are well conserved with various loops, stem loops, and pseudoknots that contribute to their function. Ribosomal RNA and protein, as the components of ribosomes, function to carry out the translation of proteins. The sequence of the 16S rRNA has alternating conserved and divergent regions that can be used to identify microorganisms. The structure of the ribosome is now known, and the rRNA is more important than ribosomal proteins in ribosome functioning. The RNA acts as a catalytic agent called a ribozyme.[27,28]

Another important group of RNAs are the tRNAs, which function as key molecules that act as a bridge between the nucleic acids and the proteins. They have a unique cloverleaf secondary structure, with the 3′ end covalently attached to the amino acid by specific aminoacyl tRNA synthetases. In the middle of the tRNA structure is the anticodon sequence that binds to a specific homologous codon in the mRNA. Therefore the codon directs the binding of a specific tRNA linked to its corresponding amino acid. The genetic code, which consists of a 64 3-base code, specifies the appropriate amino acid to be attached to the growing polypeptide chain (see Figs. 1.7 and 1.8, later in the chapter). There are several different classes of aminoacyl tRNA synthetases, but there is at least one aminoacyl tRNA synthetase for each of the 20 amino acids. There is also at least one tRNA for each amino acid; however, there can be more depending on the species.[29]

Besides the three major types of RNAs, other RNAs include nuclear, nucleolar, and cytoplasmic small RNAs, signaling RNAs, telomerase RNA, and micro-RNAs.[30] This list appears to be growing with each passing year. Some of the first characterized small RNAs, the nuclear and nucleolar small RNAs, are involved with the processing of precursor RNAs to mature RNAs, including splicing of hnRNA to mRNA and precursor rRNA to mature rRNAs. More recently a large number of microRNAs have been discovered that partly function in the regulation of translation. In addition, there are many other noncoding RNAs whose functions are just beginning to be understood.

Human Chromosome

Human double-stranded DNA that is contained in the sperm or egg is a single copy or haploid amount of DNA made up of approximately 3 billion base pairs (bp). To be more precise, the Human Genome Project consensus sequence of the human genome was 2.91×10^9 bp[31] and the first human to be sequenced, Craig Venter, had a genome size of 2.81×10^9 bp,[32] not including remaining gaps of highly repetitive sequences, many near centromeres and telomeres (see Chapter 2). The DNA in the cell is bound by many proteins to form chromatin (see Fig. 1.3). The proteins in chromatin consist of histones, which are bound in precise amounts per a length of DNA, and other proteins called nonhistone proteins that are bound more irregularly and in widely varying amounts. The histone proteins consist of eight proteins (two copies each of H2A, H2B, H3, and H4) that bind as a unit to 147 bp of DNA to make up a nucleosome, and the protein, H1, that binds between the nucleosomes (Fig. 1.4). The nucleosomes are the basic structure to which many other proteins interact and modify to regulate gene expression. For example, the access to DNA by transcription factors is controlled by proteins that remodel the histone proteins through phosphorylation, acetylation, and methylation. The

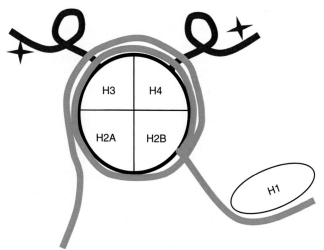

FIGURE 1.4 Schematic illustration of a nucleosome unit. A segment of DNA is wound around a nucleosome core particle consisting of an octamer of two each of the histone proteins H2A, H2B, H3, and H4. Tails with modifications (indicated by a *red star*) are shown to protrude from H3 and H4. Adjacent nucleosomes are separated by a segment of linker DNA and the linker histone, H1.

nucleosomes are condensed into filaments and even more compact structures to form a chromosome (see Fig. 1.3). There are 23 pairs of chromosomes; 22 autosomal chromosomes and 2 sex chromosomes, X and Y, with an XX pair denoting female and an XY pair denoting male. The DNA in chromosomes is continuous for each chromosome and can be as much as several hundred million base pairs in length for the largest chromosomes.

From a cytogenetic viewpoint, regions of the chromosomes can be classified by their transcriptional activity. The more condensed heterochromatin DNA is transcriptionally inactive and stains with Giemsa, a mixture of several dyes that bind to AT-rich regions of DNA. The less condensed euchromatin DNA is transcriptionally active and does not stain with Giemsa. The ends of the chromosomes, called telomeres, contain a repeat sequence, such as TTAGGG that is found in humans and shortens with age. The centromeres, at the center of most chromosomes, are important for linking sister chromatids during mitosis and contain various satellite DNAs, such as α-satellite tandem repeats (171 bp) that are over several million base pairs (Mb) in length.

Surprisingly, most of the human DNA does not code for the expression of protein. As much as 50% of human DNA consists of many types of interspersed repeat sequences, such as satellites, telomeres, microsatellites, minisatellites, short and long interspersed nuclear elements (SINES, LINES), and retrovirus elements.[31] Like other eukaryotes, human genes are in pieces with the protein-encoding regions, exons, alternating with the introns, which do not code for protein sequence and occupy more than a quarter of the human DNA.[33] Other regions around the genes, such as the promoter regions and the 3′ untranslated regions are also not translated into proteins. After all the noncoding sequences are removed, the protein-coding DNA sequence spans only approximately 1.2 to 1.5% of human DNA. Even though most human DNA is not associated with protein-producing genes, the Encyclopedia of DNA Elements (ENCODE) project has shown that much of the non–protein-encoding DNA is transcribed into noncoding RNAs, most with unknown function.

CENTRAL DOGMA OF MOLECULAR BIOLOGY

Francis Crick originated the concept of the central dogma of biology, which describes the transfer of genetic information into functional macromolecules.[34] This was generally depicted to show the movement of genetic information from DNA to RNA via transcription using RNA polymerase and further translated into protein via ribosomes and various factors. This is a simplistic version of the original concept, which took into consideration every possible transfer of information even though no evidence existed at the time. However, since the original publication a number of other postulated transfers have been described. DNA can enzymatically replicate itself by DNA polymerase, and RNA can be made into DNA using reverse transcriptase.[35] Many of these enzymes are used in molecular diagnostics assays.

Deoxyribonucleic Acid Replication

A general principle underlying the synthesis or replication of new DNA is that it uses one of the two DNA strands as a template to make a new homologous strand. This is termed semiconservative replication and was first theorized by Watson and Crick.[7] DNA replication begins at an adenine and thymine (AT)-rich structure called an origin of replication. In bacteria there is generally only one origin of replication, but in eukaryotic cells there are thousands. Since DNA can be supercoiled into more structures, a topoisomerase is required to first unwind this structure so that the DNA is accessible. A DNA helicase binds to the double-stranded DNA and separates the two strands, providing two single-stranded DNA templates. Replication progresses in a 5′ to 3′ direction; therefore one strand, the leading strand, is synthesized as one continuous strand using the 3′ to 5′ template and the other strand, called the lagging strand, is synthesized in small segments called Okazaki fragments from the 5′ to 3′ template. Because the DNA polymerase requires a primer, small RNA primers are made by a primase enzyme on the 5′ to 3′ template and the Okazaki fragments are synthesized starting from the primer. Okazaki fragments are finally linked by a ligase (Fig. 1.5).[36]

DNA polymerases of various types have been identified and they function in many different roles, the most important being the replication of new DNA and the repair of existing DNA. Using the template strand as a guide, the DNA polymerase binds a nucleotide triphosphate to the primer at a free 3′ hydroxyl group, releasing pyrophosphate. The specific nucleotide selected depends on the base on the template strand; for example, an adenine nucleotide is used if a thymine nucleotide is in the template strand. In summary, a complementary sequence is synthesized opposite the template strand. The insertion of the correct nucleotide does not always occur. Mistakes occur approximately every 100,000 nucleotides; therefore a major function of a DNA polymerase is error correction or proofreading and is accomplished by an intrinsic 3′ to 5′ exonuclease activity. DNA polymerases are important in molecular diagnostics because they are used in the polymerase chain reaction (PCR) and DNA sequencing.

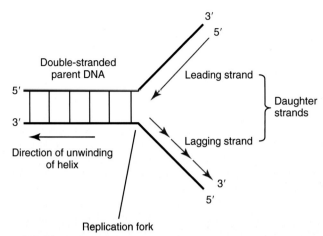

FIGURE 1.5 DNA replication. Double-stranded DNA is separated at the replication fork. The leading strand is synthesized continuously, whereas the lagging strand is synthesized discontinuously but is joined later by DNA ligase.

DNA replication is part of the cell cycle and occurs during the synthesis phase. The rest of the cell cycle is the interphase, further divided into the first growth phase (G_1) and the second growth phase (G_2), along with the DNA replication or synthesis (S) phase that lies between G_1 and G_2. The mitosis phase, which involves the splitting of one cell into two cells, occurs after the G_2 phase. Mitosis is divided into six subphases: prophase, prometaphase, metaphase, anaphase, telophase, and cytokinesis.

At important control points in the cell cycle the cell will commit significant resources to proceed further. One of these control points is between the G_1 and S phase, just before it begins DNA replication. The G_1/S boundary control point is disrupted in many cancers. It is common for neoplasms to have mutations in the retinoblastoma gene *(RB1)*, whose protein product regulates cell cycle progression from G_1 to S. Another control point is between G_2 and M, just as the cell commits to creating two cells from one.

Deoxyribonucleic Acid Repair

The integrity of DNA is damaged in a variety ways that culminate in changes or mutations in the DNA sequence. DNA bases may be damaged, removed, cross-linked or incorrectly paired with one another, and single- or double-stranded breaks may also occur.[37,38] When the cell senses that its DNA has become damaged, it stops the progression of its cell cycle and initiates DNA repair processes.[39] Cells repair these lesions by employing multiple DNA repair mechanisms that are specific for the type of DNA lesion and include base excision repair, nucleotide excision repair, mismatch repair, and homologous recombination repair.

Mechanisms

Base excision repair removes bases that are damaged by deamination, oxidation, and alkylation. Deamination of guanine, cytidine, and adenine converts them into structures that will incorrectly base pair, creating transition mutations, which are changes between similar nitrogenous bases such as a purine to a purine. A transversion mutation is a change from a purine to a pyrimidine or vice versa. DNA glycosylases, such as uracil-DNA-glycosylase, cleave the damaged base, and a 5′-deoxyribose phosphate lyase removes the nucleotide upstream of the removed base. DNA polymerase and ligase then add a new nucleotide repairing the damage. One of the inherited disorders associated with this repair process that leads to a predisposition to various neoplasms is caused by mutations in *MUTYH*, a DNA glycosylase gene.[38,40]

Nucleotide excision repair removes base modifications that change the helical structure of DNA, including bulky DNA distortions and covalently bound structures that may be created by ultraviolet radiation and certain cancer drugs. The damage is recognized by global and transcription-mediated repair processes. After the repair is initiated, the transcription factor, TFIIH, binds to a complex of proteins and makes an incision. The damaged DNA is unwound, and the gap is filled by DNA polymerase and finally sealed by DNA ligase. Mutations in the nucleotide excision repair genes cause xeroderma pigmentosum, which leaves affected individuals susceptible to specific tumors.[38,41]

Mismatch repair recognizes base incorporation errors and base damage. DNA polymerase has a 3′ to 5′ editing exonuclease with a proofreading function that is not completely effective and allows some mismatches to occur that can lead to mutations after DNA replication. The mismatched nucleotides must be repaired on the newly synthesized strand of DNA, which in prokaryotes is recognized by its unmethylated state. In eukaryotes the mechanism is different, and it is proposed that proteins associated with the replication apparatus, specifically the proliferating cell nuclear antigen protein determines the appropriate DNA strand for repair.[38] These mutations are corrected with DNA mismatch repair proteins, which identify the mismatches by their methylation patterns, excise the surrounding sequence, and then repair the excision with new sequence. Mutations in the human mismatch repair genes are associated with Lynch syndrome (hereditary nonpolyposis colorectal cancer).

Double-stand breaks are a very destructive form of DNA damage that destabilizes the genome, sometimes resulting in gross chromosomal changes, such as translocations that are frequently found in cancer. Double-stranded breaks are caused by several processes, including ionizing radiation and chemotherapy drugs, and are repaired by either homologous recombination or nonhomologous end joining.[38,41] The homologous recombination repair pathway is initiated by recognition of a double-stranded break, followed by resection using exonucleases to create a 3′ single-stranded overhang. With the assistance of many proteins, RAD51 is bound to the single-stranded DNA, which invades the intact homologous double-stranded DNA of the sister chromatid and uses it as a template for new double-stranded DNA repair.[38]

DNA repair mechanisms operate independently to repair simple lesions. However, the repair of more complex lesions involves multiple DNA processing steps regulated by the DNA damage response pathway. When single- and double-stranded DNA breaks occur, a cascade of responses is initiated that culminates in either DNA repair, stopping the cell cycle, or programmed cell death. After DNA damage has occurred, the DNA damage response pathway activates the protein kinases ATM (ataxia telangiectasia mutated) and ATR (ataxia telangiectasia and Rad3-related protein) to phosphorylate signaling proteins, such as p53, which eventually leads to cell cycle arrest at the G_1/S boundary. This gives time for the DNA repair mechanism to repair the damaged DNA; however, if the damage is too extensive, the cell initiates apoptosis or cell death.[39]

Deoxyribonucleic Acid Modification Enzymes

There are two groups of nucleases, the endonucleases that cut through the sugar-phosphate backbone and exonucleases that digest the ends of DNA. The commercially important restriction endonucleases, which bacteria have acquired to protect themselves from viral infections, are used to cleave DNA at a specific nucleotide sequence or restriction sites.[42] Several thousand restriction endonucleases have been characterized and are used extensively to manipulate DNA in molecular biology and molecular diagnostics. Recent work has described new nucleases, such as the RNA-guided engineered nuclease, CRISPR/Cas system, that can precisely cleave genomic DNA.[43]

DNA glycosylases are a family of enzymes associated with base excision repair that are used in the first step of DNA repair to remove the damaged base, without disrupting the sugar-phosphate backbone. An important member of that family, uracil DNA glycosylase, repairs the most common mutation found in humans, the spontaneous deamination of cytosine to uracil, by removing the uracil base.

Gene Structure

The structure of prokaryotic genes is straightforward; almost all of the gene sequence is used to make protein; however, this is not the case with eukaryotic genes. One of the unique hallmarks of eukaryotic genes is that the protein-coding DNA is interspersed with regions that do not code for DNA, an observation made by Richard Roberts and Phillip Sharp in 1977. A mature mRNA retains only the protein-coding sequences called exons, and the sequences between the exons are non—protein-encoding sequences called introns that are removed during mRNA maturation (Fig. 1.6).[44]

In addition to introns and exons, eukaryotic genes consist of regulatory regions, such as promoters and enhancers, and 3' regions that contain termination and polyadenylation signals. The regulation of the expression of eukaryotic genes can occur at all levels from transcription to splicing to translation to degradation; however, most gene regulation occurs at the initiation of transcription by various promoters and enhancers.[45] There are two groups of regulatory elements: one is close to the transcriptional start site and is made up of the core promoter and ancillary promoters slightly further away from the start of transcription. The other group of regulatory elements can be much further away, not only upstream but also downstream from the gene. This second group is made up of enhancers, silencers, insulators, and locus-specific control regions.[45,46] These regulatory elements contain specific sequences that bind to transcription factors that can upregulate or downregulate the expression of a gene. There are only several thousand human transcription factors, much less than the number of human genes; therefore each gene has many regulatory elements to provide the needed complexity to function in 200 different human cell types.[45]

A surprising property of human genes is that there are so few compared to less complex species. Humans have approximately 20,000 genes, many fewer than found in rice and only slightly more than found in the roundworm, *Caenorhabditis elegans*.[47-49] Recently, results from the ENCODE project have challenged the concept of "one gene, one protein."[50] Their studies show that the exon of one gene can be spliced into the exon of another gene.[51] This result, along with alternative splicing, demonstrates that one gene can make multiple proteins and is probably the reason humans have such a small number of genes.

Ribonucleic Acid Transcription and Splicing

RNA transcription involves synthesizing an RNA strand using DNA as a template. This requires many different proteins, the most important being the RNA polymerases, of which there are three types in eukaryotic cells. RNA polymerase I is specific for the rRNAs, 28S, 18S, and 5.8S, which are initially transcribed as a single primary transcript of 45S. RNA polymerase II transcribes all genes that encode proteins and the small nuclear RNA (snRNA) genes. RNA polymerase III transcribes a variety of small RNAs, including the 5S rRNA, and tRNA. Additional proteins called transcription factors function in combination to recognize and regulate transcription of different genes.[52]

The synthesis of RNA proceeds in a 5' to 3' direction using DNA as a template and a specific DNA sequence acts as a transcription start site. Transcription progresses through three phases: initiation, elongation, and termination. The initiation phase includes the binding of transcription factors to promoters upstream from the start site and includes the core promoter immediately upstream and the ancillary promoters further away. However, some of the small RNA gene promoters are in the middle of the gene. Transcription factors binding to upstream promoters act as regulators of the transcription of genes. These factors generally bind in pairs or dimers and

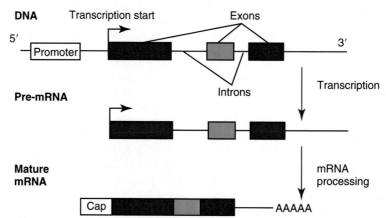

FIGURE 1.6 DNA transcription and messenger RNA processing. A gene that encodes for a protein contains a promoter region and variable numbers of introns and exons. Transcription commences at the transcription start site. Premessenger RNA or heterogeneous nuclear RNA (hnRNA) is processed by capping, polyadenylation, and intron splicing and becomes a mature messenger RNA.

have several functional domains. One functional domain of the transcription factor binds to a specific promoter DNA sequence via several structures, such as the helix-turn-helix, zinc finger, and leucine zipper structures. Another domain binds to the other transcription factor of the dimer pair, and a third domain may bind to the RNA polymerase complex that carries out transcription.[46] Even though promoters and the transcription factors binding to them are far away from the transcription initiation complex, the promoter DNA folds back on itself to allow for the transcription factors to interact with the RNA polymerase complex.[53]

Important recurring sequences are found in the core promoter. For example, the core promoter of an RNA polymerase II gene contains a TATAAA sequence, called a TATA box located upstream 25 to 40 nucleotides from the transcriptional start site. Only 20% to 30% of eukaryotic promoters contain TATA boxes, but they are highly regulated compared to those without TATA boxes that are mostly housekeeping genes.[45,54,55]

The first step in mRNA transcription is the binding of transcription factor IID (TFIID) to the TATA box, which in turn promotes the binding of other transcription factors (TFIIA, TFIIB, TFIIE, TFIIF, and TFIIH), RNA polymerase II, and proteins attached to the upstream promoter sites. To form a functional transcription complex, the promoter region's doubled-stranded DNA separates and the transcription complex moves away from the core promoter region.[45] Once started, the RNA polymerase adds nucleotides to the 3′ free hydroxyl group in a manner similar to that of DNA replication. Transcription is eventually terminated by one of several termination mechanisms. In bacteria a termination factor bound to the RNA polymerase recognizes a DNA sequence termination signal. In the case of genes transcribed by RNA polymerase II, termination is coupled with the polyadenylation step (see Fig. 1.6).

Two posttranscriptional processing events are performed on the newly formed hnRNA, one at each end of the RNA. At the 5′ end, the hnRNA is capped with a 7-methyl guanosine molecule to help protect the hnRNA from degradation. At the 3′ end, a polyadenosine (poly A) stretch is added by poly A polymerase after the RNA sequence AAUAAA is synthesized. Some transcribed mRNAs are not polyadenylated, such as histone mRNAs.[56]

Transcription initially produces an hnRNA that contains both exons and introns, which needs to be processed or spliced into mature mRNA for it to be properly translated into protein. RNA splicing involves cleavage and removal of intron RNA segments and splicing of exon RNA segments. The process uses consensus splice site sequences located at both the 5′ (GU) and 3′ (AG) ends of the intron and an internal intron sequence. Splicing requires the effort of a number of proteins and small RNAs that come together to form a spliceosome, which directs the splicing of exons and removal of introns.[57] Splicing begins with the binding of the U1 small nuclear ribonucleic protein (snRNP) to the donor splice site and the U2 snRNP to the internal intron sequence, followed by the binding of U4, U5, and U6 snRNPs, resulting in excising the intron and joining (splicing) of the ends of the two exons on either side of the excised intron (see Fig. 1.6).[57]

An important modification of the splicing process, alternative splicing, allows for the generation of different mRNAs from the same primary RNA transcript by the cutting and joining of the RNA strand at different locations. Among the types of alternative splicing are exon skipping, alternative 3′ and 5′ splice sites, and intron retention. It is estimated that 92% to 95% of all human genes are alternatively spliced.[58,59]

The movement of cellular signals from the surface of a cell to the nucleus is called signal transduction, and one of the eventual targets is the modification (eg, phosphorylation) of transcription factors, which can modulate the binding of other transcription factors to DNA and their dimerization, thereby controlling gene expression.[60] A common cascade of signaling begins with the activation of a receptor on the cell surface, such as a tyrosine kinase receptor. The tyrosine kinase receptor in the form of a dimer can be activated by binding to a hormone or growth factor, for example, which causes a dimerization and autophosphorylation of the tyrosine receptor protein kinase. This in turn activates a cytoplasmic protein, such as the guanine nucleotide exchange factor that activates the G-protein Ras, which can then modify another G-protein, Raf, which propagates the signal to a common signaling pathway, the mitogen-activated protein (MAP) kinases. The final enzyme in the pathway can then act on downstream targets, including other protein kinases, and transcriptional factors. Some mutations in the tyrosine kinase receptor or Ras protein switches them to an unregulated "on" position, which can lead to uncontrolled growth of the cell and eventually to cancer.[60]

Translation

The final phase of the transfer of information from DNA is to proteins, the structural and functional molecules that make up the majority of a living organism, such as the human body. Proteins are long single strands of various amino acids and are synthesized by a process called translation, which requires the functioning of many protein factors, tRNAs, and ribosomes.

Amino acids have a common structure consisting of a carbon atom bound to amino and carboxylic acid groups and a unique side chain. There are 20 amino acids each with a different side chain that give them their unique properties. The side chains can be divided into four types: nonpolar (hydrophobic), polar (hydrophilic uncharged), and negative and positively charged. Nonpolar (hydrophobic) amino acids include alanine, leucine, isoleucine, valine, proline, methionine, phenylalanine, and tryptophan. The uncharged polar (hydrophilic) amino acids include glycine, serine, threonine, cysteine, tyrosine, glutamine, and asparagine. The negatively charged (acidic) amino acids are aspartic acid and glutamic acid, and the positively charged (basic) amino acids are arginine, histidine, and lysine. A protein's amino acid makeup and sequence in the polypeptide chain determine the overall structure and function of the protein. Some amino acids have a more significant presence than others. For example, proline, which disrupts secondary structure, and cysteine, which can cross-link to another cysteine through disulfide bonds, can change the structure of a protein.

Protein structures are grouped into four different classes. The primary structure is the sequence of the amino acids in the protein. There are several common types of secondary structure, such as β-pleated sheets and α helixes. Proteins can be constructed with a combination of these different types of secondary structures. Tertiary structure applies to the folding of the polypeptide chain into a three-dimensional form. Quaternary structure is the structural relationship of more than one polypeptide/protein joining together, such as in immunoglobulin molecules, that contains light and heavy proteins bound together by cysteine residues.

Once proteins are synthesized, they can be modified in various ways. One of the most common modifications is phosphorylation of the amino acids serine, threonine, and tyrosine, which can regulate protein activity. Other modifications include proteolytic cleavage, such as removal of the signal transport sequence, and acetylation of the N-terminus of most eukaryotic proteins that helps to prevent degradation. Glycosylation of secreted and membrane proteins on asparagine, serine, and threonine residues and formation of disulfide bonds via cysteine cross-linking are additional modifications.

Taking into consideration these posttranslational modifications and alternatively spliced forms mentioned in an earlier section, the total number of proteins in the more than 200 human cell types is estimated to range from 250,000 to several million.[61]

The genetic code, which was deciphered in the early 1960s, is required to convert a nucleic acid sequence into an amino acid sequence.[13] It was reasoned that if there are 20 amino acids, a code of at least 3 nucleotides was necessary to have enough combinations. A 3-nucleotide code gives 64 combinations, and therefore one hallmark of the genetic code is that it is redundant, meaning that there are several codes for one amino acid. That is the case for most amino acids, but not all; for example, methionine and tryptophan have only one code. The redundancy is usually in the third base of the code. All of the 64 3-nucleotide codon possibilities code for an amino acid, except 3 that serve as stop codons (UAA, UGA, and UAG) (Fig. 1.7).

Protein synthesis or translation occurs in the cytoplasm and proceeds in three steps: initiation, elongation, and termination. The process requires tRNA and rRNA molecules, as well as ribosomes and initiation, elongation, and termination factors. One of the most important groups of molecules are the tRNAs, which are recognized by aminoacyl tRNA synthetase enzymes that attach amino acids to the 3′ end of specific tRNA molecules. Each tRNA has a 3-base sequence (anticodon) that facilitates the specific recognition and interaction with a codon in the mRNA.

The initiation step of protein synthesis is the most complex and begins with the binding of initiation factor 4E to the cap structure on the 5′ end of the mRNA and binding of polyadenosine—binding protein (PABP) to the 3′ PABP polyadenosine tail. The binding of initiation factor 4G to both initiation factor 4E and PABP circularizes the mRNA and prepares it for binding to the preinitiation complex containing the 40S ribosomal subunit, initiation factor 2, and methionine tRNA. The preinitiation complex then scans the mRNA until it finds a methionine start codon (AUG), at which point the 60S ribosomal subunit binds forming the 80S initiation complex and initiates translation elongation.[62] This is a simplistic description of the initiation process because over a dozen additional initiation and auxiliary factors are involved.

Ribosomes have at least three structural positions where tRNAs can bind, the acceptor (A), peptidyl (P), and exit (E) sites. The acceptor site binds the incoming aminoacyl-tRNA. The peptidyl site holds the peptidyl-tRNA that is covalently linked to the growing polypeptide chain, and the exit site binds to the outgoing empty tRNA that carries no amino acid.[62,63]

The first codon (AUG) always codes for methionine; therefore to initiate translation the methionine tRNA binds to the aminoacyl-tRNA binding site of the ribosome. The

		Second Letter				
		U	C	A	G	
First Letter	U	UUU UUC Phenylalanine / UUA UUG Leucine	UCU UCC UCA UCG Serine	UAU UAC Tyrosine / UAA Stop Codon / UAG Stop Codon	UGU UGC Cysteine / UGA Stop Codon / UGG Tryptophan	U C A G	Third Letter
	C	CUU CUC CUA CUG Leucine	CCU CCC CCA CCG Proline	CAU CAC Histidine / CAA CAG Glutamine	CGU CGC CGA CGG Arginine	U C A G	
	A	AUU AUC AUA Isoleucine / AUG Methionine	ACU ACC ACA ACG Threonine	AAU AAC Asparagine / AAA AAG Lysine	AGU AGC Serine / AGA AGG Arginine	U C A G	
	G	GUU GUC GUA GUG Valine	GCU GCC GCA GCG Alanine	GAU GAC Aspartic Acid / GAA GAG Glutamic Acid	GGU GGC GGA GGG Glycine	U C A G	

FIGURE 1.7 Genetic code. Translation of messenger RNA to amino acids during protein synthesis.

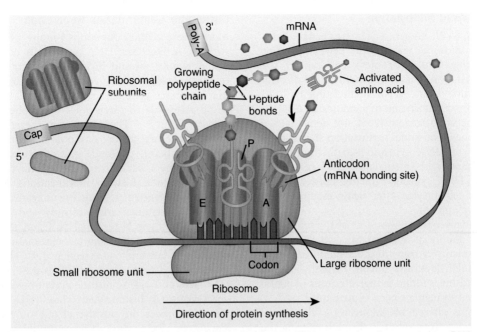

FIGURE 1.8 Translation. Shown is a ribosome bound to a messenger RNA converting the messenger RNA triplet code (codon) via a specific amino acid–bound transfer RNA containing a complementary anticodon sequence. There are three transfer RNA positions. A new amino acid–bound transfer RNA first arrives on the ribosome at the *A* or acceptor site at the front of the moving ribosome and then moves to the *P* or peptidyl site where the amino acid on the newly arrived transfer RNA combines with the growing polypeptide chain. Finally the now empty transfer RNA moves to the *E*, or exit site, where it prepares to leave the ribosome. (Modified from Huether SE, McCance KL. *Understanding pathophysiology*. 6th ed. St. Louis, Elsevier; 2017.)

tRNA specific for the next 3-base codon—for example, lysine—binds to the acceptor site of the ribosome and with the help of elongation factors (eg, eEF2), the amino acid in the peptidyl site is bound to the amino acid in the acceptor site by the formation of a peptide bond. A peptide bond is created between the amino group of one amino acid and the carboxyl group of the next amino acid through condensation releasing water. At the same time the tRNA shifts positions, with the methionine tRNA shifting to the exit site and the tRNA containing the growing chain of amino acids shifting to the peptidyl site. At the same time, the ribosome moves forward one codon and the next tRNA specific for the next codon through its anticodon binds in the acceptor site, and the process is repeated until a termination codon is reached (Fig. 1.8). Termination factors then bind and stop the translation process.[62] Protein synthesis occurs in the eukaryotic cytoplasm in the endoplasmic reticulum where multiple ribosomes called polyribosomes are involved in translating an individual mRNA.

Regulation of translation is not as extensive as that for transcription. However, there is global regulation of eukaryotic translation at the initiation step with phosphorylation of initiation factor 2B by four different protein kinases. This occurs when the cells are under stress, such as amino acid starvation or DNA damage.[64] In addition, mRNA-specific translational regulation can occur through binding to specific sequences located in the 5′ and 3′ untranslated regions. Furthermore, there are over 1000 microRNAs in humans,[65] many of which regulate transcription. The microRNA genes are transcribed as precursor RNA and then processed into a mature 22-nucleotide form by the processing enzymes Dicer and Drosha. The mature form of microRNAs can bind to specific sites on mRNA while associated with the Argonaute protein and either reversibly inhibit translation or degrade the mRNA.[62,66] For example, microRNAs Mir 15a/16-1 are deleted in chronic lymphocytic leukemia, thereby increasing Bcl2 expression and inhibiting apoptosis or cell death to prolong the life span of the cell.[67]

After proteins are synthesized there are two major processes to remove excess or damaged proteins. One process degrades the proteins ingested and uses nonspecific proteases, such as pepsin and trypsin, to digest proteins associated with foodstuff in the gut into amino acids so they can be absorbed. The second process digests extracellular and intracellular proteins by either general proteinases within lysosomes or by protein degradation via ubiquination. With the latter mechanism, proteins are tagged for degradation by binding to ubiquitin, which is recognized by a large multiprotein structure, the proteasome that degrades the ubiquinated proteins by proteolysis.[68]

EPIGENETICS

Although the original meaning of epigenetics encompassed all molecular pathways that affect the expression of genes, over time the definition has focused on the regulation of gene expression by heritable modifications that do not change the DNA sequence.[69] More recently this has been broadened to include nonheritable modifications.[70-73] Currently there are three major areas of epigenetic modifications or marks: (1) DNA methylation; (2) chromatin conformation regulation through histone modifications, including ATP-dependent remodeling enzymes and histone variants; and (3) noncoding RNAs.[74]

Deoxyribonucleic Acid Methylation

DNA methylation is a well-known epigenetic change that is important in X chromosome inactivation, gene imprinting (eg, Prader-Willi, Angelman syndromes), and cancer. The most common methylation event is the methylation of cytosine to form 5-methylcytosine. DNA methylation typically occurs at cytosines directly upstream of guanines, or CpG dinucleotides. Cytosine is both methylated and demethylated by a variety of enzymes. The initial methylation state is catalyzed by one type of DNA cytosine-5-methyltransferase, whereas the maintenance of the methylated state is performed by another type of DNA cytosine-5-methyltransferase and occurs during each cell division after being established in early embryonic development.[75]

Demethylation involves three members of the ten-eleven translocation (TET) family of dioxygenases, which catalyze the conversion of 5-methylcytosine to other modified forms, such as 5-hydroxymethylcytosine during demethylation.[76] 5-Hydroxymethylcytosine is found in high amounts in neural cells and is postulated to regulate gene expression.[76]

Gene expression is altered by methylation via several mechanisms. The most direct effect is through altering the ability of transcription factors to bind to promoters. Methylation decreases the affinity of transcription factors to a DNA promoter and enhances the binding of methylation-specific transcription factors (Fig. 1.9). Additionally, methylation compacts the chromatin structure, thus reducing the access of transcription factors to a promoter.[77] Cancer is the most common human disease associated with aberrant DNA methylation.[78] Interestingly, the overall level of 5-methylcytosine in cancer cells is 60% less than in normal cells; however, certain promoter-specific CpG islands are hypermethylated.[78] Other human diseases that are associated with methylation include lupus and many neurologic diseases.

Chromatin Conformation Regulation

Many basic cellular functions require proteins to interact with DNA. However, DNA is generally not freely accessible but is wound around histones to form nucleosomes and further condensed or compacted into heterochromatin that decreases gene expression. The cell requires the DNA to be accessible to carry out DNA replication, repair, and transcription.[74,79] The chromatin, therefore, is a very dynamic structure; at any one point in time portions of the DNA are being exposed and other portions are being covered. The mechanisms that control chromatin conformation include histone modifications, histone variants, and ATP-dependent remodeling enzymes.

Specific histones are reversibly and posttranslationally modified at their N-terminal tails and globular regions to change the chromatin from a euchromatin state to a heterochromatin state and back (see Fig. 1.9). These modifications include acetylation of lysine residues at the N-terminal tails of H2A, H3, and H4 by histone acetyltransferases (HATs) and deacetylation by histone deacetylases (HDACs). Histone acetylation removes the positive charge on the lysine residue, leaving the lysine less attracted to the negatively charged DNA phosphate backbone and thereby opening the DNA.[77]

Histone methylation of lysine and arginine residues occurs mostly on histone protein H3, but also histone protein H4, and is carried out by histone methyltransferases (HMTs) and histone demethylases (HDMs). The effect of methylation on chromatin structure ranges from active to poised to repressed. Histone lysine and arginine residues can be mono-, di-, and tri-methylated, but the positive charge is unchanged.[39,79] Histone methylation is found associated with DNA transcription, replication, and repair.

Histones are phosphorylated at serine, threonine, and tyrosine residues and are associated with DNA repair and transcription. The addition of a negatively charged phosphate

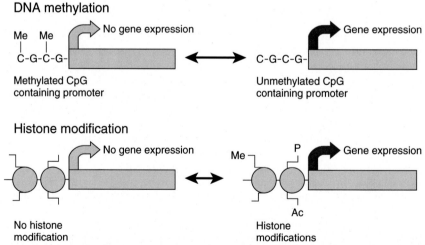

FIGURE 1.9 Epigenetics. *Top,* DNA methylation of CpG island regions indicated by *Me* in and around gene promoters is associated with loss of gene expression and silencing of the gene. When *CpG* islands are unmethylated, shown by absence of Me, gene expression is unaffected. *Bottom,* Modifications of the tails of histone proteins, such as methylation, acetylation, and phosphorylation, shown as *Me, Ac, and P,* respectively, can increase gene expression. (Modified from Zaidi SK, Young DW, Montecino M, van Wijnen AJ, Stein JL, Lian JB, et al. Bookmarking the genome: maintenance of epigenetic information. *J Biol Chem* 2011;286:18355–18361.)

group to the histone will repel the histone away from the negatively charged DNA and loosen up the chromatin structure.[80] Other modifications include poly(ADP-ribosyl)ation, ubiquitination, SUMOylation, and glycosylation.[81]

Histone variants have been known for decades, but many of their functions are not well established. Histone protein variants H3.3 and H2A.Z are the most well-known and are shown to function in regulation of gene expression.[82] Histone variant H3.3 incorporates into chromatin independent of replication and is associated with active chromatin.[83,84]

ATP-dependent remodeling enzymes use the energy from the hydrolysis of ATP to change the structure of chromatin.[84,85] ATP-dependent remodeling enzymes are grouped into four families including SWItch/Sucrose NonFermentable (SWI/SNF), imitation switch (ISWI), inositol requiring 80 (INO80), and chromodomain (CHD).[79,85]

The remodeling enzymes have similar properties, including (1) specific interaction with nucleosomes, (2) attraction to the modified histone tail residues found in nucleosomes, (3) contain an ATPase domain, (4) ATPase regulatory function, and (5) ability to interact with transcription factors and chromatin-associated proteins.[81,85] The primary role of the enzymes is to remodel the chromatin structure. The SWI/SNF proteins function in the sliding and ejecting of nucleosomes, but do not function in chromatin assembly. The IWSI family of enzymes changes the nucleosome spacing through sliding that is necessary after DNA replication. This family interacts with unmodified histone tails and functions to regulate transcription. The CHD family functions to slide and eject nucleosomes, by which it regulates transcription. The INO80 family of proteins has an insertion in the middle of its ATPase domain and functions in promoting transcription and DNA repair. A mammalian member of this family, SWR1, can exchange histones to facilitate DNA repair.[81,85-87]

Noncoding Ribonucleic Acids

Most of the expressed RNA in a cell is not translated into protein. Only the mRNAs are translated into protein, and they represent only 1% to 5% of the total RNA depending on cell type. Much of this noncoding RNA is known and includes rRNA and tRNAs. However, over the last several decades two large groups of noncoding RNAs have been discovered, the short and long noncoding RNAs. The ENCODE project tested for the expression from DNA not associated with genes by using probes that overlapped one another regardless of the location of genes. Over 80% of the human DNA could be assigned a biochemical function, although biochemical function was liberally defined.[88] Nonetheless, it was determined that the bulk of the human genome is expressed into RNA.[89]

The short noncoding RNAs consist of microRNAs, small interfering RNAs and piwi interacting RNAs.[90,91] MicroRNAs regulate gene expression by binding to a specific sequence of the mRNA and inhibiting its translation. Small interfering RNAs (siRNA) inhibit translation by also binding to a region of the mRNA, but do so by initiating the degradation of the mRNA by the associated Argonaute protein. Piwi interacting RNAs (piRNA) function in the repression of transposons and are important in the development of gametes in many multicellular eukaryotic species.

The long RNAs are arbitrarily designated to be greater than 200 nucleotides while the short RNAs are between 20 and 200 nucleotides.[92] Only recently has the extent of long noncoding RNAs been appreciated.[89] The diversity of the long noncoding RNAs is predicted to be in the hundreds of thousands in vertebrates and their expression pattern is highly regulated during the development of an organism. A well-described example of a long noncoding RNA is XIST, which associates with the Polycomb group complex 2 and inactivates the X chromosome by inducing heterochromatin formation and repressing gene expression.[93] Examples such as XIST and a similarly acting protein, HotAir, have given rise to the possibility that the noncoding regions of the human genome have important functions.[92]

The function of most noncoding RNAs is unknown, but it is speculated that coding and noncoding RNAs, referred to as competing endogenous RNAs (ceRNAs), are in competition for shared microRNA binding sites in untranslated regions of mRNAs, thereby regulating their expression. The ceRNA hypothesis proposes a new layer of regulation of gene expression that could help explain the function of the large percentage of the human genome that expresses non–protein-coding RNA.[94-96]

UNDERSTANDING OUR GENOME

Genomics is recognized as a unique field since the first free-living organisms were completely sequenced in the 1990s. With the publication of the first draft of the human genome in 2001 and the final results of the Human Genome Project in 2004, the genomics field started to impart greater influence on biomedical research and its application to medicine.[31,97] Genomics is characterized by the comprehensive nature of its collection of data and the technical development necessary to obtain, analyze, store, and make available such large amounts of data. There are also ethical, legal, and social implications of the research and clinical application of genomics.[98]

Large research projects that were initiated during the latter years of the Human Genome Project produced comprehensive biological catalogs of genetic variants, important DNA functional sequences, and expressed products from not only humans but also many other organisms.[98]

Single nucleotide variants (SNVs) are the most common DNA differences found in the human population, and they number in the millions, with each individual differing on average by 1 in 1000 nucleotides. Human SNVs (including both benign polymorphisms and causative mutations) are cataloged in the SNP database (http://www.ncbi.nlm.nih.gov/SNP).

Genome-wide association studies employ microarray tests that use large numbers of SNVs to find associations between genetic variations and diseases. DNA variants are often clustered into regions by genetic recombination during the formation of sperm and eggs that are inherited as a unit, such that a unique SNV pattern or haplotype can be passed from generation to generation. The International HapMap Project also uses SNVs to investigate haplotype associations and disease.

The 1000 Genomes Project complements the previously mentioned projects by sequencing a large number of diverse human samples from around the world. The goal is to build a comprehensive catalog of the most common human genetic variants, which includes single nucleotide variants, as well as insertions, deletions, and copy number variants that are found in the population at greater than 1%. The Exome

Aggregation Consortium (ExAC) has sequenced over 60,000 exomes to delineate common genetic variation within human exomes. The SNP database, International HapMap Project, 1000 Genomes Project, ExAC, and genome-wide association studies have helped to define genetic variability within individuals and populations to understand the basis of many genetic diseases.[98]

A more fundamental biology project is the encyclopedia of DNA elements, or ENCODE, whose goal is a catalog of the functional elements of the genomes of humans and other species. The functional elements include the genes and all their expressed RNA forms and epigenetic modifications.[51] One of the most important findings is the discovery that much of the human genome is expressed into RNA.

With the introduction of the first massively parallel DNA sequencing instrument in 2005 and subsequent instruments from 2006 onward, the current technologic era of genomics has progressed over the last decade to make significant inroads into applying genomics to patient care.[99] Along with the technologic innovation in DNA sequencing, there has been innovation in bioinformatics, which is required to manage and interpret the large amount of information generated by massively parallel DNA sequencing instruments.

Although the Human Genome Project is a significant feat, it was not the first whole genome to be sequenced. Whole genome sequencing initially focused on infectious pathogens, because of their impact on human health and also their size. The first free-living organism to be sequenced was *Haemophilus influenzae* in 1995.[100] Subsequently, many species from a cross-section of living organisms have been sequenced. The first individual human to have their whole genome sequenced was Craig Venter, who led one of the two groups that first sequenced the human genome. The second person to have their whole genome sequenced was James Watson, whose genome was the first to be sequenced by using massively parallel DNA sequencing.

An important clinical application of genomics is cancer diagnostics (see Chapters 7 and 8); however, the diversity and complexity of cancer requires a significant amount of basic biological information to interpret molecular diagnostic testing results of patient samples. The first whole genome sequencing of a cancer was an acute myeloid leukemia in 2008,[101] and many others have subsequently been sequenced. The Cancer Genome Atlas project includes large numbers of the most common cancers to identify all their associated mutations. For example, a recent study describes mutational data for 12 of the most common cancers.[102] The significant amount of basic information now available on human cancers and the availability of new therapeutics targeting specific cancer-associated genes allow the clinical use of molecular profiling in cancer patients.[103]

With the increasing use of genetic and genomic information to characterize a patient's disease, an interesting convergence of electronic medical records and genomics is emerging. The implementation of electronic medical records throughout the United States will allow for greater access to the large amount of genomic data that will be available on patients, which will eventually be a source for scientific research and discovery. The Electronic Medical Records and Genomics Network is currently developing tools and conditions under which genomic research can be pursued using electronic medical records.[104]

All of the previously discussed advances have made the field of molecular diagnostics an important and exciting area that is going to have an even greater impact on medicine in the future. As an increasing number of diseases are characterized at the molecular (eg, nucleic acid and protein) level, new therapeutics and diagnostics specifically targeting these molecular changes will continue to emerge.

POINTS TO REMEMBER

- The two strands of DNA are bound together by hydrogen bonds and stacking forces that can be broken and reformed without permanent damage to the DNA. This important property is exploited by many of the methods that are used in molecular diagnostics. This is a requirement for most of the DNA diagnostic assays.
- Even though human DNA has approximately 20,000 genes, this is far less than what would be expected given the number of proteins in a human cell. The higher number of proteins results from alternative splicing, which occurs in more than 95% of human genes.
- Only 1.2% to 1.5% of the human genome is translated into protein; however, much more of the genome is made into RNA.
- The conversion of DNA information into protein is facilitated by aminoacyl tRNA synthetases and their ability to create amino acid–specific tRNAs.
- The genetic code is redundant; the 3-base code can have 64 different combinations, but only 20 amino acids are recognized.

REFERENCES

1. Griffith F. The significance of pneumococcal types. *J Hyg* 1928;**27**:113–59.
2. Avery OT, MacLeod CM, McCarty M. Studies on the chemical nature of the substance inducing transformation of pneumococcus types: induction of transformation by a desoxyribonucleic acid fraction isolated from pneumococcus type III. *J Exp Med* 1944;**79**:137–57.
3. Hershey AD, Chase M. Independent functions of viral protein and nucleic acid in growth of bacteriophage. *J Gen Physiol* 1952;**36**:39–56.
4. Chargaff E, Zamenhof S, Green C. Composition of human desoxypentose nucleic acid. *Nature* 1950;**165**:756–7.
5. Franklin RE, Gosling RG. Molecular structure of nucleic acids: molecular configuration in sodium thymonucleate, 1953. *Ann N Y Acad Sci* 1995;**758**:16–7.
6. Wilkins MH, Stokes AR, Wilson HR. Molecular structure of deoxypentose nucleic acids. *Nature* 1953;**171**:738–40.
7. Watson JD, Crick FH. Genetical implications of the structure of deoxyribonucleic acid. *Nature* 1953;**171**:964–7.
8. Watson JD, Crick FH. Molecular structure of nucleic acids: a structure for deoxyribose nucleic acid. *Nature* 1953;**171**:737–8.
9. Lightman A. *The discoveries: great breakthroughs in 20th-century science, including the original papers*. New York: Vintage; 2006.
10. Watson JD, Crick FH. The structure of DNA. *Cold Spring Harb Symp Quant Biol* 1953;**18**:123–31.
11. Meselson M, Stahl FW. The replication of DNA. *Cold Spring Harb Symp Quant Biol* 1958;**23**:9–12.

12. Kornberg A. Biologic synthesis of deoxyribonucleic acid. *Science* 1960;**131**:1503—8.
13. Nirenberg M, Leder P, Bernfield M, et al. RNA codewords and protein synthesis. VII. On the general nature of the RNA code. *Proc Natl Acad Sci USA* 1965;**53**:1161—8.
14. Baltimore D. RNA-dependent DNA polymerase in virions of RNA tumour viruses. *Nature* 1970;**226**:1209—11.
15. Temin HM, Mizutani S. RNA-dependent DNA polymerase in virions of Rous sarcoma virus. *Nature* 1970;**226**:1211—3.
16. Berget SM, Moore C, Sharp PA. Spliced segments at the 5' terminus of adenovirus 2 late mRNA. *Proc Natl Acad Sci USA* 1977;**74**:3171—5.
17. Chow LT, Roberts JM, Lewis JB, et al. A map of cytoplasmic RNA transcripts from lytic adenovirus type 2, determined by electron microscopy of RNA: DNA hybrids. *Cell* 1977;**11**:819—36.
18. Nathans D, Smith HO. Restriction endonucleases in the analysis and restructuring of DNA molecules. *Annu Rev Biochem* 1975;**44**:273—93.
19. Jackson DA, Symons RH, Berg P. Biochemical method for inserting new genetic information into DNA of Simian Virus 40: circular SV40 DNA molecules containing lambda phage genes and the galactose operon of *Escherichia coli*. *Proc Natl Acad Sci USA* 1972;**69**:2904—9.
20. Cohen SN, Chang AC, Boyer HW, et al. Construction of biologically functional bacterial plasmids in vitro. *Proc Natl Acad Sci USA* 1973;**70**:3240—4.
21. Southern EM. Detection of specific sequences among DNA fragments separated by gel electrophoresis. *J Mol Biol* 1975;**98**:503—17.
22. Maxam AM, Gilbert W. A new method for sequencing DNA. *Proc Natl Acad Sci USA* 1977;**74**:560—4.
23. Sanger F, Nicklen S, Coulson AR. DNA sequencing with chain-terminating inhibitors. *Proc Natl Acad Sci USA* 1977;**74**:5463—7.
24. Mullis K, Faloona F, Scharf S, et al. Specific enzymatic amplification of DNA in vitro: the polymerase chain reaction. *Cold Spring Harb Symp Quant Biol* 1986;**51**(Pt 1):263—73.
25. Schena M, Shalon D, Davis RW, et al. Quantitative monitoring of gene expression patterns with a complementary DNA microarray. *Science* 1995;**270**:467—70.
26. Sharma S. Non-B DNA secondary structures and their resolution by RecQ helicases. *J Nucleic Acids* 2011;**2011**:724215.
27. Ban N, Nissen P, Hansen J, et al. The complete atomic structure of the large ribosomal subunit at 2.4 A resolution. *Science* 2000;**289**:905—20.
28. Wilson DN, Doudna Cate JH. The structure and function of the eukaryotic ribosome. *Cold Spring Harb Perspect Biol* 2012;**4**. a011536.
29. Ibba M, Soll D. Aminoacyl-tRNAs: setting the limits of the genetic code. *Genes Dev* 2004;**18**:731—8.
30. Jacquier A. The complex eukaryotic transcriptome: unexpected pervasive transcription and novel small RNAs. *Nat Rev Genet* 2009;**10**:833—44.
31. Venter JC, Adams MD, Myers EW, et al. The sequence of the human genome. *Science* 2001;**291**:1304—51.
32. Levy S, Sutton G, Ng PC, et al. The diploid genome sequence of an individual human. *PLoS Biol* 2007;**5**:e254.
33. Gazave E, Fernando O, Arcadi N. The evolution of introns in human genes. In: *Encyclopedia of Life Sciences*. John Wiley & Sons; 2008.
34. Crick FH. On protein synthesis. *Symp Soc Exp Biol* 1958;**12**:138—63.
35. Crick F. Central dogma of molecular biology. *Nature* 1970;**227**:561—3.
36. O'Donnell M, Langston L, Stillman B. Principles and concepts of DNA replication in bacteria, archaea, and eukarya. *Cold Spring Harb Perspect Biol* 2013;**5**. a010108.
37. Helleday T, Eshtad S, Nik-Zainal S. Mechanisms underlying mutational signatures in human cancers. *Nat Rev Genet* 2014;**15**:585—98.
38. Iyama T, Wilson 3rd DM. DNA repair mechanisms in dividing and non-dividing cells. *DNA Repair (Amst)* 2013;**12**:620—36.
39. Moraes MC, Neto JB, Menck CF. DNA repair mechanisms protect our genome from carcinogenesis. *Front Biosci (Landmark Ed)* 2012;**17**:1362—88.
40. Cheadle JP, Sampson JR. MUTYH-associated polyposis: from defect in base excision repair to clinical genetic testing. *DNA Repair (Amst)* 2007;**6**:274—9.
41. Jalal S, Earley JN, Turchi JJ. DNA repair: from genome maintenance to biomarker and therapeutic target. *Clin Cancer Res* 2011;**17**:6973—84.
42. Nishino T, Morikawa K. Structure and function of nucleases in DNA repair: shape, grip and blade of the DNA scissors. *Oncogene* 2002;**21**:9022—32.
43. Kim H, Kim JS. A guide to genome engineering with programmable nucleases. *Nat Rev Genet* 2014;**15**:321—34.
44. Sharp PA. The discovery of split genes and RNA splicing. *Trends Biochem Sci* 2005;**30**:279—81.
45. Maston GA, Evans SK, Green MR. Transcriptional regulatory elements in the human genome. *Annu Rev Genom Hum Genet* 2006;**7**:29—59.
46. Bansal M, Kumar A, Yella VR. Role of DNA sequence based structural features of promoters in transcription initiation and gene expression. *Curr Opin Struct Biol* 2014;**25**:77—85.
47. Clamp M, Fry B, Kamal M, et al. Distinguishing protein-coding and noncoding genes in the human genome. *Proc Natl Acad Sci USA* 2007;**104**:19428—33.
48. Pennisi E. Why do humans have so few genes? *Science* 2005;**309**:80.
49. Ezkurdia I, Juan D, Rodriguez JM, et al. Multiple evidence strands suggest that there may be as few as 19,000 human protein-coding genes. *Hum Mol Genet* 2014;**23**:5866—78.
50. Beadle G, Tatum E. The genetic control of biochemical reactions in neurospora. *Proc Natl Acad Sci USA* 1941;**27**:499—506.
51. ENCODE Project C. A user's guide to the encyclopedia of DNA elements (ENCODE). *PLoS Biol* 2011;**9**:e1001046.
52. Rowland GC, Glass RE. Conservation of RNA polymerase. *Bioessays* 1990;**12**:343—6.
53. Levine M, Cattoglio C, Tjian R. Looping back to leap forward: transcription enters a new era. *Cell* 2014;**157**:13—25.
54. Smale ST, Kadonaga JT. The RNA polymerase II core promoter. *Annu Rev Biochem* 2003;**72**:449—79.
55. Grunberg S, Hahn S. Structural insights into transcription initiation by RNA polymerase II. *Trends Biochem Sci* 2013;**38**:603—11.
56. Marzluff WF, Wagner EJ, Duronio RJ. Metabolism and regulation of canonical histone mRNAs: life without a poly(A) tail. *Nat Rev Genet* 2008;**9**:843—54.
57. Wahl MC, Will CL, Luhrmann R. The spliceosome: design principles of a dynamic RNP machine. *Cell* 2009;**136**:701—18.

58. Pan Q, Shai O, Lee LJ, et al. Deep surveying of alternative splicing complexity in the human transcriptome by high-throughput sequencing. *Nat Genet* 2008;**40**:1413–5.
59. Kornblihtt AR, Schor IE, Allo M, et al. Alternative splicing: a pivotal step between eukaryotic transcription and translation. *Nat Rev Mol Cell Biol* 2013;**14**:153–65.
60. Scott JD, Pawson T. Cell signaling in space and time: where proteins come together and when they're apart. *Science* 2009;**326**:1220–4.
61. Anonymous A. gene-centric human proteome project: HUPO—the human proteome organization. *Mol Cell Proteomics* 2010;**9**:427–9.
62. Jackson RJ, Hellen CU, Pestova TV. The mechanism of eukaryotic translation initiation and principles of its regulation. *Nat Rev Mol Cell Biol* 2010;**11**:113–27.
63. Sonenberg N, Hinnebusch AG. Regulation of translation initiation in eukaryotes: mechanisms and biological targets. *Cell* 2009;**136**:731–45.
64. Le Quesne JP, Spriggs KA, Bushell M, et al. Dysregulation of protein synthesis and disease. *J Pathol* 2010;**220**:140–51.
65. Griffiths-Jones S. miRBase: microRNA sequences and annotation. *Curr Protoc Bioinformatics* 2010;**12**(9):1–10.
66. Ha M, Kim VN. Regulation of microRNA biogenesis. *Nat Rev Mol Cell Biol* 2014;**15**:509–24.
67. Balatti V, Pekarky Y, Croce CM. Role of microRNA in chronic lymphocytic leukemia onset and progression. *J Hematol Oncol* 2015;**8**:12.
68. Reinstein E, Ciechanover A. Narrative review. Protein degradation and human diseases: the ubiquitin connection. *Ann Intern Med* 2006;**145**:676–84.
69. Russo V, Martienssen R, Riggs A, editors. *Epigenetic mechanisms of gene regulation*. New York: Cold Spring Harbor Laboratory Press; 1996. p. 1. 2-3.
70. Dupont C, Armant DR, Brenner CA. Epigenetics: definition, mechanisms and clinical perspective. *Semin Reprod Med* 2009;**27**:351–7.
71. Bird A. Perceptions of epigenetics. *Nature* 2007;**447**:396–8.
72. Feil R. Epigenetics: an emerging discipline with broad implications. *C R Biol* 2008;**331**:837–43.
73. Relton CL, Davey Smith G. Two-step epigenetic Mendelian randomization: a strategy for establishing the causal role of epigenetic processes in pathways to disease. *Int J Epidemiol* 2012;**41**:161–76.
74. Papamichos-Chronakis M, Peterson CL. Chromatin and the genome integrity network. *Nat Rev Genet* 2013;**14**:62–75.
75. Schubeler D. Function and information content of DNA methylation. *Nature* 2015;**517**:321–6.
76. Yao B, Jin P. Unlocking epigenetic codes in neurogenesis. *Genes Dev* 2014;**28**:1253–71.
77. Zhang G, Pradhan S. Mammalian epigenetic mechanisms. *IUBMB Life* 2014;**66**:240–56.
78. Portela A, Esteller M. Epigenetic modifications and human disease. *Nat Biotechnol* 2010;**28**:1057–68.
79. Mahmoud SA, Poizat C. Epigenetics and chromatin remodeling in adult cardiomyopathy. *J Pathol* 2013;**231**:147–57.
80. Zentner GE, Henikoff S. Regulation of nucleosome dynamics by histone modifications. *Nat Struct Mol Biol* 2013;**20**:259–66.
81. Nair SS, Kumar R. Chromatin remodeling in cancer: a gateway to regulate gene transcription. *Mol Oncol* 2012;**6**:611–9.
82. Volle C, Dalal Y. Histone variants: the tricksters of the chromatin world. *Curr Opin Genet Dev* 2014;**25**:8–14. 138.
83. Felsenfeld G. A brief history of epigenetics. *Cold Spring Harb Perspect Biol* 2014;**6**. a018200.
84. Luger K, Dechassa ML, Tremethick DJ. New insights into nucleosome and chromatin structure: an ordered state or a disordered affair? *Nat Rev Mol Cell Biol* 2012;**13**:436–47.
85. Clapier CR, Cairns BR. The biology of chromatin remodeling complexes. *Annu Rev Biochem* 2009;**78**:273–304.
86. Petty E, Pillus L. Balancing chromatin remodeling and histone modifications in transcription. *Trends Genet* 2013;**29**:621–9.
87. Narlikar GJ, Sundaramoorthy R, Owen-Hughes T. Mechanisms and functions of ATP-dependent chromatin-remodeling enzymes. *Cell* 2013;**154**:490–503.
88. ENCODE Project C. An integrated encyclopedia of DNA elements in the human genome. *Nature* 2012;**489**:57–74.
89. Djebali S, Davis CA, Merkel A, et al. Landscape of transcription in human cells. *Nature* 2012;**489**:101–8.
90. Castel SE, Martienssen RA. RNA interference in the nucleus: roles for small RNAs in transcription, epigenetics and beyond. *Nat Rev Genet* 2013;**14**:100–12.
91. Ipsaro JJ, Joshua-Tor L. From guide to target: molecular insights into eukaryotic RNA-interference machinery. *Nat Struct Mol Biol* 2015;**22**:20–8.
92. Mallory A, Shkumatava A. LncRNAs in vertebrates: advances and challenges. *Biochimie* 2015;**117**:3–14.
93. Mercer TR, Dinger ME, Mattick JS. Long non-coding RNAs: insights into functions. *Nat Rev Genet* 2009;**10**:155–9.
94. Salmena L, Poliseno L, Tay Y, et al. A ceRNA hypothesis: the Rosetta Stone of a hidden RNA language? *Cell* 2011;**146**:353–8.
95. Seitz H. Redefining microRNA targets. *Curr Biol* 2009;**19**:870–3.
96. Tan JY, Marques AC. The miRNA-mediated cross-talk between transcripts provides a novel layer of posttranscriptional regulation. *Adv Genet* 2014;**85**:149–99.
97. International Human Genome Sequencing C. Finishing the euchromatic sequence of the human genome. *Nature* 2004;**431**:931–45.
98. Green ED, Guyer MS. National Human Genome Research Institute. Charting a course for genomic medicine from base pairs to bedside. *Nature* 2011;**470**:204–13.
99. Wheeler DA, Wang L. From human genome to cancer genome: the first decade. *Genome Res* 2013;**23**:1054–62.
100. Fleischmann RD, Adams MD, White O, et al. Whole-genome random sequencing and assembly of *Haemophilus influenzae* Rd. *Science* 1995;**269**:496–512.
101. Ley TJ, Mardis ER, Ding L, et al. DNA sequencing of a cytogenetically normal acute myeloid leukaemia genome. *Nature* 2008;**456**:66–72.
102. Kandoth C, McLellan MD, Vandin F, et al. Mutational landscape and significance across 12 major cancer types. *Nature* 2013;**502**:333–9.
103. MacConaill LE, Van Hummelen P, Meyerson M, et al. Clinical implementation of comprehensive strategies to characterize cancer genomes: opportunities and challenges. *Cancer Discov* 2011;**1**:297–311.
104. Gottesman O, Kuivaniemi H, Tromp G, et al. The electronic medical records and genomics (eMERGE) network: past, present, and future. *Genet Med* 2013;**15**:761–71.

2

Genomes and Variants

Carl T. Wittwer and Jason Y. Park

ABSTRACT

Background

One of the defining achievements of the early 21st century is the sequencing and alignment of more than 90% of the human genome. Of course, there is not a single human genome: individuals differ from each other by about 0.1% and from other primates by about 1%. Variation comes in many different forms, including single base changes and copy number changes in large segments of DNA. Even more challenging than sequencing the whole genome is documenting and understanding the clinical significance of human sequence variation. We are still very early in our understanding of the human genome.

Content

Beginning with a historical perspective, the structure of the human genome is described in detail followed by comparison to other interesting species. Then different types of genomic variation are covered, including single base changes (substitutions, deletions, insertions), copy number variations, translocations and fusions, short tandem repeats of different size and number, and larger repetitive segments, some of which can hop around the genome as transposons. The function of different genomic elements is considered along with many different classes of RNA transcribed from the DNA. How to name all the different genes, variants, and elements is a daunting task, and accepted nomenclature is presented. Many databases are available to mine accumulated genomic information. We end with a description of basic informatics tools that provide a pipeline from the raw data of massively parallel DNA sequencing to finished sequence with annotations on the variations that are observed.

INTRODUCTION

In 1966 it was recognized that the effort to characterize human cytogenetic variation was only the tip of the iceberg in terms of our understanding of genetic detail and that many more types of variation would be revealed with advancing technology. Since the time when DNA was discovered as the major molecule for genetic inheritance, there has been a need to understand how DNA variations affect growth, development, and disease. Even after 50 years of advances in DNA technology, many types of DNA variation have yet to be identified, named, cataloged, and studied.

HUMAN GENOME

The word *genome* signifies the collection of genes in an organism and is believed to have been coined by the German botanist Hans Winkler in the 1920s.[2] The human genome encompasses all of the information needed for growth, development, and heredity. This information is copied in the nucleus of every cell in the body.[3]

Throughout the 1990s, there was an international effort to sequence the human genome. The first draft was released in 2001[4,5] followed by a more complete version in 2004.[6] The 2004 version contains 2.85 billion nucleotides (bases) and is considered 99% complete for euchromatic (actively transcribed) DNA. The overall size of the genome, including both euchromatic and heterochromatic sequences (tightly compact DNA found at centromeres and telomeres), is estimated to be 3.08 billion nucleotides. Thus, the total overall genome is only 92.5% complete. Within the 2.85 billion nucleotides of euchromatic DNA are 19,438 known genes and an additional 2188 predicted genes. The total number of nucleotides encoding protein is approximately 34 million (1.2%) of the genome. This portion of the genome encoding proteins is also known as the *exome*.

The 2004 genome contains 341 gaps in heterochromatic regions.[6] These regions could not be sequenced because of the presence of DNA that is difficult to sequence (eg, repetitive elements, GC-rich sequence) by existing technology or because no clone or template could be made for sequencing. The 2004 genome provides the reference sequence for most subsequent sequencing projects. A reference is required because commonly used DNA sequencing technologies require a scaffold on which sequence fragments are pieced together.[7] The first human reference sequences were assembled by the University of California at Santa Cruz (UCSC) and were numbered starting with "hg1" in May 2000. The National Center for Biotechnology Information (NCBI) produced its own genome builds starting in December 2001 as NCBI build 28 (equivalent to hg10 from UCSC) as the genome was further refined. This led to the publicly available 2004 version of the human genome becoming known as NCBI35/hg17. This template or reference sequence has subsequently undergone continuous improvement under the international Genome Reference Consortium (GRC),[7] producing GRCh37/hg19. In the future, only one designation will be given, such as the currently released GRCh38.[3]

Ten years after the 2004 genome publication, efforts are underway to create "platinum genomes" that address the missing information (gaps) and improve the quality of data.[8] One group has sequenced several previous gaps by using DNA from a haploid cell line.[9] Another group has combined several long-read sequencing technologies to create a de novo assembly that does not require the use of a reference genome.[10] This approach dramatically improves the mapping of sequences generated and can be combined with data from short-read sequencing instruments.[10] Thus, hybrid sequencing methods are emerging that combine the advantages of short-read sequencing for single nucleotide base accuracy with the advantages of long-read sequencing for de novo assembly without a reference genome.[10] Ongoing improvements in sequencing technology are predicted to further decrease the gaps in human genome data and reveal new mechanisms of human variation. Genomic terms and definitions used in this chapter are given in Box 2.1.

Each human cell contains two copies of the 3.08-billion-nucleotide genome divided into 46 chromosomes. Table 2.1 summarizes statistics for the human genome and the types of variations that are important in clinical diagnostics. Three quarters of human DNA is intergenic or between genes. More than 60% of this intergenic sequence consists of "parasitic" DNA regions of mostly defective transposable elements 100 to 11,000 bases in length. Between 2 and 3 million of these "retrotransposons" are present in each copy of the genome. They contribute to genetic recombination and chromosome structure and provide an evolutionary record of sequence variation and selection.

Segmental duplications constitute 5.3% of the human genome. They are more than 1 kilobase (1000 bases or Kb) in length, have a sequence identity of at least 90%, and are not transposable. Segmental duplications are common in the human genome and are prone to deletion or rearrangement (or both), often with medical consequences. Intergenic DNA also carries most of the simple sequence repeats (SSRs) present in the genome. A subset of SSRs, the short tandem repeats (STRs) have repeat units of 1 to several bases that may be repeated up to thousands of times. STRs have played a large role in genetic linkage studies and in forensic and medical identity testing. They are formed by slippage during replication and are highly polymorphic among individuals. The most common STRs are dinucleotide repeats, such as ACACAC and ATAT. On average, one STR occurs every 2000 bases.

Approximately 2% of DNA is required to maintain the structure of chromosomes and is located at chromosome centers (centromeres) and ends (telomeres), making up heterochromatic DNA. Centromeric DNA includes many tandem copies of nearly identical 171 base pair (bp) repeats encompassing 0.24 to 5.0 Mb per chromosome. Each chromosome end is capped with several Kb of the telomeric 6 base repeat TTAGGG. Although intergenic DNA does not code for protein and was originally considered "junk," much of this DNA is transcribed to RNA, producing a complex "transcriptome" network of RNA control elements whose function and mechanics are active areas of investigation.[11]

One quarter of the human genome consists of genes. There are 19,438 known genes in addition to 2188 predicted genes in the human genome. The average gene covers 27,000 bases, but only about 1300 of these bases code for amino acids. The primary RNA transcript is processed by splicing to retain exons that are interspersed throughout the gene and have a higher GC content than noncoding regions. On average, 95% of a gene is excised as introns, retaining a mean of 10.4 exons, of which on average 9.1 are translated into proteins. Exons make up only 1.9% of the total genome, with 1.2% of the genome coding for proteins. Some important genes are present in many copies, so that overall protein expression is not affected if a chance variation occurs in one copy. If extra copies of genes lose their function, they are known as pseudogenes. At least as many pseudogenes as functional genes are present in the human genome. It is important to distinguish pseudogenes from functional genes because variants in pseudogenes are seldom of clinical importance, and they often complicate DNA diagnostic assays.

POINTS TO REMEMBER

Human Genome
- Contains approximately 3 billion bases
- Protein coding nucleotides are about 1% (30 million bases)
- Noncoding sequence has important regulatory roles

VARIATIONS IN SPECIFIC POPULATIONS

Large-scale human genome sequencing projects have cast a wide net across many diverse populations. These projects have provided a wealth of knowledge of the genetic diversity that exists in humans. An alternative approach to human genetic diversity is to examine more homogenous populations. Several studies have examined the genetics of a large number of individuals from Iceland. A recent whole-genome sequencing study of 2636 Icelanders observed 20 million single nucleotide variations (SNVs) and 1.5 million insertions/deletions (indels).[12] The data from this whole genome sequencing study were combined with a previous data set of 104,220 Icelanders who had been SNV typed at 676,913 locations. By applying whole-genome sequencing data from only a small subset of individuals, the full genetics could be inferred for a larger set of more than 100,000 individuals who had only had SNV typing.

Another interesting result of the Icelandic whole-genome study was the identification of 6795 loss of function single nucleotide variants, insertions, or deletions in 4924 genes.[13] Loss of function changes (homozygous or compound heterozygous) were found in 7.7% of the individuals sequenced. In essence, this study identified a surprisingly high percentage of individuals with "knocked-out" or functionally silenced genes.

NONHUMAN GENOMES

Before the human genome was completed, other genomes of smaller size were sequenced, enabling advancements in technology and logistical organization to sequence the human genome.[14,15] Different genomes are varied in size, and the complexity can be surprising. One of the largest known

BOX 2.1 Genomic Terms and Definitions

Annotation: Biologic information attached to genomic sequence.
Annotation Track: Optional metadata in a genome browser that allows viewing of genes, exons, SNVs, repeats, etc.
Assembly: Reconstruction of short sequence reads on a scaffold of reference DNA.
Binary alignment nap (BAM): After alignment to a reference genome, the aligned data for each read produces a sequence alignment map (SAM file). The BAM file is the binary equivalent of the SAM file, and allows for efficient random access of the data.
Browser extensible data (BED): A tab delimited text file that defines the data lines in an annotation track, including the chromosome name, the starting and the ending positions.
Contig: A linear stretch of consensus sequence assembled from smaller overlapping sequence fragments.
Copy number polymorphism (CNP): A copy number variant present at more than 1% in a population.
Copy number variant (CNV): A structural variant of a large region of the genome that has been deleted or duplicated.
Deletion: A DNA sequence that is missing in one sample compared to another. Deletions may be as small as one nucleotide or as large as an entire chromosome.
De novo assembly: Formation of a contig without using a reference sequence.
FASTA File: A nucleotide sequence text file.
FASTQ file: A text output file of sequencing reads in a run, along with the quality scores of each position.
Fusion: A translocation, inversion, large deletion or large duplication resulting in a hybrid gene formed from originally separate genes.
Indel: Originally referred to a unique class of sequence variants that included both an insertion and a deletion resulting in an overall change in the number of base pairs. Today more commonly refers to either insertions or deletions or a combination thereof.
Insertion: An extra DNA sequence that is present in one sample compared with a reference sequence.
Heteroplasmy: A mixture of more than one type of mitochondrial sequence in one cell.
Intergenic: DNA sequence between genes.
Missense: A nucleotide substitution that changes a codon to the code for a different amino acid. Although these sequence changes are commonly referred to as missense "mutations," this is strictly a misnomer because missense variants may be benign and cause no disease.
Mutation: A disease-causing sequence variation. Historically, the term has been interchangeable with variant to describe any change in DNA sequence regardless of relation to disease causation. For current clinical descriptions or reporting, the use of mutation is reserved for the scenario when disease causation is known.
Nonsense: A nucleotide substitution that results in a stop codon, prematurely terminating the protein.
Nonsynonymous: Nucleotide substitutions that are predicted to change the amino acid coding. These substitutions include both missense and nonsense substitutions.
Oligonucleotide: A short single-stranded polymer of nucleic acid.
Phred score: Estimate of the error probability for a base called in DNA sequencing. It is represented as a Q-score; the higher the number, the higher the probability of a correct call.
Plasmid: An extrachromosomal ring of double-stranded, closed DNA found in bacteria.
Pseudogene: A genetic element that does not code for a functional gene product, usually because of accumulated sequence variations.
Sequence alignment map (SAM file): A file generated by alignment of sequence data to a reference genome. This file type is often converted to a BAM file to save space.
Short tandem repeat (STR): A simple sequence repeat that is 1–13 bases long.
Simple sequence repeat (SSR): A sequence from 1-500 bases that is repeated end to end. If the repeat unit is 1-13 bases, it is a microsatellite or STR. If the repeat is 14-500 bases it is a minisatellite.
Single nucleotide polymorphism (SNP): A benign single nucleotide variant (substitution, deletion, or insertion) that occurs in a population at a frequency of at least 1%.
Single nucleotide variation (SNV): A single nucleotide variant (substitution, deletion, or insertion). SNVs may be benign or may cause disease.
Structural variation: A region of DNA greater than 1000 bases in size that is inverted, translocated, inserted, or deleted.
Synonymous variant: A nucleotide change that results in no change to the predicted amino acid sequence. Although synonymous variants are typically considered to be benign since there is no protein coding change, there is the possibility of pathogenicity by changes in splicing, gene expression or mRNA stability.
Transposon: A mobile genetic element that can delete and insert itself variably into the genome.
Variant call format (VCF): After aligning all reads onto a reference sequence, variants that are different from the reference genome at a given nucleotide position are stored in a text file in a specific format.
Variation: A change in DNA sequence. It may be benign or may cause disease.

genomes is the white spruce tree *(Picea glauca)* at 26.9 billion bases. On the opposite end of the spectrum is porcine circovirus-1, a single-stranded DNA virus with a genome that is less than 2000 bases. There is overlap in genome size among eukaryotes (animals, plants, fungi), viruses, and bacteria (Table 2.2 and Fig. 2.1).

Primates

Comparison of the chimpanzee genome with the human genome shows a genome-wide difference of only 1.23%.[16] This approximate 1% difference translates to 35 million nucleotides and 5 million insertion/deletion differences. There are also differences at the level of proteins between humans

TABLE 2.1 The Human Genome and Its Sequence Variation

The Human Genome
3.08 billion base pairs in 24 chromosomes
23 chromosome pairs (46–244 million base pairs per chromosome)

75% Intergenic Sequences

Transposable elements	45%
Segmental duplications	5%
Simple sequence repeat	3%
Structural (centromeres, telomeres)	2%
Other	20%

25% Genes That Code for Proteins

Introns	23%
Exons	1.9%
• Coding segments	1.2%
• Untranslated regions	0.7%
Number of genes	19,438 known
	2,188 predicted
Average gene	27,000 base pairs
	10.4 exons
	9.1 transcribed exons
	1340 exonic bases
	446 amino acids

Sequence Variants
99.9% identity (one difference every 1250 bases between randomly selected haploid genomes)

Single-Nucleotide Variants (SNVs): Identified Every 75 Bases on Average

Noncoding	97%
Average number within a gene	126
Average number within the coding region of a gene	5

Copy Number Variants (CNVs): Involves 5%–12% of the Genome

Disease-Causing Variants

SNVs	68%
• Missense (amino acid substitution)	45%
• Nonsense (termination)	11%
• Splicing	10%
• Regulatory	2%
Small insertions or deletions (or both)	24%
Structural variants (copy number variations, inversions, translocations, rearrangements, repeats)	8%

Epigenetic Alterations
Variable initiation and alternative splicing
Cytosine methylation
Histone phosphorylation, methylation, acetylation

Data from Lander et al,[4] Venter et al,[5] and International Human Genome Sequencing Consortium.[6]

TABLE 2.2 *Homo Sapiens* in Comparison to Other Genomes

Organism/Name	Group	Size (Mb)
Human (*Homo sapiens*)	Animals	3080
White spruce tree (*Picea glauca*)	Plants	26,900
Migratory locust (*Locusta migratoria*)	Animals	5760
Mouse (*Mus musculus*)	Animals	~2500
Rat (*Rattus norvegicus*)	Animals	~2750
Apple tree (*Malus domestica*)	Plants	742
Roundworm (*Caenorhabditis elegans*)	Animals	97
Aspergillus fumigatus	Fungi	~30
Baker's yeast (*Saccharomyces cerevisiae*)	Fungi	12.3
Haemophilus influenzae	Bacteria	1.8
Human immunodeficiency virus (HIV) 1	Viruses	0.0092
Porcine circovirus-1	Viruses	0.00173

Data from the National Center for Biotechnology Information. http://www.ncbi.nlm.nih.gov/genome.

and chimpanzees: Only 29% of proteins are identical at the amino acid level, and proteins that are different only differ by an average of two amino acids.[16]

Two orangutan species have been sequenced.[17] Their genome sizes are similar to humans at approximately 3 billion bases. During evolutionary development, the number of structural rearrangements in orangutans has been less than the human and chimpanzee branches.[17] For example, the number of genome rearrangements greater than 100 Kb was 38 in the orangutan but 85 and 54 in the chimpanzee and human, respectively.

An example of non–protein-coding variation between primates is the number and types of DNA insertions. A comparison of five primate genomes (chimpanzee, gorilla, orangutan, gibbon, and macaque) identified regions of human DNA that were absent in nonhuman primates.[18] More than 200,000 human-specific DNA insertions were identified; the majority of these were less than 10 nucleotides in length and were eliminated from further study. There were 5582 genes identified that contained larger insertions; 2450 of these genes were expressed in brain tissue. Many of the human-specific insertions were transposable elements and long terminal repeats.[18]

Rodents

The mouse genome is 14% smaller than the human genome (2.5 billion bases compared with the human size of approximately 3 billion bases).[19] In comparison, the rat (*Rattus norvegicus*) genome is in between the size of the human and mouse (2.75 billion bases).[20] The number of genes is similar between the mouse and the human. About 40% of the rat, mouse, and human genomes are all in alignment. Another 30% of the rat and mouse genomes match each other but not the human genome.

Fungi

Fungi are eukaryotes and their genomes are less complex than the human genome. Common fungi that cause human disease have genome sizes of 7.5 to 30 million bases and 8 to 16 chromosomes, as well as mitochondrial genomes. Some

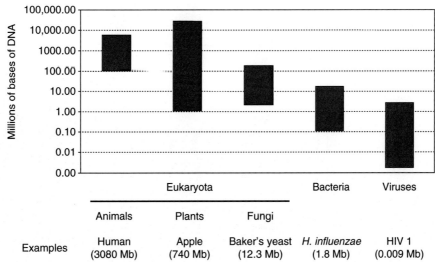

FIGURE 2.1 Range of genome sizes. Among different organisms, there is wide variation in genome size. In this plot of publicly available genomes, the *y*-axis is in megabases, and the *x*-axis lists various organisms: Eukaryota (animals, plants, fungi), bacteria, and viruses. On average, Eukaryota have larger genomes compared with bacteria and viruses; however, there are exceptions in which virus genomes are larger than bacteria or Eukaryota. The difference between the smallest and largest known genomes is more than 6 orders of magnitude. Several specific genome sizes are illustrated in Mb (megabases, million). (Data extracted from the National Center for Biotechnology Information. http://www.ncbi.nlm.nih.gov/genome.)

fungi have diploid genomes, and others have haploid genomes. Many of their genes have introns. For instance, *Aspergillus fumigatus* (a fungus that causes allergic reactions and systemic disease with a high mortality rate) has a haploid genome of about 30 million bases with more than 9900 predicted genes on eight chromosomes. Its genes are smaller than human genes, with an average length of 1400 bp and 2.8 exons per gene.

The first eukaryotic genome sequenced was *Saccharomyces cerevisiae* (baker's yeast).[15] This fungal genome has 12 million bases arranged into 16 chromosomes. In addition to the importance of yeast in baking breads and brewing alcohol, yeast is an important model organism and pathogen. With the identification of the approximately 6000 genes within the *S. cerevisiae* genome, systematic alteration of each gene or combination of genes can now be explored to examine the role of genes in yeast as well as higher organisms.

Bacteria

Bacterial genomes are considerably less complex than human or fungal genomes. Common bacteria have only one chromosome, usually a circular DNA double helix of 4 to 5 million base pairs, about 1000 times less than the amount of DNA in a human cell. About 90% of the DNA in bacteria codes for protein. There are no introns, but there are multiple small intergenic regions of repetitive sequences that are dispersed throughout the genome. *Escherichia coli*, a common bacterium in the human intestinal tract, has about 4300 genes.

In addition to the large circular chromosome that carries essential genes, bacteria also carry accessory genes in smaller circles of double-stranded DNA (dsDNA) known as plasmids. Plasmids range in size from 1000 to more than 1 million base pairs. Plasmids are important in molecular diagnosis of bacterial infections because they often encode pathogenic factors and antibiotic resistance.

The bacterial repertoire of DNA can be altered by (1) gain or loss of plasmids; (2) single-base changes, small insertions and deletions as in eukaryotic genomes; and (3) large segmental rearrangements, including inversions, deletions, and duplications. Some genes, such as those for ribosomal RNA, are present in many copies, making them good targets for molecular assays to identify the species of bacteria. In addition, the intergenic repetitive sequences serve as multiple targets for oligonucleotide probes, enabling the generation of unique DNA profiles or fingerprints for individual bacterial strains.

The first genome sequenced by random fragmentation and computational assembly was the pathogenic bacteria, *Haemophilus influenzae*.[14] The genomic DNA was fragmented into 19,687 templates inserted into plasmids and bacteriophages. A total of 24,304 sequences were successfully generated over 3 months. The sequencing data required 30 hours of computational time to be assembled. A total of 11 million bases of DNA were sequenced and used to generate the 1.8 million bases of the *H. influenzae* genome. In addition to being the first genome solved by shotgun sequencing, it was also the first bacterial genome sequenced. Multiple strains of *H. influenzae* have been subsequently sequenced. These additional genomes have revealed heterogeneity in the number of genes between different strains. Of the approximately 3000 genes identified, only 1461 are common to all strains.[21] The differences in genes among different strains may be associated with differences in the infectious pathogenicity of *H. influenzae*.[21] The success in sequencing the first *H. influenza* genome highlighted the importance and possible use of shotgun sequencing. Shotgun sequencing became an important technology for the successful completion of the human genome project. The project had begun otherwise, with the orderly "conventional" sequencing of large (150-Kb) fragments of DNA that were divided among members of the consortium and methodically sequenced.

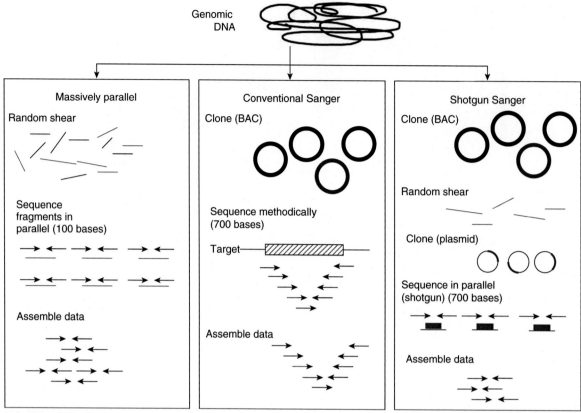

FIGURE 2.2 Genome sequencing approaches. Massively parallel sequencing *(left panel)* is the current technology used for genome sequencing, but it evolved out of earlier conventional Sanger *(middle panel)* and shotgun Sanger *(right panel)* sequencing methods. See text for details. *BAC,* Bacterial artificial chromosomes.

Fig. 2.2 contrasts different sequencing approaches. Massively parallel sequencing is the current technology used for genome sequencing, but it developed out of earlier sequencing methods. When genomic DNA is sequenced by massively parallel sequencing, the basic steps include random DNA shearing (fragmentation), sequencing in parallel reactions, and data assembly *(left panel)*. The randomly sheared fragments are end modified with oligonucleotides that aid in the identification, immobilization, and sequencing of the fragments; this step is referred to as library creation. In the case of whole-genome sequencing, this "library" of modified fragments is then sequenced. However, if only a subset of genes is of interest or if only the coding nucleotides are of interest (exome), the specific targets can be hybridized and "captured" after the library step. Targeted capture of regions of interest is a key step in exome sequencing. More than 1 million sequencing reactions occur in parallel, generating more than 100 bases of data per reaction. After sequencing, the short reads of DNA are assembled based on a reference genome (eg, GRCh38).

In comparison, the conventional sequencing of genomes was the technology used for the initiation of the Human Genome Project *(middle panel* of Fig. 2.2). This method started with the genome cloned into large molecules such as Yeast Artificial Chromosomes (YAC) and later Bacterial Artificial Chromosomes (BAC). These larger molecules, which carried genomic inserts greater than 150 Kb in size, were then divided among the members of the genome sequencing consortium and methodically sequenced in 700 base reactions. Each round of sequencing depends on the sequencing data from the prior round. The assembly of data is not as computational intensive as massively parallel or shotgun sequencing.

Finally, shotgun sequencing was key to the speedy completion of the Human Genome Project *(right panel* of Fig. 2.2). Rather than methodically sequence targets of interest, the method relied on random shearing of DNA and subcloning the fragments into plasmids. The plasmids were then sequenced in parallel (separate) reactions. The evolutionary roots of massively parallel sequencing technology can be followed back to shotgun sequencing.

Viruses

Viral genomes are considerably less complex than bacterial genomes. Common viruses that infect humans vary in size from about 5000 to 250,000 bases, or 20 to 1000 times less than the amount of nucleic acid in *E. coli*. Because viruses use the host's cellular machinery, they do not need as many genes as bacteria. Small viruses may encode only several genes, but the larger viruses can encode hundreds. The viral genome consists of either DNA or RNA, and the nucleic acid may be single stranded or double stranded, linear or circular, with one or multiple fragments or copies per viral particle. As in bacteria, there are no introns. In fact, in some viruses the exons overlap with different reading frames coding different products from the same nucleic acid sequence. Noncoding regions are usually present at the terminal ends of linear genomes. Repeat segments are often found as terminal or internal repeats and may be inverted.

Sequence alterations in viruses are common. Areas of high sequence variation may be interspersed between conserved domains. Higher frequencies of variation correlate with lower polymerase fidelity and may allow escape from antibody recognition and antiviral drugs. Common sequence variants in viruses include single base changes, insertions, and deletions. Sequence diversity within a viral species may be so great that consensus sequences for molecular typing are difficult to find.

DNA THAT CODES FOR RNA BUT NOT PROTEIN

Even though 99% of the genome does not code for protein, most of it is transcribed into noncoding RNA. At least 93% of the genome is transcribed,[11] producing more than 10 times the amount of RNA than is produced from the coding segments of genes.[22] Both strands of DNA may be transcribed, and long noncoding transcripts may overlap coding regions, producing a complex transcriptome of functional RNA molecules that may variably regulate transcription of coding regions, RNA processing, mRNA stability, translation, protein stability, and secretion. In addition to long noncoding RNA, ribosomal RNA, and transfer RNA, specific classes of noncoding RNAs include small nuclear RNAs critical for splicing, small nucleolar RNAs that modify rRNA, telomerase RNA for maintenance of telomeres, small interfering RNAs, and microRNAs (miRNAs) that regulate gene expression.[23-25] In a recent review on RNA, 54 different categories were identified.[26] Some of the more important types of RNA are listed in Table 2.3.

MicroRNAs (or miRNAs) are particularly interesting as potential markers for disease. For example, circulating miRNAs have been correlated to many different types of cancer.[27] MicroRNAs are noncoding but functional single-stranded RNAs that are 21 to 22 bases long and are expressed in a tissue-specific manner. They are initially transcribed as longer precursors that undergo two rounds of truncations as they are transported from the nucleus to the cytoplasm in the cell. The mature miRNA is then integrated into a protein complex called the *RNA-induced silencing complex*, which regulates translation of mRNA. MicroRNAs hybridize to a 6 to 8 base sequence in the 3′ untranslated region of target mRNAs and inhibit mRNA expression either by mRNA degradation if the bases are perfectly complementary or by blocking of translation if they are imperfectly complementary. Currently for humans, there are 1881 precursor miRNAs and 2588 mature miRNAs cataloged in miRBase.[28] Despite the promise of miRNAs as tumor markers, the literature is often contradictory and inconsistent with few accepted conclusions.[29]

VARIATION IN THE HUMAN GENOME

If the DNA of any two individuals is compared, on average, one difference is noted every 1250 bases (ie, approximately 99.9% of the sequence is identical between randomly chosen copies of the human genome). However, copy-number variants involve a greater amount of the genome, with 0.5% of the genome differing on average between two individuals when copy-number variants greater than 50 Kb are considered.[30] That is, between individuals, at least five times as many bases are affected by copy number changes as by small sequence differences.

Most human genetic material is present in two copies, with the exception of the unpaired sex chromosome in males and mitochondrial DNA. The presence of only single gene copies on the X and Y chromosome in males leads to well-known sex-linked disorders. In contrast, the 16,500-bp mitochondrial genome is present in multiple copies per cell, constituting about 0.3% of human DNA, depending on the tissue source. Allele fractions may vary over a wide range when all mitochondria in a cell are considered. That is, sequence variations in mitochondrial DNA are heteroplasmic, meaning that the ratio of the wild-type allele to a variant allele can vary almost continuously, sometimes resulting in a wide range of symptoms even when only one sequence variant is involved.

Any sequence change from a reference sequence is called a sequence variant or variation. Many variations do not affect human health and are benign or silent. For example, most (1) copy-number variations, (2) SNVs, and (3) STRs found between genes are seldom associated with disease.

Single Nucleotide Variants

The most common sequence variations are single base changes, also known as SNVs. More than 40 million SNVs have been described, and many new SNVs continue to be reported. Some SNVs are common in the population, with allele frequencies of 0.1 to 0.5 (ie, present in 10–50 of every 100 copies studied), but other single base changes are very rare. The vast majority of SNVs (97%) occur in noncoding

TABLE 2.3 Some Common, Interesting, and Important Types of RNA

Abbreviation	Description
mRNA	Messenger RNA is translated to protein by the ribosome.
rRNA	Ribosomal RNA is a major component of ribosomes.
tRNA	Transfer RNA pairs an amino acid with its anticodon in protein synthesis.
ncRNA	Noncoding RNA is not translated to protein.
lncRNA	Long noncoding RNA is greater than 200 bases and is not translated to protein.
hnRNA	Heterogeneous nuclear RNA is the initial RNA transcript that includes introns.
Ribozyme	RNA that has catalytic activity.
Riboswitch	RNA that switches between 2 conformations under certain conditions (ligand exposure).
Telomerase RNA	Structural part of telomerase that also provides a hexamer template.
Xist RNA	X-inactive–specific transcript RNA inactivates one X chromosome in females.
snRNA	Small nuclear RNA is found in the eukaryotic nucleus.
snoRNA	Small nucleolar RNA are intron fragments essential for pre-rRNA processing.
siRNA	Small interfering RNA can cleave perfectly complementary target RNA.
gRNA	Guide RNA pairs with a RNA target and guides proteins for cleavage and so on.
miRNA	MicroRNA affects target mRNA regulation or decay.

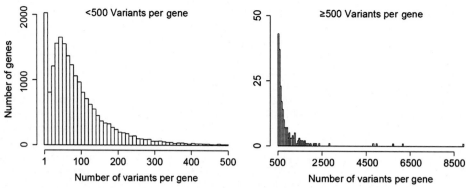

FIGURE 2.3 High number of variants per gene. In an examination of allelic heterogeneity, the Leiden Open Variation Database (LOVD) was queried for the number of unique alleles (variants) per gene. In the *left panel* are the genes with fewer than 500 variants per gene, and in the *right panel* are the genes with 500 or more variants per gene. The panels clearly demonstrate that there are currently thousands of genes with hundreds of variants and dozens of genes with thousands of variants. This high number of variants per gene is important in considering that the majority of variants are of unknown clinical significance. (From Cutting GR. Annotating DNA variants is the next major goal for human genetics. *Am J Hum Genet* 2014;94:5-10.)

regions; only 3% of SNVs are associated with exons. Similarly, most of the SNVs within introns, except for splicing and regulatory variants, are not known to affect gene function. In addition, some of the SNVs within exons are silent alterations that do not code for a change in amino acid sequence because of the redundancy in the genetic code. Still other SNVs in exons code for amino acid changes that do not affect protein function. However, some silent SNVs may affect DNA splicing, and others are of interest as genetic markers.

The international 1000 Genomes Project has a goal of sequencing approximately 2500 individuals representing populations from Europe, East Asia, South Asia, West Africa, and the Americas.[31] A report from this project on 1092 individuals from 14 populations mapped 38 million single nucleotide variants, 1.4 million short insertions and/or deletions, and more than 14,000 large deletions.[32] Individuals were found to have on average 3.6 to 3.9 million variants, of which 2300 to 2700 were nonsynonymous. Variants classified as disease causing were more than 10 per individual.

Another examination of SNVs revealed that there can be thousands of variants in each gene. These variants may or may not cause disease. In the current age of genomics, the lack of understanding of disease causation based on variant identification is referred to as an "interpretive gap."[33] In the Leiden Open Variation Database (LOVD), there are currently hundreds to thousands of variants identified for each gene (Fig. 2.3). However, the disease classification of these variants (eg, benign or pathogenic) lags far behind the ability to discover these variants.[33] Sequence alterations that are known to cause disease are often called *mutations, pathogenic variants,* or *disease-causing variants.* About 68% of known disease-causing variants involve only a single base change. Most of the remaining disease-causing variants (24%) are small insertions or deletions. The remainder (8%) includes more complex structural variations (see Table 2.1).

Most SNVs that cause disease are missense and result in an amino acid substitution; significantly fewer are nonsense variants that result in a termination codon and premature polypeptide chain termination. Approximately 10% of disease-causing variants are SNVs that affect splicing sites and result in altered concatenation of coding sequences. Finally, fewer than 2% of known disease-causing variants are SNVs that affect the regulatory efficiency of transcription by altering promoter and/or enhancer regions in introns or the stability of the RNA transcript.

Small insertion and/or deletion variants account for 24% of variants that cause disease. An insertion refers to the presence of extra bases, and deletion implies the absence of certain bases in comparison with a reference sequence. Insertions and deletions often cause a shift of the codon reading frame, resulting in altered amino acid sequence downstream of the variation—commonly followed by chain termination from a nonsense codon.

The remaining 8% of variants that relate to health and disease are mostly structural variants, including (1) duplications or deletions of entire exons or genes; (2) gene fusions, including chromosomal translocations and inversions; (3) STR expansions (eg, an increased number of trinucleotide repeats); (4) gene rearrangements (eg, rearrangements of immunoglobulin genes in B cells that are required for production of antibodies); (5) complex polymorphic loci related to health and disease (eg, human leukocyte antigens); and (6) copy number variants (CNVs).

Copy Number Variation (Gains and Losses)

Although SNVs are the most common sequence variant, CNVs cover more of the genome than SNVs. Examples of large gains or losses in genomic DNA have been known for many years with, for example, syndromic diseases. However, an examination of phenotypically normal individuals by array-based comparative genomic hybridization revealed an average of 12.4 large copy number variations per individual.[34] Some variants reached 2 million bases of DNA in phenotypically normal individuals. CNVs may be duplicated in tandem or may involve complex gains or losses of homologous sequences at multiple sites in the genome. CNV regions exist in every chromosome and involve 5% to 12% of the human genome.[30,35]

High-resolution comparative genomic hybridization has now revealed the presence of deletions across hundreds of additional individuals.[36-39] In total, these studies revealed more than 1000 unique deletions. Some deletions are in regions without known genes; however, hundreds of known or predicted genes exist at the site of the observed deletions. Interest in CNVs in relation to disease has increased recently as the extent of variation has become clear.[40]

CNVs can involve genes or contiguous sets of genes. When the normal dosage of the gene is two but more than two functional copies of a gene are present, then the gene is "amplified." If a dosage-sensitive gene, such as HER2 (ERBB2) is amplified, it usually leads to overexpression of mRNA and protein, resulting in cellular abnormalities and possible progression to diseases such as cancer. When the normal gene dosage is two and loss of one of the functional copies of the gene occurs, disorders such as mental retardation and developmental delay may result. Structural variants can be determined by cytogenetic techniques, including karyotyping, fluorescent in situ hybridization, comparative genomic hybridization, and virtual karyotyping by SNV microarrays.

Fusions

Gene fusions arise by deletions, duplications, inversions, and translocations and are commonly found in cancer.[41] Often they arise by balanced translocations, whereby a chimeric protein is created by the fusion of two coding regions. Gene fusions promote tumor proliferation by either activating an oncogenic driver or inactivating a tumor suppressor. Although translocations are rare outside of cancer, massively parallel sequencing now allows insight into the myriad of translocations that occur in both hematologic and solid tumors. By identifying gene fusions that act as primary oncogenic drivers, the hope is that targeted therapies may be available for precision treatment.

Short Tandem Repeats

Short tandem repeats are DNA motifs that are defined by 1 to several bases that are repeated many times in tandem. STRs have been implicated in more than 40 genetic diseases.[42] In the case of fragile X, an expansion of a CGG repeat results in disruption of protein expression of the FMR1 gene. For Huntington disease, an expansion of a CAG repeat results in abnormal protein expression of the HTT gene. Many massively parallel sequencing platforms use short reads of information that are less than 200 bases in length, and these short reads have made the analysis of repetitive sequences difficult. Thus, the current contribution of repetitive DNA to human variation and disease is probably underestimated. There is speculation that sequencing technology with longer reads of information in the thousands of bases will reliably detect repetitive DNA elements. In addition to advances in sequencing technology, there are promising informatics tools for characterizing repetitive DNA elements in standard massively parallel sequencing data sets.

One group proposes a "thesaurus" approach in which an extensive catalog of repetitive DNA elements is used within the existing analysis framework.[43] The catalog contains almost 3 billion entries that are representative of the variety of repetitive elements seen in a human genome. This approach was successful in detecting novel variants without extensive changes to typical analysis approaches. Massively parallel sequencing technology and informatics are limited in the amount of STR data that is sequenced and analyzed. A recent informatics tool, lobSTR, can accurately genotype STRs from massively parallel sequencing datasets.[44] When lobSTR was applied across whole-genome datasets from the 1000 Genomes Project, 700,000 STR loci were catalogued, and 350,000 STR loci were found per individual.[42] Some STR loci were common with 300,000 having a mean allele frequency of more than 1%, and 2237 were located within 20 bases of an exon–intron junction.[42] The high frequency of STR loci and their proximity to coding DNA suggest a larger role for STR variants in influencing growth, development, and disease.

Transposable Elements and Their Genetic Fossils

Transposable elements are composed of repetitive DNA that could originally facilitate homologous recombination or create deletions, duplications, inversions, and translocations.[45] Most of these elements are no longer active and are categorized as retrotransposons, including long terminal repeats (LTRs), long interspersed nuclear elements (LINEs), and short interspersed nuclear elements (SINEs). In a conservative estimate not including repeat-rich regions such as centromeres, these elements were found to comprise 30% to 50% of the total DNA in mammals. In comparison, the genomes of birds were less than 10% derived from transposable elements.[45]

In one human study, repetitive DNA including transposable elements and their nonfunctional descendants were estimated to consume 66% to 69% of the human genome.[46] In humans, active transposable elements include a subset of L1 LINEs and Alu (a type of SINE).[47] These active elements have de novo germline insertions ranging from 1 in 20 to 1 in 916 births.[47] The insertion of these elements has multiple possible effects on the transcriptional regulation of genes, including disruption of the open reading frame, creation of a novel promoter, alternative splicing, an alternative poly(A) tail, disruption of transcription factor bindings sites, and changes in small RNA regulation.[47] Common repeat sequences in the human genome are cataloged in Table 2.4. Their distribution in humans and other species is shown in Fig. 2.4.

Human Epigenetic Alterations[48]

In addition to the sequence variants considered, epigenetic alterations, including alternative splicing and methylation, affect gene expression. Even though the number of genes may be limited to less than 25,000, variable transcription initiation and exon splicing produce about 90,000 unique mRNA transcripts and protein products.

Methylation of cytosine to 5-methylcytosine occurs frequently; about 70% of CpG dinucleotides in the human genome are methylated. Although not inherited, interest in this "fifth base" has increased as correlations with cancer have been reported. CpG islands are about 1000 bases in length and are often found near the 5′ end of genes. These regions consist of clusters of CG dinucleotides that are usually not methylated in normal cells. However, CpG methylation correlates with condensed chromatin structure and promoter inactivation; an important example occurs in tumor-suppressor genes. Other epigenetic targets include nucleosome histone phosphorylation, acetylation, and methylation that can all affect gene expression.

ENCODE Project

ENCODE (Encyclopedia of DNA Elements) is a project that was initiated by the National Human Genome Research Institute in 2003 to examine all functional elements in the human

TABLE 2.4 Repeat Sequences in the Human Genome

Type	Abbreviations	Size	Copies (Thousands)	Genome (%)
Retrotransposons				
Long interspersed elements	LINEs	900 bp	850	21
	L1		516	16.9
	L2		315	3.22
	L3		37	0.31
Short interspersed elements	SINEs	100–400 bp	1500	13
	Alu	350 bp	1090	11
	MIR		393	2.2
	Ther2/MIR3		75	0.34
Long terminal repeats	LTRs	1.5–11 Kb	450	8
	ERV		112	2.89
	ERV(K)		8	0.31
	ERV(L)		83	1.44
	MaLR		240	3.65
Segmental Duplications		>1000 bp		5.3
Structural Repeat and Gene Clusters				
	Centromeres	171 bp		3-6
	Telomeres	6 bp		<0.1
	Ribosomal			0.41
DNA Transposons		80–3000 bp	300	3
Simple Sequence Repeats	SSRs	1–500 bp		3
Short Tandem Repeats	STRs	1–13 bp		
		1		0.17
		2		0.53
		3		0.10
		4		0.34
		5		0.27
		6		0.14
		7		0.09
		8		0.11
		9		0.09
		10		0.16
Processed Pseudogenes		~1300	20	1.2

Data extracted from Lander et al.,[4] Venter et al,[5] International Human Genome Sequencing Consortium,[6] Richard et al.,[70] Stultz et al.,[71] and Torrents et al.[72]

genome.[49] The functional elements defined were not only the discrete genome areas that encode a product such as protein but also any genome area that had a reproducible biologic effect on processes such as transcription or chromatin structure. These genome areas include both exons as well as non–protein-coding areas such as promoters, enhancers, and silencers. The genome areas that do not encode protein have significant contributions to human variation.[50] From a survey of 150 genome-wide association studies using SNVs to identify genes linked to disease, 465 unique disease-associated SNVs were identified.[51] Of these 465 variants, 88% ($n = 407$) were present in the regions between genes (intergenic) or within introns. These results suggest the importance of noncoding variants in disease. Thus, the initial analysis from ENCODE has revealed that genome variation in noncoding regions is significant in inherited diseases and cancer.[52]

POINTS TO REMEMBER

Variations
- Small variants such as single nucleotide changes are the easiest to correlate to human disease.
- Most of the genome is repetitive, noncoding sequences with functions just now being uncovered by massively parallel sequencing.

NOMENCLATURE

Amino acid variations were associated with human disease long before DNA variation.[53] Amino acid variants were found first because techniques for amino acid sequencing matured before DNA sequencing. Advances in DNA technology

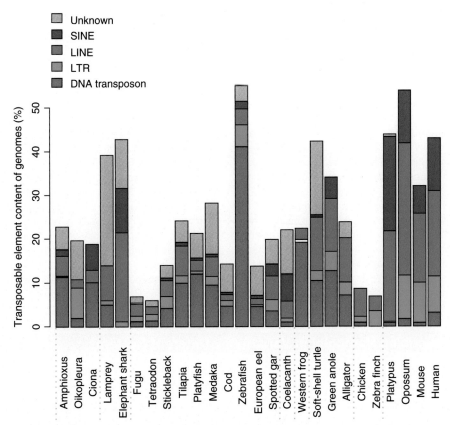

FIGURE 2.4 Transposable element diversity among species. A summary of the contribution of transposable elements to various genomes, including both retrotransposons and active transposons, is illustrated. The percentage of inactive retrotransposons (SINE, LINE, LTR) and active DNA transposons is shown for each organism. The organisms are organized into nonvertebrates (eg, Amphioxus), nonbony vertebrates (eg, lamprey), actinopterygian fish (eg, Fugu), lobe-finned fish (eg, Coelacanth), amphibians (eg, Western frog), nonbird reptiles (eg, soft-shell turtle), birds (eg, chicken), and mammals (eg, platypus). The figure clearly illustrates that in mammals, transposable elements contribute to more than 30% of the total genome. In comparison, in other organisms such as chicken and fugu, transposable elements are less than 10% of the genome. (From Chalopin D, Naville M, Plard F, et al. Comparative analysis of transposable elements highlights mobilome diversity and evolution in vertebrates. *Genome Biol Evol* 2015;7:567-580. This figure was reproduced by permission from Oxford University Press on behalf of the Society for Molecular Biology and Evolution and Dr. Jean-Nicolas Volff.)

enabled the investigation of DNA variations associated with disease. For example, the characterization of amino acid variants in the globin gene products (*HBB* and *HBA*) preceded the descriptions of DNA variation in the globin genes.

Small Variants

More than 20 years ago, the need for a database of human variants associated with disease was recognized. In 1996, the Mutation Database Initiative was sponsored by the Human Genome Organization (HUGO).[53] In 2001, the HUGO Mutation Database Initiative became the Human Genome Variation Society (HGVS).[54] In addition to the documentation and collection of variant information, the HGVS created a nomenclature system for standardizing the reporting of variations. An early recommendation proposed a hybrid model in which traditional disease alleles such as hemoglobin S for sickle cell disease and Z allele for α_1-antitrypsin deficiency would retain their historic nomenclature and new disease alleles would use the new nomenclature system.[55] The current HGVS nomenclature system has features that can be found in this early proposed system (Box 2.2).[56] In the HGVS system, disease alleles are described at the DNA level rather than as amino acid changes. Preferred terminology does not ascribe disease potential to the naming of a variant. For example, all variants are not disease-causing mutations. The preferred terms include sequence variant, CNV, and single nucleotide variant. Hemoglobin variants were initially named by a combination of letters (hemoglobin A, B, S, C, and F) and the city of discovery. Hematologists continue to use the traditional nomenclature (Table 2.5) that does not distinguish between variants from β-globin (*HBB*) and α-globin (*HBA*). In addition to the hemoglobin genes, other genes have both a traditional nomenclature system and the HGVS nomenclature. The basics of the HGVS nomenclature system for SNVs are introduced in Box 2.2.

Large Structural Variants

The determination of 46 chromosomes in humans occurred in 1956.[57] The following years were marked by increased activity in human cytogenetics, but with this rapid growth in knowledge it became apparent was that there was no coordination of how findings were named or classified.[1] Beginning in 1960, a consensus meeting of laboratories (Denver Conference) established basic guidelines for naming large

> **BOX 2.2 The Human Genome Variation Society Nomenclature for Naming Small Sequence Variants**
>
> The position of single nucleotide substitutions can be referenced to the genome (g.), amino acids in the protein (p.), or nucleotide bases in the cDNA transcript (c.). In all cases, the genome build and the sequence accession number should be specified.
>
> **For genomic coordinates,** specify the chromosome and the base position followed by the base change. For example: GRCh37/hg19 Ch17: g.3424566C>T. Genomic coordinates are the only option for all intergenic variants, as well as deep intronic variants.
>
> **When the variant is within the protein coding region of a gene** and protein coordinates are preferred, give the gene name followed by the number of the amino acid affected starting with the initiating methionine as 1. The wild-type amino acid is listed to the left of the number, and the variant amino acid is listed to the right. For example: *CYBB* p.T42R (or *CYBB* p.Tyr42Arg) for the 1 letter and 3 letter codes for amino acids, respectively.
>
> The protein coordinates best specify the phenotype, but it is a challenge to convey the results of frameshifts and the exact nucleotide change is often ambiguous.
>
> **When the variant is in or near the exons of a gene**, you can use the cDNA coordinates. In this case, give the gene name followed by the base number using the A in the ATG initiation codon as 1 and then the base change. For example: *CYBB* c.125C>G. If the base change is in an intron (eg, a splice site mutation), you count from the nearest intron–exon boundary. For example: *CYBB* c.141+2T>C or c.142-12C>T. If the variant is 3′ of the ATG initiation sequence, you count down using negative numbers (*CYBB* c.-64C>T), or if it is 5′ of the last exon, you count up (*CYBB* c.*67G>A). With c. coordinates, you know the exact base change even though it may not change the amino acid sequence.
>
> Nomenclature for insertions and deletions are also specified by the Human Genome Variation Society.

TABLE 2.5 Describing Hemoglobin Variants by Different Systems

Traditional Name	Disease Associated	Gene	Amino Acid Change (Traditional)*	Amino Acid Change (HGVS)	HGVS Nucleotide Change (mRNA Transcript)†	Genomic Coordinate (GRCh37/hg19)‡
Hemoglobin SS	Sickle cell anemia	*HBB*	Gln6Val	p.Gln7Val	c.20A>T hom (NM_000518.4)	Chr11:g.5248232
Hemoglobin CC	Hemolytic anemia	*HBB*	Glu6Lys	p.Glu7Lys	c.19G>A hom (NM_000518.4)	Chr11:g.5248233
Hemoglobin Austin	None	*HBB*	Arg40Ser	p.Arg41Ser	c.123G>T het (NM_000518.4)	Chr11:g.5247999
Hemoglobin G Philadelphia	None	*HBA2*	Asn68Lys	p.Asn69Lys	c.207C>A het (NM_000517.4)	Chr16:g.223235

*The amino acid change for hemoglobin diseases was characterized by amino acid sequencing before the advent of DNA sequencing. The first amino acid, methionine, was not included, resulting in the Gln to Val change in sickle cell anemia described as the "6" position rather than the "7" position.
† The "c." annotation is based on a reference transcript (NM number).
‡ The HGVS nucleotide position is from 5′ to 3′ on the strand. However, the genomic coordinates are not oriented to the gene. The nucleotide position based on mRNA transcript may increase while the genomic position increases or decreases, depending on the orientation of the gene on the chromosome.
Data formatted according to den Dunnen JT, Antonarakis SE. Mutation nomenclature extensions and suggestions to describe complex mutations: a discussion. *Hum Mutat* 2000;15:7-12.

chromosomal variations. The findings from multiple subsequent consensus conferences were unified in a single document, "An International System for Human Cytogenetic Nomenclature (1978)."

Some of the basic concepts of chromosomes are described in ISCN 1978: Autosomes are numbered from 1 to 22 in descending order of length. The sex chromosomes are named X and Y. The symbols p and q designated the short and long arms of the chromosomes, respectively. A chromosome band is the part of the chromosome that is clearly distinguishable from adjacent segments that are darker or lighter in appearance. G-bands are the bands resulting from Giemsa dye staining. In addition to describing the normal state of chromosomal features, ISCN 1978 considered the naming of chromosomal rearrangements such as inversions, deletions, and translocations.

In more recent times, the ISCN 2013 version introduced new features such as the term "hg" for "human genome build or assembly" and a chapter titled "Microarrays," which is devoted to naming changes identified by oligonucleotide microarrays. Of note, there is a separate consortium focused on microarrays known as ISCA (The International Standard for Cytogenomic Arrays). ISCA is focused on microarray testing quality improvement by projects such as variant databases linked to clinical data.[58]

Naming Genes

As significant as the naming of DNA variants, the naming of genes has also become standardized over the past thirty years.[59] The basic components of gene names include the gene name, which may include information on gene function, and the gene symbol, which is a short abbreviation in upper case Latin letters and Arabic numbers that are both italicized. The currently accepted gene naming system is by the HUGO Gene Nomenclature Committee (HGNC).[60] As with all standardization activities, there is a tradeoff. The more familiar historic names are established in the literature and used by practitioners in the specialty concerned with that gene. However, for a particular disease-associated gene, communication outside of the specialized field of knowledge may by difficult and lead to errors. Especially in the current era with genomic tests examining hundreds to tens of thousands of genes, a common gene naming system is necessary. In reporting specific genes, a hybrid approach that uses both the consensus nomenclature and the traditional name may be useful. A current database of recommended gene names and symbols as well as traditional names can be found on the HGNC online database. The HGNC database currently contains information on 18,990 protein-coding genes.

POINTS TO REMEMBER

Nomenclature
- The HGVS has created a systematic nomenclature for variants, including single nucleotide changes and small deletions and insertions.
- The Human Gene Nomenclature Committee provides a listing of accepted gene names.
- Within a specific field, common traditional names of genes and variants may be more familiar but communication outside of the field is best conducted with modern consensus nomenclature.

DATABASES

Databases of DNA variations may be locus or disease specific (LSDB; eg, HbVar) or general databases that seek to capture information on all variants. Some of the commonly used general databases include dbSNP (Database of Single Nucleotide Polymorphisms), OMIM (Online Mendelian Inheritance in Man), HGMD (Human Gene Mutation Database), and ClinVar (Clinical Variant). Common databases and their web accessions are given in Box 2.3.

A systematic catalog of SNVs including small insertions and deletions was created as the dbSNP in 1998 as a collaboration between the NCBI and the National Human Genome Research Institute (NHGRI).[61] The reference identifier for variants in dbSNP begins with the prefix "rs." There are separate databases for structural variants known as dbVAR (Database of Genomic Structural Variation—NCBI) and DGV (Database of Genomic Variation—European Molecular Biology Laboratory [EMBL]). The prefix is either "nsv" for structural variants originating from dbVAR of the NCBI and "esv" for structural variants originating from DGV of EMBL.[62]

OMIM is manually curated by a team of professionals located at Johns Hopkins University.[63] It was started by the geneticist Victor McKusick in the 1970s as "Mendelian Inheritance in Man," first a series of published books and later an online resource. By design, the OMIM is not a comprehensive catalog of every variant ever described with disease but rather a catalog of genes and variants representative of a disease type. As of early 2015, 5461 diseases or syndromes were described in OMIM. However, of the approximately 20,000 protein coding genes, only 3381 genes were catalogued with a known variant associated with a disease or syndrome.

In contrast to OMIM, HGMD seeks to be a comprehensive database of all variants with reported disease association.[64,65] The number of new reports of germline mutations in the literature was less than 250 per year through 1990, but throughout the 1990s, reports grew into the thousands.[64] As of 2016, there were 166,768 variants from 6905 genes cataloged in HGMD. The types of variants included 92,974 missense or nonsense; 15,168 splicing; 24,957 small deletions; 10,415 small insertions; and 12,565 gross deletions. Although HGMD is one of the largest databases, there are reports of incorrect annotations arising from database issues or

---

BOX 2.3 Human Genomic Databases

Comprehensive
NCBI (National Center for Biotechnology): ncbi.nlm.nih.gov/genome
Ensembl: ensembl.org/index.html
UCSC (University of California Santa Cruz): genome.ucsc.edu/

Genes and Disease
OMIM (Online Mendelian Inheritance in Man): ncbi.nlm.nih.gov/omim
ClinVar: ncbi.nlm.nih.gov/clinvar/
Decipher: decipher.sanger.ac.uk/index

Sequence Databanks
NCBI GenBank: ncbi.nlm.nih.gov/Genbank/
EMBL (European Molecular Biology Laboratory)-Bank: www.ebi.ac.uk/embl
DDBJ (DNA Data Bank of Japan): ddbj.nig.ac.jp/

MicroRNAs
miRBase: mirbase.org

General Variation Databases
Leiden Open Variation Database: lovd.nl/3.0/home
Human Genome Mutation Database: www.hgmd.cf.ac.uk/ac/index.php
Short Genetic Variations (dbSNP): ncbi.nlm.nih.gov/projects/SNP/
1000 Genomes: www.1000genomes.org/
Exomes (ExAC): exac.broadinstitute.org/

Specialized Variant Databases
Database of Genomic Structural Variation (dbVar): ncbi.nlm.nih.gov/dbvar
Database of Genomic Variants (DGV): dgv.tcag.ca/dgv/app/home
Retrotransposons: dbrip.brocku.ca/
Haplotypes (HapMap): hapmap.org/

Nomenclature
HUGO (Human Genome Organization) Gene names: genenames.org
HGVS (Human Genome Variation Society) Sequence variants: www.hgvs.org/mutnomen

problems with the primary literature.[66] The annotation issue is demonstrated by a report that found that 80% of the HGMD disease-causing variants had an allele frequency of more than 5% in the 1000 Genomes Project dataset; however, rare diseases would be expected to have variant frequencies much less than 5%.[66]

The chief limitation of databases such as OMIM and HGMD is that they rely on published reports. As new variants of known genes are discovered in research or clinical laboratories, they are rarely published. The recognition of the under representation of clinically significant variants resulted in the ClinVar project, which allows for the contribution of annotated variants by clinical laboratories, research laboratories, and the literature into a publicly available database.[67] The dataset that combines submitter, variant, and phenotype is given an accession number with the prefix "SCV" (Submitted Clinical Variant).

INFORMATICS

The modern era of human genome sequencing is underpinned by massively parallel sequencing. As suggested by the name of the technology, both the method as well as the amount of data is massive. Although a single human genome is approximately 3 billion bases, the amount of sequencing needed to accurately determine those bases may exceed 90 billion bases.[68] The data storage requirements are typically more than 0.5 terabytes per human genome. Fortunately, as the scale of sequencing has increased, tools have been developed to manage and analyze the information. There have been many recent reviews and evaluations of existing software tools.[69]

In brief, both publicly available tools and commercial software are assembled into what is referred to as a pipeline. In the pipeline, the information is processed in a serial manner (Fig. 2.5). First, raw sequencing data is saved in a file format such as FASTQ. The sequencing data are composed of data from millions of short sequencing reads. An alignment program uses rules that align the short sequencing reads to the reference genome. Depending on the coverage required, there may be tens to thousands of sequencing reads at a single nucleotide position. After alignment, the bases are "called" or determined by the software algorithm. The base call is dependent on a variety of factors, including the quality of the read (Q-score; Box 2.4) and the percentage of reads at the nucleotide position that are in agreement. At a single nucleotide position, there may be more than one base, which would be expected in the case of a heterozygous variant. After the bases are called for each nucleotide position, a variant call file (VCF) is generated. The VCF includes the nucleotide positions that differ from the reference genome and are tabulated by the genomic coordinate of the variant. The VCF is then examined by another set of software algorithms that query multiple databases; the examination by databases is sometimes referred to as filtering (Table 2.6). A standard filter is to further examine only variants that are predicted to change protein coding (eg, missense or nonsense variants). Other filters may be based on the frequency of specific variants in various populations. The frequency of the variant in populations is useful for determining if there is a consequence for a variant. For example, if the variant occurs in a gene

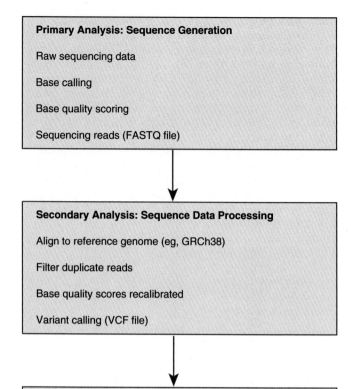

FIGURE 2.5 Bioinformatics pipeline. The analysis of data from massively parallel sequencing can be broadly considered to occur in three phases. *Primary analysis:* The raw output (eg, optical or electronic signals) from the sequencing instrument is transformed into data that describe the individual bases of DNA as well as the quality and confidence of the base call at each position. These reads of DNA are assembled into a FASTQ data file. *Secondary analysis:* The data file is then assembled onto a reference sequence. For human DNA sequencing, this is typically a reference genome such as GRCh38. If the fragments of DNA were prepared by randomly sheared fragments, then the sampling of a wide diversity of fragments improves quality and is ensured by sequencing that is from exact duplicates. When the fragments are assembled against the reference genome, the quality of each of the base calls at specific nucleotide positions can be determined. The variant at each position is then determined and reported in a single variant call file (VCF). *Tertiary analysis:* The variants are then queried against multiple databases that have information on population frequency and clinical significance. Based on these queries, the variants can be prioritized in terms of importance to the given scientific question or clinical scenario. (Modified from Oliver GR, Hart SN, Klee EW. Bioinformatics for clinical next generation sequencing. *Clin Chem* 2015;61:124-135.)

BOX 2.4 Phred Quality Score (Q-Score)[73-75]

In the 1990s, Phil Green at the University of Washington developed software to automatically read the fluorescent sequence chromatograms generated from Sanger sequencing. The original software, Phred (**Ph**il's **r**ead **ed**itor), used the following basic parameters:
1. Find the predicted location of peaks
2. Find the observed location of peaks.
3. Match predicted and observed peaks.
4. Find missing peaks.

A component of Phred was an estimator of the error probability of a base call. A quality value (Q-score) was generated from the formula:
$q = -10 \times \log_{10}(p)$
q = quality value
p = estimated error for a base call

Some representative examples of quality value (q) scores:
Q-score of 30 (Q30): The probability (p) is 1/1,000 of being incorrect.
Q-score of 20 (Q20): The probability (p) is 1/100 of being incorrect.
Q-score of 10 (Q10): The probability (p) is 1/10 of being incorrect.

Although massively parallel sequencing does not generate a Sanger sequencing type chromatogram, the convention of a Phred Q-score is still used to calculate the quality (and accuracy) of a sequenced base. Under ideal conditions, current massively parallel sequencing can achieve more than 90% of bases at Q30.

TABLE 2.6 Tertiary Analysis Databases and Software Tools*

	Annotation Source	Description	Web Address
Population frequency based	1000 Genomes Project	Whole-genome sequencing data on 2577 individuals with ancestry from East Asia, South Asia, Africa, Europe and the Americas	www.1000genomes.org
	NHLBI ESP	NHLBI Exome Sequencing Project, exome variant server. Variant data from ~6500 exomes, including 2203 African Americans and 4300 European Americans	evs.gs.washington.edu/EVS
	HapMap Project	Haplotype Mapping. Multiethnic, international database of common genetic variants	hapmap.ncbi.nlm.nih.gov
	ExAC	Exome Aggregation Consortium. Data from >60,000 individuals as part of various disease-specific and population studies	exac.broadinstitute.org
Evidence based	OMIM	Online Mendelian Inheritance in Man. Curated catalog of genes and genetic disorders. Prototypical or important variants are shown; however, the catalog is not intended to be an exhaustive compendium of all variants	www.omim.org
	LOVD	Leiden Open Variation Database. Open source database of human variation.	www.lovd.nl/3.0/home
	HGMD	Human Gene Mutation Database. Proprietary database of variants correlated with human disease	www.hgmd.cf.ac.uk/ac/index.php
	ClinVar	Clinical Variant. Human variation correlated to phenotype	www.ncbi.nlm.nih.gov/clinvar
	dbVAR	Database of Genomic Structural Variation. Genomic structural variation.	www.ncbi.nlm.nih.gov/dbvar
Prediction based	Align GVGD	Align Grantham Variation Grantham Difference. Web-based software that combines biophysical properties and protein sequence alignments to generate a score ranging from deleterious to neutral.	http://agvgd.hci.utah.edu/index.php
	ANNOVAR	Annotation of Variants including gene location (eg, exonic, splicing, UTR, intronic).	annovar.openbioinformatics.org/en/lates/
	Mutation Taster	Web-based software that generates a probability of pathogenicity based on multiple data sources.	www.mutationtaster.org/
	SIFT	Sorts Intolerant From Tolerant substitutions. Web or locally available software to predict the consequence of an amino acid change	sift.jcvi.org/
	Polyphen-2	Polymorphism Phenotyping v2. Web or locally available software to predict the consequence of an amino acid change	genetics.bwh.harvard.edu/pph2/

NHLBI, National Heart, Lung, and Blood Institute; *UTR*, untranslated region.
*Many database and software tools are available for tertiary analysis of variants, and only a few are listed here. These tools may either by publicly available or restricted for use by subscription or special permission.
Modified from Oliver GR, Hart SN, Klee EW. Bioinformatics for clinical next generation sequencing. *Clin Chem* 2015;61:124-135.

implicated in a rare disease that occurs in fewer than 1 in 100,000 individuals, then that variant should occur much less frequently than 1% in any population. In addition to population databases, there are databases that catalogue variants with clinical phenotype information or reports in the published literature. Finally, rather than an examination of databases, tools that predict the importance of variants to protein structure or function may be used. These predictive tools are not always accurate, but they may be helpful to prioritize the examination of a long queue of variants from a sequencing study.

REFERENCES

1. An international system for human cytogenetic nomenclature ISCN. Report of the standing commitee on human cytogenetic nomenclature. *Cytogenet Cell Genet 1978* 1978;**21**:309–409.
2. Lederberg J, Mccray A. 'Ome sweet' omics—a genealogical treasury of words. *The Scientist* 2001;**17**.
3. Schneider V, Church D. Genome reference consortium. NCBI Handbook. In: *National Center for Biotechnology*. 2nd ed 2013.
4. Lander ES, Linton LM, Birren B, et al. Initial sequencing and analysis of the human genome. *Nature* 2001;**409**:860–921.
5. Venter JC, Adams MD, Myers EW, et al. The sequence of the human genome. *Science* 2001;**291**:1304–51.
6. International Human Genome Sequencing Consortium. Finishing the euchromatic sequence of the human genome. *Nature* 2004;**431**:931–45.
7. Church DM, Schneider VA, Graves T, et al. Modernizing reference genome assemblies. *PLoS Biol* 2011;**9**:e1001091.
8. Callaway E. 'Platinum' genome takes on disease. *Nature* 2014;**515**:323.
9. Steinberg KM, Schneider VA, Graves-Lindsay TA, et al. Single haplotype assembly of the human genome from a hydatidiform mole. *Genome Res* 2014;**24**:2066–76.
10. Pendleton M, Sebra R, Pang AW, et al. Assembly and diploid architecture of an individual human genome via single-molecule technologies. *Nat Methods* 2015.
11. Amaral PP, Dinger ME, Mercer TR, et al. The eukaryotic genome as an RNA machine. *Science* 2008;**319**:1787–9.
12. Gudbjartsson DF, Helgason H, Gudjonsson SA, et al. Large-scale whole-genome sequencing of the Icelandic population. *Nat Genet* 2015.
13. Sulem P, Helgason H, Oddson A, et al. Identification of a large set of rare complete human knockouts. *Nat Genet* 2015.
14. Fleischmann RD, Adams MD, White O, et al. Whole-genome random sequencing and assembly of *Haemophilus influenzae* rd. *Science* 1995;**269**:496–512.
15. Goffeau A, Barrell BG, Bussey H, et al. Life with 6000 genes. *Science* 1996;**274**(546):63–7.
16. Initial sequence of the chimpanzee genome and comparison with the human genome. *Nature* 2005;**437**:69–87.
17. Locke DP, Hillier LW, Warren WC, et al. Comparative and demographic analysis of orangutan genomes. *Nature* 2011;**469**:529–33.
18. Hellen E, Kern A. The role of DNA insertions in phenotypic differentiation between humans and other primates. *Genome Biol Evol* 2015.
19. Waterston RH, Lindblad-Toh K, Birney E, et al. Initial sequencing and comparative analysis of the mouse genome. *Nature* 2002;**420**:520–62.
20. Gibbs RA, Weinstock GM, Metzker ML, et al. Genome sequence of the brown Norway rat yields insights into mammalian evolution. *Nature* 2004;**428**:493–521.
21. Hogg JS, Hu FZ, Janto B, et al. Characterization and modeling of the *Haemophilus influenzae* core and supragenomes based on the complete genomic sequences of rd and 12 clinical non-typeable strains. *Genome Biol* 2007;**8**:R103.
22. Carninci P, Kasukawa T, Katayama S, et al. The transcriptional landscape of the mammalian genome. *Science* 2005;**309**:1559–63.
23. Griffiths-Jones S, Saini HK, van Dongen S, et al. Mirbase: tools for microRNA genomics. *Nucleic Acids Res* 2008;**36**:D154–8.
24. Mendes Soares LM, Valcarcel J. The expanding transcriptome: the genome as the 'book of sand'. *EMBO J* 2006;**25**:923–31.
25. Mercer TR, Dinger ME, Mattick JS. Long non-coding RNAs: insights into functions. *Nat Rev Genet* 2009;**10**:155–9.
26. Cech TR, Steitz JA. The noncoding RNA revolution-trashing old rules to forge new ones. *Cell* 2014;**157**:77–94.
27. Di Leva G, Garofalo M, Croce CM. MicroRNAs in cancer. *Annu Rev Pathol* 2014;**9**:287–314.
28. Kozomara A, Griffiths-Jones S. Mirbase: annotating high confidence microRNAs using deep sequencing data. *Nucleic Acids Res* 2014;**42**:D68–73.
29. Witwer KW. Circulating microRNA biomarker studies: pitfalls and potential solutions. *Clin Chem* 2015;**61**:56–63.
30. McCarroll SA, Kuruvilla FG, Korn JM, et al. Integrated detection and population-genetic analysis of SNPs and copy number variation. *Nat Genet* 2008;**40**:1166–74.
31. Abecasis GR, Altshuler D, Auton A, et al. A map of human genome variation from population-scale sequencing. *Nature* 2010;**467**:1061–73.
32. Abecasis GR, Auton A, Brooks LD, et al. An integrated map of genetic variation from 1,092 human genomes. *Nature* 2012;**491**:56–65.
33. Cutting GR. Annotating DNA variants is the next major goal for human genetics. *Am J Hum Genet* 2014;**94**:5–10.
34. Iafrate AJ, Feuk L, Rivera MN, et al. Detection of large-scale variation in the human genome. *Nat Genet* 2004;**36**:949–51.
35. Redon R, Ishikawa S, Fitch KR, et al. Global variation in copy number in the human genome. *Nature* 2006;**444**:444–54.
36. Conrad DF, Andrews TD, Carter NP, et al. A high-resolution survey of deletion polymorphism in the human genome. *Nat Genet* 2006;**38**:75–81.
37. Eichler EE. Widening the spectrum of human genetic variation. *Nat Genet* 2006;**38**:9–11.
38. Hinds DA, Kloek AP, Jen M, et al. Common deletions and SNPs are in linkage disequilibrium in the human genome. *Nat Genet* 2006;**38**:82–5.
39. McCarroll SA, Hadnott TN, Perry GH, et al. Common deletion polymorphisms in the human genome. *Nat Genet* 2006;**38**:86–92.
40. Wain LV, Armour JA, Tobin MD. Genomic copy number variation, human health, and disease. *Lancet* 2009;**374**:340–50.
41. Davare MA, Tognon CE. Detecting and targetting oncogenic fusion proteins in the genomic era. *Biol Cell* 2015.
42. Willems T, Gymrek M, Highnam G, et al. The landscape of human STR variation. *Genome Res* 2014;**24**:1894–904.
43. Kerzendorfer C, Konopka T, Nijman SM. A thesaurus of genetic variation for interrogation of repetitive genomic regions. *Nucleic Acids Res* 2015.
44. Gymrek M, Golan D, Rosset S, et al. Lobstr: a short tandem repeat profiler for personal genomes. *Genome Res* 2012;**22**:1154–62.

45. Chalopin D, Naville M, Plard F, et al. Comparative analysis of transposable elements highlights mobilome diversity and evolution in vertebrates. *Genome Biol Evol* 2015;**7**:567–80.
46. de Koning AP, Gu W, Castoe TA, et al. Repetitive elements may comprise over two-thirds of the human genome. *PLoS Genet* 2011;**7**:e1002384.
47. Cowley M, Oakey RJ. Transposable elements re-wire and fine-tune the transcriptome. *PLoS Genet* 2013;**9**:e1003234.
48. Lopez J, Percharde M, Coley HM, et al. The context and potential of epigenetics in oncology. *Br J Cancer* 2009;**100**:571–7.
49. The ENCODE (encyclopedia of DNA elements) project. *Science* 2004;**306**:636–40.
50. A user's guide to the encyclopedia of DNA elements (ENCODE). *PLoS Biol* 2011;**9**:e1001046.
51. Hindorff LA, Sethupathy P, Junkins HA, et al. Potential etiologic and functional implications of genome-wide association loci for human diseases and traits. *Proc Natl Acad Sci USA* 2009;**106**:9362–7.
52. An integrated encyclopedia of DNA elements in the human genome. *Nature* 2012;**489**:57–74.
53. Cotton RG. Progress of the HUGO mutation database initiative: a brief introduction to the human mutation MDI special issue. *Hum Mutat* 2000;**15**:4–6.
54. Horaitis O, Cotton RG. The challenge of documenting mutation across the genome: the Human Genome Variation Society approach. *Hum Mutat* 2004;**23**:447–52.
55. Beaudet AL, Tsui LC. A suggested nomenclature for designating mutations. *Hum Mutat* 1993;**2**:245–8.
56. den Dunnen JT, Antonarakis SE. Mutation nomenclature extensions and suggestions to describe complex mutations: a discussion. *Hum Mutat* 2000;**15**:7–12.
57. Tjio JH, Levan A. The chromosome number of man. *Hereditas* 1956;**42**:1–6.
58. Riggs ER, Wain KE, Riethmaier D, et al. Towards a universal clinical genomics database: the 2012 international standards for cytogenomic arrays consortium meeting. *Hum Mutat* 2013;**34**:915–9.
59. Shows TB, McAlpine PJ, Boucheix C, et al. Guidelines for human gene nomenclature. An international system for human gene nomenclature (ISGN, 1987). *Cytogenet Cell Genet* 1987;**46**:11–28.
60. HGNC. HUGO Gene Nomenclature Committee. http://www.genenames.org/ [last accessed 13.07.15].
61. Sherry ST, Ward MH, Kholodov M, et al. dbSNP: the NCBI database of genetic variation. *Nucleic Acids Res* 2001;**29**:308–11.
62. MacDonald JR, Ziman R, Yuen RK, et al. The database of genomic variants: a curated collection of structural variation in the human genome. *Nucleic Acids Res* 2014;**42**:D986–92.
63. OMIM. http://omim.org/statistics/geneMap.
64. Cooper DN, Ball EV, Krawczak M. The human gene mutation database. *Nucleic Acids Res* 1998;**26**:285–7.
65. HGMD. http://www.hgmd.cf.ac.uk/ac/hahaha.php.
66. Berg JS, Adams M, Nassar N, et al. An informatics approach to analyzing the incidentalome. *Genet Med* 2013;**15**:36–44.
67. Landrum MJ, Lee JM, Riley GR, et al. Clinvar: public archive of relationships among sequence variation and human phenotype. *Nucleic Acids Res* 2014;**42**:D980–5.
68. Goldstein DB, Allen A, Keebler J, et al. Sequencing studies in human genetics: design and interpretation. *Nat Rev Genet* 2013;**14**:460–70.
69. Pabinger S, Dander A, Fischer M, et al. A survey of tools for variant analysis of next-generation genome sequencing data. *Brief Bioinform* 2014;**15**:256–78.
70. Richard GF, Kerrest A, Dujon B. Comparative genomics and molecular dynamics of DNA repeats in eukaryotes. *Microbiol Mol Biol Rev* 2008;**72**:686–727.
71. Stults DM, Killen MW, Pierce HH, et al. Genomic architecture and inheritance of human ribosomal RNA gene clusters. *Genome Res* 2008;**18**:13–8.
72. Torrents D, Suyama M, Zdobnov E, et al. A genome-wide survey of human pseudogenes. *Genome Res* 2003;**13**:2559–67.
73. Ewing B, Green P. Base-calling of automated sequencer traces using phred. II. Error probabilities. *Genome Res* 1998;**8**:186–94.
74. Ewing B, Hillier L, Wendl MC, et al. Base-calling of automated sequencer traces using phred. I. Accuracy assessment. *Genome Res* 1998;**8**:175–85.
75. Nickerson DA, Tobe VO, Taylor SL. Polyphred: automating the detection and genotyping of single nucleotide substitutions using fluorescence-based resequencing. *Nucleic Acids Res* 1997;**25**:2745–51.

Nucleic Acid Isolation

Stephanie A. Thatcher

ABSTRACT

Background
Effective isolation of nucleic acids (NAs) is important for clinical molecular methods, including polymerase chain reaction (PCR) and sequencing. Many NA sample preparation techniques (including commercial kits) are available to isolate NA for molecular detection. Solid-phase NA separation and automation are now commonly used in clinical and research laboratories.

Content
Optimal NA isolation includes lysis from diverse sources such as human cells, viruses, bacterial spores or protozoan oocysts, and purification of DNA or RNA. Techniques involve sample exposure to chemicals, enzymes, or binding matrices that reduce sample volume, variability, and complexity to achieve purity goals. The isolated NA should be efficiently separated from inhibitors of a downstream molecular assay. The best preparation method depends on the requirements for a particular application. Goals may include flexibility for multiple sample types, large batch processing, speed, or high-purity NA. Consistent results for today's molecular methods can usually be obtained by the right combination of lysis, concentration, purification, and efficiency for NA isolation.

HISTORY OF NUCLEIC ACID PREPARATION TOOLS

The function of NAs was unknown when initially isolated from eukaryotic nuclei in 1869 by chemist Friederich Miescher.[1] For more than 70 years, it was debated whether protein or NA determined heredity. Early isolations of NA used tedious methods to isolate NA away from other cellular components, often including alkaline lysis followed by acid and alcohol precipitation.[2] Researchers worked for decades with NA prepared by these chemical techniques to characterize its molecular structure and function. The four bases of DNA were initially identified in 1891–1893 by the chemist Kossel[3] from the first protein-free preparations of NA in an attempt to understand the chemistry of the nucleus. He separated "nuclein," the early descriptor for NA, into its chemical components by hydrolysis and identified phosphoric acid, the purines guanine and adenine, and a carbohydrate. He later discovered the pyrimidines thymine and cytosine from thymus NA. In 1944, Avery, MacLeod, and McCarty showed that hereditary information is contained in DNA by transforming pneumococcus organisms.[4] The structure of double-stranded DNA (dsDNA) was completed in 1953, explaining the mechanism for replication and heredity of DNA,[5-7] which finally led to acceptance that DNA carried genetic information. Subsequently, RNA was also shown to carry genetic information in 1956 by research with tobacco mosaic virus.[8,9] NA preparation techniques have improved significantly over the past 60 years in speed and complexity. Many of the procedures described in this chapter are considered routine, are used for many applications, and remain a critical component of any NA detection assay. These include clinical diagnostic procedures that use DNA or RNA to identify human genetic variants or the presence or quantity of foreign, potentially pathogenic NA.[10] NA is routinely prepared for whole-genome sequencing, mutation identification, and pathogen detection. Current NA preparation procedures have become faster, use less hazardous chemicals, and are sometimes paired with molecular analysis methods.

STEPS INVOLVED IN NA PREPARATION

Nucleic acid preparation can refer to isolation, extraction, purification, and/or separation. These terms are used interchangeably in the literature to describe preparations of NA. Here the term used is NA preparation because the type and combination of methods used may vary. Three steps are used for most samples (Fig. 3.1): (1) extracting or releasing the NA, typically from cells; (2) separating or isolating NA from other material; and (3) purifying the NA by removal of inhibitory substances. Concentrating NA can also be important for the detection of low-level analytes. Many methods of NA preparations are described here, as well as how they are used in specific applications.

Nucleic acid preparation techniques are evolving in parallel with the increased clinical utility of routine molecular testing. Methods are becoming faster, more flexible, automated and take up less space in the laboratory.[11-13] There is also a desire for flexibility, defined as the ability to isolate DNA or RNA from multiple cell or organism types (eg, human cells, viruses,

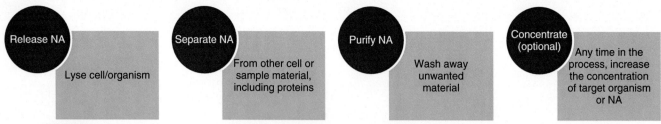

FIGURE 3.1 Basic nucleic acid (NA) preparation steps used in most isolation procedures. (Not all steps are used in all methods.)

gram-positive bacteria, bacterial spores, fungal cells, and protozoan oocysts) in many potential sample backgrounds. Because some pathogen genomes are composed of RNA and some targets for disease diagnosis are RNA, it can be important to separate both DNA and RNA from a sample. Each sample presents unique challenges; for example, blood contains many proteins and heme that can interfere with PCR, stool and other samples contain solids that can clog filters, and respiratory samples may contain RNases that can degrade RNA during preparation.[14,15]

Lysis

Release of NA from cells, nuclei, or organisms is the first step of any NA preparation. Lysis can be induced by chemical, enzymatic, or mechanical means. It is important to select an appropriate lysis technique for the target to ensure that NA is released efficiently. Sometimes lysis is all that is required, and after the NA is released, it can be analyzed.[16] Most human cells and many pathogens (especially viruses and gram-negative bacteria) require only simple chemical lysis (alkaline lysis, salt, detergents, or chaotropic agents).[17] Early methods used alkaline lysis of cells to separate NAs from proteins, with chemicals such as sodium hydroxide. Detergents also aid lysis by breaking down membranes.[18]

Enzymatic lysis, with enzymes such as proteinase K, lysozyme, or mutanolysin,[19] can increase lysis efficiency by degrading membrane or capsid proteins or attacking the peptidoglycan layer. Lytic enzymes can be bacteriolytic or yeast lysing, including lysozymes that digest the extensive peptidoglycan layer of some gram-positive bacteria.[20] For gram-negative bacteria, detergents are used first to remove the outer membrane; then enzymes are used to assist in lysis.

Lysis is a bigger challenge when working with complex sample types or a wider variety of pathogens such as bacterial spores or oocysts. Gram-positive bacteria (with a thicker peptidoglycan layer or proteinaceous spore coat), tissues, fungal cells, and protozoan oocysts are particularly resistant to some lysis techniques because of their complex cells walls. More efficient techniques should be used if target NA is low in concentration and may not be detected without efficient release of NA.[21] Resistant organisms may require more aggressive chemical, enzymatic, or physical methods. Bacterial spores,[22] fungi, yeasts, and oocysts[23] have complex coats or walls containing proteins and other complex molecules cross-linked to make them resistant to many environmental factors. Given that these organisms have very different molecules that make up their structure, it is difficult to design a universal enzymatic or chemical approach. Mechanical lysis, using external physical force, is a method for many cell types discussed later. Other physical manipulations, such as sonication, or temperature changes, such as boiling[24,25] or freeze–thaw cycles,[26] can lyse cells.

Lysis by physical means is often the best option for many hardy pathogens that are difficult to disrupt. Physical means are nonspecific methods that work with many sample types. Although it requires specialized equipment, mechanical lysis is being widely adopted because it is a rapid method that can work for many target types.[27] Mechanical lysis systems require a large input of energy and may be loud, but they do not require the addition of chemicals or enzymes that need to be removed later. NA shearing during mechanical lysis can be a concern, so care is taken to minimize lysis time, but most shearing does not impact detection because the NA fragments are larger than what is required for analysis.[28] Bead milling is a type of mechanical lysis that involves rapid motion of beads to physically break open cells.[29,30] Beads can be moved rapidly in many ways, including movement of the container. Small beads are moved rapidly within a sample, and their collision with cells physically breaks open the cells. Sonication is another mechanical method by which high-intensity sound waves can be applied to lyse cells.[31] If an analysis system targets NA from many types of cells or organisms, a broad approach such as mechanical lysis is desirable.[32] Some automated NA systems have additional external mechanical lysis options.[33,34]

> **POINTS TO REMEMBER**
>
> **Lysis Techniques**
> - Chemical: the simplest and cheapest option for easy-to-lyse cells, especially human cells and gram-negative bacteria. Solutions can include salts, chaotropic agents, strong bases (eg, sodium hydroxide), and detergents.
> - Enzymatic: a more expensive option that helps lysis for some targets, enzymes such as proteinase K, lysozyme, or mutanolysin may increase lysis by targeting proteins on the outside of the cell or organism
> - Mechanical or physical: a rigorous approach to physically disrupt the structure of any cell or organism nonspecifically that requires additional equipment. It is the most effective approach when lysing a variety of targets.

Protein Removal

In a cellular environment, NAs are surrounded by proteins called histones and many other accessory proteins. Proteins can be detrimental to either the NA preparation process or downstream analyses. Indigenous enzymes such as nucleases can break down the target NAs during the preparation method if not removed, proteases can interfere with downstream enzymatic procedures, and large amounts of proteins can interfere with specific binding of NA in some systems. Therefore, it is often important to denature or remove proteins from a sample.

Some samples contain large amounts of RNases, and RNA is particularly unstable in the presence of RNases. Some techniques used to eliminate proteins are effective against RNases, such as proteases or strong denaturing agents. Usually RNases need to be degraded or denatured immediately in the lysis reaction because RNA degradation can be very rapid. It is also important to use labware that is free from RNases when preparing RNA because RNases are prevalent on skin and in the environment.

Chemical or enzymatic techniques can be used to eliminate proteins by degradation or precipitation. Chaotropic acids such as guanidine hydrochloride or guanidinium thiocyanate protect NAs from nucleases because of their potent protein denaturing properties.[35] Chaotropic acids are also useful because they lyse bacteria and yeast.[36] These chemicals are now used in many NA preparation methods.

Detergents are added in many NA preparation methods to dissociate or remove proteins from NA preparations. Sodium dodecyl sulfate (SDS) was an early detergent used in the preparation of NA[37] to help separate NAs from proteins (including nuclear and membrane-bound proteins). Its use evolved from the observation that SDS and other surfactants disrupt bacterial and viral structures by solubilizing the proteins. Another common detergent used is Triton X-100.[38]

Unwanted proteins and enzymes can also be digested by the addition of proteases in the method. Proteases, which have many purposes, including protection from infection, are naturally occurring enzymes found in plants, animals, and microorganisms.[39] Some proteases are used in NA preparations to reduce protein background and aid in lysis by digesting membrane or capsid proteins. Chemicals and detergents can only denature proteins, but proteases actually break them into smaller molecules by cleaving peptide bonds. Most proteases, however, are too specific in their cleavage sites and not useful for NA preparations or are too difficult to produce in production quantities. An exception is the primary protease used for NA preparations, the serine protease proteinase K, originally isolated from *Engyodontium album*.[40] Its broad lysis specificity and protein degrading properties are very useful for NA preparations. Other proteases, such as the temperature-stable proteinase EA1, are also used in NA preparations.[41] Proteolysis may require an incubation step at 37° to 55°C, and the enzymes need to be removed or inactivated because they often interfere with downstream analysis.

Separation Techniques

When cell lysis is complete, NA can be isolated from other cellular or sample materials by separation methods. Proteins, polysaccharides, metals, salts, organic compounds, and dyes are examples of molecules that may need to be removed. Isolation is not required if molecules are tolerated or if a sample is relatively clean. Several techniques have been used to isolate NA from other components of the cell or sample background. The following paragraphs discuss separation methods in two broad categories: (1) liquid–liquid extraction (liquid phase separation and precipitation) and (2) liquid–solid extraction (by size exclusion or affinity separation) (Fig. 3.2). The process of isolation can also concentrate the NA, increasing the sensitivity of downstream detection methods.

Liquid phase extraction is a common method used for NA isolation that leads to a pure product. NA can be isolated from other molecules by differential solubility in immiscible liquids. The primary organic solvent is phenol,[42] usually mixed with chloroform and isoamyl alcohol.[43] Phenol denatures proteins, which stay in the organic phase while the NA is in the aqueous phase. The addition of chloroform and isoamyl alcohol helps to separate the phases and prevent foaming. After separation, NA in the aqueous phase is precipitated by ethanol to remove residual phenol for a clean, concentrated product. Initially designed to purify relatively unstable RNA vulnerable to degradation by RNases, the phenol–chloroform extraction method is very effective but also manually tedious, must be performed in a fume hood, and creates hazardous waste. Although not currently well suited for high-throughput needs, microfluidic liquid phase extraction holds promise because liquids are easy to manipulate.[44]

Precipitation of NA is a liquid phase method that achieves a clean product and concentrates NA. NA precipitates in the presence of alcohols, such as ethanol or isopropanol, and a high concentration of salt (0.1–0.5 M), such as ammonium acetate, sodium acetate, or sodium chloride.[45] Other liquids that can precipitate NA are acetone and lithium chloride. Centrifugation concentrates a NA pellet from the rest of the liquid, and the pellet is manually dried and resuspended. The resulting NA is clean, although some molecules can co-precipitate with the NA. Because manipulation of the pellet is required, this method is tedious but results in less hazardous waste than phenol–chloroform extraction.

Solid phase extraction methods are now the most common method of NA isolation for several reasons, including minimal use of hazardous chemicals, fewer and easier manual manipulations, easy adaptation to automation, and increased throughput. Solid phase extraction is often described in four basic steps: lysis, binding, washing, and eluting.

Solid phase methods rely on three principle techniques: size exclusion by gel filtration, ion exchange chromatography by charge-based reversible adsorption, and affinity chromatography by reversible surface adsorption (see Fig. 3.2). Any of these methods can be incorporated into spin filters, columns, or beads.

In gel filtration, NA molecules can be separated from smaller molecules by size through a gel matrix. Using this method, the gel matrix in the format of a spin filter or column, allows larger molecules to pass through while smaller molecules are delayed within the pores. This is useful for separating NA from smaller molecules, but molecules similar in size will separate with the NA. Sephadex[46] or derivatives are the most common matrices used.

A method called synchronous coefficient of drag alteration (SCODA) can be used to get clean concentrated NA preparation from any sample.[47] In SCODA, a rotating electrical field is used to focus NA in a spot that can be removed and purified. Advantages are that a large sample volume of almost any sample can be added and concentrated into a small volume, and NA is easily separated from any background in its movement through the gel matrix. The downside to this process is

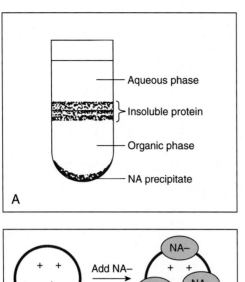
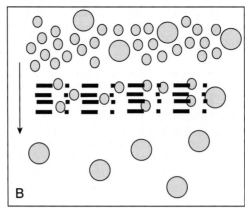
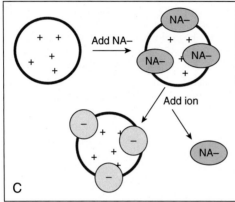
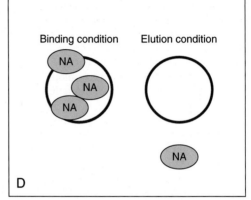

FIGURE 3.2 Nucleic acid (NA) separation methods. A, Diagram of a liquid phase separation in which the NA is soluble in a different liquid phase than proteins and impurities. After removal of the aqueous phase and precipitated protein, the NAs can be precipitated from the organic phase. B, Diagram of solid phase size exclusion with large NA molecules passing through the column faster than smaller molecules, which travel slower through smaller pores. C, Diagram of solid phase ion exchange with the negatively charged NA binding to a positively charged resin and later eluted by flooding with negatively charged ions. D, Solid phase affinity adsorption with binding of NAs under one chemical or physical condition followed by elution under another condition.

the time required (~4 hours). A similar method moves NA-bound beads through an immiscible liquid wash phase to remove contaminants.[48]

A gel electrophoresis fractionation is used for methods that require a consistent size of NAs for analysis, such as massively parallel sequencing. Gels are selected for the desired size of fragments. This size selection can be done by several semiautomated systems and provides a narrow interval of product sizes for sequencing.

In ion exchange chromatography, negatively charged NA molecules can bind selectively to surfaces with charged groups surrounded by free counter-ions. Charged NA exchange places with the ions and bind to the surface by charge. Unbound substances are washed away. NA is released by displacing it with a flood of free ions that replace the NA molecules.[49] For example, diethylaminoethyl cellulose (DEAE-C) is a common anion exchange resin that negatively charged NA will bind to. NA is released when other ions in a high salt buffer are present to exchange with the NA. Ion exchange techniques can also be used in reverse to specifically bind and separate unwanted molecules. For example, Chelex resin (Bio-Rad Laboratories) is used to separate metallic compounds and inhibitors of PCR away from NA.[50]

Affinity chromatography, using reversible adsorption of NA to surfaces such as silica, is the separation of choice for many NA preparation procedures. This technique is commonly used in automated methods. All NAs will bind to silica surfaces under specific binding conditions, especially in the presence of chaotropic salts.[51] Binding occurs when linear NA adsorbs lengthwise to silica surfaces because of complex hydrogen bond formation between the silica and NA surfaces in the presence of chaotropic salts or alcohols at high concentration and low pH (below a pH of 7).[52] Because both silica and NA surfaces are negatively charged, the binding is due to adsorption in high ionic strength conditions and hydrogen bonding that occurs as water is removed from the surfaces. The NA is released when the salt or alcohol is removed and the surfaces are hydrated. Any surface with similar NA binding properties can be used in this way, such as diatomaceous earth. Serendipitously, similar chemicals can be used for lysis and surface binding. Chaotropic salts can be used for cell lysis and binding to a silica surface.[53] In the 1990s, NA preparation became simplified using this method. Washing of the silica surface is often accomplished with alcohol. Elution occurs when binding to silica is reversed with water. Elution with small volumes can help increase the concentration of target. Unlike ion exchange, no specific chemical is required for elution using affinity chromatography.

Solid affinity isolation is flexible in that the binding surface can be created on a variety of solid surfaces. A binding filter or column is commonly used. A liquid sample passes through by centrifugation (spin filter), pressure (syringe filter), or vacuum. Kits with silica spin filters are quick to perform and do not require hazardous chemicals. A drawback is that filters can clog with thick sample types and steps are necessary to load the binding, washing, and eluting solutions. The binding surface can also be on beads or particles that mix freely with a sample to collect free NA. Glass beads or particles are the simplest silica-based surfaces. Because the beads move through the sample, clogging is not a concern. Beads or particles can then be collected by filtration of particles, centrifugation, or a magnet. Surface binding capacity is determined by the area available for binding.

Surface binding methods have been improved extensively for automation and simplicity. Paramagnetic beads coated with NA binding surfaces such as silica are used widely on automated platforms.[54,55] Paramagnetic beads do not generate a magnetic field themselves but respond to external magnetic forces; this property of paramagnetic beads is used to move the beads through solutions. Many silica paramagnetic systems are commercially available and are amenable to the use of aqueous chemistry.[56]

Paper surface binding methods, in which NAs bind to cellulose,[57] are convenient and fast. Chemically treated paper contains lysis and binding reagents that combine when the sample is applied. Lysis and binding occur in the paper, followed by washing and elution off the paper, usually in a small volume.[58]

POINTS TO REMEMBER

Primary Nucleic Acid Isolation Techniques
- Liquid phase extraction: separation of NA from unwanted proteins or other molecules by differential solubility in a mixture of immiscible liquids that separate into layers. The layer of liquid that contains the NA is processed further to isolate it, usually by precipitation.
- Solid phase size exclusion separation: a technique to isolate NA by movement through a matrix in which molecules move at different speeds depending on their size. The NA is separated by removing the appropriate fraction of eluate.
- Solid phase ion exchange chromatography: negatively charged NA can bind to a charged surface while impurities are washed away. The NA is released from the surface when it is flooded with ions that replace the NA binding so that it is released.
- Solid phase surface adsorption (or affinity chromatography): the most common method used for NA isolation, especially adsorption to silica, usually under the same chemical conditions used for chemical lysis. The surface with NA can be washed of impurities and NA eluted off with simple chemical solutions that are compatible with downstream analysis. Other methods usually result in a NA solution with chemicals that need to be removed before analysis methods such as PCR.

Concentration

Sometimes NAs are present at very low concentrations in samples. Various techniques can be used for increasing the target NA concentration during preparation. The target can be concentrated before or after lysis by either isolating target cells or by concentrating all NA during preparation.

Cell growth or selection can lead to increased target concentrations. Some pathogens can be grown to higher quantities for detection by culture, but this requires days to weeks, depending on the organism. Selective recovery of cells by centrifugation or filtration is sometimes possible, but in complex samples, too much unwanted material can clog a filter or overwhelm the system. Centrifugation methods are often used for blood. For example, red blood cells (RBCs) can be specifically lysed to allow recovery of other cells, or RBCs containing malaria can be selectively recovered by density gradient centrifugation.[59] White blood cells (WBCs) can be concentrated by centrifuging anticoagulated blood, resulting in a layer enriched in WBCs (the "buffy coat") between the RBCs and plasma.

Selection by binding is another concentration method. Some paramagnetic beads used for NA isolation also may bind to bacterial cells nonspecifically.[60] Beads with specific antibodies are used in immunomagnetic separation techniques designed to bind specific bacteria or cells as a concentration method.[61] New methods have been developed to selectively concentrate circulating cancer cells,[62] and these techniques are covered in Chapter 9.

The prepared NA product can also be concentrated after extraction if a large volume of material is produced. Alcohols such as ethanol can be used to precipitate NAs in the presence of high salts. Precipitate can be collected by centrifugation and washed to remove salt for concentrated NAs. Binding columns or dialysis can also be used to remove water and concentrate molecules in solution.

More methods are needed that can increase target concentrations in the sample before NA preparation from clinical samples, especially methods not specific to a certain organism. This requires a larger volume of sample. For example, if an analyte is present at 1 unit/mL of blood and the sensitivity of analysis requires 10 units, at least 10 mL of blood must be concentrated. This can be difficult because of the large volumes required from a patient. Concentrating techniques are used in the water testing industry, where it is much simpler to isolate organisms from a sample that is very clean and large.

Storage

Prepared NAs may be stored before they are used. The purity of the preparation, temperature, and storage buffer will determine how long they can be stored. Generally, NAs can be stored refrigerated for months or frozen for years. Highly concentrated NAs are usually more stable over time. NAs can gradually hydrolyze in acidic conditions; therefore, storage buffers are usually slightly basic. Tris is a common buffer used for NA storage. RNA is less stable and is stored at −70°C in RNase-free water. Specialized commercial products are available to increase the storage time of DNA or RNA in some sample types.

IMPACT OF SAMPLE MATRIX ON NUCLEIC ACID PREPARATION

Different sample types can contain different inhibitors of downstream analysis,[63] and extraction efficiency may vary among sample types. Preparation procedures can be designed for a specific sample background or can be flexible to handle many sample backgrounds.[64] The ability of a system to purify RNA from stool is a good measure of its sample flexibility, and several systems are able to purify RNA from stool.[65] Note that co-isolation of inhibitors is a concern with complex sample matrices, and a process control is needed to determine if inhibitor removal was effective.[66]

Human Sample Complexity

Clinical samples range from relatively clean fluids, such as urine, saliva, or buccal cells, to thick fluids, such as blood, and sputum and finally to solid materials, such as stool and tissue. The solid materials are the most challenging samples and contain portions that may sequester NA-containing cells and make it difficult to mix and extract NAs with liquid preparation solutions. Notably, even with centrifugation, solid samples can clog filters or small channels, thereby preventing liquid from flowing. Sample flexibility is enhanced by avoiding filters or small channels.

Blood

Blood is a common sample type for NA isolation, especially for human genomics or disease identification.[11] It is usually collected in an anticoagulant such as ethylenediaminetetraacetic acid (EDTA) or heparin. Specialized blood collection tubes are also available that stabilize mRNA concentrations for expression analysis. DNA is stable in refrigerated blood samples for about 1 week. For the long-term storage of DNA in blood, it should be frozen or applied to chemically treated paper for storage as a dried blood spot. Some anticoagulants, such as heparin or SPS (sodium polyanethol sulfonate used for blood culture), interfere with downstream PCR analysis if not completely removed.

DNA isolation from blood works very well, especially when recovering human DNA. These methods remove the numerous background proteins in blood samples, usually with a protease, and remove other potential inhibitors of analysis such as heme. Starting with blood plasma or serum may be appropriate for viral load testing and avoids the complexity of whole blood. However, other pathogens are present mainly in blood cells, and processing of whole blood for NA extraction is necessary. For low quantities of target in a blood sample, isolation is more complicated because a larger volume may be required. This is also true for free circulating NA (see Chapters 9 and 10).

Paraffin Embedded Tissue With Fixatives

Archived tissue is a complex solid sample that needs careful consideration, especially because sample material is limited[67] and the maintenance of sequence for genetic analysis is important. Sequence artifacts from archived tissue samples depend on the specific fixative, the fixation conditions, and storage parameters.[68-70] Fixatives such as Bouin's or acidic pretreatments for bone decalcification are known factors that degrade NAs. Formalin fixation is compatible with downstream molecular analysis and tissues stored for 1 year or less do not have many artifacts. However, NA degradation increases over time by fragmentation and formaldehyde-induced crosslinking.

Reversal of formalin (formaldehyde) cross-linking is required before DNA or RNA can be analyzed from formalin-fixed paraffin-embedded (FFPE) tissue blocks. Formalin modifies NAs with monomethylol groups, which cause NAs to cross-link by forming methylene bridges to amino groups.[71] These cross-links interfere with many processes, including reverse transcription and polymerase reactions. Heating the DNA or RNA reverses cross-linking, and most protocols use alkaline (pH 9 to 12) conditions to reverse formalin cross-linking in DNA and acidic (pH of 3–6) conditions for RNA.

Successful methods for extraction of DNA and RNA from fixed tissue samples include steps for extensive tissue and protein digestion and heating to reverse cross-linking. These methods are typically used for genetic analysis of NA from tissues.[72,73]

DNA Versus RNA

DNA and RNA may be co-prepared from the same sample. Some molecular analysis methods may target both DNA and RNA, and preparation of both is desired. Many methods will prepare both DNA and RNA from the same sample, although special effort must go into protecting RNA, if desired, from degradation, especially from RNases. Chaotropic agents are effective at removing nucleases, including RNases.

Unwanted NA, such as background RNA or single-stranded DNA (ssDNA) can be enzymatically or chemically removed from a preparation. Some methods are designed for only DNA or RNA[74] and contain steps to remove the other NA, usually by adding a nuclease specific for the undesired NA. Specific NA can also be separated by size exclusion or with liquid phase separation techniques such as phenol–chloroform.

PROCESSING THROUGHPUT

Batch Size

Sometimes many samples need to be processed at once to save time or meet demand. Some automated NA preparation systems are designed to process large batches of samples using plates. Plate-based high-throughput methods may require additional time if a full run (≥96 samples) is required before further processing. When using high-throughput NA preparation systems, measures to monitor and prevent sample cross-contamination should be implemented.

There are several automated high-throughput NA preparation systems that are designed to handle different sample types, minimize hands-on time, and shorten turn-around times. Some have combined lysis and isolation techniques for DNA or RNA (or both) that remove impurities and inhibitors.[65,66,67,75] Typically, less than 1 mL of sample is lysed with chemicals and enzymes, and NA is isolated by washing away impurities. Sometimes mechanical lysis is included as a preprocessing option. For NA separation, binding onto silica beads or spin filters is followed by magnetic separation or centrifugation. The end sample is well suited for many molecular methods. Automated methods are generally as effective as manual methods.[75] Many comparisons of methods have been published,[76,77] but they are usually limited to only a few methods and a single sample target. Most comparisons demonstrate that there are minimal differences among preparation methods.

Speed

Some applications require fast NA preparation, when time to result must be minimized or simple techniques are required for resource limited settings. Some of these methods are or could be combined with downstream analysis and are generally fast ($\leq$30 minutes). The simplest methods only dilute and lyse the sample. These are very quick for clean samples but may compromise sensitivity. Enzymatic methods using lysozyme and proteases may be adequate for bacterial cultures. Sometimes inhibitor removal is all that is needed by adding a binding resin such as Chelex to the sample and removing the supernatant. There are also simple manual kits available with paramagnetic bead binding that use minimal equipment and work with many sample types. One rapid method incorporates a binding matrix in the pipette tip that is used to manipulate the sample.[78]

SPECIFIC APPLICATIONS

Good NA preparations match the needs of analysis. No single method is optimal for every application, but methods can be selected by application. Many factors should be considered (Fig. 3.3), including how clean and concentrated the NA needs to be, the samples processed, the downstream detection method, analytical sensitivity, batch size, and preparation time.

Inherited Disorders and Pharmacogenetics

Human DNA for genetic analysis is usually extracted from blood. WBCs can be separated away from RBCs that contain no DNA. This separation reduces polymerase inhibition of downstream amplification methods by heme and hemoglobin. Large quantities of DNA can be prepared, and the amount of DNA extracted is enough to produce a visible pellet upon precipitation with ethanol.

Oncology

Cancer cells can be isolated to identify mutations and other sequence variations specific to the cancer. Cancer tissue can be processed fresh, frozen, or fixed (often formalin fixed and paraffin embedded) and may be obtained by surgical or needle biopsy. Needle biopsies are small, and cancer cells are difficult to isolate from noncancerous cells, even with careful dissection or laser capture microdissection (see Chapter 7 and Fig. 7.2). Alternatively, circulating cancer cells in the blood can be isolated and processed. Circulating cancer cells are rare, and large blood volumes processed by high-yield methods are required that can distinguish cancer cells from noncancer cells (see Chapter 9).

Fixed tissues should be prepared carefully to reduce degradation of the sample. The storage buffer, fixation procedure, and temperature can impact the quality of extracted NA. Some fixatives do not protect the integrity of NA (eg, Bouin's or B-5), but others are compatible with molecular methods

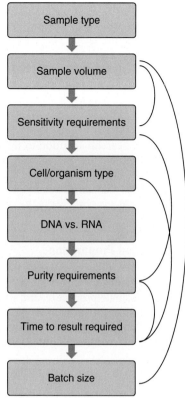

FIGURE 3.3 Factors to consider when selecting a nucleic acid (NA) preparation method. Factors at the bottom tend to be more dependent on factors at the top; the *arrows* suggest a guideline for order of decision making. *Connecting lines* refer to relationships between factors that may trigger reconsideration of a decision made above. Examples include the following: the sample volume may need to be taken into account to meet detection sensitivity requirements, a less pure product may be acceptable to save processing time, and a large batch technique might only work with small sample volumes.

(10% neutral buffered formalin and zinc-based fixatives such as acetic acid zinc formalin).

Pathogens

Organisms that cause infections or disease are present in samples at varying levels, and they may be found in a portion of the sample rather than be evenly distributed. In addition, their cell walls, outer membranes, or proteinaceous coats may be difficult to lyse. Thus, the challenges of extracting pathogen NAs from a sample are to effectively find and lyse the pathogen, which is sometimes present at low levels.

Low Concentration

Because target NAs may be present in low concentrations,[79] analytical sensitivity requirements must be considered, as well as the limit of detection of the downstream analysis. Factors include (1) starting sample volume (larger is better), (2) concentration steps that decrease the volume and increase the concentration of target material, (3) final volume of NA preparation that is used in downstream analysis, and (4) efficiency of NA recovery (yield). Large sample volumes may not be available or feasible.

Flexibility for Different Matrices

A universal method that will work for any sample type is desirable. Commercial automated platforms work with many different sample types and targets. However, there are currently no universal methods for all tissue types or matrices. Sample preparation must be adapted to the sample matrix and the targeted NAs.

RNA

Isolation of RNA requires careful attention to the removal of RNases in the sample during the preparation, especially in the final RNA isolate. Many laboratories use special plastic ware, gloves, and solutions that are RNase free to ensure that RNases, which are present everywhere in the environment, cannot degrade the RNA.

Integrated Nucleic Acid Preparation

Some detection methods contain sample NA preparation as part of the test. The benefits are that a sample can be prepared and analyzed in a single procedure, and the user skips sample handling steps that can potentially lead to sample cross-contamination. Examples are PCR systems with upstream NA preparation steps included in a single cartridge.[80,81]

Microfluidics

In-line NA preparation is frequently mentioned in research publications on microfluidic techniques for NA preparation.[82] Challenges include accepting raw complex samples, analytical sensitivity limitations, and effective lysis for hard-to-lyse organisms. If the input volume is too small, the method may not be sensitive enough for some applications. Very few microfluidic NA preparation methods are actually on the market because of the higher cost associated with developing a complex automated product.[83]

NUCLEIC ACID QUALITY AND QUANTITY

Purity

The purity and yield of NAs are primary factors for evaluating a NA method. NA yield is directly related to analytical sensitivity. The purity required depends on the target and analysis method. Sometimes purity is very important, but not every analysis method requires high-purity NA samples. Methods to measure yield and purity of NAs include ultraviolet (UV) absorbance and staining with fluorescence dyes.

Some preparation methods can lead to very clean NA. Phenol methods are still useful when a very pure product is desired. TRIzol products can be used for this purpose and are especially useful for RNA

Inhibitors

The downstream analysis dictates the extent of NA purity required. Common molecular methods such as PCR tests or loop-mediated isothermal amplification require NA of varying purity[84] and concentration. Molecules co-purified with NA can inhibit enzymatic reactions such as PCR[85] or interfere with visual real-time detection by blocking light or changing background fluorescence. Molecules that are commonly removed because of potential assay interference are nucleases, proteins, polysaccharides, salts, solvents such as alcohols, pigments such as heme, or humic acids. More extensive purification procedures may be required for samples that are destined for long-term storage.

Another approach is to make the analysis method less sensitive to interference so that the preparation procedure is less important. For example, some approaches for PCR include using forms of Taq DNA polymerase that are more resistant to inhibitors from complex sample backgrounds.[86]

It is also important to consider chemicals that the extraction procedure introduces. These chemicals or enzymes used in extraction, such as chaotropic or other salts, alcohols, or proteases, should be removed or inactivated before downstream analysis.

Yield

The yield of target NAs from a preparation can be measured. A sample may yield up to 100% of the original target NAs in the final preparation, depending on the sample and the method used. Yield is dependent on the sample background, the cell or organism, the concentration of analyte, and the method of measurement. Some material is frequently lost during processing steps. The percent yield of product can be calculated by dividing the amount recovered by the expected amount in the final sample, accounting for input and output volumes. The yield measures the extraction efficiency of a preparation procedure and affects the amount of product that should be added into a downstream analysis. Most silica binding systems, which are widely used in automated NA platforms, are similar to each other and yield human DNA above 50%.[87]

METHODS TO MEASURE NUCLEIC ACID QUALITY AND QUANTITY

The main techniques used to measure NA quantity and purity are UV absorbance and the fluorescence of stained NA in solution or in a gel matrix. These methods are good for dsDNA, and can be modified for RNA or ssDNA. Each method is differentially impacted by impurities.[88,89]

Ultraviolet Absorbance

Nucleic acids can be measured using a UV spectrophotometer at 260 nm. NA molecules absorb UV light maximally at 260 nm owing almost entirely to the constituent bases. DNA double helices have lower UV absorbance than an equivalent number of nucleotide monomers, and when DNA is denatured into single strands (ssDNA) (eg, by heat or pH), UV absorbance increases.[90] If a dsDNA preparation is pure, a 50-mg/L solution has an absorbance of 1.0 at 260 nm. RNA has a greater absorbance, so only about 30 mg/L gives an absorbance of 1.0. More precise estimates for oligonucleotides are based on dinucleotide contributions.[91] UV absorbance does not distinguish between DNA and RNA. NA concentration can be overestimated or inaccurate by the UV absorbance method when contaminants such as protein, nucleotides, single-stranded NAs, phenol, or guanidine contribute to the total absorbance.

TABLE 3.1 Comparison of Commonly Used Nucleic Acid Isolation Procedures

Isolation Method	Sample Flexibility	Lysis	Optional External Mechanical Lysis	Nucleic Acid	Sample Size	Batch Size	Time	References
Examples of Small Laboratory Automation, Good Options for Sample Flexibility								
Silica paramagnetic bead binding (or cellulose)	Many, including blood, swabs, tissue, and sputum	Chemical, enzymatic	Yes	DNA RNA	<1 mL; 2- to 10-fold concentration	6–96	20–180 min	21, 34, 64–67, 75, 76, 87
Silica spin-column binding (centrifugation)	Many, using several kits; flexible	Chemical, enzymatic	Yes	DNA RNA	Kit dependent	12		25, 33, 65–67, 77, 78
Examples of Large Laboratory Automation, Good Options for Large Batch Sizes								
Paramagnetic bead binding	Many, including swabs and tissue	Chemical	Yes	DNA RNA	≤1 mL	96		30, 76
Glass fiber plate binding vacuum	Many, including blood and tissue	Chemical, enzymatic	Yes	DNA RNA	Unknown cells; 0.5 mL	96	20–90 min	Promega, Qiagen
Examples of Simple Methods for Lysis or Inhibitor Removal								
Dilution (not sensitive)	All	None	Yes	DNA	N/A	N/A	<5 min	N/A
Mechanical lysis only	Clean samples only	Mechanical	N/A	DNA	Method dependent		<10 min	16, 31
Enzymatic lysis (lysozyme or proteinase)	Cells, tissue, food, bacteria	Enzymatic and heat	N/A	DNA RNA			5–20 min	20, 41
Bind and remove inhibitors (Chelex)	Small volume, blood cells	Boiling		DNA			30 min	50
Examples of Rapid Manual Methods with Minimal Steps for Nucleic Acid Isolation								
Tip based, porous binding matrix in tip	Liquid	Chemical	No	DNA RNA	5- to 10-fold concentration		<5 min	78
Paper based for preservation and extraction	Many: blood, cells, and tissue	Chemical	No	DNA	Spot, concentration by drying		<30 min	25, 58
Paramagnetic particles, ion exchange ligand coated on beads or surfaces	Blood, cells, forensic samples, simple, aqueous reagents	Varies, enzymatic and heat	User	DNA RNA	<1 mL		60 min	Promega, Bio-Nobile, Thermo Fisher
Examples of Methods that Can Yield Very Clean Nucleic Acid								
SCODA-gel isolation by rotating field to focused spot	Any sample type	Pre-lyse	User	DNA RNA	<10-fold concentration, ≤5 mL		4 h	84
Phenol–chloroform separation; ethanol precipitation	Cells, tissue, bacteria, yeast	Chemical		RNA	Precipitation, pure product		60 min	TRIzol Plus
Silica spin-column binding	Many, using several kits, flexible	Chemical, enzymatic	Yes	DNA RNA	Kit dependent	12		25, 33, 65–67, 77, 78

N/A, Not applicable; SCODA, synchronous coefficient of drag alteration.

The purity of a NA preparation is estimated by its ratio of absorbance at 260 and 280 nm (A_{260}/A_{280} ratio). In contrast to NAs, proteins absorb maximally at 280 nm. A pure preparation of NA should have a A_{260}/A_{280} ratio between 1.7 and 2.0 that depends on base content. Lower ratios suggest protein contamination. The A_{260}/A_{280} ratio is also affected by pH, ionic strength, and residual organic chemicals that may be introduced during extraction and purification.[92] The pH impact results from a steep change in absorbance at 280 nm with pH.

Fluorescent Staining of Nucleic Acids

Although absorbance measurements are simple and precise, they are not sensitive. Fluorescent stains that bind to NA are 1000 to 10,000 times more sensitive than absorbance measurements. They are also not as impacted by background interfering substances. The best known example of a NA dye is ethidium bromide, a positively charged, intercalating dye for dsDNA and to a lesser extent, ssDNA and RNA. Cyanine dyes are also popular stains for NAs; they have minimal fluorescence unless they are bound to NAs, thus providing very low background.[93,94] NA dyes can detect DNA and RNA in gels or in solution. Real-time PCR monitors PCR products during amplification using dyes such as SYBR Green I in solution. Each curve defines a quantification cycle (Cq) that is related to the amount of initial target. Quantification may be either absolute against a standard curve or relative against a reference target. NA dyes are also used in solution to quantify input DNA before library amplification in massively parallel sequencing. Target DNA can also be quantified by digital PCR in which many PCR partitions are run in parallel. With appropriate optics, single molecules of DNA can be visualized directly without amplification using cyanine-based NA stains.[95] Gels separate NAs by size and are visualized with fluorescent dyes with many practical applications. For additional information on gel separations, real-time PCR, digital PCR, and massively parallel sequencing, see Chapter 4.

POINTS TO REMEMBER

Common Methods to Measure Nucleic Acids
- UV absorbance: NAs are measured with a UV spectrophotometer at 260 nm. Purity of the NAs can impact the results. NA concentration may be overestimated when contaminants such as protein, nucleotides, single-stranded NAs, phenol, or guanidine contribute to the total absorbance.
- Fluorescent dye binding: Fluorescent dyes bind NA specifically (not to other molecules such as proteins) so that background material is not detected. The sensitivity of this measurement is also greater. Specific dyes can bind preferentially to dsDNA, ssDNA, or RNA.
- Gel separation and visualization: A gel matrix can be used to separate NAs by size before it is dyed. This is a good method if a NA product size needs to be measured.
- PCR and other molecular methods: A NA detection method such as PCR can be used to measure the amount of NAs by comparing them with a standard curve.

SUMMARY OF METHODS

Many methods are used to prepare NAs for molecular detection, and these are categorized in TABLE 3.1. No method serves all needs, and each is a compromise among flexibility, throughput, cost, automation, simplicity, and time to result. The needs of a central reference laboratory are different from those at the point of care. Cost reduction and large batch size are crucial to a reference laboratory, and simplicity and time to result are paramount at the point of care. Certain trends are apparent as the field moves forward toward more efficient NA preparation,[96] including (1) an increased need for less hazardous chemicals, (2) fewer preparation steps, (3) faster and more effective lysis, and (4) automation. Hard-to-lyse cells and low concentrations of NAs are challenges for any method.

CONCLUSION

Nucleic acid isolation may be required from human cells of different types or free circulating NA. When pathogens are of interest, viruses, bacteria, protozoans, and fungi must be considered. Multiplex testing may depend on simultaneous isolation of NA from some or all of these sources. Many techniques are available, requiring a considered choice of commercial products or a combination of techniques for any given application. Many NA preparation kits from companies specialize in making the process as simple as possible. NA isolation solutions are constantly improving to meet the complex needs of an evolving diagnostic world.

REFERENCES

1. Dahm R. Friedrich Miescher and the discovery of DNA. *Dev Biol* 2005;**278**:274–88.
2. Levene PA. On the preparation of nucleic acids. *J Am Chem Soc* 1900;**22**:329–31.
3. Jones ME. Albrecht Kossel, a biographical sketch. *Yale J Biol Med* 1953;**26**:80–97.
4. Avery OT, MacLeod CM, McCarty M. Studies on the chemical nature of the substance inducing transformation of Pneumococcal types. *J Exp Med* 1944;**79**:137–58.
5. Watson JD, Crick FHC. A structure for deoxyribose nucleic acid. *Nature* 1953;**171**:737–8.
6. Franklin R, Gosling RG. Molecular configuration in sodium thymonucleate. *Nature* 1953;**171**:740–1.
7. Wilkins MHF, Stokes AR, Wilson HR. Molecular structure of deoxypentose nucleic acids. *Nature* 1953;**171**:738–40.
8. Gierer A, Schramm G. Infectivity of ribonucleic acid from tobacco mosaic virus. *Nature* 1956;**177**:702–3.
9. Fraenkel-Conrat H. The role of the nucleic acid in the reconstitution of active tobacco mosaic virus. *J Am Chem Soc* 1956;**78**:882–3.
10. Niemz A, Ferguson TM, Boyle DS. Point-of-care nucleic acid testing for infectious diseases. *Trends Biotech* 2011;**29**:240–50.
11. Carpi FM, Di Pietro F, Vincenzetti S, et al. Human DNA extraction methods: patents and applications. *Recent Pat DNA Gene Seq* 2011;**5**:1–7.
12. Vomelova I, Vanickova Z, Sedo A. Methods of RNA purification. All ways (should) lead to Rome. *Folia Biol* 2009;**55**:243–51.

13. Tan SC, Yiap BC. DNA, RNA and protein extraction: the past and the present. *J Biomed Biotechnol* 2009;**2009**. 574398.
14. Rahman MM, Elaissari A. Nucleic acid sample preparation for in vitro molecular diagnosis: from conventional techniques to biotechnology. *Drug Discov Today* 2012;**17**:1199–207.
15. Thatcher SA. DNA/RNA preparation for molecular detection. *Clin Chem* 2015;**61**:89–99.
16. Folgueira L, Delgado R, Palenque E, et al. Detection of Mycobacterium tuberculosis DNA in clinical samples by using a simple lysis method and polymerase chain reaction. *J Clin Microbiol* 1993;**31**:1019–21.
17. Miller SA, Dykes DD, Polesky HF. A simple salting out procedure for extracting DNA from human nucleated cells. *Nucleic Acids Res* 1988;**16**:1215.
18. Birnboim HC, Doly J. A rapid alkaline extraction procedure for screening recombinant plasmid DNA. *Nucleic Acids Res* 1979;**7**:1513–23.
19. Yuan S, Cohen DB, Ravel J, et al. Evaluation of methods for the extraction and purification of DNA from the human microbiome. *PLoS ONE* 2012;**7**:e33865.
20. Salazar O, Asenjo JA. Enzymatic lysis of microbial cells. *Biotechnol Lett* 2007;**29**:985–94.
21. Jeddi F, Piarroux R, Mary C. Application of the NucliSENS easyMAG system for nucleic acid extraction: optimization of DNA extraction for molecular diagnosis of parasitic and fungal diseases. *Parasite* 2013;**20**:52.
22. Henriques AO, Moran CP. Structure, assembly, and function of the spore surface layers. *Annu Rev Microbiol* 2007;**61**:555–88.
23. Dumetre A, Aubert D, Puech PH, et al. Interaction forces drive the environmental transmission of pathogenic protozoa. *Appl Env Microbiol* 2012;**78**:905–12.
24. Steiner JJ, Poklemba CJ, Fjellstrom RG, et al. A rapid one-tube genomic DNA extraction process for PCR and RAPD analyses. *Nucleic Acids Res* 1995;**23**:2569–70.
25. van Tongeren SP, Degener JE, Harmsen HJM. Comparison of three rapid and easy bacterial DNA extraction methods for use with quantitative real-time PCR. *Eur J Clin Microbiol Infect Dis* 2011;**30**:1053–61.
26. Johnson DW, Pieniazek NJ, Griffin DW, et al. Development of a PCR protocol for sensitive detection of Cryptosporidium oocycsts in water samples. *Appl Environ Microbiol* 1995;**61**:3849–55.
27. Hwang KY, Kwon SH, Jung SO, et al. Miniaturized bead-beating device to automate full DNA sample preparation processes for gram-positive bacteria. *Lab Chip* 2011;**11**:3649–55.
28. Kuske CR, Banton KL, Adorada DL, et al. Small-scale DNA sample preparation method for field PCR detection of microbial cells and spores in soil. *Appl Environ Microbiol* 1998;**64**:2463–72.
29. Rantakokko-Jalava K, Jalava J. Optimal DNA isolation method for detection of bacteria in clinical specimens by broad-range PCR. *J Clin Microbiol* 2002;**40**:4211–7.
30. Halstead FD, Lee AV, Couto-Parada X, et al. Universal extraction method for gastrointestinal pathogens. *J Med Microb* 2013;**62**:1535–9.
31. Gross V, Carlson G, Kwan AT, et al. Tissue fractionation by hydrostatic pressure cycling technology: the unified sample preparation technique for systems biology studies. *J Biomol Tech* 2008;**19**:189–99.
32. Zhao J, Carmody LA, Kalikin LM, et al. Impact of enhanced Staphylococcus DNA extraction on microbial community measures in cystic fibrosis sputum. *PLoS ONE* 2012;**7**:e33127.
33. Smith B, Li N, Andersen AS, et al. Optimising bacterial DNA extraction from faecal samples: comparison of three methods. *Open Microbiol J* 2011;**5**:14–7.
34. de Boer R, Peters R, Gierveld S, et al. Improved detection of microbial DNA after bead-beating before DNA isolation. *J Microbiol Methods* 2010;**80**:209–11.
35. Chomczynski P, Sacchi N. The single-step method of RNA isolation by acid guanidinium thiocyanate-phenol-chloroform extraction: twenty-something years on. *Nat Protoc* 2006;**1**:581–5.
36. Golbang N, Burnie JP, Klapper PE, et al. Sensitive and universal method for microbial DNA extraction from blood products. *J Clin Pathol* 1996;**49**:861–3.
37. Mark AM, Butler GC. The isolation of sodium desoxyribonucleate with sodium dodecyl sulfate. *J Biol Chem* 1951;**190**:165–76.
38. Neugebauer JM. Detergents: an overview. *Methods Enzym* 1990;**182**:239–53.
39. Rao MB, Tanksale AM, Ghatge MS, et al. Molecular and biotechnological aspects of microbial proteases. *Microbiol Mol Biol Rev* 1998;**62**:597–635.
40. Ebeling W, Hennrich N, Klockow M, et al. Proteinase K from Tritiracium album Limber. *Eur J Biochem* 1974;**47**:91–7.
41. Lounsbury JA, Coult N, Miranian DC, et al. An enzyme-based DNA preparation method for application to forensic biological samples and degraded stains. *Forensic Sci Int Genet* 2012;**6**:607–15.
42. Kirby KS. A new method for the isolation of ribonucleic acids from mammalian tissues. *Biochem J* 1956;**64**:405–8.
43. Moore D, Dowhan D. Purification and concentration of DNA from aqueous solutions. *Curr Prot Mol Biol* 2002;**59**. 2.1.1-10.
44. Zhang R, Gong H, Zeng X, et al. A microfluidic liquid phase nucleic acid purification chip to selectively isolate DNA or RNA from low copy/single bacterial cells in minute sample volume followed by direct on-chip quantitative PCR assay. *Anal Chem* 2013;**85**:1484–91.
45. Eickbush TH, Moudrianakis EN. The compaction of DNA helices into either continuous supercoils or folded-fiber rods and toroids. *Cell* 1978;**13**:295–306.
46. Adell K, Ogbonna G. Rapid purification of human DNA from whole blood for potential application in clinical chemistry laboratories. *Clin Chem* 1990;**36**:261–4.
47. Broemeling DJ, Pel J, Gunn DC, et al. An instrument for automated purification of nucleic acids from contaminated forensic samples. *J Lab Autom* 2008;**13**:40–8.
48. Berry SM, Alarid ET, Beebe DJ. One-step purification of nucleic acid for gene expression analysis via immiscible filtration assisted by surface tension (IFAST). *Lab Chip* 2011;**11**:1747–53.
49. Feng G, Jiang L, Wen P, et al. A new ion-exchange adsorbent with paramagnetic properties for the separation of genomic DNA. *Analyst* 2011;**136**:4822–9.
50. Polski JM, Kimzey S, Percival RW, et al. Rapid and effective processing of blood specimens for diagnostic PCR using filter paper and Chelex-100. *J Clin Pathol* 1998;**51**:215–7.
51. Vogelstein B, Gillespie D. Preparative and analytical purification of DNA from agarose. *Proc Natl Acad Sci* 1979;**76**:615–9.
52. Melzak KA, Sherwood CS, Turner RFB, et al. Driving forces for DNA adsorption to silica in perchlorate solutions. *J Colloid Interface Sci* 1996;**181**:635–44.
53. Boom R, Sol CJA, Salimans MMM, et al. Rapid and simple method for purification of nucleic acids. *J Clin Microbiol* 1990;**28**:495–503.
54. Berensmeier S. Magnetic particles for the separation and purification of nucleic acids. *Appl Microbiol Biotechnol* 2006;**73**:495–504.
55. He J, Huang M, Wang D, et al. Magnetic separation techniques in sample preparation for biological analysis: a review. *J Pharm Biomed Anal* 2014;**101**:84–101.

56. Hourfar MK, Michelsen U, Schmidt M, et al. High-throughput purification of viral RNA based on novel aqueous chemistry for nucleic acid isolation. *Clin Chem* 2005;**51**:1217−22.
57. Martinez AW, Phillips ST, Whitesides GM, et al. Diagnostics for the developing world: microfluidic paper-based analytical devices. *Anal Chem* 2010;**82**:3−10.
58. Li Y, Yoshida H, Wang L, et al. An optimized method for elution of enteroviral RNA from a cellulose-based substrate. *J Virol Methods* 2012;**186**:62−7.
59. Trang DTX, Huy NT, Kariu T, et al. One-step concentration of malarial parasite-infected red blood cells and removal of contaminating white blood cells. *Malaria J* 2004;**3**:7.
60. Rudi K, Larsen F, Jakobsen KS. Detection of toxin-producing cyanobacteria by use of paramagnetic beads for cell concentration and DNA purification. *Appl Environ Microbiol* 1998;**64**:34−7.
61. Stevens KA, Jaykus LA. Bacterial separation and concentration from complex sample matrices: a review. *Crit Rev Microbiol* 2004;**30**:7−24.
62. Park JM, Lee JY, Lee JG, et al. Highly efficient assay of circulating tumor cells by selective sedimentation with a density gradient medium and microfiltration from whole blood. *Anal Chem* 2012;**84**:7400−7.
63. Alaeddini R. Forensic implications of PCR inhibition—a review. *Forensic Sci Int* 2012;**6**:297−305.
64. Loens K, Bergs K, Ursi D, et al. Evaluation of NucliSens easy-MAG for automated nucleic acid extraction from various clinical specimens. *J Clin Microbiol* 2007;**45**:421−5.
65. Esona MD, McDonald S, Kamili S, et al. Comparative evaluation of commercially available manual and automated nucleic acid extraction methods for rotavirus RNA detection in stools. *J Virol Methods* 2013;**194**:242−9.
66. Shulman LM, Hindiyeh M, Muhsen K, et al. Evaluation of four different systems for extraction of RNA from stool suspensions using MS-2 coliphage as an exogenous control for RT-PCR inhibition. *PLoS ONE* 2012;**7**:e39455.
67. Huijsmans CJJ, Damen J, van der Linden JC, et al. Comparative analysis of four methods to extract DNA from paraffin-embedded tissues: effect on downstream molecular applications. *BMC Res Notes* 2010;**3**:239.
68. Do H, Dobrovic A. Sequence artifacts in DNA from formalin-fixed tissues: causes and strategies for minimization. *Clin Chem* 2015;**61**:64−71.
69. Greytak SR, Engel KB, Bass BP, et al. Accuracy of molecular data generated with FFPE biospecimens: lessons from the literature. *Cancer Res* 2015;**75**:1541−7.
70. Srinivasan M, Sedmak D, Jewell S. Effect of fixatives and tissue processing on the content and integrity of nucleic acids. *Am J Path* 2002;**161**:1961−71.
71. Masuda N, Ohnishi T, Kawamoto S, et al. Analysis of chemical modification of RNA from formalin-fixed samples and optimization of molecular biology applications for such samples. *Nucleic Acids Res* 1999;**27**:4436−43.
72. Janecka A, Adamczyk A, Gasinska A. Comparison of eight commercially available kits for DNA extraction from formalin-fixed paraffin-embedded tissues. *Anal Biochem* 2015;**476**:8−10.
73. Kotorashvili A, Ramnauth A, Liu C, et al. *PLoS ONE* 2012;**7**:e34683.
74. Metcalf D, Weese JS. Evaluation of commercial kits for extraction of DNA and RNA from Clostridium difficile. *Anaerobe* 2012;**18**:608−13.
75. Dundas N, Leos NK, Mitui M, et al. Comparison of automated nucleic acid extraction methods with manual extraction. *J Mol Diagn* 2008;**10**:311−6.
76. Verheyen J, Kaiser R, Bozic M, et al. Extraction of viral nucleic acids: comparison of five automated nucleic acid extraction platforms. *J Clin Virol* 2012;**54**:255−9.
77. Shipley MA, Koehler JW, Kulesh DA, et al. Comparison of nucleic acid extraction platforms for detection of select biothreat agents for use in clinical resource limited settings. *J Microbiol Methods* 2012;**91**:179−83.
78. Chandler DP, Griesemer SB, Cooney CG, et al. Rapid, simple influenza RNA extraction from nasopharyngeal samples. *J Virol Methods* 2012;**183**:8−13.
79. Pinzani P, Salvianti F, Pzzagli M, et al. Circulating nucleic acids in cancer and pregnancy. *Methods* 2010;**50**:302−7.
80. Raja S, Ching J, Xi L, et al. Technology for automated, rapid, and quantitative PCR or reverse-transcription-PCR clinical testing. *Clin Chem* 2005;**51**:882−90.
81. Poritz MA, Blaschke AJ, Byington CL, et al. FilmArray, an automated nested multiplex PCR system for multi-pathogen detection: development and application to respiratory tract infection. *PLoS ONE* 2011;**6**:e26047.
82. Kim J, Johnson M, Hill P, et al. Microfluidic sample preparation: cell lysis and nucleic acid purification. *Integr Biol* 2009;**1**:574−86.
83. Luong JHT, Male KB, Glennon JD. Biosensor technology: technology push versus market pull. *Biotechnol Adv* 2008;**26**:492−500.
84. Ahmad F, Hashsham SA. Miniaturized nucleic acid amplification systems for rapid and point-of-care diagnostics: a review. *Anal Chim Acta* 2012;**733**:1−15.
85. Schrader C, Schielke A, Ellerbroek L, et al. PCR inhibitors — occurrence, properties and removal. *J Appl Microbiol* 2012;**113**:1014−26.
86. Baar C, d'Abbadie M, Vaisman A, et al. Molecular breeding of polymerases for resistance to environmental inhibitors. *Nucleic Acid Res* 2011;**39**:e51.
87. Schuurman T, de Boer R, Patty R, et al. Comparative evaluation of in-house manual, and commercial semi-automated and automated DNA extraction platforms in the sample preparation of human stool specimens for a Salmonella enterica 5'nuclease assay. *J Microbiol Methods* 2007;**71**:238−45.
88. Li X, Wu Y, Zhang L, et al. Comparison of three common DNA concentration measurement methods. *Anal Biochem* 2014;**451**:18−24.
89. Aranda R, Dineen SM, Craig RL, et al. Comparison and evaluation of RNA quantification methods using viral, prokaryotic, and eukaryotic RNA over a 10^4 concentration range. *Anal Biochem* 2009;**387**:122−7.
90. Blackburn GM, Gait MJ, editors. *Nucleic acids in chemistry and biology*. New York: IRL Press at Oxford University Press; 1990.
91. Borer P, Fasman GD, editors. *Handbook of biochemistry and molecular biology*. Boca Raton, Fla: CRC Press; 1975. p. 589.
92. Wilfinger WW, Mackey K, Chomczynski P. Effect of pH and ionic strength on the spectrophotometric assessment of nucleic acid purity. *Biotechniques* 1997;**22**:474−81.
93. Ahn SJ, Costa J, Emanuel JR. PicoGreen quantitation of DNA: effective evaluation of samples pre- or post-PCR. *Nucleic Acids Res* 1996;**24**:2623−5.
94. Singer VL, Jones LJ, Yue ST, et al. Characterization of PicoGreen reagent and development of a fluorescence-based solution assay for double-stranded DNA quantitation. *Anal Biochem* 1997;**249**:228−38.
95. Perkins TT, Quake SR, Smith DE, et al. Relaxation of a single DNA molecule observed by optical microscopy. *Science* 1994;**264**:822−6.
96. Dineva MA, Mahilum-Tapay L, Lee H. Sample preparation: a challenge in the development of point-of-care nucleic acid-based assays for resource-limited settings. *Analyst* 2007;**132**:1193−9.

Nucleic Acid Techniques

Carl T. Wittwer and G. Mike Makrigiorgos

ABSTRACT

Background

The expansion and power of molecular diagnostics is enabled by techniques that modify, amplify, detect, discriminate, and sequence nucleic acids (NAs). Molecular diagnostic techniques are getting faster, better, and cheaper. If these trends translate to clinical medicine, we all win.

Content

Nucleic acids are usually first purified and then amplified to provide enough NA to be easily detected and analyzed. However, not all techniques require NA purification or amplification before analysis. Although the polymerase chain reaction (PCR) remains the most common method of amplification for both research and clinical diagnostics, molecular tests that use isothermal amplification methods are now approved by the US Food and Drug Administration, and some are Clinical Laboratory Improvement Amendments waived. Radioactive detection has been replaced by fluorescence and, in some cases, electronic methods. Separation methods, particularly electrophoresis, once dominated NA analysis and are still used today. However, the advantages of closed systems that detect, identify, and quantify without separation, such as real-time PCR and melting analysis, simplify the workflow and reduce the time required. Multiplexed methods enable the physician to go beyond single target queries and can provide diagnostic answers for clinical syndromes. The specificity of NA hybridization is still central to most methods. Microarrays have great merit in research, and copy number arrays have proven clinical utility. Sequencing methods have progressed from labor-intensive base termination ladders visualized on gels, to the current workhorse of massively parallel methods that sequence during synthesis, to the promise of single molecule sequencing.

INTRODUCTION

The structure of DNA was solved in 1953,[1] with semiconservative replication demonstrated in 1958,[2] providing a conceptual background for nucleic acid (NA) replication and analysis. Restriction enzymes,[3] oligonucleotide synthesis,[4] and reverse transcriptase all become known and available around 1970.[5,6] Southern blotting, perhaps the first practical molecular diagnostic technique, appeared in 1975 using restriction enzyme digestion and size separation on agarose gels.[7] Southern blotting typically detects large structural alterations such as deletions, duplications, insertions, and rearrangements but can also detect single nucleotide variants (SNVs) if they disrupt restriction sites. Southern blotting was the first method with adequate sensitivity and specificity for DNA analysis of single-copy genes in complex genomes. Northern blotting is a parallel technique that analyzes RNA instead of DNA to size RNA transcripts and was developed in 1977.[8] However, both Southern and Northern blotting are seldom used today because they require large amounts of NAs and are also very labor intensive and time consuming. They are covered in more detail in the fifth edition of this book.[9]

DNA sequencing with chain-terminating inhibitors (dideoxynucleotides) was developed in 1977 using gels for size separation.[10] Initially sequencing was radioactive with separation on plates, but fluorescent analysis, automation, and a move to capillaries over the next 30 years increased the utility of this "gold standard" method. Originally described in 1985, the polymerase chain reaction (PCR) was greatly improved in 1988 by using heat-stable polymerases.[11] Of all molecular techniques, PCR has become the most popular method in molecular diagnostics, and many variants and improvements exist, particularly real-time PCR for quantification, which was first described in 1992,[12] with commercial instrumentation becoming available in 1997.[13,14]

DNA microarrays appeared in the 1990s with oligonucleotides attached to glass plates.[15] Applications included RNA expression profiling in 1995[16] followed by genomic arrays for SNVs and copy number comparisons. As a counterpoint to the complexity of arrays, fluorescent melting analysis, first used in 1997,[17] was upgraded in 2003 with higher resolution as a simple tool for genetic analysis.[18]

Massively parallel sequencing, first published in 2005,[19,20] continues to develop into a dominant technology because of the massive number of bases sequenced. Also called next-generation sequencing (NGS), we prefer instead the descriptive term massively parallel sequencing (MPS) rather than a temporal reference that will outdate.

Today, molecular diagnostics continues to advance as a rapidly progressing and highly competitive field, both in academics and in industry. This growth is reflected in the number of publications in molecular diagnostics in recent years

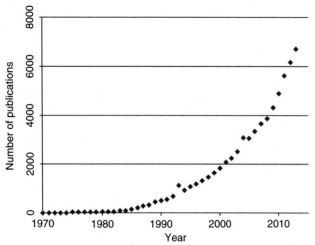

FIGURE 4.1 Molecular diagnostics publications, 1970 to 2013. (Reprinted with permission of the publisher from Chiu RWK, Lo YMD, Wittwer CT. Molecular diagnostics: a revolution in progress. *Clin Chem* 2015;61:1–3. Copyright 2015 AACC.)

(Fig. 4.1).[21] In this chapter, our intent is to focus on understanding the mechanisms of NA techniques in use today.

We begin by considering pretreatment methods for NAs and then focus on amplification techniques that are often necessary to observe or quantify NA sequences of interest. Next, the tools used to detect or visualize NAs are discussed, along with methods that allow identification, quantification, and segregation of individual NA species. Finally, we end with the extraordinary power of massively parallel sequencing. Important terms and definitions are listed in Box 4.1.

NUCLEIC ACID PREPARATION

Conventional NA testing requires (1) NA isolation of DNA, RNA, or both; (2) amplification of the NA; and (3) detection, analysis, or quantification. Sometimes one step can be eliminated or two steps can be combined depending on the sample type and quantity of the target. For example, direct amplification from blood, serum, plasma, cerebrospinal fluid, nasopharyngeal swabs, and other sources may require little or no sample preparation for PCR when high temperatures for denaturation

---

BOX 4.1 Important Terms and Definitions

Adapter: Oligonucleotides that are ligated to library fragments in order to provide consensus priming sites.

Allele-specific polymerase chain reaction (PCR): A version of PCR in which only one allele at a locus is amplified. Specificity is achieved by designing one or both PCR primers so that they overlap the site of sequence variance between alternative alleles.

Amplicon: The product of an amplification reaction, such as PCR.

Antisense RNA (asRNA): A single-stranded RNA that is complementary to a messenger RNA (mRNA) strand transcribed within a cell.

Asymmetric PCR: A version of PCR that preferentially amplifies one strand of the target DNA.

Branched-chain signal amplification: A molecular probe technique that uses branched DNA (bDNA) as a means to amplify the hybridization signal.

Conformation-sensitive gel electrophoresis (CSGE): A type of electrophoretic mutation scanning in which a segment of DNA is screened for mismatch pairing between normal and variant base pairs.

Copy number variant (CNV): A segment of DNA with copy-number differences between genomes.

Coverage: The percent of target bases that were sequenced at least a given number of times.

Denaturing gradient gel electrophoresis (DGGE): An electrophoretic method for separating DNA fragments according to their mobility under increasingly denaturing conditions (usually increasing formamide or urea concentrations). See also TGGE, SSCP and heteroduplex analysis.

Deoxyribonucleotide triphosphates (dNTPs): Usually dATP, dCTP, dGTP, and dTTP, the building blocks of DNA.

DNA library: A collection of DNA fragments with ligated adapters that will be sequenced.

Dideoxy-termination sequencing (Sanger sequencing): A method of DNA sequencing based on the selective incorporation of chain-terminating dideoxynucleotides by DNA polymerase during in vitro DNA replication.

Digital polymerase chain reaction (digital PCR or dPCR): A modification of PCR with the sample separated into many partitions so that some partitions have no template (0) and others have one or more (1).

Insert: Part of the original DNA that has been fragmented before ligation to adapters.

Fluorescent in situ hybridization (FISH): A genetic mapping technique using fluorescent tags for analysis of chromosomes and genetic abnormalities. FISH can also be referred to as chromosome painting.

gb: One billion bases or 1 gigabase (1,000,000,000 bases).

Heteroduplex: A DNA duplex with internal mismatches or loops.

Heteroduplex analysis (HDA): A type of mutation scanning in which a segment of DNA is screened by gel or capillary electrophoresis for mismatch pairing between normal and variant base pairs.

Homoduplex: A perfectly matched DNA duplex.

Hybridization: The annealing or pairing of two DNA strands.

Insertion: An extra DNA sequence that is present in one sample compared with a reference sequence.

kb: One thousand bases or 1 kilobase (1,000 bases).

Loop-mediated isothermal amplification (LAMP): A single tube technique for the amplification of DNA that uses a single temperature incubation.

Massively parallel sequencing (MPS): Sequencing of many fragments of DNA simultaneously.

Mate-pair sequence: Sequence obtained from both ends of a DNA fragment that is typically 5000 to 10,000 bases long.

mb: One million bases or 1 megabase (1,000,000 bases).

Melting curve: A measurement of the dissociation of double-stranded DNA during heating.

Multiplex ligation-dependent probe amplification (MLPA): A variation of the multiplex polymerase chain reaction that assesses the copy number of several targets by permitting multiple targets to be amplified with only a single primer pair.

Next-generation sequencing (NGS): Massively parallel sequencing. The term *massively parallel sequencing* is preferred.

BOX 4.1 Important Terms and Definitions—cont'd

Northern blot analysis: A technique for identifying specific sequences of RNA in which RNA molecules are (1) separated by electrophoresis, (2) transferred to nitrocellulose, and (3) identified with a suitable probe.

Nucleic acid analogs: Compounds that are analogous (structurally similar) to naturally occurring RNA and DNA. They are used in chemotherapy and molecular biology research.

Oligonucleotide: A short single-stranded polymer of nucleic acid.

Oligonucleotide ligation assay (OLA): A technique for determining the presence or absence of a specific nucleotide pair within a target gene, often indicating whether the gene is wild type (normal) or mutant (defective).

Paired-end sequence: Sequence from both ends of a DNA fragment typically hundreds of bases long.

Peptide nucleic acid (PNA): An artificially synthesized polymer similar to DNA or RNA with peptide instead of phosphodiester bonds. The term is somewhat of a misnomer because PNA is not an acid.

Polony: A microscopic colony of clonal temples used in massively parallel sequencing. A polony may be generated by polymerase chain reaction, bridge amplification, or isothermal amplification.

Polymerase chain reaction (PCR): An in vitro method for exponentially amplifying DNA.

Primer: An oligonucleotide that serves to initiate polymerase-catalyzed addition of nucleotides by annealing to a template strand.

Pyrosequencing: A method of DNA sequencing based on the "sequencing by synthesis" principle that relies on the detection of pyrophosphate release on nucleotide incorporation.

Read: The nucleotide sequence inferred by sequencing of a template.

Read length: Number of bases sequenced.

Real-time PCR: Observation of PCR during amplification at least once each cycle.

Restriction fragment length polymorphism (RFLP): A genetic polymorphism revealed by changes in the sizes of DNA fragments after restriction enzyme digestion and electrophoresis.

Reverse transcriptase: A polymerase that catalyzes synthesis of DNA from an RNA template.

Rolling circle amplification (RCA): A probe amplification method with a linear probe that is ligated to form a circle in the presence of template. The circle is replicated continuously by a polymerase and one or more primers.

Sanger sequencing: See dideoxy-termination sequencing.

Serial invasive signal amplification: A signal enhancing technique that combines two invasive signal amplification reactions in series in a single-tube format. The cleaved 5′-arm from the target-specific primary reaction is used to drive a secondary invasive reaction, resulting in a total signal amplification of more than seven orders of magnitude in about 4 hours.

Single nucleotide variants (SNV) chip: A type of DNA microarray that can detect single nucleotide variants, insertions, and deletions. Also known as a single nucleotide polymorphism (SNP) chip.

Single-stranded conformational polymorphism (SSCP): A gel electrophoresis technique where single-stranded DNA segments are identified by their abnormal migration patterns.

Southern blot: A method for detecting DNA sequence variants after restriction enzyme digestion and size separation by electrophoresis. Hybridization with a labeled probe reveals sequence variants that result in a change in distance between restriction sites, including (1) deletions, (2) duplications, (3) insertions, and (4) rearrangements.

Strand displacement amplification (SDA): An amplification technique that uses two types of primers, DNA polymerase, and a restriction endonuclease to exponentially produce single-stranded amplicons asynchronously.

Temperature-gradient gel electrophoresis (TGGE): A form of electrophoresis that uses temperature to denature the sample as it moves across an acrylamide gel.

Transcription-mediated amplification (TMA): An amplification method that uses RNA polymerase and DNA reverse transcriptase to produce RNA amplicon from a target nucleic acid. TMA is used to amplify both RNA and DNA.

Transcriptome: The set of all RNA molecules, including (1) mRNA, (2) rRNA, (3) tRNA, and other (4) noncoding RNA produced in one or a population of cells.

Virtual karyotyping: A technique used to identify and quantify short sequences of DNA from specific loci all over a genome to obtain information that reflects a karyotype.

Whole-genome amplification (WGA): A nonspecific amplification technique that produces an amplified product representative of the initial starting material (whole genome).

---

can make sufficient NAs available. However, quantification and sensitivity may suffer with low extraction efficiency and sample dilution. Single-molecule detection and single-molecule sequencing do not always require amplification, although whole-genome amplification is often used in single-cell analysis. Separate amplification and detection are not necessary in real-time PCR, and product melting can be seamlessly added or measured during PCR for additional analysis.[22]

DNA and RNA isolation are covered in Chapter 3. In addition to NA isolation, sometimes it is necessary to prepare samples for amplification or analysis by enzymatic or nonenzymatic means. For example, the enzyme reverse transcriptase may be needed to convert RNA to DNA. Massively parallel sequencing usually requires library preparation that may include enzymatic fragmentation with nonspecific DNAses, end repair of the generated fragments with a polymerase to produce blunt ends, 5′- phosphorylation with a kinase, A-tailing with a polymerase, and adaptor ligation with a ligase.[23] Alternatively, a transpososome complex containing a transposase and double-stranded sequencing adaptors can be used to streamline fragmentation to ligation in one step.[24] Enzymes that act on NAs are covered in Chapter 1. Nonenzymatic methods may also be used to process NAs before amplification or analysis.

Nucleic Acid Fragmentation

Boiling, acid–base treatment, sonication, mechanical shearing, and chemical cleavage can be used to cut DNA or RNA into smaller fragments to make subsequent analysis more efficient or to form libraries. Boiling is a simple way to fragment genomic DNA so that a long initial denaturation period is not necessary in PCR. Although DNA is fragmented by acid and RNA is fragmented by base, these methods are seldom used except in bulk procedures such as blotting. Sonication and acoustic shearing are common methods to fragment NA for library preparation and are used in massively parallel sequencing. Depending on the frequency and geometry of the sample and acoustic generator, fragments averaging from 100 to 20,000 bases can be produced. Hydrodynamic shearing can also be obtained by compressed air to atomize the liquid into a fine mist (nebulization), forcing the solution through a fine-gauge needle, or through a pressure cell. For chemical cleavage, several metal-ion–catalyzed chemistries can cleave single- and double-stranded NAs. Some alkylating compounds can cleave and label NAs at the same time. One example is 5-bromomethyl-fluorescein, which is catalyzed by metal ions to fragment RNA or DNA; then those fragments are simultaneously labeled with fluorescein for microarray analysis.[25] Another example of chemical treatment of DNA is the use of hydroxylamine or osmium tetroxide followed by cleavage of mismatches with piperidine as a way to detect and locate mutations.[26]

Bisulfite Treatment for Methylation Analysis

Bisulfite treatment of DNA is often used to analyze the methylation status of cytosine (C) residues in DNA. Sodium bisulfite ($NaHSO_3$),[27] optionally used together with ammonium bisulfite,[28] converts C into uracil (U), but does not affect 5-methylcytosine. The chemical process of bisulfite treatment is shown in Fig. 4.2. The process works effectively only on single-stranded DNA (ssDNA), so the sample needs to be denatured by heat, alkali, or chaotropic agents such as urea or formamide. Analysis of the bisulfite-treated DNA is usually performed after NA amplification. DNA polymerases used for amplification will recognize 5-methylcytosine as C (no change), but an unmethylated C is converted to U and will be recognized as a T (sequence change of C to T). Many methods can be used to detect and quantify the altered sequences that result from bisulfite treatment of methylated DNA, including allele-specific amplification, detection with probes, melting techniques, and sequencing.[29] One limitation of bisulfite treatment is significant degradation of DNA. Depending on the protocol, as much as 90% of the DNA can be lost. Prior affinity enrichment of methylated DNA may be used before bisulfite treatment. For example, methylated DNA can be enriched by immunoprecipitation with an antibody raised against DNA containing 5-methylcytosine.[30] Up to 90-fold enrichment of methylated DNA can be achieved by immunoprecipitation. Another method is to use a methylated DNA binding protein to capture double-stranded methylated DNA on an affinity column.[31]

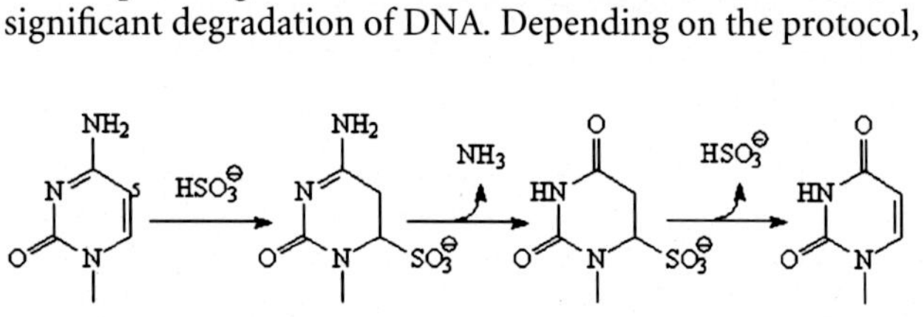

AMPLIFICATION TECHNIQUES

Molecular diagnostics requires techniques to detect extremely low concentrations of NAs in a background of complex genomic structure. Achieving sensitive detection limits is a central concern for clinical applications of NA analysis. Techniques that increase the amount of the NA target, the detection signal, or the probe are referred to as *amplification methods*. Examples of amplification methods are listed in Table 4.1. In *target amplification*, a well-defined segment of the NA (the target sequence) is copied many times by in vitro methods. Areas outside the target are not amplified.

FIGURE 4.2 Bisulfite-mediated conversion of cytosine (C) to uracil (U) occurs in three steps. The first step is the addition of bisulfite to C. This reaction occurs at acid pH. The second step is the deamination of cytidine–bisulfite (C-SO_3^-) to produce uracil–bisulfite (U-SO_3^-), which is optimal at a pH of 5 to 6. Before analysis, U-SO_3^- is converted to U by adjusting the pH to alkali. The majority of methylation on the C residue in mammalian cells occurs at the carbon 5 position (shown in the structure of C), resulting in 5-methylcytosine, which is resistant to bisulfite-mediated conversion.

TABLE 4.1 Common Amplification Techniques

Techniques	Type	Enzymes Required
Polymerase chain reaction (PCR)	Target	DNA polymerase (thermostable)
Transcription-mediated amplification (TMA)	Target	Reverse transcriptase, RNA polymerase, RNase H
Self-sustained sequence replication (3SR)		
Nucleic acid sequence–based amplification (NASBA)		
Strand displacement amplification (SDA)	Target	HincII, DNA polymerase I (5′- exo-deficient)
Loop-mediated amplification (LAMP)	Target	DNA polymerase
Helicase dependent amplification (HDA)	Target	Helicase, DNA polymerase
Recombinase polymerase amplification (RPA)	Target	Recombinase, single-strand binding protein, polymerase
Whole-genome amplification (WGA)	Target	Φ29 DNA polymerase
Multiple displacement amplification (MDA)		
Antisense RNA amplification (aRNA)	Target	T4 DNA polymerase, Klenow, S1 nuclease, T7 polymerase
Branched DNA (bDNA)	Signal	Alkaline phosphatase
Serial invasive signal amplification	Signal	Cleavase
Rolling circle amplification (RCA)	Probe	Φ29 DNA polymerase

In *signal amplification,* the amount of target stays the same, but the signal is increased by one of several methods, including sequential hybridization of branching NA structures and continuous enzyme action on substrate that may be recycled. Finally, in *probe amplification,* the probe (or a product of the probe) is amplified only in the presence of the target. Amplification techniques often can achieve more than a million-fold amplification in less than 1 hour.

> **POINTS TO REMEMBER**
>
> **Nucleic Acid Amplification Methods**
> - May amplify the target, the signal, or the probe
> - May be isothermal or cycle through different temperatures
> - Cycling speed is limited by instrumentation, not biochemistry
> - May be analyzed on gels or in real-time
> - May be qualitative, quantitative, or digital
> - Require positive and negative controls

Polymerase Chain Reaction—Target Amplification

When the amount of target NA is increased by synthetic in vitro methods, target amplification occurs. PCR[11] is the best known and most widely applied of the target amplification methods. Because of the commercial availability of thermostable DNA polymerases, kits, and instrumentation, this method has been widely adopted in research and is routinely used in the clinical laboratory.

Details of the PCR Process

Polymerase chain reaction requires (1) a thermostable DNA polymerase, (2) deoxynucleotides of each base (collectively referred to as *dNTPs*), (3) a DNA strand that includes the sequence to be amplified, and (4) a pair of oligonucleotides (referred to as *primers*) that are complementary to opposite strands that define the target sequence. In the first step, target duplexes are denatured into single strands by heat (Fig. 4.3). When the mixture is cooled, primers provided in great excess (usually over 1 million times the concentration of the initial target) specifically anneal to complementary sequences on the target. After the primers are annealed, the action of the polymerase synthesizes two new DNA strands by extending each of the two primers at their 3′ end. The primers are designed to recognize sequences of the target that are close enough together such that the polymerase extends each strand far enough to include the priming site of the other primer. Usually, the optimal temperature for polymerization (or DNA synthesis) is an intermediate temperature between the annealing temperature (at which primer hybridizes to its target) and the denaturation temperature (at which the newly generated DNA strand dissociates from its template). The second cycle also begins with denaturation, but now twice as many strands (the original genomic DNA and the extension products from the first cycle) are available for primer annealing and subsequent extension. Temperature cycling (typically) uses three temperatures: (1) a high temperature sufficient to denature the target sequence, typically 90° to 97°C but can be lower depending on the melting temperature of the product; (2) a low temperature that allows annealing of the primers to the target, typically 50° to 65°C; and (3) a third temperature optimal for polymerase extension, typically 65° to 75°C. A more complete schematic of PCR is shown in Fig. 4.4, detailing why only products of defined length are generated. The instrument that takes samples through the multiple steps of changing temperature is known as a *thermocycler*.

Repetitive thermocycling results in the exponential accumulation of the short product (consisting of primers and all intervening sequences). If the efficiency of each cycle is perfect, the number of target sequences doubles each cycle (the efficiency is 100% or 1.0). PCR efficiency depends on the primers, the temperature-cycling conditions, and any inhibitors that might limit amplification. Amplified products accumulate exponentially in the beginning cycles of PCR. At some point, however, the efficiency of amplification falls and eventually the amount of product plateaus (Fig. 4.5) as the result of exhaustion of components or competition between primer and product annealing (ie, the single strands of product are at such high concentrations that they anneal to each other rather than to the primers). The S-curve shape is similar to the logistical model for population growth. In a typical PCR reaction using 0.5 μmol/L of each primer, the maximum DNA concentration typically achieved is about 100 billion copies/μL.

With the addition of reverse transcriptase, RNA targets can be converted into cDNA and then successfully amplified. Reverse transcription and DNA amplification are most often catalyzed by two different polymerases. In one-step reverse transcriptase polymerase chain reaction (RT-PCR), both enzymes are present in a common buffer, and typically PCR primers are used for both reverse transcription and DNA amplification. Some thermostable enzymes have both DNA polymerase and reverse transcriptase activities so that both steps can be performed in the same tube with the same enzyme. In two-step RT-PCR, the reverse transcription is performed first, usually with random hexamers or a poly-dT oligonucleotide (to prime the poly-A tail of most mRNA). After reverse transcription, the second PCR step is performed on cDNA with specific primers.

After amplification, the products can be detected by various methods. Gel electrophoresis with ethidium bromide staining is a classical method that separates products by size and may

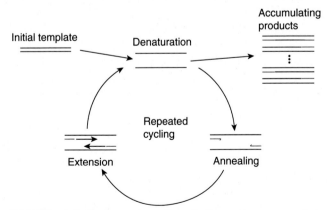

FIGURE 4.3 Simple schematic of the polymerase chain reaction. Amplification of the initial template requires denaturation by heat, allowing primers to anneal at a lower temperature, followed by primer extension at an intermediate temperature. In each cycle, a doubling of the DNA product occurs. After 20 to 40 cycles, the product accumulates more than 1 million-fold.

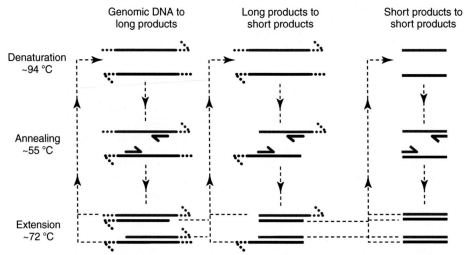

FIGURE 4.4 A more detailed schematic of polymerase chain reaction. Repetitive cycles of denaturation, annealing, and extension are paced by temperature cycling of the reaction. Two primers (indicated as short segments) anneal to opposite template strands *(long red and black lines)* to define the region to be amplified. Extension occurs from the 3′-ends *(indicated with half arrowheads)*. In each cycle, genomic DNA is denatured and annealed to primers that extend in opposite directions across the same region, producing long products of undefined length. Long products generated by extension of one of the primers anneal to the other primer during the next cycle, producing short products of defined length. Any short products present produce more short products. After n cycles, 2^n new copies of the amplified region (n long products + $[2^n - n]$ short products) are generated from each original genomic copy. A similar approach can be used to amplify RNA targets by initial reverse transcription of the RNA to produce a DNA template.

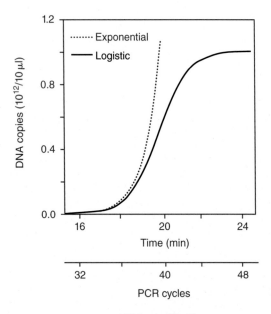

DNA amplification

FIGURE 4.5 Exponential and logistic curves for DNA amplified by polymerase chain reaction (PCR). A doubling time of 30 seconds is assumed for PCR, that is, given the equation $N_t = N_0 e^{rt}$, in which N_t is the amount of DNA at time t and N_0 is the initial amount of DNA, and r is 1.386 min^{-1} for PCR. A carrying capacity of 100 billion copies of PCR product per μL was used, assuming that the reaction is primer limited at one third the primer concentration (initially at 0.5 μmol/L, or 300 billion primer pairs per μL). Starting with only one target copy, it takes 23 minutes (46 cycles) to amplify the target to saturation when the doubling efficiency is 100%. (Modified with permission of the publisher from Wittwer CT, Kusukawa N. Real-time PCR. In: Persing DH, Tenover FC, Versalovic J, eds. *Molecular microbiology: diagnostic principles and practice*. Washington, DC: ASM Press, 2004:71–84. Copyright 2004 ASM Press.)

suffice for many applications. For fine size discrimination down to a single base, one of the primers can be fluorescently labeled and the post-PCR fragments separated in capillaries on instruments that are typically used for conventional Sanger dideoxy-termination sequencing. Alternatively, some form of hybridization assay can be used to verify or analyze the amplified product. Automated methods are always attractive, and closed-tube methods, in which amplified products are never exposed to the open environment, are particularly advantageous to avoid contamination of future reactions with the products of a prior reaction. Adding a fluorescent dye or probe before amplification allows optical monitoring to follow the reaction as it progresses (real-time PCR) or after the reaction is complete (endpoint melting) without the need to process the sample for a separate analysis step.

Polymerase Chain Reaction Kinetics

It is natural to think about PCR in terms of three events—denaturation of double-stranded target, annealing of primers to their targets, and extension of the DNA strand from the primer—that occur at three different temperatures, each requiring a certain amount of time. Indeed, it is common to perform PCR by holding the reaction mixture at three different temperatures (eg, denaturation at 94°C, annealing at 55°C, and extension at 72°C). Standard thermocyclers that use conical tubes focus on accurate temperature control of the heating block at equilibrium, not on dynamic control of the sample temperature throughout cycling. As a result, sample temperatures are not well defined during transitions, and long cycle times have become standard to ensure that samples reach target temperatures. Reproducibility between instruments and manufacturers is poor, and PCR may require an hour or more to complete 30 cycles of amplification.

The kinetics of denaturation, annealing, and extension suggest that rapid transitions between temperatures with minimal

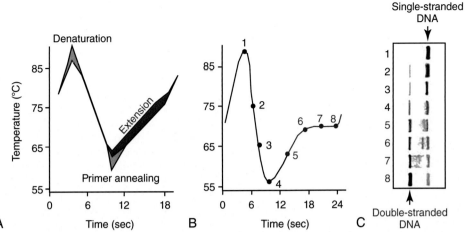

FIGURE 4.6 A visual demonstration of polymerase chain reaction (PCR) kinetics. The three phases of PCR (denaturation, annealing, and extension) occur as the temperature continuously changes (A). Toward the end of temperature cycling, the reaction contains single- and double-stranded PCR products. When different points of the cycle (B) are sampled and analyzed, the transition from denatured single-stranded DNA to double-stranded DNA appears as a continuum (C). Sampling is by snap cooling in ice water with analysis on cold agarose gels to best separate single- from double-stranded products. Progression of the extension reaction can be followed by additional bands appearing between the single- and double-stranded DNA (time points 5 to 7). (Modified with permission from Wittwer CT, Herrmann MG. Rapid thermal cycling and PCR kinetics. In: Innis M, Gelfand D, Sninsky J, eds. *PCR applications*. San Diego: Academic Press, 1999:211−229. Copyright 1999 Academic Press.)

or no pauses (temperature plateaus) provide a better paradigm of PCR amplification (Fig. 4.6). Denaturation, annealing, and extension are very rapid reactions as shown by experiments in capillaries.[32] The use of temperature "spikes" at denaturation and annealing, instead of extended temperature plateaus, allows for rapid cycling with the appropriate instrumentation.[33] The actual time required for PCR depends on the size of the product, but when it is less than 500 base pairs (bp), 30 cycles are easily completed in 15 minutes. Furthermore, rapid amplification improves specificity. Fig. 4.7 shows PCR amplification of the same product amplified at different cycling speeds. With conventional slow cycling, many nonspecific products are generated (see Fig. 4.7, A). These products disappear as the cycling time is decreased (see Fig. 4.7, B to D). In fact, amplification yield and product specificity are optimal when denaturation and annealing times are minimal. However, very rapid cycling may compromise PCR efficiency that is critical for quantitative PCR.

Requirements for denaturation, annealing, and extension in PCR have been reviewed.[34] Initial denaturation of genomic DNA may be required before PCR cycling, depending on how the DNA sample was prepared. Prior boiling of the sample or an initial denaturation step of a few seconds may be necessary to denature very longs strands of genomic DNA. During PCR, however, denaturation of the shorter products occurs very rapidly. Even for long PCR products, denaturation is complete in less than 1 second after the denaturation temperature is reached. Anything greater than a denaturation time of 0 serves only to degrade the polymerase. If longer denaturation times are required, either the sample is not reaching temperature or heat-activated polymerases are being used. Product specificity is optimal when annealing times are less than 1 second. Longer annealing times may be required if the primer concentrations are low. The required extension time for each cycle depends on the length of the PCR product. Extension is not instantaneous, although it is much faster than common practice would suggest. Extension rates of typical polymerases under in vitro conditions are 50 to 100 bases per second. Molecules are fast; people and PCR instruments are slow. Indeed, by increasing the primer and polymerase concentrations 10- to 20-fold, a 60−base pair fragment of genomic DNA was amplified with good efficiency, yield, and specificity in less than 15 seconds.[35] This was achieved with 35 cycles of PCR (each <0.5 seconds) and did not require the use of any hot start method.

Polymerase chain reaction products can be as small as 40 bp to about 40 kilobases (kb). To amplify products longer than 2 kb, mixtures of polymerases that include some 3′-exonuclease activity to edit mismatched nucleotides are usually used. Instead of separate annealing and extension temperatures, both processes can be carried out at the same temperature, resulting in two-temperature, instead of three-temperature, cycling. *Taq* DNA polymerase and other polymerases have a terminal transferase activity that may add a single unpaired nucleotide to the 3′-end of PCR product strands. In the presence of all four dNTPs, dATP is preferentially added. This means that many of the double-stranded products generated by PCR will have a protruding A at one or both 3′ ends. Although this does not influence most detection protocols, it can complicate some systems with high size resolution. On the other hand, this feature can be useful in high-efficiency cloning and library construction in massively parallel sequencing.[36]

Polymerase Chain Reaction Optimization

In addition to temperature-cycling conditions, the specificity of PCR depends on the choice of primers and the Mg^{2+} concentration. Mg^{2+} is a polymerase cofactor and also stabilizes the DNA double helix. Low concentrations of Mg^{2+} favor specificity, but high concentrations favor sensitivity. Typical Mg^{2+} concentrations in legacy PCR are rather limited (1.5−2.5 mmol/L), although greater concentrations may be needed to offset chelation of Mg^{2+} by

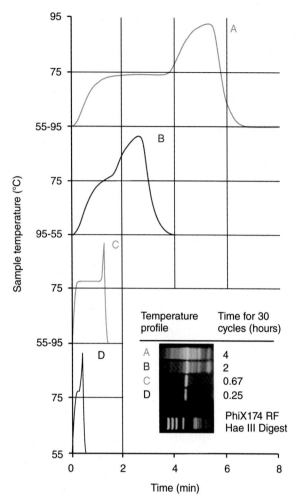

FIGURE 4.7 Rapid polymerase chain reaction improves product specificity. Human genomic DNA samples were cycled 30 times through temperature profiles *A, B, C,* and *D*. Increased specificity of amplification of the 536 bp β-globin fragment of human genomic DNA is seen with faster cycling (*C* and *D*). (Reprinted by permission of the publisher from Wittwer CT, Garling DJ. Rapid cycle DNA amplification: time and temperature optimization. *BioTechniques* 1991;10:76–83. Copyright 1991 Eaton Publishing.)

dNTPs, ethylenediaminetetraacetic acid (EDTA), or citrate in the sample.[37] Furthermore, Mg^{2+} concentrations up to 5 mmol/L are required with cycle speeds below 1 seconds.[35] Some common components of PCR buffers, such as K^+, inhibit polymerase extension 75% at 50 mmol/L.[38]

Polymerase chain reaction sensitivity and specificity can be compromised by the formation of unintended low molecular weight artifacts. This process is initiated before PCR, when the primers, template, and polymerase are all together at temperatures below the annealing temperature of the PCR primers. At low temperatures, if a primer momentarily anneals to another primer or to an unintended region, the polymerase may extend the complex. If the extension product, in turn, is primed and extended, then unintended, double-stranded products can be formed (eg, primer-dimers) that serve as amplification templates throughout the reaction. Primer-dimers can be distinguished from the intended target by their molecular weight or melting temperature, but they also influence the efficiency of the intended amplification and decrease assay sensitivity.

The formation of primer-dimers can be minimized in several ways. Almost all limit the activity of polymerase until the temperature is increased (thus the strategy is often collectively called a *hot start*). One method of hot start involves the use of antibody (or an aptamer) to bind and inactivate the polymerase at room temperature. The binding agent is released upon heating, allowing polymerase extension. Another method uses wax or paraffin to create a physical barrier between the essential components in the reaction. Finally, the polymerase, primers, or dNTPs can be chemically blocked so that extension cannot occur until activated by heat, usually requiring an initial denaturation period of 2 to 20 minutes. In an alternative approach, primers are designed to favor a self-annealing, hairpin formation at low temperatures, thus reducing the formation of primer dimers. After the temperature is raised, the primers become linearized and can bind the target DNA, resulting in polymerase extension.

Primer Design

The choice of primers often dictates the quality and success of the amplification reaction. To select primers, the sequence of the target must be known. Some guidelines for primer selection are intuitive and helpful:

1. Make sure the primers anneal to opposite DNA strands of your target with their 3′-ends directed toward each other. The shorter the distance between the primers, the smaller the PCR product and the easier it is to amplify with high efficiency and greater speed. By limiting the extension time, shorter products are amplified with greater efficiency than longer products. With rapid control of temperature cycling, short extension periods selectively amplify shorter products because longer products do not have the time to fully extend.
2. Avoid primers that anneal to themselves or to other primers and particularly avoid complementation at the 3′-end of primers.
3. Choose primers that are specific to the target. Avoid simple sequence repeats and common repeated sequences, such as Alu repeats. If your target has close relatives, design your primers so that they will anneal only to your intended target. Targets that need to be avoided include pseudogenes (for genomic DNA) and related bacterial or viral strains (for microorganisms).
4. Avoid primers that have sequence complementary to internal sequences of the intended product, especially at their 3′-ends.
5. Typically, choose primers between 18 and 25 bases that are matched in melting temperature (Tm) to each other. A primer greater than 17 bases long has a good chance of being unique in the human genome.
6. It is important to avoid sequences similar to your primers that are present in the background DNA likely to be present in your assay. The National Center for Biotechnology Information (NCBI) Basic Local Alignment Search Tool (BLAST) is commonly used to search for similar sequences.[39]
7. Mismatches are better tolerated at the 5′-end rather than the 3′-end of primers. Mismatches at the 3′-end base are most discriminatory and can be used for allele-specific amplification. For efficient amplification, try to completely match at least the first 6 bases at the 3′-end that are important for polymerase binding. Variation at the 5′-end is better tolerated and can be used to introduce restriction sites as long as melting temperature (Tm)

differences are considered. Primers can also include 5′-tails that are not homologous to the target for subsequent amplification or detection.

Many primer selection programs are available commercially and others can be freely obtained over the Internet. However, very few of the selection rules have been empirically tested.

Polymerase Chain Reaction Contamination, Inhibition, and Controls

Because PCR typically produces about 1 trillion copies of the target, a small amount of previously amplified product in a new sample reaction will result in false-positive results. Thus, minute aerosol droplets contain enough PCR product for robust amplification. PCR products can contaminate: (1) reagents, (2) pipettes, and (3) glassware. Simple laboratory precautions[40] can minimize contamination by using (1) physically separated areas for pre- and postamplification steps, (2) positive-displacement or barrier filter pipettes to minimize aerosol contamination, and (3) prior preparation and storage of individual or combined ("master mix") reaction components in small aliquots. Chemical methods can prevent contamination from affecting future reactions.[9] These include substituting U for T in PCR with subsequent degradation of the incorporated Us by uracil N-glycosylase or irradiation of DNA product with ultraviolet light. The most effective way of preventing contamination is to contain the product within a closed tube as in real-time PCR. Even with contamination precautions, a negative control or blank reaction (including all reactants except target DNA) is one of the most important controls for PCR.

One advantage of PCR is that it does not require highly purified NA to achieve a successful amplification. In practice, however, clinical samples may contain unpredictable amounts of impurities that can inhibit polymerase activity. Consequently, to ensure reliable amplification in clinical analyses, some form of NA purification is often used (see Chapter 3). The diverse nature of PCR inhibitors within clinical specimens requires demonstration that the sample (or preparation of NA purified from it) will allow amplification. Although such confirmation is automatic with genotyping assays, detection and quantification reactions typically use a control NA sequence different from the target that is added to the sample (or to NA extracted from the sample). Failure to amplify this positive control indicates that further purification of the sample is required to remove inhibitors.

Multiplex Polymerase Chain Reaction and Nested Polymerase Chain Reaction

Multiplex PCR refers to amplifying more than one target simultaneously in the same solution. This can be done by using consensus primers that bind to and amplify more than one bacterial species, for example, ribosomal sequences with high variation internal to the primers, but flanking identical sequence that can be used for consensus primer binding. Usually, however, multiple primer sets are designed for multiple targets and amplified in the same solution at the same time. Although potential primer interactions increase exponentially with multiplex PCR, it works surprising well most of the time as long as the number of PCR cycles is kept low. If lower primer concentrations are used, the annealing time may need to be increased to maintain efficient annealing. Multiplex PCR is often called "preamplification" if it is followed by an additional amplification reaction.

If PCR is performed with a pair of outer primers and then that PCR product is amplified yet again with a set of inner primers, this is referred to as *nested PCR*. Typically, the first round PCR product is diluted 1000 to 1 million times before the inner (nested) primers are added. The advantage of nested PCR is that both sensitivity and specificity increase. The disadvantage is increased potential for contamination, particularly if the first round of PCR products is handled manually for dilution and transfer. Multiplex PCR followed by nested PCR has been used in a closed-tube system for multiplex detection of infectious agents.[41]

Asymmetric Polymerase Chain Reaction

Conventional PCR uses primers that are present in equal amounts, thereby ensuring that most of the DNA products are double stranded. *Asymmetric PCR* uses different concentrations of the two primers to generate more of one strand than of the other. For instance, the use of one primer at 0.5 μmol/L and the other at 0.005 μmol/L produces mostly ssDNA. Yield of the product, however, may be low with this technique. With less extreme ratios (eg, one primer at 0.5 μmol/L and the other at 0.2 μmol/L), the yield is mostly preserved, with one strand produced in enough excess to make it readily available for probe hybridization.

One way to improve the efficiency of asymmetric PCR is to equalize the Tms of the excess and limiting primers. The lower concentration of the limiting primer results in a lower Tm than the excess primer. In "linear after the exponential" (LATE-PCR),[42] the stability of the limiting primer is raised sufficiently (typically by making it longer) to counteract the effect of its lower concentration. As a result, the two primers have a comparable ability to bind the template during the initial cycles of PCR. After exponential amplification, linear amplification provides ample template for downstream hybridization assays.

Selective Amplification of Sequence Variants by Polymerase Chain Reaction

Allele-specific PCR enables preferential amplification of one genetic allele over another by placing the 3′-end of one primer at the polymorphic site. The variant that is better matched to the primer extends more readily than the other allele. This strategy can be used to distinguish a gene from its pseudogenes and for genotyping SNVs. Allele-specific PCR can also be used to determine haplotypes.[43]

Polymerase chain reaction can also be modified to enrich for variants internal to the primer binding sites. "Co-amplification at lower denaturation temperature PCR" (COLD-PCR) enriches variants irrespective of their type or position within PCR products smaller than 200 bp.[44] Mismatch-containing sequences have a slightly lower denaturation temperature than fully homologous sequences. During PCR cycling (Fig. 4.8), a product hybridization step after denaturation enables mutation-containing product strands to bind to wild-type strands, forming *heteroduplexes*. The temperature of this step is too high for primers to bind. Next, the temperature of the reaction is raised to a critical denaturation temperature (T_c) that generates preferential denaturation of the mismatch-containing sequences. The temperature is then lowered to allow primers to bind leading to preferential replication of mutation-containing sequences. By repeating this protocol over several PCR cycles, mutation-containing strands are preferentially amplified over wild-type strands. With identification of the correct T_c to within 0.5°C, variant enrichment of 10- to 20-fold for Tm-decreasing (G:C

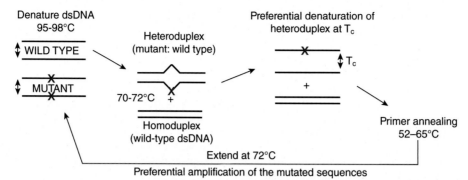

FIGURE 4.8 Schematic of COLD-PCR (co-amplification at lower denaturation temperature polymerase chain reaction). The technique enriches any sequence variant between PCR primers without requiring prior knowledge of variant type or position. Several preliminary rounds of regular PCR from genomic DNA are used to generate an initial amount of PCR product. Then a modified PCR temperature cycle is used for COLD-PCR. After DNA denaturation at 95 to 98°C, the PCR products are incubated at a temperature at which the primers do not bind (eg, 70°C for 2 to 8 minutes) for reannealing and cross-hybridization. Cross-hybridization of mutant and wild-type alleles forms mismatch-containing structures (heteroduplexes) with lower melting temperatures than the fully matched structures (homoduplexes). The temperature is next increased to a critical denaturation temperature (T_c) to preferentially denature the heteroduplex products (a single T_c is used for any mutation along the PCR product; however, different PCR products have different critical denaturation temperatures). The temperature is then reduced for primer annealing (eg, 55°C) and then increased to 72°C for primer extension, thus preferentially amplifying the variant alleles. COLD-PCR is effective for short DNA fragments 50 to 200 bp in length.

to A:T) or Tm-retaining (eg, T:A to A:T) variants and 5- to 10-fold for Tm-increasing (A:T to G:C) variants is typical. The enrichment increases if a second round of COLD-PCR is applied. Different modifications can enable higher mutation enrichment in a single step[45,46] or restrict enrichment to Tm-decreasing mutations.[47] Because during bisulfite treatment of DNA, unmethylated sequences undergo several C to T changes that decrease Tm, COLD-PCR may also be used for enriching unmethylated sequences.[48] Among various downstream detection methods, high-resolution melting[49-51] or sequencing[52-54] is most often used.

Detection Limits of Polymerase Chain Reaction

When PCR is performed under optimal conditions, a single copy of the target can be detected. In practice, however, the statistical probability of distributing at least a single copy from a dilute template solution into the PCR must be considered. The Poisson distribution indicates that if, on average, one target copy is present per tube, 37% of the tubes will have no target, 37% will have one target, and the remainder will have more than one target. If there is an average of two copies per tube, approximately 14% of the tubes will have no template and will be false negatives. About three copies on average are necessary for 95% of the tubes to include at least one copy. Therefore, the limit of detection (95% probability) of any single PCR cannot be lower than three copies per reaction. About five copies on average are necessary for 99% of the tubes to include at least one copy. This limitation of low copy analysis holds true for any amplification technique. However, digital PCR analyzes many reactions in parallel and can quantify less than one copy (on average) per reaction.

Digital Polymerase Chain Reaction

Conventional PCR averages the amplification results of many individual template molecules. Digital or single-molecule PCR is a technique that uses a dilute solution of template distributed across many reaction compartments or "partitions." Each partition either has or does not have PCR template molecules to amplify. After PCR, the partitions are scored as either positive (one or more initial templates) or negative (no initial template) for a digital readout. Thousands or even millions of partitions are typically scored. The partitions may be aqueous PCR droplets in oil or formed on chips by microfluidics.

Digital PCR can identify and quantify rare sequence variants, precisely determine copy number changes, establish the concentration of PCR standards, and determine the haplotype of variants that are on the same PCR product.[55] When properly performed, digital PCR does not require the standard curves routinely used in quantitative PCR. Digital PCR is less prone to background DNA competition because competing DNA is divided between positive and negative partitions. For example, if only 0.1% of the partitions are positive for a rare variant, 99.9% of the background DNA is in negative wells. The variant-to-background ratio is increased by a factor of 1000 in the positive wells, and better amplification of the variant can be expected. Common PCR inhibitors may not be as apparent in digital PCR because a positive threshold is reached even under conditions of moderate inhibition that would affect bulk quantitative PCR.[56]

Digital PCR results are derived from the average number of initial templates per partition, a value known as λ. Estimates of λ from experimental measurements are determined by Poisson statistics. The precision of the λ estimate increases with the number of partitions. Precision also decreases as λ gets too low (only a few partitions are positive) or too high (nearly all partitions are positive) and is optimal when λ equals 1.59.[57] Consequently, for best precision, a prior estimate of concentration is necessary. Coefficients of variation can be estimated by the Poisson distribution or determined more exactly by the binomial expansion. Single-molecule amplification and digital analysis has also been reported for other amplification methods including loop-mediated isothermal amplification[58] and recombinase polymerase amplification.[59]

Single-molecule PCR is commonly used in massively parallel sequencing methods for clonal amplification. The partitions may be minute aqueous droplets in a water-in-oil emulsion *(emulsion PCR)*,[60] PCR colonies *(polonies)* on a thin film of acrylamide gel,[61] clusters on the surface of a planar flow cell generated by bridge amplification,[62] or beads with clonally amplified template attached to their surface.[19,20,63] When amplification is observed in one of these massively parallel reactions, chances are that clonal amplification occurred from a single template molecule.

Additional Target Amplification Methods

In addition to PCR, many other methods of target amplification have been developed; some have found clinical use and are described in the next sections. These include isothermal amplification methods in which heat denaturation has been replaced by accessory proteins (helicase, recombinase) or strand displacement. These methods still resemble PCR in the products formed. Other methods do not resemble PCR, forming entirely different products based on hairpin extension or the transcription of RNA to DNA.

Transcription-Based Amplification Methods

Transcription-based amplification methods are modeled after the replication of retroviruses. These methods are known by various names, including transcription-mediated amplification (TMA),[64,65] NA sequence–based amplification (NASBA),[66] and self-sustained sequence replication (3SR) assays.[67] They amplify their target without temperature cycling (isothermally) and use the collective activities of reverse transcriptase, RNase H, and RNA polymerase. The most widely used is TMA, illustrated in Fig. 4.9. Two primers, a reverse transcriptase, and an RNA polymerase are used. The primer complementary to the RNA target has a 5′-tail that includes a promoter sequence for RNA polymerase. This primer anneals to the target RNA and is extended by the reverse transcriptase, creating an RNA–DNA duplex. The RNA strand is degraded by the RNAse H activity of the reverse transcriptase, allowing the second primer to anneal. The reverse transcriptase then extends the second primer to create double-stranded DNA (dsDNA) that includes the promoter. RNA polymerase recognizes the promoter and initiates transcription, producing 100 to 1000 copies of RNA for each DNA template. Each strand of RNA then binds and extends the second primer, forming an RNA–DNA hybrid; the RNA in the hybrid is degraded, the promoter primer binds and extends to produce dsDNA that can be transcribed, and the cycle repeats. As in PCR, all reagents are included, and amplification is exponential with completion in less than 1 hour. Unlike PCR, these methods do not require temperature cycling (except for an initial heat denaturation if a DNA template is used). They are particularly advantageous when the target is RNA (eg, human immunodeficiency virus [HIV] and hepatitis C virus [HCV] in blood bank NA testing).

Loop-Mediated Amplification Methods

Instead of producing products of a defined length, loop-mediated amplification (LAMP) produces a wide range of

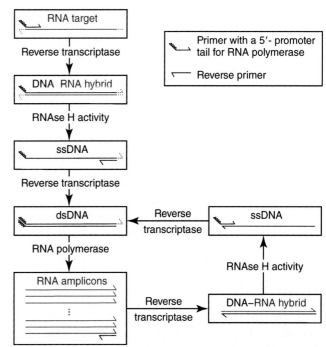

FIGURE 4.9 Schematic diagram of transcription-mediated amplification (TMA). Starting with a single-stranded RNA target, a primer with an RNA polymerase promoter on its 5′-end is extended by reverse transcriptase to form a DNA–RNA hybrid. The reverse transcriptase also has RNAse H activity that subsequently degrades the RNA strand to leave single-stranded DNA (ssDNA). A second primer then binds to the ssDNA, and extension forms double-stranded DNA (dsDNA) with the attached RNA polymerase promoter. RNA polymerase then makes 100 to 1000 copies of RNA, some of which are again primed by the second primer. Repeated cycles of reverse transcription, DNA–RNA hybrid degradation by RNAse H activity, dsDNA formation by reverse transcriptase, and further transcription by RNA polymerase exponentially produce ssRNA amplicons. Single-stranded targets are amplified isothermally, and double-stranded targets are first denatured to single strands.

different DNA structures with branches and loops. In the basic version, two strand displacement primers and two loop-forming primers recognize six segments in the target.[68] The inner two primers each include a 5′-tail that is complementary to the target sequence. After extension of the inner primers, hairpins or loops can form on each end, one of which will have a free 3′-end that can further extend. This loop formation is similar to self-probing[69] and snapback[70] primers, except that the 3′-end is not blocked. The two outer primers are used for displacement of the inner extension products to produce the starting material for cyclic amplification. The chain reaction includes both extension of the free 3′-ends and additional priming from the inner primers to the exposed single strands in the loops. The amplification results in a mixture of products with ever more loops and branching structures of increasing complexity. In another version of LAMP, allele-specific amplification with five primers and one competitive probe recognizes seven segments in the target.[71] In both versions, a variety of products are formed, and the reactions can be completed in less than 1 hour.

Strand Displacement Amplification

Similar to loop-mediated amplification, strand displacement amplification (SDA)[72] requires initial generation of starting material before the chain reaction. DNA is first heat denatured in the presence of four primers: two outer displacement primers and two inner primers with 5'-tails that includes a restriction site. An exonuclease-negative polymerase with good displacement activity is added in the presence of dCTP, dGTP, dUTP, and a modified deoxynucleotide (dATPαS), incorporating both the restriction site and the modified dATPαS into products that are ready to enter into exponential amplification. Exponential amplification occurs at 37°C by (1) nicking of the restriction enzyme site by the restriction enzyme (double-strand cutting is prevented by dATPαS); (2) extension from the nicked site with strand displacement; (3) priming of the displaced strand with the original inner primer that includes the restriction site; and (4) extension of both the primer and displaced strand, forming a new doubled-stranded product with the restriction site. Steps 1 through 4 are repeated over and over again for exponential amplification.

Variants of Polymerase Chain Reaction That Do Not Require Heat Denaturation

Variants of PCR have been developed that replace the need for heat denaturation with enzymatic separation of the double helix. These methods do not require thermal cycling and better reflect the normal DNA replication process, although the end products are the same as in the PCR. For example, helicase-dependent amplification (HDA)[73] uses the unwinding enzyme, helicase, to separate the double helix into single stands. As originally described, additional proteins were needed to stabilize the process that was performed at 37°C. Later, using a heat-stable helicase and polymerase allowed amplification at 60° to 65°C without the need for any other accessory proteins.[74] Another technique, recombinase polymerase amplification (RPA), uses a recombinase to scan dsDNA for priming sites, causing strand exchange to anneal the primers and single-stranded binding proteins to stabilize the loop structure long enough for strand displacement primer extension.[75] Two opposing primers exponentially replicate a short DNA fragment as in PCR, but the reaction is performed at 37°C without temperature cycling.

Whole-Genome and Whole-Transcriptome Amplification

Instead of specific amplification of one target to improve sensitivity, methods that amplify all genomic DNA or mRNAs are useful when the target is in short supply. For example, *multiple-displacement amplification* uses exonuclease-resistant random hexamers and a highly processive polymerase to amplify DNA nonspecifically.[76] Initial DNA denaturation is not necessary, and the reaction proceeds isothermally. Similarly, messenger RNA can be generically amplified with a poly(T) primer modified with an RNA polymerase promoter.[77] After reverse transcription, second-strand DNA synthesis and transcription, antisense RNA is produced. Both whole genome and antisense RNA amplification are also useful as NA purification methods before amplification or detection.

Signal and Probe Amplification Methods

It is not always necessary to amplify the target DNA or RNA sequence. Instead of target amplification, signal amplification or probe amplification can be used.

Branched-Chain DNA: Signal Amplification

Instead of increasing the concentration of target, signal amplification techniques use NAs to magnify the detection signal. The branched-chain DNA (bDNA) method is one of these techniques in common use. The bDNA approach hybridizes the target NA to multiple capture probes affixed to a microtiter well.[78] This is followed by hybridization to a series of "extender," "preamplifier," and "amplifier" probes. The final, highly branched amplifier probe includes multiple copies of signal-generating enzymes that act on a chemiluminescent substrate to produce light. Nucleotide analogs isoC and isoG (isomers of C and G that are complementary to each other but not to other nucleotides) are often used to increase the specificity of the signaling cascade.

Serial Invasive Amplification: Signal Amplification

When two probes overlap on one target, an "invasive" cleavage reaction can be catalyzed by certain structure-specific nucleases. The cleaved fragment, in turn, can cause invasive cleavage of a secondary probe in the shape of a hairpin. The hairpin probe can be designed as a fluorogenic indicator by using a reporter—quencher pair of dyes that are separated by cleavage. This serial sequence of events (primary invasion and cleavage followed by secondary invasion and cleavage of an indicator probe) is known as the serial invasive signal amplification reaction.[79] After DNA denaturation, cooling, and the addition of enzymes, the reaction is run at a temperature at which both the primary and secondary reactions recycle.

Rolling Circle Amplification: Probe Amplification

If a primer is annealed to a closed circle of DNA in the presence of a processive, displacing polymerase, the complement of the circle will be synthesized over and over again with displacement of the tandem repeats.[80] If two primers are used in opposite orientation, progressively more complex branches will be formed in an exponential reaction. The rolling circle can be formed by ligation of the two ends of a linear probe on template DNA. Ligation may happen directly, after polymerization through a gap, or after annealing of an additional, allele-specific oligonucleotide.

Endpoint Quantification in Amplification Assays

Molecular diagnostic assays may be qualitative (detect the presence or absence of a target or genotype identification) or quantitative (quantify the original concentration of a target sequence in the sample). When amplification is part of the assay, many variables need to be controlled carefully for accurate and precise quantification. Variation in extraction efficiency, the presence of enzyme inhibitors, lot-to-lot variation in enzyme and reagent performance, and day-to-day variation in reaction and detection conditions need to be addressed by quantitative methods. With reverse transcriptase assays, even more variables (reverse transcription efficiency and choice of reference genes) need to be considered.

Quantitative analysis at the endpoint of amplification is usually carried out with the use of calibrators with known amounts of target or a target mimic. Sample NA may be quantified by comparison with an *internal standard* of known amount that is added at the time of sample processing to control for efficiency of NA purification. These internal standards can be DNA fragments, plasmids, or RNA packaged into synthetic phage or virus particles to mimic real viruses.[81]

Real-time (continuous) analysis is simpler and more powerful for quantification than most endpoint analyses. Reactions are monitored at each cycle and initial target concentrations are typically calculated from the change (usually increase) in fluorescence with cycle number. Digital PCR is analyzed at the endpoint of amplification and provides potential advantages in precision, copy number, and rare variant analysis. Guidelines for performing and reporting quantitative real-time PCR (qPCR)[82] and digital PCR (dPCR)[83] experiments are good guides for both novice and experienced users.

DETECTION TECHNIQUES

Molecular diagnostics uses both generic and specific methods of NA detection. NAs can be quantified in bulk by optical techniques of absorbance and fluorescence (these are discussed in Chapter 3). Specific methods of detection and quantification usually use sequence-specific primers or probes with fluorescent or electronic detection.

Nucleic Acid Probes

Ultraviolet absorbance and fluorescent dyes in themselves do not discriminate between different NA sequences (ie, they are not sequence specific). Specificity in NA assays almost always comes from the hybridization of two complementary NA strands. Many reporter molecules can be covalently attached or incorporated into NA probes. Use of these probes can reveal the physical presence or location of sequences complementary to the NA portion of the probe. The first probes used in NA detection were radioactively labeled. Radioactive probes have a short shelf life limited by isotopic decay and radiolysis of the NA. This inherent instability, along with concerns of radioisotope safety and disposal, restricts the use of radioactive probes in the clinical laboratory. Although they are not discussed further here, prior versions of this chapter in earlier editions provide more detail.[9]

Indirect Detection of Hybridized Probes

The first practical example of nonradioactive probes used a biotin-labeled analog of dUTP.[84] Despite the altered steric configuration, this nucleotide is incorporated by most DNA polymerases. Other functional groups, such as digoxigenin, may also be used as affinity labels through chemical linkage to a dUTP and incorporation into polynucleotides. Alternatively, oligonucleotide probes can be labeled during synthesis with biotin or amino linkers for subsequent attachment to indicator molecules. Labels at the 5′- or 3′-end of the molecule are usually preferred because central modifications typically interfere more with hybridization.

Biotin and other affinity labels do not generate detectable signals on their own. However, they can initiate signal amplification mediated by high-affinity binding with antibodies, or in the case of biotin, with streptavidin. These binding molecules can be linked to enzymes—such as alkaline phosphatase, peroxidase, or luciferase—connecting a single target to a single enzyme. Enzyme activity is monitored by detecting catalytic turnover of enzyme substrates that result in colorimetric, fluorescent, or chemiluminescent signals.

Affinity labels can be used to capture and localize targets to an area of a solid support. For example, biotinylated probes can be affixed to a streptavidin-coated surface. After incubation with the target NA, a second probe is added, which may be directly labeled with fluorescence or conjugated through another affinity label to an enzyme. Any background or nonspecific localization of reagents results in amplification of an undesired signal along with the desired signal, and these methods usually require multiple separation and washing steps to decrease the background.

Fluorescent Labels

Advances in oligonucleotide synthesis and fluorescence detection have made fluorescently labeled probes the preferred reporter for NA analysis. Many fluorescent labels are now available, allowing color multiplexing for applications such as DNA sequencing, fragment length analysis, DNA arrays, and real-time PCR as reviewed later in this chapter. Techniques such as fluorescence polarization, fluorescence resonance energy transfer (FRET), and fluorescence quenching can provide additional detection specificity. Fluorescence polarization can be used to distinguish free from bound label, if the molecular rotation of the probe changes upon binding.[85] Molecular rotation primarily depends on the size of the molecule, so binding of a small probe onto a large target results in a polarization increase that can be measured. FRET techniques depend on the distance between two spectrally distinct fluorescent labels. Two labeled probes may be brought closer together by adjacent hybridization. Alternatively, two labels on the same probe may end up farther apart by hydrolysis or hybridization. Fluorescence quenching or augmentation does not always require FRET. For example, fluorescence may change merely by hybridization of a fluorescent oligonucleotide to its target. The effect depends on the dye and on inherent quenching from G residues in the target or probe.[86,87] Alternatively, quenching moieties can be directly incorporated into probes.[87,88]

Electrochemical Detection

Electrochemical detection of NAs is attractive for its simplicity. Hybridization events can be detected by redox indicators that recognize the DNA duplex or by other hybridization-induced changes in electrochemical parameters, such as conductivity or capacitance.[89,90] Usually, PCR amplification is performed before detection, so that many molecules are available and a bulk signal is generated to increase sensitivity. Electronic detection is also used in massively parallel sequencing, in which single nucleotide extension (SNE) from a clonally covered DNA bead produces a change in pH that is detected by complementary metal oxide semiconductor (CMOS) sensors.[63] Direct electronic sequencing of single molecules is also possible by detecting current changes that occur when a single strand of DNA passes through a nanopore.[91]

DISCRIMINATION TECHNIQUES

Nucleic acid discrimination techniques are divided into three categories: (1) electrophoretic methods that physically separate NAs based on molecular size or shape; (2) alternatives to electrophoresis that determine the size, base content, or sequence of NAs without electrophoresis, including high-performance liquid chromatography (HPLC), mass spectrometry, and pyrosequencing; and (3) hybridization assays that identify specific NAs by annealing or melting of complementary NAs. Some techniques use both electrophoresis and hybridization.

Electrophoresis

Both DNA and RNA are negatively charged and migrate toward the positively charged electrode when an electrical field is present within an appropriately buffered solution. Separation of different NAs occurs when mixtures are allowed to travel through a neutral sieving polymer under the electrical field. Separation is based primarily on molecular size, with smaller molecules traveling faster through the polymer than larger ones (Fig. 4.10). When very large molecules (≥ 50 kb) have to be separated, pulsed electrical fields are used to help move these molecules through the polymer matrix.[92] Separation also occurs based on the physical conformation, or shape, of the molecule. For instance, (1) single-stranded molecules may fold into secondary structures or they may stay as flexible linear structures; (2) linear double-stranded molecules may form heteroduplexes with mismatched bases, or they may stay in their original homoduplex forms; and (3) circular double-stranded NAs may be nicked and take a relaxed open-circular structure, or they can be in a more compact superhelical structure. Under nondenaturing conditions, each of these shapes may influence the way NA molecules travel through the electrophoretic matrix. Separation based on shape can provide useful information, but it can also confuse size-based analysis. Electrophoresis of DNA is performed under nondenaturing or denaturing conditions depending on the application.

RNA electrophoresis is commonly performed as a quality control check before transcript quantification or microarray expression analysis. RNA can be degraded easily by tissue or environmental RNAses, so it is important to assess the quality of the RNA used in these methods. Because RNA often has secondary structure, electrophoresis is usually performed under denaturing conditions to abolish these structures. Microfluidic chips with integrated microelectrophoresis channels are commercially available to rapidly assess RNA integrity by inspection of ribosomal RNA peaks (Fig. 4.11). Although specific transcripts are not detected by this method, only small amounts of starting RNA are needed.

Agarose and *polyacrylamide* are the two types of polymers commonly used in electrophoresis. Several chemical variants of the polymers are commercially available and are tailored for different separation ranges and applications. The choice of polymer and polymer concentration (usually expressed as % w/v) is dictated by (1) the size of NA to be separated, (2) the resolution that is required, and (3) how the result will be visualized and analyzed. Using various concentrations, an agarose gel can separate NA fragments as small as 20 bp to more than 10 mb (10,000 kb), including chromosomes of yeast, fungi, and parasites. However, the resolution of agarose is limited, usually to a size difference of 2% to 5%. Agarose polymers are cast in trays (sometimes commercially supplied as precast gels) and submerged in buffer. The gels are permeable to fluorescent NA–binding dyes, and results may be recorded as a photographic image of the stained gel under illumination.

Polyacrylamide polymers are suited for high-resolution separation (down to about 0.1% size differences) of short molecules (up to about 2 kb) and are the primary polymers

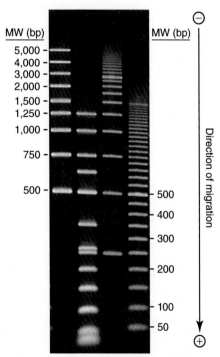

FIGURE 4.10 A photograph of multiple DNA fragments after agarose gel electrophoresis (1% w/v, SeaKem LE agarose gel) showing the separation of double-stranded DNA molecules by size. *MW,* Molecular weight. (Photograph courtesy Lonza Rockland Inc, Rockland, Maine.)

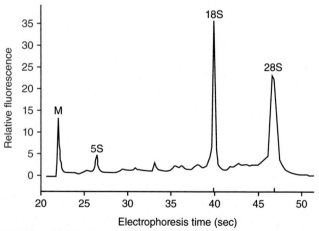

FIGURE 4.11 Microelectrophoresis of human white blood cell (WBC) RNA. After isolation of WBCs and extraction of total RNA, samples were denatured, stained with a fluorescent dye, and applied to a commercial microelectrophoresis platform for assessment of RNA quality. Prominent 18S and 28S bands of ribosomal RNA suggest the RNA is largely intact. Also indicated are a reference marker (M) and the 5S ribosomal band. Note that electrophoresis was performed in less than 1 minute.

TABLE 4.2 Commonly Used Electrophoresis-Based Techniques

Techniques Using Electrophoresis	Abbreviation	Primary Application
Polymerase chain reaction (or reverse transcriptase polymerase chain reaction) length	PCR (or RT-PCR)	Detection
Polymerase chain reaction/restriction fragment length polymorphism	PCR/RFLP	Detection
Southern blotting		Detection
Northern blotting		Detection
Dideoxy-termination sequencing (Sanger)		Sequencing
Single-nucleotide extension assay	SNE	Genotyping
Oligonucleotide ligation assay	OLA	Genotyping
Multiplex ligation-dependent probe amplification	MLPA	Quantification
Heteroduplex migration assay	HDA	Scanning
Conformation-sensitive gel electrophoresis	CSGE	Scanning
Single-strand conformation polymorphism analysis	SSCP, SSCA	Scanning
Denaturing gradient gel electrophoresis	DGGE	Scanning
Temperature gradient electrophoresis	TGGE, TGCE	Scanning
Temperature cycling capillary electrophoresis	TCCE	Scanning

for single-stranded NA separation, such as dideoxy-termination sequencing. Polyacrylamide may be used as a linear polymer solution, which is filled in capillaries *(capillary electrophoresis)*, or as cross-linked gels, which are cast between two plastic or glass plates *(slab gel electrophoresis)*. Polyacrylamide gels are permeable to fluorescent stains, and NAs can also be silver stained. In addition, the optical clarity of polyacrylamide polymers makes them ideal for visualizing emission signals from fluorescently-labeled fragments using laser-induced fluorescence detection. Table 4.2 lists common electrophoresis-based techniques.

Polymerase Chain Reaction Product Length

Polymerase chain reaction product analysis by electrophoresis is frequently used to query the product size and specificity of PCR. PCR products are visualized by staining the gel with a fluorescent DNA-binding dye, such as ethidium bromide. In some cases, the presence of an amplification product is directly diagnostic (eg, detection of sequences found only in a bacterium, virus, or fungus in a human sample). The specificity of the amplification reaction is verified by the known size of the fragment and the lack of extraneous bands. Internal negative and positive controls are used to control for potential contamination and PCR inhibitors.

Small insertions, deletions, rearrangements, and changes in the number of repeated sequences can also be detected by monitoring PCR product length on gels. Length differences may be large and easily visualized with agarose gel electrophoresis, or they may be small enough to require a denaturing polyacrylamide matrix. Fluorescent primers may be incorporated into the product during PCR to simplify detection of fragment lengths. These techniques are commonly used in the diagnosis of inherited diseases and in identity assessment.

Restriction Fragment Length Polymorphism

DNA extracted from a cell is extremely long and is usually cut into shorter fragments before electrophoresis to enhance mobility. Restriction endonucleases cut dsDNA into fragments of reproducible size; the same enzyme produces the same fragments in different specimens if they contain the same DNA sequence. If an alteration in the DNA abolishes or creates a cleavage site recognized by the enzyme (or changes the spacing between two cleavage sites), different sized fragments will be produced. These changes in fragment lengths (or polymorphisms) that result from restriction digestion are called restriction fragment length polymorphisms (RFLPs). However, restriction digestion of genomic DNA produces thousands of fragments. To be useful, specific fragments need to be visualized with probes such as in Northern and Southern blotting.[9]

Polymerase Chain Reaction/Restriction Fragment Length Polymorphisms

Many sequence alterations (eg, single base changes) do not affect the length of DNA. However, they can be amplified easily by PCR, and many can be detected as RFLPs after treatment with restriction enzymes. After PCR, the products are digested with one or more enzymes and analyzed by electrophoresis. For example, if a sample has a variant that disrupts an enzyme recognition site, it can be distinguished from a sample that does not have the variant. Such an assay will produce one uncut PCR fragment when the variant is present and two shorter fragments with normal DNA (Fig. 4.12). If the variant is present as a heterozygote (one normal and one variant copy of DNA), then one long and two shorter fragments will be observed. Usually it is possible to design the assay so that the fragments can be easily resolved by agarose electrophoresis. One variant of this method uses reverse-transcribed mRNA, which lacks the introns that would be present in the DNA. In this way, multiple exons can be analyzed in a single PCR. To detect rare variants that create a restriction site, this method can be modified by ligating an oligonucleotide to the cut site and using this oligonucleotide as one of the primers in a subsequent PCR amplification. The presence of the variant can be identified with enhanced sensitivity using either electrophoresis or real-time PCR.[93]

Conformation-Sensitive Scanning Techniques

Several electrophoretic methods can be used to scan for sequence variants after PCR amplification. For example,

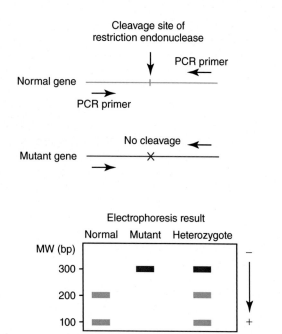

FIGURE 4.12 An example of polymerase chain reaction (PCR) restriction fragment length polymorphism (PCR-RFLP). A DNA fragment amplified by PCR carries a site (a unique sequence of generally four or more bases) that is recognized and cleaved by a restriction endonuclease. If a mutation is present, this site is altered and is no longer recognized by the enzyme. Electrophoresis reveals that the fragment from a normal specimen was indeed cut by the enzyme, generating two fragments shorter than the original length, but the fragment from a homozygous mutant was not cut, and the original length of the PCR product is preserved. In a heterozygous mutant, both the original fragment and the shorter fragments are visible. *MW*, Molecular weight.

heteroduplex analysis (HDA), also known as conformation-sensitive gel electrophoresis (CSGE),[94] reveals the presence of variants by the altered mobility of dsDNA fragments that contain one or more mismatched bases (a heteroduplex) versus one that is perfectly matched (a homoduplex). Denaturing gradient gel electrophoresis (DGGE)[95] and temperature-gradient gel electrophoresis (TGGE)[96,97] detect heteroduplexes by their lower stability. As the temperature or denaturing gradients are increased, heteroduplexes melt at lower temperatures, and eventually the strands separate, altering the gel migration. Single-strand conformation polymorphism analysis (SSCP or SSCA)[98] monitors the folding of single DNA strands produced by PCR. Electrophoretic mobility is a function of size and shape of the folded single-stranded molecules. Many of these methods, originally developed on slab gels, now have counterparts with separations by capillary electrophoresis. For example, cycling temperature capillary electrophoresis (CTCE)[99] has been proposed for pangenomic scanning for unknown point mutations. Scanning electrophoretic techniques have lost popularity as sequencing costs have decreased. Instead, direct dideoxy sequencing and massively parallel sequencing not only identify that a variant is present but also provide the variant sequence. Another alternative that does not require electrophoretic separation or expensive equipment is high-resolution melting. Additional details of scanning electrophoretic methods can be found in the previous edition of this chapter.[9]

Dideoxy-Termination Sequencing (Sanger Sequencing)

Dideoxy-termination sequencing[100] of DNA is routinely performed in the clinical laboratory. RNA can also be sequenced by first converting RNA to DNA by reverse transcriptase. NA sequence is determined and compared with a reference sequence with an error rate of approximately 0.1% (one misidentified base in 1000). Often the sequence is analyzed on both strands (sense and antisense) for even greater accuracy. Any deviation from the reference sequence is identified by using various available computer software programs to compare the sequences. Base changes, including synonymous changes that do not alter the protein sequence, as well as nonsynonymous changes resulting in altered amino acids, stop codons, and small insertions and deletions, are identified. However, larger deletions and rearrangements spanning the sequencing primer binding sites are not detected.

The most common sequencing strategy uses PCR in the first step to amplify the region of interest followed by a chain-termination reaction developed in the late 1970s.[10] This reaction generates fragments that are terminated at various lengths by incorporation of one of the four dideoxynucleotide base analogs during extension from a sequencing primer. Dideoxynucleotides lack both 3'- and 2'-hydroxyl groups on the pentose ring (Fig. 4.13). Because DNA chain elongation requires the addition of deoxynucleotides to the 3'OH group, incorporation of dideoxynucleotides terminates chain extension. The most common method for generating these terminated fragments is *cycle sequencing*, repeating the steps of annealing, chain extension-termination, and denaturation by temperature cycling, similar to PCR (Fig. 4.14). The fragments generated are tagged with a fluorescent dye (with the use of labeled primers or labeled terminator dideoxynucleotides), are separated by electrophoresis, and are detected by fluorescence as the fragments travel past a detector (Fig. 4.15). When fluorescently labeled primers are used, four tubes are needed, each with one of the four dideoxyterminators. If four colors are used, the termination reactions can be combined before electrophoresis, and only one capillary is necessary. After PCR and cycle sequencing, about 600 to 800 bases can be resolved in less than 2 hours by capillary electrophoresis, and 96 or 384 samples can be run in parallel.

RNA sequencing in the clinical laboratory is commonly used in viral genotyping such as HIV for drug resistance and HCV to establish prognosis and appropriate therapy. Sequencing is also used for bacterial speciation by analysis of ribosomal DNA, to identify mutations in many genetic diseases, and in cancer. Dideoxy sequencing, even of only one

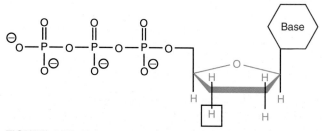

FIGURE 4.13 A dideoxynucleotide. Notice the absence of the 3'-OH that is usually present in standard deoxynucleotides. This lack of a 3'-OH prevents polymerase extension because no phosphodiester bond can form. Incorporation of a dideoxynucleotide forces termination of polymerase extension.

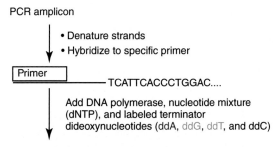

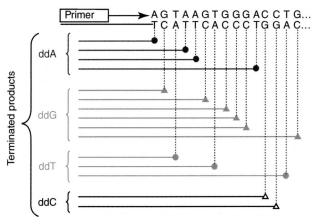

FIGURE 4.14 The dideoxy-termination reaction for sequencing. A polymerase chain reaction (PCR) product is denatured and hybridized to a specific oligonucleotide primer. As the DNA polymerase extends the primer by incorporating bases (deoxynucleotides [dNTPs]) complementary to the template, it occasionally incorporates a terminator dideoxynucleotide base analog (ddA, ddG, ddT, or ddC) that stops further extension. The result is a mixture of extended products of varying lengths. Each terminator base may be labeled with one of four different fluorescent tags (shown as different symbols in the diagram). Alternatively, the primer can carry four different fluorescent tags in individual chain-termination reactions (containing only one dideoxy nucleoside triphosphate [ddNTP]) performed in separate tubes. The original procedure incorporated a radioactive dNTP during extension, allowing monochromatic detection of the truncated fragments in four electrophoresis lanes.

gene (including exons and splice sites), in the clinical laboratory for the detection of disease-causing mutations is still an expensive proposition, with the cost of analysis proportional to the size of the gene. This is especially true for population screening (in which most samples will not have a mutation) but also for patients with symptoms of the disease. Massively parallel sequencing vastly reduces these costs and will be discussed later.

Single Nucleotide Extension

Also known as *single-base primer extension* or *minisequencing*, SNE assays[101] involve the annealing of an oligonucleotide primer to a single-stranded PCR product at a location that is immediately adjacent to, but does not include, the site of the single base variant followed by enzymatic extension of the primer in the presence of polymerase and dideoxynucleotide terminators. For example, each of the four terminators can be labeled with a unique fluorescent label so that it is possible to detect which base was incorporated. SNE assays can be multiplexed on capillary electrophoresis instruments by varying the lengths of the primers so that each SNV is resolved by size in one electrophoresis run. Many SNE detection methods are available other than electrophoresis, including (1) photometric detection on microtiter plates, (2) product capture detection systems on DNA microarrays, (3) bead hybridization assays detected by flow cytometry, (4) solution-based fluorescence polarization detection systems, and (5) mass spectrometry. SNE assays are particularly useful when the target of interest contains 5 to 50 disease-causing single base variants. SNE assays do not work well if variants are present in the primer-binding site. Nor are they usually designed to detect variants at a position other than immediately adjacent to the 3′-end of a primer.

Oligonucleotide Ligation

Another assay format frequently used for variant detection is the oligonucleotide ligation assay (OLA).[102] Two oligonucleotide probes are hybridized to adjacent sequences of amplified target DNA, with the variation site positioned at the end of one probe (Fig. 4.16). DNA ligase covalently joins the two probes only if both probes are perfectly hybridized to the target, including the polymorphic base. A probe matching the normal base and another probe matching the variant base are usually prepared. These two can be discriminated through differential electrophoretic mobility by varying the length of their 5′-tails. These tails can be noncomplementary poly A or poly C tails or penta-ethylene oxide (PEO) units. The probe hybridizing to both alleles (the *common probe*) provides the reporter molecule, usually a fluorescent label. Multiplexing is achieved by attaching different fluorescent labels to the common probes and by varying the length of the tails on the allele-specific probes. After ligation, probes for multiple variants are separated by denaturing capillary electrophoresis.

Multiplex Ligation-Dependent Probe Amplification

Multiplex ligation-dependent probe amplification (MLPA)[103] is a convenient method for relative quantification of up to 10 to 50 targets. This method is particularly useful in screening for deletions or duplications of exons, multiple exons, or whole genes. For each target, two probes are designed that hybridize adjacent to each other so that they can be ligated. The two probes have unique tails that do not hybridize to the target and that are the same between targets. After hybridization and ligation, the probes are amplified by PCR with a common primer pair (complementary to the tails). One of the primers is fluorescently labeled at its 5′-end. Because probes of different lengths are used, multiple PCR products of different sizes are produced and separated by capillary electrophoresis. Relative peak heights or areas are compared for relative quantification.

Alternatives to Electrophoresis

Electrophoresis is not the only method to determine the size, base content, and/or sequence of NAs. Some of these alternatives to electrophoresis are attractive in the clinical laboratory because of compatibility with automation and

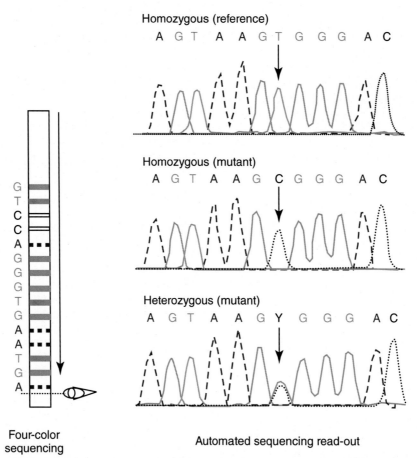

FIGURE 4.15 Schematic of dideoxy-termination DNA sequencing. Extension products (see Fig. 4.13) are labeled with different color dyes for each of the four terminator reactions and are separated by electrophoresis. The four-color strategy allows automated endpoint fluorescence detection *(eye icon)*. The direction of fragment migration is from top to bottom. The sequence is read from left to right in the electropherogram generated by the sequencer. Examples of a reference sample (homozygous T at the polymorphic site), a mutant sample (homozygous C), and a heterozygous mutant sample (T and C, reported as Y) are shown.

capacity for high throughput. Some examples include pyrosequencing, mass spectrometry, and high-pressure liquid chromatography.

Pyrosequencing

Pyrosequencing[104] is a sequencing-by-synthesis method that does not require dideoxy-termination or electrophoresis. A sequencing primer is hybridized to a single-stranded template generated by PCR. Four enzymes, a DNA polymerase, ATP sulfurylase, luciferase, and apyrase, and two substrates—adenosine 5′-phosphosulfate and luciferin—are included in the reaction mixture (Fig. 4.17). One of the four dNTPs is added to the reaction, with dATPαS substituted for dATP because it is incorporated by the polymerase but is not a luciferase substrate. If the base is complementary to the template strand, DNA polymerase catalyzes its incorporation. Each incorporation event is accompanied by release of a pyrophosphate (PPi) so that the quantity of PPi produced is equimolar to the amount of incorporated nucleotide. The release of PPi is monitored by conversion of PPi and adenosine 5′-phosphosulfate into adenosine triphosphate (ATP) by ATP sulfurylase, and ATP in turn drives conversion of luciferin into oxyluciferin, which generates visible light. The light produced is proportional to the number of nucleotides incorporated. Apyrase, which is a nucleotide-degrading enzyme, continuously degrades ATP and unincorporated dNTP. This switches off the light in preparation for the next dNTP addition. As the process is repeated by adding one dNTP at a time, the complementary DNA strand is built, and the nucleotide sequence is determined.

Mass Spectrometry

Matrix-assisted laser-desorption ionization time-of-flight (MALDI-TOF) mass spectrometry can be used to genotype sequence variants.[105,106] With mass spectrometry, no label is necessary because the alleles differ in mass. After isolation of genomic DNA, a specific DNA fragment, including the variant site, is amplified by PCR. Heat-labile alkaline phosphatase is added to the reaction to dephosphorylate any residual nucleotides, preventing future incorporation and interference with the primer extension assay. Samples are then heated to inactivate the alkaline phosphatase. An extension primer is hybridized directly or closely adjacent to the polymorphic site. Unlabeled deoxynucleotides are incorporated and terminated with a dideoxynucleotide, generating

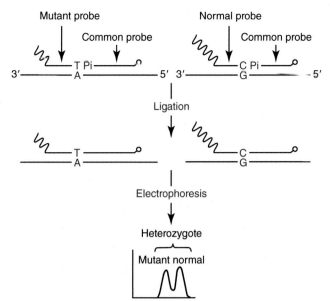

FIGURE 4.16 Oligonucleotide ligation assay of a heterozygous single nucleotide variant (SNV). A mutant probe with a 3'-T *(upper left)* hybridizes to the mutant DNA with an opposing *A*, and a normal probe with a 3'-C *(upper right)* hybridizes to the normal DNA with an opposing *G*. The probes are also attached to mobility modifying tails of different lengths *(wavy lines)*. Hybridized next to these probes is the common probe with a 5'-phosphate (Pi) and a 3'-fluorescent tag. In the presence of ligase, adjacent probes are covalently joined to generate longer probes, each with a fluorescent tag and a mobility-modifying tail. Probes that are mismatched to the target at their 3'-ends may hybridize, but they are not ligated. Electrophoresis and endpoint, laser-induced detection of ligated probes reveal the different alleles by their different mobility. Many SNVs can be analyzed in one electrophoresis assay by varying tail lengths or by using multicolor fluorescence tags.

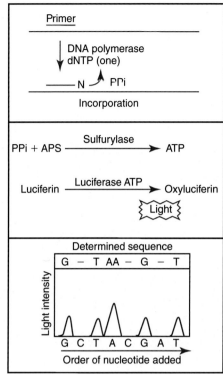

FIGURE 4.17 Schematic of pyrosequencing. Individual deoxynucleotides (dNTPs) are added one by one to the single-stranded template, a primer, and a polymerase. Pyrophosphate is generated if the dNTP is complementary to the next base on the template *(top)*. Any pyrophosphate produced reacts with adenosine-5'-phosphosulfate (APS) to produce adenosine triphosphate (ATP), which in turn generates light in the presence of luciferase *(middle)*. The sequence can be determined from the order of dTNP addition and from the intensity of light produced *(bottom)*.

allele-specific diagnostic products of different mass. Salt is removed from the sample, and approximately 10 nL is spotted onto an array coated with 3-hydroxypicolinic acid. This is placed into the MALDI-TOF instrument, which measures the mass of the extension products. After the mass is measured, the genotype is determined (Fig. 4.18). Despite the complexity, automated systems processing 384 to 1536 samples at once are available.

Another use of mass spectroscopy is infectious agent identification. After PCR, electrospray-ionization mass spectrometry determines the exact mass of the PCR product, a process known as PCR/ESI-MS. Because mass rather than sequence is determined, there is some risk of misidentification, but this can be avoided by careful choice of primers and selecting more than one target if necessary. For example, 10 bacterial and 4 viral biothreat clusters were amplified by PCR using two to four targets each.[107] The pathogenic strains within each cluster were then distinguished from near neighbors by ESI-MS.

High-Performance Liquid Chromatography

High-performance liquid chromatography is commonly used for separating and purifying oligonucleotides. Separation usually is based on ion-pair, reversed-phase chromatography and is particularly useful for purifying fluorescently labeled probes guided by absorbance and fluorescence elution profiles.

A variant of this technology is denaturing HPLC (dHPLC). dHPLC is run at a single elevated temperature to partially denature dsDNA. Similar to gel-based heteroduplex detection, dHPLC reveals the presence of heteroduplexes as additional peaks that are shifted in retention compared with homoduplexes.[108] Retention of dsDNA is governed by electrostatic interactions and an acetonitrile gradient. DNA detection is most often performed by UV absorbance, although fluorescence or mass spectroscopy can also be used. Limitations of dHPLC include sequential (one at a time) analysis and the need to analyze some samples at multiple temperatures when more than one melting domain is present.

Hybridization Assays

All hybridization assays are based on the ability of single-stranded NAs to form specific double-stranded hybrids. The process requires (1) that probe and target NAs are mixed under conditions that allow specific complementary base pairing and (2) that a method is available to detect any resulting double-stranded NAs. A *probe* indicates a NA whose identity is known, and the *target* or *sample* is a NA whose identity

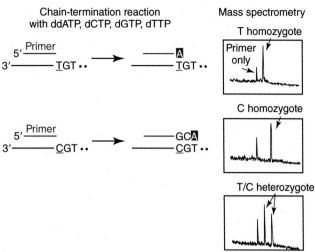

FIGURE 4.18 Sequence polymorphism analysis by mass spectrometry. The underlined base is the polymorphic site in the template (T or C). The single-stranded template is primed and extended in the presence of three deoxynucleotides (dNTPs) and one dideoxy nucleotide triphosphate (ddNTP), producing fragments of different mass depending on the sequence. The boxed "A" in this example indicates the incorporated terminator adenine base. The mass of terminated products is precisely measured by matrix-assisted laser-desorption ionization time-of-flight (MALDI-TOF) mass spectrometry.

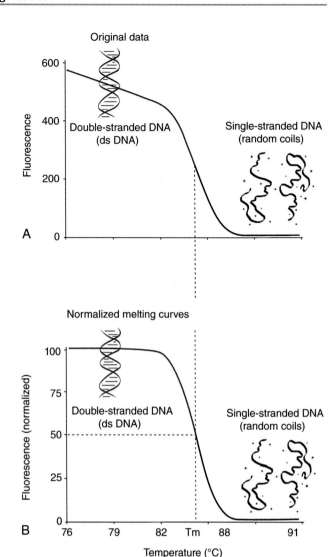

FIGURE 4.19 Fluorescence melting curve of a polymerase chain reaction (PCR) product. PCR-amplified DNA was melted in the presence of the cyanine dye, LCGreen Plus. A, Fluorescence gradually decreases as the temperature is increased until a more rapid decrease occurs near the DNA melting temperature (Tm) of the PCR product. B, The data are normalized between 0% and 100% after background removal to obtain constant fluorescence outside of the transition. (Modified with permission of the publisher from Wittwer CT, Kusukawa N. Real-time PCR and Melting Analysis. In Persing DH, Tenover FC, Tang YW, et al, eds. *Molecular microbiology: diagnostic principles and practice.* 2nd ed. Washington, DC: ASM Press; 2011:63–82. Copyright 2011 ASM Press.)

or abundance is revealed by hybridization. In some of the methods discussed here, hybridization occurs between a target in solution and a probe that is tethered to a solid surface. In *homogeneous* or *real-time* techniques, both the probes and the targets are in solution, and hybridization and detection occur without washing steps. Some of the homogeneous methods also monitor the dissociation of hybridized duplexes under controlled heating, revealing the identities of the hybridized duplexes by *melting curve* signatures.

Thermodynamics

The favored structure of DNA under physiologic conditions is an ordered double-stranded helix held together by noncovalent interactions. The duplex structure is most stable when all opposing bases are complementary, allowing for maximal hydrogen bonding and base stacking. The noncovalent binding between two DNA strands is both specific (ie, sequence dependent) and reversible. Denaturing agents (eg, high temperature, formamide, or extremes of pH) favor dissociation of the double-stranded molecule into two separate random coils (Fig. 4.19). On removal of the denaturant, single strands attempt to reform duplexes, strongly favoring interactions that maximize complementary base pairing. Because temperature is the denaturant most easily manipulated, double- to single-strand transformation is referred to as *melting*, and the temperature at which half of the DNA is melted is the *melting temperature*, or *Tm*, of the duplex. Duplexes with mismatched base pairs are less stable than those with a perfect sequence match and thus melt at lower temperatures. The reverse process, in which two complementary strands recombine to form a stable duplex molecule, is referred to as *annealing* or *hybridization*. Hybridization can occur between DNA strands, RNA strands, and strands of NA analogs (eg, peptide NAs[109]) in all combinations.

The hybridization environment defines the degree of base pair mismatch that can be tolerated in a duplex structure. Conditions of high *stringency* (low salt concentration, high formamide concentration, and high temperature) require exact base pairing. As the stringency is lowered by increasing the salt concentration, lowering the formamide concentration, or lowering temperature, more base pair mismatches can be tolerated in a duplex structure. The stringency of a hybridization reaction is determined by hybridization conditions and any washing steps designed to remove nonspecific NA. In real-time PCR and melting analysis, the hybridization solution is the buffer in which PCR occurs, and there are no washing steps.

Kinetics

The kinetics of solution-phase hybridizations is second order, being proportional to the concentrations of both hybridizing strands.[110] The rate-limiting step is nucleation, in which a small number of base pairs are formed in the correct orientation followed by rapid "zippering" of complementary sequences. In the case of a probe present in great excess to the target, hybridization proceeds as a pseudo first-order reaction, depending only on the concentration of the target. However, the time required to hybridize the probe to a given fraction of the target remains proportional to the probe concentration. For example, during most of PCR, the concentration of primers is much greater than that of the target, and the reaction rate during each annealing step depends on the concentration of available single-stranded product, but the time required to anneal primers to a certain fraction of the target is proportional to the primer concentration.

The availability of NAs for hybridization can also be an issue. In PCR, primer annealing competes with the formation of double-stranded product. As the concentration of product increases during PCR, some double-stranded product is formed before primer annealing can occur (see Fig. 4.6). Similarly, when double-stranded probes are used at high concentrations, probe self-annealing interferes with probe–target hybridization. Available hybridization sites can also be limited by the intramolecular secondary structure of the probe or target.

In addition to probe concentration and availability, the length of the probe and the NA complexity affect hybridization rates. Rates are directly proportional to the square root of the probe length and inversely proportional to complexity, defined as the total number of base pairs present in nonrepeating sequences. Mismatches up to about 10% have little effect on hybridization rates.

Hybridization rates are also influenced by many factors in the reaction environment, most notably temperature and ionic strength. Above the Tm, no stable hybrids are present, although transient complexes may form. As the temperature is lowered below the Tm, hybridization rates increase until a broad maximum occurs about 20° to 25°C below the Tm. Hybridization rates also increase with ionic strength. Divalent cations such as Mg^{2+} have a much stronger effect than monovalent cations such as Na^+ or K^+.

When the NA target or probe is immobilized on a solid support, hybridization kinetics are even more complex. Many of the preceding observations still hold true, but the rate and extent of solid-phase hybridization are lower than with solution-phase hybridization. Depending on the concentrations of the reactants, solid-phase hybridization can be nucleation limited or diffusion limited. Optimal efficiency of solid-phase hybridization is achieved under conditions that facilitate diffusion of the probe to the support and that favor hybridization over strand reassociation if double-stranded probes are used. This usually means a small volume of hybridization solution and relatively low probe concentrations. In practice, solid-phase hybridization assays are empirically designed. Time of hybridization and probe concentration are the two variables most frequently adjusted. Conditions that tend to maximize the extent of hybridization and minimize the background or nonspecific attachment of the probe are selected.

Probes

Similar to antibodies in immunoassays, probes in NA hybridization assays can be labeled in many ways to detect hybridization. Probes may be cloned (recombinant), generated by PCR, or synthesized (oligonucleotides). They may be DNA, RNA, or NA analogs and single or double stranded. Selection, purification, and labeling of probes are crucial to the success of hybridization assays.

Cloned Probes. Cloned probes consist of a known segment of DNA inserted into a plasmid vector that is propagated by growth in a bacterium. Many different plasmid vectors are available; pBR322 was one of the first in common use.[111] Some plasmids, such as the F plasmid of *Escherichia coli*, can be used to carry insert sizes that are very long (several hundred kilobases) and are called *bacterial artificial chromosomes*. The entire plasmid DNA (insert plus vector sequences) may be used as a probe, or the insert may be purified first from the vector sequences. The latter method is obviously more cumbersome but may result in reduced background. The resulting probe is a dsDNA probe, and it must be denatured before use.

Some vectors contain RNA promoter regions adjacent to the inserted DNA sequence. These regions permit generation of RNA transcripts from the DNA insert. Because only one strand is copied during RNA synthesis, single-stranded RNA probes are generated. Controlling the orientation of the insert in relation to the promoter region allows the production of transcripts in the "sense" direction (ie, same as mRNA) or the "antisense" direction (ie, complementary to mRNA).

Polymerase Chain Reaction–Generated Probes. Polymerase chain reaction–generated probes are simple to prepare.[112] During amplification, the PCR product typically is labeled with nucleotides that are fluorescent or have attached affinity labels. If desired, single-stranded probes can be obtained by amplifying with a biotin-labeled primer followed by solid-phase separation with streptavidin.

Oligonucleotide Probes. Oligonucleotide probes are even easier to obtain than PCR-generated probes. These probes are usually 15 to 45 bases of single-stranded NA that are chemically synthesized as a specified base sequence. Most commonly, they are DNA, but RNA or NA analogs can also be synthesized. Automated, efficient, and accurate methods of synthesis continue to lower the cost of production. Sequence information is now routinely available in public databases,[113] and a similarity check for probe sequence can be performed using public algorithms.[39] Probe sequences must be carefully chosen to minimize cross-hybridization with pseudogenes (eukaryotes) or related species (bacteria and viruses). The melting temperature of the probe should allow both favorable hybrid stability and discrimination against related sequences under the stringency of the assay. Oligonucleotide probes are often prepared with covalent attachment of a reporter molecule (eg, a fluorescent dye) or affinity labels that allow them to be attached to solid supports. Probes used in homogeneous (real-time) PCR are usually oligonucleotides with a fluorescent label.

Estimating Melting Temperature of Oligonucleotide Probes. Probe Tm prediction based on nearest neighbor thermodynamic parameters has improved with compilation of a unified database.[114] Consideration of all possible single-base mismatches and dangling ends further extends the usefulness of these estimates.[115] However, prediction parameters are typically determined using 1-M NaCl, far from typical assay conditions, so it is not surprising that predicted Tms are often at variance from observation. Empirical correction factors[37,116] that may include the concentrations of various cations, dNTPs, the target, and common additives may enhance prediction accuracy and are often used in software programs and websites for in silico Tm estimation. Most fluorescent dyes also stabilize duplexes,[18,117] but this increase in Tm is seldom incorporated into predictions. For these reasons, absolute Tm predictions may not be accurate with common laboratory conditions and PCR buffers. However, relative Tms (ie, the difference in Tm between two related probes, such as a probe that is matched and one that is mismatched to a single base variant) are considerably more accurate.[118]

Purity of Labeled Oligonucleotide Probes. The purity of labeled oligonucleotide probes is important for hybridization assays and critical in real-time PCR. Commercial oligonucleotides with a fluorescent label are of variable quality, and their concentration and purity should be assessed before use. Mass spectroscopy and coelution of absorbance (A_{260}) and fluorescence peaks on reversed-phase HPLC can indicate probe purity. Quantitative estimates of probe purity can also be obtained by simple absorbance measurements as previously described in the prior edition of this textbook.[9]

Examples of Hybridization Assays

Hybridization reactions can be divided into two broad categories: (1) *solid phase,* in which either probe or target is tethered to a solid support while the other is in solution, and (2) *solution phase,* in which both are in solution (Box 4.2). Several classical methods first hybridize in solution and then separate the bound from the unbound labeled probe. Exclusion chromatography and binding by hydroxyapatite, magnetic particles, or other affinity capture methods allow selective measurement of the labeled probe–target hybrid.

BOX 4.2 Hybridization Assays

Solid-Phase Hybridization
Dot blot and line probe assays
Arrays (microarrays and medium-density arrays)
Micro-bead assays
In situ hybridization
Southern and Northern blotting
Massively parallel sequencing on beads or planar flow cells

Solution Phase Hybridization
Real-time (or homogeneous) polymerase chain reaction
Melting analysis
Single-molecule sequencing
Many classical techniques

Solid Phase Hybridization

Solid-phase assays are useful because multiple samples can be processed together, which facilitates (1) control, (2) washing, and (3) separation procedures. Hybridization on a solid support is, however, less efficient than solution-hybridization, and the kinetics are slower and more difficult to predict. Both solid-phase and liquid-phase assays are used routinely in the clinical laboratory. Solid-phase assays include (1) dot blots, (2) line probes, (3) microspheres, (4) microarrays, and (5) in situ hybridization.

Dot Blot and Line Probe Assays. Conventional hybridization assays on membranes are known as dot blot or line probe assays, depending on the geometry of the individual spots. The NAs are applied with suction, forming a shape that is either round (dot) or elongated (line or slot). After immobilization, the membrane is incubated with complementary NA at a constant temperature followed by one or more washes to discriminate matched from mismatched NA. The method allows multiple simultaneous hybridizations under identical conditions.

Two general formats are used for these assays: Either multiple samples are affixed to the solid support and interrogated by a small number of probes ("sample down"), or multiple probes are attached to the support and a small number of samples are used ("probe down"). Results of dot blot and line probe assays are usually qualitative. If hybridization has occurred, a signal is generated at the specified spot, and a simple yes-or-no interpretation is given. Similar assays have been developed substituting microtiter plate wells for filters. This requires chemical modification of the plastic wells to bind short DNA probes at one end, allowing the bound probes to hybridize to sample and is more amenable to automation of washing and detection.

Medium-Density Arrays. Dot blot and line probe assays have largely been replaced by medium-density arrays that typically analyze 20 to 500 spots. Medium-density arrays are used for testing multiple mutations simultaneously in specimens for various applications, including genetic diseases, oncology, and pharmacogenetics.[119,120] These arrays do not need to be attached to a two-dimensional surface as long as their "address" can be decoded. For example, microspheres can be coded by fluorescence intensity in two different channels, and fluorescence in a third channel monitors hybridization. All channels can be read simultaneously using a flow cytometer.[121]

Many different types of medium-density and microarrays are available. The surface of an array may be glass, gold-coated piezoelectric crystals, gel pads with embedded microelectrodes, or microspheres. The bound NAs are often oligonucleotides 20 to 80 bases in length that may be conventionally synthesized and spotted onto the chip or directly synthesized in situ. Arrays can be made of expressed sequence tags (200–500 bases) or bacterial artificial chromosomes (100–200 kb). Instead of an array of probes, sample DNA or cDNA may also be bound to an array surface. After hybridization, detection may be fluorescent, electronic, or by mass spectrometry. Fluorescence is most common, and excitation may use epifluorescence, confocal, laser scanning, or surface plasmon resonance with imaging at various wavelengths, usually with a CCD camera.

Microarrays. Increasing further the density of hybridization assays, microarrays (also called DNA arrays, DNA chips, or Biochips) were introduced in the mid-1990s.[16] Compared with medium-density arrays, spot sizes in microarrays are decreased (typically to <200 microns in diameter) such that one array contains thousands to millions of spots. This dimensional change requires (1) specialized detection equipment, (2) software, and (3) informatics to analyze the data. Because of their high density, microarrays have attracted intense interest among researchers who wish to monitor the whole genome for (1) SNVs, (2) gene expression, or (3) copy number variation.

Because SNVs represent the most common genetic difference among individuals, much effort has focused on correlating SNV genotypes to phenotype and disease association. SNV microarrays have been used in many genome-wide association studies (GWAS). Microarrays that analyze human SNVs ("SNV chips") provide the technology to genotype most known human SNVs in one experiment. Nearby SNV alleles tend to cluster together as haplotypes, so disease association by haplotype simplifies the analysis. Although some valuable markers have been found by these GWAS,[122] the yield of useful disease markers obtained by these methods has been disappointing, and many difficulties remain, such as identifying adequate control populations.[123] SNV arrays can also be used to genotype SNVs of known association with disease and to assess copy number variation. Cytogenomic arrays, including SNV and comparative genomic hybridization (CGH) arrays, can analyze the entire genome, providing chromosome maps of copy number changes (large insertions and deletions) across each chromosome.

Gene expression microarrays quantify the relative amounts of different messenger RNAs in test and reference samples. An example of a two-color microarray for gene expression is shown in Fig. 4.20. Probes that hybridize to mRNA are usually directly synthesized on microarrays. Modern gene expression arrays have been used to measure the mRNA transcribed from all human genes in one experiment. They have been applied to almost every conceivable human circumstance, including (1) neoplastic, (2) inflammatory, and (3) psychiatric conditions. It is expected that application of this technology will lead to better (1) diagnosis, (2) molecular staging, (3) prognosis, and (4) therapy through understanding of disease pathogenesis. In oncology, gene expression microarrays have led to new diagnostic and prognostic markers in breast cancer,[124] bladder cancer,[125] leukemia,[126] and sarcoma,[127] among others. Even with great progress, expression arrays are used directly in only a few

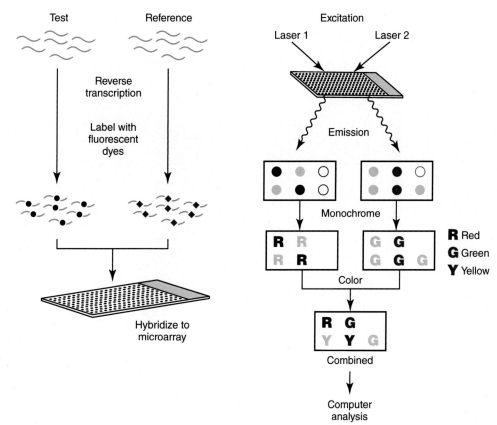

FIGURE 4.20 A two-color microarray experiment. An array of DNA oligonucleotides with sequences of messenger RNA are affixed to a glass slide. Messenger RNAs in the test and reference specimen are converted into differentially labeled cDNA by reverse transcription and incorporation of two different fluorescent dyes. The two samples are hybridized together onto the array. The array is washed, and the image is captured twice, each time with a laser of a wavelength that excites one of the dyes but not the other. The monochromatic images are then converted to two colors (green [G] for the test sample and red [R] for the reference), and the images are combined. If the abundance of cDNA is the same in each of the two samples, then the composite spot will be shown as yellow (Y). If one is in greater abundance, then that color will be preserved. Up- and downregulation of gene expression are then analyzed by software.

clinical diagnostic and prognostic tests. Most arrays are used in marker discovery projects for selection of a smaller panel of expression targets that are then analyzed by other quantitative methods, such as real-time PCR, that provide greater precision and dynamic range.

Another important clinical application of microarrays is the genome-wide analysis of deletions and duplications, referred to as copy number variants (CNVs). CNV analysis using microarrays is replacing traditional cytogenetic chromosome analysis (karyotyping) and fluorescence in situ hybridization (FISH) analysis for detection of genome-wide copy number alterations. Similar to gene expression arrays, many of the CNV arrays use two-color comparative hybridization to determine the gene dosage in a specimen compared with a normal reference genome (array comparative genomic hybridization [aCGH]). Arrays for CGH use oligonucleotide probes for very high resolution and data density. An example of CNV analysis using aCGH is shown in Fig. 4.21. SNV arrays also are used to detect copy number changes by loss of heterozygosity (this method is sometimes referred to as *virtual karyotyping*). Unlike aCGH, SNV arrays have the advantage of analyzing the specimen without the need to mix in a reference genome. SNV arrays also are able to detect copy number neutral changes caused by inversions or uniparental disomy that are not detected by aCGH methods. When a copy number change is found of potential clinical significance, it can be verified by an orthogonal method such as FISH or high-resolution melting.[128]

In Situ Hybridization. In situ hybridization is a specialized type of solid support assay with morphologically intact tissues, cells, or chromosomes affixed to a glass microscope slide to provide the matrix for hybridization. The process is analogous to immunohistochemistry, except that NA probes are used instead of antibodies. The strength of the method lies in linking morphologic evaluation with detection of specific NA sequences. When fluorescent probes are applied to

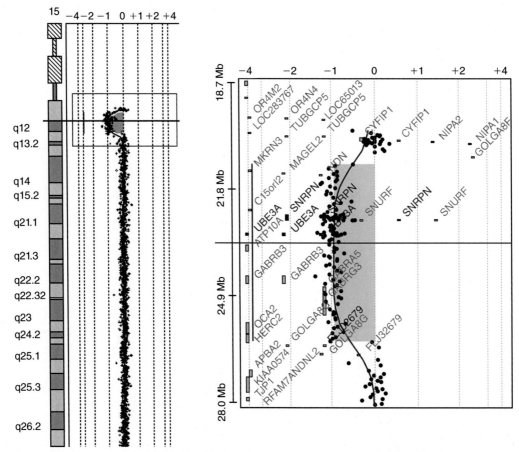

FIGURE 4.21 Copy number variation identified with a comparative genomic hybridization array made from oligonucleotides. DNA from a subject is fragmented, labeled with Cy5, and hybridized onto a microarray, together with Cy3-labeled reference DNA. On the array are nearly 44,000 oligonucleotide probes, each about 60 bases long and tiled across the whole genome at an average spacing of 75 kb. Shown on the *left panel* are results of probes on chromosome 15 (all other chromosomes are analyzed in this assay but are not shown). Each *dot* represents a specific probe to which the subject's DNA hybridizes. Their positions (0, −1, +1, and so on) reflect the dosage of the subject's DNA relative to the reference DNA. A majority of the probes line up on "0," indicating no quantitative difference compared with the reference DNA. Probes in the 15q11 to 15q13 region, however, are on the "−1" line, indicating that the subject has a deletion of that region in one of the chromosomes. A closer view of that region *(right panel)* shows that among the deleted genes are *UBE3A*, which causes Angelman syndrome, and *SNRPN*, which causes Prader-Willi syndrome. Because the method does not distinguish the methylation status of the deleted alleles, this result alone cannot determine which of the two disorders the subject has. (Courtesy Sarah South, PhD, ARUP Laboratories.)

metaphase chromosome spreads or interphase nuclei, the technique is referred to as FISH. Numeric aberrations or translocations of chromosomes can be detected rapidly. FISH can also be combined with immunohistochemistry so that information on both the amount of protein expression and the gene dosage can be found on the same slide. In situ hybridization is appropriate when localization of a target in tissue is important. However, experience in histology is necessary for accurate interpretation. In situ hybridization can provide information on the level of mRNA expression but not on the size or structure of the mRNA. As might be expected, hybridization within a tissue matrix is more variable than in solution or on well-characterized chemical surfaces.

Single-Copy Visualization. If an NA probe is labeled with many fluorescent molecules, it is possible to optically visualize a single copy of the NA target by fluorescent microscopy. One technique uses reporter probes that are labeled with a long string of multicolored fluorescent labels.[129] Several tandem color segments are placed on the reporter probe with each segment consisting of about 100 fluorophores. The combination of different color segments uniquely identifies the target. The target NA is hybridized with the probes each linked to capture probe that are (1) washed, (2) immobilized, (3) stretched, and (4) oriented on the surface of an optical slide. Each captured target is then identified by the color code of the reporter and is counted (see Fig. 50.3). Although the sensitivity of this technique is not as high as that of real-time PCR, up to 150 reporter probes have been multiplexed in one reaction. One application of this technique is direct measurement of mRNA expression in tissue specimens prepared from formalin-fixed paraffin blocks without the need for cDNA preparation or PCR.

Solution Phase Hybridization: Real-Time Polymerase Chain Reaction and Melting Analysis

Several classical hybridization methods use probe-target hybridization in solution. For example, hybrid capture uses an antibody that is specific for RNA–DNA hybrid molecules that are formed during solution-phase hybridization of a DNA sample and an unlabeled RNA probe. The assay also has been adapted to a microtiter plate format for automation of washing and detection. Solution phase hybridization has also been combined with (1) amplification, (2) detection, and (3) quantification steps all in the same tube. Such closed-tube, real-time assays do not require any additions, washing, or separation steps.

Real-time PCR and melting analysis are considered "dynamic" hybridization assays in which the formation or dissociation of the probe–target duplex (or product duplex) is monitored in real time. Data are collected throughout NA amplification rather than just at the endpoint. The technique uses fluorescent reporters and instruments to record hybridization during thermal cycling. The data obtained provide information on the (1) identity, (2) quantity, and (3) melting characteristics of the NA sample. Fluorescent dyes or probes are added to the PCR mixture before amplification. By measuring the fluorescence at each cycle of amplification, the amount of initial template present before PCR can be calculated. During the entire process, the same reaction tube is used for amplification and fluorescence monitoring, and no (1) sample transfers, (2) reagent additions, or (3) gel separation steps are required. This eliminates the risk of product contamination in subsequent reactions. Because the process is simple and fast, real-time PCR has replaced many conventional techniques in the clinical laboratory.

Real-time PCR monitors the accumulation of double-stranded PCR products with the fluorescence signal recorded once each cycle (Fig. 4.22).[12,130] If target DNA is present, fluorescence increases. How early during PCR one begins to see a signal depends on the initial amount of target DNA, and this provides a systematic method of quantification. Furthermore, when fluorescence is continuously monitored as the temperature is raised, a melting curve can be generated. Often the negative derivative of this melting curve is plotted to visually aid a person in estimating the melting temperature as peaks in the plot. Melting analysis can be used to verify the identity of the amplified product and to detect sequence variants.

Dyes and Probes. Many different fluorescence generating systems are used in real-time PCR; some of the more common ones are shown in Fig. 4.23. Many methods use probes with sequences complementary to the target. Others rely on dsDNA binding dyes and the specificity afforded by PCR primers. Some have the additional option of melting analysis to measure the melting temperature of the probe or product.

Certain cyanine dyes increase their fluorescence in the presence of dsDNA that is produced during PCR (see Fig. 4.23, *A*).[22,131] dsDNA binding dyes are commonly used for real-time quantification, particularly in the research setting when the specificity of a probe and its added cost is not needed. They also allow melting analysis of the product at the end of PCR.

Labeled primers can also be used to monitor PCR. In one system, a primer with a 5′-hairpin is labeled with a fluorophore and a quencher so that fluorescence is quenched in the hairpin conformation. When the primer straightens out during PCR, fluorescence increases.[132] If the sequence of the primer is carefully considered, the quencher moiety is not necessary.[86] Nonhairpin primers with a single label can also be used for detection and genotyping because of changes in fluorescence that occur with hybridization.[133]

One advantage of fluorescently labeled primers over dsDNA dyes is that multiplexing is possible. However, with both dsDNA dyes and labeled primers, reaction specificity depends on the primers. Any double-stranded product that is formed will be detected, including primer-dimers. Therefore, hot start techniques to increase specificity and melting curve analysis to confirm the desired product are useful.

The use of fluorescent probes in PCR adds another level of specificity to the process. Fluorescent probes that hybridize to PCR products during amplification change fluorescence by two possible mechanisms: (1) a covalent bond between two dyes is broken by hydrolysis or is made through ligation, or (2) the fluorescence change follows reversible hybridization of the probe to the target. Following this distinction, when an irreversible covalent bond is involved, the probes are called *hydrolysis probes.* When probes reversibly change fluorescence on duplex formation, they are called *hybridization probes.*

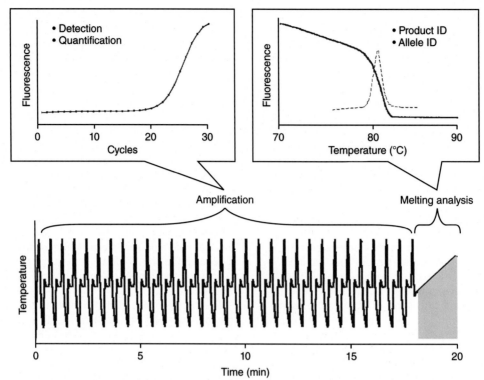

FIGURE 4.22 Real-time monitoring during amplification and melting analysis. The *bottom panel* shows a typical rapid-cycle temperature profile that is followed by a temperature ramp for melting analysis. When fluorescence is monitored once each cycle during amplification, the presence or absence and the quantification of the specific target are obtained. When fluorescence is monitored continuously through the melting phase *(shaded area)*, melting analysis can verify target identification and establishes genotype. (Modified with permission of the publisher from Wittwer CT, Kusukawa N. Real-time PCR. In: Persing DH, Tenover FC, Versalovic J, et al, eds. *Molecular microbiology: diagnostic principles and practice.* Washington, DC: ASM Press, 2004:71–84.)

One difference between the two probe types is that melting analysis is characteristic of hybridization probes, but hydrolysis probes usually have weak, if any, melting signature.

Hydrolysis probes (exonuclease probes) are synthesized with a quencher molecule positioned to quench the fluorescence of another label. If the probe is hydrolyzed between the fluorophore and the quencher during PCR, fluorescence will increase. The most common implementation uses the 5′-exonuclease activity of some DNA polymerases to hydrolyze the probe and dissociate the labels (see Fig. 4.23, *B*). This method has been simplified by putting the fluorophores on opposite ends of the probe.[134] Dual-labeled probes can also be cleaved using a DNAzyme (a DNA molecule that acts as a catalyst) generated during PCR.[135] Finally, irreversible ligation can be used for homogeneous genotyping with a fluorescent readout.[136] Hydrolysis probes generate fluorescence through changes in covalent bonds. The change in fluorescence signal is irreversible, and melting analysis of the hydrolyzed probe is seldom useful.

Hybridization probes change fluorescence upon hybridization, usually by fluorescence resonance energy transfer (FRET).[22,137] Two interacting fluorophores are typically placed on adjacent probes as dual hybridization probes (see Fig. 4.23, *C*). Only one probe with one fluorophore may be necessary if fluorescence is quenched by deoxyguanosine residues.[138] Another single-labeled probe design uses thiazole orange attached to a peptide NA.[139] In each of these designs, the fluorescence change that occurs with hybridization is reversible upon melting.

Hairpin probes (molecular beacons) typically have a fluorophore and a quencher at the 3′ and 5′-ends of a hairpin stem (see Fig. 4.23, *D*). Fluorescence increases when the distance between the quencher and the reporter increases upon target hybridization.[88] Compared with linear probes, hairpin probes discriminate mismatches with greater temperature changes.[140] Hairpin probes of different colors can be combined with melting curve analysis for highly multiplexed assays.[141]

Self-probing primers are modified at their 5′-end to include a hairpin probe with a fluorophore and quencher on opposite ends of the stem (see Fig. 4.23, *E*).[69] The hairpin loop is complementary to the primer extension product of the same strand that is generated during PCR. Through intramolecular hybridization, the quencher is separated from the fluorophore, and fluorescence increases. A blocker prevents copying the hairpin during PCR.

Highly quenched probes have a very efficient quencher at the 3′-end and a minor groove binder and fluorophore at the 5′-end (see Fig. 4.23, *F*).[142] Background fluorescence is claimed to be very low because of the combined effect of the quencher and minor groove binders. Upon hybridization, fluorescence increases because the fluorophore and quencher are separated, a process that can be reversed by melting. The minor groove binders increase probe stability, allowing

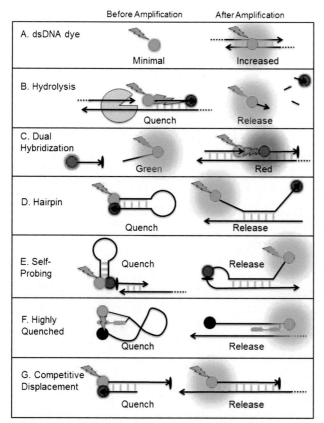

FIGURE 4.23 Common probes and dyes for real-time polymerase chain reaction (PCR). **A**, Double-stranded DNA dyes show a significant increase in fluorescence after DNA amplification. **B**, Hydrolysis probes are cleaved between a fluorescent reporter and a quencher, resulting in increased fluorescence. **C**, Dual hybridization probes. Fluorescence resonance energy transfer (FRET) is illustrated between a donor and acceptor fluorophore. The *thin vertical oval* indicates termination of the 3′-end of the probe to prevent polymerase extension. **D**, Hairpin probes are quenched in the native conformation but increase in fluorescence when hybridized. **E**, Self-probing primers retain their native, quenched conformation until they are incorporated into a double-stranded product. **F**, Highly quenched probes use a combination of minor groove binders and a very efficient quencher to limit background fluorescence. Upon hybridization, fluorescence increases. **G**, Competitive displacement probes have reporters and quenchers on opposite strands. The quenching probe is displaced by the accumulating PCR product and fluorescence increases. There are several forms of this type of probe. (Modified with permission of the publisher from Nolte FS, Wittwer CT. Nucleic acid amplification methods overview. In: Persing DH, Tenover FC, Versalovic J, et al, eds. *Molecular microbiology: diagnostic principles and practice*. 3rd ed. Washington, DC: ASM Press; 2015.)

the use of smaller probes when sequence variation of the target is high.

Competitive displacement probes come in several different varieties. In all cases, the fluorophore and quencher are on opposite strands, usually at 3′- and 5′- ends opposing each other. In its most basic form, generation of target strands during PCR displaces one labeled strand by competitive hybridization, separating fluorophore and quencher and generating fluorescence.[143] One design adds a 5′-labeled tail to one primer along with a 3′-quenching anti-primer complementary to the tail.[144] Additional designs for probes and antiprobes have recently been described that include mismatches in the antiprobe to favor displacement.[145] Another design uses partially double-stranded probes with one strand shorter than the other to also favor displacement (see Fig. 4.23, G).[146]

> **POINTS TO REMEMBER**
>
> **Fluorescent Indicators**
> - May be dyes or probes
> - Dyes are less expensive and less specific
> - Melting analysis increases specificity
> - May function by hybridization, hydrolysis, ligation, or displacement
> - Single base variants can be detected by probes or high-resolution melting
> - Enable real-time PCR in a closed system

Detection and Quantification in Real-Time PCR

When fluorescence is monitored once each cycle in the presence of a dye, the data closely follow the expected logistic shape discussed earlier (see Figs. 4.5 and 4.24, *top left*). However, with hydrolysis probes, fluorescence is cumulative and continues to increase even after the amount of product reaches a plateau (Fig. 4.24, *top middle*). In contrast, reactions monitored with hybridization probes may show a decrease in fluorescence at high cycle number[14] (Fig. 4.24, *top right*). Despite differences in the curve shape, all real-time systems follow the amount of product being produced during PCR, and this information is used for detection and quantification.

Detection. A fluorescent signal that increases during PCR and follows one of the expected curve shapes suggests that the specific target is present and was amplified. In contrast, a signal that stays at background even after 40 PCR cycles suggests that the target is absent and that no amplification occurred. Algorithms that analyze the entire curve may be more robust than simple threshold methods.[147] Positive controls (to rule out inhibitory factors) and negative controls (to rule out product contamination and nonspecific signal generation) are necessary. If the fluorescent signal is reversible with hybridization, melting analysis can verify the expected Tm of the probe or product.

Multiplex detection is possible with probes that are labeled with different colors or with probes and/or amplicons that have different melting temperatures. Examples in the clinical laboratory include probe multiplexing to detect the presence of more than one infectious organism, or to discriminate an internal control template from the target.

Quantification. Real-time PCR offers a convenient and systematic approach to quantification by monitoring the amount of product in each cycle. Some of the first clinical uses of quantitative real-time PCR were in the assessment of viral load, particularly for HIV and HCV. The clinical need for quantification is well established, and real-time methods give rapid and precise answers. However, other amplification systems, particularly transcription-based and branched DNA methods, are also used in this high-volume and highly competitive field. Additional quantitative applications of real-time PCR in clinical use are myriad and

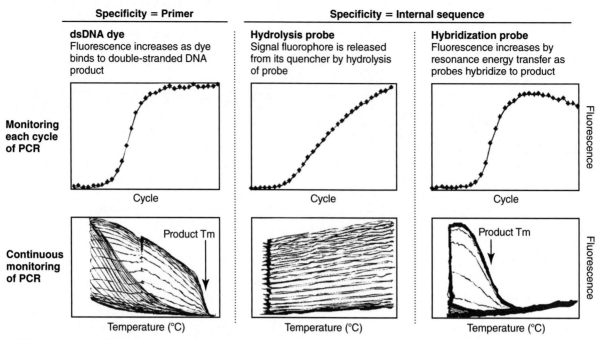

FIGURE 4.24 Monitoring polymerase chain reaction (PCR) in real time. The *top row* shows data collected once each PCR cycle, and the *bottom row* shows data collected continuously (5 times per second) during all PCR cycles. Three different reporter systems are shown. *dsDNA*, Double-stranded DNA; *Tm*, melting temperature. (Modified with permission of the publisher from Wittwer CT, Kusukawa N. Real-time PCR. In: Persing DH, Tenover FC, Versalovic J, et al, eds. *Molecular microbiology: diagnostic principles and practice.* Washington, DC: ASM Press; 2004:71–84.)

include quantification of mRNA (after reverse transcription) to monitor minimal residual disease and disease burden in leukemia with disease specific markers such as *BCR-ABL*, gene expression studies, and assessment of gene dosage in genetics and oncology.

Digital PCR has some advantages over real-time PCR, including precision and rare-allele analysis. However, one of the advantages of real-time PCR is its large dynamic range. Fig. 4.25 shows an extended range of quantification standards in a typical real-time PCR. As the initial template concentration increases, the curves shift to earlier cycles. The extent of this shift depends on the PCR efficiency (Table 4.3). The cycle at which fluorescence rises correlates inversely with the log of the initial template concentration and is the quantification cycle or Cq. This "cycle" is actually a *virtual* cycle that includes a fractional component determined by interpolation, which can be calculated by several methods. One method uses the maximum of the *second derivative* of the curve to determine Cq (Fig. 4.26). The second derivative of the amplification curve is derived from the shape of the curve and is estimated numerically with polynomials[148] without the need to adjust baselines or worry about normalizing the fluorescence values. Alternatively, in *threshold analysis,* a fluorescence level is selected that intersects with the amplification curves, and fractional cycle numbers are found by interpolation. However, when the sample fluorescence does not reach the threshold (as may happen with low copy samples), quantification is not possible.

Accuracy and Precision. The accuracy of real-time PCR quantification depends not only on the method chosen to analyze the curves but also on the quality of the quantification standards used. Purified PCR products quantified by spectrophotometry are easily obtained. When serially diluted, these calibrators can accurately quantify the amount of target in human genomic DNA.[148] Synthesized oligonucleotides, purified plasmids, and genomic DNA can also be used as calibrators. *Limiting dilution* analysis can also determine the amount of "amplifiable" DNA.[149] As previously mentioned, digital PCR is a great method to quantify PCR standards, and absolute reference standards are available for some targets.[150] Sometimes absolute quantification is not needed, and quantification relative to one or more reference genes is performed. In this case, selection of the reference genes and PCR efficiency are critical. The precision of quantitative real-time PCR depends on the initial number of template copies in the reaction. When the initial target concentration is low, imprecision is high. When this is the case, digital PCR may be a better choice.

Melting Analysis

When fluorescence is monitored continuously within each cycle of PCR, the hybridization characteristics of PCR products and probes can be observed.[34] Continuous spirals of fluorescence versus temperature are produced (see Fig. 4.24, *bottom panels*).[22] With dyes, the melting characteristics of the amplified DNA identify the product.[17] Little, if any, hybridization information is revealed with hydrolysis probes, whereas the melting of hybridization probes is readily apparent. Probe melting occurs at a characteristic temperature that can be exploited to confirm target identity and to analyze sequence variants under the probe.

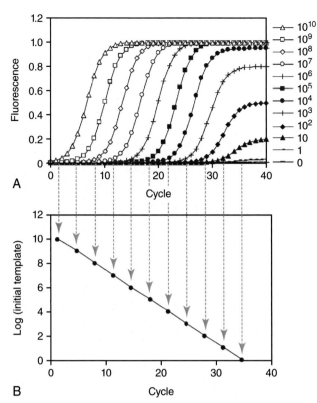

FIGURE 4.25 Quantification by real-time polymerase chain reaction. Shown are typical real-time curves for amplification reactions of varying initial target concentrations **(A)** and the log of the initial concentration plotted against the quantification cycle **(B)** as calculated by the second derivative maximum (see Fig. 4.26). (Modified with permission of the publisher from Wittwer CT, Kusukawa N. Real-time PCR. In: Persing DH, Tenover FC, Versalovic J, et al, eds. *Molecular microbiology: diagnostic principles and practice*. Washington, DC: ASM Press; 2004:71–84.)

TABLE 4.3 Correlation Between Polymerase Chain Reaction Efficiency, the Slope of the Standard Curve, and the Percentage of Polymerase Chain Reaction Product After 30 Cycles

Slope of Standard Curve*	PCR Efficiency (%)	PCR Product After 30 Cycles (% Expected)
−3.32	100	100
−3.35	99	86
−3.38	97.5	69
−3.45	95	47
−3.59	90	22
−3.74	85	10
−3.92	80	4
−4.34	70	1
−4.90	60	0.1
−5.68	50	0.02

*Assuming the log (initial template) is plotted on the *x*-axis as the independent variable and the quantification cycle is plotted on the *y*-axis as the dependent variable, the slope of the calibration curve is as follows: Slope = ΔCycle/ΔLog [initial template]. Percent PCR efficiency is calculated as $(10^{-1/\text{Slope}} - 1) \times 100$.
PCR, Polymerase chain reaction.

necessities of high-volume genomic research are different from those of a clinical reference laboratory, a medical clinic, or the STAT laboratories of the future. Many targeted genotyping techniques require complex separation or detection equipment (or both). Real-time PCR with melting curve analysis allows detection, quantification, and genotyping in

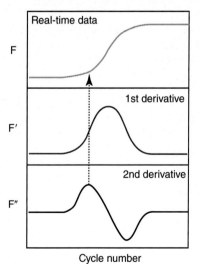

FIGURE 4.26 Estimation of the fractional cycle number for quantification. Real-time fluorescence data *(F)* from the amplification reaction are shown with the first *(F′)* and second *(F″)* derivatives. The maximum of the second derivative provides one way to determine the quantification cycle, Cq. (Modified with permission of the publisher from Wittwer CT, Kusukawa N. Real-time PCR. In: Persing DH, Tenover FC, Versalovic J, et al, eds. Molecular microbiology: diagnostic principles and practice. Washington, DC: ASM Press; 2004:71–84.)

For routine testing in the clinical laboratory, a single melting curve is usually performed at the end of PCR instead of monitoring hybridization throughout the entire PCR process (see Fig. 4.22). Genotyping is best performed in the same tube by monitoring the melting of hybridized duplexes during controlled heating, producing a melting curve signature for the duplex. Such a signature monitors melting over a range of temperatures in contrast to the single-temperature analysis of conventional hybridization techniques, such as dot blots or microarrays. The advantages of complete melting curves also apply when different homogeneous techniques are compared. Real-time amplification and melting analysis make up a powerful combination that requires only temperature control and sampling of fluorescence. When hybridization probes are used, rapid cooling maximizes formation of probe–target duplexes while minimizing formation of the duplex PCR product. Primer asymmetry and use of 5′-exonuclease–deficient polymerases can augment the probe signal.

Homogeneous Single Nucleotide Variant Typing

Many methods are available for SNV genotyping, and the method of choice depends on several factors, including turnaround time, batch size, and throughput requirements. The

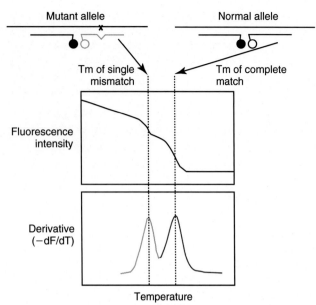

FIGURE 4.27 Melting curve single nucleotide variant (SNV) genotyping. A heterozygous specimen with an SNV under one probe is amplified and melted. Two temperature transitions are visible: one from the mutant allele that is mismatched with the probe and melts at a lower temperature and one from the normal allele that is completely matched with the probe and melts at a higher temperature. The derivative plot shows the melting temperatures (Tms) of both the mutant-probe and the normal-probe duplexes as peaks. (Modified with permission of the publisher from Bernard PS, Pritham GH, Wittwer CT. Color multiplexing hybridization probes using the apolipoprotein E locus as a model system for genotyping. *Anal Biochem* 1999; 273:221–228. Copyright 1999 Academic Press.)

less than 30 minutes (see Fig. 4.22) in a homogeneous system that does not require ancillary processing or additional equipment.

For example, a hybridization probe pair with one probe placed over a heterozygous polymorphism is shown in Fig. 4.27. In this example, the reporter probe is complementary to the normal allele. As the temperature is increased, the mismatched mutant hybrid melts first, giving the first transition, followed by the matched normal hybrid. The melting temperatures of both hybrids are easily seen in derivative plots.[148] A well-optimized probe design will provide a Tm difference of 4° to 10°C for a single base mismatch under the probe.

Single nucleotide variant genotyping by melting curve analysis can be achieved with a variety of probe and dye methods. Fig. 4.28, *A*, shows the hybridization probe pair just discussed. Virtually the same result can be achieved by using a single hybridization probe in which the fluorescent signal is quenched on the free probe, but is enhanced as it forms a hybrid with the target (see Fig. 4.28, *B*).[138] Fig. 4.28, *C*, shows genotyping with an *unlabeled probe* and a saturating DNA binding dye.[151] Both probe and amplicon melting transitions are present. Fig. 4.28, *D*, shows similar results using a *snapback primer*,[70] an unlabeled probe attached as a 5′-tail to one of the primers. The advantage of these last two methods is that they do not require fluorescently labeled probes. Finally, SNV genotyping is possible by amplicon melting with a saturating DNA binding dye (see Fig. 4.28, *E*).[152]

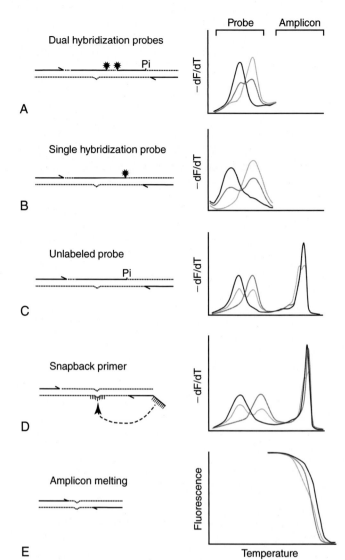

FIGURE 4.28 Five melting designs for single nucleotide variant (SNV) genotyping and corresponding melting curve results. The traditional dual hybridization probe design **(A)** uses a pair of probes: one labeled with an acceptor fluorophore and the other with a donor fluorophore. The single hybridization probe design **(B)** lacks the second probe. The unlabeled probe design **(C)** does not require a covalently attached fluorescent label, using instead a saturating DNA binding dye in solution. Snapback primers **(D)** are similar to unlabeled probes, with the probe attached to one primer as a 5′-tail. Finally, amplicon melting **(E)** uses only two regular polymerase chain reaction primers. Amplicon melting relies on high-resolution melting analysis to distinguish the small differences among genotypes. The two homozygotes are differentiated by melting temperature (Tm), and the heterozygote differs in shape because of the contribution from heteroduplexes. P_i indicates a 3′-phosphate or other blocker that prevents polymerase extension. dF/dT, Rate of fluorescence change.

The many methods of homogeneous SNV typing differ greatly in their level of complexity (Table 4.4). The number of oligonucleotides required varies from 2 to 5. The simpler techniques do not require probes at all, although some of the more complex methods require up to three labels or modifications on each probe. All of these methods use

TABLE 4.4 Comparison of Homogeneous (Closed-Tube) Genotyping Methods (in Order of Increasing Complexity)

Method	Oligonucleotides	Modifications	Comments
Amplicon melting	2	0	Simplest and least expensive
Snapback primers	2	1	Self-complementary 5′-tail
Allele-specific PCR (real time)	3	0	Requires one well for each allele
Unlabeled probes	3	1	3′-phosphate on probe
Allele-specific PCR (melting)	3	1–2	GC* clamps
Single hybridization probe	3	1–2	3′-phosphate if 5′-fluorophore
Dual hybridization probes	4	2–3	3′-phosphate if 5′-fluorophore
Hydrolysis probes	4	4	
Dual-labeled hairpin probes	4	4	
Self-probing amplicon	3–4	6	
Minor groove binder hydrolysis probe	4	6	
Serial invasive signal amplification	5	4	

*Short 5′-oligonucleotide tail of G and C bases used to modify allele melting temperatures.
PCR, Polymerase chain reaction.

fluorescence and solution hybridization. Some that use melting analysis can detect more than two alleles if present; those based on allele-specific amplification or endpoint analysis are limited to two.

The five simplest homogeneous SNV typing methods do not use fluorescently labeled probes. Amplicon melting requires only two primers and a heteroduplex-detecting DNA dye (see Fig. 4.28, *E*).[152] The snapback primer system also requires only two primers—one with a self-probing 5′-tail (see Fig. 4.28, *E*).[70] Unlabeled probe genotyping requires two primers and one 3′-blocked probe (see Fig. 4.28, *C*).[151] Allele-specific PCR requires three primers and is based on a preference by the polymerase to extend only a perfectly matched primer. Genotyping can be obtained in two wells by monitoring fluorescence at each cycle[153] or in one well by incorporating GC-clamp(s) so that alleles can be differentiated at the end of PCR by their melting temperatures.[154] Intermediate in complexity are hybridization probe melting assays. Designs with single[138] or dual hybridization[155] probes are shown in Fig. 4.28, *A* and *B*. The more complex closed-tube methods for SNV genotyping are endpoint assays. Allele-specific hydrolysis[87] and hairpin probes[156] are commonly used. Self-probing amplicons[69] and minor groove binder hydrolysis probes[157] both require three modifications on two probes for SNV typing. Finally, serial invasive signal amplification is a method of homogeneous genotyping that does not require PCR but uses the largest number of oligonucleotides for homogeneous genotyping.[79]

High-Resolution Melting Analysis

The temperature difference between genotypes in amplicon melting can be small (see Fig. 4.28, *E*), requiring high-resolution melting for accurate genotyping. High-resolution melting detects heteroduplexes with better sensitivity than gel methods and does not require any processing or separation.[158] Although legacy fluorescent melting analysis can distinguish PCR products that differed by about 1° to 2°C,[17] high-resolution melting instruments now provide precision and resolution improvement of at least 10-fold[18] and require only a few minutes.[159] Typically, melting data are normalized between 0% and 100% fluorescence, and different homozygotes are distinguished by Tm. Heterozygotes are best detected when comparing the melting curve shapes by shifting the curves along the temperature axis until they overlap. An example of heteroduplex detection and SNV genotyping by melting curve analysis is shown in Fig. 4.29 in which a PCR product melts in two domains. Domain prediction can be very useful in melting assay design.[160] Major applications of high-resolution melting include genotyping, mutation scanning, and sequence matching.[161-163] Targeted copy number assessment is a new application that may be even better than copy number determination by digital PCR.[128]

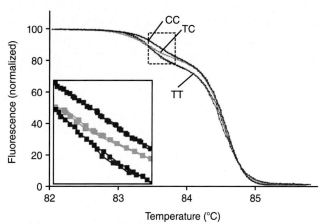

FIGURE 4.29 A single base change in a 544-bp fragment detected by melting analysis. Shown are high-resolution melting curves of polymerase chain reaction products from the gene *HTR2A* carrying a single nucleotide variant (SNV). Results are shown for six individuals, two different individuals for each of the three genotypes: wild-type homozygote (TT), mutant homozygote (CC), and heterozygote (TC). Two melting domains are present because of differing GC content. The SNV was present in the lower melting domain. The *inset* magnifies a portion of the data, showing that all three genotypes can be discriminated. (Modified by permission of the publisher from Wittwer CT, Reed GH, Gundry CN, et al. High-resolution genotyping by amplicon melting analysis using LC Green. *Clin Chem* 2003;49:853–860. Copyright AACC.)

Massively Parallel Sequencing

The need to understand the full extent of genome-wide human variation has led to the development of new massively parallel sequencing techniques.[164-166] Compared with dideoxy-termination sequencing, these techniques generate up to 1 billion more bases of sequence data in one operation at 10,000 to 100,000 times lower cost per base. These techniques continue to evolve such that throughput and cost continue to improve. Much of the progress that has been made in this technology is dependent on advances in optical data processing, bioinformatics, and overall computer power. As the cost, turnaround time, and convenience of these methods continue to decrease, they increasingly will be used in the clinical laboratory. Indeed, clinical laboratory standards for massively parallel sequencing have already appeared.[167,168] Clonal sequencing methods replicate a single DNA strand to form a clonal template in order to generate sufficient signal for detection. In contrast, single-molecule sequencing methods must be sensitive enough to detect single molecules of DNA. Characteristics of massively parallel sequencing methods are summarized in Table 4.5.

Sequencing From Clones

Clonal sequencing methods start with producing a shotgun (random) library of fragments that are typically 70 to 1000 bases in length, although some methods require 6- to 20-kb fragments. Fragmentation is usually physical or enzymatic.[23] Physical methods include sonication, acoustic shearing, and hydrodynamic shearing. Enzymatic fragmentation can be from restriction endonucleases, nonspecific DNAses, or a tranposase enzyme that simultaneously fragments and adds adapter sequences. In all cases, conditions can be modified to produce different fragment sizes.

Adapter sequences are typically added to each end of the random fragments. The primary role of these adapters is to provide common priming sites for each fragment to initiate massively parallel sequencing reactions. One primer set amplifies a massive array (beads or planer flow cell) of library inserts. Adapters also facilitate initial capture of DNA fragments onto solid surfaces and spatially restrict clonal amplification products generated from the fragments onto beads or spots on an array surface. The fragment ends typically need to be "polished" by filling in any missing bases and optionally adding a single extra A to the 3′-ends to facilitate ligation to the adapters. If multiplexing of different DNA samples is desired, a sequence "barcode" is often added as well to identify which DNA sample the clone arose from. A typical library insert with adapters and a barcode is shown in Fig. 4.30, *A*. The libraries are then partitioned according to size to select a band optimal

TABLE 4.5 Characteristics of Massively Parallel Sequencing Methods

Method	Principle	Detection	Clonal	Run Time	Output Per Run	Read Length (bp)
Synthesis	Pyrophosphate release	Chemiluminescence	Emulsion PCR	10–23 hours	40–700 mb	400–700
Synthesis	pH change	Electronic CMOS	Emulsion PCR	3–4 hours	1.5–10 gb	125–400
Synthesis	Reversible terminator	Fluorescence	Bridge amplification	2.7–12 days	15–600 gb	200–600*
Ligation	Multiple ligation events	Fluorescence	Emulsion PCR	10 days	300 gb	110
Single molecule	Zero-mode waveguide	Fluorescence	No	2 days	5 gb	10 kb
Single molecule	Conductivity	Electronic	No	Minutes to Days	Depends	5 kb

*Includes both paired end reads (sequencing from both ends).
CMOS, Complementary metal oxide semiconductor; *PCR*, polymerase chain reaction.

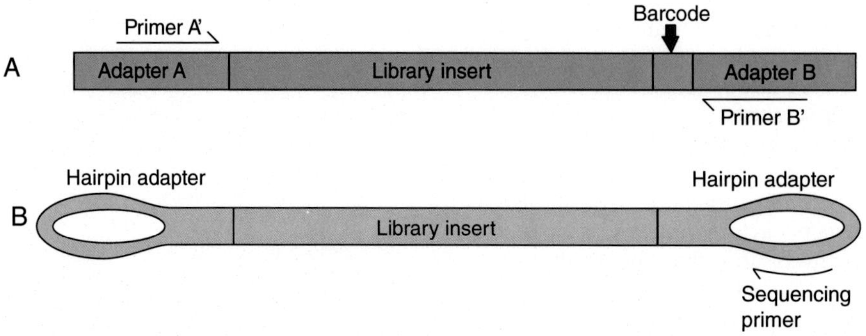

FIGURE 4.30 Diagram of different library formats used in massively parallel sequencing. **A**, Two adapters that include consensus PCR priming sites are ligated onto each end of the library inserts produced by fragmentation. If multiplexing different samples is desired, a barcode is added so that each read can be assigned to a specific sample. **B**, A library insert is bounded by hairpin adapters that allow primer binding to the single-stranded loops on each end for rolling circle amplification.

for the downstream sequencing technology. Clonal amplification is usually performed by either emulsion PCR or bridge amplification.

Emulsion Polymerase Chain Reaction. In emulsion PCR, one strand of a library element is captured on one bead and is clonally amplified inside a water-in-oil droplet, generating a bead covered with single-stranded PCR products (Fig. 4.31). The emulsion is formed by mixing beads (each covered with one primer), aqueous PCR components (including the other primer, polymerase, and dNTPs), and a mixture of oils under agitation to ideally form droplets that each contain only one bead and one library insert. The two primers are complementary to the adapters; one coats the bead surface, and one is free in solution. During emulsion PCR, all of the beads are amplified together in aqueous microdroplets dispersed in oil. The emulsion is amplified in a standard PCR thermal cycler. After PCR and denaturation, millions of copies of identical single-stranded PCR products are on each bead with each bead carrying distinct, oriented inserts flanked by the adapters. The emulsion is then broken and, after elimination of empty beads, is ready for sequencing.

Bridge Amplification. Bridge amplification generates clusters of single-stranded PCR products tethered to the surface of a planar flow cell (Fig. 4.32).[62] In contrast to the clonal bead generation of emulsion PCR, the amplification occurs on a flat surface. The primers, complementary to the adapters, are both attached to the surface, either randomly or in a fixed pattern. The library DNA is then denatured to form single strands that hybridize to the primers on the surface. After extension of the surface-bound primers, the original template strands are washed away under denaturing conditions. What follows is called bridge amplification, which is very similar to PCR except that both primers are bound to the surface, so that single strands bound to the surface must bend over to find their opposite primer, resulting in a double-stranded bridge after extension. Instead of denaturing with heat as in PCR, the flow cell is kept at 60°C, and a chemical denaturant is introduced to dissociate the two bridge strands, except now both strands are attached to the planer surface. When the flow cell is flushed with a polymerase and dNTPs under favorable extension conditions, both can find new primers to form additional bridges. The process repeats until about 1000 copies are formed. One of the bound primers can be designed to include a cleavable site (either chemical or enzymatic) so that one strand can be removed after denaturation. After capping the 3′- end of the single strands with ddNTPs (to prevent any undesired extension with the closely packed templates), the surface is ready for sequencing.

Sequencing by Synthesis

Sequencing by synthesis can be detected by (1) pyrophosphate release, (2) a pH decrease, or (3) the fluorescence of reversible terminators. The clonal amplification methods allow parallel observation of thousands to millions of strand extensions, greatly increasing the signal strength. However, the extensions must be controlled at each step because continuous strand extension would not remain synchronous between strands. This is achieved by immobilizing the clones

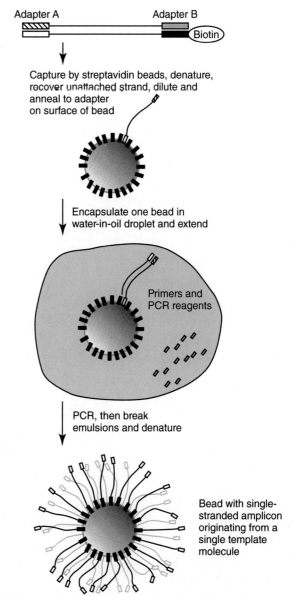

FIGURE 4.31 Emulsion polymerase chain reaction (PCR). Two adapters (*adapters A* and *B*) are randomly ligated to DNA fragments. Adapter B has a biotin on its 5′ end. Fragments with adapter B on one or both ends are captured by streptavidin beads, but fragments with only adapter A are washed away (not shown). Then the fragments are denatured, and the free strand with adapter A and adapter B at each end is collected (fragments with adapter B on both ends will not be released from the streptavidin bead). One molecule of the single-stranded template is then captured on a bead coated with adapter and is encapsulated inside a water-in-oil droplet that contains PCR reagents and primers. After PCR, the emulsion is broken, and the DNA is denatured. This generates a bead with a large number of clonal single strands tethered to it. The bead is then deposited into one of many wells on a fiber optic or onto a glass slide for sequence analysis (not shown).

into arrays so that reagents can be applied sequentially (no pun intended).

Pyrosequencing.[19] Pyrosequencing was used in the first massively parallel sequencing platform but has become less popular in recent years because of lower throughput and

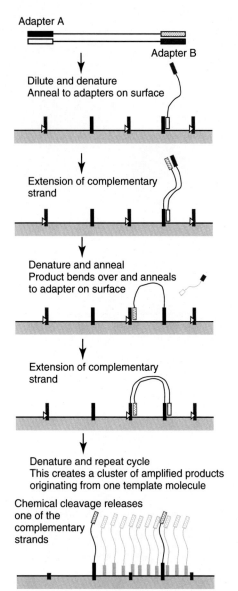

FIGURE 4.32 Bridge amplification. Two adapters (*adapters A and B*) are ligated to a DNA template. After they are diluted and denatured into single strands, the template is captured onto a flow cell surface by annealing to one of the two surface-bound primers that share sequences with adapter A or B. The polymerase reagent introduced into the flow cell extends the primer and generates the complementary strand of the template. The denaturant (usually sodium hydroxide) is introduced to the flow cell to release the original template strand. The free end of the newly synthesized strand anneals to a nearby primer by bending over, and a second round of reagent addition catalyzes the synthesis of another complementary strand. By repeating many of these cycles, a clonal cluster that consists of about 1000 copies of single-stranded template tethered to the surface is generated. This cluster is still a mixture of both complementary strands. One of the strands is selectively eliminated by treatment with periodic acid that cleaves a diol linkage present in one of the surface-bound primers *(open triangle on red primer)*. The cluster now contains only one of the template strands and is ready for sequence analysis.

higher cost. Clonal beads are fit into picoliter reaction chambers formed in etched individual fibers of a fiberoptic cable. Solutions of dATPαS, dCTP, dGTP, or dTTP are passed over chambers one at a time under conditions that favor extension. If there is a base match, the nucleotide will be incorporated, and pyrophosphate will be released. Pyrosequencing signal generation occurs by linked enzymatic reactions, leading to luciferase chemiluminescence, and is covered earlier in this chapter (including the replacement of dATP with dATPαS to prevent interference with the linked enzymatic reactions). The light produced is captured by the individual light fibers and detected on a CCD. If more than one base occurs in a row (a homopolymer stretch), multiple bases of that nucleotide will be incorporated, and the signal will be proportionately higher. As the number of identical bases increases, if becomes harder to determine their exact number.

Semiconductor Sequencing.[63] Similar to pyrosequencing detection, clonal beads are used as template. However, the beads are arrayed on semiconductor sensors modified to detect pH changes.[169] The chip does not detect light but a slight change in pH produced as a consequence of conversion of a dNTP into pyrophosphate by the many clones on the bead. Similar to pyrosequencing, homopolymer stretches can be problematic. By leveraging semiconductor technology development, the chip has rapidly increased in performance by decreasing the size of the beads and sensor wells and increasing the size of the chip. Run times are as short as 3 to 4 hours.

Reversible Terminators. After bridge amplification on a planar flow cell, one of four nucleotides is passed through the flow cell under conditions that favor extension. Unlike pyrosequencing and semiconductor sequencing, the nucleotides are fluorescent terminators so that only one base is incorporated, avoiding the problems with homopolymer stretches. Each nucleotide has a different fluorescent label so that each can be distinguished by color. Furthermore, the fluorescent terminators are reversible, which means that the blocked 3′-end can be regenerated by simple chemical means provided by the flow cell. Each cycle involves (1) adding polymerase and dNTPs as fluorescently labeled terminators under conditions that favor extension, (2) washing the flow cell, (3) imaging the fluorescence, (4) cleaving the fluorescent terminator, and (5) washing the flow cell. Output per run is up to 600 gb in 1 day.

Sequencing by Ligation

Instead of a polymerase and sequencing by synthesis, sequencing by ligation uses a ligase and a mixture of probes about eight bases long (Fig. 4.33). The template is either a clonal bead formed by emulsion PCR (similar to pyrophosphate and semiconductor sequencing) or a colony formed by isothermal amplification directly on a planer surface.[170] First, an anchor oligonucleotide is hybridized to a known portion of the immobilized template. Then a mixture of probes about eight bases long compete for ligation to the anchor. Each of the competing probes has one position with a defined base (A, T, G, or C) indicated by the color of a fluorescent label at the far end of the probe. The remaining bases of the probe may be degenerate (meaning that all four bases are used in the synthesis, creating a combination of 4^m different probes, where m is the number of degenerate base positions), or they may be universal bases (nucleotide analogs that pair with any of the four bases). The terminal fluorescent label blocks further concatenation of probes. At a stringent temperature, only the ligated anchor–probe complex remains hybridized to the template, and the defined base position is decoded by the color of the fluorescent label. The next cycle begins by stripping away the anchor–probe complex from the template and repeating the process but offset by one base by using a different anchor. Eventually, a short string of bases immediately adjacent to the anchoring site is determined. A variant of the technique uses two defined bases on each labeled probe and extends the read length by cleaving part of the probe after recording the label. Although massively parallel, the read length averages less than 75 bases.

Single-Molecule Sequencing

Single-molecule sequencing methods do not require template amplification. Base reads do not require synching with other clonal strands because there are none. Sensitive optical or electronic methods are required to detect base sequence in a single molecule.[171] If long reads and high accuracy are achieved, advantages include efficient sequence assembly, analyses of repetitive sequences, de novo sequencing, mapping of chromosomal rearrangements, and fusions. In contrast, massively parallel sequencing methods usually generate short sequence reads (30–700 bases long) that have to be aligned and analyzed to derive a consensus, then stitched together, and compared with reference genome sequences. Accurate assembly of sequence data relies on sufficient coverage and/or redundancy across the sequenced region.[172]

Real-Time Single Molecule Sequencing With Fluorescent Nucleotides.[171] Library preparation is unique for this method because fragmentation is tuned to provide about 10-kb inserts, and the adapters are designed as hairpins. The result is a double-stranded insert bounded by identical 27 base single-stranded loops on each end (see Fig. 4.30, B). Sequencing primers are annealed to the loop region and bound to a single polymerase molecule at the bottom of a zero-mode waveguide to form an active polymerase complex. The zero-mode waveguide allows single molecule detection of transient fluorescent labels that are covalently attached to the terminal phosphate of each dNTP. The four dNTPs are added to the wells with different labels that can be distinguished optically. When a fluorescent dNTP pairs with its complement near to the polymerase active site, it is in a perfect position for fluorescence detection. After being incorporated into the growing strand, the terminal fluorescent label (attached to pyrophosphate) diffuses away from the polymerase. The fluorescent signals are acquired continuously at high speed as the primer proceeds around the loop by rolling circle amplification. The process can be stopped after one loop or continued for multiple reads for error checking. Base reads of up to 40,000 to 50,000 contiguous bases have been reported.

Nanopore Sequencing.[173] Another single-molecule sequencing approach that does not require amplification and uses electronic, rather than optical, detection is nanopore sequencing. Single DNA strands are channeled through nanopores formed by immobilized proteins. Passage of

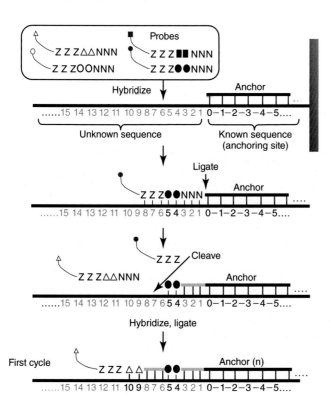

FIGURE 4.33 Sequencing by ligation (two-base encoding method). An anchor oligonucleotide is hybridized to known sequences of the template (anchoring site). Structures of octamer (8-mer) probes are shown. *N* represents a degenerate base (A, T, C, or G), *Z* represents a universal base that pairs with any base, and the defined bases *(closed circle, open circle, closed square, open triangle)* occupy the fourth and fifth base positions. The probes are color coded by one of four labels *(red),* with each color representing a set of four two-base combinations (eg, color 1 is AT, TA, CG, or GC; color 2 is AC, CA, TG, or GT; color 3 is TC, CT, AG, or GA; and color 4 is AA, TT, GG, or CC). When a probe hybridizes to the sequence adjacent to the anchoring site, ligase will connect the probe to the anchor. After the color is recorded, part of the probe is cleaved, and the label is removed together with some of the probe bases. This makes it possible for another probe to ligate to the extending complex. In fact, many rounds of ligation and cleavage can be performed, each time elongating the anchor–probe complex and providing additional two-base possibilities. After a few rounds of probe ligation and cleavage, the anchor–probe complex is stripped, and the next cycle, which is offset by one base, repeats the process. Cycles are repeated until one of the defined bases on the probe pairs with a known base on the anchoring site, allowing decoding of all of the two-base combinations. For example, in the first cycle, two-base possibilities for positions 4, 5, 9, and 10 are determined. In the second cycle, which is offset by one base (n-1), possibilities for positions 3, 4, 8, and 9 are determined. This process is repeated until the defined base on the probe pairs with the first base on the anchoring site (position 0). Because the identity of that base is known, the 0/1 base combination is decoded, which in turn decodes base position 2 and so on, until all of the two-base combinations are decoded.

individual DNA bases through the protein nanopore generates characteristic electrical signals that can reveal the identity of the base (or combination of bases) traversing the nanopore. In essence, this is similar to a nano-sized Coulter counter that quantifies base differences on a single strand of DNA rather than cell size. Kilobases of NAs can potentially be sequenced in a single read. The method is nondestructive and can discriminate methylated DNA bases as well as normal bases.[174] A number of nanopores are currently under study (α-hemolysin, *Mycobacterium smegmatis* porin A [MspA], and others), and solid-state nanopores are also being intensively investigated. Shorter and narrower nanopores that interrogate single bases would be ideal, but current material limitations allow about four bases within the nanopore at the same time.

REFERENCES

1. Watson JD, Crick FH. Molecular structure of nucleic acids; a structure for deoxyribose nucleic acid. *Nature* 1953;**171**:737–8.
2. Meselson M, Stahl FW. The replication of DNA in Escherichia coli. *Proc Natl Acad Sci USA* 1958;**44**:671–82.
3. Smith HO, Wilcox KW. A restriction enzyme from Hemophilus influenzae. I. Purification and general properties. *J Mol Biol* 1970;**51**:379–91.
4. Agarwal KL, Buchi H, Caruthers MH, et al. Total synthesis of the gene for an alanine transfer ribonucleic acid from yeast. *Nature* 1970;**227**:27–34.
5. Temin HM, Mizutani S. RNA-dependent DNA polymerase in virions of Rous sarcoma virus. *Nature* 1970;**226**:1211–3.
6. Baltimore D. RNA-dependent DNA polymerase in virions of RNA tumour viruses. *Nature* 1970;**226**:1209–11.
7. Southern EM. Detection of specific sequences among DNA fragments separated by gel electrophoresis. *J Mol Biol* 1975;**98**:503–17.
8. Alwine JC, Kemp DJ, Stark GR. Method for detection of specific RNAs in agarose gels by transfer to diazobenzyloxymethyl-paper and hybridization with DNA probes. *Proc Natl Acad Sci USA* 1977;**74**:5350–4.
9. Wittwer C, Kusukawa N. Nucleic Acid Techniques. Burtis C, Ashwood E, Bruns D, editors. Tietz Textbook of Clinical Chemistry and Molecular Diagnostics. Maryland Heights, MO: Elsevier; 2012. p. 401–442.
10. Sanger F, Nicklen S, Coulson AR. DNA sequencing with chain-terminating inhibitors. *Proc Natl Acad Sci USA* 1977;**74**:5463–7.
11. Saiki RK, Gelfand DH, Stoffel S, et al. Primer-directed enzymatic amplification of DNA with a thermostable DNA polymerase. *Science* 1988;**239**:487–91.
12. Higuchi R, Dollinger G, Walsh PS, et al. Simultaneous amplification and detection of specific DNA sequences. *Biotechnology (N Y)* 1992;**10**:413–7.
13. Heid CA, Stevens J, Livak KJ, et al. Real time quantitative PCR. *Genome Res* 1996;**6**:986–94.
14. Wittwer CT, Ririe KM, Andrew RV, et al. The LightCycler: a microvolume multisample fluorimeter with rapid temperature control. *Biotechniques* 1997;**22**:176–81.
15. Maskos U, Southern EM. Oligonucleotide hybridizations on glass supports: a novel linker for oligonucleotide synthesis and hybridization properties of oligonucleotides synthesised in situ. *Nucleic Acids Res* 1992;**20**:1679–84.
16. Schena M, Shalon D, Davis RW, et al. Quantitative monitoring of gene expression patterns with a complementary DNA microarray. *Science* 1995;**270**:467–70.
17. Ririe KM, Rasmussen RP, Wittwer CT. Product differentiation by analysis of DNA melting curves during the polymerase chain reaction. *Anal Biochem* 1997;**245**:154–60.
18. Wittwer CT, Reed GH, Gundry CN, et al. High-resolution genotyping by amplicon melting analysis using LCGreen. *Clin Chem* 2003;**49**:853–60.
19. Margulies M, Egholm M, Altman WE, et al. Genome sequencing in microfabricated high-density picolitre reactors. *Nature* 2005;**437**:376–80.
20. Shendure J, Porreca GJ, Reppas NB, et al. Accurate multiplex polony sequencing of an evolved bacterial genome. *Science* 2005;**309**:1728–32.
21. Chiu RW, Lo YM, Wittwer CT. Molecular diagnostics: a revolution in progress. *Clin Chem* 2015;**61**:1–3.
22. Wittwer CT, Herrmann MG, Moss AA, et al. Continuous fluorescence monitoring of rapid cycle DNA amplification. *Biotechniques* 1997;**22**:130–1. 4-8.
23. Head SR, Komori HK, LaMere SA, et al. Library construction for next-generation sequencing: overviews and challenges. *Biotechniques* 2014;**56**:61–4. 6, 8, passim.
24. Parkinson NJ, Maslau S, Ferneyhough B, et al. Preparation of high-quality next-generation sequencing libraries from picogram quantities of target DNA. *Genome Res* 2012;**22**:125–33.
25. Browne KA. Metal ion-catalyzed nucleic acid alkylation and fragmentation. *J Am Chem Soc* 2002;**124**:7950–62.
26. Cotton RG, Rodrigues NR, Campbell RD. Reactivity of cytosine and thymine in single-base-pair mismatches with hydroxylamine and osmium tetroxide and its application to the study of mutations. *Proc Natl Acad Sci USA* 1988;**85**:4397–401.
27. Hayatsu H, Wataya Y, Kai K, et al. Reaction of sodium bisulfite with uracil, cytosine, and their derivatives. *Biochemistry* 1970;**9**:2858–65.
28. Shiraishi M, Hayatsu H. High-speed conversion of cytosine to uracil in bisulfite genomic sequencing analysis of DNA methylation. *DNA Res* 2004;**11**:409–15.
29. Kristensen LS, Hansen LL. PCR-based methods for detecting single-locus DNA methylation biomarkers in cancer diagnostics, prognostics, and response to treatment. *Clin Chem* 2009;**55**:1471–83.
30. Weber M, Davies JJ, Wittig D, et al. Chromosome-wide and promoter-specific analyses identify sites of differential DNA methylation in normal and transformed human cells. *Nat Genet* 2005;**37**:853–62.
31. Rauch T, Pfeifer GP. Methylated-CpG island recovery assay: a new technique for the rapid detection of methylated-CpG islands in cancer. *Lab Invest* 2005;**85**:1172–80.
32. Wittwer CT, Garling DJ. Rapid cycle DNA amplification: time and temperature optimization. *Biotechniques* 1991;**10**:76–83.
33. Wittwer CT, Herrmann MG. Rapid thermal cycling and PCR kinetics. In: Innis M, Gelfand D, Sninsky J, editors. *PCR Methods Manual*. San Diego: Academic Press; 1999. p. 211–29.
34. Wittwer CT, Rasmussen RP, Ririe KM. Rapid PCR and melting analysis. In: Bustin SA, editor. *The PCR Revolution: Basic Technologies and Applications*. New York: Cambridge University Press; 2010. p. 48–69.
35. Farrar JS, Wittwer CT. Extreme PCR: efficient and specific DNA amplification in 15-60 seconds. *Clin Chem* 2015;**61**:145–53.

36. Zhou MY, Gomez-Sanchez CE. Universal TA cloning. *Curr Issues Mol Biol* 2000;**2**:1–7.
37. von Ahsen N, Wittwer CT, Schutz E. Oligonucleotide melting temperatures under PCR conditions: nearest-neighbor corrections for Mg(2+), deoxynucleotide triphosphate, and dimethyl sulfoxide concentrations with comparison to alternative empirical formulas. *Clin Chem* 2001;**47**:1956–61.
38. Montgomery JL, Wittwer CT. Influence of PCR reagents on DNA polymerase extension rates measured on real-time PCR instruments. *Clin Chem* 2014;**60**:334–40.
39. National Center for Biotechnology Information (NCBI). Basic Local Alignment Search Tool. http://blast.ncbi.nlm.nih.gov/Blast.cgi [accessed 08.15].
40. Kwok S, Higuchi R. Avoiding false positives with PCR. *Nature* 1989;**339**:237–8.
41. Poritz MA, Blaschke AJ, Byington CL, et al. FilmArray, an automated nested multiplex PCR system for multi-pathogen detection: development and application to respiratory tract infection. *PLoS ONE* 2011;**6**:e26047.
42. Sanchez JA, Pierce KE, Rice JE, et al. Linear-after-the-exponential (LATE)-PCR: an advanced method of asymmetric PCR and its uses in quantitative real-time analysis. *Proc Natl Acad Sci USA* 2004;**101**:1933–8.
43. Michalatos-Beloin S, Tishkoff SA, Bentley KL, et al. Molecular haplotyping of genetic markers 10 kb apart by allele-specific long-range PCR. *Nucleic Acids Res* 1996;**24**:4841–3.
44. Li J, Wang L, Mamon H, et al. Replacing PCR with COLD-PCR enriches variant DNA sequences and redefines the sensitivity of genetic testing. *Nat Med* 2008;**14**:579–84.
45. How-Kit A, Lebbe C, Bousard A, et al. Ultrasensitive detection and identification of BRAF V600 mutations in fresh frozen, FFPE, and plasma samples of melanoma patients by E-ice-COLD-PCR. *Anal Bioanal Chem* 2014;**406**:5513–20.
46. Milbury CA, Li J, Makrigiorgos GM. Ice-COLD-PCR enables rapid amplification and robust enrichment for low-abundance unknown DNA mutations. *Nucleic Acids Res* 2011;**39**:e2.
47. Kristensen LS, Daugaard IL, Christensen M, et al. Increased sensitivity of KRAS mutation detection by high-resolution melting analysis of COLD-PCR products. *Hum Mutat* 2010;**31**:1366–73.
48. Castellanos-Rizaldos E, Milbury CA, Karatza E, et al. COLD-PCR amplification of bisulfite-converted DNA allows the enrichment and sequencing of rare un-methylated genomic regions. *PLoS ONE* 2014;**9**:e94103.
49. Milbury CA, Li J, Makrigiorgos GM. COLD-PCR-enhanced high-resolution melting enables rapid and selective identification of low-level unknown mutations. *Clin Chem* 2009;**55**:2130–43.
50. Pritchard CC, Akagi L, Reddy PL, et al. COLD-PCR enhanced melting curve analysis improves diagnostic accuracy for KRAS mutations in colorectal carcinoma. *BMC Clin Pathol* 2010;**10**:6.
51. Carotenuto P, Roma C, Cozzolino S, et al. Detection of KRAS mutations in colorectal cancer with Fast COLD-PCR. *Int J Oncol* 2012;**40**:378–84.
52. Castellanos-Rizaldos E, Makrigiorgos GM. *Single-tube enrichment of mutations in cancer gene panels using COLD-PCR prior to targeted amplicon resequencing Clinical Chemistry 2014 in Press.* 2014.
53. Castellanos-Rizaldos E, Liu P, Milbury CA, et al. Temperature-tolerant COLD-PCR reduces temperature stringency and enables robust mutation enrichment. *Clin Chem* 2012;**58**:1130–8.
54. Murphy DM, Bejar R, Stevenson K, et al. NRAS mutations with low allele burden have independent prognostic significance for patients with lower risk myelodysplastic syndromes. *Leukemia* 2013;**27**:2077–81.
55. Huggett JF, Cowen S, Foy CA. Considerations for digital PCR as an accurate molecular diagnostic tool. *Clin Chem* 2015;**61**:79–88.
56. Dingle TC, Sedlak RH, Cook L, et al. Tolerance of droplet-digital PCR vs real-time quantitative PCR to inhibitory substances. *Clin Chem* 2013;**59**:1670–2.
57. Weaver S, Dube S, Mir A, et al. Taking qPCR to a higher level: analysis of CNV reveals the power of high throughput qPCR to enhance quantitative resolution. *Methods* 2010;**50**:271–6.
58. Nixon G, Garson JA, Grant P, et al. Comparative study of sensitivity, linearity, and resistance to inhibition of digital and nondigital polymerase chain reaction and loop mediated isothermal amplification assays for quantification of human cytomegalovirus. *Anal Chem* 2014;**86**:4387–94.
59. Shen F, Davydova EK, Du W, et al. Digital isothermal quantification of nucleic acids via simultaneous chemical initiation of recombinase polymerase amplification reactions on SlipChip. *Anal Chem* 2011;**83**:3533–40.
60. Tawfik DS, Griffiths AD. Man-made cell-like compartments for molecular evolution. *Nat Biotechnol* 1998;**16**:652–6.
61. Mitra RD, Butty VL, Shendure J, et al. Digital genotyping and haplotyping with polymerase colonies. *Proc Natl Acad Sci USA* 2003;**100**:5926–31.
62. Bentley DR, Balasubramanian S, Swerdlow HP, et al. Accurate whole human genome sequencing using reversible terminator chemistry. *Nature* 2008;**456**:53–9.
63. Merriman B, Ion Torrent R, Team D, et al. Progress in ion torrent semiconductor chip based sequencing. *Electrophoresis* 2012;**33**:3397–417.
64. Giachetti C, Linnen JM, Kolk DP, et al. Highly sensitive multiplex assay for detection of human immunodeficiency virus type 1 and hepatitis C virus RNA. *J Clin Microbiol* 2002;**40**:2408–19.
65. Hill CS. Molecular diagnostic testing for infectious diseases using TMA technology. *Expert Rev Mol Diagn* 2001;**1**:445–55.
66. Compton J. Nucleic acid sequence-based amplification. *Nature* 1991;**350**:91–2.
67. Guatelli JC, Whitfield KM, Kwoh DY, et al. Isothermal, in vitro amplification of nucleic acids by a multienzyme reaction modeled after retroviral replication. *Proc Natl Acad Sci USA* 1990;**87**:1874–8.
68. Notomi T, Okayama H, Masubuchi H, et al. Loop-mediated isothermal amplification of DNA. *Nucleic Acids Res* 2000;**28**:E63.
69. Whitcombe D, Theaker J, Guy SP, et al. Detection of PCR products using self-probing amplicons and fluorescence. *Nat Biotechnol* 1999;**17**:804–7.
70. Zhou L, Errigo RJ, Lu H, et al. Snapback primer genotyping with saturating DNA dye and melting analysis. *Clin Chem* 2008;**54**:1648–56.
71. Aomori T, Yamamoto K, Oguchi-Katayama A, et al. Rapid single-nucleotide polymorphism detection of cytochrome P450 (CYP2C9) and vitamin K epoxide reductase (VKORC1) genes for the warfarin dose adjustment by the SMart-amplification process version 2. *Clin Chem* 2009;**55**:804–12.

72. Walker GT, Linn CP, Nadeau JG. DNA detection by strand displacement amplification and fluorescence polarization with signal enhancement using a DNA binding protein. *Nucleic Acids Res* 1996;**24**:348−53.
73. Vincent M, Xu Y, Kong H. Helicase-dependent isothermal DNA amplification. *EMBO Rep* 2004;**5**:795−800.
74. An L, Tang W, Ranalli TA, et al. Characterization of a thermostable UvrD helicase and its participation in helicase-dependent amplification. *J Biol Chem* 2005;**280**:28952−8.
75. Piepenburg O, Williams CH, Stemple DL, et al. DNA detection using recombination proteins. *PLoS Biol* 2006;**4**:e204.
76. Dean FB, Hosono S, Fang L, et al. Comprehensive human genome amplification using multiple displacement amplification. *Proc Natl Acad Sci USA* 2002;**99**:5261−6.
77. Phillips J, Eberwine JH. Antisense RNA amplification: a linear amplification method for analyzing the mRNA population from single living cells. *Methods* 1996;**10**:283−8.
78. Urdea MS, Horn T, Fultz TJ, et al. Branched DNA amplification multimers for the sensitive, direct detection of human hepatitis viruses. *Nucleic Acids Symp Ser* 1991:197−200.
79. Hall JG, Eis PS, Law SM, et al. Sensitive detection of DNA polymorphisms by the serial invasive signal amplification reaction. *Proc Natl Acad Sci USA* 2000;**97**:8272−7.
80. Lizardi PM, Huang X, Zhu Z, et al. Mutation detection and single-molecule counting using isothermal rolling-circle amplification. *Nat Genet* 1998;**19**:225−32.
81. WalkerPeach CR, Winkler M, DuBois DB, et al. Ribonuclease-resistant RNA controls (Armored RNA) for reverse transcription-PCR, branched DNA, and genotyping assays for hepatitis C virus. *Clin Chem* 1999;**45**:2079−85.
82. Bustin SA, Benes V, Garson JA, et al. The MIQE guidelines: minimum information for publication of quantitative real-time PCR experiments. *Clin Chem* 2009;**55**:611−22.
83. Huggett JF, Foy CA, Benes V, et al. The digital MIQE guidelines: minimum information for publication of quantitative digital PCR experiments. *Clin Chem* 2013;**59**:892−902.
84. Langer PR, Waldrop AA, Ward DC. Enzymatic synthesis of biotin-labeled polynucleotides: novel nucleic acid affinity probes. *Proc Natl Acad Sci USA* 1981;**78**:6633−7.
85. Kwok PY. SNP genotyping with fluorescence polarization detection. *Hum Mutat* 2002;**19**:315−23.
86. Nazarenko I, Pires R, Lowe B, et al. Effect of primary and secondary structure of oligodeoxyribonucleotides on the fluorescent properties of conjugated dyes. *Nucleic Acids Res* 2002;**30**:2089−195.
87. Lee LG, Connell CR, Bloch W. Allelic discrimination by nick-translation PCR with fluorogenic probes. *Nucleic Acids Res* 1993;**21**:3761−6.
88. Tyagi S, Kramer FR. Molecular beacons: probes that fluoresce upon hybridization. *Nat Biotechnol* 1996;**14**:303−8.
89. Drummond TG, Hill MG, Barton JK. Electrochemical DNA sensors. *Nat Biotechnol* 2003;**21**:1192−9.
90. Palecek E, Bartosik M. Electrochemistry of nucleic acids. *Chem Rev* 2012;**112**:3427−81.
91. Laszlo AH, Derrington IM, Ross BC, et al. Decoding long nanopore sequencing reads of natural DNA. *Nat Biotechnol* 2014;**32**:829−33.
92. Maule J. Pulsed-field gel electrophoresis. *Mol Biotechnol* 1998;**9**:107−26.
93. Kaur M, Zhang Y, Liu WH, et al. Ligation of a primer at a mutation: a method to detect low level mutations in DNA. *Mutagenesis* 2002;**17**:365−74.
94. Highsmith Jr WE, Jin Q, Nataraj AJ, et al. Use of a DNA toolbox for the characterization of mutation scanning methods. I: construction of the toolbox and evaluation of heteroduplex analysis. *Electrophoresis* 1999;**20**:1186−94.
95. Cariello NF, Scott JK, Kat AG, et al. Resolution of a missense mutant in human genomic DNA by denaturing gradient gel electrophoresis and direct sequencing using in vitro DNA amplification: HPRT Munich. *Am J Hum Genet* 1988;**42**:726−34.
96. Riesner D, Steger G, Zimmat R, et al. Temperature-gradient gel electrophoresis of nucleic acids: analysis of conformational transitions, sequence variations, and protein-nucleic acid interactions. *Electrophoresis* 1989;**10**:377−89.
97. Li Q, Liu Z, Monroe H, et al. Integrated platform for detection of DNA sequence variants using capillary array electrophoresis. *Electrophoresis* 2002;**23**:1499−511.
98. Orita M, Iwahana H, Kanazawa H, et al. Detection of polymorphisms of human DNA by gel electrophoresis as single-strand conformation polymorphisms. *Proc Natl Acad Sci USA* 1989;**86**:2766−70.
99. Ekstrom PO, Bjorheim J, Thilly WG. Technology to accelerate pangenomic scanning for unknown point mutations in exonic sequences: cycling temperature capillary electrophoresis (CTCE). *BMC Genet* 2007;**8**:54.
100. Franca LT, Carrilho E, Kist TB. A review of DNA sequencing techniques. *Q Rev Biophys* 2002;**35**:169−200.
101. Syvanen AC. From gels to chips: "minisequencing" primer extension for analysis of point mutations and single nucleotide polymorphisms. *Hum Mutat* 1999;**13**:1−10.
102. Eggerding FA, Iovannisci DM, Brinson E, et al. Fluorescence-based oligonucleotide ligation assay for analysis of cystic fibrosis transmembrane conductance regulator gene mutations. *Hum Mutat* 1995;**5**:153−65.
103. Schouten JP, McElgunn CJ, Waaijer R, et al. Relative quantification of 40 nucleic acid sequences by multiplex ligation-dependent probe amplification. *Nucleic Acids Res* 2002;**30**:e57.
104. Ronaghi M, Uhlen M, Nyren P. A sequencing method based on real-time pyrophosphate. *Science* 1998;**281**(363):5.
105. Braun A, Little DP, Koster H. Detecting CFTR gene mutations by using primer oligo base extension and mass spectrometry. *Clin Chem* 1997;**43**:1151−8.
106. Pusch W, Wurmbach JH, Thiele H, et al. MALDI-TOF mass spectrometry-based SNP genotyping. *Pharmacogenomics* 2002;**3**:537−48.
107. Sampath R, Mulholland N, Blyn LB, et al. Comprehensive biothreat cluster identification by PCR/electrospray-ionization mass spectrometry. *PLoS ONE* 2012;**7**:e36528.
108. Xiao W, Oefner PJ. Denaturing high-performance liquid chromatography: a review. *Hum Mutat* 2001;**17**:439−74.
109. Nielsen PE. PNA Technology. *Mol Biotechnol* 2004;**26**:233−48.
110. Wetmur JG. DNA probes: applications of the principles of nucleic acid hybridization. *Crit Rev Biochem Mol Biol* 1991;**26**:227−59.
111. Bolivar F, Rodriguez RL, Greene PJ, et al. Construction and characterization of new cloning vehicles. II. A multipurpose cloning system. *Biotechnology* 1992;**24**:153−71.
112. Hopfenbeck JA, Holden JA, Wittwer CT, et al. Digoxigenin-labeled probes amplified from genomic DNA detect T-cell gene rearrangements. *Am J Clin Pathol* 1992;**97**:638−44.
113. National Center for Biotechnology Information. Genbank. http://www.ncbi.nlm.nih.gov/genbank/ [accessed 08.15].

114. SantaLucia Jr J. A unified view of polymer, dumbbell, and oligonucleotide DNA nearest-neighbor thermodynamics. *Proc Natl Acad Sci USA* 1998;**95**:1460–5.
115. SantaLucia Jr J, Hicks D. The thermodynamics of DNA structural motifs. *Annu Rev Biophys Biomol Struct* 2004;**33**:415–40.
116. Owczarzy R, Moreira BG, You Y, et al. Predicting stability of DNA duplexes in solutions containing magnesium and monovalent cations. *Biochemistry* 2008;**47**:5336–53.
117. Moreira BG, You Y, Behlke MA, et al. Effects of fluorescent dyes, quenchers, and dangling ends on DNA duplex stability. *Biochem Biophys Res Commun* 2005;**327**:473–84.
118. Lyon E. Discovering rare variants by use of melting temperature shifts seen in melting curve analysis. *Clin Chem* 2005;**51**:1331–2.
119. Heller MJ. DNA microarray technology: devices, systems, and applications. *Annu Rev Biomed Eng* 2002;**4**:129–53.
120. Holloway AJ, van Laar RK, Tothill RW, et al. Options available—from start to finish—for obtaining data from DNA microarrays II. *Nat Genet* 2002;**32**(Suppl.):481–9.
121. Kellar KL, Iannone MA. Multiplexed microsphere-based flow cytometric assays. *Exp Hematol* 2002;**30**:1227–37.
122. Grant SF, Hakonarson H. Microarray technology and applications in the arena of genome-wide association. *Clin Chem* 2008;**54**:1116–24.
123. Roeder K, Luca D. Searching for disease susceptibility variants in structured populations. *Genomics* 2009;**93**:1–4.
124. Reis-Filho JS, Pusztai L. Gene expression profiling in breast cancer: classification, prognostication, and prediction. *Lancet* 2011;**378**:1812–23.
125. Zaravinos A, Lambrou GI, Volanis D, et al. Spotlight on differentially expressed genes in urinary bladder cancer. *PLoS ONE* 2011;**6**:e18255.
126. Haferlach T, Kohlmann A, Wieczorek L, et al. Clinical utility of microarray-based gene expression profiling in the diagnosis and subclassification of leukemia: report from the International Microarray Innovations in Leukemia Study Group. *J Clin Oncol* 2010;**28**:2529–37.
127. Beck AH, West RB, van de Rijn M. Gene expression profiling for the investigation of soft tissue sarcoma pathogenesis and the identification of diagnostic, prognostic, and predictive biomarkers. *Virchows Arch* 2010;**456**:141–51.
128. Zhou L, Palais RA, Paxton CN, et al. Copy number assessment by competitive PCR with limiting deoxynucleotide triphosphates and high-resolution melting. *Clin Chem* 2015;**61**:724–33.
129. Geiss GK, Bumgarner RE, Birditt B, et al. Direct multiplexed measurement of gene expression with color-coded probe pairs. *Nat Biotechnol* 2008;**26**:317–25.
130. Higuchi R, Fockler C, Dollinger G, et al. Kinetic PCR analysis: real-time monitoring of DNA amplification reactions. *Biotechnology (N Y)* 1993;**11**:1026–30.
131. Dujols VE, Kusukawa N, McKinney JT, et al. High-resolution melting analysis for scanning and genotyping. In: Dorak MT, editor. *Real-Time PCR*. New York: Garland Science; 2006. p. 157–71.
132. Nazarenko IA, Bhatnagar SK, Hohman RJ. A closed tube format for amplification and detection of DNA based on energy transfer. *Nucleic Acids Res* 1997;**25**:2516–21.
133. Gundry CN, Vandersteen JG, Reed GH, et al. Amplicon melting analysis with labeled primers: a closed-tube method for differentiating homozygotes and heterozygotes. *Clin Chem* 2003;**49**:396–406.
134. Livak KJ, Flood SJ, Marmaro J, et al. Oligonucleotides with fluorescent dyes at opposite ends provide a quenched probe system useful for detecting PCR product and nucleic acid hybridization. *PCR Methods Appl* 1995;**4**:357–62.
135. Todd AV, Fuery CJ, Impey HL, et al. DzyNA-PCR: use of DNAzymes to detect and quantify nucleic acid sequences in a real-time fluorescent format. *Clin Chem* 2000;**46**:625–30.
136. Chen X, Livak KJ, Kwok PY. A homogeneous, ligase-mediated DNA diagnostic test. *Genome Res* 1998;**8**:549–56.
137. Lay MJ, Wittwer CT. Real-time fluorescence genotyping of factor V Leiden during rapid-cycle PCR. *Clin Chem* 1997;**43**:2262–7.
138. Crockett AO, Wittwer CT. Fluorescein-labeled oligonucleotides for real-time PCR: using the inherent quenching of deoxyguanosine nucleotides. *Anal Biochem* 2001;**290**:89–97.
139. Svanvik N, Stahlberg A, Sehlstedt U, et al. Detection of PCR products in real time using light-up probes. *Anal Biochem* 2000;**287**:179–82.
140. Bonnet G, Tyagi S, Libchaber A, et al. Thermodynamic basis of the enhanced specificity of structured DNA probes. *Proc Natl Acad Sci USA* 1999;**96**:6171–6.
141. Chakravorty S, Aladegbami B, Thoms K, et al. Rapid detection of fluoroquinolone-resistant and heteroresistant Mycobacterium tuberculosis by use of sloppy molecular beacons and dual melting-temperature codes in a real-time PCR assay. *J Clin Microbiol* 2011;**49**:932–40.
142. Lukhtanov EA, Lokhov SG, Gorn VV, et al. Novel DNA probes with low background and high hybridization-triggered fluorescence. *Nucleic Acids Res* 2007;**35**:e30.
143. Li Q, Luan G, Guo Q, et al. A new class of homogeneous nucleic acid probes based on specific displacement hybridization. *Nucleic Acids Res* 2002;**30**:E5.
144. Li J, Wang F, Mamon H, et al. Antiprimer quenching-based real-time PCR and its application to the analysis of clinical cancer samples. *Clin Chem* 2006;**52**:624–33.
145. Murray JL, Hu P, Shafer DA. Seven novel probe systems for real-time PCR provide absolute single-base discrimination, higher signaling, and generic components. *J Mol Diagn* 2014;**16**:627–38.
146. Luk KC, Devare SG, Hackett Jr JR. Partially double-stranded linear DNA probes: novel design for sensitive detection of genetically polymorphic targets. *J Virol Methods* 2007;**144**:1–11.
147. Wittwer C, Kusukawa N, Real-time PCR. In: Persing DH, Tenover FC, Relman DA, et al., editors. *Diagnostic molecular microbiology: principles and applications*. Washington: ASM Press; 2004. p. 71–84.
148. Press W, Teukolsky S, Vetterling W, et al. Salvitsky-Golay smoothing filters. In: *Numerical recipes in C: the art of scientific computing*. New York: Cambridge University Press; 1992. p. 650–5. Cambridge England.
149. Morrison TB, Weis JJ, Wittwer CT. Quantification of low-copy transcripts by continuous SYBR Green I monitoring during amplification. *Biotechniques* 1998;**24**:954–8. 60, 62.
150. Haynes RJ, Kline MC, Toman B, et al. Standard reference material 2366 for measurement of human cytomegalovirus DNA. *J Mol Diagn* 2013;**15**:177–85.
151. Zhou L, Myers AN, Vandersteen JG, et al. Closed-tube genotyping with unlabeled oligonucleotide probes and a saturating DNA dye. *Clin Chem* 2004;**50**:1328–35.
152. Liew M, Pryor R, Palais R, et al. Genotyping of single-nucleotide polymorphisms by high-resolution melting of small amplicons. *Clin Chem* 2004;**50**:1156–64.

153. Germer S, Higuchi R. Single-tube genotyping without oligonucleotide probes. *Genome Res* 1999;**9**:72–8.
154. Wang J, Chuang K, Ahluwalia M, et al. High-throughput SNP genotyping by single-tube PCR with Tm-shift primers. *Biotechniques* 2005;**39**:885–93.
155. Bernard PS, Ajioka RS, Kushner JP, et al. Homogeneous multiplex genotyping of hemochromatosis mutations with fluorescent hybridization probes. *Am J Pathol* 1998;**153**:1055–61.
156. Tyagi S, Bratu DP, Kramer FR. Multicolor molecular beacons for allele discrimination. *Nat Biotechnol* 1998;**16**:49–53.
157. de Kok JB, Wiegerinck ET, Giesendorf BA, et al. Rapid genotyping of single nucleotide polymorphisms using novel minor groove binding DNA oligonucleotides (MGB probes). *Hum Mutat* 2002;**19**:554–9.
158. Reed GH, Wittwer CT. Sensitivity and specificity of single-nucleotide polymorphism scanning by high-resolution melting analysis. *Clin Chem* 2004;**50**:1748–54.
159. Herrmann MG, Durtschi JD, Bromley LK, et al. Amplicon DNA melting analysis for mutation scanning and genotyping: cross-platform comparison of instruments and dyes. *Clin Chem* 2006;**52**:494–503.
160. Dwight Z, Palais R, Wittwer CT. uMELT: prediction of high-resolution melting curves and dynamic melting profiles of PCR products in a rich web application. *Bioinformatics* 2011;**27**:1019–20.
161. Erali M, Voelkerding KV, Wittwer CT. High resolution melting applications for clinical laboratory medicine. *Exp Mol Pathol* 2008;**85**:50–8.
162. Farrar JS, Reed GH, Wittwer CT. High resolution melting curve analysis for molecular diagnostics. In: Patrinos GP, Ansorge W, editors. *Molecular Diagnostics*. 2nd ed. London: Elsevier; 2010. p. 229–45.
163. Vossen RH, Aten E, Roos A, et al. High-resolution melting analysis (HRMA): more than just sequence variant screening. *Hum Mutat* 2009;**30**:860–6.
164. Shendure J, Ji H. Next-generation DNA sequencing. *Nat Biotechnol* 2008;**26**:1135–45.
165. Buermans HP, den Dunnen JT. Next generation sequencing technology: advances and applications. *Biochim Biophys Acta* 2014;**1842**:1932–41.
166. van Dijk EL, Auger H, Jaszczyszyn Y, et al. Ten years of next-generation sequencing technology. *Trends Genet* 2014;**30**:418–26.
167. Gargis AS, Kalman L, Berry MW, et al. Assuring the quality of next-generation sequencing in clinical laboratory practice. *Nat Biotechnol* 2012;**30**:1033–6.
168. Rehm HL, Bale SJ, Bayrak-Toydemir P, et al. ACMG clinical laboratory standards for next-generation sequencing. *Genet Med* 2013;**15**:733–47.
169. Sakurai T, Husimi Y. Real-time monitoring of DNA polymerase reactions by a micro ISFET pH sensor. *Anal Chem* 1992;**64**:1996–7.
170. Ma Z, Lee RW, Li B, et al. Isothermal amplification method for next-generation sequencing. *Proc Natl Acad Sci USA* 2013;**110**:14320–3.
171. Eid J, Fehr A, Gray J, et al. Real-time DNA sequencing from single polymerase molecules. *Science* 2009;**323**:133–8.
172. Smith DR, Quinlan AR, Peckham HE, et al. Rapid whole-genome mutational profiling using next-generation sequencing technologies. *Genome Res* 2008;**18**:1638–42.
173. Steinbock LJ, Radenovic A. The emergence of nanopores in next-generation sequencing. *Nanotechnology* 2015;**26**:074003.
174. Laszlo AH, Derrington IM, Brinkerhoff H, et al. Detection and mapping of 5-methylcytosine and 5-hydroxymethylcytosine with nanopore MspA. *Proc Natl Acad Sci USA* 2013;**110**:18904–9.

Molecular Microbiology

*Frederick S. Nolte**

ABSTRACT

Background
Nucleic acid (NA) amplification techniques are now commonly used to diagnose and manage patients with infectious diseases. The growth in the number of Food and Drug Administration—approved test kits and analyte-specific reagents has facilitated the use of this technology in clinical laboratories. Technological advances in NA amplification techniques, automation, NA sequencing, and multiplex analysis have reinvigorated the field and created new opportunities for growth. Simple, sample-in, answer-out molecular test systems are now widely available that can be deployed in a variety of laboratory and clinical settings. Molecular microbiology remains the leading area in molecular pathology in terms of both the numbers of tests performed and clinical relevance. NA-based tests have reduced the dependency of the clinical microbiology laboratory on more traditional antigen detection and culture methods and created new opportunities for the laboratory to impact patient care.

Content
This chapter reviews NA testing as it applies to specific pathogens or infectious disease syndromes, with a focus on those diseases for which NA testing is now considered the standard of care and highlights the unique challenges and opportunities that these tests present for clinical laboratories.

INTRODUCTION

Since the publication of the fifth edition of this textbook, significant changes have occurred in the practice of diagnostic molecular microbiology. Nucleic acid (NA) amplification techniques are now commonly used to diagnose and manage patients with infectious diseases. The growth in the number of Food and Drug Administration (FDA)—approved test kits and analyte-specific reagents (ASRs) has facilitated the use of this technology in clinical laboratories. Technological advances in NA amplification techniques, automation, NA sequencing, and multiplex analysis have reinvigorated the field and created new opportunities for growth. Simple, sample-in, answer-out molecular test systems are now widely available that can be deployed in a variety of laboratory and clinical settings.

Molecular microbiology remains the leading area in molecular pathology in terms of both the numbers of tests performed and clinical relevance. NA-based tests have reduced the dependency of the clinical microbiology laboratory on more traditional antigen detection and culture methods and created new opportunities for the laboratory to impact patient care. This chapter reviews NA testing as it applies to specific pathogens or infectious disease syndromes, with a focus on diseases for which NA testing is now considered the standard of care and highlights the unique challenges and opportunities that these tests present for clinical laboratories. A complete and current list of all FDA-cleared and FDA-approved microbial NA based tests can be found at http://www.fda.gov/MedicalDevices/ProductsandMedicalProcedures/InVitroDiagnostics/ucm330711.htm. Readers are directed to *Molecular Microbiology: Diagnostic Principles and Practice*, 3rd edition, for a more comprehensive and in depth examination of this dynamic and exciting discipline.[1]

VIRAL SYNDROMES

Human Immunodeficiency Virus 1

Human immunodeficiency virus 1 (HIV-1), the causative agent of the acquired immunodeficiency syndrome (AIDS), is an RNA virus belonging to the genus *Lentivirus* of the family *Retroviridae*. Replication of the virus is complex and involves reverse transcription of the RNA genome into a double-stranded DNA molecule or provirus, which is integrated into the host genome. HIV-1 enters the cell using CD4 as a receptor and CXCR4 or CCR5 as a coreceptor. In general, CCR5 coreceptors are found on macrophages, and CXCR4 coreceptors are found on T cells. Determining the cellular tropism of the virus has become important now that an antiretroviral drug targets the CCR5 coreceptor. The HIV-1 reverse transcriptase (RT) enzyme does not have proofreading capabilities, leading to the significant genetic diversity of HIV-1. Several distinct genetic subtypes or clades have been identified and are categorized into three groups: major (M), outlier (O), and N (nonmajor and nonoutlier). Recently, a new group P has been identified that is closely

---

*The author wishes to acknowledge the contributions of Aaron D. Bossler and Angela M. Caliendo who authored this chapter in the previous edition.

related to a gorilla simian immunodeficiency virus.[2] The major group is divided into nine clades (A–K) and circulating recombinant forms (CRFs), which are determined on the basis of sequence diversity within the HIV-1 *gag* and *env* genes.[3] Group M virus is found worldwide, with clade B predominating in Europe and North America, clade C in Africa and India, and clade E in much of Southeast Asia. The complex replication cycle and genetic diversity are two important factors that influence the design and interpretation of HIV-1 molecular assays.

The management of patients with HIV-1 infection was transformed by tests performed to measure the concentrations of HIV-1 RNA in blood (viral load testing) and tests for resistance to antiviral drugs. With these tools, it is possible to maximize the effectiveness of antiretroviral therapy (ART) for an individual.

The first HIV-1 viral load test was approved by the FDA in 1996 and rapidly became the standard of care for monitoring response to ART. Early studies found that patients who have higher viral loads progressed more rapidly to AIDS and death than those with low viral loads.[4-6]

Viral load testing is now widely accepted as a marker of response to ART. The guidelines for initiation of therapy based on viral load have changed as our understanding of disease progression at higher CD4 cell counts has improved. With the availability of newer more potent and less toxic drugs, ART is recommended for all HIV-infected individuals regardless of their CD4 cell count or viral load (DHHS Panel on Antiretroviral Guidelines, http://AIDSinfo.nih.gov). After treatment is initiated, viral load testing is crucial for monitoring response to therapy and should be measured in all HIV-infected individuals at the time of entry into care when therapy is initiated and at a regular interval (usually 3–4 months) while on therapy. The standard of care is to treat with a combination of antiretroviral drugs, which are classified based on their viral targets: nucleoside reverse transcriptase inhibitors (NRTIs), non-nucleoside reverse transcriptase inhibitors (NNRTIs), protease inhibitors (PIs), fusion inhibitors, integrase inhibitors (also known as integrase strand transfer inhibitors [INSTIs]), and CCR5 entry inhibitors. Current Department of Health and Human Services Panel on Antiretroviral Guidelines (http://aidsinfo.nih.gov/guidelines) recommend an initial regimen of two NRTIs and an NNRTI, a PI, or an INSTI. The pretreatment viral load values influence the treatment regimen because some regimens are less effective in patients with high viral loads. After initiation of appropriate therapy, there is typically a 2 $\log_{10}$ or greater decrease in viral load within 2 to 3 months. The goal for a patient is to achieve a viral load below the limit of detection of the most sensitive assays (20–50 copies/mL). Data have shown that the lower the absolute viral load, the better the clinical and virologic outcomes.[7,8] Guidelines recommend quantifying plasma HIV-1 RNA immediately before initiating therapy and 2 to 8 weeks later, with the goal to achieve an undetectable viral load level within 16 to 24 weeks of initiating therapy. It is important to determine early in the treatment course if there is suboptimal viral load suppression, so that factors affecting adherence can be assessed and, if needed, the regimen altered. After the initial response has been characterized, the viral load should be monitored every 3 to 4 months to ensure the response to therapy is sustained.

Of note, viral blips (defined as a detectable viral load usually <400 copies/mL) occur in successfully treated individuals and are not predictive of virologic failure.[9] For this reason, *virologic failure* is defined as a sustained viral load of more than 200 copies/mL.

Viral load testing also aids in the diagnosis of acute HIV-1 infection (the window period after infection that occurs before detectable antibody production), although the currently available viral load assays are not FDA approved for diagnostic purposes. During this period of early infection, patients typically have very high viral loads ranging from 10^5 to 10^7 copies/mL.[10] HIV RNA testing is recommended in the 2014 updated HIV diagnostic testing algorithm for patients who are reactive in a fourth-generation immunoassay but test negative in a supplemental HIV-1/HIV-2 antibody differentiation test, primarily to better detect acute HIV-1 infections.[11] However, only one HIV RNA test is FDA approved for diagnosis, the APTIMA HIV-1 qualitative assay (Hologic, San Diego, CA), and it is not widely available in clinical laboratories. Hospital laboratories would benefit from an FDA diagnostic claim for viral load testing that is done reflexively in the above diagnostic algorithm because verifying this claim at a local laboratory level is overly burdensome.

A proviral DNA test, AMPLICOR HIV-1 DNA polymerase chain reaction (PCR) (Roche Diagnostics), is available as a research use only (RUO) test that is helpful in neonates born to HIV-1–infected mothers receiving ART. Both antiretroviral agents and maternal antibody cross the placenta. Antiretroviral agents can suppress the replication of the virus in neonates, and RNA tests can be falsely negative early after birth. Maternal HIV-1 antibody can persist in neonates for up to 2 years after birth and is therefore not a reliable marker of neonatal infection.

Currently there are three commercially available, FDA-approved HIV-1 viral load assays. Two of these assays use real-time PCR, cobas AmpliPrep/cobas TaqMan HIV-1, version 2.0 (Roche Diagnostics), and m2000 RealTime System (Abbott Molecular); the third, Versant HIV-1 RNA 3.0 (Siemens), uses branched DNA signal amplification. The amplification method, gene targets, and dynamic ranges for these assays are given in Table 5.1. Newly emerging NA amplification tests (NAATs) and platforms have the potential for true point-of-care testing for HIV RNA,[12] and several (eg, Alere, BioHelix, Cepheid, and Iquum/Roche) are in commercial development. These tests may have application in detecting acute infections, confirming screening tests, and determining viral loads in resource-poor settings.

Most clinical laboratories are using one of the real-time PCR assays because of their lower limits of detection and quantification and increased dynamic range compared with older-generation PCR assays and the branched DNA assay. Genotype bias was a significant problem in some of the earlier viral load assays because the gene targets chosen were not highly conserved among the different HIV-1 subtypes. However, this problem has been addressed, and the current versions of the Roche and Abbott real-time PCR assays both accurately quantify group M, group O, and CRFs of HIV-1.[13] The intra-assay imprecision of HIV-1 viral load assays is approximately 0.12 to 0.2 $\log_{10}$ copies/mL. The biologic variation of viral load in patients not receiving therapy is

TABLE 5.1 Commercially Available Food and Drug Administration–Approved Viral Load Assays

Virus	Assay (Manufacturer)	Method	Gene Target	Dynamic Range
HIV-1	Versant 3.0 (Siemens)	Branched DNA	*pol*	75–500,000 copies/mL
	cobas Ampliprep/cobas TaqMan 2.0 (Roche)	RT-real-time PCR	*gag, LTR*	20–10,000,000 copies/mL
	RealTime (Abbott)	RT-real-time PCR	*int*	40–10,000,000 copies/mL
HCV	Versant 3.0 (Siemens)	Branched DNA	*5′ UTR*	615–7,700,000 IU/mL
	cobas Ampliprep/cobas TaqMan Test 2.0 (Roche)	RT-real-time PCR	*5′ UTR*	43–69,000,000 IU/mL
	RealTime (Abbott)	RT-real-time PCR	*5′ UTR*	12–100,000,00 IU/mL
HBV	cobas Ampliprep/cobas TaqMan Test 2.0 (Roche)	Real-time PCR	*Precore/core*	20–170,000,000 IU/mL
	RealTime (Abbott)	Real-time PCR	*Surface*	10–1,000,000,000 IU/mL
CMV	cobas Ampliprep/cobas TaqMan Test (Roche)	Real-time PCR	*UL54*	137–9,100,000 IU/mL
	artus RGQ MDx (Qiagen)	Real-time PCR	*MIE*	119–79,400,000 IU/mL

CMV, Cytomegalovirus; *HIV*, human immunodeficiency virus; *HBV*, hepatitis C virus; *HCV*, hepatitis B virus; *PCR*, polymerase chain reaction; *RT*, reverse transcriptase.

approximately 0.3 $\log_{10}$ copies/mL.[14] Consequently, changes in viral load must exceed 0.5 $\log_{10}$ copies/mL (a threefold change) to suggest a meaningful change in viral replication.

Viral load testing is routinely performed on plasma specimens, and ethylenediaminetetraacetic acid (EDTA) is the anticoagulant of choice. Acid–citrate–dextrose is also an acceptable anticoagulant, but blood anticoagulated in heparin is unacceptable for most tests. It is critical to handle clinical specimens properly to minimize the risk of RNA degradation during specimen collection and transport. Plasma should be separated within 6 hours of collection and ideally stored at −20°C, although plasma viral RNA is stable at 4°C for several days. For laboratories testing specimens collected at remote sites, sample handling can require careful attention. Special blood collection containers, or tubes, are available that contain a gel that provides a physical barrier between the plasma and cells after centrifugation. The tubes can be shipped without the need to transfer the plasma into a separate tube. Tubes should not be frozen before pouring off of the plasma because this may lead to falsely elevated viral load assays.[15,16]

Six general classes of antiretroviral drugs are used in clinical care: NRTIs, NNRTIs, PIs, fusion inhibitors, INSTIs, and CCR5 entry inhibitors. Viral resistance can occur with each of these drug classes, particularly when viral replication is not maximally suppressed during therapy. The current standard of care is to use regimens that contain a combination of drugs because resistance is less likely to occur on the complex regimens than on monotherapy.

The clinical utility of antiviral resistance testing in HIV-1–infected individuals is well established, and regularly updated guidelines for its use can be found at http://AIDSinfo.nih.gov. Current DHHS Panel on Antiretroviral Guidelines recommend that resistance testing be performed in the following situations: (1) before initiation of ART in treatment-naïve patients, (2) to guide the selection of active drugs when changing antiretroviral regimens, (3) for the management of suboptimal viral load reduction, and (4) in all pregnant women before the initiation of therapy.

HIV-1 resistance testing can be done by genotypic or phenotypic methods. Genotypic assays identify specific mutations or nucleotide changes that are associated with decreased susceptibility to an antiviral drug. Phenotypic assays are performed by the creation of a pseudoviral vector and measuring its replicative capacity in varying concentrations of drug and comparing it with replication of wild-type virus. Both genotypic and phenotypic methods are used clinically for assessing antiviral resistance in patients, with phenotypic testing usually reserved for drug-experienced patients with complicated resistance profiles.

This discussion is limited to automated sequencing methods for genotypic resistance because these are the methods used most often to inform treatment decisions. Currently available FDA-cleared assays will detect mutations in the RT and protease genes; modifications of existing assays are needed to detect resistance mutations associated with other classes of drugs such as integrase and fusion inhibitors.

The first step in genotypic assays is the isolation of HIV-1 RNA from plasma followed by reverse transcriptase polymerase chain reaction (RT-PCR) amplification and sequencing of RT and protease genes. Analysis of the results involves sequence alignment and editing, mutation identification by comparison with wild-type sequence, and interpretation of the clinical significance of the mutations identified. Most clinical laboratories performing genotypic resistance testing rely on commercial assays that provide reagents and software programs to assist with interpretation of the results. Two assays have been cleared by the FDA: the Trugene HIV-1 Genotyping Kit and OpenGene DNA Sequencing System (Siemens Healthcare Diagnostics) and the ViroSeq HIV-1 Genotyping System (Abbott Molecular).

Interpretation of genotypic resistance testing is complex. Interpretation of resistance mutations uses "rules-based" software that takes into account cross-resistance and interactions of mutations. The commercially available systems generate a summary report that lists the various mutations that have been identified in the RT and protease genes, and each drug is reported as resistant, possibly resistant, no evidence of resistance, or insufficient evidence. A comprehensive discussion of the specific mutations associated with each antiretroviral drug and the interactions of mutations is beyond the scope of this chapter but is available from a variety of sources (eg, http://www.iasusa.org, http://hivdb.stanford.edu/).

A limitation of currently used genotypic and phenotypic assays is that they can detect only those mutants that make up at least 20% of the total viral population. Regimens chosen based on resistance testing may not always be effective because the minority populations will quickly predominate in the presence of drug. Drug selection pressure is also needed for some resistance mutations to persist at detectable concentrations in the viral population; when drug therapy is discontinued, the wild-type virus may quickly predominate. For this reason, it is recommended that specimens for resistance testing be obtained while the patient is on ART. The minimum viral load required for reliable resistance testing is approximately 1000 copies/mL. Because genotyping assays are especially sensitive to RNA degradation, care must be taken to properly handle the specimen after collection.

> **POINTS TO REMEMBER**
>
> **Human Immunodeficiency Virus**
> - This RNA virus exhibits significant genetic diversity globally and within individual patients.
> - HIV RNA is the earliest marker of infection.
> - Viral load testing is a widely accepted marker of response to therapy with the goal of suppression to undetectable amounts.
> - Special care should be taken in sample processing to avoid spurious increases in HIV viral load tests.
> - HIV resistance genotyping should be performed on all therapy-naïve patients before initiation of ART and in selection of active drugs when changing therapy.

Hepatitis

Hepatitis C Virus

Approximately 3.2 million persons in the United States are living with active hepatitis C virus (HCV) infection, which is a major cause of chronic liver disease. After acute infection, 80% to 85% of individuals develop a chronic infection, and 2% to 4% of these individuals develop cirrhosis and end-stage liver disease, making end-stage liver disease secondary to HCV the most common indication for liver transplantation in the United States. Molecular tests for detection, quantification, and genotyping of HCV are standards of care for the diagnosis and management of patients with hepatitis C.

Hepatitis C virus is an RNA virus with a positive-sense, single-stranded genome of approximately 9500 nt encoding a single polyprotein of about 3000 amino acids. The long open reading frame is flanked at each end by short untranslated regions (UTRs). The genome structure is most similar to viruses of the family *Flaviviridae*, which includes many of the arthropod-borne viruses. As in other flaviviruses, the three N-terminal proteins of HCV (core, envelope 1 [E1]) and envelope 2 [E2]) are probably structural and the four C-terminal proteins (nonstructural 2, 3, 4, and 5) are thought to function in viral replication. HCV is classified within the family *Flaviviridae* in its own genus, *Hepacivirus*.

The 5' UTR is a highly conserved region of 341 nt and has a complex secondary structure. It contains an internal ribosome entry site and is important in the translation of the long open reading frame. The 3' UTR contains a short region that varies in sequence and length followed by a polypyrimidine stretch of variable length and finally a highly conserved sequence of 98 nt, which constitutes the terminus of the genome. The function of the 3' UTR is not known but is thought to be essential for viral replication.

The E1 and E2 regions of HCV are the most variable regions within the genome at both the nucleotide and amino acid levels. Two regions in E2, called hypervariable regions 1 and 2 (HVR1 and HRV2, respectively), show extreme sequence variability, which is thought to result from selective pressure by antiviral antibodies. E2 also contains the binding site for CD81, one of the putative cell receptors or coreceptors for HCV.

The nonstructural regions 2 (NS2) and 3 (NS3) contain a zinc-dependent autoprotease that cleaves the polyprotein at the NS2–NS3 junction. The aminoterminal portion of the NS3 protein also is a serine protease that cleaves the polyprotein at several sites. The carboxyterminal portion of the NS3 protein has helicase activity, which is important for HCV replication. The NS4A protein is a cofactor for NS3 serine protease. The NS5B region encodes the RNA-dependent RNA polymerase, which replicates the viral genome. A region in NS5A has been linked to interferon-α (IFN-α) response and therefore is called the IFN-α–sensitivity determining region.

The first complete HCV genome sequence was reported by Choo et al in 1991.[17] As additional genome sequences from isolates from different parts of the world were determined and compared, it was evident that HCV exists as distinct genotypes with as much as 35% sequence diversity over the whole viral genome.[18] Much of the early literature on genotyping is confusing because investigators developed and used their own classification schemes. However, a consensus nomenclature system was developed in 1994. In this system, the genotypes are numbered using Arabic numerals in order of their discovery, and the more closely related strains within some types are designated as subtypes with lowercase letters. The complex of genetic variants found within an individual isolate is termed the "quasispecies." The quasispecies results from the accumulation of mutations that occur during viral replication in the host.

The genotype and subtype assignments and nomenclature rules for HCV have recently been updated.[19] There are now 7 major genotypes and 67 subtypes of HCV recognized with another 20 provisional subtypes. HCV strains belonging to different genotypes differ at 30% to 35% of nucleotides and those that belong to the same subtype differ at fewer than 15% of nucleotides at the genome level.

Hepatitis C virus genotypes 1, 2, and 3 are found throughout the world, but there are clear differences in their distribution.[20] HCV subtypes 1a, 1b, 2a, 2b, 2c, and 3a are responsible for more than 90% of infections in North and South America, Europe, and Japan. In the United States, type 1 accounts for approximately 70% of the infections with equal distribution between subtypes 1a and 1b. Viral genotype does not correlate with disease progression.[21,22]

Detection of HCV RNA in serum or plasma by NA amplification methods is important for confirming the diagnosis of HCV, distinguishing active from resolved infection, assessing the virologic response to therapy, and screening the blood supply. These tests are incorporated into diagnostic algorithms for hepatitis C proposed by the Centers for Disease Control and Prevention (CDC),[23] American Association for the Study of Liver Diseases,[24] and National Academy of Clinical Biochemistry.[25]

The detection of HCV RNA in the plasma or serum is the earliest marker of infection, appearing 1 to 2 weeks after infection and weeks before elevation of liver enzymes or the appearance of anti-HCV antibodies. Approximately 80% of individuals infected with HCV will be chronically infected with the virus. In antibody-positive individuals, HCV RNA tests can distinguish active from resolved infections. In patients with a high pretest probability of infection, a positive serologic screening test result is usually confirmed with a test for HCV RNA rather than the recombinant immunoblot assay (RIBA). This strategy is cost-effective and more informative than using the RIBA to confirm positive antibody screening tests in a diagnostic setting.[26] However, with the discontinuation of the HCV RIBA by the manufacturer in 2012, all reactive HCV antibody screening tests should be followed by FDA-approved HCV RNA testing.[27]

Hepatitis C virus RNA testing also is helpful for the diagnosis of infection in infants born to HCV-infected mothers because of the persistence of maternal antibody and in immunocompromised or debilitated patients who may have blunted serologic responses. An HCV RNA test also should be used for patients suspected of having an acute infection and in patients with hepatitis of no identifiable cause.

Hepatitis C virus RNA tests are the most reliable means of identifying patients with active HCV infection. A negative HCV RNA test result in a serologically positive individual may indicate that the infection has resolved or that the viremia is intermittent. Up to 15% of chronically infected individuals have intermittent viremia and, as a result, a single negative HCV RNA determination may not be sufficient to exclude active infection when the index of clinical suspicion is high.[28] In these individuals a second specimen should be collected and tested.

The use of anti-HCV antibody tests to screen the blood supply has dramatically reduced the risk of transfusion-associated HCV infection in developed countries. The risk in the United States from blood that is negative for anti-HCV antibodies is less than 1 in 103,000 transfused units.[29] To drive the risk of infection from transfusion even lower, blood donor pools currently are tested for the presence of HCV RNA.[30] The serologic screening tests for HCV have a 70-day window period of seronegativity, and antigen detection tests are not yet available for blood product screening. HCV RNA testing is estimated to reduce the detection window by 25 days and reduce the number of transfused infectious units from 116 to 32 per year.[31]

Assays for the detection and quantification of HCV core antigen in serum have recently been commercially developed but are not yet FDA cleared for diagnostic use.[32-36] These tests significantly shorten the serologically silent window period using seroconversion panels, and their performance correlates closely with RNA detection tests in blood donors. However, the analytical sensitivity is less than most RNA tests, at approximately 10,000 IU/mL. The analytical sensitivity of the core antigen test is too high to be used in the monitoring of late events during and after treatment. Antigen detection may represent a cost effective alternative to HCV RNA testing to distinguish active from resolved infections in resource-poor settings.

Hepatitis C viral load testing is useful in pretreatment evaluations of patients being considered for therapy because a viral load of less 600,000 IU/mL is one of several predictors of achieving a sustained virologic response.[37,38] Other factors associated with achieving a sustained response to therapy include the absence of cirrhosis, age younger than 40 years, female gender, white race, viral genotype 2 or 3, and presence of two copies of the C allele at position rs12979860 near the gene for IFN lambda 3 (*IFNL3*, *IL28B*).[39,40]

Hepatitis C viral load does not predict disease progression and is not associated with severity of liver disease.[41] This is in sharp contrast to HIV-1, in which the viral load is the principal factor determining the rate of disease progression. Monitoring HCV viral load in untreated patients is not warranted and should be discouraged. Until recently, the standard therapy for patients with chronic HCV infection was pegylated IFN-α in combination with ribavirin administered for either 48 weeks for HCV genotype 1, 4, 5, and 6 infections or for 24 weeks for HCV genotype 2 and 3 infections. Sustained virologic response (SVR) rates were attained in 40% to 50% of patients with genotype 1 and in 80% or more of those with genotype 2 and 3 infections. SVR was defined as the absence of detectable HCV RNA in plasma or serum as determined with a test that has a limit of detection of 50 IU/mL or less and is considered a virologic cure.

The first direct-acting antivirals (DAAs) for treatment of hepatitis C were approved by the FDA in 2011. Both are NS3/4A serine protease inhibitors, boceprevir (BOC) (Merck) and telaprevir (TVR) (Vertex). These DAA agents are used in combination with pegylated IFN-α and ribavirin. Triple therapy for genotype 1 infections has led to approximately a 30% increase in SVR over the previous standard of care therapy in all patient subgroups. TVR also has activity against genotype 2 infections but not against genotype 3 infections. BOC appears to have activity against both genotypes 2 and 3 infections. However, neither drug should be used to treat patients with genotype 2 and 3 infections because SVR rates with pegylated IFN-α and ribavirin alone are much higher.[42]

The important time points for response-guided therapy are at 8, 12, and 24 weeks for BOC and 4, 12, and 24 weeks for TVR. Treatment with all three drugs should be stopped if HCV RNA is greater than 100 IU/mL at week 12 or detectable at week 24 for BOC triple therapy and if HCV RNA is greater than 1000 IU/mL at weeks 4 or 12 or detectable at week 24 with TVR.

The goal of treatment is a SVR, defined as no detectable HCV RNA in serum or plasma by a highly sensitive assay (limit of detection $\leq 10-15$ IU/mL) 6 months after the end of treatment. Patients who achieve a SVR have little or no chance of virologic relapse of their disease.

In 2013, two more potent DAAs were approved by the FDA: sofosbuvir (Gilead), a NS5B polymerase inhibitor,[43] and simeprevir (Johnson & Johnson) a second-generation protease inhibitor.[44] Sofosbuvir was approved in combination with pegylated IFN-α and ribavirin for treatment of genotypes 1 and 4 and in combination with ribavirin alone for genotypes 2 and 3. Simeprevir was approved by the FDA for treatment of genotype 1 infections in combination with pegylated IFN-α and ribavirin but only for patients with genotype 1 who have not failed therapy with first-generation protease inhibitors.

Monitoring of viral load during treatment does not affect management decisions with a sofosbuvir-based regimen because treatment failure is almost exclusively caused by relapse.[43] However, given the expense of the drugs and the potential risk of viral resistance with inappropriate use, viral load testing at week 4 and at the end of treatment (either week 12 or 24 depending on the regimen) seems prudent.

The viral load should be determined at weeks 4, 12, and 24 to assess treatment response and possible cessation of therapy in patients treated with simeprevir, pegylated IFN-α, and ribavirin. Discontinuation is warranted for patients who are unlikely to achieve a SVR based on the virologic response during treatment. If HCV RNA is greater than 25 IU/mL at week 4, the entire regimen should be discontinued. If the HCV RNA is greater than 25 IU/mL at week 12 or 24 after the simeprevir has been completed, the pegylated IFN-α and ribavirin should be discontinued.[44]

Numerous other DAAs have been developed and are currently in clinical trials. These include NS3/4A protease inhibitors, NS5B polymerase inhibitors, and inhibitors of host cell proteins required for HCV replication. The most current recommendations for all aspects of HCV treatment can be found at http://www.hcvguidelines.org.

Currently, three FDA-approved HCV viral load assays are available commercially. Two of these assays use real-time PCR, cobas AmpliPrep/cobas TaqMan, version 2.0 (Roche Diagnostics), and m2000 RealTime System (Abbott Molecular); the third uses branched DNA signal amplification, Versant 3.0 (Siemens). The amplification method, gene targets, and dynamic ranges for these assays are given in Table 5.1. The commercially available assays are calibrated against a World Health Organization (WHO) international calibration standard and report results in IU/mL. The first version of the cobas TaqMan HCV test had a genotype bias particularly against genotype 4 samples. The second version of the assay has been modified to enhance its ability to accurately quantify all of the major HCV genotypes.[45,46]

The development of the WHO first international HCV RNA standard and its acceptance by the manufacturers of HCV RNA assays as a calibrator was a significant advance in HCV RNA quantification.[47] However, despite the implementation of an international standard, HCV RNA measurements are not equivalent between the different assays.[48,49] Therefore, patients should ideally be tested with the same assay during the course of their treatment to minimize the potential for patient management errors.[50]

Although a number of baseline factors are predictive of response to treatment of chronic hepatitis C infection, HCV genotype is a strong and consistent predictor for achieving a SVR to pegylated IFN-α and ribavirin. In the large clinical trials of combination therapy with pegylated IFN-α and ribavirin, only 30% of patients infected with genotype 1 had a SVR compared with 65% of patients infected with genotypes 2 or 3.[37,38]

The determination of HCV subtypes has no clinical relevance in patients treated with IFN-α and ribavirin, but different treatment durations based on viral load kinetics are recommended for patients with different HCV genotypes. However, the emergence of resistant variants and virologic breakthrough were more common in patients infected with HCV subtype 1a than 1b when treated with TVR triple therapy.[51] HCV subtyping may play a role in helping to select treatment regimens and predict the development of resistance to DAA drugs. In addition, triple therapy with a protease inhibitor is not recommended for patients infected with genotypes 2 and 3.

Antiviral-resistance mutations that cluster around the catalytic site of the NS3/4A serine protease emerge during protease inhibitor therapy and are associated with failure and relapse.[42] Similar resistant variants are detected in both BOC- and TVR-treated patients, suggesting that cross-resistance occurs with these protease inhibitors. Also antiviral-resistant variants are found in about 5% of patients before treatment but do not appear to impact response to either protease inhibitor. Currently, there is no role for antiviral resistance genotyping at baseline or during treatment with the protease inhibitors.[52]

Several mutations in the NS3/4A protease are associated with reduced susceptibility to simeprevir. One of the most common and clinically relevant mutations is the substitution Q80K. This mutation may be present at baseline in approximately 30% of patients with genotype 1a and is associated with lower SVR rates. For patients with genotype 1a, Q80K mutation testing is recommended and patients with this variant should be offered other treatment options.[53]

A variety of laboratory-developed and commercial assays are used for HCV genotyping. The methods include NA sequencing, reverse hybridization, subtype-specific PCR, DNA fragment length polymorphism, heteroduplex mobility analysis, melting curve analysis, and serologic genotyping. Currently, there is only one FDA-approved HCV genotyping assay, the Abbott RealTime HCV Genotype II assay.[54] It uses real-time PCR and multiple hydrolysis probes to amplify and differentiate genotypes 1 to 6 and subtype 1a and 1b. The 5′ UTR is the target to identify the genotype and the NS5B is the target for subtyping genotype 1 samples. The results from this test have shown good overall agreement with direct sequencing methods. However, relatively high rates of indeterminate results and inability to distinguish all genotype 1 subtypes are limitations.

A commercially available reverse hybridization line probe assay is the most commonly used method for genotyping HCV among clinical laboratories participating in the HCV proficiency-testing surveys of the College of American Pathologists. This reverse hybridization assay was developed by Innogenetics (Fujirebio Europe) and is now marketed as the Versant HCV Genotype 2.0 Assay by Siemens. In this line probe assay (LiPA), biotinylated PCR products from the 5′ UTR and core regions of the HCV genome are hybridized under stringent conditions with oligonucleotide probes attached to a nitrocellulose strip: 19 type- and subtype-specific probes interrogate the 5′ UTR and an additional three probes interrogate the core region. The core region probes were added to provide better discrimination of subtypes 1a and 1b and genotype 6.[55] Hybridized PCR products are detected with a streptavidin–alkaline phosphatase conjugate. The pattern of reactive lines defines the genotype and in some cases the subtype. The assay discriminates among genotypes 1a, 1b, 2a/c, 2b, 3a, 3b, 3c, 3k, 4a/c/d, 4b, 4e, 4h, 5a, and 6a/b. The Versant HCV Genotype 2.0 Assay correctly identifies genotypes and distinguishes subtypes 1a and 1b when compared with sequencing but may not be able to adequately identify the other subtypes.[56] Mixed genotype infections can be recognized as unusual patterns of hybridization signals. However, the LiPA requires a considerable amount of amplicon for typing, and the assay may regularly fail when the viral load is less than 10^4 copies/mL.

Sequence analysis of amplified subgenomic sequences is the most definitive way to genotype HCV strains. Genotyping schemes based on sequencing variable genes such as E1, Core, and NS5B provide enough resolution to determine types and subtypes.[57,58] The 5' UTR is too highly conserved to discriminate all subtypes reliably. Genotyping methods targeting highly variable regions have higher failure rates because of primer mismatches and failed amplification reactions. Sequencing reactions can be performed directly on PCR products or on cloned amplicons. Mixed infections with multiple genotypes may be missed by sequence analysis. Definitive detection of mixed infections requires analysis of a large number of clones. Cloning may, however, emphasize artifactual nucleotide substitutions introduced by the DNA polymerase during amplification or by selection during the cloning procedure and is generally not practical for clinical laboratories.

A standardized direct sequencing system for HCV genotyping is commercially available (Siemens). The Trugene HCV 5'NC genotyping kit targets the 5' UTR (nt 96 to 282) and employs proprietary single-tube chemistry.[59] This method can be used with the 244-base-pair amplicon generated by either the Roche Amplicor HCV or Amplicor HCV Monitor tests as the sequencing template after a column purification step.[60] The sequencing chemistry produces bidirectional sequences. The software acquires the sequence data, and each pair of forward and reverse sequences is combined. A reference sequence library module contains approximately 200 sequences from the six major genotypes and 24 subtypes of HCV. The software automatically aligns the patient HCV sequence with the reference sequences in the library and reports type, subtype, and closest isolate determinations. The Trugene HCV 5'NC genotyping system is a reliable method for determining HCV genotypes, but similar to all approaches targeting the conserved 5' UTR, cannot reliably distinguish all HCV subtypes.[60,61]

The practice of using sequence analysis of a single subgenomic region for HCV genotyping has been challenged by the description of naturally occurring intergenotypic recombinants of two HCV genotypes.[62-65] The recombinant forms have been detected in patients in Russia (genotypes 2k and 1b), Vietnam (genotypes 2 and 6), and France (genotypes 2 and 5), as well as in experimentally infected chimpanzees (genotypes 1a and 1b).

A novel HCV genotyping method using a solid phase electrochemical array was developed by GenMark Diagnostics. The method uses sequence specific capture of a PCR amplicon from the HCV 5' UTR by surface-bound oligonucleotide capture probes formed within a preassembled monolayer with electrochemical detection using ferrocene-labeled oligonucleotide signal probes. High genotype concordance between the GenMark and LiPA HCV tests was observed; however, there were minor discrepancies in genotype 1 subtype identifications by the two tests due to differences in the regions of the HCV genome interrogated.[66]

The widespread use of tests not cleared by the FDA for HCV genotyping has placed an increased burden on clinical laboratories to verify the performance characteristics of these tests before clinical use. When validating HCV genotyping tests, laboratories should take advantage of the published evaluations and commercially available genotype panels to streamline the verification process.

The College of American Pathologists has a well-established proficiency testing program for laboratories performing tests for detection, quantification, and characterization of HCV RNA. These surveys have shown a steady improvement in the performance of laboratories over time that probably reflects progress in both the available technologies and laboratory practices.

> **POINTS TO REMEMBER**
>
> **Hepatitis C Virus**
> - There are approximately 3.2 million people living with active HCV infections in the United States. HCV is a common cause of chronic liver disease.
> - HCV viral load does not predict progression or severity of liver disease but is important in distinguishing active from resolved infections and as a widely accepted marker of response to therapy.
> - Although there are seven major HCV genotypes and numerous subtypes that differ in geographical distribution, genotype 1 infections predominate in much of the developed world and have, historically, been the most difficult to successfully treat using IFN-α and ribavirin.
> - Treatment response rates for genotype 1 infections have dramatically improved with the FDA clearance of DAA agents that target HCV proteases and polymerase.
> - Currently, there is no role for HCV resistance genotyping in guiding therapy, but that may change as the number of DAA agents increases.

Hepatitis B Virus

Hepatitis B virus (HBV) is a small enveloped DNA virus belonging to the family *Hepadnaviridae* and causes transient or persistent (chronic) infection of the liver. This family is divided into two genera, orthohepadnavirus and avihepadnavirus, which infect mammals or birds as natural hosts, respectively.[67] They possess a narrow host range determined by the initial steps of viral attachment and entry. Approximately 2 billion people have serologic evidence of hepatitis B, and of these, approximately 350 million people have chronic infections.[68] Depending on viral and host factors, the outcomes of infection with HBV range from clearance to mild or severe chronic hepatitis (CHB) to the development of cirrhosis or hepatocellular carcinoma (HCC).[69,70] The evolution of increasingly sensitive methods for HBV DNA detection and its accurate quantification has created an important role for molecular testing in routine patient care. Improved methods for sequencing and single mutation detection have led to tests that document mutational change in the viral genome. Currently, NA testing plays a critical role in the overall care of HBV-infected patients.

The HBV genome is a 3.2-kilobase relaxed circular, partially double-stranded DNA molecule. It has four partially overlapping open reading frames encoding the viral envelope (pre-S and S), nucleocapsid (precore and core), polymerase, and X proteins. After binding to hepatocytes, the virion is taken up into the cell by endocytosis and uncoated. The partially double-stranded DNA genome is converted to a covalently closed circular DNA (cccDNA) in the cell nucleus. The cccDNA is used as a template for transcription of the pregenomic RNA (pgRNA) and messenger RNA in the cell nucleus. The pgRNA moves into the host cell cytoplasm and serves as the template for translation of the HBV RT as well as the core protein by the cellular translational proteins. Concurrently, the HBV RT reverse transcribes the pgRNA to a new circular DNA molecule. Early in the replication cycle, some of the newly synthesized genomes will circulate back to the nucleus to maintain and increase the pool of cccDNA.[71]

Although HBV is a DNA virus, it replicates by an RT that lacks proofreading activity and, as a result, is prone to errors. The overlapping open reading frames of the genome limit the types of mutations that can be tolerated. However, variations in HBV sequences have been detected in almost all regions of the genome. Consequently, HBV exists as quasispecies, and different patients may be infected with different strains and genotypes.

Seven phylogenetic genotypes (A through H) of HBV have been identified, most of which have distinct geographic distribution. Genotypes are defined by intergroup divergence of greater than 8% in the complete genome nucleotide sequence. All known genotypes have been found in the United States with the prevalence of A, B, C, D, and E to G being 35%, 22%, 31%, 10%, and 2%, respectively.[72] Recent data suggest that HBV genotype plays an important role in the progression of HBV-related liver disease as well as response to IFN-α and pegylated IFN-α; however, HBV genotyping is not necessary in routine clinical practice.[73]

Serologic assays with high sensitivity, specificity, and reproducibility have been developed to detect HBV antigens and their respective antibodies. This complicated system of serologic markers is used for diagnosis of HBV infection and to define the phase of infection, degree of infectivity, prognosis, and patient's immune status. The presence of HBV DNA in the serum is a marker of viral replication in the liver and has replaced hepatitis B e antigen (HBeAg) as the most sensitive marker of viral replication. HBeAg is the extracellular form of the hepatitis B core protein. Molecular assays to quantify blood HBV DNA are useful for the initial evaluation of HBV infections, monitoring of patients with chronic infections, and assessing the efficacy of antiviral treatment.[71,73] In addition, U.S. blood donors are routinely screened for HBV DNA by qualitative tests to detect donors in the early stage of infection.[74] Antiviral resistance mutations are detected by molecular methods that identify known mutations associated with drug resistance.

The initial evaluation of patients found to have hepatitis B surface antigen (HBsAg) in serum should include routine liver tests and a variety of virologic tests, including HBV DNA testing.[73] Chronic HBV infection is a disease of variable course, and establishing baseline laboratory values at the time of diagnosis is important clinically for the tracking of disease progression over time and to evaluate candidates for liver biopsy. Monitoring disease activity in chronically HBV-infected patients is best done by measuring aminotransferase (ALT) at regular intervals in HBeAg-positive patients. However, serial HBV DNA testing is recommended in HBeAg-negative patients. The determination of serum HBV DNA (viral load) is important in the pretreatment evaluation and monitoring of therapeutic response in patients with chronic infection.[73] Currently, therapy for chronic HBV infection does not eradicate the virus and has limited long-term efficacy. The decision to treat should be based on ALT elevations; the presence of HBeAg or HBV DNA (or both); viral load of greater than 2000 IU/mL; the presence of moderate disease activity and fibrosis on liver biopsy; and virologic testing to exclude concurrent infections with hepatitis D virus (HDV), HCV, and HIV. The treatment goals for chronic hepatitis B are to achieve sustained suppression of HBV replication and to prevent further progression of liver disease. Parameters used to indicate treatment response include normalization of serum ALT, decrease in serum HBV DNA, and loss of HBeAg with or without detection of anti-HBeAg. Currently, there are eight FDA-approved therapies for chronic HBV infection: IFN-α, pegylated IFN-α2a, four nucleoside analogs (lamivudine, telbivudine, entecavir, and emtricitabine), and two nucleotide analogs (adefovir and tenofovir). Several factors predict a favorable response to IFN treatment with the most important being high ALT and low serum HBV DNA viral load, which are indirect markers of immune clearance.

Therapy usually does not eradicate the virus because the covalently closed circular form of the HBV genome is difficult to eliminate from the liver and the existence of extrahepatic reservoirs of HBV. Endpoints of treatment have traditionally been clearance of HBeAg, development of anti-HBe antibodies, and undetectable serum HBV DNA using insensitive hybridization assays with detection limits of approximately 10^6 genome copies/mL. Achieving these endpoints usually is accompanied by resolution of liver disease as evidenced by normalization of ALT and decreased inflammation on liver biopsy. The response usually is sustained at long-term follow-up. Nevertheless, most responders continue to have detectable HBV DNA when sensitive NA amplification tests are used. Responses to antiviral therapy are categorized as biochemical, virologic, or histologic and as on therapy or sustained off therapy.[73]

Several variations in the nucleotide sequence of HBV have important clinical consequences. An important mutation in the gene encoding HBsAg is a glycine-to-arginine substitution at codon 145 (G145R) in the conserved "a" determinant, which causes decreased affinity of the HBsAg for anti-HBs antibodies.[75] HBV with this mutation has been found in children of HBsAg-positive mothers who develop HBV infection despite vaccination and an adequate anti-HBs antibody response after vaccination, as well as in liver transplant recipients who have recurrent infection despite administration of HBV immune globulin.[76,77] These immune escape mutants have raised concern about vaccine efficacy and serologically silent infections. The G145R mutation has been reported in many countries and is responsible for 2% to 40% of vaccine failures. Although there is diminished binding to anti-HBs antibodies, the vast majority of S mutants can be readily detected with the current generation of HBsAg tests. Thus, an initial concern that widespread use of HBV immune globulin and vaccination would result in HBV mutants that would escape detection in the HBsAg test was unfounded.

Mutations in the basal core promoter and the precore genes affect the synthesis of HBeAg and commonly arise under immune pressure.[78] The most common basal core promoter mutation has a dual change of A to T at nt 1762 (T1762) and G to A at nt 1764 (A1764) that diminishes the amount of mRNA and hence HBeAg secretion.[79] The predominant precore mutation is a G to A change at nt 1896 (A1896), which leads to premature termination of the precore protein at codon 28, thus preventing the production of HBeAg.[80] The A1896 mutation is infrequent in North America and Western Europe but is geographically widespread. This geographic variability in frequency is related to the predominant genotypes in a geographic region because the mutation is found only in genotypes B, C, D, and E.

The A1896 mutation was first reported in patients with chronic active hepatitis or fulminant hepatitis. However, the A1896 mutation also can be present in asymptomatic carriers, and viruses with this mutation replicate no more efficiently than wild-type HBV. Thus, the pathophysiologic significance of this mutation is unclear.[81] However, the clinical picture of persistent HBV replication and active liver disease in HBeAg-negative patients appears to be increasingly prevalent, and in some regions, the A1896-mutant virus may be more prevalent than the wild-type virus.

Therapy for chronic hepatitis B requires long courses of treatment with nucleoside or nucleotide analogs. A major concern with long-term therapy is the development of antiviral resistance by mutation in one or more domains of the gene encoding the HBV polymerase. The rate at which resistant mutants are selected is related to pretreatment serum HBV DNA viral load, rapidity of viral suppression, duration of treatment, and prior antiviral exposure. The incidence of genotypic resistance also varies with the sensitivity of the methods used to detect resistance mutations and the patient population tested.

Typically, when a patient experiences a virologic breakthrough, defined as an increase in serum HBV DNA greater than 1 $\log_{10}$ above nadir after achieving a virologic response during continued treatment, HBV resistance genotyping should be performed. The standardized nomenclature of HBV antiviral resistance mutations in the polymerase is shown in Table 5.2.[71,82,83] No HBV mutations are associated with resistance to IFN-α or pegylated IFN-α2a.

There are a number of commercially available tests for quantification of HBV DNA in serum and plasma, but only two tests have United States-In Vitro Diagnostics (US-IVD) regulatory status. These tests—the cobas AmpliPrep/cobas TaqMan (Roche) and the Real-time (Abbott) HBV tests—both use real-time PCR (see Table 5.1). The others produced by Cepheid, Qiagen, and Siemens are CE (*Conformité Européenne*, meaning European Conformity) marked and RUO kits or ASRs in the United States.

TABLE 5.2 Antiviral Agents and the Hepatitis B Virus Polymerase Mutations Associated With Resistance

Antiviral Agent	Drug Class	Resistance Mutations
Lamivudine	Nucleoside analog (cytidine)	(L180M + M204V/I/S), A181V/T, S202G/I
Telbivudine	Nucleoside analog (dTTP)	M204I, A181T/V
Entecavir	Nucleoside analog (2-deoxyguanosine)	T184S/C/G/A/I/L/F/M, S202G/C/I, M250V/I/L
Emtricitabine	Nucleoside analog (cytidine)	M204V/I
Adefovir	Nucleotide analog (dATP)	A181V/T, N236T
Tenofovir	Nucleotide analog (dATP)	A194T, N2263T, A181V/T

dATP, Deoxyadenosine triphosphate; *dTTP*, deoxythymidine triphosphate.

A WHO international HBV standard was first created in 2001 in response to the recognized need to standardize HBV DNA quantification assays.[84] However, despite the availability of HBV DNA standards, the various quantitative assays usually have different conversion factors for copies to IU/mL, which may reflect their different amplification and detection chemistries. Laboratories should report HBV viral load test results in IU/mL as both $\log_{10}$ transformed and arithmetic values. HBV is included in the hepatitis viral load proficiency testing surveys available from the College of American Pathologists.

Two HBV genotyping systems are commercially available as RUO kits. Innogenetics (Fujirebio Europe) offers three different line probe assays for (1) HBV phylogenetic genotyping; (2) detection of precore mutations; and (3) detection of all relevant lamivudine, emtricitabine, telbivudine, adefovir, and entecavir resistance mutations as well as known compensatory mutations.[85,86] All assays use PCR to amplify portions of the relevant genes to produce a biotinylated product. The PCR products are denatured and hybridized to a series of informative probes immobilized on a nitrocellulose strip. The hybrids are visualized on the strip after addition of streptavidin–alkaline phosphatase and a colorimetric substrate. The mutations are identified by the colored patterns of PCR product hybridization to the probes. The line probe assays typically have better sensitivity for detection of sequence variants than direct Sanger sequencing.

The TRUGENE HBV genotyping test (Siemens) uses two fluorescently labeled DNA primers and PCR to amplify a portion of the HBsAg gene and overlapping polymerase gene, with bidirectional sequencing of these amplicons, and a software module that includes a sequence database that identifies the phylogenetic genotype (A to H) as well as mutations associated with resistance to the nucleoside and nucleotide analog drugs.[87] The sequence ladder is resolved on a polyacrylamide slab gel. The total analysis time is approximately 8 hours, including the time required for DNA extraction and purification.

The standardized nomenclature for reporting of HBV antiviral resistance mutations shown Table 5.2 should be used when resistance genotyping is performed. The inability of genotyping assays to detect minor populations of circulating HBV is a significant technical issue. In general, direct sequencing is limited to resolution of populations that are more than 20% of the viral population.

Transplant Recipients

Cytomegalovirus

Cytomegalovirus (CMV), a member of the *Herpesviridae* family, is an enveloped, double-stranded DNA virus. It has a large (240-kb) genome with approximately 95% DNA homology among different strains. CMV usually causes asymptomatic or minor infections in immunocompetent individuals but remains an important pathogen in immunocompromised individuals, including persons with AIDS, transplant recipients, and those on immune-modulating drugs. Primary infection is usually asymptomatic in immunocompetent persons, although a small percentage of individuals with CMV infection may develop a syndrome similar to mononucleosis. After primary infection, a lifelong latent infection is established that does not cause clinical symptoms. However, if an infected individual becomes immunocompromised, the virus can reactivate, leading to a wide variety of clinical syndromes.

The most severe CMV infections are seen in patients who acquire their primary infection while immunocompromised. In persons with AIDS, CMV disease rarely occurs when the CD4+ cell count is above 100 cells/mm^3; the most common clinical presentations are retinitis, esophagitis, and colitis. In transplant recipients, the occurrence and severity of CMV disease are related to the CMV serostatus of the organ donor and recipient, the type of organ transplanted, and the overall degree of immunosuppression. For example, CMV disease tends to be more severe in lung transplant recipients than in renal transplant recipients. For all types of solid organ recipients, the most severe disease occurs when CMV-seronegative recipients receive an organ from a CMV-seropositive donor and the primary CMV infection occurs while the person is immunosuppressed. In contrast, CMV-seropositive recipients of hematopoietic stem cells from CMV-seronegative donors are at highest risk of CMV diseases after hematopoietic stem cell transplant (HSCT). CMV disease can also occur in seropositive individuals, whether they receive an organ from a seropositive or seronegative donor. Clinical findings associated with CMV disease in transplant recipients are diverse and include interstitial pneumonitis; esophagitis and colitis; fever; leukopenia; and, less commonly, retinitis and encephalitis.

The diagnosis of CMV disease represents a challenge because latent infections are common. Immunocompromised individuals can have an asymptomatic, clinically insignificant, low-level, persistent infection that must be distinguished from clinically important active CMV disease. The distinction can be challenging when sensitive molecular assays are used that can detect small amounts of CMV DNA in clinical specimens.

Traditionally, the diagnosis of CMV disease relied on the detection of CMV from clinical specimens by the use of cell culture techniques in human diploid fibroblasts. Although considered the gold standard, these conventional culture methods are labor intensive and have a turnaround time (TAT) of 1 to 3 weeks. In addition, the assays lack adequate sensitivity for detecting CMV in blood specimens. The rapid shell-vial culture method can provide results in 1 to 2 days and is useful for detection of CMV in tissue, respiratory, and urine specimens. However, this method may also fail to detect CMV in blood. For many years, laboratories relied on a CMV antigenemia assay, which detects the matrix protein pp65 in circulating polymorphonuclear white blood cells. This semiquantitative assay is more rapid than culture, and the number of CMV antigen-positive cells correlates with the likelihood of CMV disease, but the assay is labor intensive, subjective, and nonstandardized and, consequently, is no longer done in most clinical laboratories.

Considering the limitations of culture, there is great interest in using NA testing for the detection and quantification of CMV DNA in blood specimens. The clinical uses of CMV molecular assays are diverse and include (1) initiation of preemptive therapy, (2) diagnosis of active CMV disease, and (3) monitoring of response to therapy. Preemptive therapy identifies a group of individuals at higher risk for developing CMV disease. For example, all members of the group would be tested for the presence of CMV DNA in their blood or plasma, and only those testing positive would be treated. Therapy is administered before development of symptoms in an attempt to prevent the onset of active disease. By contrast, with prophylactic therapy, all patients in the group are treated, without further stratification of risk, thus involving treatment of a greater number of patients. Preemptive therapy has become the standard of care for management of HSCT recipients.

Molecular assays are useful to diagnose active CMV disease because CMV DNA concentrations are higher in patients with active CMV disease than in those with asymptomatic infection.[88-91] Quantitative PCR from plasma or whole blood is now commonly used to diagnose CMV disease and monitor response to therapy. Until recently, there have been no FDA-approved CMV viral load tests, and laboratories used a variety of different laboratory-developed tests for detection and quantification of CMV DNA. Therefore, the threshold or viral load cutoff at which a diagnosis is made or preemptive therapy is begun has varied among institutions and transplant populations. Because a universal viral load cutoff for defining CMV disease has not been established, health care providers should rely on the trend of viral load values over time rather a value obtained at a single point in time for diagnosis and patient management.

After active CMV disease has been diagnosed, molecular assays are useful in monitoring response to therapy. Viral load values decrease rapidly after appropriate antiviral therapy is begun, and CMV DNA is cleared from the plasma within several weeks of initiation of therapy.[92-94] Failure of viral loads to decrease promptly should raise concerns of possible treatment failure because persistently elevated concentrations of CMV DNA during therapy indicates therapeutic resistance. Molecular assays can also identify patients at risk for relapsing CMV infection. In solid organ transplant recipients, patients with a detectable viral load after completing 14 days of ganciclovir therapy for CMV infection are at increased risk of relapse. The rate of decline in CMV

DNA after initiation of therapy can be also used to predict risk of relapse of CMV infection.[94]

Cytomegalovirus DNA concentrations are also useful in assessing the risk of developing CMV disease in persons with AIDS. Detection of CMV DNA in plasma is associated with increased risk of developing CMV disease and increased risk of death. In addition, each $\log_{10}$ increase in viral load (ie, each 10-fold increase in concentration) has been associated with a threefold increase in the risk of developing CMV disease.[95]

Currently, there are only two FDA-approved CMV viral load assays (see Table 5.1): CAP/CTM CMV test (Roche) and artus CMV RGQ MDx (Qiagen). Both are based on real-time PCR, are calibrated against the WHO CMV standard, and consequently report results in international units per milliliter. The Roche test amplifies a portion of *UL54* gene and the Qiagen test amplifies a portion of the *MIE* gene. Both assays have similar lower limits of quantification, but the Qiagen assay has a 10-fold greater dynamic range. Despite the availability of an international standard and of FDA-approved assays, there is still considerable interlaboratory variability of results in CMV proficiency testing surveys.[96]

Epstein-Barr Virus

Epstein-Barr virus (EBV) is a double-stranded DNA virus belonging to the *Herpesviridae* family. The seroprevalence of EBV is greater than 95% among adults older than the age of 40 years, and primary infection is followed by lifelong latency with reactivation of infection in immunocompromised hosts. In transplant recipients, EBV infection may cause malaise, fever, headache, and sore throat, but it is also associated with posttransplantation lymphoproliferative disease (PTLD), a significant cause of morbidity and mortality. PTLD is a spectrum of lymphocytic proliferation that ranges from benign lymphocytic hyperplasia to potentially fatal malignant lymphoma. The process is often multicentric and may involve the central nervous system (CNS), eyes, gastrointestinal (GI) tract (with bleeding and perforation), liver, spleen, lymph nodes, lungs, allograft, oropharynx, and other organs. Clinical presentations vary and include, but are not limited to, adenopathy, fever (including "fever of unknown origin"), abdominal pain, anorexia, jaundice, bowel perforation, GI bleeding, renal dysfunction, liver dysfunction, pneumothorax, pulmonary infiltrates or nodules, and weight loss.

The pathogenesis of PTLD involves the exponential proliferation of B cells as a result of uncontrolled EBV infection. Risk factors include a donor and recipient serologic mismatch (eg, donor positive/recipient negative), a high degree of immunosuppression (particularly the use of antilymphocyte therapy for rejection), and a high EBV viral load.[97] Most cases of PTLD occur during the first year after transplant, and the cumulative incidence ranges from 1% to 2% in HSCT and liver transplant recipients and up to 11% to 33% in intestinal or multiorgan transplant recipients.[98]

Treatment of EBV-related lymphoproliferative disease is challenging. After lymphoproliferative disease is established, antiviral treatment is not effective, and immunosuppression must be reduced. Murine humanized chimeric anti-CD20 monoclonal antibody (rituximab) has been helpful in some cases; some patients require chemotherapy, radiation therapy, or both. Adoptive immunotherapy using donor-derived cloned EBV-specific cytotoxic T cells may be useful for prophylaxis and treatment of lymphoproliferative disease in allogeneic HSCT and solid organ transplant recipients.

Increases in EBV viral load may be detected in patients before the development of EBV-associated PTLD[99-102]); viral loads typically decrease with effective therapy. Whereas a high EBV DNA viral load is a strong predictor for the development of PTLD, low-level EBV viral load occurs relatively frequently and may resolve without intervention.[103,104] To complicate the matter, some pediatric liver and heart transplant recipients may exhibit chronic high EBV viral loads.[105,106] Available assays lack standardization, and the optimal assay technique, sample type (ie, whole blood, lymphocytes, plasma), and sampling schedule are not defined. Nevertheless, EBV viral load assays are generally sensitive, specific, precise, linear across a wide dynamic range, rapid, reasonably inexpensive, and, overall, useful in patient care.[107] Although there are no defined "trigger points" predictive of PTLD, persistently detectable concentrations of EBV DNA by PCR (cutoffs vary between programs) typically result in a thorough evaluation for PTLD (eg, computed tomography of the chest, abdomen, and pelvis).

Epstein-Barr virus viral load testing is also indicated in transplant recipients who present with lymphadenopathy, fever, or other signs and symptoms suggestive of lymphoproliferative disease. A high EBV load should trigger the search for mass lesions or organ dysfunction pinpointing potential sites of disease, which should be biopsied.

Currently, there are no FDA-approved EBV viral load tests available. A wide variety of commercially available primers and probes for different gene targets are used in the laboratory developed tests (LDTs) deployed in clinical laboratories.[108] There is no consensus on the best target gene or specimen type (whole blood, white blood cells, or plasma); however, the first WHO international standard for EBV DNA was developed to address variation between assay results attributable to calibration (http://www.nibsc.org/documents/ifu/09-260.pdf).

The definitive diagnosis of PTLD requires biopsy. Tissues from patients with EBV-associated lymphoproliferative disease may show monoclonal, oligoclonal, or polyclonal lesions. The diagnosis of EBV-associated lymphoproliferative disease requires demonstration of EBV DNA, RNA, or protein in biopsy tissue. In situ hybridization targeting EBER1, EBER2, or both is the gold standard assay for determining whether a lymphoproliferative process is EBV related. Commercial systems for EBER in situ hybridization are available from Ventana, Leica, Dako, Invitrogen, and Biogenex.[107]

BK Virus

BK virus (BKV) is a member of family *Polyomaviridae*, which also includes JC virus (JCV) and simian virus 40 (SV40). It is an enveloped, double-stranded DNA virus that shares approximately 70% sequence homology with JCV and SV40. Seroprevalence reaches nearly 100% in early childhood, generally after an asymptomatic primary infection (although fever and nonspecific upper respiratory tract symptoms may occur).[109] Seroprevalence declines to 60% to 80% in adulthood. After primary infection, the virus can remain latent in many sites, most notably the epithelium of the urinary tract and lymphoid cells, until an immunosuppressed state allows reactivation and replication of the virus. Replication of BKV in immunocompromised hosts may be asymptomatic or cause organ dysfunction, affecting the kidney, bladder, or ureter. BKV disease in the urinary system manifests as hemorrhagic or nonhemorrhagic cystitis and with ureteric stenosis in bone marrow and solid organ transplant recipients.[110] It also causes polyomavirus-associated nephropathy (PVAN) in renal transplant recipients.[111]

Hemorrhagic cystitis (HC) is a cause of morbidity and occasional mortality in patients undergoing bone marrow transplantation.[112] The manifestations range from microscopic hematuria to severe bladder hemorrhage leading to clot retention and renal failure. Its incidence varies from 7% to 68% of bone marrow transplant recipients. Although mild HC usually resolves with supportive care, severe HC may require bladder irrigation, cystoscopy, and cauterization.[113] BKV was observed in early studies to be associated with the development of HC during bone marrow transplantation; however, later studies using sensitive PCR assays showed that BKV DNA could be detected in the blood and urine of patients with or without HC.[114-116] Recently, quantitative assays for BKV DNA in urine have demonstrated that patients with HC have higher peak BK viruria and larger total amounts of BKV excreted during bone marrow transplantation compared with asymptomatic patients.[117,118]

Although BKV was first isolated from the urine of a renal transplant recipient in 1971,[119] the association between nephropathy and the presence of BKV in renal transplant recipients was not reported until 1995.[120] BKV replication in renal allografts can lead to progressive graft dysfunction and, potentially, graft failure. Although the recognition of PVAN in renal transplant recipients coincided with the use of newer immunosuppressive drugs such as tacrolimus, sirolimus, and mycophenolate mofetil, risk factors for development of PVAN have not been elucidated.[121] The prevalence of PVAN ranges from 1% to 10% in kidney transplant recipients with loss of allograft function in about one third to half of these cases.[121] The disease appears to result from reactivation of BKV infection in the donor allograft.

The signs and symptoms of PVAN are mild and nonspecific, often with only a gradual increase in serum creatinine over weeks as the allograft loses function.[122] A definitive diagnosis of PVAN is obtained through histopathology of the biopsied kidney; the characteristic PVAN pattern includes viral cytopathic changes in epithelial cells and interstitial inflammation and fibrosis. However, these changes are not pathognomonic for PVAN, and most centers use immunohistochemical staining with antibodies specific for polyomavirus proteins or in situ hybridization to confirm the diagnosis.[121] Because of the focal nature of the nephropathy and the possibility of sampling error, a negative biopsy does not rule out PVAN. Biopsy of the kidney is an invasive procedure that is impractical for serial monitoring, early diagnosis, and clinical management of patients with PVAN. Other less invasive diagnostic methods for PVAN have also been assessed. Urine cytology may reveal renal epithelial cells with intranuclear viral inclusion bodies, termed *decoy cells*.[123] The sensitivity and specificity of decoy cells for diagnosing PVAN is 99% and 95%, respectively, but the positive predictive value varies between 27% and 90%. Quantification of BKV DNA or mRNA in urine by NA amplification methods has been proposed as a method to monitor changes in BKV replication.[124-127] However, physiological changes of urine constituents and use of different urine fractions may give rise to considerable variation in viral load values that may complicate the identification of diagnostic thresholds and quantitative cross-sectional and longitudinal studies.[111] PCR methods for detection and quantitation of BK viremia have emerged as clinically useful tools in the diagnosis and management of PVAN because viremia precedes development of nephropathy in almost all cases.[128-131]

In 2005, an expert interdisciplinary panel recommended the use of either urine cytology or NA amplification tests to screen renal transplant recipients for BK viruria every 3 months up to 2 years posttransplant or when allograft dysfunction occurs or biopsy is performed.[121] Patients with positive screening test results should have an adjunct quantitative NA amplification test performed using urine or plasma. Patients with urine DNA loads of more than 10^7 copies/mL or plasma DNA loads of more than 10^4 copies/mL that persist for more than 3 weeks have presumptive PVAN and a renal biopsy should be performed to confirm the diagnosis.

Reducing the intensity of maintenance immunosuppression is the primary intervention in patients with PVAN. No effective antiviral agents for BKV are available, but low-dose cidofovir has been used for treatment of cases not amenable or refractory to decreased maintenance immunosuppression.[121] Viral load in urine or plasma should be monitored every 2 to 4 weeks to gauge the effectiveness of the intervention.

Currently, real-time PCR is the method of choice for BKV DNA quantification because of its simplicity and wide dynamic range of 6 to 7 $\log_{10}$ virus copies/mL. Such high concentrations in plasma are uncommon but can exceed 10^{12} copies/mL. Although BK viral load tests have become a standard of care for diagnosis and monitoring of patients with PVAN, there is neither consensus in the design of PCR assays nor recognized standard reference material. As a consequence, the assays developed by different laboratories may give markedly different results, requiring individual laboratories to establish and verify their own clinical threshold values.

Polymerase chain reaction assay design is complicated by the high degree of homology between the genomes of the different human polyomaviruses. Gene targets for BKV-specific assays include coding sequences for VP1, large T antigen, and agnoprotein because these sequences are sufficiently variable among human polyomaviruses.[130]

BKV is classified into seven subtypes based on phylogenetic analysis of full-genome sequences, subtypes Ia, Ic, II, III, IV, V, and VI.[132] Hoffman et al[133] compared seven Taq-Man real-time PCR primer–probe sets in conjunction with two different reference standards to quantify BKV DNA in urine samples. They observed substantial disagreement among assays attributable both to features of the primer and probe design and to choice of reference material. The most significant source of error were primer and probe mismatches caused by BKV subtype polymorphisms, primarily among subtype III and IV isolates. However, they found less subtype bias among the seven assays for the more common subtypes Ia, V, and VI. The assay that provided the most reliable measure of all subtypes included a mixture of primers and probes that targeted both the VP1 and large T antigen sequences.

SEXUALLY TRANSMITTED INFECTIONS

Chlamydia trachomatis and *Neisseria gonorrhoeae*

Testing for *Chlamydia trachomatis* (CT) and *Neisseria gonorrhoeae* (NG) is discussed together because several of the available NA amplification tests (NAATs) for these pathogens are multiplex assays. Although both CT and NG can cause a variety of clinical infections, the focus here is on genital infections.

Detection of CT is a challenging and important public health issue. CT is a major cause of sexually transmitted infections (STIs), with an estimated 1 million cases occurring annually among sexually active adolescents and young adults in the United States.[134] More than half of the infections are asymptomatic.[135] Even when symptomatic, the diagnosis can be missed as the manifestations are protean. In men, CT infection may present as urethritis, epididymitis, prostatitis, or proctitis.[136,137] and as cervicitis, endometritis, and urethritis in women, with 10% to 40% of infections in women progressing to pelvic inflammatory disease (PID) if untreated.[138,139] Related complications include chronic pelvic pain, ectopic pregnancy, and infertility. In the United States, CT infection is the likely cause of most secondary infertility in women. In pregnant women, there is the additional risk of transmitting the infection to the newborn during labor and delivery, leading to pneumonia or conjunctivitis in the newborn.

N. gonorrhoeae infection, too, may present in various ways, and the clinical presentations overlap with those of CT. Men may have acute urethritis with discharge, epididymitis, prostatitis, and urethral strictures. In women, NG infection can produce cervicitis, which, if left untreated, can lead to PID, abscesses, or salpingitis.

Traditional methods for the diagnosis of CT infection include cell culture, antigen detection by immunofluorescence, enzyme immunoassay (EIA), and nonamplified NA probes. These traditional methods have been replaced in most laboratories by amplified NA tests, which provide greater sensitivity in detecting CT from genital specimens. For NG, which was traditionally diagnosed based on culture methods that relied on selective culture media, NA testing does not offer significant improvement in sensitivity compared with culture when culture is performed under appropriate conditions. NG is highly susceptible to extreme temperatures and desiccation, which can lead to decreased sensitivity of detection by culture, particularly when specimen transport is required before culture.[140] NA testing for NG offers a sensitive and reliable alternative to culture since it is easier to maintain the integrity of the target NA than it is to maintain the organism's viability.

In addition to high diagnostic and analytical sensitivity and specificity, NA testing offers several advantages over conventional culture and antigen detection methods for the diagnosis of CT and NG. Testing for both pathogens can be done on a single specimen, and for some multiplex assays, testing is performed in a single reaction. Unlike the infectious organism itself, the DNA and RNA of NG and CT are quite stable in commercial transport devices, thus accounting for some of the increased diagnostic sensitivity of these assays compared with culture. The stability of NA avoids the necessity of immediate transport to the laboratory, and specimens may be stored refrigerated or at room temperature before transport. Transport and storage requirements vary among tests, so it is important to refer to the package insert for specific details. An additional advantage of NA testing is the use of urine specimens, which for women allows testing to be done without the need for a pelvic examination. In men, urine offers a convenient and diagnostically sensitive alternative to collection with a urethral swab and increases the likelihood that asymptomatic men will agree to be tested.

NAATs for the detection of CT and NG from clinical specimens use a variety of specimens, including cervical and vaginal swabs, urethral swabs, and urine from both asymptomatic and symptomatic individuals. Not all assays are cleared by the FDA for use in the United States for all conditions and age ranges, and the current assays are not FDA-cleared for oropharyngeal, rectal, or conjunctival specimens. However, many of these tests have been assessed for diagnosing infections in multiple extragenital anatomic sites in men, women, and children. Current CDC guidelines for laboratory-based detection of CT and NG recommend NAATs for oropharyngeal and rectal specimens and for use in evaluating cases of adult and pediatric sexual abuse.[141] However, use of these tests outside of the FDA-approved indications requires that laboratories establish the specifications for performance characteristics according to Clinical Laboratory Improvement Amendments (CLIA) regulations. Performance characteristics vary among assays (details are available in the package inserts), but some general comments can be made. The diagnostic sensitivity of the tests varies according to the specimen type and whether the patient is asymptomatic or symptomatic. Interpretation of the results of NA testing for CT can be challenging because many studies have shown these assays to be more diagnostically sensitive than culture, which was previously used as the gold standard for clinical trials. For men, the diagnostic sensitivity of testing urine specimens is nearly equivalent to that of testing urethral swabs. A limited volume (20–50 mL) of first-passed urine is preferred because larger volumes will lead to a decreased concentration of the organism in the sample and thus reduced diagnostic sensitivity. With proper specimen collection, male urethral swabs and urine specimens have a sensitivity of nearly 100% for the detection of NG or CT infection. For women, vaginal and cervical swab specimens provide the highest sensitivity for the detection of NG and CT infection, with many studies showing a sensitivity of 90% to 95%. Vaginal swabs are preferred because they are easier to collect. Urine specimens can be used, but they generally result in a lower diagnostic sensitivity than cervical swabs (75%–85%). An alternative to urine testing in women is the use of self-collected vaginal swabs, which have been shown in some studies to have a diagnostic sensitivity that is equal to that obtained with cervical swabs; several commercial tests have been cleared for use with vaginal swabs.

Decisions regarding the selection of a specific amplification test for the detection of CT and NG should not be based solely on the cost of reagents. Other key factors to consider include test performance characteristics, such as diagnostic sensitivity and specificity, and applicability for urine and swab specimens in both symptomatic and asymptomatic individuals. Ideally, the test should include an internal control, particularly if a crude lysate is used in the assay. Other factors to consider are degree of automation, ease of use, work flow issues, and space and equipment needs.

Historically, for several of the NG assays, reduced specificity was due to presence of the gene target in nongonococcal *Neisseria* spp.[142,143] Currently, only the ProbeTec tests (Becton-Dickinson) produce false-positive results with commensal species including *Neisseria lactamica*, *Neisseria subflava*, and *Neisseria cinerea* (Table 5.3). None of the NAATs for CT have known biological false-positive results because of presence of the gene target in other organisms. Other sources of false-positive results include carryover contamination of amplified product and cross-contamination during specimen collection, transport, or processing. Concerns over these issues have led to consideration of supplemental testing for all CT- or NG-positive specimens using alternative target tests because false-positive results can have psychosocial and medicolegal ramifications.[144] However current recommendations do not advise confirming all positive NAATs unless otherwise indicated in the package insert or for tests with known cross-reactivity with commensal *Neisseria* spp.[141] False-positive results in a low-prevalence population can significantly reduce the predictive value of a positive result.

Because DNA can persist in urine samples for up to 3 weeks after completion of therapy, test of cure using NA testing is discouraged. If this must be done, then testing should be delayed for at least 3 weeks after therapy is completed to allow time for clearance of the DNA of the pathogen.

False-negative results from inhibition of amplification are a consideration for both NG and CT testing and can occur with both cervical swabs and urine specimens. Inhibition rates may vary considerably depending on the amplification and NA extraction methods used. For tests using a crude lysate (eg, ProbeTec), inhibition rates tend to be higher than those seen with the APTIMA Combo test, which uses a target capture method to purify NA. For assays that test a crude lysate, it is useful to amplify another NA sequence as an internal control (or "amplification control") to assess for inhibition of amplification. Results are reported as negative for NG or CT only when amplification of the internal control is documented.

A conserved, cryptic plasmid is found in more than 99% of strains of CT and contains the gene target for several NAATs. However, a new variant (nv) strain of CT emerged in Sweden in 2006 with a 377-base-pair deletion in the cryptic plasmid, which contained the target for several of the CT tests. This deletion led to false-negative results with some but not all of the tests that targeted the cryptic plasmid.[145] The current versions of all the NAATs for CT have been modified to detect the nv strain of CT. Obviously, tests that target the sequences contained on the cryptic plasmid will not detect the rare strain of CT that lacks the plasmid.

Performing CT and NG testing on liquid cytology media is a matter of interest because a single specimen can be used for cervical cancer screening (Papanicolaou [Pap] and human papillomavirus [HPV] tests) and for CT or NG testing.[146] The latter two tests are performed on the liquid specimen that remains after completion of the PAP and HPV testing. However, several drawbacks to this approach must be considered. The instruments used to prepare PAP tests were not designed to control for cross-contamination during processing, and this may lead to false-positive results. CT and NG testing are performed after the PAP smear and HPV testing are complete, delaying diagnosis and treatment of CT or NG infection. Moreover, the remaining specimen may be inadequate to complete CT and NG testing, thus requiring the patient to make a return visit for collection of an additional sample. Removing an aliquot for CT and NG testing before Pap testing is performed ("pre-aliquoting") may be helpful in overcoming some of these issues, provided adequate volume of sample remains for PAP and HPV testing. This approach does not completely remove the risk of cross-contamination, so specimens must be handled in a manner consistent with procedures used in molecular laboratories. In addition, not all NAATs for CT and NG have been FDA approved for use with liquid cytology media, and those that

TABLE 5.3 Amplification Methods and Target Regions for Food and Drug Administration–Cleared Nucleic Acid Amplification Tests for Detection of *Chlamydia trachomatis* and *Neisseria gonorrhoeae*

Assay (Manufacturer)	Method	*C. trachomatis* Target	*N. gonorrhoeae* Target
Abbott RealTime CT/NG (Abbott)	Real-time PCR	Two distinct regions in cryptic plasmid	*Opa* gene region
Aptima COMBO 2 assay (Hologic/Gen-Probe)	Transcription-mediated amplification	23S rRNA region	16S rRNA region
Aptima CT assay		16S rRNA region	
Aptima GC assay			Distinct 16s rRNA region
BD ProbeTec Q^X CT Amplified DNA assay	Strand displacement amplification	One region in the cryptic plasmid	
BD ProbeTec Q^X GC Amplified DNA assay			Chromosomal pilin gene-inverting protein homologue*
Xpert CT/NG test (Cepheid)	Real-time PCR	One distinct chromosomal region	Two distinct chromosomal regions
cobas CT/NG test (Roche)	Real-time PCR	One cryptic plasmid and one chromosomal region	DR-9A and DR-9B regions

*False-positive test results with some commensal *Neisseria* spp. may occur.
CT, *Chlamydia trachomatis*; *NG*, *Neisseria gonorrhoeae*; *PCR*, polymerase chain reaction.

have may not have been cleared for both types of media (Hologic PreservCyt and BD SurePath).

> **POINTS TO REMEMBER**
>
> ***Chlamydia trachomatis* and *Neisseria gonorrhoeae***
> - NAATs are the recommended test method for detection of CT and NG genital tract, oropharyngeal, and rectal infections, but they have not been cleared by the FDA for the latter two infections.
> - Routine repeat testing of NAAT-positive specimens is not recommended because this practice does not improve the positive predictive value of the test.
> - Positive reactions with nongonococcal *Neisseria* spp. have been reported with some NAATs and the use of an alternative target NAAT might be needed to avoid false-positive results for NG.
> - CT and NG DNA can persist in samples from successfully treated patients for up to 3 weeks; therefore, tests of cure using NAATs are discouraged.

Trichomonas vaginalis

Trichomoniasis is an STI caused by the protozoan *Trichomonas vaginalis*. Although *T. vaginalis* infection is not a reportable disease in the United States, it is the most prevalent nonviral STI in the United States.[147] *T. vaginalis* infection may present as vaginitis in women and urethritis in men; however, it is frequently asymptomatic. Infection with *T. vaginalis* may also cause additional adverse health outcomes, including PID in women, as well as infertility and increased incidence of HIV transmission in women and men. Current recommendations for *T. vaginalis* diagnosis and treatment can be found at http://www.cdc.gov/std/tg2015/default.htm and available diagnostic tests range from simple microscopy to NAATs.

Microscopic examination of vaginal fluid or urethral discharge for *T. vaginalis* (wet mount) in the clinic is the most commonly used test. It has low sensitivity (51%–65%) but with experienced observers can be highly specific.[148] Culture has long been considered the gold standard test, but it requires special medium and 5 days to complete. However, recent studies indicate that the sensitivity of culture may be as low as 75% to 96%.[148] Pap tests are not suitable for routine screening or diagnosis because of low sensitivity.[149] There is a single rapid antigen test for *T. vaginalis* (OSOM, Sekisui Diagnostics) that is FDA cleared for use as a point-of-care test for female patients. Test specifications include sensitivity of 82% to 95% and specificity of 97% to 100%.[150]

The Affirm VPIII Microbial Identification test is an FDA-cleared test that uses nonamplified NA probes to detect three organisms associated with vaginitis: *T. vaginalis*, *Gardnerella vaginalis*, and *Candida albicans*. Its sensitivity and specificity for detection of *T. vaginalis* are 63% and 99.9%, respectively.[151] NAATs are the most sensitive tests available for detection of *T. vaginalis*. A variety of LDT NAATs are more sensitive than the previous gold-standard test of culture but with a more rapid analysis time. Currently, there are two FDA-cleared NAATs for detection of *T. vaginalis* from female patients only. The APTIMA *Trichomonas vaginalis* assay (Hologic/Gen-Probe) detects *T. vaginalis* RNA by transcription-mediated amplification, and its sensitivity and specificity are both 95% to 100%.[148,151,152] It also offers the opportunity to test for *T. vaginalis* from the same sample submitted for CT and NG testing with their APTIMA Combo 2 assay because the test runs on the same platform. The BD Probe Tec TV Q^x Amplified DNA assay detects *T. vaginalis* using strand displacement amplification on the Viper system with performance characteristics similar to the APTIMA assay.[153] Tests for CT and NG can also be performed on the Viper system.

The laboratory diagnosis of trichomoniasis remains challenging particularly in men. Considerations for selection of diagnostic methods should include testing location, analysis time, performance characteristics, and the cost to perform the test.

Herpes Simplex Virus

Herpes simplex virus (HSV) is a double-stranded DNA virus surrounded by a lipid glycoprotein envelope. HSV persists as a latent infection in specific target cells despite the host immune response, often resulting in recurrent disease. Genital herpes is a chronic viral infection. Two serotypes of HSV have been identified, HSV-1 and HSV-2. Most cases of recurrent genital herpes in the United States are caused by HSV-2. However, an increasing proportion of anogenital herpetic infections in some populations is now attributed to HSV-1. HSV-1 is usually associated with oral lesions. The CDC estimates that 776,000 new HSV-2 infections occur each year in the United States. Most genital herpes infections are transmitted by persons unaware of their infections. Up to 90% of persons seropositive for HSV-2 antibody have not been diagnosed with genital herpes. However, many have mild or unrecognized disease, and probably most, if not all, shed virus from the genital area intermittently.

Clinical diagnosis of HSV is insensitive and nonspecific; therefore, the clinical diagnosis of genital herpes should be confirmed by laboratory testing. Many infected persons do not experience the multiple vesicular or ulcerative lesions typical of genital herpes. Both virologic and type-specific serologic tests are used to confirm the diagnosis.[154]

Cell culture and NAATs are the preferred virologic tests for persons who seek diagnosis and treatment of genital ulcers or other mucocutaneous lesions. The sensitivity of viral culture is low, especially for recurrent lesions, and declines rapidly as lesions begin to heal. NAATs for HSV DNA are increasingly used in many laboratories, and several tests are now cleared by the FDA for anogenital specimens.[155,156] NAATs are the preferred tests for detecting HSV in spinal fluid for diagnosis of HSV infection of the CNS and are discussed later in this chapter. Both culture and NAATs should determine whether the infection is due to HSV-1 or HSV-2 because recurrences and asymptomatic shedding are much less frequent for HSV-1 than for HSV-2 genital infections.[157] Failure to detect HSV by culture or NAAT does not rule out HSV infection because viral shedding is intermittent. The use of Tzanck preparations or Pap tests to detect cytologic changes produced by HSV are insensitive and nonspecific and should not be used for genital HSV infections.

Serologic tests detect type-specific and nonspecific antibodies to HSV that develop during the first several weeks to few months after infection and persist indefinitely. Type-specific serologic tests based on antigens specific for HSV-1 (gG1) and HSV-2 (gG2) are commercially available. Because almost all HSV-2 infections are sexually acquired, type-specific HSV-2 antibody indicates anogenital infection. However, the presence of HSV-1 antibody does not distinguish anogenital from orolabial infection. Type-specific HSV serologic assays might be useful in the following scenarios: (1) recurrent genital symptoms or atypical symptoms with negative HSV PCR or culture, (2) clinical diagnosis of genital herpes without laboratory confirmation, and (3) a patient whose partner has genital herpes. HSV serologic testing should be considered for persons presenting for an STI evaluation (especially for those persons with multiple sex partners), persons with HIV infection, and men who have sex with men at increased risk for HIV acquisition.[158]

Human Papillomavirus

Human papillomaviruses are small, double-stranded DNA viruses that infect squamous epithelium, subverting normal cell growth and potentially leading to squamous cell carcinoma (SCC). HPV is not a single virus but a family of more than 150 related viral genotypes that are distinguished based on sequence analysis of the L1 region of the viral genome. Anogenital HPV infections are common in both men and women. It is estimated that more than 24 million men and women in the United States are currently infected with HPV. HPV is an STI; it is most common among sexually active young women ages 15 through 25 years. In one study, cervicovaginal HPV was found in up to 43% of sexually active college women during a 3-year period.[159] Infections, however, are usually transient, and progression to cancer requires persistence of viral infection over several years. The types of HPV that are spread through sexual contact are classified as low or high risk for progression to malignancy, and there are multiple types. Infections with low-risk HPV such as types 6 and 11 can lead to benign genital warts or condyloma acuminata and have a low likelihood of progressing to malignancy. In contrast, high-risk types such as types 16, 18, and 45 are associated with development of SCC of the anogenital region and oropharynx. Currently, there are 14 high-risk (HR) HPV types recognized. The cervix is particularly affected, and worldwide, cervical SCC continues to cause significant morbidity and mortality (5% of cancer deaths).

Productive infections usually result in cytologic and histologic changes, including cellular and nuclear enlargement, nuclear hyperchromasia, and perinuclear halos (koilocytosis). These changes can be identified on a stained smear of cells collected from the cervix (the "Pap smear," developed by Dr. George Papanicolaou in the 1940s) or in a biopsy taken during colposcopy or a loop electrosurgical excision procedure. The Pap smear has been used very successfully to identify women with cervical cancer and, more important, for the detection of precursor lesions, so that biopsy or excision can be performed to remove the lesion earlier in the disease process before metastasis can occur. With the introduction of liquid cytology media and automated cytology processors, the procedure is more appropriately called the Pap test because "smears" are no longer used.

The histologic types of squamous precursor lesions are divided into three categories: mild dysplasia, or cervical intraepithelial neoplasia (CIN1); moderate dysplasia, or CIN2; and severe dysplasia, or carcinoma in situ, or CIN3. In the Bethesda System for Cytologic Classification, squamous precursor lesions are divided into low- and high-grade squamous intraepithelial lesions (LSIL and HSIL). LSIL corresponds with CIN1, and HSIL corresponds to CIN2 and CIN3. Frequently, the cytologic evaluation demonstrates mildly atypical cells that do not meet these criteria and are referred to as atypical squamous cells of undetermined significance (ASCUS); these cells may correspond to an early HPV infection. The prevalence of ASCUS on Pap tests is approximately 5% to 10%, with rates as high as 20% reported in sexually active women.

Screening for cervical cancer with cytology testing has been very effective in reducing cervical cancer in the United States. For many years, the approach was an annual Pap test. In 2000 the Hybrid Capture 2 (HC2) test (Qiagen/Digene) for detection of HR HPV types was approved by the FDA for screening women who had ASCUS detected by the Pap test to determine the need for colposcopy. At the time, the Hybrid Capture 2 test was the only FDA-approved test available. In 2003, the FDA approved expanding the use of this test to include screening preformed in conjunction with a Pap test for women over the age of 30 years, referred to as "co-testing." Co-testing allows women to extend the testing interval to 5 years if both test results are negative.[160] In 2014, the FDA approved the use of an HR HPV test (cobas HPV test, Roche) for primary cervical cancer screening for women older than the age of 25 years, without the need for a concomitant Pap test. When using the HR HPV test as the primary screening test, a Pap test is performed only when specific HR HPV types are detected (HPV-16 and -18 are excluded). Colposcopy is performed without an intervening Pap test in women who test positive for HPV-16 and -18. This algorithm was based primarily on the results of a single large FDA registration study for the cobas HPV test.[161] Interim clinical guidance is available for the use of primary HR HPV testing in cervical cancer screening.[162] However, there is still considerable debate about whether co-testing or HR HPV as a primary screening test is the optimal approach for cervical cancer screening.[163]

Four tests for the detection of HR HPV types have been cleared by the FDA for use in the United States: HC2 test, Cervista HPV HR (Hologic/Gen-Probe), cobas HPV test (Roche), and Aptima HPV test (Hologic/Gen-Probe). In addition, two different FDA-cleared tests to specifically identify HPV types 16 and 18 (Cervista) and types 16 and 18/45 (Aptima) are available. The features of these tests are compared in Table 5.4. All of these tests have been cleared by the FDA for use with ThinPrep PreservCyt liquid-based cytology medium (Hologic) but not with the other commonly used SurePath medium (Becton-Dickinson).

The HC2 test relies on hybridization of a RNA probe to the HPV DNA followed by use of an antibody for capture of the duplex (RNA-DNA) hybrids and then detection with chemiluminescent signal amplification. The test uses a pool of RNA probes spanning the entire genome that are specific for 13 HR HPV types. The specific type is not identified. The test uses a 96-well microtiter plate format and can be performed manually or with the semiautomated Rapid Capture system (Qiagen) for reagent and plate handling. It is also cleared for use on Digene specimen transport media (STM). The HC2 test has been used in several large studies and reproducibly demonstrates high sensitivity of 93% to 96%, but false-positive results occur as a result of cross-reaction with low-risk HPV types.[164]

TABLE 5.4 Features of Food and Drug Administration–Cleared High-Risk Human Papillomavirus Tests

Feature	HC2	Cervista	cobas	Aptima
Technology	Hybrid capture	Cleavase/Invader	Real-time PCR	Transcription-mediated amplification
Target(s)	Multigene	L1, E6, E7	L1	E6, E7 mRNA
LOD and clinical cutoff	5000 copies/reaction	1250–7500 copies/reaction	300–2400 copies/mL	20–240 copies/reaction
Cross reaction with low-risk types	6, 11, 40, 42, 53, 66, 67, 70, 82/82v	67, 70	None reported	26, 67, 70, 82
Internal control	None	Human histone 2 gene	Human β-globin gene	Process
HPV-16/-18 genotyping	No	Yes (separate test)	Yes (integrated)	Yes (separate test also includes type 45)
Automation	Semiautomated and automated	Semiautomated and automated	Automated (cobas 4800)	Automated (Tigris and Panther)
Sample type (volume)	ThinPrep (4 mL), sample transport medium	ThinPrep (2 mL)	ThinPrep (1 mL)	ThinPrep (1 mL)
Prealiquot required	No	No	Yes	Yes
Expanded STI menu	CT/NG	None	CT/NG, HSV 1/2	CT/NG, TV
Primary screening indication	No	No	Yes	No

CT, Chlamydia trachomatis; *HSV*, herpes simplex virus; *LOD*, limit of detection; *NG*, Neisseria gonorrhoeae; *PCR*, polymerase chain reaction; *STI*, sexually transmitted infection; *TV*, Trichomonas vaginalis.

The Cervista HPV HR assay also uses a signal amplification method that is based on cleavase/invader technology and detects the same 13 high-risk types as HC2 test plus type 66. A combination of DNA probes and invader oligonucleotides targeting the L1, E6, and E7 sequences and secondary fluorescently labeled probes are divided into three phylogenetically related reactions that are performed on 96-well microtiter plates. Unlike the HC2 assay, this assay includes an internal control with each reaction. Both assays have detection limits of around 3000 to 5000 genome copies per milliliter. The Cervista HPV HR assay has less cross-reactivity with low-risk types. Studies comparing the two assays demonstrate concordance of 82% to 88%.[165] However, the Cervista test may have poor specificity compared with other tests for HR HPV.[166,167] The Cervista HPV-16 and -18 genotyping test uses the same cleavase/invader technology.

The cobas HPV Test is the first real-time PCR method approved by the FDA for cervical cancer screening.[161] It uses a multiplexed primer and hydrolysis probe assay to individually detect both HPV types 16 and 18 simultaneously with the 12 other HR HPV types using different fluorescently labeled probes. The assay includes detection of the human β-globin gene as an internal control for extraction and amplification adequacy. The cobas 4800 system uses automated bead-based NA extraction and PCR assembly. The sensitivity and specificity is similar to the HC2 and Cervista HR HPV assays. Currently, this is the only FDA-cleared test that has an indication for primary screening.

The Aptima HPV assay targets the viral mRNA for the E6/E7 HPV genes for the 14 HR HPV types. The E6 and E7 genes of HR HPV types are known oncogenes. Proteins expressed from E6-E7 polycistronic mRNA alter cellular p53 and retinoblastoma protein functions, leading to disruption of cell-cycle check points and genome instability. Targeting the mRNA of these oncogenic elements may be a more effective approach to detect cervical disease than detection of HPV genomic DNA.[168] The APTIMA HPV Assay involves three main steps, which take place in a single tube: target capture, target amplification by transcription-mediated amplification, and detection of the amplicons by the hybridization protection assay. The assay also incorporates an internal control to monitor NA capture, amplification, and detection, as well as operator or instrument error. Unlike the internal controls used in the Cervista and Roche assays, it does not assess specimen adequacy (cellularity). An adjunctive test to detect and differentiate HPV type 16 and 18/45 based on the same principle described is also available from Hologic/Gen-Probe.

POINTS TO REMEMBER

Human Papillomavirus

- Most HPV infections in women are transient and progression to cervical cancer requires persistent infection with one of the 14 HR oncogenic types of which 16, 18, and 45 are most common.
- HR HPV testing is recommended in conjunction with the Pap test for cervical cancer screening in women older than 30 years of age (co-testing).
- Co-testing allows women to extend the testing interval to 5 years if both test results are negative because HPV DNA is more sensitive than the Pap test for detection of women with significant cervical lesions.
- Currently, there are four FDA-cleared tests for HR HPV types, each based on a different amplification method: hybrid capture, cleavase/invader, real-time PCR, and transcription mediated amplification.

RESPIRATORY TRACT INFECTIONS

Viruses

The viruses that infect the respiratory tract consist of large and diverse groups that cause disease in humans, and new ones continue to be discovered. The more common viruses that infect humans include influenza A and B, parainfluenza virus (PIV) types 1 to 4, respiratory syncytial virus (RSV), metapneumovirus, adenoviruses (>50 different types), rhinoviruses (>100 different types), and coronaviruses (4 types). The disease spectrum ranges from the common cold to severe life-threatening pneumonia. It can be difficult to differentiate the viral origin based on signs and symptoms alone, and treatment options vary depending on the viral etiology. Infection with these viruses has demonstrated the potential for global public health threats of epidemic and pandemic proportions. The 1918 influenza A pandemic, human deaths caused by infection with avian influenza A H5N1 in 1997,[169] the severe acute respiratory syndrome (SARS) coronavirus outbreak in 2003, the 2009 pandemic caused by the novel multiply-reassorted (swinelike) influenza A H1N1, and the emergence of Middle East respiratory syndrome (MERS) coronavirus in 2012 on the Arabian peninsula are all reminders of the potential threats to human health posed by novel respiratory viruses.[170] Detection of emerging respiratory viruses will require multiple modalities, but molecular methods have been crucial to their discovery and characterization and in the development of diagnostic tools.

Acute respiratory viral infection (1) is a leading cause of hospitalization and death in infants and young children; (2) contributes to problems of asthma exacerbation, otitis media, and lower respiratory tract infection; and (3) contributes to acute disease in immunocompromised and elderly patients. Rapid diagnosis aids in effective treatment (eg, with antiviral medications such as oseltamivir for influenza A virus infection) and management (eg, reduction in inappropriately prescribed antibiotics for viral infection and infection control).

Rapid antigen-based EIAs provide short TATs (minutes) but are hampered by poor diagnostic sensitivity compared with culture methods or molecular assays and low positive predictive values, especially when the prevalence is low. Direct fluorescent antibody (DFA) detection assays for viral antigens on centrifuged cellular material from nasopharyngeal swabs, aspirates, or wash specimens demonstrate greater rates of detection than the rapid antigen assays and provide results in a relatively short time frame of 2 to 4 hours. Detection rates, however, are lower for antigen detection methods than for NAATs.

Cell culture methods, although slower than antigen detection methods, have been considered the gold standard for detection of a wide range of viral pathogens. In recent years, culture methods have been optimized for detection by combining multiple cell lines and improving the TAT from weeks to days through the use of shell-vial spin amplification cultures. Here, the patient's specimen is concentrated onto cells grown on a coverslip, and fluorescent antibody detection is performed after 16 to 24 hours of incubation instead of waiting for the development of a cytopathic effect. Although this has hastened the time to detection, 1 to 2 days is still required along with significant technologist labor, and it is not quite as sensitive as molecular methods of detection.

Molecular detection of respiratory viruses offers several advantages over traditional virologic culture or antigen detection. Most important, analytical sensitivity of molecular assays, primarily using PCR or real-time PCR, is consistently better than that of traditional methods.[171-174] Results from molecular testing are more accurate, and thus the patient benefits from the most appropriate treatment decision; also, infection control practitioners can more effectively implement strategies to prevent or reduce nosocomial transmission. Molecular assays can be designed to detect a wide range of viral pathogens, including viruses that are difficult to culture.

Despite the advantages of NAATs for respiratory virus detection, their adoption by clinical laboratories was initially slow because of the limited capacity of real-time PCR assays for multiplexing. There are numerous LDTs and FDA-cleared real-time PCR assays capable of detecting one to three different viruses in single reactions, but to provide comprehensive coverage for respiratory viruses, a panel of such assays needed to be deployed, an approach not practical for most laboratories.

Currently, there are several FDA-cleared multiplexed respiratory virus panels capable of detecting up to 20 different viral targets, thus providing simplified approaches to comprehensive diagnostics for respiratory viruses.[175] See Table 5.5 for an overview of the important parameters of these respiratory virus panels. These tests are truly transformative for the laboratory in that they can replace the combination of limited multiplex NAATs, antigen detection tests, and culture-based methods that were traditionally used in clinical laboratories to detect respiratory viruses and thus dramatically increase diagnostic yield.

The xTAG Respiratory Viral Panel (RVP) v1 assay is a multiplexed RT-PCR–based assay with fluorescently color-coded microsphere (bead) hybridization for simultaneous detection and identification of 12 respiratory viruses and subtypes.[175] The multiplexed RT-PCR primers amplify conserved regions of the viruses, and the products are labeled with biotin-containing deoxynucleotides in a second target-specific primer extension reaction. The extension product has a proprietary tag sequence incorporated for hybridization to the virus-specific probe on the color-coded bead. After hybridization, phycoerythrin-conjugated streptavidin is bound by the biotin-labeled primer extension products, and the fluorescent signal is quantified on the Luminex xMAP instrument. The instrument contains two lasers: one for identification of the microbe by a color-coded bead and the other for detection of the phycoerythrin signal attached to the primer extension product. The data are recorded as mean fluorescent intensities, and the software analyzes the data and reports the positive results. The assay includes a separate lambda phage amplification control and an MS-2 bacteriophage internal control for extraction and amplification. The original version of the assay was modified to reduce the number steps and analysis time (RVP-Fast), but it does not include the parainfluenza viruses, and it is not as sensitive overall as its predecessor. Because there are a number of post-PCR processing steps in both versions of the test, care must be taken to avoid amplicon cross-contamination and false-positive results.

BioFire Diagnostics developed a PCR instrument called the FilmArray and an associated reagent pouch that together are capable of simultaneously detecting multiple organisms in

TABLE 5.5 Parameters of Different Food and Drug Administration–Cleared Respiratory Virus Panels

Parameter	Luminex xTAG RVPv1	Luminex xTAG RVP Fast	FilmArray	eSensor
Amplicon detection method	Fluorescence-labeled bead array	Fluorescence-labeled bead array	Melting curve analysis	Voltammetry
On-board sample processing	No	No	Yes	No
Post-PCR manipulation	Yes	Yes	No	Yes
Hands-on-time (min)	70	45	3	55
Throughput	High	High	Low	Moderate
Analysis time (hr)	7	4	1	6
Total time to results (hr)*	9	6	1.1	8
Complexity	High	High	Low	Moderate
Pathogens detected	ADV	ADV	ADV	ADV (B/E, C)
	INF A (H1, H3)	INF A (H1, H3)	INF A (H1, H3, 09H1)	INF A (H1, H3, 09H1)
	INF B	INF B	INF B	INF B
	MPV	MPV	MPV	MPV
	RSV (A, B)	RSV	RSV	RSV (A, B)
	RV/EV	RV/EV	RV/EV	RV
	PIV (1, 2, 3)		PIV (1, 2, 3, 4)	PIV (1, 2, 3,)
			COV (HKU1, NL63, 229E, OC43)	
			Bordetella pertussis	
			Chlamydophila pneumoniae	
			Mycoplasma pneumoniae	

*Includes the time required for nucleic acid extraction.
ADV, Adenovirus; *COV*, coronavirus *EV*, enterovirus; *INFA*, influenza A virus; *INFB*, influenza B virus; *MPV*, metapneumovirus; *PCR*, polymerase chain reaction; *PIV*, parainfluenza virus; *RSV*, respiratory syncytial virus; *RV*, rhinovirus.

the same sample. The FilmArray pouch contains freeze-dried reagents to perform NA purification; reverse transcription; and nested, multiplex PCR followed by high-resolution melting analysis. The FilmArray Respiratory Panel (RP) was designed for simultaneous detection and identification of 17 viral and 3 bacterial respiratory pathogens (see Table 5.5). The test is initiated by loading water and an unprocessed patient nasopharyngeal swab specimen mixed with lysis buffer into the FilmArray RP pouch. The pouch is then placed into the FilmArray instrument. The software has a simple interface that requires only identification of the specimen and pouch barcode to initiate a run. Multiplexed two-stage RT-PCR followed by high-resolution melting analysis of the target amplicons is used to detect each of the panel analytes.[176] Results are reported in an hour; currently, the instrument is designed to test a single sample per run, though multiple instruments can be linked. Because it is a completely closed system, false-positive results caused by amplicon cross-contamination are not an issue.

The eSensor system (GenMark Dx) uses electrochemical-detection-based DNA microarrays.[177] These microarrays are composed of a printed circuit board consisting of an array of 76 gold-plated electrodes. Each electrode is modified with a multicomponent, self-assembled monolayer that includes presynthesized oligonucleotide capture probes. NA detection is based on a sandwich assay principle. Signal and capture probes are designed with sequences complementary to immediately adjacent regions on the corresponding target DNA sequence. A three-member complex is formed between the capture probe, target sequence, and signal probe based on sequence-specific hybridization. This process brings the 5′ end of the signal probe containing electrochemically active ferrocene labels into close proximity to the electrode surface. The ferrous ion in each ferrocene group undergoes cyclic oxidation and reduction, leading to loss or gain of an electron, which is measured as current at the electrode surface using alternating-current voltammetry. Higher-order harmonic signal analysis also facilitates discrimination of ferrocene-dependent faradic current from background capacitive current.

The eSensor cartridge consists of a printed circuit board, a cover, and a microfluidic component. The microfluidic component includes a diaphragm pump and check valves in line with a serpentine channel that forms the hybridization channel above the array of electrodes. The eSensor instrument consists of a base module and up to three cartridge-processing towers, each with eight slots for cartridges. The cartridge slots operate independently of each other. The throughput of a three-tower system can reach 300 tests in 8 hours. A respiratory pathogen panel for the eSensor system that detects 14 different types and subtypes of respiratory viruses (see Table 5.5) is FDA cleared.[178] Because this test requires post-PCR manipulations of the sample, care must be taken to avoid false-positive tests caused by amplicon cross-contamination.

The Verigene system (Nanosphere) uses PCR amplification and gold nanoparticle-labeled probes to detect target NA hybridized to capture oligonucleotides arrayed on a glass slide. Silver signal amplification is then performed on the gold nanoparticle probes that are hybridized to the captured DNA targets of interest. The Verigene reader optically scans the slide for silver signal, processes the data, and produces a qualitative result. A test for detection of influenza A virus, influenza A virus subtype H3, influenza A virus subtype 2009 H1N1, influenza B, and RSV subtypes A and B is cleared by the FDA for the Verigene system.[179] The system is capable of much higher-order multiplexing and a respiratory panel that detects 13 viral and 3 bacterial targets has been developed and is available in the United States as an RUO product.

Molecular testing for respiratory viruses will likely continue to include tests designed to detect a limited number of viruses of particular importance (eg, influenza A and B viruses and RSV), as well as tests that detect a broad array of viruses because there are clinical needs for both types of tests. The use of comprehensive respiratory virus panels greatly increases diagnostic yield and the ability to detect mixed viral infections. However, the clinical significance of mixed infections is not well documented or understood. In addition, there are test options that range from simple, "sample in answer out" systems to complex tests that require multiple manual steps, meeting different niches for various clinical laboratory settings. In fact, point-of-care molecular testing for respiratory viruses is now possible with the recent development of a CLIA-waived test for influenza A and B viruses (Alere). It delivers results in 15 minutes and can be performed by nonlaboratory personnel, and its performance characteristics are similar to those of NAATs performed in laboratories.[180]

Mycobacterium tuberculosis

Mycobacterium tuberculosis causes a wide range of clinical infections, including pulmonary disease; miliary tuberculosis; meningitis; pleurisy, pericarditis, and peritonitis; GI disease; genitourinary disease, and lymphadenitis. *M. tuberculosis* infection was in steady decline in the United States with an all-time low in the late 1990s, when the number of reported cases began to increase.[181] This resurgence was related to the AIDS epidemic, homelessness, and a decreased focus on tuberculosis control programs. The infection rate continues to rise in foreign-born persons as the result of immigration from countries with a high prevalence of *M. tuberculosis* infection. This increase in *M. tuberculosis* infection has focused considerable attention on the development of assays for its rapid diagnosis; molecular methods are at the center of this effort.

Conventional tests for laboratory confirmation of tuberculosis include acid-fast bacilli (AFB) smear microscopy, which can produce results in 24 hours, and culture, which requires 2 to 6 weeks to produce results.[182,183] Although rapid and inexpensive, AFB smear microscopy is limited by its poor sensitivity (45%–80% with culture-confirmed pulmonary tuberculosis cases) and its poor positive predictive value (50%–80%) for tuberculosis in settings in which nontuberculous mycobacteria are commonly isolated.[183-185]

Compared with AFB smear microscopy, the added value of NAATs include (1) their greater positive predictive value (>95%) than AFB smear-positive specimens when nontuberculous mycobacteria are common and (2) their ability to rapidly confirm the presence of *M. tuberculosis* in 60% to 70% smear-negative, culture-positive specimens.[183-187] Compared with culture, NAATs can detect the presence of *M. tuberculosis* in specimens weeks earlier for 80% to 90% of patients suspected to have pulmonary tuberculosis ultimately confirmed by culture.[184,186,187] These advantages can impact patient care and tuberculosis control efforts, such as by avoiding unnecessary contact investigations or respiratory isolation for patients whose AFB smear-positive specimens do not contain *M. tuberculosis*.

The CDC recommends that NAATs be performed on at least one (preferably the first) respiratory specimen from each patient suspected of pulmonary tuberculosis for whom a diagnosis of tuberculosis is being considered but has not yet been established and for whom the test result would alter case management or tuberculosis control activities.[188] NAATs can also be used to inform the decision to discontinue airborne infection isolation precautions in health care settings.[189,190] NAATs do not replace the need for culture; all patients suspected of tuberculosis should have specimens collected for mycobacterial culture.[188]

Currently, two FDA-approved NAATs are available for direct detection of *M. tuberculosis* in clinical specimens: the Amplified *Mycobacterium tuberculosis* Direct test (MTD test, Hologic/Gen-Probe) and the Xpert MTB/RIF assay (Cepheid). The MTD test is based on transcription-mediated amplification of ribosomal RNA and can be used to test both AFB smear-positive and smear-negative respiratory specimens. The Xpert MTB/RIF assay uses real-time PCR to detect the DNA of *M. tuberculosis* and the mutations in the *rpoB* gene associated with rifampin resistance in sputum specimens. Rifampin resistance most often coexists with isoniazid resistance so detection of rifampin resistance serves as a marker for potentially multidrug-resistant *M. tuberculosis* strains. Similar to the other assays developed by Cepheid, the Xpert MTB/RIF assay uses a disposable cartridge that automates the NA extraction, target amplification, and amplicon detection in conjunction with the GeneXpert Instrument System. Sensitivity and specificity of the Xpert MTB/RIF assay for detection of *M. tuberculosis* appear to be comparable to other FDA-approved NAATs for this use. Sensitivity of detection of rifampin resistance was 95% and specificity 99% in a multicenter study using archived and prospective specimens from subjects suspected of having tuberculosis.[191]

Because the prevalence of rifampin resistance is low in the United States, a positive result indicating a mutation in the *rpoB* gene should be confirmed by rapid DNA sequencing for prompt reassessment of the treatment regimen and followed by growth-based drug susceptibility testing.[190] The CDC offers these services free of charge.

Bordetella pertussis

The genus *Bordetella* is composed of eight species, four of which can cause respiratory disease in humans: *B. bronchiseptica*, *B. holmesii*, *B. parapertussis*, and *B. pertussis*. Whooping cough, or pertussis, is a highly contagious respiratory disease caused by *B. pertussis*. Despite widespread childhood vaccination, more than 28,660 cases were reported in the United States in 2014. (http://www.cdc.gov/pertussis/downloads/pertuss-surv-report-2014.pdf). The reported cases represent only the "tip of the iceberg" with an estimated 800,000 to 3.3 million cases occurring in the United States annually. Although pertussis occurs most often in children younger than 1 year of age, the incidence in older children has increased substantially in recent years. Adolescents and adults, in whom immunity wanes several years after prior infection or vaccination, transmit the organism to susceptible infants. Pertussis in older children and adults is usually characterized by prolonged cough without the inspiratory whoop or posttussive vomiting that typically is observed in infants.

B. parapertussis may be responsible for up to 20% of pertussis-like disease, more often in young children.[192] Illness is generally milder than that caused by B. pertussis. B. bronchiseptica is an infrequent cause of disease in humans, usually occurring in immunocompromised individuals. Cases usually have exposure to farm animals or pets, which serve as the natural hosts for B. bronchiseptica.[193] B. holmesii is the most recently recognized species to be associated with pertussis-like illness in humans.[194] All four species play a significant role in human respiratory disease, and they should be considered in the design of NAATs for patients with pertussis-like disease.[195]

The laboratory diagnosis of pertussis has been fundamentally transformed in the past 2 decades. Culture and DFA staining of nasopharyngeal secretions are now largely replaced by NAATs in clinical laboratories. Although culture is specific for diagnosis, it is relatively insensitive. The fastidious nature and slow growth of the B. pertussis make it difficult to isolate. Although DFA staining can provide rapid results, it is neither sensitive nor specific. Serologic testing can be useful late in the disease when the organism may not be detectable by culture or NAAT and in the investigation of outbreaks, but the tests are not standardized, so the results may be difficult to interpret.

NAATs are important tools for the diagnosis of pertussis with enhanced sensitivity and rapid turnaround compared with culture and are now considered standard of care, but they can give both false-positive and false-negative results as discussed later. A variety of LDTs primarily based on real-time PCR with different performance characteristics are deployed in clinical laboratories. Currently, there are only two FDA-approved NAATs for B. pertussis, one a stand-alone test based on loop-mediated amplification (Meridian Biosciences) and the other as part of a respiratory panel (BioFire Diagnostics).

A number of different gene targets have been used in NAATs for Bordetella spp., some of which are shared among the different species.[195] Most NAATs are based on detection of multicopy insertion sequences (IS), which can increase the sensitivity of the tests. IS481 is the most validated target for B. pertussis, but it can also be found in B. holmesii and B. bronchiseptica; therefore, tests based on this target alone are of limited value, particularly when used in an outbreak setting. IS1001 is found in B. parapertussis and B. bronchiseptica but not in B. holmesii. IS1002 is found in B. pertussis and parapertussis but not in B. holmesii or bronchiseptica. Multiplex PCR targeting all three ISs may allow detection and differentiation of the major pathogens that infect humans, B. pertussis, B. parapertussis, and B. holmesii.

A number of assays based on single copy gene targets have also been described.[195] The promoter region of the pertussis toxin operon is often used in diagnostic assays. However, it is also present in B. parapertussis and B. bronchiseptica, but because of mutations in the promoter region, it is not expressed. It is not found in B. holmesii. The mutations in the promoter region found in B. parapertussis and B. bronchiseptica can be exploited in real-time PCR assays that use post-amplification analysis by melting temperature to distinguish the amplicons from the different species. Pertactin, filamentous hemagglutinin, adenylate cyclase, REC A, flagellin, and BP3385 gene sequences are also shared among the different species. BP283 and BP485 gene targets are reported to be specific for B. pertussis.[196] With the exception of the pertussis toxin gene, none of the other single gene targets has been extensively validated in diagnostic assays.

The positive predictive value of NAATs remains their biggest challenge. IS481-based tests will detect B. holmesii and B. bronchiseptica, which for clinical and epidemiologic purposes are considered biological false-positive results. Environmental contamination with B. pertussis DNA in patient clinics has been identified as a source of pseudo-outbreaks of disease.[197] The positive predictive value of NAATs for B. pertussis can be increased by amplifying gene targets not shared by other species, using multiplex assays or a two-tiered approach to confirm positives, creating an indeterminate range for assays that target multicopy ISs, segregating "clean" and "dirty" areas in the clinic and the laboratory, and testing only symptomatic patients.[195] Further guidance for health care professionals for the use and interpretation of NAATs for B. pertussis can be found on the CDC's website (http://www.cdc.gov/pertussis/clinical/diagnostic-testing/diagnosis-pcr-bestpractices.html).

BLOODSTREAM INFECTIONS

Positive Blood Culture Identification

One of the most important functions of clinical microbiology laboratories is the detection of bloodstream infections. Using conventional grow-based systems, when the blood culture system signals positive, typically within 12 to 72 hours of incubation for most pathogens, the blood culture broth is Gram stained and then subcultured to solid medium. When colonies grow on this medium, identification and antimicrobial susceptibility tests are performed. This typically takes an additional 24 to 48 hours to complete after the blood culture signals positive. Direct inoculation of conventional identification and susceptibility tests using positive blood culture broth can reduce the time required to obtain results by eliminating the subculture to solid medium, but this practice is not FDA approved for automated identification and susceptibility test systems.

A variety of NA-based tests have been developed to expedite identification of organisms in positive blood cultures. FDA-approved tests include peptide NA fluorescent in situ hybridization (PNA FISH) probes, real-time PCR assays for detection of single or limited numbers of pathogens, and high-order multiplex blood culture identification panels based on nested PCR and gold nanoparticle microarrays.[198] Matrix-assisted laser desorption ionization time-of-flight mass spectrometry (MALDI-TOF) uses proteomics rather than genomics to identify pathogens based on the mass spectrum of proteins found in the microorganisms. It has been applied to the direct identification of microorganisms from

TABLE 5.6 Key Features of Rapid Blood Culture Identification Methods

Feature	Nested Multiplex PCR FilmArray	Gold Nanoparticle Microarray	PNA FISH	MALDI-TOF MS
Inclusivity*	+++	+++	+	++++
Hands on time	2 min	5 min	5 min	30 min
Time to result	1 hr	2.5 hr	30 min	35 min
Technical complexity	+	++	++	+++
Antibiotic resistance genes (n)	Yes (3)	Yes (9)	No	No
Reagent cost	$$$$	$$$	$$	$

*Relative ability to identify common bloodstream pathogens.
MALDI-TOF MS, Matrix-assisted laser desorption ionization time-of-flight mass spectrometry; *PCR,* polymerase chain reaction; *PNA FISH,* peptide nucleic acid fluorescent in situ hybridization.

positive blood culture bottles.[199] The key features of these methods are listed in Table 5.6.

PNA FISH probes are DNA probes in which the negatively charged sugar phosphate backbone is replaced by a noncharged peptide backbone. This results in rapid binding to DNA targets because there is no electrostatic repulsion with the target.[200] PNA FISH probes are available for rapid identification of *Staphylococcus aureus* and coagulase-negative staphylococci; *Enterococcus faecalis* and other enterococci; *Escherichia coli*, *Klebsiella pneumoniae*, and *Pseudomonas aeruginosa*; and *Candida albicans*, and/or *C. parapsilosis*, *C. tropicalis* and *C. glabrata* and/or *C. krusei* (AdvanDx). The most recent protocol involves approximately 5 minutes of hands-on time and 30 minutes for results. Access to a fluorescence microscope with a special filter is required to read the stained slides.

A number of laboratory-developed NAATs for direct identification of single or a limited number of organisms directly from blood cultures have been described, but in general, they have not gained widespread acceptance in clinical laboratories. The number of commercially available assays for this application is also limited. *S. aureus* bacteremia requires prompt microbiologic diagnosis and appropriate antibiotic administration. Vancomycin is the standard treatment for suspected *S. aureus* bacteremia because in most centers, 50% or more of isolates are methicillin-resistant *S. aureus* (MRSA); however, it is less effective than methicillin for treating methicillin-susceptible *S. aureus* (MSSA) strains. Therefore, it is not surprising that many of the methods for rapid identification of bloodstream pathogens focused on the differentiation of MRSA from MSSA. Two FDA-approved real-time PCR assays for detection and differentiation of MRSA and MSSA directly from positive blood cultures are the BD GeneOhm StaphSR (BD Diagnostics) and the Xpert MRSA/SA BC (Cepheid) assays.[201,202] Each assay has limitations in accurately differentiating MRSA from MSSA largely because of assay design and selection of gene targets. See this chapter's section on antibacterial drug resistance for more details.

Two high-order multiplex assays have been approved by the FDA for identification of microorganisms from blood culture bottles, the Verigene Gram-Positive and Gram-Negative Blood Culture Tests (Nanosphere) and the FilmArray Blood Culture Identification (BCID) Panel (BioFire Diagnostics).[203-205] The organisms and antibiotic resistance genes included in each of the panels are listed in Box 5.1.

The FilmArray BCID panel uses the same technology as the respiratory panel described previously to detect 24 genus- or species-specific targets including gram-positive and gram-negative bacteria, *Candida* spp., and thee antibiotic resistance genes in approximately 1 hour.[205] This panel identifies from 80% to 90% of all positive blood cultures and provides important information about resistance to methicillin in staphylococci, vancomycin resistance in enterococci, and carbapenemase production in enteric gram-negative rods.

The Verigene BCID panels use Nanogold microarray technology to identify organisms from positive blood culture bottles without NA amplification. The gram-positive or gram-negative panel (or both) is chosen based on the results of the Gram stain that is performed when the bottle signals positive. The gram-positive panel detects 12 genus- or species-specific targets and 3 antibiotic resistance genes, and the gram-negative panel detects 8 genus- or species-specific targets and 9 antibiotic resistance genes in about 2.5 hours with minimal hands-on time. The gram-positive panel detects *mecA*, *vanA*, and *vanB* genes, and the gram-negative panel detects six different β-lactamase genes. A Verigene yeast blood culture panel is in development that will include *C. albicans*, *C. dubliniensis*, *C. glabrata*, *C. krusei*, *C. parapsilosis*, *C. tropicalis*, *C. gattii*, and *Cryptococcus neoformans*.

The FilmArray and Verigene panels provide comprehensive approaches to rapid identification of the vast majority of blood pathogens and important information about susceptibility of these pathogens to antibiotics. When coupled with active antimicrobial stewardship program interventions, the results of these tests will likely have a positive impact on the clinical outcomes of patients with sepsis.[206]

Direct Pathogen Detection

The methods discussed in the preceding section provide opportunities to expedite the identification of microorganisms when a blood culture signals positive and as such represent significant advances. However, blood cultures require 1 to 5 days of incubation before they are positive, and this timeline is inconsistent with the need to obtain rapid answers to inform treatment decisions in patients with sepsis. Direct detection of pathogens in blood without the need for culture would be ideal but presents a number of challenges. Specimen preparation, enrichment for pathogen DNA, and integration of the front-end specimen preparation with a back-end molecular analysis that identifies virtually all pathogens are major obstacles to success. Also, highly sensitive molecular methods for direct detection of microbial DNA in blood

BOX 5.1 Comparison of Organisms and Antibiotic Resistance Genes Included in the FilmArray and Verigene Blood Culture Identification Panels

FilmArray

Gram Positive
Enterococcus spp.
Listeria monocytogenes
Staphylococcus spp.
  S. aureus
Streptococcus
  S. agalactiae
  S. pneumoniae
  S. pyogenes

Gram Negative
Acinetobacter baumannii
Haemophilus influenzae
Neisseria meningitidis
Pseudomonas aeruginosa
Enterobacteriaceae
  Enterobacter cloacae complex
  Escherichia coli
  Klebsiella pneumoniae
  K. oxytoca
  Proteus spp.
  Serratia marcescens

Yeast
Candida albicans
C. glabrata
C. krusei
C. parapsilosis
C. tropicalis

Antibiotic Resistance Genes
mecA
vanA/B
bla_{KPC}

Verigene

Gram Positive
Staphylococcus spp.
  S. aureus
  S. epidermidis
  S. lugdunensis
Streptococcus spp.
  S. anginosus group
  S. agalactiae
  S. pneumoniae
  S. pyogenes
Enterococcus faecalis
E. faecium
Listeria spp.

Antibiotic Resistance Genes
mecA
vanA
vanB

Gram Negative
E. coli/Shigella
Klebsiella pneumoniae
K. oxytoca
Pseudomonas aeruginosa
Acinetobacter spp.
Proteus spp.
Citrobacter spp.
Enterobacter spp.

Antibiotic Resistance Genes
bla_{KPC}
bla_{NDM}
bla_{CTx-M}
bla_{VIM}
bla_{IMP}
bla_{OxA}

---

present significant challenges for validation given the known limitations of the sensitivity of culture, the current gold standard.

The Roche SeptiFast system has been available longer than any other method for direct detection of microorganisms in the blood.[207] It uses real-time PCR performed on a LightCycler instrument that targets the ribosomal internal transcribed spacer region. The target DNA is amplified in three parallel, multiplex, real-time PCR assays for detection of 10 gram-positive, 10 gram-negative, and 5 fungal pathogens. Melting curve analysis is used to reliably differentiate the pathogens. The assay is technically complex, requires large amounts of hands-on time, and has an analysis time of about 6 hours. Several evaluations have reported lower sensitivities and specificities in clinical settings when compared with blood cultures.[208-211]

Another molecular approach to direct detection of pathogens in blood and other body fluids has been developed by Ibis Biosciences, a subsidiary of Abbott Molecular. This method combines broad-range PCR with electrospray ionization time-of-flight mass spectrometry (PCR/ESI-MS). The technology has been described in great detail elsewhere,[212] but briefly, it works by coupling conserved-site PCR reactions that are able to amplify shared genes from diverse microorganisms to ESI-MS. Measurement of amplicon mass provides species-specific signatures that can be matched to known signatures in a database. After PCR amplification, ESI-MS analysis is performed on the amplicon mixtures, and the A, G, C, and T base compositions are compared with a database of known base compositions derived from existing sequence data. This technology accurately identifies diverse microbes from blood, other body fluids, and tissues in research settings.[213] The PCR/ESI-MS system was evaluated for the direct detection of bacteria and Candida spp. in the blood of 331 patients with suspected bloodstream infections and was found 83% sensitive and 94% specific compared with culture.[214] Replicate testing of the discrepant samples by PCR/ESI-MS resulted in increased sensitivity (91%) and specificity (99%) when confirmed infections were considered true positives. Ibis/Abbott has developed an automated and integrated platform for PCR/ES-MS analysis of clinical samples called IRIDICA. A bloodstream infection assay for this new platform that identifies up to 500 different organisms and four antibiotic resistance genes is currently in clinical trials.

Another novel technology that shows promise for direct detection of pathogens in blood is T2 magnetic resonance (T2MR)—based biosensing.[215] T2 Biosystems has developed an assay to directly identify *Candida* spp. in the blood of patients with suspected candidemia. In this assay, the *Candida* cells are lysed, their DNA is amplified by PCR, and the amplified product is detected directly in the whole-blood matrix by amplicon-induced agglomeration of superparamagnetic nanoparticles. Nanoparticle clustering yields changes in the T2 (spin-spin) relaxation time, making it detectable by magnetic resonance. A small portable T2MR instrument for rapid and precise T2 relaxation measurements has been designed for standard PCR tubes. The T2Dx instrument automates all of the steps in the assay with approximately 5 minutes of hands on time, and results are available within 3 to 5 hours. The T2 *Candida* panel is FDA approved, and the clinical trial data showed an overall sensitivity and specificity of 91.1% and 99.4%, respectively.[216] T2 Biosystems has a panel in development for direct detection of bacteria in blood.

CENTRAL NERVOUS SYSTEM

Herpes Simplex Virus

Herpes simplex virus types 1 and 2 produce various clinical syndromes involving the skin, eye, CNS, and genital tract. Although NA testing has been used to detect HSV DNA in all of these clinical manifestations, this discussion focuses on the use of HSV PCR for the diagnosis of CNS infections because NA amplification testing is widely viewed as the standard of care for diagnosis.

Herpes simplex virus causes both encephalitis and meningitis. In adults, whereas HSV encephalitis is usually attributable to infection with HSV type 1, HSV meningitis is most commonly caused by HSV type 2. HSV encephalitis is a severe infection with high morbidity and mortality; treatment with acyclovir reduces the mortality rate from approximately 70% in those with untreated infection to 19% to 28%. Neurologic impairment is common ($\approx 50\%$) in those who survive.[217] HSV encephalitis may reflect primary infection or reactivation of latent infection. HSV meningitis is usually a self-limited disease that resolves over the course of several days without therapy. In some patients, the disease may recur as a lymphocytic meningitis over a period of years.[218]

Neonatal HSV infection occurs 1 in 3500 to 1 in 5000 deliveries in the United States. It is most commonly acquired by intrapartum contact with infected maternal genital secretions and is usually HSV type 2. In newborns, three general presentations of the disease are known: skin, eye, and mouth disease, which account for approximately 45% of infections; encephalitis, which accounts for 35%; and disseminated disease, which accounts for 20%. Because disseminated disease is often associated with neurologic disease, CNS disease occurs in about 50% of newborns with neonatal HSV infection.

Herpes simplex virus encephalitis cannot be distinguished clinically from encephalitis caused by other viruses such as West Nile Virus, St. Louis encephalitis virus, and Eastern equine encephalitis virus. Historically, the gold standard for the diagnosis of HSV encephalitis required brain biopsy with identification of HSV by cell culture or immunohistochemical staining. This approach provided high sensitivity (99%) and specificity (100%), but it required an invasive procedure, and several days elapsed before results were available.

Viral culture of cerebrospinal fluid (CSF) has a sensitivity of less than 10% for the diagnosis of HSV encephalitis in adults. Tests that measure HSV antigen or antibody in CSF have diagnostic sensitivities of 75% to 85% and diagnostic specificities of 60% to 90%.[217] Because of the limitations of conventional methods, there was interest in assessing the clinical utility of PCR for the detection of HSV DNA from CSF of patients with encephalitis. The two largest studies compared HSV PCR on CSF specimens versus brain biopsy in patients with suspected HSV encephalitis.[219,220] The sensitivity and specificity of PCR were greater than 95%, and the sensitivity of HSV PCR did not decrease significantly until 5 to 7 days after the start of therapy. PCR is positive early in the course of illness, usually within the first 24 hours of symptoms, and in some individuals, HSV DNA can persist in the CSF for weeks after therapy is initiated.

The clinical utility of HSV PCR has also been established for the diagnosis of neonatal HSV infection. In one study, HSV DNA was detected in the CSF of 76% (26 of 34) of infants with CNS disease; 94% (13 of 14) of those with disseminated infection; and 24% (7 of 29) of infants with skin, eye, or mouth disease.[221] The persistence of HSV DNA in the CSF of newborns for longer than 1 week after therapy initiation is associated with a poor outcome.[222] Based on these findings, detection of HSV DNA in CSF by PCR has become the standard of care for the diagnosis of HSV encephalitis and neonatal HSV infection. In newborns with disseminated disease, HSV DNA may be detected in serum or plasma specimens and is useful diagnostically in newborns if it is not possible to do a lumbar puncture. Although the sensitivity of HSV PCR is high, it is not 100%, so a negative PCR test result may not rule out neurologic disease caused by HSV, particularly if the pretest probability is high. In this situation, it is important to consider repeat testing.

As with HSV encephalitis, HSV meningitis cannot be distinguished clinically from other viral meningitides, although recurrence of viral meningitis is a strong clue that HSV may be the etiologic agent. Unlike HSV encephalitis, HSV meningitis has not been the subject of large studies evaluating the clinical utility of PCR for diagnosis. Nonetheless, because the sensitivity of viral culture of CSF specimens is only 50%, HSV PCR of CSF is commonly used in the evaluation of meningitis and has been described as accurate in anecdotal reports.[223]

Several molecular tests for the detection of HSV DNA from genital specimens have been cleared by the FDA, but only one, the Simplexa HSV 1 and 2 Direct Kit (Focus Diagnostics), has been cleared for use with CSF specimens. Several companies provide primers and probes as ASRs, which can be used as components in LDTs.

Molecular tests are often designed to detect HSV types 1 and 2 with equal sensitivity. Distinguishing between HSV types 1 and 2 may not be necessary because the clinical management of CNS disease is the same for both infections. Primers used for the detection of HSV DNA commonly target the polymerase, glycoprotein B, glycoprotein D, or thymidine kinase genes. It is important that the primers not amplify DNA from other herpesviruses that are associated with neurologic disease; these include cytomegalovirus, varicella zoster virus, human herpes virus type 6, and Epstein-Barr virus.

Herpes simplex virus PCR assays need low detection limits (several hundred copies per milliliter of specimen) to be useful in evaluating neurologic disease. This is particularly true for the diagnosis of meningitis, in which CSF concentrations of DNA tend to be lower than those seen with encephalitis. HSV neurologic disease rarely occurs in individuals without an increased CSF white blood cell count or protein concentration. Caution should be exercised in applying this generalization to immunocompromised individuals because they may not mount a typical inflammatory response to HSV infection. Although HSV PCR of CSF specimens is clearly the gold standard for the diagnosis of neurologic disease, results should be interpreted with caution because neither sensitivity nor specificity is 100%. Test results should always be interpreted within the context of the clinical presentation of the patient. If results do not correlate with the clinical impression, repeat testing should be performed 3 to 7 days later because initial negative PCR results can occur in a small but notable number of patients with confirmed HSV encephalitis.

Enterovirus

The enteroviruses (EVs) are a diverse group of single-stranded RNA viruses belonging to the *Picornaviridae* family. Currently, human EVs are divided into seven species: EV A to D and rhinoviruses A to C. The EV species A to D contain viruses formerly referred as coxsackieviruses, EVs, polioviruses, and echoviruses. The genus Parechovirus (PeV) comprises 16 different serotypes that were originally thought to be echoviruses. Although the genomic organization is similar to EVs, the origin of PeVs is uncertain. Numerous clinical presentations are seen with EV and PeV infections, including acute aseptic meningitis, encephalitis, exanthems, conjunctivitis, acute respiratory disease, GI disease, myopericarditis, and sepsis-like syndrome in neonates. Diagnoses typically are based on clinical presentation and NAATs.

Virus culture methods have several drawbacks, including the requirement to inoculate multiple cell lines because no single cell line is optimal for all EV types, the inability to grow some EV types in cell culture, the limited diagnostic sensitivity of cell culture (65%–75%), and the long TAT of 3 to 8 days for those EVs that do grow in cell culture.[224] The long TAT for culture means that results are rarely available in a time frame to influence clinical management. NAATs offer several important advantages over cell culture, including improved sensitivity and TAT. As a result, NA testing is considered the new gold standard for the diagnosis of aseptic meningitis and neonatal sepsis syndrome caused by EV and PeV infections.

Two methods are used for the detection of enteroviral RNA from clinical specimens: RT-PCR and NA sequence–based amplification. The primers used in clinical testing most commonly target the highly conserved 5′ URT of the genome and detect polioviruses and EVs.[225] These primers do not detect parechoviruses, although these viruses can cause aseptic meningitis. In general, molecular assays have good detection limits ranging from 0.1 to 50 tissue culture infectious doses ($TCID_{50}$) per test. The assays are quite specific, but sequence similarities may allow amplification of some types of rhinoviruses.[226,227] Currently, two tests for the detection of EVs from CSF specimens have been cleared by the FDA: the NucliSENS EasyQ Enterovirus (bioMérieux) and Xpert EV (Cepheid). However, the NucliSENS Easy Q EV assay is no longer commercially available. The Xpert EV test has a sensitivity of 97% and a specificity of 100% for the diagnosis of enteroviral meningitis.[228] The Xpert test has the advantage of being very simple to perform: The specimen and reagents are added to a cartridge, which is inserted into the instrument. NA extraction, amplification, and detection are fully automated, and results are available within about 2.5 hours. The system permits random access, which allows for on-demand testing.

Nucleic acid testing for the diagnosis of enteroviral infection has been evaluated in a variety of clinical studies, with testing showing sensitivities equal to or greater than that of cell culture, a high specificity, and faster TATs than cell culture. Several studies have suggested that the use of molecular methods for the diagnosis of enteroviral infection in infants and pediatric patients can lead to an overall cost savings by reducing the use of antibiotics and imaging studies.[229-231] To maximize the benefits for patient care and cost savings, testing should be available on a daily basis.

As mentioned earlier, many EV molecular assays detect rhinoviruses, and most detect polioviruses. These two factors can lead to unexpected and misleading positive results when respiratory or stool specimens are tested. The diagnosis of EV meningitis should be based on testing of CSF specimens, and sepsis syndrome is best diagnosed in neonates by testing serum, plasma, or CSF samples.

GASTROENTERITIS

Clostridium difficile

Clostridium difficile is a gram-positive spore-forming anaerobic bacillus that is frequently found in the stool flora of healthy infants but is rarely found in the stool flora of healthy adults and children older than 12 months. The organism is acquired by ingesting spores, which survive the gastric acid barrier and germinate in the colon. Alteration of the intestinal flora with the use of antibiotics facilitates colonization of the intestinal tract. After being colonized, patients may develop symptoms of diarrhea or colitis. Most strains of *C. difficile* make two toxins: toxin A and toxin B; the regulatory proteins TcdR and TcdC control expression of the toxin A (*tcdA*) and B (*tcdB*) genes. These toxins are responsible for symptomatic disease; strains that lack these toxin genes do not cause diarrhea or colitis. Toxin B may be more important for production of disease than toxin A.[232] Detection of these toxins or of their activity is essential in diagnostic tests for *C. difficile*–associated disease. An additional toxin, the binary toxin, has been described in some strains of *C. difficile*, and recent reports have suggested that strains encoding the binary toxin (CDT) have a deletion in the *tcdC* gene, leading to overexpression of toxins A and B (ribotype 027), and are causing outbreaks of more severe disease.[233]

C. difficile is a frequent cause of antibiotic-associated diarrhea and colitis both in the hospital and community. In hospitals, the risk of infection increases with the length of hospital stay, and use of antimicrobial therapy greatly increases the likelihood of acquiring *C. difficile* colitis. *C. difficile* causes a spectrum of disease ranging from asymptomatic carrier state to fulminant, relapsing, and fatal colitis. Diarrhea may be mild to severe. Pseudomembranous colitis is a classic presentation of *C. difficile* disease, and toxic megacolon may also be seen. Although clindamycin, penicillins, and cephalosporins have commonly been associated with disease, almost all antibiotics can cause similar disease.

Various non-NA tests are available for the diagnosis of *C. difficile* infection. Culture of the organism alone is not helpful in the diagnosis because there needs to be confirmation that the organism produces toxins. The cell culture cytotoxicity test neutralization assay (CCNA), which detects the cytopathic effect of toxin B, is considered the gold standard for the diagnosis of clinically important *C. difficile* infection. The test is highly sensitive and specific but is labor intensive and technically demanding. The TAT of 1 to 3 days limits its clinical utility.[234] The most commonly used tests are EIAs and lateral flow devices that detect toxin A, toxin B, or both. Overall these tests have lower sensitivities (45%–95%) and specificities (75%–100%) than the cytotoxicity test. In general, EIAs that detect both toxins A and B are preferred because some strains may not produce toxin A. An alternative testing approach is detection of the common antigen glutamate dehydrogenase (GDH). The test does not distinguish between toxigenic and nontoxigenic strains and cannot be used alone for the diagnosis of *C. difficile* disease. A positive result needs to be confirmed with the cytotoxicity test, a toxin EIA, or a NAAT for the detection of toxin B gene. The GDH test is a useful screening test because it has a high negative predictive value. One study evaluated a two-step approach using the GDH test as the initial screen followed by a CCNA for antigen-positive specimens to confirm the presence of toxin. A negative antigen test result was more than 99% predictive of a negative CCNA.[235] A limitation of this approach is the delay in obtaining a result because of the long TAT of the CCNA test. More recently, multistep algorithms using GDH, toxin EIA, and NAATs have been deployed in clinical laboratories.[236]

In view of the limitations of traditional methods, molecular tests are a good alternative for the diagnosis of *C. difficile* infection. The first NAAT for detection of *C. difficile* in stool was approved in 2009. At the time of this writing, 15 different platforms are FDA approved and available for testing using a variety of methods, including real-time PCR, loop-mediated amplification, helicase-dependent amplification, and microarray technology. Some platforms are designed for low-volume laboratories and on-demand testing, and others are more amenable to high-volume, batch mode testing. These assays detect a variety of gene targets, including *tcdA*, *tcdB*, *cdt*, and Δ117 deletion in *tcdC*, the latter two as surrogates for identification of the ribotype 027 strain.

Although NAATs have replaced other methods in clinical laboratories for diagnosis of *C. difficile* infection and have very high negative predictive values and analytical and clinical sensitivities, there are concerns about their specificity and positive predictive value because they will detect colonization as well as infection.[236] As mentioned earlier, some laboratories have implemented multistep diagnostic test algorithms using GDH, toxin EIA, and NAAT; however, these algorithms can complicate and delay final results, may not be reimbursed, and may ultimately be less cost effective than NAAT alone. Regardless of how a laboratory chooses to deploy NAAT testing, it should be limited to only patients with diarrhea to increase the pretest probability of disease and thus help mitigate concerns about detecting patients who are asymptomatically colonized with toxigenic strains.

> **POINTS TO REMEMBER**
>
> ***Clostridium difficile***
> - *C. difficile*—associated disease spectrum ranges from mild antibiotic-associated diarrhea to life-threatening toxic megacolon and occurs both in hospitals and the community.
> - NAATs for detection of the toxin B gene of *C. difficile* have several advantages over traditional methods for diagnosis, including increased analytical and clinical sensitivity, high negative predictive value, and decreased analysis time.
> - Concerns about the specificity and positive predictive value of NAATs have led some laboratories to adopt multistep diagnostic algorithms to help mitigate the problem of detecting patients asymptomatically colonized with toxigenic strains of *C. difficile*.

Gastrointestinal Pathogen Panels

Infectious gastroenteritis (IGE) is a leading cause of global morbidity and mortality. Diarrheal disease disproportionally affects developing nations, but IGE remains a significant problem in industrialized countries as well. Each year, approximately 178.8 million cases of IGE occur in the United States, resulting in 474,000 hospitalizations and 5000 deaths.[237] IGE is associated with a diverse array of etiologic agents, including bacteria, viruses, and parasites. Clinical presentation does little to aid with a specific etiologic diagnosis because diarrhea is the predominant symptom regardless of the etiology. Accurate identification of the etiology of IGE provides important information that impacts individual patient management, infection control, and public health interventions.

Common diagnostic practice in the United States requires that providers choose among a variety of tests, including antigen detection tests, culture, ova and parasite microscopic examination, and single-target NAATs, for detection of the responsible organism or toxin. In addition, the selection of tests may be informed by patient's age, severity of disease, immunocompromised state, duration and type of diarrhea, travel history, and time of year.[238] Often the clinician is unsure of what pathogens are included in each test and consequently may miss testing for specific pathogens of interest. In the laboratory, the battery of tests required to detect all possible pathogens is laborious and expensive to maintain, can require special expertise, and may have an unacceptably long TAT. In addition, the conventional microbiologic tests have limited sensitivity for many of the major pathogens.

The application of NA amplification methods could have significant impact on the diagnosis, treatment, and understanding of the epidemiology of IGE.[239] At the time of this writing, there were five FDA-approved enteric pathogen panels. A comparison of their key features is shown in Table 5.7. The systems use a variety of different technologies and differ in the number and types of targets included in the assay and the overall platform design and throughput. The Prodesse ProGastro SSCS assay (Hologic/Gen-Probe) uses real-time PCR to detect and differentiate among *Salmonella* spp., *Shigella* spp., and *Campylobacter* spp., as well as Shiga toxin 1 (*stx1*) and Shiga toxin 2 (*stx2*) genes as indicators of Shiga-toxin producing *E. coli* in two separate master mixes.[240]

TABLE 5.7 Comparison of Different Food and Drug Administration–Approved Enteric Pathogen Panel Platforms

Feature	ProGastro SSCS	BD MAX EBP	Verigene EP	xTag GPP	FilmArray GI
Technology	Real-time PCR	Real-time PCR	PCR and gold nanoparticle microarray	PCR and bead array	Nested PCR and melting curve analysis
Automation	Separate extraction, manual PCR setup	Sample to result	Sample to result	Separate extraction, manual PCR setup, post-PCR amplicon transfer	Sample to result
Throughput	Batch (16/thermal cycler)	Batch (24)	1/run	Batch (limited by extractor)	1/run
Analysis time (hr)	3	1.5	2	4	1
Targets	5 (3 bacteria, 2 toxins)	4 (3 bacteria, 1 toxin)	9 (5 bacteria, 2 toxins, 2 viruses)	14 (8 bacteria, 3 viruses, 3 protozoa)	21 (12 bacteria, 5 viruses, 4 protozoa)
Relative cost/test	$$	$	$$$	$$$	$$$

PCR, Polymerase chain reaction.

Separate NA extraction and manual PCR setup are required. The BD MAX EBP (BD Diagnostics) uses a single real-time PCR master mix to detect essentially the same pathogens and toxins: *Salmonella* spp., *Shigella* spp. or enteroinvasive *E. coli*, *Campylobacter* spp., and *stx1/stx2*. However, the BD MAX automates all of the steps from sample preparation to target amplification and detection.[241]

Other systems have been developed to expand the panel of bacteria detected and include viral and protozoal pathogens. The Luminex xTAG GPP uses multiplex endpoint PCR and liquid bead array to detect and differentiate eight bacteria, three viruses, and three protozoa, including *Campylobacter* spp., *C. difficile* (toxins A and B), *E. coli* 0157, enterotoxigenic *E. coli*, Shiga-like toxin producing *E. coli*, *Salmonella* spp., *Shigella* spp., *Vibrio cholera*, adenovirus 40/41, norovirus GI/GII, rotavirus A, *Cryptosporidium* spp. *Entamoeba histolytica*, and *Giardia* spp.[242] This system provides for high throughput but requires a separate NA extraction step and post-PCR amplicon manipulation, which can lead to false-positive results caused by amplicon carry-over cross-contamination. It also has the longest analysis time of the available systems.

The Verigene uses multiplex PCR and a gold nanoparticle microarray to detect five bacteria, two toxins, and two viruses, including *Campylobacter* spp., *Salmonella* spp., *Shigella* spp. *Vibrio* spp., *Yersinia enterocolitica*, stx1, stx2, norovirus, and rotavirus.[243] It is a simple to use "sample in, answer out" system, but it has limited throughput because only one sample per instrument can be run at a time.

The FilmArray uses nested multiplex PCR and melting curve analysis to detect 12 bacteria, 5 viruses, and 4 protozoa, including *Campylobacter* spp., *C. difficile* toxin A/B, *Plesiomonas shigelloides*, *Salmonella* spp., *Vibrio* spp., *V. cholerae*, *Y. enterocolitica*, enteroaggregative *E. coli*, enteropathogenic *E. coli*, enterotoxigenic *E. coli*, Shiga-like toxin-producing *E. coli*, *E. coli* 0157, adenovirus F 40/41, astrovirus, norovirus GI/GII, rotavirus A, sapovirus, *Cryptosporidium* spp., *Cyclospora cayetanensis*, *E. histolytica*, and *G. lamblia*.[244] This is the most comprehensive enteric pathogen panel currently available. Similar to all of the FilmArray products, it is simple to use and provides results in about 1 hour. Its chief limitations are low throughput and high cost.

Laboratories can choose from a variety of test platforms based on whether a more focused or broader approach to IGE pathogen detection is desired. Also, the technical complexity and required throughput are important variables that may influence the approach chosen for this application. Current stool test algorithms using conventional methods typically require clinicians to consider which pathogens might be associated with the disease and choose among a variety of tests to ensure that all pathogens are covered. It is not surprising that this piecemeal approach often fails to yield positive results. The use of comprehensive pathogen panels dramatically increases diagnostic yield, but with this comes the unique challenge of interpreting the results from patients with multiple pathogens detected. In the FDA clinical trial of the FilmArray GI Panel, at least one potential pathogen was detected in 53.5% of specimens, and among these, multiple potential pathogens were detected in 32.9%.[244] Asymptomatic infections with *C. difficile*, *Cryptosporidium* spp., and *G. lamblia* are not uncommon, and some of the other IGE pathogens such as *Salmonella* spp. and norovirus can be shed for weeks after resolution of symptoms. Comprehensive panels consolidate testing platforms for agents of IGE and substantially reduce, but not completely eliminate, the need for culture because isolates are needed for epidemiologic surveillance and occasionally for antibiotic susceptibility testing.

ANTIBACTERIAL DRUG RESISTANCE

The detection of antibiotic resistance is one of the most important functions of the clinical microbiology laboratory. This has traditionally been done by phenotypic methods.

However, the delays inherent in phenotypic tests can lead to delays in appropriate therapy and adverse clinical outcomes. Molecular methods offer faster alternatives for detection of antibiotic resistance, but the genotypic approach has its own set of challenges because of the complexity of antibiotic resistance. In addition, the detection of a resistance gene may not necessarily imply phenotypic resistance if the gene is expressed at low levels or is not functional. Advances in technology and our understanding of the genetics of antibiotic resistance will likely make the use of molecular detection for antibiotic resistance more widespread in the future. This section focuses on the commonly used resistance targets, currently MRSA, vancomycin-resistant enterococci (VRE), and β-lactamases in gram-negative bacteria.

Because *S. aureus* is among the most common cause of bacterial infections in the industrialized world, particular attention has been focused on assays to rapidly differentiate MRSA from MSSA for diagnosis of infection and surveillance for infection control purposes. Molecular assays that recognize MRSA based on detection of a single target detect the junction between the staphylococcal cassette chromosome *mec* element (SCC*mec*), which carries the *mecA* resistance and other genes, and the flanking *orfX* gene.[245] This assay design has several limitations, including false-negative results caused by SCC*mec* variants and false-positive results caused by MSSA strains that carry SCC*mec* remnants lacking the *mecA* gene, sometimes referred to as "empty cassettes," or that carry SCC*mec* with a nonfunctional *mecA* gene.[245-247] An alternative approach to molecular detection of MRSA combines a *mecA* target and a second gene target specific for *S. aureus* such as *sa442*, *nuc*, *femA-femB*, *spa*, or *Idh1*.[248] At the time of this writing, there were five companies with FDA-approved assays for molecular detection of MRSA or MRSA and MSSA (BD Diagnostics, Cepheid, Elitech, Roche, and bioMérieux). In addition to these stand-alone assays, molecular detection of MRSA is incorporated into the blood culture identification panels discussed previously. These assays are intended for use in surveillance testing or to assist in the diagnosis of infections. Depending on the platform, the tests can be run on demand or in batches. Studies indicate that use of molecular methods for rapid identification of patients who are colonized with MRSA may be a cost-effective infection control strategy.[249,250]

Enterococci are commensal residents of the GI tract and female genital tract that account for about 10% of hospital-acquired infections. The vast majority of enterococcal infections are caused by *Enterococcus faecalis* and *E. faecium* and occur primarily in patients requiring long-term care. The emergence of VRE in hospitals is concerning because vancomycin is often used empirically to treat a wide variety of infections. Infection with VRE is associated with increased morbidity and mortality because of the propensity of VRE to infect patients already at high risk for comorbidity.[251]

In the United States, about 30% of enterococci are resistant to vancomycin. High-level vancomycin resistance in enterococci occurs via acquisition of mobile transposable elements carrying the *vanA* or *vanB* genes. *E. faecium* is more frequently resistant to vancomycin than *E. faecalis*, and *vanA* is more commonly found than *vanB* in resistant strains. As with MRSA, the rapid detection of VRE colonization to prevent health care–associated infections is widely recommended.[252]

Molecular assays work well for this application.[253,254] Three FDA-approved molecular assays are marketed for rapid detection of VRE from perianal and rectal swabs. The BD GeneOhm and IMDx assays detect both *vanA* and *vanB*, and the Cepheid assay detects *vanA* alone. All are based on real-time PCR, but only the Cepheid assay is designed for on-demand testing. Aside from rare reports of *vanA* in *S. aureus* and *Streptococcus* spp., detection of *vanA* is highly specific for VRE. However, *vanB* can be found in a wide variety of commensal nonenterococcal bacteria, so detection of *vanB* requires confirmation of VRE by culture. As with *mecA*, assays for detection of *vanA* and *vanB* have also been incorporated in the commercially available blood culture identification panels.

One of the greatest threats to our antibiotic formulary is the emergence of β-lactamases in gram-negative bacteria with the capabilities of hydrolyzing broad-spectrum penicillins, cephalosporins, and carbapenems. These enzymes include extended-spectrum β-lactamases (ESBLs), AmpCs, and carbapenemases. The accurate detection of these zymes is important for both treatment decisions and infection control purposes. Detection of these organisms harboring these broad-spectrum β-lactamases by phenotypic methods is imperfect.[255,256] A rapid, inexpensive, multiplex molecular assay to detect the genes encoding these enzymes would be clinically useful but presently is an unmet need. One of the biggest challenges to molecular detection is the great diversity of β-lactamases, with more than 200 described ESBLs and numerous classes of carbapenemases, including *K. pneumoniae* carbapenemase (KPC), New Delhi metallo-β-lactamase (NDM), Verona integrin-encoded metallo-β-lactamase (VIM), imipenem metallo-β-lactamase (IMP), and oxacillinase (OXA). An additional challenge is that the detection of the gene(s) does not provide information about copy number and expression, which are important to phenotypic expression of resistance to β-lactam and carbapenem antibiotics.

A number of LDTs and RUO kits have been developed for molecular detection of a variety of broad-spectrum β-lactamase genes that range from single target assays to detect KPC to highly multiplexed assays detecting multiple ESBLs, multiple AMPCs, KPC, NDM, VIM, IMP, and OXA-48.[257-262] Both the Biofire FilmArray and Verigene BCID panels include assays to detect KPC, and the Verigene panel detects genes encoding for five additional broad-spectrum β-lactamases, CTX-M, NDM, VIM, IMP, and OXA.

HUMAN MICROBIOME AND METAGENOMICS

Microbial inhabitants outnumber our own body's cells by about 10 to 1 and at a genomic level have 100-fold greater gene content than the human genome. Interest in elucidating the role of resident organisms in human health and disease has flourished over the past decade with the advent of new technologies for interrogating complex microbial communities. The microbiome is the totality of microbes, their genetic information, and the milieu in which they interact.[263] It includes bacteria, fungi, viruses and phages, and parasites, but most of the emphasis to date has been on the bacterial component of the microbiome. However, progress is being made toward defining the human virome and the role that it plays in complex microbial communities.[264]

Metagenomics refers to the concept that a collection of genes sequenced from the environment could be analyzed in a way analogous to the study of a single genome. It has been facilitated by advances in NA sequencing technology, and this technology has permitted the study of microbial communities directly in their natural environment, thus bypassing the need for isolation and cultivation of individual species. We now know that the majority of microorganisms from the human body cannot be cultured in vitro. Most taxonomic metagenomic studies have used targeted sequencing of the phylogenetically informative regions of the 16s rRNA gene from bacteria because it has long been the gold standard method for bacterial identification, and there are large sequence databases and sophisticated analysis tools available.[265] However, 16s rRNA gene sequencing does not provide enough information for comprehensive microbiome studies. To overcome the limitations of single gene-based amplicon sequencing, researchers have used whole-genome shotgun sequencing on massively parallel sequencing platforms. Whole-genome approaches permit identification and annotation of diverse sets of microbial genes that encode many different biochemical or metabolic functions, thus providing functional metagenomic information.

In 2007, the Human Microbiome Project was launched by the National Institutes of Health with the overarching goal of developing tools and resources for characterization of the human microbiome and to relate it to human health and disease.[266-268] Initial microbiome comparisons across 18 different body sites confirmed high interindividual variation with four phyla of bacteria, *Actinobacteria*, *Bacteroidetes*, *Firmicutes*, and *Proteobacteria* predominating across all body sites.[269] Additionally, the composition of the gut microbiome is most often characterized by smooth abundance gradients of key organisms and does not cluster individuals into discrete microbiome types.[268] However, the microbiome at other body sites such as the vagina can show such clustering.[266] An important point that emerged from these early studies is that although the microbial communities varied among individuals, the metabolic pathways encoded by these organisms were consistently present, forming a functional "core" to the microbiome at all body sites.[268,270,271] Although the pathways and processes of this core were consistent, the specific genes associated with these functions varied.

Alterations of the microbiome in many different disease states have been described.[263] A complete review of this topic is beyond the scope of this chapter, but some specific examples are given in Table 5.8. Establishing a causal link between microbiome changes and a specific disease often is challenging because most studies have been observational, the disease entities themselves may not be well defined, and the pathogenesis may be multifactorial.

However, it seems clear that future approaches in medical microbiology will be shaped in part by developments in metagenomics and human microbiome research. The identification of single agents of infection will be supplemented by techniques that will determine the relative composition of microbiomes in the context of different infections and other disease states. Recent evidence that recurrent *C. difficile* infections can be treated by reconstituting the normal colon microbiota in the patient by transferring feces from a normal donor is a good example of how a better understanding of changes in the microbiome composition can lead to effective treatments options.[272] Differences in the composition of the microbiome that may cause or contribute to noninfectious diseases may offer new opportunities for clinical microbiology laboratories to impact other areas of medicine. Finally, metagenomic techniques will facilitate the discovery of previously unrecognized pathogens and increase our understanding of how changes in the microbiome may contribute to infectious diseases.

TABLE 5.8 Association of Human Disease With Changes in the Microbiome

Disease	Relevant Change
Psoriasis	Increased ratio of *Firmicutes* to *Actinobacteria*[280]
Reflux esophagitis	Esophageal microbiota dominated by gram-negative anaerobes; gastric microbiota with low or absent *Helicobacter pylori*[281,282]
Obesity	Reduced ratio of *Bacteroidetes* to *Firmicutes*[283,284]
Childhood-onset asthma	Absent gastric *H. pylori* (especially cytotoxin-associated gene genotype[285,286]
Inflammatory bowel disease (colitis)	Increased *Enterobacteriaceae*[287]
Functional bowel diseases	Increased *Veillonella* spp. and *Lactobacillus* spp.[288]
Colorectal carcinoma	Increased *Fusobacterium* spp.[289,290]
Cardiovascular disease	Gut-dependent metabolism of phosphatidylcholine[291]

Modified from Cho I, Blaser MJ. The human microbiome: at the interface of health and disease. *Nat Rev Genet* 2012;13:260–270.

FUTURE DIRECTIONS

Molecular microbiology will continue to be one of the leading growth areas in laboratory medicine. The number of applications of this technology in clinical microbiology will continue to increase, and the technology will increasingly be deployed in clinical laboratories as it becomes less technically complex and thus more accessible. However, now more than ever, clinical and financial outcomes data will be needed to justify the use of this often expensive technology in an era of declining reimbursement and increased cost consciousness.

The clinical utility of molecular testing for infectious diseases is now well established, and the gap between the availability of FDA-cleared and -approved tests and clinical need is improving. However, the pending enhanced oversight of LDTs and restriction of the use of RUO and IUO reagents and systems by the FDA could limit the ability of laboratories to develop tests to meet clinical needs not met by IVD products (http://www.fda.gov/downloads/medicaldevices/deviceregulationandguidance/guidancedocuments/ucm416684.pdf).

Although considerable progress has been made in recent years, other important needs remain unmet, including the availability of international standards and traceable and commutable calibrators that can be used for assay verification and validation. These materials, when widely available, should improve agreement of the results between or among different tests and aid in the establishment of their clinical utility. Another need is for the continued development of effective proficiency testing programs that will help ensure that the results of molecular tests are reliable and reproducible among laboratories.

Digital PCR is the next advance in the evolution of quantitative PCR methods. Digital PCR has many applications, including detection and quantification of low numbers of pathogen sequences. It can provide a lower limit of detection than real-time PCR methods with better precision at very low concentrations. As opposed to relative quantification, digital PCR provides absolute quantification with no need for reference standards. Currently, digital PCR is used as a research tool, but it may find applications in clinical laboratories to resolve ambiguous results obtained with quantitative real-time PCR assays or for creating accurate viral reference standards as the technology becomes less costly.[273,274]

To a great extent, the future of molecular microbiology depends on automation. Many of the available tests are labor intensive, with much of the labor devoted to tedious sample processing methods. Several fully automated systems for molecular diagnostics have been developed for high- and midvolume laboratories, but most suffer from a limited test menu. To increase access to molecular tests, simple, affordable, fully automated, random access platforms with broad test menus are needed, particularly for laboratories that have a low- and midvolume of testing. NA testing for infectious diseases at the point of care is beginning to enter clinical practice in developed and developing countries, particularly for applications that require short TATs and in settings where a centralized laboratory approach is not feasible.[275]

The use of multiplex NA-based assays to screen at-risk patients for panels of probable pathogens remains a goal for molecular microbiology. Several such tests are currently available, but success to date has been limited by technical complexity of some systems. The development of simple, multiparametric technologies is key to providing molecular tests with the same broad diagnostic range provided by culture and other conventional methods for syndromic diagnosis of infectious diseases.

Metagenomic studies have provided new insights into the human microbiome and alterations in these communities of microorganisms have been linked to a number of disease states. With the continued decrease in the cost of massively parallel NA sequencing and the increasing availability of the necessary bioinformatics tools, it is likely that our understanding of the human microbiome will result in novel microbiome-focused diagnostics and clinical interventions. In addition, the massively parallel sequencing platforms that have enabled metagenomics will be increasingly used in epidemiological investigations[276,277] and new pathogen discovery.[278,279]

REFERENCES

1. Persing DH, Tenover FC, Hayden RT, et al. *Molecular microbiology: diagnostic principles and practice*. Washington DC. 3rd ed. ASM Press; 2016.
2. Plantier JC, Leoz M, Dickerson JE, et al. A new human immunodeficiency virus derived from gorillas. *Nat Med* 2009;15:871–2.
3. Eshleman SH, Hackett Jr J, Swanson P, et al. Performance of the Celera Diagnostics ViroSeq HIV-1 genotyping system for sequence-based analysis of diverse human immunodeficiency virus type 1 strains. *J Clin Microbiol* 2004;42:2711–7.
4. Mellors JW, Kingsley LA, Rinaldo Jr CR, et al. Quantitation of HIV-1 RNA in plasma predicts outcome after seroconversion. *Ann Intern Med* 1995;122:573–9.
5. Mellors JW, Rinaldo Jr CR, Gupta P, et al. Prognosis in HIV-1 infection predicted by the quantity of virus in plasma. *Science* 1996;272:1167–70.
6. Mellors JW, Margolick JB, Phair JP, et al. Prognostic value of HIV-1 RNA, CD4 cell count and CD4 cell count slope for progression to AIDS and death in untreated HIV-1 infection. *JAMA* 2007;297:2349–50.
7. Demeter LM, Hughes MD, Coombs RW, et al. Predictors of virologic and clinical outcomes in HIV-1-infected patients receiving concurrent treatment with indinavir, zidovudine and lamivudine. AIDS Clinical Trials Group Protocol 320. *Ann Intern Med* 2001;135:954–64.
8. Raboud JM, Montaner JS, Conway B, et al. Suppression of plasma viral load below 20 copies/mL is required to achieve a long-term response to therapy. *AIDS* 1998;12:1619–24.
9. Havlir DV, Bassett R, Levitan D, et al. Prevalence and predictive value of intermittent viremia with combination HIV therapy. *JAMA* 2001;286:171–9.
10. Kahn JO, Walker BD. Acute human immunodeficiency virus type 1 infection. *N Engl J Med* 1998;339:33–9.
11. Centers for Disease Control and Prevention and Association of Public Health Laboratories. Laboratory Testing for the Diagnosis of HIV Infection: Updated Recommendations. Available at: http://stacks.cdc.gov/view/cdc/23447; Published June 27, 2014.
12. Kumar M, Setty HG, Hewlett IK. Point of care technologies of HIV. Hindawi Publishing Corporation, Volume 2014, Article ID 497046, 20 pages. https://doi.org/10.1155/2014/497046.
13. Karasi JC, Dziezuk F, Quennery L, et al. High correlation between the Roche cobas(R) AmpliPrep/cobas(R) TaqMan(R) HIV-1, v2.0 and the Abbott m2000 RealTime HIV-1 assays for quantification of viral load in HIV-1 B and non-B subtypes. *J Clin Virol* 2011;52:181–6.
14. Saag MS, Holodniy M, Kuritzkes DR, et al. HIV viral load markers in clinical practice. *Nat Med* 1996;2:625–9.
15. Garcia-Bujalance S, Ladron de Guevara C, Gonzalez-Garcia J, et al. Elevation of viral load by PCR and use of plasma preparation tubes for quantification of human immunodeficiency virus type 1. *J Microbiol Methods* 2007;69:384–6.
16. Griffith BP, Mayo DR. Increased levels of HIV RNA detected in samples with viral loads close to the detection limit collected in plasma preparation tubes (PPT). *J Clin Virol* 2006;35:197–200.
17. Choo QL, Richman KH, Han JH, et al. Genetic organization and diversity of the hepatitis C virus. *Proc Natl Acad Sci USA* 1991;88:2451–5.
18. Okamoto H, Kurai K, Okada S, et al. Full-length sequence of a hepatitis C virus genome having poor homology to reported isolates: comparative study of four distinct genotypes. *Virology* 1992;188:331–41.
19. Smith DB, Bukh J, Kuiken C, et al. Expanded classification of Hepatitis C virus into 7 genotypes and 67 subtypes: updated Criteria and Genotype assignment web resource. *Hepatology* 2014;59:318–27.

20. Dusheiko G, Schmilovitz-Weiss H, Brown D, et al. Hepatitis C virus genotypes: an investigation of type-specific differences in geographic origin and disease. *Hepatology* 1994;**19**:13—8.
21. Benvegnu L, Pontisso P, Cavalletto D, et al. Lack of correlation between hepatitis C virus genotypes and clinical course of hepatitis C virus-related cirrhosis. *Hepatology* 1997;**25**:211—5.
22. Poynard T, Bedossa P, Opolon P. Natural history of liver fibrosis progression in patients with chronic hepatitis C. The OBSVIRC, METAVIR, CLINIVIR and DOSVIRC groups. *Lancet* 1997;**349**:825—32.
23. Centers for Disease Control and Prevention. Recommendations for prevention and control of hepatitis C virus (HCV) infection and HCV-related chronic disease. *MMWR Recomm Rep* 1998;**47**(RR—19):1—39.
24. Ghany MG, Strader DB, Thomas DL, et al. Diagnosis, management and treatment of hepatitis C: an update. *Hepatology* 2009;**49**:1335—74.
25. Dufour DR, Lott JA, Nolte FS, et al. Diagnosis and monitoring of hepatic injury. II. Recommendations for use of laboratory tests in screening, diagnosis and monitoring. *Clin Chem* 2000; **46**:2050—68.
26. Alter MJ, Kuhnert WL, Finelli L. Guidelines for laboratory testing and result reporting of antibody to hepatitis C virus. *MMWR Recomm Rep* 2003;**52**(RR03):1—16.
27. Centers for Disease Control and Prevention. Testing for HCV infection: an update of Guidance for Clinicians and Laboratorians. *MMWR Recomm Rep* 2013;**62**:362—5.
28. Fanning L, Kenny-Walsh E, Levis J, et al. Natural fluctuations of hepatitis C viral load in a homogeneous patient population: a prospective study. *Hepatology* 2000;**31**:225—9.
29. Schreiber GB, Busch MP, Kleinman SH, et al. The risk of transfusion-transmitted viral infections. The Retrovirus Epidemiology Donor Study. *N Engl J Med* 1996;**334**:1685—90.
30. Legler TJ, Riggert J, Simson G, et al. Testing of individual blood donations for HCV RNA reduces the residual risk of transfusion-transmitted HCV infection. *Transfusion* 2000;**40**: 1192—7.
31. Kolk DP, Dockter J, Linnen J, et al. Significant closure of the human immunodeficiency virus type 1 and hepatitis C virus preseroconversion detection windows with a transcription-mediated-amplification-driven assay. *J Clin Microbiol* 2002;**40**: 1761—6.
32. Icardi G, Ansaldi F, Bruzzone BM, et al. Novel approach to reduce the hepatitis C virus (HCV) window period: clinical evaluation of a new enzyme-linked immunosorbent assay for HCV core antigen. *J Clin Microbiol* 2001;**39**:3110—4.
33. Bouvier-Alias M, Patel K, Dahari H, et al. Clinical utility of total HCV core antigen quantification: a new indirect marker of HCV replication. *Hepatology* 2002;**36**:211—8.
34. Veillon P, Payan C, Picchio G, et al. Comparative evaluation of the total hepatitis C virus core antigen, branched-DNA and amplicor monitor assays in determining viremia for patients with chronic hepatitis C during interferon plus ribavirin combination therapy. *J Clin Microbiol* 2003;**41**:3213—20.
35. Mederacke I, Wedemeyer H, Ciesek S, et al. Performance and clinical utility of a novel fully automated quantitative HCV-core antigen assay. *J Clin Virol* 2009;**46**:210—5.
36. Morota K, Fujinami R, Kinukawa H, et al. A new sensitive and automated chemiluminescent microparticle immunoassay for quantitative determination of hepatitis C virus core antigen. *J Virol Methods* 2009;**157**:8—14.
37. McHutchison JG, Gordon SC, Schiff ER, et al. Interferon alfa-2b alone or in combination with ribavirin as initial treatment for chronic hepatitis C. Hepatitis Interventional Therapy Group. *N Engl J Med* 1998;**339**:1485—92.
38. Poynard T, Marcellin P, Lee SS, et al. Randomised trial of interferon alpha2b plus ribavirin for 48 weeks or for 24 weeks versus interferon alpha2b plus placebo for 48 weeks for treatment of chronic infection with hepatitis C virus. International Hepatitis Interventional Therapy Group (IHIT). *Lancet* 1998; **352**:1426—32.
39. Thompson AJ, Muir AJ, Sulkowski MS, et al. Interleukin-28B polymorphism improves viral kinetics and is the strongest pretreatment predictor of sustained virologic response in genotype 1 hepatitis C virus. *Gastroenterology* 2010; **139**:120—9.
40. Mangia A, Thompson AJ, Santoro R, et al. An IL28B polymorphism determines treatment response of hepatitis C virus genotype 2 or 3 patients who do not achieve a rapid virologic response. *Gastroenterology* 2010;**139**:821—7.
41. McCormick SE, Goodman ZD, Maydonovitch CL, et al. Evaluation of liver histology, ALT elevation and HCV RNA titer in patients with chronic hepatitis C. *Am J Gastroenterol* 1996; **91**:1516—22.
42. Ghany MG, Nelson DR, Strader DB, et al. An update on treatment of genotype 1 chronic hepatitis C virus infection: 2011 practice guideline by the American Association for the Study of Liver Diseases. *Hepatology* 2011;**54**:1433—44.
43. Lawitz E, Mangia A, Wyles D, et al. Sofosbuvir for previously untreated chronic hepatitis C. *N Engl J Med* 2013;**368**: 1878—87.
44. Fried MW, Buti M, Dore GJ, et al. Once-daily simeprevir (TMC435) with pegylated interferon and ribavirin in treatment-naïve genotype 1 Hepatitic: the randomized PILLAR study. *Hepatology* 2013;**58**:1918—29.
45. Chevaliez S, Bouvier-Alias M, Rodriguez C, et al. The cobas AmpliPrep/cobas TaqMan HCV test, version 2.0, real-time PCR assay accurately quantifies hepatitis C virus genotype 4 RNA. *J Clin Microbiol* 2013;**51**:1078—82.
46. Zitzer H, Heilek G, Truchon K, et al. Second-generation cobas AmpliPrep/cobas TaqMan HCV quantitative test for viral load monitoring: a novel dual-probe assay design. *J Clin Microbiol* 2013;**51**:571—7.
47. Saldanha J, Lelie N, Heath A. Establishment of the first international standard for nucleic acid amplification technology (NAT) assays for HCV RNA. WHO Collaborative Study Group. *Vox Sang* 1999;**76**:149—58.
48. Pyne MT, Konnick EQ, Phansalkar A, et al. Evaluation of the Abbott investigational use only RealTime hepatitis C virus (HCV) assay and comparison to the Roche TaqMan HCV analyte-specific reagent assay. *J Clin Microbiol* 2009;**47**: 2872—8.
49. Michelin BD, Muller Z, Stelzl E, et al. Evaluation of the Abbott RealTime HCV assay for quantitative detection of hepatitis C virus RNA. *J Clin Virol* 2007;**38**:96—100.
50. Laperche S, Bouchardeau F, Thibault V, et al. Multicenter trials need to use the same assay for hepatitis C virus viral load determination. *J Clin Microbiol* 2007;**45**:3788—90.
51. Sarrazin C, Kieffer TL, Bartels D, et al. Dynamic hepatitis C virus genotypic and phenotypic changes in patients treated with the protease inhibitor telaprevir. *Gastroenterology* 2007; **132**:1767—77.

52. Pawlotsky JM. Treatment failure and resistance with direct-acting antiviral drugs against hepatitis C virus. *Hepatology* 2011;**53**:1742–51.
53. Olysio Prescribing Information. Janssen Therapeutics. *Titusville NJ Reference ID 3412095*. 2013.
54. Ciotti M, Marcuccilli F, Guenci T, et al. A multicenter evaluation of the Abbott RealTime HCV Genotype II assay. *J Virol Methods* 2010;**167**:205–7.
55. Verbeeck J, Stanley MJ, Shieh J, et al. Evaluation of Versant hepatitis C virus genotype assay (LiPA) 2.0. *J Clin Microbiol* 2008;**46**:1901–6.
56. Bouchardeau F, Cantaloube JF, Chevaliez S, et al. Improvement of hepatitis C virus (HCV) genotype determination with the new version of the INNO-LiPA HCV assay. *J Clin Microbiol* 2007;**45**:1140–5.
57. Bukh J, Miller RH, Purcell RH. Genetic heterogeneity of hepatitis C virus: quasispecies and genotypes. *Semin Liver Dis* 1995;**15**:41–63.
58. Simmonds P, Holmes EC, Cha TA, et al. Classification of hepatitis C virus into six major genotypes and a series of subtypes by phylogenetic analysis of the NS-5 region. *J Gen Virol* 1993;**74**:2391–9.
59. Nolte FS, Green AM, Fiebelkorn KR, et al. Clinical evaluation of two methods for genotyping hepatitis C virus based on analysis on the 5′ noncoding region. *J Clin Microbiol* 2003;**41**:1558–64.
60. Ross RS, Viazov SO, Holtzer CD, et al. Genotyping of hepatitis C virus isolates using CLIP sequencing. *J Clin Microbiol* 2000;**38**:3581–4.
61. Halfon P, Trimoulet P, Bourliere M, et al. Hepatitis C virus genotyping based on 5′ noncoding sequence analysis (Trugene). *J Clin Microbiol* 2001;**39**:1771–3.
62. Kalinina O, Norder H, Mukomolov S, et al. A natural intergenotypic recombinant of hepatitis C virus identified in St. Petersburg. *J Virol* 2002;**76**:4034–43.
63. Noppornpanth S, Lien TX, Poovorawan Y, et al. Identification of a naturally occurring recombinant genotype 2/6 hepatitis C virus. *J Virol* 2006;**80**:7569–77.
64. Legrand-Abravanel F, Claudinon J, Nicot F, et al. New natural intergenotypic (2/5) recombinant of hepatitis C virus. *J Virol* 2007;**81**:4357–62.
65. Gao F, Nainan OV, Khudyakov Y, et al. Recombinant hepatitis C virus in experimentally infected chimpanzees. *J Gen Virol* 2007;**88**:143–7.
66. Soya SS, Steinmertz HB, Tsongalis GJ, et al. Validation of a solid-phase electrochemical array for genotyping hepatitis C virus. *Exp Mol Pathol* 2013;**95**:18–22.
67. Schaefer S, et al. Hepatitis B virus taxonomy and hepatitis B virus genotypes. *World J Gastroenterol* 2007;**13**:14–21.
68. Lavanchy D, et al. Chronic viral hepatitis as a public health issue in the world. *Best Pract Res Clin Gastroenterol* 2008;**22**:991–1008.
69. Ganem D, Prince AM, et al. Hepatitis B virus infection—natural history and clinical consequences. *N Engl J Med* 2004;**350**:1118–29.
70. McMahon BJ, et al. Epidemiology and natural history of hepatitis B. *Semin Liver Dis* 2005;**25**(Suppl. 1):3–8.
71. Horvath R, Tegtmeier GE, Hepatitis B, Viruses D. In: Versalovic J, et al., editors. *Manual of clinical microbiology*. 10th ed. Washington DC: ASM Press; 2011.
72. Chu CJ, Keeffe EB, Han SH, et al. Hepatitis B virus genotypes in the United States: results of a nationwide study. *Gastroenterology* 2003;**125**:444–51.
73. Lok AS, McMahon BJ. Chronic hepatitis B: update 2009. *Hepatology* 2009;**50**:661–2.
74. Kuhns MC, Busch MP. New strategies for blood donor screening for hepatitis B virus: nucleic acid testing versus immunoassay methods. *Mol Diagn Ther* 2006;**10**:77–91.
75. Carman WF, Zanetti AR, Karayiannis P, et al. Vaccine-induced escape mutant of hepatitis B virus. *Lancet* 1990;**336**:325–9.
76. Hsu HY, Chang MH, Ni YH, et al. Surface gene mutants of hepatitis B virus in infants who develop acute or chronic infections despite immunoprophylaxis. *Hepatology* 1997;**26**:786–91.
77. McMahon G, Ehrlich PH, Moustafa ZA, et al. Genetic alterations in the gene encoding the major HBsAg: DNA and immunological analysis of recurrent HBsAg derived from monoclonal antibody-treated liver transplant patients. *Hepatology* 1992;**15**:757–66.
78. Hillyard DR. Molecular Detection and Characterization of Hepatitis B virus. In: Persing DH, et al., editors. *Molecular microbiology: diagnostic principles and practice*. 2nd ed. Washington DC: ASM Press; 2001. p. 583.
79. Okamoto H, Tsuda F, Akahane Y, et al. Hepatitis B virus with mutations in the core promoter for an e antigen-negative phenotype in carriers with antibody to e antigen. *J Virol* 1994;**68**:8102–10.
80. Carman WF, Jacyna MR, Hadziyannis S, et al. Mutation preventing formation of hepatitis B e antigen in patients with chronic hepatitis B infection. *Lancet* 1989;**2**:588–91.
81. Miyakawa Y, Okamoto H, Mayumi M. The molecular basis of hepatitis B e antigen (HBeAg)-negative infections. *J Viral Hepat* 1997;**4**:1–8.
82. Degertekin B, Lok AS. Indications for therapy in hepatitis B. *Hepatology* 2009;**49**:S129–37.
83. Andersson KL, Chung RT. Monitoring during and after antiviral therapy for hepatitis B. *Hepatology* 2009;**49**:S166–73.
84. Saldanha J, Gerlich W, Lelie N, et al. An international collaborative study to establish a World Health Organization international standard for hepatitis B virus DNA nucleic acid amplification techniques. *Vox Sang* 2001;**80**:63–71.
85. Hussain M, Chu CJ, Sablon E, et al. Rapid and sensitive assays for determination of hepatitis B virus (HBV) genotypes and detection of HBV precore and core promoter variants. *J Clin Microbiol* 2003;**41**:3699–705.
86. Osiowy C, Giles E. Evaluation of the INNO-LiPA HBV genotyping assay for determination of hepatitis B virus genotype. *J Clin Microbiol* 2003;**41**:5473–7.
87. Gintowt AA, Germer JJ, Mitchell PS, et al. Evaluation of the MagNA Pure LC used with the TRUGENE HBV Genotyping Kit. *J Clin Virol* 2005;**34**:155–7.
88. Cope AV, Sabin C, Burroughs A, et al. Interrelationships among quantity of human cytomegalovirus (HCMV) DNA in blood, donor-recipient serostatus, and administration of methylprednisolone as risk factors for HCMV disease following liver transplantation. *J Infect Dis* 1997;**176**:1484–90.
89. Humar A, Gregson D, Caliendo AM, et al. Clinical utility of quantitative cytomegalovirus viral load determination for predicting cytomegalovirus disease in liver transplant recipients. *Transplantation* 1999;**68**:1305–11.

90. Gerna G, Baldanti F, Torsellini M, et al. Evaluation of cytomegalovirus DNAaemia versus pp65-antigenaemia cutoff for guiding preemptive therapy in transplant recipients: a randomized study. *Antivir Ther* 2007;**12**:63–72.
91. Gerna G, Lilleri D, Caldera D, et al. Validation of a DNAemia cutoff for preemptive therapy of cytomegalovirus infection in adult hematopoietic stem cell transplant recipients. *Bone Marrow Transplant* 2008;**41**:873–9.
92. Caliendo AM, St George K, Kao SY, et al. Comparison of quantitative cytomegalovirus (CMV) PCR in plasma and CMV antigenemia assay: clinical utility of the prototype AMPLICOR CMV MONITOR test in transplant recipients. *J Clin Microbiol* 2000;**38**:2122–7.
93. Caliendo AM, St. George K, Allega J, et al. Distinguishing cytomegalovirus (CMV) infection and disease with CMV nucleic acid assays. *J Clin Microbiol* 2002;**40**:1581–6.
94. Humar A, Kumar D, Boivin G, et al. Cytomegalovirus (CMV) viral load kinetics to predict recurrent disease in solid organ transplant patients with CMV disease. *J Infect Dis* 2002;**186**:829–33.
95. Spector SA, Wong R, Hsia K, et al. Plasma cytomegalovirus (CMV) DNA load predicts CMV disease and survival in AIDS patients. *J Clin Invest* 1998;**101**:497–502.
96. Hayden RT, Yan X, Wick MT, et al. Factors contributing to variability of quantitative viral PCR results in proficiency testing samples: a multivariate analysis. *J Clin Microbiol* 2012;**50**:337–45.
97. Holman CJ, Karger AB, Mullan BD, et al. Quantitative Epstein-Barr virus shedding and its correlation with the risk of post-transplant lymphoproliferative disorder. *Clin Transplant* 2012;**26**:741–7.
98. Cockfield SM. Identifying the patient at risk for post-transplant lymphoproliferative disorder. *Transpl Infect Dis* 2001;**3**:70–8.
99. Bakker NA, Verschuuren EA, Veeger NJ, et al. Quantification of Epstein-Barr virus-DNA load in lung transplant recipients: a comparison of plasma versus whole blood. *J Heart Lung Transplant* 2008;**27**:7–10.
100. Kinch AG, Oberg J, Arvidson KI, et al. Post-transplant lymphoproliferative disease and other Epstein-Barr virus diseases in allogeneic haematopoietic stem cell transplantation after introduction of monitoring of viral load by polymerase chain reaction. *Scand J Infect Dis* 2007;**39**:235–44.
101. Meerbach A, Wutzler P, Hafer R, et al. Monitoring of Epstein-Barr virus load after hematopoietic stem cell transplantation for early intervention in post-transplant lymphoproliferative disease. *J Med Virol* 2008;**80**:441–54.
102. Toyoda M, Moudgil A, Warady BA, et al. Clinical significance of peripheral blood Epstein-Barr viral load monitoring using polymerase chain reaction in renal transplant recipients. *Pediatr Transplant* 2008;**12**:778–84.
103. Aalto SM, Juvonen E, Tarkkanen J, et al. Epstein-Barr viral load and disease prediction in a large cohort of allogeneic stem cell transplant recipients. *Clin Infect Dis* 2007;**45**:1305–9.
104. Greenfield HM, Gharib MI, Turner AJ, et al. The impact of monitoring Epstein-Barr virus PCR in pediatric bone marrow transplant patients: can it successfully predict outcome and guide intervention? *Pediatr Blood Cancer* 2006;**47**:200–5.
105. Bingler MA, Feingold B, Miller SA, et al. Chronic high Epstein-Barr viral load state and risk for late-onset posttransplant lymphoproliferative disease/lymphoma in children. *Am J Transplant* 2008;**8**:442–5.
106. Green M, Soltys K, Rowe DT, et al. Chronic high Epstein-Barr viral load carriage in pediatric liver transplant recipients. *Pediatr Transplant* 2009;**13**:319–23.
107. Gulley ML, Tang W. Laboratory assays for Epstein-Barr virus-related disease. *J Mol Diagn* 2008;**10**:279–92.
108. Gulley ML, Tang W. Using Epstein Barr viral load assays to diagnose, monitor and prevent posttransplant lymphoproliferative disorder. *Clin Microbiol Rev* 2010;**23**:350–66.
109. Goudsmit J, Wertheim-van Dillen P, van Strien A, et al. The role of BK virus in acute respiratory tract disease and the presence of BKV DNA in tonsils. *J Med Virol* 1982;**10**:91–9.
110. Reploeg MD, Storch GA, Clifford DB. BK virus: a clinical review. *Clin Infect Dis* 2001;**33**:191–202.
111. Hirsch HH, Steiger J, Polyomavirus BK. *Lancet Infect Dis* 2003;**3**:611–23.
112. Sencer SF, Haake RJ, Weisdorf DJ. Hemorrhagic cystitis after bone marrow transplantation. Risk factors and complications. *Transplantation* 1993;**56**:875–9.
113. Yang CC, Hurd DD, Case LD, et al. Hemorrhagic cystitis in bone marrow transplantation. *Urology* 1994;**44**:322–8.
114. Bogdanovic G, Priftakis P, Taemmeraes B, et al. Primary BK virus (BKV) infection due to possible BKV transmission during bone marrow transplantation is not the major cause of hemorrhagic cystitis in transplanted children. *Pediatr Transplant* 1998;**2**:288–93.
115. Chan PK, Ip KW, Shiu SY, et al. Association between polyomaviruria and microscopic haematuria in bone marrow transplant recipients. *J Infect* 1994;**29**:139–46.
116. Cottler-Fox M, Lynch M, Deeg HJ, et al. Human polyomavirus: lack of relationship of viruria to prolonged or severe hemorrhagic cystitis after bone marrow transplant. *Bone Marrow Transplant* 1989;**4**:279–82.
117. Leung AY, Suen CK, Lie AK, et al. Quantification of polyoma BK viruria in hemorrhagic cystitis complicating bone marrow transplantation. *Blood* 2001;**98**:1971–8.
118. Priftakis P, Bogdanovic G, Kokhaei P, et al. BK virus (BKV) quantification in urine samples of bone marrow transplanted patients is helpful for diagnosis of hemorrhagic cystitis, although wide individual variations exist. *J Clin Virol* 2003;**26**:71–7.
119. Gardner SD, Field AM, Coleman DV, et al. New human papovavirus (B.K.) isolated from urine after renal transplantation. *Lancet* 1971;**1**:1253–7.
120. Purighalla R, Shapiro R, McCauley J, et al. BK virus infection in a kidney allograft diagnosed by needle biopsy. *Am J Kidney Dis* 1995;**26**:671–3.
121. Hirsch HH, Brennan DC, Drachenberg CB, et al. Polyomavirus-associated nephropathy in renal transplantation: interdisciplinary analyses and recommendations. *Transplantation* 2005;**79**:1277–86.
122. Ramos E, Drachenberg CB, Papadimitriou JC, et al. Clinical course of polyoma virus nephropathy in 67 renal transplant patients. *J Am Soc Nephrol* 2002;**13**:2145–51.
123. Drachenberg RC, Drachenberg CB, Papadimitriou JC, et al. Morphological spectrum of polyoma virus disease in renal allografts: diagnostic accuracy of urine cytology. *Am J Transplant* 2001;**1**:373–81.
124. Bressollette-Bodin C, Coste-Burel M, Hourmant M, et al. A prospective longitudinal study of BK virus infection in 104 renal transplant recipients. *Am J Transplant* 2005;**5**:1926–33.

125. Ding R, Medeiros M, Dadhania D, et al. Noninvasive diagnosis of BK virus nephritis by measurement of messenger RNA for BK virus VP1 in urine. *Transplantation* 2002;**74**:987−94.
126. Ramos E, Drachenberg CB, Portocarrero M, et al. BK virus nephropathy diagnosis and treatment: experience at the University of Maryland Renal Transplant Program. *Clin Transpl* 2002;**16**:143−53.
127. Randhawa P, Ho A, Shapiro R, et al. Correlates of quantitative measurement of BK polyomavirus (BKV) DNA with clinical course of BKV infection in renal transplant patients. *J Clin Microbiol* 2004;**42**:1176−80.
128. Cirocco R, Markou M, Rosen A, et al. Polyomavirus PCR monitoring in renal transplant recipients: detection in blood is associated with higher creatinine values. *Transplant Proc* 2001;**33**:1805−7.
129. Hirsch HH, Knowles W, Dickenmann M, et al. Prospective study of polyomavirus type BK replication and nephropathy in renal-transplant recipients. *N Engl J Med* 2002;**347**:488−96.
130. Limaye AP, Jerome KR, Kuhr CS, et al. Quantitation of BK virus load in serum for the diagnosis of BK virus-associated nephropathy in renal transplant recipients. *J Infect Dis* 2001;**183**:1669−72.
131. Nickeleit V, Klimkait T, Binet IF, et al. Testing for polyomavirus type BK DNA in plasma to identify renal-allograft recipients with viral nephropathy. *N Engl J Med* 2000;**342**:1309−15.
132. Nishimoto Y, Takasaka T, Hasegawa M, et al. Evolution of BK virus based on complete genome data. *J Mol Evol* 2006;**63**:341−52.
133. Hoffman NG, Cook L, Atienza E, et al. Marked variability of BK virus load measurement using quantitative real-time PCR among commonly used assays. *J Clin Microbiol* 2008;**46**:2671−80.
134. Groseclose SL, Zaidi AA, Delisle SJ, et al. Estimated incidence and prevalence of genital Chlamydia trachomatis infections in the United States. *Sex Transm Dis* 1996;**26**:339−44.
135. Stamm WE. Chlamydia trachomatis infections of the adult. In: Holmes KK, Sparling PF, Mardh P-A, et al., editors. *Sexually transmitted diseases*. 3rd ed. New York: McGraw-Hill; 1999. p. 407−22.
136. Burstein GR, Zenilman JM. Nongonococcal urethritis-a new paradigm. *Clin Infect Dis* 1999;**28**:S66−73.
137. Moss T. *International handbook of chlamydia*. Exeter, UK: Polestar Wheatons Ltd; 2001.
138. Rees E. The treatment of pelvic inflammatory disease. *Am J Obstet Gynecol* 1980;**138**:1042−7.
139. Stamm WE, Guinan ME, Johnson C, et al. Effect of treatment regimens for Neisseria gonorrhoeae on simultaneous infection with Chlamdia trachomatis. *N Engl J Med* 1984;**310**:545−9.
140. Judson FN. Gonorrhea. *Med Clin North Am* 1990;**74**:1353−66.
141. Centers for Disease Control and Prevention. Recommendations for the Laboratory-Based Detection of Chlamydia trachomatis and Neisseria gonorrhoeae. *MMWR Recomm Rep* 2014;**63**(RR−2):1−19.
142. Hagblom P, Korch C, Jonsson AB, et al. Intragenic variation by site-specific recombination in the cryptic plasmid of Neisseria gonorrhoeae. *J Bacteriol* 1986;**167**:231−7.
143. Miyada CG, Born TL. A DNA sequence for the discrimination of Neisseria gonorrhoeas from other Neisseria species. *Mol Cell Probes* 1991;**5**:327−35.
144. Centers for Disease Control and Prevention. Screening tests to detect Chlamydia trachomatis and Neisseria gonorrhoeae infections-2002. *MMWR Recomm Rep* 2002;**51**(RR−15):1−40.
145. Marions L, Rotzen-Ostlund M, Grillner L, et al. High occurrence of a new variant of Chlamydia trachomatis escaping diagnostic tests among STI clinic patients in Stockholm, Sweden. *Sex Transm Dis* 2008;**35**:61−4.
146. Bianchi A, Moret F, Desrues JM, et al. PreservCyt transport medium used for the ThinPrep Pap test is a suitable medium for detection of Chlamydia trachomatis by the cobas Amplicor CT/NG test: results of a preliminary study and future implications. *J Clin Microbiol* 2002;**40**:1749−54.
147. Satterwhite CL, Torrone E, Meites E, et al. Sexually Transmitted Infections Among US Women and Men: Prevalence and Incidence Estimates. *Sex Transm Dis* 2008;**2013**(40):187−93.
148. Nye MB, Schwebke JR, Body BA. Comparison of APTIMA Trichomonas vaginalis transcription-mediated amplification to wet mount microscopy, culture and polymerase chain reaction for diagnosis of trichomoniasis in men and women. *Am J Obstet Gynecol* 2009;**200**:188. e1−7.
149. Lobo TT, Feijo G, Carvalho SE, et al. A comparative evaluation of the Papanicolaou test for the diagnosis of trichomoniasis. *Sex Transm Dis* 2003;**30**:694−9.
150. Campbell L, Woods V, Lloyd T, et al. Evaluation of the OSOM Trichomonas rapid test versus wet preparation examination for detection of Trichomonas vaginalis vaginitis in specimens from women with a low prevalence of infection. *J Clin Microbiol* 2008;**46**:3467−9.
151. Andrea SB, Chapin KC. Comparison of Aptima Trichomonas vaginalis transcription-mediated amplification assay and BD affirm VP111 for detection of T. vaginalis in symptomatic women: performance parameters and epidemiological implications. *J Clin Microbiol* 2011;**49**:866−9.
152. Schwebke JR, Hobbs MM, Taylor SN, et al. Molecular testing for Trichomonas vaginalis in women: results from a prospective U. S. clinical trial. *J Clin Microbiol* 2011;**49**:4106−11.
153. Van Der Pol B, Williams JA, Taylor SN, et al. Detection of Trichomonas vaginalis DNA by use of self-obtained vaginal swabs with the BD ProbeTec Qx Assay on the BD Viper System. *J Clin Microbiol* 2014;**52**:885−9.
154. Scoular A. Using evidence base on genital herpes: optimizing the use of diagnostic tests and information provision. *Sex Transm Infect* 2002;**78**:160−5.
155. Van Der Pol B, Warren T, Taylor SN, et al. Type-specific identification of anogenital herpes simplex virus infections by use of a commercially available nucleic acid amplification test. *J Clin Microbiol* 2012;**50**:3466−71.
156. Kim HJ, Tong Y, Tang W, et al. A rapid and simple isothermal nucleic acid amplification test for detection of herpes simplex virus types 1 and 2. *J Clin Virol* 2011;**50**:26−30.
157. Engelberg R, Carrell D, Krantz E, et al. Natural history of genital herpes simplex virus type 1 infection. *Sex Transm Dis* 2003;**30**:174−7.
158. Song B, Dwyer DE, Mindel A. HSV type specific serology in sexual health clinics: use, benefits, and who gets tested. *Sex Transm Dis* 2004;**80**:113−7.
159. Ho G, Bierman R, Beardsley L, et al. Natural history of cervicovaginal papillomavirus infection in young women. *N Engl J Med* 1998;**338**:423−8.

160. Saslow D, Solomon D, Lawson HW, et al. American Cancer Society, American Society for Colposcopy and Cervical Pathology, and the American Society for Clinical Pathology screening guidelines for the prevention and early detection of cervical cancer. *CA Cancer J Clin* 2012;**62**:147–72.
161. Wright TC, Stoler MH, Behrens CM, et al. The ATHENA human papillomavirus study: design, methods, and baseline results. *Am J Obstet Gynecol* 2012;**206**:46. e1–11.
162. Huh WK, Ault KA, Chelmow D, et al. Use of primary high-risk human papillomavirus testing for cervical cancer screening: interim clinical guidance. *Gynecol Oncol* 2012;**136**:178–82.
163. Stoler MH, Austin M, Zhao C. Cervical cancer screeing should be done by primary HPV testing with genotyping and reflex cytology for women over the age of 25 years. *J Clin Microbiol* 2015. https://doi.org/10.1128/JCM.01087-15 [Epub ahead of print].
164. Castle PE, Solomon D, Wheeler CM, et al. Human papillomavirus genotype specificity of hybrid capture 2. *J Clin Microbiol* 2008;**46**:2595–604.
165. Youens KE, Hosler GA, Washington PJ, et al. Clinical experience with the Cervista HPV HR assay: correlation of cervical cytology and HPV status from 56,501 specimens. *J Mol Diagn* 2011;**13**:160–6.
166. Kinney W, Stoler MH, Castle PE. Patient safety and the next generation of HPV DNA tests. *Am J Clin Pathol* 2010;**134**:193–9.
167. Nolte FS, Ribeiro-Nesbitt DG. Comparison of the Aptima and Cervista tests for detection of high-risk human papillomavirus in cervical cytology specimens. *Am J Clin Pathol* 2014:561–6.
168. Czegledy JC, Losif C, Hansson BG, et al. Can a test for E6/E7 transcripts of human papillomavirus type 16 serve as a diagnostic tool for the detection of micrometastasis in cervical cancer? *Int J Cancer* 1995;**64**:211–5.
169. Snacken R, Kendal AP, Haaheim LR, et al. The next influenza pandemic: lessons from Hong Kong. *Emerg Infect Dis* 1997;**1999**(5):195–203.
170. Falsey AR, Walsh EE. Novel coronavirus and severe acute respiratory syndrome. *Lancet* 2003;**361**:1312–3.
171. Gharabaghi F, Tellier R, Cheung R, et al. Comparison of a commercial qualitative real-time RT-PCR kit with direct immunofluorescence assay (DFA) and cell culture for detection of influenza A and B in children. *J Clin Virol* 2008;**42**:190–3.
172. Kuypers J, Wright N, Ferrenberg J, et al. Comparison of real-time PCR assays with fluorescent-antibody assays for diagnosis of respiratory virus infections in children. *J Clin Microbiol* 2006;**44**:2382–8.
173. Legoff J, Kara R, Moulin F, et al. Evaluation of the one-step multiplex real-time reverse transcription-PCR ProFlu-1 assay for detection of influenza A and influenza B viruses and respiratory syncytial viruses in children. *J Clin Microbiol* 2008;**46**:789–91.
174. Weinberg GA, Erdman DD, Edwards KM, et al. Superiority of reverse-transcription polymerase chain reaction to conventional viral culture in the diagnosis of acute respiratory tract infections in children. *J Infect Dis* 2004;**189**:706–10.
175. Popowitch EB, O'Neill SS, Miller MB. Comparison of Biofire FilmArray RP, Genmark eSensor RVP, Luminex xTAG RVPv1 and Luminex RVP Fast multiplex assays for detection of respiratory viruses. *J Clin Microbiol* 2013;**51**:1528–33.
176. Poritz MA, Blasche AJ, Byington CL, et al. FilmArray, an automated nested multiplex PCR system for multipathogen detection: development and application to respiratory tract infection. *PLoS ONE* 2011;**6**:e26047.
177. Liu RH, Coty WA, Reed M, et al. Electrochemical detection-based DNA microarrays. *IVD Technol* 2008;**14**:31–8.
178. Pierce VM, Hodinka RL. Comparison of the GenMark Diagnostics eSensor Respiratory Viral Panel to real-time PCR for detection of respiratory viruses in children. *J Clin Microbiol* 2012;**50**:3458–65.
179. Alby K, Popowitch EB, Miller MB. Comparative evaluation of the Nanosphere Veirgene RV+ assay and the Simplexa Flu A/B & RSV kit for detection of influenza and respiratory syncytial viruses. *J Clin Microbiol* 2013;**51**:352–3.
180. Bell JJ, Selvarangan R. Evaluation of the Alere i Influenza A&B nucleic acid amplification test by use of respiratory specimens collected in viral transport medium. *J Clin Microbiol* 2014;**52**:3992–5.
181. Centers for Disease Control and Prevention. Tuberculosis morbidity—United States. *MMWR Recomm Rep* 1997;**1998**(47):253–7.
182. Centers for Disease Control and Prevention. National plan for reliable tuberculosis laboratory services using a systems approach: recommendations from CDC and the Association of Public Health Laboratories Task Force on Tuberculosis Laboratory Services. *MMWR Recomm Rep* 2005;**54**(RR–6):1–12.
183. American Thoracic Society; CDC; Council of Infectious Disease Society of America. Diagnostic standards and classification of tuberculosis in adults and children. *Am J Respir Crit Care Med* 2000;**161**:1376–95.
184. Moore DF, Guzman JA, Mikhail LT. Reduction in turnaround time for laboratory diagnosis of pulmonary tuberculosis by routine use of a nucleic acid amplification test. *Diagn Microbiol Infect Dis* 2005;**52**:247–54.
185. Guerra RL, Hooper NM, Baker JF, et al. Use of the Amplified Mycobacterium tuberculosis Direct Test in a public health laboratory: test performance and impact on clinical care. *Chest* 2007;**132**:946–51.
186. Dinnes J, Deeks J, Kunst H, et al. A systematic review of rapid diagnostic tests for the detection of tuberculosis infection. *Health Technol Assess* 2007;**11**:1–196.
187. Flores LL, Pai M, Colford Jr JM, et al. In-house nucleic acid amplification tests for the detection of Mycobacterium tuberculosis in sputum specimens: meta-analysis and meta-regression. *BMC Microbiol* 2005;**5**:55.
188. Centers for Disease Control and Prevention. Updated guidelines for the use of nucleic acid amplification tests in the diagnosis of tuberculosis. *MMWR Recomm Rep* 2009;**58**:7–10.
189. Marks SM, Cronin W, Venkatappa T, et al. The health-system benefits and cost-effectiveness of using Mycobacterium tuberculosis direct nucleic acid amplification testing to diagnose tuberculosis disease in the United States. *Clin Infect Dis* 2013:57532–42.
190. Centers for Disease Control and Prevention. Availability of an assay for detecting Mycobacterium tuberculosis, including rifampin-resistant strains, and considerations for its use-United States. *MMWR Recomm Rep* 2013;**2013**(62):821–4.
191. Xpert MTB/RIF assay [package insert]. *Sunnyvale CA: Cepheid*. 2013.
192. Mattoo S, Cherry JD. Molecular pathogenesis, epidemiology, and clinical manifestations of respiratory infections due to Bordetella pertussis and other Bordetella subspecies. *Clin Microbiol Rev* 2005;**18**:326–82.
193. Woofrey BF, Moody JA. Human infections associated with Bordetella bronchiseptica. *Clin Microbiol Rev* 1991;**4**:243–55.

194. Yih KW, Silva EA, Ida J, et al. Bordetella holmesii-like organisms isolated from Massachusetts patients with pertussis-like symptoms. *Emerg Infect Dis* 1999;**5**:441–3.
195. Loeffelholtz M. Towards improved accuracy of Bordetella pertussis nucleic acid amplification tests. *J Clin Microbiol* 2012;**50**:2186–90.
196. Probert WS, Ely J, Schrader K, et al. Identification and evaluation of new targets for specific detection of Bordetella pertussis by real-time PCR. *J Clin Microbiol* 2008;**46**:3228–31.
197. Mandal S, Tatti KM, Woods-Stout D, et al. Pertussis pseudo-outbreak linked to specimens contaminated by Bordetella pertussis DNA from clinic surfaces. *Pediatrics* 2012;**129**:e424.
198. Pence MA, TeKippe EM, Burnham C-A. Diagnostic assays for identification of microorganisms and antimicrobial resistance determinants directly from positive blood culture broth. *Clin Lab Med* 2013;**33**:651–84.
199. Chen JHK, Ho PL, Kwan GSW, et al. Direct bacterial identification in positive blood cultures by use of two commercial matrix-assisted laser desorption ionization-time of flight mass spectrometry systems. *J Clin Microbiol* 2013;**51**:1733–9.
200. Neilsen PE, Egholm M. An introduction to peptide nucleic acid. *Curr Issues Mol Biol* 1999;**1**:89–104.
201. Stamper PD, Cai M, Howard T, et al. Clinical validation of the molecular BD GeneOhm StaphSR assay for direct detection of Staphylococcus aureus and methicillin-resistant Staphylococcus aureus in positive blood cultures. *J Clin Microbiol* 2007;**45**:191–6.
202. Wolk DM, Struelens MJ, Pancholi P, et al. Rapid detection of Staphylococcus aureus and methicillin-resistant S. aureus (MRSA) in wound specimens and blood cultures: multicenter preclinical evaluation of the Cepheid Xpert MRSA/SA skin and soft tissue and blood culture assays. *J Clin Microbiol* 2009;**47**:823–6.
203. Buchan BW, Ginocchio CC, Manii R, et al. Multiplex identification of gram-positive bacteria and resistance determinants directly from positive blood culture broths: evaluation of an automated microarray-based nucleic acid test. *PLoS Med* 2013;**10**:e1001478.
204. Ledeboer NA, Lopansri BK, Dhiman N, et al. Identification of gram-negative bacteria and genetic resistance determinants from positive blood culture broths using Verigene gram-negative blood culture multiplex microarray-based molecular assay. *J Clin Microbiol* 2015. https://doi.org/10.1128/JCM.00581-15 [Epub ahead of print].
205. Altun O, Almuhayawi M, Ullberg M, et al. Clinical evaluation of the FilmArray blood culture identification panel in identification of bacteria and yeast from positive blood culture bottles. *J Clin Microbiol* 2013;**51**:4130–6.
206. Bauer KA, Perez KK, Forrest GN, et al. Review of rapid diagnostic tests used by antimicrobial stewardship programs. *Clin Infect Dis* 2014;**59**:S134–45.
207. Lehmann LE, Hunfeld KP, Emrich T, et al. A multiplex real-time PCR assay for detection and differentiation of 25 bacterial and fungal pathogens from whole blood. *Med Microbiol Immunol* 2008;**197**:313–24.
208. Josefson P, Strålin K, Ohlin T, et al. Evaluation of a commercial multiplex PCR test (SeptiFast) in the etiological diagnosis of community-onset bloodstream infections. *Eur J Clin Microbiol Infect Dis* 2011;**30**:1127–34.
209. Lucignano B, Ranno S, Liesenfeld O, et al. Multiplex PCR allows rapid and accurate diagnosis of bloodstream infections in newborns and children with suspected sepsis. *J Clin Microbiol* 2011;**49**:2252–8.
210. Dubská L, Vysdočilová M, Minaříková D, et al. LightCycler Septifast technology in patients with solid malignancies: clinical utility for rapid etiologic diagnosis of sepsis. *Crit Care* 2012;**16**:404–6.
211. Rath P-M, Saner F, Paul A, et al. Multiplex PCR for rapid and improved diagnosis of bloodstream infections in liver transplant recipients. *J Clin Microbiol* 2012;**50**:2069–71.
212. Ecker DJ, Sampath R, Massire C, et al. Ibis T5000: universal biosensor approach for microbiology. *Nat Rev Microbiol* 2008;**6**:553–8.
213. Wolk DM, Kaleta EJ, Wysocki VH. PCR-electrospray ionization mass spectrometry: the potential to change infectious disease diagnostics in clinical and public health laboratories. *J Mol Diagn* 2012;**14**:295–304.
214. Bacconi A, Richmond GS, Baroldi MA, et al. Improved sensitivity for molecular detection of bacterial and Candida infections in blood. *J Clin Microbiol* 2014;**52**:3164–74.
215. Neely LA, Audeh M, Phung NA, et al. T2 magnetic resonance enables nanoparticle-mediated detection of candidemia in whole blood. *Sci Transl Med* 2013;**5**. 182ra54.
216. Mylonakis E, Clancy CJ, Ostroshy-Zeichner L, et al. T2 magnetic resonance assay for rapid diagnosis of candidemia in whole blood: a clinical trial. *Clin Infect Dis* 2015;**60**:892–6.
217. Whitley R, Lakeman F. Herpes simplex virus infections of the central nervous system: therapeutic and diagnostic considerations. *Clin Infect Dis* 1995;**20**:414–20.
218. Tedder D, Ashley R, Tyler K, et al. Herpes simplex virus infection as a cause of benign recurrent lymphocytic meningitis. *Ann Intern Med* 1994;**121**:334–8.
219. Aurelius E, Johansson B, Skoldenberg B, et al. Rapid diagnosis of herpes simplex encephalitis by nested polymerase chain reaction assay of cerebrospinal fluid. *Lancet* 1991;**337**:189–92.
220. Lakeman F, Whitley RJ. Diagnosis of herpes simplex encephalitis: application of polymerase chain reaction to cerebrospinal fluid from brain-biopsied patients and correlation with disease. National Institute of Allergy and Infectious Diseases Collaborative Antiviral Study Group. *J Infect Dis* 1995;**171**:857.
221. Kimberlin DW, Lakeman FD, Arvin AM, et al. Application of the polymerase chain reaction to the diagnosis and management of neonatal herpes simplex virus disease. National Institute of Allergy and Infectious Diseases Collaborative Antiviral Study Group. *J Infect Dis* 1996;**174**:1162–7.
222. Malm G, Forsgren M. Neonatal herpes simplex virus infections: HSV DNA in cerebrospinal fluid and serum. *Arch Dis Child Fetal Neonatal Ed* 1999;**81**:F24–9.
223. Schlesinger Y, Tebas P, Gaudreault-Keener M, et al. Herpes simplex virus type 2 meningitis in the absence of genital lesions: improved recognition with use of the polymerase chain reaction. *Clin Infect Dis* 1995;**20**:842–8.
224. Chonmaitree T, Menegus MA, Powell KR. The clinical relevance of "CSF viral culture": a two-year experience with aseptic meningitis in Rochester, NY. *JAMA* 1982;**247**:1843–7.
225. Rotbart HA. Diagnosis of enteroviral meningitis with the polymerase chain reaction. *J Pediatr* 1990;**117**:85–9.

226. Bourlet T, Caro V, Minjolle S, et al. New PCR test that recognizes all human prototypes of enterovirus: application for clinical diagnosis. *J Clin Microbiol* 2003;**41**:1750—2.
227. Verstrepen WA, Bruynseels P, Mertens AH. Evaluation of a rapid real-time RT-PCR assay for detection of enterovirus RNA in cerebrospinal fluid specimens. *J Clin Virol* 2002;**25**: S39—43.
228. Kost CB, Rogers B, Oberste MS, et al. Multicenter beta trial of the GeneXpert enterovirus assay. *J Clin Microbiol* 2007;**45**: 1081—6.
229. Hamilton MS, Jackson MA, Abel D. Clinical utility of polymerase chain reaction testing for enteroviral meningitis. *Pediatr Infect Dis J* 1999;**18**:533—8.
230. Nigrovic LE, Chiang VW. Cost analysis of enteroviral polymerase chain reaction in infants with fever and cerebrospinal fluid pleocytosis. *Arch Pediatr Adolesc Med* 2000;**154**: 817—21.
231. Ramers C, Billman G, Hartin M, et al. Impact of a diagnostic cerebrospinal fluid enterovirus polymerase chain reaction test on patient management. *JAMA* 2000;**283**:2680—5.
232. Peniche AG, Savidge TC, Dann SM. Recent insights into Clostridium difficile pathogenesis. *Curr Opin Infect Dis* 2013; **26**:44753.
233. McDonald LC, Killgore GE, Thompson A, et al. An epidemic, toxin gene-variant strain of Clostridium difficile. *N Engl J Med* 2005;**353**:2433—41.
234. Wilkins TD, Lyerly DM. Clostridium difficile testing: after 20 years, still challenging. *J Clin Microbiol* 2003;**41**:531—4.
235. Ticehurst JR, Aird DZ, Dam LM, et al. Effective detection of toxigenic Clostridium difficile by a two-step algorithm including tests for antigen and cytotoxin. *J Clin Microbiol* 2006; **44**:1145—9.
236. Wilcox MH, Planche T, Fang FC. What is the role of algorithmic approaches for diagnosis of Clostridium difficile infection? *J Clin Microbiol* 2011;**48**:4347—53.
237. Scallan E, Griffin PM, Angulo FJ, et al. Foodborne illness acquired in the United States-unspecified agents. *Emerg Infect Dis* 2011;**17**:16—22.
238. Baron EJ, Miller MJ, Weinstein MP, et al. A guide to utilization of the microbiology laboratory for diagnosis of infectious diseases: 2013 recommendations by the Infectious Diseases Society of America (IDSA) and the American Society for Microbiology (ASM). *Clin Infect Dis* 2013;**57**:e22—121.
239. Reddington K, Tuite N, Minogue E, et al. A current overview of commercially available nucleic acid diagnostic approaches to detect and identify human gastroenteritis pathogens. *Biomol Detect Quantif* 2014;**1**:2—7.
240. Buchan BW, Olsen WJ, Pezewski M, et al. Clinical evaluation of a real-time PCR assay for identification of Salmonella, Shigella, Campylobacter (Campylobacter jejuni and C. coli) and Shiga toxin-producing E. coli isolates in stool specimens. *J Clin Microbiol* 2013;**51**:4001—7.
241. Harrington SM, Buchan BW, Doern C, et al. Multicenter evaluation of the BD Max enteric panel PCR assay for rapid detection of Salmonella spp., Shigella spp., Campylobacter spp. (C. jejuni and C. coli), and Shiga toxin 1 and 2 genes. *J Clin Microbiol* 2015;**53**:1639—47.
242. Navidad JF, Griswold DJ, Gradus MS, et al. Evaluation of Luminex xTAG gastrointestinal panel analyte-specific reagents for high-throughput, simultaneous detection of bacteria, viruses, and parasites of clinical and public health importance. *J Clin Microbiol* 2013;**51**:3018—24.
243. Novak SM, Bobenchik A, Cumpio J, et al. Evaluation of the Verigene EP IUO test for rapid detection of bacterial and viral causes of gastrointestinal infection. Abstract 1317 30th Clin Virol Symp Annu Meet Pan Am Soc Clin Virol.
244. Buss SN, Leber A, Chapin K, et al. Multicenter evaluation of the BioFire FilmArray gastrointestinal panel for etiologic diagnosis of infectious gastroenteritis. *J Clin Microbiol* 2015;**53**:915—25.
245. Huletsky A, Giroux R, Rossbach V, et al. New real-time PCR assay for rapid detection of methicillin-resistant Staphylococcus aureus directly from specimens containing a mixture of staphylococci. *J Clin Microbiol* 2004;**42**:1875—84.
246. Bressler AM, Williams T, Culler EE, et al. Correlation of penicillin binding protein 2a detection with oxacillin resistance in Staphylococcus aureus and discovery of a novel penicillin binding protein 2a mutation. *J Clin Microbiol* 2005:434541—4.
247. Wong H, Louie L, Lo RY, et al. Characterization of Staphylococcus aureus isolates with a partial or complete absence of staphylococcal cassette chromosome elements. *J Clin Microbiol* 2010;**48**:3525—31.
248. Carroll KC. Rapid diagnostics for methicillin-resistant Staphylococcus aureus: current status. *Mol Diagn Ther* 2008;**12**:15—24.
249. Cunningham R, Jenks P, Northwood J, et al. Effect on MRSA transmission of rapid testing of patients admitted to critical care. *J Hosp Infect* 2007;**65**:24—8.
250. Robicsek A, Beaumont JL, Paule SM, et al. Universal surveillance of methicillin-resistant Staphylococcus aureus in 3 affiliated hospitals. *Ann Intern Med* 2008;**148**:409—18.
251. Vergis EN, Hayden MK, Chow JW, et al. Determinants of vancomycin resistance and mortality rates in enterococcal bacteremia: a perspective multicenter trial. *Ann Intern Med* 2001;**135**:484—92.
252. Centers for Disease Control and Prevention. Recommendations for preventing spread of vancomycin resistance: recommendations of the Hospital Infection Control Practices Advisory Committee (HIC-PAC). *MMWR Recomm Rep* 1995;**44**:1—13.
253. Satake S, Clark N, Rimland D, et al. Detection of vancomycin-resistant enterococci in fecal samples. *J Clin Microbiol* 1997;**35**: 2325—30.
254. Paule S, Trick WE, Tenover FC, et al. Comparison of PCR assay to culture for surveillance detection of vancomycin-resistant enterococci. *J Clin Microbiol* 2003:4805—7.
255. Nordman P, Gniadkowski M, Giske CG, et al. Identification and screening of carbapenemase-producing Enterobacteriaceae. *Clin Microbiol Infect* 2012;**18**:432—8.
256. Livermore DM, Andrews JM, Hawkey PM, et al. Are susceptibility tests enough, or should laboratories still seek ESBLs and carbapenemases directly? *J Antimicrob Chemother* 2012;**67**:1569—77.
257. Cole JM, Schuetz AN, Hill CE, et al. Development and evaluation of a real-time PCR assay for detection of Klebsiella pneumoniae carbapenemase genes. *J Clin Microbiol* 2009;**47**:322—6.
258. Naas T, Cuzon G, Truong H, et al. Evaluation of a DNA microarray, the Check-Points ESBL/KPC array, for rapid detection of TEM, SHV, and CTX-M extended-spectrum beta-lactamases and KPC carbapenemases. *Antimicrob Agents Chemother* 2010;**54**:3086—92.
259. Cuzon G, Naas T, Bogaerts P, et al. Evaluation of a DNA microarray for the rapid detection of extended-spectrum beta-lactamases (TEM, SHV, and CTX-M), plasmid-mediated cephalosporinases (CMY-2-like, DHA, FOX, ACC-1, ACT/MIR, and CMY-1-like/MOX) and carbapenemases (KPC, OXA-48, VIM, IMP, and NDM). *J Antimicrob Chemother* 2012; **67**:1865—9.

260. Wintermans BB, Reuland EA, Wintermans RG, et al. The cost-effectiveness of ESBL detection: towards molecular detection methods? *Clin Microbiol Infect* 2012;**19**:662−5.
261. Kaase M, Szabados F, Wassill L, et al. Detection of carbapenemases in Enterobacteriaceae by a commercial multiplex PCR. *J Clin Microbiol* 2012;**50**:3115−8.
262. Spanu T, Fiori B, D'Inzeo T, et al. Evaluation of the new NucliSENS EasyQ KPC test for rapid detection of Klebsiella pneumoniae carbapenemase genes (blaKPC). *J Clin Microbiol* 2012;**50**:2783−5.
263. Cho I, Blaser MJ. The human microbiome: at the interface of health and disease. *Nat Rev Genet* 2012;**13**:260−70.
264. Wylie KM, Weinstock GM, Storch GA. Emerging view of the human virome. *Transl Res* 2012;**160**:283−90.
265. Petrosino JF, Highlander S, Luna RA, et al. Metagenomic pyrosequencing and microbial identification. *Clin Chem* 2009;**55**:856−66.
266. Human Microbiome Project Consortium. Evaluation of 16S rDNA-based community profiling for human microbiome research. *PLoS ONE* 2012;**7**e39315.
267. Human Microbiome Project Consortium. A framework for human microbiome research. *Nature* 2012;**486**:215−21.
268. Human Microbiome Project Consortium. Structure function and diversity of the healthy human microbiome. *Nature* 2012;**486**:207−14.
269. Huse SM, Ye Y, Zhou Y, et al. A core human microbiome as viewed through 16S rRNA sequence clusters. *PLoS ONE* 2012;**7**:e34242.
270. Abubucker S, Segata N, Goll J, et al. Metabolic reconstruction for metagenomics data and its application to the human microbiome. *PLoS Comput Biol* 2012;**8**:e1002358.
271. Cantarel BL, Lombard V, Henrissat B. Complex carbohydrate utilization by the healthy human microbiome. *PLoS ONE* 2012;**7**:e28742.
272. Taur Y, Pamer EG. Fixing the microbiota to treat Clostridium difficile infections. *Nat Med* 2014;**20**:246−7.
273. Henrich TJ, Gallien S, Li JZ, et al. Low-level detection and quantitation of cellular HIV-1 DNA and 2-LTR circles using droplet digital PCR. *J Virol Methods* 2012;**186**:68−72.
274. Hayden RT, Gu Z, Ingersoll J, et al. Comparison of droplet digital PCR to real-time PCR for quantitative detection of cytomegalovirus. *J Clin Microbiol* 2013;**51**:540−6.
275. Niemz A, Ferguson J, Boyle DS. Point-of-care nucleic acid testing for infectious diseases. *Trends Biotechnol* 2011;**29**:240−50.
276. Azarian T, Cook RL, Johnson JA, et al. Whole-genome sequencing for outbreak investigations of methicillin resistant Staphylococcus aureus in the neonatal intensive care unit: time for routine practice? *Infect Control Hosp Epidemiol* 2015;**36**:777−85.
277. Lohman NJ, Constantinidou C, Christner M, et al. A culture-independent sequence-based metagenomics approach to investigation of an outbreak of shiga-toxigenic Escherichia coli O104:H4. *JAMA* 2013;**309**:1502−10.
278. Feng H, Shuda M, Chang Y, et al. Clonal integration of a polyomavirus in human Merkel cell carcinoma. *Science* 2008;**319**:1096−100.
279. Cotton M, Watson SJ, Kellam P, et al. Transmission and evolution of the Middle East respiratory syndrome coronavirus in Saudi Arabia: a descriptive genomic study. *Lancet* 2013;**382**:1993−2002.
280. Gao Z, Tseng CH, Strober BE, et al. Substantial alterations of the cutaneous bacterial biota in psoriatic lesions. *PLoS ONE* 2008;**3**:e2719.
281. Peek Jr RM, Blaser MJ. Helicobacter pylori and gastrointestinal tract adenocarcinomas. *Nat Rev Cancer* 2002;**2**:28−37.
282. Islami F, Kamangar F. Helicobacter pylori and esophageal cancer risk: a meta-analysis. *Cancer Prev Res (Phila)* 2008;**1**:329−38.
283. Turnbaugh PJ, Ley RE, Mahowald MA, et al. An obesity-associated gut microbiome with increased capacity of energy harvest. *Nature* 2006;**444**:1027−31.
284. Ley RE, Backhed F, Turnbaugh P, et al. Obesity alters gut microbial ecology. *Proc Natl Acad Sci USA* 2005;**102**:11070−5.
285. Chen Y, Blaser MJ. Inverse associations of Helicobacter pylori with asthma and allergy. *Arch Intern Med* 2007;**167**:821−7.
286. Blaser MJ, Chen Y, Reibman J. Does Helicobacter pylori protect against asthma and allergy? *Gut* 2008;**57**:561−7.
287. Garrett WS, Gallini CA, Yatsunenko T, et al. Enterobacteriaceae act in concert with the gut microbiota to induce spontaneous and maternally transmitted colitis. *Cell Host Microbe* 2010;**8**:292−300.
288. Tana C, Umesaki Y, Imaoka A, et al. Altered profiles of intestinal microbiota and organic acids may be the origin of symptoms in irritable bowel syndrome. *Neurogastroenterol Motil* 2010;**22**:512−9.
289. Kostic AD, Gevers D, Pedamallu CS, et al. Genomic analysis identifies association of Fusobacterium with colorectal carcinoma. *Genome Res* 2012;**22**:292−8.
290. Castellarin M, Warren RL, Freeman JD, et al. Fusobacterium nucleatum infection is prevalent in human colorectal carcinoma. *Genome Res* 2012;**22**:299−306.
291. Wang Z, Klipfell E, Bennet BJ, et al. Gut flora metabolism of phosphatidylcholine promotes cardiovascular disease. *Nature* 2011:47257−63.

Genetics

Cindy L. Vnencak-Jones and D. Hunter Best

ABSTRACT

Background
The invention of polymerase chain reaction along with the chemistry of fluorescently labeled molecules, high density DNA single nucleotide polymorphism arrays for genome-wide association studies (GWAS), massively parallel sequencing (MPS) technology, the development of chromosomal microarrays, the availability of public databases, and advances in bioinformatics have revolutionized the field of human genetics. The collective use of GWAS and exome- and whole-genome DNA sequencing has facilitated disease discovery and the identification of pathogenic variants for diseases in which the underlying genetic cause was previously unknown. This knowledge has enhanced the efforts for personalized medicine, or a tailored approach to therapy based on an individual's genotype.

Content
This chapter discusses recent advances in the field using some common inherited autosomal recessive, autosomal dominant, and X-linked diseases as examples. In addition, some common mitochondrial, imprinting, and complex disorders and inherited cancers are reviewed. For each disease, information regarding the clinical phenotype, gene, protein function, treatment, and currently used clinical molecular diagnostic techniques are summarized.

DISEASES WITH MENDELIAN INHERITANCE

Autosomal Recessive Disorders

An individual with an autosomal recessive disease has inherited two abnormal alleles at a given locus by receiving one mutant allele from each carrier parent; the disease-causing gene is on one of the autosomes (1—22) and not on a sex chromosome (X or Y). Typically, the carrier parent with one abnormal allele has no clinical features of the disease yet possesses a 50% risk of donating the mutant allele to his or her offspring. Matings in which both partners are carriers of an abnormal allele have a 25% chance of producing a child with both normal alleles, a 50% chance of having a child that has received only one abnormal allele, and a 25% chance of having an affected child. The affected patient may be homozygous for a specific mutation by receiving the same mutation from each parent or may be a compound heterozygote having received a different mutation from each parent. The specific mutations present influence the clinical severity of the disease and account for variability in the expression of the disease between patients, referred to as genotype—phenotype correlation. Modifier genes and environmental factors also play a role in determining the patient's phenotype. Among pedigrees illustrating autosomal recessive disorders, males and females are equally affected, and for rare diseases, consanguinity is likely to be observed. Table 6.1 provides a list of some of the inherited autosomal recessive disorders commonly tested in clinical molecular diagnostic laboratories.

Cystic Fibrosis

Cystic fibrosis (CF) (Online Mendelian Inheritance in Man [OMIM] #219700) is one of the most common autosomal recessive diseases in people of Northern European ancestry with an estimated incidence in the United States of about 1 in 2500 and a carrier frequency of about 1 in 25.[1] Within other ethnic populations, the disease has an estimated incidence of 1 in 2300 Ashkenazi Jews, 1 in 8500 Hispanics, 1 in 17,000 African Americans, and 1 in 35,000 Asian Americans.[2] CF is a multisystem disorder characterized by progressive pulmonary disease, pancreatic insufficiency, elevated sweat electrolytes, male infertility, and a predisposition to sinonasal disease. An abnormal sweat chloride concentration is considered the gold standard for the diagnosis of CF, and a result of 60 mmol/L or greater is considered diagnostic of CF. However, some patients with disease-causing mutations may have indeterminate or borderline values of 30 to 59 mmol/L if they are younger than 6 months of age or 40 to 59 mmol/L if they are older than 6 months of age.[3,4] Although values of less than 30 mmol/L or less than 40 mmol/L, respectively make the diagnosis of CF less likely, normal measurements can be obtained in some patients. By virtue of the intricacies of this test, it is recommended that these tests only be performed at a Cystic Fibrosis Foundation—accredited care center (https://www.cff.org). Interestingly, the phenotypic expression of the disease is heterogeneous, ranging from meconium ileus and severe respiratory disease in infants to mild pulmonary symptoms and no evidence of gastrointestinal problems

TABLE 6.1 Examples of Autosomal Recessive Disorders

Disease	Gene	Location	OMIM Entry #	Incidence
α_1-Antitrypsin	SERPINA1	14q32.13	613490	1 in 5000–7000
Canavan disease	ASPA	17p13.2	271900	1 in 6400–13,400 Ashkenazi Jews (less common in other populations)
Friedreich ataxia	FXN	9q21.11	229300	1 in 25,000–50,000
Gaucher disease type I	GBA	1q22	230800	1 in 850 Ashkenazi Jews (less common in other populations)
Glycogen storage disease	G6PC	17q21.31	232200	1 in 100,000
Hereditary hemochromatosis	HFE	6p22.2	235200	1 in 200–350
Hurler syndrome: mucopolysaccharidosis type 1	IDUA	4p16.3	607014	1 in 100,000
Medium-chain acyl-coenzyme A dehydrogenase (MCAD) deficiency	ACADM	1p31.1	201450	1 in 4900–17,000
Niemann-Pick type C	NPC1	18q11.2	257220	1 in 100,000–150,000
Tay Sachs disease	HEXA	15q23	272800	1 in 3500 Ashkenazi Jews (less common in other populations)

OMIM, Online Mendelian Inheritance in Man.

in adulthood. Atypical CF patients with a nonclassic presentation may have involvement of only one organ, as in congenital bilateral absence of the vas deferens (CBAVD), pancreatitis, rhinosinusitis, or nasal polyps.[3,5,6] Variability in expression is explained by both allelic heterogeneity at the CF gene locus and genetic variation in modifier genes. Loss or decreased amounts of the disease-associated protein cause mucous accumulation and airway obstruction; recurrent infection with pathogens, such as *Pseudomonas aeruginosa, Burkholderia cepacia, Staphylococcus aureus,* and *Haemophilus influenzae;* and excessive inflammation with progressive lung damage and ultimately respiratory failure.[7-9] Because morbidity and mortality are most related to severe lung disease, daily management of these patients is to prevent bacterial lung infections through physical therapy of the chest and early or aggressive treatments to eradicate bacterial colonization. Although originally considered a fatal childhood disease, the US Cystic Fibrosis Foundation reported that in 2008, more than 45% of patients were older than 18 years of age. Furthermore, the median survival age among approximately 23,000 patients with CF who received care through one of the nationwide CF care centers was 37.4 years with the life expectancy of patients born now in their 50s. The increase in survival age is due to organ transplantation, improved nutrition, and new therapies.[10-13] The disease is complex with clinical management of most patients at one of more than 110 specialized care centers in the CF Care Center Network.[14-16] This approach provides widespread communication among health care providers who are experts in the care of patients with CF and enables monitoring of a large population of patients with respect to treatment outcome, health care, and disease-specific variables.

The severity and frequency of the disease led to an intensive search for the gene, which was eventually cloned in 1989.[17-19] The CF gene maps to chromosome 7q31.2 with 27 exons encoding a transcript of approximately 6.5 kb. The CF transmembrane conductance regulator protein (CFTR) has 1480 amino acids and is a member of the ATP-binding cassette (ABC) transporter superfamily of membrane transport proteins. CFTR consists of two transmembrane domains (TMDs), each containing six hydrophobic transmembrane sections and one hydrophilic intracellular nucleotide-binding domain (NBD).[18] The TMD/NBD segments are linked by a highly charged regulatory domain containing multiple sites for phosphorylation and activation. The molecule is unique among ABC proteins because it does not actively transport but rather serves as an ATP-gated chloride ion channel pore within the lipid bilayer, predominantly at the apical membrane of secretory epithelial cells. In addition to epithelial chloride conductance as an ion channel, CFTR mediates the passage of bicarbonate and other small ions, including sodium and potassium, from the intracellular compartment to the extracellular surface.[20-23] Whereas opening and closing of the channel requires ATP binding and hydrolysis (ATP-gated channel), channel activity requires phosphorylation of the cytoplasmic regulatory domain by protein kinase A. The wide clinical diversity of CF is based in part on the varying effects conferred on this protein from the more than 2000 mutations reported within this gene.[24]

Because CF is an autosomal recessive disorder, patients with CF must have two mutant *CFTR* alleles to develop the disease. Some mutations are "private" and unique to a family; others may be common among CF patients. Patients may be homozygous with two copies of the same mutation, or they may be compound heterozygotes with one copy of one mutation and one copy of a second mutation. More than half of all mutations are missense or frameshift mutations, with exon 14 containing the largest number of different disease-causing mutations.[24] The types of mutations and frequencies of each differ significantly among populations.[25] Mutations can be divided into six classes based on their effect on the protein[26] (Table 6.2). Class I mutations result in no functional protein production and include nonsense or frameshift mutations that cause premature truncation of the protein, splice site mutations, and exon or gene deletions or rearrangements. Class II mutations are the most common and are associated with defective processing of CFTR and the inability of the protein to reach the apical cell surface. Class II mutations cause misfolding of the fully translated CFTR protein and result from an amino acid alteration such as a deletion or a missense mutation. In the case of class I and II mutations, CFTR is not present on the apical cell membrane, and as predicted, these mutations are typically associated with a severe disease phenotype. Class III and IV mutations are generally due to missense mutations that result in full-length CFTR expression at the cell membrane. Class III mutations are

TABLE 6.2 American College of Obstetricians and Gynecologists/American College of Medical Genetics Recommended Mutation Panel for Cystic Fibrosis Carrier Screening

Mutation Frequency Among Patients With Clinically Diagnosed Cystic Fibrosis (%)

CFTR Mutation	Mutation Class	Ashkenazi Jewish	Non-Hispanic White	Hispanic White	African American	Asian American
p.Phe508del	II	31.41	72.42	54.38	44.07	38.95
p.Gly542Ter	I	7.55	2.28	5.10	1.45	0.00
p.Trp1282Ter	I	45.92	1.50	0.63	0.24	0.00
p.Gly551Asp	III	0.22	2.25	0.56	1.21	3.15
c.621+1G>T	I	0.00	1.57	0.26	1.11	0.00
p.Asn1303Lys	II	2.78	1.27	1.66	0.35	0.76
p.Arg553Ter	I	0.00	0.87	2.81	2.32	0.76
p.Ile507del	II	0.22	0.88	0.68	1.87	0.00
c.3489+10kbC>T	V	4.77	0.58	1.57	0.17	5.31
c.3120+1G>T	V	0.10	0.08	0.16	9.57	0.00
p.Arg117His	IV	0.00	0.70	0.11	0.06	0.00
c.1717-1G>T	I	0.67	0.48	0.27	0.37	0.00
c.2789+5G>A	V	0.10	0.48	0.16	0.00	0.00
p.Arg347Pro	IV	0.00	0.45	0.16	0.06	0.00
c.711+1G>T	I	0.10	0.43	0.23	0.00	0.00
p.Arg334Trp	IV	0.00	0.14	1.78	0.49	0.00
p.Arg560Thr	II	0.00	0.38	0.00	0.17	0.00
p.Arg1162Ter	I	0.00	0.23	0.58	0.66	0.00
c.3659delC	I	0.00	0.34	0.13	0.06	0.00
p.Ala455Glu	V	0.00	0.34	0.05	0.00	0.00
p.Gly85Glu	II	0.00	0.29	0.23	0.12	0.00
c.2184delA	I	0.10	0.17	0.16	0.05	0.00
c.1898+1G>A	I	0.10	0.16	0.05	0.06	0.00
Total		**94.04**	**88.29**	**71.72**	**64.46**	**48.93**

Modified from Watson MS, Cutting GR, Desnick RJ, et al. Cystic fibrosis population carrier screening: 2004 revision of American College of Medical Genetics mutation panel. *Genet Med* 2004;6:387-391.

more severe, resulting in defective regulation; class IV mutations generally cause a milder phenotype with reduced conduction of ion flow. Class V mutations are associated with reduced amounts of CFTR at the cell membrane and are most often associated with abnormal splicing and decreased amounts of normal CFTR messenger RNA (mRNA). These mutations may be associated with a severe phenotype (c.621+1G>T) or a mild phenotype (c.2789+5G>A). Last, class VI mutations cause decreased stability of CFTR in the membrane. In this chapter, the nomenclature for variants follows the Human Genome Variation Society (http://www.hgvs.org/mutnomen), and variants are specified by either the amino acid in the protein (p.) or by the base in the cDNA (c.).

The most common mutation, p.Phe508del (c.1521_1523delCTT), is a class II mutation and is detected in about 70% of CF alleles in whites of Northern European descent. This CFTR protein is misfolded and not properly processed by the endoplasmic reticulum with the majority of the protein rapidly degraded.[27] Whereas common mutations p.Gly542Ter and p.Trp1282Ter are class I mutations and cause premature translation termination and premature truncation of the protein, mutation p.Gly551Asp results in a full-length CFTR that reaches the apical membrane but improperly regulates the chloride channel.[28,29] Infertile males with only the genital form of CF—congenital bilateral absence of the vas deferens (CBAVD)—usually have a mutation associated with a severe phenotype on one allele and a mutation associated with a mild phenotype on the second allele. The most frequently reported CFTR genotype in this population is the 5T polymorphism in intron 8, c.1210-12T(5), which corresponds to a sequence of five thymidines.[3] This 5T variant is observed in about 5% of CFTR alleles and is less common than the 7T or 9T alleles, c.1210-12T(7) or c.1210-12T(9). The 5T variant affects mRNA splicing and can cause exon 9 to be deleted; without exon 9, the chloride channel is not functional.[30-32] An adjacent polymorphic TG dinucleotide sequence c.1210-34TG(9-13) regulates the efficiency of mRNA splicing, with the higher number of TG repeats c.1210-34TG(13) associated with decreased efficiency of splicing.[33] Thus, the c.1210-12T(5) c.1210-34TG(12) or c.1210-12T(5) c.1210-34TG(13) allele is more commonly associated with an abnormal phenotype than is the c.1210-12T(5) c.1210-34TG(11) allele.

Understanding the effect of each mutation on the CFTR protein is important for choosing the corrective drug therapy required for each patient.[12,13,26,34] Therapy for patients with class I or II mutations is the most challenging because CFTR in most cases is absent or in reduced amounts. Gene therapy to deliver a functional CFTR protein is one effective way to treat these patients and is also applicable to CF patients with mutations in classes other than I or II. Alternatively, for patients in this group with a nonsense stop codon such as p.Gly542Ter, the drug ataluren, an orally administered small molecule that enables readthrough of the transcript, has been attempted. Unfortunately, aminoglycoside antibiotics frequently used by patients with CF can potentially inhibit this molecule, rendering it less effective in a subset of patients. For some class II and for all class III mutations, the potentiator ivacaftor has been administered to increase the effectiveness of chloride transfer through the channel. This drug improves the clinical outcome of patients, especially for those with the common mutation p.Gly551Asp. This drug is also used in patients with mutations in class IV, V, and VI. Ivacaftor has proven more effective in some class II mutations, most notably the common p.Phe508del mutation, in combination with a second small molecule, lumacaftor. Lumacaftor, a CFTR corrector molecule, helps the misfolded p.Phe508del protein reach the cell surface. Most recently, antisense oligonucleotides have been examined as a potential therapy for some class V mutations.[35]

The CFTR genotype and clinical phenotype correlations are most closely related for pancreatic involvement rather than for pulmonary manifestations of the disease.[36,37] Most patients with two "severe" mutations, which cause a "severe" CF phenotype, usually classes I to III, have pancreatic insufficiency, but patients with one or two "mild" mutations, which cause a "mild" CF phenotype, generally classes IV to VI, have pancreatic sufficiency (PS) but have an increased risk of developing pancreatitis.[38] Furthermore, CF patients with PS generally have milder disease with longer overall survival, a later age of diagnosis, and lower sweat chloride levels. Although mutations in CFTR confer susceptibility to pancreatitis, variants in several other genes inherited in combination with or without CFTR variants are seen in patients with chronic pancreatitis.[39] In contrast, lung disease is more dependent on environmental factors and genetic modifiers.[7,38,40] Environmental modulating factors of lung disease include secondhand smoke and varying exposure to pathogens.[41] Genes that modify lung disease among patients with identical genotypes include inflammatory response genes that code for cytokines such as transforming growth factor β1 (TGF-β1) or interleukin-8 (IL8) or pathogen response genes such as mannose binding lectin 2 *(MBL2)*. Other modifying variants may include genes whose proteins are associated with tissue damage or repair, such as glutathione S-transferases *(GSTs)* or nitric oxide synthetases *(NOS1, NOS3)*. Genes that encode proteins for ion transport or cytoskeleton structure such as voltage-gated chloride channel protein or keratin 8 type 2 protein may also modulate the phenotype.[38] More recently, miRNAs have been proposed to explain the observed clinical heterogeneity among patients with the same mutation.[42]

DNA testing for the identification of CFTR mutations is performed for a variety of reasons (Box 6.1). It is performed to confirm the diagnosis of disease in patients with equivocal sweat chloride results or when an insufficient amount of sweat is collected and no results are obtained from the test. The presence of two known pathogenic CFTR mutations is diagnostic for CF. A diagnosis of CF can be considered in the patient with a CFTR-related disorder such as chronic pancreatitis, CBAVD, or sinusitis. In known CF patients, mutation analysis can be requested to help predict the prognosis based on genotype–phenotype correlations and initiate mutation-specific treatment options. At the same time, identifying the CFTR gene mutations that are present in the proband and how they segregate in the family enables preimplantation diagnosis or prenatal testing for subsequent pregnancies if desired and allows carrier or diagnostic testing for other at-risk family members. Similarly, state-sponsored newborn screening (NBS) programs have detected infants with CF and at the same time have identified at-risk family members and enabled carrier or diagnostic genetic testing. Most important, NBS allows for early diagnosis and referral to CF care centers for early intervention and management of CF-related clinical symptoms. In the United States, NBS for CF is conducted in all 50 states and the District of Columbia and is based on the measurement of immunoreactive trypsinogen (IRT) from dried blood spots to detect elevated levels of this pancreatic enzyme. However, states vary in their NBS algorithms. In some states, a positive result is followed by IRT retesting; in others, blood from the newborn is submitted for CFTR mutation analysis.[43,44] Confirmation of the diagnosis requires a sweat chloride test but may not always be effective in infants younger than 2 weeks.[43] A diagnostic conundrum occurs when sweat chloride testing is normal or intermediate and the phenotypic consequence of one or two of the CFTR variants detected is uncertain. In the United States, these children do not meet the diagnostic criteria and are classified as having CFTR-related metabolic syndrome.[45] These patients should be monitored closely and may ultimately develop symptoms of CF, albeit with a milder disease course.[46] Other patients with CF-like disease with borderline sweat chloride results and clinical features associate with CF but no CFTR gene variants detected may have complex genotypes with variants in one or more genes possibly in a contributory oligogenic fashion.[47] Some couples referred for CF carrier testing or prenatal testing of the fetus have no family history of CF but rather the presentation of hyperechogenic bowel in the fetus on routine ultrasonography.

Cystic fibrosis carrier screening may be offered as a single gene-specific test or may be included within high-throughput large carrier screening panels that detect both common CFTR gene mutations as well as targeted mutations in other genes.[48,49] Although these more extensive panels provide additional information to patients, they contain genes for inherited conditions for which population-based screening is not recommended in current practice guidelines.[50] CF carrier screening for preconception and expectant couples was first recommended in October 2001 by the American College of Obstetricians and Gynecologists (ACOG) in conjunction with the American College of Medical Genetics (ACMG). A core panel of the most common mutations includes those with an estimated prevalence of at least 0.1% of CF mutant alleles and a carrier detection rate of about 88% for non-Hispanic whites (see Table 6.2).[51] The intent of the screening panel is to identify individuals at risk for classical CF, not

---

BOX 6.1　Referrals for *CFTR* Mutation Analysis

Confirm diagnosis of cystic fibrosis
Determine prognosis
Screen patient with pancreatitis
Family member testing
Newborn screening
Preconception couples
Expectant couples
Prenatal testing—at-risk fetus
Prenatal testing—hyperechogenic bowel
Preimplantation genetic diagnosis
Infertile male with congenital bilateral absence of the vas deferens
Semen and oocyte donors

TABLE 6.3 Cystic Fibrosis Mutation Carrier Risk

	Ashkenazi Jewish	Non-Hispanic White	Hispanic White	African American	Asian American
Detection rate of ACOG/ACMG 23 mutation panel (%)	94	88	72	64	49
Estimated carrier risk in population	1/24	1/25	1/58	1/61	1/94
Estimated residual carrier risk after no mutation detected on screening panel	1/380	1/200	1/200	1/170	1/180

ACMG, American College of Medical Genetics; *ACOG*, American College of Obstetricians and Gynecologists.
Modified from ACOG committee opinion no. 486: update on carrier screening for cystic fibrosis. *Obstet Gynecol* 2011;117:1028-1031.

isolated CBAVD; thus it is recommended that the 5T/7T/9T status be reviewed and reported only in the presence of variant p.Arg117His, which can be in *cis* with either the 5T or 7T variant. This allows for distinction between p.Arg117His-5T and p.Arg117His-7T individuals and enables genetic counseling and potential prenatal testing options for individuals who are *CFTR* p.Arg117His-5T positive and are at risk of having an offspring with a classic CF phenotype if their reproductive partner is also a carrier of CF. Although CF is more common in the white and Ashkenazi Jewish populations, the standard of care recommended by the ACOG is to make CF testing available to all preconception or expectant couples, especially because it is becoming more difficult to assign a single ethnicity to a patient to best determine the carrier risk.[1] Counseling for CF carrier testing is complex and should include information about CF and CFTR-related disorders and the a priori likelihood of having a child with CF based on personal and family history and ethnicity.[52] Furthermore, it is important for the patient to understand the inability of the screening panel to detect all CF mutations and the residual risk of being a carrier despite a negative test result (Table 6.3).[1] This is especially important for patients in ethnic groups for which the *CFTR* gene mutation detection level is reduced. The genetics professional should discuss the possibility of stigmatization or anxiety associated with being a carrier of a genetic disease and how knowing this information may affect her pregnancy. After DNA testing, it is very important for the genetics professional to know relevant family history when interpreting and reporting the test results to the patient and reproductive partner to accurately assess the risk to the couple of having a child with CF.

In families at risk for a child with CF, prenatal testing of the fetus may be requested. In these prenatal cases and for all DNA testing of fetal DNA, it is the standard of care that the DNA extracted from the fetus be tested for the presence of maternal contamination because the presence of maternal DNA can interfere with interpretation of test results. This is best performed by using polymerase chain reaction (PCR) amplification for highly polymorphic short tandem repeat loci coupled with capillary electrophoresis (Fig. 6.1). In the case of prenatal *CFTR* gene mutation testing with a maternal carrier, if the fetus actually has no mutant *CFTR* allele but the extracted fetal DNA is contaminated with maternal DNA, the mutant maternal *CFTR* allele could be detected, and the fetal DNA test result could be erroneously interpreted as a carrier of a mutant *CFTR* gene. Furthermore, if the fetus inherited a paternal *CFTR* mutation and maternal DNA contaminated the fetal DNA specimen, the fetal DNA test results could show the presence of two *CFTR* gene mutations, and the fetal DNA test result could be erroneously interpreted as a compound heterozygote affected with CF.

CFTR gene mutation testing is performed using a laboratory developed test or one of a variety of commercially available platforms with kits cleared by the US Food and Drug Administration (FDA). The number of mutations detected by each assay is variable, and the median turnaround time is 14 days.[53] Furthermore, because the detection rate of the 23 mutation gene panel is lower in some ethnicities, laboratories serving such populations should consider supplementing this screening panel with additional mutation analysis.[54] Although full gene sequencing by Sanger or massively parallel sequencing (MPS) can be done to detect *CFTR* gene mutations, current practice guidelines do not recommend this methodology for CF carrier screening.[1] Rather, sequencing of the *CFTR* gene should be performed in the context of (1) confirming the diagnosis of CF in a newborn with a positive IRT and a clinical suspicion of CF but no or only one *CFTR* gene mutation was detected with the 23 common mutation panel, (2) identification of *CFTR* mutations in a CF patient in which both *CFTR* mutations were not detected with the 23 *CFTR* gene mutation panel, and (3) confirmation of CF in a patient with a *CFTR*-related disease to rule out CF. In all cases, if *CFTR* gene mutations are identified by sequencing in the proband, targeted analysis to specifically screen for that mutation in at-risk family members may be performed.

Spinal Muscular Atrophy

The spinal muscular atrophies (SMAs) are a heterogeneous group of neurodegenerative disorders characterized by progressive loss of motor neurons in the spinal cord and lower brainstem with muscle weakness and atrophy. A wide clinical spectrum is observed with variability in age of onset, motor function impairment, and inheritance patterns, with 33 contributing genes thus far.[55] SMN-related SMA or SMA5q is an autosomal recessive disorder that accounts for up to 95% of SMAs, has an incidence of 1 in 6000 to 10,000 births, and is a leading cause of death in infants. SMA5q is caused by mutations in the survival motor neuron 1 gene *(SMN1)* and is divided into four types based on clinical presentation and age of onset.[56] Type 1 (OMIM #253300) is the most common (50%); it is associated with age of onset younger than 6 months and has a median survival time of less than 1 year. These children have profound hypotonia, no control of head movement, and are unable to sit. Intercostal muscle weakness leads to respiratory failure and tongue fasciculation, dysphagia, and fatigue, making feeding difficult, worsening

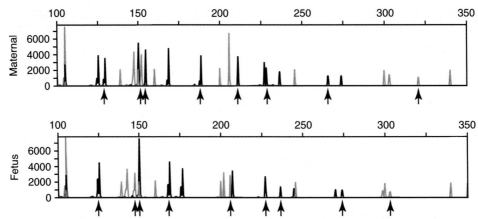

FIGURE 6.1 Electropherograms obtained after polymerase chain reaction (PCR) amplification of maternal *(upper panel)* and fetal *(lower panel)* DNA at nine independently segregating loci and one gender-specific marker. Extracted maternal and fetal DNA were amplified by PCR using a multiplex PCR assay with fluorescent labeled primers (AmpFLSTR Profiler Plus ID PCR Amplification Kit; Applied Biosystems, Foster City, California). Amplicons were detected after capillary electrophoresis on an ABI 3130*xl* Genetic Analyzer and were analyzed using GeneMapper software (Applied Biosystems). Amplicon sizes in bases are noted at the top of the figure and the relative amount of fluorescence detected for each amplicon is measured by the peak height (y-axis). Arrows in the *upper panel* denote maternal alleles absent in the fetal DNA specimen. Arrows in the *bottom panel* indicate maternal alleles inherited by the fetus.

the condition, and increasing the risk of aspiration pneumonia. A subdivision of this group into type 1 (0–6 months) and type 0 with prenatal onset and severe disease at birth has been proposed.[56] SMA type II (OMIM #253550), representing about 20% of cases, has an age of onset between 7 and 18 months with median survival into the third decade of life. These children can sit, and some can stand, although none can walk independently. SMA type III (OMIM #253400) is seen in about 30% of patients and has an age of onset after 18 months. These patients have a mild phenotype with gradually progressive disease but a normal life expectancy. SMA type IV (OMIM #271150) is rare and is the mildest of all forms. This type was initially characterized by muscle weakness in the second or third decade of life. These patients are ambulatory and have a normal lifespan.

The gene for SMA, *SMN1*, was mapped to chromosome 5q11.2-13.3 in 1990 and cloned in 1995.[57-59] This gene contains 9 exons (numbered 1, 2a, 2b, and 3–8) spanning about 28,000 bases and encodes a 1.7 kb mRNA transcript producing a 38-kDa protein composed of 294 amino acids that is ubiquitously expressed in the nucleus and cytoplasm.[60] The SMN protein is one of nine core proteins in a multiprotein complex that also contains the proteins Unrip and Gemins2-8.[61] This complex is enriched in the nucleus in size and number to form Gems.[62] The SMN-GEMINs is essential for the cytoplasmic assembly of small ribonucleoproteins (snRNPs) into the spliceosome, critical components for pre-mRNA splicing.[63] SMN is essential during embryogenesis, evidenced by embryonic lethality in *SMN1* knockout mice.[64] In the majority of cases, SMA results from homozygous deletions of *SMN1*. Some SMA cases result from gene conversion of *SMN1*, and in 2% to 5% of cases, patients are compound heterozygotes with an *SMN1* deletion on one allele and a pathogenic loss-of-function or point mutation on the second allele.[65,66] In rare cases, a point mutation in *SMN1* is present on both alleles.

SMN1 is contained within a large inverted repeat sequence that contains the highly homologous *SMN2* gene. *SMN2* differs from *SMN1* by only five bases; it lies in the opposite orientation and is centromeric to *SMN1*.[58,67] Although the five bases that differ between *SMN1* and *SMN2* do not affect the amino acid sequence of the protein, a C-to-T transition in exon 7 of *SMN2* corresponding to codon 280 causes alternative splicing and the deletion of exon 7 in 90% of *SMN2* transcripts.[68] Without exon 7, SMN is unstable and is unable to efficiently oligomerize to form the SMN complex that drives snRNP assembly.[69] Thus, even though most patients with SMA have intact *SMN2* genes, they still have disease because most *SMN2* transcripts do not contain exon 7 (SMAΔ7). Thus the pathogenesis of the disease in most patients results from no *SMN1* transcripts, few full length *SMN2* transcripts, limited functional SMN protein, reduced snRNP assembly, and ultimately aberrant mRNA splicing.

The *SMN2* gene is in part a modifier of the severity of SMA, and its effect is based on the *SMN2* gene copy number.[65,70,71] Some *SMN1* deletion haplotypes also contain an *SMN2* deletion, but other *SMN1* deletion haplotypes may have two or even three copies of *SMN2*. In patients with milder forms of the disease, three or four copies of *SMN2* may be present.[65] Increased *SMN2* copies result in production of more *SMN2* transcripts, some of which translate to functional SMN protein, thereby providing some normal SMN protein for required cellular functions. Prior et al described three unrelated individuals with *SMN1* homozygous deletions and 1 or 2 *SMN2* genes but an unexpected mild phenotype.[72] A point mutation in exon 7 of *SMN2* in these patients created an exonic splicing enhancer element and increased levels of full-length *SMN2* transcript to regain increased cellular levels of SMN protein despite the homozygous loss of *SMN1*. Although *SMN2* copy number or transcripts from *SMN2* are known to influence SMA severity, other uncharacterized factors appear to be contributory. Males with identical biallelic *SMN1* deletions and identical *SMN2* copy numbers appear to be more severely affected than females, and the SMA phenotype is variable even within families whose members share identical genotypes.[73]

Treatment for SMA is supportive for the management of respiratory insufficiency, nutritional deficiency, and orthopedic needs.[74] Because *SMN2* serves to modify the SMA phenotype, it can provide a therapeutic target for treatment by correcting aberrant splicing of *SMN2* using antisense oligonucleotides (ASOs) to allow for full-length transcription of SMN2 and a functional SMN protein. This approach has been successful in mice and more recently in pigs.[75,76] Clinical trials using ASOs are under way. Clinical trials using administration of small molecules to protect motor neuron loss or prevent RNA degradation are also in progress, and emerging therapies using gene or stem cell therapy are planned.[74,77,78]

DNA testing for SMA is performed using a variety of techniques.[79] Diagnostic or carrier testing for SMA can be complicated by (1) the polymorphic nature of the *SMN* locus, with alleles containing varying copy numbers of *SMN1* and/or *SMN2* genes; (2) the degree of homology between *SMN1* and *SMN2*; (3) a small percentage of affected alleles with point mutations rather than deletions within *SMN1*; and (4) a 2% rate of de novo cases, which most frequently occur during paternal meiosis.[65,80] A common diagnostic assay for SMA includes PCR amplification coupled with restriction endonuclease digestion with *DraI* and gel electrophoresis.[81] Only amplicons derived from *SMN2* will contain the restriction site; those generated from *SMN1* do not. Because most SMA-affected patients have *SMN1* deletions involving exon 7, no bands corresponding to *SMN1* are observed. Although this assay is simple and robust, it is not quantitative and cannot determine the *SMN2* copy number, which is required for prognosis, nor can this assay detect other *SMN1* pathogenic mutations. *SMN2* copy number determination requires a quantitative method of analysis such as quantitative PCR or multiplex ligation-dependent probe amplification.

Although population-based carrier screening for SMA has been endorsed by the American College of Medical Genetics and Genomics, the committee on Genetics in the ACOG supports SMA testing in patients with a family history of SMA but does not recommend population-based preconception or prenatal testing at this time.[82,83] With an early age of onset for most SMA patients and the prospects of therapeutic treatment of this disease, NBS for SMA would allow early identification of patients and enable timely treatment intervention, thereby minimizing the severity of the disease. However, no state NBS programs currently include SMA.

Nonsyndromic Hearing Loss and Deafness

More than 100 genetic loci have been linked to nonsyndromic hearing loss and deafness, and most demonstrate an autosomal recessive mode of inheritance.[84] The most common autosomal recessive nonsyndromic hearing loss and deafness locus is DFNB1 (DeaFNess autosomal recessive [B] locus 1), which in most cases is associated with congenital, nonprogressive moderate to profound impairment and no other clinical phenotypic findings.[85] DFNB1 mutations occur in the gene gap junction protein beta-2 *(GJB2)* encoding the protein connexin 26 (OMIM #220290).[86] Also mapped to this region on chromosome 13q12.11 is the gap junction protein beta-6 *(GJB6)* gene encoding the protein connexin 30, which is also associated with DFNB1 autosomal recessive nonsyndromic hearing loss and deafness (OMIM #612645).

The incidence of newborn or prelingual hearing loss is about two to three per 1000 births in the United States, and although the etiology is heterogeneous, including cytomegalovirus infection and other environmental causes, the primary cause is genetic.[87] Because the presentation of deafness at this age is considered relatively common and early intervention in these patients is associated with improved clinical outcomes, newborn hearing screening programs are mandated across the United States. Newborns identified with a sensorineural loss in either ear are referred for audiologic confirmatory testing and, if a sensorineural loss is confirmed, are further referred for genetic evaluation, including a physical examination and pre- and postnatal history. If there is a strong suspicion of a genetic basis for hearing loss, DNA testing is ordered to make or confirm a diagnosis.[88]

The first linkage of a gene to autosomal recessive nonsyndromic hearing loss on chromosome 13q was by Guilford and colleagues in 1994.[89] In 1996, the connexin 26 gene *(GJB2)* was mapped to 13q11-q12, and the following year, mutations in this gene were identified as disease-causing in Pakistani families with profound deafness.[90,91] The *GJB2* gene encodes a member of the other gap junction protein genes with gap junction family of connexin proteins.[92] *GJB2* is flanked by other gap junction proteins with gap junction protein beta-6 *(GJB6)* positioned 5′ to *GJB2* and gap junction protein alpha-3 *(GJA3)* located 3′ to *GJB2*. Common to other connexin genes, although the 5510-bp gene has two exons, only the second exon contains coding sequences for this 26-kDa, 226-amino-acid protein. More than 20 genes encode the connexin proteins, which are expressed throughout the body, most notably in the skin, nervous tissue, heart, muscle, and ear. Each protein has four TMDs connected by two extracellular loops and one intracellular cytoplasmic loop, with the amino and carboxyl termini located in the cytoplasm.[93,94] Connexin proteins oligomerize to form a hexameric connexon or hemichannel of identical (homomeric) or different (heteromeric) connexin proteins dependent on the tissue.[92,95] The connexon formed in the plasma membrane of one cell aligns with the connexon from the plasma membrane of the adjacent cell to form gap junction channels in the extracellular space. These channels allow for the exchange of ions and small molecules between adjacent cells. Connexin 26 and connexin 30 are widely expressed in epithelial cells and interspersed hair cells of the cochlea as well as in the connective tissue, and to a lesser extent, connexins 29, 31, 43, and 45 have also been detected.[87,96,97] In the cochlea, normal hearing requires properly functioning gap junctions for the movement and homeostasis of potassium ions between these cells.

More than 200 *GJB2* gene mutations have been described with most associated with a loss of normal function of the connexin 26 protein.[98] The different *GJB2* mutations result in varying effects on connexin 26 and ultimately on gap function. Mutations can affect the proper formation of the gap junction, a loss of function of the gap junction, or a loss of permeability of selected ions through the gap junction, or mutations can cause a gain of function with abnormal opening and increased gap junction activity.[97] The most common mutation, c.35delG, has a carrier frequency as high as 3% to 4% in some white populations and the highest worldwide carrier rate of 1.5%.[99,100] This is a frameshift mutation resulting in premature termination of the protein, and its relative frequency in multiple populations suggests an ancestral founder mutation. Other common *GJB2* frameshift founder mutations include c.167delT and c.235delC, which are common in the Ashkenazi Jewish and Asian populations, respectively.[101,102] The c.35delG mutation can be homozygous or present with another *GJB2* gene mutation on the second allele. In addition to *GJB2* c.35delG, a less common mutation that is frequently included in a first-tier genetic screening test for nonsyndromic autosomal recessive hearing loss is mutation *GJB6*-D13S1830.[103-105] *GJB6*-D13S1830 is a 342-kb deletion encompassing a portion of the *GJB6* gene and the 5′

regulatory sequences of *GJB2*, thus disrupting normal expression of *GJB2*. This mutation is most commonly associated with *GJB2* c.35delG in compound heterozygous patients with one copy of each mutation, or this mutation can be detected in a homozygous state in which it is present on both alleles of the patient.

Although most *GJB2* gene mutations cause a loss of function and demonstrate autosomal recessive inheritance, some mutations in the first extracellular domain of the protein in a heterozygous state have a dominant-negative effect on connexin 26.[106] In these cases, the production and subsequent incorporation of a mutant protein into the hexameric connexon structure results in abnormal function of the gap junction. These autosomal dominant nonsyndromic hearing loss *GJB2* mutations (OMIM#601544) and similar dominant *GJB6* gene mutations (OMIM#604418) define DFNA3 (DeaFNess autosomal dominant [A] locus 3).[107] Gap junctions are also important in the epidermis for intercellular communication, and connexin 26 as well as connexin proteins 30, 30.3, 31, and 43 are widely expressed here and are important for growth and differentiation of keratinocytes.[94] Some *GJB2* gene mutations demonstrating autosomal dominant inheritance patterns are associated with syndromic hearing loss and characteristic skin diseases such as Bart-Pumphrey syndrome (OMIM#149200), Vohwinkel syndrome (OMIM#124500), and others.[94,108]

Genetic testing to identify the pathogenic variant associated with hearing loss is important to families for diagnosis, determining recurrence risks, enabling subsequent targeted mutation analysis for at-risk family members, and determining the likely degree of hearing impairment for the child (ie, mild, moderate, severe or profound).[85] However, despite the presence of the same mutation within family members, modifier genes and environmental factors influence the phenotype such that the degree of hearing impairment may be different between siblings.[85] After a thorough examination of the newborn by the clinical geneticist and a review of family and patient medical history, nonsyndromic autosomal recessive hearing loss may be suspected. Because up to 50% of these patients have *GJB2* gene mutations including the common 342 kb deletion *GJB6*-D13S1830, full-gene *GJB2* Sanger sequencing and *GJB6*-D13S1830 deletion mutation analysis is often performed (Fig. 6.2).[105] Because the etiology of hearing loss is heterogeneous, if no mutation is detected or if the clinical suspicion of a *GJB2* gene mutation is not high, MPS with targeted panels containing genes known to cause hearing loss is suggested.[86,109,110] Although MPS targeted panels increase the diagnostic yield and the likelihood of finding the pathogenic disease-causing mutation(s), some patients will not have disease variants identified because their pathogenic mutation(s) are in genes not included in the panel. In these cases, and if clinically indicated, whole-exome sequencing can be done to identify pathogenic variants in novel candidate genes segregating in the family.[111]

Regardless of the etiology of the hearing loss, these patients and their families require a multidisciplinary team of both health care professionals to manage the clinical needs of the patient and family support services to assist them in adjusting to these new and challenging circumstances.[88] Treatment for patients with hearing loss is dependent on the degree of impairment, including hearing aids or cochlear implantation. Early cochlear implantation surgery results in significant speech perception and language advantages.[112]

Autosomal Dominant Diseases

In autosomal dominant disorders, a single abnormal allele is sufficient to cause disease despite the presence of a normal allele. An individual with an autosomal dominant disease may have inherited an abnormal allele from an affected parent, or the mutant allele may have arisen de novo as a new mutation during gametogenesis in an unaffected parent. The disease-causing gene is on one of the autosomes (1–22) and is not on a sex chromosome (X or Y). An affected individual has a 50% risk of donating the mutant allele to each offspring. Different mutations within the gene may have varying effects on the protein, causing variability in clinical expression between patients who have pathogenic mutations. In some instances, known mutant gene carriers have no clinical symptoms of the disease, a phenomenon referred to as *reduced penetrance*; however, they still possess a 50% risk of donating the mutant allele to each offspring. Differences in phenotypic expression of the disease between patients who share identical gene mutations are commonly explained by the effects of modifier genes or environmental influences (or both). Among pedigrees illustrating autosomal dominant inheritance, both males and females are affected, and male-to-male transmission is observed (unlike X-linked inheritance). Table 6.4 provides a list of some of the inherited autosomal dominant disorders commonly tested in clinical molecular diagnostic laboratories.

Huntington Disease

Huntington disease (HD; OMIM#143100) is an autosomal dominant, late-onset neurodegenerative disorder with an incidence of about 3 to 10 per 100,000 in most populations but may be as high as 10 to 15 in some populations of Western European origin. First described by George Huntington in 1872,[113] this progressive disease is characterized by choreic movement, cognitive decline, and ultimately dementia and psychiatric disturbances.[114,115] The mean age of onset is between 35 and 44 years, but subtle changes in personality may be evident before diagnosis. Approximately 25% of patients first display symptoms after the age of 50 years, and about 5% to 10% of patients have juvenile HD with the age of onset before 20 years.[116] The median survival time is 15 to 20 years after the onset of symptoms. Early in the disease, primary symptoms include cognitive deficits; clumsiness; and

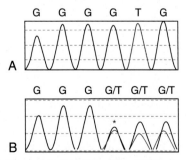

FIGURE 6.2 Sanger sequencing of the *GJB2* gene illustrating a wild-type sequence **(A)** and on a patient with the common c.35delG mutation **(B**, *asterisk*). This mutation results in a frameshift and premature truncation of the protein.

TABLE 6.4 Examples of Autosomal Dominant Disorders

Disease	Gene(s)	Location	OMIM Entry #	Incidence
Achondroplasia	FGFR3	4p16.3	100800	1 in 26,000–28,000
CHARGE syndrome	CHD7	8q12.1-q12.2	214800	1 in 10,000
Familial hypercholesterolemia	LDLR	19p13.2	143890	1 in 200–500
Hereditary hemorrhagic telangiectasia	ACVRL1	9q34.11	187300	1 in 10,000
	ENG	12q13.13		
	GDF2	10q11.22		
	SMAD4	18q21.2		
Long QT syndrome	Numerous	—	192500	1 in 3000–7000
Myotonic dystrophy type 1	DMPK	19q13.32	160900	1 in 20,000
Neurofibromatosis type 1	NF1	17q11.2	162200	1 in 3000
Polycystic kidney disease	PKD1	16p13.3	173900	1 in 400–1000
	PKD2	4q22.1	613095	
Retinoblastoma	RB1	13q14.2	180200	1 in 15,000–20,000
Tuberous sclerosis	TSC1	9q34.13	191100	1 in 5,800
	TSC2	16p13.3	613254	

OMIM, Online Mendelian Inheritance in Man.

mood disturbances such as depression, anxiety, irritability, and apathy.[114,116] The next stage of the disease is associated with slurred speech (dysarthria), impairment of voluntary movements, hyperreflexia, chorea, gait abnormalities, and behavioral disturbances such as intermittent explosiveness and aggression. As the disease advances, bradykinesia, rigidity, dementia, dystonia, dysphagia, severe weight loss, sleep disturbances, and incontinence occur. The HD phenotype is initially caused by selective loss of the medium spiny neurons in the striatum, but in later stages of the disease, there are cortical atrophy and widespread degeneration.

In 1983, Gusella and associates reported linkage between DNA marker D4S10, on the short arm of chromosome 4, and HD, based on studies of a large kindred in Venezuela.[117] Through an international collaborative effort, 10 years after its initial localization, the HD gene was cloned.[118] The molecular basis for HD was determined to be an expansion of a glutamine-encoding CAG trinucleotide repeat in exon 1 of the HD gene, *HTT*. This was confirmed in a worldwide study by the identification of expanded CAG-repeat alleles in HD patients from 565 families, representing 43 national or ethnic groups.[119] In this initial international study, the median CAG-repeat length was reported to be 44 in affected patients and 18 in control participants. Normal CAG repeat lengths range from 10 to 26, repeats of 27 to 35 are considered intermediate or "mutable," repeats of 36 to 39 are considered HD alleles associated with reduced penetrance of the disease, and repeats of 40 or greater are diagnostic of HD (Fig. 6.3).

HD is one of many trinucleotide repeat expansion diseases associated with neurologic and neuromuscular dysfunction.[120,121] In HD, the number of CAG repeats is inversely correlated with the age at onset of the disease. Patients with onset as early as 2 years of life have a repeat number approaching 100 or greater, and late-onset-disease patients have repeat numbers of 36 to 39.[122,123] Although the CAG-repeat number accounts for the majority of variance in age of onset for HD ($\approx$70%), the remainder of the variance comes from modifier genes and environmental factors. Analysis of the large Venezuelan HD kindreds encompassing 18,149 individuals spanning 10 generations indicates that

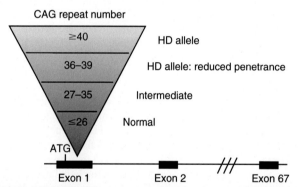

FIGURE 6.3 Schematic representation of the polyglutamine-encoding CAG repeat in exon 1 of the HD gene and associated alleles. A CAG-repeat number of 26 or less is considered normal. CAG-repeat numbers of 27 to 35 are intermediate, and although they are not associated with an abnormal phenotype, these alleles are susceptible to meiotic expansion to an HD allele. CAG repeats of 36 to 39 are considered HD alleles with reduced penetrance, indicating that both unaffected and affected patients have been reported with alleles of this size. CAG repeats of 40 or more are associated with HD with complete penetrance.

40% of the nongenetic variance is due to heritable modifier genes and 60% to environmental factors.[124] The CAG repeat number and polyglutamine length in the normal huntingtin protein (htt) is not a modifier of age of onset for HD.[125] Genes that encode proteins that interact with htt regulate the expression of *HTT*, dysregulate energy metabolism in the cell, and alter neurotransmitter receptor function have all been considered as modifier genes, but no correlation with age of onset or penetrance has been found.[126]

Pathogenic mutations in *HTT* causing HD in the absence of a family history occur from expansion of a CAG-intermediate allele, which occurs almost exclusively through paternal transmission, although maternal intermediate allele expansion has been reported.[127,128] Intermediate alleles are present in about 1% of the population. The instability of these alleles may be influenced by flanking DNA sequences, which may enhance the formation of hairpin loop structures and cause replication slippage.[129,130] Studies in the HD mouse model have proposed that variations in mismatch repair genes may modify somatic CAG expansion and disease progression.[131,132] Single sperm analysis studies have demonstrated 11% instability (9% expansions and 2.5% contractions) in CAG repeats of 30 compared with 0.6% instability (contractions only) seen in average-sized alleles of 15 to 18 repeats, indicating that CAG instability increases as the repeat number increases.[133] CAG repeats of 36 showed 53% instability, and CAG repeats from 38 to 51 had instability ranging from 92% to 99%. In addition to the CAG repeat number, *cis*-elements may play a role in CAG-repeat instability.[134] In families with HD, the onset of symptoms occurs at a progressively younger age in successive generations, a pattern referred to as *anticipation*. Anticipation is explained by meiotic expansion of the unstable CAG repeat during transmission by the affected parent, resulting in an even higher CAG-repeat number in the offspring and an earlier age of onset. In addition, although 69% of affected father–child pairs show expansion, only 32% of affected mother–child pairs demonstrate expansion. Furthermore, less than 2% of maternal expansions result in a change of more than five repeats, but up to 21% of paternal transmissions increase by more than 7 repeats.[135] For this reason, the affected parent in most cases of juvenile-onset HD is the father. However, the largest reported CAG-repeat number of approximately 130 occurred via maternal expansion of 70 CAG repeats.[136]

The *HTT* gene contains 67 exons and encodes a novel protein, huntingtin (htt), with 3144 amino acids and a molecular mass of approximately 350 kDa.[137] Htt is ubiquitously expressed in both neural and nonneural tissue with highest levels in the brain.[138,139] Htt is required for neuronal development with the complete absence of htt being lethal in mice.[140] Htt is predominantly localized in the cytoplasm, but lesser amounts are also found in the nucleus.[141,142] The structure of htt includes the CAG encoded polyglutamine repeat, a proline-rich domain, and 3 HEAT repeats, which form a rodlike helical scaffold to which other components can attach and are involved in intracellular transport and a nuclear localization domain at the C terminal end. The HEAT acronym is derived from four proteins found to also contain this amino acid sequence and unique structure: **H**untingtin, **E**longation factor 3, regulatory **A** subunit of protein phosphatase 2A and **T**OR1. In neurons, htt is associated with synaptic vesicles and microtubules and is abundant in dendrites and nerve terminals. Huntingtin interacts with more than 200 proteins, including those involved in intracellular trafficking and signaling, cytoskeletal organization, endocytosis, and transcription regulation.[143] Mutant htt with expanded polyglutamine tracks are effectively transcribed and translated, but as a result of the increase in glutamine residues, the protein is misfolded.[144] This abnormal folding conveys a gain of function and an increased amyloid-like ability to form protein aggregates, sequestering other cellular proteins preventing their normal function, thereby disrupting normal protein homeostasis and conferring neurotoxicity. Interestingly, patients with two expanded CAG-repeat alleles do not have more severe disease than heterozygotes.[119]

HD remains an incurable complex disease requiring a multidisciplinary approach and medication to treat the motor, psychiatric, and cognitive symptoms of the disease.[145,146] Current therapeutic strategies include the use of antisense-oligonucleotides specific for HD-associated single nucleotide polymorphisms (SNPs), or RNA interference (RNAi) technology for CAG-expanded allele inhibition.[147-149]

DNA testing for HD is performed using PCR to determine the CAG-repeat number.[150] The most common method includes PCR with fluorescently labeled primers coupled with capillary electrophoresis (Fig. 6.4).[151] Technical standards and guidelines for HD diagnostic testing for clinical laboratories have been developed by the American College of Medical Genetics and Genomics.[150]

Many ethical issues are associated with HD testing, primarily as they relate to presymptomatic testing (Box 6.2). The first policy statement on ethical issues related to predictive genetic testing for HD was adopted in 1989 at a joint meeting with representatives from the International Huntington Association and the World Federation of Neurology.[152] At that time, the gene had not yet been cloned, and predictive testing was performed using linkage studies and the at-risk patient was quoted only the likelihood of inheriting the mutant allele. These tests were less than perfect and provided, at best, results in only 60% to 75% of families. Moreover, the possibility of recombination resulted in an inaccurate carrier assessment.[153] In other families, living affected members were not available, or markers were not informative. After the gene had been cloned and direct mutation analysis was possible, risk assessments were reversed in a small percentage of patients.[154]

However, after the gene was cloned, direct mutation analysis for CAG repeat number was implemented, and guidelines for predictive testing using direct mutation analysis were established.[155,156] The approach used for presymptomatic HD testing has become the model for late-onset single-gene disorders, and similar formats have been applied to other late-onset inherited diseases.[157,158] This model includes a multidisciplinary team involving neurology, psychiatry, and genetics specialists; pretest evaluation and counseling; individual support from family members or close friends; and posttest follow-up sessions.[159] Predictive testing should be performed for patients 18 years of age or older and only with informed consent. Informed consent implies that the patient has been thoroughly counseled and clearly understands both the advantages and the disadvantages of knowing the results. An advantage of having this test is the removal of uncertainty regarding whether patients have or have not inherited the mutant allele and thus a feeling of relief for those who have not inherited a mutant *HTT* allele. This knowledge can help patients plan their career goals and personal affairs involving marriage, children, and long-term care insurance. Disadvantages of knowing this information include, but are not limited to, (1) the feeling of "survivor's guilt" in those who learn that they have not inherited a mutant allele while other family members have, (2) fear from learning that they have inherited a mutant *HTT* allele and will develop this incurable disease, (3) potential risk of discrimination in employment, (4) concern for passing this gene on to their offspring, and (5) uncertainty of developing the disease if a mutant *HTT* allele with 36 to 39 CAG repeats is identified.

When at all possible, the patient should be accompanied by a trusted friend or loved one throughout the counseling

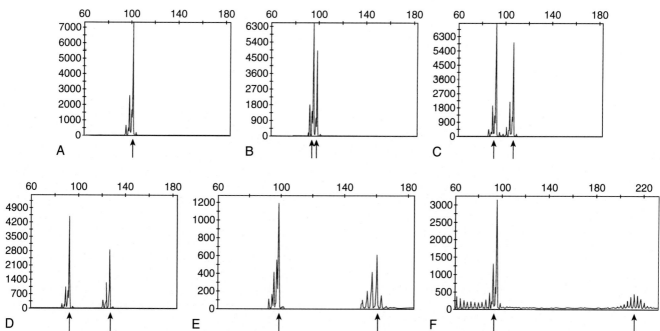

FIGURE 6.4 Electropherograms representing various patterns observed in patients referred for Huntington disease (HD) testing. The polyglutamine-encoding CAG repeat in exon 1 is amplified by polymerase chain reaction (PCR) using flanking oligonucleotide primers, one of which is labeled with a fluorescent dye. Amplicon sizes in bases are noted at the top of the figure, and the relative amount of fluorescence detected for each amplicon is measured by the peak height (y-axis). *Arrows* indicate the predominant amplicons observed for each patient with the size of the amplicons corresponding to the CAG repeat number. Additional smaller peaks not indicated by arrows represent "stutter" peaks and result from strand slippage during PCR of repetitive sequences. **A,** Patient 1 has amplicons 101 bp in length corresponding to 20 CAG repeats on both HD alleles. **B,** Patient 2 has 95- and 98-bp amplicons, corresponding to CAG repeats of 18 and 19. **C,** Patient 3 has 92- and 107-bp amplicons representing CAG repeats of 17 and 22, respectively. The diagnosis of HD can be ruled out in these three patients. **D,** Patient 4 has 92- and 128-bp amplicons corresponding to CAG repeats of 17 and 29. The results would not support a diagnosis of HD. However, a CAG repeat of 29 is considered an intermediate allele and can undergo meiotic expansion to an HD allele during gamete formation. **E,** Patient 5 has CAG repeats of 19 and 39, as depicted by amplicons 98 and 158 bp in length. In a symptomatic patient, these results would support the diagnosis of HD. However, in a presymptomatic patient, the phenotype of an HD allele with reduced penetrance cannot be predicted with certainty. **F,** Patient 6 has CAG repeats corresponding to 18 and 57 with amplicons 95 and 212 bp in length. These results confirm the diagnosis of HD in this patient. Genetic counseling regarding the implications of HD DNA findings is indicated.

BOX 6.2 Ethical Issues Associated With Presymptomatic DNA Testing for Huntington Disease

Patients must be 18 years of age or older.
The decision to proceed with testing must be voluntary and informed.
Genetic counseling regarding the benefits and pitfalls of testing is required.
A support partner is needed for the patient for counseling and the testing process.
Diagnosis of Huntington disease in the family should be confirmed by DNA testing before presymptomatic testing.
Psychiatric assessment of patient is necessary before testing.
Follow-up genetic counseling is recommended after delivery of results.
Prenatal testing of fetuses is controversial; preimplantation diagnosis is available.

---

and testing procedure. This person can provide stability to the patient by being able to intimately speak to the patient about the situation and discuss the information shared at the counseling sessions. Most important, as a part of this process, the partner will be present when the results of testing are revealed and can provide comfort and support as needed both then and in the following days, weeks, or months. Ultimately, however, it is the patient's decision to proceed with this testing and to accept both the benefits and pitfalls of knowing this information. The patient's decision to proceed must be his or hers, without coercion from family members, clinicians, friends, or employers.

For at-risk individuals requesting mutation studies, their mental stability should be considered for the safety of the patient; a psychiatric assessment is often part of the testing protocol because HD test results can precipitate depression. Results of the psychiatric evaluation can influence the timing of the DNA test, postponing it until such time when the patient is considered mentally able to deal with the possibly devastating news. Suicidal tendencies are present in both at-risk patients and those with HD.[160,161]

Because HD is a delayed-onset disease, as the asymptomatic at-risk patient ages, the risk of testing positive with an expanded CAG repeat decreases.[162] Thus if the patient elects not to have predictive testing, the genetic counselor can provide information regarding the probability that an HD mutation exists, which, based on the individual's age, may provide some comfort to the patient. Most individuals who seek presymptomatic testing have a mean age of about 40 years and are more often female.[163] Some individuals may begin the multistep counseling process and then withdraw from the study without receiving test results. If possible, before presymptomatic mutation testing, the diagnosis of HD should be confirmed in an affected member of the family to be certain that the disease segregating in the family is HD. A normal *HTT* CAG repeat number does not rule out a different, dominantly inherited neurodegenerative disease in the family for which the patient likely retains a risk of development.

Counseling in families with no prior family history of HD can be challenging when the expanded CAG repeat appears to be de novo. Multiple factors can explain this phenomenon, including (1) expansion of an intermediate-sized allele (27–35), (2) premature death of an adult asymptomatic yet affected individual with a reduced or full penetrance allele, (3) misdiagnosis of HD in other family members, (4) alternative paternity, and (5) undisclosed adoption. Without understanding the mechanism of a new mutation in the affected individual, a priori risk assessment for HD in at risk family members may be inaccurate. Thus, counseling for the risk of HD is only prudent after accurate DNA testing.

Prenatal testing for HD, another complicated issue associated with this disease, may not be provided in all laboratories that perform routine HD testing for several reasons. Ethical issues include possible pregnancy termination for a late-onset disorder, presymptomatic testing of the child if the parents choose not to terminate, and technical issues that may compromise testing of chorionic villus and amniocentesis samples. As an alternative for prenatal testing, preimplantation genetic diagnosis (PGD) can be performed.[164] In PGD, in vitro fertilization is used to produce embryos that then undergo a single cell biopsy for genetic analysis. After PCR testing has been used to determine the *HTT* CAG-repeat numbers, embryos with normal HD alleles are implanted. This method, which combines direct mutation analysis and PGD, eliminates the necessity for subsequent prenatal testing to determine the HD status of the fetus. Exclusion testing whereby only embryos who have not inherited an affected allele are implanted without disclosing information regarding the *HTT* CAG repeat numbers in all tested embryos is an option for some families that do not wish to know the genotype of the asymptomatic at-risk parent.[165,166]

Marfan Syndrome

Marfan syndrome (MFS; OMIM #154700) is a relatively common autosomal dominant multisystem connective tissue disorder with primary manifestations involving the ocular, musculoskeletal, and cardiovascular systems with an estimated worldwide incidence of about 1 in 5000.[167,168] The most common ocular feature of MFS is myopia, but other associated features include unilateral or bilateral ectopia lentis (60%), or retinal detachment. In addition, patients with MFS are at increased risk for glaucoma and cataracts at a younger age as compared with the general population. Characteristic facial features may include a long narrow face, deep-set and downward-slanting eyes, flat cheek bones, and micrognathia. Skeletal abnormalities arise from bone overgrowth and joint hypermobility. MFS patients are tall with frequent clinical findings including pectus excavatum or pectus carinatum, caused by overgrowth of the ribs, which can interfere with pulmonary function and require surgery. Patients typically have an arm span-to-height ratio greater than 1.05 and a reduced upper body-to-lower extremity ratio. Scoliosis is present in about half of patients; it can be mild to severe and is progressive. Morbidity and early mortality are linked to cardiovascular manifestations of the disorder, which are characterized by progressive dilation of the aortic root, predisposition to aortic dissection, mitral and tricuspid valve prolapse with or without regurgitation, and dilation of the proximal pulmonary artery. Wide phenotypic variability is observed, with some patients presenting as neonates with severe and progressive disease that is sometimes fatal, but others can remain undiagnosed until adulthood. An early diagnosis is associated with an improved long-term outcome.[169,170] Unfortunately, for undiagnosed patients, presentation may be sudden premature death caused by aortic dissection or rupture.[171,172]

The diagnosis of MFS is based on family history of the disease; however, as many as 25% of MFS arise from a new mutation with no family history of disease.[173] In the absence of a documented family history of MFS, currently the clinical diagnosis of MFS is made using the revised Ghent diagnostics nosology with most of the emphasis on the presence of an aortic root aneurysm and ectopia lentis.[168] In the absence of one or both of these, a mutation in the gene associated with MFS or manifestations in other MFS-related organ systems are required. Some MFS features may be isolated findings and not associated with MFS, and some overlap with other genetic syndromes.[167,168] Because most MFS clinical manifestations increase with age, these criteria may make diagnosis in the pediatric patient more difficult, and often these patients may carry a diagnosis of "potential" MFS and require periodic follow-up visits for reevaluation.[167]

MFS is associated with mutations in the fibrillin-1 gene (*FBN1*) mapped to chromosome 15q21.1.[173-175] The gene spans 237,414 bp, is composed of 65 exons, and encodes a 10 kb mRNA, prefibrillin-1.[176] *FBN1* is ubiquitously expressed in connective tissue. The 320-kDa, 2871-amino-acid extracellular glycoprotein, fibrillin-1, self-assembles into macroaggregates to serve as the primary structural component of 10-nm-diameter microfibrils located throughout the basal lamina in both elastic and nonelastic tissue.[176-178] In elastin-expressing tissue, such as blood vessels, lung, and skin, microfibrils make up the scaffold for elastin assembly within the extracellular matrix (ECM). In nonelastic tissue, including the ciliary zonule of the eye and of basement membranes, they have an anchoring function and provide tensile strength.[179] In most cases in which the mutation causes abnormal fibrillin-1 protein, a dominant negative effect occurs when fibrillin-1 is incorporated into the microfibril, resulting in functionally inferior connective tissue. In other cases, disease results from reduced protein production or haploinsufficiency. Fibrillin-1 contains several motifs, including 47 cysteine-rich epidermal growth factor-like (EGF) domains, most of which bind calcium.[180] These domains are interspersed with 7 TGF-β binding protein-like (TB) domains, and there are two hybrid domains with sequences similar to both the EGF and TB motifs.[180] Latent TGF-β binding proteins interact with fibrillin-1 and together bind TGF-β to inactivate and thereby prevent signaling through the SMAD 2/3 pathway.[177,179,181,182] Dysregulation of TGF-β signaling affects the development of vascular smooth muscle and the integrity of the ECM.

Mutations in FBN1 are heterogeneous with almost 2000 reported throughout the gene.[183] Point mutations represent the largest category at 66%; 20% are small deletions or insertions, 11% are splice site mutations, and the remainder represent large deletions or duplications. In all cases, increased TGF-β signaling is observed. Phenotype–genotype studies have shown that neonatal MFS or early onset and severe MFS is associated with mutations spanning exons 24 to 32.[184,185] These patients also have an increased likelihood of developing ectopia lentis, ascending aortic dilation, mitral valve anomalies, scoliosis, and a reduced life expectancy. This region of the protein contains the longest stretch of EGF-like domains and is thought to be important for microfibril biogenesis. In-frame missense mutations are more often associated with more severe complications of MFS than are mutations predicted to cause premature truncation of the protein, and ectopia lentis is more frequently seen in patients with a missense mutation resulting in cysteine than with other missense mutations.[184] Furthermore, because wide phenotypic variability is observed even for individuals harboring the same recurrent mutation, modifier genes likely play a role in the MFS phenotype.[184] Interestingly, the phenotypic expression of MFS may be related in part to varying degrees of expression between the normal and mutant allele, and these differences may also be tissue specific.[186] The lack of consistent genotype–phenotype correlations provides little prognostic value for individual patient management. Although variability in expression is observed, FBN1 mutations are considered highly penetrant.

Management of patients with MFS is similar to that of patients with other inherited multisystem disorders, involving a team approach with specialists in many areas of medicine including a cardiologist, ophthalmologist, orthopedist, and geneticist. Because the pathogenesis of MFS is due to dysregulation of TGF-β and increased TGF-β signaling, one pharmacologic treatment for MFS is the drug losartan, an angiotensin II type 1 receptor inhibitor used to inhibit excessive TGF-β signaling and slow aortic growth.[187] More recent studies suggest that this therapy may be more effective in patients with FBN1 haploinsufficiency as opposed to dominant negative FBN1 mutations.[188] Other successful treatments for aortic aneurysms associated with MFS include β-blockers and statins or tetracycline to inhibit TGF-β signaling through inhibition of matrix metalloproteinase-2 and -9 and ERK inhibitors to inhibit ERK signaling.[187] Clinical trials investigating the best therapy to treat patients with MFS and other aortopathies are ongoing. Annual ophthalmologic and transthoracic echocardiographic imaging is important for monitoring, and prophylactic surgery for aortic root replacement is recommended when the aortic root reaches a critical diameter of 5.5 cm.[189,190] Lifestyle changes to limit physical activity to low-impact sports in order to prevent physical exhaustion are recommended to prevent high blood pressure and aortic wall stress.

FBN1 is one of the largest human genes, and although it is labor intensive and costly, DNA sequencing is the gold standard for genetic testing for MFS. This may be most effective in patients when the clinical diagnosis of MFS is likely.[191] After a FBN1 mutation has been identified within a family, predictive targeted FBN1 mutation specific testing for at-risk family members can be performed and enables early diagnosis and proper management in identified mutation positive family members. However, despite extensive FBN1 analysis, 7% to 30% of MFS patients will have no mutation detected. These cases could reflect patients with FBN1 mutations that are contained within regions of the gene that cannot be detected by current screening techniques.[192] Alternatively, these may be patients who have a phenotype suspicious for MFS but who do not meet the strict Ghent diagnostic criteria. MPS with a targeted panel of genes, which includes FBN1 as well as other candidate genes associated with thoracic aortic aneurysms or aortic dissections, may be an efficient screening method. Associated syndromes and genes may include Loeys-Dietz (SMAD3, TGFBR1, TGFBR2) Ehlers-Danlos type IV (COL3A1), or genes associated with thoracic aortic aneurysm and dissections (ACAT2, MYH11, MYLK, TGFBR1, and TGFBR2).[193] Collectively, these disorders are described as aortopathies. Aortopathies are a common cause of morbidity and mortality in the United States. In fact, it is reported that aneurysm of the aorta is responsible for 1% to 2% of all deaths in the Western world.[194] MPS-based testing using an aortopathy gene panel is rapidly becoming the first-line diagnostic test for individuals who present with a phenotype that could fit any one of multiple disorders (eg, MFS and Loeys-Dietz syndrome).

Multiple Endocrine Neoplasia

Multiple endocrine neoplasia (MEN) is an autosomal dominant disorder characterized by the presence of tumors in two or more endocrine glands. There are two major types of MEN disease: type 1 (MEN1, which is also known as Wermer syndrome) and type 2 (MEN2, which is also known as Sipple syndrome). Both MEN1 and MEN2 are clinically distinct and should be considered as independent disorders with separate genetic causes.

Multiple endocrine neoplasia type 1 (OMIM #131100) is a relatively common disorder with an estimated incidence ranging from 1 in 10,000 to 30,000.[195,196] MEN1 is characterized by the presence of any of many (>20) tumor types, and a clinical diagnosis requires that an individual have at least two endocrine tumors that are parathyroid, pituitary, or gastroenteropancreatic in nature.[196] Parathyroid tumors are the most common tumors observed in MEN1 and occur in approximately 95% of patients.[197] Consequently, hyperparathyroidism resulting from overproduction of hormones by parathyroid tumors is a common early manifestation.[196,198] Parathyroid tumors also often cause hypercalcemia, which ultimately results in a multitude of medical issues (ie, depression, nausea, vomiting, kidney stones, and hypertension, among others).[196] Pancreatic islet cell tumors are the second most common tumors observed in MEN1 with approximately 40% of patients developing a neoplasm of this area.[197,198] Pancreatic tumors are of particular importance in MEN1 because gastrinomas (Zollinger-Ellison syndrome) are the most common cause of morbidity in patients with this disease.[197] Pituitary tumors occur in 30% of patients with MEN1 and represent the third most prevalent tumor type. Of note, a multitude of non–endocrine-associated tumors (carcinoid, adrenocorticoid, facial angiofibromas, lipomas, meningiomas, among others) commonly occur in patients with MEN1.[196-198]

Using a combination of linkage analysis and deletion mapping techniques, investigators in several groups were able to identify a region on chromosome 11 (11q13) likely to contain the gene responsible for MEN1.[199-208] In 1997, the gene causative for MEN1 was identified and named multiple endocrine neoplasia I (*MEN1*).[209,210] The *MEN1* gene contains ten exons that encode a 610 amino acid protein known as menin.[209] Menin is a ubiquitously expressed protein that can localize to either the nucleus or cytoplasm.[211,212] However, other than clearly defined nuclear localization signals, menin lacks functional domains homologous to those observed in other proteins, making it difficult to predict how it functions.[213] Menin protein interactions suggest that it functions in transcriptional regulation, cell division and proliferation, and genome stability.[213-216] Based on loss of heterozygosity (LOH) patterns observed during gene identification, *MEN1* was considered a likely tumor suppressor gene. Several studies have since shown that overexpression of menin in vitro results in suppression of cellular proliferation.[217-219] Furthermore, the loss of menin results in immortalization of cells.[220] Taken together, these data support a tumor suppressor role for menin, but its specific function is not yet known.

To date, more than 500 unique disease-causing mutations have been described in the *MEN1* gene.[216] Mutations are located in all coding exons of the gene, and there are no significant mutational hot spots. The bulk of mutations reported in *MEN1* are those that lead to truncated forms of the menin protein (eg, frameshift, nonsense, gross deletions).[216] Most known mutations in *MEN1* result in the loss of nuclear localization, and loss of function appears to be the mechanism of disease.[196,216] However, there are no clear genotype–phenotype correlations in patients with MEN1. *MEN1* missense mutations, which are unlikely to be loss of function, have been reported in individuals with familial isolated hyperparathyroidism (FIHP).[221-223] There have also been several truncating mutations reported in FIHP that also occur in classic MEN1 disease.[216] It is therefore not possible to predict the course of disease based on genotype alone.

Molecular diagnosis of MEN1 typically involves Sanger sequencing of the entire coding region of the *MEN1* gene. This method detects disease-causing mutations in 80% to 90% of familial cases and 65% of isolated cases.[196,224,225] Symptomatic individuals with negative Sanger sequencing results should be screened for gross deletions and duplications. Testing for large deletions and duplications (typically performed by multiplex ligation-dependent probe amplification [MLPA]) detects mutations in an additional 1% to 4% of patients.[196,216,226-230] Combining both techniques, a causative mutation is found in approximately 95% of familial MEN1 cases. After a causative germline mutation is identified in a family, other at-risk family members should be offered targeted testing for the identified mutation as soon as possible because the disease may begin to show manifestations as early as 5 years of age.[231] Because MEN1 is autosomal dominant, the risk of MEN1 in the child of an affected individual is 50%. Approximately 10% of mutations identified in *MEN1* are de novo, so recurrence risk is much lower in families where the proband is the child of individuals in which no *MEN1* gene mutation has been identified and correct parentage has been confirmed.[216] However, all children of an individual with a de novo change have a 50% chance of inheriting the causative mutation.

Treatment of MEN1 disease is largely driven by the presentation of disease in the individual patient. Surgical intervention is recommended to remove all functional tumors as well as those that are greater than 4 cm in size or demonstrate rapid growth.[231] In some cases in which pancreatic tumors become metastatic (or are inoperable), chemotherapy may be used.[231] Individuals with hyperparathyroidism may undergo a subtotal or total parathyroidectomy.[231] Thymectomy may be performed prophylactically to prevent the development of carcinoid tumors, but in most cases, it is performed after tumor development.[196,231] Presymptomatic individuals with a known disease-causing mutation in *MEN1* should undergo routine clinical screening for early detection of cancer.

Multiple endocrine neoplasia type 2 is divided into three phenotypically distinct subtypes: MEN2A (OMIM #171400), MEN2B (OMIM #162300), and familial medullary thyroid carcinoma (FMTC, OMIM #155240). Unlike MEN1, tumors associated with MEN2 disease are highly malignant and life threatening. The MEN2A and MEN2B subtypes are both characterized by medullary thyroid carcinoma (MTC) and pheochromocytoma but also have unique distinguishing features. MEN2A is the most common form of MEN2 disease and accounts for approximately 55% of patients. It is characterized by MTC with pheochromocytomas in 50% of patients and parathyroid adenoma in approximately 20% of patients.[197,232] A clinical diagnosis of MEN2A requires the presence of both MTC and pheochromocytoma or a parathyroid adenoma (or parathyroid hyperplasia) in a single individual.[232] MEN2B is much less common, representing only 5% to 10% of MEN2 patients.[197,198] It is characterized by MTC with pheochromocytoma but very little risk for parathyroid adenoma. MEN2B patients also commonly have mucosal neuromas of the lips or tongue, marfanoid habitus, distinctive facies, intestinal autonomic ganglion dysfunction, and medullated corneal fibers.[197,232] A clinical diagnosis of MEN2B typically requires the presence of most of these features in addition to MTC. FMTC accounts for 35% of MEN2 cases and is characterized by MTC in the absence of other malignancies.[197,198] FMTC is clinically diagnosed in families with four or more individuals affected with MTC alone.[232] Interestingly, the onset of disease varies significantly between the subtypes of this disorder. The onset of MTC is typically observed in early adulthood in MEN2A, in early childhood in MEN2B, and often in middle age in FMTC patients.[232]

Using a combination of linkage analysis and gene mapping techniques the gene causative for MEN2A was located in a 480-kilobase region of chromosome 10q11.2.[233-236] The proto-oncogene *RET* was later identified as causative for both MEN2A and FMTC[236,237] and MEN2B.[238,239] *RET* is a proto-oncogene containing 21 exons that encode an 1114-amino-acid protein called RET. RET is a receptor tyrosine kinase for members of the glial cell line—derived neurotrophic factor family (GDNF) of signaling molecules.[240-243] RET contains three functional domains: an extracellular ligand-binding domain, a TMD, and a cytoplasmic tyrosine kinase domain.[244] It is involved in several signaling pathways during development that control proliferation, differentiation, survival, and migration of enteric nervous system progenitor cells.[244] Not surprisingly, *RET* gene mutations that result in MEN2 disease are activating in nature.[245] Interestingly, mutations that inactivate the *RET* gene result in Hirschsprung disease (OMIM #142623), a disorder characterized by the absence of neuronal ganglion cells in the large intestine.[246,247]

More than 100 variants have been described in the *RET* gene in association with the MEN2 phenotype with several mutation hotspots.[248] In fact, mutations at *RET* codon 634 account for more than 85% of familial MEN2A, and mutations in three cysteine residues located at codons 609, 618, and 620 account for 50% of FMTC.[197,232] Similarly, a single mutation resulting in a change from methionine to threonine at codon 918 (p.Met918Thr) in *RET* exon 16 accounts for 95% of MEN2B patients.[198] Because the bulk of MEN2-associated *RET* gene mutations are limited to exons 10, 11, and 13 to 16, molecular testing is typically limited to Sanger sequencing of these regions. When a diagnosis of MEN2B is suspected, targeted testing for the p.Met918Thr mutation is often performed first. All subtypes of MEN2 are inherited in an autosomal dominant manner, so any individual found to have a disease-causing *RET* mutation has a 50% chance of passing the disease-causing allele to each of his or her offspring. Five percent of MEN2A-related and 50% of MEN2B-related *RET* gene mutations are caused by a de novo change.[232] In these cases, the children of an individual with a de novo mutation also have a 50% chance of inheriting the disease.

Treatment of MEN2 is dependent on the disease presentation. Individuals with MTC typically undergo thyroidectomy and lymph node dissection.[232] Patients with pheochromocytoma have laparoscopic adrenalectomy.[198] Presymptomatic individuals with a disease-causing *RET* mutation should be screened regularly to detect early manifestations of disease. These individuals may also elect to have prophylactic thyroidectomy to prevent MTC.[232]

X-Linked Diseases

In X-linked diseases, the mutant allele resides on the X chromosome. In X-linked recessive diseases, females are heterozygous carriers of the disease with one normal and one mutant allele and are typically not affected. Males receiving the mutant allele from their mothers are considered hemizygous with one mutant allele and no normal allele. All daughters of affected males are carriers of a mutant allele. A carrier female has a 25% chance of transmitting her normal allele to a son, a 25% chance of having an affected son, a 25% chance of having a daughter who carries the mutant allele, and a 25% chance of having a daughter who receives her normal allele. In the absence of a family history, an affected male can have a mutant allele that arose de novo as a new mutation during gametogenesis in the formation of the egg. Roughly one third of all cases of X-linked disorders represent de novo new mutations with the absence of a family history. In these cases, the mother is not a carrier of a mutant allele and is not at risk for having subsequent affected children. In pedigrees associated with X-linked recessive conditions, typically only males are affected, and male-to-male transmission of the disease is not seen. In less frequent, X-linked dominant diseases, one copy of the mutant allele is sufficient to cause disease despite the presence of a normal allele. In these disease processes, females are affected, and in males with only a single mutant allele, these diseases are often lethal. Table 6.5 provides a list of some of the inherited X-linked disorders commonly tested in clinical molecular diagnostic laboratories.

Duchenne Muscular Dystrophy

Duchenne muscular dystrophy (DMD; OMIM #310200) is a fatal X-linked recessive disorder characterized by progressive skeletal muscle wasting. The incidence of DMD is about 1 in 3500 male births, making it the most common severe neuromuscular disease in humans. Classic DMD presents in early childhood, with delayed motor skills or an abnormal gait. This is followed by progressive muscle weakness, calf hypertrophy, and grossly elevated serum creatine kinase (>10× normal) caused by degenerating muscle fibers. A muscle biopsy will show variation in fiber size, necrosis, inflammation, fibrosis, and fiber regeneration and may be required for confirmation of disease in about 5% of patients in whom

TABLE 6.5 Examples of X-Linked Disorders

Disease	Gene	Location	OMIM Entry #	Incidence
Fabry disease	GLA	Xq22.1	301500	1 in 50,000 males
Hemophilia A	HEMA	Xq28	306700	1 in 4000–5000 males
Hemophilia B	HEMB	Xq27.1	306900	1 in 20,000 males
Hunter syndrome: mucopolysaccharidosis type II	IDS	Xq28	309900	1 in 100,000 males
Incontinentia pigmenti	IP	Xq28	308300	1 in 1,000,000 females
Lesch-Nyhan syndrome	HRPT1	Xq26.2-26.3	300322	1 in 380,000 males
Menkes disease	ATP7A	Xq21.1	309400	1 in 100,000 males
Ornithine transcarbamylase deficiency	OTC	Xp11.4	300461	1 in 14,000 males
Severe combined immunodeficiency	IL2RG	Xq13.1	300400	1 in 50,000–100,000 males
Wiskott Aldrich syndrome	WAS	Xp11.23	301000	1 in 100,000 males

OMIM, Online Mendelian Inheritance in Man.

no DNA mutation is identified. Immunohistochemistry (IHC) staining using antibodies directed against different epitopes of the DMD encoded protein, dystrophin, shows complete or almost complete absence of carboxy-terminal antigens in the majority of DMD patients. Progressive weakness initially affects the lower extremities, causing most DMD patients to require wheelchairs between 10 and 15 years of age. Continual degeneration and regeneration and inflammation of muscle eventually lead to the replacement of muscle tissue by adipose and connective tissue and progressive disease. Scoliosis is common and affects respiratory function. Chronic respiratory insufficiency develops in all patients. Cardiac disease is most commonly dilated cardiomyopathy (DCM) caused by cardiac fibrosis or rhythm and conduction abnormalities. Cardiorespiratory failure is the primary cause of death.[249] In some patients, however, only the heart is affected, causing DMD-associated X-linked DCM (OMIM #302045).[250] Additionally, many patients with DMD exhibit lower IQs and nonprogressive cognitive impairment.[251] Clinical management of patients with DMD is complex, requiring a multidisciplinary team approach.[252,253] Glucocorticoids are used to slow muscle weakness, angiotensin-converting enzyme inhibitors, β-blockers and diuretics for cardiac disease, and noninvasive ventilation for respiratory care. In addition to health care professionals in these respective areas, team members in the areas of psychosocial, gastrointestinal, pain, and speech and language are required. Although elevated serum creatine kinase (CK) levels can be measured from blood spots, NBS for DMD, enabling early identification of affected children and potential better outcomes from early intervention has not been implemented in the United States. This may be in part due to false-positive elevated CK levels as a result of the birthing process and the lack of evidence to support early intervention in DMD patients.[254] A two-tiered pilot program performing both CK and DNA analysis on dried blood spots was successfully implemented in Ohio, screening 37,649 newborn males and identifying 6 affected patients.[254] As novel therapies for DMD continue to develop and the efficacy of targeted treatment for early intervention is validated, NBS for DMD may be adopted.[255]

Because DMD is an X-linked recessive disorder, most carrier females are asymptomatic. Similar to other X-linked diseases, the varying degree of clinical manifestations among carrier females depends on the degree of inactivation of the X chromosome harboring the mutant *DMD* gene in various tissues where the DMD protein is expressed.[256] Up to as many as 20% of carriers can display some symptoms of DMD and Becker muscular dystrophy (BMD). Most frequently observed is muscle weakness with elevated serum CK levels or cardiac involvement, including DCM or left ventricle dilation. Females with severe disease most often result from skewed lyonization in carrier females or an X-autosome translocation involving the *DMD* gene.[257-261]

Cytogenetic abnormalities in DMD patients and DNA linkage studies localized the DMD locus to chromosome Xp21.[258,259,262-264] By mixing DNA enhanced for X-linked genes from a 49, XXXXY cell line with DNA from a patient with DMD and a cytogenetic deletion in Xp21, Kunkel and colleagues cleverly used subtraction hybridization to clone the DNA corresponding to the patient's deletion.[265] During hybridization, Xp21 sequences from the cell line had no complementary sequences with which to anneal in the patient's DNA; thus, they were available for cloning. The *DMD* gene (Xp21.2) is complex and is one of the largest genes in the human genome, spanning 2.4 megabases. DMD contains 79 exons, representing less than 1% of the gene, and encodes a 14 kb mRNA.[266] The gene has multiple tissue-specific promoters that transcribe various full-length dystrophin isoforms differing in their amino terminal sequences. The full-length protein product, dystrophin, contains 3685 amino acids; has a molecular weight of 427 kDa; and contains four distinct domains, including an actin-binding domain, a central rod domain with spectrin-like repeats, a cysteine-rich domain, and a unique COOH-terminal domain. Dystrophin is predominantly expressed in skeletal, cardiac, and smooth muscle. Additional dystrophin isoforms transcribed from four internal promoters and splice variants are found in nonmuscle organs throughout the body.[267] Dystrophin is a rodlike cytoskeletal protein and is a critical component of the dystrophin-associated protein complex (DAPC).[268]

The DAPC interacts with a host of cytoskeletal, transmembrane, extracellular, trafficking, and intracellular signaling proteins. In skeletal muscle, DAPC plays a structural role by connecting the actin cytoskeleton to the ECM, stabilizing the sarcolemma of muscle fibers during repeated cycles of contraction and relaxation, and transmitting force generated in the muscle sarcomeres to the ECM.[269] The DAPC is also important for Ca^{2+} homeostasis. In the absence of normal dystrophin and the DAPC, the sarcolemmal integrity is compromised, allowing an influx of calcium, immune cells, and cytokines to occur, causing the activation of proteases and the breakdown of the ECM. Secondary morphologic findings including autophagy, necrosis, and fibrosis are associated with progressive muscle wasting.[269,270] Dysregulation of matrix metalloproteinases important for normal muscle repair may also contribute to the pathogenesis in muscle tissue. In addition, there is abnormal signaling of nuclear factor-kappa β, mitogen-activated protein kinases (MAPK), and phosphatidylinositol 3 kinase/AKT pathways.[269]

DMD gene mutations are heterogeneous.[271] Intragenic deletions encompassing one or more exons represent 70% of mutations and affect the translational reading frame of the protein leading to a truncated and nonfunctional protein. Duplications of one or more exons are observed in about 5% of patients. Point mutations account for the majority of remaining mutations, but small insertions, deletions, or splice site mutations are also detected.

Becker muscular dystrophy (OMIM #300376) is a milder and less common form of muscular dystrophy with an estimated incidence of 1 in 18,500 births. BMD is an allelic variation of DMD that is caused by different mutations within the *DMD* gene that result in either reduced protein levels or a partially functional protein. As such, BMD is associated with a milder phenotype, with only half of BMD patients displaying symptoms of disease by 10 years of age with the mean age of death in the mid-40s.[271-276] About 85% of BMD patients have deletions of one or more exons; 5% to 10% have duplications involving one or more exons; and 5% to 10% have a small insertion, deletion, or point mutation. The BMD phenotype is variable and is associated with the type of mutation and the resulting effect on the corresponding structural characteristics of dystrophin.[277] Patients with deletions involving the distal rod domain of dystrophin (exons 45 to 60) show the mild BMD phenotype and in some cases remain free of symptoms until their 50s. However, BMD patients with deletions involving the amino-terminal domain of dystrophin (exons 1–9) have a more severe BMD phenotype with an earlier age of onset and more rapid progression of disease.

Patients with DCM may have a mutation at the dystrophin locus, resulting in *DMD*-associated X-linked DCM.[250,278-282] DCM is characterized by dilation of the left ventricle and reduced left ventricular systolic function and is a rapidly progressive, fatal disease with an onset of symptoms early in the third decade of life.[283] The lack of the functional dystrophin isoform in the heart results from altered tissue-specific transcription or alternative splicing from mutations involving the promoter region or exon 1, splice sites, or specific exonic duplications and deletions.[283] Although mutations in *DMD* can cause DCM, mutations in other genes are also associated with this phenotype.[284]

Because of the tremendous size and diversity of mutations within the *DMD* gene, DNA testing for DMD presents a challenge for clinical laboratories. DNA testing confirms the diagnosis, identifies the pathogenic mutations segregating in the family, and enables targeted analysis for carrier testing of at-risk females as well as prenatal or preimplantation testing as requested. Deletion and duplication testing for all 79 exons is most frequently performed by multiplex ligation-dependent probe amplification assay (Fig. 6.5).[285] This method determines whether each exon is present, deleted, or doubled compared with control DNA. Alternatively, microarray-based comparative genomic hybridization can be used for deletion and duplication screening.[286] Sanger sequence analysis is typically performed for the remaining 30% of mutations.[286] However more recently, targeted MPS has been used for DMD analysis.[286,287] In about 2% of DMD and BMD patients, the DNA mutation is not identified. Although a muscle biopsy can be used to confirm the diagnosis, in the absence of the pathogenic mutation, targeted mutational analysis is not possible for additional family

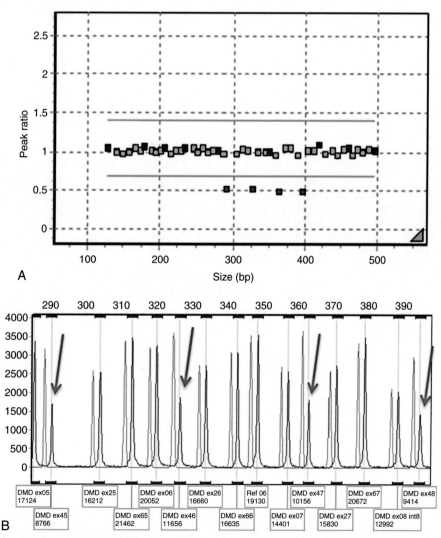

FIGURE 6.5 Deletion/duplication analysis of the *DMD* gene in a female patient by multiplex ligation-dependent probe amplification (MLPA). **A,** Wild-type (normal) alleles demonstrate a peak ratio of about 1. In this example, the patient is positive for a deletion of exons 45 to 48 *(red squares)*. Note that the deleted exons show a peak ratio of approximately 0.5. **B,** Electropherogram data showing the reduction in peak height (y-axis, relative fluorescence) for exons 45 to 48 *(arrows)* in one haplotype *(blue peaks)* relative to the other control haplotype *(red)*.

members, and risk assessment for carrier females can be complex.[288] Particularly difficult are sporadic cases of DMD or BMD in which no other family member with DMD or BMD is known and no mutation in the affected individual is detected. Generally, one third of sporadic cases are thought to represent a new mutation in the mother's gamete from which that individual was derived. Thus, neither the mother nor female siblings would be carriers, and the risk to the mother for a second affected son would be considered minimal. Alternatively, a female in these families could be a carrier who, although clinically asymptomatic, would be at risk of having an affected child. Furthermore, carrier assessment can be complicated by the phenomenon of germline mosaicism, in which no DMD gene mutation is present in lymphocyte DNA but a DMD gene mutation is present in germline tissue.[288] These mutations occur during mitosis in germline proliferation and explain the report of multiple affected children of women whose lymphocyte DNA contains no DMD mutation. If no mutation within the family is known, linkage analysis using intragenic and DMD flanking polymorphic markers can be used for carrier, prenatal, and preimplantation genetic testing.[262,289]

Although no curative treatment for DMD is available, innovative therapies are emerging.[268,290] Recombinant adeno-associated virus vectors have been used to restore 90% of strength in mdx mice by delivery of shortened dystrophin gene constructs. Suppression of nonsense mutations causing premature truncation of the dystrophin protein generated by point mutations in about 10% to 15% of patients can be accomplished with some success using the drug ataluren. Alternatively, for the majority of patients with gene deletions or duplications, exon skipping using antisense oligonucleotide therapy targeting specific exons has been successful for exon 51 using the drugs drisapersen or eteplirsen.[268,291] Clinical trials for these and other drugs are ongoing.[68,290-292] Alternative therapeutic approaches and new targets linked to the pathogenesis of the disease are being evaluated including use of the patient's stem cells, gene editing by homologous recombination, nonhomologous end joining, and increased expression of dystrophin similar proteins.[293-296]

Fragile X Syndrome

Fragile X syndrome (OMIM #300624) is one of the most commonly inherited forms of intellectual disability, with an estimated incidence of approximately 1 in 4000 males and approximately 1 in 5000 to 8000 females.[297] The name of the condition reflects the cytogenetic abnormality of a breakpoint or fragile site in the X chromosome. The clinical syndrome was first described by Martin and Bell in 1943 in a family with sex-linked mental retardation in both males and females yet who had no dysmorphic features.[298] The disease was later redefined by Lubs, who noted the presence of a marker X chromosome in leukocytes of males incubated in cell culture media depleted of folate and thymidine and that segregated with mental retardation within the family.[299]

The chromosomal locus for this fragile site was later localized to Xq27.3.[300] Common clinical features associated with fragile X syndrome are intellectual disability, delayed motor and speech development, macroorchidism, long face, prominent forehead and jaw, large ears, flat feet, and abnormal behavioral characteristics (eg, hyperactivity, hand flapping, temper tantrums, persevering speech patterns, poor eye contact, and autism spectrum disorder [ASD]).[301] Fragile X syndrome represents about 5% of patients with ASD.[301] These features often are less frequent and milder in affected females than in affected males because of random X inactivation of the abnormal fragile X gene in females and the expression of the normal gene in roughly half of their tissues. The primary molecular basis of fragile X syndrome includes expansion of the 5'UTR (untranslated region) CGG repeat sequence that is coupled with hypermethylation and histone deacetylation of this region and the adjacent CpG island in the promoter region of the FMR1 gene.[302-306] This results in transcriptional silencing of the gene and no production of the associated FMR1 protein, FMRP. Males with full expansion alleles but incomplete methylation, methylation mosaic males, or males with FMR1 alleles differing in size within or between tissues may have a more positive clinical outcome.[307]

As a sex-linked disease, fragile X syndrome has a complicated inheritance pattern. Affected females are heterozygous for the mutation, and unaffected males can transmit the mutation through the family. For this reason, Sherman and colleagues proposed that fragile X syndrome was an X-linked dominant disorder with reduced penetrance (79% for males and 35% for females), but the penetrance of the disease appeared to increase in subsequent generations within a family.[308,309] The mechanism of this "Sherman paradox" was resolved when the gene causing fragile X syndrome, FMR1 (Fragile X mental retardation 1), was cloned in 1991.[302,310-313] FMR1 was the first gene discovered to cause disease through expansion of an unstable trinucleotide repeat sequence. The gene spans 38 kb, with 17 exons, and encodes a 4.4-kb mRNA transcript that contains 190 bp of the 5'UTR.[314] FMR1 mRNA is expressed in neural and nonneural tissues during embryonic development and throughout life.[315] Interestingly, multiple mRNA splice variants have been identified in humans, mice, and fruit flies, although the roles of their associated FMRP isoforms are not clear.[316] The primary FMR1 protein, FMRP, is a 71-kDa transacting RNA-binding protein with multiple domains including two chromatin-binding domains (Agenet 1 and 2); a nuclear localization signal; a nuclear export signal domain; and three RNA-binding domains, KH1, KH2, and a RGG box.[316] FMRP is most abundant in the testes and the brain, correlating with the two most prominent features of this disease, intellectual disability and macro-orchidism. In the neurons, FMRP shuttles mRNAs from the nucleus to the cytoplasm, but it predominantly resides in postsynaptic spaces of dendritic spines, where it is associated with polyribosomes and plays a role in the regulation of translation of mRNAs important for synaptic plasticity.[316] Several models have been proposed to explain the process by which FMRP inhibits mRNA translation, including (1) blockage of translation initiation, (2) stalling of the polyribosome during translation, and (3) repression of translation via the RNA interference pathway.[316] Thus, loss of function of FMRP in fragile X patients results in abnormal translation profiles and altered synapse structure and signaling. FMRP mRNA targets are large in number yet are specific, with binding occurring only to mRNA transcripts with specific motifs.[317] Most well-characterized is FMRPs normal steady-state repression of translationally upregulated mRNAs in response to stimulation of group 1 metabotrophic glutamate receptors (mGluRs). Upon mGluR activation, FMRP is dephosphorylated, translation inhibition ceases, mRNA translation is enabled, and long-term depression of synaptic transmission occurs. In the absence of FMRP in fragile X patients, there is constitutive translation of these mRNAs in the absence of mGluR activation and excessive and prolonged synaptic long-term depression. However, clinical trials using GluR antagonists for patients with fragile X have been largely unsuccessful.[318] In addition, FMRP associates with mRNAs encoding for subunits of the neurotransmitter gamma-aminobutyric acid (GABA) receptor.[319,320] *Fmr1* knockout mice display reduced $GABA_A$ receptors, associated altered GABA concentrations, abnormal $GABA_A$ receptor signaling, and overall decreased GABAergic input. Because these aberrations are believed to play a role in the pathogenesis of fragile X syndrome, clinical trials targeting this pathway are underway.

FMR1 alleles contain blocks of CGG repeats usually 7 to 13 repeats in length, which can be interspersed with single AGG repeats.[321,322] Allelic diversity results from the variable numbers and lengths of these CGG-repeat blocks. Normal alleles have 5 to 44 repeats; gray zone, intermediate, or borderline alleles have 45 to 54 repeats; premutation alleles have 55 to 200; and full mutation expansion alleles contain more than 200 repeats (Fig. 6.6).[323] Individuals with a normal number of CGG repeats do not have fragile X syndrome, nor are they at risk of having an affected child. Individuals with 45 to 54 repeats represent alleles in the upper range of normal. These individuals do not have fragile X syndrome, yet some families may have a slightly increased risk of repeat instability and expansion to a *FMR1* premutation allele in their offspring. Premutation alleles are unstable and can expand to a full mutation allele and are associated with a risk of an offspring with fragile X syndrome. However, this expansion is largely confined to maternal transmission. Alternatively, premutation alleles can remain stable or increase to a larger premutation allele. Less commonly, a premutation allele can contract to a smaller premutation allele, a gray zone allele, or even to a normal allele.[324] The risk of CGG expansion from a premutation to a full mutation allele is dependent on both cis- and trans-acting factors, including the number of pure uninterrupted CGG repeats, the number and position of interspersed AGG repeats, maternal age, haplotype background, and less well-characterized heritable factors.[325-330] As the CGG-repeat length increases, the risk of expansion from a premutation to a full mutation in premutation carrier females increases. A carrier woman with a premutation CGG repeat length of 55 to 59 has about a 5% risk of expansion to a full mutation, a woman with a repeat length of 70 to 79 has a 31% risk of expansion, and a woman with a CGG-repeat length greater than 100 has close to a 100% chance of expansion.[331] Although CGG repeat length is the best predictor of maternal premutation expansion, AGG interruptions reduce the risk of transmission for CGG repeat lengths below 100. Nolin et al examined 457 maternal transmissions, including intermediate (45–54) and small premutation alleles (55–69) and reported that 97% of premutation alleles with 0 AGG repeats were unstable, displaying an increase in CGG repeat number compared with only 19% of alleles with 2 interspersed AGG repeats. Only 9 of these 457 transmissions resulted in a full mutation, which originated from a CGG repeat with no interspersed AGG repeats. This study supports the low frequency of full mutation expansion from CGG repeat numbers below 65 and the relevance of AGG repeats for reducing the risk of expansion.[329] Thus, the presence of AGG repeat numbers within CGG repeats can be incorporated into risk assessment for the premutation carrier patient at risk of having a child with fragile X syndrome.[327,329]

Most CGG expansion occurs before zygote formation. CGG expansion that occurs after zygote formation will result in mosaicism with the presence of cells containing either a premutation or full mutation *FMR1* allele. CGG repeat

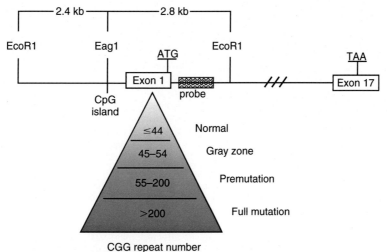

FIGURE 6.6 Schematic representation of the CGG repeat in exon 1 of *FMR1* and associated alleles. A CGG-repeat number less than or equal to 44 is normal. A CGG-repeat number of 45 to 54 is in the gray zone and can expand to a premutation in some families. A CGG-repeat number of 55 to 200 is an unstable premutation allele and is prone to expand to a full mutation allele during female meiosis. A CGG-repeat number in excess of 200 is considered a full mutation and is diagnostic of fragile X syndrome. Restriction endonuclease *EcoRI* and methylation sensitive restriction enzyme *EagI* are used to digest genomic DNA before hybridization with a ^{32}P-labeled probe (see Figure 6.7).

expansion occurs during DNA replication through either the incorporation of looped DNA intermediates on the nascent leading or lagging strand or stalling and restarting of the replication fork.[325,332] Alternatively, trinucleotide repeat expansion may occur at the time of DNA repair of single-strand breaks during the excision of damaged DNA.

Premutation carriers do not have fragile X syndrome. Although *FMR1* is overexpressed in these individuals, FMRP is significantly diminished, indicating reduced translation efficiency.[316] The incidence of a premutation allele in females is estimated to be between 1 in 130 to 250 females.[333] About 20% of premutation carrier females have fragile X associated primary ovarian insufficiency (FXPOI) with cessation of menstrual periods before 40 years of age.[334] Interestingly, women with premutation alleles between 70 and 90 repeats have the highest risk of developing FXPOI. Toxicity from increased expression of premutation *FMR1* mRNA contributes to these clinical symptoms. As expected, full mutation carrier females with no *FMR1* mRNA do not develop ovarian dysfunction, suggesting that decreased or absent FMRP is not contributory to this phenotype. Recently, reports suggest that premutation carrier females are at an increased risk for other medical, reproductive, psychiatric, and cognitive features compared with noncarrier control participants.[335]

Premutation carrier males are less common and are estimated at 1 in 250–810 males.[333] Although premutation carrier males do not have fragile X syndrome, they do have neurodevelopmental problems, including higher rates of attention deficit disorder, shyness, social deficits, and ASDs.[333] As adults, about one third of premutation carrier males older than the age of 50 years exhibit fragile X associated tremor and ataxia syndrome (FXTAS) with 17% of males between the ages of 50 and 59 years and as many as 75% of males older than 80 years of age affected.[336] FXTAS is a neurodegenerative syndrome characterized by progressive intention tremor, cerebellar gait ataxia, parkinsonism, neuropathy, cognitive decline, psychiatric features, and generalized brain atrophy.[337] Although predominantly in premutation carrier males, FXTAS can be seen in up to 16% of premutation carrier females. The lower incidence of FXTAS in females may be explained by the presence of a second X chromosome with a normal *FMR1* gene expressed in approximately half of all cells. The pathogenesis of FXTAS likely results from multiple factors, including (1) increased expression of *FMR1* mRNA, resulting in a gain-of-function through sequestration of specific proteins and a loss of their normal function; (2) translation of unique (CGG) proteins without an AUG start site; or (3) antisense transcription from the *FMR1* locus and decreased production of FMRP.[338]

DNA testing for fragile X syndrome can be performed using Southern blot analysis to detect the 5'UTR CGG repeat expansion as well as hypermethylation of this region, yet PCR analysis is required to accurately determine the precise CGG-repeat number (Fig. 6.7). However, because this is a CG-rich sequence, large premutation and full mutation alleles were at one time difficult to amplify by PCR. These regions can now be successfully amplified using three primers, including the typical forward and reverse primer specific to this unique area within the genome as well as a third oligonucleotide complementary to the CGG repeat itself (Fig. 6.8).[339,340] The methylation status of the CGG repeat

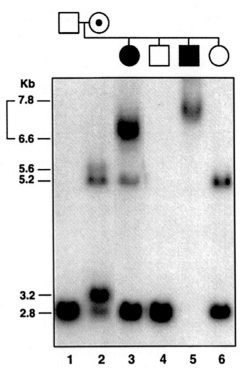

FIGURE 6.7 Southern blot analysis for the diagnosis of fragile X syndrome. Patient DNA is simultaneously digested with restriction endonucleases *EcoR*1 and *Eag*1, blotted to a nylon membrane, and hybridized with a ^{32}P-labeled probe adjacent to exon 1 of *FMR1* (see Figure 6.6). *Eag*1 is a methylation-sensitive restriction endonuclease that will not cleave the recognition sequence if the cytosine in the sequence is methylated. A normal male DNA pattern is seen in the father *(lane 1)* with a CGG-repeat number of 20 on his single X chromosome and in the brother *(lane 4)*, who has a CGG repeat of 22. This generates a band about 2.8 kb in length, corresponding to *Eag*1-*EcoR*1 fragments (see Figure 6.6). A normal female DNA pattern with a CGG-repeat number of 20 on one X chromosome and a CGG-repeat number of 22 on her second X chromosome *(lane 6)* generates two bands: one at about 2.8 kb and a second at 5.2 kb. *EcoR*1-*EcoR*1 fragments 5.2 kb in length result from methylated DNA sequences characteristic of the lyonized chromosome in each cell that is not digested with restriction endonuclease *Eag*1. DNA in *lane 2* is characteristic of a premutation carrier female with one normal allele having a CGG-repeat number of 22 (band at about 2.8 kb) and a second premutation allele with a CGG repeat number of 90 (band at about 3.0 kb). In premutation carrier females, the lyonized (methylated) cells of the X chromosome with the premutation allele is larger than 5.2 kb because of the increased CGG-repeat number and in this case is about 5.4 kb in length. *Lane 3* is diagnostic of a female with fragile X syndrome with one normal allele and one full mutation allele that is completely methylated and transcriptionally silenced. The full mutation allele arose from expansion of the maternal 90 CGG premutation allele *(lane 2)* during meiosis and the normal allele with a CGG repeat number of 20 she inherited from her father *(lane 1)*. The banding pattern observed in *lane 5* is that of an affected male with fragile X syndrome, illustrating the typical expanded allele that is fully methylated in all cells.

in a full mutation allele can also be assessed using PCR.[341] These advances can alleviate the need for Southern blot analysis and reduce both the time required to perform the test and the overall turnaround time.

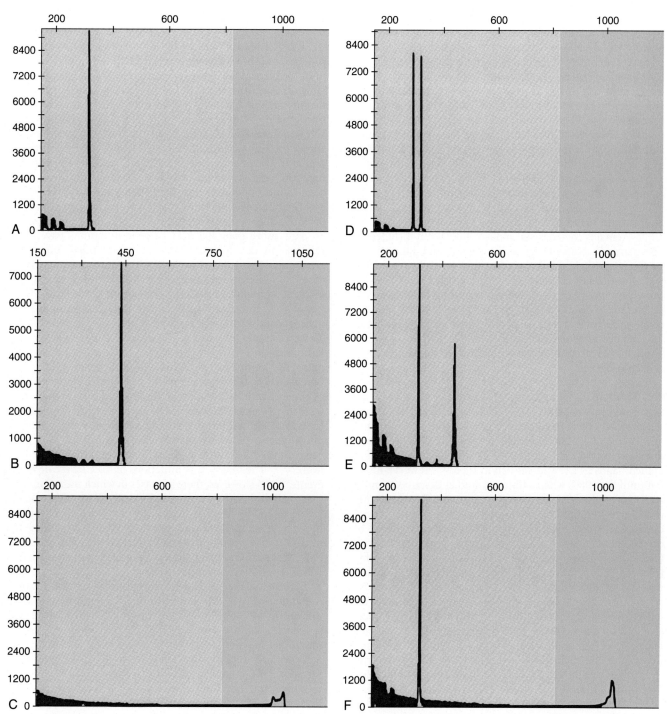

FIGURE 6.8 Electropherograms of polymerase chain reaction (PCR) products of the CGG repeat number in exon 1 of the *FMR1* gene. Amplicon sizes in bases are noted at the top of the figure and the relative amount of fluorescence detected for each amplicon is measured by the peak height (y-axis). Results were generated using the AmplideX kit (Asuragen, Inc). PCR products were amplified by using a primer pair flanking the CGG repeat of the *FMR1* gene and a 15-bp oligonucleotide as a primer within the CGG repeat itself. **A,** Subject 1 is a normal male with a CGG repeat number of 29 corresponding to an amplicon of 316 bp in length. **B,** Subject 2 represents a premutation carrier male with a CGG repeat number of 67 and an amplicon 430 bp in length. **C,** Subject 3 is the characteristic full mutation pattern obtained from a male with fragile X syndrome with more than 200 CGG repeats. **D,** Subject 4 is a pattern from a normal female with CGG repeats of 20 and 30 corresponding to amplicons at 289 bp and 319 bp, respectively. **E,** Subject 5 is a premutation female with CGG repeats of 29 and 74 corresponding to amplicons of 316 bp and 454 bp, respectively. **F,** Subject 6 is characteristic of a female affected with fragile X syndrome with both a normal *FMR1* allele with 31 CGG repeats corresponding to an amplicon of 322 bp, and a second *FMR1* allele with a full mutation with CGG repeats greater than 200 *(pink shaded area)*.

FMR1 DNA testing for fragile X syndrome is often requested for children with (1) the fragile X phenotype, (2) developmental delay, (3) intellectual disability, (4) ASD, or (5) family members of individuals with a diagnosis of fragile X syndrome. Fragile X DNA testing is performed for carrier testing in at-risk pregnant or preconception female patients with a family history of fragile X or intellectual disability of unknown cause. Prenatal testing can be performed on chorionic villi tissue or cultured amniocytes for at-risk pregnancies. Preimplantation diagnosis for fragile X syndrome has been reported but can be complicated by ovarian dysfunction in premutation carrier females.[342] *FMR1* premutation allele testing is performed on patients with clinical suspicion of FXPOI or FXTAS. NBS for fragile X syndrome, although feasible from blood spots as demonstrated from several pilot studies, remains controversial and currently is not included in NBS panels.[343] NBS enables early detection and intervention for the affected child but also identifies at-risk family members who did not consent to this test and may not wish to know personal health information regarding their own risk for late-onset disorders such as FXPOI and FXTAS. Educational resources and counseling programs regarding the medical implications of the various *FMR1* alleles identified through NBS need to be established before implementation of such programs.

Rett Syndrome

Rett syndrome (OMIM #312750) is an X-linked dominant cause of inherited intellectual disability with an estimated incidence of 1 in 10,000 female births.[344] This disorder was first described in a cohort of patients with identical wringing of their hands by Dr. Andreas Rett in 1966.[345] However, it was not until the 1980s when additional studies described similar patients that the syndrome was given its name. Rett syndrome is characterized by progression in stages with initial normal development up to age 18 months followed by a period of developmental inactivity and overall signs of failure to thrive (eg, microcephaly, weight loss).[346] This stage is quickly followed by a period of rapid developmental regression and the loss of purposeful hand movement. It is during this period that patients exhibit signs of autism and profound intellectual disability. Rett syndrome is also marked by progressive motor deterioration that typically results in patients being wheelchair bound by their teens.[346] Although some patients have unexplained death,[347] many patients survive into the sixth or seventh decade of life. It should be noted that Rett syndrome is almost exclusively observed in females because *MECP2* mutations in males are typically embryonic lethal.

Identification of the genetic cause of Rett syndrome was initially difficult because traditional linkage methods were ineffective because of the sporadic nature of the disease.[346] However, using exclusion mapping methods, it was determined that the causative gene for Rett syndrome was located at chromosome Xq28.[348-352] Systematic analysis of this region identified mutations in the methyl-CpG binding protein 2 (*MECP2*) gene as causative for Rett syndrome in 1999.[353] The *MECP2* gene is composed of four exons that produce two different protein isoforms.[346,354] The protein produced by the *MECP2* gene is a chromosome binding protein expressed in all tissues that specifically targets 5-methyl cytosine residues.[355] MECP2 contains three functional domains: a methyl-CpG binding domain (MBD),[356] a transcriptional repression domain (TRD),[357] and a C-terminal domain (CTD).[358] The MBD specifically binds to methylated cytosine residues and shows a preference for binding to CpG dinucleotide sequences that are adjacent to A/T-rich motifs.[359] Downstream from the MBD is the TRD, which is critical in the interaction of MECP2 with histone deacetylases and other transcription corepressors.[346,353,354,358,360,361] Finally, the CTD enables MECP2 binding to the nucleosome core and allows the protein to bind to naked DNA.[346,354] All of these domains are essential for the MECP2 protein to properly function, and mutations in each of these domains have been found in patients with Rett syndrome.

MECP2 gene mutations are identified in approximately 95% of patients with a classic Rett syndrome phenotype.[346] To date, more than 400 *MECP2* gene mutations have been reported in the literature, and newly identified mutations are regularly added to the RettBASE online mutation database.[362,363] Described pathogenic *MECP2* gene mutations include nonsense, missense, frameshift, and large deletions.[362,363] Interestingly, mutations at eight common residues account for approximately 70% of all cases.[346,354] These common mutations all occur in CpG dinucleotides and include p.Arg106, p.Arg133, p.Thr158, p.Arg168, p.Arg255, p.Arg270, p.Arg294, and p.Arg306 (Fig. 6.9). Because there is a large degree of phenotypic heterogeneity in patients with Rett syndrome, several studies have been performed to correlate genotype with phenotype. Truncating (nonsense or frameshift) mutations cause a more severe phenotype than missense mutations, and mutations that occur in the CTD typically result in milder disease presentation.[346,364] The majority of mutations in *MECP2* are de novo events.[365,366] However, there are cases in which mutations are maternally inherited from an individual who is either unaffected or mildly affected because of skewed X-inactivation. It is therefore important to establish whether a pathogenic mutation is de novo or maternally derived because the recurrence risk in future pregnancies of mutation-positive females with skewed X-inactivation approaches 50%.

Because of the heterogeneous nature of disease-causing mutations in *MECP2*, clinical testing for confirmation of a diagnosis of Rett syndrome typically involves DNA sequencing of the entire *MECP2* gene coding region. If mutations are not identified by sequencing analysis, large deletion and duplication analysis (typically MLPA) is performed. This combination detects approximately 95% of causative mutations in patients with a clinical diagnosis of Rett syndrome. Recently, mutations in the *CDKL5* and *FOXG1* genes have been shown to cause variant forms of Rett syndrome and should therefore be considered for

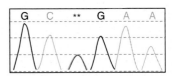

FIGURE 6.9 Sanger sequencing of the *MECP2* gene illustrating a common pathogenic mutation (c.808C>T) at p.Arg270 (asterisk). This C>T transition mutation results in a stop codon (TGA) and premature truncation of the protein.

follow-up testing in individuals not found to have mutations in *MECP2*.[367-373] Rett syndrome treatment is largely based on the manifestations of disease and therefore should be tailored to each individual patient.

Complex Diseases

A complex or multifactorial inheritance pattern suggests interaction of one or more genes in combination with lifestyles and one or more environmental factors. Multifactorial diseases can be prevalent in some families with several affected family members, but the disease does not follow typical Mendelian inheritance patterns. A disease may present in multiple family members because of the sharing of similar disease-predisposing alleles and often sharing of similar daily habits, routines, and diet. The specific genes, lifestyle habits, and environmental factors and their respective contribution in predisposition to disease vary among diseases and are difficult to elucidate. Twin studies are often used to determine the relative importance of each component. Among twins who were raised together, a greater concordance of disease among monozygotic (MZ) twins (who share all of their genes) than among dizygotic (DZ) twins (who share 50% of their genes) provides strong evidence of a genetic component of the disease. Conversely, disease concordance of less than 100% in MZ twins is strong evidence that nongenetic factors play a role in the disease process. Large genome-wide association studies (GWAS) are used to identify genes and genetic variants that play a role in the pathogenesis of common complex diseases. Examples of complex adult-onset diseases, include type 1 diabetes, rheumatoid arthritis, multiple sclerosis, osteoporosis, Parkinson disease, Alzheimer disease, hypertension, atrial fibrillation, alcoholism, schizophrenia, depression, obesity, and thrombophilia.

Thrombophilia

Thrombophilia is defined as an abnormality of hemostasis with a predisposition to thrombosis. A common complication of venous thromboembolism (VTE) is the development of a deep vein thrombosis (DVT) or a more serious, and potentially fatal, pulmonary embolism. Thrombophilia (OMIM #188050) is a multifactorial disorder resulting from the interaction of genetic, lifestyle, and environmental factors. Risk factors include the use of oral contraceptives, hormone replacement therapy, trauma, obesity, malignancy, surgery, immobility, pregnancy, and advanced age.[374,375]

Protein products of many genes are involved in the anticoagulation and coagulation pathway to regulate hemostasis. Hypercoagulability, or an alteration in the coagulation pathway that predisposes to thrombosis, can be caused by variants in genes encoding proteins involved in the coagulation pathway.[376] Although familial thrombophilia can be attributed to mutations in genes encoding protein C, *PROC* (OMIM #176860); protein S, *PROS1* (OMIM #612336) or antithrombin III, *SERPINC1* (OMIM #107300), 50% to 60% of familial thrombophilia is associated with variants in genes encoding coagulation factor V, *FV* (OMIM #188055) or factor II, *F2* (OMIM #176930). In 1993, Dahlbäck reported that familial thrombophilia caused resistance to activated protein C (APC).[377] In 1994, Bertina and colleagues reported linkage of a common G-to-A base substitution at nucleotide 1691 (c.1691G>A) in exon 10 of the *FV* gene with the APC resistance phenotype.[378] This nucleotide change results in an arginine-to-glutamine substitution in the FV protein at codon 506 (p.Arg506Gln) and is commonly referred to as FV Leiden, named for the Dutch city where it was discovered.[378] The c.1691G>A substitution is common in the white population of Northern European descent with a reported frequency of about 3% to 5%, but it is absent in other populations, including those in Africa and Southeast Asia.

The *FV* gene is localized to chromosome 1q24.2, is about 70 kb, contains 25 exons, and encodes a 330 kDa protein.[379] In the coagulation pathway, FV is converted to an activated form, FVa, by thrombin. FVa is a cofactor for activated factor X, FXa, and is required for the conversion of prothrombin (F2) to thrombin (F2a). Thrombin is essential for the last step of the coagulation cascade by catalyzing the conversion of fibrinogen to fibrin for clot formation. Activated FV is converted to an inactive form by APC. The arginine residue at codon 506 is one of three peptide bonds (Arg306, Arg506, and Arg679) cleaved by APC to inactivate FV and decrease the affinity to FXa, thereby reducing the conversion of prothrombin to thrombin.[380,381] Substitution of a glutamine residue at this site prolongs APC inactivation of FVa by approximately 10-fold, thereby shifting the balance of hemostasis to favor coagulation and increasing thrombin production.[382]

Heterozygous *FV* gene c.1691G>A carriers have a lifelong 7.9-fold increased relative risk of venous thrombosis compared with an increased relative risk for homozygotes as high as 80-fold.[375,383] However, FV Leiden does not confer increased risk for arterial thrombosis.[384] The mean age of onset of symptoms associated with thrombosis is 44 years for heterozygotes and 31 years for homozygotes.[383] *FV* c.1691G>A carriers represent about 25% of patients with idiopathic VTE, 30% to 50% of patients with recurrent VTE, 20% to 60% of oral contraceptive–associated VTE, 20% to 40% of pregnancy-associated VTE, and 8% to 30% of patients with pregnancy loss.[385] Although *FV* gene mutations are considered dominant (heterozygous mutations carriers can be symptomatic), many heterozygous carriers remain asymptomatic because thrombophilia is a complex disease resulting from the interaction of genetic, lifestyle, and environmental factors.[383,386]

Several years later also in Leiden, Poort and coworkers described a genetic variant in the 3′ untranslated region of the *F2* gene present in 18% of patients with a documented family history of venous thrombosis.[387] The *F2* gene maps to chromosome 11p11.2, is 21 kb in length, contains 14 exons, and encodes a 70-kDa protein.[388] The *F2* 3′ variant that segregated with disease, c.*97G>A, results in increased levels of plasma coagulation F2, prothrombin, and a 2.8-fold lifelong increased risk of venous thrombosis.[387] This variant allele is largely confined to white populations at a frequency of 1% to 2% and is rare in other populations. FII requires activation and conversion to thrombin by FXa and FVa to catalyze the conversion of fibrinogen to fibrin, the last step of blood clot formation. The G-to-A substitution does not alter the coding region of the protein; rather, it enhances prothrombin mRNA stability and ultimately results in increased production of both prothrombin and thrombin.[389]

The risk of venous thrombosis is increased 16-fold in *F2* c.*97 G>A carriers using oral contraceptives, and the risk of cerebral vein thrombosis increases 149-fold.[390,391] In the white population, this base substitution is present in 6.2% of patients with venous thrombosis compared with 2.3% of control participants.

Inherited together, *FV* c.1691G>A and *F2* c.*97G>A convey an increased relative risk for a VTE event, an example of additive genetic effects associated with a complex disease.[392-394] Metaanalysis shows a higher risk for severe preeclampsia in women with *FV* or *F2* variants, with odds ratios of 1.9 and 2.01, respectively.[395]

Venous thromboembolism most often is classified as "provoked" with one or more predisposing risk factors; however, in 25% to 50% of cases, it is "unprovoked" with the precipitating cause not determined. The initial treatment of patients with VTE uses both heparin and vitamin K antagonists (eg, warfarin).[396] After several days, the heparin is discontinued, and warfarin therapy is continued. For "provoked" VTE, therapy is discontinued after 3 months for distal DVT and 6 months for proximal DVT or PE. Therapy may be altered based on the type of thrombotic event, the type of precipitating event, or the absence of a triggering cause. The recurrence risk is higher in "unprovoked" VTE and in patients with cancer, suggesting prolonged or indefinite anticoagulant therapy.[397] D-dimer levels may identify patients for continued anticoagulation therapy.[398,399] Antithrombic agents may be most advantageous in the management of pregnancy VTE and associated thrombophilia.[400] To avoid the risk of bleeding with sometimes fatal complications, treatment is regularly monitored using prothrombin time reported as the international normalized ratio (INR) ideally achieving an INR of 2.0 to 3.0. Because of the challenges with warfarin therapy, new oral anticoagulants have been developed and effectively used to prevent recurrence of VTE. These include FIIa and FXa inhibitors, and most recently a FIXa inhibitor that does not require regular laboratory monitoring.[401] These drugs carry a lower risk of bleeding complications and appear to provide similar efficacy compared to standard treatment.[401-406]

Another genetic risk factor for predisposition to thrombophilia is a common 1-bp guanine deletion/insertion (4G/5G) polymorphism at c.-817dupG in the promoter region of the *SERPINE1* gene located on chromosome 7q22.1 (OMIM#173360).[407,408] This gene is approximately 12.2 kb in length with 9 exons and encodes a mature 379-amino-acid protein with a molecular weight of about 42.7 kDa.[409-411] This protein, plasminogen activator inhibitor-1 (PAI-1), is a member of the serine protease inhibitor family and is released by endothelial cells to block the degradation of fibrin clots. Increased levels of PAI-1 can be associated with thrombophilia, and the *SERPINE1* 4G allele is associated with higher transcription levels than the 5G allele.[407,412,413] Testing for common variants in the 5,10-methylenetetrahydroflate reductase *(MTHFR)* gene, c.665C>T and c.1286A>C, once thought to convey a slight risk for thrombophilia, are not as frequently used for routine clinical evaluation of this disease.[414]

Many DNA testing platforms have been used for the detection of *FV* c.1691G>A and *F2* c.*97G>A variants, including Invader technology, PCR coupled with restriction-endonuclease digestion and gel electrophoresis, or real-time

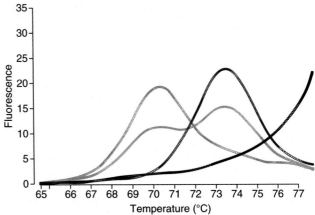

FIGURE 6.10 Interrogation of the common c.*97G>A mutation in the *F2* gene by melting curve analysis. A perfectly matched probe is more tightly bound to the polymerase chain reaction (PCR) product and therefore will melt at a higher temperature than when there is a nucleotide mismatch between the probe and the PCR product. In this example, the probe perfectly matches the wild-type allele *(red)*. When the *F2* c.*97G>A mutation is heterozygous, the probe binds to the mutant allele imperfectly (with a lower melting temperature) and to the wild-type allele perfectly, resulting in the presence of two peaks *(pink)*. When a patient is homozygous for the c.*97G>A mutation, only an imperfect match is possible, resulting a single peak at the lower temperature *(gray)*. A no-template control is always included to rule out potential DNA contamination *(black)*.

PCR followed by melting curve analysis with fluorescence resonance energy transfer probes (Fig. 6.10). The most common testing methods for the 4G/5G *SERPINE1* polymorphism involve the use of PCR and melting curve analysis or PCR coupled with capillary electrophoresis and fragment analysis. Any testing platform is acceptable for clinical use as long as the procedure has been properly validated in a CLIA-certified laboratory that follows appropriate regulatory and quality assurance guidelines.

DNA testing for factor V may be requested when a patient tests positive on the functional activated protein C resistance assay to both confirm the diagnosis and distinguish between *FV* c.1691G>A heterozygotes and homozygotes.[415] At many hospitals, screening for *FV* abnormalities using the functional activated protein C resistance assay is the preferred method because it is both cost effective and can be automated.[416] However, DNA testing should be ordered in place of the functional assay for patients taking FIIa or FVa oral inhibitors (eg, argatroban, dabigatran, bivalirudin, or rivaroxaban) because these can interfere with the functional assay causing a falsely normal result. Similarly, inaccurate results can be obtained in patients with lupus anticoagulants.[415] The clinical utility of knowing the *FV* or *F2* genotype has been debated because it may not affect the clinical management of patients.[417-419] Yet testing for *FV* and *F2* variants is common in clinical practice, and most experts believe that testing is appropriate for targeted patients, including those with (1) venous thrombosis or pulmonary embolism before 50 years of age; (2) venous thrombosis at an unusual site (hepatic, mesenteric, portal, or cerebral veins); (3) recurrent VTE; (4) VTE and a strong family history of thrombotic disease; (5) VTE in pregnancy, postpartum, or associated with contraceptive use or hormone replacement therapy; or (6) unexplained recurrent pregnancy loss.[420-423] However, routine screening before hormone use is not routinely performed, nor is it considered cost-effective. Modifications to patient management after knowledge of this information may involve length of treatment with anticoagulants and management of other procoagulation risk factors. First-degree relatives of *FV* or *F2* variant positive patients are at a higher risk of related thrombotic events if the proband was younger and experienced an "unprovoked" thrombophilic event. DNA testing for *FV* or *F2* variants is not recommended for the general population, prenatal carrier screening or NBS. In addition, variants at two other factor V arginine-cleavage sites have been reported, but these variants are rare and are not part of routine *FV* DNA testing.[424] Although *FV* or *F2* variants are present in 50% to 60% of families with inherited thrombophilia, defects in protein C, protein S, and antithrombin are detected in 10% to 15% of families with inherited venous thrombosis. These less common coagulation deficiencies are typically diagnosed through immunologic or functional assays that do not involve DNA testing. Furthermore, unlike *FV* and *F2* genes, no single common variants in *PROC*, *PROS1*, or *SERPINEC1* have been identified.

Hereditary Pancreatitis

Hereditary pancreatitis (HP; OMIM# 167800) occurs in approximately 1 in 300,000 individuals in the Western world.[425] HP traditionally refers to the occurrence of pancreatitis in two or more individuals in two or more generations of a family.[426] Additionally, HP can refer to individuals with disease-causing germline mutations in any of the genes associated with the disorder.[426] Individuals affected with HP are typically characterized by acute pancreatitis (disease with a duration of a few days to a few weeks) that eventually progresses to chronic pancreatitis (disease with a duration >6 months) over a period of years.[426] The onset of symptoms is difficult to predict, but most individuals with HP are first affected with acute pancreatitis by age 10 years and chronic disease by age 20 years.[427] Given the inflammatory nature of the disease, HP patients are also at higher risk for pancreatic cancer than the general population and should be routinely screened.[428] The complex nature of disease (and variable penetrance) indicates that although genetics is a risk factor for disease, several environmental factors (eg, diet, tobacco use, alcohol consumption) also contribute to disease manifestations.[427] Mutations in the protease, serine, 1 *(PRSS1)* gene (also known as cationic trypsinogen, located at chromosome 7q35) are the most common genetic cause of pancreatitis, accounting for 60% to 100% of HP families.[426,427] Cationic trypsinogen is the main isoform of trypsinogen in human pancreatic juice and is involved in facilitating zymogen activation.[429-431] Mutations in the *PRSS1* gene cause a gain of function and typically result in the constitutive autoactivation of the cationic trypsinogen protein.[432-436] These mutations are inherited in an autosomal dominant manner with variable penetrance.[425] In fact, several studies conducted using large HP patient cohorts have reported that the penetrance of *PRSS1* mutations ranges between 40% and 96%[437-444]; therefore, predicting disease progression in presymptomatic individuals who carry pathogenic *PRSS1* gene mutations is not possible because symptoms and onset are extremely variable even among affected members of the same family.

Several additional genetic risk factors for pancreatitis have been identified.[445-447] In 1998, mutations in the *CFTR* gene (located at chromosome 7q31.2) were reported as a genetic cause of pancreatitis.[447] As discussed in detail earlier, *CFTR* gene mutations are causative for the multisystem autosomal recessive disorder CF. One of the most common features of CF is altered pancreatic function, so the discovery of genetic mutations in the *CFTR* gene in patients with isolated pancreatic disease is not surprising. However, unlike individuals with classic CF disease who carry two severe *CFTR* gene mutations, pancreatitis patients often carry a severe pathogenic *CFTR* gene mutation on one chromosome and a mild mutation on the other (or a mild mutation on both chromosomes).[447] About 30% of patients with pancreatitis remained idiopathic and did not have mutations in *PRSS1* or *CFTR*. In 2000, autosomal recessive mutations in the serine protease inhibitor, Kazal type 1 *(SPINK1)* gene (located at chromosome 5q32), were reported as an additional cause of HP.[446] In this study, 96 patients with chronic pancreatitis were tested for mutations in the *SPINK1* gene, and 23% of the patients tested positive for mutations.[446] Interestingly, a single mutation in codon 34 (p.Asn34Ser) was the most common mutation identified.[446] The *SPINK1* gene produces a protein (SPINK1) that prevents premature trypsinogen activation through the inhibition of trypsin activity in the pancreas.[448] Therefore, *SPINK1* mutations may cause a decrease in the ability of SPINK1 protein to inhibit trypsin.[446] However, functional studies have been inconclusive and the exact mechanism by which *SPINK1* mutations cause pancreatitis remains unknown.

Mutations in the chymotrypsin C *(CTRC)* gene (located at chromosome 1p36.21) are the most recently described genetic cause of HP.[445] An international team of investigators screened a large cohort of German pancreatitis patients (n = 901) for mutations in the *CTRC* gene.[445] *CTRC* mutations were identified in 3.3% of this patient group, and subsequent studies in an Indian population identified *CTRC* variants in 14.1% of affected individuals.[445] Given the overrepresentation of *CTRC* variants in patients with pancreatitis (compared with control populations), the authors were able to conclude that mutations in the *CTRC* gene increase the risk for pancreatitis.[445] The *CTRC* gene produces a digestive enzyme called chymotrypsin C (CTRC) that promotes proteolytic inactivation of trypsinogen and trypsin and is essential for limiting intrapancreatic trypsinogen activation.[449,450] Functional studies indicate that mutations in *CTRC* are loss of function, resulting in diminished secretion ability, impaired catalytic activity, and an inability to degrade trypsin.[450] Similar to *SPINK1*, mutations in the *CTRC* gene are autosomal recessive and require the presence of two pathogenic mutations to cause disease.

DNA testing for pancreatitis is most often performed stepwise with Sanger sequencing of the *PRSS1* gene as the first step. The identification of a single pathogenic *PRSS1* gene mutation confirms a diagnosis of HP in symptomatic patients and indicates a substantial risk for the development of disease in an asymptomatic individual. If *PRSS1* analysis is negative, Sanger sequencing of *CFTR*, *SPINK1*, and *CTRC* can be considered. All three of these genes are inherited in an autosomal recessive manner, and therefore identifying two pathogenic mutations is required to confirm a diagnosis in a symptomatic individual. The identification of a single pathogenic mutation may increase the risk for pancreatitis but, by itself, is not causative for disease. Comprehensive testing of all four HP-associated genes can be performed as the initial test in HP patients. However, this may increase the complexity of the results obtained (eg, variants of uncertain significance), so patients should be counseled carefully before ordering such a panel.

Even with the advances over the past decade in our understanding of the genetics responsible for HP, cases remain that are not explained by mutations in a single gene. These individuals appear to have complex disease that involves a number of pathogenic mutations in several genes (ie, *CFTR* and *SPINK1*) suggesting that gene–gene interactions govern their disease.[426] In fact, as many as one third of pancreatitis cases (acute and chronic) result from complex inheritance patterns.[426,451] In some cases, environmental factors may play a more significant role than genetics in the development of this disease.[451]

Treatment for HP varies between acute and chronic disease. Pain management is the primary focus for patients with acute pancreatitis.[426] Patients with acute disease are also counseled to abstain from activities that will predispose them to future attacks (eg, smoking, alcohol consumption).[426,427] Patients with chronic pancreatitis may be treated with pancreatic enzyme replacement therapy to aid with digestion.[427] In some severe cases, patients may undergo total pancreatectomy with islet autotransplantation.[426] Additional treatment should be based on manifestations of disease such as diabetes mellitus.

Hereditary Breast and Ovarian Cancer

Breast cancer is the most common cancer in women with about 1 in 8 developing invasive cancer during their lifetime.[452] Approximately 10% of breast cancers appear to be familial, exhibiting dominant inheritance. Inherited breast cancer is associated with multiple cases of breast or ovarian cancer within the family, an early age of disease onset, bilateral disease or multiple cancers in the same breast, and an increased prevalence of male breast cancer. Epithelial carcinoma of the ovary is the fifth leading cause of death in women.[453]

Most hereditary breast and ovarian cancer (HBOC) is caused by mutations in *BRCA1* (OMIM #113705) or *BRCA2* (OMIM #600185). In addition to a predisposition to breast and ovarian cancer, germline mutation carriers are at an increased risk for pancreas cancer. and male carriers are at an increased risk for prostate and male breast cancer.[454] The incidence of *BRCA1/BRCA2* mutations in the United States is estimated to be between 1 in 300 to 500.[455] However, the combined frequency of two founder mutations in *BRCA1* (185delAG, 5382insC) and one in *BRCA2* (6174delT) in the Ashkenazi Jewish population is as high as almost 3%.[456]

Although dominantly inherited, *BRCA1/BRCA2* mutation carriers do not possess a 100% certainty of developing cancer. Rather, as tumor suppressor genes, inactivation of both alleles is required for tumorigenesis. In a patient who has inherited a mutant *BRCA1* or *BRCA2* allele, the second normal allele is mutated and most often deleted via somatic mutation in cancer cells. By the age of 70 years, *BRCA1* mutation carriers have a 44% to 75% risk of developing breast cancer, an 83% chance of developing contralateral breast cancer after an initial cancer is found, and a 43% to 76% possibility of developing ovarian cancer.[457] Slightly lower frequencies are observed for *BRCA2* mutation carriers who have a 41% to 70% risk of developing breast cancer, a 62% chance of developing contralateral breast cancer, and a 7.4% to 34% possibility of developing ovarian cancer by 70 years of age. Penetrance of the disease in *BRCA1/BRCA2* mutation carriers is determined by both other genes and lifestyle and environmental factors. Identified susceptibility factors influencing development of disease include parity, body mass index, age at menarche, menopause and first full-term pregnancy, breastfeeding, smoking, and oral contraceptive usage.[458,459] Additional genetic susceptibility factors include variants in multiple genes or genetic regions such as *FGFR2*, *TOX3*, *MAP3K1*, *LSP1*, *SLC4A7*, 2q35, and 5p12 for *BRCA2* germline mutation carriers and variants in *TOX3* and genetic factors located at 2q35 for *BRCA1* germline mutation carriers.[460] Understanding susceptibility factors, both environmental and genetic, aid in appropriate counseling for patients by either increasing or decreasing their overall risks for the development of cancer and help guide their clinical management decisions for surgical interventions. However, to prevent disease in *BRCA1/BRCA2* mutation carrier women, population-based testing for all women as routine medical care has been proposed.[461] This screening would identify germline mutation carriers and enable appropriate clinical management to reduce their lifetime cancer risks.

In a large study, Mayaddat and colleagues compared the pathologic characteristics between *BRCA1* and *BRCA2* mutation–positive breast and ovarian tumors.[462] *BRCA1*-mutated breast tumors were most frequently invasive ductal carcinomas (80%), grade 3 (77%) with the majority of tumors (69%) triple negative (TN) for the expression of estrogen (ER), progestin (PR) and *HER2*. *BRCA2* mutation positive breast tumors likewise were predominantly invasive ductal carcinomas (83%) but with relatively equal numbers of grade 2 (43%) and grade 3 (50%) observed. However, unlike *BRCA1* mutation–positive tumors, only 16% were TN. Medullary carcinomas were more frequent in *BRCA1* mutation–positive tumors (9.4%) compared with *BRCA2* (2.2%), and invasive lobular was seen more frequently in *BRCA2* mutation positive tumors (8.4%) compared with *BRCA1* mutated tumors (2.2%). Interestingly, *BRCA1*-mutated TN tumors were highest in patients with an earlier age of onset, but TN tumors were more common in tumors from *BRCA2*-positive carriers with an older age of onset. Both *BRCA1* and *BRCA2* mutation–positive ovarian tumors were morphologically similar, with most serous (66% and 70%) and grade 3 (77% and 73%), respectively. Overall, about 15% of ovarian cancers are associated with germline mutations in *BRCA1/BRCA2*, and up to 25% of high-grade serous tumors are positive for germline mutations in one of these two genes.[463] Although the median age for ovarian cancer is close to 60 years, it is about 10 years earlier in patients with a genetic predisposition to this disease.

Using DNA linkage studies, Hall and coworkers mapped the gene for early-onset familial breast cancer to chromosome 17q21.[464] In 1994, *BRCA1* was cloned; it was later confirmed by several other investigators as the susceptibility gene in breast and ovarian cancer kindreds.[465-467] The *BRCA1* gene on chromosome 17q21.31 spans 80 kb and is composed of 24 exons; 22 encode the 7.8-kb mRNA that is translated into a protein of 1863 amino acids. Exon 11, encoding 60% of the protein, is alternatively spliced in a number of tissues. In families not linked to *BRCA1*, a second susceptibility locus at chromosome 13q12-13 was proposed, and in 1995, *BRCA2* was cloned.[468-470] The *BRCA2* gene spans 70 kb, contains 27 exons, encodes an 11.5-kb mRNA, and is translated into a protein of 3418 amino acids.

BRCA1 is a widely expressed multifunctional protein that maintains genomic stability in the repair of double-strand breaks in DNA via the homologous recombination pathway and cell cycle checkpoint control.[471,472] BRCA1 also plays a role in transcription regulation, chromatin architecture, apoptosis, mRNA splicing, and ubiquitination of multiple proteins. Although a distinctly different protein, BRCA2 is also important in the repair of double-strand DNA breaks and in transcription regulation.[473,474] Because BRCA1/BRCA2-deficient tumors have aberrant DNA repair pathways, they are more sensitive to treatment options causing DNA damage such as platinum-based chemotherapies cisplatin and carboplatin.[473] Clinical trials using PARP (poly ADP ribose polymerase) inhibitors for the treatment of patients with BRCA1/ BRCA2-deficient tumors have been effective.[475] PARP1 is required for base excision repair and repair of DNA single-strand breaks. When PARP1 inhibition occurs in BRCA1/BRCA2 tumors, both double- and single-strand DNA breaks are not repaired, and cell death occurs. However, similar to other targeted therapy, resistant clones develop, and the effectiveness of therapy diminishes.

Variants in *BRCA1* and *BRCA2* are heterogeneous and are located throughout each gene with founder mutations identified in many different ethnic populations.[476] Most variants represent loss-of-function alleles, with more than 75% of the reported variants as deletions, nonsense mutations, or insertions, with deletions representing the majority. Many pathogenic missense mutations occur at critical BRCA1- or BRCA2-protein binding sites that are required for normal function. Although the majority of mutations detected are obviously pathogenic (eg, nonsense or frameshift mutations causing premature truncation of the protein), some detected gene variations may be family specific and may not result in an obvious biologic functional change to the protein based on in silico analysis.[477] These variants of unknown significance (VUS) are most often missense, splice site, or small in-frame insertions or deletions. These may be as common as 7% to 15% of the reported BRCA variants in individuals of European ancestry and can be even higher for patients of other ethnicities for which common variants have not been well characterized.[478] Segregation studies confirming linkage of VUS with disease can be performed if archived tissue or DNA from multiple family members is available.[477] Unfortunately, in many situations, DNA from an adequate number of family members to perform these studies is not possible. The clinical significance for some VUS may become clear by sharing data between laboratories; however, at some testing sites, these databases remain proprietary.[479]

In families in which the clinical history is suggestive of hereditary breast and ovarian cancer, DNA tests for *BRCA1* and *BRCA2* may be requested. Testing should be considered for a woman if (1) there are at least three cases in her family of breast, ovarian, or pancreatic cancer at an early age or aggressive prostate cancer; (2) she has breast or epithelial ovarian cancer and is younger than 45 years; (3) both breast and ovarian cancers are present in her or in a family member; (4) she has breast or ovarian cancer and is of Ashkenazi Jewish descent; (5) she has a male relative with breast cancer; (6) she has a family member with a known *BRCA1* or *BRCA2* mutation; or (7) she has breast cancer and a family history of *BRCA1*- or *BRCA2*-related tumors.[480] In addition, *BRCA1* and *BRCA2* testing should be performed on any man with breast cancer. Nevertheless, *BRCA1* and *BRCA2* mutations can be found in families that do not meet these criteria, and mutations may not be found even when expected.

Before testing, a genetic professional should discuss the likelihood of a *BRCA1* or *BRCA2* mutation, the types of variants that may be identified, and that offspring and other family members may also be at risk for having a mutation.[481] Mathematical models are available to determine the likelihood of *BRCA1* or *BRCA2* mutations, and these models can be used to assist in counseling.[482,483] After a *BRCA* mutation has been detected, targeted mutation analysis can be performed in presymptomatic family members at risk for inheriting the mutation. Counseling for presymptomatic patients may include discussing psychological issues involving the fear of cancer or medical procedures and cancer surveillance and potential risk-reducing surgery options if a mutation is identified. Management of presymptomatic *BRCA1* or *BRCA2* mutation–positive women is complex. The National Comprehensive Cancer Network Clinical Practice Guidelines has established surgical and surveillance guidelines to decrease the risk of disease in these women.[480] The patient may wish to have prophylactic mastectomy, salpingo-oophorectomy, or both. In the United States, it is estimated that these elective procedures are performed at a frequency of 20% to 49% and 37% to 60%, respectively.[478] Alternatively, she may choose to use increased surveillance and prevention strategies for the early detection of breast and ovarian cancer. Surveillance for breast cancer in *BRCA1* or *BRCA2* mutation–positive individuals should include annual mammography and breast magnetic resonance imaging beginning at 25 years of age. Surveillance for ovarian cancer should include transvaginal ultrasonography and CA-125 measurement every 6 months beginning at age 35 or 10 years earlier than the earliest age of onset in the family. Clinical management for patients with a *BRCA1* or *BRCA2* VUS is challenging, and the use of risk-reducing surgeries is appropriately lower.[478] In these cases, because the VUS may ultimately be reclassified as benign, counseling is recommended depending on personal and family medical history rather than solely on *BRCA* mutations.

For presymptomatic patients with a family history but no prior DNA testing for breast or ovarian cancer, the genetic professional should explain before testing the possibility of a VUS result and continued anxiety and uncertainty. In addition, the patient should understand the possibility of false-negative results because (1) not all possible *BRCA1* or *BRCA2* gene mutations are tested or (2) there may be possible mutations in another breast cancer susceptibility gene.

Because only 25% of familial breast and/or ovarian cancer is due to *BRCA* mutations, MPS analysis for a panel of breast- and/or ovarian-related cancer predisposing genes may be more appropriate and more cost effective (Table 6.6).[484-489] This strategy reduces the possibility of a false-negative result by increasing the mutation detection rate by a small percentage. Appropriate patient surveillance and management depend on the mutant gene identified.[490] Because founder gene mutations are observed for some susceptibility genes, a specific gene test or cancer panel may be requested based on the ethnicity of the patient and the presence of that gene in a panel. If multiple cancers are reported in the family in addition to breast or ovarian cancer, a more comprehensive cancer susceptibility panel may be advised as a first-tier testing strategy.[488] The genetic professional must be ever mindful of the insurance coverage for the patient and the financial resources available to the patient for testing to be sure to maximize efficacy of the testing yet minimize unnecessary costs to the patient and carefully determine the appropriate first-tier test to order, whether it is a single gene test, small targeted cancer-specific panel, or a larger comprehensive cancer panel. Regardless of the MPS cancer panel chosen, pathogenic mutations are detected in only about 30% of cases suggestive of familial breast cancer.[486]

One limitation of this personalized medicine approach and increased clinical utilization of MPS panels is the increase in the number of rare missense variants detected, many of which are VUS. These are detected at a rate of about 0.008 variants per 1000 bases of exonic DNA that is sequenced.[491] Thus, while providing more comprehensive testing to patients, the lack of certainty regarding clinical relevance of VUS will remain challenging for health care providers in the care and management of patients and will likely be disconcerting to patients. In addition to the identification of the problematic VUS, expanded genetic testing also reveals variants in cancer susceptibility genes that are less well characterized than *BRCA1* and *BRCA2*. Many of these genes demonstrate reduced penetrance, making it difficult to translate the identification of a variant to a calculable cancer risk, and for many, clinical recommendations and guidelines have not yet been established.[492]

Inherited Colon Cancer

Colorectal cancer (CRC) is the third most common cancer in the United States with close to 150,000 new cases each year and about 50,000 deaths annually. About 3% to 5% of CRC cases are associated with inherited mutations linked to highly penetrant colon cancer syndromes.[493] In as many as one third of cases, familial clustering is observed, thereby suggesting the involvement of less penetrant susceptibility genes and environmental factors. Environmental risk factors include obesity; lack of exercise; moderate to heavy alcohol use; smoking; increased consumption of red or processed meats; and reduced whole-grain fiber, fruits, and vegetables.[453] About 85% of carcinomas arise from transformation of the normal mucosa to adenomas and then to carcinomas, and carcinoma arises through the serrated polyp pathway in 15%.[494] The molecular basis of CRC is a complex multistep process that involves genetic and epigenetic alterations. The various molecular pathways of disease convey different clinical features, prognosis, treatment plans, and pathological findings.

The microsatellite instability (MSI) pathway represents about 15% to 20% of CRC cases and is characterized by inherited, inactivating mutations in genes involved in DNA mismatch repair (MMR) or by acquired epigenetic silencing of these genes. MMR is a ubiquitous DNA repair process that occurs in all dividing cells. Because of a dysfunctional MMR system, DNA replication errors within microsatellite repeats or short tandem repeat sequences of 1 to 6 base pairs in length remain uncorrected and accumulate throughout the genome. Expansion or contraction of the microsatellite-repeat number in noncoding areas of the genome is of little significance; however, the predisposition to CRC results when changes occur within coding microsatellites of the genome and specifically within targeted genes whose protein products are involved in cell growth (*TGFβR2*), apoptosis (*BAX*), or DNA repair (*MSH6*).[495,496]

The chromosomal instability (CIN) pathway is observed in 75% to 80% of CRC cases and was first proposed more than 25 years ago.[497] The CIN pathway is characterized by inherited or somatic mutations causing inactivation of the tumor suppressor gene *APC*, resulting in activation of the Wnt signaling pathway coupled with the acquired loss of chromosomal material as tumorigenesis progresses. These karyotypic changes most frequently involve the chromosomal regions of 5q, 18q and 17p encompassing tumor suppressor genes *APC*, *DCC*, *SMAD2*, SMAD4, *TP53*, and adjacent DNA sequences. Activating mutations in codons 12, 13, and 61 of the *KRAS* proto-oncogene on chromosome 12p12.1 is also common.[494] Collectively, this molecular profile results in unrestrained cell growth, proliferation, and loss of apoptosis. The tumors derived from the CIN pathway are microsatellite stable because CIN and MSI pathways are considered mutually exclusive.

The CpG island methylator phenotype (CIMP) is considered a subset of the MSI pathway by which CRC may arise.[498] As a subset of MSI tumors, these tumors are expectedly CIN negative, and CpG island hypermethylation occurs in a characteristic pattern of specific genes. These epigenetic changes cause silencing of the respective genes with no transcription; therefore, ultimately no translation of the associated gene product.[499] CIMP tumors can be further classified based on the degree of methylation and the number of genes in which hypermethylation is observed.[500] CIMP tumors typically display MSI because the MMR gene, *MLH1*, is hypermethylated.

Lynch Syndrome

The most common CRC susceptibility syndrome is Lynch syndrome (LS; OMIM #120435), also referred to as hereditary nonpolyposis colorectal cancer (HNPCC), which represents about 3% to 5% of all CRC cases. This disorder has an incidence of about 1 in 500 and is named after Dr. Henry Lynch's observation of an autosomal dominant predisposition to early-onset CRC with stomach and endometrial tumors in two large Midwestern kindreds.[501] LS is a heterogeneous disorder caused by one of multiple genes, manifests a variable age of onset, demonstrates reduced penetrance with lifetime risks of CRC of 50% to 70%, and confers increased risks for other associated tumors.[502] CRC in LS patients is distinguished by a few polyps that possess an accelerated transformation potential to carcinoma in as little as

TABLE 6.6 Common Inherited Breast, Gynecologic, and Gastrointestinal Cancer Susceptibility Genes

Gene	Location	Name	Function	Associated Cancer	Disease
APC	5q22.2	Adenomatous polyposis coli	Control of cell proliferation	Colon, small bowel, thyroid, liver, pancreas	Familial adenomatous polyposis
ATM	11q22.3	Ataxia-telangiectasia mutated	Cell cycle control	Breast, ovarian, gastric, hematologic	Ataxia-telangiectasia
AXIN2	17q24.1	Axin 2	Assumed WNT signaling pathway regulator	Colon	Oligodontia-colorectal cancer syndrome
BARD1	2q35	BRCA1-associated ring domain 1	DNA repair, apoptosis, cell cycle arrest	Breast, ovarian, brain	Familial breast cancer
BMPR1A	10q23.2	Bone morphogenetic protein receptor, type IA	Cell signaling, proliferation, and differentiation	Colon, stomach, pancreas	Familial juvenile polyposis
BRCA1	17q21.31	Breast cancer 1, early onset	DNA repair	Breast, ovarian, prostate, pancreas	Hereditary breast and ovarian cancer
BRCA2	13q13.1	Breast cancer 2, early onset	DNA repair	Breast, ovarian, prostate, pancreas, brain, kidney, gastric	Hereditary breast and ovarian cancer
BRIP1	17q23.2	BRCA1-interacting protein C-terminal helicase 1	DNA helicase, DNA repair	Breast, ovarian, hematologic	Fanconi anemia type J, familial breast cancer
CDH1	16q22.1	E-cadherin	Cell signaling, adhesion, and proliferation	Gastric, breast, ovarian, endometrium, prostate	Hereditary diffuse gastric cancer
CDKN2A	9p21.3	Cyclin-dependent kinase inhibitor 2A	Cell cycle control	Pancreas, skin	Pancreatic cancer/melanoma syndrome
CHEK2	22q12.1	Checkpoint kinase 2	Cell cycle control	Breast, prostate, colon, bone	Li-Fraumeni syndrome
EPCAM	2p21	Epithelial cell adhesion molecule	Cell adhesion, signaling, proliferation, differentiation, and migration	Colon, endometrium, ovary, stomach, small bowel, hepatobiliary tract, urinary tract, brain, pancreas, sebaceous	Lynch syndrome
GREM1	15q13.3	Gremlin 1	Control of cell proliferation	Colon	Hereditary mixed polyposis syndrome
MLH1	3p22.3	MutL homolog 1	DNA mismatch repair	Colon, endometrium, ovary, stomach, small bowel, hepatobiliary tract, urinary tract, brain, pancreas, sebaceous	Lynch syndrome
MSH2	2p21	MutS homolog 2	DNA mismatch repair	Colon, endometrium, ovary, stomach, small bowel, hepatobiliary tract, urinary tract, brain, pancreas, sebaceous	Lynch syndrome
MSH6	2p16.3	MutS homolog 6	DNA mismatch repair	Colon, endometrium, ovary, stomach, small bowel, hepatobiliary tract, urinary tract, brain, pancreas, sebaceous	Lynch syndrome
MUTYH	1p34.1	Mut Y homolog	DNA repair	Colon	MUTYH-associated polyposis
POLD1	19q13.33	Polymerase (DNA-directed), delta 1, catalytic subunit	DNA replication and repair	Colon, endometrium	CRC-polymerase proofreading-associated polyposis syndrome
POLE	12q24.33	Polymerase (DNA-directed), epsilon, catalytic subunit	DNA replication and repair	Colon	CRC-polymerase proofreading-associated polyposis syndrome

Continued

TABLE 6.6 Common Inherited Breast, Gynecologic, and Gastrointestinal Cancer Susceptibility Genes—cont'd

Gene	Location	Name	Function	Associated Cancer	Disease
PALB2	16p12.2	Partner and localizer of BRCA2	DNA repair	Breast, pancreas	Familial breast cancer; Fanconi anemia type N
PMS2	7p22.1	PMS2 postmeiotic segregation increased 2	DNA mismatch repair	Colon, endometrium, ovary, stomach, small bowel, hepatobiliary tract, urinary tract, brain, pancreas, sebaceous	Lynch syndrome
PTEN	10q23.31	Phosphatase and tension homolog	Cell cycle control	Breast, thyroid, renal, endometrium, colon, skin, CNS	PTEN hamartoma tumor syndrome
RAD51C	17q22	RAD51 paralog C	DNA repair	Breast, ovarian	Familial breast and ovarian cancer
RAD51D	17q12	RAD51 paralog D	DNA repair	Breast, ovarian	Familial breast and ovarian cancer
SMAD4	18q21.2	SMAD family member 4	Cell signaling and control of proliferation	Colon, stomach, pancreas	Familial juvenile polyposis
STK11	19p13.3	Serine/threonine kinase 11	Cell signaling and control of proliferation	Breast, colon, ovary, stomach, lung, pancreas	Peutz-Jeghers syndrome
TP53	17p13	Tumor protein p53	DNA repair; cell cycle control	Breast, brain, renal, adrenal, hematologic	Li-Fraumeni syndrome

CNS, Central nervous system; *CRC,* colorectal cancer; *PTEN,* phosphatase and tensin homolog.

2 to 3 years. This predisposing event is linked to germline mutations in MMR genes that are associated with the MSI pathway. Both CRC tumors in LS families and sporadic tumors with MSI, more commonly occur in the proximal part of the colon (right sided). These tumors are typically associated with a better prognosis but, in the absence of MMR proteins, have a poor response to adjuvant 5-FU based chemotherapy.[503] Histologically, these tumors display infiltrating lymphocytes, are mucinous with signet ring cells, and are poorly differentiated.

The first MMR gene was mapped to chromosome 2p15-16 using large LS kindreds.[504] Simultaneously, MSI was noted in a subset of sporadic CRC.[505,506] The MMR genes associated with LS include *MSH2* (2p15-16), *MSH6* (2p15-16), *MLH1* (3p21), and *PMS2* (7p22).[507] The majority of LS mutations are observed in *MSH2* (40%) and *MLH1* (50%).[496] A small percentage of patients with LS (1%–3%) have germline deletions affecting the epithelial cell adhesion molecule *(EPCAM)* gene (2p21), resulting in silencing of the downstream *MSH2* gene and no translation of the associated MSH2 protein.

Lynch syndrome–related *MMR* gene mutations are diverse and are located throughout these genes. Almost all errors made during DNA replication are repaired through the proofreading 3′-to-5′ exonuclease activity of DNA polymerase. Uncorrected errors of mismatched bases between the two strands are repaired before cell division by the MMR proteins, a process that is critical for maintaining genomic stability.[503] In addition to providing the repair of a mismatched base–base pair, the MMR system repairs "loop outs" from small insertions or deletions, unmatched bases that can occur during replication of a microsatellite or small repetitive sequences. The repair process includes three steps: (1) recognition of the mismatch, (2) excision, and (3) resynthesis to restore the correct sequence. In the MMR process, the MLH1 and PMS2 proteins dimerize facilitating the binding of other proteins involved in MMR, and the MSH2 protein forms a heterodimer with the *MSH6* gene product to identify mismatches.[507] When the normal function of MMR proteins is altered through DNA mutations, mismatched bases generated through DNA replication are not fixed, leading to strands of DNA with repeats of different lengths. In LS, a germline mutation in an MMR gene is inherited, causing one allele to be nonfunctional. In the tumor tissue of these patients, the second allele is inactivated through a somatic mutation or is deleted, a phenomena referred to as LOH (Fig. 6.11). Uncorrected somatic replication errors thus accumulate in noncoding and insignificant locations throughout the genome but also in the coding regions of genes involved in cell growth, signaling, and DNA repair. Approximately 15% to 20% of all CRC display MSI. In non-LS cases, MSI is attributed to epigenetic silencing of *MLH1* expression through biallelic methylation.[508]

To assist clinicians in identifying patients in LS kindreds, several criteria have been adopted, including the Amsterdam criteria first developed in 1990 and the more inclusive Bethesda criteria developed in 1998.[509,510] These recommendations have further evolved to maximize the identification of index cases in these families.[511,512] MSI testing should be performed on CRC tumor tissue if one of the following criteria is met: (1) age younger than 50 years; (2) regardless of age, synchronous or metachronous CRC or the presence of other LS-associated tumors, including the endometrium, ovary, stomach, hepatobiliary system, small bowel, ureter, renal, pelvis, and brain; (3) age younger than 60 with histology demonstrating a typical MSI pattern of tumor-infiltrating lymphocytes, Crohn-like lymphocytic reaction, mucinous–signet ring differentiation, or a medullary growth pattern; (4) the patient has one or more first-degree relatives with a

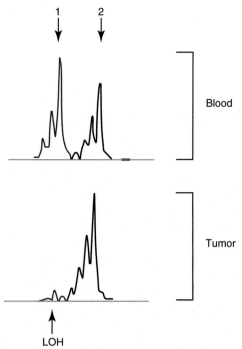

FIGURE 6.11 Electropherograms illustrating loss of heterozygosity (LOH) in tumor DNA. Patient DNA is extracted from peripheral blood and tumor tissue. Polymerase chain reaction of a microsatellite repeat locus contained within the chromosomal region thought to be deleted during tumorigenesis is performed. One of the primers is labeled with a fluorescent dye, and amplicons are subjected to capillary electrophoresis. Relative amplicon size is on the x-axis, and relative fluorescence is on the y-axis. Constitutive DNA from the patient's blood illustrates heterozygosity for this marker as alleles 1 and 2. In DNA from the tumor, a single peak suggesting a homozygous pattern is observed. Thus, LOH is present in the tumor DNA. The absence of one of the amplicons signifies the loss of the second allele and indicates loss of chromosomal material at this particular locus.

LS-related tumor that was diagnosed before age 50 years; or (5) there are two or more first- or second-degree relatives with CRC or LS-related tumors, regardless of their age.[512] Despite these revised guidelines, some index cases are still not identified, so universal screening for all CRC in patients younger than 70 years of age has been recommended.[513] Equally important is screening for LS in patients with endometrial cancer (EC) because approximately 3% of patients with EC have an MMR gene mutation. The lifetime risk of EC in LS women is 25% to 60% with an average age at diagnosis between 48 and 62 years.[514,515] The lifetime risk for developing ovarian cancer is 1% to 3% with 10% to 15% of hereditary ovarian cancer caused by LS.[516] Synchronous tumors of the endometrium and ovary may be observed in as many as 20% of women with LS.[517] In many women with LS, EC often precedes CRC as their initial malignancy. Identification of their increased risks of LS-related tumors can be followed by increased surveillance, counseling, and testing for at-risk family members. For these reasons, universal screening for all newly diagnosed EC has been proposed because a significant percentage of LS patients are missed when relying on MMR-associated tumor morphology or patient indices such as age (<50 years) and history.[518-520]

Molecular testing for MSI can identify defects in MMR genes through germline or somatic changes, and IHC testing is performed to detect expression of MMR proteins.[521] MSI testing by PCR may be the preferred method in some facilities. Alternatively, IHC testing may be performed if MSI testing is not readily available or if there is a limited amount of tumor tissue. In some settings, however, both molecular MSI testing and IHC testing are performed concurrently to account for possible false-negative results that may be obtained from either methodology. At many institutions, a multidisciplinary team develops and agrees on a standard protocol for universal screening of suspected LS patients. This practice, however, is not well adopted by many community hospitals.[522] Because mononucleotide microsatellite repeats are more susceptible to MMR errors, MSI testing is most often performed using a multiplex PCR for several mononucleotide loci and requires normal adjacent tissue for comparison (Fig. 6.12). More recently, MSI has been detected with MPS.[523] MSI is characterized by the expansion or contraction of DNA sequences through the insertion or deletion of repeated sequences. If MSI is detected at two or more of five loci, or more than 30% of the loci analyzed, the tumor has a "high" frequency of MSI (MSI-H). If MSI is detected at one locus, or less than 30%, the tumor has a "low" frequency of MSI (MSI-L). If MSI is not detected at any locus, the tumor is considered to be microsatellite stable (MSS). MSI-L or MSS results greatly reduce the likelihood of LS.

If MSI-H is detected, it is necessary to determine whether the MSI results from an inherited inactivating germline MMR gene mutation or from the more frequent somatic CpG methylation of *MLH1* seen commonly in sporadic CRC. Because somatic CpG methylation of *MLH1* is frequently associated with a somatic *BRAF* proto-oncogene p.Val600Glu mutation, this test is often performed as a reflex test on tumor tissue DNA after MSI-H results are obtained (Fig. 6.13).[524] If a somatic *BRAF* p.Val600Glu mutation is detected, LS is unlikely. If, however, no *BRAF* gene mutation is detected, a germline MMR gene mutation is more likely, and the testing algorithm continues with *MLH1* promoter hypermethylation studies of tumor tissue DNA. If epigenetic biallelic *MLH1* gene promoter hypermethylation is detected, a sporadic CRC tumor is the diagnosis. Conversely, the lack of *MLH1* promoter hypermethylation indicates the possibility of LS, and DNA sequence analysis of MMR genes from the patient's peripheral blood lymphocytes to identify a germline MMR gene mutation is performed.[507]

The presence of a germline MMR mutation confirms the diagnosis of LS. If IHC testing has been performed, the IHC pattern obtained may direct which MMR gene to sequence. For example, whereas the loss of MSH2 expression would suggest an inactivating mutation in the *MSH2* gene, the loss of expression of MSH6 would indicate a mutation in the *MSH6* gene. Thus gene-specific, *MSH2* or *MSH6*, Sanger DNA sequencing can be requested to search for an unknown mutation. However, IHC-directed gene testing is not always concordant, and IHC is not always performed. Therefore, germline testing on DNA extracted from a peripheral

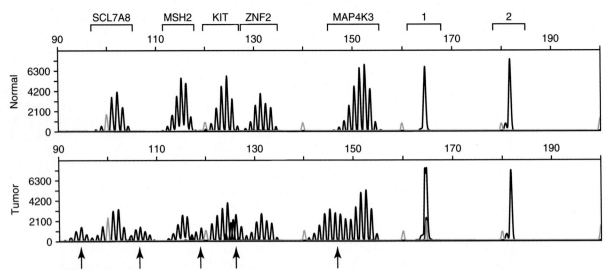

FIGURE 6.12 Electropherograms illustrating microsatellite instability (MSI) in tumor DNA. Patient DNA is extracted from peripheral blood or normal tissue and is compared with DNA extracted from tumor tissue. DNA is amplified in a multiplex polymerase chain reaction (PCR) assay using fluorescent labeled primers corresponding to five mononucleotide loci (*SCL7A8, MSH2, KIT, ZNF2,* and *MAP4K3*) and two pentanucleotide loci (1 and 2) (Promega Corporation, Madison, WI). Amplicon sizes in bases are noted at the top of the figure, and the relative amount of fluorescence detected for each amplicon is measured by the peak height (y-axis). Multiple peaks seen at each locus represent "stutter" peaks and result from strand slippage during PCR of repetitive sequences. *Arrows* denote a shift in product size at five of five mononucleotide loci, indicating microsatellite instability. Identical patterns between normal and tumor DNA at the polymorphic pentanucleotide markers suggest that normal and tumor DNA are derived from the same individual.

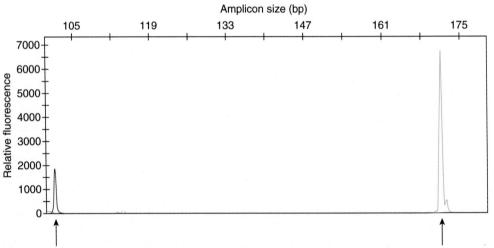

FIGURE 6.13 Detection of *BRAF* mutation c.1799T>A (p.Val600Glu) by allele-specific polymerase chain reaction (PCR). Amplicon sizes in bases are noted at the top of the figure, and the relative amount of fluorescence detected for each amplicon is measured by the peak height (y-axis). *BRAF* mutation c.1799T>A can be rapidly detected using allele-specific PCR coupled with capillary electrophoresis. DNA was extracted from a paraffin-embedded MSI-H (microsatellite instability–high) colon tumor. An internal control 171-bp fragment was amplified using a primer pair flanking exon 15 with the 3′ primer fluorescently labeled. In the same reaction, a 101-bp allele specific amplicon was generated with the same 5′ forward primer and a second 3′-fluorescently labeled primer specific for *BRAF* c.1799T>A.[524] The *right arrow* points to the internal control, and the *left arrow* confirms the presence of mutation *BRAF* c.1799T>A (p.Val600Glu) in the tumor, thereby suggesting that the MSI-H phenotype observed in the colon tumor is associated with sporadic colon cancer and not to mutations in an *MMR* gene associated with Lynch syndrome.

blood specimen for a MPS LS specific panel with all four MMR genes and *EPCAM* may be considered. After a specific mutation has been identified in a patient with LS, directed Sanger DNA sequencing analysis may be used to screen at-risk family members for the specific mutation that is segregating in the family.

Both pre- and posttest genetic counseling are indicated. Before testing, the genetics professional should review the family history, assess the psychosocial needs of the patient, review risk counseling, and discuss genetic testing and possible outcomes. Prediction models can be used to determine the probability of identifying an MMR gene mutation.[525] Posttest counseling includes disclosing the results and implications to the patient and his or her family members. The genetics professional should provide a management and surveillance plan for the patient based on the MMR gene involved and the identified mutation.[526]

Lynch syndrome—associated gene mutations are heterogeneous; most are nonsense or frameshift mutations expected to result in premature truncation of the protein and loss of function.[527] The majority of germline mutations occur in *MLH1* and *MSH2* and to a lesser degree in the *MSH6* gene. Mutations in the *PMS2* gene are the least common. Whether Sanger- or MPS-based DNA sequencing analysis is used, an inherent problem is the identification of sequence VUS, which are most often missense or possible splice variants and for which no significant biologic function can be determined. Thus, it is difficult to assign a pathologic role to these variants without functional assays to confirm their relevance in the disease process. Because of this, the International Society for Gastrointestinal Hereditary Tumours (InSiGHT) developed a database with close to 2500 reported constitutional variants identified in LS patients.[527] With a multidisciplinary approach, this group systematically applied a standard classification scheme for the database variants and determined the clinical recommendations for each.[528] This comprehensive approach assists in risk assessments for specific gene mutations and facilitates appropriate care and management for families. Challenging however, are "private" variants segregating within a single family and the associated uncertainty of characterizing those genotype—phenotype correlations and predictive risk assessments for specific mutations based on the clinical findings in just one family.

Overall genotype—phenotype correlations for MMR genes indicate that individuals with *MLH1/MSH2* gene mutations have the earliest age of onset and highest risk for developing CRC by age 70 years.[502,529] *MSH6* gene mutations convey the lowest lifetime risk of CRC to female carriers (10%), yet *MSH6* gene mutation—positive carriers have the highest risk of extracolorectal cancer by age 70 years. *PMS2* gene mutation carriers have a lower penetrance of disease and a lower risk of extracolorectal cancer.

If an MMR gene mutation in a patient is identified, at-risk adult family members can pursue presymptomatic testing with appropriate genetic counseling. Similar to presymptomatic testing for other adult-onset disorders, the counseling session should include verification of the family history and discussion of the clinical course of the disease, including risk of developing the disease and issues associated with disease management. Discussions should be incorporated into the session, including how the patient will act on both positive and negative results, feelings of survivor guilt or stigmatization, and the possibility of discrimination for insurance and employment. If a germline mutation is detected in the presymptomatic patient, a colonoscopy should be performed every 1 or 2 years beginning at age 20 to 25 years or 10 years younger than the youngest age of diagnosis in the family.[530] Surveillance for EC and ovarian cancer for at-risk women with LS should include transvaginal ultrasonography and endometrial biopsy every 1 to 2 years beginning at 30 to 35 years of age.[530] Furthermore, prophylactic hysterectomy and bilateral salpingo-oophorectomy can be considered for women who have completed childbearing. Alternatively, if genetic testing is not pursued, relatives should begin an intensive screening program with colonoscopy every 1 to 2 years, starting between 20 and 30 years of age, and then annually after age 40 years. If no pathogenic MMR germline mutation is detected in the proband despite MSI and IHC testing suggestive of a MMR germline mutation and LS, presymptomatic DNA testing for family members is not possible. These mutation-negative families or Lynch-like families may have (1) a MMR VUS that may eventually be reclassified, (2) a MMR gene mutation in a region of the genome that is not screened, (3) a structural variant not detected by current methodologies, or (4) a germline mutation in other yet to be determined MMR genes or genes that regulate MMR genes.[519] Although detection of a mutation in a family that meets LS criteria is not always possible, heightened surveillance of the proband and at-risk family members should be implemented.

In addition, *MLH1* and *MSH2* mutations have been identified in about 70% of patients with Muir-Torre syndrome (MTS) at frequencies of 10% and 90%, respectively. MTS is characterized by at least one sebaceous gland neoplasia, adenoma or carcinoma, and one or more visceral tumors, which most frequently are colorectal or endometrial, although other LS-associated tumors may be seen. MTS is considered a subtype of LS with a mean age at diagnosis around 50 years.[531] Because sebaceous neoplasms may be sporadic, LS confirmation is recommended. In rare cases, patients with biallelic germline mutations in an MMR gene result in constitutional mismatch repair deficiency syndrome (OMIM #276300) with early-onset CRC and childhood hematologic or brain tumors.

Familial Adenomatous Polyposis

Familial adenomatous polyposis (FAP; OMIM #175100) is an autosomal dominant disorder with an incidence of about 1 in 8000 to 15,000 characterized by hundreds to thousands of adenomatous polyps both in the colon and rectum. This disorder has a high penetrance, conveying a lifetime risk of CRC in untreated mutation carriers of close to 100%. FAP accounts for 1% of CRC cases observed in the United States. Interestingly, in about 20% of cases, no family history exists, and the disease is the result of a new mutation. Polyps appear during the second decade of life in about half of the patients.[532] CRC ultimately develops approximately 10 to 15 years after the onset of polyposis, with the median age of CRC about 40 years of age without surgical intervention.[532,533] It is the sheer number of polyps in these patients that increases the likelihood that one will progress to cancer. For this reason, close surveillance of these patients is indicated, and colectomy is advised when multiple adenomas are observed or if high-grade histologic findings are reported. Furthermore, patients with FAP have an increased risk of developing other malignancies, including carcinoma of the small bowel, most often in the duodenum or periampulla; papillary thyroid carcinoma; hepatoblastoma; pancreatic cancer; and brain tumors.[534] In addition, these patients have a 7% to 20% lifetime risk of developing benign extraintestinal manifestations of the disease, including adrenal or desmoid tumors, osteomas, congenital hypertrophy of the retinal epithelium, and dental abnormalities.[534] Desmoid tumors in vital areas have been reported as a cause of death in as many as 21% of patients with FAP.[535] Thus, close clinical surveillance of FAP patients and at-risk family members is critical to reducing CRC and CRC-associated mortality and APC-associated complications.

Familial adenomatous polyposis is caused by germline mutations in the adenomatous polyposis coli (APC) gene, which was cloned in 1991.[536-538] The APC gene encodes 8535 base pairs contained within 15 exons and produces a protein of 2843 amino acids and a molecular weight of about 310 kDa. The APC protein is a multidomain, multifunctional protein with multiple binding partners that play key roles in regulating the β-catenin level in the Wnt signaling pathway.[539,540] This protein is also involved in other cellular processes, including cell adhesion and migration, DNA repair, apoptosis, FAK/Src signaling, microtubule assembly, and chromosome segregation.[540,541] However, the main tumor suppressor function of APC appears to be its regulation of the proto-oncoprotein β-catenin and canonical Wnt signaling. In this process, APC together with glycogen synthase kinase-3β and axin regulate the amount of β-catenin through the phosphorylation of cytoplasmic β-catenin for ubiquitin-dependent degradation. In the absence of functional APC, β-catenin is unregulated and accumulates in the nucleus, leading to ligand independent constitutive Wnt signaling and transcription mediated by lymphoid enhancer-binding factor/T-cell factor (LEF/TCF) transcription factors, resulting in the upregulation of genes involved in proliferation.[542] For this reason, APC is commonly referred to as a "gatekeeper" of tumor progression.[540,541,543] Similar to autosomal dominant inherited mutations in other tumor suppressor genes, the normal allele must also be inactivated to manifest the disease phenotype. Therapeutic approaches targeting the multifaceted functions of APC have been reported.[540]

Sanger and MPS DNA sequencing studies have identified well over 1500 germline mutations in APC, most of which result in truncated proteins because of small frameshift or nonsense mutations.[528] Gross rearrangements have been identified using MLPA.[544,545] Despite the heterogeneous nature of these mutations, two hot spots occur with 10% and 15% of all mutations found specifically at codons 1061 and 1309, respectively. Additionally, about 33% of mutations occur between these two sites. Genotype–phenotype correlations exist for some APC mutations. The AAAAG deletion at codon 1309 is associated with a younger age of onset.[546] Classic FAP is primarily observed from mutations occurring between exon 5 and the 5′ portion of exon 15, the largest of all exons spanning 6.5 kb. Patients with FAP who have mutations between codons 1250 and 1464 or truncating mutations between codons 1403 and 1578 have a severe phenotype and are at increased risk for extracolonic disease.[547,548] Attenuated FAP (AFAP) is seen in about 8% to 10% of patients with FAP and is associated with fewer colonic polyps (10–100) and an older age of onset for both the development of polyps and cancer.[549-552] These patients typically have truncating mutations at the extreme 5′ end of the gene (codons 1–163) or mutations at the carboxyl-terminal end of the gene (codons 1860–1987). Intra- and interfamilial phenotypic variability exists even in the presence of identical mutations and may be explained by a modifier gene or genes.[553,554]

If APC is suspected, full gene sequence analysis is recommended coupled with deletion and duplication analysis.[507,544] If the mutation within the family can be identified, at-risk family members as young as 10 to 12 years of age may be referred for genetic counseling and presymptomatic DNA testing.[555] Although DNA testing on an asymptomatic minor is usually not endorsed by the genetic community, in this scenario, early identification of the mutation in these patients will clearly affect their clinical management because intense screening programs and possible prophylactic colectomy may be initiated as early as the second decade of life. If a mutation is detected in the family, endoscopy should be performed every 2 years and if adenomas are detected, colonoscopy should be performed every year until the colectomy is performed.[556] Annual thyroid and abdominal ultrasonography for detection of thyroid malignancy and desmoid tumors may also be included in the surveillance screening. Family members who test negative for the mutation do not have increased risk of CRC, can avoid these intensive screening programs, and should follow the screening programs for the general population. If the mutation in the family cannot be identified, screening with sigmoidoscopy is recommended every 2 years, starting as early as age 10 to 12 years. Furthermore, when repeatedly negative sigmoidoscopy results are obtained, the frequency of such examinations can be reduced in each subsequent decade of life, and frequent surveillance may be discontinued at age 50 years.[556] Because AFAP is associated with a later age of onset and most adenomas are in the right part of the colon, colonoscopy is the recommended screening method beginning at 18 to 20 years of age.

In 2002, APC mutation-negative adenomatous polyposis patients were found to have biallelic mutations in the base excision repair gene MUTYH, a disorder known as MYH-associated polyposis (MAP; OMIM #604933).[557,558] This gene contains 16 exons, is located on chromosome 1p34.3-p32.1, and encodes an adenine-specific DNA glycosylase involved in DNA base excision repair (BER). This protein removes adenine when it is inappropriately paired with guanine, cytosine, or oxidatively damaged DNA containing 8-oxo-7,8-dihydroguanine. If not repaired, these mispairings can result in G:C to T:A transversion mutations in the APC gene and sometimes the KRAS gene as well.[557,559] MUTYH gene mutation–positive patients have CRC at a mean age of 51.7 ± 9.5 years.[560] Incomplete penetrance is observed, and the polyp burden is variable with typical patients having between 10 and 100 polyps. Close surveillance of these patients is recommended, similar to that provided for AFAP, with screening beginning as early as 18 to 20 years of age and colonoscopy performed every 2 years. In addition, although this is an autosomal recessive condition, heterozygous carriers with only one MUTYH gene mutation are at increased risk for CRC, thus indicating that close clinical surveillance may be indicated in this population as well.[560,561]

Mutations in the MUTYH gene are heterogeneous and scattered throughout the gene with more than 300 unique variants reported.[528] Most variants cause amino acid substitutions in the gene, and ethnic-specific common mutations are observed.[562] Patients can be homozygous for the same mutation or compound heterozygotes with two different mutations. DNA testing for MUTYH gene mutations can be targeted to common mutations or known mutations segregating in the family. Alternatively, full gene sequencing of the *MUTYH* gene can be requested. If *MUTYH* DNA sequence analysis results are negative, *APC* testing may be offered.[507]

Clinical presentation and personal and family history should be carefully reviewed when determining the most appropriate molecular testing to recommend. In cases of higher disease certainty, gene-specific Sanger DNA sequencing is more cost effective. However, in some families, MPS using a targeted comprehensive panel of susceptibility genes for colon cancer may afford a higher likelihood for identification of the disease gene segregating in the family in a faster and more cost-effective manner (see Table 6.6).[563]

> **POINTS TO REMEMBER**
>
> **Inherited Cancer**
> - Disease can result from the inheritance of a single activating mutation in a proto-oncogene.
> - Disease can result from the inheritance of a single loss-of-function mutation in a tumor suppressor or mismatch repair gene followed by somatic inactivation of the second allele.
> - Age of onset and type of tumor(s) observed within family members are variable.
> - MPS-based disease specific tumor panels are currently most often used to identify the gene mutation segregating in the family.
> - Routine clinical surveillance screening is initiated in mutation-positive asymptomatic family members.
> - Genetic counseling is an important component of patient care and management.

MITOCHONDRIAL DNA DISEASES

Mitochondria are organelles ubiquitous to the cytoplasm of all eukaryotic cells of animals, higher plants, and some microorganisms. Mitochondria generate energy for cellular processes by producing ATP through oxidative phosphorylation (OXPHOS); they are important in maintaining both calcium homeostasis and various intracellular signaling cascades, including apoptosis.[564-566] The matrix of the mitochondrion is surrounded by a cardiolipin-rich inner membrane, and both are enclosed by a second outer membrane. Within the matrix are copies of mitochondrial DNA (mtDNA). Each mitochondria contains between about 2 and 10 copies of mtDNA, so with hundreds of mitochondria per cell, an estimated 10^3 to 10^4 copies of mtDNA exist within each cell, with brain, skeletal, and cardiac muscle having particularly high concentrations. Alterations in mtDNA copy number or mutations in mtDNA are associated with both inherited and acquired diseases.[567-569] The mitochondrial genome is composed of a double-stranded, circular DNA molecule containing 16,569 base pairs that encodes 37 genes, including two ribosomal RNAs (rRNA), 22 transfer RNAs (tRNA), and 13 subunits required for the OXPHOS system, with 7 belonging to complex I, 1 to complex III, 3 to complex IV, and 2 to complex V.[570] Most subunits involved in the OXPHOS system are nuclear encoded, as are several nuclear gene products that regulate mitochondrial gene expression. The mitochondrial genetic code is slightly different from the universal code. For example, in mtDNA, TGA codes for tryptophan rather than a termination codon, and all mitochondrial-translated mRNA contains codons requiring only 22 mitochondrial-encoded tRNA molecules for translation rather than the 31 predicted by Crick's wobble hypothesis.[571,572] The high copy number of mtDNA per cell coupled with a small genome and highly polymorphic sequence variations among individuals makes mtDNA sequence analysis an ideal tool for forensic studies.[573,574]

Mitochondria-related diseases have an incidence of 1 in 1000 to 5000 and can result from mutations in nuclear DNA (85%—90%) or, as first reported in 1988, from mutations in the mitochondrial genome (10%—15%).[575-577] Mutations in mtDNA occur at a higher rate than nuclear DNA, probably because of differences in chromatin structure, lack of DNA repair machinery, and the continual generation of reactive oxygen species. Mitochondrial genetics is different from Mendelian genetics in several aspects. First, all mtDNA is maternally inherited, with mature oocytes having the highest mtDNA copy number per cell at 10^5 and with sperm having the lowest mtDNA copy number per cell at 10^2. After fertilization, sperm mtDNA is selectively degraded so that only maternal mtDNA remains. Thus, if a mother is carrying an mtDNA mutation, it will be transmitted to all of her children, but only her daughters can transmit the disease to their offspring. Although this is considered the rule, paternal mtDNA inheritance has been reported and may result from incomplete degradation of sperm mtDNA in early embryogenesis.[578,579] If a mtDNA mutation arises, it will exist among a population of normal mtDNA. This coexistence of normal and mutant mtDNA copies within the same cell is referred to as *heteroplasmy* and is the second unique feature of mitochondrial genetics. Third, during cell division, the proportions of normal and mutant mtDNA can shift as mitochondria, and their accompanying genomes are partitioned into daughter cells. Thus, in development and differentiation, the proportions of normal and mutant mtDNA can vary among cells and tissues within the body. Last, the percentage of mutant mtDNA required within a cell, tissue, or organ system to produce a deleterious phenotype is referred to as the *threshold effect*. The threshold for disease varies among people, energetic requirements for tissue, and the mtDNA mutation. Genetic counseling for families with mtDNA disorders is complicated by an inability to accurately predict phenotype caused by heteroplasmy and the threshold effect.

Two types of mtDNA mutations exist: those that affect mitochondrial protein synthesis (tRNA and rRNA genes) and those within the protein-encoding genes themselves.[580,581] Traditionally, testing for mtDNA mutations was performed by direct Sanger sequencing or by targeted mutation testing for specific disease-related mutations. Over the past several years, most clinical laboratories have moved to MPS-based testing for the detection of mtDNA alterations. Recently, the Mitochondrial Medicine Society released a consensus statement on the diagnosis of mitochondrial diseases.[582] This statement indicates that MPS-based testing of the entire mitochondrial genome is now the preferred method for the diagnosis of a suspected mitochondrial disorder and should be considered the first-line DNA-based test (as opposed to targeted mutation analysis).[582] MPS-based testing for mtDNA mutations usually involves a preliminary long-range PCR to amplify the entire mitochondrial genome.[583-585] After this step is complete, the amplified mtDNA is processed, sequenced, and then analyzed.[584] mtDNA mutations identified by MPS-based methods are usually confirmed by a secondary method (eg, Sanger sequencing). Using MPS-based testing and subsequent mutational confirmation techniques, investigators have been able to reproducibly detect heteroplasmy levels of approximately 10%.[583,584] Because of the limitations in the detection of heteroplasmy using MPS-based methods, some laboratories are continuing to use PCR-based methods capable of detecting lower levels of heteroplasmy to target specific disease-causing mtDNA mutations. Although mtDNA mutations are now associated with a significant number of inherited diseases, acquired mtDNA deletions are associated with the aging process, and mitochondrial dysfunction is associated with neurodegenerative diseases and cancer.[568] Many somatic mtDNA mutations occur via damage by oxygen free radicals produced as byproducts of aerobic metabolism.[586,587]

Clinical treatment of mitochondrial disorders is largely supportive in nature. Several treatment modalities are currently under investigation (eg, antioxidant therapy, gene therapy, stimulation of mitochondrial biosynthesis, among others).[588] However, these potential treatments remain largely experimental.

Leber Hereditary Optic Neuropathy
Leber hereditary optic neuropathy (LHON; OMIN #53500), the most common mitochondrial disease, was the first linked to maternal inheritance through a mutation in the mtDNA.[575] LHON is characterized by acute or subacute bilateral loss of central vision caused by focal degeneration of the retinal ganglion cell (RGC) layer and, in some individuals, impairment of optic nerve function.[589] The specific nature of the disease in terms of RGC degeneration is unknown, but it could be caused by differences in superoxide regulation.[590] Age of onset is typically in the second to fourth decade of life, and after initial symptoms, both eyes are usually affected within 6 months. Approximately 50% of males and only 10% of females who possess an LHON mtDNA mutation will develop disease.[589] In addition, yet to be defined environmental factors, nuclear-encoded modifier genes that affect mtDNA expression, mtDNA products, or mitochondrial metabolism may modify the phenotypic expression of LHON. The explanation for differences in rates between genders has not been determined but could be related to genes on the X chromosome.[591-593] It has also been suggested that sex hormones may provide a protective effect in females. Experiments using LHON cybrid cell lines indicate that the presence of estrogens results in more efficient oxidative phosphorylation suggesting that hormones explain gender differences in LHON.[594] Genetic counseling in LHON is complicated because the amount of mutant mtDNA transmitted by heteroplasmic females is not predictable. Furthermore, genetic testing does not predict which individuals will develop visual symptoms.[595] LHON can be confused with autosomal dominant optic atrophy (OMIN #165500), which shares a similar ocular phenotype but results from mutations in the gene OPA1 (3q28-29). It is interesting to note that OPA1 is a nuclear-encoded mitochondrial protein required for mitochondrial fusion, maintenance of cristae architecture, and regulation of apoptosis.[566]

Leber hereditary optic neuropathy is a disorder caused by OXPHOS deficiency. Although many mutations have been associated with this disease, mtDNA mutations m.3460G>A, m.11778G>A, and m.14484T>C represent 90% of those identified.[596] Mutation m.11778G>A was the first described, is the most common, and accounts for at least 50% of cases. In most affected individuals, LHON mutations appear to be homoplasmic, with only mutant mtDNA detected, but in 15% of cases, the mutations are heteroplasmic, with a mixture of both normal and mutant mtDNA detected.[597,598] Each of the common mutations affects a subunit of the nicotinamide adenine dinucleotide: ubiquinone oxidoreductase in complex I of the OXPHOS pathway. The mechanism by which these mutations cause the LHON phenotype is not well understood.[599]

Leigh Syndrome
Leigh syndrome (OMIM #256000), or subacute necrotizing encephalopathy, is a progressive neurodegenerative disorder that most often leads to death before the age of 5 years. In contrast to LHON, most patients present within the first year of life with hypotonia, failure to thrive, psychomotor regression, ocular movement abnormalities, ataxia, and brainstem and basal ganglia dysfunction caused by severe dysfunction of mitochondrial energy metabolism. The clinical phenotype for Leigh syndrome is variable in patients with the same pathogenic mtDNA mutation and results from differences in the percentages of mutant mtDNA among organs and tissues within an individual.[600,601] However, it appears that heteroplasmy alone cannot explain the differences in phenotypic presentation because individuals with high levels of the common Leigh syndrome mutation m.8993T>C present with other disease manifestations.[602]

Leigh syndrome exhibits extensive genetic heterogeneity, with disease-causing mutations identified in both nuclear-encoded genes and mtDNA, making both Mendelian and maternal patterns of inheritance possible for this syndrome.[602] Mutations in more than 35 genes have been described to cause Leigh syndrome.[602] The most common mitochondrial-encoded mutation associated with Leigh syndrome is seen in the *MT-ATPase 6* gene (complex V) with a T>G transversion mutation at nucleotide m.8993. The most common nuclear-encoded mutation associated with Leigh syndrome is in the *SURF1* gene (9q34), which encodes a cytochrome oxidase assembly factor. Regardless of which gene is involved, the overall prognosis of these patients is generally poor, and treatment of mitochondrial disease is in its infancy.[588] Because of the lethality of Leigh syndrome, PGD can be considered and has been successfully performed with the implantation of a disease-free embryo.[603]

Mitochondrial Encephalomyopathy, Lactic Acidosis, and Stroke-Like Episodes
Mitochondrial encephalomyopathy, lactic acidosis, and stroke-like episodes (MELAS; OMIM #540000) is a multisystem disorder characterized by generalized tonic-clonic seizures, recurrent headaches and vomiting, hearing loss, exercise intolerance, and proximal limb weakness.[604] Manifestations of MELAS routinely appear in early childhood, and the disease is commonly fatal by young adulthood.[605] As with other mitochondrial disorders, phenotypic presentation in MELAS

is widely variable among individuals. In fact, it is not possible to predict the course of disease even among individuals with the same mutation or in the same family.[567,605,606]

The disorder is primarily caused by mutations in the mitochondrial tRNA encoding gene *MT-TL1*.[607] A single *MT-TL1* A>G transition mutation located at nucleotide m.3243 is responsible for disease in approximately 80% of MELAS patients.[604] An additional two *MT-TL1* point mutations (m.3271T>C and m.3252A>G) are responsible for disease in approximately 12% of patients.[604] Mutations in a second mitochondrial gene (*MT-ND5*) that encodes the NADH-ubiquinone oxidoreductase subunit 5 are also a relatively common cause of MELAS.[608] Although several causative mutations have been described in the *MT-ND5* gene, a single point mutation m.13513G>A is by far the most common.[609] There are also rare cases in which causative mutations have been identified in other mitochondrial (and some nuclear) genes.[604]

Myoclonic Epilepsy Associated With Ragged Red Fibers

Myoclonic epilepsy associated with ragged red fibers (MERRF; OMIM #545000) is characterized by myoclonus, epilepsy, ataxia, dementia, muscle weakness, hearing loss, short stature, and optic atrophy.[610] MERRF patients commonly undergo a period of normal development before showing manifestations of disease in childhood.[610] The "ragged red fibers" denote the hallmark finding of frayed muscle fibers observed in muscle biopsies from these patients. Although the clinical presentation of MERRF can vary widely, there are four key characteristics required for diagnosis: myoclonus, ataxia, generalized epilepsy, and ragged red fibers observed in the muscle biopsy.[610]

The disorder is caused by mutations in mitochondrial tRNA genes that result in altered translational efficiency.[568,611] Mutations in several tRNA genes have been described in this disorder, but the most frequently observed alteration (m.8344A>G) occurs in the *MT-TK* gene that encodes the tRNA^{Lys}.[612] A single point *MT-TK* point mutation (m.8344A>G) is causative in approximately 80% of MERRF patients. Three additional *MT-TK* gene mutations (m.8356T>C, m.8361G>A, and m.8363G>A) are responsible for disease in an additional 10%.[610] Rare mutations in the *MT-TI*, *MT-TP*, *MT-TF*, and *MT-TL1* genes also result in MERRF. There is much heterogeneity in the presentation of MERRF patients. In fact, a large percentage of patients with the common m.8344A>G mutation do not exhibit all four of the hallmark findings of MERRF,[613] suggesting further refinement of diagnostic criteria in the future.

Kearns-Sayre Syndrome

Kearns-Sayre Syndrome (KSS; OMIM #530000) is a progressive multisystem disorder with onset occurring before age 20 years. KSS is defined by the presence of progressive external ophthalmoplegia and pigmentary retinopathy and at least one additional hallmark finding (cardiac conduction block, cerebrospinal fluid protein concentrations of >100 mg/dL, or cerebellar ataxia).[568,569] Additional clinical findings of KSS may include ptosis, hearing loss, short stature, limb weakness, dementia, hypoparathyroidism, and others.[569] The disease often results in early adulthood death. Unlike most other mitochondrial disorders, KSS usually results from de novo alterations that likely occur in the maternal oocyte during germline development or embryogenesis.[569,614]

In the late 1980s, large deletions in mtDNA were identified as the cause of KSS.[576,615] The size of these deletions varies among individuals, but a common deletion approximately 4.9 kb in size is found in approximately 30% of KSS patients.[615] Studies indicate that regardless of the deletion size, the removal of critical tRNAs needed for protein synthesis results in the disease phenotype.[569] Additionally, the variable presence of deleted mtDNA in specific tissues results in the clinical phenotype of patients.[568] Patients with KSS often have partially deleted mtDNAs in all tissues examined, which is likely to explain the multisystem involvement of this disorder.[568]

Clinical DNA testing for KSS involves the use of deletion and duplication analysis performed by any number of testing modalities (eg, CGH microarray, quantitative PCR, MLPA). Deletion and duplication testing detects disease-causing mutations in 90% of patients affected with KSS.[569] Patients with a KSS phenotype that test negative for deletions may pursue MPS-based mitochondrial genome sequencing because rare point mutations can cause a KSS-like phenotype.

IMPRINTING

Imprinting refers to the differential marking or "imprinting" of specific paternally and maternally inherited alleles during gametogenesis, resulting in differential expression of those genes. Such imprints on the DNA during gametogenesis must be maintained through DNA replication in the somatic cells of the offspring, must be reversible from generation to generation, and must influence transcription. DNA methylation is the primary mechanism for genomic imprinting. The number of imprinted genes in the human genome is estimated to be fewer than 200, and most are clustered around imprinting control centers. Alterations in normal imprinting patterns can result in disease.[616]

Prader-Willi and Angelman Syndromes

Prader-Willi syndrome (PWS; OMIM #176270) is a complex multisystem, neurogenetic disorder with an incidence of 1 in 10,000–30,000. PWS along with Angelman syndrome (AS), were the first reported human disorders resulting from imprinting. Prenatally, fetuses with PWS exhibit diminished movement, peculiar fetal position, and often polyhydramnios.[617] At birth, dysmorphic features, small hands and feet, and hypogonadism are observed, and the child has persistent hypotonia that results in poor feeding and failure to thrive.[618] Development is delayed for both motor skills and language, and this delay continues throughout life, with a mean IQ of 60 to 70. Early in childhood, a unique and characteristic insatiable appetite that is hypothalamic in origin presents; obesity ensues with associated complications and is the major cause of morbidity, mortality, and sleep disorders.[619] In addition, patients have short stature and abnormal body composition characteristic of growth hormone deficiency. This aspect of the disorder can be treated with exogenous growth hormone, although many other aspects of this disease are difficult to manage and require a multidisciplinary approach.[620,621] Children with PWS can develop behavioral disorders, and psychiatric disorders can be present in up to 10% of young adults.

Angelman syndrome (OMIM #105830) is a neurogenetic disorder with a similar incidence in the population as PWS. The AS clinical phenotype includes intellectual disability (IQ <40), inappropriate bouts of laughter, absence of speech, gait ataxia, progressive microcephaly, dysmorphic facial features, and epilepsy.[622] Because these patients demonstrate bursts of laughter and smiling, AS is sometimes referred to as the "happy puppet syndrome." Unique electroencephalographic patterns are seen in most individuals with AS younger than 2 years of age and can be helpful in diagnosing the condition. As many as 6% of patients who display both intellectual disability and epilepsy may have AS. Some of the phenotypic features associated with AS can be nonspecific or can occur separately in other syndromes or nonsyndromic conditions; thus, a constellation of findings with associated laughter, unique smiling, and happy demeanor of people with AS helps in the diagnosis.

Apparent from the characteristic physical findings, PWS and AS are clinically distinct syndromes, yet each results from different genetic alterations involving an imprinted segment within 8 million bases on chromosome 15q11.2-q13. The genes at both ends of this region have biparental expression, but they flank genes that demonstrate exclusively paternal or maternal expression. The expression of either paternal or maternal genes is controlled by an imprinting center (IC) with the imprinting "reset" during gametogenesis between parent and offspring of a different gender.[618] If paternally expressed genes in this region are missing, defective, or epigenetically silenced through DNA methylation and only an inactive, nonexpressed maternal allele remains, PWS is observed. Conversely, if maternally expressed genes in this region are not functional and only the inactive paternal allele remains, the clinical phenotype will be AS.

The most 5' gene in the paternally expressed region is *MKRN3* that encodes makorin ring finger protein 3, with several zinc finger motifs and no introns.[623] Adjacent to *MKRN3* are the structurally similar and intronless genes *MAGEL2* (alias *NDNL1*) coding melanoma antigen family L2 and *NDN*, a melanoma antigen family member, which is involved in terminal differentiation of neurons.[624,625] This region also contains the locus *SNRPN-SNURF*, small nuclear ribonucleoprotein polypeptide N *(SNRPN)* that is involved in mRNA processing and splicing and *SNURF* (*SNRPN* upstream reading frame), which is found in the nucleus, contains 71 amino acids, may bind RNA, and has a C-terminal motif similar to ubiquitin. Last are multiple C/D box snoRNA genes, which are noncoding RNA molecules that modify both rRNA and snRNA by the methylation of the ribose 2'-hydroxyl group.[626,627] Sahoo et al were the first to link the loss of paternal snoRNA genes to PWS.[628] The mechanism by which the absence of these genes results in the pathogenesis of PWS is not clearly understood.

Loss of normally expressed paternal genes on chromosome 15q11-q13 resulting in PWS can occur by several mechanisms (Table 6.7).[619] Most commonly, PWS (65%–75%) results from a de novo deletion after unequal homologous recombination involving one of two common centromeric breakpoints (BP1 or BP2) and one of multiple telomeric breakpoints (BP3–BP6) on the paternal allele.[629] This renders the zygote monosomic for these genes, and the zygote possesses only the maternal copy of this region. Alternatively,

TABLE 6.7 Molecular Mechanisms for Prader-Willi and Angelman Syndromes

Molecular Mechanism of Disease	Angelman Syndrome (Frequency)	Prader-Willi Syndrome (Frequency)
Deletion of 15q11-13	Loss of maternal allele (70%–75%)	Loss of paternal allele (65%–75%)
Uniparental disomy	Two paternal chromosomes (3%–7%)	Two maternal chromosomes (20%–30%)
Imprinting center defect epimutations or microdeletions	Maternal allele (2%–3%)	Paternal allele (2%)
UBE3A mutation	10%	Not applicable
Rearrangement involving 15q11-13	Maternal allele (<1%)	Paternal allele (<1%)
Cause not identified	10%	Rare

20% to 30% of cases of PWS are caused by uniparental disomy (UPD). In the case of PWS, although two copies of the genes located in 15q11.2-q13 exist, both are maternal in origin and in most cases arise from meiosis I nondisjunction followed by postzygotic mitotic loss of the third, paternally derived chromosome 15 via a process referred to as *trisomy rescue*. This mechanism rescues the zygote from trisomy 15, a condition that is incompatible with life.[630] Although the fetus is genetically complete with two chromosome 15s (disomy), both chromosomes are received from the mother (uniparental), and no expression of paternally expressed genes occurs in this imprinted region. Not surprisingly, maternal age has been reported to be significantly higher in PWS patients resulting from UPD caused by maternal meiosis I nondisjunction than in PWS patients resulting from a de novo deletion.[631] Some cases of PWS result from microdeletions encompassing the paternal IC or, in 2% of cases, from abnormal methylation at this site.[632] A mutation involving the IC prevents this *cis*-acting control center from resetting the imprint in the germline. These mutations will result in PWS because if they are present on the maternal chromosome of phenotypically normal fathers, they will be transmitted to offspring because now the paternal chromosome will maintain the maternal imprint and will be silenced. Finally, fewer than 1% of PWS cases are caused by chromosomal rearrangements disrupting the genes in the 15q11.2-q13 region.[633] Some genotype–phenotype correlations have been noted. Patients with PWS with UPD are more likely to have psychotic episodes, compulsive behaviors (eg, skin picking), and ASDs than are PWS patients with deletions.[634,635]

The maternally expressed gene *UBE3A* is telomeric to the snoRNA genes and oriented in the opposite orientation on chromosome 15q11.2-q13.[636,637] The *UBE3A* gene encompasses 120 kb; contains 16 exons; and encodes E3 ubiquitin protein ligase, which is involved in the ubiquitin proteasome degradation pathway.[638] Three protein isoforms are produced from this gene by alternative splicing, and they differ at their N-termini.[639] Interestingly, in the brain, only the maternal *UBE3A* allele is expressed; however, both alleles are expressed in other tissues.[640] A second gene, *ATP10A*, is upstream from *UBE3A* and is also preferentially expressed only from the maternal allele.

Similar to PWS, most patients with AS (70%–75%) have a deletion of the critical 15q1.21-q13 region. However, unlike PWS, in AS, the disease-causing deletion occurs on the maternal allele. In 3% to 7% of AS patients, the syndrome is attributed to UPD from the inheritance of two paternally derived chromosome 15s; as a consequence, there is no transcription of *UBE3A*. An IC defect has been described in 2% to 3% of patients with AS ; a chromosome rearrangement has been reported in fewer than1%; and in 10% of cases, a *UBE3A* mutation has been detected.[632,641,642] Most of the *UBE3A* gene mutations are frameshift or nonsense mutations and result in loss of function.[643] Using the AS mouse model, a possible treatment for this disease could be increasing expression of the normal paternal allele and silencing the mutant transcript with antisense oligonucleotides.[644] In about 10% of AS cases, the molecular basis of the disease has not been determined. It is possible that these patients are misdiagnosed with AS and rather are similar to AS but clinically and molecularly separate from those with AS.[645] Genotype–phenotype associations are known, including the fact that patients with AS with deletions are more likely to have hypopigmentation of skin, eye, and hair or microcephaly and are more likely to be severely affected.[646] In contrast, patients with AS arising from UPD have normal head circumference and are more often mildly affected.

Diagnostic testing for individuals suspected of having AS or PWS can involve a variety of laboratory techniques and testing algorithms. Methylation-specific PCR (msPCR) coupled with gel electrophoresis is one cost-effective approach (Fig. 6.14). In methylation-specific PCR, genomic DNA is treated with sodium bisulfite to convert unmethylated cytosine residues to uracil without altering the methylated cytosine residues (those silenced in the 15q11.2-13 region). Subsequent PCR reactions use oligonucleotide primers specific to DNA strands that contain uracil (from unmethylated cytosines) or cytosine (from methylated cytosines).[647,648] Methylation-specific PCR (msPCR) provides a rapid and reliable diagnostic test for PWS or AS. Fewer than 1% of PWS cases and about 20% of AS cases are not detected by this assay. Additionally, msPCR coupled with melting curve analysis or methylation-specific MLPA can be used.[649-651] Although msPCR is frequently the first tier of testing and can be used to diagnose PWS or AS, it cannot determine the molecular basis of disease. If the msPCR result is positive, chromosomal microarray analysis will identify chromosome deletions, and in UPD cases, the parental origin of both chromosomes can be determined.[652-654] Alternatively, if the msPCR result is positive, fluorescent in situ hybridization (FISH) studies will detect deletions, and in UPD cases, molecular testing using polymorphic short tandem repeat sequences coupled with capillary electrophoresis will detect UPD.[655] Patients with PWS and patients with a chromosomal rearrangement disrupting the genes in this area will not be identified by these testing methods nor will AS patients with a *UBE3A* mutation.

FIGURE 6.14 Methylation-specific polymerase chain reaction (PCR) for the diagnosis of Prader-Willi syndrome (PWS) and Angelman syndrome (AS). Extracted DNA is treated with sodium bisulfate before amplification using multiplex PCR and oligonucleotide primers specific for modified DNA. PCR products are subjected to gel electrophoresis. Normal individuals show two amplicons representing their methylated maternal allele and unmethylated paternal allele. PWS patients show only the maternal allele, and AS patients show only the paternal allele. Patient DNA with patterns diagnostic of AS (*lanes 1* and *5*) and PWS (*lanes 2* and *6*) and patients with normal methylation patterns (*lanes 3* and *4*) are shown. Normal control DNA patterns and a negative control reaction in which no template DNA was added are indicated in *lanes 7 and 8*, respectively. No amplification products are observed in unmodified normal control DNA (*lane 9*), illustrating the specificity of PCR primers prepared specifically for sodium bisulfate–modified DNA. (Courtesy Jack Tarleton, PhD, Director of Genetics Laboratory, Fullerton Genetics Center, Mission Health System, Asheville, North Carolina.)

Rather, in these cases, a routine karyotype or DNA sequence analysis is required for diagnosis. Testing for PWS and AS is critical because knowledge regarding the molecular mechanism of disease is important for accurately determining recurrence risk to the family.[656] For example, although mutations causing AS can arise de novo (eg, UPD with a <1% recurrence risk), other AS-causing mutations can be silently transmitted through several generations. If a *UBE3A* mutation arose de novo on a paternal allele transmitted to a son, the son could transmit the mutation to a son or daughter to produce a normal phenotype. However, although this son could transmit the silenced *UBE3A* mutation to his offspring, his sister could donate her mutated *UBE3A* paternally derived allele to her offspring, and the child would have AS. The recurrence risk for her to have another affected child in this case would be 50%.

POINTS TO REMEMBER

Inherited Diseases
- Mitochondrial and imprinting disorders follow non-Mendelian patterns of inheritance.
- Complex diseases result from the contribution of both genetic and environmental factors and do not follow Mendelian inheritance patterns.
- For most genes, pathogenic mutations are located throughout the gene and are heterogeneous in nature.
- For some disease genes, the type of mutation and effect on the encoded protein can predict the clinical phenotype and identify targeted therapy for patient care.
- Diagnostic DNA testing is complicated by genetic and allelic heterogeneity.
- Genetic counseling is an important component of patient care and management.

EXPANDED CARRIER SCREENING

Carrier screening refers to the use of genetic testing to determine individuals who are at risk of having a child affected with an autosomal recessive disorder. Carrier screening for disorders has long been a mainstay of genetic laboratories and has increased dramatically over the past decade. Carrier screening for some common lethal disorders (eg, CF) is considered the standard of care in prenatal patients regardless of ethnicity.[2] For some ethnic groups, such as Ashkenazi Jews, the carrier frequency for several lethal disorders is relatively high.[657] As a result, screening for many of these disorders (eg, Tay-Sachs, Canavan) has long been recommended to Ashkenazi Jewish couples during preconception and prenatal counseling.[657]

Conventional carrier screening used mutation panels (eg, the ACMG *CFTR* 23 mutation panel) that identify the majority of mutation carriers for a single disorder. Over the past several years, carrier screening (particularly in preconception/prenatal care) has shifted toward new testing platforms that simultaneously screen for hundreds of common disease-associated point mutations. With the advent of these new technologies many laboratories are offering expanded carrier screening (ECS) panels. Initially, ECS panels simply increased the number of mutations or variants that were tested in a single gene (eg, *CFTR*). More recently, ECS panels have broadened to include known mutations or variants for a multitude (>100 in some cases) of inherited disorders in a single test.[658] The clear advantage of ECS panels is that they provide a cost-effective method to screen for multiple genetic disorders. In many cases, the cost of an ECS panel is less than screening for a single gene by traditional means. ECS panels also provide patients with carrier status data on additional disorders not routinely included on traditional carrier screening panels. However, ECS panels have come under criticism because of several drawbacks.

Traditionally, mutations included on carrier screening panels were selected based on confirmed pathogenicity and the carrier frequency of the mutation.[51] With the advent of ECS panels, some laboratories have included variants with reduced penetrance or mild clinical effects.[659,660] Some expanded CF screening panels include variants that are known to have variable clinical impact and in some cases result in no discernible phenotype.[659] The inclusion of such variants in a genetic screening assay can be confusing and lead individuals to make reproductive decisions without a complete understanding of the information provided.[659,661] Some disorders that have been selected for inclusion on commercially available ECS panels do not meet generally accepted criteria for carrier screening.[661,662] For example, some disorders are rare and have a reported incidence of less than 1 per million births.[49,662] Not surprisingly, the targeted mutations in many of these rare disorders only account for a small fraction of the mutations capable of causing disease. A negative screening result in an individual with a family history for a rare disorder may give a false sense of security. Because of this low sensitivity, if a mutation is identified in one partner for a rare disorder, that individual's reproductive partner likely will undergo full gene sequencing for the causative gene, thus increasing the costs of screening dramatically.[662] ECS panels have also been criticized for including mutations (and functional polymorphisms) of variable penetrance that are very common in certain ethnic groups or society at large (eg, factor V Leiden, *HFE*-related hemochromatosis, MTHFR deficiency).[661] Typically, disorders with such high frequency in the population are not recommended for carrier screening because their clinical implications are uncertain. The identification of these mutations or variants in the context of preconception (or prenatal) counseling are especially controversial because fetal testing for some of these disorders (eg, MTHFR deficiency) is not routinely offered in the United States.[661] The clinical validity of some of the variants that are included on prenatal screening panels is unclear.

Another aspect of ECS panels that merits mention is the inclusion of adult onset disorders. Some disorders (eg, familial Mediterranean fever, α_1-antitrypsin deficiency, *GJB2*-related nonsyndromic hearing loss, atypical CF) have variable ages of onset and disease manifestations. Carrier screening for mutations that cause these disorders in individuals of reproductive age can provide an unexpected diagnosis of a disease that they have not yet developed. In one recent study, 78 of 23,453 individuals tested were identified as either compound heterozygous or homozygous for disease-causing mutations.[49] Of the patients identified, only three patients reported a previous diagnosis or history of disease.[49] These data illustrate the complex counseling-related issues that ECS panels have created. Because traditional carrier screening has focused on severe, disease-causing mutations, the likelihood of diagnosing an asymptomatic individual with disease was very low. The newer ECS panels require that all patients undergoing such screening be properly advised as to the potential testing outcomes.

To address the issues associated with ECS testing, the ACMG released a position statement regarding prenatal and preconception ECS that outlined criteria for inclusion of diseases or mutations on a carrier screening panel, including (1) a clearly defined clinical association; (2) most at-risk patients would choose fetal testing to aid in preconception or prenatal decision making; (3) a clearly defined residual risk for individuals that test negative; and (4) for any adult-onset disorder that may affect the individual being tested, pretest counseling and consent should be performed.[663] Recently, several professional societies released a joint statement regarding the use of ECS panels.[50] These statements can guide physicians in offering, consenting, and counseling patients about ECS panels and their results.[50]

The clinical use of ECS panels has only just begun. As with any new technology and the increase in data that it provides, unforeseen ethical issues can arise. As the use of ECS panels becomes widespread and professional organizations (eg, the ACMG) develop formal guidelines, many of the issues outlined here will be resolved. However, preconception and prenatal genetic counseling will continue to be important in helping patients to understand the clinical implications of their screening results.

MASSIVELY PARALLEL SEQUENCING

Massively parallel sequencing (also referred to as next-generation sequencing) is a high-throughput DNA sequencing technology capable of generating data on a genomic scale in a short period of days (see Chapter 4 on nucleic acid techniques). Not since the introduction of PCR in the 1980s has a technology revolutionized the field of molecular diagnostics like MPS has in the past few years. At the most basic level, MPS uses similar concepts to traditional capillary-based Sanger sequencing in that fluorescently labeled dNTPs are used to determine the template sequence. However, the distinct advantage of MPS is its ability to perform simultaneous sequencing reactions on millions of target sequences at a vastly lower cost per base than traditional Sanger sequencing. There are several different MPS methods, but most have similar sample preparation workflows. Typically, DNA fragmentation is followed by insert selection, library formation, and clonal amplification. Then sequencing by synthesis signals is acquired by measuring pyrophosphate release, generation of hydrogen ions, or the fluorescence of reversible terminators. This massively parallel clonal sequencing of library inserts generates gigabases of sequencing data at a cost that is feasible for clinical testing. One drawback to such massive data generation is that bioinformatics filtering processes are required to efficiently interpret the numerous variants identified. Data filtering in MPS typically involves the utilization of publicly available variant databases (eg, dbSNP; Exome Aggregation Consortium [ExAC], National Heart, Lung, and Blood Institute Exome Sequencing Project) to eliminate variants that occur at high frequency in the general population and therefore are likely benign. After the common variants have been filtered, locus-specific mutation databases and in silico prediction programs (eg, SIFT and PolyPhen2) can aid in the interpretation of potential disease-causing variants. Variants are then classified as pathogenic, likely pathogenic, uncertain significance, likely benign, or benign.[664] Data filtering systems are commercially available to help in the interpretation of MPS data, but many laboratories have chosen to develop their own software pipelines internally.

One of the most significant advantages of MPS for diagnostic testing is the ability to target all genes known to be associated with a specific diagnosis (or phenotype) in a single test that is comparable in price and turnaround time to that of Sanger sequencing for a single-gene analysis. These targeted gene panels are the most commonly ordered clinical tests using MPS technology. Before development of MPS, testing patients for causative mutations in multiple genes that cause a single syndrome was a very expensive and time-consuming process. For example, retinitis pigmentosa (RP, OMIM #268000) is a group of inherited degenerative ocular disorders that affects 1 in 3000 to 7000 people.[665] Locus heterogeneity is a hallmark of RP because mutations in more than 60 genes have been reported to cause this disease. Using traditional Sanger sequencing to determine the underlying molecular alteration would be cost prohibitive for most patients with RP, yet an MPS RP gene panel is cost effective and timely. RP is one example of how MPS implementation has advanced our ability to provide a molecular diagnosis for a genetically heterogeneous disorder. Table 6.8 lists some commonly ordered MPS-based gene panels. MPS also enables the sequencing of very large genes at a reasonable cost (eg, the *DMD* gene that causes DMD and BMD). The flexibility of MPS selection and library preparation can

TABLE 6.8 Commonly Ordered Massively Parallel Sequencing–Based Gene Panels

Panel Name	Included Disorders*	Targeted Patient Group
Aortopathy disorders	Ehlers-Danlos syndrome (I, II, and IV), Loeys-Deitz syndrome, Marfan syndrome	Individuals with disease affecting any aortic section
Breast and ovarian hereditary cancer	Hereditary breast and ovarian cancer syndrome	Individuals with a strong family history of breast and ovarian cancer; individuals with early onset of breast or ovarian cancer
Cardiomyopathy disorders	Dilated cardiomyopathy, hypertrophic cardiomyopathy, long QT syndrome	Individuals with a suspected diagnosis of a hereditary cardiomyopathy disorder
Expanded carrier screening	Numerous (>100)	Individuals planning pregnancy or prenatal reproductive partners
Hearing loss	Keratitis–ichthyosis–deafness syndrome, nonsyndromic hearing loss, Usher syndrome	Individuals with a suspected diagnosis of either syndromic or nonsyndromic hearing loss
Hereditary endocrine cancer	Multiple endocrine neoplasia type 1, multiple endocrine neoplasia type 2, Von Hippel-Lindau disease	Individuals with a family history of endocrine cancer; individuals with a personal history of endocrine cancer
Hereditary gastrointestinal cancer	Familial adenomatous polyposis, juvenile polyposis syndrome, Lynch syndrome	Individuals with a personal or family history of gastrointestinal cancer
Noonan spectrum disorders	Cardiofaciocutaneous syndrome, Costello syndrome, Noonan syndrome	Individuals with a suspected diagnosis of Noonan syndrome or a related disorder
Periodic fever syndromes	Familial Mediterranean fever, Majeed syndrome, Muckle-Wells syndrome	Individuals with a suspected diagnosis of a periodic fever syndrome
X-linked intellectual disability	Rett syndrome, Duchenne muscular dystrophy, ornithine transcarbamylase deficiency	Individuals with intellectual disability inherited in an X-linked manner

*This is a selected list of commonly included disorders and is not comprehensive.

provide a comprehensive test of more than 100 genes (eg, an X-Linked Intellectual Disabilities Panel) or a more targeted panel that analyzes a handful of genes associated with a specific phenotype (eg, a hereditary gastrointestinal cancer panel). Over the next few years, it is likely that MPS will be widely implemented in diagnostic laboratories, Sanger sequencing assays will decline in use, and sequencing technology will continue to evolve.

WHOLE-EXOME SEQUENCING

Whole-exome sequencing (WES) refers to an MPS-based DNA sequencing method that specifically targets the coding regions (exons) and directly adjacent intronic regions for the majority of the approximately 20,000 genes known to exist in the human genome. Although the exome only accounts for approximately 1% of the human genome, mutations in gene encoding regions are responsible for the vast majority of human inherited diseases. On average, WES is able to identify an underlying genetic alteration in approximately 25% to 50% of patients (depending on the inclusion criteria).[666-668] This makes WES an effective diagnostic tool for patients with a phenotype that suggests an inherited disorder that does not fit the clinical characteristics of previously described syndromes.[666,669]

Most exome sequencing performed in clinical laboratories currently uses a hybridization based capture method of tagged (biotinylated) probes targeted to specific areas of the fragmented template DNA.[670] These probes are bound to magnetic beads allowing a simple washing process to separate targeted DNA regions from the excess unwanted (intronic) DNA.[670] After this enrichment process, the DNA is ready to be sequenced. Several exome capture kits are commercially available, making WES easily performed in most molecular diagnostic laboratories. Often, the limiting factor in implementing WES is the large amount of data produced; therefore, a well-defined bioinformatics workflow is critical for the timely reporting of WES results. Several data analysis programs are commercially available to aid in the interpretation of WES data, and many publications describe informatics workflows.[671]

Clinical WES sequencing is often performed on both the symptomatic proband and their (typically asymptomatic) parents (often referred to as a trio). Sequencing parents helps with interpretation of sequence variants that are identified in the proband. For example, if a potentially pathogenic variant is identified in the proband but not observed in the parental samples, that variant is likely de novo (assuming confirmed paternity) and potentially causative. Likewise, if a potentially pathogenic variant is identified in a gene that is dominantly inherited but is also identified in an asymptomatic parent, it is less likely to cause the patient's phenotype. Parental samples can also be used to establish phase in the detection of two pathogenic mutations in a gene inherited in an autosomal recessive manner. Use of familial samples to aid interpretation of WES results is a powerful tool capable of dramatically increasing clinical sensitivity of the test and should be pursued in all patients undergoing WES.

WES has been implemented widely in clinical diagnostic laboratories in the United States, and thousands of patients have been tested by this method. WES should be considered in patients with a phenotype suspected to be caused by a mutation in a single gene when known single-gene disorders have been eliminated.[669] Careful consideration should always be given to the patient's presentation before determining whether or not WES is the appropriate test for a given phenotype. Specifically, a detailed family history, a systematic characterization of the patient's phenotype, and a careful literature review is recommended before ordering WES.[669] Obtaining this information can help determine if the patient is actually affected by a previously described, but rare, syndrome with a known genetic cause that should be ruled out before proceeding to WES.[669] In many cases, a single gene test or a targeted MPS-based gene panel with multiple genes may be the appropriate first test to order. To aid clinicians in determining which molecular testing protocol is best for their patients, algorithms have been developed.[672,673] These testing algorithms suggest that individuals with multiple nonspecific clinical findings or with a clinical presentation associated with marked genetic heterogeneity (eg, intellectual disability) are good candidates for WES.[672,673] Individuals with distinctive clinical features, family history of a specific disorder, or indicative findings for specific disorders should be counseled to pursue either single-gene or MPS-based gene panel testing.[672,673] Using these testing guidelines, approximately 50% of patients received a genetic diagnosis using "traditional" methods of diagnosis.[673] Single-gene testing is therefore not obsolete because of WES but continues to be the appropriate diagnostic tool in many clinical cases.

Counseling for WES is highly complex because issues relating to test results must be considered. The risk of identifying a VUS exists in all sequencing-based genetic tests. However, the risk for the identification of a VUS increases dramatically for WES. Patient counseling for WES should always include a discussion of VUSs because these are likely to appear on any WES report even though the clinical implications of these findings are unclear.[672] Patients should be counseled on the possibility that incidental findings may include the identification of pathogenic mutations in clinically actionable genes (eg, *BRCA1*) that are unrelated to the patient's current phenotype. The return of incidental findings (IFs) in WES is a controversial topic and has resulted in the ACMG formalizing recommendations for which IFs should be returned to the patient.[674] The ACMG recommendations provide a list of 56 genes that represent the "minimum" IFs that should be reported to the clinician when a pathogenic variant is identified regardless of the clinical indication for testing.[674] The gene list generated by the ACMG was developed to include genes for conditions that are verifiable by other diagnostic methods and that cause highly penetrant disorders that would likely benefit from medical intervention.[674] The release of the ACMG recommendations was met by criticism because the guidelines were seen by some to violate existing ethical norms in genetic testing and the patient's right to autonomy by suggesting that IFs in the 56-gene list should be returned regardless of patient preference.[674,675] In response to criticism from its members, the ACMG released a statement in April 2014 revising its recommendations to allow patients to opt out of receiving incidental findings. Even after this revision, debate continues among members of the ACMG and clinical geneticists regarding how IFs should be returned to patients.[676] Recommendations on the return of IFs likely will continue to evolve as WES genomic testing becomes more commonplace in clinics.[677]

Over the past several years, WES has proven to be an invaluable research tool in the discovery of the underlying genetic alterations for many Mendelian disorders.[678-680] In some cases, these discoveries identified the first known genetic cause of a disorder (eg, Miller syndrome, Kabuki syndrome), and in others, additional genes were discovered to cause an already well-defined phenotype (eg, RP, nonsyndromic hearing loss, osteogenesis imperfecta, intellectual disability, and many others).[681-687] WES studies also have elucidated alternative phenotypes caused by mutations in genes already known to cause a genetic disorder.[688-690] The benefits of these discoveries in the clinical diagnosis and treatment of patients cannot be overstated. Identification of the underlying molecular alteration (and molecular pathogenesis) that results in a specific disease is the first step in developing treatment modalities. Because of the discoveries made by WES, the next decade will see a vast improvement in the treatment of many inherited disorders. Within the next decade, as WES technology improves and WES costs decline, the underlying cause for the vast majority of inherited genetic disorders may become known.

CYTOGENOMICS

The term *cytogenomics* describes the application of molecular techniques to cytogenetics. In a broad sense, this applies to FISH. In FISH analysis, a fluorescently labeled DNA molecule serves as a "probe" and is hybridized to metaphase chromosomes or interphase nuclei, and the fluorescent probes are visualized using a fluorescent microscope.[691] In a more narrow sense, cytogenomics applies to the use of chromosome microarray (CMA) technology, including array comparative genomic hybridization (aCGH) and SNP arrays.[692] SNP and aCGH arrays have revolutionized the field of cytogenetics as important tools for both clinical diagnosis and disease discovery.[693] However, these technologies do not identify balanced translocations or inversions, both of which require routine karyotyping, FISH analysis, or both.

Instrumentation and associated kits for aCGH and SNP arrays are commercially available. Platforms and methods vary, including the probe size, spacing between the probes on the array, copy number resolutions, and probe sensitivity.[692,694] In aCGH, the patient and control DNA are labeled with two different fluorochromes and co-hybridized to the array, but with a SNP array, no control DNA is used; rather, the fluorescent signals are measured against a reference pattern.[692] Both aCGH and SNP arrays can detect copy number variants (CNVs). In addition, SNP arrays provide the genotype and determine if the patient is homozygous (AA, BB) or heterozygous (AB) for each SNP present on the array. SNP genotype analysis allows long stretches of homologous DNA sequences, also referred to as regions of homology (ROHs) or regions with an absence of heterozygosity (AOHs), to be identified. These segments of DNA are important to identify UPD in the diagnosis of imprinting disorders.[695,696] However, detection of an ROH or AOH could represent an incidental or unexpected finding of parental relatedness or consanguinity. Depending on the degree of homozygosity, these findings could suggest incestral mating. Standards and guidelines for laboratory reporting of incidental findings suggestive of consanguinity have been developed by the ACMG, and care must be taken by clinicians in communicating these results to patients.[697,698]

CMA's clinical utility is in the diagnosis and management of patients referred for multiple congenital anomalies, developmental delay or intellectual disability, and ASDs.[699-701] The ACMG recommends CMA as a first-tier genetic test for patients with these conditions, and practice standards and guidelines for these applications have been established.[702] Interestingly, CMA analysis can identify CNVs of clinical significance in as many as 21% of patients with ASDs.[703] Common ASD CNV hot spots are known, and a database of previously reported CNVs with a documented association with ASDs is available for reference.[703-705] FISH, MLPA, or real-time PCR studies are often used for confirmation of novel aberrant CNV findings to prevent false-positive reporting (Fig. 6.15). Public databases (eg, Database of Genomic Variants,[706] Ensembl,[707] National Center for Biotechnology Information[708]) are used to determine the genes that are contained within the CNV and that may be either lost or gained. Generally, the larger the CNV and the more genes contained within the DNA region of interest, the more the variant is likely to have a clinical consequence and the more likely a deletion is to be pathogenic compared with duplications.[704] After careful review of various databases, peer-reviewed publications, and the clinical findings of the patient, the significance of the CNV is reported using guidelines established by the ACMG.[709] Results may be pathogenic or benign or may be reported as uncertain clinical significance. Furthermore, a variant of uncertain clinical significance can be further classified as likely pathogenic or likely benign.[709] Appropriate literature should be referenced, and the variant should be reported according to standard CNV nomenclature. To ascertain the significance of any variant with uncertain clinical significance, parental specimens should be requested (Fig. 6.16).

Chromosome microarray is also appropriate for prenatal testing when abnormal ultrasound findings are evident or for additional information when a normal karyotype is unexpected.[710,711] CMA testing can be performed on DNA extracted from cultured amniocytes or chorionic villi tissue. However, CMA testing in prenatal cases can be especially challenging if a variant of uncertain clinical significance is identified because it may be difficult to predict the postnatal effect.[710] CMA analysis on DNA extracted from representative tissue of products of conception can be useful in determining the etiology of the pregnancy loss.[712]

In addition to using aCGH and SNP arrays to detect constitutional or germline changes associated with disease, these arrays are also used on hematologic and solid tumors.[713,714] In somatic tissue, the detection of copy number changes can define regions of DNA and specific genes involved in the pathogenesis of neoplastic processes. In addition, the ability of SNP arrays to determine genotype enables

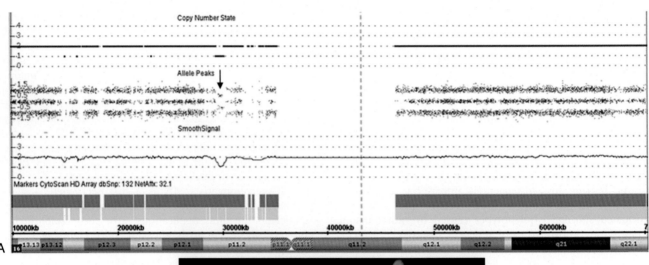
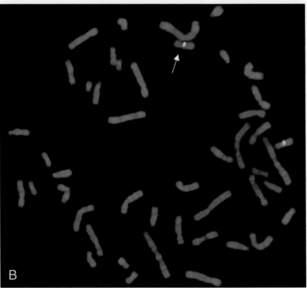

FIGURE 6.15 Loss of 16p11.2 detected by microarray analysis with confirmatory testing performed by fluorescent in situ hybridization (FISH) analysis. **A,** DNA from a patient referred for autism was analyzed using the Affymetrix CytoScan platform. The copy number state *(top)*, allele peaks *(middle)*, and smooth signal pattern of the copy number *(lower rows)* are shown. A partial ideogram of chromosome 16 is shown to localize the data to the corresponding chromosomal bands. A 751-kb loss of nucleotides 29,427,215 to 30,177,916, localized to chromosome band 16p11.2 as denoted by the *arrow,* is shown. Given that only 1 copy of DNA sequence is present at this site, the allele peak single nucleotide polymorphism pattern for this region shows only two lines (AA and BB), and the remaining portions of the chromosome have the characteristic three-line pattern (AA, AB, and BB). The repeat sequences localized to the pericentromeric region of chromosome 16 are excluded from the array (absence of *bars*) because they are not uniquely localized to chromosome 16 and could cross hybridize to other chromosomes. **B,** FISH results obtained using a red probe (PR11-301D18) specific for band 16p11.2 including nucleotides 29,776,142-29,961,746 and an aqua pericentromeric control probe (D16Z3) for chromosome 16 to assess probe hybridization efficiency and identify both chromosomes 16. After hybridization, the metaphase spreads showed a normal pattern for the control probe (one signal on each chromosome 16), but an abnormal pattern for the test probe (one probe signal on a structurally normal chromosome 16, with no signal observed on a structurally abnormal chromosome 16 *[arrow]*). Based on the banding pattern (reverse DAPI and GTG [not shown]) and morphology of the chromosomes, this loss resulted from an interstitial deletion [del(16)(p11.2p11.2)(PR11-301D18-,D16Z3+)] that was not detected by routine karyotype analysis. (Courtesy Dr. Colleen Jackson-Cook, Director Cytogenetics Laboratory, Department of Pathology, Virginia Commonwealth University Health System).

the identification of copy neutral LOH. Similar to constitutional UPD seen with imprinting disorders, copy neutral LOH is a somatic event and indicates two copies of the same chromosomal region. This may involve part of the chromosome or the entire chromosome, and most often, the region involved harbors a "driver" mutation in a particular gene that promotes growth and proliferation for the neoplastic process.[713]

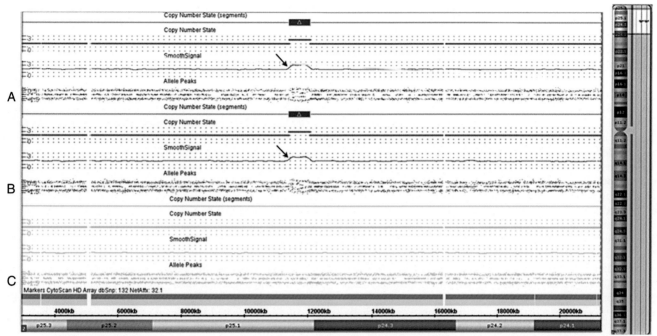

FIGURE 6.16 Detection of inherited copy number gain at chromosome region 2p25.1 by microarray analysis. DNA from a 7-year-old male patient referred for developmental delay was analyzed using an Affymetrix CytoScan platform. A copy number variant (CNV) of uncertain clinical significance representing a 654-kb copy number gain (three copies involving nucleotides 11,410,614–12,064,438) was identified in the patient *(row a)*. The *arrow* denotes the gain of DNA to three copies in the smooth signal and copy number state plots; with four "bands" (AAA, AAB, ABB, and BBB) being observed in the allele peak pattern. A partial ideogram of chromosome 2 is shown horizontally to show the location of the CNV detected in the patient and the location of the CNV relative to the entire chromosome number 2 as seen on the vertical orientation. Maternal *(row b)* and paternal *(row c)* specimens were requested for analysis to determine if the CNV observed in the patient was inherited or represented a de novo event. The results indicate the presence of no CNVs in the paternal DNA but show that the CNV in the patient is of maternal origin. Furthermore, the CNV in the maternal sample is identical to that observed in the patient and demonstrates that the CNV was stably inherited, thereby suggesting that this CNV is benign. The data shown for each family member includes the copy number state *(top rows of each case)*, smooth signal of copy number values *(middle rows of each case)*, and allele peaks *(lower rows of each case)*. (Courtesy Dr. Colleen Jackson-Cook, Director Cytogenetics Laboratory, Department of Pathology, Virginia Commonwealth University Health System.)

POINTS TO REMEMBER

Clinical Utility of Molecular Methods
- Targeted PCR amplification is most useful in the identification of common pathogenic point mutations.
- msPCR is capable of determining the methylation status of DNA and is often used in the diagnosis of imprinting disorders.
- Full-gene Sanger sequencing is typically used to identify pathogenic point mutations and small insertion or deletion mutations in disorders associated with a single disease-causing gene.
- MPS is the ideal technology to use for the identification of a pathogenic point mutation or a small insertion or deletion mutation in disorders caused by mutations in any of a number of genes.
- MLPA is most useful in the detection of large (exon level or bigger) deletions or duplications in three or fewer disease-causing genes.
- Array-based technology is most useful in the simultaneous detection of large deletions or duplications in numerous genes.

REPORTING OF TEST RESULTS

As the preceding pages show, DNA testing for inherited diseases is complex, and thoroughly conveying genetic test results is important. Results must be presented so they can be easily and accurately understood by a professional whose expertise may not be genetics because in many instances, primary care providers communicate test results to the patient. With the increasing clinical demand for genetic testing and the increasing numbers of laboratories performing such tests, uniformity in communicating these complex results to referring clinicians is important, and failure to include pertinent information in these reports constitutes a deficiency in the molecular pathology laboratory inspection checklist of the College of American Pathologists (CAP).[715] A comprehensive genetic report should include the patient's name, medical record number or birth date, sex of the patient, ethnicity of the patient (if relevant), type of specimen and date received, specimen's laboratory identification number, laboratory test requested, name and address of laboratory performing the test, name and address of referring health care professional or hospital, date of the report, analytic interpretation of the results using standard nomenclature for all variants identified, detailed description of the method used (citing literature if needed), and sensitivity and specificity of the assay (eg, number of variants analyzed, percentage of variants not detected, possibility of genetic heterogeneity, chance of genetic recombination). All sequence variants are classified as one of the following: pathogenic, likely pathogenic, uncertain significance, likely benign, or benign.[664,716] In silico tools such as PolyPhen2, SIFT, and MutationTaster should be listed if they were used to determine significance of the variant.[716] The DNA and protein change, if applicable, should be listed using guidelines of the Human Genome Variation Society.[717] The laboratory should include the reference sequence and genome build and provide genomic coordinates for the variant. The report must also include a clinical interpretation of the findings as applicable. Although preparation of the clinical interpretation can be labor intensive, this section is vital to most genetic reports and is important for describing the clinical significance of the results as they apply specifically to the patient and his or her family. This section should include a brief clinical history of patient (indicating the reason for testing) and may discuss recurrence risk, genotype—phenotype correlation or penetrance, associated disease or carrier risk calculations for other members of his or her family, and citations of literature as needed. Importantly, a statement that genetic counseling for the patient is indicated must be included.

Furthermore, because many assays performed in clinical DNA laboratories are laboratory-developed tests or procedures that have been developed, designed, or validated by the laboratory and are not approved by the FDA, reports must include a disclaimer to state this fact. Class I analyte-specific reagents may be purchased from a vendor and sold for a specific test, or they may be independently purchased by the laboratory and assembled into a laboratory-designed test. An example of the disclaimer would state: "This test was developed and its performance characteristics determined by [laboratory name]. It has not been cleared or approved by the U.S. Food and Drug Administration." In addition, the CAP recommends inclusion of these additional statements: "The FDA does not require this test to go through premarket FDA review. This test is used for clinical purposes. It should not be regarded as investigational or for research. This laboratory is certified under the Clinical Laboratory Improvement Amendments (CLIA) as qualified to perform high-complexity clinical laboratory testing."[715] Last, reports should be reviewed and signed by the laboratory director or a qualified designee.

LABORATORY REGULATION

Regulatory oversight of clinical laboratories is essential to maintaining consistency across testing centers. All molecular genetic laboratories offering clinical testing should be CLIA certified and be actively participating in proficiency testing. In most cases, molecular laboratories are accredited by the CAP, which is considered to be the gold standard. CAP accreditation requires that laboratories undergo biannual inspection by an outside team of laboratory scientists using a specified checklist of requirements.[715] Maintaining accreditation requires that any deficiencies identified during a CAP inspection must be corrected. Laboratories are required to perform proficiency testing, which covers the scope of the tests performed in the laboratory. CAP provides proficiency testing samples or packets for a number of commonly ordered tests (eg, HD, fragile X). CAP also provides method-based proficiency testing to verify that a clinical laboratory using a general method (eg, Sanger sequencing) reports results consistent with those of other clinical laboratories. For tests that are offered clinically but are not covered by CAP proficiency testing, other means of confirming test accuracy must be pursued. This can involve sample exchanges with other laboratories that offer a similar clinical test or can simply be internal proficiency testing whereby a sample is randomly selected and is anonymously retested (among other patient samples) to confirm that the same results are obtained. Regardless of the method used, adequate records of proficiency testing results must be kept, and any discrepancies among results must be investigated and addressed. More information on the CAP accreditation and proficiency testing process can be found at http://www.cap.org.

Over the past several years, regulation of molecular genetic testing has become the focus of government agencies. The rise of companies that offer direct-to-consumer (DTC) genetic testing with no involvement of medical professionals or counselors elicited significant concerns regarding the accuracy and clinical validity of some genetic tests. In 2006, the Government Accountability Office (GAO) investigated the activities of multiple DTC genetic testing companies and found that many of the tests being offered were medically unsound or of no practical use. The GAO findings increased the concerns that many agencies had previously voiced regarding DTC genetic testing and complex genetic testing in general. In 2011, the National Institutes of Health started a voluntary registry of genetic tests available in the US. Subsequently, the FDA announced that they would also start reviewing all (non—FDA-approved) laboratory-developed tests. These actions may mark the first steps in what could be an increase in government oversight of genetic testing.

REFERENCES

1. ACOG committee opinion no. 486: Update on carrier screening for cystic fibrosis. *Obstet Gynecol* 2011;**117**:1028—31.
2. ACOG committee opinion. Number 325, December. Update on carrier screening for cystic fibrosis. *Obstet Gynecol* 2005; **2005**(106):1465—8.
3. Chillon M, Casals T, Mercier B, et al. Mutations in the cystic fibrosis gene in patients with congenital absence of the vas deferens. *N Engl J Med* 1995;**332**:1475—80.
4. Highsmith WE, Burch LH, Zhou Z, et al. A novel mutation in the cystic fibrosis gene in patients with pulmonary disease but normal sweat chloride concentrations. *N Engl J Med* 1994;**331**: 974—80.
5. Paranjape SM, Zeitlin PL. Atypical cystic fibrosis and CFTR-related diseases. *Clin Rev Allergy Immunol* 2008;**35**: 116—23.
6. Berkhout MC, van Rooden CJ, Rijntjes E, et al. Sinonasal manifestations of cystic fibrosis: a correlation between genotype and phenotype? *J Cyst Fibros* 2014;**13**:442—8.
7. Saiman L, Siegel J. Infection control in cystic fibrosis. *Clin Microbiol Rev* 2004;**17**:57—71.
8. Collawn JF, Matalon S. CFTR and lung homeostasis. *Am J Physiol Lung Cell Mol Physiol* 2014;**307**:L917—23.

9. Stoltz DA, Meyerholz DK, Welsh MJ. Origins of cystic fibrosis lung disease. *N Engl J Med* 2015;**372**:351–62.
10. Aurora P, Carby M, Sweet S. Selection of cystic fibrosis patients for lung transplantation. *Curr Opin Pulm Med* 2008;**14**:589–94.
11. Dodge JA, Turck D. Cystic fibrosis: nutritional consequences and management. *Best Pract Res Clin Gastroenterol* 2006;**20**:531–46.
12. Bell SC, De Boeck K, Amaral MD. New pharmacological approaches for cystic fibrosis: promises, progress, pitfalls. *Pharmacol Ther* 2015;**145**:19–34.
13. Pettit RS, Fellner C. CFTR modulators for the treatment of cystic fibrosis. *P T* 2014;**39**:500–11.
14. Grosse SD, Schechter MS, Kulkarni R, et al. Models of comprehensive multidisciplinary care for individuals in the United States with genetic disorders. *Pediatrics* 2009;**123**:407–12.
15. Chopra R, Paul L, Manickam R, et al. Efficacy and adverse effects of drugs used to treat adult cystic fibrosis. *Expert Opin Drug Saf* 2015;**14**:401–11.
16. Cystic fibrosis foundation. http://www.cff.org; [accessed 01.04.2015].
17. Kerem B, Rommens JM, Buchanan JA, et al. Identification of the cystic fibrosis gene: genetic analysis. *Science* 1989;**245**:1073–80.
18. Riordan JR, Rommens JM, Kerem B, et al. Identification of the cystic fibrosis gene: cloning and characterization of complementary DNA. *Science* 1989;**245**:1066–73.
19. Rommens JM, Iannuzzi MC, Kerem B, et al. Identification of the cystic fibrosis gene: chromosome walking and jumping. *Science* 1989;**245**:1059–65.
20. Schwiebert EM, Egan ME, Hwang TH, et al. CFTR regulates outwardly rectifying chloride channels through an autocrine mechanism involving ATP. *Cell* 1995;**81**:1063–73.
21. Stutts MJ, Canessa CM, Olsen JC, et al. CFTR as a camp-dependent regulator of sodium channels. *Science* 1995;**269**:847–50.
22. Jih KY, Hwang TC. Nonequilibrium gating of CFTR on an equilibrium theme. *Physiology (Bethesda)* 2012;**27**:351–61.
23. Linsdell P. Functional architecture of the CFTR chloride channel. *Mol Membr Biol* 2014;**31**:1–16.
24. The sick kids cystic fibrosis mutation database. http://www.genet.sickkids.on.ca/cftr; [accessed 01.04.2015].
25. De Boeck K, Zolin A, Cuppens H, et al. The relative frequency of CFTR mutation classes in European patients with cystic fibrosis. *J Cyst Fibros* 2014;**13**:403–9.
26. Boyle MP, De Boeck K. A new era in the treatment of cystic fibrosis: correction of the underlying CFTR defect. *Lancet Respir Med* 2013;**1**:158–63.
27. Cheng SH, Gregory RJ, Marshall J, et al. Defective intracellular transport and processing of CFTR is the molecular basis of most cystic fibrosis. *Cell* 1990;**63**:827–34.
28. Welsh MJ, Smith AE. Molecular mechanisms of CFTR chloride channel dysfunction in cystic fibrosis. *Cell* 1993;**73**:1251–4.
29. Hamosh A, Rosenstein BJ, Cutting GR. CFTR nonsense mutations G542X and W1282X associated with severe reduction of CFTR mRNA in nasal epithelial cells. *Hum Mol Genet* 1992;**1**:542–4.
30. Chu CS, Trapnell BC, Curristin S, et al. Genetic basis of variable exon 9 skipping in cystic fibrosis transmembrane conductance regulator mRNA. *Nat Genet* 1993;**3**:151–6.
31. Strong TV, Wilkinson DJ, Mansoura MK, et al. Expression of an abundant alternatively spliced form of the cystic fibrosis transmembrane conductance regulator (CFTR) gene is not associated with a camp-activated chloride conductance. *Hum Mol Genet* 1993;**2**:225–30.
32. Delaney SJ, Rich DP, Thomson SA, et al. Cystic fibrosis transmembrane conductance regulator splice variants are not conserved and fail to produce chloride channels. *Nat Genet* 1993;**4**:426–31.
33. Groman JD, Hefferon TW, Casals T, et al. Variation in a repeat sequence determines whether a common variant of the cystic fibrosis transmembrane conductance regulator gene is pathogenic or benign. *Am J Hum Genet* 2004;**74**:176–9.
34. Trescott L, Holcomb J, Spellmon N, et al. Targeting the root cause of cystic fibrosis. *Curr Drug Targets* 2014.
35. Siva K, Covello G, Denti MA. Exon-skipping antisense oligonucleotides to correct missplicing in neurogenetic diseases. *Nucleic Acid Ther* 2014;**24**:69–86.
36. Burke W, Aitken ML, Chen SH, et al. Variable severity of pulmonary disease in adults with identical cystic fibrosis mutations. *Chest* 1992;**102**:506–9.
37. Durno C, Corey M, Zielenski J, et al. Genotype and phenotype correlations in patients with cystic fibrosis and pancreatitis. *Gastroenterology* 2002;**123**:1857–64.
38. Guillot L, Beucher J, Tabary O, et al. Lung disease modifier genes in cystic fibrosis. *Int J Biochem Cell Biol* 2014;**52**:83–93.
39. Ravi Kanth V, Nageshwar Reddy D. Genetics of acute and chronic pancreatitis: an update. *World J Gastrointest Pathophysiol* 2014;**5**:427–37.
40. Collaco JM, Vanscoy L, Bremer L, et al. Interactions between secondhand smoke and genes that affect cystic fibrosis lung disease. *JAMA* 2008;**299**:417–24.
41. Hampton TH, Green DM, Cutting GR, et al. The microbiome in pediatric cystic fibrosis patients: the role of shared environment suggests a window of intervention. *Microbiome* 2014;**2**:14.
42. Sonneville F, Ruffin M, Guillot L, et al. New insights about miRNAs in cystic fibrosis. *Am J Pathol* 2015;**185**:897–908.
43. Parad RB, Comeau AM, Dorkin HL, et al. Sweat testing infants detected by cystic fibrosis newborn screening. *J Pediatr* 2005;**147**:S69–72.
44. Sontag MK, Hammond KB, Zielenski J, et al. Two-tiered immunoreactive trypsinogen-based newborn screening for cystic fibrosis in Colorado: screening efficacy and diagnostic outcomes. *J Pediatr* 2005;**147**:S83–8.
45. Borowitz D, Parad RB, Sharp JK, et al. Cystic fibrosis foundation practice guidelines for the management of infants with cystic fibrosis transmembrane conductance regulator-related metabolic syndrome during the first two years of life and beyond. *J Pediatr* 2009;**155**:S106–16.
46. Ren CL, Desai H, Platt M, et al. Clinical outcomes in infants with cystic fibrosis transmembrane conductance regulator (CFTR) related metabolic syndrome. *Pediatr Pulmonol* 2011;**46**:1079–84.
47. Ramos MD, Trujillano D, Olivar R, et al. Extensive sequence analysis of CFTR, SCNN1A, SCNN1B, SCNN1G and SERPINA1 suggests an oligogenic basis for cystic fibrosis-like phenotypes. *Clin Genet* 2014;**86**:91–5.

48. Langlois S, Benn P, Wilkins-Haug L. Current controversies in prenatal diagnosis 4: Pre-conception expanded carrier screening should replace all current prenatal screening for specific single gene disorders. *Prenat Diagn* 2015;**35**:23–8.
49. Lazarin GA, Haque IS, Nazareth S, et al. An empirical estimate of carrier frequencies for 400+ causal mendelian variants: results from an ethnically diverse clinical sample of 23,453 individuals. *Genet Med* 2013;**15**:178–86.
50. Edwards JG, Feldman G, Goldberg J, et al. Expanded carrier screening in reproductive medicine—points to consider: a joint statement of the American College of Medical Genetics and Genomics, American College of Obstetricians and Gynecologists, National Society of Genetic Counselors, Perinatal Quality Foundation, and Society for Maternal-Fetal Medicine. *Obstet Gynecol* 2015;**125**:653–62.
51. Watson MS, Cutting GR, Desnick RJ, et al. Cystic fibrosis population carrier screening: 2004 revision of American College of Medical Genetics mutation panel. *Genet Med* 2004;**6**:387–91.
52. Langfelder-Schwind E, Karczeski B, Strecker MN, et al. Molecular testing for cystic fibrosis carrier status practice guidelines: recommendations of the National Society of Genetic Counselors. *J Genet Couns* 2014;**23**:5–15.
53. Lyon E, Schrijver I, Weck KE, et al. Molecular genetic testing for cystic fibrosis: laboratory performance on the College of American Pathologists external proficiency surveys. *Genet Med* 2015;**17**:219–25.
54. Zvereff VV, Faruki H, Edwards M, et al. Cystic fibrosis carrier screening in a North American population. *Genet Med* 2014;**16**:539–46.
55. Farrar MA, Kiernan MC. The genetics of spinal muscular atrophy: progress and challenges. *Neurother* 2014.
56. Arnold WD, Mo X, Kolb SJ, et al. The motor neuron response to SMN1 deficiency in spinal muscular atrophy. *Muscle Nerve* 2014;**50**:457–8.
57. Brzustowicz LM, Lehner T, Castilla LH, et al. Genetic mapping of chronic childhood-onset spinal muscular atrophy to chromosome 5q11.2-13.3. *Nature* 1990;**344**:540–1.
58. Lefebvre S, Burglen L, Reboullet S, et al. Identification and characterization of a spinal muscular atrophy-determining gene. *Cell* 1995;**80**:155–65.
59. Melki J, Abdelhak S, Sheth P, et al. Gene for chronic proximal spinal muscular atrophies maps to chromosome 5q. *Nature* 1990;**344**:767–8.
60. Liu Q, Dreyfuss G. A novel nuclear structure containing the survival of motor neurons protein. *EMBO J* 1996;**15**:3555–65.
61. Ding G, Zhang Z, Chopp M, et al. MRI evaluation of BBB disruption after adjuvant AcSDKP treatment of stroke with tPA in rat. *Neuroscience* 2014;**271**:1–8.
62. Battle DJ, Kasim M, Yong J, et al. The SMN complex: an assembly machine for RNPs. *Cold Spring Harb Symp Quant Biol* 2006;**71**:313–20.
63. Pellizzoni L, Yong J, Dreyfuss G. Essential role for the SMN complex in the specificity of SNRNP assembly. *Science* 2002;**298**:1775–9.
64. Schrank B, Gotz R, Gunnersen JM, et al. Inactivation of the survival motor neuron gene, a candidate gene for human spinal muscular atrophy, leads to massive cell death in early mouse embryos. *Proc Natl Acad Sci USA* 1997;**94**:9920–5.
65. Prior TW, Russman BS. Spinal muscular atrophy. In: Pagon RA, Adam MP, Ardinger HH, et al., editors. *Genereviews(r), vol.* Seattle (WA); 1993.
66. Jedrzejowska M, Gos M, Zimowski JG, et al. Novel point mutations in survival motor neuron 1 gene expand the spectrum of phenotypes observed in spinal muscular atrophy patients. *Neuromuscul Disord* 2014;**24**:617–23.
67. Parsons DW, McAndrew PE, Iannaccone ST, et al. Intragenic telAMN mutations: frequency, distribution, evidence of a founder effect, and modification of the spinal muscular atrophy phenotype by cenSMN copy number. *Am J Hum Genet* 1998;**63**:1712–23.
68. Lorson CL, Hahnen E, Androphy EJ, et al. A single nucleotide in the SMN gene regulates splicing and is responsible for spinal muscular atrophy. *Proc Natl Acad Sci USA* 1999;**96**:6307–11.
69. Seng CO, Magee C, Young PJ, et al. The SMN structure reveals its crucial role in SNRNP assembly. *Hum Mol Genet* 2015;**24**:2138–46.
70. Taylor JE, Thomas NH, Lewis CM, et al. Correlation of SMNt and SNMc gene copy number with age of onset and survival in spinal muscular atrophy. *Eur J Hum Genet* 1998;**6**:467–74.
71. Mailman MD, Heinz JW, Papp AC, et al. Molecular analysis of spinal muscular atrophy and modification of the phenotype by SMN2. *Genet Med* 2002;**4**:20–6.
72. Prior TW, Krainer AR, Hua Y, et al. A positive modifier of spinal muscular atrophy in the SMN2 gene. *Am J Hum Genet* 2009;**85**:408–13.
73. Jedrzejowska M, Borkowska J, Zimowski J, et al. Unaffected patients with a homozygous absence of the SMN1 gene. *Eur J Hum Genet* 2008;**16**:930–4.
74. Castro D, Iannaccone ST. Spinal muscular atrophy: therapeutic strategies. *Curr Treat Options Neurol* 2014;**16**:316.
75. Yee JK, Lin RJ. Antisense oligonucleotides shed new light on the pathogenesis and treatment of spinal muscular atrophy. *Mol Ther* 2012;**20**:8–10.
76. Duque SI, Arnold WD, Odermatt P, et al. A large animal model of spinal muscular atrophy and correction of phenotype. *Ann Neurol* 2015;**77**:399–414.
77. Frattini E, Ruggieri M, Salani S, et al. Pluripotent stem cell-based models of spinal muscular atrophy. *Mol Cell Neurosci* 2015;**64**:44–50.
78. Zanetta C, Nizzardo M, Simone C, et al. Molecular therapeutic strategies for spinal muscular atrophies: current and future clinical trials. *Clin Ther* 2014;**36**:128–40.
79. Nurputra DK, Lai PS, Harahap NI, et al. Spinal muscular atrophy: from gene discovery to clinical trials. *Ann Hum Genet* 2013;**77**:435–63.
80. Prior TW. Spinal muscular atrophy: newborn and carrier screening. *Obstet Gynecol Clin North Am* 2010;**37**:23–36. Table of Contents.
81. van der Steege G, Grootscholten PM, van der Vlies P, et al. PCR-based DNA test to confirm clinical diagnosis of autosomal recessive spinal muscular atrophy. *Lancet* 1995;**345**:985–6.
82. Prior TW. Carrier screening for spinal muscular atrophy. *Genet Med* 2008;**10**:840–2.
83. ACOG committee opinion no. 432: Spinal muscular atrophy. *Obstet Gynecol* 2009;**113**:1194–6.
84. The hereditary hearing loss page. http://hereditaryhearingloss.org; [accessed 19.04.2015].

85. Smith RJH, Van Camp G. Nonsyndromic hearing loss and deafness, DFNB1. In: Pagon RA, Adam MP, Ardinger HH, et al., editors. *Genereviews(r), vol. Seattle (WA)*; 1993.
86. Duman D, Tekin M. Autosomal recessive nonsyndromic deafness genes: a review. *Front Biosci (Landmark Ed)* 2012;**17**:2213–36.
87. Morton CC, Nance WE. Newborn hearing screening—a silent revolution. *N Engl J Med* 2006;**354**:2151–64.
88. Alford RL, Arnos KS, Fox M, et al. American College of Medical Genetics and Genomics guideline for the clinical evaluation and etiologic diagnosis of hearing loss. *Genet Med* 2014;**16**:347–55.
89. Guilford P, Ben Arab S, Blanchard S, et al. A non-syndrome form of neurosensory, recessive deafness maps to the pericentromeric region of chromosome 13q. *Nat Genet* 1994;**6**:24–8.
90. Mignon C, Fromaget C, Mattei MG, et al. Assignment of connexin 26 (GJB2) and 46 (GJA3) genes to human chromosome 13q11—>q12 and mouse chromosome 14D1-E1 by in situ hybridization. *Cytogenet Cell Genet* 1996;**72**:185–6.
91. Kelsell DP, Dunlop J, Stevens HP, et al. Connexin 26 mutations in hereditary non-syndromic sensorineural deafness. *Nature* 1997;**387**:80–3.
92. Segretain D, Falk MM. Regulation of connexin biosynthesis, assembly, gap junction formation, and removal. *Biochim Biophys Acta* 2004;**1662**:3–21.
93. Milks LC, Kumar NM, Houghten R, et al. Topology of the 32-kd liver gap junction protein determined by site-directed antibody localizations. *EMBO J* 1988;**7**:2967–75.
94. Lee JR, White TW. Connexin-26 mutations in deafness and skin disease. *Expert Rev Mol Med* 2009;**11**:e35.
95. Kumar NM, Gilula NB. The gap junction communication channel. *Cell* 1996;**84**:381–8.
96. Jagger DJ, Forge A. Compartmentalized and signal-selective gap junctional coupling in the hearing cochlea. *J Neurosci* 2006;**26**:1260–8.
97. Hoang Dinh E, Ahmad S, Chang Q, et al. Diverse deafness mechanisms of connexin mutations revealed by studies using in vitro approaches and mouse models. *Brain Res* 2009;**1277**:52–69.
98. Leiden open variation database (gjb2). https://research.cchmc.org/LOVD2/home.php?select_db=GJB2; [accessed 19.05.2015].
99. Petersen MB, Willems PJ. Non-syndromic, autosomal-recessive deafness. *Clin Genet* 2006;**69**:371–92.
100. Chan DK, Chang KW. GJB2-associated hearing loss: systematic review of worldwide prevalence, genotype, and auditory phenotype. *Laryngoscope* 2014;**124**:E34–53.
101. Morell RJ, Kim HJ, Hood LJ, et al. Mutations in the connexin 26 gene (GJB2) among Ashkenazi Jews with nonsyndromic recessive deafness. *N Engl J Med* 1998;**339**:1500–5.
102. Abe S, Usami S, Shinkawa H, et al. Prevalent connexin 26 gene (GJB2) mutations in Japanese. *J Med Genet* 2000;**37**:41–3.
103. del Castillo I, Villamar M, Moreno-Pelayo MA, et al. A deletion involving the connexin 30 gene in nonsyndromic hearing impairment. *N Engl J Med* 2002;**346**:243–9.
104. Rodriguez-Paris J, Tamayo ML, Gelvez N, et al. Allele-specific impairment of GJB2 expression by GJB6 deletion del(GJB6-D13S1854). *PLoS ONE* 2011;**6**:e21665.
105. Brown KK, Rehm HL. Molecular diagnosis of hearing loss. *Curr Protoc Hum Genet* 2012. Chapter 9:Unit 9 16.
106. Denoyelle F, Lina-Granade G, Plauchu H, et al. Connexin 26 gene linked to a dominant deafness. *Nature* 1998;**393**:319–20.
107. Smith RJH, Ranum PT, Van Camp G. Nonsyndromic hearing loss and deafness, DFNA3. In: Pagon RA, Adam MP, Ardinger HH, et al., editors. *Genereviews(r), vol. Seattle (WA)*; 1993.
108. Avshalumova L, Fabrikant J, Koriakos A. Overview of skin diseases linked to connexin gene mutations. *Int J Dermatol* 2014;**53**:192–205.
109. Schrauwen I, Sommen M, Corneveaux JJ, et al. A sensitive and specific diagnostic test for hearing loss using a microdroplet PCR-based approach and next generation sequencing. *Am J Med Genet A* 2013;**161A**:145–52.
110. Vona B, Muller T, Nanda I, et al. Targeted next-generation sequencing of deafness genes in hearing-impaired individuals uncovers informative mutations. *Genet Med* 2014;**16**:945–53.
111. Xing G, Yao J, Wu B, et al. Identification of OSBP12 as a novel candidate gene for progressive nonsyndromic hearing loss by whole-exome sequencing. *Genet Med* 2015;**17**:210–8.
112. Black J, Hickson L, Black B, et al. Prognostic indicators in paediatric cochlear implant surgery: a systematic literature review. *Cochlear Implants Int* 2011;**12**:67–93.
113. Huntington G. On chorea. *Med Surg Rep* 1872;**26**:317–21.
114. Dayalu P, Albin RL. Huntington disease: pathogenesis and treatment. *Neurol Clin* 2015;**33**:101–14.
115. Clabough EB. Huntington's disease: the past, present, and future search for disease modifiers. *Yale J Biol Med* 2013;**86**:217–33.
116. Warby SC, Graham RK, Hayden MR, et al. Huntington disease. In: Pagon RA, Adam MP, Ardinger HH, editors. *Genereviews(r), vol. Seattle (WA)*; 1993.
117. Gusella JF, Wexler NS, Conneally PM, et al. A polymorphic DNA marker genetically linked to Huntington's disease. *Nature* 1983;**306**:234–8.
118. The Huntington's disease collaborative research group. A novel gene containing a trinucleotide repeat that is expanded and unstable on Huntington's disease chromosomes. *Cell* 1993;**72**:971–83.
119. Kremer B, Goldberg P, Andrew SE, et al. A worldwide study of the Huntington's disease mutation. The sensitivity and specificity of measuring CAG repeats. *N Engl J Med* 1994;**330**:1401–6.
120. Budworth H, McMurray CT. A brief history of triplet repeat diseases. *Methods Mol Biol* 2013;**1010**:3–17.
121. Iyer RR, Pluciennik A, Napierala M, et al. DNA triplet repeat expansion and mismatch repair. *Annu Rev Biochem* 2015.
122. Andrew SE, Goldberg YP, Kremer B, et al. The relationship between trinucleotide (CAG) repeat length and clinical features of Huntington's disease. *Nat Genet* 1993;**4**:398–403.
123. Brinkman RR, Mezei MM, Theilmann J, et al. The likelihood of being affected with Huntington disease by a particular age, for a specific CAG size. *Am J Hum Genet* 1997;**60**:1202–10.
124. Wexler NS, Lorimer J, Porter J, et al. Venezuelan kindreds reveal that genetic and environmental factors modulate Huntington's disease age of onset. *Proc Natl Acad Sci USA* 2004;**101**:3498–503.
125. Lee JM, Ramos EM, Lee JH, et al. CAG repeat expansion in Huntington disease determines age at onset in a fully dominant fashion. *Neurology* 2012;**78**:690–5.
126. Arning L, Epplen JT. Genetic modifiers in Huntington's disease: fiction or fact? *Neurogenetics* 2013;**14**:171–2.

127. Goldberg YP, Kremer B, Andrew SE, et al. Molecular analysis of new mutations for Huntington's disease: intermediate alleles and sex of origin effects. *Nat Genet* 1993;**5**:174–9.
128. Semaka A, Kay C, Belfroid RD, et al. A new mutation for Huntington disease following maternal transmission of an intermediate allele. *Eur J Med Genet* 2015;**58**:28–30.
129. Chong SS, Almqvist E, Telenius H, et al. Contribution of DNA sequence and CAG size to mutation frequencies of intermediate alleles for Huntington disease: evidence from single sperm analyses. *Hum Mol Genet* 1997;**6**:301–9.
130. Goldberg YP, McMurray CT, Zeisler J, et al. Increased instability of intermediate alleles in families with sporadic Huntington disease compared to similar sized intermediate alleles in the general population. *Hum Mol Genet* 1995;**4**:1911–8.
131. Pinto RM, Dragileva E, Kirby A, et al. Mismatch repair genes MLH1 and MLH3 modify CAG instability in Huntington's disease mice: genome-wide and candidate approaches. *PLoS Genet* 2013;**9**:e1003930.
132. Tome S, Manley K, Simard JP, et al. MSH3 polymorphisms and protein levels affect CAG repeat instability in Huntington's disease mice. *PLoS Genet* 2013;**9**:e1003280.
133. Leeflang EP, Zhang L, Tavare S, et al. Single sperm analysis of the trinucleotide repeats in the Huntington's disease gene: quantification of the mutation frequency spectrum. *Hum Mol Genet* 1995;**4**:1519–26.
134. Warby SC, Montpetit A, Hayden AR, et al. CAG expansion in the Huntington disease gene is associated with a specific and targetable predisposing haplogroup. *Am J Hum Genet* 2009;**84**:351–66.
135. Kremer B, Almqvist E, Theilmann J, et al. Sex-dependent mechanisms for expansions and contractions of the CAG repeat on affected Huntington disease chromosomes. *Am J Hum Genet* 1995;**57**:343–50.
136. Nahhas FA, Garbern J, Krajewski KM, et al. Juvenile onset Huntington disease resulting from a very large maternal expansion. *Am J Med Genet A* 2005;**137A**:328–31.
137. Hoogeveen AT, Willemsen R, Meyer N, et al. Characterization and localization of the Huntington disease gene product. *Hum Mol Genet* 1993;**2**:2069–73.
138. De Rooij KE, Dorsman JC, Smoor MA, et al. Subcellular localization of the Huntington's disease gene product in cell lines by immunofluorescence and biochemical subcellular fractionation. *Hum Mol Genet* 1996;**5**:1093–9.
139. Gutekunst CA, Levey AI, Heilman CJ, et al. Identification and localization of Huntingtin in brain and human lymphoblastoid cell lines with anti-fusion protein antibodies. *Proc Natl Acad Sci USA* 1995;**92**:8710–4.
140. Zeitlin S, Liu JP, Chapman DL, et al. Increased apoptosis and early embryonic lethality in mice nullizygous for the Huntington's disease gene homologue. *Nat Genet* 1995;**11**:155–63.
141. Velier J, Kim M, Schwarz C, et al. Wild-type and mutant huntingtins function in vesicle trafficking in the secretory and endocytic pathways. *Exp Neurol* 1998;**152**:34–40.
142. Trottier Y, Devys D, Imbert G, et al. Cellular localization of the Huntington's disease protein and discrimination of the normal and mutated form. *Nat Genet* 1995;**10**:104–10.
143. Li SH, Li XJ. Huntingtin-protein interactions and the pathogenesis of Huntington's disease. *Trends Genet* 2004;**20**:146–54.
144. Schilling G, Sharp AH, Loev SJ, et al. Expression of the Huntington's disease (IT15) protein product in HD patients. *Hum Mol Genet* 1995;**4**:1365–71.
145. Videnovic A. Treatment of Huntington disease. *Curr Treat Options Neurol* 2013;**15**:424–38.
146. Zielonka D, Mielcarek M, Landwehrmeyer GB. Update on Huntington's disease: advances in care and emerging therapeutic options. *Parkinsonism Relat Disord* 2015;**21**:169–78.
147. Kay C, Skotte NH, Southwell AL, et al. Personalized gene silencing therapeutics for Huntington disease. *Clin Genet* 2014;**86**:29–36.
148. Skotte NH, Southwell AL, Ostergaard ME, et al. Allele-specific suppression of mutant huntingtin using antisense oligonucleotides: providing a therapeutic option for all Huntington disease patients. *PLoS ONE* 2014;**9**:e107434.
149. Hu J, Liu J, Yu D, et al. Exploring the effect of sequence length and composition on allele-selective inhibition of human huntingtin expression by single-stranded silencing RNAs. *Nucleic Acid Ther* 2014;**24**:199–209.
150. Bean L, Bayrak-Toydemir P. American College of Medical Genetics and Genomics standards and guidelines for clinical genetics laboratories, 2014 edition: Technical standards and guidelines for Huntington disease. *Genet Med* 2014;**16**:e2.
151. Vnencak-Jones CL. Fluorescence PCR and Genescan analysis for the detection of CAG repeat expansions associated with Huntington's disease. *Methods Mol Biol* 2003;**217**:101–8.
152. Went L. Ethical issues policy statement on Huntington's disease molecular genetics predictive test. International Huntington Association and World Federation of Neurology. *J Med Genet* 1990;**27**:34–8.
153. Hayden MR, Bloch M, Fahy M. Predictive testing for Huntington's disease using linked DNA markers. *N Engl J Med* 1988;**319**:583–4.
154. Almqvist E, Adam S, Bloch M, et al. Risk reversals in predictive testing for Huntington disease. *Am J Hum Genet* 1997;**61**:945–52.
155. International Huntington Association and the World Federation of Neurology Research Group on Huntington's Chorea. Guidelines for the molecular genetics predictive test in Huntington's disease. *J Med Genet* 1994;**31**:555–9.
156. International Huntington Association (IHA) and the World Federation of Neurology (WFN) Research Group on Huntington's Chorea. Guidelines for the molecular genetics predictive test in Huntington's disease. *Neurology* 1994;**44**:1533–6.
157. Ulm JE, Vnencak-Jones CL, Bosque P. Research on familial Creutzfeldt-Jakob disease (FCJD) resulting in presymptomatic testing: implications for the human genome project. *J Genet Couns* 1993;**2**:9–15.
158. Goizet C, Lesca G, Durr A. Presymptomatic testing in Huntington's disease and autosomal dominant cerebellar ataxias. *Neurology* 2002;**59**:1330–6.
159. MacLeod R, Tibben A, Frontali M, et al. Recommendations for the predictive genetic test in Huntington's disease. *Clin Genet* 2013;**83**:221–31.
160. Paulsen JS, Hoth KF, Nehl C, et al. Critical periods of suicide risk in Huntington's disease. *Am J Psychiatry* 2005;**162**:725–31.
161. Booij SJ, Tibben A, Engberts DP, et al. Thinking about the end of life: a common issue for patients with Huntington's disease. *J Neurol* 2014;**261**:2184–91.

162. Harper PS, Newcombe RG. Age at onset and life table risks in genetic counselling for Huntington's disease. *J Med Genet* 1992;**29**:239–42.
163. Trembath MK, Tassicker RJ, Collins VR, et al. Fifteen years of experience in predictive testing for Huntington disease at a single testing center in Victoria, Australia. *Genet Med* 2006;**8**:673–80.
164. Tur-Kaspa I, Jeelani R, Doraiswamy PM. Preimplantation genetic diagnosis for inherited neurological disorders. *Nat Rev Neurol* 2014;**10**:417–24.
165. Sermon K, De Rijcke M, Lissens W, et al. Preimplantation genetic diagnosis for Huntington's disease with exclusion testing. *Eur J Hum Genet* 2002;**10**:591–8.
166. van Rij MC, de Koning Gans PA, van Belzen MJ, et al. The uptake and outcome of prenatal and pre-implantation genetic diagnosis for Huntington's disease in the Netherlands (1998-2008). *Clin Genet* 2014;**85**:87–95.
167. Tinkle BT, Saal HM. Health supervision for children with Marfan syndrome. *Pediatrics* 2013;**132**:e1059–72.
168. Loeys BL, Dietz HC, Braverman AC, et al. The revised Ghent nosology for the Marfan syndrome. *J Med Genet* 2010;**47**:476–85.
169. Willis L, Roosevelt GE, Yetman AT. Comparison of clinical characteristics and frequency of adverse outcomes in patients with Marfan syndrome diagnosed in adulthood versus childhood. *Pediatr Cardiol* 2009;**30**:289–92.
170. Amado M, Calado MA, Ferreira R, et al. Neonatal Marfan syndrome: a successful early multidisciplinary approach. *BMJ Case Rep* 2014:2014.
171. Dolynchuk KN. Plasma fibronectin levels as predictive of survival in major thermal injury. *Burns* 1991;**17**:185–7.
172. Ripperger T, Troger HD, Schmidtke J. The genetic message of a sudden, unexpected death due to thoracic aortic dissection. *Forensic Sci Int* 2009;**187**:1–5.
173. Dietz HC, Cutting GR, Pyeritz RE, et al. Marfan syndrome caused by a recurrent de novo missense mutation in the fibrillin gene. *Nature* 1991;**352**:337–9.
174. Hollister DW, Godfrey M, Sakai LY, et al. Immunohistologic abnormalities of the microfibrillar-fiber system in the Marfan syndrome. *N Engl J Med* 1990;**323**:152–9.
175. Kainulainen K, Pulkkinen L, Savolainen A, et al. Location on chromosome 15 of the gene defect causing Marfan syndrome. *N Engl J Med* 1990;**323**:935–9.
176. Biery NJ, Eldadah ZA, Moore CS, et al. Revised genomic organization of FBN1 and significance for regulated gene expression. *Genomics* 1999;**56**:70–7.
177. Hubmacher D, El-Hallous EI, Nelea V, et al. Biogenesis of extracellular microfibrils: multimerization of the fibrillin-1 c terminus into bead-like structures enables self-assembly. *Proc Natl Acad Sci USA* 2008;**105**:6548–53.
178. Halper J, Kjaer M. Basic components of connective tissues and extracellular matrix: elastin, fibrillin, fibulins, fibrinogen, fibronectin, laminin, tenascins and thrombospondins. *Adv Exp Med Biol* 2014;**802**:31–47.
179. Ramirez F, Dietz HC. Fibrillin-rich microfibrils: structural determinants of morphogenetic and homeostatic events. *J Cell Physiol* 2007;**213**:326–30.
180. Jensen SA, Robertson IB, Handford PA. Dissecting the fibrillin microfibril: structural insights into organization and function. *Structure* 2012;**20**:215–25.
181. Goumans MJ, Liu Z, ten Dijke P. TGF-beta signaling in vascular biology and dysfunction. *Cell Res* 2009;**19**:116–27.
182. Kumar A, Agarwal S. Marfan syndrome: an eyesight of syndrome. *Meta Gene* 2014;**2**:96–105.
183. Collod-Beroud G, Le Bourdelles S, Ades L, et al. Update of the UMD-FBN1 mutation database and creation of an FBN1 polymorphism database. *Hum Mutat* 2003;**22**:199–208.
184. Faivre L, Collod-Beroud G, Loeys BL, et al. Effect of mutation type and location on clinical outcome in 1,013 probands with Marfan syndrome or related phenotypes and FBN1 mutations: an international study. *Am J Hum Genet* 2007;**81**:454–66.
185. Sipek Jr A, Grodecka L, Baxova A, et al. Novel FBN1 gene mutation and maternal germinal mosaicism as the cause of neonatal form of Marfan syndrome. *Am J Med Genet A* 2014;**164A**:1559–64.
186. Aubart M, Gross MS, Hanna N, et al. The clinical presentation of Marfan syndrome is modulated by expression of wild-type FBN1 allele. *Hum Mol Genet* 2015;**24**:2764–70.
187. Attenhofer Jost CH, Greutmann M, Connolly HM, et al. Medical treatment of aortic aneurysms in Marfan syndrome and other heritable conditions. *Curr Cardiol Rev* 2014;**10**:161–71.
188. Franken R, den Hartog AW, Radonic T, et al. Beneficial outcome of losartan therapy depends on type of FBN1 mutation in Marfan syndrome. *Circ Cardiovasc Genet* 2015;**8**:383–8.
189. Goldfinger JZ, Halperin JL, Marin ML, et al. Thoracic aortic aneurysm and dissection. *J Am Coll Cardiol* 2014;**64**:1725–39.
190. Radke RM, Baumgartner H. Diagnosis and treatment of Marfan syndrome: an update. *Heart* 2014;**100**:1382–91.
191. Lerner-Ellis JP, Aldubayan SH, Hernandez AL, et al. The spectrum of FBN1, TGFbetaR1, TGFbetaR2 and ACTA2 variants in 594 individuals with suspected Marfan syndrome, Loeys-Dietz syndrome or thoracic aortic aneurysms and dissections (TAAD). *Mol Genet Metab* 2014;**112**:171–6.
192. Gillis E, Kempers M, Salemink S, et al. An FBN1 deep intronic mutation in a familial case of Marfan syndrome: an explanation for genetically unsolved cases? *Hum Mutat* 2014;**35**:571–4.
193. Lebo MS, Baxter SM. New molecular genetic tests in the diagnosis of heart disease. *Clin Lab Med* 2014;**34**:137–56. vii–viii.
194. Lindsay ME, Dietz HC. Lessons on the pathogenesis of aneurysm from heritable conditions. *Nature* 2011;**473**:308–16.
195. Trump D, Farren B, Wooding C, et al. Clinical studies of multiple endocrine neoplasia type 1 (MEN1). *QJM* 1996;**89**:653–69.
196. Giusti F, Marini F, Brandi ML, et al. Multiple endocrine neoplasia type 1. In: Pagon RA, Adam MP, Ardinger HH, editors. *Genereviews(r)*, vol. Seattle (WA); 1993.
197. Thakker RV. Multiple endocrine neoplasia. *Medicine (Baltimore)* 2013;**41**:562–5.
198. Walls GV. Multiple endocrine neoplasia (MEN) syndromes. *Semin Pediatr Surg* 2014;**23**:96–101.
199. Larsson C, Skogseid B, Oberg K, et al. Multiple endocrine neoplasia type 1 gene maps to chromosome 11 and is lost in insulinoma. *Nature* 1988;**332**:85–7.
200. Nakamura Y, Larsson C, Julier C, et al. Localization of the genetic defect in multiple endocrine neoplasia type 1 within a small region of chromosome 11. *Am J Hum Genet* 1989;**44**:751–5.
201. Bystrom C, Larsson C, Blomberg C, et al. Localization of the MEN1 gene to a small region within chromosome 11q13 by deletion mapping in tumors. *Proc Natl Acad Sci USA* 1990;**87**:1968–72.

202. Courseaux A, Grosgeorge J, Gaudray P, et al. Definition of the minimal MEN1 candidate area based on a 5-mb integrated map of proximal 11q13. The European Consortium on MEN1, (GENEM 1; groupe d'etude des neoplasies endocriniennes multiples de type 1). *Genomics* 1996;**37**:354–65.
203. Pang JT, Lloyd SE, Wooding C, et al. Genetic mapping studies of 40 loci and 23 cosmids in chromosome 11p13-11q13, and exclusion of mu-calpain as the multiple endocrine neoplasia type 1 gene. *Hum Genet* 1996;**97**:732–41.
204. Thakker RV, Bouloux P, Wooding C, et al. Association of parathyroid tumors in multiple endocrine neoplasia type 1 with loss of alleles on chromosome 11. *N Engl J Med* 1989;**321**:218–24.
205. Thakker RV, Wooding C, Pang JT, et al. Linkage analysis of 7 polymorphic markers at chromosome 11p11.2-11q13 in 27 multiple endocrine neoplasia type 1 families. *Ann Hum Genet* 1993;**57**:17–25.
206. Friedman E, Sakaguchi K, Bale AE, et al. Clonality of parathyroid tumors in familial multiple endocrine neoplasia type 1. *N Engl J Med* 1989;**321**:213–8.
207. Emmert-Buck MR, Lubensky IA, Dong Q, et al. Localization of the multiple endocrine neoplasia type 1 (MEN1) gene based on tumor loss of heterozygosity analysis. *Cancer Res* 1997;**57**:1855–8.
208. Debelenko LV, Emmert-Buck MR, Manickam P, et al. Haplotype analysis defines a minimal interval for the multiple endocrine neoplasia type 1 (MEN1) gene. *Cancer Res* 1997;**57**:1039–42.
209. Chandrasekharappa SC, Guru SC, Manickam P, et al. Positional cloning of the gene for multiple endocrine neoplasia-type 1. *Science* 1997;**276**:404–7.
210. Lemmens I, Van de Ven WJ, Kas K, et al. Identification of the multiple endocrine neoplasia type 1 (MEN1) gene. The European Consortium on MEN1. *Hum Mol Genet* 1997;**6**:1177–83.
211. Guru SC, Goldsmith PK, Burns AL, et al. Menin, the product of the MEN1 gene, is a nuclear protein. *Proc Natl Acad Sci USA* 1998;**95**:1630–4.
212. Huang SC, Zhuang Z, Weil RJ, et al. Nuclear/cytoplasmic localization of the multiple endocrine neoplasia type 1 gene product, menin. *Lab Invest* 1999;**79**:301–10.
213. Matkar S, Thiel A, Hua X. Menin: a scaffold protein that controls gene expression and cell signaling. *Trends Biochem Sci* 2013;**38**:394–402.
214. Balogh K, Racz K, Patocs A, et al. Menin and its interacting proteins: elucidation of menin function. *Trends Endocrinol Metab* 2006;**17**:357–64.
215. Poisson A, Zablewska B, Gaudray P. Menin interacting proteins as clues toward the understanding of multiple endocrine neoplasia type 1. *Cancer Lett* 2003;**189**:1–10.
216. Lemos MC, Thakker RV. Multiple endocrine neoplasia type 1 (MEN1): analysis of 1336 mutations reported in the first decade following identification of the gene. *Hum Mutat* 2008;**29**:22–32.
217. Stalberg P, Grimfjard P, Santesson M, et al. Transfection of the multiple endocrine neoplasia type 1 gene to a human endocrine pancreatic tumor cell line inhibits cell growth and affects expression of junD, delta-like protein 1/preadipocyte factor-1, proliferating cell nuclear antigen, and QM/Jif-1. *J Clin Endocrinol Metab* 2004;**89**:2326–37.
218. Kim YS, Burns AL, Goldsmith PK, et al. Stable overexpression of MEN1 suppresses tumorigenicity of Ras. *Oncogene* 1999;**18**:5936–42.
219. Yumita W, Ikeo Y, Yamauchi K, et al. Suppression of insulin-induced AP-1 transactivation by menin accompanies inhibition of c-fos induction. *Int J Cancer* 2003;**103**:738–44.
220. Lin SY, Elledge SJ. Multiple tumor suppressor pathways negatively regulate telomerase. *Cell* 2003;**113**:881–9.
221. Miedlich S, Lohmann T, Schneyer U, et al. Familial isolated primary hyperparathyroidism—a multiple endocrine neoplasia type 1 variant? *Eur J Endocrinol* 2001;**145**:155–60.
222. Warner J, Epstein M, Sweet A, et al. Genetic testing in familial isolated hyperparathyroidism: unexpected results and their implications. *J Med Genet* 2004;**41**:155–60.
223. Hannan FM, Nesbit MA, Christie PT, et al. Familial isolated primary hyperparathyroidism caused by mutations of the MEN1 gene. *Nat Clin Pract Endocrinol Metab* 2008;**4**:53–8.
224. Brandi ML, Gagel RF, Angeli A, et al. Guidelines for diagnosis and therapy of MEN type 1 and type 2. *J Clin Endocrinol Metab* 2001;**86**:5658–71.
225. Guo SS, Sawicki MP. Molecular and genetic mechanisms of tumorigenesis in multiple endocrine neoplasia type-1. *Mol Endocrinol* 2001;**15**:1653–64.
226. Tham E, Grandell U, Lindgren E, et al. Clinical testing for mutations in the MEN1 gene in Sweden: a report on 200 unrelated cases. *J Clin Endocrinol Metab* 2007;**92**:3389–95.
227. Bergman L, Teh B, Cardinal J, et al. Identification of MEN1 gene mutations in families with MEN1 and related disorders. *Br J Cancer* 2000;**83**:1009–14.
228. Cavaco BM, Domingues R, Bacelar MC, et al. Mutational analysis of Portuguese families with multiple endocrine neoplasia type 1 reveals large germline deletions. *Clin Endocrinol (Oxf)* 2002;**56**:465–73.
229. Ellard S, Hattersley AT, Brewer CM, et al. Detection of an MEN1 gene mutation depends on clinical features and supports current referral criteria for diagnostic molecular genetic testing. *Clin Endocrinol (Oxf)* 2005;**62**:169–75.
230. Klein RD, Salih S, Bessoni J, et al. Clinical testing for multiple endocrine neoplasia type 1 in a DNA diagnostic laboratory. *Genet Med* 2005;**7**:131–8.
231. Thakker RV, Newey PJ, Walls GV, et al. Clinical practice guidelines for multiple endocrine neoplasia type 1 (MEN1). *J Clin Endocrinol Metab* 2012;**97**:2990–3011.
232. Moline J, Eng C. Multiple endocrine neoplasia type 2. In: Pagon RA, Adam MP, Ardinger HH, et al., editors. *Genereviews(r), vol. Seattle (WA)*; 1993.
233. Gardner E, Papi L, Easton DF, et al. Genetic linkage studies map the multiple endocrine neoplasia type 2 loci to a small interval on chromosome 10q11.2. *Hum Mol Genet* 1993;**2**:241–6.
234. Mole SE, Mulligan LM, Healey CS, et al. Localisation of the gene for multiple endocrine neoplasia type 2a to a 480 kb region in chromosome band 10q11.2. *Hum Mol Genet* 1993;**2**:247–52.
235. Mathew CG, Chin KS, Easton DF, et al. A linked genetic marker for multiple endocrine neoplasia type 2a on chromosome 10. *Nature* 1987;**328**:527–8.
236. Simpson NE, Kidd KK, Goodfellow PJ, et al. Assignment of multiple endocrine neoplasia type 2a to chromosome 10 by linkage. *Nature* 1987;**328**:528–30.

237. Mulligan LM, Kwok JB, Healey CS, et al. Germ-line mutations of the RET proto-oncogene in multiple endocrine neoplasia type 2a. *Nature* 1993;**363**:458—60.
238. Hofstra RM, Landsvater RM, Ceccherini I, et al. A mutation in the RET proto-oncogene associated with multiple endocrine neoplasia type 2b and sporadic medullary thyroid carcinoma. *Nature* 1994;**367**:375—6.
239. Carlson KM, Dou S, Chi D, et al. Single missense mutation in the tyrosine kinase catalytic domain of the RET protooncogene is associated with multiple endocrine neoplasia type 2b. *Proc Natl Acad Sci USA* 1994;**91**:1579—83.
240. Knowles PP, Murray-Rust J, Kjaer S, et al. Structure and chemical inhibition of the RET tyrosine kinase domain. *J Biol Chem* 2006;**281**:33577—87.
241. Durbec P, Marcos-Gutierrez CV, Kilkenny C, et al. GDNF signalling through the RET receptor tyrosine kinase. *Nature* 1996;**381**:789—93.
242. Robertson K, Mason I. The GDNF-RET signalling partnership. *Trends Genet* 1997;**13**:1—3.
243. Trupp M, Arenas E, Fainzilber M, et al. Functional receptor for GDNF encoded by the c-RET proto-oncogene. *Nature* 1996;**381**:785—9.
244. Krampitz GW, Norton JA. RET gene mutations (genotype and phenotype) of multiple endocrine neoplasia type 2 and familial medullary thyroid carcinoma. *Cancer* 2014;**120**:1920—31.
245. Santoro M, Carlomagno F, Romano A, et al. Activation of RET as a dominant transforming gene by germline mutations of MEN2A and MEN2B. *Science* 1995;**267**:381—3.
246. Edery P, Lyonnet S, Mulligan LM, et al. Mutations of the RET proto-oncogene in Hirschsprung's disease. *Nature* 1994;**367**:378—80.
247. Romeo G, Ronchetto P, Luo Y, et al. Point mutations affecting the tyrosine kinase domain of the RET proto-oncogene in Hirschsprung's disease. *Nature* 1994;**367**:377—8.
248. Margraf RL, Crockett DK, Krautscheid PM, et al. Multiple endocrine neoplasia type 2 RET protooncogene database: repository of MEN2-associated RET sequence variation and reference for genotype/phenotype correlations. *Hum Mutat* 2009;**30**:548—56.
249. Mosqueira M, Zeiger U, Forderer M, et al. Cardiac and respiratory dysfunction in Duchenne muscular dystrophy and the role of second messengers. *Med Res Rev* 2013;**33**:1174—213.
250. Diegoli M, Grasso M, Favalli V, et al. Diagnostic work-up and risk stratification in X-linked dilated cardiomyopathies caused by dystrophin defects. *J Am Coll Cardiol* 2011;**58**:925—34.
251. Wingeier K, Giger E, Strozzi S, et al. Neuropsychological impairments and the impact of dystrophin mutations on general cognitive functioning of patients with Duchenne muscular dystrophy. *J Clin Neurosci* 2011;**18**:90—5.
252. Bushby K, Finkel R, Birnkrant DJ, et al. Diagnosis and management of Duchenne muscular dystrophy, part 1: Diagnosis, and pharmacological and psychosocial management. *Lancet Neurol* 2010;**9**:77—93.
253. Bushby K, Finkel R, Birnkrant DJ, et al. Diagnosis and management of Duchenne muscular dystrophy, part 2: Implementation of multidisciplinary care. *Lancet Neurol* 2010;**9**:177—89.
254. Mendell JR, Shilling C, Leslie ND, et al. Evidence-based path to newborn screening for Duchenne muscular dystrophy. *Ann Neurol* 2012;**71**:304—13.
255. Scully MA, Farrell PM, Ciafaloni E, et al. Cystic fibrosis newborn screening: a model for neuromuscular disease screening? *Ann Neurol* 2015;**77**:189—97.
256. Hoogerwaard EM, Bakker E, Ippel PF, et al. Signs and symptoms of Duchenne muscular dystrophy and Becker muscular dystrophy among carriers in the Netherlands: a cohort study. *Lancet* 1999;**353**:2116—9.
257. Richards CS, Watkins SC, Hoffman EP, et al. Skewed X inactivation in a female MZ twin results in Duchenne muscular dystrophy. *Am J Hum Genet* 1990;**46**:672—81.
258. Verellen-Dumoulin C, Freund M, De Meyer R, et al. Expression of an X-linked muscular dystrophy in a female due to translocation involving Xp21 and non-random inactivation of the normal X chromosome. *Hum Genet* 1984;**67**:115—9.
259. Zatz M, Vianna-Morgante AM, Campos P, et al. Translocation (X;6) in a female with Duchenne muscular dystrophy: implications for the localisation of the DMD locus. *J Med Genet* 1981;**18**:442—7.
260. Juan-Mateu J, Rodriguez MJ, Nascimento A, et al. Prognostic value of X-chromosome inactivation in symptomatic female carriers of dystrophinopathy. *Orphanet J Rare Dis* 2012;**7**:82.
261. Soltanzadeh P, Friez MJ, Dunn D, et al. Clinical and genetic characterization of manifesting carriers of DMD mutations. *Neuromuscul Disord* 2010;**20**:499—504.
262. Bakker E, Hofker MH, Goor N, et al. Prenatal diagnosis and carrier detection of Duchenne muscular dystrophy with closely linked RFLPs. *Lancet* 1985;**1**:655—8.
263. Davies KE, Pearson PL, Harper PS, et al. Linkage analysis of two cloned DNA sequences flanking the Duchenne muscular dystrophy locus on the short arm of the human X chromosome. *Nucleic Acids Res* 1983;**11**:2303—12.
264. Francke U, Ochs HD, de Martinville B, et al. Minor Xp21 chromosome deletion in a male associated with expression of Duchenne muscular dystrophy, chronic granulomatous disease, retinitis pigmentosa, and Mcleod syndrome. *Am J Hum Genet* 1985;**37**:250—67.
265. Kunkel LM, Monaco AP, Middlesworth W, et al. Specific cloning of DNA fragments absent from the DNA of a male patient with an X chromosome deletion. *Proc Natl Acad Sci USA* 1985;**82**:4778—82.
266. Hoffman EP, Brown Jr RH, Kunkel LM. Dystrophin: the protein product of the Duchenne muscular dystrophy locus. *Cell* 1987;**51**:919—28.
267. Blake DJ, Weir A, Newey SE, et al. Function and genetics of dystrophin and dystrophin-related proteins in muscle. *Physiol Rev* 2002;**82**:291—329.
268. Gintjee TJ, Magh AS, Bertoni C. High throughput screening in Duchenne muscular dystrophy: from drug discovery to functional genomics. *Biology (Basel)* 2014;**3**:752—80.
269. Walsh CM. Grand challenges in cell death and survival: apoptosis vs. necroptosis. *Front Cell Dev Biol* 2014;**2**:3.
270. Kharraz Y, Guerra J, Pessina P, et al. Understanding the process of fibrosis in Duchenne muscular dystrophy. *Biomed Res Int* 2014;**2014**:965631.
271. Aartsma-Rus A, Van Deutekom JC, Fokkema IF, et al. Entries in the Leiden Duchenne muscular dystrophy mutation database: an overview of mutation types and paradoxical cases that confirm the reading-frame rule. *Muscle Nerve* 2006;**34**:135—44.
272. Yazaki M, Yoshida K, Nakamura A, et al. Clinical characteristics of aged Becker muscular dystrophy patients with onset after 30 years. *Eur Neurol* 1999;**42**:145—9.

273. Bushby KM, Gardner-Medwin D, Nicholson LV, et al. The clinical, genetic and dystrophin characteristics of Becker muscular dystrophy. Correlation of phenotype with genetic and protein abnormalities. *J Neurol* 1993;**240**:105–12.
274. Comi GP, Prelle A, Bresolin N, et al. Clinical variability in Becker muscular dystrophy. Genetic, biochemical and immunohistochemical correlates. *Brain* 1994;**117**(Pt 1):1–14.
275. Hoffman EP, Kunkel LM. Dystrophin abnormalities in Duchenne/Becker muscular dystrophy. *Neuron* 1989;**2**: 1019–29.
276. Takeshima Y, Yagi M, Okizuka Y, et al. Mutation spectrum of the dystrophin gene in 442 Duchenne/Becker muscular dystrophy cases from one Japanese referral center. *J Hum Genet* 2010;**55**:379–88.
277. Nicolas A, Raguenes-Nicol C, Ben Yaou R, et al. Becker muscular dystrophy severity is linked to the structure of dystrophin. *Hum Mol Genet* 2015;**24**:1267–79.
278. Feng J, Yan JY, Buzin CH, et al. Comprehensive mutation scanning of the dystrophin gene in patients with non-syndromic X-linked dilated cardiomyopathy. *J Am Coll Cardiol* 2002;**40**:1120–4.
279. Ferlini A, Sewry C, Melis MA, et al. X-linked dilated cardiomyopathy and the dystrophin gene. *Neuromuscul Disord* 1999; **9**:339–46.
280. Milasin J, Muntoni F, Severini GM, et al. A point mutation in the 5' splice site of the dystrophin gene first intron responsible for X-linked dilated cardiomyopathy. *Hum Mol Genet* 1996;**5**: 73–9.
281. Muntoni F, Cau M, Ganau A, et al. Brief report: deletion of the dystrophin muscle-promoter region associated with X-linked dilated cardiomyopathy. *N Engl J Med* 1993;**329**:921–5.
282. Ortiz-Lopez R, Li H, Su J, et al. Evidence for a dystrophin missense mutation as a cause of X-linked dilated cardiomyopathy. *Circulation* 1997;**95**:2434–40.
283. Finsterer J, Stollberger C. The heart in human dystrophinopathies. *Cardiology* 2003;**99**:1–19.
284. Mestroni L, Brun F, Spezzacatene A, et al. Genetic causes of dilated cardiomyopathy. *Prog Pediatr Cardiol* 2014;**37**:13–8.
285. Lalic T, Vossen RH, Coffa J, et al. Deletion and duplication screening in the DMD gene using MLPA. *Eur J Hum Genet* 2005;**13**:1231–4.
286. Nallamilli BR, Ankala A, Hegde M. Molecular diagnosis of Duchenne muscular dystrophy. *Curr Protoc Hum Genet* 2014; **83**. 9 25 1–9 9.
287. Wang Y, Yang Y, Liu J, et al. Whole dystrophin gene analysis by next-generation sequencing: a comprehensive genetic diagnosis of Duchenne and Becker muscular dystrophy. *Mol Genet Genomics* 2014;**289**:1013–21.
288. Grimm T, Kress W, Meng G, et al. Risk assessment and genetic counseling in families with Duchenne muscular dystrophy. *Acta Myol* 2012;**31**:179–83.
289. Ye Y, Yu P, Yong J, et al. Preimplantational genetic diagnosis and mutation detection in a family with duplication mutation of DMD gene. *Gynecol Obstet Invest* 2014;**78**:272–8.
290. Seto JT, Bengtsson NE, Chamberlain JS. Therapy of genetic disorders—novel therapies for Duchenne muscular dystrophy. *Curr Pediatr Rep* 2014;**2**:102–12.
291. Aoki Y, Yokota T, Wood MJ. Development of multiexon skipping antisense oligonucleotide therapy for Duchenne muscular dystrophy. *Biomed Res Int* 2013;**2013**:402369.
292. Hoffman EP, Connor EM. Orphan drug development in muscular dystrophy: update on two large clinical trials of dystrophin rescue therapies. *Discov Med* 2013;**16**:233–9.
293. Jarmin S, Kymalainen H, Popplewell L, et al. New developments in the use of gene therapy to treat Duchenne muscular dystrophy. *Expert Opin Biol Ther* 2014;**14**: 209–30.
294. Bertoni C. Emerging gene editing strategies for Duchenne muscular dystrophy targeting stem cells. *Front Physiol* 2014;**5**: 148.
295. Darras BT, Miller DT, Urion DK. Dystrophinopathies. In: Pagon RA, Adam MP, Ardinger HH, et al., editors. *Genereviews(r), vol. Seattle (WA)*; 1993.
296. Kornegay JN, Spurney CF, Nghiem PP, et al. Pharmacologic management of Duchenne muscular dystrophy: target identification and preclinical trials. *ILAR J* 2014;**55**:119–49.
297. Hill MK, Archibald AD, Cohen J, et al. A systematic review of population screening for fragile X syndrome. *Genet Med* 2010; **12**:396–410.
298. Martin JP, Bell J. A pedigree of mental defect showing sex-linkage. *J Neurol Psychiatry* 1943;**6**:154–7.
299. Lubs HA. A marker X chromosome. *Am J Hum Genet* 1969;**21**: 231–44.
300. Harrison CJ, Jack EM, Allen TD, et al. The fragile X: a scanning electron microscope study. *J Med Genet* 1983;**20**:280–5.
301. Hersh JH, Saul RA. Health supervision for children with fragile X syndrome. *Pediatrics* 2011;**127**:994–1006.
302. Bell MV, Hirst MC, Nakahori Y, et al. Physical mapping across the fragile X: hypermethylation and clinical expression of the fragile X syndrome. *Cell* 1991;**64**:861–6.
303. Coffee B, Zhang F, Warren ST, et al. Acetylated histones are associated with FMR1 in normal but not fragile X-syndrome cells. *Nat Genet* 1999;**22**:98–101.
304. Pieretti M, Zhang FP, Fu YH, et al. Absence of expression of the FMR-1 gene in fragile X syndrome. *Cell* 1991;**66**:817–22.
305. Pietrobono R, Tabolacci E, Zalfa F, et al. Molecular dissection of the events leading to inactivation of the FMR1 gene. *Hum Mol Genet* 2005;**14**:267–77.
306. Sutcliffe JS, Nelson DL, Zhang F, et al. DNA methylation represses FMR-1 transcription in fragile X syndrome. *Hum Mol Genet* 1992;**1**:397–400.
307. Pretto D, Yrigollen CM, Tang HT, et al. Clinical and molecular implications of mosaicism in FMR1 full mutations. *Front Genet* 2014;**5**:318.
308. Sherman SL, Jacobs PA, Morton NE, et al. Further segregation analysis of the fragile X syndrome with special reference to transmitting males. *Hum Genet* 1985;**69**:289–99.
309. Sherman SL, Morton NE, Jacobs PA, et al. The marker (X) syndrome: a cytogenetic and genetic analysis. *Ann Hum Genet* 1984;**48**:21–37.
310. Kremer EJ, Pritchard M, Lynch M, et al. Mapping of DNA instability at the fragile X to a trinucleotide repeat sequence p(CGG)n. *Science* 1991;**252**:1711–4.
311. Oberle I, Rousseau F, Heitz D, et al. Instability of a 550-base pair DNA segment and abnormal methylation in fragile X syndrome. *Science* 1991;**252**:1097–102.
312. Verkerk AJ, Pieretti M, Sutcliffe JS, et al. Identification of a gene (FMR-1) containing a CGG repeat coincident with a breakpoint cluster region exhibiting length variation in fragile X syndrome. *Cell* 1991;**65**:905–14.

313. Yu S, Pritchard M, Kremer E, et al. Fragile X genotype characterized by an unstable region of DNA. *Science* 1991;**252**:1179–81.
314. Ennis S, Murray A, Brightwell G, et al. Closely linked cis-acting modifier of expansion of the CGG repeat in high risk FMR1 haplotypes. *Hum Mutat* 2007;**28**:1216–24.
315. Lim JH, Booker AB, Fallon JR. Regulating fragile X gene transcription in the brain and beyond. *J Cell Physiol* 2005;**205**:170–5.
316. Santoro MR, Bray SM, Warren ST. Molecular mechanisms of fragile X syndrome: a twenty-year perspective. *Annu Rev Pathol* 2012;**7**:219–45.
317. Suhl JA, Chopra P, Anderson BR, et al. Analysis of FMRP mRNA target datasets reveals highly associated mRNAs mediated by g-quadruplex structures formed via clustered WGGA sequences. *Hum Mol Genet* 2014;**23**:5479–91.
318. Scharf SH, Jaeschke G, Wettstein JG, et al. Metabotropic glutamate receptor 5 as drug target for fragile X syndrome. *Curr Opin Pharmacol* 2015;**20**:124–34.
319. Lozano R, Hare EB, Hagerman RJ. Modulation of the GABAergic pathway for the treatment of fragile X syndrome. *Neuropsychiatr Dis Treat* 2014;**10**:1769–79.
320. Braat S, Kooy RF. Insights into GABAergic system deficits in fragile X syndrome lead to clinical trials. *Neuropharmacology* 2015;**88**:48–54.
321. Kunst CB, Warren ST. Cryptic and polar variation of the fragile X repeat could result in predisposing normal alleles. *Cell* 1994;**77**:853–61.
322. Snow K, Doud LK, Hagerman R, et al. Analysis of a CGG sequence at the FMR-1 locus in fragile X families and in the general population. *Am J Hum Genet* 1993;**53**:1217–28.
323. Monaghan KG, Lyon E, Spector EB. ACMG standards and guidelines for fragile X testing: a revision to the disease-specific supplements to the standards and guidelines for clinical genetics laboratories of the American College of Medical Genetics and Genomics. *Genet Med* 2013;**15**:575–86.
324. Alfaro MP, Cohen M, Vnencak-Jones CL. Maternal FMR1 premutation allele expansion and contraction in fraternal twins. *Am J Med Genet A* 2013;**161A**:2620–5.
325. McMurray CT. Mechanisms of trinucleotide repeat instability during human development. *Nat Rev Genet* 2010;**11**:786–99.
326. Gerhardt J, Zaninovic N, Zhan Q, et al. Cis-acting DNA sequence at a replication origin promotes repeat expansion to fragile X full mutation. *J Cell Biol* 2014;**206**:599–607.
327. Yrigollen CM, Durbin-Johnson B, Gane L, et al. AGG interruptions within the maternal FMR1 gene reduce the risk of offspring with fragile X syndrome. *Genet Med* 2012;**14**:729–36.
328. Yrigollen CM, Martorell L, Durbin-Johnson B, et al. AGG interruptions and maternal age affect FMR1 CGG repeat allele stability during transmission. *J Neurodev Disord* 2014;**6**:24.
329. Nolin SL, Sah S, Glicksman A, et al. Fragile X AGG analysis provides new risk predictions for 45-69 repeat alleles. *Am J Med Genet A* 2013;**161A**:771–8.
330. Nolin SL, Glicksman A, Ding X, et al. Fragile X analysis of 1112 prenatal samples from 1991 to 2010. *Prenat Diagn* 2011;**31**:925–31.
331. Nolin SL, Brown WT, Glicksman A, et al. Expansion of the fragile X CGG repeat in females with premutation or intermediate alleles. *Am J Hum Genet* 2003;**72**:454–64.
332. Mirkin SM. Expandable DNA repeats and human disease. *Nature* 2007;**447**:932–40.
333. Tassone F, Hagerman PJ, Hagerman RJ. Fragile X premutation. *J Neurodev Disord* 2014;**6**:22.
334. Sherman SL, Curnow EC, Easley CA, et al. Use of model systems to understand the etiology of fragile X-associated primary ovarian insufficiency (FXPOI). *J Neurodev Disord* 2014;**6**:26.
335. Wheeler AC, Bailey Jr DB, Berry-Kravis E, et al. Associated features in females with an FMR1 premutation. *J Neurodev Disord* 2014;**6**:30.
336. Besterman AD, Wilke SA, Mulligan TE, et al. Towards an understanding of neuropsychiatric manifestations in fragile X premutation carriers. *Future Neurol* 2014;**9**:227–39.
337. Hagerman RJ, Leehey M, Heinrichs W, et al. Intention tremor, Parkinsonism, and generalized brain atrophy in male carriers of fragile X. *Neurology* 2001;**57**:127–30.
338. Sellier C, Usdin K, Pastori C, et al. The multiple molecular facets of fragile X-associated tremor/ataxia syndrome. *J Neurodev Disord* 2014;**6**:23.
339. Lyon E, Laver T, Yu P, et al. A simple, high-throughput assay for fragile X expanded alleles using triple repeat primed PCR and capillary electrophoresis. *J Mol Diagn* 2010;**12**:505–11.
340. Filipovic-Sadic S, Sah S, Chen L, et al. A novel FMR1 PCR method for the routine detection of low abundance expanded alleles and full mutations in fragile X syndrome. *Clin Chem* 2010;**56**:399–408.
341. Chen L, Hadd AG, Sah S, et al. High-resolution methylation polymerase chain reaction for fragile X analysis: evidence for novel FMR1 methylation patterns undetected in Southern blot analyses. *Genet Med* 2011;**13**:528–38.
342. Tsafrir A, Altarescu G, Margalioth E, et al. PGD for fragile X syndrome: ovarian function is the main determinant of success. *Hum Reprod* 2010;**25**:2629–36.
343. Tassone F. Newborn screening for fragile X syndrome. *JAMA Neurol* 2014;**71**:355–9.
344. Laurvick CL, de Klerk N, Bower C, et al. Rett syndrome in Australia: a review of the epidemiology. *J Pediatr* 2006;**148**:347–52.
345. Rett A. On a unusual brain atrophy syndrome in hyperammonemia in childhood. *Wien Med Wochenschr* 1966;**116**:723–6.
346. Chahrour M, Zoghbi HY. The story of Rett syndrome: from clinic to neurobiology. *Neuron* 2007;**56**:422–37.
347. Sekul EA, Moak JP, Schultz RJ, et al. Electrocardiographic findings in Rett syndrome: an explanation for sudden death? *J Pediatr* 1994;**125**:80–2.
348. Archidiacono N, Lerone M, Rocchi M, et al. Rett syndrome: exclusion mapping following the hypothesis of germinal mosaicism for new X-linked mutations. *Hum Genet* 1991;**86**:604–6.
349. Curtis AR, Headland S, Lindsay S, et al. X chromosome linkage studies in familial Rett syndrome. *Hum Genet* 1993;**90**:551–5.
350. Ellison KA, Fill CP, Terwilliger J, et al. Examination of X chromosome markers in Rett syndrome: exclusion mapping with a novel variation on multilocus linkage analysis. *Am J Hum Genet* 1992;**50**:278–87.
351. Schanen NC, Dahle EJ, Capozzoli F, et al. A new Rett syndrome family consistent with X-linked inheritance expands the X chromosome exclusion map. *Am J Hum Genet* 1997;**61**:634–41.

352. Sirianni N, Naidu S, Pereira J, et al. Rett syndrome: confirmation of X-linked dominant inheritance, and localization of the gene to Xq28. *Am J Hum Genet* 1998;**63**:1552–8.
353. Amir RE, Van den Veyver IB, Wan M, et al. Rett syndrome is caused by mutations in X-linked MECP2, encoding methyl-CpG-binding protein 2. *Nat Genet* 1999;**23**:185–8.
354. Matijevic T, Knezevic J, Slavica M, et al. Rett syndrome: from the gene to the disease. *Eur Neurol* 2009;**61**:3–10.
355. Lewis JD, Meehan RR, Henzel WJ, et al. Purification, sequence, and cellular localization of a novel chromosomal protein that binds to methylated DNA. *Cell* 1992;**69**:905–14.
356. Nan X, Meehan RR, Bird A. Dissection of the methyl-CpG binding domain from the chromosomal protein MECP2. *Nucleic Acids Res* 1993;**21**:4886–92.
357. Nan X, Campoy FJ, Bird A. MECP2 is a transcriptional repressor with abundant binding sites in genomic chromatin. *Cell* 1997;**88**:471–81.
358. Adams VH, McBryant SJ, Wade PA, et al. Intrinsic disorder and autonomous domain function in the multifunctional nuclear protein, MECP2. *J Biol Chem* 2007;**282**:15057–64.
359. Klose RJ, Sarraf SA, Schmiedeberg L, et al. DNA binding selectivity of MECP2 due to a requirement for A/T sequences adjacent to methyl-CpG. *Mol Cell* 2005;**19**:667–78.
360. Jones PL, Veenstra GJ, Wade PA, et al. Methylated DNA and MECP2 recruit histone deacetylase to repress transcription. *Nat Genet* 1998;**19**:187–91.
361. Nan X, Ng HH, Johnson CA, et al. Transcriptional repression by the methyl-CpG-binding protein MECP2 involves a histone deacetylase complex. *Nature* 1998;**393**:386–9.
362. Christodoulou J, Grimm A, Maher T, et al. Rettbase: the IRSA MECP2 variation database-A new mutation database in evolution. *Hum Mutat* 2003;**21**:466–72.
363. Rettbase: Rettsyndrome.Org variation database. http://mecp2.chw.edu.au; [accessed 08.04.2015].
364. Smeets E, Terhal P, Casaer P, et al. Rett syndrome in females with CTS hot spot deletions: a disorder profile. *Am J Med Genet A* 2005;**132A**:117–20.
365. Trappe R, Laccone F, Cobilanschi J, et al. MECP2 mutations in sporadic cases of Rett syndrome are almost exclusively of paternal origin. *Am J Hum Genet* 2001;**68**:1093–101.
366. Wan M, Lee SS, Zhang X, et al. Rett syndrome and beyond: recurrent spontaneous and familial MECP2 mutations at CpG hotspots. *Am J Hum Genet* 1999;**65**:1520–9.
367. Ariani F, Hayek G, Rondinella D, et al. FOXG1 is responsible for the congenital variant of Rett syndrome. *Am J Hum Genet* 2008;**83**:89–93.
368. Bahi-Buisson N, Nectoux J, Girard B, et al. Revisiting the phenotype associated with FOXG1 mutations: two novel cases of congenital Rett variant. *Neurogenetics* 2010;**11**:241–9.
369. Evans JC, Archer HL, Colley JP, et al. Early onset seizures and Rett-like features associated with mutations in CDKL5. *Eur J Hum Genet* 2005;**13**:1113–20.
370. Mencarelli MA, Spanhol-Rosseto A, Artuso R, et al. Novel FOXG1 mutations associated with the congenital variant of Rett syndrome. *J Med Genet* 2010;**47**:49–53.
371. Scala E, Ariani F, Mari F, et al. CDKL5/STK9 is mutated in Rett syndrome variant with infantile spasms. *J Med Genet* 2005;**42**:103–7.
372. Tao J, Van Esch H, Hagedorn-Greiwe M, et al. Mutations in the X-linked cyclin-dependent kinase-like 5 (CDKL5/STK9) gene are associated with severe neurodevelopmental retardation. *Am J Hum Genet* 2004;**75**:1149–54.
373. Weaving LS, Christodoulou J, Williamson SL, et al. Mutations of CDKL5 cause a severe neurodevelopmental disorder with infantile spasms and mental retardation. *Am J Hum Genet* 2004;**75**:1079–93.
374. Heit JA, Silverstein MD, Mohr DN, et al. Risk factors for deep vein thrombosis and pulmonary embolism: a population-based case-control study. *Arch Intern Med* 2000;**160**:809–15.
375. Liem TK, Deloughery TG. First episode and recurrent venous thromboembolism: who is identifiably at risk? *Semin Vasc Surg* 2008;**21**:132–8.
376. Martinelli I, De Stefano V, Mannucci PM. Inherited risk factors for venous thromboembolism. *Nat Rev Cardiol* 2014;**11**:140–56.
377. Dahlback B, Carlsson M, Svensson PJ. Familial thrombophilia due to a previously unrecognized mechanism characterized by poor anticoagulant response to activated protein C: prediction of a cofactor to activated protein C. *Proc Natl Acad Sci USA* 1993;**90**:1004–8.
378. Bertina RM, Koeleman BP, Koster T, et al. Mutation in blood coagulation factor V associated with resistance to activated protein C. *Nature* 1994;**369**:64–7.
379. Cripe LD, Moore KD, Kane WH. Structure of the gene for human coagulation factor V. *Biochemistry* 1992;**31**:3777–85.
380. Dahlback B. Inherited thrombophilia: resistance to activated protein C as a pathogenic factor of venous thromboembolism. *Blood* 1995;**85**:607–14.
381. Kalafatis M, Bertina RM, Rand MD, et al. Characterization of the molecular defect in factor V R506Q. *J Biol Chem* 1995;**270**:4053–7.
382. Heeb MJ, Kojima Y, Greengard JS, et al. Activated protein C resistance: molecular mechanisms based on studies using purified Gln506-factor V. *Blood* 1995;**85**:3405–11.
383. Rosendaal FR, Koster T, Vandenbroucke JP, et al. High risk of thrombosis in patients homozygous for factor V Leiden (activated protein C resistance). *Blood* 1995;**85**:1504–8.
384. Juul K, Tybjaerg-Hansen A, Steffensen R, et al. Factor V Leiden: the Copenhagen city heart study and 2 meta-analyses. *Blood* 2002;**100**:3–10.
385. Kujovich JL. Factor V Leiden thrombophilia. *Genet Med* 2011;**13**:1–16.
386. Greengard JS, Eichinger S, Griffin JH, et al. Brief report: variability of thrombosis among homozygous siblings with resistance to activated protein C due to an Arg−>Gln mutation in the gene for factor V. *N Engl J Med* 1994;**331**:1559–62.
387. Poort SR, Rosendaal FR, Reitsma PH, et al. A common genetic variation in the 3'-untranslated region of the prothrombin gene is associated with elevated plasma prothrombin levels and an increase in venous thrombosis. *Blood* 1996;**88**:3698–703.
388. Degen SJ, Davie EW. Nucleotide sequence of the gene for human prothrombin. *Biochemistry* 1987;**26**:6165–77.
389. Gehring NH, Frede U, Neu-Yilik G, et al. Increased efficiency of mRNA 3' end formation: a new genetic mechanism contributing to hereditary thrombophilia. *Nat Genet* 2001;**28**:389–92.
390. Martinelli I, Sacchi E, Landi G, et al. High risk of cerebral-vein thrombosis in carriers of a prothrombin-gene mutation and in users of oral contraceptives. *N Engl J Med* 1998;**338**:1793–7.
391. Martinelli I, Taioli E, Bucciarelli P, et al. Interaction between the G20210A mutation of the prothrombin gene and oral contraceptive use in deep vein thrombosis. *Arterioscler Thromb Vasc Biol* 1999;**19**:700–3.

392. Simsek E, Yesilyurt A, Pinarli F, et al. Combined genetic mutations have remarkable effect on deep venous thrombosis and/or pulmonary embolism occurence. *Gene* 2014;**536**:171–6.
393. Lenicek Krleza J, Jakovljevic G, Bronic A, et al. Contraception-related deep venous thrombosis and pulmonary embolism in a 17-year-old girl heterozygous for factor V Leiden, prothrombin G20210A mutation, MTHFR C677T and homozygous for PAI-1 mutation: report of a family with multiple genetic risk factors and review of the literature. *Pathophysiol Haemost Thromb* 2010;**37**:24–9.
394. ACOG practice bulletin no. 138: Inherited thrombophilias in pregnancy. *Obstet Gynecol* 2013;**122**:706–17.
395. Fong FM, Sahemey MK, Hamedi G, et al. Maternal genotype and severe preeclampsia: a huge review. *Am J Epidemiol* 2014;**180**:335–45.
396. Dahlback B. Advances in understanding pathogenic mechanisms of thrombophilic disorders. *Blood* 2008;**112**:19–27.
397. Howard LS, Hughes RJ. Nice guideline: management of venous thromboembolic diseases and role of thrombophilia testing. *Thorax* 2013;**68**:391–3.
398. Kearon C, Spencer FA, O'Keeffe D, et al. D-dimer testing to select patients with a first unprovoked venous thromboembolism who can stop anticoagulant therapy: a cohort study. *Ann Intern Med* 2015;**162**:27–34.
399. Bruinstroop E, Klok FA, Van De Ree MA, et al. Elevated D-dimer levels predict recurrence in patients with idiopathic venous thromboembolism: a meta-analysis. *J Thromb Haemost* 2009;**7**:611–8.
400. Bates SM, Greer IA, Pabinger I, et al. Venous thromboembolism, thrombophilia, antithrombotic therapy, and pregnancy: American College of Chest Physicians evidence-based clinical practice guidelines (8th edition). *Chest* 2008;**133**. 844S–86S.
401. Ahrens I, Peter K, Lip GY, et al. Development and clinical applications of novel oral anticoagulants. Part I. Clinically approved drugs. *Discov Med* 2012;**13**:433–43.
402. Schulman S, Kakkar AK, Goldhaber SZ, et al. Treatment of acute venous thromboembolism with dabigatran or warfarin and pooled analysis. *Circulation* 2014;**129**:764–72.
403. Bauersachs R, Berkowitz SD, Brenner B, et al. Oral rivaroxaban for symptomatic venous thromboembolism. *N Engl J Med* 2010;**363**:2499–510.
404. Prins MH, Lensing AW, Bauersachs R, et al. Oral rivaroxaban versus standard therapy for the treatment of symptomatic venous thromboembolism: a pooled analysis of the EINSTEIN-DVT and PE randomized studies. *Thromb J* 2013;**11**:21.
405. Buller HR, Prins MH, Lensin AW, et al. Oral rivaroxaban for the treatment of symptomatic pulmonary embolism. *N Engl J Med* 2012;**366**:1287–97.
406. Buller HR, Decousus H, Grosso MA, et al. Edoxaban versus warfarin for the treatment of symptomatic venous thromboembolism. *N Engl J Med* 2013;**369**:1406–15.
407. Dawson SJ, Wiman B, Hamsten A, et al. The two allele sequences of a common polymorphism in the promoter of the plasminogen activator inhibitor-1 (PAI-1) gene respond differently to interleukin-1 in HEPG2 cells. *J Biol Chem* 1993;**268**:10739–45.
408. Gohil R, Peck G, Sharma P. The genetics of venous thromboembolism. A meta-analysis involving approximately 120,000 cases and 180,000 controls. *Thromb Haemost* 2009;**102**:360–70.
409. Loskutoff DJ, Linders M, Keijer J, et al. Structure of the human plasminogen activator inhibitor 1 gene: nonrandom distribution of introns. *Biochemistry* 1987;**26**:3763–8.
410. Ginsburg D, Zeheb R, Yang AY, et al. cDNA cloning of human plasminogen activator-inhibitor from endothelial cells. *J Clin Invest* 1986;**78**:1673–80.
411. Klinger KW, Winqvist R, Riccio A, et al. Plasminogen activator inhibitor type 1 gene is located at region q21.3-q22 of chromosome 7 and genetically linked with cystic fibrosis. *Proc Natl Acad Sci USA* 1987;**84**:8548–52.
412. Nilsson IM, Ljungner H, Tengborn L. Two different mechanisms in patients with venous thrombosis and defective fibrinolysis: low concentration of plasminogen activator or increased concentration of plasminogen activator inhibitor. *Br Med J (Clin Res Ed)* 1985;**290**:1453–6.
413. Engesser L, Brommer EJ, Kluft C, et al. Elevated plasminogen activator inhibitor (PAI), a cause of thrombophilia? A study in 203 patients with familial or sporadic venous thrombophilia. *Thromb Haemost* 1989;**62**:673–80.
414. Hickey SE, Curry CJ, Toriello HV. ACMG practice guideline: lack of evidence for MTHFR polymorphism testing. *Genet Med* 2013;**15**:153–6.
415. Kadauke S, Khor B, Van Cott EM. Activated protein C resistance testing for factor V Leiden. *Am J Hematol* 2014;**89**:1147–50.
416. Pruller F, Weiss EC, Raggam RB, et al. Activated protein C resistance assay and factor V. *Leiden. N Engl J Med* 2014;**371**:685–6.
417. Laberge AM, Psaty BM, Hindorff LA, et al. Use of factor V Leiden genetic testing in practice and impact on management. *Genet Med* 2009;**11**:750–6.
418. Smith TW, Pi D, Hudoba M, et al. Reducing inpatient heritable thrombophilia testing using a clinical decision-making tool. *J Clin Pathol* 2014;**67**:345–9.
419. Middeldorp S, van Hylckama Vlieg A. Does thrombophilia testing help in the clinical management of patients? *Br J Haematol* 2008;**143**:321–35.
420. De Stefano V, Rossi E. Testing for inherited thrombophilia and consequences for antithrombotic prophylaxis in patients with venous thromboembolism and their relatives. A review of the guidelines from scientific societies and working groups. *Thromb Haemost* 2013;**110**:697–705.
421. Baglin T, Gray E, Greaves M, et al. Clinical guidelines for testing for heritable thrombophilia. *Br J Haematol* 2010;**149**:209–20.
422. Pietropolli A, Giuliani E, Bruno V, et al. Plasminogen activator inhibitor-1, factor V, factor II and methylenetetrahydrofolate reductase polymorphisms in women with recurrent miscarriage. *J Obstet Gynaecol* 2014;**34**:229–34.
423. Rott H. Prevention and treatment of venous thromboembolism during HRT: current perspectives. *Int J Gen Med* 2014;**7**:433–40.
424. Chan WP, Lee CK, Kwong YL, et al. A novel mutation of Arg306 of factor V gene in Hong Kong Chinese. *Blood* 1998;**91**:1135–9.
425. Rebours V, Levy P, Ruszniewski P. An overview of hereditary pancreatitis. *Dig Liver Dis* 2012;**44**:8–15.
426. LaRusch J, Solomon S, Whitcomb DC. Pancreatitis overview. In: Pagon RA, Adam MP, Ardinger HH, et al., editors. *Genereviews(r), vol. Seattle (WA)*; 1993.
427. Solomon S, Whitcomb DC, LaRusch J. PRSS1-related hereditary pancreatitis. In: Pagon RA, Adam MP, Ardinger HH, et al., editors. *Genereviews(r), vol. Seattle (WA)*; 1993.

428. Weiss FU. Pancreatic cancer risk in hereditary pancreatitis. *Front Physiol* 2014;5:70.
429. Teich N, Rosendahl J, Toth M, et al. Mutations of human cationic trypsinogen (PRSS1) and chronic pancreatitis. *Hum Mutat* 2006;27:721–30.
430. Guy O, Lombardo D, Bartelt DC, et al. Two human trypsinogens. Purification, molecular properties, and N-terminal sequences. *Biochemistry* 1978;17:1669–75.
431. Rinderknecht H, Renner IG, Carmack C. Trypsinogen variants in pancreatic juice of healthy volunteers, chronic alcoholics, and patients with pancreatitis and cancer of the pancreas. *Gut* 1979;20:886–91.
432. Sahin-Toth M, Toth M. Gain-of-function mutations associated with hereditary pancreatitis enhance autoactivation of human cationic trypsinogen. *Biochem Biophys Res Commun* 2000;278:286–9.
433. Sahin-Toth M. Human cationic trypsinogen. Role of Asn-21 in zymogen activation and implications in hereditary pancreatitis. *J Biol Chem* 2000;275:22750–5.
434. Szilagyi L, Kenesi E, Katona G, et al. Comparative in vitro studies on native and recombinant human cationic trypsins. Cathepsin B is a possible pathological activator of trypsinogen in pancreatitis. *J Biol Chem* 2001;276:24574–80.
435. Teich N, Ockenga J, Hoffmeister A, et al. Chronic pancreatitis associated with an activation peptide mutation that facilitates trypsin activation. *Gastroenterology* 2000;119:461–5.
436. Chen JM, Kukor Z, Le Marechal C, et al. Evolution of trypsinogen activation peptides. *Mol Biol Evol* 2003;20:1767–77.
437. Comfort MW, Steinberg AG. Pedigree of a family with hereditary chronic relapsing pancreatitis. *Gastroenterology* 1952;21:54–63.
438. Le Bodic L, Schnee M, Georgelin T, et al. An exceptional genealogy for hereditary chronic pancreatitis. *Dig Dis Sci* 1996;41:1504–10.
439. Keim V, Bauer N, Teich N, et al. Clinical characterization of patients with hereditary pancreatitis and mutations in the cationic trypsinogen gene. *Am J Med* 2001;111:622–6.
440. Rebours V, Boutron-Ruault MC, Schnee M, et al. The natural history of hereditary pancreatitis: a national series. *Gut* 2009;58:97–103.
441. Joergensen M, Brusgaard K, Cruger DG, et al. Incidence, prevalence, etiology, and prognosis of first-time chronic pancreatitis in young patients: a nationwide cohort study. *Dig Dis Sci* 2010;55:2988–98.
442. Sibert JR. Hereditary pancreatitis in England and Wales. *J Med Genet* 1978;15:189–201.
443. de las Heras-Castano G, Castro-Senosiain B, Fontalba A, et al. Hereditary pancreatitis: clinical features and inheritance characteristics of the R122C mutation in the cationic trypsinogen gene (PRSS1) in six Spanish families. *JOP* 2009;10:249–55.
444. Howes N, Lerch MM, Greenhalf W, et al. Clinical and genetic characteristics of hereditary pancreatitis in Europe. *Clin Gastroenterol Hepatol* 2004;2:252–61.
445. Rosendahl J, Witt H, Szmola R, et al. Chymotrypsin C (CTRC) variants that diminish activity or secretion are associated with chronic pancreatitis. *Nat Genet* 2008;40:78–82.
446. Witt H, Luck W, Hennies HC, et al. Mutations in the gene encoding the serine protease inhibitor, Kazal type 1 are associated with chronic pancreatitis. *Nat Genet* 2000;25:213–6.
447. Sharer N, Schwarz M, Malone G, et al. Mutations of the cystic fibrosis gene in patients with chronic pancreatitis. *N Engl J Med* 1998;339:645–52.
448. Masamune A. Genetics of pancreatitis: the 2014 update. *Tohoku J Exp Med* 2014;232:69–77.
449. Szmola R, Sahin-Toth M, Chymotrypsin C. (caldecrin) promotes degradation of human cationic trypsin: identity with Rinderknecht's enzyme Y. *Proc Natl Acad Sci USA* 2007;104:11227–32.
450. Beer S, Zhou J, Szabo A, et al. Comprehensive functional analysis of chymotrypsin C (CTRC) variants reveals distinct loss-of-function mechanisms associated with pancreatitis risk. *Gut* 2013;62:1616–24.
451. Whitcomb DC. Framework for interpretation of genetic variations in pancreatitis patients. *Front Physiol* 2012;3:440.
452. Edwards BK, Noone AM, Mariotto AB, et al. Annual report to the nation on the status of cancer, 1975-2010, featuring prevalence of comorbidity and impact on survival among persons with lung, colorectal, breast, or prostate cancer. *Cancer* 2014;120:1290–314.
453. Society AC. *Cancer facts & figures 2015*. Atlanta: American Cancer Society; 2015.
454. Mersch J, Jackson MA, Park M, et al. Cancers associated with BRCA1 and BRCA2 mutations other than breast and ovarian. *Cancer* 2015;121:269–75.
455. Gabai-Kapara E, Lahad A, Kaufman B, et al. Population-based screening for breast and ovarian cancer risk due to BRCA1 and BRCA2. *Proc Natl Acad Sci USA* 2014;111:14205–10.
456. Roa BB, Boyd AA, Volcik K, et al. Ashkenazi Jewish population frequencies for common mutations in BRCA1 and BRCA2. *Nat Genet* 1996;14:185–7.
457. Mavaddat N, Peock S, Frost D, et al. Cancer risks for BRCA1 and BRCA2 mutation carriers: results from prospective analysis of embrace. *J Natl Cancer Inst* 2013;105:812–22.
458. Friebel TM, Domchek SM, Rebbeck TR. Modifiers of cancer risk in BRCA1 and BRCA2 mutation carriers: systematic review and meta-analysis. *J Natl Cancer Inst* 2014;106. dju091.
459. Prosperi MC, Ingham SL, Howell A, et al. Can multiple SNP testing in BRCA2 and BRCA1 female carriers be used to improve risk prediction models in conjunction with clinical assessment? *BMC Med Inform Decis Mak* 2014;14:87.
460. Antoniou AC, Beesley J, McGuffog L, et al. Common breast cancer susceptibility alleles and the risk of breast cancer for BRCA1 and BRCA2 mutation carriers: implications for risk prediction. *Cancer Res* 2010;70:9742–54.
461. King MC, Levy-Lahad E, Lahad A. Population-based screening for BRCA1 and BRCA2: 2014 Lasker award. *JAMA* 2014;312:1091–2.
462. Mavaddat N, Barrowdale D, Andrulis IL, et al. Pathology of breast and ovarian cancers among BRCA1 and BRCA2 mutation carriers: results from the Consortium of Investigators of Modifiers of BRCA1/2 (CIMBA). *Cancer Epidemiol Biomarkers Prev* 2012;21:134–47.
463. Jayson GC, Kohn EC, Kitchener HC, et al. Ovarian cancer. *Lancet* 2014;384:1376–88.
464. Hall JM, Lee MK, Newman B, et al. Linkage of early-onset familial breast cancer to chromosome 17q21. *Science* 1990;250:1684–9.
465. Castilla LH, Couch FJ, Erdos MR, et al. Mutations in the BRCA1 gene in families with early-onset breast and ovarian cancer. *Nat Genet* 1994;8:387–91.

466. Friedman LS, Ostermeyer EA, Szabo CI, et al. Confirmation of BRCA1 by analysis of germline mutations linked to breast and ovarian cancer in ten families. *Nat Genet* 1994;**8**:399–404.
467. Miki Y, Swensen J, Shattuck-Eidens D, et al. A strong candidate for the breast and ovarian cancer susceptibility gene BRCA1. *Science* 1994;**266**:66–71.
468. Tavtigian SV, Simard J, Rommens J, et al. The complete BRCA2 gene and mutations in chromosome 13q-linked kindreds. *Nat Genet* 1996;**12**:333–7.
469. Wooster R, Bignell G, Lancaster J, et al. Identification of the breast cancer susceptibility gene BRCA2. *Nature* 1995;**378**:789–92.
470. Wooster R, Neuhausen SL, Mangion J, et al. Localization of a breast cancer susceptibility gene, BRCA2, to chromosome 13q12-13. *Science* 1994;**265**:2088–90.
471. Gudmundsdottir K, Ashworth A. The roles of BRCA1 and BRCA2 and associated proteins in the maintenance of genomic stability. *Oncogene* 2006;**25**:5864–74.
472. Savage KI, Harkin DP. BRCA1, a 'complex' protein involved in the maintenance of genomic stability. *FEBS J* 2015;**282**:630–46.
473. Paul A, Paul S. The breast cancer susceptibility genes (BRCA) in breast and ovarian cancers. *Front Biosci (Landmark Ed)* 2014;**19**:605–18.
474. Lee H. Cycling with BRCA2 from DNA repair to mitosis. *Exp Cell Res* 2014;**329**:78–84.
475. Rosen EM, Pishvaian MJ. Targeting the BRCA1/2 tumor suppressors. *Curr Drug Targets* 2014;**15**:17–31.
476. University of Utah/ARUP laboraotories BRCA mutation database. http://arup.utah.edu/database/BRCA/; [accessed 01.06.2015].
477. Santos C, Peixoto A, Rocha P, et al. Pathogenicity evaluation of BRCA1 and BRCA2 unclassified variants identified in Portuguese breast/ovarian cancer families. *J Mol Diagn* 2014;**16**:324–34.
478. Garcia C, Lyon L, Littell RD, et al. Comparison of risk management strategies between women testing positive for a BRCA variant of unknown significance and women with known BRCA deleterious mutations. *Genet Med* 2014;**16**:896–902.
479. Evans BJ. Mining the human genome after association for molecular pathology v. Myriad genetics. *Genet Med* 2014;**16**:504–9.
480. Daly MB, Axilbund JE, Buys S, et al. Genetic/familial high-risk assessment: breast and ovarian. *J Natl Compr Canc Netw* 2010;**8**:562–94.
481. Cragun D, Camperlengo L, Robinson E, et al. Differences in BRCA counseling and testing practices based on ordering provider type. *Genet Med* 2015;**17**:51–7.
482. Antoniou AC, Hardy R, Walker L, et al. Predicting the likelihood of carrying a BRCA1 or BRCA2 mutation: validation of Boadicea, BRCApro, Ibis, Myriad and the Manchester scoring system using data from UK genetics clinics. *J Med Genet* 2008;**45**:425–31.
483. Panchal SM, Ennis M, Canon S, et al. Selecting a BRCA risk assessment model for use in a familial cancer clinic. *BMC Med Genet* 2008;**9**:116.
484. Yiannakopoulou E. Etiology of familial breast cancer with undetected BRCA1 and BRCA2 mutations: clinical implications. *Cell Oncol (Dordr)* 2014;**37**:1–8.
485. Tung N, Battelli C, Allen B, et al. Frequency of mutations in individuals with breast cancer referred for BRCA1 and BRCA2 testing using next-generation sequencing with a 25-gene panel. *Cancer* 2015;**121**:25–33.
486. Shiovitz S, Korde LA. Genetics of breast cancer: a topic in evolution. *Ann Oncol* 2015.
487. LaDuca H, Stuenkel AJ, Dolinsky JS, et al. Utilization of multigene panels in hereditary cancer predisposition testing: analysis of more than 2,000 patients. *Genet Med* 2014;**16**:830–7.
488. Hall MJ, Forman AD, Pilarski R, et al. Gene panel testing for inherited cancer risk. *J Natl Compr Canc Netw* 2014;**12**:1339–46.
489. Hampel H, Bennett RL, Buchanan A, et al. A practice guideline from the American College of Medical Genetics and Genomics and the National Society of Genetic Counselors: referral indications for cancer predisposition assessment. *Genet Med* 2015;**17**:70–87.
490. King-Spohn K, Pilarski R. Beyond BRCA1 and BRCA2. *Curr Probl Cancer* 2014;**38**:235–48.
491. Shirts BH, Jacobson A, Jarvik GP, et al. Large numbers of individuals are required to classify and define risk for rare variants in known cancer risk genes. *Genet Med* 2014;**16**:529–34.
492. Domchek SM, Nathanson KL. Panel testing for inherited susceptibility to breast, ovarian, and colorectal cancer. *Genet Med* 2014;**16**:827–9.
493. Stoffel EM, Kastrinos F. Familial colorectal cancer, beyond Lynch syndrome. *Clin Gastroenterol Hepatol* 2014;**12**:1059–68.
494. Bhalla A, Zulfiqar M, Weindel M, et al. Molecular diagnostics in colorectal carcinoma. *Clin Lab Med* 2013;**33**:835–59.
495. Woerner SM, Kloor M, Mueller A, et al. Microsatellite instability of selective target genes in HNPCC-associated colon adenomas. *Oncogene* 2005;**24**:2525–35.
496. Lynch HT, de la Chapelle A. Hereditary colorectal cancer. *N Engl J Med* 2003;**348**:919–32.
497. Fearon ER, Vogelstein B. A genetic model for colorectal tumorigenesis. *Cell* 1990;**61**:759–67.
498. Toyota M, Ahuja N, Ohe-Toyota M, et al. CpG island methylator phenotype in colorectal cancer. *Proc Natl Acad Sci USA* 1999;**96**:8681–6.
499. Wong JJ, Hawkins NJ, Ward RL. Colorectal cancer: a model for epigenetic tumorigenesis. *Gut* 2007;**56**:140–8.
500. Ogino S, Goel A. Molecular classification and correlates in colorectal cancer. *J Mol Diagn* 2008;**10**:13–27.
501. Lynch HT, Shaw MW, Magnuson CW, et al. Hereditary factors in cancer. Study of two large midwestern kindreds. *Arch Intern Med* 1966;**117**:206–12.
502. Giardiello FM, Allen JI, Axilbund JE, et al. Guidelines on genetic evaluation and management of Lynch syndrome: a consensus statement by the US multi-society task force on colorectal cancer. *Am J Gastroenterol* 2014;**109**:1159–79.
503. Guillotin D, Martin SA. Exploiting DNA mismatch repair deficiency as a therapeutic strategy. *Exp Cell Res* 2014;**329**:110–5.
504. Peltomaki P, Aaltonen LA, Sistonen P, et al. Genetic mapping of a locus predisposing to human colorectal cancer. *Science* 1993;**260**:810–2.

505. Thibodeau SN, Bren G, Schaid D. Microsatellite instability in cancer of the proximal colon. *Science* 1993;**260**:816–9.
506. Ionov Y, Peinado MA, Malkhosyan S, et al. Ubiquitous somatic mutations in simple repeated sequences reveal a new mechanism for colonic carcinogenesis. *Nature* 1993;**363**:558–61.
507. Hegde M, Ferber M, Mao R, et al. ACMG technical standards and guidelines for genetic testing for inherited colorectal cancer (Lynch syndrome, familial adenomatous polyposis, and MYH-associated polyposis). *Genet Med* 2014;**16**:101–16.
508. Kuismanen SA, Holmberg MT, Salovaara R, et al. Genetic and epigenetic modification of MLH1 accounts for a major share of microsatellite-unstable colorectal cancers. *Am J Pathol* 2000;**156**:1773–9.
509. Boland CR, Thibodeau SN, Hamilton SR, et al. A National Cancer Institute workshop on microsatellite instability for cancer detection and familial predisposition: development of international criteria for the determination of microsatellite instability in colorectal cancer. *Cancer Res* 1998;**58**:5248–57.
510. Vasen HF, Mecklin JP, Khan PM, et al. The International Collaborative Group on Hereditary Non-polyposis Colorectal Cancer (ICG-HNPCC). *Dis Colon Rectum* 1991;**34**:424–5.
511. Vasen HF, Watson P, Mecklin JP, et al. New clinical criteria for hereditary nonpolyposis colorectal cancer (HNPCC, Lynch syndrome) proposed by the International Collaborative Group on HNPCC. *Gastroenterology* 1999;**116**:1453–6.
512. Umar A, Boland CR, Terdiman JP, et al. Revised Bethesda guidelines for hereditary nonpolyposis colorectal cancer (Lynch syndrome) and microsatellite instability. *J Natl Cancer Inst* 2004;**96**:261–8.
513. Stoffel EM, Mangu PB, Gruber SB, et al. Hereditary colorectal cancer syndromes: American Society of Clinical Oncology clinical practice guideline endorsement of the familial risk-colorectal cancer: European Society for Medical Oncology clinical practice guidelines. *J Clin Oncol* 2015;**33**:209–17.
514. Hampel H, Frankel W, Panescu J, et al. Screening for Lynch syndrome (hereditary nonpolyposis colorectal cancer) among endometrial cancer patients. *Cancer Res* 2006;**66**:7810–7.
515. Kohlmann W, Gruber SB. Lynch syndrome. In: Pagon RA, Adam MP, Ardinger HH, et al., editors. *Genereviews(r)*, vol. Seattle (WA); 1993.
516. Nakamura K, Banno K, Yanokura M, et al. Features of ovarian cancer in Lynch syndrome (review). *Mol Clin Oncol* 2014;**2**:909–16.
517. Lynch HT, Casey MJ. Prophylactic surgery prevents endometrial and ovarian cancer in Lynch syndrome. *Nat Clin Pract Oncol* 2007;**4**:672–3.
518. Mills AM, Liou S, Ford JM, et al. Lynch syndrome screening should be considered for all patients with newly diagnosed endometrial cancer. *Am J Surg Pathol* 2014;**38**:1501–9.
519. Djordjevic B, Broaddus RR. Role of the clinical pathology laboratory in the evaluation of endometrial carcinomas for Lynch syndrome. *Semin Diagn Pathol* 2014;**31**:195–204.
520. Moline J, Mahdi H, Yang B, et al. Implementation of tumor testing for Lynch syndrome in endometrial cancers at a large academic medical center. *Gynecol Oncol* 2013;**130**:121–6.
521. Samowitz WS. Evaluation of colorectal cancers for Lynch syndrome: practical molecular diagnostics for surgical pathologists. *Mod Pathol* 2015;**28**(Suppl. 1):S109–13.
522. Beamer LC, Grant ML, Espenschied CR, et al. Reflex immunohistochemistry and microsatellite instability testing of colorectal tumors for Lynch syndrome among US cancer programs and follow-up of abnormal results. *J Clin Oncol* 2012;**30**:1058–63.
523. Salipante SJ, Scroggins SM, Hampel HL, et al. Microsatellite instability detection by next generation sequencing. *Clin Chem* 2014;**60**:1192–9.
524. Cushman-Vokoun AM, Stover DG, Zhao Z, et al. Clinical utility of KRAS and BRAF mutations in a cohort of patients with colorectal neoplasms submitted for microsatellite instability testing. *Clin Colorectal Cancer* 2013;**12**:168–78.
525. Win AK, Macinnis RJ, Dowty JG, et al. Criteria and prediction models for mismatch repair gene mutations: a review. *J Med Genet* 2013;**50**:785–93.
526. Weissman SM, Burt R, Church J, et al. Identification of individuals at risk for Lynch syndrome using targeted evaluations and genetic testing: National Society of Genetic Counselors and the Collaborative Group of the Americas on inherited colorectal cancer joint practice guideline. *J Genet Couns* 2012;**21**:484–93.
527. Thompson BA, Spurdle AB, Plazzer JP, et al. Application of a 5-tiered scheme for standardized classification of 2,360 unique mismatch repair gene variants in the INSIGHT locus-specific database. *Nat Genet* 2014;**46**:107–15.
528. International Society for Gastrointestinal Hereditary Tumours. http://insight-group.org/; [accessed 28.05.2015].
529. ten Broeke SW, Brohet RM, Tops CM, et al. Lynch syndrome caused by germline PMS2 mutations: delineating the cancer risk. *J Clin Oncol* 2015;**33**:319–25.
530. Lindor NM, Petersen GM, Hadley DW, et al. Recommendations for the care of individuals with an inherited predisposition to Lynch syndrome: a systematic review. *JAMA* 2006;**296**:1507–17.
531. Bhaijee F, Brown AS. Muir-Torre syndrome. *Arch Pathol Lab Med* 2014;**138**:1685–9.
532. Petersen GM, Slack J, Nakamura Y. Screening guidelines and premorbid diagnosis of familial adenomatous polyposis using linkage. *Gastroenterology* 1991;**100**:1658–64.
533. Bisgaard ML, Fenger K, Bulow S, et al. Familial adenomatous polyposis (FAP): frequency, penetrance, and mutation rate. *Hum Mutat* 1994;**3**:121–5.
534. Groen EJ, Roos A, Muntinghe FL, et al. Extra-intestinal manifestations of familial adenomatous polyposis. *Ann Surg Oncol* 2008;**15**:2439–50.
535. Jones IT, Jagelman DG, Fazio VW, et al. Desmoid tumors in familial polyposis coli. *Ann Surg* 1986;**204**:94–7.
536. Groden J, Thliveris A, Samowitz W, et al. Identification and characterization of the familial adenomatous polyposis coli gene. *Cell* 1991;**66**:589–600.
537. Nishisho I, Nakamura Y, Miyoshi Y, et al. Mutations of chromosome 5q21 genes in FAP and colorectal cancer patients. *Science* 1991;**253**:665–9.
538. Kinzler KW, Nilbert MC, Su LK, et al. Identification of FAP locus genes from chromosome 5q21. *Science* 1991;**253**:661–5.
539. Vogelstein B, Kinzler KW. Cancer genes and the pathways they control. *Nat Med* 2004;**10**:789–99.
540. Lesko AC, Goss KH, Prosperi JR. Exploiting APC function as a novel cancer therapy. *Curr Drug Targets* 2014;**15**:90–102.

541. Aoki K, Taketo MM. Adenomatous polyposis coli (APC): a multi-functional tumor suppressor gene. *J Cell Sci* 2007;**120**: 3327–35.
542. Polakis P. The oncogenic activation of beta-catenin. *Curr Opin Genet Dev* 1999;**9**:15–21.
543. Katoh M. Wnt signaling pathway and stem cell signaling network. *Clin Cancer Res* 2007;**13**:4042–5.
544. Kerr SE, Thomas CB, Thibodeau SN, et al. APC germline mutations in individuals being evaluated for familial adenomatous polyposis: a review of the Mayo Clinic experience with 1591 consecutive tests. *J Mol Diagn* 2013;**15**:31–43.
545. Jasperson KW, Burt RW. APC-associated polyposis conditions. In: Pagon RA, Adam MP, Ardinger HH, et al., editors. *Genereviews(r)*, vol. Seattle (WA); 1993.
546. Caspari R, Friedl W, Mandl M, et al. Familial adenomatous polyposis: mutation at codon 1309 and early onset of colon cancer. *Lancet* 1994;**343**:629–32.
547. Dobbie Z, Spycher M, Mary JL, et al. Correlation between the development of extracolonic manifestations in FAP patients and mutations beyond codon 1403 in the APC gene. *J Med Genet* 1996;**33**:274–80.
548. Nagase H, Miyoshi Y, Horii A, et al. Correlation between the location of germ-line mutations in the APC gene and the number of colorectal polyps in familial adenomatous polyposis patients. *Cancer Res* 1992;**52**:4055–7.
549. Brensinger JD, Laken SJ, Luce MC, et al. Variable phenotype of familial adenomatous polyposis in pedigrees with 3' mutation in the APC gene. *Gut* 1998;**43**:548–52.
550. Friedl W, Meuschel S, Caspari R, et al. Attenuated familial adenomatous polyposis due to a mutation in the 3' part of the APC gene. A clue for understanding the function of the APC protein. *Hum Genet* 1996;**97**:579–84.
551. Spirio L, Olschwang S, Groden J, et al. Alleles of the APC gene: an attenuated form of familial polyposis. *Cell* 1993;**75**: 951–7.
552. Soravia C, Berk T, Madlensky L, et al. Genotype-phenotype correlations in attenuated adenomatous polyposis coli. *Am J Hum Genet* 1998;**62**:1290–301.
553. Crabtree MD, Tomlinson IP, Hodgson SV, et al. Explaining variation in familial adenomatous polyposis: relationship between genotype and phenotype and evidence for modifier genes. *Gut* 2002;**51**:420–3.
554. Giardiello FM, Krush AJ, Petersen GM, et al. Phenotypic variability of familial adenomatous polyposis in 11 unrelated families with identical APC gene mutation. *Gastroenterology* 1994;**106**:1542–7.
555. Giardiello FM, Brensinger JD, Petersen GM. AGA technical review on hereditary colorectal cancer and genetic testing. *Gastroenterology* 2001;**121**:198–213.
556. Vasen HF, Moslein G, Alonso A, et al. Guidelines for the clinical management of familial adenomatous polyposis (FAP). *Gut* 2008;**57**:704–13.
557. Al-Tassan N, Chmiel NH, Maynard J, et al. Inherited variants of MYH associated with somatic G: C—>T: A mutations in colorectal tumors. *Nat Genet* 2002;**30**:227–32.
558. Sampson JR, Jones S, Dolwani S, et al. MUTYH (MYH) and colorectal cancer. *Biochem Soc Trans* 2005;**33**:679–83.
559. Jones S, Lambert S, Williams GT, et al. Increased frequency of the k-RAS G12C mutation in MYH polyposis colorectal adenomas. *Br J Cancer* 2004;**90**:1591–3.
560. Cleary SP, Cotterchio M, Jenkins MA, et al. Germline MUTY human homologue mutations and colorectal cancer: a multisite case-control study. *Gastroenterology* 2009;**136**:1251–60.
561. Olschwang S, Blanche H, de Moncuit C, et al. Similar colorectal cancer risk in patients with monoallelic and biallelic mutations in the MYH gene identified in a population with adenomatous polyposis. *Genet Test* 2007;**11**:315–20.
562. Yamaguchi S, Ogata H, Katsumata D, et al. MUTYH-associated colorectal cancer and adenomatous polyposis. *Surg Today* 2014;**44**:593–600.
563. Chubb D, Broderick P, Frampton M, et al. Genetic diagnosis of high-penetrance susceptibility for colorectal cancer (CRC) is achievable for a high proportion of familial CRC by exome sequencing. *J Clin Oncol* 2015;**33**:426–32.
564. Feissner RF, Skalska J, Gaum WE, et al. Crosstalk signaling between mitochondrial Ca2+ and ROS. *Front Biosci (Landmark Ed)* 2009;**14**:1197–218.
565. Petit E, Oliver L, Vallette FM. The mitochondrial outer membrane protein import machinery: a new player in apoptosis? *Front Biosci (Landmark* 2009;**14**:3563–70.
566. Soubannier V, McBride HM. Positioning mitochondrial plasticity within cellular signaling cascades. *Biochim Biophys Acta* 2009;**1793**:154–70.
567. Davis RL, Sue CM. The genetics of mitochondrial disease. *Semin Neurol* 2011;**31**:519–30.
568. Schon EA, DiMauro S, Hirano M. Human mitochondrial DNA: roles of inherited and somatic mutations. *Nat Rev Genet* 2012;**13**:878–90.
569. DiMauro S, Hirano M. Mitochondrial DNA deletion syndromes. In: Pagon RA, Adam MP, Ardinger HH, et al., editors. *Genereviews(r)*, vol. Seattle (WA); 1993.
570. Anderson S, Bankier AT, Barrell BG, et al. Sequence and organization of the human mitochondrial genome. *Nature* 1981;**290**:457–65.
571. Barrell BG, Anderson S, Bankier AT, et al. Different pattern of codon recognition by mammalian mitochondrial tRNAs. *Proc Natl Acad Sci USA* 1980;**77**:3164–6.
572. Barrell BG, Bankier AT, Drouin J. A different genetic code in human mitochondria. *Nature* 1979;**282**:189–94.
573. Tzen CY, Wu TY, Liu HF. Sequence polymorphism in the coding region of mitochondrial genome encompassing position 8389-8865. *Forensic Sci Int* 2001;**120**:204–9.
574. Melton T, Clifford S, Kayser M, et al. Diversity and heterogeneity in mitochondrial DNA of North American populations. *J Forensic Sci* 2001;**46**:46–52.
575. Wallace DC, Singh G, Lott MT, et al. Mitochondrial DNA mutation associated with Leber's hereditary optic neuropathy. *Science* 1988;**242**:1427–30.
576. Zeviani M, Moraes CT, DiMauro S, et al. Deletions of mitochondrial DNA in Kearns-Sayre syndrome. *Neurology* 1988;**38**: 1339–46.
577. Holt IJ, Cooper JM, Morgan-Hughes JA, et al. Deletions of muscle mitochondrial DNA. *Lancet* 1988;**1**:1462.
578. Schwartz M, Vissing J. Paternal inheritance of mitochondrial DNA. *N Engl J Med* 2002;**347**:576–80.
579. Schwartz M, Vissing J. New patterns of inheritance in mitochondrial disease. *Biochem Biophys Res Commun* 2003; **310**:247–51.
580. Di Donato S. Multisystem manifestations of mitochondrial disorders. *J Neurol* 2009;**256**:693–710.

581. Wong LJ. Pathogenic mitochondrial DNA mutations in protein-coding genes. *Muscle Nerve* 2007;**36**:279—93.
582. Parikh S, Goldstein A, Koenig MK, et al. Diagnosis and management of mitochondrial disease: a consensus statement from the Mitochondrial Medicine Society. *Genet Med* 2014.
583. Dames S, Chou LS, Xiao Y, et al. The development of next-generation sequencing assays for the mitochondrial genome and 108 nuclear genes associated with mitochondrial disorders. *J Mol Diagn* 2013;**15**:526—34.
584. Dames S, Eilbeck K, Mao R. A high-throughput next-generation sequencing assay for the mitochondrial genome. *Methods Mol Biol* 2015;**1264**:77—88.
585. Wong LJ. Next generation molecular diagnosis of mitochondrial disorders. *Mitochondrion* 2013;**13**:379—87.
586. Richter C, Park JW, Ames BN. Normal oxidative damage to mitochondrial and nuclear DNA is extensive. *Proc Natl Acad Sci USA* 1988;**85**:6465—7.
587. Ames BN, Shigenaga MK, Hagen TM. Oxidants, antioxidants, and the degenerative diseases of aging. *Proc Natl Acad Sci USA* 1993;**90**:7915—22.
588. Rahman S. Emerging aspects of treatment in mitochondrial disorders. *J Inherit Metab Dis* 2015.
589. Yu-Wai-Man P, Griffiths PG, Hudson G, et al. Inherited mitochondrial optic neuropathies. *J Med Genet* 2009;**46**:145—58.
590. Hoegger MJ, Lieven CJ, Levin LA. Differential production of superoxide by neuronal mitochondria. *BMC Neurosci* 2008;**9**:4.
591. Shankar SP, Fingert JH, Carelli V, et al. Evidence for a novel X-linked modifier locus for Leber hereditary optic neuropathy. *Ophthalmic Genet* 2008;**29**:17—24.
592. Ji Y, Jia X, Li S, et al. Evaluation of the X-linked modifier loci for Leber hereditary optic neuropathy with the G11778A mutation in Chinese. *Mol Vis* 2010;**16**:416—24.
593. Hudson G, Keers S, Yu Wai Man P, et al. Identification of an X-chromosomal locus and haplotype modulating the phenotype of a mitochondrial DNA disorder. *Am J Hum Genet* 2005;**77**:1086—91.
594. Giordano C, Montopoli M, Perli E, et al. Oestrogens ameliorate mitochondrial dysfunction in Leber's hereditary optic neuropathy. *Brain* 2011;**134**:220—34.
595. Huoponen K, Puomila A, Savontaus ML, et al. Genetic counseling in Leber hereditary optic neuropathy (LHON). *Acta Ophthalmol Scand* 2002;**80**:38—43.
596. Yu-Wai-Man P, Chinnery PF. Leber hereditary optic neuropathy. In: Pagon RA, Adam MP, Ardinger HH, et al., editors. *Genereviews(r), vol.* Seattle (WA); 1993.
597. Smith KH, Johns DR, Heher KL, et al. Heteroplasmy in Leber's hereditary optic neuropathy. *Arch Ophthalmol* 1993;**111**:1486—90.
598. Vilkki J, Savontaus ML, Nikoskelainen EK. Segregation of mitochondrial genomes in a heteroplasmic lineage with Leber hereditary optic neuroretinopathy. *Am J Hum Genet* 1990;**47**:95—100.
599. Newman NJ. From genotype to phenotype in Leber hereditary optic neuropathy: still more questions than answers. *J Neuroophthalmol* 2002;**22**:257—61.
600. Tsao CY, Herman G, Boue DR, et al. Leigh disease with mitochondrial DNA A8344G mutation: case report and brief review. *J Child Neurol* 2003;**18**:62—4.
601. Sgarbi G, Baracca A, Lenaz G, et al. Inefficient coupling between proton transport and ATP synthesis may be the pathogenic mechanism for NARP and Leigh syndrome resulting from the T8993G mutation in mtDNA. *Biochem J* 2006;**395**:493—500.
602. Ruhoy IS, Saneto RP. The genetics of Leigh syndrome and its implications for clinical practice and risk management. *Appl Clin Genet* 2014;**7**:221—34.
603. Unsal E, Aktas Y, Uner O, et al. Successful application of preimplantation genetic diagnosis for Leigh syndrome. *Fertil Steril* 2008;**90**:2017. e11—13.
604. DiMauro S, Hirano M. MELAS. In: Pagon RA, Adam MP, Ardinger HH, et al., editors. *Genereviews(r), vol.* Seattle (WA); 1993.
605. Kaufmann P, Engelstad K, Wei Y, et al. Natural history of MELAS associated with mitochondrial DNA m.3243A>G genotype. *Neurology* 2011;**77**:1965—71.
606. Crimmins D, Morris JG, Walker GL, et al. Mitochondrial encephalomyopathy: variable clinical expression within a single kindred. *J Neurol Neurosurg Psychiatry* 1993;**56**:900—5.
607. Goto Y, Nonaka I, Horai S. A mutation in the tRNA(Leu)(UUR) gene associated with the MELAS subgroup of mitochondrial encephalomyopathies. *Nature* 1990;**348**:651—3.
608. Santorelli FM, Tanji K, Kulikova R, et al. Identification of a novel mutation in the mtDNA ND5 gene associated with MELAS. *Biochem Biophys Res Commun* 1997;**238**:326—8.
609. Shanske S, Coku J, Lu J, et al. The G13513A mutation in the ND5 gene of mitochondrial DNA as a common cause of MELAS or Leigh syndrome: evidence from 12 cases. *Arch Neurol* 2008;**65**:368—72.
610. DiMauro S, Hirano M. MERRF. In: Pagon RA, Adam MP, Ardinger HH, et al., editors. *Genereviews(r), vol.* Seattle (WA); 1993.
611. Suzuki T, Nagao A. Human mitochondrial diseases caused by lack of taurine modification in mitochondrial tRNAs. *Wiley Interdiscip Rev RNA* 2011;**2**:376—86.
612. Shoffner JM, Lott MT, Lezza AM, et al. Myoclonic epilepsy and ragged-red fiber disease (MERRF) is associated with a mitochondrial DNA tRNA(Lys) mutation. *Cell* 1990;**61**:931—7.
613. Mancuso M, Orsucci D, Angelini C, et al. Phenotypic heterogeneity of the 8344A>G mtDNA "MERRF" mutation. *Neurology* 2013;**80**:2049—54.
614. DiMauro S, Schon EA. Mitochondrial respiratory-chain diseases. *N Engl J Med* 2003;**348**:2656—68.
615. Moraes CT, DiMauro S, Zeviani M, et al. Mitochondrial DNA deletions in progressive external ophthalmoplegia and Kearns-Sayre syndrome. *N Engl J Med* 1989;**320**:1293—9.
616. Ishida M, Moore GE. The role of imprinted genes in humans. *Mol Aspects Med* 2013;**34**:826—40.
617. Bigi N, Faure JM, Coubes C, et al. Prader-Willi syndrome: is there a recognizable fetal phenotype? *Prenat Diagn* 2008;**28**:796—9.
618. Driscoll DJ, Miller JL, Schwartz S, et al. Prader-Willi syndrome. In: Pagon RA, Adam MP, Ardinger HH, et al., editors. *Genereviews(r), vol.* Seattle (WA); 1993.
619. Butler MG. Prader-Willi syndrome: obesity due to genomic imprinting. *Curr Genomics* 2011;**12**:204—15.
620. Goldstone AP, Holland AJ, Hauffa BP, et al. Recommendations for the diagnosis and management of Prader-Willi syndrome. *J Clin Endocrinol Metab* 2008;**93**:4183—97.
621. Lindgren AC, Lindberg A. Growth hormone treatment completely normalizes adult height and improves body composition in Prader-Willi syndrome: experience from KIGS (Pfizer International Growth Database). *Horm Res* 2008;**70**:182—7.
622. Dagli AI, Mueller J, Williams CA. Angelman syndrome. In: Pagon RA, Adam MP, Ardinger HH, et al., editors. *Genereviews(r), vol.* Seattle (WA); 1993.

623. Jong MT, Gray TA, Ji Y, et al. A novel imprinted gene, encoding a ring zinc-finger protein, and overlapping antisense transcript in the Prader-Willi syndrome critical region. *Hum Mol Genet* 1999;**8**:783–93.

624. Lee S, Kozlov S, Hernandez L, et al. Expression and imprinting of MAGEL2 suggest a role in Prader-Willi syndrome and the homologous murine imprinting phenotype. *Hum Mol Genet* 2000;**9**:1813–9.

625. Nakada Y, Taniura H, Uetsuki T, et al. The human chromosomal gene for necdin, a neuronal growth suppressor, in the Prader-Willi syndrome deletion region. *Gene* 1998;**213**:65–72.

626. Girardot M, Cavaille J, Feil R. Small regulatory RNAs controlled by genomic imprinting and their contribution to human disease. *Epigenetics* 2012;**7**:1341–8.

627. Kiss T. Small nucleolar rnas: an abundant group of non-coding RNAs with diverse cellular functions. *Cell* 2002;**109**:145–8.

628. Sahoo T, del Gaudio D, German JR, et al. Prader-Willi phenotype caused by paternal deficiency for the HBII-85 C/D box small nucleolar RNA cluster. *Nat Genet* 2008;**40**:719–21.

629. Kitsiou-Tzeli S, Tzetis M, Sofocleous C, et al. De novo interstitial duplication of the 15q11.2-q14 PWS/AS region of maternal origin: clinical description, array CGH analysis, and review of the literature. *Am J Med Genet A* 2010;**152A**:1925–32.

630. Robinson WP, Christian SL, Kuchinka BD, et al. Somatic segregation errors predominantly contribute to the gain or loss of a paternal chromosome leading to uniparental disomy for chromosome 15. *Clin Genet* 2000;**57**:349–58.

631. Matsubara K, Murakami N, Nagai T, et al. Maternal age effect on the development of Prader-Willi syndrome resulting from udp(15)mat through meiosis 1 errors. *J Hum Genet* 2011;**56**:566–71.

632. Buiting K, Saitoh S, Gross S, et al. Inherited microdeletions in the Angelman and Prader-Willi syndromes define an imprinting centre on human chromosome 15. *Nat Genet* 1995;**9**:395–400.

633. Sun Y, Nicholls RD, Butler MG, et al. Breakage in the SNRPN locus in a balanced 46,XY,t(15;19) Prader-Willi syndrome patient. *Hum Mol Genet* 1996;**5**:517–24.

634. Dykens EM, Roof E. Behavior in Prader-Willi syndrome: relationship to genetic subtypes and age. *J Child Psychol Psychiatry* 2008;**49**:1001–8.

635. Veltman MW, Craig EE, Bolton PF. Autism spectrum disorders in Prader-Willi and Angelman syndromes: a systematic review. *Psychiatr Genet* 2005;**15**:243–54.

636. Kishino T, Lalande M, Wagstaff J. UBE3A/E6-AP mutations cause Angelman syndrome. *Nat Genet* 1997;**15**:70–3.

637. Meguro M, Kashiwagi A, Mitsuya K, et al. A novel maternally expressed gene, ATP10C, encodes a putative aminophospholipid translocase associated with Angelman syndrome. *Nat Genet* 2001;**28**:19–20.

638. Matentzoglu K, Scheffner M. Ubiquitin ligase E6-AP and its role in human disease. *Biochem Soc Trans* 2008;**36**:797–801.

639. Yamamoto Y, Huibregtse JM, Howley PM. The human E6-AP gene (EBE3A) encodes three potential protein isoforms generated by differential splicing. *Genomics* 1997;**41**:263–6.

640. Vu TH, Hoffman AR. Imprinting of the Angelman syndrome gene, UBE3A, is restricted to brain. *Nat Genet* 1997;**17**:12–3.

641. Horsthemke B, Dittrich B, Buiting K. Imprinting mutations on human chromosome 15. *Hum Mutat* 1997;**10**:329–37.

642. Fang P, Lev-Lehman E, Tsai TF, et al. The spectrum of mutations in UBE3A causing Angelman syndrome. *Hum Mol Genet* 1999;**8**:129–35.

643. Sadikovic B, Fernandes P, Zhang VW, et al. Mutation update for UBE3A variants in Angelman syndrome. *Hum Mutat* 2014;**35**:1407–17.

644. Meng L, Ward AJ, Chun S, et al. Towards a therapy for Angelman syndrome by targeting a long non-coding RNA. *Nature* 2015;**518**:409–12.

645. Tan WH, Bird LM, Thibert RL, et al. If not Angelman, what is it? A review of Angelman-like syndromes. *Am J Med Genet A* 2014;**164A**:975–92.

646. Lossie AC, Whitney MM, Amidon D, et al. Distinct phenotypes distinguish the molecular classes of Angelman syndrome. *J Med Genet* 2001;**38**:834–45.

647. Kubota T, Das S, Christian SL, et al. Methylation-specific PCR simplifies imprinting analysis. *Nat Genet* 1997;**16**:16–7.

648. Velinov M, Jenkins EC. PCR-based strategies for the diagnosis of Prader-Willi/Angelman syndromes. *Methods Mol Biol* 2003;**217**:209–16.

649. Henkhaus RS, Kim SJ, Kimonis VE, et al. Methylation-specific multiplex ligation-dependent probe amplification and identification of deletion genetic subtypes in Prader-Willi syndrome. *Genet Test Mol Biomarkers* 2012;**16**:178–86.

650. Wang W, Law HY, Chong SS. Detection and discrimination between deletional and non-deletional Prader-Willi and Angelman syndromes by methylation-specific PCR and quantitative melting curve analysis. *J Mol Diagn* 2009;**11**:446–9.

651. Hung CC, Lin SY, Lin SP, et al. Quantitative and qualitative analyses of the SNRPN gene using real-time PCR with melting curve analysis. *J Mol Diagn* 2011;**13**:609–13.

652. Altug-Teber O, Dufke A, Poths S, et al. A rapid microarray based whole genome analysis for detection of uniparental disomy. *Hum Mutat* 2005;**26**:153–9.

653. Khan WA, Knoll JH, Rogan PK. Context-based FISH localization of genomic rearrangements within chromosome 15q11.2q13 duplicons. *Mol Cytogenet* 2011;**4**:15.

654. Utine GE, Haliloglu G, Volkan-Salanci B, et al. Etiological yield of SNP microarrays in idiopathic intellectual disability. *Eur J Paediatr Neurol* 2014;**18**:327–37.

655. Giardina E, Peconi C, Cascella R, et al. A multiplex molecular assay for the detection of uniparental disomy for human chromosome 15. *Electrophoresis* 2008;**29**:4775–9.

656. Ramsden SC, Clayton-Smith J, Birch R, et al. Practice guidelines for the molecular analysis of Prader-Willi and Angelman syndromes. *BMC Med Genet* 2010;**11**:70.

657. ACOG committee opinion no. 442: Preconception and prenatal carrier screening for genetic diseases in individuals of Eastern European Jewish descent. *Obstet Gynecol* 2009;**114**:950–3.

658. Srinivasan BS, Evans EA, Flannick J, et al. A universal carrier test for the long tail of Mendelian disease. *Reprod Biomed Online* 2010;**21**:537–51.

659. Strom CM, Redman JB, Peng M. The dangers of including nonclassical cystic fibrosis variants in population-based screening panels: p.L997F, further genotype/phenotype correlation data. *Genet Med* 2011;**13**:1042–4.

660. Grody WW. Expanded carrier screening and the law of unintended consequences: from cystic fibrosis to fragile X. *Genet Med* 2011;**13**:996–7.

661. Wienke S, Brown K, Farmer M, et al. Expanded carrier screening panels—does bigger mean better? *J Community Genet* 2014;**5**:191—8.
662. Stoll K, Resta R. Considering the cost of expanded carrier screening panels. *Genet Med* 2013;**15**:318—9.
663. Grody WW, Thompson BH, Gregg AR, et al. ACMG position statement on prenatal/preconception expanded carrier screening. *Genet Med* 2013;**15**:482—3.
664. Richards S, Aziz N, Bale S, et al. Standards and guidelines for the interpretation of sequence variants: a joint consensus recommendation of the American College of Medical Genetics and Genomics and the Association for Molecular Pathology. *Genet Med* 2015;**17**:405—23.
665. Fahim AT, Daiger SP, Weleber RG. Retinitis pigmentosa overview. In: Pagon RA, Adam MP, Ardinger HH, et al., editors. *Genereviews(r), vol. Seattle (WA)*; 1993.
666. Yang Y, Muzny DM, Reid JG, et al. Clinical whole-exome sequencing for the diagnosis of Mendelian disorders. *N Engl J Med* 2013;**369**:1502—11.
667. Yang Y, Muzny DM, Xia F, et al. Molecular findings among patients referred for clinical whole-exome sequencing. *JAMA* 2014;**312**:1870—9.
668. Need AC, Shashi V, Hitomi Y, et al. Clinical application of exome sequencing in undiagnosed genetic conditions. *J Med Genet* 2012;**49**:353—61.
669. Biesecker LG, Green RC. Diagnostic clinical genome and exome sequencing. *N Engl J Med* 2014;**371**:1170.
670. Teer JK, Mullikin JC. Exome sequencing: the sweet spot before whole genomes. *Hum Mol Genet* 2010;**19**:R145—51.
671. Bao R, Huang L, Andrade J, et al. Review of current methods, applications, and data management for the bioinformatics analysis of whole exome sequencing. *Cancer Inform* 2014;**13**:67—82.
672. Xue Y, Ankala A, Wilcox WR, et al. Solving the molecular diagnostic testing conundrum for Mendelian disorders in the era of next-generation sequencing: single-gene, gene panel, or exome/genome sequencing. *Genet Med* 2014.
673. Shashi V, McConkie-Rosell A, Rosell B, et al. The utility of the traditional medical genetics diagnostic evaluation in the context of next-generation sequencing for undiagnosed genetic disorders. *Genet Med* 2014;**16**:176—82.
674. Green RC, Berg JS, Grody WW, et al. ACMG recommendations for reporting of incidental findings in clinical exome and genome sequencing. *Genet Med* 2013;**15**:565—74.
675. Burke W, Antommaria AH, Bennett R, et al. Recommendations for returning genomic incidental findings? We need to talk! *Genet Med* 2013;**15**:854—9.
676. Yu JH, Harrell TM, Jamal SM, et al. Attitudes of genetics professionals toward the return of incidental results from exome and whole-genome sequencing. *Am J Hum Genet* 2014;**95**:77—84.
677. Hegde M, Bale S, Bayrak-Toydemir P, et al. Reporting incidental findings in genomic scale clinical sequencing—a clinical laboratory perspective: a report of the Association for Molecular Pathology. *J Mol Diagn* 2015;**17**:107—17.
678. Gilissen C, Hoischen A, Brunner HG, et al. Unlocking Mendelian disease using exome sequencing. *Genome Biol* 2011;**12**:228.
679. Rabbani B, Tekin M, Mahdieh N. The promise of whole-exome sequencing in medical genetics. *J Hum Genet* 2014;**59**:5—15.
680. Bamshad MJ, Ng SB, Bigham AW, et al. Exome sequencing as a tool for Mendelian disease gene discovery. *Nat Rev Genet* 2011;**12**:745—55.
681. Ng SB, Bigham AW, Buckingham KJ, et al. Exome sequencing identifies MLL2 mutations as a cause of Kabuki syndrome. *Nat Genet* 2010;**42**:790—3.
682. Ng SB, Buckingham KJ, Lee C, et al. Exome sequencing identifies the cause of a Mendelian disorder. *Nat Genet* 2010;**42**:30—5.
683. Walsh T, Shahin H, Elkan-Miller T, et al. Whole exome sequencing and homozygosity mapping identify mutation in the cell polarity protein GPSM2 as the cause of nonsyndromic hearing loss DFNB82. *Am J Hum Genet* 2010;**87**:90—4.
684. Zuchner S, Dallman J, Wen R, et al. Whole-exome sequencing links a variant in DHDDS to retinitis pigmentosa. *Am J Hum Genet* 2011;**88**:201—6.
685. Becker J, Semler O, Gilissen C, et al. Exome sequencing identifies truncating mutations in human SERPINF1 in autosomal-recessive osteogenesis imperfecta. *Am J Hum Genet* 2011;**88**:362—71.
686. Schuurs-Hoeijmakers JH, Oh EC, Vissers LE, et al. Recurrent de novo mutations in PACS1 cause defective cranial-neural-crest migration and define a recognizable intellectual-disability syndrome. *Am J Hum Genet* 2012;**91**:1122—7.
687. Basel-Vanagaite L, Dallapiccola B, Ramirez-Solis R, et al. Deficiency for the ubiquitin ligase UBE3B in a blepharophimosis-ptosis-intellectual-disability syndrome. *Am J Hum Genet* 2012;**91**:998—1010.
688. Norton N, Li D, Rieder MJ, et al. Genome-wide studies of copy number variation and exome sequencing identify rare variants in BAG3 as a cause of dilated cardiomyopathy. *Am J Hum Genet* 2011;**88**:273—82.
689. Rehman AU, Santos-Cortez RL, Morell RJ, et al. Mutations in TBC1D24, a gene associated with epilepsy, also cause nonsyndromic deafness DFN86. *Am J Hum Genet* 2014;**94**:144—52.
690. Sawyer SL, Tian L, Kahkonen M, et al. Biallelic mutations in BRCA1 cause a new Fanconi anemia subtype. *Cancer Discov* 2015;**5**:135—42.
691. Gijsbers AC, Ruivenkamp CA. Molecular karyotyping: from microscope to SNP arrays. *Horm Res Paediatr* 2011;**76**:208—13.
692. Keren B, Le Caignec C. Oligonucleotide microarrays in constitutional genetic diagnosis. *Expert Rev Mol Diagn* 2011;**11**:521—32.
693. Schaaf CP, Wiszniewska J, Beaudet AL. Copy number and SNP arrays in clinical diagnostics. *Annu Rev Genomics Hum Genet* 2011;**12**:25—51.
694. Haraksingh RR, Abyzov A, Gerstein M, et al. Genome-wide mapping of copy number variation in humans: comparative analysis of high resolution array platforms. *PLoS ONE* 2011;**6**:e27859.
695. Kearney HM, Kearney JB, Conlin LK. Diagnostic implications of excessive homozygosity detected by SNP-based microarrays: consanguinity, uniparental disomy, and recessive single-gene mutations. *Clin Lab Med* 2011;**31**:595—613. ix.
696. Wang JC, Ross L, Mahon LW, et al. Regions of homozygosity identified by oligonucleotide SNP arrays: evaluating the incidence and clinical utility. *Eur J Hum Genet* 2015;**23**:663—71.
697. Rehder CW, David KL, Hirsch B, et al. American College of Medical Genetics and Genomics: standards and guidelines for documenting suspected consanguinity as an incidental finding of genomic testing. *Genet Med* 2013;**15**:150—2.

698. Delgado F, Tabor HK, Chow PM, et al. Single-nucleotide polymorphism arrays and unexpected consanguinity: considerations for clinicians when returning results to families. *Genet Med* 2015;**17**.400–4.
699. Coulter ME, Miller DT, Harris DJ, et al. Chromosomal microarray testing influences medical management. *Genet Med* 2011;**13**:770–6.
700. Riggs ER, Wain KE, Riethmaier D, et al. Chromosomal microarray impacts clinical management. *Clin Genet* 2014;**85**:147–53.
701. Ellison JW, Ravnan JB, Rosenfeld JA, et al. Clinical utility of chromosomal microarray analysis. *Pediatrics* 2012;**130**:e1085–95.
702. South ST, Lee C, Lamb AN, et al. ACMG standards and guidelines for constitutional cytogenomic microarray analysis, including postnatal and prenatal applications: revision 2013. *Genet Med* 2013;**15**:901–9.
703. Schaefer GB, Mendelsohn NJ. Clinical genetics evaluation in identifying the etiology of autism spectrum disorders: 2013 guideline revisions. *Genet Med* 2013;**15**:399–407.
704. Heil KM, Schaaf CP. The genetics of autism spectrum disorders—a guide for clinicians. *Curr Psychiatry Rep* 2013;**15**:334.
705. Sfari gene. http://gene.sfari.org; [accessed 04.06.2015].
706. Database of genomic variants. http://dgv.tcag.ca/dgv/app/home; [accessed 04.06.2015).
707. Ensembl genome browzer. http://www.ensembl.org; [accessed 04.06.2015].
708. National Center for Biotechnology Information. http://www.ncbi.nlm.nih.gov/; [accessed 03.06.2015].
709. Kearney HM, Thorland EC, Brown KK, et al. American College of Medical Genetics standards and guidelines for interpretation and reporting of postnatal constitutional copy number variants. *Genet Med* 2011;**13**:680–5.
710. Brady PD, Delle Chiaie B, Christenhusz G, et al. A prospective study of the clinical utility of prenatal chromosomal microarray analysis in fetuses with ultrasound abnormalities and an exploration of a framework for reporting unclassified variants and risk factors. *Genet Med* 2014;**16**:469–76.
711. Wei Y, Xu F, Li P. Technology-driven and evidence-based genomic analysis for integrated pediatric and prenatal genetics evaluation. *J Genet Genomics* 2013;**40**:1–14.
712. Bug S, Solfrank B, Schmitz F, et al. Diagnostic utility of novel combined arrays for genome-wide simultaneous detection of aneuploidy and uniparental isodisomy in losses of pregnancy. *Mol Cytogenet* 2014;**7**:43.
713. Sato-Otsubo A, Sanada M, Ogawa S. Single-nucleotide polymorphism array karyotyping in clinical practice: where, when, and how? *Semin Oncol* 2012;**39**:13–25.
714. Jackson-Cook C, Ponnala S. Application of chromosomal microarray. In: Idowu MO, Dumur CI, Garrett CT, editors. *Molecular oncology testing for solid tumors: a pragmatic approach, vol.* New York: Springer International Publishing; 2015.
715. College of American Pathologists. Molecular pathology checklist. http://www.cap.org; [accessed 12.05.2015].
716. Bahcall OG. Genetic testing: ACMG guides on the interpretation of sequence variants. *Nat Rev Genet* 2015;**16**:256–7.
717. Human Genome Variation Society (HGVS). http://hgvs.org/mutnomen; [accessed 12.05.2015].

7

Solid Tumor Genomics

Elaine R. Mardis

ABSTRACT

Background
Since the initial report of the finished human genome reference sequence in 2004,[1] cancer genetics research has focused on using this reference as a template for characterizing the somatic genomic alterations that underlie cancer's development and the germline genomic alterations that underlie human susceptibility to develop cancer. Because technology has enabled a transition from polymerase chain reaction (PCR)—based mutation discovery, to microarray-based copy number detection, to massively parallel sequencing (MPS) assays that provide a comprehensive somatic landscape of the cancer genome, our understanding of cancer genome alterations has increased in scope while becoming increasingly refined.[2]

Content
Modern day cancer diagnostic assays have been devised based on the cumulative knowledge gained from large-scale cancer genomics discovery efforts. These studies of tens to thousands of cancers across many tissues of origin have catalogued the genomic landscape of human cancers by MPS and often included data from RNA expression and DNA methylation. This chapter outlines aspects of these molecular assays of solid tissue malignancies in the clinical setting that have been enabled by research-based discovery work over the past approximately 20 years.

CONSIDERATIONS FOR SOLID TUMOR GENOMICS

Sampling and Preservation Methods

Solid tumors can be sampled by a variety of procedures that yield different amounts of tumor cells from which DNA, RNA, or both, can be isolated for diagnostic assays. In the ideal setting, the number of assays required for pathological evaluation of the tumor would dictate the amount of sampling done; however, the typical scenario involves a limited amount of a sample that restricts the comprehensiveness of assays that can be performed. This scenario has become even more common with the introduction of nucleic acid—based assays to the diagnostic repertoire because they require tissue that must first satisfy the standard diagnostics of microscopy-based pathology. As will be discussed, the increasing use of MPS in genomic assays of solid tumors has somewhat ameliorated this dilemma.

Briefly, solid tumors can be sampled using either fine needle aspiration or core biopsy procedures.[3] These are often done in advance of surgery or in patients with nonresectable tumors as determined by imaging, to provide a sample for pathology-based examination and diagnosis. Fine needle aspiration (21–25 gauge needle) generates a minimal amount of tissue and can consist of a few tumor cells in the fluid that is co-aspirated into the needle.[4,5] Core biopsies (18–21 gauge needle) can obtain intact and solid cores from the tumor mass. If imaging such as ultrasound or computed tomography scanning is used to guide the biopsy needle to the tumor mass, several passes are made for better sampling. Alternatively, for patients with resectable cancer without plans for neoadjuvant therapy, the tumor mass is removed at surgery and may be preserved either as a bulk resection sample or sampled with core needle biopsies that are preserved for pathology-based assays according to the clinical protocol being followed.

Once the tumor biopsy or resected tumor is obtained, the tissue can be preserved by several methods. Historically, pathology of solid tumors has focused on preservation methods that stabilize proteins and other cellular structural components, preserving tissue structure for microscopic visualization in the presence of specific staining or immunohistochemistry. Hence, tissue fixation by soaking in formalin or formaldehyde was developed to stabilize the cellular proteins. Subsequent embedding in paraffin wax is performed to create an impervious, room temperature stable substrate. Once the paraffinized tissue has solidified into a block, thin sections can be cut for subsequent characterization and diagnosis by staining and microscopic examination. Although this approach is facile and preserves tissues for long-term storage at room temperature, formalin causes crosslinking with cytosine residues in DNA and RNA. Subsequent oxidation results in breaks in the nucleic acid sugar-phosphate backbone, thereby degrading the nucleic acids. The degradation is time-dependent, however, such that preserved tissues that have been formalin-fixed and paraffin-embedded (FFPE) for less than 3 years are, in general, equivalent to fresh frozen tissues in terms of yield and quality of nucleic acids. Because diagnostic assays of solid tumors have begun to include nucleic acids, alternative preservatives and preservation methods are now being used. These methods include flash

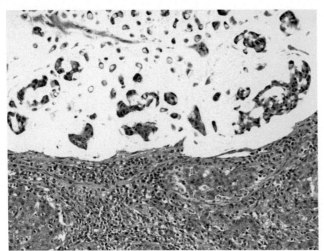

FIGURE 7.1 Illustration of an Hematoxylin and Eosin *(H&E)* Stained Tumor Section (Invasive Colon Adenocarcinoma). This adenocarcinoma has both mucinous *(top)* and medullary *(bottom)* histologic features. In addition, tumor-infiltrating lymphocytes are prominent in the portion with medullary features. These histologic features are frequently found in colon cancers that develop in patients with Lynch syndrome. Lynch syndrome is an autosomal dominant disorder caused by a DNA mismatch repair gene defect (magnification 200×, H&E stain).

freezing tissue at −80°C (dry ice acetone bath) or immersion in nucleic acid stabilizers. Stabilizer examples include freezing tissues in optimal cutting temperature (OCT) compound (Tissue-Plus OCT, Thermo Fisher Scientific), and preserving nucleic acids in tissues by addition of liquid reagents (PAXgene tissue STABILIZER, Qiagen or RNAlater, Thermo Fisher Scientific).[6]

Staining and Selection of Tumor Cells

Several staining methods have been developed to examine the preserved needle biopsy or resection materials obtained in solid tumor sampling. The most basic assay is referred to as an Hematoxylin and Eosin stain (H&E) and readily identifies tumor and normal cells in a tissue section under a light microscope (Fig. 7.1). H&E staining is not only used to diagnose cancer, but also to enumerate cancerous cells in a tissue section to provide an estimate of the percentage of tumor nuclei present. Because solid tumors are a mixture of tumor cells and various normal cells, this estimate of tumor "cellularity" indicates how tumor rich the sample is relative to other needle biopsies or other portions of the bulk tumor. Tumor features such as necrotic areas may be also identified in the microscopic evaluation. In sections with evident necrosis, a process of macro-dissection to cut away the non-necrotic sections for subsequent nucleic acid isolation can be pursued. In tumor types such as prostate and pancreatic adenocarcinomas, there is a low proportion of tumor nuclei present relative to the surrounding normal cells (eg, stroma and immune cells). In these tumor types, a laser capture microdissection (LCM) instrument can be used to isolate tumor cells from surrounding normal cells before nucleic acid isolation (Fig. 7.2). The LCM imaging system produces an image of the tumor section, and an operator uses software to identify the tumor cells for harvest. A membrane is placed adjacent to the tumor section, and the LCM fires an infrared laser pulse at each tumor cell identified for harvest, thereby affixing it to the membrane. Subsequent cutting of tissue

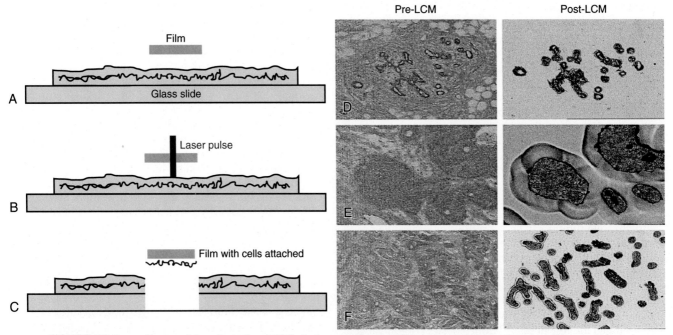

FIGURE 7.2 Laser Capture Microdissection (LCM). **A**, An optically clear film is layered over a tissue section on a glass slide that is viewed through a microscope to locate the cells of interest. **B**, A laser pulse then affixes the film to the tissue of interest *(black)* that can then be removed from the slide for further processing **(C)**. LCM can also remove the cells that are not of interest from the slide. For example, selection of residual normal **(D)** breast glands, **(E)** carcinoma in situ, or **(F)** infiltrating ductal carcinoma of the breast is possible. (Photomicrographs reprinted with permission from Palmer-Toy DE, Sarracino DA, Sgroi D, LeVangie R, Leopold PE. Direct acquisition of matrix-assisted laser desorption/ionization time-of-flight mass spectra from laser capture microdissected tissues. *Clin Chem* 2000;46:1513–1516. Copyright AACC Press.)

and membrane is performed by a ultraviolet laser, completing a laser-induced forward transfer for cellular isolation. Once the desired number of tumor cells is harvested, the membrane goes through a series of processing steps to isolate DNA or RNA (or both) from the captured cells.

Another staining-based method to identify tumor-specific antigens is immunohistochemistry; in this technique, a protein- or protein-epitope–specific antibody is coupled to an enzyme (eg, horseradish peroxidase) to identify tumor cells in a tissue section expressing that protein. Examples of immunohistochemistry stains include those for estrogen and progesterone receptors, and the cellular proliferation marker Ki-67 (MIB-1).

GENOMIC ANALYSIS OF SOLID TUMORS

Background

Although we have known for many decades that cancer's origins lie in changes to the cellular genome, only recently have technologies become available to profile the mutations in genes. Initial gene cloning efforts in the early 1980s identified the chromosomal locations and sequences of many oncogenes and tumor suppressors in the human genome.[7-9] The decoding of the human genome, coupled with technological advances, opened the door to genome-wide studies of cancer. For example, learning the sequences of human genes enabled the construction of microarrays to query RNA expression in tumors (described in Chapter 4). By using advanced bioinformatic analysis approaches (eg, clustering algorithms),[10] similarities and differences in gene expression across tumors from a single tissue site (eg, lung adenocarcinoma) were revealed.[11] Clusters of gene expression, when correlated with other pathology-based categories, revealed subtypes within a given tissue site, such as the intrinsic subtypes of breast cancer.[12] Similarly, single nucleotide polymorphism–based microarrays could be used to identify gross-scale chromosomal aberrations by comparing normalized signal strength between tumor and normal DNA.[13] As explored in the following, conducting research-based inquiries of cancer genomes with these tools provided the means by which the clinical usefulness (diagnosis, prognosis, treatment decision) of the resulting classifications was demonstrated, thereby providing a rationale for the approach to translate them into clinical laboratory assays. More recently, the introduction of MPS platforms and associated methods for exploring cancer genes, genomes, and transcriptomes (expressed RNAs) have supplanted foundational technologies such as capillary sequencing and microarrays. MPS methods have the potential to produce more quantitative data while simultaneously expanding the identification of somatic alterations. The first application of MPS to decode a cancer genome was published in 2008[14] when Ley and colleagues sequenced and compared the whole genome sequence (WGS) data from a patient with FAB M1 (normal karyotype) acute myeloid leukemia (AML) to a matched normal skin sample. The second WGS case published[15] was also on a single patient AML sample, and it revealed an unexpected mutation in the gene *IDH1* (a metabolic enzyme in the glycolytic pathway). This *IDH1* mutation was then characterized as a recurrent mutation in a panel of 188 AML samples and associated with a poor prognosis. From these single patient beginnings, the field of cancer genomics has exploded to characterize tens of thousands of cases by 2015.

> **POINTS TO REMEMBER**
>
> **Cancer Genomics**
> - Gene expression microarrays were initially used to characterize large numbers of human cancers.
> - Cancer gene cloning efforts and the completed Human Genome Sequence enabled the use of PCR and sequencing efforts to catalog cancer mutations.
> - MPS enables the identification of somatic alterations in targeted gene panels, in all known genes, or the whole genome.
> - The scale of massively parallel cancer sequencing has permitted large studies of DNA and RNA from the major solid tissue malignancies.

Large-Scale Discovery Efforts

One consequence of rapid, high-throughput and inexpensive sequencing has been the production of large data sets that explore the mutational, transcriptional, and methylation landscape of thousands of tumors representing major human cancer types. Examples of these large-scale discovery projects include The Cancer Genome Atlas (http://cancergenome.nih.gov), the International Cancer Genome Consortium (https://icgc.org), the Pediatric Cancer Genome Project (http://www.pediatriccancergenomeproject.org/site), and numerous other government and privately funded cancer genomics efforts. The resulting tumor type-specific "omics" catalogues revealed several unexpected discoveries, including the fact that many mutated genes occur in multiple tumor types[16] and that different tissue sites have widely different mutational burdens.[17] These efforts also identified new classes of mutated genes that had previously not been considered as contributing to cancer development. Included in these new classes of genes were proteins involved in the spliceosome complex (U2AF3, SF3B1), proteins involved in cellular metabolism (IDH1, IDH2), and proteins that contribute to DNA packaging in histones (H3.3, ARID1A, and ARID1B). Not surprisingly, the greatly expanded characterization of the cancer mutational landscape afforded by these large-scale projects revealed that most known cancer-associated genes had a multitude of previously unknown mutations, some of which were recurrent. Due to the magnitude of these efforts in a relatively short time span, the overwhelming number of mutations discovered by genomic methods have yet to be functionally studied to determine their effects on the resulting protein.

Once significant progress was made in the individual cancer-specific studies, for example, meta- or "pan can" analyses of these catalogues were pursued, further reinforcing the similarities and differences between tissue sites by virtue of integration across various "omic" data sets. The similarities, in particular, challenge long-held notions about tissue site specificity of therapeutic approaches and introduce the concept that oncogenic "drivers" present in cancers from different tissues could respond to similar targeted therapeutic interventions. These metaanalyses further reinforced the complexity of cancer in that each tumor arises due to a unique and intricate interplay of molecular events. Although some occur more commonly than others, the impact on protein function and amount ultimately manifests in activated or repressed cellular pathways. Hence, cancer is a disease of pathway alterations, rather than a specific gene and/or protein alteration, and therapeutic decision-making should occur using this framework.

Pre-Massively Parallel Sequencing Approaches to "Hotspot" Characterization

The discovery of oncogenes and their activating mutations (also known as "hot spot" or "gain of function" mutations) emerged from gene cloning efforts in the 1980s and 1990s. The description of PCR by Kary Mullis and Fred Faloona[18] tremendously facilitated the selective amplification and sequencing of these genes from DNA isolates of solid tumor blocks. The concomitant development of an automated fluorescent Sanger sequencing instrument by Leroy Hood's group[19] and its commercialization ultimately led to a combined PCR and sequencing assay to detect these hot spot mutations in a few days' time. Automated software to identify variants in capillary electropherograms further facilitated rapid analysis and interpretation of sequencing results (see also the Phred Quality Score, Box 2-5 in Chapter 2). The clinical usefulness of hotspot characterization to identify a drug target in solid tumors was initiated by three studies in lung adenocarcinoma published in 2004.[20-22] These results demonstrated that patients who responded to a new class of targeted therapies called tyrosine kinase inhibitors carried mutations in the tyrosine kinase domain of the epidermal growth factor receptor (EGFR) gene. The correlation of these mutations to treatment response initiated a paradigm in which the clinical assay for hotspot mutations became the companion diagnostic for the targeted therapy.

RNA-Based Approaches

The development and use of microarrays to query gene expression and correlate the analyzed data to clinical features has been widely used in the research setting to identify clinically relevant subtypes within specific malignancies. Samples that group into a specific subtype were evaluated in the context of correlative data types such as patient outcome (eg, overall survival, disease-specific survival, or progression-free survival) or treatment, or both. Correlative gene expression analyses can subtype disease and determine treatment as tested in clinical trials. An example of the translation of gene expression assays to clinical tests is the US Food and Drug Administration (FDA) —approved Mammaprint microarray diagnostic from Agendia (www.agendia.com) that originated from a 70-gene breast cancer expression profiling test developed in the Netherlands.[23] Results of this assay indicate whether node negative estrogen receptor positive (ER+) or negative (ER−) patients are at low risk or high risk for disease recurrence. When combined with other clinical risk factors, the Mammaprint score contributes to determining whether a patient will benefit from adjuvant chemotherapy.

The expression of genes previously determined to predict outcomes or characterize subtypes can also be compared with reverse transcriptase quantitative PCR (RT-qPCR). This method is described in Chapter 4, and begins with total RNA isolated from tumor cells, converted to DNA, and subjected to gene-specific amplification by PCR in a specialized instrument that acquires fluorescence at each amplification cycle. Such "real-time" data are then analyzed for each gene to calculate its expression relative to those of previously selected calibrating genes (reference genes). Typically, reference genes are specific to the tissue and/or tumor type of interest and are selected because their expression is constant across multiple samples. By analyzing the qPCR-based gene expression levels of specific subtype-determining genes from banked samples with known outcomes, a classifier algorithm can be derived that returns a risk of recurrence (ROR) score. Oncologists may use the ROR score in combination with other pathology assay results to determine which treatment the patient will receive. An example of such an assay is the Oncotype Dx breast cancer risk of recurrence assay used to classify ROR for women with node-negative ER+ breast cancer.[24]

A third type of RNA quantitation uses a specialized instrument that detects and identifies specific gene transcripts based on combinations of fluorophores ("barcodes") with which they are labeled. In this system, shown in Fig. 7.3, an initial hybridization step permits gene-specific "capture" probes to hybridize to the extracted RNA from the tumor sample. A second "reporter" probe that contains the fluorescent barcode for that gene also hybridizes near the capture probe hybridization site on the transcript. The resulting mixture is placed onto a silicon substrate derivatized with streptavidin, and the biotin-labeled capture probes immobilize each labeled transcript on the substrate. The application of an electric current across the substrate linearizes the RNAs and orders the fluorophores. A scanning step then excites the fluorescent dyes and detects their emission wavelengths across an x-y coordinate grid. The scanning step enumerates each barcoded molecule on the silicon substrate, thereby quantitating, and through software-based decoding, identifying each transcript of interest and its absolute expression level. This approach was commercialized by Nanostring, Inc. and currently forms the basis of a clinical assay for breast cancer risk of recurrence (Prosigna).

MASSIVE PARALLEL SEQUENCING OF SOLID TUMOR SAMPLES

Clinical assays of solid tumors using MPS to identify somatic mutations in multiple genes ("gene panels") or in all annotated genes (the "exome") have been developed for several reasons, including: (1) large-scale, MPS-based discovery efforts have identified mutated genes in the major adult and pediatric cancers; (2) specific genes and/or mutations have been studied in clinically annotated sample sets, and their mutational status correlates with outcome (and is therefore prognostic); and (3) novel small molecule- and antibody-based targeted therapies have been developed to address specific somatic alterations, and therefore, the mutational signature of a patient's somatic genome can identify targeted therapies based on predicted gene—drug interactions.

Although testing multiple genes from a single tumor DNA isolate could be accomplished using combinations of individual clinical assays to evaluate individual gene and/or hotspot mutation status, often tumor material is limited due to the size of a biopsy and the need to use the material for other more conventional pathology assays such as H&E and immunohistochemistry. Hence, MPS-based assays make more efficient use of precious diagnostic material, especially when so

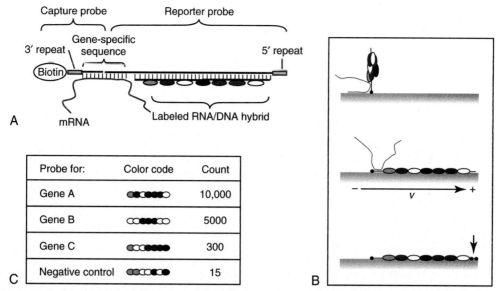

FIGURE 7.3 Single Molecule Detection of Messenger RNA (mRNA). **A**, A pair of probes (one capture probe and one reporter probe) hybridizes to the RNA target in solution through gene-specific probe sequences. The reporter probe has seven color segments, each segment is made of approximately 900 RNA bases that are labeled with approximately 100 hundred fluorophores of one type. The labeled portion of the probe is a DNA/RNA hybrid that can be observed as a approximate 3-nm fluorescent spot. **B**, (top) The target complex is immobilized on a streptavidin-coated slide through the biotin moieties on the capture probe. (Middle) An electrical current is applied, and the complex is stretched. (Bottom) The reporter probe is immobilized in extended form by biotin-labeled oligonucleotides complementary to its 5′ repeat sequence. **C**, The color code of the probe fluors is read by an epi-fluorescent microscope, and each unique probe is counted. Normally, a number of negative control probes are present in the hybridization solution to establish nonspecific background counts.

many genes that are potentially mutated can provide important information regarding patient care.

Borrowing from the design of PCR-based hotspot mutation assays, MPS assays often study only the tumor-derived DNA. This is an economic decision because performing the same assay on the matched normal tissue (adjacent nonmalignant or peripheral blood monocytic cell–derived DNA from blood) essentially doubles the cost of the assay. However, there is a fundamental difference between a hotspot assay and an MPS assay; namely, the breadth of genomic territory studied by MPS assays is much larger. As such, variants that lie outside of known hotspots will be identified and may still contribute to the patient's prognostic or therapeutic diagnosis. Further, as shown by focused and large-scale cancer genomics studies, known germline susceptibility loci, such as *BRCA1*, *BRCA2*, *TP53*, and *APC*, are also mutated in the somatic genome. Hence, comparing tumor to normal gene sequences may have important consequences for understanding whether variants that are identified are inherited or acquired.[25] Obviously, determining somatic versus germline origin is important for the patient because the former may help to identify a therapeutic path and the latter will ethically and responsibly involve genetic counseling and enable possible further testing of parents, children, and siblings.

Due to the heterogeneity of solid tumors and the increasing knowledge that certain mutations in genes, if present at low levels in the tumor cell population, can predict early onset of acquired resistance to targeted therapies,[26] MPS-based assays must provide a high depth of tumor sequence data coverage for each gene in the assay. Coverage is defined as the number of sequencing reads that align uniquely at a given gene locus, where each read providing coverage depth must be from a unique DNA fragment. High coverage can be defined as 300- to 1000-fold read depth and can be challenging to obtain when MPS libraries are constructed from low amounts of input DNA because there are fewer unique molecules available for ligation to the adapters in the library construction step.[27] The issue of uniqueness is exacerbated because the use of PCR to amplify the completed library can lead to PCR-based "jackpotting" (preferential amplification of certain sequences). This, in turn, can bias the representation in the completed library. In general, with 500-fold or higher coverage, detecting mutations can be accomplished even from low percentage tumor cellularity samples (10% to 20%), but detecting low prevalence mutations becomes less likely with decreasing tumor cell percentages. High coverage is also important to attain because not all probes capture complementary sequences with the same avidity. When this deficiency is combined with representation bias from MPS processes downstream of capture, lowered coverage at certain sites may result. Hence, higher coverage overall may reduce or eliminate low and/or no coverage sites, thereby reducing the incidence of false negative results or type II errors, wherein a mutation is missed due to lack of coverage in the tumor or the normal gene.

Whole Genome Sequencing

Ideally, the most unbiased characterization of each cancer sample by MPS-based methods should be pursued to optimize the likelihood of identifying all druggable targets in the genome. This is best accomplished using WGS, which can deliver the totality of point mutations, small insertions and deletions, structural variants (including copy number alterations), and other large chromosomal rearrangements in a tumor-to-normal genome comparison. Different technologies for performing WGS are covered in Chapter 4. However, at this time, WGS is rarely performed as a clinical assay that can inform solid tumor therapeutic interventions due to the following limitations. First, the genome is quite large (3 gb), and although new MPS instrumentation can produce the data rapidly (eg, the Illumina HiSeq X produces 30-fold coverage for 16 human genomes in 3 days), the effort to analyze a whole genome comparison of tumor tissue to matched normal tissue and compile a clinical report is in conflict with the time frame needed to inform the oncologist's decision-making process. Second, due to the need for high coverage of the tumor genome, generating the necessary 100-fold or higher WGS coverage needed for confident variant detection remains cost prohibitive. Third, most of the somatic variants will be uninterpretable in terms of their impact on tumor progression because they will lie outside of genes. Fourth, many variants in the coding sequences will not be interpretable in terms of their impact on the protein's function as a potential driver of disease development (ie, gain or loss of function). However, the latter issue is not unique to WGS assays, but reflects on all MPS assay results in terms of the difficulty of interpreting variants. Due to these limitations, among others, it is presently rare (circa 2015) for whole genome comparisons of tumor tissue to normal tissue to be assayed in the clinical diagnostics setting.

Selective Enrichment by Hybridization Capture

Locus-selective or hybridization capture methods have been developed that use synthetic probes designed to hybridize the human loci and/or genes of interest as a means of selectively isolating sequences from a whole human genome library. MPS of the captured sequences is followed by alignment of the reads to the human genome reference sequence, identifying the loci of interest for mutational analysis by applying a Browser Extensible Data format file[28] to delineate the specific genomic regions that should be examined for variants in the tumor tissue compared with normal tissue, and reporting any mutations that are identified. Depending upon the assay, either a select set of genes or hotspots, or the entire annotated gene set of the human (the exome) can be captured. There are several commercially available exome reagents, including "clinical" exome reagents that offer enhanced hybridization times of 2 to 4 hours (compared with approximately 72 hours for research reagents).

As the number of loci assayed by this approach increase, so do both the analysis time and the complexity of the mutational signature. The decision to include drug-targetable genes that carry novel, non-canonical mutations with untested drug interactions is one challenge that can arise in the clinical interpretation of these assays.

Although there is a desire to develop assays with small numbers of genes and/or hotspot loci that will (1) keep cost per assay low, (2) speed the interpretation of data, and (3) decrease the detection of non-canonical mutations, the ability of gene-selective hybridization capture to subset the genome does have a lower limit. In particular, when targeting loci that total less than approximately 300 kb, inefficiencies of selective hybridization capture predominate, requiring increased sequencing coverage to achieve the desired coverage of the targeted genes. In particular, the lower the target space for hybrid capture, the larger the effect of so-called "off target" reads. These spurious hybridization events result in the capture of nontarget sequences by capture probes and effectively decrease the overall coverage achieved from on-target reads.

Multiplex Polymerase Chain Reaction Approach

Because of the lower limit described for hybrid capture, one alternative that has been used in clinical assays of solid tumors is multiplex PCR. In this type of assay, careful design of PCR primer pairs that target the selected loci for mutation detection achieves the coincident amplification of all loci in a single reaction. Multiplex PCR has the distinct advantages of: (1) selectively amplifying smaller target regions of the genome below that which is optimal for selective hybridization capture; (2) requiring only 5 to 10 ng of DNA for the PCR; and (3) focusing mutation detection on a relatively small portion of the genome in an efficient manner. These assays are also relatively inexpensive to perform, and by post-PCR addition of DNA barcoded linkers for MPS, they can be multiplexed with other patients onto a single MPS instrument run. There are obvious limitations to multiplex PCR that include: (1) the ability to design PCR primers with similar T_m, uniqueness, and lack of interprimer complementarity to diminish amplification artifacts (eg, primer dimers, hybrid amplicons); and (2) the presence of pseudogenes and highly conserved regions of gene families may preclude unique amplification of specific gene sequences. PCR primers are often designed with FFPE-associated degradation in mind to produce small products of less than 100 bp; therefore, these require a large number of primers to cover a large gene, which further complicates the design process and/or complicates obtaining complete coverage from the resulting amplicons. Despite the limitations, multiplex PCR assays are widely used in clinical testing of solid tumors. The challenges of overcoming these limitations can be circumvented by purchasing prequalified multiplex PCR primer sets from commercial vendors, many of which frequently include tested medically relevant genes for different types of clinical questions, including cancer mutation hotspots.

RNA Sequencing of Solid Tumor Isolates

RNA isolates from cancer samples can be assayed using MPS technologies (RNA-seq) by simply converting the RNA to DNA via RT before library construction. There are several nuances to this general statement that require further detail, as follows. First, RNA can be reliably amplified if the starting quantities are low, such as isolates from a fine needle aspiration or core biopsy. Second, coding RNAs can be selected from a total RNA population by implementing a polyA selection step. This selection works because messenger RNAs (mRNAs) are polyadenylated at their 3' ends and thereby can hybridize to superparamagnetic beads with covalently attached polyT sequences. Typically, mRNA represents only 2% of the total RNA population, so the resulting yields will be quite small if one begins with a limited total RNA isolate. In practice, polyA selection is not advised if the starting amount of total RNA is less than 2 μg. Third, because the number of genes transcribed in any given tissue and/or tumor is quite variable in the number and range of expression levels, the notion of coverage as presented for DNA assays does not typically apply to RNA-seq. Rather, a set number of sequencing reads are produced, aligned to the genome, and evaluated. As discussed, FFPE preservation can cause fractionation of the RNA in a tumor sample, so an initial evaluation of the quantity and intactness of the RNA isolate is advised when there are sufficient quantities obtained to permit an initial quality control (QC). This evaluation can be accomplished using a variety of sensitive devices such as the Agilent Bioanalyzer with an RNA Picochip or Agilent TapeStation. As a general rule, RNA that is degraded to an average length of 300 bp or less will not perform well in RNA-seq library construction.

There are valid reasons to assay the transcribed genes in a solid tumor sample in the clinic, including: (1) DNA sequencing assays do not reflect higher order changes in the tumor genome, such as methylation or chromatin packaging that can influence the amount of RNA produced from a gene; (2) in high mutation load tumors such as melanomas and lung squamous or adenocarcinomas, RNA-seq can indicate which of the thousands of mutations identified from DNA sequencing assays are actually being transcribed and could therefore be targeted with a specific therapy; (3) alternatively spliced transcript (isoform) expression can only be detected at the level of RNA and may be important in a diagnostic, prognostic, or therapeutic context; and (4) predicted gene fusions from DNA assays can be verified by the presence of the predicted fusion transcript in the RNA-seq data. Despite these advantages, RNA-seq is rarely pursued in the clinical setting for several reasons. Adding another assay increases the cost of clinical testing and adds time to the delivery of information back to the ordering physician. Furthermore, RNA is an unstable molecule, and preservation techniques such as FFPE degrade RNA more noticeably than DNA, especially if the temperature of the paraffin bath is above 65°C. RNA can be reliably amplified before sequencing when only limited amounts are available from a clinical sample. Alternatively, limited quantities of RNA or degraded RNA from FFPE-preserved tissue can be sequenced by subjecting the MPS library to an intermediate selective hybridization capture step with the same types of reagents used for DNA.[29] Perhaps most daunting is the bioinformatic analysis of RNA-seq data, because there are many different algorithms to evaluate the data once reads are aligned to the human genome reference, depending upon whether straightforward determination of gene expression is the goal, or identifying fusion transcripts, alternative transcript expression, or other types of specialized inquiry are pursued.[30] Despite the challenges, there is increasing recognition that DNA assays alone do not provide a complete picture of the drug-specific vulnerabilities of a solid tumor.

Detection of Cancer Mutations in Body Fluids

Cancer cells are characterized by their rapid growth and proliferation, which infers that they turn over rapidly. In so doing, the cells release DNA into bodily fluids that can be assayed by sequencing and other nucleic acid detection techniques. Chapter 9 describes high-sensitivity, blood-based detection of cancer-specific mutations from circulating tumor cells and circulating tumor DNA. Also referred to as a "liquid biopsy," these sources provide tumor DNA for either diagnosis or therapeutic response monitoring that requires only a blood draw and specific processing steps to obtain the analyte without biopsy. In addition to blood-based tests, several commercially available noninvasive assays use either feces or urine as the analyte and survey a single marker or small number of markers for evidence of cancer-specific DNA or RNA. Hence, these are screening or diagnostic assays and are primarily used for detection of cancer.

One such screening assay is marketed as Cologuard by Exact Sciences and uses stool DNA to detect 11 distinct colorectal cancer biomarkers that are a combination of methylation markers (NDRG4 and BMP3) and mutations (eg, KRAS), as well as a hemoglobin immunoassay. This assay was compared in a clinical trial of 9989 participants to a standard fecal immunochemical test (FIT) in which 0.7% of participants had a colon cancer diagnosis and 7.6% had advanced precancerous lesions by colonoscopy.[31] The assay had 92.3% sensitivity to detect colorectal cancer compared with the 73.8% sensitivity of the FIT assay. Sensitivity for advanced precancerous lesions was 42.4% compared with 23.8% for FIT. However, the multitarget stool assay had a lower specificity compared with the FIT test (89.8% compared with 96.4% for FIT) in patients with negative colonoscopy. As a result, Cologuard is an FDA-approved assay in which a positive test result is suggestive of a follow-up colonoscopy to confirm the presence of cancer in the rectum or colon.

There are two available urine-based noninvasive tests that detect prostate cancer, both of which detect specific RNA markers. These tests use urine obtained subsequent to a digital rectal examination of the prostate, which is sufficient mechanical stimulation to release prostate cancer cells into the urine. The Progensa assay from Hologic gained FDA approval in 2012. This quantitative real-time PCR assay evaluates the expression of two genes, prostate specific antigen *(PSA)* and *PCA3*, both of which are overexpressed in prostate cancer. The data analysis compares *PCA3* to *PSA* expression as a ratio that generates a PCA3 score. This score helps determine the need for repeat prostate biopsies in men who have had a previous negative biopsy. In a clinical trial, the Progensa assay had a negative predictive value of 90%. Following a positive assay by Progensa, prostate biopsy is needed for a definitive diagnosis.

The Mi prostate score test is a laboratory-developed test by the University of Michigan that quantifies both *PCA3* and the *TMPRSS2:ERG* fusion transcript (http://www.mlabs.umich.edu/files/pdfs/MiPS_FAQ.pdf). The *TMPRSS2:ERG* gene fusion occurs in 47% of prostate cancers as a known driver event. This test was validated in a cohort of 1225 patients with no history of prostate cancer[32] and was shown to enhance the usefulness of serum PSA for predicting prostate cancer risk and clinically relevant cancer on biopsy.

> **POINTS TO REMEMBER**
>
> **Cancer Genomics Assays**
> - DNA and RNA can be isolated from cancer cells for further study.
> - There are multiple technology platforms for directed gene assays of mutations or RNA expression levels.
> - MPS assays of DNA permit a broad survey of genetic variation and can detect both germline and somatic alterations.
> - Targeted gene assays can detect specific mutations from DNA or quantify both expression and fusion transcripts in RNA.

INTERPRETING SOMATIC ALTERATIONS IN A CLINICAL CONTEXT

The use of MPS-based solid tumor diagnostics is increasing across clinical laboratories, both commercial and academic, largely due to the increasing knowledge of somatic alterations in cancers, to the increasing numbers of targeted therapies designed to interact with known cancer-relevant alterations, and to the demands of patients and oncologists to add this information to the evidence being considered in determining the treatment regimen for the patient. Progress in this area is confounded by the need to relate somatic alterations to other clinical data, such as the diagnosis of a specific tumor type or subtype, the range of possible patient outcomes, and therapeutic response. Additional studies of the functional impact of a somatic alteration, how that functional impact may contribute to the onset or progression of the cancer, or how and/or whether the alteration interacts with a variety of potential therapies based on putative drug–gene interactions are often needed. Drug studies typically involve preclinical testing in cell lines or other disease models, followed by clinical trials that require 1 or more years to compare patient outcomes when patients carrying the gene alteration(s) are treated with a specific therapy in comparison to patients treated with the standard of care. In essence, there is often a significant delay between the research-based discovery of a somatic alteration and when its full clinical implication can be established. The clinical sequencing pipeline and how to determine which of the identified mutations in a cancer biopsy should be evaluated further in terms of their prognostic, diagnostic, or therapeutic value have been described.[33,34] There are also publicly available curated cancer mutation databases that permit a look-up of known cancer genes and mutations that offer prognostic, diagnostic, and therapeutic information, including TARGET (www.broadinstitute.org/cancer/cga/target) and CIViC (www.civicdb.org).

Prognostic Interpretation

Understanding the prognostic or outcome-related value of a specific gene alteration depends upon a number of factors, such as the prevalence of the disease, the time to establish outcome, the specific outcome measure, and the uniformity of patient care in each group. The time to establish outcome can vary widely; for example, patients with lung squamous cell carcinoma or glioblastoma typically present with metastatic disease or progression in 6 to 18 months after a primary diagnosis, whereas ER+ breast cancer may relapse after 15 to 20 years. The defined outcome may be overall survival, disease-specific survival, complete pathologic response, or other responses. Some examples of prognostic mutations and their ability to assort to different disease subtypes are seen in gliomas,[35] colorectal,[36] endometrial,[37] and ovarian[38] cancers.

Diagnostic Interpretation

Multigene capture panels that detect alterations in known cancer susceptibility genes such as TP53, APC, BRCA1/2, and others in germline DNA can be used to identify inherited cancer susceptibility. These assays are discussed in Chapter 6. Diagnostic characterization of solid tumor DNA based on somatic alterations is taking on increased significance due to the availability of targeted therapies that correspond to these

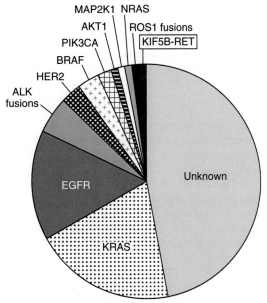

FIGURE 7.4 Molecular Subsets of Lung Adenocarcinoma. The clinically important driver mutations and fusions identified to date from molecular studies of lung adenocarcinoma. The width of the chart wedge for each driver event corresponds to its percentage distribution in the disease. (Reprinted by permission from Macmillan Publishers Ltd. From Pao W, Hutchinson KE. Chipping away at the lung cancer genome. *Nat Med* 2012;18:349–351.)

somatic alterations. Non-small cell lung cancers, especially adenocarcinomas, perhaps provide the richest example of sequencing-based diagnosis and therapeutic option guidance among the solid tumors. In 2012, the spectrum of genomic alterations in genes that drove the development of non-small cell lung adenocarcinomas encompassed approximately 50% of the disease, of which approximately 40% were therapeutically targeted (Fig. 7.4).[39] Increasingly, a gene panel assay is pursued to sequence these genes or gene fusions, and can indicate one or several targeted therapies that represent the first line of therapy. This paradigm shift represents a remarkable transition in medical practice over just a 10-year period from when lung adenocarcinoma mutations that predicted response to tyrosine kinase inhibitors were first described, and is indicative of the directions in diagnostic pathology for other solid cancers. However, there are several challenges to this paradigm that merit further discussion.

Challenges of Variants of Unknown Significance

As somatic gene panel assays expand to generate sequence data across the length of each gene rather than from focused hotspots, more variants are identified. In particular, the resulting breadth of variant discovery permits the detection of variants for which no clear indication of response for a given therapeutic indication might be available. These variants of unknown significance (VUS) raise issues with regard to whether the result should be reported because the therapeutic indication is uncertain. Although basic science experiments may eventually characterize the functional impact of all variants in all cancer-related genes, most molecular pathology groups that are conducting selective hybridization capture panel assays of cancers are instead developing local databases to catalogue VUS in an effort to initially provide a record of whether the variant has been seen in previously completed assays. Over time, data mining of electronic medical records for patients with variants, targeted therapy, and response data recorded may transition VUS to therapeutic indications.

Challenges of Off-Label Indications

The current process for clinical trials of new therapies in cancer toward FDA approval rests on a paradigm in which trials compare each new therapy in a specific tumor type to the standard of care therapy. Hence, cancer therapies are tested and approved in a disease site—specific manner. By contrast, cancer genomics efforts have demonstrated that known cancer-initiating gene alterations occur in multiple disease and/or tissue sites.[16,40] This fact, in turn, raises the possibility that a given targeted therapy might be effective for multiple solid tumor types that all carry the same somatic alteration. There is support for multiple tissue—site therapeutic effectiveness, including the FDA-approved use of the kinase inhibitor Gleevec (imatinib) in both chronic myeloid leukemia (a blood cancer) targeting *ABL* kinase and in gastrointestinal stromal tumors targeting *KIT*.[41] Another example is EGFR-targeting therapies with FDA approval for use in lung adenocarcinomas, head and neck squamous cell carcinomas, and colorectal cancers with somatically altered *EGFR*.[42-44] These approved tissue sites each required a separate series of clinical trials to gain FDA approval. However, in solid tumor assays with gene panels, off-label indications will arise due to the shared nature of cancer-initiating mutations across different tissue sites. This challenge occurs when a gene with a functional mutation is detected that indicates a therapy not yet approved in the specific tissue site of origin. Current changes in clinical trial design are attempting to address this issue by creating either a trial design such as National Cancer Institute's Molecular Analysis for Therapy Choice (MATCH),[45] which assigns patients with a druggable mutation and/or gene to a specific therapy (basket trial) or a design that studies only a single tissue site but permits multiple therapies to be assigned depending upon the mutation identified (umbrella trial).[46]

Importance of Tissue Biology

A further confounder of the shared nature of cancer-initiating mutations that can defeat a seemingly clear therapeutic indication is that the accompanying tissue biology may render the therapy ineffective in a particular tissue. One such example is the presence of point mutations in *BRAF* with corresponding targeted therapies shown by clinical trials to provide therapeutic benefit in melanomas.[47] However, a clinical trial of patients with colorectal adenocarcinomas carrying the same functional *BRAF* V600E mutation was stopped due to a lack of therapeutic benefit. Subsequent research from Bernard's laboratory showed definitively this lack of benefit in colorectal cancer was due to up-regulation of other tyrosine kinases that compensated for the inhibition of mutated *BRAF*.[48] Hence, tissue biology must be considered in the prediction of therapeutic response. Importantly, this study also demonstrated that a combination of BRAF and EGFR inhibitors, which were predicted to alleviate the compensatory mechanism, rendered a therapeutic benefit.[49]

SUMMARY AND CONCLUDING REMARKS

Solid tumor gene-based assays are in a dramatic transition to multigene testing. This trend is driven by (1) the increasing use of selective hybridization capture or other targeting approaches coupled with MPS, (2) the discovery of somatic alterations in the genomes of cancer cells, and (3) the emerging number of targeted therapies that address specific cancer-initiating genes or gene families. The comprehensive nature of multigene assays creates specific challenges, mainly due to the impedance mismatch between cancer mutation discovery and the functional characterization of those alterations that either contribute to the development and progression of cancer or do not contribute to it. The need to accelerate testing of new therapeutics across multiple tissue sites at once is slowly being addressed by new clinical trial designs to identify the most effective therapies for a given gene, alteration, and tissue site. Ultimately, the widespread use of multigene testing for cancer diagnosis, prognosis, and therapeutic decision-making rests in the demonstration of clinical usefulness, which is established for some tissue sites, but as yet, not for all.

REFERENCES

1. International Human Genome Sequencing C. Finishing the euchromatic sequence of the human genome. *Nature* 2004;**431**: 931–45.
2. Koboldt DC, Steinberg KM, Larson DE, et al. The next-generation sequencing revolution and its impact on genomics. *Cell* 2013;**155**:27–38.
3. Renshaw AA, Pinnar N. Comparison of thyroid fine-needle aspiration and core needle biopsy. *Am J Clin Pathol* 2007;**128**: 370–4.
4. de Biase D, Visani M, Baccarini P, et al. Next generation sequencing improves the accuracy of KRAS mutation analysis in endoscopic ultrasound fine needle aspiration pancreatic lesions. *PLoS ONE* 2014;**9**:e87651.
5. Marshall D, Laberge JM, Firetag B, et al. The changing face of percutaneous image-guided biopsy: molecular profiling and genomic analysis in current practice. *J Vasc Interv Radiol* 2013; **24**:1094–103.
6. Staff S, Kujala P, Karhu R, et al. Preservation of nucleic acids and tissue morphology in paraffin-embedded clinical samples: comparison of five molecular fixatives. *J Clin Pathol* 2013;**66**: 807–10.
7. Friend SH, Bernards R, Rogelj S, et al. A human DNA segment with properties of the gene that predisposes to retinoblastoma and osteosarcoma. *Nature* 1986;**323**:643–6.
8. McGrath JP, Capon DJ, Smith DH, et al. Structure and organization of the human KRAS proto-oncogene and a related processed pseudogene. *Nature* 1983;**304**:501–6.
9. Popescu NC, Amsbaugh SC, DiPaolo JA, et al. Chromosomal localization of three human RAS genes by in situ molecular hybridization. *Somat Cell Mol Genet* 1985;**11**:149–55.
10. Eisen MB, Spellman PT, Brown PO, et al. Cluster analysis and display of genome-wide expression patterns. *Proc Natl Acad Sci USA* 1998;**95**:14863–8.
11. Welsh JB, Zarrinkar PP, Sapinoso LM, et al. Analysis of gene expression profiles in normal and neoplastic ovarian tissue samples identifies candidate molecular markers of epithelial ovarian cancer. *Proc Natl Acad Sci USA* 2001;**98**:1176–81.

12. Sorlie T, Perou CM, Tibshirani R, et al. Gene expression patterns of breast carcinomas distinguish tumor subclasses with clinical implications. *Proc Natl Acad Sci USA* 2001;**98**:10869–74.
13. Pinkel D, Segraves R, Sudar D, et al. High resolution analysis of DNA copy number variation using comparative genomic hybridization to microarrays. *Nat Genet* 1998;**20**:207–11.
14. Ley TJ, Mardis ER, Ding L, et al. DNA sequencing of a cytogenetically normal acute myeloid leukaemia genome. *Nature* 2008;**456**:66–72.
15. Mardis ER, Ding L, Dooling DJ, et al. Recurring mutations found by sequencing an acute myeloid leukemia genome. *N Engl J Med* 2009;**361**:1058–66.
16. Kandoth C, McLellan MD, Vandin F, et al. Mutational landscape and significance across 12 major cancer types. *Nature* 2013;**502**: 333–9.
17. Lawrence MS, Stojanov P, Polak P, et al. Mutational heterogeneity in cancer and the search for new cancer-associated genes. *Nature* 2013;**499**:214–8.
18. Saiki RK, Scharf S, Faloona F, et al. Enzymatic amplification of beta-globin genomic sequences and restriction site analysis for diagnosis of sickle cell anemia. *Science* 1985;**230**:1350–4.
19. Smith LM, Sanders JZ, Kaiser RJ, et al. Fluorescence detection in automated DNA sequence analysis. *Nature* 1986;**321**:674–9.
20. Paez JG, Janne PA, Lee JC, et al. EGFR mutations in lung cancer: correlation with clinical response to gefitinib therapy. *Science* 2004;**304**:1497–500.
21. Lynch TJ, Bell DW, Sordella R, et al. Activating mutations in the epidermal growth factor receptor underlying responsiveness of non-small-cell lung cancer to gefitinib. *N Engl J Med* 2004;**350**: 2129–39.
22. Pao W, Miller V, Zakowski M, et al. EGF receptor gene mutations are common in lung cancers from "never smokers" and are associated with sensitivity of tumors to gefitinib and erlotinib. *Proc Natl Acad Sci USA* 2004;**101**:13306–11.
23. van 't Veer LJ, Dai H, van de Vijver MJ, et al. Gene expression profiling predicts clinical outcome of breast cancer. *Nature* 2002; **415**:530–6.
24. Paik S, Shak S, Tang G, et al. A multigene assay to predict recurrence of tamoxifen-treated, node-negative breast cancer. *N Engl J Med* 2004;**351**:2817–26.
25. Jones S, Anagnostou V, Lytle K, et al. Personalized genomic analyses for cancer mutation discovery and interpretation. *Sci Transl Med* 2015;**7**. 283ra53.
26. Engelman JA, Zejnullahu K, Mitsudomi T, et al. MET amplification leads to gefitinib resistance in lung cancer by activating ERBB3 signaling. *Science* 2007;**316**:1039–43.
27. Head SR, Komori HK, LaMere SA, et al. Library construction for next-generation sequencing: overviews and challenges. *Biotechniques* 2014;**56**:61–4. 6, 8. passim.
28. Quinlan AR, Hall IM. Bedtools: a flexible suite of utilities for comparing genomic features. *Bioinformatics* 2010;**26**:841–2.
29. Cabanski CR, Magrini V, Griffith M, et al. cDNA hybrid capture improves transcriptome analysis on low-input and archived samples. *J Mol Diagn* 2014;**16**:440–51.
30. Auer PL, Srivastava S, Doerge RW. Differential expression—the next generation and beyond. *Brief Funct Genomics* 2012;**11**:57–62.
31. Imperiale TF, Ransohoff DF, Itzkowitz SH, et al. Multitarget stool DNA testing for colorectal-cancer screening. *N Engl J Med* 2014;**370**:1287–97.
32. Tomlins SA, Aubin SM, Siddiqui J, et al. Urine TMPRSS2:ERG fusion transcript stratifies prostate cancer risk in men with elevated serum PSA. *Sci Transl Med* 2011;**3**. 94ra72.
33. Van Allen EM, Wagle N, Stojanov P, et al. Whole-exome sequencing and clinical interpretation of formalin-fixed, paraffin-embedded tumor samples to guide precision cancer medicine. *Nat Med* 2014;**20**:682–8.
34. Andre F, Mardis E, Salm M, et al. Prioritizing targets for precision cancer medicine. *Ann Oncol* 2014;**25**:2295–303.
35. Beiko J, Suki D, Hess KR, et al. IDH1 mutant malignant astrocytomas are more amenable to surgical resection and have a survival benefit associated with maximal surgical resection. *Neuro Oncol* 2014;**16**:81–91.
36. Popovici V, Budinska E, Bosman FT, et al. Context-dependent interpretation of the prognostic value of BRAF and KRAS mutations in colorectal cancer. *BMC Cancer* 2013;**13**:439.
37. Cancer Genome Atlas Research Network, Kandoth C, Schultz N, et al. Integrated genomic characterization of endometrial carcinoma. *Nature* 2013;**497**:67–73.
38. Nodin B, Zendehrokh N, Sundstrom M, et al. Clinicopathological correlates and prognostic significance of KRAS mutation status in a pooled prospective cohort of epithelial ovarian cancer. *Diagn Pathol* 2013;**8**:106.
39. Pao W, Hutchinson KE. Chipping away at the lung cancer genome. *Nat Med* 2012;**18**:349–51.
40. Ciriello G, Miller ML, Aksoy BA, et al. Emerging landscape of oncogenic signatures across human cancers. *Nat Genet* 2013;**45**: 1127–33.
41. Lasota J, Corless CL, Heinrich MC, et al. Clinicopathologic profile of gastrointestinal stromal tumors (GISTs) with primary KIT exon 13 or exon 17 mutations: a multicenter study on 54 cases. *Mod Pathol* 2008;**21**:476–84.
42. Janne PA, Johnson BE. Effect of epidermal growth factor receptor tyrosine kinase domain mutations on the outcome of patients with non-small cell lung cancer treated with epidermal growth factor receptor tyrosine kinase inhibitors. *Clin Cancer Res* 2006;**12**. 4416s–20s.
43. Lee CC, Shiao HY, Wang WC, et al. Small-molecule EGFR tyrosine kinase inhibitors for the treatment of cancer. *Expert Opin Investig Drugs* 2014;**23**:1333–48.
44. Dietel M, Johrens K, Laffert M, et al. Predictive molecular pathology and its role in targeted cancer therapy: a review focusing on clinical relevance. *Cancer Gene Ther* 2013;**20**: 211–21.
45. American Association for Cancer Research. NCI prepares to launch MATCH trial. *Cancer Discov* 2015;**5**:685.
46. Redig AJ, Janne PA. Basket trials and the evolution of clinical trial design in an era of genomic medicine. *J Clin Oncol* 2015;**33**: 975–7.
47. Sosman JA, Kim KB, Schuchter L, et al. Survival in BRAF V600-mutant advanced melanoma treated with vemurafenib. *N Engl J Med* 2012;**366**:707–14.
48. Prahallad A, Sun C, Huang S, et al. Unresponsiveness of colon cancer to BRAF(V600E) inhibition through feedback activation of EGFR. *Nature* 2012;**483**:100–3.
49. Sun C, Wang L, Huang S, et al. Reversible and adaptive resistance to BRAF(V600E) inhibition in melanoma. *Nature* 2014;**508**:118–22.

Genetic Aspects of Hematopoietic Malignancies

Todd W. Kelley and Jay L. Patel

ABSTRACT

Background

The field of hematology has long been at the forefront of using genetic testing to improve clinical outcomes. There have been many benefits for patients with hematopoietic malignancies, including improved diagnostic accuracy, prognostic stratification, and identification of new therapeutic targets. Molecular methods such as polymerase chain reaction, dideoxy-termination sequencing, and fluorescence in situ hybridization have been a routine part of the practice of hematopathology for many years. However, like other areas of oncology, the last few years have seen further advances in the understanding of the genetic basis of these neoplasms, primarily due to the influence of new sequencing technologies. Massively parallel sequencing has made routine what was just a few years ago unthinkable.

Content

In this chapter we cover the genetic abnormalities common in hematopoietic malignancies, including many of the newest discoveries. We approach hematopoietic malignancies from a laboratory standpoint starting with structural chromosomal abnormalities and translocations and moving to smaller-scale genetic changes found in single genes and finally to epigenetic changes. We compare and contrast the various laboratory methods used to query these abnormalities and highlight the utility of new and advancing technologies and platforms, including array-based methods and massively parallel sequencing. Finally, the chapter ends with a discussion of lymphoid clonality testing, an area of hematology testing that is also benefiting from the influence of modern sequencing technology.

RECURRENT TRANSLOCATIONS AND STRUCTURAL CHROMOSOMAL ABNORMALITIES

Many hematopoietic malignancies harbor underlying chromosomal abnormalities, including balanced and unbalanced translocations and large-scale structural changes such as deletions or duplications. Some of these abnormalities are characteristic of certain disease subtypes and are thus repeatedly, or recurrently, found in different individuals with the same disease. Recurrent genetic abnormalities of all types are found across the spectrum of hematopoietic malignancies. This theme is revisited throughout this chapter.

Chromosomal abnormalities may be detectable by conventional cytogenetics or may require more specialized molecular techniques. The specificity and utility of these findings are variable and highly context-dependent. While there are some genetic lesions that are disease defining, most others have utility in diagnosis, prognosis, and clinical management. Proper diagnosis and classification of acute myelogenous leukemia (AML) and acute lymphoblastic leukemia (ALL), according to the WHO classification, require karyotyping and/or fluorescent in situ hybridization (FISH) studies, in conjunction with morphologic, immunophenotypic, and clinical correlation.[1] The workup of myeloid neoplasms such as myelodysplastic syndromes (MDS) and myeloproliferative neoplasms (MPNs) also often includes these types of ancillary studies. Chromosomal analysis requires bone marrow aspirate samples that are amenable to cytogenetic testing.[2]

A hallmark of certain subtypes of mature B-cell lymphoproliferative disorders is the presence of a balanced translocation involving *IGH* on chromosome 14q32 with a proto-oncogene—for example, *BCL2* in the case of follicular lymphoma. The latter is thereby placed under the influence of *IGH* enhancer elements, resulting in dysregulated expression and lymphomagenesis. The mechanistic details underlying the development of recurrent translocations in hematolymphoid neoplasms continue to be the subject of investigation.

The normal process of lymphocyte antigen receptor gene rearrangement and sequence remodeling, which are necessary for proper B- and T-cell function, is vulnerable to mistakes. Antigen receptor rearrangement occurs in a sequential fashion involving recombination of *IGH*-V, D, and J gene segments mediated by an enzyme complex in which the nucleases RAG1 and RAG2 introduce double-strand DNA breaks adjacent to specific recombination signal sequences (described in more detail in the "Clonality Testing in Hematopoietic Malignancies" section). The presence of recombination signal sequences adjacent to *IGH-BCL2* breakpoints in follicular lymphoma provides evidence for the involvement of failed V(D)J recombination in translocation events involving *IGH*.[3] The precise mechanism of DNA breakage at oncogene loci is less clear but may involve the ability of RAG1 and RAG2 proteins to additionally act as transposases, with the ability to catalyze excision and subsequent integration of a DNA fragment into another molecule via a transesterification reaction.[4] However, aberrant V(D)J recombination–associated events do not account for the entirety of oncogenic translocations leading to lymphoma development. Consider that while the number of B- and T-lymphocytes produced by the human body is roughly equal and both undergo V(D)J recombination, the clear majority of lymphoid neoplasms are of B-cell lineage. This discrepancy is explained in part by the fact that B cells undergo secondary antibody diversification by somatic hypermutation and class switch recombination after migration into the germinal centers of peripheral lymphoid tissue; these processes inherently involve high mutational rates and provide additional opportunities for pathogenic events to manifest during B-cell development. *IGH* breakpoints in most cases of sporadic Burkitt lymphoma, for example, occur 3' to the switch region adjacent to the mu constant segment (C_μ), consistent with a mechanism involving abnormal class switch recombination.[5] T-cell development, in comparison, does not undergo similar mutation-prone events.

Recent studies of mutation events in B cells during somatic hypermutation and class switch recombination have centered on the enzymatic role of activation-induced cytidine deaminase (AID). AID is highly expressed in the germinal center microenvironment, where it normally introduces mutations into variable and class switching regions of immunoglobulin genes to promote increased diversity of the immunoglobulin antigen-binding repertoire. AID is transcription-dependent and acts by deamination of cytidine to uracil in single-stranded DNA targets, which is propagated as a uracil-guanine mismatch and replicated as a cytidine-thymine transition. Alternatively, the uracil residue may be removed by uracil DNA glycolase to create an abasic site, which is recognized by nucleases or proofreading DNA polymerases.[6] Aberrant targeting of AID at the genomic level contributes to genomic instability in B cells and appears to occur preferentially in highly transcribed super-enhancer domains that are also linked to other promoters and enhancers to form a regulatory cluster.[7] Further, AID initiation occurs in association with focal regions of target genes in which sense and antisense transcription converge.[8] These recent discoveries provide a framework for understanding the role of AID in nonimmunoglobulin genes, including *MYC*, *BCL6*, and *PAX5*. Aberrant AID activity confers a predisposition to the development of mutations and chromosomal aberrations found in many B-cell lymphomas.

Lymphoid Disorders

Precursor lymphoid neoplasms are primarily diseases of children and adolescents and are composed of lymphoblasts committed to B or T cell lineage as defined by immunophenotypic features. B-lymphoblastic leukemia (B-ALL) is significantly more common than its T cell–derived counterpart and usually presents with cytopenias accompanying florid peripheral blood and bone marrow disease. T-lymphoblastic leukemia/lymphoma (T-ALL) may manifest as a nodal disease, frequently in association with a mediastinal mass, and shows variable bone marrow involvement. Classification of B lymphoblastic leukemia is based in part on detection of certain recurrent genetic abnormalities that often confer a specific clinicopathologic phenotype and carry prognostic implications (Table 8.1). For detection of chromosomal abnormalities, routine karyotyping along with FISH analysis is the clinical approach used by most laboratories. While T-lymphoblastic leukemia/lymphoma is not currently subclassified according to genetic findings, an abnormal karyotype is found in the majority of cases.

Mature lymphoid neoplasms are diagnosed largely based on morphology and immunophenotypic features. Cytogenetic and/or FISH studies are typically not required in order to appropriately diagnose and subclassify most cases due to the use of surrogate markers. For example, overexpression of cyclin D1 by immunohistochemistry is often used as presumptive evidence of the presence of t(11;14)(q13;q32); *IGH-CCND1* in the case of mantle cell lymphomas if the overall features are otherwise consistent.

B-Lymphoblastic Leukemia/Lymphoma

B-lymphoblastic leukemia with t(v;11q23); *KMT2A* (previously known as *MLL*) rearrangements (see Fig. 8.1 for an explanation of the general format used to represent chromosomal abnormalities) are characterized by a translocation between *KMT2A* on 11q23 and any one of a large number of potential fusion partners, the most common of which is *AFF1* on 4q21. These translocations confer high-risk disease due to aberrant expression of the epigenetic regulator *KMT2A*. *KMT2A* rearrangements represent the most likely recurrent chromosomal abnormality in infantile B-lymphoblastic leukemia and may occur in utero.[9] Patients are typically younger than 2 years of age and characteristically present with marked leukocytosis and central nervous system involvement. *KMT2A* rearrangements in lymphoblastic leukemia are typically detected by FISH studies using a break-apart probe spanning the 11q23 region. This FISH strategy employs locus-specific probes designed to flank a known breakpoint region. When a translocation is present, the probes are separated (they "break apart") and only primary colors are observed (eg, red and green), whereas normal cells demonstrate the secondary color (eg, yellow).

B-lymphoblastic leukemia with t(9;22)(q34;q11.2) is a high-risk disease characterized by production of a fusion protein with constitutively active ABL1 tyrosine kinase activity (*BCR-ABL1*; known as the Philadelphia chromosome). This is the most common recurrent genetic abnormality in adult B-lymphoblastic leukemia patients and confers poor prognosis among all age groups. Patients may benefit from tyrosine kinase inhibitor (TKI) therapy in addition to traditional high-dose chemotherapy. While t(9;22)(q34;q11.2) is reliably detected by FISH, the type of *BCR-ABL1* fusions present should be confirmed by quantitative reverse transcription-PCR for purposes of ongoing monitoring during treatment and minimal residual disease detection. The p190 kDa fusion protein (e1a2 transcript) is seen in pediatric disease, while adults may demonstrate either the p190 kDa form or the larger p210 kDa form (e13a2 [b2a2] or e14a2 [b3a2] transcript). The latter is also seen in almost all cases of chronic myelogenous leukemia (CML). *BCR-ABL1* fusion transcripts are described in more detail later in this section.

The balanced cryptic (not identifiable by routine karyotype analysis) translocation t(12;21)(p13;q22); *ETV6-RUNX1* requires FISH for detection and is the most common recurrent abnormality in childhood B-ALL, accounting for 25% of cases. Its incidence decreases with age, and *ETV6-RUNX1* is only

TABLE 8.1 Select Recurrent Chromosomal Abnormalities in Lymphoid Malignancies

Genetic Abnormality	Disease	Incidence	Prognosis	Clinical Notes
t(v;11q23); KMT2A rearranged	B-ALL	5%	Poor	Most common form of infant ALL
t(9;22)(q34;q11.2); BCR-ABL1	B-ALL	5% (pediatric) 25% (adult)	Poor	p190 (minor) isoform typical in children, p210 (major) common in adults, quantitative RT-PCR for monitoring
t(12;21)(p13;q22); ETV6-RUNX1	B-ALL	25% (pediatric) 3% (adults)	Very good	Balanced cryptic translocation, detection requires FISH
t(5;14)(q31;q32); IL3-IGH	B-ALL	1%	Intermediate	Eosinophilia
(1;19)(q23;p13.3); TCF3-PBX1	B-ALL	5%	Intermediate	—
Hyperdiploidy (>50 chromosomes)	B-ALL	25%	Very good	"Triple trisomy" of chromosomes 4, 10, and 17 particularly favorable
Hypodiploidy (24–44 chromosomes)	B-ALL	1%	Variable	Prognosis depends on degree of chromosome loss, near haploid patients fare particularly poorly
iAMP21	B-ALL	2%	Poor	Detectable by FISH for RUNX1 on chromosome 21
t(1;14)(p32;q11); TRD-TAL1	T-ALL	3%	Intermediate	Difficult to detect due to cryptic deletion at 1p32
t(1;7)(p32q35); TRB-TAL1	T-ALL	1%	Intermediate	—
t(10;14)(q24;q11); TRD-TLX1	T-ALL	4%	Good	—
t(8;14)(q24;q32); IGH-MYC	BL	75%	Good	Aggressive disease but curable, IGH-MYC rare in DLBCL
t(2;8)(p12;q24); IGK-MYC	BL	15%	Good	Aggressive disease but curable
t(8;22)(q24;q11); IGL-MYC	BL	10%	Good	Aggressive disease but curable
t(14;18)(q32;q21); IGH-BCL2	FL, DLBCL	90% (FL) 20%-30% (DLBCL)	Variable	Prognosis in FL depends on grade and clinical stage. Aggressive in combination with MYC rearrangement —"double-hit."
t(v;3q27); BCL6 rearrangement	FL, DLBCL	5%-10% (FL) 30% (DLBCL)	Variable	Aggressive disease when in combination with MYC rearrangement —"double-hit"
t(v;8q24); MYC rearrangement	DLBCL	10%	Poor	Common in plasmablastic lymphoma
t(11;14)(q13;q32); IGH-CCND1	MCL, MM	100%	Poor	FISH preferred for detection, association with lymphoplasmacytic morphology in MM
t(11;18)(q21;q21); BIRC3-MALT1	EMZL	40% (lung) 30% (stomach)	Good	Nonresponsive to H. pylori–directed antibiotic therapy, low likelihood of DLBCL transformation
del17p13; TP53 deletion	MM	10% to 15%	Poor	Associated with disease progression
t(2;5)(p23;q35.1); NPM1-ALK	ALK+ ALCL	85%	Good	Indirect evidence provided by immunohistochemistry
6p25.3; DUSP22 rearrangement	ALK- ALCL	30%	Good	Prognosis similar to ALK+ ALCL, FISH for detection
3q27; TP63 rearrangement	ALK- ALCL	8%	Very poor	FISH for detection

Incidence is estimated with respect to disease category.
B-ALL, B-lymphoblastic leukemia/lymphoma; T-ALL, T-lymphoblastic leukemia/lymphoma; BL, Burkitt lymphoma; DLBCL, diffuse large B cell lymphoma; FL, follicular lymphoma; MCL, mantle cell lymphoma; MM, multiple myeloma; EMZL, extranodal marginal zone lymphoma; ALCL, anaplastic large cell lymphoma.

rarely seen in adult B-ALL. The event appears to occur early in leukemogenesis and results in an abnormal fusion of the transcription factor ETV6 and the DNA-binding domain of RUNX1. The fusion protein interferes with the normal function of RUNX1, a factor that is critical for hematopoietic cell differentiation. Patients with B-ALL and ETV6-RUNX1 have an excellent prognosis and achieve cure in greater than 90% of cases.

In contrast to mature B-cell malignancies, B-ALL rarely demonstrates translocations involving immunoglobulin loci. An exception is B-ALL with t(5;14)(q31;q32); IL3-IGH. The cytokine interleukin-3 (IL-3) is constitutively overexpressed as a result of being brought under the control of the IGH enhancer. One consequence is variable secondary eosinophilia, potentially leading to end-organ damage to sensitive tissues such as cardiac muscle. Otherwise, the clinical characteristics and prognosis associated with this rare disease are similar to B-ALL in general.

B-lymphoblastic leukemia with t(1;19)(q23;p13.3); TCF3-PBX1 accounts for approximately 6% of cases in children and

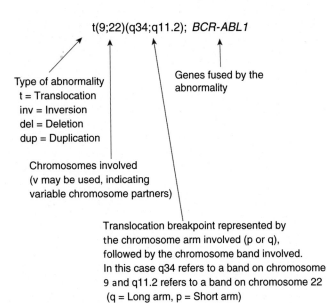

FIGURE 8.1 Format used to represent chromosomal abnormalities.

less than 5% in adults. The genetic abnormality is a translocation between the transcription factors *TCF3* and *PBX1*, resulting in creation of a leukemogenic fusion gene in which the DNA binding domain of *TCF3* is replaced with that of *PBX1*. This results in constitutive transcriptional activity of genes regulated by the PBX protein family. The prognosis of *TCF3-PBX1*–positive B-ALL is similar to other subtypes with comparable risk factors.

Numerical chromosomal abnormalities are relatively common in B-lymphoblastic leukemia and define additional subtypes of the disease. Precise definitions vary, but generally cases with greater than 50 chromosomes are referred to as hyperdiploid, while those with less than 45 chromosomes are designated hypodiploid. Concurrent structural abnormalities may be encountered but are uncommon. B-ALL with hyperdiploidy is seen in 25% of childhood disease and decreases in incidence with age. The prognosis is very good, and achievement of cure is highly likely with standard therapy. The number of chromosomes may be less important than the specific chromosomes involved. Patients with three copies of chromosomes 4, 10, and 17, so-called "triple-trisomy ALL," have an excellent prognosis. Hypodiploid B-ALL is seen in less than 5% of patients overall. Prognosis is generally poor and appears to be correlated to the degree of chromosome loss, with near-diploid (44 chromosomes) and high-hypodiploid (40 to 43 chromosomes) patients faring relatively well. Low-hypodiploid (32 to 39 chromosomes) cases show a high frequency of *TP53* mutations (often germline) along with *IKZF2* and *RB1* abnormalities.[10] Near-haploid (24 to 31 chromosomes) B-ALL is associated with *RAS* mutations and other alterations targeting receptor tyrosine kinase signaling.

Intrachromosomal amplification of chromosome 21 (iAMP21) occurs as a primary genetic event in approximately 2% of childhood B lymphoblastic leukemia and arises from the effect of "gene dosage" due to copy-number alterations of potentially hundreds of linked genes. The iAMP21 abnormality in B-ALL is associated with older age at presentation (median age is 9 years) and with low white blood cell counts. FISH testing with probes to the *RUNX1* locus demonstrates the presence of five or more signals on a single chromosome. Recent data show that B-ALL patients with iAMP21 fare poorly when treated according to standard-risk protocols and benefit from high-risk ALL therapy.[11]

T-Lymphoblastic Leukemia/Lymphoma

The most frequent nonrandom cytogenetic abnormalities seen in T-lymphoblastic leukemia/lymphoma are translocations involving T cell receptor (TCR) loci on chromosomes 7 (*TRA* and *TRD*) and 14 (*TRB* and *TRG*) in association with various partner genes. Many of these rearrangements involve dysregulation of T cell–specific cellular transcription factors, resulting in disruption of normal maturation or uncontrolled cellular proliferation. One frequent genetic target is *TAL1*, which is rearranged in approximately 60% of cases of T-ALL. Other genes implicated in T-ALL–related translocations include *TLX1*, *TLX3*, *MYC*, *LMO1*, *LMO2*, and *LYL1*.[12] The prognostic significance of these rearrangements is generally not well established, but *TLX1* abnormalities appear to correlate with improved clinical outcome, while *LYL1* and *TAL1* rearrangements appear to be less favorable. The Philadelphia chromosome, t(9;22)(q34;q11.2), is rarely detected in T-ALL but confers a poor prognosis.[13] Chromosomal deletions also occur in T-ALL, the most important of which involves the short arm of chromosome 9 and is seen in approximately 30% of cases. This results in loss of the important tumor suppressor gene *CDKN2A* and compromised cell cycle control. Notably, cryptic abnormalities are frequently found in T-ALL. Activating mutations in *NOTCH1*, which encodes a protein critical for early T cell development, have been demonstrated in the majority of T-ALL patients and appear to be associated with a favorable outcome. Increased NOTCH1 signaling can arise from point mutations, insertions, deletions, or, rarely, translocations, and appears to play an important role in leukemogenesis.[14]

Burkitt Lymphoma

Burkitt lymphoma was the first lymphoma shown to harbor a recurrent genetic abnormality, and in many ways it serves as an archetype for the study of mature B cell non-Hodgkin lymphomas.[15] Translocations involving the immunoglobulin locus and the *MYC* oncogene on 8q24 are primary genetic events in most cases of Burkitt lymphoma. The typical finding is t(8;14)(q24;q32), resulting in formation of a derivative chromosome 14 in which *MYC* is brought under control of the *IGH* locus, leading to constitutive *MYC* expression. Variant translocations involving *MYC* and the immunoglobulin light chain loci at 2p12 (*IGK*) and 22q11 (*IGL*) are seen in 5% to 10% of cases. MYC overexpression contributes to genomic instability through various mechanisms, including disruption of DNA double-strand break repair pathways, as well as both activation and repression of transcriptional activity. *IG-MYC* translocations are not specific for Burkitt lymphoma and occur as secondary events in a minor subset of other aggressive B cell non-Hodgkin lymphomas, as well as, rarely, in plasma cell myeloma and B-lymphoblastic leukemia. Burkitt lymphoma is perhaps most reliably defined by gene expression profiling with a molecular signature distinct from diffuse large B cell lymphoma, but such testing remains unavailable in most clinical settings.[16]

Follicular Lymphoma

A translocation involving *IGH* at 14q32 and *BCL2* on 18q21 is present in up to 90% of cases of follicular lymphoma. *BCL2* is thereby placed under the influence of the *IGH* promoter, and the result is overexpression of the antiapoptotic protein bcl-2. *IGH-BCL2* translocations are also found in approximately 20% to 30% of cases of diffuse large B-cell lymphoma (DLBCL). Surprisingly, t(14;18) can sometimes be detected in histologically benign lymph nodes or tonsillar tissue (usually at low levels) in otherwise normal individuals and is therefore not definitively diagnostic of lymphoma in isolation.[17] Additional genetic aberrations are present in most cases of follicular lymphoma, including a variety of chromosomal gains and losses.[18] Notably, 10% to 15% of follicular lymphoma, particularly high-grade cases, lack t(14;18). Approximately 80% to 90% of the breakpoints on 18q21 are located either in the major breakpoint region (MBR) found within the 3' untranslated region of exon 3 of *BCL2* or in the minor cluster region (MCR) found 3' further downstream of exon 3 and are amenable to detection by multiplex PCR using consensus primer sets (Fig. 8.2). Less common breakpoints (10%), many of which are found upstream of exon one of *BCL2*, are not targeted by typical PCR assays.[19] For this reason, FISH is often used to detect t(14;18).

Diffuse Large B-Cell Lymphoma

Diffuse large B-cell lymphoma (DLBCL) is a genetically heterogeneous group of large B cell lymphomas that display a variety of underlying chromosomal abnormalities, including t(14;18)(q32;q21). In addition to *IGH-BCL2* rearrangement, which may signify evolution from preexisting follicular lymphoma, rearrangements of the transcriptional regulator *BCL6* on 3q27 are common. The prevalence of *BCL6* somatic mutations and translocations is explained by the fact that *BCL6* is one of several other genes expressed in the germinal center that are known to normally undergo the process of somatic hypermutation, as does the variable region of *IGH*.[20] *MYC* rearrangements are also present in approximately 10% of patients. A subset of high-grade B cell lymphoma patients with variable histology harbor concurrent rearrangements of *MYC* in combination with *BCL2* or *BCL6*, a phenomenon referred to as "double-hit" lymphoma.[21] Rarely, all three abnormalities may be observed, in which case the designation "triple-hit" B cell lymphoma may be used. In either situation, the clinical phenotype is particularly aggressive. Due to the aggressive nature of the disease and a propensity for involvement of the central nervous system, prompt recognition of these patients is warranted for purposes of therapeutic decision making and prognostic stratification. This requires FISH testing at the time of initial diagnosis.

Mantle Cell Lymphoma

Mantle cell lymphoma is characterized by t(11;14)(q13;q32); *IGH-CCND1* in nearly all cases. The juxtaposition of the *CCND1* gene, encoding cyclin D1, with the *IGH* enhancer is a primary genetic event and results in increased progression through the cell cycle due to cyclin D1 protein overexpression.[22] The latter is detectable by immunohistochemistry, which often renders direct molecular genetic demonstration of *IGH-CCND1* unnecessary for diagnosis in most settings. However, clinical scenarios often arise in which mantle cell lymphoma is in the differential diagnosis but cyclin D1 immunohistochemical evaluation is not feasible, perhaps due to specimen limitations (eg, peripheral blood). In these situations, alternative testing is required. FISH represents the most sensitive testing strategy and is capable of detecting nearly 100% of translocations, assuming that cells harboring the translocation are present above established sensitivity thresholds.[23] Notably, t(11;14)(q13;q32) is also seen in 5% to 10% of plasma cell myeloma patients. Numerous breakpoints involving the *CCND1* locus may be encountered in mantle cell lymphoma and span a 350-kilobase region on 11q13. Approximately 40% of these breakpoints are clustered in a 1-kilobase segment referred to as the major translocation cluster (MTC) found 110 kb downstream of the *CCND1* locus (Fig. 8.3). Most PCR-based assays do not interrogate *CCND1* breakpoints outside the MTC. Translocations with breakpoints outside of this region are therefore not detected, and the sensitivity of these assays is only 40% to 50%.

Extranodal Marginal Zone Lymphoma (MALT Lymphoma)

Several recurrent translocations have been described in extranodal marginal zone lymphoma (MALT lymphoma), and their incidence varies with the anatomic site of disease. The t(11;18)(q21;q21) occurs mostly in gastric and pulmonary MALT lymphoma and results in *BIRC3-MALT1* fusion.[24] Detection is important because patients with t(11;18) are less likely to respond to antibiotic therapy directed at *H. pylori* and only rarely progress to large-cell lymphoma.[25] Three additional translocations—t(14;18)(p14;q32), t(1;14)(p22;q32), and t(3;14)(p22q32)—are seen with relatively low incidence but reinforce the paradigm of proto-oncogenes (*MALT1*, *BCL10*, and *FOXP1*, respectively) under the control of the *IGH* enhancer complex in B cell lymphoproliferative disorders.

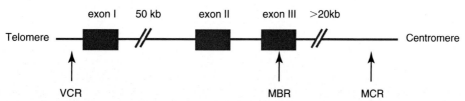

FIGURE 8.2 Schematic depiction of the architecture of *BCL2* on chromosome 18q21 with exons depicted as rectangles. *IGH-BCL2* translocation breakpoints occur most often within the major breakpoint cluster region (MBR, 50% to 60%) or the minor breakpoint cluster region (MCR, 20% to 25%). In approximately 10% of translocations, the *BCL2* breakpoint occurs in the variable cluster region (VCR) found upstream of exon 1 and will not be detected by most PCR assays.

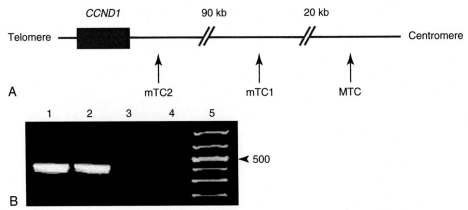

FIGURE 8.3 A, Schematic representation of the organization of *CCND1* on 11q13 and the range of breakpoint locations observed in *IGH-CCND1* translocation. The rectangle represents the *CCND1* gene. Various breakpoints may be observed and approximately 40% cluster in the major translocation cluster (MTC), the target of most PCR assays. Most of the remaining breakpoints occur in the minor translocation cluster 1 or 2 regions (mTC1 or mTC2). Note the relatively large distances involved because translocation breakpoints may span a region upward of 350 kb downstream of the *CCND1* locus. **B,** Example of an *IGH-CCND1* PCR assay with primers targeting the MTC followed by agarose gel electrophoresis. Lane 1 is a positive control, lane 2 is the patient sample, lane 3 is a negative control, and lane 4 is a no-template control with lane 5 showing the molecular size marker. The patient sample shows a strong discrete band at approximately 450 bp, comparable to the positive control. Note that the size of the PCR product may vary according to the specific *CCND1* breakpoint involved.

Multiple Myeloma

Multiple myeloma is a clinically and genetically heterogeneous clonal disorder of terminally differentiated plasma cells. Incorporation of clinicopathologic and radiographic findings, including bone marrow evaluation, is often necessary for definitive diagnosis. Karyotypic and/or FISH abnormalities are detectable in a large proportion of cases and include a variety of aberrations. The most common abnormalities appear to be early events and include either trisomies of various chromosomes resulting in hyperdiploidy or translocations involving the *IGH* locus at 14q32. *IGH* translocation partners include, in decreasing frequency, *CCND1* (11q13), *FGFR3/MMSET* (4p16.3), *C-MAF* (16q23), *CCND3* (6p21), and *MAFB* (20q11). Monosomy 13 or deletion of 13q14 is also observed with regularity in plasma cell myeloma (40%), particularly in its leukemic form (70%). Disease progression is associated with acquisition of various additional genetic abnormalities, perhaps the most important of which is deletion of the tumor suppressor *TP53* at 17p13.[26] *TP53* deletion signals high-risk disease with significantly decreased overall survival and additionally may serve as a marker of extramedullary involvement.[27]

Anaplastic Large Cell Lymphoma

Examples of recurrent chromosomal abnormalities are less common in mature T cell lymphomas. Anaplastic lymphoma kinase (ALK)–positive anaplastic large cell lymphoma (ALCL), the hallmark of which is translocation of the tyrosine kinase receptor *ALK*, is one exception. The most common translocation, t(2;5)(p23;q35.1), accounts for 85% of cases and involves fusion of *ALK* to nucleophosmin *(NPM1)*, resulting in nuclear and cytoplasmic expression of ALK. Numerous, less-frequent variant translocation partners have also been described. *ALK* rearrangements are not disease-specific and are seen in a subset of non–small cell lung cancers. *NPM1-ALK* and variant fusions can be detected by FISH and other means, but this is usually unnecessary because normal postnatal human tissues, except for rare central nervous system constituents, lack ALK expression. Therefore the presence of an *ALK* translocation can be inferred quickly and cost effectively by the immunohistochemical detection of ALK protein expression in the neoplastic cells. Notably, ALK inhibitor therapies are in development, and treatment with these agents may lead to the acquisition of *ALK* kinase domain activating mutations conferring drug resistance.[28]

ALK-negative anaplastic large cell lymphoma is a T cell lymphoma that shares essentially identical morphologic and immunohistochemical features with ALK-positive ALCL, but it occurs in older patients and lacks both *ALK* rearrangements and ALK protein expression. ALK-negative ALCL has a worse prognosis relative to ALK-positive ALCL, possibly justifying a more aggressive therapeutic approach. Recent studies have revealed genetic heterogeneity in these cases. Patients with ALK-negative ALCL and rearrangement of *DUSP22* on 6p25.3 (demonstrated by FISH) fare relatively well, with overall survival resembling ALK-positive disease.[29] Rearrangements of *TP63* on 3q27 appear to be mutually exclusive with regard to *DUSP22* and imply a very poor prognosis.

Myeloid Disorders

Myeloid malignancies are a relatively diverse group of diseases that arise as a consequence of a variety of genetic aberrations. Generally they are classified in the WHO system as myeloproliferative neoplasms (MPN), myelodysplastic syndromes (MDS), myelodysplastic/myeloproliferative overlap neoplasms (MDS/MPN), and acute myeloid leukemia (AML). They are further subclassified through the synthesis of cytologic, morphologic, clinical, and genetic findings (Table 8.2). A variety of laboratory tests are important for genetic evaluation, including cytogenetic karyotyping, FISH, array-based genotyping (eg, single nucleotide polymorphism arrays), and molecular techniques such as PCR and sequencing, including both dideoxy-termination

TABLE 8.2 Selected Recurrent Chromosomal Abnormalities in Myeloid Neoplasms

Genetic Abnormality	Disease	Incidence	Prognosis	Clinical Notes
t(8;21)(q22,q22); RUNX1-RUNX1T1	AML	10%	Good	Core binding factor leukemia, favorable prognosis partly negated by concurrent KIT mutation (20%-25%)
inv(16)(q13.1;q22); CBFB-MYH11	AML	10%	Good	Core binding factor leukemia, favorable prognosis negated by concurrent KIT mutation (30%)
t(15;17)(q24;q12); PML-RARA	AML	7%	Good	Hematologic emergency (risk of DIC), FISH is test of choice for diagnostic confirmation, responsive to ATRA, monitor with quantitative RT-PCR
t(9;11)(p22;q23); MLLT3-KMT2A	AML	5%	Poor	More common in pediatric AML (12%), monocytic differentiation, gingival hypertrophy, risk of DIC
t(6;9)(p23q34); DEK-NUP214	AML, MDS	1%	Very poor	Basophilia, may arise de novo or in the setting of MDS, frequent FLT3 ITD (70%)
inv3(q21;q26.2); RPN1-EVI1	AML, MDS	2%	Very poor	Normal or increased platelet count, may arise de novo or in the setting of MDS, occasional FLT3 ITD (13%)
t(1;22)(p13;q13); RBM15-MKL1	AML	<1%	Good	Pediatric disease associated with Down syndrome, megakaryoblastic differentiation, somatic GATA1 mutations
del(5q) or monosomy 5	MDS	10%	Good	Excellent response to lenalidomide when del(5q) is sole abnormality, TP53 mutations confer poor response
del(7q) or monosomy 7	MDS	10%	Intermediate	Poor prognosis with monosomy 7
del(11q)	MDS	3%	Very good	—
del(12p)	MDS	12%	Good	—
del(20q)	MDS	8%	Good	Insufficient to diagnose MDS as a sole abnormality
i(17q) or t(17p); TP53 deletion	MDS	5%	Intermediate	Acquired (pseudo) Pelger-Huet anomaly
Trisomy 8	MDS	10%	Intermediate	Common but nonspecific, seen in other myeloid neoplasms
Normal	MDS	50%	Good	Somatic mutations by massively parallel sequencing
Complex (>3 unrelated defects)	MDS	7%	Very poor	High risk of evolution to AML
t(9;22)(q34;q11.2); BCR-ABL1	CML	100%	Good	Response to TKI therapy monitored by quantitative RT-PCR, myeloid or B lymphoblastic blast phase possible
del(4q12); FIP1L1-PDGFRA	CEL	Rare	Unknown	May present as AML or T-LBL, cryptic deletion detectable by FISH for CHIC2, responsive to TKI
t(5;12)(q33;p12); ETV6-PDGFRB	CMML	Rare	Unknown	Eosinophilia, responsive to TKI
t(v;8p11); FGFR1 rearranged	Variable	Rare	Very poor	Stem cell disease with lymphomatous presentation common, unresponsive to TKI

Incidence is estimated with respect to disease category.
AML, Acute myeloid leukemia; MDS, myelodysplastic syndrome; CML, chronic myelogeneous leukemia; CEL, chronic eosinophilic leukemia; CMML, chronic myelomonocytic leukemia; T-LBL, T-lymphoblastic lymphoma; DIC, disseminated intravascular coagulation; ATRA, all-trans-retinoic acid; ITD, internal tandem duplication; TKI, tyrosine kinase inhibitors.

(Sanger) sequencing and massively parallel sequencing. Gene mutations are detectable only by molecular techniques. Such mutations and the techniques used to detect them are the focus of the "Gene Mutations in Hematopoietic Malignancies" section.

Acute Myeloid Leukemias

Core binding factor–associated acute myeloid leukemias (AMLs) are unified by molecular pathophysiology and include t(8;21)(q22;q22); *RUNX1-RUNX1T1* and inv(16)(q13.1;q22); *CBFB-MYH11*. The core binding factor is a heterodimeric protein composed of alpha and beta subunits encoded by *RUNX1* and *CBFB*, respectively, which is normally involved in regulation of hematopoiesis. The *RUNX1-RUNX1T1* and *CBFB-MYH11* abnormalities disrupt this function, resulting in the impairment of differentiation leading to the accumulation of immature myeloid blasts. A finding of either t(8;21)(q22;q22); *RUNX1-RUNX1T1* or inv(16)(q13.1;q22); *CBFB-MYH11* is diagnostic of acute leukemia, regardless of blast count. Patients with either of these two AML subtypes respond well to cytarabine-based consolidation chemotherapy.[30] Importantly, prognosis is adversely affected by activating *KIT* gene mutations in exon 8 or 17, which are present in up to 30% of cases.[31] FISH is preferred for detection at diagnosis, while quantitative reverse transcription PCR is highly sensitive and ideally suited for disease monitoring. Minimal residual disease testing has clear prognostic significance in the setting of core binding factor AML.[32] *RUNX1-RUNX1T1* fusion transcripts are readily detectable by reverse transcription-PCR (RT-PCR) utilizing relatively simple primer sets due to the clustering of breakpoints at a limited number of intronic sites.[33] Three dominant *CBFB-MYH11* fusion transcripts (types A, D, and E) corresponding to different breakpoints account for over 95% of inv(16)-positive AML cases and are also amenable to detection by RT-PCR.

The presence of the balanced reciprocal translocation t(15;17)(q24;q12); *PML-RARA* is diagnostic of acute promyelocytic leukemia. The chimeric fusion protein resulting from the translocation mediates an arrest in myeloid differentiation at the promyelocyte stage. Timely recognition is required due to the high risk of disseminated intravascular coagulation (DIC) seen in these patients and the associated clinical ramifications. Prompt DIC prophylaxis may prevent a bleeding diathesis that otherwise could be fatal. Morphologic and flow cytometric studies allow for a presumptive diagnosis in most cases, but genetic confirmation of *PML-RARA* is nevertheless required. FISH is the test of choice at diagnosis due to high sensitivity and typically rapid turnaround time. Rare cryptic *PML-RARA* fusions have been documented, which are not detectable by routine karyotyping.[34] The disease is responsive to therapy with all-*trans*-retinoic acid, which drives the terminal differentiation of the neoplastic cells and is used in combination with anthracyclines or arsenic trioxide to induce durable remission in 80% to 90% of patients.[35] Variant *RARA* fusion partners may be seen on occasion, the most important of which is *ZBTB16* at 11q23. Recognition is important due to the lack of therapeutic response to all-*trans*-retinoic acid.[36] Assessment of response to therapy by a quantitative RT-PCR—based monitoring test has emerged as standard practice and appears to improve clinical outcome.[37] Three *PML-RARA* fusion transcripts (bcr1, bcr2, and bcr3) may be encountered corresponding to breakpoints at different regions of *PML* at 15q24 (Fig. 8.4). *RARA* breakpoints are clustered within intron 2 in a 15 kb region on 17q12. The bcr3 fusion results in a relatively short transcript and may be associated with the presence of *FLT3* mutations.

Translocations involving the *KMT2A* gene and various partner genes occur with some frequency in AML cases, most commonly in children. Over 80 *KMT2A* translocation partners have been described. In AML, the most common is t(9;11)(p22;q23); *MLLT3-KMT2A*, which is routinely detectable by FISH. This disease is often associated with extramedullary manifestations, and patients classically

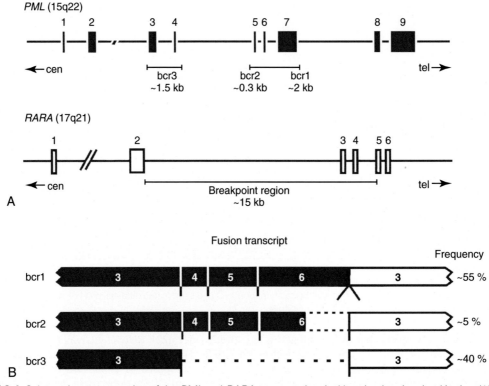

FIGURE 8.4 Schematic representation of the *PML* and *RARA* genes and typical breakpoints involved in the t(15;17)(q24;q12) diagnostic of acute promyelocytic leukemia. **A,** *PML* exons are shown as red rectangles, and *RARA* exons are in white. Translocation breakpoints in *PML* are found in one of three breakpoint cluster regions (bcr1, bcr2, and bcr3), while *RARA* breakpoints are clustered in a single intronic region. **B,** The configuration of the *PML-RARA* fusion transcripts corresponding to different *PML* breakpoint cluster regions along with their relative frequency is shown. *cen,* Centromere; *tel,* telomere.

present with gingival hyperplasia or cutaneous lesions due to tissue infiltration by the leukemia cells. Due to the heterogeneity of the translocation breakpoints, a widely applicable quantitative RT-PCR assay for disease monitoring is difficult to design, but testing can be performed on those patients with more common breakpoints.[38]

A subset of cases of acute myeloid leukemia may arise from a preexisting myelodysplastic syndrome (MDS), and therefore some genetic features common to both diseases are observed. Two examples are t(6;9)(p23q34); *DEK-NUP214* and inv(3)(q21;q26.2); *RPN1-EVI1*.[39,40] Both are associated with multilineage dysplasia and occur in 1% to 2% of AML patients overall. As expected, prognosis is generally poor. *FLT3* mutations are frequently detected in AML with *DEK-NUP214*, which has a tendency to affect younger adults and children, and it is unclear whether the negative prognostic implication is independent of *FLT3* status. *FLT3* mutations are relatively uncommon in AML with inv(3) in which *RPN1* acts as an enhancer of the oncogene *EVI1* to drive cellular proliferation possibly in collaboration with *RAS*-pathway mutations.[41]

A particularly rare form of AML occurs in infants and children with Down syndrome and involves fusion of *RBM15* to *MKL1* as a result of t(1;22)(p13;q13). The fusion gene's precise role in leukemogenesis is unclear, but transcriptional activation and modulation of chromatin organization are likely mechanisms.[42] Somatic N-terminal truncating *GATA1* mutations are common in patients with Down syndrome–associated AML but do not affect prognosis. *GATA1* mutations are also seen in individuals with Down syndrome who develop transient abnormal myelopoiesis (TAM), a disorder which may mimic AML but usually resolves in the first 6 months of life.[43] These two conditions are not readily distinguishable, especially in early infancy, and careful clinical correlation is necessary.

Chronic Myelogeneous Leukemia

The discovery in 1960 of a recurrent chromosomal abnormality in patients with a disease then known as chronic granulocytic leukemia by Peter Nowell and David Hungerford was a landmark observation. This was followed 13 years later by Janet Rowley's demonstration of a consistent reciprocal translocation between 9q34 and 22q11.2, resulting in a der(22q) (now known as the Philadelphia chromosome) found in virtually all cases of chronic myelogenous leukemia (CML).[44,45] Another 10 years would pass before the precise breakpoints were cloned, and it was shown that the translocation results in juxtaposition of the *ABL1* proto-oncogene on 9q34 to *BCR* on 22q11.2 with formation of a novel *BCR-ABL1* fusion gene.[46] This results in constitutive ABL1 protein tyrosine kinase activity and dysregulated cellular proliferation. Documentation of t(9;22)(q34;q11.2); *BCR-ABL1* fusion is necessary for diagnosis and is readily demonstrated by metaphase cytogenetics, FISH and/or RT-PCR.

Distinct *BCR-ABL1* fusion transcripts, which correspond to variably sized fusion proteins, are encountered based on the translocation breakpoints (Fig. 8.5). The *ABL1* breakpoint is largely conserved at a location upstream of exon 2. The bulk of *ABL1* is therefore fused to BCR and preserves the ABL1 kinase domain. The major breakpoint cluster region (M-bcr) is the location of the majority of *BCR* breakpoints in CML, and up to half of the *BCR* breakpoints in Philadelphia chromosome–positive (Ph+) adult B-ALL. MBR breakpoints occur between exons 13 or 15, resulting in a fusion transcript consisting of e13a2 (b2a2) or e14a2 (b3a2), both of which encode the p210 kDa *BCR-ABL1* isoform typical of CML.

Less frequently, BCR breakpoints occur in the minor breakpoint cluster region (m-bcr) between *BCR* exon 1 and exon 2, resulting in a shorter e1a2 fusion transcript encoding the p190 kDa fusion protein seen rarely in CML but commonly in pediatric Ph+ B-ALL. At initial diagnosis, CML patients may harbor low levels of e1a2 transcript detectable by RT-PCR in addition to the M-bcr transcript. This phenomenon is likely due to alternative splicing, and it is important to note that for purposes of disease monitoring, only the M-bcr transcript should be followed by RT-PCR. Occasionally, e1a2 is the exclusive transcript detected in CML and is associated with monocytic proliferation resembling chronic myelomonocytic leukemia. Rare cases of CML harbor *BCR* breakpoints that occur in the mu breakpoint cluster region (μ-BCR) at exon 19 (c3), resulting in an e19a2 (c3a2) fusion. The corresponding *BCR-ABL1* protein is larger (p230 kDa) and is characteristically seen in association with marked neutrophilic maturation that may mimic chronic neutrophilic leukemia. Notably, most RT-PCR assays are not designed to detect the e19a2 fusion transcript, so FISH studies should be recommended if clinical suspicion for CML persists despite a negative RT-PCR result.

Despite dramatic improvement in long-term survival, treatment of CML patients with TKIs does not typically result in cure. This is evidenced by the fact that low levels of *BCR-ABL1* persist even in patients who achieve a major molecular response. Over time, point mutations in the *BCR-ABL1* kinase domain may develop in a subset of patients and confer resistance to the TKI.[47] This resistance may be overcome by changing therapy to a different TKI. Documentation of a *BCR-ABL1* kinase domain mutation in a patient with suboptimal or failed response to first-line TKI therapy may therefore be indicated and can be accomplished by Sanger sequencing and, increasingly, by massively parallel sequencing. The latter strategy may offer additional testing benefits,[48] as will be discussed in the "Gene mutations in hematopoietic malignancies" section of this chapter.

Myelodysplastic Syndromes

The most common chromosomal abnormalities observed in myelodysplastic syndromes (MDSs) take the form of unbalanced structural chromosomal deletions or gains (Table 8.2). Deletion of the long arm (q arm) or outright loss of chromosome 5 or 7 is seen relatively frequently. Balanced abnormalities are rare in MDS, but they do occur and show some overlap with recurrent translocations found in AML. These findings provide evidence of clonality that is particularly useful when the differential diagnosis in a cytopenic patient includes reactive conditions that must be excluded. Indeed, several chromosomal aberrations can provide presumptive evidence of MDS in the setting of persistent cytopenias, even in the absence of sufficient morphologic evidence of dysplasia.[1] In addition, many of these cytogenetic abnormalities have established prognostic significance and are an integral component of the widely applied International

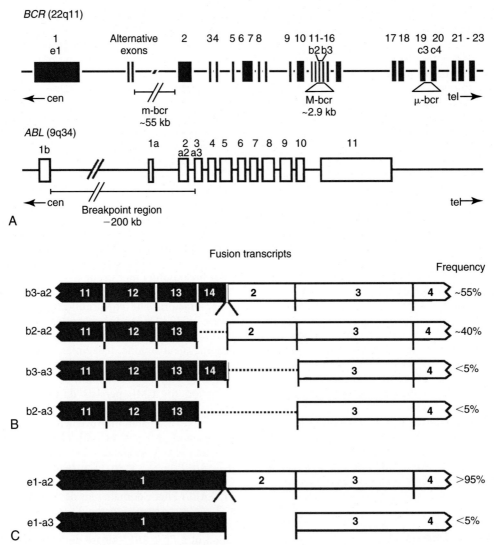

FIGURE 8.5 Schematic representation of the *BCR* and *ABL1* genes and typical breakpoints regions involved in the t(9;22)(q34;q11.2) characteristic of chronic myelogenous leukemia (CML) and a subset of B-lymphoblastic leukemia (B-ALL). **A,** *BCR* exons are shown as red rectangles, and *ABL* exons are in white. *BCR* translocation breakpoints are found in one of three locations: the major breakpoint cluster region (M-bcr), the minor breakpoint cluster region, and the mu breakpoint cluster region (µ-BCR). *ABL* breakpoints are clustered in a single conserved region upstream of exon a2 (or rarely exon a3). **B,** Configuration and relative frequency of *BCR-ABL1* fusion transcripts involving the M-bcr encountered in CML. The e13a2 (b2a2) or e14a2 (b3a2) transcripts encode a p210 kDa protein and are seen in most cases of CML along with a proportion of adult B-ALL. **C,** Configuration and relative frequency of *BCR-ABL1* fusion transcripts involving the m-bcr commonly encountered in B-lymphoblastic leukemia. The e1a2 transcript encodes a p190 kDa protein and is rarely seen in CML but quite frequently in pediatric B-ALL. *cen,* Centromere; *tel,* telomere.

Prognostic Scoring System for MDS.[49] For example, a complex cytogenetic profile (greater than three unrelated defects) is a harbinger of evolution to AML and portends a very poor prognosis. Copy number alterations or copy number neutral loss of heterozygosity (eg, *TP53*) also occur in MDSs and are detectable by array methods.[50,51]

Myeloid and Lymphoid Neoplasms With Eosinophilia and Abnormalities of PDGFRA, PDGFRB, or FGFR1

A rare but distinct group of myeloid and lymphoid neoplasms demonstrate variable clinical presentations but are unified by eosinophilia and gene fusions involving the receptor protein tyrosine kinases *PDGFRA, PDGFRB,* or *FGFR1* (see Table 8.2).[52] Patients with *PDGFRA*-associated disease most often present with a myeloproliferative neoplasm resembling chronic eosinophilic syndrome, but a range of presentations, including acute myeloid leukemia and T-lymphoblastic leukemia/lymphoma, are possible. A *FIP1L1-PDGFRA* fusion is formed as a result of a cryptic deletion at chromosome 4q12. FISH is well suited for the detection of this abnormality, and probes targeting *FIP1L1, PDGFRA,* and *CHIC2* (also located at chromosome 4q12) are often employed. Fusion of the *FIP1L1* and *PDGFRA* loci results in loss of the *CHIC2* locus and creates a signal pattern detectable by FISH. An *ETV6-PDGRFB* fusion resulting from t(5;12)(q33;p12) is the most commonly observed fusion involving the *PDGFRB* gene and typically occurs in the context of a chronic myelomonocytic leukemia-like disorder, often with eosinophilia. *PDGRFB-*

associated translocations are detectable by routine karyotype. Prompt recognition of these diseases is crucial because end-organ damage due to eosinophilia may result in significant morbidity. In addition, most patients with *PDGFRA* and *PDGFRB* abnormalities are highly sensitive to treatment with the TKI imatinib.[53] *FGFR1*-related myeloid and lymphoid neoplasms are characteristically heterogeneous and may manifest as chronic eosinophilic leukemia, acute myeloid leukemia, or T- or B-lymphoblastic leukemia/lymphoma, among other possibilities. Various fusion partners have been reported, notably including *BCR*, but rearrangement of *FGFR1* at 8p11 is a constant finding. This is readily detectable by karyotyping. Patients with this abnormality have a very poor prognosis disease that is unresponsive to TKI therapy.

Test Applications

Cytogenetics

Conventional cytogenetics refers to visual analysis of a karyotype composed of a complete set of metaphase-arrested chromosomes typically stained by Giemsa (G-banding). G-banding highlights light and dark zones corresponding to A-T-rich and G-C-rich regions of chromosomes. This decades-old technique remains very important for the comprehensive bone marrow evaluation of patients with acute leukemia or a myelodysplastic syndrome. G-banded karyotyping enabled the discovery of many of the recurring translocations characteristic of specific hematologic diseases. Karyotyping provides a genome-scale perspective and is well suited for uncovering large-scale chromosomal aberrations (resolution greater than 5 Mb) including reciprocal translocations, large deletions, and aneuploidy. Cryptic rearrangements and small insertions or deletions are likely to be missed by conventional karyotype and require more sensitive techniques for detection. Karyotyping is a labor-intensive process that requires considerable expertise for performance and interpretation. In addition, because cell growth in culture is required, turnaround times often range from 3 to 7 days, making this technique inappropriate for urgent clinical scenarios such as detection of t(15;17)(q24;q12); *PML-RARA*.

Fluorescence in Situ Hybridization

Fluorescence in situ hybridization (FISH) continues to be a powerful tool in the modern cytogenetics laboratory due to its ability to overcome several of the limitations of conventional G-banded karyotype. This technique utilizes fluorescently labeled DNA probes designed to hybridize to specific sequences on metaphase or interphase chromosomal preparations. Resolution is improved to approximately 2 Mb. FISH allows detection of cryptic chromosomal abnormalities such as translocations or deletions, which may not be identifiable by routine karyotype. The proliferation of FISH testing has been facilitated by the commercial availability of extensive libraries of probes targeting various recurrent abnormalities. A variety of specimen types can be used, importantly including paraffin-embedded formalin fixed samples, because FISH assays do not require viable cells or growth in culture. Break-apart FISH probes allow for identification of gene rearrangements without a priori knowledge of the translocation partner, of which there may be numerous possibilities. For example, an *MLL* break-apart probe (*KMT2A* gene) spanning the appropriate 11q23 region can effectively confirm or exclude the presence of an *MLL* rearrangement without having to query any of the dozens of potential MLL translocation partners. Global assessment of copy number changes is not possible by FISH because only a limited number of probes are utilized.

Single Nucleotide Polymorphism Array

Gene amplification is found in a variety of hematolymphoid neoplasms and may contribute to disease pathogenesis.[54] Array-based strategies offer a method by which to identify potentially important copy number changes that are not detectable by conventional karyotyping or FISH studies. Such arrays are based on genome-wide analysis of common single nucleotide polymorphisms (SNPs) using genomic DNA prepared from both tumor and normal reference samples. Copy number changes and copy neutral loss of heterozygosity are detectable by this method, but balanced translocations are not. Therefore this technique is best utilized as an adjunct tool in combination with conventional cytogenetics and/or FISH testing.

Polymerase Chain Reaction

Conventional polymerase chain reaction (PCR) assays targeting recurrent translocations are occasionally employed in the diagnostic workup of hematopoietic malignancies when evidence of a particular gene fusion (eg, *IGH-BCL2*) is sought. Depending on the context, test sensitivity may be limited due to marked heterogeneity in translocation breakpoints and the fact that abnormal mRNA fusion transcripts may not be produced by a particular abnormality (eg, *IGH-BCL2*). For this reason, FISH is often a preferable alternative for identification of lymphoma-associated translocations involving the *IGH* locus. On the other hand, RT-PCR allows for detection of abnormal mRNA fusion transcripts where applicable, such as in the setting of AML-associated translocations t(8;21)(q22;q22); *RUNX1-RUNX1T1*, t(15;17)(q22;q12); *PML-RARA*, inv(16)(p13.1;q22); *CBFB-MYH11* or in the t(9;22)(q34;q11.2); *BCR-ABL1*—positive leukemias. In RT-PCR assays, mRNA acts as a template for transcription of complementary DNA (cDNA) that is utilized as a template for subsequent PCR reactions. These reactions utilize assay-specific primers spanning the breakpoint of interest and thus will only yield a PCR product in the context of abnormal fusion transcripts.

The advent of technologies allowing real-time detection of PCR products has greatly facilitated the quantitative analysis of pathologic gene fusion transcripts seen in specific diseases. Although several technologies are available (see Chapter 4), oligonucleotide hydrolysis probes are commonly used for quantitative RT-PCR assays. The most important clinical application of quantitative RT-PCR is in minimal residual disease testing or disease monitoring. An important example is in the clinical diagnosis and management of chronic myelogenous leukemia (Fig. 8.6). Routine molecular monitoring of response to TKI therapy in CML is the standard of care. Testing should be performed at diagnosis and periodically throughout the course of treatment. A three-log or greater reduction in *BCR-ABL1* transcript levels within 18 months of initiation of therapy is designated a major molecular response (MMR) and portends a high likelihood of a favorable long-term outcome.[55] Quantitative PCR assays, therefore, should be designed and validated with clinically important levels of sensitivity in mind.

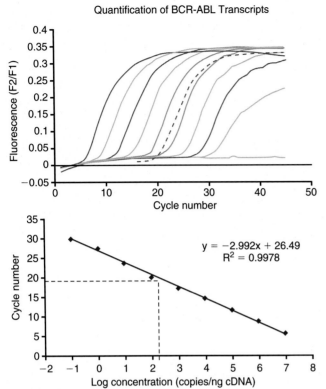

FIGURE 8.6 An example of quantitative reverse transcription PCR (RT-PCR) for enumeration of *BCR-ABL1* transcripts. The upper panel shows amplification curves for standards containing serial log dilutions of *BCR-ABL1* fusion transcripts ranging from 10^0 to 10^9 copies *(solid lines)* and an unknown patient sample *(dotted line)*. The lower panel shows the standard curve for the reaction that plots cycle number versus the log of the template concentration. The cycle threshold is then used to quantitate the *BCR-ABL1* transcript copy number (CN) in the patient sample. For normalization purposes, total *ABL1* transcripts can be analyzed separately in an identical fashion to generate a ratio of *BCR-ABL1* CN to *ABL1* CN referred to as the *normalized copy number* (NCN). The patient NCN can then be converted to the international standard scale (IS).

POINTS TO REMEMBER

Recurrent Translocations and Structural Chromosomal Abnormalities

- Recurrent genetic abnormalities are those that are repeatedly found in different patients with the same or similar diseases.
- Recurrent translocations are common in mature B-cell non-Hodgkin lymphoma and prototypically involve immunoglobulin gene loci.
- Detection of balanced translocations and other abnormalities is crucial in the diagnosis and prognostication of acute leukemia and mature myeloid malignancies.
- Preferred detection methods vary depending on the specific genetic abnormality and clinical context (eg, FISH for diagnosis vs. quantitative RT-PCR for disease monitoring).
- Massively parallel sequencing may allow for more efficient detection of multiple translocations or large-scale genomic alterations, but technical challenges exist.

GENE MUTATIONS IN HEMATOPOIETIC MALIGNANCIES

This chapter is organized by genetic mechanism (translocations, single-gene mutations) and test methodology. For example, cytogenetics and FISH are used for translocations and copy number changes. SNP arrays unveil more cryptic cytogenetic abnormalities, including copy neutral loss of heterozygosity, and molecular studies are employed for targeted detection of abnormal gene fusions and single-gene abnormalities. However, in terms of biology and pathophysiology, such distinctions are artificial. Many different types of genetic abnormalities may be present in a neoplasm, in many cases acting in concert to promote the development and progression of the disease. This section covers gene mutations, including insertions, deletions, and single-nucleotide variants (SNVs), that occur across the spectrum of hematopoietic malignancies and that are not detectable by karyotyping or FISH-based assays but rather by molecular techniques such as PCR and sequencing. In some disorders, such mutations may be the only detectable genetic abnormality and are likely sufficient in and of themselves to promote development of the neoplasm (eg, *BRAF* Val600Glu and hairy cell leukemia). In other cases multiple single-gene mutations are present and may occur along with structural chromosomal abnormalities and/or translocations detectable by classical karyotyping or FISH (eg, myelodysplastic syndromes and acute myeloid leukemia) and with epigenetic changes that must be queried by yet additional testing modalities. In addition, copy number changes may be present and typically require additional, array-based technologies for detection. A comparison of these common testing strategies used to evaluate structural and focal genetic abnormalities in hematopoietic neoplasms is summarized in Table 8.3.

Mechanisms

Mutations can result from exposure to mutagenic agents (ie, ionizing radiation or chemotherapeutic agents) or randomly during the process of DNA replication during cell division. Thus mutation rates at any particular nucleotide position depend on a variety of factors, including nucleotide stability due to genome context (eg, homopolymer tracts or repetitive sequences) or chemical modification (eg, methylation), the error rate of DNA synthesis, and the accuracy and effectiveness of DNA repair mechanisms. Mutations may therefore leave a signature with clues about the original mechanism that resulted in the error.[56]

Mutations that correlate most closely with aging include C to T transition mutations that arise in CpG dinucleotides due to spontaneous deamination of 5-methylcytosine to thymine. Indeed, there is a positive correlation between the number of such mutations and the age of the patient at the time of cancer diagnosis.[56] Chemotherapeutic alkylating agents such as cyclophosphamide also have a characteristic mutation signature involving C to T mutations.[57] Interestingly, and perhaps not unexpectedly, there is also a positive correlation between the number of divisions of normal tissue stem cells and the rate of malignancies among tissue types.[58] This finding suggests that the majority of malignant processes, including those involving the hematopoietic system, are derived from

TABLE 8.3 Testing Strategies Commonly Used to Evaluate Genetic Abnormalities in Hematopoietic Neoplasms

Testing Methodology	Pan-Genomic or Targeted	Sensitivity	Detection of Single Nucleotide Variants (SNVs)	Detection of Small Insertions and Deletions (<10² Base Pairs)	Detection of Copy Number Changes	Detection of Balanced Chromosomal Translocations	Detection of CN-LOH	Cost
Karyotyping by metaphase cytogenetics	Pan-genomic	Low	No	No	Yes	Yes	No	Moderate
FISH	Targeted	Low	No	No	Yes	Yes	No	Low
SNP arrays	Pan-genomic	Low	No	No	Yes	No	Yes	Moderate
Quantitative PCR techniques	Targeted	High	Yes	No	No	Yes	No	Low
Sanger sequencing	Targeted	Low	Yes	Yes	No	No	No	Low
Massively parallel sequencing mutation panels	Targeted	High	Yes	Yes	Yes/No*	Yes/No*	Yes/No*	Moderate
Whole genome sequencing	Pan-genomic	Low	Yes	Yes	Yes/No*	Yes/No*	Yes/No*	High

*Some bioinformatics pipelines have the capacity to detect copy number changes and chromosomal rearrangements if designed appropriately.
CN-LOH, Copy neutral loss of heterozygosity.
Modified from Nybakken GE, Bagg A. The genetic basis and expanding role of molecular analysis in the diagnosis, prognosis, and therapeutic design for myelodysplastic syndromes. J Mol Diagn 2014;16:145–58.

the evolution of chance mutations that occur during the process of DNA replication and not from some environmental or toxic insult. Mutations that arise in a stem cell may be noncontributory to the process of malignant transformation. However, they remain a component of the malignant clone and are detectable if queried in an appropriate manner—for example, by whole genome sequencing. These have come to be known as "passenger" mutations and can sometimes be segregated from true disease "driver" mutations that occur in coding regions, at splice sites, or in RNA genes because they are not recurrent over large disease-specific patient cohorts. As more data are generated, the differentiation between passenger and driver will become easier.

Myeloid Diseases

Two landmark studies with very similar findings published in late 2014 offer important insight into the development of myeloid malignancies.[59,60] Both of these studies described an age-associated rate of mutation in genes known to be involved in myeloid cancers in older individuals who had no evidence of a hematopoietic malignancy. Both studies found that the mutations were primarily in the same three genes: *ASXL1*, *TET2*, and *DNMT3A*. Patients with mutations had an elevated risk for the subsequent development of a hematologic neoplasm.[59] The most common somatic variants were the C to T transition mutations that are typically associated with aging.[59] It has long been known that clonal hematopoiesis is a feature that is detectable in a subset of older individuals, and these studies provide further support for this observation. The implication of these studies is that these mutations are a part of a founding clone and, in and of themselves, are not sufficient to promote the development of an overt hematological malignancy but are predisposing to additional genetic abnormalities that eventually lead to clinical disease.

In myeloid cancers recurrent driver mutations tend to affect discrete cellular pathways and regulatory mechanisms. Although many genes tend to be recurrently mutated in specific myeloid malignancies, there is a large degree of overlap among the various disease entities and there are few, if any, true disease-defining mutations (Table 8.4). Generally the affected pathways may be classified into those involving signal transduction (eg, *JAK2*, *KIT*, *CSF3R*, *MPL*, *FLT3*, *CALR*, *PTPN11*, *NRAS/KRAS*), epigenetic modifiers (eg, *TET2*, *IDH1*, *IDH2*, *EZH2*, *DNMT3A*), the RNA splicing complex known as the spliceosome (*SF3B1*, *SRSF2*, *ZRSR2*), the cohesion complex involved in sister chromatid separation during meiosis and mitosis (*STAG2*, *RAD21*), myeloid transcription factors (*CEBPA*, *RUNX1*), and chromatin modifiers (*ASXL1*, *EZH2*). Some of these genes are more associated with specific histologically defined myeloid malignancies—for example, *FLT3* mutations and AML or *JAK2* mutations—and myeloproliferative neoplasms such as polycythemia vera, while others are found more widely across the spectrum of myeloid diseases—for example, *TET2* mutations. The order in which mutations arise also appears to have consequences with respect to the overall clinical and pathophysiologic features of a disease process. For example, patients with myeloproliferative neoplasms who acquire *JAK2* mutations prior to acquiring a *TET2* mutation are more likely to have erythrocytosis and have increased sensitivity to *JAK2* inhibitor therapies versus those patients who acquired *JAK2* mutations after a *TET2* mutation.[61]

Signal Transduction Pathways: Disease-Specific Signaling Mutations

Signal transduction pathways define cellular communication networks that allow cells to communicate with other cells and with the extracellular environment via cell surface receptors. These receptors are often linked to tyrosine and serine/threonine kinases that serve to propagate the message, ultimately to the nucleus where modulation of gene expression results in some change in cellular activity. Important pathways in

TABLE 8.4 Recurrently Mutated Genes in Myeloid Malignancies

Gene	MPN	MDS	MDS/MPN	De novo AML	Secondary AML	Effect*
JAK2	++	−	+	−	−	Gain
MPL	+	−	−	−	−	Gain
CALR	++	−	+	−	−	Gain
FLT3	−	−	−	++	−	Gain
NPM1	−	−	+	++	−	Gain
CEBPA	−	−	−	+	−	Loss
RUNX1	−	+	++	+	−	Loss
KIT	+	−	−	+	−	Gain
CSF3R	+	−	+	−	−	Gain
DNMT3A	+	+	+	++	−	Loss
TET2	+	++	++	++	+	Loss
IDH1/2	+	+	+	++	+	Gain
SF3B1	−	+	+	−	+	Unknown
SRSF2	−	+	++	+	++	Unknown
STAG2	−	+	−	−	++	Loss
ASXL1	++	++	++	+	++	Unknown
EZH2	+	+	+	−	++	Loss
TP53	+	+	+	+	+	Loss

*Gain = gain of function. Loss = loss of function. − = rare (<2% to 3%). + = moderate frequency overall or high frequency in certain small defined subsets (eg, KIT mutations and systemic mastocytosis). ++ = high frequency overall (>~15%).

myeloid malignancies include growth factor signaling pathways linked to JAK-STAT signaling and other cell surface receptors linked to the RAS-ERK (mitogen activated protein kinase; MAPK) pathway, among others. Thus, the mutations that occur in signaling pathways tend to affect tyrosine kinase mediated signal transduction and often result in constitutively active tyrosine kinases. Activating mutations like these tend to result from recurrent so-called "hot spots" often in regulatory or enzymatic (ie, tyrosine kinase) domains of these important signaling molecules.

JAK2

JAK2 is a protein tyrosine kinase that is mutated in MPNs, including nearly all cases of polycythemia vera and approximately half of cases of essential thrombocythemia and primary myelofibrosis. The most common JAK2 mutation is c.1849G>T, p.Val617Phe, which occurs in the autoinhibitory pseudokinase (JH2) domain of JAK2 and imparts dysregulated kinase activity. The JAK2 protein is a prominent component of JAK-STAT dependent growth factor signaling pathways, including those downstream of the erythropoietin (EPO) receptor (EPOR). In MPNs, this abnormality results in JAK2-mutated progenitor cells that are abnormally EPO independent. The phenomenon of EPO-independent progenitor cells in patients with polycythemia vera was well recognized many years prior to the discovery of the presence of activating JAK2 mutations.[62] Frequently, homozygous JAK2 mutations are present in neoplastic cells in MPNs and may be a result of loss of heterozygosity (LOH) due to mitotic recombination.[63] In fact, LOH had originally been observed at the JAK2 locus at chromosome 9p prior to the identification of JAK2 mutations. The observation of recurrent LOH at chromosome 9p prompted the original analysis of the genes located in this region, including JAK2.[64] The detection of a JAK2 p.Val617Phe mutation is important in the workup of a patient suspected of having a myeloproliferative neoplasm and allows neoplasia to be distinguished from benign reactive-type disorders in the differential diagnosis.

When the JAK2 p.Val617Phe mutation was discovered, there was an initial hope that therapeutic inhibition of JAK2 would produce results similar to those seen with the ABL1 kinase inhibitor imatinib in CML patients. However, the experience with JAK2 inhibitors has not been the panacea that was initially expected. The first JAK2 inhibitor approved in the United State was ruxolitinib, a selective oral inhibitor of JAK1 and JAK2. A series of trials in patients with myelofibrosis, including primary myelofibrosis, postessential thrombocythemia myelofibrosis, and postpolycythemia vera myelofibrosis showed that ruxolitinib reduced patient spleen size and improved the overall burden of symptoms but had uncertain benefits on survival.[65] In a more recent study in patients with polycythemia vera who had an inadequate response to standard treatment with hydroxyurea, ruxolitinib treatment demonstrated clear benefits in terms of disease symptoms and promoted a modest mean reduction in the JAK2 p.Val617-Phe mutant allele burden of −12.2% at 32 weeks of treatment.[66] Overall, it is clear that JAK2 inhibitor therapy markedly improves the quality of life of MPN patients, but, in contrast to the success of imatinib and other TKIs in CML patients, JAK2 inhibition may not improve overall survival rates in patients with MPNs.

KIT

KIT is a receptor protein tyrosine kinase expressed by hematopoietic progenitor cells. KIT mutations occur primarily in the tyrosine kinase domain. They are found in most cases of systemic mastocytosis, a MPN that manifests primarily as a mast cell expansion. Patients present with symptoms such as flushing and diarrhea from increased mast cells and their associated secretory products. KIT mutations are also found in a subset of cases of core binding factor (CBF) AML, which collectively includes cases with t(8;21)(q22;q22); RUNX1-RUNX1T1 or inv(16)(p13.1;q22); CBFB-MYH11 abnormalities, which both involve the same transcription factor complex. The most common KIT mutation in both AML and systemic mastocytosis is the KIT c. 2447A>T, p.Asp816Val mutation, which results in constitutive protein tyrosine kinase activity. KIT mutations may also be present in exon 8, the extracellular domain, and these are commonly in-frame insertions or deletions that also lead to constitutive activation of KIT receptor signaling. Cases of CBF-AML with KIT mutations have a higher risk of relapse, and these patients are considered intermediate risk compared to CBF AML cases without KIT mutations, which are considered low risk.[31] Because of this, testing for KIT mutations is the standard of care in those patients with CBF-AML and inv(16)(p13.1;q22); CBFB-MYH11 or t(8; 21)(q22;q22); RUNX1-RUNX1T1 abnormalities. KIT mutations are also seen in nonhematopoietic malignancies such as melanoma and gastrointestinal stromal tumors (GIST), among others. Certain patients with KIT-mutated tumors—for example, GISTs—respond quite well to treatment with the TKI imatinib.[67] The small subset of systemic mastocytosis patients who lack a KIT p.Asp816Val mutation may also respond to imatinib, while those with the mutation are resistant.[68]

CSF3R

Activating mutations in the receptor for colony stimulating factor 3 (CSF3R) are present in most cases of chronic neutrophilic leukemia, a rare BCR-ABL1-negative expansion of mature granulocytes in blood and bone marrow without features of dysplasia.[69,70] CSF3R mutations have also been described in patients with atypical chronic myeloid leukemia, an MDS/MPN neoplasm with morphologic features distinct from those of chronic neutrophilic leukemia. However, some degree of controversy exists concerning the presence of CSF3R mutations in atypical chronic myeloid leukemia, and if strict criteria from the World Health Organization (WHO) classification system are applied, then CSF3R mutations appear much more specific for chronic neutrophilic leukemia.[70] Most mutations occur at codon 618 (p. Thr618Ile; sometimes annotated as p.Thr595Ile) in the membrane proximal extracellular domain and result in hyperactive signaling through the JAK-STAT pathway. Other membrane proximal mutations occur as well, albeit with lower frequencies. However, a significant subset of cases also have frameshift or nonsense mutations in the cytoplasmic domain of CSF3R that result in truncation of the CSF3R protein with enhanced signaling through other pathways such as those involving SRC kinases. These are identical to the CSF3R truncation mutations that are sometimes observed in patients with severe congenital neutropenia. This syndrome is frequently due to germline mutations in ELANE. The subsequent acquisition of somatic CSF3R mutations in these patients appears to herald progression to MDS or AML, an outcome for which these individuals are known to be at increased risk. Patients with membrane proximal *CSF3R* mutations and resultant hyperactive JAK-STAT signaling may respond best to treatment with a JAK inhibitor such as ruxolitinib, while those with truncating mutations appear to respond best to treatment with a SRC kinase inhibitor such as dasatinib.[71] In a small subset of cases of chronic neutrophilic leukemia and atypical chronic myeloid leukemia, *CSF3R* mutations may cooperate with mutations in the oncogene *SETBP1* to promote the development of the diseases. Stronger associations exist between *SETBP1* mutations and *ASXL1* mutations, which are very frequently found together in patients with MDS/MPN neoplasms.[72] Recurrent cooperating mutation pairs have been identified across the spectrum of myeloid malignancies and may yield insight into the resultant pathophysiology of the neoplastic cells.

FLT3

FLT3 encodes a receptor protein tyrosine kinase involved in the regulation of hematopoiesis. *FLT3* is one of the most frequently mutated genes in adult de novo AML, along with *NPM1* and *DNMT3A*,[73] and is associated with cytogenetically normal AML (CN-AML). In de novo AML, *FLT3* mutations are present in approximately 25% to 30% of cases.[73] *FLT3* mutations are also found frequently in AML with t(15;17)(q22;q12); *PML-RARA* and in AML with t(6;9)(p23;q34); *DEK-NUP214*.[74] Mutations in *FLT3* are most frequently internal tandem duplication (*FLT3*-ITD) mutations. These are in-frame insertions from three to hundreds of base pairs in length in the juxtamembrane domain (exon 14) that result in constitutive receptor activation. Point mutations also occur in exon 20 in the FLT3 tyrosine kinase domain (FLT3-TKD). These occur most commonly at position D835 (D835Y>D835H/V/E) and also result in constitutive receptor activation.[75] In spite of their apparent biological similarities, *FLT3*-ITD and *FLT3*–TKD mutations do not carry the same clinical significance. *FLT3*-ITD mutations are associated with poor outcome due to a very high likelihood of relapse after first remission. In addition, those patients with the highest *FLT3*-ITD mutant allele burdens (often referred to as a *FLT3*-ITD allelic ratio) do worse than those with lower levels, so testing strategies may be designed to assess mutant allele frequencies as well.

Typically *FLT* allelic ratios are derived by capillary electrophoresis of amplicons from a PCR that spans the region. This is accomplished via a comparison of wild-type and mutant product relative fluorescence (Fig. 8.7). *FLT3*-TKD mutations are found in less than 10% of cases and do not appear to affect prognosis.[75] However, *FLT3*-TKD mutations may be a source of TKI resistance because certain point mutations interfere with TKI binding. *FLT3* mutations are known to be unstable. Patients without evidence of *FLT3*-ITD mutation at diagnosis may relapse with a clone carrying *FLT3*-ITD mutation and vice versa. This underscores the fact that leukemia clones present at relapse may differ substantially from the dominant clone that was initially present at diagnosis. The unstable nature of *FLT3* mutations in AML makes them less suitable for use as a minimal residual disease marker.

As monotherapy, FLT3 inhibition has shown only limited success in early phase clinical trials that have been conducted thus far in AML patients. Therefore current strategies are primarily focused on the addition of FLT3 inhibitors to the commonly used chemotherapeutic regimens. Combination therapies are being explored using the FLT3 inhibitors midostaurin, quizartinib, and others with some fairly promising early results.[76] However, to date, the optimum treatment for *FLT3*-ITD mutated AML remains early allogeneic stem cell transplantation during first remission before the almost inevitable relapse occurs.

CALR

In late 2013, two landmark studies were published showing that the majority of patients with essential thrombocythemia or primary myelofibrosis who lack *JAK2 p.Val617Phe* mutations have mutations in *CALR*.[77,78] *CALR* and *JAK2* mutations are mutually exclusive in patients with these diseases, and most of the remaining small subset of patients with essential thrombocythemia or primary myelofibrosis without *JAK2* or *CALR* mutations have mutations in *MPL*, the gene that encodes the receptor for thrombopoietin (TPO). A small subset of individuals lack *JAK2*, *CALR*, and *MPL* mutations and are known as "triple negative." *CALR* mutations are not seen in polycythemia vera patients, making the detection of *CALR* mutations useful for disease subclassification.

CALR encodes a protein, calreticulin, which appears to have multiple functions, including as a calcium-binding

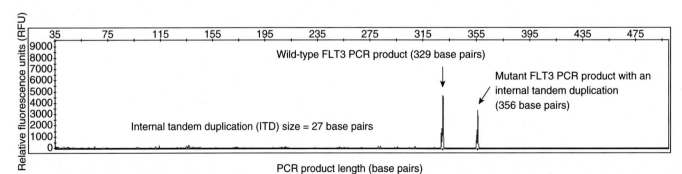

FIGURE 8.7 An example of results from a test for detection of *FLT3* internal tandem duplication (ITD) mutations. In this strategy, a DNA sample is subjected to PCR amplification using primers flanking the region of exon 14 where the insertions occur. The resulting amplicons are resolved by capillary electrophoresis. In this example, the wild-type *FLT3* PCR product is evident (329 base pairs) along with an abnormal *FLT3* amplicon of 356 base pairs representing an in-frame *FLT3* insertion (an internal tandem duplication) of 27 base pairs. A negative test result would demonstrate only the wild-type PCR fragment of 329 base pairs.

endoplasmic reticulum (ER) chaperone. Calreticulin has a C-terminal acidic domain with calcium binding motifs and an ER retention motif (the amino sequence is Lys-Asp-Glu-Leu). The *CALR* gene has 9 exons. *CALR* mutations are recurrent frameshifts in exon 9, most commonly a 52–base pair deletion (known as a type 1 variant) or a 5–base pair insertion (known as a type 2 variant). All *CALR* mutations demonstrate the same abnormal reading frame, shifted by 1 base pair. The mutations result in the production of an abnormal protein with a C-terminus that lacks the ER retention motif and that gains a novel 36 amino acid sequence with a basic, rather than acidic, charge.[77,79] There have been reports of *CALR* mutated myelofibrosis patients responding to treatment with JAK2 inhibitors, suggesting that JAK-STAT signaling is also important in *CALR*-mutated patients.[80] Individuals with primary myelofibrosis with *CALR* mutations appear to have a better prognosis than those with primary myelofibrosis with a *JAK2 p.Val617Phe* mutation or an *MPL* mutation.[81] Interestingly, primary myelofibrosis patients in the "triple negative" subgroup (negative for mutations in *CALR*, *JAK2*, and *MPL*), have the worst outcome.[81]

Mutations in Epigenetic Regulators

Mutations in genes that function in pathways involved in the epigenetic regulation of gene expression are extremely common and important across the spectrum of myeloid malignancies. Mutations result in aberrant function of several related mechanisms of epigenetic regulation. These include DNA methylation, DNA hydroxymethylation, histone modification via methylation and acetylation, metabolic pathways, and others.[82] The end result is abnormal epigenetic regulation and concomitant altered gene expression patterns that result in abnormal cellular differentiation and ultimately lead to the clinical manifestations of disease.

DNA methylation of the C-5 position of cytosine to form 5-methylcytosine in the context of CpG dinucleotides is a fundamentally important mechanism in the regulation of gene expression. Such CpG motifs are clustered in the genome into so-called CpG islands that tend to occur in gene promoter regions. The methylation of CpG islands within promoters results in repression of the expression of the associated downstream genes. In myeloid malignancies, there are recurrent and abnormal methylation patterns that have been observed as a result of aberrant function of genes involved in gene methylation.[83,84] Normally, CpG islands tend to be relatively hypomethylated. Conversely, in the context of many cancers, including myeloid malignancies, CpG islands are frequently found to be hypermethylated, particularly those that are associated with tumor suppressor genes. This mechanism serves to suppress their expression. Mutations in epigenetic regulators are common in AML, MDS, and MDS/MPN neoplasms, but are also occasionally seen in patients with MPNs. Four of the most commonly mutated genes in myeloid malignancies are involved in epigenetic regulatory functions: *DNMT3A*, *TET2*, *IDH1*, and *IDH2*.

DNMT3A

The *DNMT3A* gene encodes a DNA methyltransferase that participates in the de novo methylation of cytosines within CpG dinucleotides. Other closely related genes include *DNMT1*, which stabilizes methylation, and the methyltransferase *DNMT3B*. *DNMT3A* mutations are present in about 10% of cases of MDS and in 20% to 25% of cases of cytogenetically normal AML (CN-AML). The majority of mutations occur in a hotspot at Arg882 (approximately 60%) in the C-terminal methyltransferase domain, while frame-shift, nonsense, and splice-site mutations predicted to inactivate the gene account for most of the remainder of mutations identified.[84] Mutations at R882 result in reduced DNA methylation compared to the wild-type protein,[85] and the most common mutation at this codon, p.Arg882His, binds and inhibits wild-type *DNMT3A* in a dominant negative manner.[86] Thus the majority of mutations found in *DNMT3A* appear to suppress methyltransferase activity either through reduced expression of functional *DNMT3A* or by expression of a dominant negative form that binds and inactivates wild-type *DNMT3A* protein. In terms of prognostic significance, the findings for patients with *DNMT3A* mutations have been mixed. Many, but not all, studies have found a negative impact of mutated *DNMT3A* on outcome in AML patients.[84,87]

TET2

The *TET2* (ten-eleven translocation gene 2) gene encodes an alpha-ketoglutarate dependent 5-methylcytosine dioxygenase with an important role in demethylation of 5-methylcytosine. In myeloid neoplasms, acquired somatic mutations of *TET2* often result in loss of function of the TET2 protein and include frameshift, nonsense, and missense mutations throughout the coding sequence.[88] Somatic *TET2* mutations are found in patients with MPN, AML, and MDS/MPN neoplasms, such as chronic myelomonocytic leukemia.[83] The activity of wild-type TET2 protein results in the conversion of 5-methylcytosine to 5-hydroxymethylcytosine. The latter is subsequently converted to unmodified cytosine. In individuals with *TET2* loss-of-function mutations, there is a measurable loss of 5-hydroxymethylcytosine in genomic DNA from bone marrow cells, supporting a model where TET2 plays a critical role in the overall regulation of gene expression through the formation of 5-hydroxymethylcytosine.[89] Loss of heterozygosity has been frequently reported at the *TET2* locus at chromosome 4q24, including copy number neutral loss of heterozygosity as well as deletions, and this may result in homozygous or hemizygous mutations.[88] There is also evidence of multiple (biallelic) *TET2* mutations in patients with myeloid neoplasms, suggesting there is a selective advantage when both *TET2* alleles are mutated and inactivated in a neoplastic clone.[83] The observations of biallelic *TET2* loss-of-function mutations and loss of heterozygosity at chromosome 4q24 in *TET2*-mutated patients are evidence that *TET2* is a classic tumor suppressor gene.

IDH1 and IDH2

Mutations in the *IDH1* (isocitrate dehydrogenase 1) gene were originally detected in a whole genome sequencing study performed on a group of patients with AML.[90] In subsequent studies, mutations in *IDH2* (isocitrate dehydrogenase 2), the mitochondrial homologue of *IDH1*, were also found in other myeloid malignancies.[91] The mutations center on key arginine that resides in the catalytic sites of *IDH1* (Arg132) and *IDH2* (Arg140 and Arg172). In the Krebs cycle, isocitrate dehydrogenase coverts isocitrate to alpha-ketoglutarate. Mutations at Arg132 in *IDH1* or Arg140/Arg172 in *IDH2* confer novel (neomorphic) enzymatic activity that results in

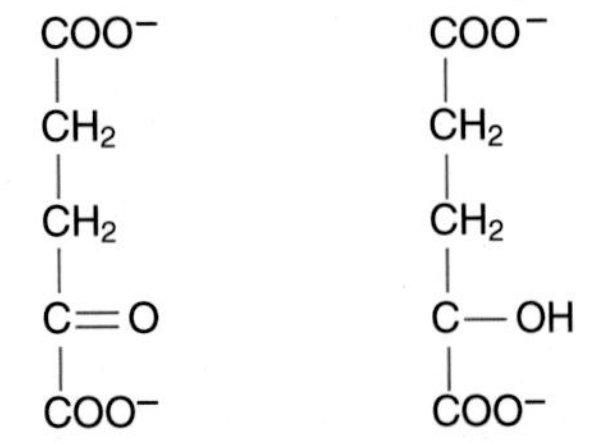

$$
\begin{array}{cc}
\text{COO}^- & \text{COO}^- \\
| & | \\
\text{CH}_2 & \text{CH}_2 \\
| & | \\
\text{CH}_2 & \text{CH}_2 \\
| & | \\
\text{C}=\text{O} & \text{C}-\text{OH} \\
| & | \\
\text{COO}^- & \text{COO}^- \\
\text{alpha-ketoglutarate} & \text{2-hydroxyglutarate}
\end{array}
$$

FIGURE 8.8 Structures of alpha-ketoglutarate and 2-hydroxyglutarate.

the formation of the oncometabolite 2-hydroxyglutarate rather than alpha-ketoglutarate (see Fig. 8.8 for a comparison of the structures of alpha-ketoglutarate and 2-hydroxyglutarate).[92,93] 2-Hydroxyglutarate accumulates and competitively inhibits alpha-ketoglutarate-dependent dioxygenases, including TET2, in affected individuals.[94] Thus IDH1/IDH2 mutations impair TET2 enzymatic function. As such, mutations in IDH1/IDH2 and TET2 tend to be mutually exclusive in myeloid malignancies.

Cases with TET2 or IDH1/IDH2 mutations display similar epigenetic abnormalities characterized by histone and DNA hypermethylation.[94] IDH1/IDH2 mutations are found most frequently in patients with cytogenetically normal AML but are also seen to a lesser degree in cases of MDS and in MDS/MPN neoplasms.[95] Selective inhibitors of mutated IDH1/IDH2 reverse the abnormal histone and DNA hypermethylation, and thus inhibition of mutant forms of isocitrate dehydrogenase is potentially a promising pharmacotherapeutic strategy.[96] An oral inhibitor of mutant IDH2 appears particularly effective at inducing differentiation and ultimately clearance of leukemic blasts in IDH2-mutated AML patients and may obviate the need for highly toxic chemotherapy regimens in these patients.[96] IDH1 and IDH2 mutations have also been identified in nonmyeloid hematopoietic malignancies, including T cell lymphoma and in nonhematopoietic tumors, including many gliomas.

Mutations in the RNA Spliceosome

The RNA spliceosome is a large multimeric protein/RNA complex that functions to remove introns and ligate exons from immature precursor messenger RNA (mRNA), resulting in the formation of mature mRNA transcripts. For many genes, multiple possible normal splice variants may be created with sometimes differing functions. Mutations in components of the spliceosome are quite common in myeloid malignancies, particularly in cases of MDS or AML that developed from MDS or MPN (so-called secondary AML). Spliceosome mutations are uncommon in de novo AML and MPNs that have not progressed to AML. Spliceosome mutations have also been identified in lymphoid malignancies, primarily in chronic lymphocytic leukemia and almost always in the gene SF3B1. In myeloid cancers, the most commonly mutated spliceosome components include SF3B1, SRSF2, U2AF1, and ZRSR2.[97] Mutations in SF3B1 are found in the majority of cases of MDS with ring sideroblasts.[98,99] Ring sideroblasts are abnormal erythroblasts with iron-containing mitochondria present in a ring pattern that are visible on an iron-stained bone marrow aspirate sample. This observation highlights an important genotype-phenotype association. Although the mechanistic reason for the presence of ring sideroblasts in cases with SF3B1 mutations is unclear, it correlates to SF3B1 haploinsufficiency.[100] Cases of MDS with SF3B1 mutations have a better prognosis with lower risk of AML progression compared to cases of MDS without SF3B1 mutations.[99] Conversely, MDS patients with mutations in SRSF2 demonstrate adverse outcome with more rapid progression to AML.[101] Almost half of cases of chronic myelomonocytic leukemia demonstrate mutations in SRSF2, and some have mutations in ZRSR2.[102,103] Mutations in SRSF2 most commonly are missense variants affecting codon Pro95 and include p.Pro95His, p.Pro95Leu, and p.Pro95Arg.[102] Mutations in SF3B1 are most commonly missense variants affecting codon Lys700 (p.Lys700Glu), as well as missense variants affecting codons Glu622, Arg625, His662, and Lys666.[98]

As these patterns suggest, spliceosome mutations do not appear to be simple loss-of-function mutations but rather are hypothesized to change the nature of the splice variants that are generated. Missplicing of pre-mRNA can change the ratio of physiologic splice variants or may lead to the generation of new, abnormal splice variants lacking certain exons, a process known as exon skipping. In the context of U2AF1 mutations in myeloid neoplasms, distinct patterns of missplicing were observed in functionally related genes, including those involved in regulation of the cell cycle.[104] However, the identity and nature of most splice variants that result from spliceosome mutations are unknown. Spliceosome mutations are not generally seen in pairs but do coexist with nonspliceosomal mutations, commonly with TET2.[105]

Mutations in the Cohesin Complex

The cohesin complex is a multimeric protein structure involved in critical genetic regulatory mechanisms, including the segregation of sister chromatids during cell division and the regulation of gene expression. It is composed of four core subunits encoded by the genes: SMC1, SMC3, STAG2, and RAD21. Overall, somatic cohesin mutations are seen in approximately 12% of patients with myeloid malignancies and appear most commonly in AML.[106,107] Cohesin mutations are mutually exclusive but may cooccur with other mutations common in AML, including in NPM1, RUNX1, and RAS.[107] Mutations in STAG2 are often nonsense or frameshift variants that likely result in loss of function, while missense mutations are typical for the other cohesion components.[108] The functional role for cohesin mutations in leukemogenesis is unclear but likely involves alterations in gene expression patterns. MDS patients with cohesin mutations have a worse overall survival compared to MDS patients without cohesin mutations.[107]

Mutations in Myeloid Transcription Factors and Chromatin Modifiers

The transcription factor CCAAT/enhancer binding protein-alpha (CEBPA) is critical for expression of myeloid-specific genes and for the repression of proliferation, both of which are necessary for the differentiation of myeloid lineage cells. CEBPA expression is highest in differentiated myeloid cells and lowest in hematopoietic progenitor cells.[109] CEBPA is an intronless gene approximately 1 kilobase in length with a basic leucine zipper domain (bZIP), which is important for

DNA binding at promoter regions, and two N-terminal transactivation (TAD) domains, which are important for protein-protein binding. *CEBPA* can be alternatively transcribed into a long isoform (p42) and a short isoform (p30). The latter lacks the two N-terminal transactivation domains. These two isoforms are translated into proteins with slightly different functions.[110] In AML patients, the mutations are often biallelic in nature. Frequently, individuals with biallelic somatic *CEBPA* mutations demonstrate an N-terminal mutation at one of the TAD domains in one allele and a second bZIP domain mutation in the other allele.[111] *CEBPA* mutations are most commonly seen in normal karyotype AML patients and are associated with good prognosis, but only in those patients with biallelic mutations representing approximately 5% of cases.[112] Germline *CEBPA* mutations are a cause of congenital forms of AML that arise when a second somatic *CEBPA* mutation is acquired in an individual with a single germline mutation.[113]

RUNX1 (runt-related transcription factor -1) is a heterodimeric transcription factor important for hematopoietic cell differentiation and maintenance. Translocations involving the *RUNX1* gene are relatively common in myeloid malignancies and include t(8;21)(q22;q22); *RUNX1-RUNX1T1* and t(3;21)(q26;q22); *EVI1-RUNX1*. In ALL, t(12;21)(p13; q22); *ETV6-RUNX1* is found. *RUNX1* copy number alterations and gene mutations also contribute to the pathogenesis of hematopoietic malignancies. The RUNX1 protein contains an evolutionarily conserved runt-domain, which is important for DNA binding and heterodimerization. Mutations involving *RUNX1* predominantly occur in the runt domain and disrupt *RUNX1* function and include frameshift, nonsense, and missense variants. *RUNX1* mutations are observed in patients with AML, MDS, and chronic myelomonocytic leukemia. Patients with AML with mutated *RUNX1* have shorter survival and tend to display resistance to chemotherapy.[114]

Polycomb group proteins are critical, evolutionarily conserved epigenetic regulators first discovered in *Drosophila melanogaster* (fruit flies). They are involved in histone modification and chromatin remodeling, and their activity results in gene silencing. They function within multiprotein transcriptional-repressive complexes. Two of the major complexes are known as polycomb repressive complex 1 and 2 (PRC1 and PRC2). PRC1 and PRC2 act together but have distinct biochemical roles. *ASXL1* is one polycomb group gene commonly mutated in myeloid malignancies. Mutations in exon 12 of *ASXL1* are found across the spectrum of myeloid malignancies with frequencies from approximately 5% in de novo AML to almost 40% in chronic myelomonocytic leukemia.[115] *ASXL1* mutations are also common in individuals with primary myelofibrosis and are associated with worse outcome in these patients.[116] In fact, *ASXL1* mutations appear to predict worse outcome across all myeloid neoplasms.[115,117,118] The exact biologic functions of *ASXL1* remain unclear, but it may regulate histone H3 lysine 27 (H3K27) methylation status, which influences chromatin structure. It also interacts with a variety of other polycomb complex proteins, including *EZH2* and *SUZ12*. *ASXL1* may be responsible for recruitment and localization of this complex to certain important regions of the genome such as *HOXA* genes.[119] *ASXL1* mutations appear to be primarily localized to the last exon (exon 12) and result in production of an abnormal truncated protein product. It is unclear if these mutations are simple loss-of-function mutations, if they are dominant negative, or if loss of the C-terminus results in a gain-of-function by the truncated protein. The most frequently encountered mutation is a duplication of a nucleotide in a homopolymer track consisting of an 8 base pair guanine repeat (c.1943dupG), resulting in a frameshift initially thought to be a PCR artifact,[120] but that is now accepted as a true somatic variant located in a mutation hot-spot. *ASXL1* mutations are commonly seen together with mutations in *SETBP1* in patients with MDS/MPN.[72]

EZH2 and *SUZ12* are a part of the polycomb repressive complex 2 (PRC2), the activity of which is modified by *ASXL1*. PRC2 is involved in the repression of gene expression via trimethylation of histone H3K27. *EZH2* is a H3K27 histone methyltransferase that is also found mutated in myeloid malignancies. *EZH2* mutations appear to be true loss-of-function variants resulting from deletion, nonsense, and frameshift-type mutations. Such mutations result in enhanced expression of genes epigenetically regulated by *EZH2*.[121] *EZH2* mutations are most common in patients with primary myelofibrosis.[122,123] Patients with primary myelofibrosis with *EZH2* mutations have higher leukocyte counts, higher blast counts, larger spleens, and reduced overall survival.[124] The functional consequence of mutated *EZH2* in myeloid cancers stands in contrast to the observation of *EZH2* mutations in B-cell lymphoma, where they are primarily gain-of-function variants (see below). This contrast highlights the importance of lineage context in the pathophysiology of these complex genetic regulatory pathways.

TP53

The tumor suppressor p53 regulates a variety of critical anticancer functions and is nicknamed the "guardian of the genome." Its functions include regulation of cell cycle entry and exit at various checkpoints, maintenance of cellular senescence, modulation of autophagy, and control of apoptosis.[125] Patients with germline *TP53* mutations (Li-Fraumeni syndrome) have a very high risk for the development of cancers, often sarcoma or breast cancer. In general, dysregulation of p53-dependent signaling networks is a common feature of aggressive malignancies, including those of the hematopoietic system. Genetic abnormalities that result in loss of p53 activity include chromosomal deletions of the *TP53* gene locus on chromosome 17 (17p), loss of function mutations in the *TP53* gene, and epigenetic suppression of *TP53* expression. The presence of del17p or *TP53* mutations is strongly associated with poor prognosis in cancer patients.

In myeloid malignancies, *TP53* mutations are seen most frequently in patients with AML, MDS, and primary myelofibrosis. *TP53* mutations are loss-of-function variants, including insertions, deletions, nonsense mutations, and splice site mutations that result in the production of truncated protein products. In AML, *TP53* mutations are commonly associated with complex (three or more) karyotypic abnormalities.[126] AML patients with both complex karyotypes and *TP53* mutations have an extremely poor prognosis.[126] *TP53* mutations are commonly seen in a subtype of myeloid cancer that develops following exposure to cytotoxic chemotherapy for another malignancy called therapy-related AML (tAML) or therapy-related MDS (tMDS), depending on the blast count at diagnosis, although this distinction has little prognostic significance. The myeloid clones that expand following chemotherapy represent rare hematopoietic stem and progenitor cells (HSPCs) with age-related *TP53* mutations, rather than newly induced *TP53* mutated clones.[127]

NPM1

NPM1 (nucleophosmin) is a multifunctional nucleolar phosphoprotein that continuously shuttles between the nucleus and the cytoplasm. Mutations in the *NPM1* gene are quite common in AML patients and are seen in approximately 50% to 60% of cases of normal karyotype AML, making it one of the most common genetic abnormalities in this disease. Mutations in *NPM1* are found in exon 11 (formerly known as exon 12) and are characteristically tetranucleotide insertions. The most common insertion is a TCTG duplication termed *type A*. The mutation results in abnormal accumulation of mutated *NPM1* protein in the cytoplasm. The finding of abnormal cytoplasmic staining for *NPM1* by immunohistochemistry is a surrogate method that may be used to detect *NPM1* mutations in formalin fixed paraffin embedded tissue.[128] The mutations are almost always heterozygous and appear to act in a dominant negative fashion. They are usually stable at relapse (ie, present at relapse if they were present at diagnosis) and can be used as a marker for the presence of minimal residual disease in posttreatment follow-up samples.[129] *NPM1* mutations are associated with favorable prognosis in acute myeloid leukemia patients who do not have coexisting *FLT3*-internal tandem duplication mutations.[130]

Lymphoid Diseases

Whereas myeloid neoplasms tend to have overlapping mutation profiles with fewer disease-specific mutations, the somatic mutations found in lymphoid disorders are slightly more disease specific, although significant overlap still exists, particularly in T cell lymphoproliferative disorders. Due to this, the discussion of mutations in lymphoid neoplasms is structured by disease rather than by pathway. In addition, the pathophysiology of certain lymphoproliferative disorders is driven much more by the presence of specific translocations rather than single-gene mutations—for example, follicular lymphoma and the t(14;18)(q32;q21); *IGH-BCL2* translocation. The focus in this section is on those disorders where single-gene mutations, rather than balanced translocations, are of central importance with clearcut diagnostic, therapeutic, and/or prognostic significance.

Mature B-Cell Lymphoproliferative Disorders

The spectrum of B-cell lymphoproliferative disorders runs the gamut from those that are relatively indolent and typically incurable, such as chronic lymphocytic leukemia (CLL), to those with a much higher proliferative rate that can be quite aggressive but that may also be cured in certain circumstances, including diffuse large B-cell lymphoma (DLBCL) and Burkitt lymphoma. New therapeutic options are needed for many of these disorders, and the somatic mutations discussed are the targets for much of these efforts. The morphologic and histopathologic subclassification of these diseases are not discussed in detail here.

One theme that is applicable to many types of B cell lymphoproliferative disorders is that they often have dysregulated activation of the B cell receptor (BCR) signaling pathway and thus may be susceptible to inhibitors of BCR signaling, including those targeting Bruton's tyrosine kinase (BTK) or phosphatidylinositide 3-kinase (PI3K). In addition to somatic mutations, other mechanisms may promote inappropriate BCR signaling, including chronic antigenic stimulation by certain microorganisms—for example, *Helicobacter pylori* infection and the development of gastric extranodal marginal zone B cell lymphoma of mucosa-associated lymphoid tissue (MALT lymphoma). Thus both intrinsic and extrinsic activators of the BCR pathway may promote the development of these diseases.

Hairy Cell Leukemia

Hairy cell leukemia is an indolent B cell lymphoproliferative disorder typically diagnosed based on immunophenotypic findings assessed by flow cytometry and/or immunohistochemistry along with characteristic neoplastic cells with "hairy" cytoplasmic projections. Although the diagnosis is typically straightforward based on the clinical and laboratory findings, there are occasional cases that can cause diagnostic difficulty. For these cases, identification of the presence of the *BRAF* p.Val600Glu mutation is diagnostically important. *BRAF* p.Val600Glu mutations are present in nearly all cases of classic hairy cell leukemia and absent in other B cell lymphoproliferative disorders, including those that may cause diagnostic confusion with hairy cell leukemia.[131] The mutation promotes constitutive signaling through the MEK-ERK pathway, downstream of RAS, as demonstrated by high levels of phosphorylated (activated) ERK in the leukemia cells.[132] BRAF inhibitor therapies are available and have been used with some success in other tumors with BRAF-activating mutations such as melanoma. In patients with hairy cell leukemia, combined BRAF and MEK inhibition appears effective at inducing apoptosis in the leukemic cells and thus offers another treatment option for hairy cell leukemia patients who fail or are intolerant to other therapies currently in use.[133]

Lymphoplasmacytic Lymphoma

Lymphoplasmacytic lymphoma is a low-grade B cell lymphoproliferative disorder with plasmacytoid morphology and commonly associated with an immunoglobulin-M (IgM) monoclonal protein and hyperviscosity due to increased serum protein levels. The combination of an IgM monoclonal protein and hyperviscosity constitutes a clinical syndrome known as Waldenstrom macroglobulinemia. Most cases of lymphoplasmacytic lymphoma (approximately 90%) carry an activating mutation (usually, p.Leu265Pro) in *MYD88* (myeloid differentiation factor 88) that encodes an adapter protein in the Toll-like receptor (TLR) signaling pathway that affects the innate immune response to bacterial pathogens.[134] Wild-type MYD88 protein promotes the assembly of a multimeric complex with the serine-threonine kinase interleukin-1 receptor–associated kinase 4 (IRAK-4). The complex activates IRAK-1 and IRAK-2, ultimately leading to activation of nuclear factor-kappa B (NF-κβ). *MYD88* p.Leu265Pro mutations occur in an important domain (Toll-Interleukin 1 receptor; TIR) that functions in oligomerization/homodimerization with other TIR domains. The presence of *MYD88* p.Leu265Pro enhances the formation of MYD88-dependent signaling complexes, resulting in abnormal activation of NF-KB signaling. This appears to require signaling through BTK.[135] In support of a role for BTK in *MYD88* p.Leu265Pro signaling, a therapeutic inhibitor of BTK (ibrutinib) induces killing of lymphoplasmacytic lymphoma cells in vitro.[135] This observation served as the rationale for testing the efficacy of ibrutinib in patients with lymphoplasmacytic lymphoma, and early results have been encouraging.[136]

Diffuse Large B Cell Lymphoma

Diffuse large B cell lymphoma (DLBCL) is an aggressive lymphoma of large B cells that may be classified into various subtypes based on histopathologic and clinical findings. The subtypes include primary DLBCL of the CNS, T cell/histiocyte rich large B cell lymphoma, Epstein-Barr virus (EBV) positive DLBCL, and DLBCL, not otherwise specified (NOS). DLBCL, NOS may be further segregated into activated B cell–type (ABC) and germinal center B cell (GCB)–type based on gene expression profiling or, more commonly in clinical practice, by expression of surrogate markers by immunohistochemistry.[137] Here again, BCR signaling may be important in the disease, and BCR pathway mutations appear particularly relevant to the pathogenesis of ABC-DLBCL. The B cell receptor comprises surface immunoglobulin (Ig) associated with the signaling molecules CD79a and CD79b, which contain signaling motifs in their intracellular domains called immunoreceptor tyrosine–based activating motifs (ITAMs). ITAMs are paired canonical tyrosine–containing sequences (YXXL/I; where X is any amino acid) separated by six to eight amino acids. Somatic mutations affecting the ITAMs of CD79a and CD79b have been identified in ABC-DLBCL cell lines and patient samples. The mutations increase expression of surface B cell receptors and attenuate negative regulatory mechanisms, thus enhancing BCR signaling.[138] The mutations that occur in the first tyrosine residue of the ITAM result in the maintenance of surface B cell receptor expression, even under chronic active stimulating conditions, that otherwise, in normal cells, leads to downregulation of receptor expression and concomitant decreased cellular activation.[138] CD79B mutations, along with mutations in MYD88, are more frequent in ABC-type DLBCL, whereas mutations in EZH2 are more frequent in GCB-type DLBCL.[139] In DLBCL, EZH2 mutations (primarily at codon Y641 and at codon A677) result in a gain of function, and samples harboring the mutations display increased levels of histone H3K27 trimethylation.[140,141] As pointed out previously, this is in contrast to the loss-of-function variants observed in myeloid malignancies. EZH2 inhibitor drugs are being tested in early-stage clinical trials in patients with EZH2 mutated B-cell lymphomas.

Chronic Lymphocytic Leukemia

Chronic lymphocytic leukemia (CLL) is a B cell lymphoproliferative disorder of small, mature B cells that have an immunophenotype characterized by expression of pan B cell markers along with CD23, aberrant expression of CD5, and a restricted pattern of immunoglobulin light chain expression. Many prognostic markers have been described, most of which are not discussed in detail here. A few of the more important prognostic markers include IGH variable segment somatic hypermutation status (unmutated correlates with poor outcome), ZAP70 protein expression (positivity correlates with poor outcome), and cytogenetic findings (del17p and del11q correlate with poor outcome). Numerous single-gene mutations have recently been identified in CLL, some of which appear to have important prognostic significance. These include mutations in NOTCH1, MYD88, BIRC3, TP53, and SF3B1. In terms of prognostic significance, mutations in TP53 and SF3B1 are independently associated with poor outcome.[142] The poor prognostic outcome observed in SF3B1-mutated CLL patients is in contrast to SF3B1-mutated MDS patients who appear to have a more favorable outcome. The mechanistic role of SF3B1 mutations in CLL is unclear but likely involves the production of abnormal mRNA splice variants.

Mature T and NK Cell Lymphoproliferative Disorders

Similar to mature B cell lymphoproliferative disorders, T and NK cell lymphoproliferative disorders include indolent disorders such as T cell large granular lymphocyte (LGL) leukemia, which can be difficult to distinguish from a reactive process, to aggressive disorders such as peripheral T cell lymphoma. It also includes neoplasms in which viral infection plays a significant etiologic role. These include adult T cell leukemia/lymphoma, in which human T cell leukemia virus type-1 (HTLV-1) infection is causative, and extranodal NK/T cell lymphoma, where EBV is universally present in a clonally integrated state in the neoplastic cells. In terms of somatic single-gene mutations and affected pathways, aberrant JAK-STAT pathway signal transduction is very commonly involved in the pathogenesis.

T cell LGL leukemia is a rare clonal disorder of cytotoxic T cells that typically does not progress but that can be associated with severe cytopenias that may themselves be a cause of significant morbidity. A significant subset of these patients (approximately 40%) demonstrate acquired somatic mutations in the signal transducer and activator of transcription 3 (STAT3) gene, which encodes a transcription factor important for cytokine and growth factor signaling that is downstream of JAK kinases in various signaling pathways.[143] STAT3 mutations are located in an important signaling domain called the src homology 2 (SH2) domain that promotes homodimerization by mediating binding to tyrosine phosphorylated STAT3 monomers. STAT3 mutations lead to inappropriate STAT3 dimerization, phosphorylation, and translocation to the nucleus and increased transcription of STAT3 gene targets in the leukemia cells. Other LGL leukemia patients with wild-type STAT3 may demonstrate similar SH2 domain mutations in the STAT5 gene or in genes that influence STAT mediated signaling.[144,145] Dysregulated JAK-STAT signaling is critically important across the spectrum of T-cell lymphoproliferative disorders. Gain-of-function mutations in JAK2 and STAT3 have been observed in angioimmunoblastic T cell lymphoma, an aggressive peripheral T cell lymphoma primarily seen in older individuals.[146] Similarly, mutually exclusive gain-of-function mutations in JAK1, JAK3, and STAT5B have been identified in T cell prolymphocytic leukemia (T-PLL), an aggressive T-cell leukemia that frequently demonstrates translocations involving TCL1 or MTCP1.[147] T-PLL may also demonstrate mutations in IL2RG, encoding the common gamma chain important for signaling from the IL2 and other cytokine receptors. IL2RG mutations in T-PLL also upregulate of JAK-STAT signaling in the leukemia cells.[147] Thus T-PLL appears to arise as a result of the combined effects of the common translocations acting in concert with gene mutations that activate JAK-STAT signaling.

B/T Lymphoblastic Leukemia/Lymphoma

Lymphoblastic leukemia is a neoplasm of immature lymphoid progenitor cells typically displaying evidence of B- or T-lymphoid differentiation. As discussed, B-lymphoblastic leukemias are subclassified on the presence of discrete recurrent cytogenetic abnormalities such as translocations or hypodiploidy or hyperdiploidy. As outlined in the "Recurrent Translocations and Structural Chromosomal Abnormalities" section, the detection of these abnormalities has prognostic significance. Copy number changes and gene mutations not detectable by karyotyping or FISH are also present in cases of lymphoblastic leukemia, and these may provide additional clinically important information. The commonly affected genes are those important for lymphoid differentiation, maturation, and signaling, including *IKZF1*, *NOTCH1*, *PAX5*, and *PTEN1*. However, copy number changes at these loci, rather than gene mutations, appear to be the most common mechanism. *IKZF1* codes for a transcription factor that is important in the development of lymphoid lineage cells and is a tumor suppressor that is frequently deleted in cases of B-lymphoblastic leukemia, including most cases of B-lymphoblastic leukemia with t(9;22)(q34;q11.2); *BCR-ABL1*. *IKZF* deletions include complete loss due to structural abnormalities of chromosome 7 (the gene is located at chromosome 7p), kilobase scale deletions involving *IKZF* (often exons 4-7), and, finally, frameshift, missense, and nonsense mutations.[148] The genetic abnormalities affecting *IKZF* highlight the complexity of testing that is required for analysis. The different types of abnormalities affecting *IKZF* collectively require metaphase cytogenetics, FISH, array-based copy number analysis, and molecular techniques for comprehensive coverage. Detection may be clinically important because *IKZF* abnormalities are associated with poor prognosis of patients with B-lymphoblastic leukemia.

Acquired Drug Resistance Mutations

Patients with either CML or Ph+ lymphoblastic leukemia are treated with ABL1 TKIs. These include the first-generation TKI (imatinib), second-generation TKIs (dasatinib and nilotinib), or third-generation TKIs (bosutinib and ponatinib). A subset of patients may develop acquired mutations in the tyrosine kinase domain of *ABL1* that impede drug binding. When this occurs, patients will manifest signs of resistance typically evident in rising levels of *BCR-ABL1* transcripts detected by a routine quantitative RT-PCR. Many mutations have been identified, and dozens are recurrent across the spectrum of patients treated with TKIs. The identification of one or more mutations is important for clinical treatment because the sensitivity of *BCR-ABL1* to the available TKIs differs depending on which mutation is present. For example, patients who develop imatinib resistance due to the emergence of a clone harboring an *ABL1* p.Tyr253His mutation would also be relatively resistant to treatment with the second-generation TKI nilotinib but sensitive to treatment with the second-generation TKI dasatinib. However, mutations at the so-called gatekeeper residue Thr315, most commonly a p.Thr315Ile mutation, result in resistance to all but one of the currently approved therapies (ponatinib).

In a subset of patients, particularly those who have been treated with multiple sequential TKIs, more than one mutation may arise in the same clone (ie, on the same *BCR-ABL1* allele). These are commonly referred to as compound mutations, and they may impart resistance profiles that differ substantially from those seen when either mutation alone is present. Certain compound mutations, such as p.Glu255Val with p.Thr315Ile, display resistance to ponatinib, whereas either mutation occurring alone does not.[149] The compound mutations that have been identified in patients tend to center around a dozen or so key residues that are frequently found mutated alone.[149] Thus the typical sequence of events in the development of a compound mutation is likely similar in most patients. First, the acquisition and outgrowth of an initial single mutation (X) in a key residue occurs. This is followed by the acquisition of a second mutation (Y) in a clone harboring mutation X, thereby generating a new clone with two mutations (X + Y). At this time, both clone X and clone X + Y coexist in the patient, although clone X + Y may quickly dominate due to enhanced resistance properties. Strategies for analyzing and evaluating the complex clonal architecture in patients with *BCR-ABL1*-positive leukemias who have developed resistance mutations are often imperfect. Sanger sequencing is not typically useful because the data generated consist of a mixture of sequences and clonal relationships are mostly lost. Other techniques, such as massively parallel sequencing, offer more promise because the sequence data comprises single-molecule sequences and not mixtures, and therefore the clonal architecture is retained.[48]

Acquired resistance mutations also occur in patients with CLL who have undergone therapy with the BTK inhibitor, ibrutinib. A *BTK* p.Cys481Ser mutation has been seen in multiple patients. This mutation decreases the affinity of ibrutinib for BTK and reduces its ability to inhibit BTK tyrosine kinase activity. It also renders inhibition reversible instead of the irreversible inhibition seen with wild-type *BTK*.[150] Mutations have also been observed in PLCγ2 (*PLCG2*), a downstream effector of BTK. These mutations likely circumvent ibrutinib-induced inhibition of BTK by leading to constitutive activation of PLCγ2 through gain of function and BTK-independent activation of B cell receptor signaling pathways.[150] The mutations include *PLCG2* p.Arg665Trp and p.Leu845Phe.

Laboratory Techniques for the Detection of Single-Gene Variants

The clinically important variants outlined in this section include highly recurrent single nucleotide variants such as *JAK2* c.1849G>T, p.Val617Phe or *IDH1* c.394C>T, p.Arg132Cys, and more complex insertions or deletions that may result in a frameshift and lead to loss of function and that are not as recurrent. When deciding which type of testing strategy to use for a particular gene, many factors must be considered (see Table 8.3). These include the nature of the targeted variants (ie, single nucleotide variants vs. insertions and deletions); performance characteristics, including test sensitivity and precision, cost, throughput, and ease of interpretation; and the type of result desired (qualitative vs. quantitative). For example, in the context of a test for a recurrent single nucleotide variant that is used as a marker of response to a targeted therapy, it would be desirable to employ a strategy that is highly sensitive and quantitative with a very narrow focus—for example, allele-specific PCR. This strategy would likely result in the ability to detect specific single nucleotide variants down to an allele frequency of less than 1%. On the other hand, the strategy chosen for detection of loss-of-function mutations that result from various insertions or deletions occurring throughout a single or perhaps in multiple exons of a gene would likely be relatively insensitive but have much broader coverage. An example would be a test such as Sanger sequencing that typically has a comparatively poor sensitivity of approximately 10% to 20% variant allele frequency. Sensitivity is critically important for somatic mutation testing in hematopoietic disorders because there is almost always a variable degree of normal background polyclonal hematopoiesis that serves to dilute signals from the abnormal variant(s). Mutation-specific antibodies have also been developed for certain common, recurrent missense mutations that result in the production of an abnormal protein product. Many are quite specific when used in the appropriate context. For example, antibodies to IDH1 p.Arg132His are available and may be less expensive than molecular methods.[151]

Massively Parallel Sequencing for Somatic Variants in Hematopoietic Malignancies

Massively parallel sequencing (also known as next-generation sequencing or NGS) as a testing strategy is becoming more commonplace in the clinical hematology laboratory and offers numerous advantages to traditional techniques. Currently, panel-based testing, rather than whole exome or whole genome sequencing, is most frequently utilized in the clinical laboratory. Such panels combine a relatively small number of gene targets, at least compared to the breadth of sequence covered by whole genome or whole exome sequencing. Panels are composed of genes that are recurrently mutated in a particular disease or class of disease. For example, a panel comprising a few dozen genes frequently mutated in myeloid malignancies is a strategy currently employed by a variety of clinical laboratories. To enhance the representation of specific gene targets for sequencing, PCR- or hybrid capture–based enrichment is commonly employed. The end result is very high coverage (usually greater than 1000×) of a clinically important set of targets. This degree of coverage depth yields test sensitivities that may be as low as 1% to 2% variant allele frequency, far better than what is possible by Sanger sequencing but not as good as what is achievable by targeted allele-specific PCR assays or digital droplet PCR.

In general, the workflow for massively parallel sequencing panels is similar, regardless of the exact methods or platforms used (see Chapters 2 and 4). In brief, it consists of (1) DNA extraction; (2) sequencing library preparation; (3) target enrichment by PCR or hybrid capture; (4) hybridization of library fragments to a solid surface (ie, flow cell); (5) clonal amplification of library fragments; (6) massively parallel sequencing; (7) data analysis, including variant identification and annotation; and (8) variant interpretation. The variant identification and annotation step is a computationally intensive process requiring a bioinformatics pipeline employing data analysis algorithms to sort and align the generated sequencing data (reads) to a reference sequence and to identify variants (see Chapter 2). A variant is simply a position or positions within the generated sequence that differ from a reference sequence. Typically, the majority of the identified variants represent benign germline polymorphisms unrelated to the disease process. However, a subset may represent true clinically important somatic mutations present only in cancer cells and/or clinically important germline abnormalities associated with a cancer predisposition. Finally, some variants may represent sequencing error and need to be identified as such.

Currently, there is debate about the necessity of comparing the results obtained from massively parallel sequencing of tumor tissue to results obtained from paired "normal" tissues from the same patient obtained at the same time. The rationale is that this allows for the distinction between variants that are truly somatic (pathogenic) and thus not present in normal tissue and those that are germline (likely nonpathogenic) and thus present in both normal and tumor tissue. This comparison may also help in the identification of systematic errors that may be introduced during various stages of the testing process. For example, PCR or sequencing errors tend to occur at highly repetitive sequences that may also be a so-called "hot spots" for somatic variants. In a patient with a solid tumor, a concurrent skin biopsy obtained from normal skin, a buccal swab, or peripheral blood white cells could be used as the paired normal tissue. However, in patients with hematopoietic malignancies, particularly bone marrow–based myeloid or lymphoid disorders, it may be difficult to obtain normal tissue that is not contaminated with granulocytes or lymphocytes that may be derived from the abnormal clone and thus harbor the same somatic variants. It may be necessary to resort to other strategies such as fluorescence activated cell sorting (FACS) to obtain a population of cells that are not part of the malignant clone. Sorted T cell populations have been used as paired normal controls in the context of B cell malignancies such as chronic lymphocytic leukemia.[152] The sequencing of paired normal tissue alongside tumor DNA reduces the rate of false-positive findings in clinically actionable genes that could otherwise result in the use of targeted therapies that are unlikely to be effective.[153] However, sequencing of paired normal tissue essentially doubles the cost and effort involved in testing. Thorough test validation to identify systematic sequencing errors and a careful, measured, and evidence-based approach to variant interpretation, including tiered reporting of variants into separate classes containing those variants of known clinical significance and those variants of unknown clinical significance (so-called "VUSs"), may mitigate the need for sequencing paired normal tissue. This matter remains unsettled, and the arguments for and against will continue to evolve.

> **POINTS TO REMEMBER**
>
> **Gene Mutations in Hematopoietic Malignancies**
> - Massively parallel sequencing is revolutionizing the laboratory-based evaluation of hematopoietic malignancies by allowing for the simultaneous evaluation of dozens to hundreds of genes in the context of testing panels.
> - In myeloid malignancies, gene mutations tend to affect a discrete series of cellular pathways involved in cell signaling and genetic regulation.
> - In B cell lymphoproliferative disorders, gene mutations tend to affect B cell receptor signaling pathways.
> - In T cell lymphoproliferative disorders, gene mutations tend to affect JAK-STAT signaling pathways.

ROLE OF GENETIC REGULATORY MECHANISMS IN HEMATOPOIETIC MALIGNANCIES

The regulation of gene expression involves a host of interacting pathways, a number of which are affected by the genetic abnormalities that occur in hematopoietic malignancies, including those of both lymphoid and myeloid differentiation. As outlined above, these include methylation of cytosine bases in CpG islands by DNA methyltransferases, modulation of the activity of transcriptional enhancers and repressors, chemical modification of histones, and thus chromatin structure by a host of changes, including methylation, acetylation, phosphorylation, and ubiquitination. Abnormal patterns of expression of certain microRNAs, often due to structural chromosomal abnormalities or epigenetic changes, are another mechanism for regulation of gene expression and appear quite important in the pathophysiology of cancer, including various hematopoietic malignancies.[154] The end result of aberrant epigenetics is the development of abnormal gene expression patterns, often leading to a stem cell–like state and a lack of terminal differentiation. The abnormal gene expression pattern can be identified and may itself be of diagnostic or prognostic significance.

Methylation Patterns

Methylation is regulated by DNA methyltransferases, including *DNMT1*, *DNMT3A*, and *DNMT3B*. As outlined above, *DNMT3A* is frequently mutated in myeloid malignancies, but the presence of *DNMT3A* mutations does not necessarily correlate with changes in the global methylation state,[84] although the presence of recurrent *DNMT3A* missense mutations appears to disrupt enzymatic function.[85] It is likely that more focal changes in methylation, and therefore focal changes in gene expression, contribute to the pathophysiology of *DNMT3A*-mutated AML. In fact, hypermethylation of CpG islands is commonly observed in cancer cells. Drugs, such as 5-azacitidine, that reduce methylation by inhibiting methyltransferases, so-called hypomethylating agents, can be effective in certain myeloid malignancies such as MDS. DNA hypermethylation has been associated with poor risk and progression of MDS to AML.[155]

Measurement of DNA methylation status can be performed using chromatographic methods such as high-performance liquid chromatography (HPLC) to very precisely measure overall levels of methylated and unmethylated cytosine. However, this method does not give any information about local CpG methylation. To measure more focal changes, which may be more informative, bisulfite sequencing can be performed. Bisulfite treatment is used to convert unmethylated cytosine to uracil, while methylcytosine remains unconverted. Uracil is then converted to thymine when PCR is amplified and sequenced, usually by pyrosequencing. In this way the methylation status of individual cytosines can be assessed.

Micro RNA Expression Patterns

The discovery of small noncoding regulatory RNA (microRNA; miRNA) represented an entirely new and unappreciated paradigm for the regulation of gene expression. Mature, processed miRNA bind to the 3' untranslated region of target mRNA sequences, where there is sufficient, but usually imperfect, sequence complementarity. The binding of miRNA to target mRNA sequences leads to decreased translation through various pathways mediated by inhibition, inactivation, or degradation of the bound mRNA molecule. Various miRNAs have been associated with the pathophysiology of hematological malignancies and may add prognostic information but are not currently part of the routine clinical workup of patients with hematological malignancies.

Gene Expression Profiling

Gene expression profiling, wherein the expression levels of a series of mRNA transcripts are measured and compared to standards using microarray technology, is becoming more widely employed in the clinical hematology laboratory, although it is still not a common test. One use for gene expression profiling in the clinical laboratory is "cell of origin" testing to determine the differentiation state of cases of diffuse large B cell lymphoma, either activated B cell–type or germinal center B cell–type. Patients with activated B cell–type diffuse large B cell lymphoma appear to have a worse clinical outcome compared to those with germinal center B cell–type lymphoma when treated with standard front-line therapy with rituximab, cyclophosphamide, doxorubicin, vincristine, and prednisone (known as R-CHOP therapy).[156] In the future this distinction is likely to become even more clinically important because the two profiles may be associated with distinct outcomes in the setting of combination therapies incorporating the kinase inhibitor ibrutinib as well as those incorporating the proteasome inhibitor bortezomib.[157,158] Gene expression profiling is considered the gold standard for making this distinction, although in clinical practice most of the testing is done via the surrogate method of immunohistochemistry. However, a significant number of cases are misclassified by immunohistochemistry, potentially leading to inappropriate treatment decisions.[156] The gene expression profiling test is performed using a Bayesian algorithm to assign a profile to either of the two classes, or as unclassifiable, based on statistical probability.[159] Activated B cell–type diffuse large B cell lymphoma tends to display relative overexpression of genes associated with plasmacytic differentiation, while germinal center types demonstrate higher expression of genes seen in normal germinal center B cells.

A number of expression profiles have been published in other hematopoietic malignancies with claimed prognostic implications. However, for the most part none are in widespread use in clinical practice. Multiple myeloma, a systemic disease caused by the expansion of clonal plasma cells and the overproduction of clonal immunoglobulin heavy and/or light chains, is one promising area where gene expression profiling appears to provide clinically useful information. One 70-gene panel has been used in a number of studies and predicts prognosis based on a score derived from the profiling performed on purified bone marrow plasma cells with high analytical reproducibility and low variation.[160]

CLONALITY TESTING IN HEMATOPOIETIC MALIGNANCIES

Lymphoid Clonality Testing

Cancers, including those of hematopoietic origin, are composed of a population of cells derived from a parental cell that has undergone neoplastic transformation. The daughter cells are therefore "monoclonal" by definition and share important biologic characteristics that may be exploited

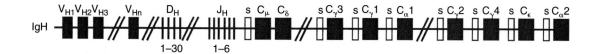

FIGURE 8.9 Schematic representation of immunoglobulin heavy chain *(IGH)*, kappa light chain *(IGK)*, and lambda light chain *(IGL)* genetic loci in germline configuration. *IGH* contains diversity gene segments, whereas *IGK* and *IGL* do not. Switch regions (s) involved in class switch recombination are indicated adjacent to constant (C) gene segments. *V,* Variable; *D,* diversity; *J,* joining.

by molecular genetic testing strategies. Putative lymphoproliferative disorders are particularly amenable to clonality testing by molecular methods due to the nature of the role of lymphocytes in immunity and the fact that they express antigen receptors. In the context of adaptive immune function, B-lymphocytes mediate the humoral immune response via the production of immunoglobulins, while T-lymphocytes mediate the cellular immune response. The specificity of these processes requires lymphocytes to be capable of producing an almost unlimited array of antigen receptors. This is accomplished in large part by recombination of germline immunoglobulin (Ig) and T cell receptor (TCR) gene segments in B and T cells, respectively, which results in highly diverse Ig and TCR antigen binding repertoires (Figs. 8.9 and 8.10).[161,162] Clonality has historically been less readily demonstrable in myeloid malignancies but can be shown indirectly in female patients by studying the pattern of inactivation of genes encoded on the X chromosome. However, given the progress of modern genetic testing strategies, it is much easier to demonstrate evidence of a clonal genetic abnormality in myeloid malignancies rather than rely on a surrogate marker of clonality such as X-inactivation patterns.

Although not required in most instances, evidence of monoclonality may aid in the diagnosis of hematologic malignancy in challenging cases such as those presenting with unusual clinical or morphologic features. However, clonality assays are fraught with potential for misinterpretation and should be treated as ancillary diagnostic studies. The recognition of small clonal hematopoietic cell populations, of both myeloid and lymphoid origin, in the peripheral blood of hematologically normal individuals further highlights the need for a comprehensive diagnostic approach.[59,163]

V(D)J Recombination in Normal B- and T-Lymphocytes

The genes encoding the immunoglobulin (Ig) and TCR proteins are the targets for DNA-based lymphoid clonality assays. The Ig and TCR gene loci, with some exceptions, are composed of a series of variable (V), diversity (D), joining (J), and constant (C) regions (Table 8.5). The V, D, and J segments recombine during early lymphocyte development in a process known as V(D)J recombination (Fig. 8.11). In B cells, recombination takes place in the bone marrow, whereas T cell maturation occurs primarily in the thymus. The result is that each B or T cell harbors a distinct productive V(D)J rearrangement that is unique in sequence and usually also in length. Monoclonal lymphoid populations, having been derived from a single progenitor cell of origin in most instances, can be expected to demonstrate identical rearrangements, a fact that is exploited for the purposes of clonality testing.

The recombination events are choreographed by an enzyme complex in which the nucleases RAG1 and RAG2 introduce double-strand DNA breaks adjacent to recombination signal sequences.[164] Recombination signal sequences are present downstream of V-gene segments, at each side of D-gene segments, and upstream of J-gene segments. D to J rearrangement is followed by a V to D-J rearrangement. Immunoglobulin light chains kappa *(IGK)* and lambda *(IGL)*, as well as the TCR alpha *(TRA)* and gamma *(TRG)*, lack diversity segments and proceed directly with V to J rearrangement. The imprecise nature of the joining reactions in V(D)J recombination, which are subject to deletion and random insertion of nucleotides, also contributes to the variability of functional rearrangements. Through these mechanisms, phenomenal diversity of Ig and TCR molecules (estimated $>10^{12}$) is achieved. B cells further diversify the repertoire of rearranged antigen receptor molecules by somatic hypermutation, a process that occurs in the germinal center of lymph nodes.

Antigen receptor rearrangements normally occur in a hierarchical sequence. In B cells, *IGH* rearranges, followed by *IGK*. If recombination of *IGK* fails to produce a productive rearrangement, a kappa-deleting element located near the constant region undergoes rearrangement and results in its deletion. *IGL* is the last immunoglobulin gene to participate in recombination and is a less desirable target for clonality investigation. The high frequency of detectable *IGH* and/or *IGK* monoclonal rearrangements in malignant B-cell proliferations (>95%) typically obviates the need for analysis of *IGL*.

T cells begin with rearrangement of the gene encoding TCR delta *(TRD)*, followed by *TRG*, TCR beta *(TRB)*, and finally TCR alpha *(TRA)*. Productive *TRD* rearrangement results in surface expression of the gamma-delta TCR that

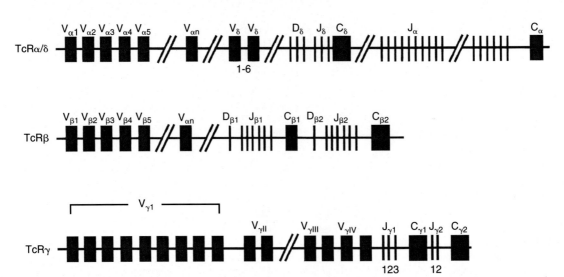

FIGURE 8.10 Schematic representation of T cell antigen receptor genes *TRA*, *TRD*, *TRB*, and *TRG* in germline configuration. *TRB* and *TRD* loci contain diversity gene segments (D), whereas *TRD* and *TRG* do not. Note that *TRD* is located in between the V and J segments of *TRA* and is deleted in the process of productive TRA rearrangement. *V*, Variable; *D*, diversity; *J*, joining; *C*, constant; TcRα/δ, TRA/TRD; TcRβ, TRB; TcRγ, TRG.

TABLE 8.5 Chromosomal Location and Number of V, D, and J Segments Potentially Available for Rearrangement of Immunoglobulin and T Cell Receptor Loci

| | RECOMBINATION HIERARCHY | | | | | | |
| | B Cells → | | | T Cells → | | | |
Genetic Locus	IGH	IGK	IGL	TRD	TRG	TRB	TRA
Chromosomal location	14q32.33	2p11.2	22q11.2	14q11.2	7p14	7q34	14q11.2
V gene segments	123–129	76	73–74	8	12–15	67	54
D gene segments	27	—	—	3	—	2	—
J gene segments	9	5	7–11	4	5	14	61

Data from Lefranc MP, Giudicelli V, Duroux P, Jabado-Michaloud J, Folch G, Aouinti S, et al. IMGT, the international ImMunoGeneTics information system 25 years on. Nucleic Acids Res 2015;43:D413–22.

defines a minor subset of T cells that are preferentially found in mucocutaneous sites. In most T cells, V to J rearrangement of *TRA* results in deletion of *TRD* due to its location between *TRA* V- and J-gene segments. Expression of the common alpha-beta TCR generally follows. The ubiquitous presence of *TRG* rearrangements in T cells, regardless of the type of TCR expressed, lends itself to utilization in clonality studies.

Identification of Clonal Rearrangements in the Workup of Lymphoid Malignancies

The diagnosis of lymphoid malignancies requires a comprehensive approach, including careful consideration of clinical findings, patient history, histopathology, and immunophenotype. The preponderance of lymphoid proliferations can be effectively separated into benign and malignant diagnostic categories based on this strategy. However, a definitive diagnosis may prove difficult in a minor subset of cases (5% to 10%). In such instances, clonality assays may provide additional diagnostic support. Knowledge of the underlying biology and the normal pattern of maturation is essential for interpretation of the clonality results. The finding of a monoclonal immunoglobulin or T cell receptor gene rearrangement may support a diagnosis of lymphoma, particularly in the setting of an atypical lymphoid proliferation showing subtle or otherwise challenging morphologic and immunophenotypic features. Monoclonal immunoglobulin or T cell receptor gene rearrangements are detectable in greater than 95% of cases of mature B cell and T cell neoplasms, respectively.[165]

Monoclonal rearrangements of Ig and TCR genes may also be identified in the context of certain reactive processes. For example, T cell clonality testing is occasionally "positive" in patients with viral infection, immune dysfunction, posttransplantation, or even in certain normal individuals.[166,167] Prominent B cell clones may be found in histologically reactive lymph nodes (particularly with HIV infection or in pediatric patients) and in autoimmune states such as rheumatoid arthritis.[168] A small population of monoclonal B cells, termed *monoclonal B cell lymphocytosis*, is found with surprising

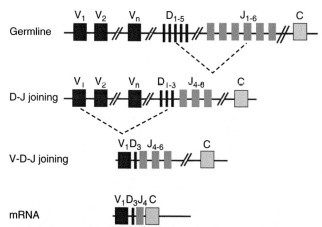

FIGURE 8.11 Schematic depiction of VDJ recombination using the immunoglobulin heavy chain locus (IGH) as an example. The process is initiated by D to J segment joining, after which V to D-J rearrangement occurs. Antigen receptor loci that lack diversity segments proceed directly with V to J rearrangement. Following RNA splicing, the IGH VDJ transcript is juxtaposed to constant (C) region genes to complete the VDJC transcript. T cell antigen receptor gene loci undergo rearrangement via an analogous process.

frequency in otherwise healthy adults.[169] Evidence of clonality, therefore, does not equate to malignancy. Further, antigen receptor rearrangements are not helpful for lineage determination due to lineage infidelity, a phenomenon particularly well illustrated by precursor lymphoid malignancies.[170] Ig rearrangements may be detectable in up to 15% of cases of T-lymphoblastic leukemia, while up to 40% of cases of B-lymphoblastic leukemia show monoclonal TCR rearrangement. Lineage infidelity is less common, but it can occur in mature T and B cell lymphomas as well. B cell clonality studies cannot distinguish between plasmablastic lymphoma and plasma cell myeloma, a clinically important distinction in which both neoplasms would be expected to harbor monoclonal Ig rearrangements. Coexisting clonal Ig and TCR rearrangements are frequently encountered in patients with angioimmunoblastic T cell lymphoma.[171] Such limitations require that clonality testing always be interpreted within the larger clinical context.

Conversely, lack of evidence of clonality does not necessarily mean benign. For example, a negative TCR gene rearrangement study does not exclude the possibility of an NK cell-derived malignancy in which the neoplastic cells would exhibit germline TCR. The limit of detection of lymphocyte clonality assays, in terms of proportion of clonal cells required in a given sample, varies by testing methodology and ranges from 1% to 10%. Scarcity of malignant cells, often in association with an exuberant cytokine-driven reactive cellular background, is a hallmark of some B cell malignancies, including Hodgkin lymphoma and T cell/histiocyte-rich large B cell lymphoma. Unless specialized techniques are employed to enrich for tumor cells (eg, laser microdissection), attempts at B cell clonality testing in such situations are likely to yield a false-negative result.

Myeloid Clonality Testing

From a historical perspective, clonality investigation of myeloid proliferations was inherently more difficult due to the lack of accessibility of specific genetic markers. To this end, X-chromosome inactivation analysis can provide an indirect method of clonality assessment in most female patients. Females normally have two copies of the X chromosome—one derived maternally and the other paternally. One of these copies is randomly inactivated during embryonic development in a process known as lyonization. When either the maternal or paternal X-chromosome is preferentially inactivated, a nonrandom or skewed pattern of inactivation is usually demonstrable and may serve as evidence of clonality.

The human androgen receptor gene (AR) is the most commonly used target of X-chromosome inactivation assays. The AR locus contains a hypervariable CAG short tandem repeat that is most often (but not always) heterozygous and in close proximity to cleavage sites of methylation-sensitive restriction enzymes (eg, HpaII and HhaI).[172] Because X-chromosome inactivation is associated with hypermethylation of cytosine residues, active (unmethylated) alleles are digested by the enzymes and can be distinguished by PCR and subsequent fragment analysis from inactive (methylated) alleles, which are not cut. The ratio of methylated and unmethylated alleles is then assessed and used to determine clonality based on skewing from the theoretically expected normal distribution (50:50). This assay can also determine clonality in NK cell proliferations that, like their myeloid counterparts, lack immunoglobulin or T cell receptor rearrangements.[173]

While capable of providing evidence of clonality in a subset of patients, X-chromosome inactivation studies are seldom performed in today's clinical laboratory. In addition to the obvious limitation with regard to patient gender, the method is subject to other important limitations that may complicate interpretation. One limitation is nonrandom X-chromosome inactivation, which occurs with some frequency in older females and results in skewing of lyonization ratios, potentially mimicking clonality.[174] Diagnosis of myeloid neoplasms is typically accomplished through comprehensive clinical and morphologic evaluation in conjunction with selective use of molecular genetic studies and conventional cytogenetics/FISH. The discovery of mutually exclusive genetic abnormalities involving JAK2, CALR, and MPL in the vast majority of myeloproliferative neoplasms, for example, has rendered X-chromosome inactivation-based clonality studies obsolete with respect to diagnosis of these malignancies. The emergence of massively parallel sequencing allows efficient mutation analysis of gene panels that, in addition to providing prognostic and/or predictive information, may also provide evidence of clonality in patients with morphologically equivocal or otherwise diagnostically challenging myeloid neoplasms.

Test Applications
Southern Blotting

At one time, Southern blot–based lymphoid clonality assays were recognized as the "gold standard" for specificity. However, this method suffers from a number of practical limitations that have contributed to its decline. Southern blot–based clonality assays utilize multiple restriction endonucleases (eg, EcoRI, BamHI, and HindIII) carefully chosen in order to digest DNA at specific sites within the corresponding antigen receptor genes. The resulting fragments from each enzymatic digestion undergo separation by gel electrophoresis, followed by transfer and immobilization to a membrane. Complementary labeled probes, usually targeted to the *IGH* or *TRG* J-segment, are hybridized to the membrane. The hallmark of a monoclonal rearrangement is the detection of distinct, novel fragments (visualized as distinct bands on the Southern blot) that differ from the pattern seen in the germline, nonrearranged controls indicating the presence of a dominant (clonal) rearrangement in the sample. Polyclonal rearrangements in a normal T or B cell population will yield a smear and no distinct bands, indicating that no dominant rearrangement is present.

The principal shortcomings of clonality assessment by Southern blotting relate to the time-consuming, labor-intensive nature of the method, limited sensitivity, and restrictive specimen requirements. Formalin-fixed paraffin embedded tissue, a common clinical sample type in lymphoma diagnosis, provides DNA of insufficient quality for Southern blot– based testing due to DNA degradation and cross-linking. PCR-based clonality testing has largely overcome these limitations without significant sacrifices in performance.

Polymerase Chain Reaction

PCR-based clonality test methods seek to amplify DNA across the V(D)J junction of rearranged Ig and TCR loci in order to capture the repertoire of unique rearrangements. Upon analysis by capillary electrophoresis, the lengths of the PCR products typically follow a Gaussian distribution due to junction size heterogeneity. A clonal population of B or T cells results in the overabundance of a unique V(D)J junction sequence evidenced by a spike emerging from the Gaussian background of polyclonal fragments (Fig. 8.12). Typically, a peak has to be at least twofold higher than the Gaussian background to be called clonal. To amplify all possible V and J combinations of a locus, multiplexing strategies have been devised to limit the number of required PCR reactions.[19] In addition, V-segment consensus primers are targeted toward framework sequences that define the Ig domain structure and are conserved between V-segments. Despite these efforts, sources of possible false negatives and false positives remain. High levels of somatic hypermutation observed in some B cell lymphoma subtypes such as follicular lymphoma may result in V gene segment alterations that could compromise consensus primer recognition. Multiple V(D)J fragments of different sequences but identical length can lead to false clonal peaks by capillary electrophoresis, especially when studying less complex loci such as *TRG*.[175]

Massively Parallel Sequencing

Massively parallel sequencing strategies allow for sequencing the entire repertoire of junction sequences for an Ig or TCR locus. Typically 100,000 to 400,000 reads are obtained that can be attributed to 15,000 to 30,000 unique rearrangements in a typical clinical sample.[176] The number of sequencing reads obtained for a particular V(D)J rearrangement is proportional to the relative size of the clone harboring that particular rearrangement. Similar to capillary electrophoresis, the analysis is performed on the repertoire of PCR-amplified rearrangements. However, the use of a single-tube multiplexed PCR amplification strategy is highly desirable in order to preserve the relative abundance of the different V and J elements present in the original repertoire.[176] Clonality is determined based on the relative

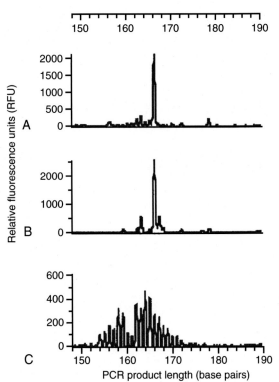

FIGURE 8.12 An example of T cell clonality testing by PCR of the *TRG* locus. The PCR products are resolved by capillary electrophoresis. **A** and **B** each show a dominant peak with amplitude at least five times background, representing monoclonal rearrangements of the same size (166 bp). **C** illustrates a typical polyclonal pattern, with a Gaussian distribution of rearrangements.

abundance of a particular sequence compared to the background of polyclonal sequences. Using massively parallel sequencing (Fig. 8.13), different rearrangements that result in PCR products of identical length can be resolved based on sequence data potentially resulting in a lower false-positive rate. In addition, identification of the junction sequence present in neoplastic cells constitutes a very sensitive marker for subsequent use in minimal residual disease detection at very low levels (0.1% to 0.01% of cells) over the course of treatment, possibly allowing early prediction of relapse.[176]

POINTS TO REMEMBER

Clonality Testing in Hematopoietic Malignancies
- The unique biology of B and T cell antigen receptor gene rearrangement allows for molecular genetic assessment of clonality.
- Failure to interpret clonality assay results within the appropriate clinicopathologic context may lead to diagnostic error.
- Monoclonal T cell receptor and immunoglobulin gene rearrangements may be demonstrable in benign clinical settings.
- Clonality assays employing massively parallel sequencing may be particularly useful for monitoring a previously documented clone.

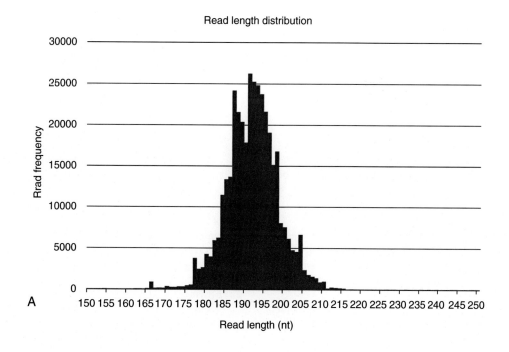

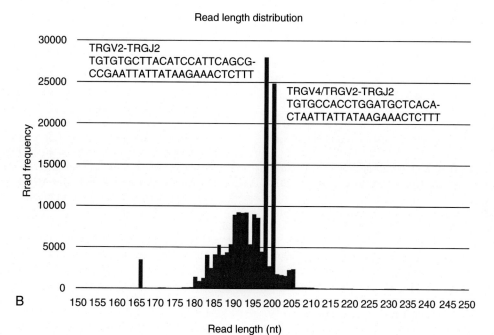

FIGURE 8.13 Examples of T cell clonality analysis by massively parallel sequencing of *TRG* rearrangements. **A,** A polyclonal (normal) pattern. Individual read clusters representing distinct rearrangements each represent 1% or less of total reads, indicating that no dominant clonal rearrangements are present. The histogram shows a Gaussian distribution when read frequencies are graphed according to length (total number of reads: 499,238). **B,** A monoclonal pattern. The top two sequence clusters, with V-J segments and complete TRG junctional sequences shown, represent 12.1% and 10.5% of total reads and show a sixfold difference over the background. The histogram shows a non-Gaussian distribution of read frequencies graphed according to length and illustrates the biallelic nature of the monoclonal rearrangement (total number of reads: 358,674). *nt,* Nucleotides.

CONCLUSION

The spectrum of genetic abnormalities present in hematopoietic malignancies requires the use of many complex testing strategies for comprehensive detection. These testing strategies include detection of translocations, copy number changes, gene mutations, and epigenetic abnormalities. As the number and type of clinically actionable genetic variants continue to grow, the laboratory evaluation of these patients will likely become a more multidisciplinary effort. However, traditional clinical, laboratory, and pathologic methods remain central to patient diagnosis and management and are unlikely to be supplanted in the near future by genetic testing.

REFERENCES

1. Vardiman JW, Thiele J, Arber DA, et al. The 2008 revision of the World Health Organization (WHO) classification of myeloid neoplasms and acute leukemia: Rationale and important changes. *Blood* 2009;**114**:937–51.
2. McNeil N, Ried T. Novel molecular cytogenetic techniques for identifying complex chromosomal rearrangements: Technology and applications in molecular medicine. *Expert Rev Mol Med* 2000;**2000**:1–14.
3. Jager U, Bocskor S, Le T, et al. Follicular lymphomas' BCL-2/IGH junctions contain templated nucleotide insertions: Novel insights into the mechanism of t(14;18) translocation. *Blood* 2000;**95**:3520–9.
4. Kuppers R, Dalla-Favera R. Mechanisms of chromosomal translocations in B cell lymphomas. *Oncogene* 2001;**20**:5580–94.
5. Hecht JL, Aster JC. Molecular biology of Burkitt's lymphoma. *J Clin Oncol* 2000;**18**:3707–21.
6. Pasqualucci L, Bhagat G, Jankovic M, et al. AID is required for germinal center-derived lymphomagenesis. *Nat Genet* 2008;**40**: 108–12.
7. Qian J, Wang Q, Dose M, et al. B cell super-enhancers and regulatory clusters recruit AID tumorigenic activity. *Cell* 2014; **159**:1524–37.
8. Meng FL, Du Z, Federation A, et al. Convergent transcription at intragenic super-enhancers targets aid-initiated genomic instability. *Cell* 2014;**159**:1538–48.
9. Bueno C, Montes R, Catalina P, et al. Insights into the cellular origin and etiology of the infant pro-B acute lymphoblastic leukemia with MLL-AF4 rearrangement. *Leukemia* 2011;**25**: 400–10.
10. Holmfeldt L, Wei L, Diaz-Flores E, et al. The genomic landscape of hypodiploid acute lymphoblastic leukemia. *Nat Genet* 2013;**45**:242–52.
11. Harrison CJ, Moorman AV, Schwab C, et al. An international study of intrachromosomal amplification of chromosome 21 (iAMP21): Cytogenetic characterization and outcome. *Leukemia* 2014;**28**:1015–21.
12. Harrison CJ. Cytogenetics of paediatric and adolescent acute lymphoblastic leukaemia. *Br J Haematol* 2009;**144**:147–56.
13. Killick S, Matutes E, Powles RL, et al. Outcome of biphenotypic acute leukemia. *Haematologica* 1999;**84**:699–706.
14. Weng AP, Ferrando AA, Lee W, et al. Activating mutations of NOTCH1 in human T cell acute lymphoblastic leukemia. *Science* 2004;**306**:269–71.
15. Manolov G, Manolova Y. Marker band in one chromosome 14 from Burkitt lymphomas. *Nature* 1972;**237**:33–4.
16. Dave SS, Fu K, Wright GW, et al. Molecular diagnosis of Burkitt's lymphoma. *N Engl J Med* 2006;**354**:2431–42.
17. Limpens J, de Jong D, van Krieken JH, et al. BCL-2/JH rearrangements in benign lymphoid tissues with follicular hyperplasia. *Oncogene* 1991;**6**:2271–6.
18. Cheung KJ, Shah SP, Steidl C, et al. Genome-wide profiling of follicular lymphoma by array comparative genomic hybridization reveals prognostically significant DNA copy number imbalances. *Blood* 2009;**113**:137–48.
19. van Dongen JJ, Langerak AW, Bruggemann M, et al. Design and standardization of PCR primers and protocols for detection of clonal immunoglobulin and T-cell receptor gene recombinations in suspect lymphoproliferations: Report of the BIOMED-2 concerted action BMH4-CT98-3936. *Leukemia* 2003;**17**:2257–317.
20. Pasqualucci L, Neumeister P, Goossens T, et al. Hypermutation of multiple proto-oncogenes in B-cell diffuse large-cell lymphomas. *Nature* 2001;**412**:341–6.
21. Petrich AM, Nabhan C, Smith SM. MYC-associated and double-hit lymphomas: A review of pathobiology, prognosis, and therapeutic approaches. *Cancer* 2014;**120**:3884–95.
22. Fernandez V, Hartmann E, Ott G, et al. Pathogenesis of mantle-cell lymphoma: All oncogenic roads lead to dysregulation of cell cycle and DNA damage response pathways. *J Clin Oncol* 2005;**23**:6364–9.
23. Monteil M, Callanan M, Dascalescu C, et al. Molecular diagnosis of t(11;14) in mantle cell lymphoma using two-colour interphase fluorescence in situ hybridization. *Br J Haematol* 1996;**93**:656–60.
24. Isaacson PG, Du MQ. MALT lymphoma: From morphology to molecules. *Nat Rev Cancer* 2004;**4**:644–53.
25. Liu H, Ruskon-Fourmestraux A, Lavergne-Slove A, et al. Resistance of t(11;18) positive gastric mucosa-associated lymphoid tissue lymphoma to Helicobacter pylori eradication therapy. *Lancet* 2001;**357**:39–40.
26. Fonseca R, Bergsagel PL, Drach J, et al. International myeloma working group molecular classification of multiple myeloma: Spotlight review. *Leukemia* 2009;**23**:2210–21.
27. Billecke L, Murga Penas EM, May AM, et al. Cytogenetics of extramedullary manifestations in multiple myeloma. *Br J Haematol* 2013;**161**:87–94.
28. Ceccon M, Mologni L, Bisson W, et al. Crizotinib-resistant NPM-ALK mutants confer differential sensitivity to unrelated ALK inhibitors. *Mol Cancer Res* 2013;**11**:122–32.
29. Parrilla Castellar ER, Jaffe ES, Said JW, et al. ALK-negative anaplastic large cell lymphoma is a genetically heterogeneous disease with widely disparate clinical outcomes. *Blood* 2014; **124**:1473–80.
30. Borthakur G, Kantarjian H, Wang X, et al. Treatment of core-binding-factor in acute myelogenous leukemia with fludarabine, cytarabine, and granulocyte colony-stimulating factor results in improved event-free survival. *Cancer* 2008;**113**: 3181–5.
31. Paschka P, Marcucci G, Ruppert AS, et al. Adverse prognostic significance of KIT mutations in adult acute myeloid leukemia with inv(16) and t(8;21): A Cancer and Leukemia Group B study. *J Clin Oncol* 2006;**24**:3904–11.
32. Jourdan E, Boissel N, Chevret S, et al. Prospective evaluation of gene mutations and minimal residual disease in patients with core binding factor acute myeloid leukemia. *Blood* 2013;**121**: 2213–23.
33. Zhang Y, Strissel P, Strick R, et al. Genomic DNA breakpoints in AML1/RUNX1 and ETO cluster with topoisomerase II DNA cleavage and DNase I hypersensitive sites in t(8;21) leukemia. *Proc Natl Acad Sci USA* 2002;**99**:3070–5.

34. Hiorns LR, Min T, Swansbury GJ, et al. Interstitial insertion of retinoic acid receptor-alpha gene in acute promyelocytic leukemia with normal chromosomes 15 and 17. *Blood* 1994;**83**: 2946—51.
35. Lo-Coco F, Avvisati G, Vignetti M, et al. Retinoic acid and arsenic trioxide for acute promyelocytic leukemia. *N Engl J Med* 2013;**369**:111—21.
36. Wells RA, Hummel JL, De Koven A, et al. A new variant translocation in acute promyelocytic leukaemia: Molecular characterization and clinical correlation. *Leukemia* 1996;**10**: 735—40.
37. Kayser S, Schlenk RF, Grimwade D, et al. Minimal residual disease-directed therapy in acute myeloid leukemia. *Blood* 2015;**125**:2331—5.
38. Abildgaard L, Ommen HB, Lausen B, et al. A novel RT-qPCR assay for quantification of the MLL-MLLT3 fusion transcript in acute myeloid leukaemia. *Eur J Haematol* 2013;**91**:394—8.
39. Secker-Walker LM, Mehta A, Bain B. Abnormalities of 3q21 and 3q26 in myeloid malignancy: A united kingdom cancer cytogenetic group study. *Br J Haematol* 1995;**91**:490—501.
40. Alsabeh R, Brynes RK, Slovak ML, et al. Acute myeloid leukemia with t(6;9) (p23;q34): Association with myelodysplasia, basophilia, and initial CD34 negative immunophenotype. *Am J Clin Pathol* 1997;**107**:430—7.
41. Lavallee VP, Gendron P, Lemieux S, et al. EVI1-rearranged acute myeloid leukemias are characterized by distinct molecular alterations. *Blood* 2015;**125**:140—3.
42. Mercher T, Busson-Le Coniat M, Nguyen Khac F, et al. Recurrence of OTT-MAL fusion in t(1;22) of infant AML-M7. *Genes Chromosomes Cancer* 2002;**33**:22—8.
43. Roberts I, Izraeli S. Haematopoietic development and leukaemia in Down syndrome. *Br J Haematol* 2014;**167**:587—99.
44. Nowell PC, Hungerford DA. Chromosome studies on normal and leukemic human leukocytes. *J Natl Cancer Inst* 1960;**25**:85—109.
45. Rowley JD. Letter: A new consistent chromosomal abnormality in chronic myelogenous leukaemia identified by quinacrine fluorescence and giemsa staining. *Nature* 1973;**243**:290—3.
46. Heisterkamp N, Stephenson JR, Groffen J, et al. Localization of the c-ABl oncogene adjacent to a translocation break point in chronic myelocytic leukaemia. *Nature* 1983;**306**: 239—42.
47. Soverini S, Hochhaus A, Nicolini FE, et al. BCR-ABL kinase domain mutation analysis in chronic myeloid leukemia patients treated with tyrosine kinase inhibitors: Recommendations from an expert panel on behalf of European LeukemiaNet. *Blood* 2011;**118**:1208—15.
48. Machova Polakova K, Kulvait V, Benesova A, et al. Next-generation deep sequencing improves detection of BCR-ABL1 kinase domain mutations emerging under tyrosine kinase inhibitor treatment of chronic myeloid leukemia patients in chronic phase. *J Cancer Res Clin Oncol* 2015;**141**:887—99.
49. Greenberg PL, Tuechler H, Schanz J, et al. Revised international prognostic scoring system for myelodysplastic syndromes. *Blood* 2012;**120**:2454—65.
50. Starczynowski DT, Vercauteren S, Sung S, et al. Copy number alterations at polymorphic loci may be acquired somatically in patients with myelodysplastic syndromes. *Leuk Res* 2011;**35**: 444—7.
51. Jasek M, Gondek LP, Bejanyan N, et al. TP53 mutations in myeloid malignancies are either homozygous or hemizygous due to copy number-neutral loss of heterozygosity or deletion of 17p. *Leukemia* 2010;**24**:216—9.
52. Gotlib J. World Health Organization-defined eosinophilic disorders: 2014 update on diagnosis, risk stratification, and management. *Am J Hematol* 2014;**89**:325—37.
53. Barraco D, Carobolante F, Candoni A, et al. Complete and long-lasting cytologic and molecular remission of FIP1L1-PDGFRA-positive acute eosinophil myeloid leukaemia, treated with low-dose imatinib monotherapy. *Eur J Haematol* 2014;**92**:541—5.
54. Bouska A, McKeithan TW, Deffenbacher KE, et al. Genome-wide copy-number analyses reveal genomic abnormalities involved in transformation of follicular lymphoma. *Blood* 2014;**123**:1681—90.
55. Press RD, Willis SG, Laudadio J, et al. Determining the rise in BCR-ABL RNA that optimally predicts a kinase domain mutation in patients with chronic myeloid leukemia on imatinib. *Blood* 2009;**114**:2598—605.
56. Alexandrov LB, Nik-Zainal S, Wedge DC, et al. Signatures of mutational processes in human cancer. *Nature* 2013;**500**:415—21.
57. Tomita-Mitchell A, Kat AG, Marcelino LA, et al. Mismatch repair deficient human cells: Spontaneous and MNNG-induced mutational spectra in the HPRT gene. *Mutat Res* 2000;**450**:125—38.
58. Tomasetti C, Vogelstein B. Cancer etiology. Variation in cancer risk among tissues can be explained by the number of stem cell divisions. *Science* 2015;**347**:78—81.
59. Jaiswal S, Fontanillas P, Flannick J, et al. Age-related clonal hematopoiesis associated with adverse outcomes. *N Engl J Med* 2014;**371**:2488—98.
60. Genovese G, Kahler AK, Handsaker RE, et al. Clonal hematopoiesis and blood-cancer risk inferred from blood DNA sequence. *N Engl J Med* 2014;**371**:2477—87.
61. Ortmann CA, Kent DG, Nangalia J, et al. Effect of mutation order on myeloproliferative neoplasms. *N Engl J Med* 2015;**372**:601—12.
62. Eridani S, Dudley JM, Sawyer BM, et al. Erythropoietic colonies in a serum-free system: Results in primary proliferative polycythaemia and thrombocythaemia. *Br J Haematol* 1987;**67**: 387—91.
63. Levine RL, Wadleigh M, Cools J, et al. Activating mutation in the tyrosine kinase JAK2 in polycythemia vera, essential thrombocythemia, and myeloid metaplasia with myelofibrosis. *Cancer Cell* 2005;**7**:387—97.
64. Kralovics R, Guan Y, Prchal JT. Acquired uniparental disomy of chromosome 9p is a frequent stem cell defect in polycythemia vera. *Exp Hematol* 2002;**30**:229—36.
65. Sonbol MB, Firwana B, Zarzour A, et al. Comprehensive review of jak inhibitors in myeloproliferative neoplasms. *Ther Adv Hematol* 2013;**4**:15—35.
66. Vannucchi AM, Kiladjian JJ, Griesshammer M, et al. Ruxolitinib versus standard therapy for the treatment of polycythemia vera. *N Engl J Med* 2015;**372**:426—35.
67. Demetri GD, von Mehren M, Blanke CD, et al. Efficacy and safety of Imatinib mesylate in advanced gastrointestinal stromal tumors. *N Engl J Med* 2002;**347**:472—80.
68. Zermati Y, De Sepulveda P, Feger F, et al. Effect of tyrosine kinase inhibitor STI571 on the kinase activity of wild-type and various mutated c-KIT receptors found in mast cell neoplasms. *Oncogene* 2003;**22**:660—4.
69. Maxson JE, Gotlib J, Pollyea DA, et al. Oncogenic CSF3R mutations in chronic neutrophilic leukemia and atypical CML. *N Engl J Med* 2013;**368**:1781—90.
70. Pardanani A, Lasho TL, Laborde RR, et al. CSF3R T618I is a highly prevalent and specific mutation in chronic neutrophilic leukemia. *Leukemia* 2013;**27**:1870—3.

71. Gotlib J, Maxson JE, George TI, et al. The new genetics of chronic neutrophilic leukemia and atypical CML: Implications for diagnosis and treatment. *Blood* 2013;**122**: 1707–11.
72. Meggendorfer M, Bacher U, Alpermann T, et al. SETBP1 mutations occur in 9% of MDS/MPN and in 4% of MPN cases and are strongly associated with atypical CML, monosomy 7, isochromosome i(17)(q10), ASXL1 and CBL mutations. *Leukemia* 2013;**27**:1852–60.
73. Cancer Genome Atlas Research N. Genomic and epigenomic landscapes of adult de novo acute myeloid leukemia. *N Engl J Med* 2013;**368**:2059–74.
74. Oyarzo MP, Lin P, Glassman A, et al. Acute myeloid leukemia with t(6;9)(p23;q34) is associated with dysplasia and a high frequency of FLT3 gene mutations. *Am J Clin Pathol* 2004;**122**: 348–58.
75. Bacher U, Haferlach C, Kern W, et al. Prognostic relevance of FLT3-TKD mutations in AML: The combination matters—An analysis of 3082 patients. *Blood* 2008;**111**:2527–37.
76. Pratz KW, Luger SM. Will FLT3 inhibitors fulfill their promise in acute meyloid leukemia? *Curr Opin Hematol* 2014; **21**:72–8.
77. Nangalia J, Massie CE, Baxter EJ, et al. Somatic CALR mutations in myeloproliferative neoplasms with nonmutated JAK2. *N Engl J Med* 2013;**369**:2391–405.
78. Klampfl T, Gisslinger H, Harutyunyan AS, et al. Somatic mutations of calreticulin in myeloproliferative neoplasms. *N Engl J Med* 2013;**369**:2379–90.
79. Shivarov V, Ivanova M, Tiu RV. Mutated calreticulin retains structurally disordered c terminus that cannot bind Ca(2+): Some mechanistic and therapeutic implications. *Blood Cancer J* 2014;**4**:e185.
80. Passamonti F, Caramazza D, Maffioli M. JAK inhibitor in CALR-mutant myelofibrosis. *N Engl J Med* 2014;**370**:1168–9.
81. Tefferi A, Lasho TL, Finke CM, et al. CALR vs JAK2 vs MPL-mutated or triple-negative myelofibrosis: Clinical, cytogenetic and molecular comparisons. *Leukemia* 2014;**28**: 1472–7.
82. Shih AH, Abdel-Wahab O, Patel JP, et al. The role of mutations in epigenetic regulators in myeloid malignancies. *Nat Rev Cancer* 2012;**12**:599–612.
83. Delhommeau F, Dupont S, Della Valle V, et al. Mutation in TET2 in myeloid cancers. *N Engl J Med* 2009;**360**:2289–301.
84. Ley TJ, Ding L, Walter MJ, et al. DNMT3A mutations in acute myeloid leukemia. *N Engl J Med* 2010;**363**:2424–33.
85. Holz-Schietinger C, Matje DM, Reich NO. Mutations in DNA methyltransferase (DNMT3A) observed in acute myeloid leukemia patients disrupt processive methylation. *J Biol Chem* 2012;**287**:30941–51.
86. Russler-Germain DA, Spencer DH, Young MA, et al. The R882H DNMT3A mutation associated with AML dominantly inhibits wild-type DNMT3A by blocking its ability to form active tetramers. *Cancer Cell* 2014;**25**: 442–54.
87. Marcucci G, Metzeler KH, Schwind S, et al. Age-related prognostic impact of different types of DNMT3A mutations in adults with primary cytogenetically normal acute myeloid leukemia. *J Clin Oncol* 2012;**30**:742–50.
88. Jankowska AM, Szpurka H, Tiu RV, et al. Loss of heterozygosity 4q24 and TET2 mutations associated with myelodysplastic/myeloproliferative neoplasms. *Blood* 2009;**113**:6403–10.
89. Ko M, Huang Y, Jankowska AM, et al. Impaired hydroxylation of 5-methylcytosine in myeloid cancers with mutant TET2. *Nature* 2010;**468**:839–43.
90. Mardis ER, Ding L, Dooling DJ, et al. Recurring mutations found by sequencing an acute myeloid leukemia genome. *N Engl J Med* 2009;**361**:1058–66.
91. Green A, Beer P. Somatic mutations of IDH1 and IDH2 in the leukemic transformation of myeloproliferative neoplasms. *N Engl J Med* 2010;**362**:369–70.
92. Ward PS, Patel J, Wise DR, et al. The common feature of leukemia-associated IDH1 and IDH2 mutations is a neomorphic enzyme activity converting alpha-ketoglutarate to 2-hydroxyglutarate. *Cancer Cell* 2010;**17**:225–34.
93. Dang L, White DW, Gross S, et al. Cancer-associated IDH1 mutations produce 2-hydroxyglutarate. *Nature* 2010;**465**:966.
94. Figueroa ME, Abdel-Wahab O, Lu C, et al. Leukemic IDH1 and IDH2 mutations result in a hypermethylation phenotype, disrupt TET2 function, and impair hematopoietic differentiation. *Cancer Cell* 2010;**18**:553–67.
95. Kosmider O, Gelsi-Boyer V, Slama L, et al. Mutations of IDH1 and IDH2 genes in early and accelerated phases of myelodysplastic syndromes and mds/myeloproliferative neoplasms. *Leukemia* 2010;**24**:1094–6.
96. Kernytsky A, Wang F, Hansen E, et al. IDH2 mutation-induced histone and DNA hypermethylation is progressively reversed by small-molecule inhibition. *Blood* 2015;**125**:296–303.
97. Abdel-Wahab O, Levine R. The spliceosome as an indicted conspirator in myeloid malignancies. *Cancer Cell* 2011;**20**: 420–3.
98. Papaemmanuil E, Cazzola M, Boultwood J, et al. Somatic SF3B1 mutation in myelodysplasia with ring sideroblasts. *N Engl J Med* 2011;**365**:1384–95.
99. Malcovati L, Papaemmanuil E, Bowen DT, et al. Clinical significance of SF3B1 mutations in myelodysplastic syndromes and myelodysplastic/myeloproliferative neoplasms. *Blood* 2011; **118**:6239–46.
100. Visconte V, Rogers HJ, Singh J, et al. SF3B1 haploinsufficiency leads to formation of ring sideroblasts in myelodysplastic syndromes. *Blood* 2012;**120**:3173–86.
101. Thol F, Kade S, Schlarmann C, et al. Frequency and prognostic impact of mutations in SRSF2, U2AF1, and ZRSR2 in patients with myelodysplastic syndromes. *Blood* 2012;**119**:3578–84.
102. Meggendorfer M, Roller A, Haferlach T, et al. SRSF2 mutations in 275 cases with chronic myelomonocytic leukemia (CMML). *Blood* 2012;**120**:3080–8.
103. Yoshida K, Sanada M, Shiraishi Y, et al. Frequent pathway mutations of splicing machinery in myelodysplasia. *Nature* 2011;**478**:64–9.
104. Przychodzen B, Jerez A, Guinta K, et al. Patterns of missplicing due to somatic U2AF1 mutations in myeloid neoplasms. *Blood* 2013;**122**:999–1006.
105. Malcovati L, Papaemmanuil E, Ambaglio I, et al. Driver somatic mutations identify distinct disease entities within myeloid neoplasms with myelodysplasia. *Blood* 2014;**124**: 1513–21.
106. Kon A, Shih LY, Minamino M, et al. Recurrent mutations in multiple components of the cohesin complex in myeloid neoplasms. *Nat Genet* 2013;**45**:1232–7.
107. Thota S, Viny AD, Makishima H, et al. Genetic alterations of the cohesin complex genes in myeloid malignancies. *Blood* 2014;**124**:1790–8.

108. Losada A. Cohesin in cancer: Chromosome segregation and beyond. *Nat Rev Cancer* 2014;**14**:389–93.
109. Zhang P, Nelson E, Radomska HS, et al. Induction of granulocytic differentiation by 2 pathways. *Blood* 2002;**99**:4406–12.
110. Roe JS, Vakoc CR. C/EBPalpha: Critical at the origin of leukemic transformation. *J Exp Med* 2014;**211**:1–4.
111. Szankasi P, Ho AK, Bahler DW, et al. Combined testing for ccaat/enhancer-binding protein alpha (CEBPA) mutations and promoter methylation in acute myeloid leukemia demonstrates shared phenotypic features. *Leuk Res* 2011;**35**:200–7.
112. Wouters BJ, Lowenberg B, Erpelinck-Verschueren CA, et al. Double CEBPA mutations, but not single CEBPA mutations, define a subgroup of acute myeloid leukemia with a distinctive gene expression profile that is uniquely associated with a favorable outcome. *Blood* 2009;**113**:3088–91.
113. Debeljak M, Kitanovski L, Pajic T, et al. Concordant acute myeloblastic leukemia in monozygotic twins with germline and shared somatic mutations in the gene for CCAAT-enhancer-binding protein alpha with 13 years difference at onset. *Haematologica* 2013;**98**:e73–4.
114. Gaidzik VI, Bullinger L, Schlenk RF, et al. RUNX1 mutations in acute myeloid leukemia: Results from a comprehensive genetic and clinical analysis from the AML study group. *J Clin Oncol* 2011;**29**:1364–72.
115. Gelsi-Boyer V, Brecqueville M, Devillier R, et al. Mutations in ASXL1 are associated with poor prognosis across the spectrum of malignant myeloid diseases. *J Hematol Oncol* 2012;**5**:12.
116. Tefferi A, Guglielmelli P, Lasho TL, et al. CALR and ASXL1 mutations-based molecular prognostication in primary myelofibrosis: An international study of 570 patients. *Leukemia* 2014;**28**:1494–500.
117. Gelsi-Boyer V, Trouplin V, Roquain J, et al. ASXL1 mutation is associated with poor prognosis and acute transformation in chronic myelomonocytic leukaemia. *Br J Haematol* 2010;**151**:365–75.
118. Schnittger S, Eder C, Jeromin S, et al. ASXL1 exon 12 mutations are frequent in AML with intermediate risk karyotype and are independently associated with an adverse outcome. *Leukemia* 2013;**27**:82–91.
119. Abdel-Wahab O, Adli M, LaFave LM, et al. ASXL1 mutations promote myeloid transformation through loss of PRC2-mediated gene repression. *Cancer Cell* 2012;**22**:180–93.
120. Abdel-Wahab O, Kilpivaara O, Patel J, et al. The most commonly reported variant in ASXL1 (c.1934dupG; p.Gly646Trpfsx12) is not a somatic alteration. *Leukemia* 2010;**24**:1656–7.
121. Khan SN, Jankowska AM, Mahfouz R, et al. Multiple mechanisms deregulate EZH2 and histone H3 lysine 27 epigenetic changes in myeloid malignancies. *Leukemia* 2013;**27**:1301–9.
122. Ernst T, Chase AJ, Score J, et al. Inactivating mutations of the histone methyltransferase gene EZH2 in myeloid disorders. *Nat Genet* 2010;**42**:722–6.
123. Nangalia J, Green TR. The evolving genomic landscape of myeloproliferative neoplasms. *Hematology Am Soc Hematol Educ Program 2014* 2014:287–96.
124. Guglielmelli P, Biamonte F, Score J, et al. EZH2 mutational status predicts poor survival in myelofibrosis. *Blood* 2011;**118**:5227–34.
125. Zilfou JT, Lowe SW. Tumor suppressive functions of p53. *Cold Spring Harb Perspect Biol* 2009;**1**. a001883.
126. Rucker FG, Schlenk RF, Bullinger L, et al. TP53 alterations in acute myeloid leukemia with complex karyotype correlate with specific copy number alterations, monosomal karyotype, and dismal outcome. *Blood* 2012;**119**:2114–21.
127. Wong TN, Ramsingh G, Young AL, et al. Role of TP53 mutations in the origin and evolution of therapy-related acute myeloid leukaemia. *Nature* 2015;**518**:552–5.
128. Falini B, Mecucci C, Tiacci E, et al. Cytoplasmic nucleophosmin in acute myelogenous leukemia with a normal karyotype. *N Engl J Med* 2005;**352**:254–66.
129. Falini B, Nicoletti I, Martelli MF, et al. Acute myeloid leukemia carrying cytoplasmic/mutated nucleophosmin (NPMc+ AML): Biologic and clinical features. *Blood* 2007;**109**:874–85.
130. Falini B, Sportoletti P, Martelli MP. Acute myeloid leukemia with mutated NPM1: Diagnosis, prognosis and therapeutic perspectives. *Curr Opin Oncol* 2009;**21**:573–81.
131. Tiacci E, Trifonov V, Schiavoni G, et al. BRAF mutations in hairy-cell leukemia. *N Engl J Med* 2011;**364**:2305–15.
132. Tiacci E, Schiavoni G, Martelli MP, et al. Constant activation of the RAF-MEK-ERK pathway as a diagnostic and therapeutic target in hairy cell leukemia. *Haematologica* 2013;**98**:635–9.
133. Pettirossi V, Santi A, Imperi E, et al. BRAF inhibitors reverse the unique molecular signature and phenotype of hairy cell leukemia and exert potent antileukemic activity. *Blood* 2015;**125**:1207–16.
134. Treon SP, Xu L, Yang G, et al. MYD88 L265P somatic mutation in Waldenstrom's macroglobulinemia. *N Engl J Med* 2012;**367**:826–33.
135. Yang G, Zhou Y, Liu X, et al. A mutation in MYD88 (L265P) supports the survival of lymphoplasmacytic cells by activation of Bruton tyrosine kinase in waldenstrom macroglobulinemia. *Blood* 2013;**122**:1222–32.
136. Treon SP, Tripsas CK, Meid K, et al. Ibrutinib in previously treated Waldenstrom's macroglobulinemia. *N Engl J Med* 2015;**372**:1430–40.
137. Hans CP, Weisenburger DD, Greiner TC, et al. Confirmation of the molecular classification of diffuse large B-cell lymphoma by immunohistochemistry using a tissue microarray. *Blood* 2004;**103**:275–82.
138. Davis RE, Ngo VN, Lenz G, et al. Chronic active B-cell-receptor signalling in diffuse large B-cell lymphoma. *Nature* 2010;**463**:88–92.
139. Zhang J, Grubor V, Love CL, et al. Genetic heterogeneity of diffuse large B-cell lymphoma. *Proc Natl Acad Sci USA* 2013;**110**:1398–403.
140. McCabe MT, Graves AP, Ganji G, et al. Mutation of A677 in histone methyltransferase EZH2 in human B-cell lymphoma promotes hypertrimethylation of histone H3 on lysine 27 (H3K27). *Proc Natl Acad Sci USA* 2012;**109**:2989–94.
141. Yap DB, Chu J, Berg T, et al. Somatic mutations at EZH2 Y641 act dominantly through a mechanism of selectively altered PRC2 catalytic activity, to increase H3K27 trimethylation. *Blood* 2011;**117**:2451–9.
142. Baliakas P, Hadzidimitriou A, Sutton LA, et al. Recurrent mutations refine prognosis in chronic lymphocytic leukemia. *Leukemia* 2015;**29**:329–36.
143. Koskela HL, Eldfors S, Ellonen P, et al. Somatic STAT3 mutations in large granular lymphocytic leukemia. *N Engl J Med* 2012;**366**:1905–13.

144. Jerez A, Clemente MJ, Makishima H, et al. STAT3 mutations unify the pathogenesis of chronic lymphoproliferative disorders of NK cells and T-cell large granular lymphocyte leukemia. *Blood* 2012;**120**:3048–57.
145. Andersson EI, Rajala HL, Eldfors S, et al. Novel somatic mutations in large granular lymphocytic leukemia affecting the STAT-pathway and T-cell activation. *Blood Cancer J* 2013;**3**:e168.
146. Odejide O, Weigert O, Lane AA, et al. A targeted mutational landscape of angioimmunoblastic T-cell lymphoma. *Blood* 2014;**123**:1293–6.
147. Kiel MJ, Velusamy T, Rolland D, et al. Integrated genomic sequencing reveals mutational landscape of T-cell prolymphocytic leukemia. *Blood* 2014;**124**:1460–72.
148. Kastner P, Dupuis A, Gaub MP, et al. Function of Ikaros as a tumor suppressor in B cell acute lymphoblastic leukemia. *Am J Blood Res* 2013;**3**:1–13.
149. Zabriskie MS, Eide CA, Tantravahi SK, et al. BCR-ABL1 compound mutations combining key kinase domain positions confer clinical resistance to Ponatinib in Ph chromosome-positive leukemia. *Cancer Cell* 2014;**26**:428–42.
150. Woyach JA, Furman RR, Liu TM, et al. Resistance mechanisms for the Bruton's tyrosine kinase inhibitor Ibrutinib. *N Engl J Med* 2014;**370**:2286–94.
151. Byers R, Hornick JL, Tholouli E, et al. Detection of IDH1 R132H mutation in acute myeloid leukemia by mutation-specific immunohistochemistry. *Appl Immunohistochem Mol Morphol* 2012;**20**:37–40.
152. Sutton LA, Ljungstrom V, Mansouri L, et al. Targeted next-generation sequencing in chronic lymphocytic leukemia: A high-throughput yet tailored approach will facilitate implementation in a clinical setting. *Haematologica* 2015;**100**: 370–6.
153. Jones S, Anagnostou V, Lytle K, et al. Personalized genomic analyses for cancer mutation discovery and interpretation. *Sci Transl Med* 2015;**7**. 283ra53.
154. Calin GA, Croce CM. Chromosomal rearrangements and micrornas: A new cancer link with clinical implications. *J Clin Invest* 2007;**117**:2059–66.
155. Jiang Y, Dunbar A, Gondek LP, et al. Aberrant DNA methylation is a dominant mechanism in MDS progression to AML. *Blood* 2009;**113**:1315–25.
156. Gutierrez-Garcia G, Cardesa-Salzmann T, Climent F, et al. Gene-expression profiling and not immunophenotypic algorithms predicts prognosis in patients with diffuse large B-cell lymphoma treated with immunochemotherapy. *Blood* 2011;**117**:4836–43.
157. Ruan J, Martin P, Furman RR, et al. Bortezomib plus CHOP-Rituximab for previously untreated diffuse large B-cell lymphoma and mantle cell lymphoma. *J Clin Oncol* 2011;**29**: 690–7.
158. Younes A, Thieblemont C, Morschhauser F, et al. Combination of ibrutinib with rituximab, cyclophosphamide, doxorubicin, vincristine, and prednisone (R-CHOP) for treatment-naive patients with CD20-positive B-cell non-hodgkin lymphoma: A non-randomised, phase 1b study. *Lancet Oncol* 2014;**15**: 1019–26.
159. Wright G, Tan B, Rosenwald A, et al. A gene expression-based method to diagnose clinically distinct subgroups of diffuse large B cell lymphoma. *Proc Natl Acad Sci USA* 2003;**100**:9991–6.
160. van Laar R, Flinchum R, Brown N, et al. Translating a gene expression signature for multiple myeloma prognosis into a robust high-throughput assay for clinical use. *BMC Med Genomics* 2014;**7**:25.
161. Tonegawa S. Somatic generation of antibody diversity. *Nature* 1983;**302**:575–81.
162. Davis MM, Bjorkman PJ. T-cell antigen receptor genes and T-cell recognition. *Nature* 1988;**334**:395–402.
163. Rawstron AC, Bennett FL, O'Connor SJ, et al. Monoclonal B-cell lymphocytosis and chronic lymphocytic leukemia. *N Engl J Med* 2008;**359**:575–83.
164. Hiom K, Gellert M. A stable RAG1-RAG2-DNA complex that is active in V(D)J cleavage. *Cell* 1997;**88**:65–72.
165. Rizvi MA, Evens AM, Tallman MS, et al. T-cell non-Hodgkin lymphoma. *Blood* 2006;**107**:1255–64.
166. Ibrahim HA, Menasce LP, Pomplun S, et al. Presence of monoclonal T-cell populations in B-cell post-transplant lymphoproliferative disorders. *Mod Pathol* 2011;**24**:232–40.
167. Degauque N, Boeffard F, Foucher Y, et al. The blood of healthy individuals exhibits CD8 T cells with a highly altered TCR Vb repertoire but with an unmodified phenotype. *PLoS ONE* 2011;**6**:e21240.
168. Kussick SJ, Kalnoski M, Braziel RM, et al. Prominent clonal B-cell populations identified by flow cytometry in histologically reactive lymphoid proliferations. *Am J Clin Pathol* 2004;**121**:464–72.
169. Shim YK, Rachel JM, Ghia P, et al. Monoclonal B-cell lymphocytosis in healthy blood donors: An unexpectedly common finding. *Blood* 2014;**123**:1319–26.
170. van der Velden VH, Bruggemann M, Hoogeveen PG, et al. TCRB gene rearrangements in childhood and adult precursor-B-all: Frequency, applicability as MRD-PCR target, and stability between diagnosis and relapse. *Leukemia* 2004;**18**: 1971–80.
171. Tan BT, Warnke RA, Arber DA. The frequency of B- and T-cell gene rearrangements and Epstein-Barr virus in T-cell lymphomas: A comparison between angioimmunoblastic T-cell lymphoma and peripheral T-cell lymphoma, unspecified with and without associated B-cell proliferations. *J Mol Diagn* 2006;**8**:466–75.
172. Allen RC, Zoghbi HY, Moseley AB, et al. Methylation of HpaII and HhaI sites near the polymorphic CAG repeat in the human androgen-receptor gene correlates with X chromosome inactivation. *Am J Hum Genet* 1992;**51**:1229–39.
173. Boudewijns M, van Dongen JJ, Langerak AW. The human androgen receptor X-chromosome inactivation assay for clonality diagnostics of natural killer cell proliferations. *J Mol Diagn* 2007;**9**:337–44.
174. Busque L, Mio R, Mattioli J, et al. Nonrandom X-inactivation patterns in normal females: Lyonization ratios vary with age. *Blood* 1996;**88**:59–65.
175. Delfau MH, Hance AJ, Lecossier D, et al. Restricted diversity of V gamma 9-Jp rearrangements in unstimulated human gamma/delta T lymphocytes. *Eur J Immunol* 1992;**22**:2437–43.
176. Schumacher JA, Duncavage EJ, Mosbruger TL, et al. A comparison of deep sequencing of TCRG rearrangements vs traditional capillary electrophoresis for assessment of clonality in T-cell lymphoproliferative disorders. *Am J Clin Pathol* 2014; **141**:348–59.

Circulating Tumor Cells and Circulating Tumor DNA

*Evi Lianidou and Dave Hoon**

ABSTRACT

Background
Classic tissue biopsies or surgical resections are invasive procedures that capture only a single snapshot in the evolution of cancer. In contrast, a blood-based test or "liquid biopsy" has the potential to characterize the evolution of a solid tumor in real time by extracting molecular information from circulating tumor cells (CTCs), circulating tumor DNA (ctDNA), circulating miRNAs, or exosomes. Molecular characterization of CTCs and ctDNA holds considerable promise for the identification of therapeutic targets and resistance mechanisms and for real-time monitoring of the efficacy of systemic therapies. The major potential advantage of CTC and ctDNA analysis is that they are minimally invasive and can be serially repeated.

Content
This overview is focused on the diagnostic, prognostic, and predictive value of CTCs and ctDNA in cancer patients. It includes key studies in different cancers and incorporates the latest advances in genome-wide analysis of ctDNA. Focus includes (1) CTC isolation, enumeration, and detection systems; (2) clinical applications of CTC; (3) different forms of ctDNA; (4) ctDNA isolation and detection systems; (5) clinical applications of ctDNA; (6) quality control and standardization of liquid biopsy assays; (7) the potential of liquid biopsy in the clinical laboratory; and (8) the potential of the molecular characterization of CTCs and ctDNA analysis as a liquid biopsy for individualized therapy. With respect to the clinical laboratory, the development of targeted molecular assays as companion diagnostics, for disease monitoring, and even for early cancer detection are all potential possibilities at various stages of development.

Cancer genomes are not static but change over the course of therapy. The term *liquid biopsy* refers to blood-based cancer testing and often involves detailed molecular analysis of the tumor from circulating genetic material in the peripheral blood. This genetic information is derived mainly from circulating tumor cells (CTCs), circulating tumor DNA (ctDNA), circulating miRNAs, and exosomes (Fig. 9.1). Liquid biopsy offers a simple and noninvasive insight into a patient's cancer. Blood-based targeted molecular assays have potential utility as companion diagnostics, in disease monitoring, and even early cancer detection.[1-3]

Currently the most promising and readily applicable role for liquid biopsy is the profiling of CTCs or ctDNA as a way to monitor patients in the course of therapy—particularly by using novel technologies for a better and earlier indication of either response or emerging resistance to a particular treatment. The next logical step is to couple this ability with analysis of the genomic landscape of CTCs or ctDNA to better understand the mechanisms of evolving resistance and hopefully to guide treatment strategies to overcome resistance. The utility of liquid biopsy is not just limited to being a mirror of tissue biopsy, but it is a potential tool that can detect unique and impactful information about a patient's cancer that tissue testing cannot. Just a few years ago, the liquid biopsy approach was limited to research studies, but it is now entering prospective clinical trials as a companion diagnostic for evaluation. It can be used for patients whose tumors are hard to access or when the site of the primary tumor is unknown. It may in the future enable decisions for activating genomic targeted therapies in patients who have failed treatment on a particular drug regimen (Fig. 9.2).

Potentially, liquid biopsy may aid in the investigation of the evolution of subclonal cancer cell populations. Liquid biopsy may be a minimally invasive method for determining dominant clones to direct targeted therapies against. There is hope that this approach can illuminate strategies to combine drugs that affect the dominant mutated populations and also inhibit other subclonal populations from expanding. This approach may impact the definition of minimal residual disease because it can change the clinician's ability to predict the risk of recurrence in early-stage cancer patients whose tumors have been surgically removed. The most exciting potential clinical application of blood-based cancer testing is early cancer detection.

---

*This work was supported by the CANCER-ID project (E.L.) and by the Dr. Miriam and Sheldon G. Adelson Medical Research Foundation (AMRF; D.H.), Weil Family Foundation, and Gonda Foundation (D.H.). We would also like to thank Ms. Anupam Singh (MSc student) and Mrs. Nousha Javanmardi for their editorial assistance. We would like to thank Ms. Cleo Parisi (PhD student) for her assistance in organizing the references and Mr. Ilias Agelidis (MSc student) for his assistance in designing Figs. 9.1 and 9.6.

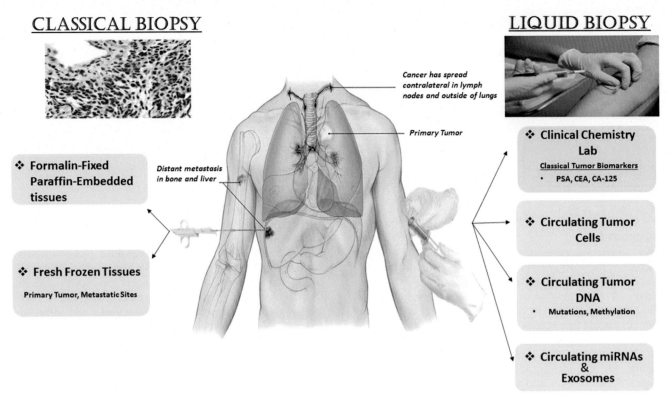

FIGURE 9.1 Classic versus liquid biopsy approaches.

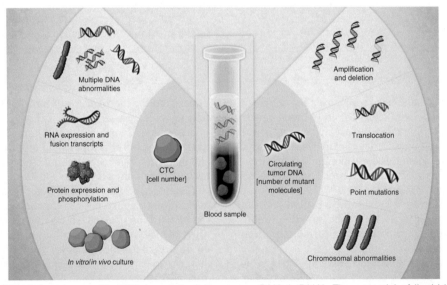

FIGURE 9.2 Circulating tumor cells (CTCs) and circulating tumor DNA (ctDNA). The potential of liquid biopsy for the evaluation and management of cancer patients from CTCs includes nucleic acid analysis, protein expression and phosphorylation, and in vivo and in vitro cultures. Analysis of ctDNA is limited to DNA. (From Haber DA, Velculescu VE. Blood-based analyses of cancer: circulating tumor cells and circulating tumor DNA. *Cancer Discov* 2014;4:650–61. Copyright 2014 American Association for Cancer Research.)

Liquid biopsy as a diagnostic, prognostic, and theranostic tool is appealing because it is minimally invasive and easily performed in a serial manner. However, there are several barriers to the routine clinical use of liquid biopsy: (1) numerous technologies are available for the detection of circulating cancer biomarkers, (2) more and more biomarkers for evaluation of CTCs and ctDNA are required, (3) well-designed comparison studies between liquid biopsy (CTCs and/or ctDNA) with conventional biopsy samples are still needed, (4) it is still difficult to control the preanalytical phase to obtain robust and reproducible results, (5) currently available techniques are costly, and (6) the turnaround time is currently too slow for maximal clinical utility.

> **POINTS TO REMEMBER**
>
> **Liquid Biopsy**
> - The liquid biopsy approach extracts molecular information from the tumor by detailed analysis of circulating tumor-derived genetic material in the bloodstream. The sources of this material are circulating tumor cells (CTCs), circulating tumor DNA (ctDNA), circulating miRNAs, and exosomes.
> - Liquid biopsy can provide detailed information on tumor genome evolution over time through conventional peripheral blood sampling that can be used for serial monitoring of a patient.

CIRCULATING TUMOR CELLS

Circulating Tumor Cells: Historical Background

The presence of circulating tumor cells (CTCs) was first reported in 1869 by Thomas Ashworth (Fig. 9.3).[4] In 2005, the clinical importance of disseminated tumor cells (DTCs) in the bone marrow of breast cancer patients was shown.[5] However, analysis of DTCs in bone marrow is invasive and thus difficult to repeat. CTCs are rare cells that originate from primary and metastatic tumors that have managed to get into the circulation and that may extravasate to different organs. Only a small fraction of CTCs will develop into metastasis.[6] CTCs are a major player in the liquid biopsy approach and may provide real-time information on a patient's disease status. Cancer metastasis is the main cause of cancer-related death, and dissemination of tumor cells through the blood circulation is an important intermediate step that also exemplifies the switch from localized to systemic disease.[7]

Many advances have been made in the detection and molecular characterization of CTCs. The presence of CTCs in peripheral blood has been linked to worse prognosis and early relapse in various types of solid cancers.[8] The FDA has cleared the CellSearch system for breast (2004), colorectal (2008), and prostate (2008) cancer based on the critical role that CTCs play in the metastatic spread of carcinomas.[9] Detection of CTCs is correlated with decreased progression-free survival (PFS) and overall survival (OS) in both operable breast cancer and metastatic breast cancer.

CTCs are targets for understanding tumor biology and tumor cell dissemination in humans. Their molecular characterization offers an exciting approach to understanding resistance to established therapies and elucidating the complex biology of metastasis.[10] Further research on the molecular characterization of CTCs should contribute to a better understanding of the biology of metastatic development in cancer patients and the identification of novel therapeutic targets, especially after elucidating the relationship of CTCs to cancer stem cells. This approach may provide individualized targeted treatments and spare breast cancer patients unnecessary and ineffective therapies.[11]

CTCs are rare, and the amount of available sample is limited, presenting formidable analytical and technical challenges. Recent technical advancements in CTC detection and characterization include multiplex reverse transcription quantitative PCR (RT-qPCR) methods, image-based approaches, and microfilter and microchip devices for their isolation. However, direct comparison of different methods for detecting CTCs in blood from patients with breast cancer has revealed a substantial variation in the detection rates[12,13] There is a lack of standardization in reference material, which hampers the implementation of CTC measurement in clinical routine practice. The potential of CTC analysis is now widely recognized, although many challenges remain(Fig. 9.4).[14]

Circulating Tumor Cells: Analytical Techniques
General Overview

The isolation and further analysis of highly pure CTCs are difficult because these cells are extremely rare in the peripheral blood.[15] Typically, CTCs are coisolated with a background of peripheral blood mononuclear cells. The combination of high-throughput and automated CTC isolation technologies with generally accepted and validated downstream molecular assays is necessary for the routine use of CTC-based diagnostics in the clinical management of cancer patients.

CTC analysis includes isolation/enrichment, detection, enumeration, and molecular characterization. The main analytical systems used are described below, and an outline is presented in Fig. 9.5. Recent advances in this area, as described in recently published reviews, can further supplement this information.[7,16-19]

The main strategies for CTC isolation/enrichment are based on separation by density, size, and/or electrical charge, or by immunomagnetic isolation of specific proteins on their cell surface. A variety of microfluidic and filtration devices has been developed and are currently under evaluation for the isolation and enrichment of CTCs, including in vivo capture and isolation of viable single CTCs. Detection systems for CTC analysis are typically protein- and image-based, although an increasing number of molecular assays are now performed on nucleic acids. Reliable single CTC isolation

FIGURE 9.3 Historical medical journal article by Thomas Ashworth from 1869 describing circulating tumor cells (I) blood smear from a normal donor, (II) section of tumor from the patient, and (III) blood smear of patient. The larger cells in the blood smear of the patient are circulating tumor cells (From Ashworth TR. A case of cancer in which cells similar to those in the tumours were seen in the blood after death. *Med J Australia* 1869;14:146–47.)

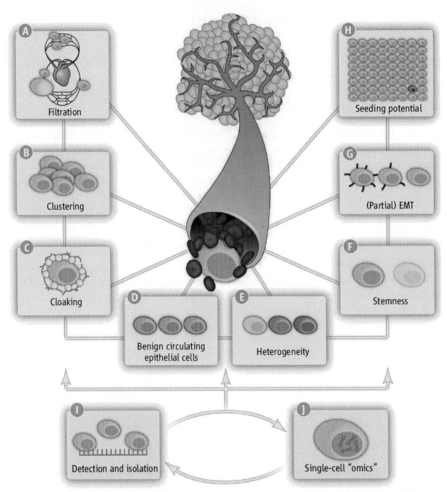

FIGURE 9.4 Challenges in CTC research. Biophysical factors that may diminish the detection of CTCs include (A) filtration of large CTCs in smaller capillaries, (B) clustering of tumor cells that lodge in capillaries, and (C) cloaking of CTCs by platelets or coagulation factors. Biological factors that complicate the detection and isolation of clinically relevant populations of CTCs that rely on epithelial markers include (D) the presence of benign circulating epithelial cells, (E) the large heterogeneity among CTCs, (F) the possible stemness of a subpopulation of CTCs, (G) the (partial) epithelial-mesenchymal transition (EMT) that some CTCs undergo during dissemination, and (H) the unclear seeding potential of detected CTCs. Future research should use technologies focused on (I) improving the detection and isolation of CTCs and (J) single-cell "omics." (From Plaks V, Koopman D, Werb Z. Circulating tumor cells. *Science* 2013;341:1186–88. Copyright 2013 American Association for the Advancement of Science.)

followed by massive parallel sequencing (see Chapters 4 and 7) may open new frontiers for the management of patients.

The CellSearch System

The CellSearch system (Janssen Diagnostics, Raritan, N.J.) is considered the gold standard for detecting CTCs in the clinical setting because it was used in many clinical studies correlating CTC enumeration and prognosis. Epithelial CTCs are detected and enumerated in blood against a backdrop of millions of leukocytes and rely on epithelial cell adhesion molecule (EpCAM)–based immunomagnetic separation. This system has been cleared by the US Food and Drug Administration for prognostication and disease monitoring in patients with metastatic breast (2004), colon (2008), and prostate (2008) cancer and is still the only FDA-cleared system for clinical CTC measurement in metastatic cancer.[20]

In CellSearch, immunofluorescence is used for CTC detection based on specific markers for CTCs such as cytoplasmic epithelial cytokeratins (8, 18, and 19), leukocytes with CD45, and cell viability as indicated by 4′,6-diamidino-2-phenylindole (DAPI) staining. This system is based on a combination of positive immunomagnetic enrichment of CTCs and automated digital microscopy. More specifically (Fig. 9.6), the CTC enumeration steps are (1) CTCs are labeled with magnetic beads coated with an antibody against EpCAM, an epithelial marker expressed by the cell membrane of the majority (but not all) of CTCs; (2) CTCs are immunomagnetically captured and concentrated; and (3) CTCs are stained with phycoerythrin (PE)-labeled anti-CK antibodies (CK-PE). Cell nuclei are then fluorescently stained with DAPI. In addition, leukocytes are stained with allophycocyanin (APC)-labeled anti-CD45 antibodies (CD45-APC), and cellular fixation is performed; and (4) concentrated, stained tumor cells are examined by fluorescence microscopy to assess labeling by PE, DAPI, and APC, thereby distinguishing tumor cells from leukocytes (Fig. 9.7). By using these

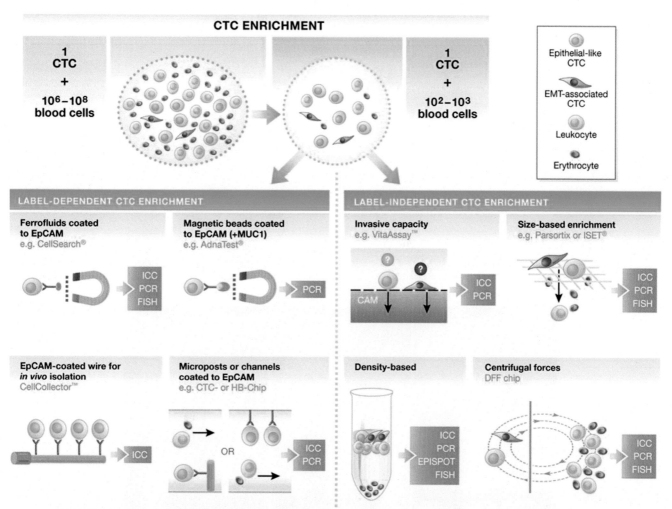

FIGURE 9.5 Overview of CTC enrichment and identification systems. A cartoon of CTC enrichment is shown at the top with depletion of RBCs and leukocytes, leaving mostly epithelial-like CTCs and epithelial mesenchymal transition (EMT)—associated CTCs. Enrichments of 10^4 to 10^5 can be achieved. Enrichment strategies for CTCs can be separated into label-dependent and label-independent techniques. Among label-dependent techniques, immunomagnetic assays targeting EpCAM (epithelial cell adhesion molecule) and sometimes MUC-1 (a cell-surface glycoprotein) are the most common. The antibodies can be attached to ferrofluids (CellSearch, Janssen Diagnostics, Raritan, N.J.), magnetic coated beads (AdnaTest, Qiagen Hanover, Germany), wires (CellCollector, Gilupi GmbH, Potsdam, Germany), microposts (CTC-Chip,[82] or herringbone channels (HB-Chip).[467] Label-independent methods include invasion into a cell adhesion matrix (CAM) (Vita-Assay, Vitatex, Stony Brook, N.Y.), enrichment by size (Parsortix, ANGLE, Guildford, UK) and ISET,[72] enrichment by density (Ficoll gradients), and vortex flows in curvilinear channels known as Dean flow fractionation (DFF).[468] A combination of different enrichment strategies is also practicable. Captured tumor cells are ready for molecular characterization by immunocytochemistry (ICC), using antibodies for tumor-specific markers, or by PCR approaches targeting tumor-specific mRNA or DNA sequences. Additionally, fluorescence in situ hybridization (FISH) can be used for the detection of tumor-specific gene aberrations, or tumor-specific proteins released by CTCs can be detected with labeled antibodies on immobilized membranes by "epithelial immunoSPOT" assays (EPISPOT).[469] (Modified from Joosse SA, Gorges TM, Pantel K. Biology, detection, and clinical implications of circulating tumor cells. *EMBO Mol Med.* 2014;7:1–11. Copyright 2014 Wiley-VCH Verlag.)

fluorescent labels, CD45-APC fluorescence indicates the presence of leukocytes, while CK-PE indicates the presence of epithelial cells. A subcomponent of the CellSearch, the CellSpotter Analyzer, is a semiautomated fluorescence microscope that is used to enumerate the fluorescently labeled cells that are immunomagnetically selected and aligned. This system additionally allows for the detection of a fourth protein biomarker of choice, depending on the type of cancer being investigated—for example, human epidermal growth factor receptor 2 (HER-2) in breast cancer. Moreover, downstream molecular characterization of CTCs is possible because DNA can be isolated from cells captured in the CellSearch cartridge.[21] However, this application of CellSeach is not FDA-cleared.

The main advantages of the CellSearch system are its ease of use and high reproducibility. The preanalytical phase in this system is controlled by collecting blood in special tubes containing a cell preservative that enables sample transportation and stability. All analytical steps are well controlled and automated, similar to a standardized biochemical analyzer.

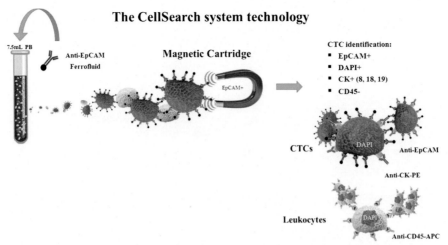

FIGURE 9.6 Positive immunomagnetic isolation and staining of CTC on the CellSearch system. Most nucleated cells in the blood are white blood cells (WBC). CTCs are enriched by a magnetic cartridge and are identified by being EpCAM+, CK+, and CD45−. After isolation, they can be enumerated and/or characterized.

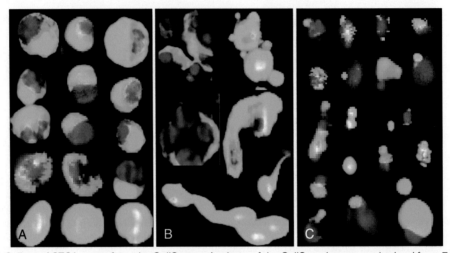

FIGURE 9.7 Gallery of CTC images from the CellSpotter Analyzer of the CellSearch system obtained from 7.5 mL of blood from cancer patients. EpCAM is stained *light gray*, and nuclear DNA is stained *dark gray*. Images that have no nuclear DNA are not viable cells and are not counted. Panel A shows examples of typical intact CTCs, panel B shows examples of intact CTCs present as clusters or with odd shapes that are present less frequently, and panel C provides examples of CTC fragments and apoptotic CTCs. Images presented in panel C are not included in CTC counts but are frequently observed in CTC analysis of carcinoma patients. A full color version of this figure is available online at ExpertConsult.com.

The blood tubes are placed into an automated sample preparation system that uses ferro-fluids loaded with an antibody against EpCAM to capture CTCs (Fig. 9.8). The CellSearch System can enumerate CTCs in 7.5 mL of peripheral blood and can detect down to one cell. The analytical accuracy, precision, linearity, and diagnostic specificity of this system were first validated in 2004.[21] A prospective multicenter validation study at three independent laboratories showed that CTC counts remained stable for 72 hours after blood sample collection, even at room temperature.[22] Interassay variability was below 5% for 299 analyzed control samples, and the mean recovery rate of tumor cells spiked into normal blood at the 4- to 12-cell level was 82% and 80%, respectively. Within-run precision was evaluated in two different centers by eight replicates of the same spiked samples, and CVs were 18.4% to 26% for low cell numbers and 2.1% to 11.6% for high numbers of spiked cells.

The majority of evidence supporting the use of CTCs in clinical decision making has been related to epithelial CTC enumeration using the CellSearch system. However, the system has several limitations. The main drawback of CellSearch is that it depends entirely on the expression of the epithelial marker EpCAM on CTCs. This means that CTCs that do not express EpCAM because of epithelial mesenchymal transition (EMT) are not detected, as is the case for a subpopulation of highly aggressive breast cancer stem cells.[23] Another limitation is that gene expression cannot be performed on CTCs isolated with CellSearch because the cell preservative used inhibits downstream RNA analysis.

Leon Terstappen's group, which developed the CellSearch system, evaluated different ways to express changes in CTC counts with respect to overall survival. They clarified that CTC measurements on a patient at successive times may show a decline in CTCs. However, for low CTC numbers,

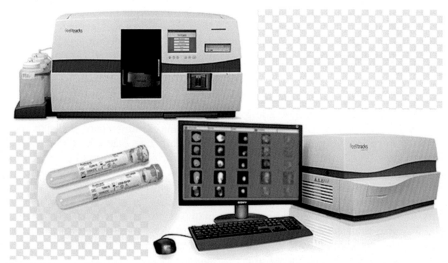

FIGURE 9.8 The CellSearch CTC detection and enumeration system. See text for discussion (Image reproduced by permission of Janssen Diagnostics LLC from *https://www.cellsearchctc.com/product-systems-overview/cellsearch-system-overview.*)

this decline may result from either a true decline in the number of CTCs in the patient's blood, or it may arise from a Poisson sampling error. They propose that a static CTC cutoff is the best method to determine whether a therapy is effective, and they provide a lookup table from which the significance of a change in CTCs can be derived.[24]

An automated CTC counting algorithm to eliminate the variability in manual counting of CTCs has been validated on prostate, breast, and colorectal cancer CTC images. By using this approach, CTCs were identified and morphological features extracted from images stored by the CellSearch system. The automated CTC counts were strongly correlated with clinical outcomes in metastatic breast, colorectal, and prostate cancers.[25] Recent advances in the classification of CTCs captured by the CellSearch system and the implications of their features and numbers have been recently reviewed.[26]

Since its introduction, CellSearch has been widely used in many clinical labs worldwide, and numerous clinical studies have been performed.[27] The enumeration of CTCs continues to hold great promise in guiding treatment decisions in breast cancer patients. However, guidelines on how to use CTC enumeration by CellSearch in clinical decisions in both primary and metastatic breast cancer are still lacking.[28]

Isolation/Enrichment Systems for CTC Analysis

CTCs can be infrequent, depending on the cancer stage. One CTC may be present among 10^6 to 10^8 peripheral blood cells. CTC isolation and enrichment from peripheral blood are thus extremely challenging and very demanding. It is not only that these cells circulate at very low numbers, but they are heterogeneous even within the same patient. Highly standardized and robust isolation protocols are necessary for downstream CTC analysis and molecular characterization. Toward this goal, a lot of effort has focused on developing novel technologies for the isolation and enrichment of CTCs from peripheral blood. These can be categorized into two main systems based on the different properties of CTCs that distinguish them from surrounding normal hematopoietic cells: (1) label-independent systems based on physical properties, such as size, density, electric charge, and deformability that may utilize microfluidics, and (2) label-dependent systems based on biological properties of CTCs, such as expression of specific proteins on their cell surface. These include immunomagnetic bead separation systems with positive and/or negative selection. Approaches using both physical and biologic properties are common. In vivo methods for increasing CTC yield have also been developed.

The most important systems for the isolation/enrichment of CTCs are discussed in the following text. In many cases these technologies are complementary to one another because they target different properties of CTCs and may define unique CTC populations.

Label-Dependent CTC Isolation/Enrichment Systems

Positive Selection. Positive selection is the most widely used CTC isolation/enrichment system. This approach captures CTCs through specific monoclonal antibodies against epithelial cell surface markers that are expressed on CTCs but are absent from normal leukocytes. CTCs can be tagged using antibody-conjugated magnetic microbeads (diameter: 0.5–5 μm) or nanoparticles (diameter: 50–250 nm) that bind to a specific surface antigen. Intracellular antigens like cytokeratins can also be used as targets.[29,30] Immunomagnetic assays require a short incubation (~30 minutes) for antigen/antibody binding that couples the cells to magnetic beads, followed by isolation of the cells using a magnetic field (see Fig. 9.5).

Various antigens have been exploited for the positive immunomagnetic isolation of CTCs. Among these, EpCAM is the most common. This approach is well established in terms of proven clinical significance of the captured cells. However, capture with EpCAM has the disadvantage of missing some cells that are undergoing EMT. We also now know that CTCs are highly heterogeneous, but one approach to partially overcome this issue is to use a "cocktail" of antibodies that targets multiple antigens.[29,31-33] Along these lines, several organ- or tumor-specific markers, such as carcinoembryonic antigen (CEA), epidermal growth factor receptor (EGFR), prostate-specific antigen, HER-2, cell surface–associated mucin 1 (MUC-1), ephrin receptor B4 (EphB4), insulin-like growth factor 1 receptor (IGF-1R), cadherin-11 (CAD11), and tumor-associated glycoprotein 72 (TAG-72), have been used to isolate CTCs.[34,35] Cell surface vimentin on CTCs is a good marker for epithelial cancers undergoing EMT. By using a specific monoclonal antibody against vimentin, EMT-CTCs were detected in patients undergoing post-surgery adjuvant chemotherapy for metastatic colon cancer.[36] These isolated EMT-CTCs were characterized further using EMT-specific markers, fluorescent in situ hybridization, and single-cell mutation analysis. This antibody exhibited high specificity and sensitivity toward different epithelial cancer cells and was used to detect and enumerate EMT-CTCs from patients. The number of EMT-CTCs detected correlated with the therapeutic outcome of the disease. According to these results, cell surface vimentin is a promising marker for the isolation of EMT-CTCs from a wide variety of tumor types.[36]

Another example of positive selection is the MagSweeper, an immunomagnetic cell separator.[37,38] In this device, magnetic beads are coated with an antibody targeting epithelial cell surface markers. These immunomagnetic beads are added into blood samples, and cancer cells are attached to the beads. This device gently enriches target cells and eliminates cells that are not bound to magnetic particles by using centrifugal forces. The isolated cells are easily accessible and can be extracted individually based on their physical characteristics to deplete any cells nonspecifically bound to beads.[37,38] The same group has recently developed the magnetic sifter, a miniature microfluidic chip with a dense array of magnetic pores; tumor cells are labeled with magnetic nanoparticles and are captured from whole blood with high efficiency.[39] The use of isolation technologies that take advantage of magnetic fields may lead the way to routine preparation and characterization of liquid biopsies from cancer patients.

Negative Selection. This isolation approach is completely independent of the phenotype of CTCs and is based on the depletion of noncancerous peripheral blood cells by first lysing red blood cells (RBCs) and further using specific markers for white blood cells (WBCs) like CD45 or CD61 to magnetically remove WBCs from the sample.[40-42] Another variation of CTC enrichment by negative depletion is the commercially available RosetteSep system (StemCell technologies, Canada), which uses a mixture of antibodies that link RBCs to one another and to WBCs. In this way, the majority of the erythrocytes and leukocytes in the peripheral blood sample are depleted, thereby negatively enriching the sample with tumor cells if present. This approach is independent of the expression of EpCAM on CTCs and has better recoveries than the conventional density gradient method for the isolation of CTCs from normal leukocytes in blood.[43]

Flow Cytometry–Based Technologies. Multiparameter flow cytometry was firstly evaluated for the detection of human tumor cells in experimental mouse models of breast cancer.[44] This methodology was later adapted for the analysis of tumor dissemination to bone marrow and lymph nodes, in combination with laser scanning cytometry.[45] It is feasible to isolate CTCs from peripheral blood with high purity through fluorescence-activated cell sorting and profile them downstream via gene expression analysis.[46-48] Recently, cell sorting was used to isolate CTCs from CellSearch isolates and perform downstream high-resolution genomic profiling at the single-cell level.[49]

Label-Independent CTC Isolation/Enrichment Systems

Systems Based on Density. Density gradient centrifugation using commercially available reagents (eg, Ficoll, GE Healthcare Life Sciences, Pittsburgh, Penn.) is one of the most widely used approaches for CTC enrichment.[50,51] It is based on the lower density of mononuclear cells (including CTCs) compared to RBCs and polymorphonuclear leukocytes.[52] The OncoQuick system (Greiner Bio-One, Germany) is an improved version of this approach that uses a porous membrane placed on top of the gradient media to prevent mixing.[53] Experiments performed in cell lines have shown CTC recovery rates of 70% to 90%.[54] Although simple and inexpensive, the OncoQuick system has a relatively low yield and enrichment when compared to the CellSearch system because in the same group of 61 patients, at least one CTC was detected in 23% with OncoQuick and in 54% with CellSearch.[55]

Systems Based on CTC Size. Size-based isolation systems are independent of tumor markers and separate CTCs (that are usually larger) from smaller leukocytes. Several different size–based systems have been developed and include membrane filters, microfluidic chips, and hydrodynamic methods.[56,57] The first CTC filtration device was described by Vona and colleagues in 2000.[58] Since then, filtration devices have been improved, and many downstream applications have been developed. Filters used for CTC isolation/enrichment are often disposable porous membranes (usually polycarbonate) containing numerous randomly distributed 7- to 8-μm-diameter holes that allow blood constituents to cross but capture the larger CTCs.[58,59] Specific microfabrication techniques have been applied to build microfilters with controlled pore distribution, size, and geometry,[60-62] and different materials like polycarbonate,[58,59,63-66] parylene C,[67] nickel,[60] and, most recently, silicon[68] have been tested for fabrication of these membranes. Live CTCs isolated in this way can be further cultured in vitro for downstream studies.

Specifically designed polycarbonate membrane filter kits are usually accompanied by specific syringes or pumps so the pressure on the filter is optimized to keep fragile CTCs intact. Commercially available filtering devices for CTC isolation include ScreenCell (ScreenCell Inc., France), the ISET system (isolation by size of epithelial tumor cells, RareCells, France), and the CellSieve system (CREATV MicroTech, Potomac, Md.). Another filter-based microdevice that is both a capture and analysis platform capable of performing multiplexed imaging and genetic analysis has the potential to enable routine CTC analysis in the clinical setting.[67]

ScreenCell filters allow the isolation of live, spiked cancer cells or paraformaldehyde-fixed CTCs.[59] The performance of this technology for CTC capture was evaluated in 76 patients and has potential as a diagnostic blood test for lung cancer.[69] The ISET filtration system allows for cytomorphological, immunocytological, and genetic characterization of CTCs. This isolation approach is sensitive and can detect one single tumor cell added to 1 mL of blood. It is also simple and fast, consisting of only one filtration step.[58] Furthermore, ISET-isolated tumor cells can be subsequently recovered from the filter and evaluated by immunohistochemistry and/or molecular-genetic analyses.[70] In non–small cell lung cancer, this system can detect CTCs with a hybrid (epithelial/mesenchymal) phenotype, isolate anaplastic lymphoma receptor tyrosine kinase (ALK)–positive CTCs,[71] and isolate ROS-rearranged CTCs.[72] By using the CellSieve filtering system, very large cells (25 to 300 microns), called cancer-associated macrophage-like cells, were recently identified in the peripheral blood of cancer patients.[73]

CTC isolation by size filtration is simple and reliable; filtration and staining are easy to perform, and the method is rapid. CTCs can be identified using classic cytopathologic criteria. No complex instrumentation or specific training is needed. However, false-positive results can be obtained by using filtration devices. This is due to the presence of normal epithelial cells, which are collected by an intradermic needle and can be misinterpreted as CTCs. Moreover, endothelial cells or rare hematologic cells such as megakaryocytes may also be present on the filter. For this reason, downstream cytomorphological, immunocytochemical, and molecular characterization are important to rule out cells that are not CTCs.[74]

Separation Based on the Electric Charge of CTCs. CTCs are different from peripheral blood mononuclear cells with respect to morphology and dielectric properties. Dielectrophoresis (DEP) is a technology for CTC isolation based on electrical properties of cancer cells. Dielectric properties (polarizability) of cells are dependent on cell diameter, membrane area, density, conductivity, and volume. Depending on their phenotype and morphology, different cells have different dielectric properties, which is the principle employed for electro-kinetic isolation of CTCs.[75]

DEP can precisely manipulate and collect single cells. The DEPArray (Silicon Biosystems, Menarini group, IT) is a microfluidic device consisting of 320 × 320 arrayed electrodes generating over 10,000 spherical DEP cages that can guide single CTCs to predetermined spatial coordinates. This lab-on-a-chip platform is completely independent of the expression of antibodies on the cell surface and can be used for quick, arrayed, software-guided binding of individual CTCs. This system has been evaluated in the clinical setting with a goal toward showing superior performance in single-cell CTC analysis (Fig. 9.9).[76] Dielectrophoretic manipulation and isolation of single CTCs at 100% purity has been achieved from preenriched blood samples.[77] The single-use, microfluidic cartridge contains an array of individually controllable electrodes, each with embedded sensors. This circuitry enables the creation of DEP cages around cells. After imaging, individual cells of interest are gently moved to specific locations on the cartridge—for example, for cell–cell interaction studies—or into the holding chamber for isolation and recovery.[77,78] This system can be used in combination with other CTC platforms, such as the CellSearch, for downstream single CTC molecular characterization.[77,79,80]

Microfluidic Systems for CTC Isolation/Enrichment. The main advantages of microfluidic systems for CTC isolation/enrichment are their simplicity and the potential to be fully automated, unlike traditional affinity-based CTC isolation

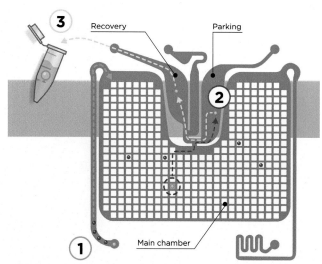

FIGURE 9.9 The DEPArray Cartridge for the isolation of single CTCs (Menarini Silicon Biosystems). The system is based on dielectrophoresis. The sequential steps are (1) inject, trap, and image cells; (2) move the cells of interest into the parking chamber; and (3) move the individual or multiple cells into the recovery chamber. (Image reproduced by permission of Menarini Silicon Biosystems from http://cdn2.hubspot.net/hubfs/304284/Brochures/Silicon_Biosystems_DEPArray_Brochure_2015.pdf?t=1466668511037.)

techniques. The main disadvantage up to now is the long time needed to run each sample and the low capacity in terms of the volume of peripheral blood that can be processed. These systems are based on a combination of precisely defined topography of microstructures (traps) with laminar flow in microchannels.[7,81]

The first specific microfluidic device for CTC isolation was described in 2007. This "CTC-Chip" was a silicon chamber etched with 78,000 microposts that were coated with an anti-EpCAM antibody.[82] Captured CTCs attached to microposts were visualized by staining with antibodies against cytokeratin or tissue-specific markers. Using this design, approximately 60% of cancer cells have been recovered from spiked whole blood samples. Due to the high rate of collisions enabled by the dense array of microposts, varying expressions of EpCAM by target cells do not affect the recovery rate. However, in this first device, many false-positive results were detected in a healthy individual's blood.[82] The same group has further developed a simpler device, the "HB-chip," which is characterized by a herringbone structure and reliable coating of the inner surface with antibodies. Moreover, the chambers are made out of transparent materials that enhance high-resolution imaging, including the use of transmitted light microscopy.[83] The latest development of this group is the "CTC-iChip," another microfluidic CTC capture platform that is capable of sorting rare CTCs from whole blood at 10^7 cells/s (Fig. 9.10). This device can isolate CTCs using strategies that are either label-dependent or label-independent and thus is applicable to virtually all cancers. The CTC-iChip was recently evaluated in an expanded set of both epithelial and nonepithelial cancers, including lung, prostate, pancreas, breast, and melanoma. CTC sorting in solution allows high-quality,

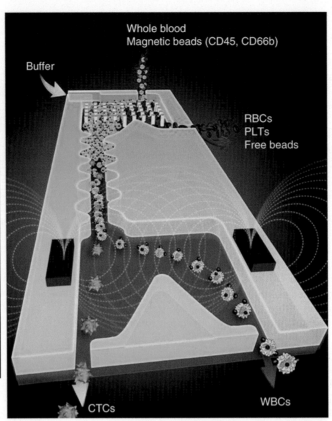

FIGURE 9.10 The CTC-iChip schematic. The CTC-iChip is composed of two separate microfluidic devices that house three different microfluidic components engineered for inline operation: deterministic lateral displacement to remove nucleated cells from whole blood by size-based deflection by using a specially designed array of posts performed in CTC-iChip1, inertial focusing to line up cells to prepare for precise magnetic separation, and magnetophoresis for sensitive separation of bead-labeled WBCs and unlabeled CTCs, which are performed in CTC-iChip2. *PLTs,* Platelets. (From Karabacak NM, Spuhler PS, Fachin F, Lim EJ, Pai V, Ozkumur E, et al. Microfluidic, marker-free isolation of circulating tumor cells from blood samples. *Nat Protoc.* 2014;9:694–710. Copyright 2014 Nature Publishing Group.)

clinically standardized, morphological and immunohistochemical analyses, as well as RNA-based single-cell molecular characterization.[84]

Most microfluidic devices are reliant on three-dimensional structures that limit the characterization of cells on the chip. Recently, Nagrath's group isolated CTCs from blood samples from patients with pancreatic, breast, and lung cancers by using functionalized graphene oxide nanosheets on a patterned gold surface.[85] This method allows for CTC capture with high sensitivity. A variety of highly sophisticated microfluidic devices have been developed by this group,[86,87] some of which have enabled further culturing of isolated cells on the chip. Recently, this group also developed an in situ capture and culture methodology for ex vivo expansion of CTCs using a three-dimensional co-culture model, simulating a tumor microenvironment to support tumor development. They managed to successfully expand CTCs isolated from early-stage lung cancer patients and characterize them further by both Sanger sequencing and massively parallel sequencing, as summarized in Fig. 9.11.[88]

A multiplexed microfluidic chip was recently developed and clinically validated for the ultra-high-throughput, low-cost, and label-free enrichment of CTCs.[89] Retrieved cells were unlabeled and viable, enabling potential propagation and real-time downstream analysis using massively parallel sequencing or proteomic analysis (Fig. 9.12).[89] In a separate study, HER-2 was used as an alternative to EpCAM in a novel microfluidic device for the isolation of CTCs from peripheral blood of patients with HER-2-expressing solid tumors.[90] In summary, microfluidic devices for CTC isolation have been described, and this number is increasing continuously.[57,91-98]

In Vivo Systems for CTC Isolation/Enrichment. The recent development of in vivo systems for CTC isolation from whole blood adds another dimension to the field of CTC isolation (Fig. 9.13). These systems aim to overcome the limitation of small blood sample volumes inherent to the ex-vivo CTC isolation techniques previously described.

Leukapheresis is a laboratory procedure in which WBCs or peripheral blood stem cells are separated from blood. During leukapheresis, a patient's blood is passed through a machine that removes the WBCs or peripheral blood stem cells and then returns the balance of the blood back to the patient (Fig. 9.13A). Eifler and colleagues first showed that isolation of CTCs via leukapheresis was feasible.[99] A recent study that screened leukapheresis products generated from up to 25 L of processed blood per patient demonstrated that CTCs can be detected in more than 90% of nonmetastatic breast cancer patients.[100]

Another novel in vivo CTC isolation system is the CellCollector (Gilupi GmbH, Germany) (Fig. 9.13B). CellCollector is a nanowire that is coated at its tip (2 cm of length) with pure gold, to which chimeric antibodies directed to EpCAM are

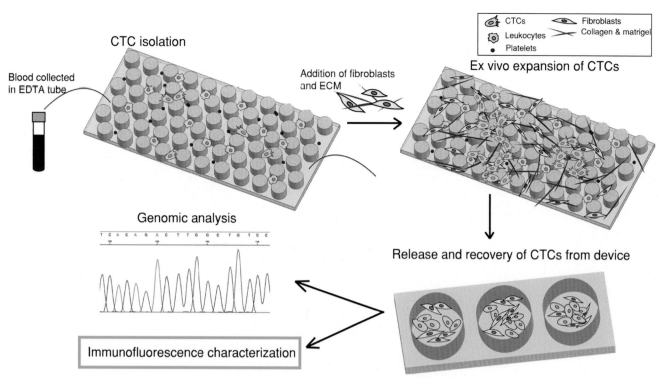

FIGURE 9.11 Expansion of CTCs from early-stage lung cancer patients using microfluidics. The first step is to capture CTCs by flow of a patient blood sample through a CTC-capture chip. The second step is to introduce fibroblasts and extracellular matrix (ECM) to the same chip to establish a co-culture environment for ex vivo expansion of CTCs. The third step is to release and recover CTCs from the device, and the fourth step is downstream characterization. (From Zhang Z, Shiratsuchi H, Lin J, Chen G, Reddy RM, Azizi E, Fouladdel et al. Expansion of CTCs from early stage lung cancer patients using a microfluidic co-culture model. *Oncotarget* 2014;5:12383−97. Copyright 2014 Impact Journals, LLC.)

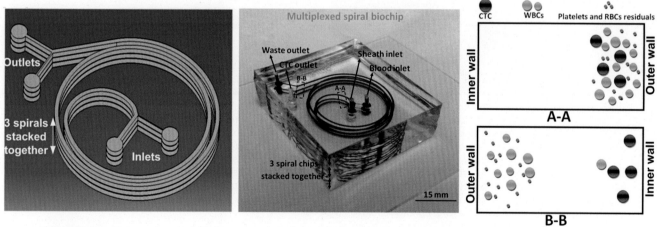

FIGURE 9.12 The design *(left)* and a photograph *(middle)* of a parallel spiral microfluidic device for separating CTCs. Blood sample and sheath fluid are pumped through the device using two separate syringe pumps. Under the influence of inertial lift and Dean drag forces in the fluid flow, CTCs focus near the microchannel inner wall, while WBCs and platelets go through one Dean cycle and migrate toward the outer wall, thus achieving separation. A full color version of this figure is available online at ExpertConsult.com. (Modified from Khoo BL, Warkiani ME, Tan DS, Bhagat AA, Irwin D, Lau DP, et al. Clinical validation of an ultra high-throughput spiral microfluidics for the detection and enrichment of viable circulating tumor cells. *PLoS One.* 2014;9:e99409.)

covalently attached. The CellCollector is placed into the antecubital vein of a cancer patient for in vivo binding of rare CTCs from the patient's entire circulating blood pool of several liters.[101] More CTCs to support multiple downstream assays are available as compared to traditional in vitro approaches with samples of 7 to 15 mL of blood. This approach increases the chances of isolating rare tumor cells, especially in the early stages of the disease.

Photoacoustic flow cytometry can also detect CTCs in vivo (Fig. 9.13C). A high−pulse repetition rate diode laser shines

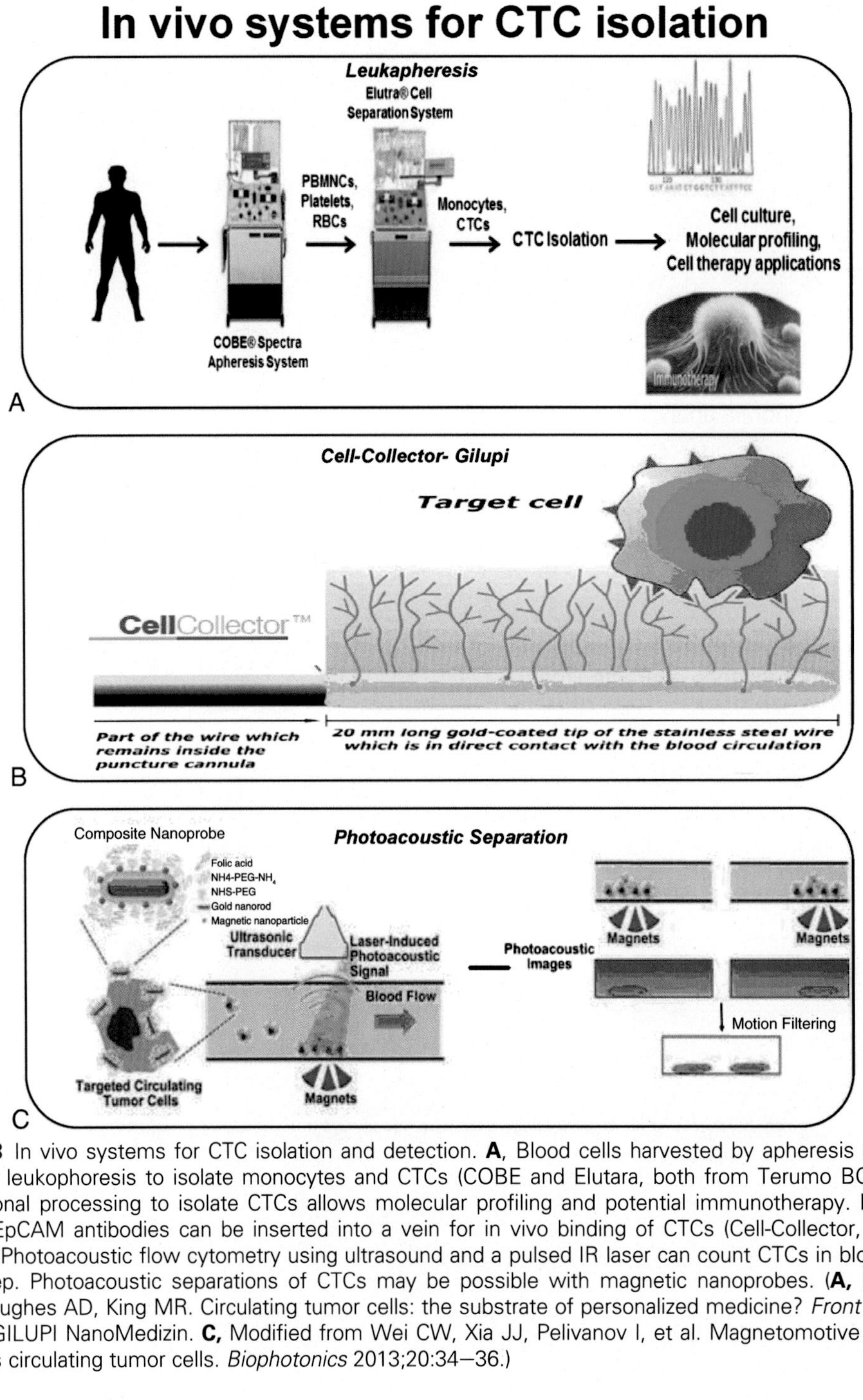

FIGURE 9.13 In vivo systems for CTC isolation and detection. **A**, Blood cells harvested by apheresis can be further processed by leukophoresis to isolate monocytes and CTCs (COBE and Elutara, both from Terumo BCT, Lakewood, Colo.). Additional processing to isolate CTCs allows molecular profiling and potential immunotherapy. **B**, A nanowire coated with EpCAM antibodies can be inserted into a vein for in vivo binding of CTCs (Cell-Collector, Gilupi GmbH, Germany). **C**, Photoacoustic flow cytometry using ultrasound and a pulsed IR laser can count CTCs in blood vessels up to 3 mm deep. Photoacoustic separations of CTCs may be possible with magnetic nanoprobes. (**A**, Modified from Greene BT, Hughes AD, King MR. Circulating tumor cells: the substrate of personalized medicine? *Front Oncol* 2012;2: 69. **B,** From GILUPI NanoMedizin. **C,** Modified from Wei CW, Xia JJ, Pelivanov I, et al. Magnetomotive photoacoustic imaging spots circulating tumor cells. *Biophotonics* 2013;20:34–36.)

light through the skin into a vessel that is up to 3 mm deep to detect acoustic vibrations that result from the absorption of laser light by target nanoparticles.[102] By using this technology, circulating melanoma cells were detected in blood.[103] In vivo isolation technologies for CTC detection are now being evaluated in many centers; their future use will depend on the results of these clinical evaluation studies.

CTC Analysis: Detection and Molecular Characterization

CTC detection and molecular characterization are achieved by immunofluorescence; molecular nucleic acid analysis, including RT-qPCR and multiplex RT-qPCR; and detection of tumor-specific proteins released by CTCs. An outline of the main approaches for CTC detection is presented in Fig. 9.14.

Image-Based Approaches. Detection of epithelial CTCs by immunofluorescence using anti-CK antibodies is currently the most validated and standardized approach and also allows for morphological interpretation of positive events. However, detection of CTC by classical immunofluorescence, typically done by trained pathologists through visual observation of stained cytokeratin-positive epithelial CTCs, is time-consuming and may take days if many samples are to be analyzed. Additional image-based modalities include detecting CTCs with multiple antibodies (eg, against cytokeratin, Her-2, and

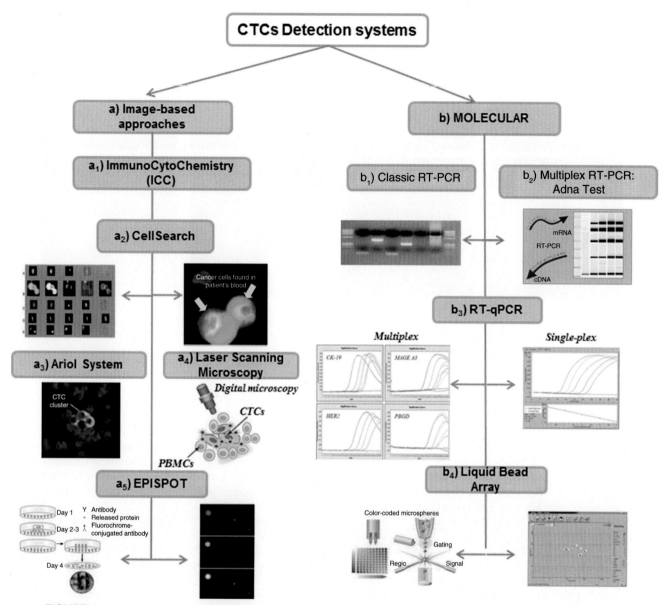

FIGURE 9.14 Main approaches for CTC detection and molecular characterization. **A**, Image-based approaches: (A_1) classic immunocytochemistry (ICC), (A_2) CellSearch system (FDA cleared), (A_3) digital image capture and analysis (Ariol system, Leica Biosystems, Buffalo Grove, Ill.), (A_4) laser-scanning cytometry, and (A_5) EPISPOT assay (detects tumor-specific proteins released by CTCs). **B**, Molecular assays based on nucleic acid analysis: (B_1) classic reverse transcriptase PCR (RT-PCR), (B_2) multiplex RT-PCR—for example, the AdnaTest for breast cancer (Qiagen Hannover, Germany), (B_3) singleplex or multiplex RT-qPCR, and (B_4) liquid bead array. (Modified with permission of the publisher from Lianidou ES, Markou A. Circulating tumor cells in breast cancer: detection systems, molecular characterization, and future challenges. *Clin Chem.* 2011;57:1242–55. Copyright 2011 American Association of Clinical Chemistry.)

the stem cell antigens ALDH1, CD44, and CD24) labeled with different fluorochromes (DyLight Technology, ThermoFisher) and spectral image analysis to separate the different emissions.[104] Another technology is the laser scanning cytometer, a fast and quantitative automated microscopic procedure for screening with up to 10,000-fold enrichment. However, large numbers of apparent CTCs are detected even in healthy donors, possibly because of nonspecific binding of antibodies.[105] A third example of an image analysis method is the Ariol high-throughout automated image analysis system (Leica Biosystems, Germany) that has been widely used for high-definition imaging of CTCs (Fig. 9.15). The high-definition circulating tumor cell assay is a fluid-phase biopsy approach that identifies CTCs by high-definition imaging without any surface protein-based enrichment at a high enough definition to satisfy diagnostic pathology image quality requirements. This system has been evaluated in various types of cancers, such as lung, colorectal, and prostate.[106-109]

Molecular Assays. Molecular assays for CTC detection and molecular characterization take advantage of the sensitivity and specificity of PCR and the high throughput of massively parallel sequencing (MPS). Molecular assays have been widely applied for CTC detection and characterization and can interrogate CTCs at the single-cell level. CTC

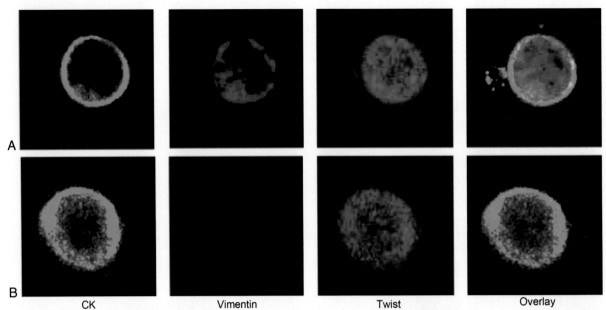

FIGURE 9.15 Digital image capture and analysis of CTCs. Representative confocal laser-scanning photomicrographs of CTC cytospins after negative immunomagnetic separation in a patient with metastatic breast cancer showing coexpression of cytokeratin, Twist and vimentin (Ariol system, Leica Biosystems, Buffalo Grove, Ill.). Cells were triple-stained for cytokeratin (CK), vimentin, and Twist. Original magnification is 600×. **A,** A CTC expressing CK, Twist, and vimentin. **B,** A CTC expressing CK and Twist, but not vimentin. A full color version of this figure is available online at ExpertConsult.com. (From Kallergi G, Papadaki MA, Politaki E, et al. Epithelial to mesenchymal transition markers expressed in circulating tumour cells of early and metastatic breast cancer patients. *Breast Cancer Res.* 2011;13:R59.)

molecular assays include RT-qPCR, multiplex RT-qPCR, methylation-specific PCR, allele-specific PCR, fluorescent in situ hybridization, array comparative genomic hybridization (arrayCGH), and MPS technologies.

PCR assays are very sensitive and can analyze both RNA and DNA. PCR assays can be designed in silico, automated, and subjected to internal and external quality control systems.[19,110] In silico design of PCR assays using databases and specific software programs can avoid cross-reactions with nontarget genes. Molecular assays can be quantitative, high throughput, and easy to perform and usually require a very small sample amount for analysis. Especially when multiplex PCR is used, many targets can be evaluated from the same sample, thus enabling a multiparametric approach for the often limited CTC sample. In contrast to imaging approaches, measurements obtained by molecular assays are objective and quantifiable. Molecular assays are relatively low cost, high throughput, and amenable to quantifiable quality control. Another advantage is that many molecular methods can be automated. Fully automated systems for RNA and DNA isolation and downstream PCR analysis are already used in routine molecular in vitro diagnostics labs.

The main disadvantages of molecular assays are preanalytical issues concerning CTC stability during sample shipment and storage, a problem that CellSearch has solved for capture, staining, and image analysis. However, molecular assays currently require immediate handling of blood samples for CTC isolation and downstream analysis, a fact that hinders long-distance shipment of samples. Moreover, PCR assays require specially designed lab areas to avoid end-product contamination. Separate areas are usually needed for RNA/DNA isolation, setting up the reactions, and amplification and analysis. Unless single-cell analysis is performed, molecular assays provide only bulk information on a sample. For example, in mRNA analysis, it is not known whether transcripts are coexpressed in the same cell or derived from different cell populations.

PCR-Based Assays. PCR-based assays can be highly specific and sensitive. RT-qPCR assays targeting specific expression of genes in cancer cells and not in peripheral blood mononuclear cells are especially sensitive, with the ability to detect one cancer cell in the presence of more than 10^6 leukocytes.[111] However, this requires the detection of mRNA markers that are specifically expressed in CTCs and not expressed in leukocytes. In all cases, cutoffs should be estimated based on the expression values of these markers in a significant number of healthy donor samples that should be analyzed in exactly the same way as patient samples. RT-qPCR assays are high throughput and easy to perform because they are based on the isolation of total RNA from viable CTCs and subsequent RT-PCR amplification of tumor- or epithelial-specific targets.

Cytokeratins are not only used as biomarkers in imaging but are also used as biomarkers in RT-PCR as well. *KRT19* (the gene for CK-19) is an especially specific and sensitive marker for CTCs, provided that the primers used are well designed to avoid false-positive results due to the presence of the *CK-19* pseudogene. Using a highly specific EpCAM-independent RT-qPCR assay,[112,113] *KRT19* was a molecular marker of prognostic significance for OS and PFS in early breast cancer before, during, and after chemotherapy.[114-116]

Other targets can also be used for CTC detection. Smirnov and colleagues studied the expression profile of many genes on CTCs and observed that *AGR2, S100A14, S100A16,*

FABP1, and other specific genes could be used for the detection of CTCs in the peripheral blood of advanced cancer patients.[117] Since then, many research groups have described the use of RT-PCR to characterize single or multiple targets of CTCs in cancer patients.

Another advantage of RT-qPCR is the flexibility it offers, especially in multiplexing. Multiplexing minimizes time, cost, and conserves available nucleic acid. Only a few transcripts provide adequate sensitivity individually, but combinations of transcripts may produce better sensitivity for CTC detection. AdnaGen (Qiagen, Hamburg) is commercializing a number of molecular diagnostic kits for different types of cancer based on RT-PCR. These kits use positive immunomagnetic isolation of CTC, followed by RT-PCR and electrophoretic detection with a Bioanalyzer (Agilent Technologies) instrument.[118,119] A quantitative gene expression profiling methodology based on RT-qPCR requires only one breast cancer CTC and amplifies a set of genes with no or minor expression by leukocytes.[120] Several mRNA markers may be useful for RT-PCR–based detection of CTCs. Quantification of these mRNAs is essential to distinguish normal expression in blood cells from that due to the presence of CTCs. A technical protocol to measure the expression of thousands of genes in CTCs captured with magnetic beads linked to EpCAM and the carcinoma-associated mucin, MUC-1, was recently reported. The expression profiles were reproducible and the technology suitable for prospective studies to assess the clinical utility of CTCs.[121]

Another approach using multiplex PCR is the liquid bead array (Luminex) that is automated and often present in clinical labs for a variety of applications.[122] A sensitive and specific CTC gene expression assay using multiplexed PCR coupled to a liquid bead array has been developed and validated.[123] Five established CTC genes—the HER-2 gene *ERRB2*, the mammaglobin A gene *SCGB2A2*, the *CK-19* gene *KRT19*, the melanoma antigen family A1 gene *MAGEA1*, the twist basic helix-loop-helix family 1 gene *TWIST1*, and one reference gene *HMBS*—were simultaneously amplified and detected in the same reaction using only 1 μL of cDNA from CTCs. Up to 100 genes in CTCs can be studied.[123]

Fluorescence in Situ Hybridization and RNA in Situ Hybridization. Fluorescence in situ hybridization (FISH) is widely used in the CTC field to verify *HER-2* amplification status and detect the presence of genomic rearrangements, such as *ALK* in lung cancer, and the androgen receptor in prostate cancer. In most cases FISH analysis has followed CellSearch selection and staining. After enumeration of cytokeratin+, CD45−, nucleated cells, the cells are fixed in the cartridge to maintain their original position and hybridized to FISH probes. Next, fluorescence images of the FISH probes are acquired. Heterogeneity of chromosomal abnormalities is observed between CTCs of different patients and among CTCs of the same patient.[124,125]

RNA in situ hybridization is also used for CTC detection; a system called CTCscope detects a multitude of tumor-specific markers from single CTCs. Breast cancer CTC transcripts of eight epithelial markers and three EMT markers have been evaluated.[126] RNA in situ hybridization can also characterize circulating brain tumor cells at the single-cell level isolated with microfluidics from patients with glioblastoma.[127] Both these techniques are mainly used in the research setting.

Array comparative genomic hybridization (arrayCGH) has also been used to detect DNA copy number changes in single CTCs. Multiplex FISH in combination with array CGH revealed that occult disseminated cells are characterized by very complex numerical and structural aberrations. Array CGH in micrometastatic cells allows for high-resolution assessment of copy number changes, pinpointing commonly gained or lost regions to narrow down the regions critically involved in metastasis.[128] Using a combination of whole genome amplification (WGA) technology and high-resolution oligonucleotide arrayCGH, reliable detection of numerical genomic alterations as small as 0.1 Mb in a single cell is possible. Analysis of single cells from well-characterized cell lines and single normal cells confirmed the stringent quantitative nature of the amplification and hybridization protocol (Fig. 9.16).[129] Comprehensive genomic profiling of CTCs can be achieved by combining arrayCGH and MPS technology. An integrated process to isolate, quantify, and sequence whole exomes of CTCs suggests that mapping of greater than 99.9955 of the CTC exome is possible,[130] indicating the clinical potential for CTC genomics in the future.

Protein Analyses on CTCs (EPISPOT Assay). The epithelial immunoSPOT (EPISPOT, a specific type of ELISPOT or enzyme-linked immunoSPOT) assay was developed to detect tumor-specific proteins released by viable CTCs. This assay detects proteins secreted/released/shed from single epithelial cancer cells. Cells are cultured for a short time on a membrane coated with antibodies that capture the secreted/released/shed proteins, with subsequent detection by secondary antibodies labeled with fluorochromes. This assay has been used in many types of solid cancers, such as breast, colorectal, prostate, and others.[75]

Single-Cell Analysis of CTCs. CTCs are highly heterogeneous, even within the same individual.[131] This has been verified by using reliable single-CTC isolation followed by MPS to clearly reveal differences among CTCs. This powerful combination offers a new dimension in CTC molecular characterization. Using the DEPArray system (see Fig. 9.9) for isolation of single CTCs, Peeters and colleagues have obtained reliable gene expression profiles from single cells and groups of up to 10 cells.[77]

MPS technologies reveal that intratumor heterogeneity may reflect tumor evolution and adaptation that hinder personalized medicine strategies developed from results on single-tumor biopsy samples.[132,133] Analysis of CTCs reduces this complexity because CTCs are present in blood in low cell numbers that provide a better picture for individualized treatments because they represent both cells that have disseminated from the primary tumor and cells from secondary metastases. Heitzer and colleagues performed the first comprehensive genomic profiling of CTCs in patients with stage IV colorectal carcinoma using arrayCGH and MPS. Mutations in driver genes of the primary tumor and metastasis were also detected in the corresponding CTCs. Mutations that appeared exclusively in CTCs were later found by additional deep sequencing in metastases from the same patient.[134]

The importance of characterizing single CTCs to investigate their molecular heterogeneity is clear. Polzer and colleagues combined enrichment and isolation of pure CTCs with a WGA method for single cells.[135] Focusing on

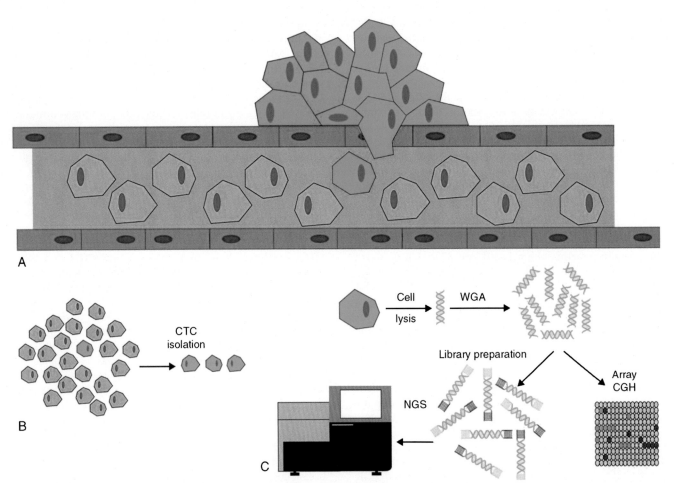

FIGURE 9.16 Workflow of CTC analysis by array comparative genomic hybridization and massively parallel sequencing. **A,** CTCs *(light blue cells)* are rare cells in the circulation; the vast majority of nucleated cells are normal blood cells *(orange)*. **B,** CTCs are isolated by one of many methods. **C,** After cell lysis, DNA is accessible for whole genome amplification (WGA). The WGA products can be analyzed for copy number changes on an array platform by comparative genomic hybridization (arrayCGH). Alternatively, libraries can be prepared and subjected to massively parallel sequencing (also known as next-generation sequencing, or NGS). By NGS, both copy number changes and mutations within genes can be detected. (From Heitzer E, Auer M, Ulz P, Geigl JB, Speicher MR. Circulating tumor cells and DNA as liquid biopsies. *Genome Med.* 2013;5:73. Copyright 2013 BioMed Central.)

metastatic breast cancer, both molecular heterogeneity between single CTCs and between CTCs and the primary tumors were identified. Some individual CTCs were resistant to HER2-targeted therapies, suggesting ongoing microevolution at late-stage disease relevant to personalized treatment decisions and acquired drug resistance.[135] Toward this direction, the heterogeneity of *PIK3CA* mutation status within single CTCs isolated from individual metastatic breast cancer patients was recently studied by combining the CellSearch and DEPArray technologies.[80] *TP53* mutations were also found in single CTCs using this method.[79]

CTCs: "A Window to Metastasis"

In 1889, in the very first issue of *Lancet*, Steve Paget described "the seed and soil hypothesis," in which "metastasis depends on the cross talk between selected cancer cells (the seed) and specific organ microenvironments (the soil)," a hypothesis revisited many years later by Fidler.[136] Detailing the mechanism of metastasis remains a very hot topic in cancer research today.[14,137] According to the parallel progression model recently proposed by Klein,[6,138] parallel, independent metastases may arise from early disseminated tumor cells (Fig. 9.17). Analysis of disease course, tumor growth rates, autopsy studies, clinical trials, and molecular genetic analyses of primary and disseminated tumor cells all contribute to our understanding of systemic cancer.[6,138] Molecular characterization of CTCs provides a level of detail not previously possible and may be the key to our further understanding of metastatic progression.

CTC biology can be viewed as a "window to metastasis" because CTCs play a critical role in the metastatic spread of carcinomas (see Figs. 9.17 and 9.18).[7,8] If CTCs are effectively targeted or kept in a dormant state, the cancer may be prevented from progressing to metastatic disease. Molecular characterization of CTCs from patients may be the shortest path to determine which and when patients might relapse and to identify specific mechanisms to target these cells. Dormancy gene signatures that identify individuals with dormant disease have also been explored.[139,140] In contrast, CTCs with stemness and EMT features display enhanced

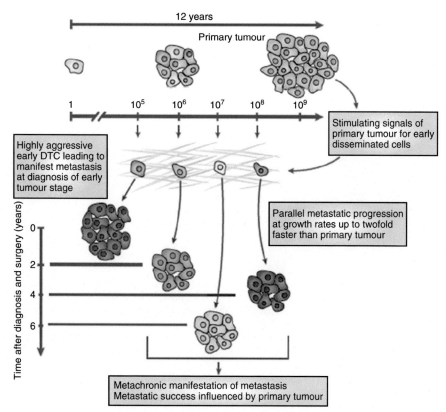

FIGURE 9.17 The parallel progression model of tumor metastasis. In the parallel progression model, several waves of disseminated tumor cells may originate before diagnosis and may progress in parallel at different rates in different organs. (From Klein CA. Parallel progression of primary tumours and metastases. *Nat Rev Cancer*. 2009;9:302–12.)

malignant and metastatic potential. The role of CTCs in treatment failure and disease progression is likely explained by their biological processes, such as EMT, stemness features, dormancy, and heterogeneity.[141]

Epithelial and Mesenchymal Transitions of CTCs

Epithelial-mesenchymal transition (EMT) is an essential process in the metastatic cascade.[23,142-144] This biological process is highly associated with an invasive phenotype and enables detachment of tumor cells from the primary site and migration. The reverse process of mesenchymal epithelial transition (MET) might play a crucial role in the further steps of metastasis when CTCs seed distant organs and establish metastasis (Fig. 9.18). The mechanisms and the interplay of EMT and MET have been intensively studied, but only limited data suggest the existence of the EMT process in CTCs. It is now clear that CTCs from metastatic breast cancer patients exhibit heterogeneous epithelial and mesenchymal phenotypes and display higher frequencies of partial or full-blown mesenchymal phenotype than carcinoma cells within primary tumors. Mesenchymal-like CTCs are also elevated in patients who are refractory to therapy.[142]

Currently, most systems that detect CTCs, including the CellSearch system, are based on the expression of the epithelial marker EpCAM and do not specifically identify CTC subtypes with EMT. Only recently, EMT-related markers have been applied in CTC studies. Three EMT markers (*TWIST1*, *AKT2*, and *PI3Kα*) and the stem cell marker *ALDH1* were evaluated in CTCs from 502 primary breast cancer patients by a multiplex RT-PCR assay.[145] A subset of CTCs showed EMT and stem cell characteristics. The expression levels of EMT-inducing transcription factors (*TWIST1*, *SNAIL1*, *SLUG*, *ZEB1*, and *FOXC2*) were also determined in CTCs from primary breast cancer patients.[146] In another study, rare primary tumor cells simultaneously expressed mesenchymal and epithelial markers, but mesenchymal cells were highly enriched in CTCs, and serial monitoring suggested an association of mesenchymal CTCs with disease progression.[147] Mesenchymal CTCs occurred as both single cells and multicellular clusters, expressed known EMT regulators, including transforming growth factor (TGF)-β pathway components and the *FOXC1* transcription factor. These data support a role for EMT in the blood-borne dissemination of human breast cancer.[147] When the EMT phenotype of CTCs was studied through the expression of two important EMT-connected genes—namely, *VIM* and *SNAIL*—using cytokeratin-negative CTCs from nonmetastatic breast cancer patients, the simultaneous detection of both EGFR and EMT markers may improve prognostic or predictive information.[148] A new assay based on triple immunofluorescence examines the coexpression of ALDH1 (a stemness marker) and TWIST (an EMT marker) of single CTCs in patients with breast cancer. A differential expression pattern for these markers in CTCs was observed both in early and metastatic disease. CTCs expressing high ALDH1 along with nuclear TWIST were more frequently found in patients with metastatic breast cancer, suggesting that CTCs undergoing EMT may prevail during disease progression.[149]

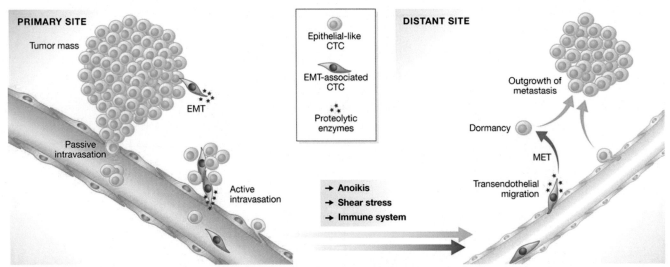

FIGURE 9.18 The metastatic cascade. Tumor cells may enter the bloodstream passively or actively via biological events caused by epithelial mesenchymal transition or centrosome amplification. Disseminating tumor cells must overcome several hurdles, including shear stress in the bloodstream, the immune system both inside and outside of the blood circulation, and anoikis (programmed cell death induced by a cell detaching from the extracellular matrix). Once at a distant site, tumor cells may extravasate, undergo mesenchymal epithelial transition, and grow locally to become a metastasis or remain in dormancy. (From Joosse SA, Gorges TM, Pantel K. Biology, detection, and clinical implications of circulating tumor cells. *EMBO Mol Med.* 2014;7:1–11.)

Gorges and colleagues reported low CTC numbers in patients with late metastatic cancers.[150] These results prompted the search for new markers, including those for mesenchymal-like subpopulations. Plastin-3 is an EMT marker in CTCs in colorectal cancer. Aberrant expression of plastin-3 was associated with increased CTCs and poor prognosis in colorectal cancer and may be involved in the regulation of EMT.[151] Cell-surface vimentin is specifically expressed on the surface of CTCs from epithelial cancers undergoing EMT, and the number of CTCs undergoing EMT correlates with disease outcome.

Cytokeratins are widely used for the identification of CTCs by immunocytochemistry, but even these established markers might be modulated during EMT. Breast cancer cells display a complex pattern of cytokeratin expression with potential biological relevance. Individual cytokeratin antibodies may recognize only certain cytokeratins, and important subsets of biologically relevant CTCs in cancer patients may be missed.[152] EMT and MET transitions are central to the metastatic potential of CTCs; the elucidation of their role in the clinical outcome of cancer patients can be achieved by CTC analysis in single cells.

CTCs and Cancer Stem Cells

Stemness is the ability of cells to self-renew and differentiate into cancer cells.[153] There is substantial evidence that many cancers are driven by a population of cells that display stem cell properties. These cells, called cancer stem cells (CSCs) or tumor initiating cells, not only drive tumor initiation and growth but also mediate tumor metastasis and therapeutic resistance. There is in vitro and clinical evidence that CSCs mediate metastasis and treatment resistance in breast cancer. Novel strategies to isolate CTCs that contain CSCs and the use of patient-derived xenograft models in preclinical breast cancer research have been developed to study the biology of CSCs.[154] Therapeutic resistance, underlying tumor recurrence, and the lack of curative treatments in metastatic disease raise the question as to whether conventional anticancer therapies target the right cells. Indeed, these treatments might miss CSCs that are resistant to many current cancer treatments, including chemotherapy and radiation therapy.[155] Emerging data suggest that the remarkable clinical efficacy of HER-2 targeting agents may be related to their ability to target the breast CSC population. In breast cancers that do not display HER-2 gene amplification, HER-2 is selectively expressed in the CSC population. This expression is regulated by the tumor microenvironment, suggesting that novel and effective adjuvant therapies may need to target the CSC population.[156,157]

EMT induction not only allows for cancer cells to disseminate from the primary tumor but also promotes their self-renewal capability.[158] Breast CSCs have elevated tumorigenicity required for metastatic outgrowth, while EMT may promote CSC character and endows breast cancer cells with enhanced invasive and migratory potential.[159] Emerging evidence indicates that CSCs and EMT cooperate to produce CTCs that are highly competent for metastasis.[160] CTCs with both CSC and EMT characteristics have been identified in the bloodstream of patients with metastatic disease.[118,145] Furthermore, the expression of stemness and EMT markers in CTCs is associated with resistance to conventional anticancer therapies and treatment failure.[158] Some subsets of CTCs have a putative breast cancer stem cell phenotype and express EMT markers. The first evidence of the existence of this putative stemlike phenotype within disseminated breast cancer cells in the bone marrow in early breast cancer patients was shown in 2006.[160] The expression of cancer stem cell markers such as CD44, CD24, or ALDH1, both by molecular assays[118] and imaging,[161] has also been shown in CTCs.[118,161,162]

CTCs: Clinical Significance, Molecular Characterization, and Its Impact on Individualized Treatment of Cancer Patients

Breast Cancer

The clinical significance of CTCs has been extensively evaluated in patients with breast cancer. Many clinical studies have shown that CTC detection is associated with OS and PFS both in early and metastatic breast cancer. A comprehensive meta-analysis of published literature on the prognostic relevance of CTC clearly indicates that the detection of CTCs is a reliable prognostic factor in patients both with early-stage and metastatic breast cancer.[163]

At the moment, many clinical studies are evaluating the potential of CTC testing in the routine management of breast cancer patients.[164] A number of prospective interventional studies are designed to demonstrate that CTC enumeration/characterization may improve the management of breast cancer patients and aim to assess CTC-guided hormone therapy versus chemotherapy decisions in M1 patients, changes in CTC counts during treatment in metastatic patients, and anti–HER-2 treatments in HER-2–negative breast cancer patients selected on the basis of CTCs detection/characterization. The results of these trials will be very important for CTC implementation in the routine management of breast cancer patients.[165]

Clinical Significance of CTCs in Metastatic Breast Cancer. In a seminal paper in 2004, Cristofanilli and colleagues demonstrated that CTCs are an independent prognostic factor for PFS and OS in patients with metastatic breast cancer (MBC).[9] The CellSearch system was used with a cutoff of 5 CTCs/7.5 mL of peripheral blood. This paper led to the FDA clearance of the CellSearch assay in MBC. Many clinical studies have since verified the importance of CTC enumeration in MBC.[162,166-170] It is now clear that MBC patients who present basal counts of 5 CTCs/7.5 mL of blood or greater have a poor prognosis and that enumeration of CTCs during treatment predicts progression of disease earlier than conventional imaging tests.[162] Changing chemotherapy and switching to an alternate cytotoxic therapy was not effective in prolonging OS in patients with persistent increase in CTCs.[171] The independent prognostic effect of CTC count by CellSearch on PFS and OS was further confirmed across 20 studies at 17 European centers that included 1944 eligible patients.[172] CTC analysis may predict the effect of treatment earlier than imaging.[173-175] The prognostic significance of CTC detection is also supported by numerous clinical studies performed on systems other than the CellSearch. RT-PCR assays, especially for the epithelial marker cytokeratin-19 (*CK-19*) alone or in combination with other transcripts, can identify CTCs in MBC.[12] Before the initiation of front-line treatment in patients with MBC, the median PFS and OS were significantly shorter when *CK-19* mRNA-positive CTCs were detected compared with patients who were negative for *CK-19* mRNA.[176] The presence of baseline *CK-19* mRNA-positive CTCs was associated with poor prognosis, and a decrease in mammaglobin mRNA in CTCs may correlate to therapeutic response.[177] The detection of viable CTCs that excrete *CK-19* correlates with OS using the EPISPOT assay.[178]

Clinical Significance of CTCs in Early Breast Cancer. In early breast cancer, CTC numbers are low, and molecular assays are most successful for detection. Nested RT-PCR for *CK-19* expression in the peripheral blood of node-negative breast cancer patients was first shown to be of prognostic value in 2002.[179] Detection of *CK-19* transcripts by using an EpCAM-independent RT-qPCR assay in peripheral blood of early breast cancer patients[112,113] was an independent prognostic factor for disease-free survival (DFS) and OS before,[114] during,[115] and after[116] chemotherapy. Before administration of adjuvant chemotherapy, CTC detection based on *CK-19* positivity by RT-qPCR predicted poor clinical outcome mainly in patients with ER-negative, triple-negative, and HER-2–positive early-stage breast cancer.[180] By using the same EpCAM-independent assay, persistent CTC detection during the first 5 years of follow-up indicated resistance to chemotherapy and hormonal therapy and predicted late relapses in patients with operable breast cancer.[181] Elimination of these *CK-19* mRNA-positive CTCs during adjuvant chemotherapy reflects successful treatment.[182]

CellSearch has also been used in early disease stages, but it requires more than 7.5 mL of blood. By using CellSearch, detection of one or more CTCs in 7.5 mL of blood before neoadjuvant chemotherapy accurately predicted OS.[183] In another CellSearch study, chemonaive patients with nonmetastatic breast cancer at the time of definitive surgery were studied. After a median follow-up of 35 months, the presence of one or more CTCs predicted early recurrence and decreased OS.[184] The REMAGUS02 neoadjuvant study was the first to report the significance of CellSearch CTC detection on distant metastasis-free survival and OS. The detection of CTCs was independently associated with a significantly worse outcome but mainly during the first 3 to 4 years of follow-up.[185] In the German SUCCESS trial, the presence of CTCs was associated with poor DFS, distant DFS, breast cancer–specific survival, and OS, and the prognosis was worse in patients with at least 5 CTCs/30 mL blood. CTCs were confirmed as independent prognostic markers by multivariable analysis for DFS and OS.[186] These findings may change the clinical management of early breast cancer because they clearly indicate the metastatic potential of CTC early in the disease.

Molecular Characterization of CTCs in Breast Cancer and Its Impact on Individualized Treatment. The main goal of adjuvant therapy is to prevent distant recurrences by targeting residual disseminated tumor cells. However, almost 70% of deaths in patients with early breast cancer occur after 5 years, showing that residual disease can be dormant for very long periods. Differences between primary tumors and CTCs could be of crucial importance for therapeutic decisions. In breast cancer, molecular characterization of CTCs may help identify therapeutic targets and resistance mechanisms and to stratify breast cancer patients.[11,187-189] Molecular characterization of CTCs may help to identify new druggable targets in MBC patients.[162] The application of extremely powerful MPS technologies to single CTCs may find a place in the management of patients. However, it is currently a technical challenge for the clinical lab to perform these tests on CTCs on a regular basis, under standardized conditions, with robust and accredited methodologies.

HER-2. In breast cancer, anti–HER-2 therapies are prescribed according to the HER-2 status of the primary tumor, as assessed by immunohistochemistry or FISH.

However, a continuously growing body of evidence indicates that CTC HER-2 status can be different from that of the corresponding primary tumor. Moreover, CTC HER-2 status can change over time, particularly during disease recurrence or progression.[110,119,190-194] By using CellSearch and an automated algorithm to evaluate CTC HER-2 expression, heterogeneity even within the same patient, is the rule.[25] Many research groups have shown that HER-2–positive CTCs can be detected in patients with HER-2–negative primary tumors.[190,193-197]

The HER-2 status of CTCs was correlated to the clinical response of HER-2–targeted therapies in the first "liquid biopsy trial" completed in 2012. Georgoulias and colleagues investigated the effect of trastuzumab in a small number of patients who were HER-2 negative in the primary tumor but were positive for CK(+)/HER2(+) CTCs as evaluated by immunofluorescence. These patients were randomized into two groups, and one group received trastuzumab, while the other received a placebo. After trastuzumab administration, 75% of the women became *CK-19* mRNA-negative, the risk of disease recurrence was reduced, and the median DFS was longer.[198] Similarly, a multicenter phase II trial evaluated the activity of lapatinib in MBC patients with HER-2–negative primary tumors and HER-2–positive CTCs using the CellSearch.[199] The TREAT-CTC trial, initiated in 2012, is a randomized phase II trial for patients with HER-2 negative primary breast cancer but HER-2–positive CTCs. This trial is specifically designed to test the efficacy of trastuzumab in HER-2–negative early breast cancer.

Endocrine Treatment. Endocrine treatment is the preferred systemic treatment in MBC patients who have had an estrogen receptor (ER)–positive primary tumor or metastatic lesions. However, 20% of these patients do not benefit from this therapy and demonstrate further metastatic progress. A possible explanation for failure of endocrine therapy might be the heterogeneity of ER expression in CTCs. Similar to the HER-2 story, there is a growing body of evidence that hormone receptor status can change over time, especially during disease recurrence or progression in breast cancer patients.[110,119,190-194] In this context, reevaluation of hormone receptor status by molecular characterization of CTCs is a strategy with potential clinical application. Optimal individualized treatment could be selected by characterizing ER status in CTCs and comparing it to the primary tumor.[200] The commercially available AdnaTest Breast Cancer kit (AdnaGen, Qiagen, Germany) can detect EpCAM, MUC-1, and HER-2 transcripts in CTCs; it was found that a major proportion of CTCs in MBC patients showed EMT and tumor stem cell characteristics.[118] Interestingly, when ER and progesterone receptor (PR) CTC expression was assessed by RT-PCR, CTCs were mostly triple-negative regardless of the ER, PR, and HER-2 status of the primary tumor.[119] CTCs frequently lack ER expression in MBC patients with ER-positive primary tumors and show a considerable intrapatient heterogeneity, which may reflect a mechanism to escape endocrine therapy. Single-cell analysis based on WGA and MPS did not support a role for *ESR1* mutations.[201] In nonmetastatic breast cancer patients, the expression of estrogen, progesterone, and epidermal growth factor receptor (EGFR) by immunofluorescence experiments revealed heterogeneous expression of these hormonal receptors in single CTCs in the same patient.[202]

EMT and Stem Cell Markers on CTCs. EMT and tumor stem cell characteristics were detected in CTCs from metastatic breast cancer patients.[118] CTCs expressing TWIST and vimentin were identified in patients with both metastatic and early breast cancer.[203] Subpopulations of CTCs with putative stem cell progenitor phenotypes in patients with MBC was shown by triple-marker immunofluorescence microscopy.[161] Serial monitoring of CTCs in patients with breast cancer reveals simultaneous expression of mesenchymal and epithelial markers. Mesenchymal cells expressing known EMT regulators, including transforming growth factor (TGF)-β pathway components and the FOXC1 transcription factor, were associated with disease progression.[147]

Apoptosis and Clinical Dormancy on CTCs. CTCs and DTCs can enter a state of dormancy and become resistant to targeted or conventional therapies.[139] Our understanding of the biology of dormant DTCs and CTCs is currently limited. Adjuvant chemotherapy reduced both the number of CTCs per patient and the number of proliferating CTCs.[146] Apoptotic CTCs were detected in patients with breast cancer irrespective of their clinical status, although more were detected in early cancer compared to metastatic cancer.[204] Apoptotic CTCs were more frequently encountered during follow-up in those patients who remained disease-free compared to those with subsequent late relapse.[205] These results indicate that the detection of CTCs that survive despite adjuvant therapy implies that CTC elimination should be attempted using agents targeting their distinctive molecular characteristics.

DNA Methylation in CTCs. Recent studies have shown that DNA methylation silences many key tumor and metastasis suppressor genes in CTCs, including those that suppress growth and proliferation, invasiveness, epithelial differentiation, and stemness like *CST6*, *BRMS1*, and *SOX17*.[206-208]

Overall, CTC analysis is highly correlative to systemic disease spreading and disease outcome in breast cancer patients. Monitoring of CTCs can predict early subclinical metastasis recurrence and monitor progression during and after treatment. Many clinical studies support the potential utility of CTC detection in breast cancer.

Prostate Cancer. In prostate cancer, CTCs have been extensively studied and validated as a prognostic tool.[209] In advanced prostate cancer, CTC enumeration on the CellSearch system is FDA-cleared as a prognostic test of survival at baseline and posttreatment. Integration into routine clinical practice is ongoing.[211] The main CTC studies in advanced and localized prostate cancer highlight the important gains as well as the challenges posed by various approaches and their implications for advancing prostate cancer management.[210,211]

Clinical Significance of CTCs in Metastatic Prostate Cancer. In 2001, CTC numbers were first quantified in the circulation of patients with metastatic prostate cancer. This change in CTC numbers was of prognostic significance.[212] Later, CTC enumeration before therapy was also related to prognosis. The shedding of CTCs into the circulation was an intrinsic property of the tumor, distinct from the extent of the disease.[213] In 2008, the FDA cleared the CellSearch assay for the enumeration of CTCs in castration-resistant prostate cancer (CRPC) based on data presented by de Bono and colleagues,[214] who showed that CTC enumeration had prognostic value as an independent predictor of OS in CRPC. This study was followed by many others, confirming that CTC enumeration can be used to monitor disease status.[215,216] In patients with metastatic hormone-sensitive prostate cancer, correlation of prostate-specific antigen, Gleason score, and TNM stage with CTC counts showed that CTCs could correctly stage prostate cancer and assess prognosis.[217,218] According to the results of a prospective phase III trial, baseline CTC counts were prognostic and could be used as an early metric to help redirect and optimize therapy.[219]

Clinical Significance of CTCs in Early Prostate Cancer. In early-stage prostate cancer, posttreatment reduction in CTCs may indicate radiation therapy response[220] and can assist in the decision between systemic or local treatment.[221] Recent trials in patients with CRPC are incorporating the detection of CTCs, imaging, and patient-reported outcomes to improve future drug development and patient management.[222]

Molecular Characterization of CTCs in Prostate Cancer and Its Impact on Individualized Treatment. Advances in the understanding of prostate cancer signaling pathways have led to the development and subsequent approval of multiple novel therapies, especially for metastatic castration-resistant prostate cancer. The androgen receptor (AR) is a key target in prostate cancer, and many current therapies for metastatic CRPC target AR signaling. For example, abiraterone and enzalutamide are novel endocrine treatments that abrogate AR signaling in CRPC, but resistance to these therapies is also common. CRPC therapeutic intervention after clinical progression may be achieved through the characterization of AR activity. Biopsies of bone metastases can also assess AR activity but are highly invasive. On the other hand, the molecular characterization of CTCs offers a minimally invasive approach to study late-stage disease. However, patient benefit is variable with these agents, so development of other predictive biomarkers is very important. In addition to CTC enumeration by CellSearch, molecular profiling may better reflect tumor evolution in an individual during the pressure of systemic therapies.[223]

Pretreatment detection of androgen receptor splice variant-7 (AR-V7) in CTCs from CRPC patients is associated with resistance to enzalutamide and abiraterone.[224] Such androgen ablation therapy induces expression of constitutively active AR splice variants that drive disease progression. Recently, the same group confirmed these results by conducting CTC-based AR-V7 analysis at baseline, during therapy, and at progression.[225] Nuclear AR expression in CTCs of CRPC patients treated with enzalutamide and abiraterone was recently evaluated in real-time using the CellSearch system and an automated algorithm to identify CTCs.[226] Large intrapatient heterogeneity of CTC AR expression was observed, including AR-positive CD45-negative CTCs that were CK-negative. The number of these CK-19–negative cells correlated with traditional CTCs and associated with an even worse outcome on univariate analysis. This is important because these events are completely missed by using standard CellSearch detection.[226] A recent study, based on fluorescence-activated cell sorting for CTC isolation and downstream molecular characterization of AR expression, revealed a very high interpatient and intrapatient heterogeneity of the expression and localization of the AR. Increased AR expression and nuclear localization were associated with elevated coexpression of a marker of proliferation, Ki-67, consistent with a continued role for AR expression in castration-resistant disease. However, despite this heterogeneity, it was clear that CTCs from patients with prior exposure to abiraterone had increased AR expression compared to CTCs from patients who were abiraterone naive. Thus the evaluation of AR expression in CTCs is critical for the management of patients with advanced prostate cancer.[227]

Taxanes are a standard of care therapy for CRPC. According to a recent study, two clinically relevant AR splice variants, AR-V567 and AR-V7, differentially associated with microtubules and dynein motor protein, resulting in differential taxane sensitivity in vitro and in vivo. Androgen receptor variants that accumulate in CRPC cells utilize distinct pathways of nuclear import that affect the antitumor efficacy of taxanes, suggesting a mechanistic rationale to customize treatments for patients with CRPC that might improve outcomes.[228]

Clinical trials that evaluate the efficacy of drugs in CRPC need new clinical endpoints that are valid surrogates for survival. In a clinical trial of abiraterone plus prednisone versus prednisone alone for patients with metastatic CRPC, Scher and colleagues evaluated CTC enumeration as a surrogate outcome. They developed a biomarker panel that includes CTC number and LDH concentration as a surrogate for survival at the individual patient level.[229]

Although these findings would require large-scale prospective validation before routine clinical practice, they do show the strong potential of CTC analysis for the management of prostate cancer. Serial CTC AR-V7 expression testing may soon be implemented in clinical practice. This should also provide further insights into tumor evolution.

Overall, CTC analysis is highly correlative to systemic disease spread and disease outcome in prostate cancer patients. Recent clinical studies support the potential utility of CTC molecular characterization, especially in respect to androgen receptor mutations, splice variants, and response to targeted therapies in prostate cancer.

Colorectal Cancer

A metaanalysis of 12 studies revealed prognostic value of CTCs and DTCs in patients with metastatic colorectal cancer.[230] CTC number was an independent predictor of cancer recurrence in six out of nine studies that examined the detection of postoperative CTCs in CRC.[231] The prognostic significance of CTCs in colorectal cancer has been recently reviewed.[232]

Clinical Significance of CTCs in Metastatic Colorectal Cancer. The CellSearch assay was cleared by the FDA for metastatic CRC in 2008.[233] In advanced colorectal cancer, CTC enumeration before and during treatment independently predicts PFS and OS and provides additional information beyond CT imaging,[234-236] while surgical resection of metastases immediately decreases CTC levels.[237] CTC numbers are higher in the mesenteric venous blood compartments of patients with CRC, and viable CTCs are trapped in the liver, a finding that may possibly explain the high rates of liver metastasis in this type of cancer.[238] Six CTC markers (tissue specific and EMT transcripts) used in metastatic CRC patients identified therapy-refractory patients not detected by standard image techniques. Patients with increased CTCs numbers, even when classified as responders by computed tomography, showed significantly shorter survival times.[239] The presence of CTCs in stage III colon cancer patients undergoing curative resection followed by mFOLFOX chemotherapy, as determined by telomerase reverse transcriptase, CK-19, CK-20, and CEA transcripts, was an independent predictor of postchemotherapeutic relapse and strongly correlated with DFS and OS.[240]

Clinical Significance of CTCs in Early Colorectal Cancer. CTC studies in nonmetastatic CRC are more limited compared to metastatic CRC, due to the very low number of CTCs. It was shown that preoperative CTC detection is an independent prognostic marker in nonmetastatic CRC,[241] and the presence of CTCs correlates with reduced DFS in patients with nonmetastatic CRC.[242] CTC detection might help in the selection of high-risk stage II CRC candidates for adjuvant chemotherapy.[243] CEA, CK, and CD133 expression were used to evaluate the clinical significance of CTCs as a prognostic factor for OS and DFS in patients with colorectal cancer after curative surgery. In patients with Duke's stage B and C CRC who required adjuvant chemotherapy, detection of CEA/CK/CD133 mRNA in CTCs predicted risk of recurrence and poor prognosis.[244]

Molecular Characterization of CTCs in Colorectal Cancer and Its Impact on Individualized Treatment. Patients with benign inflammatory diseases of the colon can harbor viable circulating epithelial cells that are detected as "CTCs" on the CellSearch system because they are of epithelial origin and CD45 negative.[245] Hence, further molecular characterization of CTCs is important in colorectal cancer. A considerable portion of viable CTCs are trapped in the liver, as shown by their enumeration in the peripheral and mesenteric blood, using both the CellSearch and EPISPOT assays,[246] potentially explaining the high incidence of liver metastasis in colon cancer. Anti-EGFR therapy in metastatic colorectal cancer may select for *KRAS* and *BRAF* mutations. However, the occurrence of these mutations in metastatic colorectal cancer may vary among primary tumors, CTCs, and metastatic tumors.[247,248] Using the CellSearch system, the expression of *EGFR*, *EGFR* gene amplification, *KRAS*, *BRAF*, and *PIK3CA* mutations were evaluated in single CTCs of patients with metastatic colorectal cancer,[249] and the concordance between *KRAS* mutations in CTCs and primary tumors was 50%.[78] *APC*, *KRAS*, and *PIK3CA* mutations that were found in CTCs were also present at subclonal levels in the primary tumors and metastases from the same patients.[134] When *KRAS* mutations were investigated in CTCs from patients with metastatic CRC throughout the course of the disease and compared to the corresponding primary tumors, CTCs exhibited different *KRAS* mutations during treatment.[250] Plastin-3 is a marker for CTCs undergoing EMT and is associated with colorectal cancer prognosis, particularly in patients with Duke's B and C stage tumors.[251] Patients with CTC positivity at baseline had a shorter median PFS compared to patients with no CTCs, and a significant correlation was also found between CTC detection during treatment and radiographic findings at 6 months.[236] CTCs are promising markers for the evaluation and prediction of treatment responses in rectal cancer patients, superior to CEA. CTC number correlates with treatment outcome, but not serum CEA.[252]

In conclusion, CTC analysis highly correlates to systemic disease spread and disease outcome in colorectal cancer patients. Many clinical studies support the potential utility of CTC detection in colorectal cancer.

Lung Cancer

Lung cancer biopsies are difficult to obtain, while a liquid biopsy of CTCs is relatively easy. However, CTCs in lung cancer are difficult to detect because they seldom have epithelial characteristics. Detection methods that rely on epithelial markers to identify CTCs may not be effective. Even so, there is evidence that in lung cancer CTC numbers are prognostic and that CTCs counted before and after treatment mirror treatment response. In patients with molecularly defined subtypes of non–small cell lung cancer, CTCs demonstrate the same molecular changes as the cancer cells of the tumor.[253] Chronic obstructive pulmonary disease is a risk factor for lung cancer, and monitoring CTCs in these patients may allow for early diagnosis.[254]

Clinical Significance of CTCs in Non–Small Cell Lung Cancer. The presence of EpCAM/MUC-1 mRNA-positive CTCs in non–small cell lung cancer (NSCLC) patients preoperatively and postoperatively revealed shortened DFS and OS.[255] In NSCLC patients undergoing surgery, the presence and the number of CTCs were associated with worse survival.[256] By using the ISET filtration system and immunofluorescence, hybrid CTCs with an EMT phenotype were detected.[257] CTCs are detectable in patients with untreated stage III or IV NSCLC and are prognostic according to another single-center prospective study.[258] Advanced-stage NSCLC patients with elevated CEA had higher numbers of CTCs.[259] Using the ISET filtration system and filter-adapted fluorescence in situ hybridization, genetic alterations of tyrosine kinase oncogenes were examined in CTCs. Specifically, CTCs from four *ROS1*-rearranged NSCLC patients were followed during treatment with the *ROS1*-inhibitor crizotinib.[72]

Clinical Significance of CTCs in Small Cell Lung Cancer. Small cell lung cancer (SCLC) accounts for 15% to 20% of lung cancer cases. It is very aggressive and characterized by early dissemination and dismal prognosis. In most cases, SCLC is not operable, and it is difficult to obtain biopsies to investigate its biology and therapeutic options. In this type of cancer, CTCs circulate in high numbers and are readily accessible through a single blood draw, which can be easily repeated for follow-up over time. A research team that was headed by Dive showed that CTCs in SCLC are tumorigenic in immunocompromised mice; the resultant CTC-derived explants mirror the patient's response to platinum and etoposide chemotherapy and can be used for the selection of appropriate therapies.[260] The same group showed that in patients with SCLC undergoing standard treatment, CTCs and CTC clusters, called circulating tumor microemboli (CTM), can be detected and are independent prognostic factors.[261] Evaluating the presence of CTM also improved diagnostic accuracy in NSCLC patients based on clinical and imaging data.[262] The change in CTC count after the first cycle of chemotherapy, evaluated by using the CellSearch system, provided useful prognostic information in SCLC[263] and was the strongest response predictor for chemotherapy and survival.[261]

Molecular Characterization of CTCs in Lung Cancer and Its Impact on Individualized Treatment. CTCs isolated from early-stage lung cancer patients have been successfully expanded through cell culture. After CTC isolation with an in situ capture and culture methodology, ex vivo expansion of CTCs is performed using a three-dimensional co-culture model. These expanded lung CTCs carried mutations of *TP53* identical to those observed in the matched primary tumors.[264] Mutations were first detected in CTCs of NSCLC patients in 2008.[265] In this study, NSCLC patients with EpCAM-positive CTCs carrying *EGFR* mutations had faster disease progression than CTCs who lacked mutations. In another study, a mutation panel for six genes (*EGFR, KRAS, BRAF, NRAS, AKT1,* and *PIK3CA*) revealed only one *EGFR* mutation (exon 19 deletion) in the 38 patient samples analyzed.[266] In some of these studies, the analytical sensitivity may not be high enough to detect all variants. The CellSearch System coupled with MPS may be the most sensitive and specific diagnostic tool for CTC evaluation.[267] Crizotinib is an effective molecular treatment for *ALK* rearrangement–positive NSCLC. The companion diagnostic test for *ALK* rearrangements in NSCLC for crizotinib treatment is currently done by tumor biopsy or fine-needle aspiration. By using a filtration technique and FISH, Pailler and colleagues successfully managed to detect *ALK* rearrangements in CTCs of NSCLC patients, enabling both diagnostic testing and monitoring of crizotinib treatment.[268] CellSearch technology was recently adapted to identify tumor cells in malignant pleural effusions. The pleural fluid CellSearch assay may complement traditional cytology in the diagnosis of malignant effusions.[269]

Cutaneous Melanoma

Cutaneous melanoma is a cancer of the skin derived from transformation of melanocytes. Melanoma is mainly a disease of developed countries, with the highest incidence in Australia, North America, and Europe.[270] Unfortunately, melanoma is among the fastest-growing cancers in Western societies. A unique characteristic of primary melanoma is that small lesions more than 2.5 mm deep can be highly metastatic with very poor prognosis compared to similar-size tumors of other cancers, so early detection of CTCs is very important. Melanoma often metastasizes to regional lymph nodes and to distant organs.[270,271] Melanoma metastasis may occur to almost any organ, but brain, liver, and lung are the most common, so CTCs are particularly interesting due to their aggressive ability to metastasize systemically. Recently, the approval of targeted therapies such as *BRAF* mutation inhibitors (vemurafenib, dabrafenib), the MEK inhibitor trametinib, and immune checkpoint inhibitors PDL-1 (pembrolizumab), PD1 (nivolumab), and ipilimumab alone and in combinations have improved OS and DFS considerably.[272-274]

Clinical Significance of CTCs in Cutaneous Melanoma. The systemic spread of melanoma as viewed through CTCs is in essence real-time monitoring of metastasis as it occurs.[275,276] The only approved blood biomarker for melanoma is LDH in American Joint Cancer Committee stage III/IV patients. Because melanoma is highly malignant, it is not surprising that studies have found CTC analysis clinically important. CTC analysis of melanoma patients started in the early 1990s and has progressed through the years, along with advancements in molecular assays.[277-279] Circulating melanoma cells have differentiation lineage and tumor-related gene expression patterns that are not found in peripheral blood leukocytes. RT-qPCR can be performed on blood after lysis of RBCs and mononuclear cell preparation. Unlike most other RT-qPCR CTC assays, analysis of melanoma cells does not require antibody capture through cell surface antigens. There are only a few cell-surface melanoma-associated antigens for targeting antibodies.[280-282] Therefore molecular assays that target melanoma transcripts are often used, including *MART-1, MAGE-A3, PAX3,* and ganglioside *GM2/GD2 glycosyltransferase (GalNAc-T)*.[278,283-285]

Melanoma is heterogeneous in genomic aberrations and transcriptome expression. Therefore it is important to use multiple markers to assess CTCs to improve sensitivity.[281,283,286-288] New approaches to detect CTCs include the inertial focusing spiral microfluidics CTChip in the ClearCell FX system (Clearbridge BioMedics, Singapore). It can separate CTCs by size and mass, followed by RT-qPCR or immunohistochemistry. The system is rapid, and the yield of CTCs from peripheral blood leukocytes is higher than with other methods.[98]

Multiple marker RT-qPCR assays allow monitoring both early- and advanced-stage patients during treatment; several well-annotated studies with long-term follow-up have been published.[284,285,287,289,290] CTC monitoring post- and pretreatment also predicts OS in surgically resected disease-free stage III/IV patients. A recent multicenter international phase III trial demonstrated that CTC analysis predicts the outcome of melanoma patients with positive sentinel lymph nodes after their resection with disease-free status.[290] Multimarker RT-qPCR analysis may be helpful in identifying patients who have high risk of systemic disease progression after surgery or therapy.

Molecular Characterization of Isolated CTCs From Melanoma Patients. Melanoma CTCs are unique in their associated antigens, mRNA expression patterns, and genomic aberrations.[280,281,286,291-293] Recently, an in-depth genomic analysis was performed on paired primary tumors and CTCs whereby copy number aberrations and loss of heterozygosity were analyzed. CTCs were captured by several antiganglioside cell surface human IgM monoclonal antibodies and then subjected to a genome-wide SNP array (Array 6.0, Affymetrix, Santa Clara, Calif.).[282] IgM provides better capture of CTCs than IgG. Greater than 90% of SNPs were concordant between the primary tumor and isolated CTCs. Several frequent copy number aberrations were identified and validated in a separate cohort of patients with advanced-stage melanoma. These studies indicate the presence of many unexplored key copy number variants in CTCs that can be used to monitor patients' progression. Other groups have reported known genomic mutations in *BRAF* and *KIT,* using isolated melanoma CTCs.[294,295]

Overall CTC analysis is highly correlated to systemic disease spread and disease outcome in melanoma patients. With the availability of new and improved therapeutics in melanoma, the monitoring of CTCs may be important to assess early subclinical recurrence and progression both during and after treatment. Phase II and III clinical trials support the potential utility of melanoma patient CTC detection and analysis.

Ovarian Cancer

Molecular assays are most commonly used for the characterization of CTCs in ovarian cancer. A panel of six genes for the PCR-based detection of CTCs in endometrial, cervical, and ovarian cancers detected CTCs in 44% of cervical, 64% of endometrial, and 19% of ovarian cancers.[296] This same group later identified additional markers for CTCs in patients with epithelial ovarian cancer and evaluated their impact on clinical outcome.[297] Aktas and colleagues investigated CTCs in the blood of 122 ovarian cancer patients at primary diagnosis and after platinum-based chemotherapy by using immunomagnetic enrichment and multiplex RT-PCR (AdnaTest, Qiagen, Germany). They reported that CTCs correlated with shorter OS before surgery and after chemotherapy.[298] When ovarian cancer is studied on the CellSearch system, some studies show that elevated CTCs impart an unfavorable prognosis,[299] while others find no correlation.[300] Even if CTCs are associated with poor outcomes in ovarian cancer, clinical implementation will require uniform methodology and prospective validation.[301] By using a cell adhesion matrix for functional enrichment and identification, the presence of CTCs was correlated with shorter OS and PFS and had a better positive predictive value than CA-125.[302]

Platinum resistance constitutes one of the most recognized clinical challenges for ovarian cancer. Molecular CTC analysis in ovarian cancer correlates with platinum resistance. Although the immunohistochemistry of ERCC1 protein in primary tumors does not predict platinum resistance, ERCC1 (+) CTCs do predict platinum resistance at primary diagnosis of ovarian cancer.[303] A recent metaanalysis of eight studies, including 1184 patients, showed that patients with ERCC1(+) CTCs had significantly shorter OS and DFS than patients with ERCC1(−) CTCs.[304]

Pancreatic Cancer

Pancreatic cancers frequently spread to the liver, lung, and skeletal system, suggesting that pancreatic tumor cells must be able to intravasate and travel through the circulation to distant organs. The presence of CTCs correlates with an unfavorable outcome in pancreatic cancer.[305,306] However, as stated by Gall and colleagues, CTCs are rare in pancreatic cancer, and it is unclear whether CTCs actually contribute toward tumor invasiveness and spread in such an aggressive cancer.[307] A recent metaanalysis including nine studies with a total of 623 pancreatic cancer patients showed that 268 CTC-positive patients had poorer PFS and OS compared to 355 CTC-negative patients.[308] Larger studies, as well as characterization of the CTC population, are required to achieve further insight into the clinical implications of CTC analysis in pancreatic cancer.

Head and Neck Cancer

Head and neck squamous cell carcinoma (HNSCC) is the sixth most common cancer and causes high morbidity due to the lack of early detection. A recent comprehensive review details studies over the past 5 years on the detection of CTCs in HNSCC.[309] When CTCs from locally advanced NHSCC are enriched by the CellSearch system, CTCs are detected in only a low fraction of cases.[310] A metaanalysis of eight studies and 433 patients with HNSCC concluded that the presence of CTCs portends a poor prognosis compared to patients without CTCs.[311] In patients with HNSCC undergoing surgical intervention, patients with no detectable CTCs had a higher probability of longer DFS.[312,313] In both of these studies, CTCs were isolated only by negative enrichment that is not dependent on the expression of surface epithelial markers. Another prospective multicentric study evaluated the role of CTC detection in locally advanced head and neck cancer; a decrease in the CTC number or their absence throughout the treatment was related to nonprogressive disease.[314] Current staging methods for squamous cell carcinomas of the oral cavity need to be improved to predict the risk to individual patients. This can be achieved by counting bone marrow DTCs and peripheral blood CTCs that predict relapse with higher sensitivity than routine staging procedures.[315] The persistence of CTCs after upfront tumor surgery may be useful for the identification of patients who benefit from treatment intensification in locally advanced HNSCC.[316]

Hepatocellular Carcinoma

CTC analysis in hepatocellular carcinoma (HCC) is a new field.[163] A large variation of CTCs with epithelial, mesenchymal, liver-specific, and mixed characteristics, including different size ranges, is observed among patient groups and is associated with therapeutic outcome.[317] Frequent EpCAM+ CTCs in intermediate or advanced HCC are seen.[318] In HCC patients undergoing curative resection, stem cell-like phenotypes have been observed in EpCAM+ CTCs. Preoperative CTC numbers predicted tumor recurrence in HCC patients after surgery, especially in patient subgroups with α-fetoprotein concentrations of up to 400 ng/mL.[319]

Bladder Cancer

CTCs may be used as a noninvasive, real-time tool for the stratification of early-stage bladder cancer patients according to individual risk of progression.[320] The potential prognostic value of CTCs in patients with advanced nonmetastatic urothelial carcinoma of the bladder was shown in a recent clinical study, where CTC-positive patients had significantly higher risks of disease recurrence, as well as cancer-specific and overall mortality.[321] CellSearch was also used to detect and evaluate prospectively the biological significance of CTC in patients with nonmetastatic, advanced bladder cancer; according to this study, the presence of CTCs may be predictive for early systemic disease because CTCs were detected in 30% of patients with nonmetastatic disease.[322] The prognosis of T1G3 bladder cancer is highly variable and unpredictable from clinical and pathological prognostic factors. When survivin-expressing CTCs were evaluated in patients with T1G3 bladder tumors, the presence of CTCs was an independent prognostic factor for DFS.[323] However, in metastatic urothelial carcinoma, CTCs could not predict extravesical disease and were not a clinically useful parameter for directing therapeutic decisions.[324]

CHAPTER 9 Circulating Tumor Cells and Circulating Tumor DNA

POINTS TO REMEMBER

CTCs
- CTC enumeration tests in metastatic breast, colorectal, and prostate cancer are cleared by the FDA.
- A plethora of analytical systems are available for CTC isolation, detection, and molecular characterization that are currently under analytical and clinical evaluation.
- CTC molecular characterization may be translated into individualized targeted treatments.
- Single-cell analysis of CTCs holds considerable promise for the identification of therapeutic targets and resistance mechanisms in CTCs, as well as for real-time monitoring of the efficacy of systemic therapies.
- Quality control and standardization of CTC isolation, detection, and molecular characterization are required for the incorporation of CTC analysis into routine clinical laboratory practice.

CIRCULATING TUMOR DNA

Circulating tumor DNA (ctDNA) isolation, detection, and analysis have significantly improved over the past 2 decades. Several comprehensive reviews have been published.[325-331] Key studies are reviewed in this chapter. Different forms of ctDNA in several cancer types and their roles in clinical oncology also are examined.

Origin and Function of ctDNA

Free circulating DNA can be found in both cancer patients and noncancer patients with various pathologies, including trauma, inflammatory diseases, and autoimmune diseases. This chapter focuses on the circulating DNA in plasma or serum of cancer patients (ctDNA). Most ctDNA is from cancer cells, although some may be from the tumor microenvironment. ctDNA can be actively or passively released by tumor cells and enter different fluid compartments of the body such as lymphatics, urine, blood, semen, saliva, and cerebral spinal fluid.[332,333] ctDNA may also be from CTCs that are disrupted in the bloodstream. ctDNA may have physiological functions that influence normal cells, particularly those adjacent to the tumor. Single- and double-stranded DNA bind to toll-like receptors (TLRs) expressed on the surface of tumor cells and tumor-infiltrating cells.[334] ctDNA can bind to TLRs on leukocytes and activate various specific signal transduction pathways that could alter host-immune responses to tumor cells in the tumor microenvironment. Activation of TLRs can initiate signal transduction pathways that result in cytokine release and other functional changes in the targeted binding cells. The release of ctDNA may have direct effects on tumor microenvironment immune cells, as well as distant organ sites. This may be an important ctDNA physiological influence in the host. ctDNA also may act through horizontal gene transfection into normal cells in the tumor microenvironment.[337] Cellular transformation and tumorigenesis have been implicated with oncogene-containing DNA transfected in vivo.[336-338]

Types of ctDNA

Multiple forms of ctDNA exist in the blood of cancer patients (Figs. 9.19 and 9.20). The most frequent ctDNA results reported are mutations. Tumor mutations in several cancers have drawn more attention because of the availability of

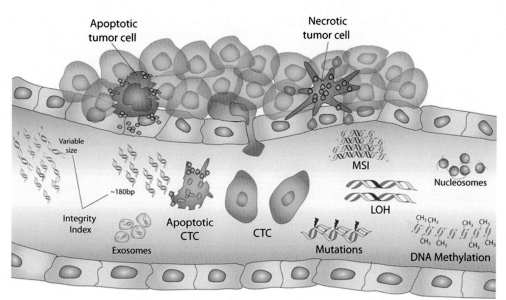

FIGURE 9.19 Schematic representation of circulating tumor DNA (ctDNA) in the systemic circulation. ctDNA arises from apoptosis or necrosis of primary and metastatic solid tumors and from degradation of CTCs. In the circulation, ctDNA is of variable size, can be quantified by an integrity index, and is often enriched with multiples of 180 bp fragments that wind around nucleosomes. In the circulation, ctDNA may be contained within vesicles, attached to nucleosomes or free in solution. Analysis of ctDNA can identify tumor mutations, microsatellite instability (MSI), loss of heterozygosity (LOH), and methylation patterns. Apoptosis of normal cells also results in circulating DNA that dilutes the variant fraction of abnormal DNA.

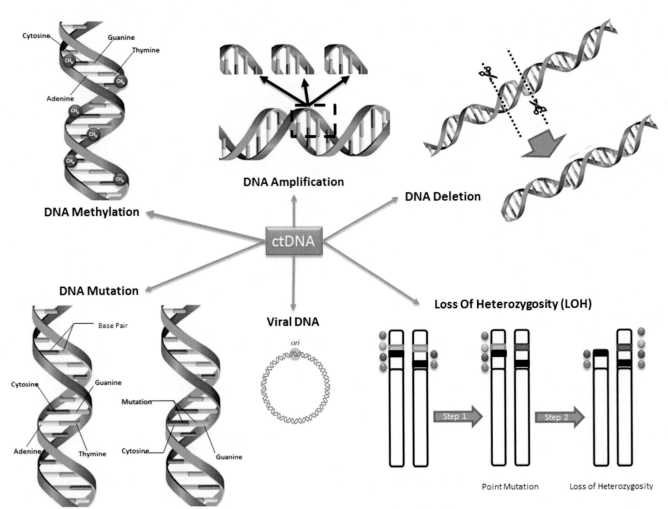

FIGURE 9.20 Examples of genomic variation revealed by analysis of ctDNA in blood. Variants include abnormal DNA methylation patterns, copy number variations (amplifications, deletions, insertions), point mutations, and loss of heterozygosity. Nonhuman DNA, including viral and bacterial DNA, may also be found in the circulation.

respective targeted therapies that are FDA approved in clinical oncology. However, in addition to mutations (single-nucleotide variants, small insertions and small deletions), there are additional types of aberrations in individual cancers with potential utility.[339] These potential targets include methylation of gene promoter regions of coding and noncoding genomic sequences, microsatellite loss of heterozygosity, DNA integrity, gene fusion, copy number variation, and cancer viral DNA.[339]

These different types of ctDNA vary among individual cancer types, although some genomic aberrations can overlap, particularly with cancers of similar embryonic or carcinogenesis origin. The clinical utility of one type of ctDNA in a specific cancer may not be the same for other cancers. For example, *BRAF* mutations are found in several cancers, including melanoma, colorectal, and lung, but they are only partially responsive to therapy with limited long-term response. BRAF inhibitors in melanoma have the highest response overall. Thus ctDNA BRAF mutations may potentially be useful for several types of solid tumor cancers. Several solid tumor cancers have viral etiology, whereby viral specific ctDNA can be used for detection (see the viral section below for more details).

Isolation of ctDNA

The amount and quality of ctDNA are very important for reproducible and accurate assays. Multiple techniques have been used for the isolation of ctDNA, including laboratory developed methods and commercial kits. Unfortunately, there are currently no set standards for isolation or quantification. Isolation of ctDNA is logistically laborious, and there are issues with reproducibility among and within runs.[340] More accurate, robust, and consistent isolation of ctDNA is needed, particularly with small volumes of serum/plasma. Also, these procedures must be standardized to simplify performance across multiple centers and become established laboratory regulations. Both commercial kits and in-house approaches for ctDNA isolation may involve proprietary chemicals and reagents that require centrifugation, bead hybridization, and/or column separations. Yields from different assays can vary considerably.

Quantification of ctDNA

There are many assays for quantification of specific ctDNA fragments. Usually, qPCR assays that incorporate hydrolysis or other types of probes, peptide or locked nucleic acid clamping, and digital PCR are used. However, because only small amounts of ctDNA may be recovered from plasma or serum, reproducible quantification is difficult, accuracy may be low, and false-negative ctDNA results may occur. The issue then becomes how much ctDNA to use in individual assays, and that requires overall quantification of ctDNA. Quantification methods for purified ctDNA include UV spectrophotometry, electrophoresis, double-stranded (eg, PicoGreen) or single-stranded (eg, OligoGreen) DNA assays (Thermo-Fisher), qPCR, and digital PCR. Isolated ctDNA can vary in purity and be double-stranded or single-stranded, and it may be complexed to histones, all of which can affect quantification. UV spectrophotometry can assess protein contamination, but it is typically not sensitive enough for ctDNA quantification. Electrophoresis also lacks sensitivity, and quantification is difficult. If enough ctDNA is available, multiple assays can be performed to increase confidence, such as digital PCR, qPCR, and/or PicoGreen assays. ctDNA quantification for downstream assays remains an area that needs improvement.

Massive Parallel Sequencing

Massive parallel sequencing (MPS) can analyze the full sequence of multiple genes at once. MPS provides wider coverage of genes and their aberrations with verification of the ctDNA detection accuracy. MPS analysis of ctDNA continues to improve,[341-345] but it may not be as sensitive as qPCR assays for specific gene targets that have been well optimized for low copy detection. Nevertheless, the development of more sensitive MPS techniques and targeted qPCR with reduced costs will continue to make ctDNA more useful in companion diagnostics.

MPS provides much wider genomic coverage than specific qPCR assays. For example, in an exome analysis of ctDNA, thousands of DNA genes are analyzed for all sequence variants that may be present. The MPS approach allows analysis of many genomic aberrations, including deletions, copy number changes, and mutations, providing a comprehensive profile of ctDNA. Although MPS requires optimization, only one gene panel can cover many tumors, in contrast to multiple qPCR assays that target specific genes and must be optimized for each tumor type. As MPS becomes more cost efficient, targeted gene panels and whole exon sequencing of ctDNA may reveal many specific variants to guide targeted therapies in clinical trials.

The current issues with MPS are the high costs and laborious preparations of libraries from small amounts of extracted DNA from serum or plasma. Bioinformatics also remains a problem, with many different methods for analyzing and reporting results. Reporting needs to be better standardized for easy clinical interpretation. Digital single-molecule sequencing with digital informatics is a new approach that allows for targeted gene sequencing at high sensitivity and specificity for ctDNA analysis.[346-350]

Plasma Versus Serum for ctDNA

ctDNA has been assessed both in plasma and serum of cancer patients. Fetal circulating DNA testing has focused on plasma with CLIA approval using stabilizing DNA blood tubes (cell-free DNA BCT, Streck, Omaha, Neb.) for consistent collection and preservation at room temperature for up to 72 hours before extraction. In general, serum has more ctDNA than plasma and has the advantage of no interference from preservatives. However, plasma is preferred over serum because clotting in serum causes trapped leukocytes to release DNA that can interfere in ctDNA detection. Collection of plasma for ctDNA requires expedient centrifugation and filtration of contaminant cells such as leukocytes and CTCs prior to cryopreservation to prevent cells from breaking up and releasing DNA.

The field of ctDNA analysis takes advantage of well-developed assays designed for tumor tissues. However, the extracted amount of ctDNA can be quite low, and, depending on the particular assay, low copy numbers of ctDNA may not be detectable. Uncertain detection limits plague ctDNA analysis. Low copy number ctDNA is particularly problematic in methylation assays when bisulphite treatment is required. On the other hand, ctDNA assays may be easier to interpret than tumor assays because of limited tumor-derived DNA background.

Early assays required large volumes of blood, varying from 0.5 to 10 mL, but improvements in sensitivity and efficiency have lowered the amounts needed. Most current ctDNA assays require less than 2 mL of plasma or serum as the extraction efficiency of methods and laboratories improve. With the demand of various clinical blood tests in cancer patients, lower blood requirements are more practical and also better tolerated by patients.

ctDNA is also present in body fluids other than blood. Urine is an important fluid to detect ctDNA for urogenital cancers.[351-353] However, degradation of DNA in urine is a concern. Cerebrospinal fluid (CSF) is also a potential source for detection of brain tumor ctDNA mutations.[332] CSF in general has less proteins, lipids, and cells compared to blood, which allows for easier extraction of ctDNA. However, the amount of ctDNA in CSF is far less than in blood. ctDNA can also be monitored in pleural and peritoneal fluids for cancer-related progression. These latter fluids can provide a large source of cell-free DNA for very comprehensive analysis of genomic aberrations before and after treatment.

Tumor-Specific ctDNA Mutations

Tumor-specific mutations in ctDNA are most commonly studied in cancer patients. Tumor mutations in ctDNA become increasingly important when the mutations have approved targeted therapies as companion diagnostics and are associated with therapeutic response.[354-376] ctDNA mutations can easily be assessed with high sensitivity and specificity by qPCR or MPS.[346,377] TABLE 9.1 lists the mutations detected in ctDNA for different cancers. Most mutations can now be detected in ctDNA as well as tumor tissue. Using digital sequencing (Guardent Health, Redwood City, Calif.), complete exomes of genes for ctDNA mutations and amplifications have been reported.[350,366,378]

In NSCLC, mutations in *EGFR* include deletions in exon 19, affecting the amino acid motif ELREA (delE746-A750) and an amino acid substitution (L858R) in exon 21.[379-383] These genomic aberrations result in constitutively activated EGFR tyrosine kinase. These EGFR activation changes are responsive to EGFR tyrosine kinase inhibitor (TKI) therapies (gefitnib, erlotinib) in up to 78% of patients, as shown in

TABLE 9.1 Detection of Circulating Tumor DNA (ctDNA) Mutations in Various Cancers

Cancer	Gene	References
Breast	PIK3CA, BMI1, SMC4, FANCD2, MED1, ATM, PDGFRA, GAS6, TP53, ESR1, PTEN, AKT1, IDH2 SMAD4	355, 356, 358, 364, 366, 369, 370, 458
Ovarian	RB1, ZEB2, MTOR, CES4A, BUB1, PARP8	458
Non–small cell lung cancer	KRAS, EGFR, TP53, NFKB1	359, 364, 366, 368, 376, 379–385, 458
Melanoma	BRAFV600, TP53, E2H2	346, 358, 371
Pancreatic	KRAS, TP53, APC, SMAD4, FBXW7	349, 357
Hepatocellular carcinoma	TP53	459
Colorectal cancer and other gastrointestinal malignancies	KRAS, WTKRAS, NRAS, BRAF, EGFR, cKIT, PIK3CA	347, 354, 356, 363, 365, 372–375, 378
Cervical	KRAS, TP53	460

multiple studies. Recently, detection of *EGFR* mutations in ctDNA has mirrored the *EGFR* mutation status in associated lung tumors.[378,384] In lung cancer patients treated with erlotinib, the presence of *EGFR* L858R ctDNA was associated with a decrease in OS; however, exon 19 deletions detected in ctDNA were associated with increased OS. When serial peripheral blood sampling for ctDNA was assessed, acquired drug resistance correlated with the appearance of *EGFR* T290M.[366] Anti-EGFR drug therapies can be monitored by assessing different ctDNA mutations.[385] Currently, these companion ctDNA diagnostics are being verified in the clinic. Assessment of ctDNA mutations in blood during treatment and follow-up provide information to the clinician about new mutations that were not present in the original tumor. New strategies for initiating targeted therapies based on ctDNA analysis (Fig. 9.21) may replace biopsy of metastases for new gene mutations arising during follow-up.

Gene Amplification in ctDNA

Amplification of *ERBB2*, *BRAF*, and *MET* have been observed in various cancers.[386-389] More types of tumors with these amplifications and more types of amplifications in various tumors will likely be found as techniques to assess gene amplification in ctDNA develop. For example, androgen receptor amplification in prostate cancer can be detected from ctDNA and may be a resistance mechanism for patients on enzalutamide and abiraterone.[389] TABLE 9.2 lists amplifications detected in different cancers. As MPS becomes more sensitive for the detection of gene amplification, it will be easier to detect amplifications from ctDNA assays.

Copy Number Variation in ctDNA

MPS can measure copy number variants (CNVs) in ctDNA.[344,359,390-392] This new strategy for ctDNA analysis is currently very costly. There are multiple CNVs on specific chromosome arms that are frequent for individual types of cancers. CNVs may cover multiple genes or noncoding sequences and can be valuable biomarkers of tumor progression. For example, 9p1q deletions are found frequently in melanoma and other solid tumors. The main technical issues are, again, low amounts of ctDNA and bioinformatics analysis.

Microsatellite Instability and Loss of Heterozygosity in ctDNA

Microsatellite instability revealed as a loss of heterogeneity (LOH) is common in many cancers and occurs in specific chromosome regions.[393-400] LOH of tumors revealed by ctDNA can correlate with clinical outcome. However, technical issues in assessing LOH of specific chromosome loci can make some assays difficult to interpret and others uninformative. To assess LOH, multiple primer sets to specific loci have to be run to obtain an accurate profile. PCR products are conventionally evaluated by gel electrophoresis with variable sensitivity. Recently, more accurate assays performed by capillary electrophoresis arrays with discrimination down to a few base pairs have appeared. ctDNA LOH analysis now has strong prognostic and diagnostic significance in many different cancer types (TABLE 9.3). It remains to be seen if ctDNA LOH can be assessed through MPS.

Methylated ctDNA

Methylated ctDNA is detectable in cancer patients and has utility in both early detection and treatment monitoring. Typically, gene promoter regions contain CpG sites that are hypermethylated or hypomethylated to control transcription. In cancer, the promoter regions of tumor suppressor genes may be hypermethylated to limit transcription, and oncogene transcription may be increased by hypomethylation. In the blood of cancer patients, methylated CpG ctDNA can be detected and used for prognosis and as markers of treatment response.[208,285,401-427] Cancer genes that are known to be controlled by methylation are listed in TABLE 9.4. However, some promoter CpG sites can be nonspecifically hypermethylated and not involved in gene transcription. Therefore, one has to carefully select the regulatory CpG sites in the promoter region for analysis. More than one site can be regulatory, and methylation can vary throughout the gene promoter. This can cause difficulty in interpreting methylated ctDNA results. During tumor progression, promoters typically become hypermethylated. Apparent hypomethylated CpG sites in ctDNA may result from normal cell shedding of hypomethylated DNA. Normal cell regulatory genes can also change in methylation status and be used for methylated ctDNA analysis. An example is estrogen receptor (ER)

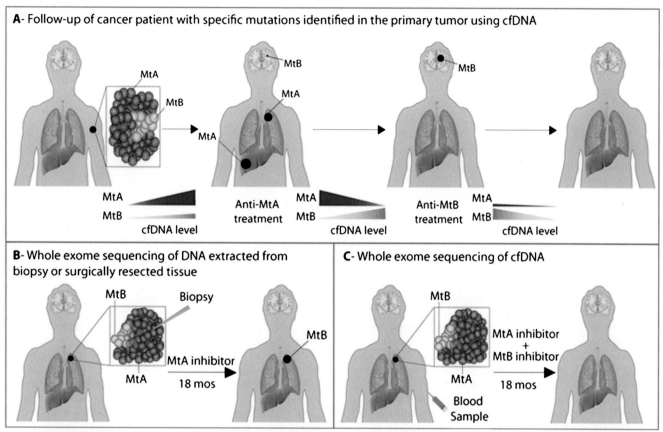

FIGURE 9.21 Examples of ctDNA analysis in cancer patients with multiple mutations. **A,** A melanoma patient presents a heterogeneous primary tumor with both mutation profile A (MtA) and mutation profile B (MtB). Initial circulating free DNA (cfDNA) assessment indicates a higher level of MtA as a consequence of being the predominant tumor subclone. After starting with anti-MtA therapy, MtA clones decrease, while MtB may increase. If the MtB burden increases, anti-mtB treatment can be initiated, potentially clearing both clones by personalized treatment. **B,** A resected tumor of a patient with non–small cell lung carcinoma is analyzed, revealing the targetable mutation L858R in *EGFR* but missing a minor clone harboring *BRAF* V600E. After 1 year of anti-EGFR treatment, the patient relapses with a tumor harboring *BRAF* V600E. **C,** Whole genome sequencing of cfDNA identifies both L858 *EGFR* and an underrepresented subclone with a V600E *BRAF* mutation. In this diagnostic setting, both targetable mutations could be considered for treatment selection and potential cure.

TABLE 9.2 Cell-Free DNA Copy Number Variants (CNVs) and Amplification in Various Cancers

Cancer	CNV	References
Prostate	*AR, CYP17A1* CNV	342, 387
Breast	*HER2* amp, *ERBB2, CDK6*	384, 387
Esophageal	*CCND1* amp	386
Liver	Multiple chromosomes	390
Colorectal	*ERBB2, CDK6, EGFR, BRAF, MYC, SMO*	347, 387

TABLE 9.3 Circulating Tumor DNA (ctDNA) Loss of Heterozygosity (LOH) Sites in Various Cancers

Cancer	LOH Site	References
Non–small cell lung cancer	D9S286, D9S942, GATA49D12, D13S170	461
Breast	D8S321, D1S228, D14S51, D14S62, D17S855	395, 397, 399, 400
Melanoma	12q22-23	393, 396, 398
Colorectal	D18S59, D18S1140, D18S976	462
Ovarian	D10S1765, D13S218	463
Prostate	D9S171, D10S591, D8S261, D6S286, D8S262, D18S70	408

TABLE 9.4 Circulating Tumor DNA (ctDNA) Methylation in Various Cancers

Cancer	Methylation Gene	References
Colorectal	MGMT, P16, RAR-ß2, RASSFIA, APC, SOX17, SEPT9	208, 405–407, 416
Breast	APC, RASSF1, DAPESR1, SOX17, PCDH10, PAX 5 RARß2, APC, GSTP1	411, 417, 418, 424, 427
Gastric	XAF1, CKIT, PDGFRA	363, 412
Melanoma	AIM1, LINE1, MGMT, ESR1, RAR-ß2, RASF1	285, 401–405
Lung	RAR-ß2, PAX5, MGMT, DCC, BRMS1	415, 420, 422–424
Prostate	RASSF1A, GSTP1, RAR-ß2, ARHGAP6	408, 410, 422, 426
Thyroid (papillary thyroid carcinoma)	BRAF (V600E)	369
Liver	LINE1, RASSF1	409, 428
Ovary	KRAS, RASSF1, CALCA, EP300, H1C1, PAX5	420, 421, 460
Brain	MGMT	425

methylation, which increases in advanced melanoma patients and can be detected in ctDNA.[403]

Recently, methylated SEPT9 ctDNA has been detected in early-stage colon cancer, specifically in asymptomatic average-risk individuals undergoing screening.[405–407] These studies included multiple colorectal cancer patients and healthy cohorts and are a good example of the potential for ctDNA analysis to identify early-stage cancer and serve as a cancer screening test. The assay still needs improvement before widespread clinical use because it requires up to 15 mL of blood, and the sensitivity for early-stage colorectal cancer and precancerous adenomas is not optimal. Interestingly, sensitivity of SEPT9 ctDNA detection is not accurate in late-stage CRC.[405,406]

Some methylation ctDNA biomarkers occur in more than one type of cancer. Some common methylated ctDNA markers found in multiple types of cancer include *RASSF1*, *ESR1*, *CDKN2A*, *RARB*, and *MGMT* (see TABLE 9.4). This limits the specificity of methylation ctDNA biomarkers for early detection in patients without cancer. Nevertheless, methylated ctDNA biomarkers have potential utility in patients with a specific type of cancer. Identification of specific-functional hypermethylated CpG sites in the promoter region of genes may lead to better specificity. One of the drawbacks of methylated ctDNA is that promoter regions of genes can be hypermethylated in both benign and malignant cancers.

There are several types of assays that detect methylated ctDNA. These assays include quantitative methylation-specific PCR, mass spectrometry, pyrosequencing, and array hybridization. Each technique has pros and cons, but all require bisulphite conversion that is laborious and destroys a large fraction of the rare ctDNA during the process. The reliance on bisulphite treatment limits the sensitivity of methylation ctDNA assays until new approaches for methylation detection are developed. Typically, real-time PCR assays are more quantitative and sensitive than the other methods.

Noncoding ctDNA

With whole genome sequencing, repetitive noncoding sequences such as long interspersed nucleotide elements (LINEs) and short interspersed nucleotide elements (SINEs) are also potential biomarkers in multiple cancers.[331,339,402,428,429] There are many variants in the LINE family, such as the *LINE1* subfamily consisting of 6–8 kb CpG-rich DNA elements. Their function is not completely understood, but some act as retrotransposons and, when hypomethylated, increase tumor genomic instability. The human genome contains over 5 million copies of *LINE1*, representing almost 17% of the entire genome and are commonly found in ctDNA. The hypomethylated *LINE1* status in tumor tissues can be assessed as hypomethylated ctDNA. In normal tissue, the *LINE1* subfamily is usually hypermethylated, but in cancers such as melanoma, breast cancer, hepatocellular carcinoma, colorectal cancer, and prostate cancer, *LINE1* elements are hypomethylated and activated. SINEs, mostly *ALU* sequences, are assessed in ctDNA by a DNA integrity index that is described below.

Viral ctDNA

Several cancers have an etiology strongly linked to viral DNA integration. In these cancers, ctDNA can be tested for specific cell-free viral DNA (TABLE 9.5). The advantage of assessing viral ctDNA is the high specificity and detection sensitivity because of the unique viral sequences. The most notable and extensive assessment for clinical utility of viral ctDNA is Epstein Barr virus (EBV) detection in nasopharyngeal cancer (NPC).[430,431] NPC is one of the most common malignancies in southern China and Southeast Asia. Patients with metastatic disease have a poor prognosis. Plasma EBV DNA in NPC patients is a well-established ctDNA biomarker. Studies have shown its utility in assessing clinically concealed residual NPC after curative-intent radiotherapy. Furthermore, EBV

TABLE 9.5 Viral Circulating Tumor DNA (ctDNA) in Various Cancers

Cancer	Virus	References
Nasopharyngeal carcinoma	Epstein Barr virus	430–434
Liver	Hepatitis B virus	437, 459
Head and neck	Human papilloma virus	435, 436, 438

ctDNA is associated with treatment outcome prediction.[432] Plasma EBV ctDNA is useful in screening early-stage NPC and is relatively low in cost.[433] Radiotherapy and combination chemoradiotherapy are major treatment modalities for NPC, whereby in the setting of these treatments, plasma EBV ctDNA levels can help assess treatment response and outcome.[434] EBV ctDNA is one of the best examples of the clinical utility of ctDNA biomarkers.

Additional examples of viral ctDNA include human papilloma virus (HPV) for head and neck cancer[435,436] and cervical cancer, and hepatitis virus B for hepatocellular carcinoma.[437] In cervical cancer patients, the specificity of HPV ctDNA detection can be improved by analyzing the cell–viral DNA junction sequences in individual patients.[438] This involves assessing the specific cell genomic locus that the viral DNA integrates into. Although most current viral ctDNA assays are based on quantitative PCR, MPS assays have great potential in the future because they can cover larger DNA or RNA sequences. In the future, as new associations of cancer with virus etiology are found, new ctDNA assays may be developed to improve monitoring these viral-related cancers

ctDNA Integrity

Another approach to evaluating ctDNA is to examine its size, often referred to as "integrity." DNA fragment size can be used to determine the origin of cell-free DNA from either apoptotic or necrotic cells.[439-441] Programmed cell death under normal conditions is accompanied by an organized degradation of the DNA, which results in small and uniform DNA fragments (approximately 140–200 bp). In contrast, cell necrosis, which is a frequent event in solid tumors, generates a wide spectrum of DNA fragment lengths, including those greater than 200 bp as a consequence of digestion. The increased DNA integrity of cell-free DNA in different cancers[442] can be quantified as an integrity index that compares long to short cell-free DNA.

The DNA integrity index may be calculated in different ways. AC electrokinetics can separate cell-free DNA fragments by size (Biological Dynamics, Calif.). However, ctDNA integrity is usually assessed by qPCR of different fragment sizes, often from noncoding repeats such as *ALU*.[331,429,443] *ALU* sequences are approximately 300 bp in length and are found on all human chromosomes, being the most abundant transposable elements of the human genome. Umetani and colleagues developed a method using qPCR to directly measure the integrity of cell-free DNA without DNA purification.[441] The integrity index was calculated as the ratio between a 115-bp (small) amplicon and a 247-bp (large) amplicon of *ALU*. Another DNA integrity index distinguished between lymph node–negative and lymph node–positive breast cancer patients.[444] In this breast cancer study, the integrity index was derived from the quantification cycle of qPCR for two amplicons (one of 400 bp and the other 100 bp). DNA integrity was calculated as the ratio of the 400 bp fragment to total DNA and was clinically useful for detecting breast cancer progression.

The index of integrity has been successfully applied to detect multiple malignancies including breast, colorectal, esophagus, head and neck, kidney, melanoma, nasopharyngeal, periampullary, and prostate cancer (TABLE 9.6).[444-452] Persistence of a high DNA integrity index after radiotherapy is associated with poor prognosis in nasopharyngeal carcinoma patients.[453] The DNA integrity index is also associated with response to preoperative chemotherapy in rectal carcinoma patients.[454]

DNA integrity analysis is a ctDNA biomarker with applications in many cancers, although the specific calculations and cutoffs may vary between cancer types. Isolation methods may also bias enrichment of different size fragments. As new methods become available for quantification of a larger size range of ctDNA, these assays may increase in utility.

TABLE 9.6 Circulating Tumor DNA (ctDNA) Integrity in Various Cancers

Cancer	Integrity Sites	References
Breast	ALU	441, 449
Melanoma	Nonspecific	450
Colorectal	ALU, LINE1	429, 441, 452, 454
Esophageal	Nonspecific	451
Head and neck	Nonspecific	445, 448
Renal	Nonspecific	446
Bladder	Nonspecific	464
Brain	ALU	465
Prostate	Nonspecific	447
Ovarian	Nonspecific	466
Liver	Nonspecific	409

Half-Life of ctDNA

The half-life of ctDNA is important in the assessment of cancer patients. Understanding the rapid clearance and/or degradation from circulation is needed to interpret patient results. Unfortunately, it has been difficult to determine the actual half-life of ctDNA in cancer patients. In fetal cell–free DNA analysis, clearance from maternal blood occurs within 30 minutes.[455] However, in cancer patients, we do not know how long ctDNA survives in the circulation or the kinetics of its clearance. Long-fragment ctDNA (>250 bp) is cleared more rapidly than shorter fragments. Important factors may include degradation by enzymes in blood, binding to blood lipids/proteins, binding to nucleoprotein, uptake by leukocytes in blood and tissues, and clearance by normal physiological mechanisms in the liver and kidney. Tumors located in highly vascular organs or with active angiogenesis will have higher access and release into the systemic circulation. In addition, ctDNA can be taken up by cells in the surrounding microenvironment, as well as be incorporated into vesicles of cancer cells and released as exosomes.[456,457] Although not strictly free ctDNA, exosomes are an additional source of ctDNA. This topic is difficult to address in patients because many physiologic events can control ctDNA half-life in blood. Further studies are needed in patients who are disease free by surgical resection and assessed for ctDNA before and several years after surgery.

ctDNA Detection by Single-Marker Versus Multimarker Approach Assays

The approach of multimarker versus single-marker assays is clearly essential in most cancers. Because of tumor heterogeneity and continual genomic/epigenomic changes during tumor progression, multiple-marker assays are often necessary for diagnosis and monitoring. Furthermore, therapeutics can modify the molecular profile of tumors, particularly when drug resistance develops. Most patients with distant metastasis have molecular heterogeneity between clinical and subclinical metastasis. Multiple molecular markers and platforms may be necessary to study different genomic/epigenomic aberrations and/or multiple targets within an aberration. Combination drug therapy is often required because monotherapy in many advanced-stage cancers is not sufficient to control long-term tumor progression in melanoma, breast cancer, and colorectal cancer. More information is needed on multiple events occurring before, during, and after treatment. Because signal pathways can change during tumor progression and resistance, more complete profiling of specific gene aberrations may be needed (eg, therapies that target tyrosine kinase activation where resistance development frequently occurs). Cancer genomic changes are complex. It is not surprising that multiple markers and assays are needed for adequate study and understanding.[339] Several studies have been conducted whereby multiple markers have shown better utility than single markers, particularly in melanoma. As we know, cancer is quite heterogeneous, so in genomic changes it is more important to have a panel of markers for efficient detection. For example, very few cancers have a single mutation at higher than 50% frequency.

ctDNA in Clinical Oncology Trials

ctDNA testing is now incorporated into many clinical phase II and III trials, particularly for targeted therapeutics to mutations and fusions. To date, the most reported ctDNA companion diagnostics is *EGFR* mutation analysis in ctDNA, particularly in NSCLC.[368] Many clinical cancer trials have shown the potential utility of ctDNA *EGFR* mutations in monitoring treatment response. As more targeted therapies to particular mutations or fusions are developed, more companion ctDNA assays will likely be incorporated into clinical trials. Companion ctDNA diagnostics can either be specific to the targeted gene or to other tumor biomarkers that correlate to patient response. The merit of MPS ctDNA analysis is that tumor heterogeneity with changing mutations can be monitored in real time and associated with emerging drug resistance. This approach identifies new mutations as they arise in the tumor during treatment or follow-up that may provide guidance for decisions to stop treatment and/or initiate new targeted drug(s). Noninvasive ctDNA assays are a welcome replacement for repeated tumor sampling by core biopsies or fine-needle aspirations.

Future Challenges for ctDNA Analysis

The future of ctDNA lies in the hands of investigators in academia and the pharmaceutical and biotechnology industry to validate assays for CLIA and FDA approval for clinical diagnostics. ctDNA assays provide a unique opportunity to better monitor cancer patients with gene-targeted therapies. However, continued validation in well-defined multicenter clinical trials is necessary. Its technical success depends on the robust, consistent, easy, and efficient isolation of ctDNA in small amounts of blood.

In addition to technical improvements, quantification and reporting methods that allow oncologists to make decisions in patient treatment are critical. Standards for ctDNA will be essential to set quality controls, consistent quantification, and reporting among laboratories. Until regulatory groups start implementing these standards, clinical implementation will remain chaotic. The complexity inherent in cancer has resulted in many approaches to ctDNA measurement. MPS will also generate new challenges and opportunities for ctDNA analysis.

> **POINTS TO REMEMBER**
>
> **ctDNA**
> - ctDNA can identify DNA changes in the primary tumor, including mutations, loss of microsatellite heterozygosity, methylation of gene promoter regions, DNA integrity, amplification, copy number variation, and the presence of viral DNA.
> - ctDNA assays can be performed on both serum and plasma.
> - There are multiple types of detection assays for ctDNA. Some are highly dependent on the form of ctDNA. Most are based on quantitative PCR.
> - Massively parallel sequencing (MPS) in recent years has elevated the level of ctDNA analysis to include whole genome analysis of mutations and copy number variations.
> - ctDNA mutation analysis can be successful companion diagnostics for targeted therapies (eg, EGFR mutations in NSCLC treatment).

COMBINED DISCUSSION OF CTC AND ctDNA

Comparison Between CTCs and ctDNA

Both CTC and ctDNA technology and applications have greatly advanced in the last decade. Because liquid biopsy of cancer patients can evaluate either CTC or ctDNA, what are the merits of each? Both assays provide some overlapping information as well as unique information for specific cancers. CTC detection provides real-time information on tumor spread. If CTCs are sampled repetitively over days to weeks and are increasing in number, this may indicate that the tumor is progressing and active metastasis is occurring. Assessing the presence of CTCs in blood may be important before, during, and after therapy because their presence suggests potential metastasis and colonization in distant sites. Continued development of better markers for CTCs that predict survival and establishment of metastasis is important. Molecular characterization of CTCs in addition to quantification will provide more guidance for treatment decisions. Markers expressed by CTCs may allow prediction of what organ site is likely to be colonized.

ctDNA detection indicates there is tumor present, but unlike CTCs, it does not indicate if it is spreading through the circulation. Consistent detection or increasing concentrations over time suggest that the tumor is active (progressing or rapidly renewing cell turnover). ctDNA assays require less blood, typically less than 2 mL, whereas most CTC assays require 7 to 10 mL. When the demand for other blood tests is high, the additional volume required by CTC assays can be a concern. Assays that require less blood will have better patient compliance, particularly for repetitive phlebotomy during treatment and follow-up. TABLE 9.7 compares the relative merits of CTC and ctDNA analyses.

TABLE 9.7 Comparison Between Circulating Tumor DNA (ctDNA) and Circulating Tumor Cells (CTCs)

ctDNA	CTCs
Minimal amount of blood (<2 mL)	Larger blood draw (>7 mL)
Logistically easier to assay due to purity of ctDNA	Logistically difficult to assay due to small number of cells and purity
More informative	Less informative for the sample used
Easily quantifiable due to high-sensitivity molecular assays	Difficult to quantify due to heterogeneity and cell recovery
Easier to assess tumor heterogeneity	Tumor heterogeneity is difficult to assess
Identifies presence of tumor	Identifies presence of tumor
Cannot identify metastasis occurring	Can identify metastasis occurring
Highly diagnostic	Diagnosis not dependable clinically
Prognostic	Prognostic
Can identify tumor resistance during therapy	Poor in identifying tumor resistance during therapy

TABLE 9.8 Comparison Between Circulating Tumor (ctDNA) and Solid Tumor Biopsy

ctDNA	Tumor Biopsy
Repetitive analysis	Minimal biopsy material obtainable
Minimally invasive	Highly invasive
More representative genomic information	Variation from biopsy sampling error
Can be repeated as needed	Difficult to repeat
Can be assessed from small blood draw	Not practical in all anatomical sites
Can monitor tumor heterogeneity	Limited assessment of tumor heterogeneity
Better patient compliance	Poor patient compliance
Cost-effective	Expensive

Quality Control and Logistic Issues of CTCs and ctDNA

CTC and ctDNA assays need better quality control and standardization to provide more consistent reporting and comparisons. There are many different ways to detect and procure CTCs, as well as different downstream assays. The CTC field has attracted many new companies in the last 5 years. This is important and supportive for continued advancement, but each company tends to brand its specific techniques and reports, resulting in confusion and poor standardization among companies. The development of quality controls and reporting standards is needed to allow comparisons and determine clinical utility. As better isolation methods develop, downstream analysis has improved and become more informative. Blood CTC analysis can be a tremendous advantage over the necessary sampling error of tumor biopsies. However, because CTCs are heterogeneous, all detected CTCs may not be relevant in assessing metastatic potential. Another issue for CTCs is the volume of blood needed (>5 mL) to accurately assess CTCs, particularly in early-stage cancer patients.

Quality control issues in ctDNA must be improved before wider clinical use. Blood tube collection, transport to a reference or hospital laboratory, ctDNA reference standards, isolation, reproducibility, robustness, accuracy, and reporting format are all issues that need to be addressed. There are currently several clinically validated ctDNA assays available to oncologists.

Prenatal plasma cell–free DNA testing has led the field of circulating nucleic acid assessment and has shown how efficient and useful such testing can be (see Chapter 10). ctDNA assays need to model this approach for cross-validated comparisons of different approaches. Similar to currently accepted blood cancer biomarkers, multicenter validation and reporting must be performed for ctDNA assay acceptance and wide use.

Liquid Biopsy Versus Tumor Biopsy

A major objective in the field is to determine if CTC and ctDNA analysis can replace tumor biopsies. Liquid biopsies are minimally invasive and can be performed repetitively without the complications of tumor biopsy. Tumor biopsies cannot always be performed because of the tumor site location, and repetitive samples may not be sufficient or representative of the tumor. Repeat ctDNA molecular analysis may be more consistent than tumor biopsies because of tumor sampling error. CTC analysis of genomic aberrations can now be performed with the same potential quality as a tumor biopsy such as a fine-needle aspiration. While initial diagnosis is best performed by tumor biopsy and conventional histopathology, liquid biopsies may in the future be used once validated for targeted therapy stratification and monitoring patients. With the advent of MPS, ctDNA and CTC assays have become more informative and offer better homogeneity than a solid tumor for sequencing analysis. As liquid biopsies become better validated, they may replace some tumor biopsies. ctDNA and CTC analysis may enable more rapid decisions for treatment stratification and treatment modifications during therapy (TABLE 9.8). A newly formed European consortium, CANCER-ID (http://www.cancer-id.eu/), consisting of 33 partners from 13 countries, aims to establish standard procedures for clinical validation of liquid biopsy biomarkers and to evaluate the clinical utility of ctDNA and CTC in several types of carcinomas, including breast and lung. The objective of the CANCER-ID consortium is to have multicenter prospective trials to evaluate techniques for assessing CTC and ctDNA in different laboratories using specimens from different clinical sites.

CTCs and ctDNA Combined as Companion Blood Biomarkers

The combination of both CTCs and ctDNA as a liquid biopsy may augment the accurate assessment of cancer patients.

Both can be used together to provide informative patient results.[285] In addition, the use of liquid biopsy (CTC and/or ctDNA) may be combined with imaging and blood protein/analyte tumor biomarkers to give a more comprehensive evaluation of the patient status. The real challenge is to combine all the information into an interpretable result for the clinician to enable decisions on treatment.

CONCLUSIONS

The development of both CTC and ctDNA liquid biopsies provides many new opportunities for implementation over the next 5 years. These assays, upon validation, may provide better clinical management in treatment monitoring and understanding of tumor progression for many different types of cancers. As new biosensors, molecular procedures, and molecular devices develop in both CTC and ctDNA analysis, the sensitivity and specificity, as well as logistics, will improve. With the resource of the cancer gene atlas (TCGA NIH USA: http://cancergenome.nih.gov/), there is a significant amount of sequencing data of multiple cancer types that can be translated into new CTC and ctDNA targets in the future.

REFERENCES

1. Alix-Panabières C, Pantel K. Clinical applications of circulating tumor cells and circulating tumor DNA as liquid biopsy. *Cancer Discov* 2016;**6**:479–91.
2. Heitzer E, Auer M, Ulz P, et al. Circulating tumor cells and DNA as liquid biopsies. *Genome Med* 2013;**5**:73.
3. Lianidou ES. Molecular characterization of circulating tumor cells: holy Grail for personalized cancer treatment? *Clin Chem* 2014;**60**:1249–51.
4. Ashworth T. A case of cancer in which cells similar to those in the tumours were seen in the blood after death. *Med J Aust* 1869:146–7.
5. Braun S, Vogl FD, Naume B, et al. A pooled analysis of bone marrow micrometastasis in breast cancer. *N Engl J Med* 2005;**353**:793–802.
6. Klein CA. Parallel progression of primary tumours and metastases. *Nat Rev Cancer* 2009;**9**:302–12.
7. Joosse SA, Gorges TM, Pantel K. Biology, detection, and clinical implications of circulating tumor cells. *EMBO Mol Med* 2015;**7**:1–11.
8. Pantel K, Alix-Panabieres C, Riethdorf S. Cancer micrometastases. *Nat Rev Clin Oncol* 2009;**6**:339–51.
9. Cristofanilli M, Budd GT, Ellis MJ, et al. Circulating tumor cells, disease progression, and survival in metastatic breast cancer. *N Engl J Med* 2004;**351**:781–91.
10. Lianidou ES, Mavroudis D, Georgoulias V. Clinical challenges in the molecular characterization of circulating tumour cells in breast cancer. *Br J Cancer* 2013;**108**:2426–32.
11. Lianidou ES, Markou A, Strati A. Molecular characterization of circulating tumor cells in breast cancer: challenges and promises for individualized cancer treatment. *Cancer Metastasis Rev* 2012;**31**:663–71.
12. Van der Auwera I, Peeters D, Benoy IH, et al. Circulating tumour cell detection: a direct comparison between the CellSearch™ System, the AdnaTest and CK-19/mammaglobin RT-PCR in patients with metastatic breast cancer. *Br J Cancer* 2010;**102**:276–84.
13. Strati A, Kasimir-Bauer S, Markou A, et al. Comparison of three molecular assays for the detection and molecular characterization of circulating tumor cells in breast cancer. *Breast Cancer Res* 2013;**15**:R20.
14. Plaks V, Koopman CD, Werb Z. Cancer. Circulating tumor cells. *Science* 2013;**341**:1186–8.
15. Tibbe AG, Miller MC, Terstappen LW. Statistical considerations for enumeration of circulating tumor cells. *Cytometry A* 2007;**71**:154–62.
16. Lowes LE, Allan AL. Recent advances in the molecular characterization of circulating tumor cells. *Cancers (Basel)* 2014;**6**:595–624.
17. Alix-Panabieres C, Pierga JY. Circulating tumor cells: liquid biopsy. *Bull Cancer* 2014;**101**:17–23.
18. Broersen LH, van Pelt GW, Tollenaar RA, et al. Clinical application of circulating tumor cells in breast cancer. *Cell Oncol (Dordr)* 2014;**37**:9–15.
19. Lianidou ES, Markou A. Circulating tumor cells in breast cancer: detection systems, molecular characterization, and future challenges. *Clin Chem* 2011;**57**:1242–55.
20. Miller MC, Doyle GV, Terstappen LW. Significance of circulating tumor cells detected by the CellSearch™ system in patients with metastatic breast colorectal and prostate cancer. *J Oncol* 2010;**2010**:617421.
21. Allard WJ, Matera J, Miller MC, et al. Tumor cells circulate in the peripheral blood of all major carcinomas but not in healthy subjects or patients with nonmalignant diseases. *Clin Cancer Res* 2004;**10**:6897–904.
22. Riethdorf S, Fritsche H, Muller V, et al. Detection of circulating tumor cells in peripheral blood of patients with metastatic breast cancer: a validation study of the CellSearch™ system. *Clin Cancer Res* 2007;**13**:920–8.
23. Bednarz-Knoll N, Alix-Panabieres C, Pantel K. Plasticity of disseminating cancer cells in patients with epithelial malignancies. *Cancer Metastasis Rev* 2012;**31**:673–87.
24. Coumans FA, Ligthart ST, Terstappen LW. Interpretation of changes in circulating tumor cell counts. *Transl Oncol* 2012;**5**:486–91.
25. Ligthart ST, Coumans FA, Bidard FC, et al. Circulating tumor cells count and morphological features in breast, colorectal and prostate cancer. *PLoS ONE* 2013;**8**:e67148.
26. Barradas AM, Terstappen LW. Towards the biological understanding of CTC: capture technologies, definitions and potential to create metastasis. *Cancers (Basel)* 2013;**5**:1619–42.
27. Truini A, Alama A, Dal Bello MG, et al. Clinical applications of circulating tumor cells in lung cancer patients by CellSearch™ system. *Front Oncol* 2014;**4**:242.
28. Beije N, Jager A, Sleijfer S. Circulating tumor cell enumeration by the CellSearch™ system: the clinician's guide to breast cancer treatment? *Cancer Treat Rev* 2015;**41**:144–50.
29. Deng G, Herrler M, Burgess D, et al. Enrichment with anti-cytokeratin alone or combined with anti-EpCAM antibodies significantly increases the sensitivity for circulating tumor cell detection in metastatic breast cancer patients. *Breast Cancer Res* 2008;**10**:R69.
30. Pamme N. On-chip bioanalysis with magnetic particles. *Curr Opin Chem Biol* 2012;**16**:436–43.
31. Ghazani AA, Castro CM, Gorbatov R, et al. Sensitive and direct detection of circulating tumor cells by multimarker micro-nuclear magnetic resonance. *Neoplasia* 2012;**14**:388–95.

32. Hager G, Cacsire-Castillo TD, Schiebel I, et al. The use of a panel of monoclonal antibodies to enrich circulating breast cancer cells facilitates their detection. *Gynecol Oncol* 2005;**98**: 211–6.
33. Kruger W, Datta C, Badbaran A, et al. Immunomagnetic tumor cell selection—Implications for the detection of disseminated cancer cells. *Transfusion* 2000;**40**:1489–93.
34. Lacroix M. Significance, detection and markers of disseminated breast cancer cells. *Endocr Relat Cancer* 2006;**13**:1033–67.
35. Pantel K, Brakenhoff RH, Brandt B. Detection, clinical relevance and specific biological properties of disseminating tumour cells. *Nat Rev Cancer* 2008;**8**:329–40.
36. Satelli A, Brownlee Z, Mitra A, et al. Circulating tumor cell enumeration with a combination of epithelial cell adhesion molecule- and cell-surface vimentin-based methods for monitoring breast cancer therapeutic response. *Clin Chem* 2015;**61**:259–66.
37. Talasaz AH, Powell AA, Huber DE, et al. Isolating highly enriched populations of circulating epithelial cells and other rare cells from blood using a magnetic sweeper device. *Proc Natl Acad Sci USA* 2009;**106**:3970–5.
38. Powell AA, Talasaz AH, Zhang H, et al. Single cell profiling of circulating tumor cells: transcriptional heterogeneity and diversity from breast cancer cell lines. *PLoS ONE* 2012;**7**: e33788.
39. Earhart CM, Hughes CE, Gaster RS, et al. Isolation and mutational analysis of circulating tumor cells from lung cancer patients with magnetic sifters and biochips. *Lab Chip* 2014;**14**: 78–88.
40. Liu Z, Fusi A, Klopocki E, et al. Negative enrichment by immunomagnetic nanobeads for unbiased characterization of circulating tumor cells from peripheral blood of cancer patients. *J Transl Med* 2011;**9**:70.
41. Yang L, Lang JC, Balasubramanian P, et al. Optimization of an enrichment process for circulating tumor cells from the blood of head and neck cancer patients through depletion of normal cells. *Biotechnol Bioeng* 2009;**102**:521–34.
42. Zborowski M, Chalmers JJ. Rare cell separation and analysis by magnetic sorting. *Anal Chem* 2011;**83**:8050–6.
43. Naume B, Borgen E, Tossvik S, et al. Detection of isolated tumor cells in peripheral blood and in BM: evaluation of a new enrichment method. *Cytotherapy* 2004;**6**:244–52.
44. Allan AL, Vantyghem SA, Tuck AB, et al. Detection and quantification of circulating tumor cells in mouse models of human breast cancer using immunomagnetic enrichment and multiparameter flow cytometry. *Cytometry A* 2005;**65**:4–14.
45. Goodale D, Phay C, Postenka CO, et al. Characterization of tumor cell dissemination patterns in preclinical models of cancer metastasis using flow cytometry and laser scanning cytometry. *Cytometry A* 2009;**75**:344–55.
46. Lustberg MB, Balasubramanian P, Miller B, et al. Heterogeneous atypical cell populations are present in blood of metastatic breast cancer patients. *Breast Cancer Res* 2014;**16**:R23.
47. Lang JE, Scott JH, Wolf DM, et al. Expression profiling of circulating tumor cells in metastatic breast cancer. *Breast Cancer Res Treat* 2015;**149**:121–31.
48. Lu Y, Liang H, Yu T, et al. Isolation and characterization of living circulating tumor cells in patients by immunomagnetic negative enrichment coupled with flow cytometry. *Cancer* 2015;**121**:3036–45.
49. Neves RP, Raba K, Schmidt O, et al. Genomic high-resolution profiling of single CKpos/CD45neg flow-sorting purified circulating tumor cells from patients with metastatic breast cancer. *Clin Chem* 2014;**60**:1290–7.
50. Yang ZF, Ngai P, Ho DW, et al. Identification of local and circulating cancer stem cells in human liver cancer. *Hepatology* 2008;**47**:919–28.
51. Lalmahomed ZS, Kraan J, Gratama JW, et al. Circulating tumor cells and sample size: the more, the better. *J Clin Oncol* 2010;**28**: e288–9.
52. Mostert B, Sleijfer S, Foekens JA, et al. Circulating tumor cells (CTCs): detection methods and their clinical relevance in breast cancer. *Cancer Treat Rev* 2009;**35**:463–74.
53. Rosenberg R, Gertler R, Friederichs J, et al. Comparison of two density gradient centrifugation systems for the enrichment of disseminated tumor cells in blood. *Cytometry* 2002;**49**:150–8.
54. Wu LJ, Pan YD, Pei XY, et al. Capturing circulating tumor cells of hepatocellular carcinoma. *Cancer Lett* 2012;**326**:17–22.
55. Balic M, Dandachi N, Hofmann G, et al. Comparison of two methods for enumerating circulating tumor cells in carcinoma patients. *Cytometry B Clin Cytom* 2005;**68**:25–30.
56. Mohamed H, Murray M, Turner JN, et al. Isolation of tumor cells using size and deformation. *J Chromatogr A* 2009;**1216**: 8289–95.
57. Tan SJ, Lakshmi RL, Chen P, et al. Versatile label free biochip for the detection of circulating tumor cells from peripheral blood in cancer patients. *Biosens Bioelectron* 2010;**26**:1701–5.
58. Vona G, Sabile A, Louha M, et al. Isolation by size of epithelial tumor cells: a new method for the immunomorphological and molecular characterization of circulating tumor cells. *Am J Pathol* 2000;**156**:57–63.
59. Desitter I, Guerrouahen BS, Benali-Furet N, et al. A new device for rapid isolation by size and characterization of rare circulating tumor cells. *Anticancer Res* 2011;**31**:427–41.
60. Hosokawa M, Hayata T, Fukuda Y, et al. Size-selective microcavity array for rapid and efficient detection of circulating tumor cells. *Anal Chem* 2010;**82**:6629–35.
61. Zheng S, Lin H, Liu JQ, et al. Membrane microfilter device for selective capture, electrolysis and genomic analysis of human circulating tumor cells. *J Chromatogr A* 2007;**1162**:154–61.
62. Zheng S, Lin HK, Lu B, et al. 3D microfilter device for viable circulating tumor cell (CTC) enrichment from blood. *Biomed Microdevices* 2011;**13**:203–13.
63. Rostagno P, Moll JL, Bisconte JC, et al. Detection of rare circulating breast cancer cells by filtration cytometry and identification by DNA content: sensitivity in an experimental model. *Anticancer Res* 1997;**17**:2481–5.
64. Pinzani P, Salvadori B, Simi L, et al. Isolation by size of epithelial tumor cells in peripheral blood of patients with breast cancer: correlation with real-time reverse transcriptase-polymerase chain reaction results and feasibility of molecular analysis by laser microdissection. *Hum Pathol* 2006;**37**:711–8.
65. Hou JM, Krebs M, Ward T, et al. Circulating tumor cells as a window on metastasis biology in lung cancer. *Am J Pathol* 2011; **178**:989–96.
66. De Giorgi V, Pinzani P, Salvianti F, et al. Application of a filtration- and isolation-by-size technique for the detection of circulating tumor cells in cutaneous melanoma. *J Invest Dermatol* 2010;**130**:2440–7.

67. Lin HK, Zheng S, Williams AJ, et al. Portable filter-based microdevice for detection and characterization of circulating tumor cells. *Clin Cancer Res* 2010;**16**:5011–8.
68. Lim LS, Hu M, Huang MC, et al. Microsieve lab-chip device for rapid enumeration and fluorescence in situ hybridization of circulating tumor cells. *Lab Chip* 2012;**12**:4388–96.
69. Freidin MB, Tay A, Freydina DV, et al. An assessment of diagnostic performance of a filter-based antibody-independent peripheral blood circulating tumour cell capture paired with cytomorphologic criteria for the diagnosis of cancer. *Lung Cancer* 2014;**85**:182–5.
70. Vona G, Beroud C, Benachi A, et al. Enrichment, immuno-morphological, and genetic characterization of fetal cells circulating in maternal blood. *Am J Pathol* 2002;**160**:51–8.
71. Faugeroux V, Pailler E, Auger N, et al. Clinical utility of circulating tumor cells in alk-positive non-small-cell lung cancer. *Front Oncol* 2014;**4**:281.
72. Pailler E, Auger N, Lindsay CR, et al. High level of chromosomal instability in circulating tumor cells of ROS1-rearranged non-small-cell lung cancer. *Ann Oncol* 2015;**26**:1408–15.
73. Adams DL, Martin SS, Alpaugh RK, et al. Circulating giant macrophages as a potential biomarker of solid tumors. *Proc Natl Acad Sci USA* 2014;**111**:3514–9.
74. El-Heliebi A, Kroneis T, Zohrer E, et al. Are morphological criteria sufficient for the identification of circulating tumor cells in renal cancer? *J Transl Med* 2013;**11**:214.
75. Broche LM, Bhadal N, Lewis MP, et al. Early detection of oral cancer - Is dielectrophoresis the answer? *Oral Oncol* 2007;**43**(2):199–203.
76. Abonnenc M, Manaresi N, Borgatti M, et al. Programmable interactions of functionalized single bioparticles in a dielectrophoresis-based microarray chip. *Anal Chem* 2013;**85**:8219–24.
77. Peeters DJ, De LB, Van den Eynden GG, et al. Semiautomated isolation and molecular characterisation of single or highly purified tumour cells from CellSearch™ enriched blood samples using dielectrophoretic cell sorting. *Br J Cancer* 2013;**108**:1358–67.
78. Fabbri F, Carloni S, Zoli W, et al. Detection and recovery of circulating colon cancer cells using a dielectrophoresis-based device: kRAS mutation status in pure CTCs. *Cancer Lett* 2013;**335**:225–31.
79. Fernandez SV, Bingham C, Fittipaldi P, et al. TP53 mutations detected in circulating tumor cells present in the blood of metastatic triple negative breast cancer patients. *Breast Cancer Res* 2014;**16**:445.
80. Pestrin M, Salvianti F, Galardi F, et al. Heterogeneity of PIK3CA mutational status at the single cell level in circulating tumor cells from metastatic breast cancer patients. *Mol Oncol* 2015;**9**:749–57.
81. Bhagat AA, Bow H, Hou HW, et al. Microfluidics for cell separation. *Med Biol Eng Comput* 2010;**48**:999–1014.
82. Nagrath S, Sequist LV, Maheswaran S, et al. Isolation of rare circulating tumour cells in cancer patients by microchip technology. *Nature* 2007;**450**:1235–9.
83. Stott SL, Hsu CH, Tsukrov DI, et al. Isolation of circulating tumor cells using a microvortex-generating herringbone-chip. *Proc Natl Acad Sci USA* 2010;**107**:18392–7.
84. Ozkumur E, Shah AM, Ciciliano JC, et al. Inertial focusing for tumor antigen-dependent and -independent sorting of rare circulating tumor cells. *Sci Transl Med* 2013;**5**. 179ra47.
85. Yoon HJ, Kim TH, Zhang Z, et al. Sensitive capture of circulating tumour cells by functionalized graphene oxide nanosheets. *Nat Nanotechnol* 2013;**8**:735–41.
86. Kim TH, Yoon HJ, Stella P, et al. Cascaded spiral microfluidic device for deterministic and high purity continuous separation of circulating tumor cells. *Biomicrofluidics* 2014;**8**:064117.
87. Murlidhar V, Zeinali M, Grabauskiene S, et al. A radial flow microfluidic device for ultra-high-throughput affinity-based isolation of circulating tumor cells. *Small* 2014;**10**:4895–904.
88. Zhang Z, Nagrath S. Microfluidics and cancer: are we there yet? *Biomed Microdevices* 2013;**15**:595–609.
89. Khoo BL, Warkiani ME, Tan DS, et al. Clinical validation of an ultra high-throughput spiral microfluidics for the detection and enrichment of viable circulating tumor cells. *PLoS ONE* 2014;**9**:e99409.
90. Galletti G, Sung MS, Vahdat LT, et al. Isolation of breast cancer and gastric cancer circulating tumor cells by use of an anti HER2-based microfluidic device. *Lab Chip* 2014;**14**:147–56.
91. McFaul SM, Lin BK, Ma H. Cell separation based on size and deformability using microfluidic funnel ratchets. *Lab Chip* 2012;**12**:2369–76.
92. Hughes AD, Mattison J, Western LT, et al. Microtube device for selectin-mediated capture of viable circulating tumor cells from blood. *Clin Chem* 2012;**58**:846–53.
93. Kirby BJ, Jodari M, Loftus MS, et al. Functional characterization of circulating tumor cells with a prostate-cancer-specific microfluidic device. *PLoS ONE* 2012;**7**:e35976.
94. Harb W, Fan A, Tran T, et al. Mutational analysis of circulating tumor cells using a novel microfluidic collection device and qPCR assay. *Transl Oncol* 2013;**6**:528–38.
95. Warkiani ME, Guan G, Luan KB, et al. Slanted spiral microfluidics for the ultra-fast, label-free isolation of circulating tumor cells. *Lab Chip* 2014;**14**:128–37.
96. Autebert J, Coudert B, Champ J, et al. High purity microfluidic sorting and analysis of circulating tumor cells: towards routine mutation detection. *Lab Chip* 2015;**15**:2090–101.
97. Dieguez L, Winter MA, Pocock KJ, et al. Efficient microfluidic negative enrichment of circulating tumor cells in blood using roughened PDMS. *Analyst* 2015;**140**:3565–72.
98. Lim C, Hoon D. Circulating tumor cells: cancer's deadly couriers. *Phys Today* 2014;**67**:26–30.
99. Eifler RL, Lind J, Falkenhagen D, et al. Enrichment of circulating tumor cells from a large blood volume using leukapheresis and elutriation: proof of concept. *Cytometry B Clin Cytom* 2011;**80**:100–11.
100. Fischer JC, Niederacher D, Topp SA, et al. Diagnostic leukapheresis enables reliable detection of circulating tumor cells of nonmetastatic cancer patients. *Proc Natl Acad Sci USA* 2013;**110**:16580–5.
101. Saucedo-Zeni N, Mewes S, Niestroj R, et al. A novel method for the in vivo isolation of circulating tumor cells from peripheral blood of cancer patients using a functionalized and structured medical wire. *Int J Oncol* 2012;**41**:1241–50.
102. Zharov VP, Galanzha EI, Shashkov EV, et al. Photoacoustic flow cytometry: principle and application for real-time detection of circulating single nanoparticles, pathogens, and contrast dyes in vivo. *J Biomed Opt* 2007;**12**:051503.
103. Nedosekin DA, Sarimollaoglu M, Ye JH, et al. In vivo ultra-fast photoacoustic flow cytometry of circulating human melanoma cells using near-infrared high-pulse rate lasers. *Cytometry A* 2011;**79**:825–33.

104. Balic M, Rapp N, Stanzer S, et al. Novel immunofluorescence protocol for multimarker assessment of putative disseminating breast cancer stem cells. *Appl Immunohistochem Mol Morphol* 2011;**19**:33—40.
105. Pachmann K, Camara O, Kavallaris A, et al. Monitoring the response of circulating epithelial tumor cells to adjuvant chemotherapy in breast cancer allows detection of patients at risk of early relapse. *J Clin Oncol* 2008;**26**:1208—15.
106. Dago AE, Stepansky A, Carlsson A, et al. Rapid phenotypic and genomic change in response to therapeutic pressure in prostate cancer inferred by high content analysis of single circulating tumor cells. *PLoS ONE* 2014;**9**:e101777.
107. Bethel K, Luttgen MS, Damani S, et al. Fluid phase biopsy for detection and characterization of circulating endothelial cells in myocardial infarction. *Phys Biol* 2014;**11**:016002.
108. Wendel M, Bazhenova L, Boshuizen R, et al. Fluid biopsy for circulating tumor cell identification in patients with early- and late-stage non-small cell lung cancer: a glimpse into lung cancer biology. *Phys Biol* 2012;**9**:016005.
109. Marrinucci D, Bethel K, Kolatkar A, et al. Fluid biopsy in patients with metastatic prostate, pancreatic and breast cancers. *Phys Biol* 2012;**9**:016003.
110. Sieuwerts AM, Mostert B, Bolt-de VJ, et al. mRNA and microRNA expression profiles in circulating tumor cells and primary tumors of metastatic breast cancer patients. *Clin Cancer Res* 2011;**17**:3600—18.
111. Stathopoulou A, Ntoulia M, Perraki M, et al. A highly specific real-time RT-PCR method for the quantitative determination of CK-19 mRNA positive cells in peripheral blood of patients with operable breast cancer. *Int J Cancer* 2006;**119**:1654—9.
112. Stathopoulou A, Ntoulia M, Perraki M, et al. A highly specific real-time RT-PCR method for the quantitative determination of CK-19 mRNA positive cells in peripheral blood of patients with operable breast cancer. *Int J Cancer* 2006;**119**:1654—9.
113. Stathopoulou A, Gizi A, Perraki M, et al. Real-time quantification of CK-19 mRNA-positive cells in peripheral blood of breast cancer patients using the lightcycler system. *Clin Cancer Res* 2003;**9**:5145—51.
114. Xenidis N, Perraki M, Kafousi M, et al. Predictive and prognostic value of peripheral blood cytokeratin-19 mRNA-positive cells detected by real-time polymerase chain reaction in node-negative breast cancer patients. *J Clin Oncol* 2006;**24**:3756—62.
115. Xenidis N, Markos V, Apostolaki S, et al. Clinical relevance of circulating CK-19 mRNA-positive cells detected during the adjuvant tamoxifen treatment in patients with early breast cancer. *Ann Oncol* 2007;**18**:1623—31.
116. Xenidis N, Ignatiadis M, Apostolaki S, et al. Cytokeratin-19 mRNA-positive circulating tumor cells after adjuvant chemotherapy in patients with early breast cancer. *J Clin Oncol* 2009;**27**:2177—84.
117. Smirnov DA, Zweitzig DR, Foulk BW, et al. Global gene expression profiling of circulating tumor cells. *Cancer Res* 2005;**65**:4993—7.
118. Aktas B, Tewes M, Fehm T, et al. Stem cell and epithelial-mesenchymal transition markers are frequently overexpressed in circulating tumor cells of metastatic breast cancer patients. *Breast Cancer Res* 2009;**11**:R46.
119. Fehm T, Hoffmann O, Aktas B, et al. Detection and characterization of circulating tumor cells in blood of primary breast cancer patients by RT-PCR and comparison to status of bone marrow disseminated cells. *Breast Cancer Res* 2009;**11**:R59.
120. Sieuwerts AM, Kraan J, Bolt-de VJ, et al. Molecular characterization of circulating tumor cells in large quantities of contaminating leukocytes by a multiplex real-time PCR. *Breast Cancer Res Treat* 2009;**118**:455—68.
121. Fina E, Callari M, Reduzzi C, et al. Gene expression profiling of circulating tumor cells in breast cancer. *Clin Chem* 2015;**61**:278—89.
122. Dunbar SA. Applications of Luminex xMAP technology for rapid, high-throughput multiplexed nucleic acid detection. *Clin Chim Acta* 2006;**363**:71—82.
123. Markou A, Strati A, Malamos N, et al. Molecular characterization of circulating tumor cells in breast cancer by a liquid bead array hybridization assay. *Clin Chem* 2011;**57**:421—30.
124. Leversha MA, Han J, Asgari Z, et al. Fluorescence in situ hybridization analysis of circulating tumor cells in metastatic prostate cancer. *Clin Cancer Res* 2009;**15**:2091—7.
125. Swennenhuis JF, Tibbe AG, Levink R, et al. Characterization of circulating tumor cells by fluorescence in situ hybridization. *Cytometry A* 2009;**75**:520—7.
126. Payne RE, Wang F, Su N, et al. Viable circulating tumour cell detection using multiplex RNA in situ hybridisation predicts progression-free survival in metastatic breast cancer patients. *Br J Cancer* 2012;**106**:1790—7.
127. Sullivan JP, Nahed BV, Madden MW, et al. Brain tumor cells in circulation are enriched for mesenchymal gene expression. *Cancer Discov* 2014;**4**:1299—309.
128. Kraus J, Pantel K, Pinkel D, et al. High-resolution genomic profiling of occult micrometastatic tumor cells. *Genes Chromosomes Cancer* 2003;**36**:159—66.
129. Czyz ZT, Hoffmann M, Schlimok G, et al. Reliable single cell array CGH for clinical samples. *PLoS ONE* 2014;**9**:e85907.
130. Lohr JG, Adalsteinsson VA, Cibulskis K, et al. Whole-exome sequencing of circulating tumor cells provides a window into metastatic prostate cancer. *Nat Biotechnol* 2014;**32**:479—84.
131. Powell AA, Talasaz AH, Zhang H, et al. Single cell profiling of circulating tumor cells: transcriptional heterogeneity and diversity from breast cancer cell lines. *PLoS ONE* 2012;**7**:e33788.
132. Yap TA, Gerlinger M, Futreal PA, et al. Intratumor heterogeneity: seeing the wood for the trees. *Sci Transl Med* 2012;**4**. 127ps10.
133. Wang Y, Waters J, Leung ML, et al. Clonal evolution in breast cancer revealed by single nucleus genome sequencing. *Nature* 2014;**512**:155—60.
134. Heitzer E, Auer M, Gasch C, et al. Complex tumor genomes inferred from single circulating tumor cells by array-CGH and next-generation sequencing. *Cancer Res* 2013;**73**:2965—75.
135. Polzer B, Medoro G, Pasch S, et al. Molecular profiling of single circulating tumor cells with diagnostic intention. *EMBO Mol Med* 2014;**6**:1371—86.
136. Fidler IJ. The pathogenesis of cancer metastasis: the "seed and soil" hypothesis revisited. *Nat Rev Cancer* 2003;**3**:453—8.
137. Kaiser J. Medicine. Cancer's circulation problem. *Science* 2010;**327**:1072—4.
138. Polzer B, Klein CA. Metastasis awakening: the challenges of targeting minimal residual cancer. *Nat Med* 2013;**19**:274—5.
139. Aguirre-Ghiso JA, Bragado P, Sosa MS. Metastasis awakening: targeting dormant cancer. *Nat Med* 2013;**19**:276—7.
140. Sosa MS, Bragado P, Aguirre-Ghiso JA. Mechanisms of disseminated cancer cell dormancy: an awakening field. *Nat Rev Cancer* 2014;**14**:611—22.

141. Mego M, Mani SA, Cristofanilli M. Molecular mechanisms of metastasis in breast cancer—clinical applications. *Nat Rev Clin Oncol* 2010;**7**:693—701.
142. Thiery JP, Lim CT. Tumor dissemination: an EMT affair. *Cancer Cell* 2013;**23**:272—3.
143. Lim J, Thiery JP. Epithelial-mesenchymal transitions: insights from development. *Development* 2012;**139**:3471—86.
144. Barriere G, Fici P, Gallerani G, et al. Circulating tumor cells and epithelial, mesenchymal and stemness markers: characterization of cell subpopulations. *Ann Transl Med* 2014;**2**:109.
145. Kasimir-Bauer S, Hoffmann O, Wallwiener D, et al. Expression of stem cell and epithelial-mesenchymal transition markers in primary breast cancer patients with circulating tumor cells. *Breast Cancer Res* 2012;**14**:R15.
146. Mego M, Mani SA, Lee BN, et al. Expression of epithelial-mesenchymal transition-inducing transcription factors in primary breast cancer: the effect of neoadjuvant therapy. *Int J Cancer* 2012;**130**:808—16.
147. Yu M, Bardia A, Wittner BS, et al. Circulating breast tumor cells exhibit dynamic changes in epithelial and mesenchymal composition. *Science* 2013;**339**:580—4.
148. Serrano MJ, Ortega FG, Alvarez-Cubero MJ, et al. EMT and EGFR in CTCs cytokeratin negative non-metastatic breast cancer. *Oncotarget* 2014;**5**:7486—97.
149. Papadaki MA, Kallergi G, Zafeiriou Z, et al. Co-expression of putative stemness and epithelial-to-mesenchymal transition markers on single circulating tumour cells from patients with early and metastatic breast cancer. *BMC Cancer* 2014;**14**:651.
150. Gorges TM, Tinhofer I, Drosch M, et al. Circulating tumour cells escape from EpCAM-based detection due to epithelial-to-mesenchymal transition. *BMC Cancer* 2012;**12**:178.
151. Sugimachi K, Yokobori T, Iinuma H, et al. Aberrant expression of plastin-3 via copy number gain induces the epithelial-mesenchymal transition in circulating colorectal cancer cells. *Ann Surg Oncol* 2014;**21**:3680—90.
152. Joosse SA, Hannemann J, Spotter J, et al. Changes in keratin expression during metastatic progression of breast cancer: impact on the detection of circulating tumor cells. *Clin Cancer Res* 2012;**18**:993—1003.
153. Gupta PB, Chaffer CL, Weinberg RA. Cancer stem cells: mirage or reality? *Nat Med* 2009;**15**:1010—2.
154. Luo M, Clouthier SG, Deol Y, et al. Breast cancer stem cells: current advances and clinical implications. *Methods Mol Biol* 2015;**1293**:1—49.
155. Brooks MD, Burness ML, Wicha MS. Therapeutic implications of cellular heterogeneity and plasticity in breast cancer. *Cell Stem Cell* 2015;**17**:260—71.
156. Korkaya H, Paulson A, Iovino F, et al. HER2 regulates the mammary stem/progenitor cell population driving tumorigenesis and invasion. *Oncogene* 2008;**27**:6120—30.
157. Korkaya H, Wicha MS. HER2 and breast cancer stem cells: more than meets the eye. *Cancer Res* 2013;**73**:3489—93.
158. Bonnomet A, Brysse A, Tachsidis A, et al. Epithelial-to-mesenchymal transitions and circulating tumor cells. *J Mammary Gland Biol Neoplasia* 2010;**15**:261—73.
159. Charpentier M, Martin S. Interplay of stem cell characteristics, EMT, and microtentacles in circulating breast tumor cells. *Cancers (Basel)* 2013;**5**:1545—65.
160. Balic M, Lin H, Young L, et al. Most early disseminated cancer cells detected in bone marrow of breast cancer patients have a putative breast cancer stem cell phenotype. *Clin Cancer Res* 2006;**12**:5615—21.
161. Theodoropoulos PA, Polioudaki H, Agelaki S, et al. Circulating tumor cells with a putative stem cell phenotype in peripheral blood of patients with breast cancer. *Cancer Lett* 2010;**288**:99—106.
162. Giordano A, Gao H, Anfossi S, et al. Epithelial-mesenchymal transition and stem cell markers in patients with HER2-positive metastatic breast cancer. *Mol Cancer Ther* 2012;**11**:2526—34.
163. Zhang Y, Li J, Cao L, et al. Circulating tumor cells in hepatocellular carcinoma: detection techniques, clinical implications, and future perspectives. *Semin Oncol* 2012;**39**:449—60.
164. Bidard FC, Hajage D, Bachelot T, et al. Assessment of circulating tumor cells and serum markers for progression-free survival prediction in metastatic breast cancer: a prospective observational study. *Breast Cancer Res* 2012;**14**:R29.
165. Bidard FC, Fehm T, Ignatiadis M, et al. Clinical application of circulating tumor cells in breast cancer: overview of the current interventional trials. *Cancer Metastasis Rev* 2013;**32**:179—88.
166. Pierga JY, Hajage D, Bachelot T, et al. High independent prognostic and predictive value of circulating tumor cells compared with serum tumor markers in a large prospective trial in first-line chemotherapy for metastatic breast cancer patients. *Ann Oncol* 2012;**23**:618—24.
167. Giordano A, Egleston BL, Hajage D, et al. Establishment and validation of circulating tumor cell-based prognostic nomograms in first-line metastatic breast cancer patients. *Clin Cancer Res* 2013;**19**:1596—602.
168. Wallwiener M, Hartkopf AD, Baccelli I, et al. The prognostic impact of circulating tumor cells in subtypes of metastatic breast cancer. *Breast Cancer Res Treat* 2013;**137**:503—10.
169. Giuliano M, Giordano A, Jackson S, et al. Circulating tumor cells as early predictors of metastatic spread in breast cancer patients with limited metastatic dissemination. *Breast Cancer Res* 2014;**16**:440.
170. Muller V, Riethdorf S, Rack B, et al. Prognostic impact of circulating tumor cells assessed with the CellSearch™ System and AdnaTest Breast in metastatic breast cancer patients: the DETECT study. *Breast Cancer Res* 2012;**14**:R118.
171. Smerage JB, Barlow WE, Hortobagyi GN, et al. Circulating tumor cells and response to chemotherapy in metastatic breast cancer: sWOG S0500. *J Clin Oncol* 2014;**32**:3483—9.
172. Bidard FC, Peeters DJ, Fehm T, et al. Clinical validity of circulating tumour cells in patients with metastatic breast cancer: a pooled analysis of individual patient data. *Lancet Oncol* 2014;**15**:406—14.
173. Budd GT, Cristofanilli M, Ellis MJ, et al. Circulating tumor cells versus imaging—predicting overall survival in metastatic breast cancer. *Clin Cancer Res* 2006;**12**:6403—9.
174. Nakamura S, Yagata H, Ohno S, et al. Multi-center study evaluating circulating tumor cells as a surrogate for response to treatment and overall survival in metastatic breast cancer. *Breast Cancer* 2010;**17**:199—204.
175. Liu MC, Shields PG, Warren RD, et al. Circulating tumor cells: a useful predictor of treatment efficacy in metastatic breast cancer. *J Clin Oncol* 2009;**27**:5153—9.
176. Androulakis N, Agelaki S, Perraki M, et al. Clinical relevance of circulating CK-19mRNA-positive tumour cells before front-line treatment in patients with metastatic breast cancer. *Br J Cancer* 2012;**106**:1917—25.
177. Reinholz MM, Kitzmann KA, Tenner K, et al. Cytokeratin-19 and mammaglobin gene expression in circulating tumor cells from metastatic breast cancer patients enrolled in North

Central Cancer Treatment Group trials, N0234/336/436/437. *Clin Cancer Res* 2011;**17**:7183–93.
178. Ramirez JM, Fehm T, Orsini M, et al. Prognostic relevance of viable circulating tumor cells detected by EPISPOT in metastatic breast cancer patients. *Clin Chem* 2014;**60**:214–21.
179. Stathopoulou A, Vlachonikolis I, Mavroudis D, et al. Molecular detection of cytokeratin-19-positive cells in the peripheral blood of patients with operable breast cancer: evaluation of their prognostic significance. *J Clin Oncol* 2002;**20**:3404–12.
180. Ignatiadis M, Xenidis N, Perraki M, et al. Different prognostic value of cytokeratin-19 mRNA positive circulating tumor cells according to estrogen receptor and HER2 status in early-stage breast cancer. *J Clin Oncol* 2007;**25**:5194–202.
181. Saloustros E, Perraki M, Apostolaki S, et al. Cytokeratin-19 mRNA-positive circulating tumor cells during follow-up of patients with operable breast cancer: prognostic relevance for late relapse. *Breast Cancer Res* 2011;**13**:R60.
182. Xenidis N, Perraki M, Apostolaki S, et al. Differential effect of adjuvant taxane-based and taxane-free chemotherapy regimens on the CK-19 mRNA-positive circulating tumour cells in patients with early breast cancer. *Br J Cancer* 2013;**108**:549–56.
183. Pierga JY, Bidard FC, Mathiot C, et al. Circulating tumor cell detection predicts early metastatic relapse after neoadjuvant chemotherapy in large operable and locally advanced breast cancer in a phase II randomized trial. *Clin Cancer Res* 2008;**14**:7004–10.
184. Lucci A, Hall CS, Lodhi AK, et al. Circulating tumour cells in non-metastatic breast cancer: a prospective study. *Lancet Oncol* 2012;**13**:688–95.
185. Bidard FC, Belin L, Delaloge S, et al. Time-dependent prognostic impact of circulating tumor cells detection in non-metastatic breast cancer: 70-month analysis of the REMAGUS02 Study. *Int J Breast Cancer* 2013;**2013**:130470.
186. Rack B, Schindlbeck C, Juckstock J, et al. Circulating tumor cells predict survival in early average-to-high risk breast cancer patients. *J Natl Cancer Inst* 2014:106.
187. Nadal R, Lorente JA, Rosell R, et al. Relevance of molecular characterization of circulating tumor cells in breast cancer in the era of targeted therapies. *Expert Rev Mol Diagn* 2013;**13**:295–307.
188. Becker TM, Caixeiro NJ, Lim SH, et al. New frontiers in circulating tumor cell analysis: a reference guide for biomolecular profiling toward translational clinical use. *Int J Cancer* 2014;**134**:2523–33.
189. Turner N, Pestrin M, Galardi F, et al. Can biomarker assessment on circulating tumor cells help direct therapy in metastatic breast cancer? *Cancers (Basel)* 2014;**6**:684–707.
190. Fehm T, Muller V, Aktas B, et al. HER2 status of circulating tumor cells in patients with metastatic breast cancer: a prospective, multicenter trial. *Breast Cancer Res Treat* 2010;**124**:403–12.
191. Tewes M, Aktas B, Welt A, et al. Molecular profiling and predictive value of circulating tumor cells in patients with metastatic breast cancer: an option for monitoring response to breast cancer related therapies. *Breast Cancer Res Treat* 2009;**115**:581–90.
192. Riethdorf S, Muller V, Zhang L, et al. Detection and HER2 expression of circulating tumor cells: prospective monitoring in breast cancer patients treated in the neoadjuvant Gepar-Quattro trial. *Clin Cancer Res* 2010;**16**:2634–45.
193. Ignatiadis M, Rothe F, Chaboteaux C, et al. HER2-positive circulating tumor cells in breast cancer. *PLoS ONE* 2011;**6**:e15624.
194. Punnoose EA, Atwal SK, Spoerke JM, et al. Molecular biomarker analyses using circulating tumor cells. *PLoS ONE* 2010;**5**:e12517.
195. Pestrin M, Bessi S, Galardi F, et al. Correlation of HER2 status between primary tumors and corresponding circulating tumor cells in advanced breast cancer patients. *Breast Cancer Res Treat* 2009;**118**:523–30.
196. Flores LM, Kindelberger DW, Ligon AH, et al. Improving the yield of circulating tumour cells facilitates molecular characterisation and recognition of discordant HER2 amplification in breast cancer. *Br J Cancer* 2010;**102**:1495–502.
197. Krishnamurthy S, Bischoff F, Ann MJ, et al. Discordance in HER2 gene amplification in circulating and disseminated tumor cells in patients with operable breast cancer. *Cancer Med* 2013;**2**:226–33.
198. Georgoulias V, Bozionelou V, Agelaki S, et al. Trastuzumab decreases the incidence of clinical relapses in patients with early breast cancer presenting chemotherapy-resistant CK-19mRNA-positive circulating tumor cells: results of a randomized phase II study. *Ann Oncol* 2012;**23**:1744–50.
199. Pestrin M, Bessi S, Puglisi F, et al. Final results of a multicenter phase II clinical trial evaluating the activity of single-agent lapatinib in patients with HER2-negative metastatic breast cancer and HER2-positive circulating tumor cells. A proof-of-concept study. *Breast Cancer Res Treat* 2012;**134**:283–9.
200. Rack B, Bock C, Andergassen U, et al. Hormone receptor status, erbB2 expression and cancer stem cell characteristics of circulating tumor cells in breast cancer patients. *Histol Histopathol* 2012;**27**:855–64.
201. Babayan A, Hannemann J, Spotter J, et al. Heterogeneity of estrogen receptor expression in circulating tumor cells from metastatic breast cancer patients. *PLoS ONE* 2013;**8**:e75038.
202. Nadal R, Fernandez A, Sanchez-Rovira P, et al. Biomarkers characterization of circulating tumour cells in breast cancer patients. *Breast Cancer Res* 2012;**14**:R71.
203. Kallergi G, Papadaki MA, Politaki E, et al. Epithelial to mesenchymal transition markers expressed in circulating tumour cells of early and metastatic breast cancer patients. *Breast Cancer Res* 2011;**13**:R59.
204. Kallergi G, Konstantinidis G, Markomanolaki H, et al. Apoptotic circulating tumor cells in early and metastatic breast cancer patients. *Mol Cancer Ther* 2013;**12**:1886–95.
205. Spiliotaki M, Mavroudis D, Kapranou K, et al. Evaluation of proliferation and apoptosis markers in circulating tumor cells of women with early breast cancer who are candidates for tumor dormancy. *Breast Cancer Res* 2014;**16**:485.
206. Chimonidou M, Strati A, Tzitzira A, et al. DNA methylation of tumor suppressor and metastasis suppressor genes in circulating tumor cells. *Clin Chem* 2011;**57**:1169–77.
207. Chimonidou M, Kallergi G, Georgoulias V, et al. Breast cancer metastasis suppressor-1 promoter methylation in primary breast tumors and corresponding circulating tumor cells. *Mol Cancer Res* 2013;**11**:1248–57.
208. Chimonidou M, Strati A, Malamos N, et al. SOX17 promoter methylation in circulating tumor cells and matched cell-free DNA isolated from plasma of patients with breast cancer. *Clin Chem* 2013;**59**:270–9.

209. Schilling D, Todenhofer T, Hennenlotter J, et al. Isolated, disseminated and circulating tumour cells in prostate cancer. *Nat Rev Urol* 2012;**9**:448–63.
210. Hu B, Rochefort H, Goldkorn A. Circulating tumor cells in prostate cancer. *Cancers (Basel)* 2013;**5**:1676–90.
211. Li J, Gregory SG, Garcia-Blanco MA, et al. Using circulating tumor cells to inform on prostate cancer biology and clinical utility. *Crit Rev Clin Lab Sci* 2015:1–20.
212. Moreno JG, O'Hara SM, Gross S, et al. Changes in circulating carcinoma cells in patients with metastatic prostate cancer correlate with disease status. *Urology* 2001;**58**:386–92.
213. Danila DC, Heller G, Gignac GA, et al. Circulating tumor cell number and prognosis in progressive castration-resistant prostate cancer. *Clin Cancer Res* 2007;**13**:7053–8.
214. de Bono JS, Scher HI, Montgomery RB, et al. Circulating tumor cells predict survival benefit from treatment in metastatic castration-resistant prostate cancer. *Clin Cancer Res* 2008;**14**: 6302–9.
215. Scher HI, Jia X, de Bono JS, et al. Circulating tumour cells as prognostic markers in progressive, castration-resistant prostate cancer: a reanalysis of IMMC38 trial data. *Lancet Oncol* 2009; **10**:233–9.
216. Olmos D, Arkenau HT, Ang JE, et al. Circulating tumour cell (CTC) counts as intermediate end points in castration-resistant prostate cancer (CRPC): a single-centre experience. *Ann Oncol* 2009;**20**:27–33.
217. Resel FL, San Jose ML, Galante RI, et al. Prognostic significance of circulating tumor cell count in patients with metastatic hormone-sensitive prostate cancer. *Urology* 2012;**80**:1328–32.
218. Saad F, Pantel K. The current role of circulating tumor cells in the diagnosis and management of bone metastases in advanced prostate cancer. *Future Oncol* 2012;**8**:321–31.
219. Goldkorn A, Ely B, Quinn DI, et al. Circulating tumor cell counts are prognostic of overall survival in SWOG S0421: a phase III trial of docetaxel with or without atrasentan for metastatic castration-resistant prostate cancer. *J Clin Oncol* 2014;**32**:1136–42.
220. Lowes LE, Lock M, Rodrigues G, et al. Circulating tumour cells in prostate cancer patients receiving salvage radiotherapy. *Clin Transl Oncol* 2012;**14**:150–6.
221. Doyen J, Alix-Panabieres C, Hofman P, et al. Circulating tumor cells in prostate cancer: a potential surrogate marker of survival. *Crit Rev Oncol Hematol* 2012;**81**:241–56.
222. Scher HI, Morris MJ, Larson S, et al. Validation and clinical utility of prostate cancer biomarkers. *Nat Rev Clin Oncol* 2013; **10**:225–34.
223. Danila DC, Fleisher M, Scher HI. Circulating tumor cells as biomarkers in prostate cancer. *Clin Cancer Res* 2011;**17**: 3903–12.
224. Antonarakis ES, Lu C, Wang H, et al. AR-V7 and resistance to enzalutamide and abiraterone in prostate cancer. *N Engl J Med* 2014;**371**:1028–38.
225. Nakazawa M, Lu C, Chen Y, et al. Serial blood-based analysis of AR-V7 in men with advanced prostate cancer. *Ann Oncol* 2015; **26**(9):1859–65.
226. Crespo M, van Dalum G, Ferraldeschi R, et al. Androgen receptor expression in circulating tumour cells from castration-resistant prostate cancer patients treated with novel endocrine agents. *Br J Cancer* 2015;**112**:1166–74.
227. Reyes EE, VanderWeele DJ, Isikbay M, et al. Quantitative characterization of androgen receptor protein expression and cellular localization in circulating tumor cells from patients with metastatic castration-resistant prostate cancer. *J Transl Med* 2014;**12**:313.
228. Thadani-Mulero M, Portella L, Sun S, et al. Androgen receptor splice variants determine taxane sensitivity in prostate cancer. *Cancer Res* 2014;**74**:2270–82.
229. Scher HI, Heller G, Molina A, et al. Circulating tumor cell biomarker panel as an individual-level surrogate for survival in metastatic castration-resistant prostate cancer. *J Clin Oncol* 2015;**33**:1348–55.
230. Groot KB, Rahbari NN, Buchler MW, et al. Circulating tumor cells and prognosis of patients with resectable colorectal liver metastases or widespread metastatic colorectal cancer: a meta-analysis. *Ann Surg Oncol* 2013;**20**:2156–65.
231. Peach G, Kim C, Zacharakis E, et al. Prognostic significance of circulating tumour cells following surgical resection of colorectal cancers: a systematic review. *Br J Cancer* 2010;**102**:1327–34.
232. Akagi Y, Kinugasa T, Adachi Y, et al. Prognostic significance of isolated tumor cells in patients with colorectal cancer in recent 10-year studies. *Mol Clin Oncol* 2013;**1**:582–92.
233. Cohen SJ, Punt CJ, Iannotti N, et al. Relationship of circulating tumor cells to tumor response, progression-free survival, and overall survival in patients with metastatic colorectal cancer. *J Clin Oncol* 2008;**26**:3213–21.
234. Tol J, Koopman M, Miller MC, et al. Circulating tumour cells early predict progression-free and overall survival in advanced colorectal cancer patients treated with chemotherapy and targeted agents. *Ann Oncol* 2010;**21**:1006–12.
235. Matsusaka S, Suenaga M, Mishima Y, et al. Circulating endothelial cells predict for response to bevacizumab-based chemotherapy in metastatic colorectal cancer. *Cancer Chemother Pharmacol* 2011;**68**:763–8.
236. de Albuquerque A, Kubisch I, Stolzel U, et al. Prognostic and predictive value of circulating tumor cell analysis in colorectal cancer patients. *J Transl Med* 2012;**10**:222.
237. Jiao LR, Apostolopoulos C, Jacob J, et al. Unique localization of circulating tumor cells in patients with hepatic metastases. *J Clin Oncol* 2009;**27**:6160–5.
238. Rahbari NN, Bork U, Kircher A, et al. Compartmental differences of circulating tumor cells in colorectal cancer. *Ann Surg Oncol* 2012;**19**:2195–202.
239. Barbazan J, Muinelo-Romay L, Vieito M, et al. A multimarker panel for circulating tumor cells detection predicts patient outcome and therapy response in metastatic colorectal cancer. *Int J Cancer* 2014;**135**:2633–43.
240. Lu CY, Tsai HL, Uen YH, et al. Circulating tumor cells as a surrogate marker for determining clinical outcome to mFOLFOX chemotherapy in patients with stage III colon cancer. *Br J Cancer* 2013;**108**:791–7.
241. Bork U, Rahbari NN, Scholch S, et al. Circulating tumour cells and outcome in non-metastatic colorectal cancer: a prospective study. *Br J Cancer* 2015;**112**:1306–13.
242. Thorsteinsson M, Soletormos G, Jess P. Low number of detectable circulating tumor cells in non-metastatic colon cancer. *Anticancer Res* 2011;**31**:613–7.
243. Gazzaniga P, Gianni W, Raimondi C, et al. Circulating tumor cells in high-risk nonmetastatic colorectal cancer. *Tumour Biol* 2013;**34**:2507–9.
244. Iinuma H, Watanabe T, Mimori K, et al. Clinical significance of circulating tumor cells, including cancer stem-like cells, in peripheral blood for recurrence and prognosis in patients with Dukes' stage B and C colorectal cancer. *J Clin Oncol* 2011;**29**: 1547–55.

245. Pantel K, Deneve E, Nocca D, et al. Circulating epithelial cells in patients with benign colon diseases. *Clin Chem* 2012;**58**: 936–40.
246. Deneve E, Riethdorf S, Ramos J, et al. Capture of viable circulating tumor cells in the liver of colorectal cancer patients. *Clin Chem* 2013;**59**:1384–92.
247. Mostert B, Jiang Y, Sieuwerts AM, et al. KRAS and BRAF mutation status in circulating colorectal tumor cells and their correlation with primary and metastatic tumor tissue. *Int J Cancer* 2013;**133**:130–41.
248. Mohamed Suhaimi NA, Foong YM, Lee DY, et al. Non-invasive sensitive detection of KRAS and BRAF mutation in circulating tumor cells of colorectal cancer patients. *Mol Oncol* 2015;**9**: 850–60.
249. Gasch C, Bauernhofer T, Pichler M, et al. Heterogeneity of epidermal growth factor receptor status and mutations of KRAS/PIK3CA in circulating tumor cells of patients with colorectal cancer. *Clin Chem* 2013;**59**:252–60.
250. Kalikaki A, Politaki H, Souglakos J, et al. KRAS genotypic changes of circulating tumor cells during treatment of patients with metastatic colorectal cancer. *PLoS ONE* 2014;**9**:e104902.
251. Yokobori T, Iinuma H, Shimamura T, et al. Plastin3 is a novel marker for circulating tumor cells undergoing the epithelial-mesenchymal transition and is associated with colorectal cancer prognosis. *Cancer Res* 2013;**73**:2059–69.
252. Sun W, Huang T, Li G, et al. The advantage of circulating tumor cells over serum carcinoembryonic antigen for predicting treatment responses in rectal cancer. *Future Oncol* 2013;**9**:1489–500.
253. Fusi A, Metcalf R, Krebs M, et al. Clinical utility of circulating tumour cell detection in non-small-cell lung cancer. *Curr Treat Options Oncol* 2013;**14**:610–22.
254. Ilie M, Hofman V, Long-Mira E, et al. "Sentinel" circulating tumor cells allow early diagnosis of lung cancer in patients with chronic obstructive pulmonary disease. *PLoS ONE* 2014;**9**: e111597.
255. Zhu WF, Li J, Yu LC, et al. Prognostic value of EpCAM/MUC1 mRNA-positive cells in non-small cell lung cancer patients. *Tumour Biol* 2014;**35**:1211–9.
256. Hofman V, Bonnetaud C, Ilie MI, et al. Preoperative circulating tumor cell detection using the isolation by size of epithelial tumor cell method for patients with lung cancer is a new prognostic biomarker. *Clin Cancer Res* 2011;**17**:827–35.
257. Lecharpentier A, Vielh P, Perez-Moreno P, et al. Detection of circulating tumour cells with a hybrid (epithelial/mesen-chymal) phenotype in patients with metastatic non-small cell lung cancer. *Br J Cancer* 2011;**105**:1338–41.
258. Krebs MG, Sloane R, Priest L, et al. Evaluation and prognostic significance of circulating tumor cells in patients with non-small-cell lung cancer. *J Clin Oncol* 2011;**29**:1556–63.
259. Chen X, Wang X, He H, et al. Combination of circulating tumor cells with serum carcinoembryonic antigen enhances clinical prediction of non-small cell lung cancer. *PLoS ONE* 2015;**10**:e0126276.
260. Hodgkinson CL, Morrow CJ, Li Y, et al. Tumorigenicity and genetic profiling of circulating tumor cells in small-cell lung cancer. *Nat Med* 2014;**20**:897–903.
261. Hiltermann TJ, Pore MM, van den Berg A, et al. Circulating tumor cells in small-cell lung cancer: a predictive and prognostic factor. *Ann Oncol* 2012;**23**:2937–42.
262. Carlsson A, Nair VS, Luttgen MS, et al. Circulating tumor microemboli diagnostics for patients with non-small-cell lung cancer. *J Thorac Oncol* 2014;**9**:1111–9.
263. Normanno N, Rossi A, Morabito A, et al. Prognostic value of circulating tumor cells' reduction in patients with extensive small-cell lung cancer. *Lung Cancer* 2014;**85**:314–9.
264. Zhang Z, Shiratsuchi H, Lin J, et al. Expansion of CTCs from early stage lung cancer patients using a microfluidic co-culture model. *Oncotarget* 2014;**5**:12383–97.
265. Maheswaran S, Sequist LV, Nagrath S, et al. Detection of mutations in EGFR in circulating lung-cancer cells. *N Engl J Med* 2008;**359**:366–77.
266. Punnoose EA, Atwal S, Liu W, et al. Evaluation of circulating tumor cells and circulating tumor DNA in non-small cell lung cancer: association with clinical endpoints in a phase II clinical trial of pertuzumab and erlotinib. *Clin Cancer Res* 2012;**18**: 2391–401.
267. Marchetti A, Del GM, Felicioni L, et al. Assessment of EGFR mutations in circulating tumor cell preparations from NSCLC patients by next generation sequencing: toward a real-time liquid biopsy for treatment. *PLoS ONE* 2014;**9**: e103883.
268. Pailler E, Adam J, Barthelemy A, et al. Detection of circulating tumor cells harboring a unique ALK rearrangement in ALK-positive non-small-cell lung cancer. *J Clin Oncol* 2013;**31**: 2273–81.
269. Schwed Lustgarten DE, Thompson J, Yu G, et al. Use of circulating tumor cell technology (CELLSEARCH™) for the diagnosis of malignant pleural effusions. *Ann Am Thorac Soc* 2013;**10**:582–9.
270. Gershenwald J, Soong S-J, Thompson J. Prognostic factors and natural history of melanoma. In: Balch C, editor. *Cutaneous Melanoma*. 5th ed. St. Louis: Quality Medical Pub; 2009. p. 921.
271. Edge S, Compton C, Fritz A, et al. Melanoma of the skin. In: Edge S, editor. *AJCC Cancer Staging Manual*. 7th ed. New York: Springer; 2010. p. 648.
272. Dossett LA, Kudchadkar RR, Zager JS. BRAF and MEK inhibition in melanoma. *Expert Opin Drug Saf* 2015;**14**:559–70.
273. Eggermont AM, Maio M, Robert C. Immune checkpoint inhibitors in melanoma provide the cornerstones for curative therapies. *Semin Oncol* 2015;**42**:429–35.
274. Queirolo P, Picasso V, Spagnolo F. Combined BRAF and MEK inhibition for the treatment of BRAF-mutated metastatic melanoma. *Cancer Treat Rev* 2015;**41**:519–26.
275. Hoon DS. Are circulating tumor cells an independent prognostic factor in patients with high-risk melanoma? *Nat Clin Pract Oncol* 2004;**1**:74–5.
276. Mocellin S, Hoon D, Ambrosi A, et al. The prognostic value of circulating tumor cells in patients with melanoma: a systematic review and meta-analysis. *Clin Cancer Res* 2006;**12**:4605–13.
277. Kuo CT, Bostick PJ, Irie RF, et al. Assessment of messenger RNA of beta 1→4-N-acetylgalactosaminyl-transferase as a molecular marker for metastatic melanoma. *Clin Cancer Res* 1998;**4**:411–8.
278. Hoon DS, Wang Y, Dale PS, et al. Detection of occult melanoma cells in blood with a multiple-marker polymerase chain reaction assay. *J Clin Oncol* 1995;**13**:2109–16.
279. Hoon DS, Kuo CT, Wascher RA, et al. Molecular detection of metastatic melanoma cells in cerebrospinal fluid in melanoma patients. *J Invest Dermatol* 2001;**117**:375–8.

280. Kitago M, Koyanagi K, Nakamura T, et al. mRNA expression and BRAF mutation in circulating melanoma cells isolated from peripheral blood with high molecular weight melanoma-associated antigen-specific monoclonal antibody beads. *Clin Chem* 2009;**55**:757−64.
281. Goto Y, Koyanagi K, Narita N, et al. Aberrant fatty acid-binding protein-7 gene expression in cutaneous malignant melanoma. *J Invest Dermatol* 2010;**130**:221−9.
282. Chiu CG, Nakamura Y, Chong KK, et al. Genome-wide characterization of circulating tumor cells identifies novel prognostic genomic alterations in systemic melanoma metastasis. *Clin Chem* 2014;**60**:873−85.
283. Koyanagi K, Kuo C, Nakagawa T, et al. Multimarker quantitative real-time PCR detection of circulating melanoma cells in peripheral blood: relation to disease stage in melanoma patients. *Clin Chem* 2005;**51**:981−8.
284. Koyanagi K, O'Day SJ, Gonzalez R, et al. Serial monitoring of circulating melanoma cells during neoadjuvant biochemotherapy for stage III melanoma: outcome prediction in a multicenter trial. *J Clin Oncol* 2005;**23**:8057−64.
285. Koyanagi K, Mori T, O'Day SJ, et al. Association of circulating tumor cells with serum tumor-related methylated DNA in peripheral blood of melanoma patients. *Cancer Res* 2006;**66**:6111−7.
286. Miyashiro I, Kuo C, Huynh K, et al. Molecular strategy for detecting metastatic cancers with use of multiple tumor-specific MAGE-A genes. *Clin Chem* 2001;**47**:505−12.
287. Koyanagi K, O'Day SJ, Gonzalez R, et al. Microphthalmia transcription factor as a molecular marker for circulating tumor cell detection in blood of melanoma patients. *Clin Cancer Res* 2006;**12**:1137−43.
288. Gray ES, Reid AL, Bowyer S, et al. Circulating melanoma cell subpopulations: their heterogeneity and differential responses to treatment. *J Invest Dermatol* 2015;**135**(8):2040−8.
289. Koyanagi K, O'Day SJ, Boasberg P, et al. Serial monitoring of circulating tumor cells predicts outcome of induction biochemotherapy plus maintenance biotherapy for metastatic melanoma. *Clin Cancer Res* 2010;**16**:2402−8.
290. Hoshimoto S, Shingai T, Morton DL, et al. Association between circulating tumor cells and prognosis in patients with stage III melanoma with sentinel lymph node metastasis in a phase III international multicenter trial. *J Clin Oncol* 2012;**30**:3819−26.
291. Hoshimoto S, Faries MB, Morton DL, et al. Assessment of prognostic circulating tumor cells in a phase III trial of adjuvant immunotherapy after complete resection of stage IV melanoma. *Ann Surg* 2012;**255**:357−62.
292. Khoja L, Lorigan P, Zhou C, et al. Biomarker utility of circulating tumor cells in metastatic cutaneous melanoma. *J Invest Dermatol* 2013;**133**:1582−90.
293. Zhang Z, Irie RF, Chi DD, et al. Cellular immuno-PCR. Detection of a carbohydrate tumor marker. *Am J Pathol* 1998;**152**:1427−32.
294. Sakaizawa K, Goto Y, Kiniwa Y, et al. Mutation analysis of BRAF and KIT in circulating melanoma cells at the single cell level. *Br J Cancer* 2012;**106**:939−46.
295. Reid AL, Freeman JB, Millward M, et al. Detection of BRAF-V600E and V600K in melanoma circulating tumour cells by droplet digital PCR. *Clin Biochem* 2014;**48**:999−1002.
296. Obermayr E, Sanchez-Cabo F, Tea MK, et al. Assessment of a six gene panel for the molecular detection of circulating tumor cells in the blood of female cancer patients. *BMC Cancer* 2010;**10**:666.
297. Obermayr E, Castillo-Tong DC, Pils D, et al. Molecular characterization of circulating tumor cells in patients with ovarian cancer improves their prognostic significance—A study of the OVCAD consortium. *Gynecol Oncol* 2013;**128**:15−21.
298. Aktas B, Kasimir-Bauer S, Heubner M, et al. Molecular profiling and prognostic relevance of circulating tumor cells in the blood of ovarian cancer patients at primary diagnosis and after platinum-based chemotherapy. *Int J Gynecol Cancer* 2011;**21**:822−30.
299. Poveda A, Kaye SB, McCormack R, et al. Circulating tumor cells predict progression free survival and overall survival in patients with relapsed/recurrent advanced ovarian cancer. *Gynecol Oncol* 2011;**122**:567−72.
300. Liu JF, Kindelberger D, Doyle C, et al. Predictive value of circulating tumor cells (CTCs) in newly diagnosed and recurrent ovarian cancer patients. *Gynecol Oncol* 2013;**131**:352−6.
301. Romero-Laorden N, Olmos D, Fehm T, et al. Circulating and disseminated tumor cells in ovarian cancer: a systematic review. *Gynecol Oncol* 2014;**133**:632−9.
302. Pearl ML, Zhao Q, Yang J, et al. Prognostic analysis of invasive circulating tumor cells (iCTCs) in epithelial ovarian cancer. *Gynecol Oncol* 2014;**134**:581−90.
303. Kuhlmann JD, Wimberger P, Bankfalvi A, et al. ERCC1-positive circulating tumor cells in the blood of ovarian cancer patients as a predictive biomarker for platinum resistance. *Clin Chem* 2014;**60**:1282−9.
304. Zeng L, Liang X, Liu Q, et al. The predictive value of circulating tumor cells in ovarian cancer: a meta analysis. *Int J Gynecol Cancer* 2015 [epub ahead of print].
305. Bidard FC, Huguet F, Louvet C, et al. Circulating tumor cells in locally advanced pancreatic adenocarcinoma: the ancillary CirCe 07 study to the LAP 07 trial. *Ann Oncol* 2013;**24**:2057−61.
306. Tjensvoll K, Nordgard O, Smaaland R. Circulating tumor cells in pancreatic cancer patients: methods of detection and clinical implications. *Int J Cancer* 2014;**134**:1−8.
307. Gall TM, Frampton AE, Krell J, et al. Is the detection of circulating tumor cells in locally advanced pancreatic cancer a useful prognostic marker? *Expert Rev Mol Diagn* 2013;**13**:793−6.
308. Han L, Chen W, Zhao Q. Prognostic value of circulating tumor cells in patients with pancreatic cancer: a meta-analysis. *Tumour Biol* 2014;**35**:2473−80.
309. Kulasinghe A, Perry C, Jovanovic L, et al. Circulating tumour cells in metastatic head and neck cancers. *Int J Cancer* 2015;**136**:2515−23.
310. Bozec A, Ilie M, Dassonville O, et al. Significance of circulating tumor cell detection using the CellSearch™ system in patients with locally advanced head and neck squamous cell carcinoma. *Eur Arch Otorhinolaryngol* 2013;**270**:2745−9.
311. Wang Z, Cui K, Xue Y, et al. Prognostic value of circulating tumor cells in patients with squamous cell carcinoma of the head and neck: a systematic review and meta-analysis. *Med Oncol* 2015;**32**:164.
312. Jatana KR, Balasubramanian P, Lang JC, et al. Significance of circulating tumor cells in patients with squamous cell carcinoma of the head and neck: initial results. *Arch Otolaryngol Head Neck Surg* 2010;**136**:1274−9.
313. Balasubramanian P, Lang JC, Jatana KR, et al. Multiparameter analysis, including EMT markers, on negatively enriched blood samples from patients with squamous cell carcinoma of the head and neck. *PLoS ONE* 2012;**7**:e42048.

314. Buglione M, Grisanti S, Almici C, et al. Circulating tumour cells in locally advanced head and neck cancer: preliminary report about their possible role in predicting response to non-surgical treatment and survival. *Eur J Cancer* 2012;**48**: 3019–26.
315. Grobe A, Blessmann M, Hanken H, et al. Prognostic relevance of circulating tumor cells in blood and disseminated tumor cells in bone marrow of patients with squamous cell carcinoma of the oral cavity. *Clin Cancer Res* 2014;**20**:425–33.
316. Tinhofer I, Konschak R, Stromberger C, et al. Detection of circulating tumor cells for prediction of recurrence after adjuvant chemoradiation in locally advanced squamous cell carcinoma of the head and neck. *Ann Oncol* 2014;**25**: 2042–7.
317. Nel I, Baba HA, Ertle J, et al. Individual profiling of circulating tumor cell composition and therapeutic outcome in patients with hepatocellular carcinoma. *Transl Oncol* 2013;**6**:420–8.
318. Schulze K, Gasch C, Staufer K, et al. Presence of EpCAM-positive circulating tumor cells as biomarker for systemic disease strongly correlates to survival in patients with hepatocellular carcinoma. *Int J Cancer* 2013;**133**:2165–71.
319. Sun YF, Xu Y, Yang XR, et al. Circulating stem cell-like epithelial cell adhesion molecule-positive tumor cells indicate poor prognosis of hepatocellular carcinoma after curative resection. *Hepatology* 2013;**57**:1458–68.
320. Raimondi C, Gradilone A, Naso G, et al. Clinical utility of circulating tumor cell counting through CellSearch™: the dilemma of a concept suspended in limbo. *Onco Targets Ther* 2014;**7**:619–25.
321. Rink M, Chun FK, Dahlem R, et al. Prognostic role and HER2 expression of circulating tumor cells in peripheral blood of patients prior to radical cystectomy: a prospective study. *Eur Urol* 2012;**61**:810–7.
322. Rink M, Chun FK, Minner S, et al. Detection of circulating tumour cells in peripheral blood of patients with advanced non-metastatic bladder cancer. *BJU Int* 2011;**107**:1668–75.
323. Gradilone A, Petracca A, Nicolazzo C, et al. Prognostic significance of survivin-expressing circulating tumour cells in T1G3 bladder cancer. *BJU Int* 2010;**106**:710–5.
324. Guzzo TJ, McNeil BK, Bivalacqua TJ, et al. The presence of circulating tumor cells does not predict extravesical disease in bladder cancer patients prior to radical cystectomy. *Urol Oncol* 2012;**30**:44–8.
325. Schwarzenbach H, Hoon DSB, Pantel K. Cell-free nucleic acids as biomarkers in cancer patients. *Nat Rev Cancer* 2011;**11**: 426–37.
326. Fleischhacker M, Schmidt B. Circulating nucleic acids (CNAs) and cancer–a survey. *Biochim Biophys Acta* 2007;**1775**:181–232.
327. Jung K, Fleischhacker M, Rabien A. Cell-free DNA in the blood as a solid tumor biomarker - A critical appraisal of the literature. *Clin Chim Acta* 2010;**411**:1611–24.
328. van der Vaart M, Pretorius PJ. Circulating DNA. Its origin and fluctuation. *Ann N Y Acad Sci* 2008;**1137**:18–26.
329. Anker P, Mulcahy H, Chen XQ, et al. Detection of circulating tumour DNA in the blood (plasma/serum) of cancer patients. *Cancer Metastasis Rev* 1999;**18**:65–73.
330. Stroun M, Anker P, Lyautey J, et al. Isolation and characterization of DNA from the plasma of cancer patients. *Eur J Cancer Clin Oncol* 1987;**23**:707–12.
331. Sunami E, Vu AT, Nguyen SL, et al. Quantification of LINE1 in circulating DNA as a molecular biomarker of breast cancer. *Ann N Y Acad Sci* 2008;**1137**:171–4.
332. Pan W, Gu W, Nagpal S, et al. Brain tumor mutations detected in cerebral spinal fluid. *Clin Chem* 2015;**61**:514–22.
333. Sidransky D, Von Eschenbach A, Tsai YC, et al. Identification of p53 gene mutations in bladder cancers and urine samples. *Science* 1991;**252**:706–9.
334. Kawai T, Akira S. Toll-like receptors and their crosstalk with other innate receptors in infection and immunity. *Immunity* 2011;**34**:637–50.
335. Reference deleted in review.
336. Bergsmedh A, Szeles A, Henriksson M, et al. Horizontal transfer of oncogenes by uptake of apoptotic bodies. *Proc Natl Acad Sci USA* 2001;**98**:6407–11.
337. Gahan PB. Circulating DNA: intracellular and intraorgan messenger? *Ann N Y Acad Sci* 2006;**1075**:21–33.
338. Garcia-Olmo DC, Dominguez C, Garcia-Arranz M, et al. Cell-free nucleic acids circulating in the plasma of colorectal cancer patients induce the oncogenic transformation of susceptible cultured cells. *Cancer Res* 2010;**70**:560–7.
339. Marzese DM, Hirose H, Hoon DS. Diagnostic and prognostic value of circulating tumor-related DNA in cancer patients. *Expert Rev Mol Diagn* 2013;**13**:827–44.
340. Fleischhacker M, Schmidt B, Weickmann S, et al. Methods for isolation of cell-free plasma DNA strongly affect DNA yield. *Clin Chim Acta* 2011;**412**:2085–8.
341. Roschewski M, Dunleavy K, Pittaluga S, et al. Circulating tumour DNA and CT monitoring in patients with untreated diffuse large B-cell lymphoma: a correlative biomarker study. *Lancet Oncol* 2015;**16**:541–9.
342. Breitbach S, Tug S, Helmig S, et al. Direct quantification of cell-free, circulating DNA from unpurified plasma. *PLoS ONE* 2014;**9**:e87838.
343. Newman AM, Bratman SV, To J, et al. An ultrasensitive method for quantitating circulating tumor DNA with broad patient coverage. *Nat Med* 2014;**20**:548–54.
344. Salvi S, Casadio V, Conteduca V, et al. Circulating cell-free AR and CYP17A1 copy number variations may associate with outcome of metastatic castration-resistant prostate cancer patients treated with abiraterone. *Br J Cancer* 2015;**112**:1717–24.
345. Vaca-Paniagua F, Oliver J, Nogueira da Costa A, et al. Targeted deep DNA methylation analysis of circulating cell-free DNA in plasma using massively parallel semiconductor sequencing. *Epigenomics* 2015;**7**:353–62.
346. Hoon DSB, Huang S, Sebisanovic D, et al. Identification of multiple informative genomic mutations by deep sequencing of circulating cell-free tumor DNA in plasma of metastatic melanoma patients. *ASCO Meeting Abstracts* 2014;**32**:9018.
347. Morelli MP, Overman MJ, Vilar Sanchez E, et al. Frequency of concurrent gene mutations and copy number alterations in circulating cell-free DNA (cfDNA) from refractory metastatic CRC patients. *ASCO Meeting Abstracts* 2014;**32**:11117.
348. Talasaz A, Mortimer S, Sebisanovic D, et al. Use of the GUARDANT360 noninvasive tumor sequencing assay on 300 patients across colorectal, melanoma, lung, breast, and prostate cancers and its clinical utility. *ASCO Meeting Abstracts* 2014;**32**: e22041.
349. Zill OA, Greene C, Sebisanovic D, et al. Cell-free dna next-generation sequencing in pancreatobiliary carcinomas. *Cancer Discov* 2015;**5**(10):1040–8.
350. Lanman RB, Mortimer SA, Zill OA, et al. Analytical and clinical validation of a digital sequencing panel for quantitative, highly accurate evaluation of cell-free circulating tumor DNA. *PLoS ONE* 2015;**10**:e0140712.

351. Miyake M, Sugano K, Sugino H, et al. Fibroblast growth factor receptor 3 mutation in voided urine is a useful diagnostic marker and significant indicator of tumor recurrence in non-muscle invasive bladder cancer. *Cancer Sci* 2010;**101**:250–8.
352. Rieger-Christ KM, Mourtzinos A, Lee PJ, et al. Identification of fibroblast growth factor receptor 3 mutations in urine sediment DNA samples complements cytology in bladder tumor detection. *Cancer* 2003;**98**:737–44.
353. van Kessel KE, Kompier LC, de Bekker-Grob EW, et al. FGFR3 mutation analysis in voided urine samples to decrease cystoscopies and cost in nonmuscle invasive bladder cancer surveillance: a comparison of 3 strategies. *J Urol* 2013;**189**:1676–81.
354. Bettegowda C, Sausen M, Leary RJ, et al. Detection of circulating tumor DNA in early- and late-stage human malignancies. *Sci Transl Med* 2014;**6**. 224ra24.
355. Board RE, Wardley AM, Dixon JM, et al. Detection of PIK3CA mutations in circulating free DNA in patients with breast cancer. *Breast Cancer Res Treat* 2010;**120**:461–7.
356. Braig F, Marz M, Schieferdecker A, et al. Epidermal growth factor receptor mutation mediates cross-resistance to panitumumab and cetuximab in gastrointestinal cancer. *Oncotarget* 2015;**6**:12035–47.
357. Castells A, Puig P, Mora J, et al. K-ras mutations in DNA extracted from the plasma of patients with pancreatic carcinoma: diagnostic utility and prognostic significance. *J Clin Oncol* 1999;**17**:578–84.
358. Daniotti M, Vallacchi V, Rivoltini L, et al. Detection of mutated BRAFV600E variant in circulating DNA of stage III-IV melanoma patients. *Int J Cancer* 2007;**120**:2439–44.
359. Dawson SJ, Tsui DW, Murtaza M, et al. Analysis of circulating tumor DNA to monitor metastatic breast cancer. *N Engl J Med* 2013;**368**:1199–209.
360. Del Re M, Vasile E, Falcone A, et al. Molecular analysis of cell-free circulating DNA for the diagnosis of somatic mutations associated with resistance to tyrosine kinase inhibitors in non-small-cell lung cancer. *Expert Rev Mol Diagn* 2014;**14**:453–68.
361. Diehl F, Schmidt K, Choti MA, et al. Circulating mutant DNA to assess tumor dynamics. *Nat Med* 2008;**14**:985–90.
362. Guttery DS, Page K, Hills A, et al. Noninvasive detection of activating Estrogen Receptor 1 (ESR1) mutations in estrogen receptor-positive metastatic breast cancer. *Clin Chem* 2015;**61**(7):974–82.
363. Maier J, Lange T, Kerle I, et al. Detection of mutant free circulating tumor DNA in the plasma of patients with gastrointestinal stromal tumor harboring activating mutations of CKIT or PDGFRA. *Clin Cancer Res* 2013;**19**:4854–67.
364. Mok T, Wu YL, Lee JS, et al. Detection and dynamic changes of EGFR mutations from circulating tumor DNA as a predictor of survival outcomes in NSCLC patients treated with first-line intercalated erlotinib and chemotherapy. *Clin Cancer Res* 2015;**21**(14):3196–203.
365. Morelli MP, Overman MJ, Dasari A, et al. Characterizing the patterns of clonal selection in circulating tumor DNA from patients with colorectal cancer refractory to anti-EGFR treatment. *Ann Oncol* 2015;**26**:731–6.
366. Murtaza M, Dawson SJ, Tsui DW, et al. Non-invasive analysis of acquired resistance to cancer therapy by sequencing of plasma DNA. *Nature* 2013;**497**:108–12.
367. Oshiro C, Kagara N, Naoi Y, et al. PIK3CA mutations in serum DNA are predictive of recurrence in primary breast cancer patients. *Breast Cancer Res Treat* 2015;**150**:299–307.
368. Oxnard GR, Paweletz CP, Kuang Y, et al. Noninvasive detection of response and resistance in EGFR-mutant lung cancer using quantitative next-generation genotyping of cell-free plasma DNA. *Clin Cancer Res* 2014;**20**:1698–705.
369. Pupilli C, Pinzani P, Salvianti F, et al. Circulating BRAFV600E in the diagnosis and follow-up of differentiated papillary thyroid carcinoma. *J Clin Endocrinol Metab* 2013;**98**:3359–65.
370. Rothe F, Laes JF, Lambrechts D, et al. Plasma circulating tumor DNA as an alternative to metastatic biopsies for mutational analysis in breast cancer. *Ann Oncol* 2014;**25**:1959–65.
371. Shinozaki M, O'Day SJ, Kitago M, et al. Utility of circulating B-RAF DNA mutation in serum for monitoring melanoma patients receiving biochemotherapy. *Clin Cancer Res* 2007;**13**:2068–74.
372. Sorenson GD. Detection of mutated KRAS2 sequences as tumor markers in plasma/serum of patients with gastrointestinal cancer. *Clin Cancer Res* 2000;**6**:2129–37.
373. Spindler KL, Pallisgaard N, Vogelius I, et al. Quantitative cell-free DNA, KRAS, and BRAF mutations in plasma from patients with metastatic colorectal cancer during treatment with cetuximab and irinotecan. *Clin Cancer Res* 2012;**18**:1177–85.
374. Taly V, Pekin D, Benhaim L, et al. Multiplex picodroplet digital PCR to detect KRAS mutations in circulating DNA from the plasma of colorectal cancer patients. *Clin Chem* 2013;**59**:1722–31.
375. Wong AL, Lim JS, Sinha A, et al. Tumour pharmacodynamics and circulating cell free DNA in patients with refractory colorectal carcinoma treated with regorafenib. *J Transl Med* 2015;**13**:57.
376. Yung TK, Chan KC, Mok TS, et al. Single-molecule detection of epidermal growth factor receptor mutations in plasma by microfluidics digital PCR in non-small cell lung cancer patients. *Clin Cancer Res* 2009;**15**:2076–84.
377. Austin LK, Fortina P, Sebisanovic D, et al. Circulating tumor DNA (ctDNA) as a molecular monitoring tool in metastatic breast cancer (MBC). *ASCO Meeting Abstracts* 2014;**32**:11093.
378. Karachaliou N, Mayo-de-Las-Casas C, Molina-Vila MA, et al. Real-time liquid biopsies become a reality in cancer treatment. *Ann Transl Med* 2015;**3**:36.
379. Fukuoka M, Wu YL, Thongprasert S, et al. Biomarker analyses and final overall survival results from a phase III, randomized, open-label, first-line study of gefitinib versus carboplatin/paclitaxel in clinically selected patients with advanced non-small-cell lung cancer in Asia (IPASS). *J Clin Oncol* 2011;**29**:2866–74.
380. Rosell R, Carcereny E, Gervais R, et al. Erlotinib versus standard chemotherapy as first-line treatment for European patients with advanced EGFR mutation-positive non-small-cell lung cancer (EURTAC): a multicentre, open-label, randomised phase 3 trial. *Lancet Oncol* 2012;**13**:239–46.
381. Seto T, Kato T, Nishio M, et al. Erlotinib alone or with bevacizumab as first-line therapy in patients with advanced non-squamous non-small-cell lung cancer harbouring EGFR mutations (JO25567): an open-label, randomised, multicentre, phase 2 study. *Lancet Oncol* 2014;**15**:1236–44.
382. Zhou C, Wu YL, Chen G, et al. Erlotinib versus chemotherapy as first-line treatment for patients with advanced EGFR mutation-positive non-small-cell lung cancer (OPTIMAL, CTONG-0802): a multicentre, open-label, randomised, phase 3 study. *Lancet Oncol* 2011;**12**:735–42.

383. Morgensztern D, Politi K, Herbst RS. Egfr mutations in non—small-cell lung cancer: find, divide, and conquer. *JAMA Oncol* 2015;**1**:146—8.
384. Rosell R, Moran T, Queralt C, et al. Screening for epidermal growth factor receptor mutations in lung cancer. *N Engl J Med* 2009;**361**:958—67.
385. Kimura H, Suminoe M, Kasahara K, et al. Evaluation of epidermal growth factor receptor mutation status in serum DNA as a predictor of response to gefitinib (IRESSA). *Br J Cancer* 2007;**97**:778—84.
386. Bechmann T, Andersen RF, Pallisgaard N, et al. Plasma HER2 amplification in cell-free DNA during neoadjuvant chemotherapy in breast cancer. *J Cancer Res Clin Oncol* 2013;**139**:995—1003.
387. Leary RJ, Sausen M, Kinde I, et al. Detection of chromosomal alterations in the circulation of cancer patients with whole-genome sequencing. *Sci Transl Med* 2012;**4**:162ra54.
388. Komatsu S, Ichikawa D, Hirajima S, et al. Clinical impact of predicting CCND1 amplification using plasma DNA in superficial esophageal squamous cell carcinoma. *Dig Dis Sci* 2014;**59**:1152—9.
389. Azad AA, Volik SV, Wyatt AW, et al. Androgen receptor gene aberrations in circulating cell-free DNA: biomarkers of therapeutic resistance in castration-resistant prostate cancer. *Clin Cancer Res* 2015;**21**:2315—24.
390. Chan KC, Jiang P, Zheng YW, et al. Cancer genome scanning in plasma: detection of tumor-associated copy number aberrations, single-nucleotide variants, and tumoral heterogeneity by massively parallel sequencing. *Clin Chem* 2013;**59**:211—24.
391. Shaw JA, Page K, Blighe K, et al. Genomic analysis of circulating cell-free DNA infers breast cancer dormancy. *Genome Res* 2012;**22**:220—31.
392. Belic J, Koch M, Ulz P, et al. Rapid identification of plasma DNA samples with increased ctDNA levels by a modified FAST-SeqS approach. *Clin Chem* 2015;**61**:838—49.
393. Fujimoto A, O'Day SJ, Taback B, et al. Allelic imbalance on 12q22-23 in serum circulating DNA of melanoma patients predicts disease outcome. *Cancer Res* 2004;**64**:4085—8.
394. Fujiwara Y, Chi DD, Wang H, et al. Plasma DNA microsatellites as tumor-specific markers and indicators of tumor progression in melanoma patients. *Cancer Res* 1999;**59**:1567—71.
395. Schwarzenbach H, Eichelser C, Kropidlowski J, et al. Loss of heterozygosity at tumor suppressor genes detectable on fractionated circulating cell-free tumor DNA as indicator of breast cancer progression. *Clin Cancer Res* 2012;**18**:5719—30.
396. Taback B, O'Day SJ, Boasberg PD, et al. Circulating DNA microsatellites: molecular determinants of response to biochemotherapy in patients with metastatic melanoma. *J Natl Cancer Inst* 2004;**96**:152—6.
397. Wang Q, Larson PS, Schlechter BL, et al. Loss of heterozygosity in serial plasma DNA samples during follow-up of women with breast cancer. *Int J Cancer* 2003;**106**:923—9.
398. Fujimoto A, Takeuchi H, Taback B, et al. Allelic imbalance of 12q22-23 associated with APAF-1 locus correlates with poor disease outcome in cutaneous melanoma. *Cancer Res* 2004;**64**:2245—50.
399. Taback B, Fujiwara Y, Wang HJ, et al. Prognostic significance of circulating microsatellite markers in the plasma of melanoma patients. *Cancer Res* 2001;**61**:5723—6.
400. Taback B, Giuliano AE, Hansen NM, et al. Detection of tumor-specific genetic alterations in bone marrow from early-stage breast cancer patients. *Cancer Res* 2003;**63**:1884—7.
401. Hoon DS, Spugnardi M, Kuo C, et al. Profiling epigenetic inactivation of tumor suppressor genes in tumors and plasma from cutaneous melanoma patients. *Oncogene* 2004;**23**:4014—22.
402. Hoshimoto S, Kuo CT, Chong KK, et al. AIM1 and LINE-1 epigenetic aberrations in tumor and serum relate to melanoma progression and disease outcome. *J Invest Dermatol* 2012;**132**:1689—97.
403. Mori T, Martinez SR, O'Day SJ, et al. Estrogen receptor-alpha methylation predicts melanoma progression. *Cancer Res* 2006;**66**:6692—8.
404. Mori T, O'Day SJ, Umetani N, et al. Predictive utility of circulating methylated DNA in serum of melanoma patients receiving biochemotherapy. *J Clin Oncol* 2005;**23**:9351—8.
405. Church TR, Wandell M, Lofton-Day C, et al. Prospective evaluation of methylated SEPT9 in plasma for detection of asymptomatic colorectal cancer. *Gut* 2014;**63**:317—25.
406. deVos T, Tetzner R, Model F, et al. Circulating methylated SEPT9 DNA in plasma is a biomarker for colorectal cancer. *Clin Chem* 2009;**55**:1337—46.
407. Ladabaum U, Allen J, Wandell M, et al. Colorectal cancer screening with blood-based biomarkers: cost-effectiveness of methylated septin 9 DNA versus current strategies. *Cancer Epidemiol Biomarkers Prev* 2013;**22**:1567—76.
408. Sunami E, Shinozaki M, Higano CS, et al. Multimarker circulating DNA assay for assessing blood of prostate cancer patients. *Clin Chem* 2009;**55**:559—67.
409. Chan KC, Lai PB, Mok TS, et al. Quantitative analysis of circulating methylated DNA as a biomarker for hepatocellular carcinoma. *Clin Chem* 2008;**54**:1528—36.
410. Cortese R, Kwan A, Lalonde E, et al. Epigenetic markers of prostate cancer in plasma circulating DNA. *Hum Mol Genet* 2012;**21**:3619—31.
411. Liggett TE, Melnikov AA, Marks JR, et al. Methylation patterns in cell-free plasma DNA reflect removal of the primary tumor and drug treatment of breast cancer patients. *Int J Cancer* 2011;**128**:492—9.
412. Ling ZQ, Lv P, Lu XX, et al. Circulating methylated DNA indicates poor prognosis for gastric cancer. *PLoS ONE* 2013;**8**:e67195.
413. Misawa A, Tanaka S, Yagyu S, et al. RASSF1A hypermethylation in pretreatment serum DNA of neuroblastoma patients: a prognostic marker. *Br J Cancer* 2009;**100**:399—404.
414. Wallner M, Herbst A, Behrens A, et al. Methylation of serum DNA is an independent prognostic marker in colorectal cancer. *Clin Cancer Res* 2006;**12**:7347—52.
415. Belinsky SA, Klinge DM, Dekker JD, et al. Gene promoter methylation in plasma and sputum increases with lung cancer risk. *Clin Cancer Res* 2005;**11**:6505—11.
416. Cassinotti E, Melson J, Liggett T, et al. DNA methylation patterns in blood of patients with colorectal cancer and adenomatous colorectal polyps. *Int J Cancer* 2012;**131**:1153—7.
417. Danese E, Minicozzi AM, Benati M, et al. Epigenetic alteration: new insights moving from tissue to plasma—The example of PCDH10 promoter methylation in colorectal cancer. *Br J Cancer* 2013;**109**:807—13.

418. Hoque MO, Feng Q, Toure P, et al. Detection of aberrant methylation of four genes in plasma DNA for the detection of breast cancer. *J Clin Oncol* 2006;**24**:4262–9.
419. Liggett T, Melnikov A, Yi QL, et al. Differential methylation of cell-free circulating DNA among patients with pancreatic cancer versus chronic pancreatitis. *Cancer* 2010;**116**:1674–80.
420. Liggett TE, Melnikov A, Yi Q, et al. Distinctive DNA methylation patterns of cell-free plasma DNA in women with malignant ovarian tumors. *Gynecol Oncol* 2011;**120**:113–20.
421. Melnikov A, Scholtens D, Godwin A, et al. Differential methylation profile of ovarian cancer in tissues and plasma. *J Mol Diagn* 2009;**11**:60–5.
422. Ostrow KL, Hoque MO, Loyo M, et al. Molecular analysis of plasma DNA for the early detection of lung cancer by quantitative methylation-specific PCR. *Clin Cancer Res* 2010;**16**:3463–72.
423. Ponomaryova AA, Rykova EY, Cherdyntseva NV, et al. RARbeta2 gene methylation level in the circulating DNA from blood of patients with lung cancer. *Eur J Cancer Prev* 2011;**20**:453–5.
424. Balgkouranidou I, Chimonidou M, Milaki G, et al. Breast cancer metastasis suppressor-1 promoter methylation in cell-free DNA provides prognostic information in non-small cell lung cancer. *Br J Cancer* 2014;**110**:2054–62.
425. Barault L, Amatu A, Bleeker FE, et al. Digital PCR quantification of MGMT methylation refines prediction of clinical benefit from alkylating agents in glioblastoma and metastatic colorectal cancer. *Ann Oncol* 2015;**26**(9):1994–9.
426. Mahon KL, Qu W, Devaney J, et al. Methylated glutathione S-transferase 1 (mGSTP1) is a potential plasma free DNA epigenetic marker of prognosis and response to chemotherapy in castrate-resistant prostate cancer. *Br J Cancer* 2014;**111**:1802–9.
427. Martinez-Galan J, Torres-Torres B, Nunez MI, et al. ESR1 gene promoter region methylation in free circulating DNA and its correlation with estrogen receptor protein expression in tumor tissue in breast cancer patients. *BMC Cancer* 2014;**14**:59.
428. Tangkijvanich P, Hourpai N, Rattanatanyong P, et al. Serum LINE-1 hypomethylation as a potential prognostic marker for hepatocellular carcinoma. *Clin Chim Acta* 2007;**379**:127–33.
429. Mead R, Duku M, Bhandari P, et al. Circulating tumour markers can define patients with normal colons, benign polyps, and cancers. *Br J Cancer* 2011;**105**:239–45.
430. Leung SF, Zee B, Ma BB, et al. Plasma Epstein-Barr viral deoxyribonucleic acid quantitation complements tumor-node-metastasis staging prognostication in nasopharyngeal carcinoma. *J Clin Oncol* 2006;**24**:5414–8.
431. Lin JC, Wang WY, Chen KY, et al. Quantification of plasma Epstein-Barr virus DNA in patients with advanced nasopharyngeal carcinoma. *N Engl J Med* 2004;**350**:2461–70.
432. Leung SF, Chan KC, Ma BB, et al. Plasma Epstein-Barr viral DNA load at midpoint of radiotherapy course predicts outcome in advanced-stage nasopharyngeal carcinoma. *Ann Oncol* 2014;**25**:1204–8.
433. Chan KC, Hung EC, Woo JK, et al. Early detection of nasopharyngeal carcinoma by plasma Epstein-Barr virus DNA analysis in a surveillance program. *Cancer* 2013;**119**:1838–44.
434. Wang WY, Twu CW, Chen HH, et al. Plasma EBV DNA clearance rate as a novel prognostic marker for metastatic/recurrent nasopharyngeal carcinoma. *Clin Cancer Res* 2010;**16**:1016–24.
435. Capone RB, Pai SI, Koch WM, et al. Detection and quantitation of human papillomavirus (HPV) DNA in the sera of patients with HPV-associated head and neck squamous cell carcinoma. *Clin Cancer Res* 2000;**6**:4171–5.
436. Dahlstrom KR, Li G, Hussey CS, et al. Circulating human papillomavirus DNA as a marker for disease extent and recurrence among patients with oropharyngeal cancer. *Cancer* 2015;**121**(19):3455–64.
437. Ren XD, Lin SY, Wang X, et al. Rapid and sensitive detection of hepatitis B virus 1762T/1764A double mutation from hepatocellular carcinomas using LNA-mediated PCR clamping and hybridization probes. *J Virol Methods* 2009;**158**:24–9.
438. Campitelli M, Jeannot E, Peter M, et al. Human papillomavirus mutational insertion: specific marker of circulating tumor DNA in cervical cancer patients. *PLoS ONE* 2012;**7**:e43393.
439. Jahr S, Hentze H, Englisch S, et al. DNA fragments in the blood plasma of cancer patients: quantitations and evidence for their origin from apoptotic and necrotic cells. *Cancer Res* 2001;**61**:1659–65.
440. Jiang P, Chan CW, Chan KC, et al. Lengthening and shortening of plasma DNA in hepatocellular carcinoma patients. *Proc Natl Acad Sci USA* 2015;**112**:E1317–25.
441. Umetani N, Giuliano AE, Hiramatsu SH, et al. Prediction of breast tumor progression by integrity of free circulating DNA in serum. *J Clin Oncol* 2006;**24**:4270–6.
442. Wang BG, Huang H-Y, Chen Y-C, et al. Increased plasma DNA integrity in cancer patients. *Cancer Res* 2003;**63**:3966–8.
443. Deininger P. Alu elements: know the SINEs. *Genome Biol* 2011;**12**:236.
444. Umetani N. Prediction of breast tumor progression by integrity of free circulating DNA in serum. *J Clin Oncol* 2006;**24**:4270–6.
445. Chan KC, Leung SF, Yeung SW, et al. Persistent aberrations in circulating DNA integrity after radiotherapy are associated with poor prognosis in nasopharyngeal carcinoma patients. *Clin Cancer Res* 2008;**14**:4141–5.
446. Gang F, Guorong L, An Z, et al. Prediction of clear cell renal cell carcinoma by integrity of cell-free DNA in serum. *Urology* 2010;**75**:262–5.
447. Hanley R, Rieger-Christ KM, Canes D, et al. DNA integrity assay: a plasma-based screening tool for the detection of prostate cancer. *Clin Cancer Res* 2006;**12**:4569–74.
448. Jiang WW, Zahurak M, Goldenberg D, et al. Increased plasma DNA integrity index in head and neck cancer patients. *Int J Cancer* 2006;**119**:2673–6.
449. Lehner J, Stotzer OJ, Fersching D, et al. Circulating plasma DNA and DNA integrity in breast cancer patients undergoing neoadjuvant chemotherapy. *Clin Chim Acta* 2013;**425**:206–11.
450. Pinzani P, Salvianti F, Zaccara S, et al. Circulating cell-free DNA in plasma of melanoma patients: qualitative and quantitative considerations. *Clin Chim Acta* 2011;**412**:2141–5.
451. Tomita H, Ichikawa D, Ikoma D, et al. Quantification of circulating plasma DNA fragments as tumor markers in patients with esophageal cancer. *Anticancer Res* 2007;**27**:2737–41.
452. Umetani N. Increased integrity of free circulating DNA in sera of patients with colorectal or periampullary cancer: direct quantitative PCR for ALU repeats. *Clin Chem* 2006;**52**:1062–9.
453. Chan KC, Leung SF, Yeung SW, et al. Persistent aberrations in circulating DNA integrity after radiotherapy are associated

with poor prognosis in nasopharyngeal carcinoma patients. *Clin Cancer Res* 2008;**14**:4141–5.
454. Agostini M, Pucciarelli S, Enzo MV, et al. Circulating cell-free DNA: a promising marker of pathologic tumor response in rectal cancer patients receiving preoperative chemoradiotherapy. *Ann Surg Oncol* 2011;**18**:2461–8.
455. Lo YM, Zhang J, Leung TN, et al. Rapid clearance of fetal DNA from maternal plasma. *Am J Hum Genet* 1999;**64**:218–24.
456. Azmi AS, Bao B, Sarkar FH. Exosomes in cancer development, metastasis, and drug resistance: a comprehensive review. *Cancer Metastasis Rev* 2013;**32**:623–42.
457. Thakur BK, Zhang H, Becker A, et al. Double-stranded DNA in exosomes: a novel biomarker in cancer detection. *Cell Res* 2014; **24**:766–9.
458. Esposito A, Bardelli A, Criscitiello C, et al. Monitoring tumor-derived cell-free DNA in patients with solid tumors: clinical perspectives and research opportunities. *Cancer Treat Rev* 2014;**40**:648–55.
459. Hosny G, Farahat N, Tayel H, et al. Ser-249 TP53 and CTNNB1 mutations in circulating free DNA of Egyptian patients with hepatocellular carcinoma versus chronic liver diseases. *Cancer Lett* 2008;**264**:201–8.
460. Dobrzycka B, Terlikowski SJ, Mazurek A, et al. Circulating free DNA, p53 antibody and mutations of KRAS gene in endometrial cancer. *Int J Cancer* 2010;**127**:612–21.
461. Gahan PB. Circulating nucleic acids in plasma and serum: diagnosis and prognosis in cancer. *EPMA J* 2010;**1**:503–12.
462. Schwarzenbach H, Goekkurt E, Pantel K, et al. Molecular analysis of the polymorphisms of thymidylate synthase on cell-free circulating DNA in blood of patients with advanced colorectal carcinoma. *Int J Cancer* 2010;**127**:881–8.
463. Kuhlmann JD, Schwarzenbach H, Wimberger P, et al. LOH at 6q and 10q in fractionated circulating DNA of ovarian cancer patients is predictive for tumor cell spread and overall survival. *BMC Cancer* 2012;**12**:325.
464. Ellinger J, Bastian PJ, Ellinger N, et al. Apoptotic DNA fragments in serum of patients with muscle invasive bladder cancer: a prognostic entity. *Cancer Lett* 2008;**264**:274–80.
465. Shi W, Lv C, Qi J, et al. Prognostic value of free DNA quantification in serum and cerebrospinal fluid in glioma patients. *J Mol Neurosci* 2012;**46**:470–5.
466. Salani R, Davidson B, Fiegl M, et al. Measurement of cyclin E genomic copy number and strand length in cell-free DNA distinguish malignant versus benign effusions. *Clin Cancer Res* 2007;**13**:5805–9.
467. Stott SL, Hsu CH, Tsukrov DI, et al. Isolation of circulating tumor cells using a microvortex-generating herringbone-chip. *Proc Natl Acad Sci USA* 2010;**107**:18392–7.
468. Khoo BL, Warkiani ME, Tan DS, et al. Clinical validation of an ultra high-throughput spiral microfluidics for the detection and enrichment of viable circulating tumor cells. *PLoS ONE* 2014;**9**:e99409.
469. Ramirez JM, Fehm T, Orsini M, et al. Prognostic relevance of viable circulating tumor cells detected by EPISPOT in metastatic breast cancer patients. *Clin Chem* 2014;**60**:214–21.

Circulating Nucleic Acids for Prenatal Diagnostics

Rossa W.K. Chiu and Y.M. Dennis Lo

ABSTRACT

Background
Prenatal diagnosis is an important part of prenatal care for many patients. Traditional methods of sampling fetal genetic material for a definitive diagnosis, such as chorionic villus sampling and amniocentesis, are invasive and expose the fetus to the risk of spontaneous miscarriage. The discovery of the presence of fetal cell–free DNA in the circulation of pregnant women led to the possibility of noninvasive genetic and chromosomal assessment of the fetus through the sampling of maternal peripheral blood. The clinical introduction of cell-free fetal DNA (cffDNA) tests has led to a substantial reduction in the number of invasive procedures performed worldwide.

Content
This chapter describes the biological properties of circulating cffDNA, the applications that have been developed, and their clinical uses, and it highlights some analytical features that require careful consideration. Circulating cffDNA is derived from cell turnover of the placenta. This DNA is highly fragmented, with the majority of fragments less than 200 bp in length. Circulating cffDNA exists with a substantial background of maternal DNA. Circulating cffDNA can be detected from early pregnancy onward, and it rapidly disappears from maternal circulation following delivery of the newborn. The analysis of cffDNA is now clinically used for the assessment of sex-linked diseases, fetal blood group incompatibility, fetal chromosomal aneuploidies, and some single-gene diseases. Analytical protocols are designed to maximize the yield of fetal DNA by minimizing maternal DNA contamination and by preserving the abundance of short DNA molecules. The inclusion of internal positive controls for the presence of fetal DNA or the measurement of fetal DNA fraction is an important quality control parameter.

BRIEF OVERVIEW OF THE EARLY DEVELOPMENTS OF PRENATAL GENETIC DIAGNOSTICS

Prenatal diagnosis is an important part of prenatal care for many patients. It encompasses both diagnostic and screening tests that detect or exclude morphological, structural, functional, chromosomal, and molecular defects in a fetus.[1] Amniocentesis was first introduced in 1952 for the prenatal assessment of fetal hemolytic disease.[2] This was followed by karyotyping of amniotic fluid cells in 1966,[3] and then ultrasonography for fetal structural abnormalities in the 1970s.[4,5] Later, maternal serum biochemistry testing was shown to be of value in the screening of neural tube defects[6,7] and fetal aneuploidies.[8] In the early 1980s, chorionic villus sampling (CVS) became available as an alternative to amniocentesis for prenatal genetic assessment.[9] For many years, amniocentesis and CVS were the key approaches for providing fetal genetic material used in prenatal testing.

The main disadvantage of amniocentesis and CVS is the procedural-related risk of fetal miscarriage. The fetal loss rate associated with the performance of these invasive procedures is about 0.5%.[10] Consequently, much effort has been devoted to the development of noninvasive approaches to identify high-risk pregnant women.

STRATEGIES TO MITIGATE RISKS OF INVASIVE PRENATAL DIAGNOSIS

The risk for Down syndrome, with an incidence rate of 1 in 800 pregnancies,[11] is one of the predominant reasons for women seeking prenatal diagnosis. Strategies have been devised to identify high-risk pregnancies by the combined assessment of maternal age, serum biochemical markers, and ultrasonographic findings. The purpose of this assessment is to risk stratify pregnancies where the chance of having an affected fetus is higher than the chance of a procedure-related fetal loss. Different combinations of screening strategies have been practiced, with different levels of specificity and sensitivity.[12]

Maternal Age
The probability of giving birth to an infant with Down syndrome increases with advancing maternal age.[13] The risk of giving birth to an affected infant at term is estimated to be less than 1 in 1000 at a maternal age of 29 years and younger, but it increases to 1 in 385 at 35 years of age.[14] Hence, prior to the development of more elaborate prenatal screening strategies, it was customary to offer prenatal diagnosis to women aged 35 years or older. However, because a significant

proportion of women become pregnant before 35 years of age, maternal age alone would only identify 51% of Down syndrome–affected pregnancies at a 14% false-positive rate.[15]

Serum Biochemistry Screening

The combination of maternal age assessment with maternal serum screening of various biomarkers between 15 and 22 weeks of gestation was later developed as a second trimester screening protocol to identify high-risk pregnancies. This screening strategy is referred to as the "triple test," and the serum biomarkers include alpha-fetoprotein, human chorionic gonadotropin, and unconjugated estriol. When the analytical cutoff values are set to give a 5% false-positive rate, the detection sensitivity for Down syndrome is 70%.[16] Testing maternal serum inhibin A and the triple test markers during the second trimester, termed the "quadruple test," provided a detection sensitivity of 75% at a false-positive rate of 5%.[15]

First Trimester Screening

The triple test is used during the second trimester; during the first trimester, alternative Down syndrome screening strategies have been developed. Free β-human chorionic gonadotropin and pregnancy-associated plasma protein A are used for Down syndrome screening in the first trimester.[17,18] Down syndrome is associated with an increase in fetal nuchal translucency measured by first trimester ultrasound.[19] Subsequently, the combination of first trimester biochemical markers, fetal nuchal translucency, and maternal age assessments came to be known as the "first trimester combined test." As a false-positive rate of 5%, the test could detect 95% of Down syndrome fetuses.[20]

The approaches described above have been incorporated into many prenatal screening programs. However, the main disadvantage of these tests lies in their high false-positive rates. Most of the test cutoff values used to identify those deemed to be at high risk had false-positive rates of 5%. This meant that 1 in every 20 women would be labeled as high risk and would need to face the decision of whether or not to undergo an invasive diagnostic procedure. Because the average Down syndrome risk is 1 in 800, this meant that a substantial number of women undergoing an invasive diagnostic procedure did not carry an affected fetus. Therefore there was a need to identify new screening methods that had lower false-positive rates and improved detection rates.

NONINVASIVE FETAL DNA ANALYSIS

The above prenatal screening methods are based on the detection of prenatal phenotypic features that tend to be associated with Down syndrome. It had been reasoned that to improve the sensitivity and specificity of prenatal screening, methods directed at the detection of the core pathology of the condition are needed—for example, trisomy 21 for Down syndrome or the fetal mutations for single-gene diseases. To this end, noninvasive methods have been developed to provide access to fetal DNA for analysis.

Fetal Cells in Maternal Blood

Over a century ago, a German pathologist, Schmorl,[21] observed the presence of trophoblasts in the lung tissues of women who died of preeclampsia. The existence of such circulating fetal cells was later confirmed by molecular techniques based on the detection of chromosome Y DNA sequences in the blood of women pregnant with male fetuses.[22] The idea of noninvasive prenatal diagnosis based on maternal blood sampling began to emerge. It was subsequently realized that intact fetal cells are present in maternal circulation rarely, with about just one cell per milliliter of maternal blood.[23] Protocols have since been developed to isolate and enrich for these rare fetal cells, including fluorescence activated cell sorting[24] and magnetic activated cell sorting.[25] To date, the identification and isolation of intact fetal cells in maternal circulation have remained difficult.[26] However, once isolated, they may be amenable to single-cell whole-genome analysis for the potential identification of fetal genetic and genomic abnormalities.[27] Therefore with further improvement in technologies, circulating fetal cell–based noninvasive prenatal diagnostics may be a clinically viable option in the future.

Cell-Free Fetal DNA in Maternal Plasma

Instead of intact cells, Lo and colleagues[28] searched for fetal DNA in the cell-free portion of maternal blood—namely, plasma and serum. Chromosome Y DNA was detected in the plasma and serum of women who were pregnant with male fetuses but not in women with female fetuses. Since this first report in 1997, much research investigated the properties and potential utilities of circulating cell-free fetal DNA (cffDNA). cffDNA analysis is currently used for routine prenatal testing.

THE BIOLOGY OF CIRCULATING CELL-FREE FETAL NUCLEIC ACIDS IN MATERNAL PLASMA

Every milliliter of maternal plasma contains thousands of genome-equivalents of cell-free DNA, with the fetus contributing a minor proportion.[29] In other words, the majority of the cell-free DNA in maternal plasma is derived from the mother. The median amount of cffDNA as a proportion of the total DNA in maternal plasma, termed the "fetal fraction," is around 10% in the first and second trimesters and around 20% during the third trimester.[30] The absolute concentration of cffDNA increases with gestational age, probably as a consequence of the increase in placental tissue mass. Nonetheless, its abundance in the maternal circulation far exceeds that of intact fetal cells.

> **POINTS TO REMEMBER**
>
> **Biological Properties of cffDNA**
> - cffDNA is a by-product of placental cell turnover.
> - cffDNA coexists in maternal plasma with a major background of maternal DNA.
> - cffDNA is highly fragmented and is generally less than 200 bp long.
> - cffDNA has rapid clearance kinetics and an apparent half-life of 1 hour.
> - cffDNA is detectable in early pregnancy. It is most reliably detected from the late first trimester onward.

Fetal DNA molecules are detectable in maternal plasma quite early in pregnancy. Depending on the analytical platform used, cffDNA is detected from around the 10th week

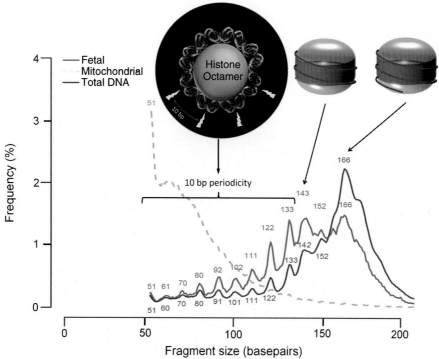

FIGURE 10.1 Size distribution of fetal, mitochondrial, and total DNA. Numbers denote the DNA size in bps at the peaks. Schematic illustrations of the structural organization of a nucleosome are shown above the graph. From left to right, DNA double helix wound around a nucleosomal core unit with the sites for nuclease cleavage shown; a nucleosome core unit (blue) with approximately 146 bp of DNA (red strand) wound around it; and a nucleosomal core unit with an approximately 20-bp linker intact.

of gestation.[31] cffDNA is cleared from maternal plasma very rapidly after delivery.[32] Using sensitive methods to measure cffDNA serially after delivery, the half-life of cffDNA in the postpartum maternal serum is about an hour.[33] No cffDNA could be detected in maternal plasma 1 day after delivery. These observations reveal that fetal DNA molecules in maternal plasma have a high turnover rate with efficient clearance mechanisms. Renal clearance studies performed by serial monitoring of cffDNA in maternal plasma and maternal urine after delivery show that renal excretion is only a minor component of fetal DNA clearance.[33]

Despite the rapid clearance kinetics, cffDNA amounts to some 10% to 20% of the total DNA in maternal plasma; this suggests that substantial amounts of fetal DNA may be released by a tissue source at any point in time. Two lines of evidence demonstrate the placenta as the predominant tissue that releases fetal DNA into the maternal circulation: first, epigenetic markers specific to the placenta are detectable in maternal plasma;[34,35] second, chromosomal abnormalities confined to the placenta are detectable in maternal plasma.[36] On the other hand, the main contributors of cell-free DNA from the mother are maternal hematological cells.[35,37]

Cell-free DNA is a metabolic by-product of cell death and is therefore present in the circulation as short fragments in a "cell-free" form. Using paired-end massively parallel sequencing, Lo and colleagues studied the size profile of maternal plasma DNA in a high-resolution manner (Fig. 10.1).[38] By noting the genomic coordinates of the outermost ends of each plasma DNA molecule, the size of the fragments was determined. The cell-free DNA molecules are generally shorter than 200 bp. The most frequently represented size in plasma is 166 bp in length. 166 bp corresponds to the length of DNA that is wound around a histone core with a linker. This characteristic size reflected that cell-free DNA is mainly derived from mononucleosomes. Interestingly, cffDNA molecules are somewhat shorter, and the most frequently represented size is 142 bp in length.[38] This shorter length corresponds to the length of DNA wound around a histone core without a linker. This implies that the fetal DNA molecules may have undergone further steps of degradation compared to maternal DNA in maternal plasma. In addition to these dominant peaks, there are smaller amounts of cell-free DNA that are successively shorter in 10-bp increments. This observation suggests that DNase participates in the degradation of cell-free DNA because DNase cutting sites are located every 10 bases around the nucleosomal DNA.

DIAGNOSTIC APPLICATIONS OF CIRCULATING CFFDNA

Cell-free fetal DNA is a source of fetal genetic material that may be sampled noninvasively by maternal phlebotomy. Its relative high abundance in maternal plasma from early pregnancy, with no postpartum persistence, has facilitated the development of a number of applications for noninvasive prenatal assessment.

> **POINTS TO REMEMBER**
>
> **Diagnostic Applications of cffDNA**
> - Fetal sex assessment for the management of fetal sex–associated diseases.
> - Fetal *RHD* genotyping for the management of rhesus D incompatibility.
> - Fetal chromosomal aneuploidy screening, especially trisomies 21, 13, and 18.
> - Fetal single-gene disease detection, including paternally or maternally transmitted autosomal dominant diseases, autosomal recessive diseases, and sex-linked diseases.

Fetal Sex Assessment for Sex-Associated Disorders

The first report of cffDNA was based on the detection of male fetal DNA sequences in the plasma of women pregnant with male fetuses.[28] This noninvasive test for fetal sex assessment was immediately useful for clinical purposes. The accurate assessment of fetal sex is useful for the prenatal management of diseases with sex-linked patterns of inheritance and conditions such as congenital adrenal hyperplasia, where the disease manifestation differs between male and female offspring. For sex-linked genetic diseases, such as hemophilia and Duchenne muscular dystrophy, invasive prenatal diagnostic procedures could be avoided if the noninvasive fetal sex assessment suggested a female fetus. For congenital adrenal hyperplasia secondary to 21-hydroxylase deficiency, female fetuses are at risk of virilization. Thus steroid therapy and further prenatal genetic assessment may be avoided if the noninvasive fetal sex assessment suggested a male fetus. In general, the specificity for male cffDNA detection approaches 99%.[31] In terms of sensitivity, Devaney and colleagues showed that higher sensitivities could be reached by using later gestational ages, more replicate analyses, and higher-sensitivity analytical approaches.[31] For example, real-time PCR provides higher sensitivity than conventional PCR.

Fetal Rhesus D Status Determination

Rhesus (Rh) D incompatibility occurs when an RhD-negative women is pregnant with a RhD-positive fetus. RhD-negative blood cells lack the RhD antigen. Therefore when the RhD-negative maternal immune cells are presented with the RhD antigen of the RhD-positive fetal blood, alloimmunization occurs. Upon the next pregnancy with an RhD-positive fetus, the sensitized woman and the anti-RhD antibodies may cause destruction of the fetal tissues, causing hemolytic disease of the newborn. However, such risks do not exist if the woman is pregnant with an RhD-negative fetus. Therefore prenatal RhD genotyping of the fetus is useful in the management of RhD-negative pregnant women. The great majority of RhD-negative individuals lack the RhD gene, *RHD*, due to gene deletion. Therefore one could noninvasively assess the fetal RhD status by detecting the presence of *RHD* in the plasma of RhD-negative pregnant women.[39,40] Unlike conventional methods, such as amniocentesis or CVS, noninvasive methods are free from the risk of inducing fetomaternal hemorrhage and further sensitization. In fact, the analysis of cffDNA for noninvasive fetal RhD genotyping has been globally implemented for clinical use. In addition, the test has also been used as the basis for rationalizing the administration of prophylactic anti-D immunoglobulin only to pregnancies involving an RhD-positive fetus.[41] Such an approach may minimize the unnecessary use of the scarce and expensive anti-D immunoglobulin, as well as reduce the need to unnecessarily expose the pregnant woman to the anti-D blood product.

Fetal Chromosomal Aneuploidy Screening

Down syndrome is one of the key reasons for couples to consider prenatal diagnosis. Thus there has been a longstanding interest to develop noninvasive tests for Down syndrome assessment based on cffDNA analysis. Down syndrome is typically caused by the presence of an additional dose of chromosome 21—namely, trisomy 21—in the genome of affected individuals. Therefore the key to achieving noninvasive prenatal detection of Down syndrome is to provide evidence that increased copies of chromosome 21 are present in maternal plasma. The majority of the DNA in maternal plasma, including DNA from chromosome 21, originates from the mother who has a normal amount of chromosome 21. If the fetus has trisomy 21, it would contribute additional amounts of chromosome 21 into maternal plasma. Therefore the additional amount of cell-free chromosome 21 DNA molecules in maternal plasma is dependent on the fetal fraction (ie, the percent of fetal DNA in the background of maternal DNA in the plasma). The higher the fetal fraction, the easier the identification of trisomy 21—associated changes in maternal plasma DNA analysis.

Methodological Approaches

To precisely quantify and detect this small additional amount of chromosome 21 DNA, most protocols utilize massively parallel sequencing (see Chapter 4). One approach is based on random or shotgun sequencing of maternal plasma DNA.[42] The rationale is that among the cell-free DNA fragments in maternal plasma, if one sequences a random fraction of all the molecules, the relative amount of DNA obtained from each segment across the genome should reflect the relative DNA contribution of that segment in the genome of the tested individual. To determine the relative amount of DNA from each segment—say, each chromosome—one could determine the number of DNA molecules sequenced from that chromosome as a proportion of all molecules sequenced from the sample. The relative amount, or genomic representation, is then compared with the expected amount for the same chromosome among a control group of samples representing euploid pregnancies (Fig. 10.2). If a sample shows a genomic representation of chromosome 21 that is significantly different (eg, more than 3 standard deviations) from the control group, the amount of chromosome 21 is considered elevated and therefore suggestive of trisomy 21.

Because this approach is based on random whole-genome sequencing, it could in principle be applied to other chromosomal aneuploidies, such as trisomy 18 and trisomy 13, and the sex chromosome aneuploidies, such as Turner syndrome, 45 X0, and Klinefelter syndrome, 47 XXY.[43,44] The protocol could be repurposed for the detection of microdeletion and microduplication syndromes.[45,46] In fact, molecular karyotyping at Mb-level of resolution covering most parts of the genome appears possible.[45,46] Indeed, there is great versatility in the random whole-genome sequencing approach. As described in the section on analytical aspects, the successful implementation of these protocols relies on the precise quantification of the genomic segment of interest. For example, the signal-to-noise ratio for relative quantification of a

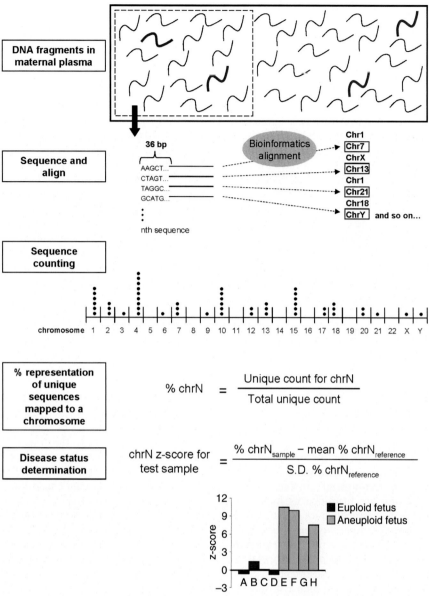

FIGURE 10.2 Schematic illustration of the procedural framework for using massively parallel genomic sequencing for the noninvasive prenatal detection of fetal chromosomal aneuploidy. Fetal DNA (thick red fragments) circulates in maternal plasma as a minor population among a high background of maternal DNA (black fragments). A sample containing representative DNA molecules in maternal plasma is obtained. Plasma DNA molecules are sequenced, and the chromosomal origin of each molecule is identified through mapping to the human reference genome by bioinformatic analysis. The number of sequences mapped to each chromosome is counted and then expressed as a percentage of all unique sequences generated for the sample, termed %chrN for chromosome N. Z-scores for each chromosome and each test sample are calculated using the formula shown. The z-score of a potentially aneuploid chromosome is expected to be higher for pregnancies with an aneuploid fetus (cases E to H) than those with a euploid fetus (cases A to D). (From Chiu et al. Noninvasive prenatal diagnosis of fetal chromosomal aneuploidy by massively parallel genomic sequencing of DNA in maternal plasma. *Proc Natl Acad Sci U S A* 2008;105:20458–63.)

genomic region is partly governed by the sequencing depth.[46,47] Thus the sequencing depth may need to be adjusted if the analysis is intended for the detection of sub-chromosomal changes instead of whole-chromosome aneuploidies. In addition, target capture of cell-free DNA originating from chromosomal regions of interest followed by targeted analysis could similarly achieve relative genomic representation assessment.[48]

To detect the presence of extra or missing copies of chromosomes, allelic ratio approaches have also been developed. Such approaches take advantage of polymorphic differences between homologous chromosomes between the mother and the fetus.[49] In principle, such polymorphic loci may include loci where the mother is homozygous and the fetus is heterozygous, or when the mother is heterozygous and the fetus is homozygous. The rationale is that when the fetus has aneuploidy, the ratio between alleles on that chromosome will be skewed. However, the extent of skewing between the alleles is dependent on the configuration of the polymorphic markers between the fetus and the mother as well as the fetal fraction. To implement this method, the allelic information of each cell-free DNA fragment needs to be compared to relative amounts of the homologous chromosomes.

Clinical Implementation

cffDNA-based noninvasive prenatal screening for fetal chromosomal aneuploidy became clinically available in 2011.[47] Within 3 years, service availability extended to over 60 countries.[50] In general, cffDNA-based noninvasive prenatal screening achieved detection rates of about 99% for Down syndrome with less than 1% false-positive rates.[51,52] Depending on the specific protocol used, the detection rates for trisomy 18 and trisomy 13 are about or greater than 90% with less than 1% false-positive rates.[51,52] In other words, cffDNA-based prenatal screening achieved better detection sensitivities and specificities than those screening modalities based on maternal serum biochemistry and/or fetal ultrasonography. Consequently, there are now many professional guidelines and recommendations supporting the use of cffDNA-based prenatal assessment for fetal chromosomal aneuploidy screening.[52-55] However, it is noteworthy that there are false-positive cases (see discussion below). Therefore cffDNA assessment is a screening procedure and not a definitive diagnosis. All clinical guidelines recommend that cffDNA tests with positive findings suggestive of chromosomal aneuploidy should be confirmed by tests on fetal genetic material collected by conventional invasive methods such as chorionic villus sampling or amniocentesis.

While the performance profile of the cffDNA-based tests is quite attractive for screening, the direct costs are higher than that of maternal serum biochemistry screening. To balance the overall costs, various modalities for incorporating the cffDNA-based tests into prenatal screening programs have evolved.[51] One option is to recommend cffDNA testing only to pregnant women who are identified as high risk and who would otherwise be recommended for conventional invasive prenatal diagnosis. For example, women who receive a high-risk score upon having undergone the first trimester combined screening test may consider the option of cffDNA testing. In this scenario, the cffDNA assessment is only performed for the 5% of pregnancies with high-risk scores. With their high-specificity profile, the cffDNA tests should be able to further identify 99% of the unaffected pregnancies. As a result, the number of pregnancies recommended for invasive testing would be reduced to only those patients with a positive cffDNA test result. In practice, since the implementation of cffDNA testing in the clinical setting, the reduction in invasive prenatal diagnostic procedures is between 26% and 69%.[56-58]

The two-step screening approach described above has the advantage of reducing the high false-positive rate of the conventional prenatal screening program. However, it does not raise the aneuploidy detection rates that are limited by the detection performance of first-tier screening tests. Consequently, some groups have proposed a "contingent approach" where the threshold to label a pregnancy as "high risk" by the conventional screening tests is relaxed to include a greater proportion of the population (ie, 10% of all pregnancies).[51] Theoretically, this would increase the detection rates for fetal chromosomal aneuploidies without an increase in the overall false-positive rate in the context of a two-tier screening program.

Recently, evidence has emerged that cffDNA tests for chromosomal aneuploidies have similar detection sensitivities and specificities among high- as well as average-risk pregnant women.[59-61] This has led to discussions on applying the cffDNA tests as a primary screening test.[55,62] Bianchi and colleagues[62] reported that the positive predictive values of the cffDNA sequencing tests were substantially higher than that of maternal serum biochemistry–based screening. For trisomy 21 detection, the positive predictive value of cffDNA sequencing was 45.5%, while that of conventional screening was 4.2%. For trisomy 18 detection, the positive predictive value of cffDNA sequencing was 40.0% and 8.3% for conventional screening. In this study, the negative predictive values for trisomy 21 and trisomy 18 detection were 100% for both cffDNA sequencing and conventional screening. If the cost of cffDNA testing could be substantially reduced, cost-benefit studies have identified it as the preferred primary screen for aneuploidies.[63] In this regard, there has been a recent report that single-molecule sequencing may be applied to cffDNA analysis.[64] Such a development may lead to a reduction in costs and the potential for point-of-care noninvasive prenatal testing.

Discordant Results

While cffDNA tests demonstrate high sensitivities and specificities, there are false-negative and false-positive results. Some of the false-negative cases are a result of low fetal DNA fraction.[65] In these cases, the proportion of fetal DNA in the sample is too low to produce a statistically significant change in the genomic representation, even in the presence of chromosome aneuploidy. Other false-negative cases are due to the mosaic nature of some chromosomal abnormalities. Mosaicism refers to the situation when only a proportion of the fetal cells harbor the chromosomal abnormality. Because the proportion of fetal cells that are contributing the DNA from the affected chromosome is reduced, the ability of the analytical protocol to detect the abnormality is also reduced.[65] Interestingly, some of the false-negative cases are a result of the absence of the chromosomal aneuploidy in the placental tissue.[66] In other words, the chromosomal aneuploidy is present in the fetus proper but not in the placenta or is present at an exceedingly low proportion of the placental cells. cffDNA in maternal plasma is mainly placental DNA, and this discrepancy between aneuploidy in the placenta and the fetus may result in false-negative test results.

> **POINTS TO REMEMBER**
>
> **Reasons for Discordant cffDNA Results**
> - False-negative results due to low fetal fraction, mosaicism, and absence of the abnormality in the placenta
> - False-positive results due to statistical reasons, confined placental mosaicism, and maternal DNA abnormalities
> - Incidental findings could be due to occult maternal malignancy or diseases with plasma DNA abnormalities, such as systemic lupus erythematosus.

False-positivity can be statistical. For example, the chromosome 21 DNA amount is considered elevated when it is 3 S.D. above that of a control population. However, 0.01% of the control population falls beyond 3 S.D. in a one-tailed normal distribution. Therefore the choice of a cutoff value for aneuploidy detection influences the theoretical false-positive rate of the test. A relatively common biological reason for "false-positive" results is confined placental mosaicism.[51] Confined placental mosaicism refers to chromosomal aneuploidy in the placenta but not the fetus proper; one report showed that placental mosaicism occurred in 2% of cases by chorionic villus sampling.[67] Because cffDNA is placental DNA, placental mosaicism may cause false-positive test results reflecting the state of the placenta, not the fetus. In fact, only 13% of mosaic chorionic villus abnormalities are detected in amniocytes.[67] Finally, another reason for "false-positive" fetal aneuploidy detection relates to aneuploidies of the mother. This is especially the case for subclinical mosaic sex chromosome aneuploidies. It has been reported that 8.6% of the sex chromosome aneuploidies detected by cffDNA testing occur when the maternal blood cell DNA showed the same finding.[68] This most commonly occurs with monosomy X (45, X0) and triple X (47, XXX). Consequently, some centers offer reflex confirmatory testing for maternal DNA when the cffDNA test suggests the presence of sex chromosome aneuploidy.[68]

Other non-pregnancy-related diseases may confound the use of cffDNA testing. For example, malignant tumors release cell-free DNA (see Chapter 9).[69] Occult malignancies have been suspected in pregnant women after cffDNA testing.[70] The suspicion arises when multiple chromosomal aneuploidies are detected in the same sample, or the cell-free DNA chromosomal copy number aberration shows a magnitude that is substantially larger than that expected for the measured fetal fraction. Other conditions, such as systemic lupus erythematosus, are also associated with abnormalities in the cell-free DNA profile.[71] If these abnormalities preexist in the plasma of a woman, the cffDNA test interpretation may become more challenging during pregnancy.

In summary, cffDNA results for chromosomal aneuploidy screening may, in rare instances, be inaccurate. Positive results require confirmation by definitive invasive testing. The obstetric history, other obstetric findings, and ultrasound features should be taken into account for the interpretation of the cffDNA tests.

Single-Gene Diseases

Besides fetal chromosomal aneuploidies, many prenatal programs address the screening and diagnosis of single-gene diseases, such as cystic fibrosis, sickle cell anemia, and thalassemias. Noninvasive prenatal diagnosis of autosomal dominant diseases of paternal origin may be achieved in a similar manner to fetal rhesus D genotyping. For example, when a paternal mutation is detected in the plasma of a mother known not to share the same mutation as the father, this may imply that the fetus has inherited the paternal mutation.[72] Maternally inherited mutations, on the other hand, are more challenging to diagnose by cffDNA analysis because cffDNA is surrounded by maternal DNA molecules that harbor the mutation. In view of the maternal DNA interference, the fetal inheritance of the maternal mutation can be assessed by quantifying the ratio between the mutant and the normal alleles in the sample—namely, the relative mutation dosage approach (Fig. 10.3).[73] For example, for a person with a heterozygous

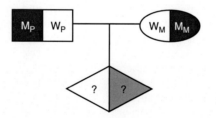

Condition	Approach
Autosomal dominant traits or mutations	
Paternally inherited	Qualitative detection of M_P
Maternally inherited	Quantitative comparison between M_M and W. If M_M = W, fetus has inherited M_M. If M_M < W, fetus has not inherited M_M.
Autosomal recessive conditions or diseases	
When M_P and M_M are identical	Quantitative comparison between M and W. If M = W, fetus is heterozygous. If M > W, fetus is homozygous for the mutation. If M < W, fetus has not inherited the mutation.
When M_P and M_M are different	Assess which paternal allele has been transmitted to the fetus by qualitative detection of M_P or W_P by a polymorphism that distinguishes W_P from M_P, M_M, and W_M. Assess which maternal allele has been transmitted to the fetus by quantitative comparison between M_M and W. If M_M = W, fetus has inherited M_M. If M_M < W, fetus has not inherited M_M.

FIGURE 10.3 Approaches to noninvasive prenatal diagnosis of monogenic diseases by maternal plasma DNA analysis. *M*, Mutant allele; *W*, wild-type allele; M_P, paternally inherited mutant allele; W_P, paternally inherited wild-type allele; M_M, maternally inherited mutant allele; W_M, maternally inherited wild-type allele.

mutation, there should be equal amounts of mutant and normal alleles among the cell-free DNA molecules. When the person is pregnant with a heterozygous fetus, the relative amounts between the mutant and normal alleles remain equal. If the fetus has not inherited the maternal mutation and is homozygous for the normal allele, there should be a slight overrepresentation in the normal allele compared with the mutant allele. Finally, if the fetus is homozygous for the mutation (maternal mutation and the same mutation from the father), there would be a slight overrepresentation of the mutant allele when compared with the normal allele. On the other hand, to detect compound heterozygous mutations, a combination approach could be used (ie, direct detection of the paternal mutation combined with the allelic ratio assessment for the maternal mutation in maternal plasma) (see Fig. 10.3).

Protocols that are based on digital PCR[73] or sequencing have been developed for quantifying amounts of the maternal mutant allele and normal allele in maternal plasma. Digital PCR protocols for the noninvasive assessment of beta-thalassemia, hemophilia, and sickle cell disease exist.[73-75] Sequencing-based protocols, such as those for beta-thalassemia[76] and congenital adrenal hyperplasia,[77] use targeted capture of the disease locus. To render the protocols even more cost effective, haplotype-based analyses have been developed, termed *relative haplotype dosage analysis* (RHDO).[38,76,77] In the RHDO method, haplotypes of the mutant and normal alleles are known. During interpretation of the maternal plasma DNA sequencing results, the number of DNA molecules detected that cover SNP alleles belonging to the inheritance block that contains the mutation are counted. The counts from each consecutive informative SNP allele are combined. The same analysis is performed for the homologous allele that does not contain the mutation. The total number of DNA molecules that originate from the haplotype block of the allele containing the mutation are compared with the count for the haplotype block of the allele not containing the mutation. Based on a statistical comparison, an interpretation is made regarding which haplotype block the fetus has likely inherited—namely, the mutant or the normal allele.[38,77] It is envisioned that in the future, targeted sequence analysis covering genomic loci of clinical importance may be the approach of choice to deliver noninvasive prenatal diagnosis of single-gene diseases. To realize this, more convenient methods to generate long-range haplotype information are needed.

NONINVASIVE FETAL 'OMICS

Many advances have occurred in cffDNA analysis. Massively parallel sequencing[38] has enabled an approach to determine the fetal genome noninvasively. Studies have demonstrated that the entire fetal genome is represented in maternal plasma. The fetal genome assembly requires information regarding the paternal genotypes as well as the maternal haplotypes across the genome. Paternal-specific alleles as well as overrepresented maternal haplotype segments that are detected in maternal plasma form the basis for assembling the paternal and maternal portions of the fetal genome. Therefore fetal genotype determination for any disease loci is theoretically possible.

Besides the fetal genome, the transcriptome[78] and the DNA methylome of the placenta[35] may be determined directly from maternal plasma analysis. These developments are particularly important because they offer the means to monitor placental function. Placental dysfunction occurs or is suspected to occur in some pregnancy-associated diseases, such as preeclampsia, preterm labor, and intrauterine growth retardation. The ability to noninvasively monitor the health of the placenta by molecular means, which was previously not possible, means that the utility of maternal plasma cell-free nucleic acid analysis could be extended beyond the assessment of fetal genetic and chromosomal diseases.

ANALYTICAL ASPECTS

Cell-free nucleic acids exist in plasma in the form of short fragments and are present in low abundance. In addition, the key to successful analysis of cffDNA lies in the preservation and maximization of the fetal fraction in the maternal blood sample. Thus attention to a number of preanalytical and analytical details, as outlined below, is important to ensure the quality of the analysis.

Maternal Blood Sample Collection and Processing

One advantage of cffDNA analysis for prenatal screening is that, unlike maternal serum biochemistry testing, the use of the cffDNA tests is not restricted to a specific gestational period. This is because the cffDNA tests target the detection of the core pathologies of the fetus—for example, inherited mutations or chromosomal aneuploidies. These pathologies exist throughout the pregnancy. However, during the earliest stage of pregnancy, there may not be a sufficient amount of fetal DNA in the maternal circulation. While cffDNA has been detected in maternal plasma in individual pregnant women from as early as 18 days of pregnancy,[79] they become quantitatively sufficient for most testing purposes from about the 11th week of pregnancy. Between 9 to 13 weeks of gestation, 2% of pregnancies show fetal DNA fractions that are below 5% and are prone to potential false-negativity.[47] Therefore some cffDNA testing programs specify the minimum gestational age for samples that are acceptable.[80]

As will be discussed, maximizing the fetal DNA fraction in the maternal sample is a key factor to many cffDNA tests. With such a consideration, maternal plasma is preferred over maternal serum. Clotting of the maternal blood cells results in the release of more maternal DNA into serum as compared to plasma.[29] While the absolute fetal DNA amounts in plasma and serum are similar, the fetal DNA fraction is significantly reduced in serum. Even for plasma, different anticoagulants show varying degrees of efficacy in suppressing the rise in maternal background DNA over time. EDTA is more effective than heparin and citrate in maintaining a relatively stable total plasma DNA concentration and fetal DNA fraction up to about 24 hours after phlebotomy.[81] Some companies now manufacture blood collection tubes containing proprietary cell-stabilizing reagents that are effective in maintaining the plasma DNA concentration and fetal fraction for up to 14 days.[82] The availability of these tubes has helped to facilitate the shipping of maternal blood samples across long distances for cffDNA testing.

After the maternal peripheral blood samples are collected, plasma is harvested via centrifugation. The goal of centrifugation is to minimize the residual maternal blood cells in the sample. Furthermore, it is not recommended to use hemolyzed samples for cffDNA analysis. Reduction in maternal DNA contamination results in maximization of the fetal fraction. Therefore two-step centrifugation protocols are recommended.[83] The plasma supernatant is carefully harvested after the first centrifugation with care not to disturb the underlying maternal blood cell layer. With the second centrifugation, the remaining cells in the form of a cell pellet are separated from the final supernatant. The resultant plasma can be stored frozen until further analysis.

Circulating Cell-Free Fetal DNA Analysis

A significant amount of cffDNA has a length shorter than 100 bp.[38] For plasma DNA extraction, a protocol that is efficient in preserving the small DNA molecules is advantageous.[84] For the same reason, PCR assays intended for cffDNA or placental RNA detection should be designed to amplify shorter products, preferably less than 100 bp.[30,85,86] The fact that circulating fetal DNA molecules are generally shorter than maternal DNA can be used to enhance the detection of fetal DNA. Researchers have used size-selection methods to increase the proportional representation of short DNA within the sample or dataset to improve the detection of fetal DNA.[73,87,88] In addition, the detection of short DNA from maternal plasma can be used to detect fetal disorders.[89] The rationale is that when the fetus has an aneuploidy, such as trisomy 21, there should be an additional dose of short DNA from chromosome 21. Therefore the overall size profile of the chromosome 21 DNA molecules would be shorter than that of other nonaneuploid chromosomes. If the fetus has monosomy X, it is then expected that there would be less short DNA from chromosome X, and thus the overall size profile of chromosome X would be longer than the other nonaneuploid chromosomes.

Besides the need for maximizing the fetal DNA fraction, there are circumstances where it may be advantageous to employ measures to minimize the maternal DNA interference. Unless the fetal DNA sequence of interest is absent in the maternal genome—for example, chromosome Y or RHD in rhesus-negative women—the need to minimize the background maternal DNA interference is particularly relevant when one aims to directly detect a fetal-specific sequence in maternal plasma. For example, for the detection of fetal-specific point mutations in maternal plasma, the large background of wild-type DNA often results in nonspecificity of the assay.[90] Researchers have used a number of different approaches to minimize background maternal DNA. Minisequencing or primer extension assays have been designed that only allow the extension of the fetal DNA sequence but not the maternal DNA sequence.[90] Peptide nucleic acid clamping has been used to suppress the maternal DNA from interacting with the detection reagents of the fetal-specific assay.[91] Restriction enzyme cutting has been used to remove the maternal DNA sequence from the sample.[92] Single-molecule or digital PCR is conducted when the average amount of template DNA per reaction well is less than one molecule. Therefore the fetal DNA sequence is separated from the maternal counterpart sequence in the reaction environment and therefore is less prone to nonspecific amplification of the maternal DNA sequence.[73]

Measuring the Fetal DNA Fraction

Analysis of the fetal DNA fraction is an important quality control parameter for cffDNA analysis. For assays that aim to detect the presence of a fetal-specific DNA sequence, such as the presence of chromosome Y sequences for determining male fetal sex and the presence of RHD to determine a rhesus D positive fetus, the inclusion of an internal fetal DNA control is preferable. The internal fetal DNA control is particularly useful when the assay reveals that the target sequence is negative. For example, when the chromosome Y assay shows negative results, it may either mean the presence of the female fetus or the lack of fetal DNA in the sample. When the RHD assay is negative, it may mean the fetus is rhesus D negative or the sample lacked fetal DNA. The positive detection of an internal fetal DNA control in such situations would exclude the scenario of lack of fetal DNA. Thus the report for the female fetus or rhesus D negative fetus could be issued with confidence.

There are several different methods for detecting the presence of fetal DNA. Some laboratories detect the fetal chromosome Y sequence as the indicator of the presence of fetal DNA, in the case of male fetuses. Some laboratories include the analysis of a panel of polymorphisms with the target sequence analysis. The panel aims to detect alleles that are not present in the maternal DNA or that show a minor contribution as an indicator of the presence of fetal DNA in the sample. Another approach is to detect placental-specific methylation signatures in maternal plasma. These are gene loci where the methylation status is opposite in the placenta as compared to the maternal blood cells. Assays specific for the detection of the placental form of the gene are used for maternal plasma analysis. The assay is designed to only detect the placental and not the maternal form of the DNA sequences. For example, the RASSF1A gene was shown to be hypermethylated in the placenta and not the maternal blood cells.[92] The presence of the methylated form of RASSF1A in maternal plasma suggests the presence of fetal/placental DNA.

There are a number of cffDNA tests where ensuring the presence of a minimum amount of fetal DNA in the sample—namely, the fetal DNA fraction—is an important quality control parameter.[65] For example, the noninvasive detection of fetal chromosomal aneuploidies by maternal plasma DNA sequencing and the assessment of the fetal inheritance of the maternal allele by relative mutation dosage or relative haplotype dosage all rely on the sample containing a certain minimum amount of fetal DNA.[65,73,77] The statistics developed to determine the presence or absence of a statistically significant difference in chromosome dosage or allelic ratio is dependent on the sample containing at least a certain amount of fetal DNA. Therefore for the tests based on massively parallel sequencing, the proportion of chromosome Y sequences is often used as a fetal fraction measurement for chromosome Y–positive samples.[61] A panel of placenta-specific DNA methylation markers has also been developed to quantify the fetal DNA.[93] Recently, the proportion of short DNA in a sample has also been shown as a reasonable measure of the fetal DNA fraction.[89]

Massively Parallel Maternal Plasma DNA Sequencing

The advantage of massively parallel sequencing–based cffDNA assessment is that many maternal plasma DNA molecules can be analyzed per sample, regardless of whether the DNA fragment is originating from the fetus or the mother. For quantitative sequencing applications, such as for fetal chromosomal aneuploidy detection or relative haplotype dosage analysis, the key to success is to maximize the signal-to-noise ratio of the sequencing data covering the genetic abnormality of interest. Trying to maximize the fetal fraction improves the chance of detecting the fetal abnormality.[65]

Increasing the sequencing depth either for random or targeted sequencing improves the precision, and thus reduces the noise, for genomic representation or allelic ratio assessments.[47] The size of the genomic locus of interest also matters. At the same sequencing depth, larger loci, such as whole chromosome aneuploidies, are much easier to detect than subchromosomal aneuploidies.[45,46] Thus protocols for the detection of subchromosomal aneuploidies require higher sequencing depths. The current massively parallel sequencing protocols also suffer from a certain extent of GC bias. To reduce the imprecision surrounding the quantitative measurement of cffDNA by sequencing, GC normalization steps are typically included in the bioinformatics analysis of the data. This is especially important for the detection of aneuploidies in GC-rich chromosomes, such as chromosomes 13 and 18.

On the other hand, current massively parallel sequencing protocols have a sequencing error rate that cannot be ignored. This has an impact on the detection of single-nucleotide variants of the fetus, such as point mutations, polymorphic alleles, and de novo mutations. Because cffDNA is the minor species of DNA in maternal plasma, high-depth sequencing is needed to detect fetal single-nucleotide variants. Yet, the higher the total number of bases sequenced, the higher the chance for sequencing errors.[94] Thus extra care is needed in designing protocols for the detection of fetal single-nucleotide variants by sequencing. For example, targeted capture of loci of interest followed by sequencing is one feasible option. Targeted sequencing allows a high depth to be achieved without sequencing a high total number of bases, thereby reducing the number of sequencing errors detected and improving the signal-to-noise ratio.

> **POINTS TO REMEMBER**
>
> **Analytical Factors to Be Mindful of**
> - EDTA plasma is preferred over serum.
> - When delayed blood processing is expected, collection of blood into tubes containing cell-stabilizing agents is recommended.
> - Avoid hemolysis.
> - Harvest the plasma using protocols to remove as much of the maternal blood cells as possible.
> - Use protocols that preserve the short cell-free DNA molecules.
> - Design assays to maximize the chance of detecting the short cffDNA molecules.
> - Consider approaches to further minimize the effect of the maternal DNA interference.
> - Include an internal control to indicate the presence of fetal DNA or measure the fetal DNA fraction.
> - Maximize the signal-to-noise ratio of the massively parallel sequencing protocols used for cffDNA analysis.

CONCLUSION

Cell-free nucleic acids in the maternal circulation are a reliable and noninvasive source of fetal genetic material. Knowledge about their biological properties has translated into useful information for guiding the design of the preanalytical and analytical approaches relevant for cffDNA analysis. cffDNA analysis is useful for the prenatal assessment of fetal chromosomal aneuploidies and genetic diseases. Its effectiveness in some of these areas has led to a major reduction in the number of amniocenteses performed worldwide, causing a paradigm shift in prenatal diagnosis. cffDNA analysis has the potential to detail the entire fetal genome noninvasively before the birth of the child. Nonetheless, the vast amount of information that one may be able to access before the birth of the child has raised some potential concerns and spurred research interests in studying the ethical, legal, and social implications of such technologies.[95] It remains to be seen how noninvasive prenatal diagnostics will continue to develop and how it will contribute to improving and maintaining fetal and maternal health.

REFERENCES

1. Papp C, Papp Z. Chorionic villus sampling and amniocentesis: what are the risks in current practice? *Curr Opin Obstet Gynecol* 2003;**15**:159–65.
2. Bevis DC. The antenatal prediction of haemolytic disease of the newborn. *Lancet* 1952;**1**:395–8.
3. Steele MW, Breg Jr WR. Chromosome analysis of human amniotic-fluid cells. *Lancet* 1966;**1**:383–5.
4. Campbell S, Johnstone FD, Holt EM, et al. Anencephaly: early ultrasonic diagnosis and active management. *Lancet* 1972;**2**:1226–7.
5. Hobbins JC, Grannum PA, Berkowitz RL, et al. Ultrasound in the diagnosis of congenital anomalies. *Am J Obstet Gynecol* 1979;**134**:331–45.
6. Norgaard-Pedersen B, Schultz-Larsen P. Reference interval for maternal serum alpha-fetoprotein. *Clin Genet* 1976;**9**:374–7.
7. Brock DJ, Scrimgeour JB, Steven J, et al. Maternal plasma alpha-fetoprotein screening for fetal neural tube defects. *Br J Obstet Gynaecol* 1978;**85**:575–81.
8. Merkatz IR, Nitowsky HM, Macri JN, et al. An association between low maternal serum alpha-fetoprotein and fetal chromosomal abnormalities. *Am J Obstet Gynecol* 1984;**148**:886–94.
9. Williamson R, Eskdale J, Coleman DV, et al. Direct gene analysis of chorionic villi: a possible technique for first-trimester antenatal diagnosis of haemoglobinopathies. *Lancet* 1981;**2**:1125–7.
10. Mennuti MT, Driscoll DA. Screening for Down's syndrome—Too many choices? *N Engl J Med* 2003;**349**:1471–3.
11. Haddow JE, Palomaki GE. Prental screening for Down syndrome. In: Simpson JL, Elias S, editors. *Essentials of prenatal diagnosis, vol.* New York: Churchill Livingstone; 1993. p. 185–220.
12. Wald NJ, Watt HC, Hackshaw AK. Integrated screening for Down's syndrome on the basis of tests performed during the first and second trimesters. *N Engl J Med* 1999;**341**:461–7.
13. Akesson HO, Forssman H. A study of maternal age in Down's syndrome. *Ann Hum Genet* 1966;**29**:271–6.
14. Hook EB. Rates of chromosome abnormalities at different maternal ages. *Obstet Gynecol* 1981;**58**:282–5.
15. Wald NJ, Huttly WJ, Hackshaw AK. Antenatal screening for Down's syndrome with the quadruple test. *Lancet* 2003;**361**:835–6.
16. Cuckle H. Established markers in second trimester maternal serum. *Early Hum Dev* 1996;**47**(Suppl.):S27–9.
17. Brizot ML, Snijders RJ, Bersinger NA, et al. Maternal serum pregnancy-associated plasma protein a and fetal nuchal translucency thickness for the prediction of fetal trisomies in early pregnancy. *Obstet Gynecol* 1994;**84**:918–22.
18. Brizot ML, Snijders RJ, Butler J, et al. Maternal serum hCG and fetal nuchal translucency thickness for the prediction of fetal trisomies in the first trimester of pregnancy. *Br J Obstet Gynaecol* 1995;**102**:127–32.
19. Nicolaides KH, Azar G, Byrne D, et al. Fetal nuchal translucency: ultrasound screening for chromosomal defects in first trimester of pregnancy. *BMJ* 1992;**304**:867–9.

20. Wright D, Spencer K, Kagan KK, et al. First-trimester combined screening for trisomy 21 at 7-14 weeks' gestation. *Ultrasound Obstet Gynecol* 2010;**36**:404–11.
21. Schmorl G. *Pathologisch-anatomische untersuchungen ueber publereklampsie*. Leipzig: Vogel; 1893.
22. Lo YMD, Patel P, Wainscoat JS, et al. Prenatal sex determination by DNA amplification from maternal peripheral blood. *Lancet* 1989;**2**:1363–5.
23. Bianchi DW, Williams JM, Sullivan LM, et al. PCR quantitation of fetal cells in maternal blood in normal and aneuploid pregnancies. *Am J Hum Genet* 1997;**61**:822–9.
24. Bianchi DW, Flint AF, Pizzimenti MF, et al. Isolation of fetal DNA from nucleated erythrocytes in maternal blood. *Proc Natl Acad Sci USA* 1990;**87**:3279–83.
25. Ganshirt-Ahlert D, Burschyk M, Garritsen HS, et al. Magnetic cell sorting and the transferrin receptor as potential means of prenatal diagnosis from maternal blood. *Am J Obstet Gynecol* 1992;**166**:1350–5.
26. Bianchi DW, Simpson JL, Jackson LG, et al. Fetal gender and aneuploidy detection using fetal cells in maternal blood: analysis of NIFTY-I data. National Institute of Child Health and Development Fetal Cell Isolation study. *Prenat Diagn* 2002;**22**:609–15.
27. Hua R, Barrett AN, Tan TZ, et al. Detection of aneuploidy from single fetal nucleated red blood cells using whole genome sequencing. *Prenat Diagn* 2015;**35**:637–44.
28. Lo YMD, Corbetta N, Chamberlain PF, et al. Presence of fetal DNA in maternal plasma and serum. *Lancet* 1997;**350**:485–7.
29. Lo YMD, Tein MS, Lau TK, et al. Quantitative analysis of fetal DNA in maternal plasma and serum: implications for noninvasive prenatal diagnosis. *Am J Hum Genet* 1998;**62**:768–75.
30. Lun FMF, Chiu RWK, Chan KCA, et al. Microfluidics digital PCR reveals a higher than expected fraction of fetal DNA in maternal plasma. *Clin Chem* 2008;**54**:1664–72.
31. Devaney SA, Palomaki GE, Scott JA, et al. Noninvasive fetal sex determination using cell-free fetal DNA: a systematic review and meta-analysis. *JAMA* 2011;**306**:627–36.
32. Lo YMD, Zhang J, Leung TN, et al. Rapid clearance of fetal DNA from maternal plasma. *Am J Hum Genet* 1999;**64**:218–24.
33. Yu SCY, Lee SW, Jiang P, et al. High-resolution profiling of fetal DNA clearance from maternal plasma by massively parallel sequencing. *Clin Chem* 2013;**59**:1228–37.
34. Chim SSC, Tong YK, Chiu RWK, et al. Detection of the placental epigenetic signature of the maspin gene in maternal plasma. *Proc Natl Acad Sci USA* 2005;**102**:14753–8.
35. Lun FMF, Chiu RWK, Sun K, et al. Noninvasive prenatal methylomic analysis by genomewide bisulfite sequencing of maternal plasma DNA. *Clin Chem* 2013;**59**:1583–94.
36. Masuzaki H, Miura K, Yoshiura KI, et al. Detection of cell free placental DNA in maternal plasma: direct evidence from three cases of confined placental mosaicism. *J Med Genet* 2004;**41**:289–92.
37. Lui YYN, Chik KW, Chiu RWK, et al. Predominant hematopoietic origin of cell-free DNA in plasma and serum after sex-mismatched bone marrow transplantation. *Clin Chem* 2002;**48**:421–7.
38. Lo YMD, Chan KCA, Sun H, et al. Maternal plasma DNA sequencing reveals the genome-wide genetic and mutational profile of the fetus. *Sci Transl Med* 2010;**2**. 61ra91.
39. Lo YMD, Hjelm NM, Fidler C, et al. Prenatal diagnosis of fetal RhD status by molecular analysis of maternal plasma. *N Engl J Med* 1998;**339**:1734–8.
40. Faas BH, Beuling EA, Christiaens GC, et al. Detection of fetal RhD-specific sequences in maternal plasma [letter]. *Lancet* 1998;**352**:1196.
41. van der Schoot CE, Hahn S, Chitty LS. Non-invasive prenatal diagnosis and determination of fetal Rh status. *Semin Fetal Neonatal Med* 2008;**13**:63–8.
42. Chiu RWK, Chan KCA, Gao Y, et al. Noninvasive prenatal diagnosis of fetal chromosomal aneuploidy by massively parallel genomic sequencing of DNA in maternal plasma. *Proc Natl Acad Sci USA* 2008;**105**:20458–63.
43. Chen EZ, Chiu RWK, Sun H, et al. Noninvasive prenatal diagnosis of fetal trisomy 18 and trisomy 13 by maternal plasma DNA sequencing. *PLoS ONE* 2011;**6**:e21791.
44. Bianchi DW, Platt LD, Goldberg JD, et al. Genome-wide fetal aneuploidy detection by maternal plasma DNA sequencing. *Obstet Gynecol* 2012;**119**:890–901.
45. Srinivasan A, Bianchi DW, Huang H, et al. Noninvasive detection of fetal subchromosome abnormalities via deep sequencing of maternal plasma. *Am J Hum Genet* 2013;**92**:167–76.
46. Yu SCY, Jiang P, Choy KW, et al. Noninvasive prenatal molecular karyotyping from maternal plasma. *PLoS ONE* 2013;**8**:e60968.
47. Chiu RWK, Akolekar R, Zheng YW, et al. Non-invasive prenatal assessment of trisomy 21 by multiplexed maternal plasma DNA sequencing: large scale validity study. *BMJ* 2011;**342**:c7401.
48. Sparks AB, Struble CA, Wang ET, et al. Noninvasive prenatal detection and selective analysis of cell-free DNA obtained from maternal blood: evaluation for trisomy 21 and trisomy 18. *Am J Obstet Gynecol* 2012;**206**:319 e1–9.
49. Zimmermann B, Hill M, Gemelos G, et al. Noninvasive prenatal aneuploidy testing of chromosomes 13, 18, 21, x, and y, using targeted sequencing of polymorphic loci. *Prenat Diagn* 2012;**32**:1233–41.
50. Chandrasekharan S, Minear MA, Hung A, et al. Noninvasive prenatal testing goes global. *Sci Transl Med* 2014;**6**. 231fs15.
51. Cuckle H, Benn P, Pergament E. Cell-free DNA screening for fetal aneuploidy as a clinical service. *Clin Biochem* 2015. https://doi.org/10.1016/j.clinbiochem.2015.02.011 [Epub Feb 27].
52. Dondorp W, de Wert G, Bombard Y, et al. Non-invasive prenatal testing for aneuploidy and beyond: challenges of responsible innovation in prenatal screening. *Eur J Hum Genet* 2015. https://doi.org/10.1038/ejhg.2015.57.
53. ISPD, editor. Position statement from the aneuploidy screening committee on behalf of the board of the international society for prenatal diagnosis, vol. 2012.
54. Benn P, Borell A, Chiu R, et al. Position statement from the aneuploidy screening committee on behalf of the board of the international society for prenatal diagnosis. *Prenat Diagn* 2013;**33**:622–9.
55. Benn P, Borell A, Chiu RW, et al. Position statement from the chromosome abnormality screening committee on behalf of the board of the international society for prenatal diagnosis. *Prenat Diagn* 2015;**35**:725–34.
56. Larion S, Warsof SL, Romary L, et al. Association of combined first-trimester screen and noninvasive prenatal testing on diagnostic procedures. *Obstet Gynecol* 2014;**123**:1303–10.
57. Robson SJ, Hui L. National decline in invasive prenatal diagnostic procedures in association with uptake of combined first trimester and cell-free DNA aneuploidy screening. *Aust N Z J Obstet Gynaecol* 2015;**55**:507–10.
58. Chan YM, Leung WC, Chan WP, et al. Women's uptake of noninvasive DNA testing following a high-risk screening test for trisomy 21 within a publicly funded healthcare system: findings from a retrospective review. *Prenat Diagn* 2015;**35**:342–7.
59. Nicolaides KH, Syngelaki A, Ashoor G, et al. Noninvasive prenatal testing for fetal trisomies in a routinely screened first-trimester population. *Am J Obstet Gynecol* 2012;**207**:374 e1–6.

60. Fairbrother G, Burigo J, Sharon T, et al. Prenatal screening for fetal aneuploidies with cell-free DNA in the general pregnancy population: a cost-effectiveness analysis. *J Matern Fetal Neonatal Med* 2015:1—5.
61. Hudecova I, Sahota D, Heung MMS, et al. Maternal plasma fetal DNA fractions in pregnancies with low and high risks for fetal chromosomal aneuploidies. *PLoS ONE* 2014;**9**:e88484.
62. Bianchi DW, Parker RL, Wentworth J, et al. DNA sequencing versus standard prenatal aneuploidy screening. *N Engl J Med* 2014;**370**:799—808.
63. Neyt M, Hulstaert F, Gyselaers W. Introducing the non-invasive prenatal test for trisomy 21 in Belgium: a cost-consequences analysis. *BMJ Open* 2014;**4**:e005922.
64. Cheng SH, Jiang P, Sun K, et al. Noninvasive prenatal testing by nanopore sequencing of maternal plasma DNA: feasibility assessment. *Clin Chem* 2015;**61**:1305—6.
65. Canick JA, Palomaki GE, Kloza EM, et al. The impact of maternal plasma DNA fetal fraction on next generation sequencing tests for common fetal aneuploidies. *Prenat Diagn* 2013;**33**:667—74.
66. Hochstenbach R, Page-Christiaens GC, van Oppen AC, et al. Unexplained false negative results in noninvasive prenatal testing: two cases involving trisomies 13 and 18. *Case Rep Genet* 2015;**2015**:926545.
67. Malvestiti F, Agrati C, Grimi B, et al. Interpreting mosaicism in chorionic villi: results of a monocentric series of 1001 mosaics in chorionic villi with follow-up amniocentesis. *Prenat Diagn* 2015. https://doi.org/10.1002/pd.4656 [Epub Jul 25].
68. Wang Y, Chen Y, Tian F, et al. Maternal mosaicism is a significant contributor to discordant sex chromosomal aneuploidies associated with noninvasive prenatal testing. *Clin Chem* 2014;**60**:251—9.
69. Chan KCA, Jiang P, Zheng YW, et al. Cancer genome scanning in plasma: detection of tumor-associated copy number aberrations, single-nucleotide variants, and tumoral heterogeneity by massively parallel sequencing. *Clin Chem* 2013;**59**:211—24.
70. Bianchi DW, Chudova D, Sehnert AJ, et al. Noninvasive prenatal testing and incidental detection of occult maternal malignancies. *JAMA* 2015;**314**:162—9.
71. Chan RWY, Jiang P, Peng X, et al. Plasma DNA aberrations in systemic lupus erythematosus revealed by genomic and methylomic sequencing. *Proc Natl Acad Sci USA* 2014;**111**:E5302—11.
72. Chitty LS, Mason S, Barrett AN, et al. Non-invasive prenatal diagnosis of achondroplasia and thanatophoric dysplasia: next-generation sequencing allows for a safer, more accurate, and comprehensive approach. *Prenat Diagn* 2015;**35**:656—62.
73. Lun FMF, Tsui NBY, Chan KCA, et al. Noninvasive prenatal diagnosis of monogenic diseases by digital size selection and relative mutation dosage on DNA in maternal plasma. *Proc Natl Acad Sci USA* 2008;**105**:19920—5.
74. Tsui NBY, Kadir RA, Chan KCA, et al. Noninvasive prenatal diagnosis of hemophilia by microfluidics digital PCR analysis of maternal plasma DNA. *Blood* 2011;**117**:3684—91.
75. Barrett AN, McDonnell TC, Chan KC, et al. Digital PCR analysis of maternal plasma for noninvasive detection of sickle cell anemia. *Clin Chem* 2012;**58**:1026—32.
76. Lam KWG, Jiang P, Liao GJW, et al. Noninvasive prenatal diagnosis of monogenic diseases by targeted massively parallel sequencing of maternal plasma: application to beta-thalassemia. *Clin Chem* 2012;**58**:1467—75.
77. New MI, Tong YK, Yuen T, et al. Noninvasive prenatal diagnosis of congenital adrenal hyperplasia using cell-free fetal DNA in maternal plasma. *J Clin Endocrinol Metab* 2014;**99**:E1022—30.
78. Tsui NBY, Jiang P, Wong YF, et al. Maternal plasma RNA sequencing for genome-wide transcriptomic profiling and identification of pregnancy-associated transcripts. *Clin Chem* 2014;**60**:954—62.
79. Guibert J, Benachi A, Grebille AG, et al. Kinetics of SRY gene appearance in maternal serum: detection by real time PCR in early pregnancy after assisted reproductive technique. *Hum Reprod* 2003;**18**:1733—6.
80. Canick JA, Saller Jr DN. Maternal serum screening for aneuploidy and open fetal defects. *Obstet Gynecol Clin North Am* 1993;**20**:443—54.
81. Lam NYL, Rainer TH, Chiu RWK, et al. EDTA is a better anticoagulant than heparin or citrate for delayed blood processing for plasma DNA analysis. *Clin Chem* 2004;**50**:256—7.
82. Fernando MR, Chen K, Norton S, et al. A new methodology to preserve the original proportion and integrity of cell-free fetal DNA in maternal plasma during sample processing and storage. *Prenat Diagn* 2010;**30**:418—24.
83. Chiu RWK, Poon LLM, Lau TK, et al. Effects of blood-processing protocols on fetal and total DNA quantification in maternal plasma. *Clin Chem* 2001;**47**:1607—13.
84. Holmberg RC, Gindlesperger A, Stokes T, et al. Akonni TruTip((r)) and Qiagen((r)) methods for extraction of fetal circulating DNA—evaluation by real-time and digital PCR. *PLoS ONE* 2013;**8**:e73068.
85. Chan KCA, Zhang J, Hui AB, et al. Size distributions of maternal and fetal DNA in maternal plasma. *Clin Chem* 2004;**50**:88—92.
86. Tsui NBY, Ng EKO, Lo YMD. Molecular analysis of circulating RNA in plasma. In: Lo YMD, Chiu RWK, Chan KCA, editors. *Clinical applications of PCR, vol.* 2nd ed. Humana Press; 2006.
87. Li Y, Di Naro E, Vitucci A, et al. Detection of paternally inherited fetal point mutations for beta-thalassemia using size-fractionated cell-free DNA in maternal plasma. *JAMA* 2005;**293**:843—9.
88. Fan HC, Blumenfeld YJ, Chitkara U, et al. Analysis of the size distributions of fetal and maternal cell-free DNA by paired-end sequencing. *Clin Chem* 2010;**56**:1279—86.
89. Yu SCYJ, Chan KCA, Zheng YW, et al. Size-based molecular diagnostics using plasma DNA for noninvasive prenatal testing. *Proc Natl Acad Sci USA* 2014;**111**:8583—8.
90. Ding C, Chiu RWK, Lau TK, et al. MS analysis of single-nucleotide differences in circulating nucleic acids: application to noninvasive prenatal diagnosis. *Proc Natl Acad Sci USA* 2004;**101**:10762—7.
91. Galbiati S, Restagno G, Foglieni B, et al. Different approaches for noninvasive prenatal diagnosis of genetic diseases based on PNA-mediated enriched PCR. *Ann N Y Acad Sci* 2006;**1075**:137—43.
92. Chan KCA, Ding C, Gerovassili A, et al. Hypermethylated RASSF1A in maternal plasma: a universal fetal DNA marker that improves the reliability of noninvasive prenatal diagnosis. *Clin Chem* 2006;**52**:2211—8.
93. Nygren AO, Dean J, Jensen TJ, et al. Quantification of fetal DNA by use of methylation-based DNA discrimination. *Clin Chem* 2010;**56**:1627—35.
94. Kitzman JO, Snyder MW, Ventura M, et al. Noninvasive whole-genome sequencing of a human fetus. *Sci Transl Med* 2012;**4**. 137ra76.
95. Greely HT. Get ready for the flood of fetal gene screening. *Nature* 2011;**469**:289—91.

Pharmacogenetics

Gwendolyn A. McMillin, Mia Wadelius and Victoria M. Pratt

ABSTRACT

Background

Pharmacogenetics describes how genes influence drug response. Genes can impact either the pharmacokinetics or pharmacodynamics of a drug to influence the dose required and associated therapeutic or toxic effects. Pharmacogenetic testing performed before drug administration may guide the selection of drugs and drug dosing. Posttherapeutic pharmacogenetic testing can explain an adverse drug reaction, including therapeutic failure.

Content

This chapter reviews pharmacokinetics and pharmacodynamics, the two major processes involved in drug response, and describes how genes that encode for proteins involved in these processes influence drug response. Important nongenetic factors that influence drug response, such as drug formulation differences, drug—drug and food—drug interactions, and clinical status are discussed, along with appropriate specimens and analytical strategies for performing, reporting, and interpreting pharmacogenetic testing results. In addition, specific gene—drug examples are described in detail relative to the nomenclature of genetic variants and allele assignments, genotype-phenotype predictions, clinical applications, and associated guidance for dosing. Specific examples include *ABCB1, CFTR, CYP2C9, CYP2C19, CYP2D6, CYP3A4/5, DPYD, G6PD, HLA-B, NATs, SLCO1B1, TPMT, UGT1A1,* and *VKORC1.*

PRINCIPLES OF PHARMACOGENETICS

The term pharmacogenetics comes from merging the terms pharmacology and genetics. Pharmacogenetics can predict and/or explain how individuals respond to drugs, and it is a prominent component of personalized, precision medicine. Initiatives exist around the globe to improve patient care with pharmacogenetics. For example, US President Barack Obama announced his "Precision Medicine Initiative" at his January 2015 State of the Union Address. This initiative will promote "patient-powered research" and "provide clinicians with new tools, knowledge, and therapies to select which treatments will work best for which patients," including pharmacogenetics.[1] Related work associating drug response with many genes, and ultimately the whole genome, is known as pharmacogenomics, although the terms are commonly used interchangeably. Pharmacogenetics and pharmacogenomics can apply to the human germline genome, tumor genomes (eg, somatic mutations), and pathogen genomes (eg, viral genomes). The goal of this chapter is to explain concepts and provide examples of human pharmacogenetics, with an emphasis on germline variants. Targeted variants of several genes and specific applications to drugs are described.

Drug Response

For simplicity, the term drug(s) is used throughout this chapter to reflect any xenobiotic (foreign compound absorbed by the human body) that is capable of evoking a physiologic or behavioral response. Response to drugs depends on many variables, such as drug formulation, route of administration, age, gender, clinical status (eg, kidney function, liver function, protein status), comedications, and genetics. Most drugs are selected and initially dosed according to drug labeling, clinical experience, and institutional protocols that stem from population-based dosage and dose frequency recommendations. Many of the nongenetic variables affecting drug response are measurable and are currently applied to drug therapy decisions, but dose optimization still largely relies on trial and error. Minimizing this process of trial and error with pharmacogenetics is proposed to improve the efficacy of drugs and prevent up to 60% of adverse drug reactions.[2,3]

Pharmacogenetic testing is designed to predict specific aspects of the two major processes upon which drug response is based: pharmacokinetics and pharmacodynamics. Pharmacokinetics describes how the body acts on a drug, often called "ADME," referring to absorption, distribution, metabolism, and elimination. Genes that encode drug metabolizing enzymes and transport proteins are involved in pharmacokinetics. Pharmacodynamics describes how the body responds to drugs, both desirable (eg, therapeutic) and undesirable (eg, therapeutic failure and/or toxicity). Genes that encode for mechanistic proteins such as enzymes, receptors, and ion channels are involved in pharmacodynamics. Sometimes drugs cause adverse effects due to mechanisms that are unrelated to the intended use of the drug. For example, carbamazepine, a drug used to treat seizures and neuropathic pain largely by inhibiting voltage gated sodium channels, can stimulate the immune system, leading to a severe cutaneous adverse reaction that can be life-threatening.[4,5] This adverse reaction is attributed to the presence of the *HLA-B*15:02* allele, representing variation in the genes that code for the human leukocyte antigen (HLA) system, and is unrelated to the mechanisms responsible for managing seizures and pain. The adverse reaction is also unrelated to the pharmacokinetics of carbamazepine. Pretherapeutic testing to detect this allele in people being considered for carbamazepine therapy is recommended for vulnerable populations.[6,7]

There are often many genes associated with the pharmacokinetics and pharmacodynamics of a drug. Selecting drugs and drug dosing for an individual should consider the clinically significant genes involved in pharmacokinetics and pharmacodynamics, in combination with relevant nongenetic factors. After drug and dose are selected, response to the drug should be monitored. If the response is desirable (therapeutic), the pharmacokinetics and pharmacodynamics are appropriate. Therapeutic failure may occur if the concentrations of active drug are insufficient (eg, pharmacokinetic variability) or if the physiology to elicit the response to the drug is absent or impaired (eg, pharmacodynamic variability). If the response is not optimal or is undesirable, therapy may need to be adjusted. Dose of a drug is adjusted based on results of clinical measurements (eg, blood pressure, for antihypertensive drugs), therapeutic drug monitoring, or monitoring biochemical markers of response.

The goal of therapeutic drug monitoring is to adjust the dose to achieve blood concentrations of active drug that fall within an established therapeutic range, at particular times after dose administration. This practice of dose adjustment is common to immunosuppressants such as tacrolimus.[8] Thus blood samples are collected at specific times after administration of tacrolimus, ideally after the concentrations of drug have achieved steady state. The dose is adjusted to achieve concentrations that consistently fall within the therapeutic range selected for the patient population and clinical indication, noting that pharmacokinetic variation can contribute to differences in time to achieve a steady-state drug concentration. Pretherapeutic pharmacogenetic testing can guide the initial dose of tacrolimus and predict an altered time to steady state.[9]

An example of a biochemical marker of response is the international normalized ratio (INR) that is calculated from prothrombin time (PT) (Box 11.1) and is used to guide and adjust doses of the common anticoagulant drug warfarin. As with tacrolimus, blood samples are collected at specific times after administration of warfarin. The dose is adjusted until the INR consistently falls within a target range, selected for the patient population and clinical indication.[10] The target INR range is set to 2.0 to 3.0 for most indications.[11] Genetic testing can help predict a therapeutic dose, thus making warfarin a drug of pharmacogenetic interest.[12]

Adverse drug reactions (therapeutic failure or toxicity) that are dose-dependent are classified as "type A" reactions. Such scenarios can be managed by adjusting the dose of a drug based on clinical or laboratory monitoring of drug concentrations and/or biomarkers. Inappropriate dosing of tacrolimus or warfarin can lead to type A adverse reactions. Adverse drug reactions may also occur independently of dose and are classified as "type B" reactions. As such, monitoring and adjusting dose will be unsuccessful. In this case, an alternate drug is required. The carbamazepine-induced hypersensitivity example illustrates a type B reaction.[13] Pretherapeutic pharmacogenetic testing may predict vulnerability to type A and type B adverse reactions if these drugs are administered.

Likelihood of a desirable (therapeutic) response can be predicted through pharmacogenetics as well. The US Food and Drug Administration (FDA) designates in vitro diagnostic (IVD) devices that identify candidates for a specific therapeutic product as *companion diagnostic* devices. The labeling for both the IVD diagnostic and the companion therapeutic product stipulate the pretherapeutic use of the test to determine whether a person is likely to respond to the therapeutic product.[14,15] Nearly all currently approved companion diagnostics are for use in treating cancer and are designed to detect drug targets (pharmacodynamics), be they proteins or somatic gene mutations, in tumor tissue.[16] Presence of the drug target qualifies a patient to receive the corresponding companion drug. Absence of the drug target suggests that a patient should not receive the companion drug and should be considered for an alternate therapy. Examples of approved companion diagnostics for cancer treatment are shown in Table 11.1. These tests have demonstrated clinical validity and have improved patient outcomes for select drugs, such as for prescribing cetuximab in colorectal carcinoma based on *KRAS* mutation results.[17] This approach to personalized care is also prominent in drug development efforts and clinical trials for new drugs and new drug indications.[18]

BOX 11.1 International Normalized Ratio (INR)

INR = [(PT result for the patient)/(PT result for a normal control)]^{ISI}

Established by the World Health Organization and the International Committee on Thrombosis and Hemostasis.

Standardizes reporting of the prothrombin time (PT) test, a common method of evaluating how many seconds it takes for a person's blood to clot.

Includes an international sensitivity index (ISI) that compensates for variability in laboratory methods (usually 1.0 to 2.0).

A typical therapeutic range of the INR for a patient treated with warfarin is 2.0 to 3.0.

POINTS TO REMEMBER

Comparing Phenotype ("State") and Genotype ("Trait") Strategies for Pharmacogenetic Testing

- The drug response phenotype displays the current response to a drug, which reflects genetics but also considers real-time drug–drug and food–drug interactions, protein expression, blood transfusions, solid organ transplantation, and so on.
- Phenotype testing may require collection of multiple blood and/or urine samples, collected at specific times after administration of a drug, coupled to targeted testing to detect the drug and drug metabolites concentrations/ratios.
- Phenotype testing may require rapid processing of specimens due to poor analyte (eg, enzyme) stability.
- Genotype testing can be performed anytime, but it reflects only specific genetic variants or alleles that the test is designed to detect.
- The specific relationship between phenotype and genotype may not be known.

TABLE 11.1 Examples of Companion Diagnostics

Cancer Indication	Analytical Target	Companion Trade Drug* (Generic)	Example Devices* (Manufacturer)
Breast	ERBB2 (Her2/Neu) gene expression	Herceptin (trastuzumab)	INFORM HER2 Dual ISH DNA Probe Cocktail (Ventana Medical; Tucson, Ariz.)
Colorectal	KRAS somatic mutations	Erbitux (cetuximab), Vectibix (panitumumab)	Therascreen KRAS RGQ PCR kit (Qiagen; Hilden, Germany)
Gastrointestinal	c-Kit protein	Gleevec/Glivec (Imatinib mesylate)	c-Kit pharmDx (Dako; Carpinteria, Calif.)
Lung	ALK gene rearrangements	Xalkori (crizotinib)	Vysis ALK Break Apart FISH Probe Kit (Abbott Molecular; Abbott Park, Ill.)
Melanoma	BRAF gene mutation (V600E)	Zelboraf (vemurafenib)	BRAF V600 Mutation Test (Roche Molecular; Pleasanton, Calif.)

*This table provides examples of drug-device pairs, and is not intended to be comprehensive. Consult the FDA for a complete and contemporary list of approved companion diagnostic devices.

Pharmacokinetics

The first recognized pharmacogenetic findings described pharmacokinetic variation, specifically interindividual differences in drug metabolism. The early focus on drug metabolism reflects the fact that measurements of drug and drug metabolite concentrations in biological fluids predate detailed understanding of genetics and development of most biomarkers that describe pharmacodynamics. Distinct metabolic phenotypes were characterized and subsequently found to cluster within families. For example, the metabolic phenotype for N-acetyltransferase (NAT) and isoniazid, a drug used to treat tuberculosis, was recognized in the 1950s.[19,20] Population studies revealed a bimodal distribution in plasma and urine concentrations of the N-acetylated isoniazid metabolite, which correlated with the phenotype. An example of the phenotypic differences observed with metabolic ratios in urine is shown in Fig. 11.1. The concentration of the parent drug was also correlated with the prevalence of toxic symptoms, including hepatotoxicity and a painful, progressive peripheral neuropathy that affected up to one-third of white and African American patients. Such testing defines the phenotypic "state" rather than the genetic "trait" and can require collection of several biological specimens, which can be costly. Phenotype testing may look at protein function as well, such as direct testing of enzyme activity. Phenotype testing is not routinely performed today when informative gene-based testing is available.

When thinking about the role of drug metabolism and associated genes in pharmacogenetics, one must consider whether the drug administered is active or inactive. Parent drugs are administered as either active drug or as inactive *prodrugs*. A prodrug is a compound that requires metabolism to be converted to an active drug. Many of the drugs discussed in this chapter are classified as a prodrug or active drug in Table 11.2. Both prodrugs and active drugs are usually metabolized by many enzymes, producing both active and inactive metabolites. Fig. 11.2 illustrates the common metabolic relationship for prodrugs, active drugs, and metabolites (active and/or inactive). Primary active metabolites are identified for the specific drug examples shown in Table 11.2.

Metabolic phenotypes, whether determined directly or predicted by genotype, typically include extensive (normal), intermediate, and poor metabolizers. For some enzymes, an ultra-rapid metabolizer phenotype is also described. The

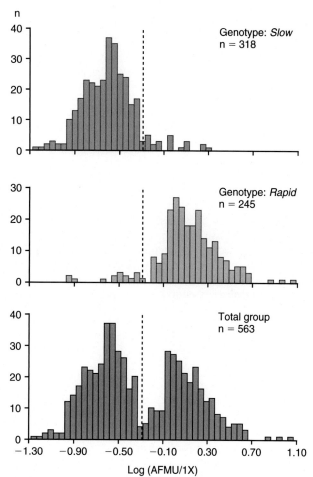

FIGURE 11.1 Histograms representing N-acetyl-transferase (NAT2) phenotype data in urine, obtained by the caffeine test. Urine was collected 5 hours after administration of caffeine, and the concentration of caffeine metabolites was determined: 5-acetylamino-6-formylamino-3-methyluracil (AFMU) and l-methylxanthine (1X). The antimode *(dashed line)* discriminates the slow metabolizers from the rapid metabolizers. (From Cascorbi I, Drakoulis N, Brockmoller J, et al. Arylamine N-acetyltransferase [NAT2] mutations and their allelic linkage in unrelated Caucasian individuals: correlation with phenotypic activity. *Am J Hum Genet* 1995;57:581–92. Reproduced with permission from the University of Chicago Press.)

TABLE 11.2 Examples of Prodrugs, Active Drugs, and Active Metabolites

COMMON FORMULATION			
Prodrug	Active Drug	Generic Drug Name	Example of an Active Metabolite*
	X	Abacavir	
	X	Amitriptyline	Nortriptyline
	X	Carbamazepine	Carbamazepine-10, 11-epoxide
X		Clopidogrel	2-oxo-clopidogrel hydrolysis products
X		Codeine	Morphine
X		5-Fluorouracil	Fluorodeoxyuridine monophosphate
X		Irinotecan	SN-38
	X	Ivacaftor	
X		6-Mercaptopurine	Thioguanine nucleotides
	X	Rasburicase	
X		Simvastatin	Simvastatin acid
	X	Tacrolimus	13-O-desmethyl-tacrolimus
X		Tamoxifen	Endoxifen
	X	Warfarin	

*Note that some active metabolites of prodrugs are available independently as active drugs.

extensive metabolizer is consistent with normal metabolic function and expression for the enzymes evaluated. Genetic variants can affect the function or expression of the enzymes and, consequently, the predicted phenotype. Two copies of nonfunctional (loss-of-function) alleles predicts the *poor metabolizer* phenotype. Such persons should seek therapeutic products that do not require the drug metabolizing enzymes coded by the affected gene or consider dose adjustment. The *intermediate metabolizer* phenotype predicts reduced metabolic activity as compared to the extensive metabolizer phenotype. When increased activity or expression is predicted, or when more than two copies of a functional gene are present (eg, gene duplication events), the ultra-rapid metabolizer phenotype is expected. The specific phenotype prediction is dependent on the nomenclature for each gene and on the specific combinations of variants detected.

Multiple drug metabolizing enzymes may be involved in activation and inactivation of a drug. The composite drug metabolizer phenotype must consider the impact of all known metabolic pathways. The phenotype may also be affected by nongenetic factors such as drug–drug interactions. Depending on the scenario, the dose of a drug may be adjusted to compensate for the predicted phenotype. Therefore clinical use of pharmacogenetic testing can guide both drug and dose selection by understanding how the genetic variation affects pharmacokinetics.

Drug transporter proteins may also affect pharmacokinetics of a drug by preventing or enhancing the transport of drug molecules across membranes. Drug transport proteins may affect absorption, distribution, and/or elimination of a drug. A loss-of-function variant in a gene that codes for a drug transport protein may prevent drug absorption, leading to therapeutic failure, or may prevent elimination of a drug, leading to drug accumulation that can contribute to a type A adverse reaction. The composite roles of both drug transporter proteins and drug metabolizing enzymes will affect overall pharmacokinetics of a drug and the associated phenotype for a person.[21]

An example pharmacokinetic pathway for the analgesic opioid drug codeine is shown in Fig. 11.3.[22] Codeine is a prodrug, and the active drug is morphine. Bioactivation of codeine occurs primarily through the drug metabolizing enzyme cytochrome P450 2D6 (CYP2D6). Codeine is also inactivated by reactions mediated by cytochrome P450 3A4

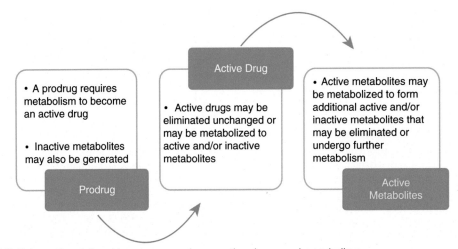

FIGURE 11.2 Schematic relationships among prodrugs, active drugs, and metabolites.

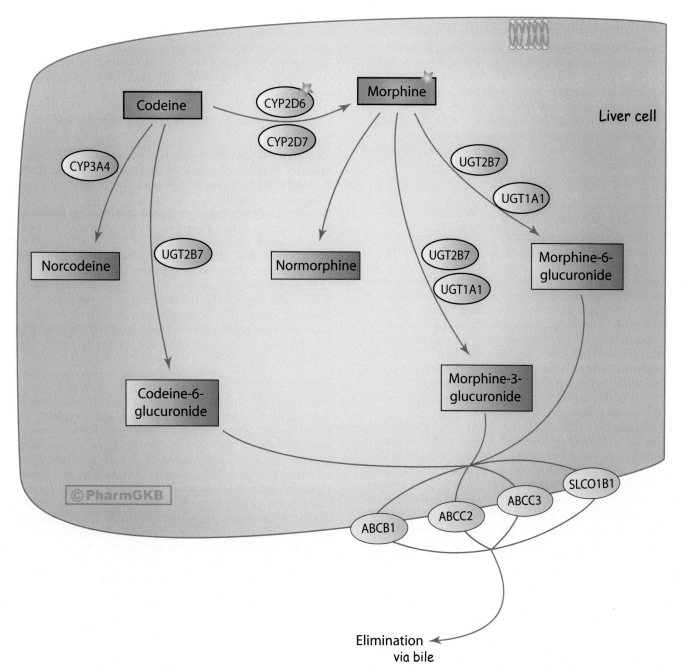

FIGURE 11.3 Schematic illustration of codeine pharmacokinetics. Purple boxes indicate codeine and metabolites, and the blue ovals are genes that code for drug-metabolizing enzymes. Codeine is the prodrug. Morphine is the primary active metabolite, formed through a reaction mediated by CYP2D6. The orange stars are to indicate the importance of this metabolic activation. All other metabolites are thought to have little to no pharmacological activity. (From Thorn CF, Klein TE, Altman RB. Codeine and morphine pathway. *Pharmacogenet Genomics* 2009;19:556–8. Figure was accessed May 21, 2015 and was reproduced with permission from the copyright holder, PharmGKB and Stanford University. Link to original pathway [https://www.pharmgkb.org/pathway/PA146123006].)

(CYP3A4) and uridine diphosphate glucuronic acid glucuronosyltransferase 2B7 enzyme (UGT2B7). The UGT family of drug metabolizing enzymes mediates the transferase reaction necessary to form glucuronide metabolites that are in general more water-soluble than the nonglucuronidated forms, which promotes drug elimination. Morphine is inactivated by reactions mediated by several UGT enzymes. The various glucuronide metabolites are transported out of the liver cell for elimination by drug transporters, including ABCB1, ABCC2, ABCC3, and SLCO1B1. This example demonstrates that several drug metabolizing enzymes and drug transporters may be involved in the metabolism and elimination of a drug.

> **POINTS TO REMEMBER**
>
> **Phenotypes Described for Drug Metabolizing Enzymes**
> - Ultra-rapid or rapid metabolizer: more than normal enzyme activity and/or gene expression is expected/observed.
> - Extensive metabolizer: normal enzyme activity and/or gene expression is expected/observed.
> - Intermediate metabolizer: less than normal enzyme activity and/or gene expression is expected/observed.
> - Poor metabolizer: little or no enzyme activity is expected/observed.

Pharmacodynamics

Pharmacogenetics of pharmacodynamics can predict desirable and undesirable responses to a drug. The range of pharmacodynamic targets is vast, including receptors, ion channels, other signaling proteins, enzymes, and the immune system. Pharmacodynamic responses can be elicited by any active component of a drug, including a range of active metabolites. The overall response phenotype represents a composite of how all pharmacodynamic and pharmacokinetic processes function, representing both genetic and nongenetic factors. Thus achieving appropriate concentrations of active drug and/or metabolites (pharmacokinetics) does not ensure but may increase the likelihood that a person will respond to the drug.

An example pharmacodynamic pathway for the anticoagulant drug warfarin is illustrated in Fig. 11.4.[23] As shown, warfarin is a racemic mixture of R- and S-stereoisomers. The primary pharmacodynamic target of warfarin is vitamin K epoxide reductase complex 1 (VKORC1). Other proteins involved in the pharmacodynamics of warfarin include epoxide hydrolase 1 (EPHX1), cytochrome P450 4F2 (CYP4F2), gamma-glutamyl carboxylase (GGXC), calumenin (CALU), and the various blood clotting factors that are activated by vitamin K.[23,24] Variants in *VKORC1* are well recognized to affect the sensitivity to warfarin, which is discussed in detail later in this chapter.

Implementation of Pharmacogenetics

Pharmacogenetic testing is intended to predict and/or explain discrete aspects of pharmacokinetics and pharmacodynamics to guide drug and dose selection. Once a drug is initiated, the response is monitored and dosing optimized with clinical and/or laboratory tools. It is not practical, medically indicated, or cost-effective to apply pharmacogenetics to every drug therapy situation. The pharmacogenetic tests that have proven most successful are the ones that produce actionable results. Many resources for labeling information, gene–drug associations, and clinical consensus guidelines are maintained and updated electronically.[25] In general, these guidelines promote implementation of pharmacogenetic testing when patient outcomes are improved, when specific dose or dosing strategies are available, and when alternative therapeutic choices are available for an indication. Some specific gene–drug relationships for which the FDA labeling includes pharmacogenetic information are shown in Table 11.3. These gene–drug examples apply to medical disciplines that include oncology, psychiatry, neurology, infectious disease, pain management, and cardiology. Other areas of medicine can also benefit from pharmacogenetic testing, and the number of clinically useful targets will only increase as comprehensive pharmacogenetic testing becomes more widely available with analytical techniques such as massively parallel sequencing.

Logistics of Pharmacogenetics in a Clinical Setting
Specimens

Most pharmacogenetic testing revolves around DNA that is extracted from blood, saliva, buccal cells, or other specimens from which DNA can be obtained. Specimens for DNA testing do not require special timing or patient preparation in most situations. Saliva or buccal cells may be preferred due to the noninvasive nature of collection but may not produce sufficient quantity or quality of DNA for all applications.[26] That said, these specimen types may be requested for patients who have recently received a blood transfusion or bone marrow transplant in order to minimize the likelihood of chimerism affecting the accuracy of the results.[27,28] Phenotyping assays are typically performed with blood or urine and may require special patient preparation as well as coordination of specimen collection relative to the timing of drug administration.

Analytical Strategies

The analytical strategy used in pharmacogenetic testing depends on a variety of factors, such as the complexity of the gene; the extent, frequency, and type of genetic variation;, and the time needed for return of the results. Most genotyping assays cannot detect all variants that have been identified in a gene. If the need for a rapid time to result is clinically indicated, an assay may be designed to limit complexity, such as through targeted detection of a small number of the most common and clinically relevant variants. Commercially available IVDs are available for some pharmacogenetic applications and may reduce complexity of testing and data analysis in order to reduce the time to result.[29,30] However, most pharmacogenetic testing is based on laboratory-developed approaches performed at a central or reference laboratory. Clinical laboratories that offer pharmacogenetic testing can be found through the voluntary National Institutes of Health Genetic Testing Registry.[31]

Targeted Testing

Most pharmacogenetic testing involves targeted genes and variants. Laboratory services may include a single gene or multiple gene panels where known variants are interrogated. Targeted genotyping will not detect any variant or alleles that are not directly interrogated, so a negative genotyping result does not rule out the possibility that a patient carries another variant not detected by the assay.

Whole Exome/Genome

Exome sequencing is only able to identify those variants near to and including the coding regions of genes. This approach cannot identify intergenic differences, including structural and noncoding variants, which can be found using other methods such as whole genome sequencing. Presently, whole genome sequencing is not practical for pharmacogenetics applications due to the high costs and time associated with

CHAPTER 11 Pharmacogenetics 301

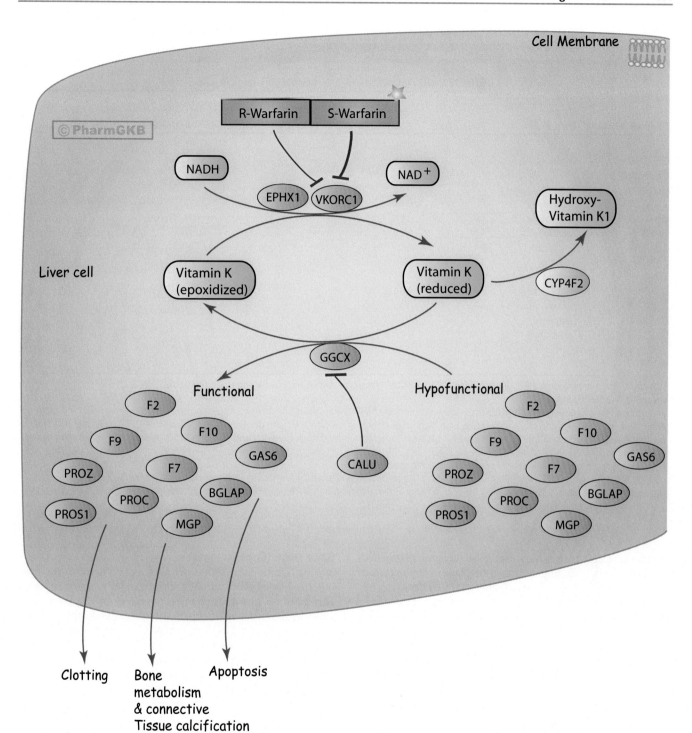

FIGURE 11.4 Schematic illustration of warfarin pharmacodynamics. *Purple boxes* show the two enantiomers of warfarin. *Green rounded boxes* are cofactors important for reductive metabolism. *Blue ovals* are genes that code for enzymes or coagulation factors. Warfarin is administered as a racemic mixture of R- and S-enantiomers, with S-warfarin being the more potent inhibitor of vitamin K epoxide reductase (VKORC1) in the vitamin K cycle. Other enzymes involved in the vitamin K cycle are epoxide hydrolase 1 (EPHX1) and gamma-glutamyl carboxylase (GGCX). The vitamin K cycle is further regulated by calumenin (CALU) that inhibits GGCX and by oxidation of reduced vitamin K by cytochrome P450 4F2. The vitamin K cycle activates the coagulation factors F2, F7, F9, and F10 and proteins C, S1, and Z into their functional forms. Other vitamin K–dependent proteins activated are the apoptotic growth arrest specific 6 (GAS6), the bone gamma-carboxyglutamate (Gla) protein (BGLAP) that regulates bone remodeling, and the matrix Gla protein (MGP) that inhibits osteogenic factors. (From Thorn CF, Klein TE, Altman RB. Codeine and morphine pathway. *Pharmacogenet Genomics* 2009;19:556–8. Figure was accessed May 21, 2015 and was reproduced with permission from the copyright holder, PharmGKB and Stanford University. Link to original pathway [https://www.pharmgkb.org/pathway/PA145011114].)

TABLE 11.3 Drug-Gene Examples With Published Guidelines, FDA Labeling Comments, and Graded Level of Evidence for Clinical Implementation

Gene or Allele	Generic Drug Name	GRADES, GUIDELINES*			COMMENT IN FDA LABELING*			
		PharmGKB	CPIC	DPWG	Testing Required	Testing Advised	Actionable	Not Available
HLA-B*57:01	Abacavir	1A	Yes	Yes		X		
CYP2D6 CYP2C19	Amitriptyline	1A	Yes	Yes			X	
HLA-B*15:02	Carbamazepine	1A	Yes	No	X†			
CYP2C19	Clopidogrel	1A	Yes	Yes		X		
CYP2D6	Codeine	1A	Yes	Yes			X	
DPYD	Fluorouracil	1A	Yes	Yes			X	
UGT1A1	Irinotecan	2A	No	Yes			X	
CFTR	Ivacaftor	1A	Yes	No	X			
TPMT	Mercaptopurine	1A	Yes	Yes		X		
G6PD	Rasburicase	1A	Yes	No	X			
SLCO1B1	Simvastatin	1A	Yes	No				X
CYP3A4	Tacrolimus	1A	Yes	Yes				X
CYP2D6	Tamoxifen	2A	No	Yes				X
CYP2C9 VKORC1	Warfarin	1A	Yes	No			X	

PharmGKB, Pharmacogenomics Knowledge Base; *CPIC,* Clinical Pharmacogenetics Implementation Consortium; *DPWG,* Dutch Pharmacogenomics Working Group.

*Levels of evidence (grades), status of guidelines, and comments in FDA labeling are provided as examples but are time-sensitive and therefore may change. The PharmGKB grade of 1A indicates a high level of evidence to support the gene-drug association, available CPIC guidelines, and/or widespread clinical implementation. The PharmGKB grade of 2A indicates moderate evidence and lack of a CPIC guideline or widespread clinical implementation.

† In patients with ancestry in genetically at-risk populations, mainly Asians.

Data from Klein TE, Chang JT, Cho MK, et al. Integrating genotype and phenotype information: an overview of the PharmGKB project. Pharmacogenetics Research Network and Knowledge Base. *Pharmacogenomics J* 2001;1:167–70.

sequencing. Current massively parallel sequencing platforms have limitations, including decreased coverage in regions with high GC content (eg, the 5′ end of genes), limited detection of copy number variants, limited detection of insertions/deletions, and interference from pseudogenes. These technical limitations are expected to improve, if not resolve, over time. However, rare and novel variants will likely be of uncertain clinical significance, causing difficulty in interpretation. Novel combinations of variants may also be identified by sequencing, which will necessitate evolution of nomenclature and may affect phenotype predictions.

Genome-Wide Association Studies

A genome-wide association study (GWAS) is an approach that involves rapidly scanning genetic markers across the complete sets of DNA, or genomes, of many people in order to find genetic variations associated with a particular disease or pharmacogenetic response. If certain genetic variations are found more frequently in people with a pharmacogenetic response compared to people without a response, the variations are said to be "associated" with the response. To date, one of the successes of the GWAS project is identification of a predictive common variant near the *IFNL3 (IL28B)* gene. The variant rs12979860 is linked to peginterferon alfa-2a and ribavirin response in individuals infected with HCV type 1.[32]

Gene Dose (Copy Number)

Normally, human DNA has two copies of each gene, one copy inherited from each parent for autosomal regions on each chromosome. However, many genetic regions display a variation in the number of copies and are termed *copy number variants* (CNVs).[33] CNVs can range in size from one kilobase to several megabases due to deletion, insertion, inversion, duplication, or complex recombination. CNVs in some pharmacogenetic genes (pharmacogenes) play a clear role in drug efficacy and toxicity.[34] One example is the human CYP2D locus that contains two pseudo-genes, *CYP2D7* and *CYP2D8*, which are closely located and evolutionarily related to *CYP2D6*.[35] The presence of highly homologous gene units in *CYP2D6* and *CYP2D7* facilitates homologous crossovers and formation of large gene conversions, deletions, duplications, and multiplications.

Haplotyping

Many of the pharmacogenes exhibit combinations of variants—for example, single nucleotide variants (SNVs) and insertion-deletions (indels) within a gene that are inherited together and are referred to as a haplotype. Many nomenclature systems have been developed to describe pharmacogenetic haplotypes. In the most commonly used nomenclature system, combinations of pharmacogenetic sequence variants are designated by star (*) alleles, where *1 is designated as normal (commonly referred to as wild-type or fully functional) and numbered star alleles are assigned as new variants are identified.[36] Pragmatically, the *1 allele is assigned by default when none of the targeted variant alleles are detected. Therefore the true accuracy of a *1 allele designation depends on whether the assay detects all possible functional variants.

Reporting

The results of clinical pharmacogenetic testing should be reported using current recommended standard nomenclature. Pharmacogenetic nomenclature is constantly evolving, and laboratories or users of laboratory services might not be familiar with most current nomenclature. Therefore test results should be provided in current recommended standard nomenclature, which should include clarifications of commonly used terms and should indicate the genotypes detected.

In general, which variants/alleles should be interrogated and reported for the pharmacogenes is not standardized. In addition, some assays use different combinations of variants to define or infer the haplotypes that the assay detects, which ultimately can lead to discrepancies in star allele genotypes between platforms. Laboratories are required by the Clinical Laboratory Improvement Amendment (CLIA) Program to include interpretation of the test results on the test report. Clinical pharmacogenetic laboratories often provide an interpreted phenotype (eg, extensive metabolizer, intermediate metabolizer, poor metabolizer, ultra-rapid metabolizer) based on the genotype results. Laboratories that are accredited by the College of American Pathologists are required to include a summary of methods and the variants that can be detected on the report, but overall, the lack of consensus for pharmacogenetic nomenclature (eg, *alleles, rs#, or HGVS, diplotype vs. haplotype, etc.) and variable assay designs add to the complexity of analyzing and reporting results from pharmacogenetic assays.

Clinical Interpretation

Pharmacogenetic testing is clinically useful only when information is sufficient to allow clinical interpretation of the results. This information must be derived from in vivo human studies. Many examples of this type of information can be found in the peer-reviewed literature. One limitation of existing literature, however, is that most data are based on retrospective studies, and there is no single printed source in which all of this information has been collated. The Pharmacogenetics and Pharmacogenomics Knowledge Base (PharmGKB) is a publicly available Internet research tool developed at Stanford University with funding from the National Institutes of Health and is part of the Pharmacogenetics Research Network, a nationwide collaborative research consortium.[37] Its aim is to aid researchers in understanding how genetic variation among individuals contributes to differences in reactions to drugs. This regularly updated database is an excellent source of genetic and clinical information.

Many ongoing clinical trials designed to study the efficacy and toxicity of new pharmaceutical products or new indications for previously developed pharmaceuticals have employed pharmacogenetic testing. Thus it is anticipated that pharmacogenetic guidelines, and possibly new companion diagnostics that would be simultaneously released to market with the pharmaceutical, will be available in the coming years. In addition, drugs that were previously removed from development because of adverse drug reactions may be reconsidered if a genetic test can be demonstrated to identify individuals at high risk for adverse drug reactions, who could then avoid use of that drug. For each of the major genes discussed in this chapter, Table 11.3 provides a list of common drugs with pharmacogenetic associations that have resulted in several FDA-revised drug labels. Because many other drug–gene combinations are recognized that may lead to additional drug label revisions, the discussion provided here is in no way comprehensive. In addition, many of the specific genes discussed later have applications to other areas of medicine, some of which are mentioned briefly. The European Medicines Agency has included pharmacogenetic information in drug labeling as well. While not all European countries participate, the European Union has promoted standardization in labeling of pharmaceuticals and included pharmacogenetic information in the summary of product characteristics. Table 11.4 provides examples of the sections in a European drug label that may include pharmacogenetic information and examples of genes that are targeted in drug labels.[38] The specific examples discussed below can be used as a guide for translating the principles of pharmacogenetics to additional gene–drug pairs in this continually evolving field of laboratory medicine.

SPECIFIC EXAMPLES OF PHARMACOGENE ASSOCIATIONS

Associations With Pharmacokinetics

Drug Metabolizing Enzymes

Drug metabolizing enzymes are often categorized as Phase I or Phase II (Box 11.2). Phase I reactions typically convert the parent compound into a more polar metabolite by introducing or removing a single functional group. Examples include oxidation, reduction, and hydrolysis. The metabolites may or may not be active. Most Phase I reactions are oxidative and are mediated by cytochrome P450 enzymes (CYPs). CYPs are heme-containing enzymes that are synthesized from a superfamily of CYP genes and are classified into families, and further into subfamilies, based on amino acid homology. Using *CYP2D6* as an example, the naming convention for CYPs is shown in Fig. 11.5, illustrating that the core name of the gene and protein is shared, identifying the family, subfamily, and polypeptide, also referred to as the isozyme. The variant allele designations follow the name of the protein. When referring to a gene, the name should be italicized, whereas the name of the associated the protein should not.

Genetic variants of CYPs have been associated with changes in enzyme activity, stability, and/or substrate affinity that can lead to clinically significant phenotypes. Alleles for all CYP genes are described according to international consensus, based on star nomenclature, often followed by a letter to describe an allele subtype.[39,40] Subtypes do not usually influence the clinical phenotype but could be important for haplotype assignment or to support research studies.[41] Common alleles for CYP2D6 are shown in

TABLE 11.4 Recommendations for Pharmacogenetics in the Summary of Product Characteristics Sections of European Union Drug Labeling for Germline Variation, by Gene

Section	Section Title	Recommendation	Gene Examples
4.1	Therapeutic indications	State when a product indication depends on a particular genotype, phenotype, or expression of a gene	HLA-B*57:01, CFTR
4.2	Posology and method of administration	State dose adjustment recommendations linked to a particular genotype	TPMT, CFTR
4.3	Contraindications	State contraindications linked to a particular genotype	DPYD, G6PD
4.4	Special warnings and precautions for use	State when adverse drug reactions, including therapeutic failure, are linked to a specific genotype or phenotype	HLA-B*57:01, TPMT, CFTR, CYP2C19, UGT1A1
4.5	Interaction with other medicinal products	State when interactions with other medicinal products depend on specific genotype or phenotype	CYP2D6, CYP2C19, UGT1A1
4.8	Undesirable effects	State any clinically relevant differences in adverse drug reactions, including therapeutic failure, that are linked to a specific genotype	HLA-B*57:01, CFTR, G6PD
5.1	Pharmacodynamic properties	State relevant clinical studies that show a difference in benefit or risk, depending on a specific genotype or phenotype	HLA-B*57:01, CFTR, G6PD
5.2	Pharmacokinetic properties	State variations in metabolism and associated quantitative terms, if clinically relevant	CYP2D6, TPMT, DPYD, CYP2C19, UGT1A1

Data from Ehmann F, Caneva L, Prasad K, et al. Pharmacogenomic information in drug labels: European Medicines Agency perspective. *Pharmacogenomics J* 2015;15:201–10.

BOX 11.2 Categorization of Drug Metabolizing Enzyme Reactions

- Phase I reactions that introduce or remove functional groups on a drug and/or drug metabolite usually increase the polarity of the compound and may change its pharmacological activity and/or kinetics. Typical reactions include oxidation, reduction, and hydrolysis. Examples of drug-metabolizing enzymes responsible for Phase I reactions include cytochrome P450s (CYPs) and dihydropyrimidine dehydrogenase (DPD).
- Phase II reactions conjugate a drug and/or drug metabolite with a chemical moiety that frequently leads to detoxification/inactivation of the compound, and may promote elimination. Phase II drug metabolizing enzymes are typically transferases, such as N-acetyltransferases (NATs), UDP-glucuronosyltransferases (UGTs), and thiopurine-S-methyl transferase (TPMT).

Table 11.5 with the predicted functional consequences. While definition of an allele may involve detection of many variants, a common SNV is frequently chosen as the primary analytical target used to detect each allele (bolded), but in many cases detection of other variants is required to accurately classify an allele. For example, the *CYP2D6* c.100C>T variant (rs1065852) is present in three of the alleles shown in Table 11.5. Misclassification of *CYP2D6*4* or *CYP2D6*36* as *CYP2D6*10* would incorrectly predict the phenotype. Such misclassifications based on discordant analytical results or interpretation of results has been observed in proficiency testing administered by the College of American Pathologists.[42] The assignment of phenotype predictions from the *CYP2D6* genotype is described in Table 11.6. An alternate approach to *CYP2D6* phenotype characterization is assignment of *activity scores*.[43,44] The range of activity scores observed in the extensive metabolizer category is relatively wide because the score reflects the actual alleles represented and may or may not be more predictive of actual phenotype.

Variants in the genes that code for Phase II enzymes can also lead to changes in enzyme activity, stability, and/or substrate affinity and contribute to pharmacokinetic variation and adverse drug effects.[45] These enzymes are not typically induced or inhibited to the same degree as CYPs. However,

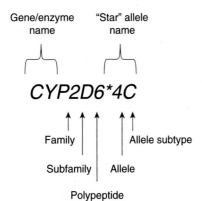

FIGURE 11.5 Description of the nomenclature for the *CYP2D6*4C* allele.

TABLE 11.5 Examples of Common Star (*) Allele Definitions for *CYP2D6*

Allele	Nucleotide Changes (cDNA)[†]	Effect	Enzyme Function
CYP2D6*1	None		Normal
CYP2D6xN	Gene amplifiers (multiple copies)	Depends on allele	Increased for functional alleles; no effect for nonfunctional alleles
CYP2D6*2A	−1584C>G; −1235A>G; −740C>T; −678G>A; *CYP2D7* gene conversion in intron 1; 1661G>C; 2850C>T; 4180G>C	R296C; S486T	Normal
CYP2D6*3A	**2549delA**	Frame shift	Nonfunctional
CYP2D6*4A	100C>T; 974C>A; 984A>G; 997C>G; 1661G>C; **1846G>A;** 4180G>C	P34S; L91M; H94R; splicing defect; S486T	Nonfunctional
CYP2D6*5	Gene deletion		Nonfunctional
CYP2D6*10A	**100C>T;** 1661G>C; 4180G>C	**P34S;** S486T	Decreased function
CYP2D6*17	**1023C>T;** 1661G>C; **2850C>T;** 4180G>C	T107I; R296C; S486T	Decreased function
CYP2D6*36 (single)	−1426C>T; −1235A>G; −1000G>A; **100C>T;** 310G>T; 843T>G; 1039C>T; 1661G>C; 2097A>G; 3384A>C; 3582A>G; gene conversion to *CYP2D7* in exon 9; 4180G>C	P34S; P469A; T470A; H478S; G479R; F481V; A482S; S486T	Nonfunctional
CYP2D6*41	−1235A>G; −740C>T; −678G>A; *CYP2D7* gene conversion in intron 1; 1661G>C; 2850C>T; **2988G>A;** 4180G>C	R296C; splicing defect; S486T	Decreased function

[†] When known, **bold** nucleotide variants represent the major alterations responsible for the effect.
Data from Sim SC, Ingelman-Sundberg M. The Human Cytochrome P450 (CYP) Allele Nomenclature website: A peer-reviewed database of CYP variants and their associated effects. *Hum Genomics* 2010;4:278–81.

TABLE 11.6 Summarized Assignment of CYP2D6 Phenotype

Predicted Function of Star (*) Allele	Star Allele (Determined From Genotype)	Diplotype	Predicted Phenotype Based on Diplotype	Predicted Activity Score Based on Diplotype
Increased function		More than two copies of normal functional alleles	Ultra-rapid metabolizer (~2% of people*)	>2.0
Normal function	*1, *2, *33, *35	Two copies of functional or decreased function alleles One functional allele and one decreased or nonfunctional allele	Extensive metabolizer (~80% of people*)	1.0–2.0
Decreased function	*9, *10, *17, *29, *36, *41	One decreased function allele and one nonfunctional allele	Intermediate metabolizer (~10% of people*)	0.5
Nonfunctional	*3, *4, *5, *6, *7, *8, *11, *12, *13, *14, *15, *16, *18, *19, *20, *21, *36, *38, *40, *42, *44, *68, *92, *100, *101	Two or more copies of nonfunctional alleles	Poor metabolizer (~8% of people*)	0

*Actual prevalence is population-dependent; percentages shown here represent published estimates for the general population.
Data from Kroetz DL, Yee SW, Giacomini KM. The pharmacogenomics of membrane transporters project: research at the interface of genomics and transporter pharmacology. *Clin Pharmacol Ther* 2010;87:109–16.

> **BOX 11.3 Definition of Drug Inhibitors According to the US Food and Drug Administration**
>
> **Strong Inhibitors**
> A fivefold or greater increase in the area under the curve (AUC), wherein the AUC refers to a plot of drug concentration in blood versus time after drug administration *or* a greater than 80% decrease in clearance (CL) of a drug from the blood, expressed in terms of volume cleared per unit of time.
>
> **Moderate Inhibitors**
> A greater than twofold but less than fivefold increase in AUC *or* 50% to 80% decrease in CL.
>
> **Weak Inhibitors**
> A 1.25-fold or greater but less than twofold increase in AUC *or* 20% to 50% decrease in CL.

exhausting the substrates or cofactors for transfer, such as glutathione or acetyl-CoA, will prevent the corresponding transferase reactions from occurring. Like the CYPs, the naming convention for Phase II enzymes is typically based on star alleles.[46-49]

Drug–drug or food–drug interactions often occur at the level of drug metabolism and are of particular concern under circumstances of polypharmacy. Drugs can be classified as inducers, as well as strong, moderate, or weak inhibitors (Box 11.3). Some drugs may be both substrates and inducers or inhibitors for the same enzyme.[50,51] A well-recognized food–drug interaction is inhibition of CYP3A4 by grapefruit juice.[52] For a person who is genetically an intermediate, extensive, or ultra-rapid metabolizer, the metabolic phenotype may be modified by drug–drug interactions. Consideration of drug–drug interactions in the activity score for an individual will improve the accuracy of a genotype-based phenotype prediction.[43] For example, the genotype-based activity scores for each phenotype in Table 11.6 may be adjusted to improve accuracy of a phenotype prediction by multiplying the activity score by an inhibitor score. The inhibitor score would be 0, 0.5, or 1, depending on whether prescribed drugs are classified as strong, moderate, or weak inhibitors, respectively. For a person who is genetically a poor metabolizer, drug–drug and/or food–drug interactions are not expected to impact drug and dose selections.

Cytochrome P450 2D6 (CYP2D6). The CYP2D6 enzyme is known to metabolize hundreds of drugs and environmental toxins, and it is implicated in many adverse drug events. As such, CYP2D6 is the subject of more than 100 FDA-issued public health advisories and labeling revisions and is included in dozens of clinical practice guidelines.[51,53] Many of the drug labeling revisions and consensus guidelines focus on neuropsychiatric medications, of which 84% (27 of 34 drugs as of this writing) include CYP2D6 as an important biomarker.[54]

The CYP2D6 polypeptide is comprised of 497 amino acids. The CYP2D6 gene contains 4408 bases and is located on chromosome 22q13.2, near two pseudogenes (*CYP2D7* and *CYP2D8*) that exhibit greater than 90% homology. More than 100 allelic variants have been described in the CYP2D6 gene.[40,55,56] The CYP2D6 gene is challenging to genotype because of the presence of pseudogenes, the sheer number of genetic variants described, the complexity of the genetic variants, and the need for identification of gene dose, specifically gene duplications and deletions.[57] The nomenclature of *CYP2D6* and examples of some common *CYP2D6* alleles are shown in Table 11.5, with associated phenotype predictions in Table 11.6. Allele frequencies vary among populations. The most common variant allele in whites is *CYP2D6*4* (18%), whereas the most common variant alleles in African Americans, Middle Eastern, and East Asian populations are *CYP2D6*17* (18%), *CYP2D6*41* (22%), and *CYP2D6*10* (42%), respectively.[55]

Tamoxifen Application. Tamoxifen is an antiestrogenic prodrug widely used to treat and prevent breast cancer. Tamoxifen mediates its therapeutic effects through modulation of estrogen receptors (ERs), leading to suppression of estrogen-mediated cell proliferation. Therefore hormone-sensitive breast tumors (ER-positive) are most likely to respond to tamoxifen. A metaanalysis of the Early Breast Cancer Trialist Collaborative Group study at the 15-year follow-up period for ER-positive breast cancer showed that 5 years of tamoxifen therapy reduced the recurrence rate by 50% and the mortality rate by 33%.[58] However, the success of tamoxifen therapy is variable, and 30% to 45% of tamoxifen patients relapse or die from recurrent cancer.[59]

As shown in Fig. 11.6, tamoxifen is extensively metabolized by many Phase I and Phase II enzymes.[60] The lack of efficacy with tamoxifen can be explained in part by interindividual differences in the metabolic activation of tamoxifen. Tamoxifen is a prodrug and must be metabolized to active metabolites to elicit the desired therapeutic effect. The most potent antiestrogenic metabolites are 4-hydroxy tamoxifen and endoxifen (4-hydroxy, *N*-desmethyl tamoxifen), each shown to exhibit approximately 100-fold greater affinity for the estrogen receptor than the parent drug tamoxifen. The concentrations of these metabolites compared with the concentrations of *N*-desmethyl tamoxifen, an inactive metabolite, and tamoxifen itself, are dependent on CYP2D6 phenotype. A therapeutic range for endoxifen in blood has not been well established. Because the amounts of endoxifen are greater than the amounts of 4-hydroxy tamoxifen, endoxifen is largely credited with producing antiestrogenic effects of tamoxifen.[61] Considerable evidence suggests that CYP2D6 is a major route for production of endoxifen. Thus patients with impaired CYP2D6 produce less endoxifen than patients with a CYP2D6 extensive metabolizer phenotype. In addition, a poor metabolizer phenotype and lower endoxifen concentrations are observed in extensive metabolizers who are coprescribed known strong inhibitors of CYP2D6, such as fluoxetine or paroxetine.[43,62] Patients treated with CYP2D6 strong inhibitors and patients who are known CYP2D6 poor metabolizers also exhibit fewer of the common antiestrogenic adverse drug reactions such as hot flashes.[61,63,64]

Pretherapeutic pharmacogenetic testing for *CYP2D6* variants could suggest whether a patient is a good candidate for tamoxifen, and the Dutch Pharmacogenetics Working Group has released guidelines.[60,65] An alternate antiestrogen drug, such as an aromatase inhibitor, is suggested for postmenopausal women who are poor or intermediate metabolizers. The International Tamoxifen Pharmacogenomics

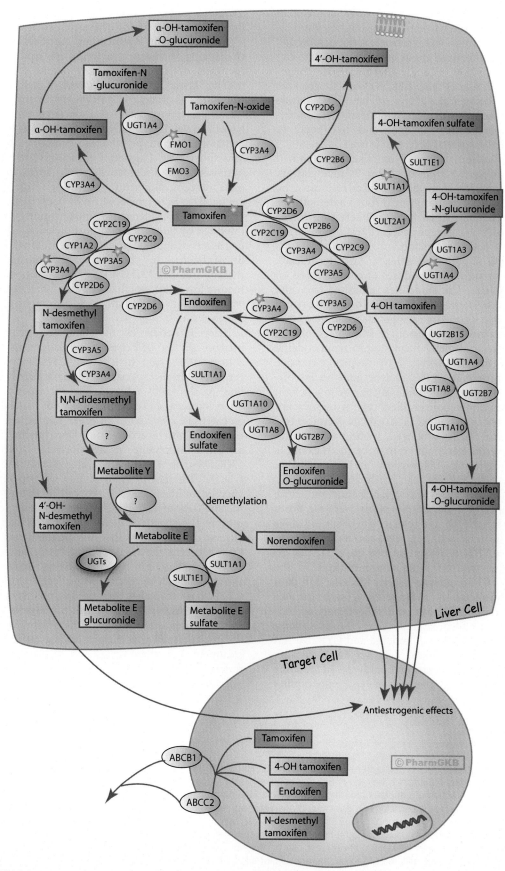

FIGURE 11.6 Schematic illustration of tamoxifen pharmacokinetics. *Purple boxes* indicate tamoxifen and metabolites, and the *blue ovals* are genes that code for drug-metabolizing enzymes in the liver cell. The *blue ovals* in the target cell indicate the genes that code for p-glycoprotein, which mediates transport of tamoxifen and metabolites out of the cell. *Yellow stars* indicate the primary pathways. Tamoxifen is extensively metabolized, but the primary pathways are

Consortium performed a metaanalysis wherein CYP2D6 poor metabolizers trended toward poor invasive disease-free survival, but statistical significance was not achieved.[66] The considerable controversy about outcomes from tamoxifen therapy and CYP2D6 has hindered widespread pretherapeutic testing. The role of other pharmacokinetic genes as well as the pharmacodynamics of tamoxifen must be considered to optimize therapy. Correlation of metabolite concentrations with therapeutic outcomes and likelihood of adverse drug reactions would be useful as well.[67,68]

Codeine Application. Codeine, an alkaloid obtained from opium or prepared from morphine by methylation, is a widely used analgesic medication. Codeine must be activated by conversion to morphine through a demethylation reaction mediated primarily by CYP2D6 to produce analgesia (see Fig. 11.3). Therefore a CYP2D6 poor metabolizer would not activate codeine and should avoid this drug due to lack of predicted efficacy. An intermediate metabolizer may require higher doses of codeine than an extensive metabolizer. The guidelines published from the Clinical Pharmacogenetics Implementation Consortium (CPIC) suggest that alternative analgesia should be sought if a CYP2D6 intermediate metabolizer fails to respond to codeine. An ultra-rapid metabolizer should avoid codeine due to the potential for toxicity.[69] In a typical extensive metabolizer, approximately 10% of the codeine is converted to morphine. Administration of codeine to a person with the ultra-rapid metabolizer phenotype is a significant safety concern because higher than expected concentrations of morphine can be produced, leading to a risk of unintentional overdose and opioid toxicity. In an example case report, respiratory depression and coma were observed in a 62-year-old man with the ultra-rapid metabolizer phenotype who was given a moderate dose of codeine. Concentrations of morphine were 80 times higher than expected.[70] Opioid toxicity can occur in babies when nursing women with an ultra-rapid metabolizer phenotype for CYP2D6 are prescribed codeine, due to higher than expected morphine concentrations in the breast milk.[71,72] Children who receive codeine for pain control after tonsillectomy and adenoidectomy are also vulnerable to unintentional but potentially deadly opioid-induced respiratory depression.[73] For these reasons, pharmacogenetic association of codeine and CYP2D6 has been incorporated into clinical decision support tools administered through the electronic health record for some institutions.[74,75] A similar pharmacogenetic vulnerability to opioid-induced toxicity exists for other opioid prodrugs that utilize CYP2D6 for activation, such as tramadol.[69]

Antidepressant Application. Antidepressant dosing is challenging because several weeks are required to assess efficacy, and optimizing dose may require several months of trial and error.[76] Many antidepressant medications available today are substrates of CYP2D6, and many are also inhibitors. Case reports of CYP2D6-related toxicity and death have been published, such as the death of a child with a poor metabolizer genotype who was prescribed the antidepressant fluoxetine.[77] Fluoxetine is a selective serotonin reuptake inhibitor that is demethylated to form the only active metabolite, norfluoxetine. Although several CYP enzymes are involved in the metabolism of fluoxetine, metabolism by CYP2D6 is considered the major route. CYP2D6 is also inhibited by fluoxetine and norfluoxetine, which are classified as strong inhibitors that lead to prolongation in the half-life of the drug and the metabolite within the first weeks of therapy (see Box 11.3). A CYP2D6 poor metabolizer would require lower doses of fluoxetine and might be best managed with another drug that does not require CYP2D6 for metabolism.[78,79]

Tricyclic antidepressants such as nortriptyline are also metabolized by CYP2D6. The relationship between drug concentrations and toxicity for CYP2D6 phenotypes and genotypes has been extensively characterized for nortriptyline. As shown in Table 11.7, lower doses of nortriptyline are recommended for a CYP2D6 poor metabolizer than for an extensive metabolizer in order to produce similar serum concentrations of active drug. Nortriptyline, which is also an active metabolite of amitriptyline, is hydroxylated by CYP2D6 to form an inactive metabolite (10-hydroxy-nortriptyline). Clearance of amitriptyline, nortriptyline, and other tricyclic antidepressants is reduced by at least 50% in poor metabolizers of CYP2D6.[80] Tricyclic antidepressants have a narrow therapeutic index and are associated with severe adverse drug reactions that may be life-threatening. Despite these concerns, the drugs are attractive because response rates are higher than for many other classes of antidepressants, and they are available in generic (cost-effective) formulations. The CYP2D6 genotype has been used to pretherapeutically triage patients as high-risk or low-risk for adverse drug reactions and response to amitriptyline and nortriptyline. For example, the proportion of patients with adverse drug reactions during treatment with these drugs was 12.1% in those with two functional CYP2D6 alleles versus 76.5% in carriers with one functional allele.[81,82]

4-hydroxylation and N-demethylation, mediated by CYP3A4, CYP3A5, and CYP2D6, producing the active metabolites 4-hydroxytamoxifen (4-OH tamoxifen) and endoxifen. Endoxifen occurs at approximately 10 times higher concentrations than 4-OH tamoxifen and is thereby believed to be responsible for most of the antiestrogenic effects of tamoxifen. These active metabolites are inactivated by conjugative reactions, the most common of which is mediated by SULT1A1. N-desmethyl tamoxifen is the most prevalent metabolite and is inactive. Tamoxifen is also metabolized to tamoxifen-N-oxide by flavin monooxygenases, primarily FMO1, which is reversed to regenerate tamoxifen by CYP3A4. Tamoxifen and metabolites exert therapeutic effects through interaction with estrogen receptors expressed on the target (breast cancer) cells. However, the effects can be inadequate if p-glycoprotein transporters expel the tamoxifen and metabolites from the cell, leading to subtherapeutic intracellular concentrations. (From Klein DJ, Thorn CF, Desta Z, et al. PharmGKB summary: tamoxifen pathway, pharmacokinetics. *Pharmacogenet Genomics* 2013;23:643–7. Figure was accessed May 21, 2015 and was reproduced with permission from the copyright holder, PharmGKB and Stanford University. Link to original pathway [https://www.pharmgkb.org/pathway/PA145011114].)

TABLE 11.7 Example Dose Adjustments for Antidepressants, Based on CYP2D6 or CYP2C19 Metabolizer Phenotype

Antidepressant	CYP2C19	CYP2D6	METABOLIZER PHENOTYPE			
			Ultra-Rapid	Extensive	Intermediate	Poor
Amitriptyline	X		141%	105%	87%	70%
		X	138%	114%	90%	67%
Citalopram	X		130%	107%	87%	59%
Desipramine		X	207%	136%	76%	25%
Doxepine		X	204%	131%	77%	34%
Fluvoxamine		X	147%	117%	89%	68%
Nortriptyline		X	195%	133%	72%	50%
Paroxetine		X	169%	125%	81%	51%
Trimipramine	X		154%	107%	76%	45%
		X	147%	118%	88%	59%
Venlafaxine	X		124%	107%	89%	44%
		X	172%	109%	92%	77%

Data from Stingl J, Viviani R. Polymorphism in CYP2D6 and CYP2C19, members of the cytochrome P450 mixed-function oxidase system, in the metabolism of psychotropic drugs. *J Intern Med* 2015;277:167–77.

The CPIC has recommended dosing for tricyclic antidepressants based on CYP2D6 phenotype or activity scores. Poor and ultra-rapid metabolizers should avoid these drugs due to risk of toxicity and lack of efficacy, respectively. Intermediate metabolizers should consider a 25% dose reduction.[83] Recommendations for adjustments from standard dosing for these drugs, for serotonin selective reuptake inhibitors,[84] and for other antidepressants have been published as well. Examples of recommended dose adjustments (see Table 11.7), range from 75% dose reduction (eg, desipramine for a CYP2D6 poor metabolizer) to more than twice the recommended dose (eg, desipramine for a CYP2D6 ultra-rapid metabolizer). The role of CYP2C19 phenotype, discussed below, is also relevant to drug and dose selection decisions for many antidepressants.[65,84,85] Therapeutic drug monitoring to evaluate the ratios of active drugs and drug metabolites will both identify and help manage phenotypic variation and optimize dose of antidepressants.[50,54,82,86]

Cytochrome P450 2C9 (CYP2C9). The CYP2C9 enzyme is a member of the CYP2C family, which includes CYP2C8, CYP2C9, CYP2C18, and CYP2C19. CYP2C9 is coded by the gene of the same name (*CYP2C9*), which is primarily expressed in the liver. Concentrations of CYP2C9 enzyme in the liver are exceeded only by those of CYP3A4.[87,88] CYP2C9 is associated with the metabolism of 15% to 20% of all drugs. Examples of drugs metabolized by CYP2C9 include vitamin K–dependent anticoagulants (eg, warfarin), anticonvulsants, nonsteroidal antiinflammatory drugs, antidiabetic agents, cholesterol-lowering drugs, angiotensin receptor blockers, and drugs used to treat infection. Genotype-based dosing guidelines have been proposed for some CYP2C9 substrates.[89] CYP2C9 is subject to induction and inhibition through drug–drug interactions, which will affect the phenotype. For example, CYP2C9 is induced by rifampicin, which will significantly increase the clearance of drugs eliminated by CYP2C9. Common inhibitors of CYP2C9 that reduce clearance of drugs include amiodarone and fluconazole.[90,91] Coadministration of CYP2C9 substrates and inducers or inhibitors can lead to life-threatening adverse drug reactions, based on whether CYP2C9 activates or inactivates a drug.[92,93]

The *CYP2C9* gene is clustered in a 500-kb region on chromosome 10q24 containing CYP genes in the order *CYP2C8-CYP2C9-CYP2C19-CYP2C18*.[94] At least 60 allelic variants have been described for this gene.[39] The most common variant in whites is *CYP2C9*2* (c.430C>T, rs1799853), with a minor allele frequency of 13%, which leads to an amino acid change (p.R144C) that decreases enzyme activity to 12% in homozygotes.[95,96] *CYP2C9*2* has a lower allele frequency in individuals of African and Native Indian ancestry (3%) and is not detected in Asians. The second most common variant in whites is *CYP2C9*3* (c.1075A>C, rs1057910), with a minor allele frequency of 7%, which leads to an amino acid change (p.I359L) that decreases enzyme activity to 5% in homozygotes.[95,97] The allele frequency of *CYP2C9*3* is also lower in people of Asian, African, and Native Indian ancestry, with allele frequencies of 4%, 1%, and 6%, respectively. A noteworthy low-activity variant with allele frequencies up to 12% in African populations is *CYP2C9*8* (rs7900194, c.449G>A, p.R150H), which is discussed in the warfarin application in the pharmacodynamics section (VKORC1) later in this chapter.[98,99] CYP2C9 is also involved in the CPIC guideline for phenytoin that is discussed in the Human Leukocyte Antigen (HLA) association. Several other low-activity variants of *CYP2C9*, with allele frequencies in the range of 1% to 3% in different populations exist.[39] Many promoter variants are known as well, but their clinical significance has yet to be determined.[100] Numerous protocols have been published for detecting variants in this gene. Most focus on detection of the *CYP2C9*2* and *CYP2C9*3* alleles, some are FDA-cleared IVDs, and comprehensive multiplexed assays have also been described.[12,101,102]

Cytochrome P450 2C19 (CYP2C19). The CYP2C19 enzyme is involved in the metabolism of a number of therapeutic drugs such as citalopram, diazepam, omeprazole, and clopidogrel.[50,51] The CYP2C19 gene contains nine exons and eight introns, and exhibits substantial variability resulting in poor, intermediate, extensive, and ultra-rapid metabolizer phenotypes. The CYP2C19 metabolic phenotype is also vulnerable to drug–drug interactions.[103] Although at least 30 allelic variants of CYP2C19 have been described, CYP2C19*2 and CYP2C19*3 account for more than 90% of poor metabolizers.[104] The poor metabolizer phenotype occurs in 2% to 5% of white and black Zimbabwean Shona populations, and 10% to 23% of Asian populations. This phenotype results from the presence of two nonfunctional alleles. The most common is CYP2C19*2 c.681G>A (rs4244285), resulting in a splicing defect that eliminates enzyme activity, and accounts for more than 75% of all poor metabolizers.[105] The second most common CYP2C19 allele (CYP2C19*3, c.636G>A, rs4986893) associated with the poor metabolizer phenotype results in a premature stop codon and no active enzyme product.[106] The CYP2C19*3 is most common in Asians, with allele frequencies reported from 6% to 10% in East Asians, 13.3% in Polynesians, and less than 1% among whites.[107,108] A single copy of CYP2C19*2 or CYP2C19*3 is associated with an intermediate metabolizer phenotype, which includes up to 45% of patients.[109]

The CYP2C19*17 has allele frequencies of approximately 20% in whites and African Americans, approximately 10% in Hispanics, and only 3% in Asians. The allele frequency in Central Europeans is 34%.[104,109] This allele arises from a promoter variant in the gene (c.-806C>T, rs12248560) and is associated with the ultra-rapid metabolizer phenotype for CYP2C19 due to increased expression. Depending on the population, 5% to 30% of patients are classified as ultra-rapid metabolizers for CYP2C19 because one copy of CYP2C19*17 is sufficient to classify a person as a ultra-rapid metabolizer.[109] However, the phenotype prediction is altered when multiple variants are detected in cis. For example, the variant associated with the CYP2C19*4 allele (c.1A>G, rs28399504) may occur in cis with the variant associated with the CYP2C19*17 allele. Because increased expression of a nonfunctional allele would not increase enzymatic activity, the predicted phenotype of this variant combination is an intermediate metabolizer.[107] This haplotype is referred to as the CYP2C19*4B allele and is observed most commonly in Ashkenazi and Sephardic Jewish populations, with a frequency of 2%.[110] The variant associated with the CYP2C19*17 allele is also recognized to be in cis with the variant associated with CYP2C19*2 and also predicts an intermediate metabolizer phenotype.[109]

Clopidogrel Application. Clopidogrel inhibits platelet aggregation and is widely used in combination with aspirin to reduce the incidence of thrombotic and ischemic events in patients with coronary artery disease, acute coronary syndrome, and/or after percutaneous coronary intervention with stenting. Yet, many patients do not achieve adequate platelet inhibition, and rates of drug resistance are estimated at 30%.[111] One explanation for drug resistance is based on the fact that clopidogrel is a prodrug that requires metabolism to the active metabolite. The active metabolite irreversibly inhibits the platelet adenosine diphosphate receptor, P2Y12. Carriers of the CYP2C19*2 and/or CYP2C19*3 alleles have reduced formation of the active metabolite of clopidogrel, demonstrate reduction in clopidogrel-induced platelet inhibition, and have a higher incidence of major thrombotic events.[112-114] The CPIC guidelines recommend that alternative drugs, such as prasugrel or ticagrelor, be sought for CYP2C19 intermediate and poor metabolizers. Prasugrel and ticagrelor do not require bioactivation by CYP2C19. No dose adjustment for clopidogrel is recommended for CYP2C19 ultra-rapid or extensive metabolizers.[109]

Antidepressant Application. A common reason for genotyping CYP2C19 is to explain inappropriate response to antidepressant medications.[54,78] CYP2C19 is the primary enzyme responsible for converting amitriptyline to its active metabolite nortriptyline. Monitoring serum or plasma concentrations of both amitriptyline and nortriptyline as a sum has been historically used to guide amitriptyline and other tricyclic antidepressant doses.[115,116] Amitriptyline, clomipramine, doxepin, imipramine, and trimipramine undergo demethylation, primarily mediated by CYP2C19, to form active metabolites. Both the parent drugs and the active metabolites are inactivated by hydroxylation that is mediated primarily by CYP2D6. As mentioned previously, the CPIC has published dosing recommendations for tricyclic antidepressants that consider phenotypes for both genes independently and in combination.[83] When variants are present in both CYP2D6 and CYP2C19, the recommendation is to avoid tricyclic antidepressants. In the absence of CYP2D6 and CYP2C19 variants, standard dosing is appropriate. A CYP2C19 ultra-rapid metabolizer should consider alternate drug and/or therapeutic drug monitoring of parent and metabolite concentrations to optimize dose regardless of CYP2D6 genotype, and a CYP2C19 poor metabolizer may consider a 50% dose reduction and guide dosing with therapeutic drug monitoring if the CYP2D6 phenotype is normal.[83]

Other antidepressants, such as citalopram or venlafaxine, may be affected by CYP2C19 phenotype as well. Examples of dose adjustment recommendations based on CYP2C19 phenotype are shown in Table 11.7.[85] Clinical practice guidelines have also been published, and drug labels for serotonin selective reuptake inhibitors and other neuroleptic drugs include pharmacogenetic information.[78,84,117] However, pharmacogenetic testing to guide antidepressant therapy has been criticized and is not currently routine.

Cytochrome P4503A Family. The CYP3A family is found on chromosome 7q21.3-q22.1 and includes four genes that code for enzymes with the same name: CYP3A4, CYP3A5, CYP3A7, and CYP3A43. There are also four pseudogenes. CYP3A7 is expressed during the fetal period, and CYP3A43 is not significantly involved in drug metabolism. In contrast, CYP3A4 and CYP3A5 enzymes are responsible for the metabolism of approximately half of all currently used drugs. Variants of CYP3A4 are rare, but the CYP3A5*3 allele, a splicing defect (rs776746, c.6986A>G), is common in many populations. CYP3A5 is expressed in only 10% to 30% of whites and Asians, correlating well with the observed high-allele frequencies for the loss-of-function CYP3A5*3 allele. Thus 70% to 90% of whites and Asians are poor metabolizers for CYP3A5.[118] The CYP3A5*6 (rs10264272, c.14690G>A) and CYP3A5*7 (rs41303343, c.27131_27132insT) allele frequencies approach 20% in African populations but are rare in whites, Asians, and Middle Eastern populations. These nonfunctional alleles also predict a poor metabolizer phenotype. Approximately 50% of African Americans are predicted to be poor metabolizers.

As with other CYPs, drug–drug interactions are a common concern. Drugs such as clarithromycin and itraconazole can produce a poor metabolizer phenotype through inhibition of CYP3A4 and CYP3A5 enzymes.[119] Grapefruit juice is also an inhibitor of these enzymes.[57] The CYP3A family is inducible by drugs including carbamazepine, phenytoin, and the commonly used herbal, St. John's wort.[120]

Tacrolimus Application. The immunosuppressant drug, tacrolimus, is commonly used to prevent rejection in solid organ transplantation. This drug has a narrow therapeutic index and is monitored frequently to achieve and maintain therapeutic blood concentrations. Acute rejection may result if blood concentrations are insufficient, and life-threatening toxicity can ensue with excess dosing. Predicting a target dose for a patient pretherapeutically could reduce dosing errors, particularly in the critical early posttransplant period.

Tacrolimus is inactivated by demethylation and hydroxylation, mediated primarily by CYP3A4 and CYP3A5. A metaanalysis of 21 studies demonstrated that the risk of transplant rejection is significantly higher in patients with the *CYP3A5 *1/*1* or *CYP3A5 *1/*3* genotypes.[121] The *CYP3A5 *3/*3* genotype predicts lower dose requirements than the extensive metabolizer (eg, *CYP3A5 *1/*1*) or intermediate metabolizer phenotypes (eg, *CYP3A5 *1/*3*). The CPIC guideline suggests standard doses for poor metabolizers and 1.5 to 2 times higher doses for extensive and intermediate metabolizers, followed by therapeutic drug monitoring to optimize dose.[9] These recommendations apply to pediatric patients as well as adults and are irrespective of the type of organ transplant. However, in liver transplantation, the *CYP3A5* genotype of the organ donor should be considered, as the relative contribution of donor and recipient genotypes to metabolism of tacrolimus is not known.

Dihydropyrimidine Dehydrogenase. The rate-limiting step of 5-fluorouracil (5-FU) catabolism is dihydropyrimidine dehydrogenase (DPD) conversion of 5-FU to dihydrofluorouracil (DHFU).[122] The main mechanism of 5-FU activation is believed to be conversion to fluorodeoxyuridine monophosphate (FdUMP), which inhibits the enzyme thymidylate synthase, an important part of the folate-homocysteine cycle and purine and pyrimidine synthesis. The resultant damage from increased base excision repair causes DNA fragmentation and ultimately cell death. In addition, the fluorouridine triphosphate (FUTP) metabolite can be incorporated into RNA in place of uridine triphosphate (UTP), thus interfering with RNA processing and protein synthesis. Approximately 10% to 40% of individuals who receive 5-FU develop severe toxicity such as neutropenia, nausea, vomiting, severe diarrhea, stomatitis, mucositis, hand-foot syndrome, and neuropathy.[123,124] The FDA label for the fluoropyrimidines indicates that variants in the gene that codes for DPD (*DPYD*) are associated with increased risk for adverse and potentially toxic events, and therefore the drug is contraindicated in patients with known DPD deficiency. However, genetic testing or screening of DPD activity is not mentioned in the drug label, partly because variants do not always result in toxicity, and treatment regimens are not standardized across studies.[125]

Over 13 variant alleles in *DPYD* (chromosome 1p21.3) have been identified.[126] The star (*) allele nomenclature is used to describe *DPYD* haplotypes, although most alleles are defined by a single variant. The most common variant is *DPYD*2* (c.1905+1G>A, rs3918290), which results in the skipping of exon 14, a 165 base pair segment encoding amino acid residues 581 to 635 of the DPYD protein.[127] The **2* variant is reported at frequencies of 0% to 3.5% in white populations.[41]

Dihydropyrimidine dehydrogenase (DPD) deficiency is an autosomal recessive disorder that is characterized by a wide range of severity, with neurological problems in some individuals and no signs or symptoms in others. In individuals with severe DPD deficiency, the disorder becomes apparent in infancy with recurrent seizures, intellectual disability, microcephaly, hypertonia, delayed development of motor skills such as walking, and autistic behaviors that affect communication and social interaction. More than 50 mutations in the *DPYD* gene have been identified in people with DPD deficiency. It is estimated that 3% to 5% of the white population have partial DPD deficiency and 0.2% have complete DPD deficiency.[128]

Genetic testing for *DPYD* typically involves targeted genotyping of the *DPYD*2* (often referred to as *2A) decreased activity allele; however, many other variant alleles have been identified.[41,129] Full gene sequencing will detect all *DPYD* variants, but rare or novel variants will likely be of uncertain clinical significance. In addition, other genes may influence responses to 5-FU, including *ABCB1*, *MTHFR*, and *TYMS*, which will not be detected by *DPYD* genetic testing.[130,131] Alternatives to *DPYD* genotyping that assess DPD enzyme activity directly include dihydrouracil/uracil ratio determination in plasma, the uracil breath test method, and measurement of DPD activity in peripheral mononuclear cells.[132]

Fluoropyrimidine Application. Fluoropyrimidines (ie, 5-fluorouracil, capecitabine, tegafur) are widely used for the treatment of several solid tumors, including breast and colorectal cancer, and typically are administered in combination with other antineoplastic agents.[131] Both capecitabine and tegafur are inactive prodrugs that are metabolized to 5-FU.

Evidence for the clinical utility of *DPYD* genotype-directed fluoropyrimidine dosing is based on prospective studies, retrospective genetic studies, case studies of patients with severe toxicity, and metaanalyses.[128,130,133] Together, these data suggest that patients heterozygous for loss-of-function *DPYD* alleles have significantly reduced 5-FU clearances, ranging from 40% to 80% less than the clearances in patients without these variants.[134]

The available literature on *DPYD* and fluoropyrimidine response prompted CPIC guidelines that recommend a 50% reduction in starting dose for patients who are heterozygous for a nonfunctional *DPYD* variant and an alternate therapy for those with two nonfunctional *DPYD* variants (ie, homozygous or compound heterozygous), despite the lack of a prospective randomized clinical trial directly evaluating the utility of *DPYD* genotyping.[125] These guidelines are summarized in Table 11.8.

N-Acetyltransferases (NAT1 and NAT2). The *N*-acetyltransferase (NAT) polymorphism is one of the earliest pharmacogenetic targets recognized and characterized. NATs are Phase II enzymes that catalyze the transfer of an acetyl moiety from acetyl-CoA to homocyclic and heterocyclic arylamines and hydrazines. Substrates include drugs, carcinogens, toxicants,

TABLE 11.8 Clinical Pharmacogenetics Implementation Consortium (CPIC) Recommended Dosing for the Drug/Gene Pairs: Fluoropyrimidine/*DPYD*, Rasburicase/*G6PD*, and Thiopurine/*TPMT*

Phenotype/Genotype	Examples of Diplotypes	Implications for Outcome	Dosing Recommendations
DPYD			
Homozygous wild-type or normal, high DPD enzyme activity (two or more functional *1 alleles)	*1/*1	"Normal" risk for fluoropyrimidine toxicity	Use label-recommended dosage and administration
Heterozygous or intermediate DPD enzyme activity (one functional allele *1, plus one nonfunctional allele)	*1/*2; *1/*13	Decreased DPD enzyme activity (leukocyte DPD activity at 30% to 70% of the normal population) and increased risk for severe or even fatal drug toxicity when treated with fluoropyrimidines	Start with at least a 50% reduction in starting dose followed by titration of dose based on toxicity or pharmacokinetic test (if available)
Homozygous or compound heterozygous variant, DPD deficiency, at risk for toxicity with drug exposure (2 nonfunctional alleles)	*2/*2; *2/*13; *13/*13	Complete DPD enzyme deficiency and increased risk for severe or even fatal drug toxicity when treated with fluoropyrimidines	Select alternate drug; avoid use of fluoropyrimidines
G6PD			
Normal. A male carrying a nondeficient (class IV) allele or a female carrying two nondeficient (class IV) alleles.	Male: B, Sao Boria. Female: B/B, B/Sao Boria	Low or reduced risk of hemolytic anemia	Use label-recommended dosage and administration
Deficient or Deficient with CNSHA. A male carrying a class I, II, or III allele, a female carrying two deficient class I–III alleles.	Male: A-, Orissa, Kalyan-Kerala, Mediterranean, Canton, Chatham, Bangkok, Villeurbanne. Female: A/A, A-/Orissa, Orissa/Kalyan-Kerala, Mediterranean/Mediterranean, Chatham/Mediterranean, Canton/Viangchan, Bangkok/Bangkok, Bangkok/Villeurbanne.	At risk of acute hemolytic anemia	Select alternate drug; avoid use of rasburicase

Variable. A female carrying one nondeficient (class IV) and one deficient (class I–III variants) allele	B/A−, B/Mediterranean, B/Bangkok.	Unknown risk of hemolytic anemia	G6PD enzyme activity must be measured to determine phenotype or rasburicase use should be avoided; consider allopurinol
TPMT			
Homozygous wild-type or normal, high activity (two functional *1 alleles)	*1/*1	"Normal" (low) concentrations of thioguanine nucleotide metabolites. Note that thioguanine nucleotide metabolite concentrations with thioguanine are 5 to 10 times higher than after mercaptopurine or azathioprine	Use label-recommended dosage and administration. Adjust based on degree of myelosuppression and disease-specific guidelines. Allow 2–4 weeks to reach steady state after each dose adjustment.
Heterozygote or intermediate activity (one functional allele—*1, plus one nonfunctional allele)	*1/*2, *1/*3A, *1/*3B, *1/*3C, *1/*4	Moderate to high concentrations of thioguanine nucleotide metabolites	Reduce dose by 30% to 50%, and adjust based on degree of myelosuppression and disease-specific guidelines. Allow 2–4 weeks to reach steady state after each dose adjustment.
Homozygous or compound heterozygous variant, mutant, low, or deficient activity (two nonfunctional alleles)	*3A/*3A, *2/*3A, *3C/*3A, *3C/*4, *3C/*2, *3A/*4	Extremely high concentrations of thioguanine nucleotide metabolites; fatal toxicity possible without dose decrease	Reduce daily dose by 10-fold and dose thrice weekly instead of daily; adjust based on degree of myelosuppression and disease-specific guidelines. Allow 4–6 weeks to reach steady state after each dose adjustment. For nonmalignant conditions, consider nonthiopurine alternative

Fluoropyrimidines and *DPYD* data from Caudle KE, Thorn CF, Klein TE, et al. Clinical Pharmacogenetics Implementation Consortium guidelines for dihydropyrimidine dehydrogenase genotype and fluoropyrimidine dosing. *Clin Pharmacol Ther* 2013;94:640–5; Rasburicase and *G6PD* data from Relling MV, McDonagh EM, Chang T, et al. Clinical Pharmacogenetics Implementation Consortium (CPIC) guidelines for rasburicase therapy in the context of G6PD deficiency genotype. *Clin Pharmacol Ther* 2014;96:169–74; and Thiopurines and *TPMT* data from Relling MV, Gardner EE, Sandborn WJ, et al. Clinical Pharmacogenetics Implementation Consortium guidelines for thiopurine methyltransferase genotype and thiopurine dosing. *Clin Pharmacol Ther* 2011;89:387–9.

and possibly endogenous compounds. There are two forms of NAT, with 81% amino acid sequence identity, coded by genes of the same names (*NAT1* and *NAT2*). The NAT1 enzyme is unstable and therefore more difficult to study than the NAT2 enzyme, but both enzymes have affinity for most substrates. Slow metabolizer phenotypes, which may affect up to 90% of some populations, are manifested by changes in protein expression, protein stability, and/or enzyme kinetics.

Three *NAT* genes are mapped to chromosome 8p22. The *NAT1* and *NAT2* genes share 87% nucleotide sequence identity; the third gene, *NATP*, is thought to be a noncoding pseudogene. Consensus nomenclature has been published wherein *NAT1*3* and *NAT2*4* are considered the normal alleles, and *NAT2*5*, *6*, *7*, *13*, and *14* alleles are thought to account for more than 99% of slow acetylator phenotypes.[49,135] NAT2 slow acetylators are common in many populations, including approximately 83% of Egyptians; 40% to 60% of whites, Europeans, and African Americans; 10% to 30% of Asians; and 5% of Canadian Eskimos.[136]

Clinical Applications. Genotyping can predict NAT phenotype quite well, with concordance of 90% to 100% for *NAT2*. However, neither genotyping nor phenotyping methods are widely available due to limited clinical utility. For many NAT drug substrates, the acetylator status is managed through nongenetic approaches. For example, the antiarrhythmic drug procainamide is routinely monitored along with the active metabolite *N*-acetylprocainamide in timed blood samples. The dose of procainamide is adjusted based on the parent/metabolite ratio. For isoniazid, the neuropathic adverse drug reactions associated with isoniazid are linked to pyridoxine deficiency and can be avoided by coadministration of pyridoxine to all patients. Also, isoniazid dosing intervals were changed from once a week to twice per week to compensate for the fact that rapid acetylators are less likely to respond to the conventionally administered dose.

NAT status is associated with risk of immunologic disorders such as rheumatoid arthritis and systemic lupus erythematosus and several cancers, particularly bladder, lung, gastric, and colorectal. Those risks could be associated with environmental and occupational toxicants that are NAT substrates. As such, *NAT2* genotyping, along with testing for other risk factors, may be important for individuals at high risk of environmental or occupational exposure to NAT substrates.[137] Glucose-6-phosphate dehydrogenase (G6PD) is one of the most common enzymatic defects in humans, where defects in G6PD, triggered by oxidative stress, lead to neonatal jaundice and hemolytic anemia. G6PD catalyzes the first step of the pentose phosphate pathway, resulting in the production of nicotinamide adenine dinucleotide phosphate (NADPH).[138] NADPH is essential to cope with oxidative stress in red blood cells because red blood cells do not have mitochondria.

More than 400 disease causing mutations in the *G6PD* gene have been identified, and most are missense mutations that affect protein stability.[139] The nomenclature for *G6PD* variants is based on the location of where the variant was first described. As the underlying DNA defects have been characterized, many variants that were thought to be unique have been found to be identical on sequence analysis. This finding should not be surprising because biochemical characterizations are not very accurate, particularly when dealing with unstable enzymes such as G6PD. All *G6PD* variants are broadly divided into five classes according to the resulting level of enzyme activity, with class I being the most severely dysfunctional and class V having the highest enzyme activity (see Table 11.8).

A pathological disorder linked to ingestion of fava beans (Vicia faba), commonly described as favism, was later identified as G6PD deficiency. G6PD deficiency is an X-linked disorder that affects over 400 million people worldwide and approximately 1 in 10 African American males in the United States.[140] It occurs most frequently in the malarial endemic regions of Africa, Asia, and the Mediterranean due to the protection it provides against malaria infection.[141]

Of note, different racial and ethnic groups have predominant founder mutations such as the G6PD Mediterranean (c.563C>T) variant, which has important implications when considering genetic testing (see below). The G6PD enzyme catalyzes the first step in the pentose phosphate pathway, which produces antioxidants to protect cells against oxidative stress.[142] Triggers that heighten oxidative stress in red blood cells result in hemolytic anemia and symptom onset in patients with G6PD deficiency.[140]

Without enough functional G6PD, red blood cells are unable to protect themselves from the damaging effects of reactive oxygen species and subsequent hemolysis. Factors such as infections, certain drugs, and ingesting fava beans can increase the levels of reactive oxygen species, thus causing red blood cells to undergo hemolysis faster than the body can replace them. The loss of red blood cells causes the signs and symptoms of hemolytic anemia such as dark urine, enlarged spleen, fatigue, rapid heart rate, shortness of breath and jaundice, which are the characteristic features of G6PD deficiency.[140]

G6PD variants that result in enzyme deficiency confer a G6PD-deficient phenotype in hemizygous males and homozygous or compound heterozygous females. It is difficult to biochemically diagnose G6PD deficiency in heterozygous females due to random X chromosome inactivation as the red cells are mosaic with two populations: one G6PD normal and one G6PD deficient. Furthermore, in females with 50% enzymatic activity, approximately half of the mosaic red cells with G6PD deficiency are prone to hemolysis, although it is generally milder than in affected males. Targeted *G6PD* genotyping can establish a molecular diagnosis of G6PD deficiency; however, prediction of drug response can be difficult without testing G6PD enzyme activity levels.

Genetic testing for *G6PD* typically involves targeted genotyping of a panel of *G6PD*-deficient alleles; however, many *G6PD* variants have been identified.[142] In addition, the US National Newborn Screening Program routinely tests for G6PD deficiency by genotyping in some states with a panel of five variants, followed by confirmatory enzyme activity testing.[143]

Targeted *G6PD* genotyping will not detect any alleles that are not directly interrogated, so a negative genotyping result does not rule out the possibility that a patient carries another *G6PD* variant. Full gene sequencing will detect all *G6PD* variants. In addition, other genes may influence responses to rasburicase, which will not be detected by *G6PD* genetic testing. Alternatives to *G6PD* genotyping that assess G6PD enzyme activity directly are available and can be used to confirm a diagnosis of G6PD deficiency.

Rasburicase Application. Rasburicase is a drug approved by the FDA for prophylaxis and treatment of hyperuricemia during chemotherapy in adults and children with lymphoma, leukemia, and solid tumors. When chemotherapy is administered, cancer cells are destroyed, releasing large amounts of uric acid into the blood. Rasburicase is a recombinant urate oxidase enzyme that works by breaking down uric acid to allantoin and hydrogen peroxide, which is eliminated from the body by the kidneys. The pegylated form of urate oxidase, pegloticase, is also FDA approved for the treatment of refractory gout.[144] Notably, both rasburicase and pegloticase carry an FDA boxed warning and are contraindicated for use in patients with known G6PD deficiency due to mutations in the G6PD gene on chromosome Xq28.[142] Additionally, the CPIC guidelines (see Table 11.8) recommend an alternative therapy (eg, allopurinol) for G6PD-deficient patients.[145]

The limited data reported from cost-effectiveness studies on G6PD genotyping suggest that G6PD screening may be cost effective.[146] A number of adverse reactions, such as drug-induced hemolytic anemia, have been reported for several medications among patients with G6PD deficiency. As such, individuals with a diagnosis of G6PD deficiency should select drug treatments that avoid hemolytic anemia and other adverse phenotypes.

Thiopurine S-Methyltransferase. Thiopurine S-methyltransferase (TPMT) is a Phase II metabolic enzyme that catalyzes the inactivation of thiopurine drugs (ie, azathioprine, mercaptopurine, and thioguanine) by S-methylation, thus preventing the formation of thioguanine nucleotides (TGNs). These drugs are analogs of the nucleic acid guanine and are incorporated into RNA and DNA by phosphodiester linkages, ultimately inhibiting several metabolic pathways and inducing apoptosis. In addition, mercaptopurines are metabolized to methyl-thioinosine monophosphate, which inhibits de novo purine synthesis and cell proliferation, thus adding another mechanism of cytotoxicity. However, approximately 10% of the population have intermediate levels of TPMT activity, and 0.3% have low or undetectable enzyme activity, which results in significantly increased risks for thioguanine nucleotides toxicity and life-threatening myelosuppression.[147,148]

The phenotypic nomenclature is usually described as normal or high function for individuals where no variant is identified, intermediate function where one nonfunctional allele is identified, and low or deficient where two nonfunctional alleles are identified. Over 30 variant alleles of the TPMT gene on chromosome 6p22.3 have been identified, many of which are missense mutations associated with decreased in vitro activity.[46] The star (*) allele nomenclature is used to describe TPMT haplotypes, although most alleles are defined by a single variant. The only exception is *3. TPMT*3A contains two missense variants in cis, p.Ala154Thr (c.460G>A; rs1800460) and p.Tyr240Cys (c.719A>G; rs1142345), while TPMT*3B contains only p.Ala154Thr (c.460G>A; rs1800460) and TPMT*3C contains only p.Tyr240Cys (c.719A>G; rs1142345). TPMT*3A is the most common variant allele associated with low TPMT activity in whites (frequency ~5%), and the assumed diplotype is *1/*3A when both variants are present and the individual is of white descent, although *3B/*3C cannot be ruled out. If *3B/*3C is suspected, phenotypic testing should be used to distinguish between *1/*3A and *3B/*3C. TPMT*3C is the most common variant allele in East Asian and African American populations, with a frequency of approximately 2%. TPMT*2 (c.238G>C; rs1800462) was the first variant allele described, but it occurs at approximately 1% allele frequency in all populations studied.[147]

Thiopurines are used for the treatment of childhood acute lymphoblastic leukemia, autoimmune diseases, inflammatory bowel diseases, lupus, and transplantation. Specifically, mercaptopurine and azathioprine are used for nonmalignant immunologic disorders—mercaptopurine for lymphoid malignancies and thioguanine for myeloid leukemias. Thiopurines are inactive precursors that are metabolized by hypoxanthine guanine phosphoribosyl transferase to active thioguanine nucleotides, which are inactivated by thiopurine methyltransferase (TPMT).[149] With conventional doses of thiopurines, individuals who inherit two loss-of-function TPMT alleles universally experience severe myelosuppression, a high proportion of heterozygous individuals show moderate to severe myelosuppression, and individuals in whom no variant is detected have lower levels of TGN metabolites and a low risk of myelosuppression.[147,150,151]

Genetic testing for TPMT typically involves targeted genotyping of the TPMT*2, *3A, *3B, and *3C alleles.[152,153] DNA-based assays that interrogate the three common variants, as described above, detect 80% to 95% of low and intermediate enzyme activity in individuals in the white, African American, and Asian populations.[154] Other genes may influence responses to thiopurines, including ITPA, which will not be detected by TPMT genetic testing.[130,131] Trinucleotide repeat variants in the promoter region have been described that may explain the 1% to 2% of whites who demonstrate the ultra-rapid metabolizer phenotype.[155,156]

Alternatives to TPMT genotyping are available and include testing TPMT enzyme activity directly and/or TPMT metabolite levels. TPMT enzyme activity depends on stable enzyme activity between the times of blood collection and analytical testing. This approach therefore is challenged by storage and stability concerns. In addition, because the TPMT enzyme is expressed in red blood cells, testing is limited to patients who have not received a blood transfusion over the weeks previous to TPMT testing and who have healthy red blood cells at the time of testing (often not the case at the time acute lymphoid leukemia is diagnosed).

Thiopurine Application. Available data suggest that patients with reduced or nonfunctional TPMT alleles are at high risk for bone marrow toxicity from thiopurine therapy and require significant dose reduction. The available evidence was utilized to inform the CPIC guidelines on dose reduction among TPMT variant carriers when genotype data are available.[147] These guidelines are summarized in Table 11.8. Cost-effectiveness studies on TPMT genotype-directed thiopurine dosing have been reported, but conflicting results are common.[157-159]

UDP-Glucuronyltransferase 1A1. The uridine diphosphate (UDP) glucuronosyl transferase (UGT) family comprises 117 members that can be divided among UGT1, UGT2,

UGT3, and UGT4. The UGT1 and UGT2 families are most efficient at glucuronidation, and the UGT1 family is of most interest clinically and is best studied. The primary goal of glucuronidation is to increase the water solubility of a compound, which typically inactivates that compound and promotes its elimination. An important endogenous substrate for UGT1A1 is bilirubin. Impairment of UGT1A1 leads to accumulation of bilirubin. The UGT family is responsible for the glucuronidation of hundreds of compounds, including hormones, flavonoids and environmental mutagens, and pharmaceutical drugs. Most of the UGTs are expressed in the liver as well as in other types of tissues, such as intestinal, stomach, and breast tissue.

Irinotecan is used to treat metastatic colorectal cancer, typically given in combination with other anticancer agents (eg, 5-FU, leucovorin). It is also used in combination with cisplatin for the treatment of extensive small cell lung cancer. Irinotecan works by binding to the topoisomerase I-DNA complex and preventing DNA replication, and thus causes double-strand DNA breakage and cell death. The active form of irinotecan is SN-38, which is converted to SN-38 glucuronic acid (SN-38G), an inactive metabolite that is subsequently eliminated through the intestines.[23] Impaired elimination of SN-38 can result in severe toxicities, including myelosuppresion, diarrhea, and neutropenia, which are associated with variation in the UGT1A1 gene.[160]

Over 100 variant alleles of the UGT1A1 gene on chromosome 2q37.1 have been identified, and the star (*) allele nomenclature is used to describe UGT1A1 haplotypes. One of the most clinically relevant variant UGT1A1 alleles is *28, which is a promoter polymorphism comprised of seven thymine-adenine (TA) dinucleotide repeats [$(TA)_7TAA$], compared to the normal UGT1A1*1 allele with six TA repeats [$(TA)_6TAA$].[161] Very important, the length of the TA repeat sequence is inversely correlated with UGT1A1 expression and activity.[162]

The UGT1A1*28 allele and other UGT1A1 missense variants have been implicated in Gilbert syndrome, which is an autosomal recessive disorder characterized by elevated unconjugated hyperbilirubinemia.[163] This mild disorder does not indicate liver damage but affects the metabolism of several substances, which can present with jaundice among affected individuals and/or with mild abdominal pain or nausea triggered by fasting or infections. Given that patients with Gilbert syndrome have normal liver function tests and typically require no treatment, a correct diagnosis is essential to avoid unnecessary testing.

Mutations in the UGT1A1 gene that lead to complete absence of UGT1A1 activity have been associated with the severe hyperbilirubinemia seen in Crigler-Najjar syndrome. Crigler-Najjar syndrome is divided into two types. Type 1 is the more severe form where affected individuals can die in childhood due to kernicterus, although with proper treatment they may survive longer. Type 2 is less severe, and affected individuals are less likely to develop kernicterus; most survive into adulthood.

Genetic testing for UGT1A1 typically involves targeted testing of the UGT1A1*28 TA repeat polymorphism. The most common assay is a laboratory-developed test involving fluorescent PCR amplification and size separation by capillary electrophoresis. Of note, this assay will also detect the five [$(TA)_5TAA$] and eight [$(TA)_8TAA$] repeat alleles. The five TA repeat allele is assumed to maintain efficient transcription, while the uncommon eight TA repeat allele indicates irinotecan sensitivity similar to the seven TA repeat allele (*28). The UGT1A1*28 allele is common in white populations and populations of African origin (allele frequencies of 26% to 56%). The UGT1A1*6 is defined by UGT1A1 211G>A (rs4148323, Arg71Gly) and is found almost exclusively in Asian populations, with a frequency of 13% to 25%.[164] This variant allele is also associated with hyperbilirubinemia. Other genes may influence irinotecan toxicity risk, including CYP3A4, which will not be detected by UGT1A1 genetic testing.[165]

Irinotecan Application. UGT1A1*28 heterozygotes and homozygotes have an approximate 25% and 70% reduction in enzyme activity, respectively, and individuals who are homozygous for UGT1A1*28 are at increased risk for myelosuppression, diarrhea, and neutropenia due to the buildup of SN-38.[162] Although the US FDA package insert for irinotecan does include information on toxicity risk due to UGT1A1*28 and the availability of testing, clinical UGT1A1 genetic testing is not required prior to treatment. Instead, it is recommended that a loading dose be administered and subsequent dosing based on symptoms (ie, granulocyte count and treatment-related diarrhea).

The Evaluation of Genomic Applications in Practice and Prevention Working Group found that the evidence was insufficient to recommend for or against the routine use of UGT1A1 genotyping in patients with metastatic colorectal cancer who are treated with irinotecan.[166] The Royal Dutch Pharmacists Association—Pharmacogenetics Working Group (KNMP-PWG) has evaluated irinotecan dosing based on UGT1A1 genotype and recommended dose reduction for *28 homozygous patients receiving more than 250 mg/m^2.[65] Cost-effectiveness studies on UGT1A1 genotype-directed irinotecan dosing indicate that UGT1A1*28 testing may be cost effective, but only if irinotecan dose reduction in homozygotes does not reduce efficacy.[167,168]

Drug Transporters

Drug transporters are membrane proteins that mediate the influx and efflux of chemicals using active and passive mechanisms.[169] They can be divided into two major families: the adenosine triphosphate binding cassette family and the solute carrier family.[170] Transporters have been shown to affect the absorption, tissue distribution, and elimination of drugs when expressed in the membranes of cells important for pharmacokinetics.

Adenosine Triphosphate (ATP)-Binding Cassette Transporters.
The (ATP)-binding cassette (ABC) family includes several transporter and channel proteins that translocate substrates across the cell membrane.[171,172] P-glycoprotein 1 (P-gp) is one of many ABC transporters responsible for cellular homeostasis. It is encoded by the multidrug resistance gene (ABCB1), which as its name indicates is responsible for the influx and efflux of multiple drugs. Numerous common coding variants in ABCB1 have been studied for their potential influence on drug exposure and adverse effects. Genetic associations with clinical phenotypes have, however, largely been inconsistent.[173] So far, no adjustments in drug dosing have been

recommended for individuals carrying sequence variants of *ABCB1*. Current clinical considerations for *ABCB1* are therefore related to its possible role in multidrug resistance and drug–drug interactions.[174]

Solute Carrier Organic Anion-Transporters. Uptake into the liver is an important step for the effect and elimination of many drugs. The solute carrier organic anion-transporting polypeptide 1B1 (OATP1B1) is one of the main influx transporters expressed on the basolateral membrane of human hepatocytes.[169,170,175,176] Examples of OATP1B1 substrates are HMG-CoA reductase inhibitors (statins), drugs acting on the angiotensin system, a number of antiinfective drugs, and certain anticancer, antiinflammatory, and antihistamine drugs.[177] OATP1B1 also transports endogenous compounds such as bile acids, thyroid hormones, and estrogens.

OATP1B1 is a 691 amino acid glycoprotein encoded by the solute carrier organic anion transporter family member 1B1 (*SLCO1B1*) gene.[178] *SLCO1B1* consists of 15 exons spanning over 108 kb on chromosome 12. More than 40 variants have been described in *SLCO1B1*. Of these, two variants have been well characterized: a polymorphism commonly named 388A>G (rs2306283, c.492A>G, p.N130D) and a low function variant referred to as 521T>C (rs4149056, c.625T>C, p.V174A). These two variants are in partial linkage disequilibrium (LD) and lead to important haplotypes that contain neither, either, or both polymorphisms.[179] The most common haplotype in Europeans, *1A, contains neither variant. The most common haplotype in Asians and Africans, *1B, contains 388G alone. The variant 521C has an allele frequency of 15%, 13%, and 3% in Europeans, Asians, and Africans, respectively. It is present on three haplotypes: alone on *5, together with 388G on *15, and together with 388G and the upstream variant -11187G>A (rs4149015) on *17.[180]

Many studies on the functional consequences of *SLCO1B1* haplotypes have been performed.[177] The *1B haplotype (388G) has yielded contradictory results, perhaps due to substrate-specific effects or different experimental conditions. In contrast, the *5, *15, and *17 haplotypes that all have the 521C variant in common have consistently been associated with decreased OATP1B1 transport activity. Convincing clinical relevance of 521C has been shown especially for simvastatin used for cholesterol reduction and for high-dose methotrexate in children with acute lymphoblastic leukemia.[179,181] In addition, many pharmaceutical compounds are known to inhibit OATP1B1 transport activity. Strong inhibitors include atorvastatin, rifamycin antibiotics, the immunosuppressants cyclosporine and tacrolimus, and the HIV drugs ritonavir, lopinavir, and nelfinavir.[177]

Simvastatin Application. Simvastatin was the first HMG-CoA reductase inhibitor to come onto the market, and it is still among the most commonly used for the treatment of hypercholesterolemia. This drug is discussed in a CPIC guideline.[179] Statins are safe and effective in the majority of patients, but they are associated with muscle toxicity in approximately 10% of patients.[182] The clinical spectrum of muscle toxicity ranges from mild forms with asymptomatic elevations of creatine kinase (CK) or pain without evidence of muscle degradation (myalgia), to pain with evidence of muscle degradation (myopathy), and finally to severe muscle damage with acute kidney injury (rhabdomyolysis).[183]

Homozygous carriers of the *SLCO1B1* 521C variant are exposed to higher levels of the active simvastatin acid than subjects homozygous for the T allele due to decreased hepatic uptake.[184] Because statin-induced myopathy is a dose-dependent adverse drug reaction, it was reasonable to believe that the 521C variant would increase the risk of myopathy during treatment with simvastatin. This was verified in a whole genome scan on patients with simvastatin-induced myopathy compared to patients who tolerated the drug.[185] In this study high-dose simvastatin (80 mg/day) increased the odds ratio for myopathy by 4.5 per copy of the minor C allele. An association between the 512C variant and statin-intolerance has since been replicated in several studies.[186-188] Even though this relationship is well established, the *SLCO1B1* 512C variant does not, however, explain all simvastatin myopathy. The possibility that other genes could contribute is currently being explored in international exome sequencing studies.[183]

Associations With Pharmacodynamics
Vitamin K Epoxide Reductase Complex 1

Vitamin K epoxide reductase (VKOR) performs the rate-limiting step in the vitamin K cycle: conversion of vitamin K epoxide to vitamin K.[189] Coagulation factors II, VII, IX, and X are vitamin K dependent. Impairment of VKOR limits activation of these factors and explains the anticoagulant effect of warfarin. Although the activity of VKOR has been recognized since the 1970s, the VKORC1 gene was not cloned and characterized until 2004.[190,191] In 2005, several independent laboratories identified common *VKORC1* variants associated with warfarin sensitivity, based on low-dose requirements.[192-194] Subsequently, a multitude of studies have dealt with the influence of *VKORC1* genetic variants on warfarin dose requirements.[195]

The VKORC1 gene is located on chromosome 16 and contains 5126 base pairs arranged in three exons.[191] *VKORC1* variants are associated most commonly with warfarin sensitivity, but also with warfarin resistance, and they are implicated in multiple coagulation factor deficiency disorders. Common variants associated with warfarin sensitivity are noncoding polymorphisms that are thought to act through differential expression of the VKOR protein.[196] There are population differences in the frequency of the warfarin-sensitive haplotype: approximately 10% for African Americans, 40% for whites, and 90% for Asians. Due to high LD in the region, it has not been resolved which polymorphism mediates the effect, but a study on liver expression quantitative trait loci (eQTL) points toward rs2303222 downstream of *VKORC1*.[95,197] Most clinical testing currently available detects the upstream *VKORC1* variant c.-1639G>A (rs9923231).[12] This variant may be used to classify the *VKORC1*2* haplotype. Typically, this variant is tested along with *CYP2C9*2* (rs1799853) and *CYP2C9*3* (rs1057910). Additional variants in these and other genes, such as *CYP4F2*, appear to have little importance for warfarin dosing in whites but may be of importance in other populations.[198-200]

Warfarin Application. Warfarin is an oral anticoagulant used to treat deep vein thrombosis and to prevent thromboembolic disease in atrial fibrillation, recurrent stroke, heart valve prosthesis, and after surgery (FDA, Coumadin label). Warfarin is widely used, but its dominance is being challenged by novel non-vitamin-K antagonist oral anticoagulants such as dabigatran, rivaroxaban, apixaban, and edoxaban.[201] Individual response to warfarin is highly variable and is monitored by INR (see Box 11.1).[202] A 10% increase in time out of the target INR range is associated with a higher risk of mortality, ischemic stroke, and other thromboembolic events.[203]

The individual response to warfarin is influenced by clinical factors such as age, gender, body size, diet, comorbidities, and concomitant interacting medications, and by variation in genes involved in its pharmacokinetics and pharmacodynamics.[201] The main pharmacokinetic gene involved is CYP2C9, which encodes an enzyme important for the metabolism of warfarin.[204] Carriers of CYP2C9 variant alleles are sensitive to warfarin, and are likely to be overdosed by a standard dose of warfarin.[205] The pharmacodynamic gene implicated is VKORC1 that encodes the drug target of warfarin.[189] Common genetic variants of VKORC1 are associated with warfarin sensitivity and account for up to a third of the variance in dose in whites.[206] Patients who are sensitive to warfarin have been shown to spend a higher proportion of time above target range in the first 90 days of treatment and to have an increased risk of bleeding compared to normal responders.[207]

Several genotype-guided warfarin dose algorithms have been developed to predict individual warfarin dose requirements.[208] These algorithms usually include the CYP2C9 alleles *2 and *3, a VKORC1 genetic variant, and clinical factors such as age, body size, and interacting drugs. Dose revision algorithms also include previous warfarin doses and existing INR data. Examples of warfarin dose algorithms are shown in Table 11.9. The performance of these algorithms varies between populations, partly due to different frequencies of the included variables.[95] For example, the aggregated allele frequencies of *2 and *3 range from 4% in Asians and Africans to 20% in whites (see CYP2C9 above) and therefore explain a larger percentage of the variance in dose in whites.

In 2009, the International Warfarin Pharmacogenetics Consortium predicted that dosing guided by CYP2C9 and VKORC1 would improve the management of warfarin, particularly for warfarin-sensitive patients.[209,210] At least 10 randomized clinical trials have been performed to compare genotype-guided dosing with standard or clinically guided dosing.[201] The two largest randomized clinical trials completed so far are the European EU-PACT and the American COAG.[211,212] Both trials incorporated the CYP2C9 alleles *2 and *3 and VKORC1 -1639 G>A and were adequately powered to evaluate the primary outcome time in therapeutic INR range. In EU-PACT, the time in the therapeutic INR range improved with genotype-guided dosing, but in COAG there was no difference between the two treatment arms. In COAG, African Americans actually fared worse with genotype-guided than with clinical dosing. These contradictory findings have led to much debate.[201] Differences in dosing strategies and allele frequencies could explain part of the divergent results. Thus EU-PACT compared a genotype-guided loading dose algorithm with a standard loading dose,[213] while COAG compared a genotype-guided maintenance dose algorithm with a clinical maintenance dose algorithm.[214] Also, CYP2C9 variants that are prevalent in people of African descent were not accounted for in COAG that had 27% African American patients. It has subsequently been shown that to avoid dosing error in African Americans, it is necessary to include CYP2C9 variants such as *8.[98,99] Furthermore, VKORC1 explains less of the variability in dose in African Americans, partly due to a low allele frequency of -1639A.[95] In conclusion, clinical utility studies have not consistently demonstrated that genotype-guided dosing improves the management of warfarin. In spite of

TABLE 11.9 Example Warfarin Dosing Prediction Models

Type of Algorithm (Reference)	Common Input Variables	Body Size Input	Other Input Variables	Dose Calculator Reference
Maintenance dose by IWPC[209]	CYP2C9*2, CYP2C9*3, VKORC1 rs9923231, Age	Height, weight	Race, enzyme inducer, amiodarone	www.nejm.org/doi/full/10.1056/NEJMoa0809329
Maintenance dose by Gage et al.[235]		BSA	Race, amiodarone, smoking, VTE, target INR	www.warfarindosing.org
Dose initiation by Avery et al.[213]		Height, weight	Amiodarone	Not available
Dose revision by Lenzini et al.[206]		BSA	Race, amiodarone, fluvastatin, stroke, diabetes, target INR, previous INR, 3 previous warfarin doses	www.warfarindosing.org
PK-PD model by Hamberg et al.[236]		Weight	Base INR, target INR, dose interval*	www.warfarindoserevision.com

IWPC, International Warfarin Pharmacogenetics Consortium; *BSA*, Body surface area; INR, International Normalized Ratio; *VTE*, venous thromboembolism; *PK-PD*, a pharmacokinetics-pharmacodynamics–based pharmacometric model that can be used for the prediction of maintenance dose, dose initiation, and dose revision and for the prediction of INR over time. See original references for complete details.
*After dose initiation, add warfarin doses, INR observations, time of dosing, and time of blood sampling for INR.

the FDA and European labeling efforts, there are at present no clear recommendations as to whether pharmacogenetic testing should be applied routinely at the initiation of warfarin.

Cystic Fibrosis Transmembrane Conductance Regulator

Cystic fibrosis (CF) is one of the most common lethal autosomal recessive diseases in whites, with an estimated incidence of 1 in 2500 to 3300 live births. Approximately 30,000 children and adults in the United States are affected, and approximately 1000 individuals are newly diagnosed annually, the majority less than 1 year old.[215] There is, however, wide variability in clinical presentation, severity, and the rate of disease progression among patients, which can be influenced by the underlying cystic fibrosis transmembrane conductance regulator (CFTR) genotype, as well as other genetic modifiers and environmental factors. The incidence of CF is around 70,000 cases worldwide, although it may be largely underdiagnosed in parts of Asia, Africa, and Latin America.[216]

The gene mutated in CF, *CFTR*, spans approximately 230 kilobases on chromosome 7q31.2 and contains 27 coding exons. The protein CFTR is in the ATP-binding cassette family of transporter proteins, containing five domains: two membrane-spanning domains, a regulatory domain, and two nucleotide-binding domains that interact with ATP. Mutations that lead to an abnormal CFTR protein cause defective electrolyte transport in apical membrane epithelial cells, resulting in complex multisystem disease that can affect the respiratory tract, pancreas, intestine, male genital tract, hepatobiliary system, and exocrine system.[215]

Since the identification of the *CFTR* gene, more than 1800 mutations have been identified. Human Genome Variation Society (HGVS) nomenclature is used to describe mutations in *CTFR*. A relatively small number of mutations account for the majority of CF alleles. The most frequent mutation is c.1521_1523delCTT, commonly known as deltaF508. This mutation is a three-base pair deletion that occurs on approximately 70% of CF chromosomes worldwide; thus approximately half of CF patients are homozygous for deltaF508. The mutations and their frequency vary by ethnic group. The American College of Medical Genetics and the American Congress of Obstetricians and Gynecologists recommended a panethnic panel of 23 mutations (which was revised from an original 25 mutations) that occur at a frequency of greater than 0.1% in individuals with CF from any of the major US ethnic groups, plus reflex testing for four additional sequence variants under specified conditions.[217-219]

Ivacaftor Application. Traditionally, the treatment of CF has focused on ameliorating symptoms (eg, fighting infection, thinning mucus, and reducing inflammation), rather than directly targeting the genetic cause. Ivacaftor (VX-770, Kalydeco) is the first FDA-approved therapeutic developed to target a specific *CFTR* defect by improving the transport of sodium through the CFTR ion channel. It was originally indicated in CF patients 6 years and older who have at least one G551D variant (c.1652G>A). The indication section of the FDA-approved ivacaftor drug label has now been amended to include additional CFTR gating defect variants. In all these variants, CFTR is trafficked correctly to the epithelial cell surface, but once there, the protein cannot transport chloride through the channel. Use with CYP3A inducers (eg, rifampin, St. John's wort) substantially decreases exposure of ivacaftor, which may diminish effectiveness. Therefore coadministration is not recommended by the CPIC.[220]

Human Leukocyte Antigen Complex

The human leukocyte antigen (HLA) complex is a cornerstone for the immune system because of its involvement in identification of foreign proteins. HLA is the human version of the major histocompatibility complex, a gene family common to many species. The HLA proteins produced from the associated genes are expressed on the surface of nearly all cells, where they bind to peptides that are exported from the cell. These peptides are thereby captured and displayed to the circulating cells of the immune system. If the immune system recognizes the peptides as foreign (eg, viral or bacterial peptides), it responds by triggering the infected cell to self-destruct.

The HLA locus consists of more than 200 genes clustered on chromosome 6 that encode cell surface proteins involved in presenting intracellular antigens to the immune system. Genes in this complex are categorized into three groups: class I, class II, and class III. The class I genes include *HLA-A*, *HLA-B*, and *HLA-C*. It is the *HLA-B* gene that is best characterized from the perspective of pharmacogenetics, based on a high degree of polymorphism and the association of variants with drug hypersensitivity reactions. The World Health Organization recognized over 3000 HLA-B alleles in humans as of July 2015 and standardized the nomenclature. The variant alleles within the class I genes are classified by a letter and up to four sets of digits separated by a colon; closely related alleles are numbered together.[221] For example *HLA-B*57:01* differs from *HLA-B*57:03* by only two base substitutions. *HLA-B* gene variants are associated with risk of disease, particularly immune-mediated inflammatory disease and drug-associated hypersensitivity.

Severe cutaneous reaction syndromes, such as Stevens-Johnson syndrome and toxic epidermal necrolysis, are classified as type B adverse drug reactions and lead to extremely painful rashes that sometimes are life-threatening. The diagnosis of Stevens-Johnson syndrome versus toxic epidermal necrolysis is based on the amount of skin affected, wherein toxic epidermal necrolysis affects more than 30% of the body surface area and therefore is more serious. Overall, drug-induced cutaneous reaction syndromes occur rarely (1 to 6 per 10,000), but the incidence is approximately 10-fold greater in Asian populations, and reactions are characteristically observed with certain drugs. The mortality rates are 5% to 10% for Stevens-Johnson syndrome and up to 40% for toxic epidermal necrolysis, and pretherapeutic testing has been recommended for select applications.[222,223]

Clinical Applications. The best-characterized HLA-drug associations for which clinical consensus guidelines are published today are *HLA-B*15:02* with carbamazepine and phenytoin,[6,224] *HLA-B*58:01* with allopurinol,[225] and *HLA-B*57:01* with abacavir.[226,227] The first well-characterized association was between HLA-B*57:01 and abacavir. This association was identified in 2002 during a prospective, double-blinded randomized study performed with nearly 2000 HIV patients from 19 countries in 2006. The

TABLE 11.10 Clinical Pharmacogenetics Implementation Consortium (CPIC) Recommended Dosing for Carbamazepine and Phenytoin With *HLA-B*15:02* and Abacavir With *HLA-B*57:01*

Phenotype/Genotype	Implications for Outcome	Dosing Recommendations
HLA-B*15:02		
Absence of *15:02 alleles (may be reported as "negative" on a genotyping test)	Low or reduced risk of carbamazepine and phenytoin associated hypersensitivity	Use label-recommended dosage and administration
Presence of at least one *15:02 allele (may be reported as "positive" on a genotyping test)	Increased risk of carbamazepine or and phenytoin hypersensitivity	Select alternative drug; avoid use of carbamazepine and phenytoin.
HLA-B*57:01		
Absence of *57:01 alleles (may be reported as "negative" on a genotyping test)	Low or reduced risk abacavir hypersensitivity.	Use label-recommended dosage and administration
Presence of at least one *57:01 allele (may be reported as "positive" on a genotyping test)	Increased risk of abacavir hypersensitivity	Select alternate drug; avoid use of abacavir.

See original references for complete details.
HLA-B*15:02 data from Leckband SG, Kelsoe JR, Dunnenberger HM, et al. Clinical Pharmacogenetics Implementation Consortium guidelines for HLA-B genotype and carbamazepine dosing. Clin *Pharmacol Ther* 2013;94:324—8; Caudle KE, Rettie AE, Whirl-Carrillo M, et al. Clinical Pharmacogenetics Implementation Consortium Guidelines for CYP2C9 and HLA-B genotypes and phenytoin dosing. *Clin Pharmacol Ther* 2014;96: 542—8; and HLA-B*57:01 data from Martin MA, Klein TE, Dong BJ, et al. Clinical Pharmacogenetics Implementation Consortium guidelines for HLA-B genotype and abacavir dosing. *Clin Pharmacol Ther* 2012;91:734—8.

prevalence of the variant allele was 5.6%, and screening for the allele eliminated the hypersensitivity reaction, suggesting a negative predictive value of 100% and positive predictive value of 47.9%.[228] This association was embraced by the clinical community such that 2008 guidelines from the Office of AIDS Research Advisory Council recommended preemptive screening. The FDA also revised the labeling for abacavir to reflect the association. For carbamazepine, a similar clinical study was published in 2011 wherein nearly 5000 patients were evaluated in Taiwan. A drug hypersensitivity was not observed in any patients who screened negative for the HLA-15:02, suggesting 100% negative predictive value in that application as well.[229]

A summary of CPIC guidelines for selection of carbamazepine or abacavir based on HLA allele detection is shown in Table 11.10. Note that a contraindication for carbamazepine would also suggest a contraindication to phenytoin because phenytoin-associated cutaneous reactions also occur in people who carry the *HLA-B*15:02*.[224] The *HLA-B*15:02* association with cutaneous adverse drug reactions has been best studied in parts of Asia where the allele is most common.[229,230] Detection of a high-risk allele before therapy is initiated could allow drug avoidance and substantially reduced risk of the associated adverse drug reaction.

Genetic testing for HLA variants is highly variable among laboratories, ranging from complete gene sequencing, to identification of SNVs that are in LD with high-risk haplotypes. For example, an SNV in *HCP5* (rs2395029) is a surrogate marker for HLA-B*57:01, with 99.9% predictive value for the presence of the allele.[231,232] However, *HCP5* occurs within a region of copy number variation, and the SNV of interest may be deleted. As such, false-negative results can exist using the SNV approach.[233] Phenotyping through detection of the B57 and B58 serotypes or skin patch testing can also identify hypersensitivity risks.[223,228,234]

FUTURE DIRECTIONS

Any gene with variation has potential pharmacogenetic implications. The future success of pharmacogenetics depends on understanding (1) the interrelationship of the various proteins involved in pharmacokinetics and pharmacodynamics, (2) the balance between major versus minor pathways, and (3) the mechanisms of action of specific drugs. Also important is an understanding of haplotype relationships, the possible implications of heterozygote variants, and gene dose. Standardization of test content and interpretation is needed. Although massively parallel sequencing will increase the complexity of pharmacogenetics, it will also improve our understanding and the utility of genotype—phenotype relationships. Additional clinical studies, particularly prospective and outcome-related studies, are needed to guide implementation of pharmacogenetic findings in the clinic. Finally, closer integration of pharmacy professionals with clinical laboratories and clinicians and increased availability of cost-effective commercial testing will improve the success of pharmacogenetic applications.

REFERENCES

1. Obama. <http://www.whitehouse.gov/the-press-office/2015/01/30/fact-sheet-president-obama-s-precision-medicine-initiative>.
2. Abul-Husn NS, Owusu Obeng A, Sanderson SC, et al. Implementation and utilization of genetic testing in personalized medicine. *Pharmgenomics Pers Med* 2014;7:227—40.

3. Arnaout R, Buck TP, Roulette P, et al. Predicting the cost and pace of pharmacogenomic advances: an evidence-based study. *Clin Chem* 2013;**59**:649–57.
4. Krasowski MD, McMillin GA. Advances in anti-epileptic drug testing. *Clin Chim Acta* 2014;**436**:224–36.
5. Davies JA. Mechanisms of action of antiepileptic drugs. *Seizure* 1995;**4**:267–71.
6. Leckband SG, Kelsoe JR, Dunnenberger HM, et al. Clinical Pharmacogenetics Implementation Consortium guidelines for HLA-B genotype and carbamazepine dosing. *Clin Pharmacol Ther* 2013;**94**:324–8.
7. Tangamornsuksan W, Chaiyakunapruk N, Somkrua R, et al. Relationship between the HLA-B*1502 allele and carbamazepine-induced Stevens-Johnson syndrome and toxic epidermal necrolysis: a systematic review and meta-analysis. *JAMA Dermatol* 2013;**149**:1025–32.
8. Oellerich M, Armstrong VW, Schutz E, et al. Therapeutic drug monitoring of cyclosporine and tacrolimus. Update on Lake Louise Consensus Conference on cyclosporin and tacrolimus. *Clin Biochem* 1998;**31**:309–16.
9. Birdwell KA, Decker B, Barbarino JM, et al. Clinical pharmacogenetics implementation consortium (CPIC) guidelines for CYP3A5 genotype and tacrolimus dosing. *Clin Pharmacol Ther* 2015.
10. Pruthi RK. Review of the American College of Chest Physicians 2012 Guidelines for Anticoagulation Therapy and Prevention of Thrombosis. *Semin Hematol* 2013;**50**:251–8.
11. Hirsh J, Dalen J, Anderson DR, et al. Oral anticoagulants: mechanism of action, clinical effectiveness, and optimal therapeutic range. *Chest* 2001;**119**:8S–21S.
12. Johnson JA, Gong L, Whirl-Carrillo M, et al. Clinical Pharmacogenetics Implementation Consortium Guidelines for CYP2C9 and VKORC1 genotypes and warfarin dosing. *Clin Pharmacol Ther* 2011;**90**:625–9.
13. Egan LJ. Mechanisms of drug toxicity or intolerance. *Dig Dis* 2011;**29**:172–6.
14. Agarwal A, Ressler D, Snyder G. The current and future state of companion diagnostics. *Pharmgenomics Pers Med* 2015;**8**:99–110.
15. McCormack RT, Armstrong J, Leonard D. Codevelopment of genome-based therapeutics and companion diagnostics: insights from an Institute of Medicine roundtable. *JAMA* 2014;**311**:1395–6.
16. FDA. List of Cleared or Approved Companion Diagnostic Devices (In Vitro and Imaging Tools). <http://www.fda.gov/medicaldevices/productsandmedicalprocedures/invitrodiagnostics/ucm301431.htm>.
17. Evaluation of Genomic Applications in P. Prevention Working G. Recommendations from the EGAPP Working Group: Can testing of tumor tissue for mutations in EGFR pathway downstream effector genes in patients with metastatic colorectal cancer improve health outcomes by guiding decisions regarding anti-EGFR therapy? *Genet Med* 2013;**15**:517–27.
18. Maitland ML, Schilsky RL. Clinical trials in the era of personalized oncology. *CA Cancer J Clin* 2011;**61**:365–81.
19. Bonicke R, Reif W. Enzymatic inactivation of isonicotinic acid hydrazide in humans and animals. *Naunyn Schmiedebergs Arch Exp Pathol Pharmakol* 1953;**220**:321–33.
20. Weber WW, Hein DW. Clinical pharmacokinetics of isoniazid. *Clin Pharmacokinet* 1979;**4**:401–22.
21. Shi S, Li Y. Interplay of drug-metabolizing enzymes and transporters in drug absorption and disposition. *Curr Drug Metab* 2015;**15**:915–41.
22. Thorn CF, Klein TE, Altman RB. Codeine and morphine pathway. *Pharmacogenet Genomics* 2009;**19**:556–8.
23. Whirl-Carrillo M, McDonagh EM, Hebert JM, et al. Pharmacogenomics knowledge for personalized medicine. *Clin Pharmacol Ther* 2012;**92**:414–7.
24. Wadelius M, Pirmohamed M. Pharmacogenetics of warfarin: current status and future challenges. *Pharmacogenomics J* 2007;**7**:99–111.
25. Zhang G, Zhang Y, Ling Y, et al. Web resources for pharmacogenomics. *Genomics Proteomics Bioinformatics* 2015;**13**:51–4.
26. Rogers NL, Cole SA, Lan HC, et al. New saliva DNA collection method compared to buccal cell collection techniques for epidemiological studies. *Am J Hum Biol* 2007;**19**:319–26.
27. Sanchez R, Lee TH, Wen L, et al. Absence of transfusion-associated microchimerism in pediatric and adult recipients of leuko-reduced and gamma-irradiated blood components. *Transfusion* 2012;**52**:936–45.
28. Themeli M, Waterhouse M, Finke J, et al. DNA chimerism and its consequences after allogeneic hematopoietic cell transplantation. *Chimerism* 2011;**2**:25–8.
29. King CR, Porche-Sorbet RM, Gage BF, et al. Performance of commercial platforms for rapid genotyping of polymorphisms affecting warfarin dose. *Am J Clin Pathol* 2008;**129**:876–83.
30. Roberts JD, Wells GA, Le May MR, et al. Point-of-care genetic testing for personalisation of antiplatelet treatment (RAPID GENE): a prospective, randomised, proof-of-concept trial. *Lancet* 2012;**379**:1705–11.
31. Rubinstein WS, Maglott DR, Lee JM, et al. The NIH genetic testing registry: a new, centralized database of genetic tests to enable access to comprehensive information and improve transparency. *Nucleic Acids Res* 2013;**41**:D925–35.
32. Ge D, Fellay J, Thompson AJ, et al. Genetic variation in IL28B predicts hepatitis C treatment-induced viral clearance. *Nature* 2009;**461**:399–401.
33. Weinshilboum R. Inheritance and drug response. *N Engl J Med* 2003;**348**:529–37.
34. He Y, Hoskins JM, McLeod HL. Copy number variants in pharmacogenetic genes. *Trends Mol Med* 2011;**17**:244–51.
35. Kimura S, Umeno M, Skoda RC, et al. The human debrisoquine 4-hydroxylase (CYP2D) locus: sequence and identification of the polymorphic CYP2D6 gene, a related gene, and a pseudogene. *Am J Hum Genet* 1989;**45**:889–904.
36. Robarge JD, Li L, Desta Z, et al. The star-allele nomenclature: retooling for translational genomics. *Clin Pharmacol Ther* 2007;**82**:244–8.
37. Klein TE, Chang JT, Cho MK, et al. Integrating genotype and phenotype information: an overview of the PharmGKB project. Pharmacogenetics Research Network and Knowledge Base. *Pharmacogenomics J* 2001;**1**:167–70.
38. Ehmann F, Caneva L, Prasad K, et al. Pharmacogenomic information in drug labels: European Medicines Agency perspective. *Pharmacogenomics J* 2015;**15**:201–10.
39. Sim SC. The Human Cytochrome P450 (CYP) Allele Nomenclature Database. <http://www.cypalleles.ki.se/>.
40. Sim SC, Ingelman-Sundberg M. The Human Cytochrome P450 (CYP) Allele Nomenclature website: A peer-reviewed

database of CYP variants and their associated effects. *Hum Genomics* 2010;**4**:278–81.
41. McLeod HL, Collie-Duguid ES, Vreken P, et al. Nomenclature for human DPYD alleles. *Pharmacogenetics* 1998;**8**:455–9.
42. Wu AH. Genotype and phenotype concordance for pharmacogenetic tests through proficiency survey testing. *Arch Pathol Lab Med* 2013;**137**:1232–6.
43. Borges S, Desta Z, Jin Y, et al. Composite functional genetic and comedication CYP2D6 activity score in predicting tamoxifen drug exposure among breast cancer patients. *J Clin Pharmacol* 2010;**50**:450–8.
44. Gaedigk A, Simon SD, Pearce RE, et al. The CYP2D6 activity score: translating genotype information into a qualitative measure of phenotype. *Clin Pharmacol Ther* 2008;**83**:234–42.
45. Sim SC, Kacevska M, Ingelman-Sundberg M. Pharmacogenomics of drug-metabolizing enzymes: a recent update on clinical implications and endogenous effects. *Pharmacogenomics J* 2013;**13**:1–11.
46. Appell ML, Berg J, Duley J, et al. Nomenclature for alleles of the thiopurine methyltransferase gene. *Pharmacogenet Genomics* 2013;**23**:242–8.
47. Garte S, Crosti F. A nomenclature system for metabolic gene polymorphisms. *IARC Sci Publ* 1999;5–12.
48. Mackenzie PI, Bock KW, Burchell B, et al. Nomenclature update for the mammalian UDP glycosyltransferase (UGT) gene superfamily. *Pharmacogenet Genomics* 2005;**15**:677–85.
49. Vatsis KP, Weber WW, Bell DA, et al. Nomenclature for N-acetyltransferases. *Pharmacogenetics* 1995;**5**:1–17.
50. Samer CF, Lorenzini KI, Rollason V, et al. Applications of CYP450 testing in the clinical setting. *Mol Diagn Ther* 2013;**17**:165–84.
51. Rendic S, Di Carlo FJ. Human cytochrome P450 enzymes: a status report summarizing their reactions, substrates, inducers, and inhibitors. *Drug Metab Rev* 1997;**29**:413–580.
52. Lilja JJ, Kivisto KT, Neuvonen PJ. Duration of effect of grapefruit juice on the pharmacokinetics of the CYP3A4 substrate simvastatin. *Clin Pharmacol Ther* 2000;**68**:384–90.
53. Gough AC, Miles JS, Spurr NK, et al. Identification of the primary gene defect at the cytochrome P450 CYP2D locus. *Nature* 1990;**347**:773–6.
54. Drozda K, Muller DJ, Bishop JR. Pharmacogenomic testing for neuropsychiatric drugs: current status of drug labeling, guidelines for using genetic information, and test options. *Pharmacotherapy* 2014;**34**:166–84.
55. Gaedigk A. Complexities of CYP2D6 gene analysis and interpretation. *Int Rev Psychiatry* 2013;**25**:534–53.
56. Ingelman-Sundberg M. Genetic polymorphisms of cytochrome P450 2D6 (CYP2D6): clinical consequences, evolutionary aspects and functional diversity. *Pharmacogenomics J* 2005;**5**:6–13.
57. Hosono N, Kato M, Kiyotani K, et al. CYP2D6 genotyping for functional-gene dosage analysis by allele copy number detection. *Clin Chem* 2009;**55**:1546–54.
58. Early Breast Cancer Trialists' Collaborative Group (EBCTCG). Effects of chemotherapy and hormonal therapy for early breast cancer on recurrence and 15-year survival: an overview of the randomised trials. *Lancet* 2005;**365**:1687–717.
59. Osborne CK. Tamoxifen in the treatment of breast cancer. *N Engl J Med* 1998;**339**:1609–18.
60. Klein DJ, Thorn CF, Desta Z, et al. PharmGKB summary: tamoxifen pathway, pharmacokinetics. *Pharmacogenet Genomics* 2013;**23**:643–7.
61. Brauch H, Murdter TE, Eichelbaum M, et al. Pharmacogenomics of tamoxifen therapy. *Clin Chem* 2009;**55**:1770–82.
62. Saladores P, Murdter T, Eccles D, et al. Tamoxifen metabolism predicts drug concentrations and outcome in premenopausal patients with early breast cancer. *Pharmacogenomics J* 2015;**15**:84–94.
63. Goetz MP, Rae JM, Suman VJ, et al. Pharmacogenetics of tamoxifen biotransformation is associated with clinical outcomes of efficacy and hot flashes. *J Clin Oncol* 2005;**23**:9312–8.
64. Mortimer JE, Flatt SW, Parker BA, et al. Tamoxifen, hot flashes and recurrence in breast cancer. *Breast Cancer Res Treat* 2008;**108**:421–6.
65. Swen JJ, Nijenhuis M, de Boer A, et al. Pharmacogenetics: from bench to byte—an update of guidelines. *Clin Pharmacol Ther* 2011;**89**:662–73.
66. Province MA, Goetz MP, Brauch H, et al. CYP2D6 genotype and adjuvant tamoxifen: meta-analysis of heterogeneous study populations. *Clin Pharmacol Ther* 2014;**95**:216–27.
67. Hertz DL, McLeod HL, Irvin Jr WJ. Tamoxifen and CYP2D6: a contradiction of data. *Oncologist* 2012;**17**:620–30.
68. Lum DW, Perel P, Hingorani AD, et al. CYP2D6 genotype and tamoxifen response for breast cancer: a systematic review and meta-analysis. *PLoS ONE* 2013;**8**:e76648.
69. Crews KR, Gaedigk A, Dunnenberger HM, et al. Clinical Pharmacogenetics Implementation Consortium guidelines for cytochrome P450 2D6 genotype and codeine therapy: 2014 update. *Clin Pharmacol Ther* 2014;**95**:376–82.
70. Gasche Y, Daali Y, Fathi M, et al. Codeine intoxication associated with ultrarapid CYP2D6 metabolism. *N Engl J Med* 2004;**351**:2827–31.
71. Madadi P, Shirazi F, Walter FG, et al. Establishing causality of CNS depression in breastfed infants following maternal codeine use. *Paediatr Drugs* 2008;**10**:399–404.
72. Willmann S, Edginton AN, Coboeken K, et al. Risk to the breast-fed neonate from codeine treatment to the mother: a quantitative mechanistic modeling study. *Clin Pharmacol Ther* 2009;**86**:634–43.
73. Whittaker MR. Opioid use and the risk of respiratory depression and death in the pediatric population. *J Pediatr Pharmacol Ther* 2013;**18**:269–76.
74. Bell GC, Crews KR, Wilkinson MR, et al. Development and use of active clinical decision support for preemptive pharmacogenomics. *J Am Med Inform Assoc* 2014;**21**:e93–9.
75. Hoffman JM, Haidar CE, Wilkinson MR, et al. PG4KDS: a model for the clinical implementation of pre-emptive pharmacogenetics. *Am J Med Genet C Semin Med Genet* 2014;**166C**:45–55.
76. Mann JJ. The medical management of depression. *N Engl J Med* 2005;**353**:1819–34.
77. Sallee FR, DeVane CL, Ferrell RE. Fluoxetine-related death in a child with cytochrome P-450 2D6 genetic deficiency. *J Child Adolesc Psychopharmacol* 2000;**10**:27–34.
78. de Leon J, Armstrong SC, Cozza KL. Clinical guidelines for psychiatrists for the use of pharmacogenetic testing for CYP450 2D6 and CYP450 2C19. *Psychosomatics* 2006;**47**:75–85.
79. Kirchheiner J, Nickchen K, Bauer M, et al. Pharmacogenetics of antidepressants and antipsychotics: the contribution of allelic variations to the phenotype of drug response. *Mol Psychiatry* 2004;**9**:442–73.
80. Tomalik-Scharte D, Lazar A, Fuhr U, et al. The clinical role of genetic polymorphisms in drug-metabolizing enzymes. *Pharmacogenomics J* 2008;**8**:4–15.

81. Spear BB, Heath-Chiozzi M, Huff J. Clinical application of pharmacogenetics. *Trends Mol Med* 2001;7:201–4.
82. Steimer W, Zopf K, von Amelunxen S, et al. Amitriptyline or not, that is the question: pharmacogenetic testing of CYP2D6 and CYP2C19 identifies patients with low or high risk for side effects in amitriptyline therapy. *Clin Chem* 2005;51:376–85.
83. Hicks JK, Swen JJ, Thorn CF, et al. Clinical Pharmacogenetics Implementation Consortium guideline for CYP2D6 and CYP2C19 genotypes and dosing of tricyclic antidepressants. *Clin Pharmacol Ther* 2013;93:402–8.
84. Hicks JK, Bishop JR, Sangkuhl K, et al. Clinical Pharmacogenetics Implementation Consortium (CPIC) guideline for CYP2D6 and CYP2C19 genotypes and dosing of selective serotonin reuptake inhibitors. *Clin Pharmacol Ther* 2015;98: 127–34.
85. Stingl J, Viviani R. Polymorphism in CYP2D6 and CYP2C19, members of the cytochrome P450 mixed-function oxidase system, in the metabolism of psychotropic drugs. *J Intern Med* 2015;277:167–77.
86. van der Weide J, van Baalen-Benedek EH, Kootstra-Ros JE. Metabolic ratios of psychotropics as indication of cytochrome P450 2D6/2C19 genotype. *Ther Drug Monit* 2005;27: 478–83.
87. Soars MG, Gelboin HV, Krausz KW, et al. A comparison of relative abundance, activity factor and inhibitory monoclonal antibody approaches in the characterization of human CYP enzymology. *Br J Clin Pharmacol* 2003;55:175–81.
88. Rettie AE, Jones JP. Clinical and toxicological relevance of CYP2C9: drug-drug interactions and pharmacogenetics. *Annu Rev Pharmacol Toxicol* 2005;45:477–94.
89. PharmGKB. The Pharmacogenomics Knowledgebase. <https://www.pharmgkb.org/>.
90. Vormfelde SV, Brockmoller J, Bauer S, et al. Relative impact of genotype and enzyme induction on the metabolic capacity of CYP2C9 in healthy volunteers. *Clin Pharmacol Ther* 2009;86: 54–61.
91. Kanebratt KP, Diczfalusy U, Backstrom T, et al. Cytochrome P450 induction by rifampicin in healthy subjects: determination using the Karolinska cocktail and the endogenous CYP3A4 marker 4beta-hydroxycholesterol. *Clin Pharmacol Ther* 2008; 84:589–94.
92. Siddoway LA. Amiodarone: Guidelines for use and monitoring. *Am Fam Physician* 2003;68:2189–96.
93. Lu Y, Won KA, Nelson BJ, et al. Characteristics of the amiodarone-warfarin interaction during long-term follow-up. *Am J Health Syst Pharm* 2008;65:947–52.
94. Walton R, Kimber M, Rockett K, et al. Haplotype block structure of the cytochrome P450 CYP2C gene cluster on chromosome 10. *Nat Genet* 2005;37:915–6.
95. Limdi NA, Wadelius M, Cavallari L, et al. Warfarin pharmacogenetics: a single VKORC1 polymorphism is predictive of dose across 3 racial groups. *Blood* 2010;115:3827–34.
96. Rettie AE, Wienkers LC, Gonzalez FJ, et al. Impaired (S)-warfarin metabolism catalysed by the R144C allelic variant of CYP2C9. *Pharmacogenetics* 1994;4:39–42.
97. Gaedigk A, Casley WL, Tyndale RF, et al. Cytochrome P4502C9 (CYP2C9) allele frequencies in Canadian Native Indian and Inuit populations. *Can J Physiol Pharmacol* 2001;79:841–7.
98. Drozda K, Wong S, Patel SR, et al. Poor warfarin dose prediction with pharmacogenetic algorithms that exclude genotypes important for African Americans. *Pharmacogenet Genomics* 2015;25:73–81.
99. Nagai R, Ohara M, Cavallari LH, et al. Factors influencing pharmacokinetics of warfarin in African-Americans: implications for pharmacogenetic dosing algorithms. *Pharmacogenomics* 2015;16:217–25.
100. Kramer MA, Rettie AE, Rieder MJ, et al. Novel CYP2C9 promoter variants and assessment of their impact on gene expression. *Mol Pharmacol* 2008;73:1751–60.
101. Pickering JW, McMillin GA, Gedge F, et al. Flow cytometric assay for genotyping cytochrome p450 2C9 and 2C19: comparison with a microelectronic DNA array. *Am J Pharmacogenomics* 2004;4:199–207.
102. Lyon E, McMillin G, Melis R. *Pharmacogenetic testing for warfarin sensitivity. Clinics in laboratory medicine*. Philadephia, PA: WB Saunders; 2009. p. 525–37.
103. Flockhart DA. Drug interactions and the cytochrome P450 system. The role of cytochrome P450 2C19. *Clin Pharmacokinet* 1995;29(Suppl. 1):45–52.
104. Strom CM, Goos D, Crossley B, et al. Testing for variants in CYP2C19: population frequencies and testing experience in a clinical laboratory. *Genet Med* 2012;14:95–100.
105. de Morais SM, Wilkinson GR, Blaisdell J, et al. The major genetic defect responsible for the polymorphism of S-mephenytoin metabolism in humans. *J Biol Chem* 1994;269: 15419–22.
106. Mizutani T. PM frequencies of major CYPs in Asians and Caucasians. *Drug Metab Rev* 2003;35:99–106.
107. Scott SA, Martis S, Peter I, et al. Identification of CYP2C19*4B: pharmacogenetic implications for drug metabolism including clopidogrel responsiveness. *Pharmacogenomics J* 2012;12: 297–305.
108. Yin OQ, Tomlinson B, Chow AH, et al. Omeprazole as a CYP2C19 marker in Chinese subjects: assessment of its gene-dose effect and intrasubject variability. *J Clin Pharmacol* 2004; 44:582–9.
109. Scott SA, Sangkuhl K, Stein CM, et al. Clinical Pharmacogenetics Implementation Consortium guidelines for CYP2C19 genotype and clopidogrel therapy: 2013 update. *Clin Pharmacol Ther* 2013;94:317–23.
110. Scott SA, Martis S, Peter I, et al. Identification of CYP2C19*4B: pharmacogenetic implications for drug metabolism including clopidogrel responsiveness. *Pharmacogenomics J* 2012;12: 297–305.
111. Gurbel PA, Bliden KP, Hiatt BL, et al. Clopidogrel for coronary stenting: response variability, drug resistance, and the effect of pretreatment platelet reactivity. *Circulation* 2003;107: 2908–13.
112. Brandt JT, Close SL, Iturria SJ, et al. Common polymorphisms of CYP2C19 and CYP2C9 affect the pharmacokinetic and pharmacodynamic response to clopidogrel but not prasugrel. *J Thromb Haemost* 2007;5:2429–36.
113. Mega JL, Close SL, Wiviott SD, et al. Cytochrome p-450 polymorphisms and response to clopidogrel. *N Engl J Med* 2009;360:354–62.
114. Simon T, Verstuyft C, Mary-Krause M, et al. Genetic determinants of response to clopidogrel and cardiovascular events. *N Engl J Med* 2009;360:363–75.
115. Linder MW, Keck Jr PE. Standards of laboratory practice: antidepressant drug monitoring. National Academy of Clinical Biochemistry. *Clin Chem* 1998;44:1073–84.
116. Perry PJ, Pfohl BM, Holstad SG. The relationship between antidepressant response and tricyclic antidepressant plasma concentrations. A retrospective analysis of the literature using

logistic regression analysis. *Clin Pharmacokinet* 1987;**13**: 381–92.
117. Wang B, Canestaro WJ, Choudhry NK. Clinical evidence supporting pharmacogenomic biomarker testing provided in US Food and Drug Administration drug labels. *JAMA Intern Med* 2014;**174**:1938–44.
118. Kuehl P, Zhang J, Lin Y, et al. Sequence diversity in CYP3A promoters and characterization of the genetic basis of polymorphic CYP3A5 expression. *Nat Genet* 2001;**27**:383–91.
119. Back D, Else L. The importance of drug-drug interactions in the DAA era. *Dig Liver Dis* 2013;**45**(Suppl. 5):S343–8.
120. Nowack R. Review article: Cytochrome P450 enzyme, and transport protein mediated herb-drug interactions in renal transplant patients: grapefruit juice, St. John's Wort—and beyond! *Nephrology (Carlton)* 2008;**13**:337–47.
121. Rojas L, Neumann I, Herrero MJ, et al. Effect of CYP3A5*3 on kidney transplant recipients treated with tacrolimus: a systematic review and meta-analysis of observational studies. *Pharmacogenomics J* 2015;**15**:38–48.
122. van Kuilenburg AB, Meinsma R, Zonnenberg BA, et al. Dihydropyrimidinase deficiency and severe 5-fluorouracil toxicity. *Clin Cancer Res* 2003;**9**:4363–7.
123. Amstutz U, Farese S, Aebi S, et al. Dihydropyrimidine dehydrogenase gene variation and severe 5-fluorouracil toxicity: a haplotype assessment. *Pharmacogenomics* 2009;**10**:931–44.
124. Van Kuilenburg AB, Vreken P, Abeling NG, et al. Genotype and phenotype in patients with dihydropyrimidine dehydrogenase deficiency. *Hum Genet* 1999;**104**:1–9.
125. Caudle KE, Thorn CF, Klein TE, et al. Clinical Pharmacogenetics Implementation Consortium guidelines for dihydropyrimidine dehydrogenase genotype and fluoropyrimidine dosing. *Clin Pharmacol Ther* 2013;**94**:640–5.
126. Collie-Duguid ES, Etienne MC, Milano G, et al. Known variant DPYD alleles do not explain DPD deficiency in cancer patients. *Pharmacogenetics* 2000;**10**:217–23.
127. Meinsma R, Fernandez-Salguero P, Van Kuilenburg AB, et al. Human polymorphism in drug metabolism: mutation in the dihydropyrimidine dehydrogenase gene results in exon skipping and thymine uracilurea. *DNA Cell Biol* 1995;**14**:1–6.
128. Morel A, Boisdron-Celle M, Fey L, et al. Clinical relevance of different dihydropyrimidine dehydrogenase gene single nucleotide polymorphisms on 5-fluorouracil tolerance. *Mol Cancer Ther* 2006;**5**:2895–904.
129. Saif MW, Ezzeldin H, Vance K, et al. DPYD*2A mutation: the most common mutation associated with DPD deficiency. *Cancer Chemother Pharmacol* 2007;**60**:503–7.
130. Schwab M, Zanger UM, Marx C, et al. Role of genetic and nongenetic factors for fluorouracil treatment-related severe toxicity: a prospective clinical trial by the German 5-FU Toxicity Study Group. *J Clin Oncol* 2008;**26**:2131–8.
131. Thorn CF, Marsh S, Carrillo MW, et al. PharmGKB summary: fluoropyrimidine pathways. *Pharmacogenet Genomics* 2011;**21**: 237–42.
132. van Staveren MC, Guchelaar HJ, van Kuilenburg AB, et al. Evaluation of predictive tests for screening for dihydropyrimidine dehydrogenase deficiency. *Pharmacogenomics J* 2013; **13**:389–95.
133. Terrazzino S, Cargnin S, Del Re M, et al. DPYD IVS14+1G>A and 2846A>T genotyping for the prediction of severe fluoropyrimidine-related toxicity: a meta-analysis. *Pharmacogenomics* 2013;**14**:1255–72.
134. Deenen MJ, Tol J, Burylo AM, et al. Relationship between single nucleotide polymorphisms and haplotypes in DPYD and toxicity and efficacy of capecitabine in advanced colorectal cancer. *Clin Cancer Res* 2011;**17**:3455–68.
135. Rodrigues-Lima F, Dairou J, Laurieri N, et al. Pharmacogenomics, biochemistry, toxicology, microbiology and cancer research in one go. *Pharmacogenomics* 2011;**12**:1091–3.
136. Gross M, Kruisselbrink T, Anderson K, et al. Distribution and concordance of N-acetyltransferase genotype and phenotype in an American population. *Cancer Epidemiol Biomarkers Prev* 1999;**8**:683–92.
137. Peluso ME, Munnia A, Srivatanakul P, et al. DNA adducts and combinations of multiple lung cancer at-risk alleles in environmentally exposed and smoking subjects. *Environ Mol Mutagen* 2013;**54**:375–83.
138. Cappellini MD, Fiorelli G. Glucose-6-phosphate dehydrogenase deficiency. *Lancet* 2008;**371**:64–74.
139. Mason PJ, Bautista JM, Gilsanz F. G6PD deficiency: the genotype-phenotype association. *Blood Rev* 2007;**21**:267–83.
140. Frank JE. Diagnosis and management of G6PD deficiency. *Am Fam Physician* 2005;**72**:1277–82.
141. Nkhoma ET, Poole C, Vannappagari V, et al. The global prevalence of glucose-6-phosphate dehydrogenase deficiency: a systematic review and meta-analysis. *Blood Cells Mol Dis* 2009; **42**:267–78.
142. McDonagh EM, Thorn CF, Bautista JM, et al. PharmGKB summary: very important pharmacogene information for G6PD. *Pharmacogenet Genomics* 2012;**22**:219–28.
143. Lin Z, Fontaine JM, Freer DE, et al. Alternative DNA-based newborn screening for glucose-6-phosphate dehydrogenase deficiency. *Mol Genet Metab* 2005;**86**:212–9.
144. Sundy JS, Baraf HS, Yood RA, et al. Efficacy and tolerability of pegloticase for the treatment of chronic gout in patients refractory to conventional treatment: two randomized controlled trials. *JAMA* 2011;**306**:711–20.
145. Relling MV, McDonagh EM, Chang T, et al. Clinical Pharmacogenetics Implementation Consortium (CPIC) guidelines for rasburicase therapy in the context of G6PD deficiency genotype. *Clin Pharmacol Ther* 2014;**96**:169–74.
146. Khneisser I, Adib SM, Loiselet J, et al. Cost-benefit analysis of G6PD screening in Lebanese newborn males. *J Med Liban* 2007;**55**:129–32.
147. Relling MV, Gardner EE, Sandborn WJ, et al. Clinical Pharmacogenetics Implementation Consortium guidelines for thiopurine methyltransferase genotype and thiopurine dosing. *Clin Pharmacol Ther* 2011;**89**:387–91.
148. Weinshilboum RM, Sladek SL. Mercaptopurine pharmacogenetics: monogenic inheritance of erythrocyte thiopurine methyltransferase activity. *Am J Hum Genet* 1980;**32**:651–62.
149. Evans WE. Pharmacogenetics of thiopurine S-methyltransferase and thiopurine therapy. *Ther Drug Monit* 2004;**26**:186–91.
150. Black AJ, McLeod HL, Capell HA, et al. Thiopurine methyltransferase genotype predicts therapy-limiting severe toxicity from azathioprine. *Ann Intern Med* 1998;**129**:716–8.
151. Relling MV, Hancock ML, Rivera GK, et al. Mercaptopurine therapy intolerance and heterozygosity at the thiopurine S-methyltransferase gene locus. *J Natl Cancer Inst* 1999;**91**: 2001–8.
152. Evans WE, Hon YY, Bomgaars L, et al. Preponderance of thiopurine S-methyltransferase deficiency and heterozygosity

152. among patients intolerant to mercaptopurine or azathioprine. *J Clin Oncol* 2001;**19**:2293–301.
153. Yates CR, Krynetski EY, Loennechen T, et al. Molecular diagnosis of thiopurine S-methyltransferase deficiency: genetic basis for azathioprine and mercaptopurine intolerance. *Ann Intern Med* 1997;**126**:608–14.
154. Zhou S. Clinical pharmacogenomics of thiopurine S-methyltransferase. *Curr Clin Pharmacol* 2006;**1**:119–28.
155. Roberts RL, Gearry RB, Bland MV, et al. Trinucleotide repeat variants in the promoter of the thiopurine S-methyltransferase gene of patients exhibiting ultra-high enzyme activity. *Pharmacogenet Genomics* 2008;**18**:434–8.
156. Zukic B, Radmilovic M, Stojiljkovic M, et al. Functional analysis of the role of the TPMT gene promoter VNTR polymorphism in TPMT gene transcription. *Pharmacogenomics* 2010;**11**:547–57.
157. Compagni A, Bartoli S, Buehrlen B, et al. Avoiding adverse drug reactions by pharmacogenetic testing: a systematic review of the economic evidence in the case of TPMT and AZA-induced side effects. *Int J Technol Assess Health Care* 2008; **24**:294–302.
158. Donnan JR, Ungar WJ, Mathews M, et al. A cost effectiveness analysis of thiopurine methyltransferase testing for guiding 6-mercaptopurine dosing in children with acute lymphoblastic leukemia. *Pediatr Blood Cancer* 2011;**57**:231–9.
159. Thompson AJ, Newman WG, Elliott RA, et al. The cost-effectiveness of a pharmacogenetic test: a trial-based evaluation of TPMT genotyping for azathioprine. *Value Health* 2014;**17**: 22–33.
160. Marsh S, Hoskins JM. Irinotecan pharmacogenomics. *Pharmacogenomics* 2010;**11**:1003–10.
161. Perera MA, Innocenti F, Ratain MJ. Pharmacogenetic testing for uridine diphosphate glucuronosyltransferase 1A1 polymorphisms: are we there yet? *Pharmacotherapy* 2008;**28**: 755–68.
162. Zhang D, Zhang D, Cui D, et al. Characterization of the UDP glucuronosyltransferase activity of human liver microsomes genotyped for the UGT1A1*28 polymorphism. *Drug Metab Dispos* 2007;**35**:2270–80.
163. Strassburg CP. Pharmacogenetics of Gilbert's syndrome. *Pharmacogenomics* 2008;**9**:703–15.
164. Boyd MA, Srasuebkul P, Ruxrungtham K, et al. Relationship between hyperbilirubinaemia and UDP-glucuronosyltransferase 1A1 (UGT1A1) polymorphism in adult HIV-infected Thai patients treated with indinavir. *Pharmacogenet Genomics* 2006;**16**:321–9.
165. van der Bol JM, Mathijssen RH, Creemers GJ, et al. A CYP3A4 phenotype-based dosing algorithm for individualized treatment of irinotecan. *Clin Cancer Res* 2010;**16**:736–42.
166. Evaluation of Genomic Applications in P. Prevention Working G. Recommendations from the EGAPP Working Group: Can UGT1A1 genotyping reduce morbidity and mortality in patients with metastatic colorectal cancer treated with irinotecan? *Genet Med* 2009;**11**:15–20.
167. Gold HT, Hall MJ, Blinder V, et al. Cost effectiveness of pharmacogenetic testing for uridine diphosphate glucuronosyltransferase 1A1 before irinotecan administration for metastatic colorectal cancer. *Cancer* 2009;**115**:3858–67.
168. Pichereau S, Le Louarn A, Lecomte T, et al. Cost-effectiveness of UGT1A1*28 genotyping in preventing severe neutropenia following FOLFIRI therapy in colorectal cancer. *J Pharm Pharm Sci* 2010;**13**:615–25.
169. Klaassen CD, Aleksunes LM. Xenobiotic, bile acid, and cholesterol transporters: function and regulation. *Pharmacol Rev* 2010;**62**:1–96.
170. International Transporter C, Giacomini KM, Huang SM, et al. Membrane transporters in drug development. *Nat Rev Drug Discov* 2010;**9**:215–36.
171. Borst P, Elferink RO. Mammalian ABC transporters in health and disease. *Annu Rev Biochem* 2002;**71**:537–92.
172. Jones PM, George AM. The ABC transporter structure and mechanism: perspectives on recent research. *Cell Mol Life Sci* 2004;**61**:682–99.
173. Leschziner GD, Andrew T, Pirmohamed M, et al. ABCB1 genotype and PGP expression, function and therapeutic drug response: a critical review and recommendations for future research. *Pharmacogenomics J* 2007;**7**:154–79.
174. Aszalos A. Drug-drug interactions affected by the transporter protein, P-glycoprotein (ABCB1, MDR1) II. Clinical aspects. *Drug Discov Today* 2007;**12**:838–43.
175. Fahrmayr C, Fromm MF, Konig J. Hepatic OATP and OCT uptake transporters: their role for drug-drug interactions and pharmacogenetic aspects. *Drug Metab Rev* 2010;**42**:380–401.
176. Kalliokoski A, Niemi M. Impact of OATP transporters on pharmacokinetics. *Br J Pharmacol* 2009;**158**:693–705.
177. Niemi M, Pasanen MK, Neuvonen PJ. Organic anion transporting polypeptide 1B1: a genetically polymorphic transporter of major importance for hepatic drug uptake. *Pharmacol Rev* 2011;**63**:157–81.
178. Abe T, Kakyo M, Tokui T, et al. Identification of a novel gene family encoding human liver-specific organic anion transporter LST-1. *J Biol Chem* 1999;**274**:17159–63.
179. Ramsey LB, Johnson SG, Caudle KE, et al. The clinical pharmacogenetics implementation consortium guideline for SLCO1B1 and simvastatin-induced myopathy: 2014 update. *Clin Pharmacol Ther* 2014;**96**:423–8.
180. Niemi M, Schaeffeler E, Lang T, et al. High plasma pravastatin concentrations are associated with single nucleotide polymorphisms and haplotypes of organic anion transporting polypeptide-C (OATP-C, SLCO1B1). *Pharmacogenetics* 2004; **14**:429–40.
181. Ramsey LB, Panetta JC, Smith C, et al. Genome-wide study of methotrexate clearance replicates SLCO1B1. *Blood* 2013;**121**: 898–904.
182. Sathasivam S. Statin induced myotoxicity. *Eur J Intern Med* 2012;**23**:317–24.
183. Alfirevic A, Neely D, Armitage J, et al. Phenotype standardization for statin-induced myotoxicity. *Clin Pharmacol Ther* 2014;**96**:470–6.
184. Pasanen MK, Neuvonen M, Neuvonen PJ, et al. SLCO1B1 polymorphism markedly affects the pharmacokinetics of simvastatin acid. *Pharmacogenet Genomics* 2006;**16**:873–9.
185. Group SC, Link E, Parish S, et al. SLCO1B1 variants and statin-induced myopathy—a genomewide study. *N Engl J Med* 2008; **359**:789–99.
186. Donnelly LA, Doney AS, Tavendale R, et al. Common nonsynonymous substitutions in SLCO1B1 predispose to statin intolerance in routinely treated individuals with type 2 diabetes: a go-DARTS study. *Clin Pharmacol Ther* 2011;**89**:210–6.
187. Heart Protection Study Collaborative G, Bulbulia R, Bowman L, et al. Effects on 11-year mortality and morbidity of lowering LDL cholesterol with simvastatin for about 5 years in 20,536 high-risk individuals: a randomised controlled trial. *Lancet* 2011;**378**:2013–20.

188. Voora D, Shah SH, Spasojevic I, et al. The SLCO1B1*5 genetic variant is associated with statin-induced side effects. *J Am Coll Cardiol* 2009;**54**:1609–16.
189. Garcia AA, Reitsma PH. VKORC1 and the vitamin K cycle. *Vitam Horm* 2008;**78**:23–33.
190. Li T, Chang CY, Jin DY, et al. Identification of the gene for vitamin K epoxide reductase. *Nature* 2004;**427**:541–4.
191. Rost S, Fregin A, Ivaskevicius V, et al. Mutations in VKORC1 cause warfarin resistance and multiple coagulation factor deficiency type 2. *Nature* 2004;**427**:537–41.
192. D'Andrea G, D'Ambrosio RL, Di Perna P, et al. A polymorphism in the VKORC1 gene is associated with an interindividual variability in the dose-anticoagulant effect of warfarin. *Blood* 2005;**105**:645–9.
193. Rieder MJ, Reiner AP, Gage BF, et al. Effect of VKORC1 haplotypes on transcriptional regulation and warfarin dose. *N Engl J Med* 2005;**352**:2285–93.
194. Wadelius M, Chen LY, Downes K, et al. Common VKORC1 and GGCX polymorphisms associated with warfarin dose. *Pharmacogenomics J* 2005;**5**:262–70.
195. Jorgensen AL, FitzGerald RJ, Oyee J, et al. Influence of CYP2C9 and VKORC1 on patient response to warfarin: a systematic review and meta-analysis. *PLoS ONE* 2012;**7**:e44064.
196. Wang D, Chen H, Momary KM, et al. Regulatory polymorphism in vitamin K epoxide reductase complex subunit 1 (VKORC1) affects gene expression and warfarin dose requirement. *Blood* 2008;**112**:1013–21.
197. Innocenti F, Cooper GM, Stanaway IB, et al. Identification, replication, and functional fine-mapping of expression quantitative trait loci in primary human liver tissue. *PLoS Genet* 2011;**7**:e1002078.
198. Bress A, Patel SR, Perera MA, et al. Effect of NQO1 and CYP4F2 genotypes on warfarin dose requirements in Hispanic-Americans and African-Americans. *Pharmacogenomics* 2012;**13**:1925–35.
199. Daneshjou R, Gamazon ER, Burkley B, et al. Genetic variant in folate homeostasis is associated with lower warfarin dose in African Americans. *Blood* 2014;**124**:2298–305.
200. Perera MA, Cavallari LH, Limdi NA, et al. Genetic variants associated with warfarin dose in African-American individuals: a genome-wide association study. *Lancet* 2013;**382**:790–6.
201. Pirmohamed M, Kamali F, Daly AK, et al. Oral anticoagulation: a critique of recent advances and controversies. *Trends Pharmacol Sci* 2015;**36**:153–63.
202. van den Besselaar AM. Standardization of the prothrombin time in oral anticoagulant control. *Haemostasis* 1985;**15**:271–7.
203. Jones M, McEwan P, Morgan CL, et al. Evaluation of the pattern of treatment, level of anticoagulation control, and outcome of treatment with warfarin in patients with non-valvar atrial fibrillation: a record linkage study in a large British population. *Heart* 2005;**91**:472–7.
204. Wittkowsky AK. Warfarin and other coumarin derivatives: pharmacokinetics, pharmacodynamics, and drug interactions. *Semin Vasc Med* 2003;**3**:221–30.
205. Rettie AE, Korzekwa KR, Kunze KL, et al. Hydroxylation of warfarin by human cDNA-expressed cytochrome P-450: a role for P-4502C9 in the etiology of (S)-warfarin-drug interactions. *Chem Res Toxicol* 1992;**5**:54–9.
206. Lenzini P, Wadelius M, Kimmel S, et al. Integration of genetic, clinical, and INR data to refine warfarin dosing. *Clin Pharmacol Ther* 2010;**87**:572–8.
207. Mega JL, Walker JR, Ruff CT, et al. Genetics and the clinical response to warfarin and edoxaban: findings from the randomised, double-blind ENGAGE AF-TIMI 48 trial. *Lancet* 2015.
208. Eriksson N, Wadelius M. Prediction of warfarin dose: why, when and how? *Pharmacogenomics* 2012;**13**:429–40.
209. International Warfarin Pharmacogenetics C, Klein TE, Altman RB, et al. Estimation of the warfarin dose with clinical and pharmacogenetic data. *N Engl J Med* 2009;**360**:753–64.
210. Francis B, Lane S, Pirmohamed M, et al. A review of a priori regression models for warfarin maintenance dose prediction. *PLoS ONE* 2014;**9**:e114896.
211. Pirmohamed M, Burnside G, Eriksson N, et al. A randomized trial of genotype-guided dosing of warfarin. *N Engl J Med* 2013;**369**:2294–303.
212. Kimmel SE, French B, Kasner SE, et al. A pharmacogenetic versus a clinical algorithm for warfarin dosing. *N Engl J Med* 2013;**369**:2283–93.
213. Avery PJ, Jorgensen A, Hamberg AK, et al. A proposal for an individualized pharmacogenetics-based warfarin initiation dose regimen for patients commencing anticoagulation therapy. *Clin Pharmacol Ther* 2011;**90**:701–6.
214. Kimmel SE, French B, Anderson JL, et al. Rationale and design of the Clarification of Optimal Anticoagulation through Genetics trial. *Am Heart J* 2013;**166**:435–41.
215. Moskowitz SM, Chmiel JF, Sternen DL, et al. CFTR-Related Disorders. In: Pagon RA, Bird TC, Dolan CR, Stephens K, ed. GeneReviews. <www.ncbinlmnihgov/sites/GeneTests/>.
216. WHO. The Molecular Genetic Epidemiology of Cystic Fibrosis. 2004.
217. Watson MS, Cutting GR, Desnick RJ, et al. Cystic fibrosis population carrier screening: 2004 revision of American College of Medical Genetics mutation panel. *Genet Med* 2004;**6**:387–91.
218. Grody WW, Cutting GR, Klinger KW, et al. Laboratory standards and guidelines for population-based cystic fibrosis carrier screening. *Genet Med* 2001;**3**:149–54.
219. Amos J, Feldman G, Grody WW, et al. American College of Medical Genetics standards and guidelines for clinical genetics laboratories. ACMG. <www.acmg.net>.
220. Clancy JP, Johnson SG, Yee SW, et al. Clinical Pharmacogenetics Implementation Consortium (CPIC) guidelines for ivacaftor therapy in the context of CFTR genotype. *Clin Pharmacol Ther* 2014;**95**:592–7.
221. Marsh S. Nomenclature of HLA Alleles. <http://hla.alleles.org>.
222. Chung WH, Hung SI, Chen YT. Human leukocyte antigens and drug hypersensitivity. *Curr Opin Allergy Clin Immunol* 2007;**7**:317–23.
223. Agundez JA, Esguevillas G, Amo G, et al. Pharmacogenomics testing for type B adverse drug reactions to anti-infective drugs: the example of hypersensitivity to abacavir. *Recent Pat Antiinfect Drug Discov* 2015.
224. Caudle KE, Rettie AE, Whirl-Carrillo M, et al. Clinical pharmacogenetics implementation consortium guidelines for CYP2C9 and HLA-B genotypes and phenytoin dosing. *Clin Pharmacol Ther* 2014;**96**:542–8.
225. Hershfield MS, Callaghan JT, Tassaneeyakul W, et al. Clinical Pharmacogenetics Implementation Consortium guidelines for human leukocyte antigen-B genotype and allopurinol dosing. *Clin Pharmacol Ther* 2013;**93**:153–8.
226. Martin MA, Hoffman JM, Freimuth RR, et al. Clinical Pharmacogenetics Implementation Consortium guidelines for

HLA-B genotype and abacavir dosing: 2014 update. *Clin Pharmacol Ther* 2014;**95**:499—500.
227. Martin MA, Klein TE, Dong BJ, et al. Clinical pharmacogenetics implementation consortium guidelines for HLA-B genotype and abacavir dosing. *Clin Pharmacol Ther* 2012;**91**.734—8.
228. Mallal S, Phillips E, Carosi G, et al. HLA-B*5701 screening for hypersensitivity to abacavir. *N Engl J Med* 2008;**358**: 568—79.
229. Chen P, Lin JJ, Lu CS, et al. Carbamazepine-induced toxic effects and HLA-B*1502 screening in Taiwan. *N Engl J Med* 2011;**364**:1126—33.
230. Chang CC, Too CL, Murad S, et al. Association of HLA-B*1502 allele with carbamazepine-induced toxic epidermal necrolysis and Stevens-Johnson syndrome in the multi-ethnic Malaysian population. *Int J Dermatol* 2011;**50**:221—4.
231. Rodriguez-Novoa S, Cuenca L, Morello J, et al. Use of the HCP5 single nucleotide polymorphism to predict hypersensitivity reactions to abacavir: correlation with HLA-B*5701. *J Antimicrob Chemother* 2010;**65**:1567—9.
232. Sanchez-Giron F, Villegas-Torres B, Jaramillo-Villafuerte K, et al. Association of the genetic marker for abacavir hypersensitivity HLA-B*5701 with HCP5 rs2395029 in Mexican Mestizos. *Pharmacogenomics* 2011;**12**:809—14.
233. Melis R, Lewis T, Millson A, et al. Copy number variation and incomplete linkage disequilibrium interfere with the HCP5 genotyping assay for abacavir hypersensitivity. *Genet Test Mol Biomarkers* 2012;**16**:1111—4.
234. Kostenko L, Kjer-Nielsen L, Nicholson I, et al. Rapid screening for the detection of HLA-B57 and HLA-B58 in prevention of drug hypersensitivity. *Tissue Antigens* 2011;**78**:11—20.
235. Gage BF, Eby C, Johnson JA, et al. Use of pharmacogenetic and clinical factors to predict the therapeutic dose of warfarin. *Clin Pharmacol Ther* 2008;**84**:326—31.
236. Hamberg AK, Hellman J, Dahlberg J, et al. A Bayesian decision support tool for efficient dose individualization of warfarin in adults and children. *BMC Med Inform Decis Mak* 2015;**15**:7.

12

Identity Testing

Victor W. Weedn, Katherine B. Gettings and Daniele S. Podini

ABSTRACT

Background
DNA has sufficient variation among individuals to discriminate all individuals who have ever lived on the earth or will in the foreseeable future. Furthermore, DNA persists as an identification taggant over the lifetime of the individual and can be obtained from all tissues and fluids—even trace amounts. Accordingly, DNA identity testing has become routine in forensic laboratories around the world. This genetic testing is also useful for parentage and kinship relationship testing as well as certain other clinical applications.

Content
This chapter describes the technologies and methods used in forensic science, parentage, and clinical laboratories for individual identification and genetic relationship testing. Virtually all such testing is based upon capillary electrophoresis (CE) with laser induced fluorescence instrumental analysis following sample collection, DNA extraction, quantification, and polymerase chain reaction (PCR) amplification. Short tandem repeat (STR) testing, using the Federal Bureau of Investigation's (FBI's) combined DNA index system (CODIS) loci, has become routine throughout the United States. An expanded set of 20 STRs will soon replace the original 13 loci. Y-STR analysis, mitochondrial DNA sequencing, and single nucleotide polymorphisms (SNP) tests are important ancillary tests. Massively parallel sequencing and rapid DNA testing will offer new technologic capabilities. Forensic testing must comply with chain-of-custody and other rigorous standards and must be capable of withstanding courtroom scrutiny. The same identity tests are useful for parentage testing, as well as in other clinical applications, such as sample verification, prenatal validation testing, bone marrow transplantation, and loss of heterozygosity (LOH) tumor testing.

Identity testing began with the use of serologic methods to identify variations in proteins that differ among individuals (ie, blood group and type, serum protein isoenzymes). Technical advances in genetic analysis methods allowed the field to move to direct typing of DNA. Genetic variation among individuals is extensive, with about one sequence difference for every 400 to 1250 nucleotides on autosomal chromosomes. This variation is found in coding regions and even more abundantly in noncoding regions. Variants of a genetic locus in a population are referred to as *alleles*.

DNA testing has revolutionized criminalistics. Only fingerprint evidence can sometimes rival the ability of DNA as trace evidence left at a scene to identify a perpetrator. Any nucleated cell, regardless of the tissue or body fluid of origin, can be an informative source of DNA. As a general rule, other trace evidence merely links an article, instrument, or material to a scene. The origin of DNA-based identity testing is generally traced to a 1985 article in *Nature* by Alec Jeffreys.[1] He coined the term "DNA fingerprint" and suggested that the hybridization of DNA probes to polymorphic genetic loci could be used for human identification and thus exploited for forensic purposes. Jeffreys first applied his techniques to civil and criminal cases in England. In the United States, DNA-based identity testing was introduced via commercial laboratories and later by the FBI. Today over 400 forensic DNA typing laboratories have been established in the United States, along with many other DNA laboratories around the globe.[2] Forensic DNA testing is also used to identify decomposed unknown human remains through kinship analysis and can be used to identify victims of mass disaster and servicemen who have perished in past conflicts.[3-6]

Forensic testing differs from clinical laboratory testing in several ways: (1) the forensic question is usually one of identity rather than one of presence or absence of a trait or analyte quantification, as is done in most clinical laboratory analyses; (2) specimens received by forensic laboratories are much more diverse than the typical blood, fluid, and tissue samples handled by clinical laboratories; (3) clinical samples are collected under controlled circumstances, while biologic evidence is exposed to the environment, potentially resulting in degradation, modification, and inhibition; (4) forensic samples may include DNA mixtures from multiple donors, contamination, and nonhuman sources; (5) evidentiary material cannot be replenished and may be present in only trace amounts—for example, testing may consume the sample, rendering complete or repeat testing impossible; and (6) forensic identity testing is scrutinized in a judicial environment, requiring complete accounting for chain of custody following its collection and strict validation of procedures (Box 12.1).

BOX 12.1 Historical Context

1985 Alec Jeffreys's article appears in *Nature*.
1986-87 Forensic Science Associates, Lifecodes, Cellmark begin casework.
1986 *Pennsylvania v. Pestinikas* first use of DNA profiling in court using ALA DQ-α.
1987 Tommy Lee Andrews becomes the first person convicted in the United States.
1988 TWGDAM established.
1989 *NY v. Castro* first successful defense challenge to DNA evidence.
1989 First TWGDAM guidelines published.
1990 *US v. Yee* unsuccessful defense challenge in federal court.
1991 Caskey describes STRs for DNA typing.
1992 NRC I report published.
1994 DNA Identification Act authorizes FBI database using CODIS software.
1995 DNA Advisory Board established to review standards.
1996 NRC II report published.
1997 13 CODIS core loci selected.
1998 FBI's QAS standards for DNA testing labs published.
1999 FBI's QAS standards for DNA databasing labs published.
2011 FBI announces consideration of additional CODIS core loci.
2017 Expanded CODIS core loci to be implemented.

SHORT TANDEM REPEATS AND AMELOGENIN

STR or microsatellite loci consist of DNA sequence motifs that have core repeats of two to seven base pairs.[7,8] Examples include the dinucleotide 5′ CACACACA 3′ and the tetranucleotide 5′ AGATAGATAGATAGAT 3′ (Fig. 12.1). Thousands of STRs are scattered throughout the genome.[9] Because they are flanked by unique sequences, each can be specifically amplified with the PCR for analysis.[10] In populations of individuals, multiple alleles may be present based on differences in the numbers of repeated motifs at a locus.[8] STRs have many characteristics that make them ideal for identity testing: (1) they are relatively short sequences, less than 400 nucleotides, thus still amplifiable in moderately degraded DNA; (2) they can be amplified with PCR in multiplex and analyzed on automated fluorescent CE platforms; (3) alleles can be assigned in a definitive manner following analysis; (4) STR loci are almost always transmitted in families in a Mendelian fashion[11]; (5) the loci may have 10 or more alleles with high power of discrimination, making them highly informative; and (6) extensive information is available about allele frequencies in many human populations for STRs commonly used in identity testing.[12]

Forensic STR testing is based on PCR,[13] which is inherently sensitive, allowing routine analysis of subnanogram quantities of genomic DNA and often successful testing down to 100 picograms of sample (one human diploid cell contains approximately 6 pg of DNA).[14] Commercially available STR systems employ predominantly tetrameric repeat loci, although a few trimeric and pentameric loci are also in use.[15-17] Fragments can be labeled during PCR amplification with fluorescently tagged primers using different dyes facilitating multiplexing.

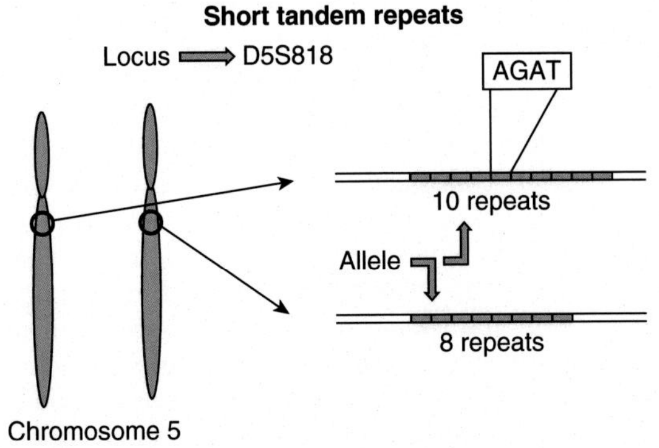

The National Institute of Justice provided funding for the initial application of STRs to forensics. STRs were used in forensic casework during the first Persian Gulf War and were widely adopted for testing by forensic laboratories in the United Kingdom and the United States in the mid to late 1990s.

In 1997, the FBI established the CODIS, which blends forensic DNA analysis and computer technology into an effective tool for investigating violent crimes.[18,19] CODIS enables federal, state, and local crime laboratories to exchange and compare DNA profiles electronically, thereby linking crimes to one another, and to convicted offenders. A panel of forensic scientists was convened to select a panel of 13 STR loci for use in the National DNA Index System (NDIS). These 13 core loci have become the standard for casework and data banking; they are now routinely used in crime laboratories globally and typically yield discriminatory values of one in 10^{12} to one in 10^{21}.[20-23]

In April 2011, the FBI launched an effort with the CODIS Core Loci Working Group to select additional markers that could be implemented in the core CODIS loci. The purpose of this effort was threefold: to reduce the likelihood of adventitious matches given the constant increase of profiles stored in NDIS, to increase international compatibility because some non-CODIS loci are popular in other countries (for example, Europe), and to increase the power of discrimination to aid in criminal and missing person cases.[24] The working group selected an additional seven markers, for a total of 20 STR loci, to be implemented into CODIS. The FBI intends to require CODIS laboratories to implement the new 20 CODIS Core Loci by January 1, 2017.[25] More detailed information on the new 20 CODIS Core Loci can be found in TABLE 12.1.

Amelogenin is a low molecular weight protein found in tooth enamel. The Amelogenin gene is useful as a sex marker and is usually included with the CODIS loci. A specific region of the Amelogenin gene on the X chromosome is six base pairs shorter than its homolog on chromosome Y, allowing the distinction between individuals with 46XY and 46XX

FIGURE 12.1 Example of a tetrameric STR locus located on chromosome 5. The allele designation (genotype) is based on the number of times the four base pair unit (in this example, AGAT) is repeated.

TABLE 12.1 List of the New 20 CODIS Core STR Loci With Corresponding Chromosome Locations

Locus	Chromosome Location
Original 13 Core Loci	
TPOX	2p23-2per
D3S1358	3p21.31
FGA	4q28
D5S818	5q21-31
CSF1PO	5q33.3-34
D7S820	7q11.21-22
D8S1179	8q24.13
TH01	11p15.5
vWA	12p13.31
D13S317	13q22-31
D16S539	16q24.1
D18S51	18q21.33
D21S11	21q11.2-q21
Amelogenin*	X: p22.1-22.3 Y: p11.2
Newly Added 7 Core Loci	
D1S1656	1q42.2
D2S441	2p14
D2S1338	2q35
D10S1248	10q26.3
D12S391	12p13.2
D19S433	19q12
D22S1045	22q12.3

*Amelogenin is not a short tandem repeat (STR) but a locus used to determine sex that is located both on the X and Y chromosomes.

karyotypes. By using primers that encompass this region, males will display Amelogenin locus heterozygosity, while females will exhibit homozygosity, with some rare exceptions.[26-28]

POINTS TO REMEMBER

Short Tandem Repeats (STRs)
- STRs are the most common DNA identity tests.
- STRs in use today are generally composed of four base repeats.
- Alleles are designated by the number of repeats.
- FBI CODIS loci are used for national database purposes.

THE ANALYTICAL PROCESS

Sample Collection

Most nonforensic laboratories perform routine analyses of pristine samples collected in defined ways. Forensic identity testing must contend with much greater variability in sample types, conditions, and purity. Samples for identity testing can be any biologic specimen that contains DNA. Unlike blood or buccal swab samples obtained from an individual for a reference sample, parentage testing, or clinical purposes that are obtained under controlled conditions, evidentiary DNA specimens may be swabs or scrapings of stains, smears, or body orifices; may be composed of various body fluids or human remains; may be contaminated by DNA from other sources; and are exposed to various environments for varying lengths of time. Reference specimens, when not of the person themselves, resemble evidentiary specimens and may include paraffin-embedded tissue blocks, newborn screening bloodstain archives, cigarette butts, shavers, toothbrushes, and clothing. Although subject to degradation over time in the presence of enzymes, acidic or basic conditions, and high temperatures, DNA is a remarkably stable molecule that can be recovered and successfully analyzed from solutions, surfaces, and cells, sometimes decades or centuries after it was deposited.

Extraction

Different samples require different extraction methods. The sample diversity that forensic laboratories encounter requires that multiple extraction methods be validated within a laboratory. All methods have in common cell lysis, DNA nuclease inhibition, and DNA purification, yet the manner in which these objectives are achieved may vary significantly. Postcoital swabs from sexual assault evidence, for example, require a specific process called *differential extraction*. This method differentially lyses the more fragile nonspermatic content of the sample (ie, female epithelial cells) from the less fragile encapsulated sperm heads.[29] Furthermore, as the amount of samples submitted to laboratories continues to increase, automation and the ability to process a variety of sample types are attractive qualities when selecting an extraction method. Examples of useful methods include, but are not limited to, the QIAGEN Investigator kit series (QIAGEN; Redwood City, California),[30] PrepFiler (ThermoFisher Scientific; Rochester, New York),[31] ChargeSwitch (ThermoFisher Scientific),[32] ZyGEM (ZyGEM; Hamilton, New Zealand),[33] and DNA IQ (Promega; Sunnyvale, California).[34]

Quantification

In the extraction process, all sources of DNA present in the sample will be recovered in the final product. In fact, if the sample contains bacteria, plant, fungi, animal blood, or tissue, the DNA will be extracted and present in the processed sample. To address this potential issue, the FBI's quality assurance standards mandate that, as part of the analytical process, human specific DNA quantitation is performed on the sample. The main purpose of the quantitation step is to determine the appropriate amount of sample to input in the following PCR amplification step. Excessive template will cause excessive amount of PCR product to be loaded on the genetic analyzer, which in turn causes problems with signal saturation and artifacts that can hinder interpretation. Conversely insufficient input DNA will translate into allele dropout and total or partial absence of the profile. However, newer STR kits (discussed below) are substantially more robust and less sensitive to the amount of input DNA.

Several quantification methods have been used in the past by the forensic community, including UV absorbance, PicoGreen (ThermoFisher Scientific), AluQuant (Promega), slot blots, end-point PCR, and others.[35] Currently, the most popular methods are based on real-time quantitative PCR (qPCR). The advantage of the PCR methods is that they provide information on how amplifiable the DNA is, rather than just its quantity. In fact, DNA may be degraded, or the sample

TABLE 12.2 Most Common Real Time PCR Quantification Kits Used in Crime Laboratories

Manufacturer	Kits	Target
Applied Biosystems	Quantifiler Human DNA Quantification Kit	Single-copy autosomal chromosome target
	Quantifiler Y DNA Quantification Kit	Single-copy Y chromosome target
	Quantifiler Duo DNA Quantification Kit	Single-copy autosomal and single copy Y chromosome targets
	Quantifiler HP DNA Quantification Kit	Multicopy autosomal target and DNA degradation detection
	Quantifiler Trio DNA Quantification Kit	Multicopy autosomal and Y chromosome targets, and DNA degradation detection
Promega	Plexor HY System	Multicopy autosomal and Y chromosome targets
Qiagen	Investigator Quantiplex Kit	Multicopy autosomal target

Applied Biosystems' assays are based on TaqMan chemistry. The Plexor System works by measuring a reduction in fluorescent signal during amplification by site-specific incorporation of a fluorescent quencher opposite a complementary fluorescently tagged modified base. The Investigator Quantiplex Kit uses Scorpion primers and rapid polymerase chain reaction (PCR) chemistry.

TABLE 12.3 List of Some Commercially Available STR PCR Kits

Manufacturer	Kit	Number of Loci
Autosomal STR Kits*		
ThermoFisher	AmpFℓSTR Cofiler	6
	AmpFℓSTR Profiler Plus	9
	AmpFℓSTR SGM Plus	10
	AmpFℓSTR SEFiler	11
	AmpFℓSTR Identifiler[†,‡]	15
	AmpFℓSTR Identifiler Plus[†]	15
	AmpFℓSTR Minifiler	8
	GlobalFiler[‡,§,ǁ]	21
Promega	PowerPlex 1.1	8
	PowerPlex ES	8
	PowerPlex S5	4
	PowerPlex 16[†]	15
	PowerPlex ESX and ESI Fast Systems	16
	PowerPlex 18D System[†,‡]	17
	PowerPlex 21 System	20
	PowerPlex Fusion System[§]	23
	PowerPlex Fusion 6C System[§,ǁ]	26
Qiagen	Investigator 24plex QS Kit[§,¶]	22
	Investigator 24plex GO! Kit[‡,§]	22
Y Chromosome STR Kits		
ThermoFisher	AmpFlSTR YFiler	17
	Yfiler Plus[ǁ,#]	27
Promega	PowerPlex Y	12
	PowerPlex Y23 System	23

*Number of loci excludes Amelogenin.
[†] Targets all 13 original core loci.
[‡] Available in the "direct PCR" version of the kit for reference samples.
[§] Targets all 20 New Core STR loci and includes one or two Y chromosome markers.
[ǁ] Uses 6 dye chemistry.
[¶] Includes a quality sensor to detect presence of PCR inhibitors.
[#] Contains seven rapidly mutating Y-STR loci.
PCR, Polymerase chain reaction; STR, short tandem repeat.

may contain PCR inhibitors that are not detected by non-PCR methods. Furthermore, multiplex PCR combined with multicolor detection allows for simultaneous detection of autosomal DNA, Y chromosome, and an internal positive control for inhibition. These features provide useful information in potentially mixed (male/female DNA ratio) and inhibited samples.[36,37] Furthermore, recent commercialized kits also provide information on the level of degradation of the DNA, which can be useful in selecting the multiplex STR kit most likely to be successful with the sample at hand. TABLE 12.2 lists popular qPCR kits on the market for DNA quantification.

Polymerase Chain Reaction Amplification

Since the start of fluorescent DNA technology, manufacturers have been developing, for the forensic community, multiplex STR kits that allow typing of several loci in a single assay.[38-41] With the purpose of maximizing the amount of information that can be obtained from a minimal amount of sample, manufacturers have been able to progressively increase sensitivity and the number of markers typed in a single reaction. Using fluorescent dyes during PCR, fragments are specifically labeled for subsequent detection and genotyping. The most recent products include the use of six different dyes and enable multiplex amplification and genotyping of all the 20 CODIS Core Loci plus Amelogenin, together with other autosomal or Y chromosome markers, totaling up to 27 regions from as little as 100 picograms of template DNA.[42,43] Some kits have been developed specifically for direct amplification from reference samples such as buccal swabs or blood punches, rendering prior DNA extraction unnecessary, while others are optimized for forensic samples purposely designed to be resistant to high levels of PCR inhibitors. Volume reduction is also possible; proportionally scaling down reagents increases the number of samples that can be processed with a single kit. Amplification volumes have been reduced down to 2 μL for single-source reference samples.[44] TABLE 12.3 lists some of the most popular STR PCR kits available on the world's market. The list also includes kits specifically targeting STRs on the Y chromosome (discussed later in this chapter).

Capillary Electrophoresis and Data Analysis

After PCR amplification, STR alleles are separated and detected by CE. The most popular instruments used in crime labs are the ThermoFisher/ABI Genetic Analyzer series (310, 3100, 3130, 3130*xl*, 3500, 3500*xl*, 3730, 3730*xl*). These instruments (except the 310, which is a single-capillary instrument) run multiple samples simultaneously (the 3500*xl* and the 3730*xl* can process up to 24 and 96 respectively) and are able to separate alleles based on their size, because smaller fragments move faster through the capillary, and based on their color, because different fluorescent dyes are attached to PCR fragments for loci that have similar size during amplification (Fig. 12.2).[45] The latest version of these instruments allow for the detection of up to six different dyes rather than five dyes as in the earlier versions. This feature can increase the number of loci coamplified in the recently expanded STR kits discussed previously. An internal size standard, labeled with a specific dye, is also added to the sample and is used to precisely size the alleles. This latter step is performed after the electrophoresis run using dedicated software, which also generates genotypes by comparing the size of the fragments in questioned samples to those obtained by running an allelic ladder. Popular software used in crime labs include GeneMapper IDX (Applied Biosystems, ThermoFisher Scientific), GeneMarker HID (SoftGenetics; State College, Pennsylvania), and OSIRIS (open source product from NCBI).

The final genotype is simply the list of the loci in the kit and the corresponding allele call—that is, the number of times the STR motif is repeated in the allele. Procedures vary from lab to lab, but generally after the analyst reviews the allele calls, a final genotype is determined. Only then is a profile compared to a

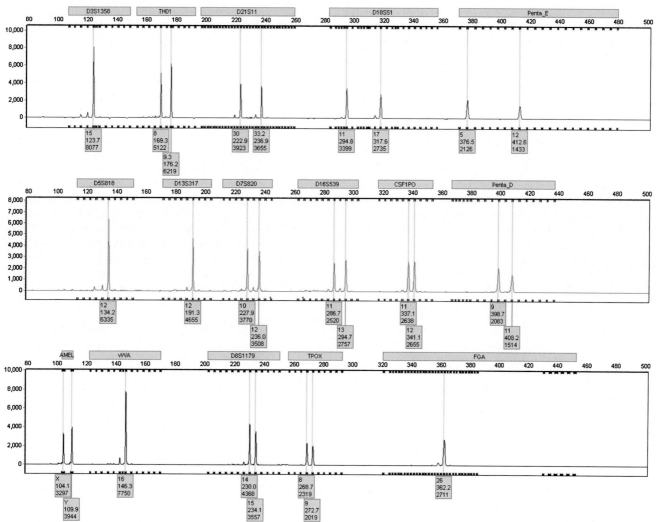

FIGURE 12.2 Electropherogram of a sample amplified with PowerPlex 16HS, run on an ABI 3130 Genetic Analyzer and analyzed with GeneMarker HID software. The PowerPlex 16 kit includes the 13 core CODIS loci, Amelogenin, and proprietary Penta-D and Penta-E loci. Each locus is labeled in a gray box above each lane. Each lane represents a different fluorophore color (a fourth color not shown is used for an internal control). The x-axis is time, which corresponds to the size of the allele. The y-axis is fluorescent intensity, measured in relative fluorescent units (RFU), which corresponds to the amount of PCR product in the sample. Each peak represents an allele of the locus indicated in the gray box over the peak. The tick marks represent the range of common alleles for each locus. The box below each peak has three numbers that correspond to the allele call (i.e., number of STR repeats in the allele), the approximate size of the fragment in nucleotides, and the RFU value, respectively. Homozygous loci have only one peak, while heterozygous loci have two. The genotype of this profile is D3S1358 15; TH01 8,9.3; D21S11 30,33.2; ... Penta E 5,12; ... Amel X,Y; ... FGA 26.

reference, which can be a suspect, a victim, or a database. To avoid bias, known reference testing is conducted only after the allele call on the evidence sample is complete.

Y—SHORT TANDEM REPEATS

When traditional autosomal STRs are analyzed on vaginal swabs from sexual assault cases, the female epithelial cells will swamp the relatively few male spermatozoa in the sample, resulting in a mixture at best and no male contribution at worst. In these cases, Y chromosomal markers, typically Y-STRs, found in the nonrecombinant section of the Y chromosome, can be amplified to generate a male-specific profile that can be used for comparison.[46-48] Y-STRs can also be used to clarify distant paternal kindred relationships, as in the case of President Thomas Jefferson's lineage.

Y-STRs are genetically linked, so the product rule does not apply. Instead, multiple Y-STR results are treated as a single set or *haplotype*, resulting in discriminatory power that is significantly less than that of a panel of independently segregating autosomal STR loci. Furthermore, paternally related males cannot be distinguished because they inherited the same Y chromosome (unless a mutation occurs). Discriminatory values can be increased by using a large panel of Y chromosome markers in conjunction with a large database of typed individuals.[49]

The most recent generation of Y-STR kits are able to simultaneously amplify up to 27 Y-STR loci in a single PCR reaction and are able to generate a full profile in a 1 : 4000 male to female DNA mixture.[49a] Yfiler Plus (ThermoFisher Scientific) also includes rapidly mutating loci that significantly increase the chances of distinguishing paternally related individuals. With this assay, approximately 7 out of 10 pairs of brothers can be distinguished (see TABLE 12.3).[50]

MITOCHONDRIAL DNA

Mitochondrial genomes are circular double-stranded DNA molecules that are 16,569 bp long and are present as one or more copies within all mitochondria of a cell. Thus mitochondrial DNA (mtDNA) is present in hundreds to thousands of copies per cell. mtDNA, unlike nuclear DNA, does not undergo meiosis and does not participate in genetic recombination events. mtDNA remains stable over generations, except for the acquisition of mutations at a rate 10 to 20 times that of nuclear DNA.

mtDNA is maternally inherited, so siblings will share the same DNA as their mother but not that of the father. The normal state of mitochondria is generally thought to be one of *homoplasmy*, in which all the mtDNA has the same sequence. However, because of mutational events, a state of *heteroplasmy*, in which more than one mtDNA sequence is present in the same tissue, may exist. High-level heteroplasmy is generally on the order of 30% of the mtDNA sequence before it is reported when analyzed with conventional methods. In high-quality sequences it can be detected down to the 10% to 15% range. Although often undetectable with Sanger sequencing, low-level heteroplasmy is common, and massive parallel sequencing (MPS) has offered new insight because it allows detection down to approximately 1%.[51] Heteroplasmy appears to be somewhat tissue-specific rather than uniform throughout the body. Thus two shed hairs may show discrepant mtDNA sequences.

In the human mitochondrial genome, only about 1200 bases near the transcription origin (15971-579), known as the *displacement loop (D-loop)* or the *control region,* are noncoding. This D-loop consists of two hypervariable regions that contain the polymorphisms conventionally used for identity (HVI: 16024–16365; HVII: 73–340), although additional polymorphisms are present in the coding regions. mtDNA polymorphisms are typically identified for forensic testing via Sanger sequencing of hypervariable regions. This method is expensive, labor intensive, and highly sensitive to contamination. The implementation of MPS technology, described below, is likely to provide forensic investigators access to the entire mtDNA genome, thus significantly increasing the discrimination power of its analysis.

The mtDNA sequence obtained from a specimen (*mtDNA haplotype*) is compared with a reference sequence (revised Cambridge reference sequence ([rCRS]; www.mitomap.org/MITOMAP), and the "profile" of the sample consists of the recorded differences between the sample and the rCRS. The Scientific Working Group on DNA Analysis Methods' (SWGDAM) Interpretation Guidelines for Mitochondrial DNA Analysis by Forensic DNA Testing Laboratories state that when comparing sequences, if the samples differ at two or more nucleotide positions (excluding length heteroplasmy, like the homopolymorphic "C-Stretch" in HVI[52,53]), they can be excluded as coming from the same source or maternal lineage. The comparison should be reported as inconclusive if samples differ at a single nucleotide position. Finally, two samples cannot be excluded as originating from the same source or maternal lineage if they have the same sequence or are concordant (sharing a common DNA base at every nucleotide position, including common length variants).

As in the case with Y chromosome STRs, because mtDNA polymorphisms are linked, individual polymorphism frequencies cannot be multiplied together to generate a likelihood of identity. Instead, the *mtDNA haplotype* identified in a sample is compared with those deposited in a database to derive a frequency statistic using the counting method. Thus the size of the database becomes crucial: the larger the database, the greater the significance of a match. The European DNA Profiling group has developed an mtDNA Population Database (EMPOP) that can be used to determine the frequency of the haplotype in various populations. EMPOP is constantly updated and, as of May 2015, contains over 34,600 control sequences and will soon be upgraded to store full mtDNA genome sequences.[54] The FBI also has a reference database with over 4000 control region sequences.

mtDNA is useful primarily for identity testing in four contexts. First, a sample may be available that contains mitochondrial but not nuclear DNA. For example, shed hairs that do not have roots generally contain only mtDNA. Second, when severely decomposed skeletal remains are recovered, DNA is substantially degraded; the high copy number and small size of mtDNA make it more likely to yield a result than nuclear DNA. Third, mtDNA analysis may become essential when only a distant relative is available for a reference specimen. An example is the identification of the remains of Tsar Nicholas II, from which mtDNA was recovered and compared to that of his sister's descendant three generations down in the family tree.[55] Nuclear DNA requires samples from multiple close kindred and would not have been useful in this case. Fourth, in database searches of autosomal STR data from unidentified human remains or missing person relatives, the algorithms for the search often produce several potential matches, and mitochondrial DNA analysis is needed to identify the true match.

> **POINTS TO REMEMBER**
>
> **Mitochondrial DNA (mtDNA)**
> - mtDNA is maternally inherited.
> - mtDNA is present in cells at a high copy number.
> - mtDNA is typically sequenced in the two hypervariable regions.

TOUCH SAMPLES AND DNA MIXTURES

Most identity testing performed today relies on the PCR. PCR testing is inherently sensitive, allowing routine analysis of nanogram quantities of genomic DNA and often successful testing of picogram quantities. By increasing the number of PCR cycles and the quantity of Taq polymerase, it is possible to maximize the sensitivity of PCR down to the single-cell level. This approach has been referred to as low copy number or low template DNA, which in some cases included performing the analysis in replicates and generating a consensus profile to be used for comparison.[56-58]

Very small amounts of DNA can be detected, but because of the small quantities, stochastic effects often can appear as imbalanced (or a complete dropout of) sister alleles. DNA degradation and secondary transfer of skin cells and free DNA, from one touched surface to another, further complicate the analysis of touched samples, which often result in mixtures.[58-61] In the absence of a clear major contributor to a touch DNA mixture, electropherogram interpretation can be challenging even for experienced DNA analysts, particularly if there are more than two contributors. To increase consistency in data interpretation and eliminate human subjectivity, laboratories can use dedicated probabilistic software.[62-65]

When processing degraded DNA samples, the sensitivity of the analysis can be enhanced by the use of mini-STRs, which essentially are the same tandem repeats used in the commercial kits described previously, but the flanking PCR primers are moved closer to the tandem repeats. This results in amplification of smaller fragments, which are more likely to remain intact (thus amplifiable) when the DNA is partially degraded.[66,67]

SINGLE NUCLEOTIDE POLYMORPHISMS

Four categories of SNPs can be useful to the forensic community: individual identification SNPs (IISNPs), ancestry informative SNPs (AISNPs or AIMs), phenotype informative SNPs (PISNPs or PIMs), and lineage informative SNPs (LISNPs).[68] IISNPs are SNPs that collectively give very low probabilities of two individuals having the same multilocus genotype. Their mutation rate is very low compared to STRs (important for parentage testing), and if typed with PCR-based methods, they have the advantage of requiring a very small target that is likely to amplify even in highly degraded DNA. On the other hand, being biallelic, a greater number of loci (~70) are necessary to reach the same discrimination power of 20 STRs, and and they may be of limited use in mixed samples. AIMs have alleles that have very high frequencies in some populations and are rare or absent in others. PIMs are found in genes (or expression control regions) coding for proteins that play an important role in determining individual physical characteristics, such as hair, skin, and eye color, and skull morphology. AIMs and PIMs can provide useful information on the physical characteristics of an individual—a molecular eyewitness that can help investigators, prioritize suspect processing, interview suspects, and determine the relevance of a piece of evidence to a crime.[69] LISNPs are sets of tightly linked SNPs that function as multiallelic markers that can serve to identify relatives with higher probabilities than simple biallelic SNPs and have the potential to help with resolving DNA mixtures when typed with MPS methods (see later).

Although forensic SNP analysis offers great potential for human identification, the absence of a reference database and the over 12 million STR profiles in CODIS make it unlikely that SNPs will replace STRs anytime soon.[70] Yet, as MPS methods become more sensitive and cost-effective (discussed below), forensic SNP analysis likely will be implemented in crime labs to support cases where STR profiles from crime scenes do not match any of the suspects with no hits in CODIS.

QUALITY ASSURANCE

In general, forensic science laboratories are unregulated, but this may change as federal legislation in this area is being considered. This legislation would mandate certification of forensic science analysts and accreditation of forensic science providers. A few states have forensic science advisory or regulatory boards and require accreditation (New York, Virginia, Texas, North Carolina, Montana). Thus accreditation of the labs is currently voluntary for most labs; nonetheless, the vast majority of crime labs are accredited.

Although not every crime laboratory is regulated, effectively all governmental forensic DNA laboratories in the United States are regulated because of the federal *DNA Identification Act of 1994*,[71] which gave the FBI regulatory oversight of DNA profiles entered into the national database. Specifically, all laboratories that submit DNA results to the NDIS using CODIS software are required to comply with the FBI's *Quality Assurance Standards for Forensic DNA Testing Laboratories* and *Quality Assurance Standards for Convicted Offender DNA Databasing Laboratories* (the FBI QAS).[72] All testing requires both technical and administrative reviews. Laboratories are required to be accredited and have annual audits. All personnel must meet minimum educational requirements and be proficiency tested at least twice a year.

The FBI standards arose from and continue to be guided by advisory bodies from the forensic DNA community. The FBI created the Technical Working Group on DNA Analysis Methods (TWGDAM) in 1988. TWGDAM issued guidelines for forensic DNA analysis. TWGDAM was later renamed as SWGDAM and continued to develop guidelines for the community. When the 1994 DNA Identification Act gave the FBI direct regulatory authority, the FBI created a DNA Advisory Board, which used the TWGDAM guidelines as the starting point and then recommended standards to the attorney general. SWGDAM continued to review and make recommendations thereafter. In 2014, the Department of Justice, in a collaborative agreement with the Department of Commerce (DOC), gave over the administration of the Scientific Working Groups to the National Institute of Standards and Technology (NIST), with the exception of SWGDAM and a few others. NIST replaced the SWGs with an Organization of Scientific Area Committees (OSAC). SWGDAM was retained for its statutory duty to oversee the FBI QAS, but the NIST OSAC includes two DNA subcommittees.[73]

Forensic science laboratories are generally accredited by the American Society of Crime Laboratory Directors/Laboratory Accreditation Board, ANSI-ASQ National Accreditation Board (formerly Forensic Quality Services), and the American Association for Laboratory Accreditation. These accreditation programs are based upon the International Organization for Standardization (ISO) 17025 standards, which incorporate more than 500 separate standards and are the basis for conformity assessments. Accreditation programs issue *Supplemental Standards* as community-specific consensus standards, which supplement the ISO 17025 standards.

Proficiency test providers for forensic laboratories in the United States include the Collaborative Testing Service, the College of American Pathologists, Orchid Cellmark, and Forensic Assurance. Standard reference materials from the NIST are available for PCR-Based Profiling DNA Standard (SRM 2391b), Mitochondrial DNA Sequencing (SRM 2392, 2392-I, and 2394), Human DNA Quantitation (SRM 2372), and Y-chromosome testing (SRM 2395). Standards require annual NIST-traceable comparisons.[74]

STATISTICAL INTERPRETATION

Discriminatory power is the ability of an identity testing system to distinguish an individual or group from the rest of the population. The power of discrimination of a locus or testing system should not be confused with accuracy. ABO blood group typing is accurate but poorly discriminating in that this locus results in only a few phenotypes of generally high frequency in populations. Current identity testing systems that employ the 13 core CODIS loci have powers of discrimination that exceed 1×10^{-14}, making it very unlikely that any unrelated individual on earth other than a nonexcluded suspect or his or her identical twin could be the source of an evidence sample. However, likelihoods of this magnitude should be viewed with knowledge that a variety of potential problems extraneous to the testing technology, involving sample collection and labeling and test interpretation and reporting, may lead to an erroneous result.

In the early days of DNA identity testing, significant challenges were launched regarding the interpretation of DNA typing results.[75] Questions included whether loci exhibit Hardy-Weinberg equilibrium and whether allele frequencies vary significantly among ethnic groups. Current forensic genetic systems have not demonstrated significant deviation from Hardy-Weinberg equilibrium. These systems show greater intragroup allelic diversity than between-group diversity. A National Research Council panel was created to address these issues.[76] The resultant report (so-called NRC I) introduced a *ceiling principle* to ensure the conservativism of the frequency estimates, which itself generated considerable controversy. This led to NRC II, which articulated the current standards of statistical analysis.[77] Statistical formulae are routinely applied to European American, African American, and Hispanic population databases. Specialized databases also exist for Native American Indian and for Pacific and African populations. Current statistical formulae assume that populations are substructured or inbred and correct for this occurrence ($\theta = 0.01$ for the US general population; $\theta = 0.03$ for isolated populations). In a substructured population, mates are not chosen randomly from the population. Instead, a person will choose a mate of similar religious, ethnic, or geographic origin, such as by marrying within his or her own religion. In a substructured population, an individual is more likely to marry a distant relative such as a second, third, or fourth cousin because selection of a mate is limited, leading to an inbred population.

Identity testing frequently excludes the tested person with almost absolute reliability. Exclusion results can indicate that the sampled DNA did not originate from the suspect or that an alleged man did not father a child. The exclusion is based on the presence of alleles at a locus that make it impossible for him or her to be a contributor to the tested sample. For example, if the person has alleles j and k at the autosomal locus L, then, in the absence of mutations, it is not possible for him to be the major contributor to an evidence sample with the alleles m and n at L or to be the father of a child with the alleles m and p at L. In practice, laboratory protocols require that exclusion be based on incompatible results for at least two loci to rule out mutation events or other sources of error. In this context, "impossible" implies a situation in which samples have been collected correctly and have not been mislabeled, testing has been performed accurately, and results have been interpreted and reported appropriately.

If a tested individual has genotypes at all the tested loci that match the genotypes found in an evidence sample—an *inclusion*—then the tested person cannot be excluded as the contributor to the sample. If the tested loci independently segregate during meiosis, the overall probability that a random person matches the DNA profile can be calculated by multiplying the genotype population frequencies for each locus (so-called *Product Rule*).[77] When several loci are tested and each has many possible alleles, it becomes extremely likely that an individual whose genotypes match those found in the evidence sample is the person who contributed the DNA to the sample. This random match probability (RMP) is calculated using published allele frequency databases.[19,23,78] Most crime laboratories in the United States report population phenotypic frequencies for European American, African American, and Hispanic populations.

Outside of the United States and some labs in the United States report likelihood ratios (LR). The LR is the probability of the evidence if it originated from the suspect (prosecution hypothesis) divided by the probability of the evidence if the sample originated from someone other than the suspect (defense hypothesis). For good-quality single-source samples (without the possibility for drop-ins and dropouts), the LR is 1/genotype frequency, so LRs greater than 1 favor the prosecution and those less than 1 favor the defense. LR can be developed as a frequentist or Bayesian statistic. Using the Bayesian approach, a subjective assessment of the prior probability that the suspect is in fact the source is used to obtain the posterior probability. Generally, forensic scientists make no determination or assumption of the prior probability. However, in determining a probability of paternity, the prior probability that the putative father is in fact the father is generally assumed to be 0.5 (ie, an equal probability that he is or is not the father). In cases where the quality of the data is good for single-source samples and when the interpretation is done by considering peak heights as discrete variables, the LR is the inverse of the RMP. A common verbal scale asserts that LR values of 100 are moderately supportive of the prosecution's hypothesis and values greater than 1000 are strong support.[79]

For genetically linked systems such as mitochondrial DNA testing or Y chromosome STRs in which the loci do not independently segregate during meiosis, inclusion statistics are calculated using the observed frequency of the haplotype in the relevant population frequency database (the so-called *Counting Method*).

Many times, forensic evidence samples contain a mixture of DNA because they are collected in the real world and not under controlled laboratory conditions. In a mixed sample, if the individual constituents of the mixture cannot be separated, a forensic laboratory will calculate an LR or a combined probability of exclusion, which determines the percentage of the population who could not contribute to the mixture. Mixture interpretation is difficult and problematic and is conducted variously and not by one single consensus method. When questions arise regarding assumptions that must be made during the calculation of inclusion probabilities, laboratories generally choose the conservative option that favors the accused individual.

> **POINTS TO REMEMBER**
> **Statistical Interpretation**
> - Autosomal STR loci are genetically independent; STR genotype frequencies can be multiplied.
> - Y-STRs and mtDNA are genetically linked; haplotype statistics are based on counting.

RAPID DNA

With conventional methods the analysis of a reference sample such as a buccal swab takes about 8 hours, with PCR being the longest part of the process (around 3 hours). In 2008, Dr. Peter Vallone demonstrated that with the use of fast-processing polymerases and rapid cycling thermocyclers, the amplification time could be reduced to 30 minutes.[80] This study sparked efforts by the industry, supported by the federal government, to develop all-in-one systems that can process a reference sample in less than 2 hours, from extraction to profile, without human interaction.[81] In 2010, the FBI established a Rapid DNA Program Office to direct the development and integration of Rapid DNA technology for use by law enforcement. Several government agencies are involved in this effort, including the National Institute of Justice, the NIST, the Department of Homeland Security, and other federal agencies. The goal is to develop a fourth tier of NDIS called RDIS (Rapid DNA Index System) that will integrate rapid DNA analysis systems in the booking process at police stations to rapidly determine if an individual's profile is in CODIS (parallel to fingerprint processing).

There are currently two major players in the industry of rapid DNA technology: NetBio's DNAscan Rapid DNA Analysis System (NetBio; Waltham, Massachusetts)[80] and IntegenX's RapidHIT System (IntegenX; Pleasanton, California).[82] Both systems are standalone machines that can generate DNA profiles from buccal swabs in less than 2 hours and can be operated by individuals with limited scientific background after minimal training. Although both instruments are undergoing validation in selected crime labs, several issues need to be resolved for RDIS to be established. For example, the DNA Identification Act of 1994 requires that the DNA records be generated by accredited laboratories following the FBI Director's Quality Assurance Standards (QAS). Consequently, specific legislation is necessary before DNA records generated by Rapid DNA instruments outside accredited laboratories can be considered NDIS compatible. Furthermore, validation and certification issues for Rapid DNA analysis instruments must be addressed prior to implementation of the booking process. Thus, at this time, it is difficult to predict when rapid DNA testing will become a reality.

MASSIVELY PARALLEL SEQUENCING

MPS can simultaneously sequence many thousands of genomic regions at the same time, which offers the forensic community the opportunity to sequence all markers of forensic interest within a single workflow.[83] Streamlined, high-throughput approaches will be needed to bring the cost of MPS closer to the level of current forensic technologies and to facilitate implementation. Additionally, significant foundational work is needed in technical areas such as bioinformatics, characterizing sequencing error rates (ie, background noise), evaluating sample bar code demultiplexing, and determining what controls should be included in each sequencing run. Other considerations such as data storage policies, nomenclature, and the ethics of genotyping/reporting variants that may be clinically relevant, also remain to be addressed. The advantages of using MPS for the previously described forensic marker types are outlined below.

Short Tandem Repeats

STR analysis has traditionally been performed by size-based DNA separations using gel electrophoresis or CE. However, analysis of PCR product length alone fails to capture the potential internal sequence variation that may exist in many STR loci detected via base composition mass spectrometry[84,85] or through full sequence analysis.[86,87] MPS offers the potential for forensic laboratories to routinely obtain full sequence data from STR loci as opposed to length-based genotypes. Some forensic STR loci will demonstrate double or even triple the number of alleles by sequence compared to the current system of length-based alleles. In addition, sequencing the nearby flanking regions may reveal SNPs that will further effectively increase the number of alleles. These gains will increase statistical discrimination and may improve the ability to interpret mixtures. Finally, traditional CE assay design requires a size range of amplicons from approximately 100 bp to 500 bp because the markers must be separated by size. MPS enables reduction in size of STR amplicons because loci will not require size differentiation. This is expected to offer improved amplification performance in degraded DNA samples, akin to the smaller loci in mini-STR CE assays. However, STRs are among the most difficult regions to interrogate by MPS technologies due to the difficulty of accurate sequence alignment of repetitive DNA sequences.

Mitochondrial DNA

Forensic laboratories offering mtDNA analysis have historically used Sanger sequencing, which is significantly limited in throughput and typically only applied to the approximately 1200 bases in the control region of the mtDNA genome. Due to background noise levels with CE analysis, this method also has limited sensitivity (approximately 10% to 15%) in detecting heteroplasmy and/or mixtures. Finally, when a mixture is detected via Sanger sequencing, the data are not interpretable. MPS offers the ability to sequence the full mtDNA genomes of multiple individuals in a single sequencing run.[88] In addition, the sensitivity of MPS in detecting heteroplasmies and mixtures is improved.[89] Finally, due to the clonal nature of MPS, samples composed of multiple individuals may become interpretable. It is likely that mtDNA analysis will be the first area of forensic DNA analysis for which MPS replaces current methodology.

Single-Nucleotide Polymorphisms

Current CE-based methods for genotyping SNPs are labor intensive and low-throughput, resulting in a barrier to forensic laboratory implementation. However, studies have shown advantages in using SNP markers with degraded DNA samples that may be encountered in forensic casework and disaster victim identification,[90-92] and publications in forensic and clinical journals have characterized SNPs suitable for human identification,[93,94] ancestry determination,[69,95] and phenotype prediction.[96,97] MPS offers the ability to genotype hundreds of SNPs alongside STRs, and this may provide the impetus for incorporating this marker type into routine forensic workflows.

LEGAL ISSUES

Early challenges to the practice and interpretation of DNA-based forensic identification have faded as the public, attorneys, and judges have become more knowledgeable about this technology. In fact, no court has rejected DNA evidence based upon invalid scientific theory. The most common challenges today involve issues regarding sample collection, preservation of the evidence, chain-of-custody documentation, and validation studies. Judicial scrutiny based upon compliance with national standards promulgated by the NIST OSAC and quality assurance will likely also increase. New applications of DNA-based testing to assess ethnicity, infer phenotypic features from evidence, release the names of arrestees or potential family members of convicted offenders, or aid in the interpretation of mixtures and touch evidence testing are likely to cause controversy in the future.[98] Current practice, often mandated by biologic evidence retention laws,[99] involves splitting biologic samples and retaining a portion of the sample to permit the defense to conduct its own DNA testing; however, this is not always possible due to the minute sample that may be available. State and federal laws have been promulgated to permit and pay for postconviction DNA testing, permit appeals based upon *actual innocence* as determined through DNA testing, extend statutes of limitations for such cases, and allow "John Doe" warrants based upon a DNA profile without a name. Statutes also have mandated convicted offender DNA collections and, more recently, arrestee samples. In *Maryland v. King*, the US Supreme Court ruled that "when officers make an arrest supported by probable cause to hold for a serious offense and bring the suspect to the station to be detained in custody, taking and analyzing a cheek swab of the arrestee's DNA is, like fingerprinting and photographing, a legitimate police booking procedure that is reasonable under the Fourth Amendment."[100]

PARENTAGE AND KINSHIP

Commercial forensic identity STR testing kits have been adopted by the parentage testing community as the standard method for their use. Thus many of the same considerations expressed above apply to parentage testing. Questions regarding the parentage of minor and adult children arise frequently in modern society. Generally, paternity is at issue—that is, whether a particular man is the father of a child. Parentage testing allows an individual to be excluded or not excluded as the parent of a child. If not excluded, a calculation is performed that gives the likelihood of observing the allele sharing between the child and the alleged father at all loci tested compared to the likelihood of that match occurring if the true biological father is a random man from the same racial population. The same methods can be applied to the search for a mother, a situation that sometimes arises when an adopted person is trying to find his or her biological parents. Court-ordered or privately sought parentage testing is usually performed to facilitate decisions regarding the financial support of a child. However, individuals may wish to establish parentage for other reasons—for example, the settlement of an estate. Laboratories that perform other types of identification tests for purposes unrelated to parentage (eg, HLA typing for transplantation) should be aware of the possibility of inadvertently uncovering "false paternity"; informed consent forms often declare that such findings will not be disclosed to protect family harmony.

Standard parentage testing involves genotyping of several polymorphic loci in samples from a trio consisting of the mother, child, and putative father. Programs generally perform testing on peripheral blood samples or buccal swabs. Buccal swabs have virtually replaced blood samples because they offer a noninvasive means of sample collection, which is especially convenient when testing minors. The identity of the family references should be verified using picture identification cards, and chain-of-custody should be maintained.

Alleles in the child that are not found in the mother are assumed to be contributed by the biologic father, *obligatory paternal alleles*. If the alleged father is the biologic father, then his profile is expected to contain the obligatory paternal alleles. Most laboratories require the absence of an obligatory allele for at least *three* loci[100a] in a tested man to exclude him as the father due to the possibility of a technical problem or mutation at one or, rarely, two of the examined loci. The soundest approach, though, would be to calculate the LR factoring in the probability of mutation at the excluded loci.

If the putative father is not excluded, then the likelihood that he is truly the father, rather than a random individual who is not excluded, can be calculated.[101-103] A number of assumptions underlie accurate calculations of the likeliness of paternity. One of these assumptions is that the population of individuals to whom the comparison is being made is unrelated to both the tested man and the mother. When this is not the case, and it is known that another alleged father is a close relative of the tested man—for example, a brother or his father—then additional calculations are necessary in cases of nonexclusion. Tested individuals must be properly identified, testing must be accurate, and allele frequencies in relevant populations must be well characterized.

The discriminatory power of STRs for parentage is not as strong as for identity testing because only half of the autosomal alleles will be inherited from a parent. However, like identity testing, the power of a given locus depends on the number and individual frequency of the alleles found at the locus. A single STR locus typically has a 30% to 60% likelihood of excluding a wrongly accused man. If several STRs that independently segregate during meiosis are studied, a cumulative probability of exclusion can be calculated that is based on the product of exclusion power for each tested locus. The use of commercial 13 CODIS core loci STR kits generally results in at least a 99.9% probability of exclusion of a falsely accused man.

Although STRs are usually transmitted in a faithful Mendelian fashion, mutations can occur. A child and a tested man may be encountered who share an allele at all but one of the tested loci. The possibility that the man is truly the father and that a mutation has occurred at the mismatched locus should then be entertained. Most of the loci in the expanded STR kits will be particularly useful to recognize mutational events because additional loci can be tested to ensure that no more exclusionary loci are identified. The more complex loci such as SE33 have a higher mutation rate, thus they are not used for parentage testing. It is particularly important in cases where one genetic inconsistency is discovered that the mother of the child also participate in the testing because the knowledge of her genetic contribution to the child's genotype will often reveal additional exclusionary loci. The paternity index (PI) and the probability of paternity are two closely related values that express the likelihood that the tested man is truly the father, rather than another man who by chance shares alleles at a tested locus. Calculation of the PI takes into account allele frequencies at the locus in relevant populations. If multiple independent loci are analyzed, a cumulative PI (CPI) based on the products of individual loci can be calculated. Government entities typically require that the CPI be greater than 100 and that the probability of paternity be greater than 99% as conclusive evidence that the tested man is the biologic father of the child. Other evidence bearing on the probability that a man is the father before testing is performed can be integrated with test results using Bayesian analysis to calculate a posterior probability of paternity. The prior probability that an accused man is the father is typically set at 50%, but a range of likelihoods given different prior probabilities (eg, 10% to 50%) can be calculated for comparison.

In many cases, the standard paternity trio may not all be available for testing. If a sample cannot be obtained from the mother, it may still be possible to exclude a tested man or to calculate his probability of paternity. For example, the finding of locus L genotypes 1,3 for the child and 4,5 for the tested man is inconsistent with the hypothesis that he is the father (barring mutation), whether or not we know the mother's genotype. Similarly, when the accused man is not available, testing performed for individuals related to him may be used to calculate the likelihood that he is the biological father of the child in question. Y-chromosomal markers are useful to sort paternal lineages, and mtDNA is useful to sort maternal lineages.

Recommendations and standards for paternity testing are formulated by government agencies or professional organizations (eg, the Standards for Parentage Testing Laboratories of the American Association of Blood Banks [AABB]). The AABB administers a laboratory inspection and accreditation program.[104] ASHI and CAP also publish relevant standards. CAP and AABB jointly offer a proficiency testing survey.

OTHER CLINICAL APPLICATIONS OF DNA IDENTITY MARKERS

Genetic identity markers, as used in forensic applications, have also found application in several clinical applications.

Verification of Sample

Identity testing can be used to confirm the identity of a clinical or anatomic pathology laboratory patient specimen.[105] Occasionally, questions arise regarding the identity of specimens in the clinical laboratory. DNA prepared from a peripheral blood or buccal mucosa specimen can be compared with the pathology specimen to confirm if it is derived from the same patient. Such testing has revealed accidental sample switches and "floaters" on microscopic biopsy specimens. The availability of identity testing in situations such as these can be a significant benefit for involved patients and health care providers.

Prenatal Testing

In prenatal testing for inherited disorders, chorionic villus sampling (CVS) performed early in pregnancy is sometimes employed as a source of fetal DNA for testing. Testing for a disorder such as cystic fibrosis may reveal that the genotypes for the mother and the fetus are identical—for example, that they are both heterozygous carriers of a cystic fibrosis mutation. This result is entirely consistent with the usual segregation of chromosomes during meiosis. However, it is possible that the CVS material was not derived largely from the fetus but consists of decidual cells primarily from the mother. If the father is a carrier of a cystic fibrosis mutation, this may result in failure to diagnose cystic fibrosis in a fetus. Identity testing can be employed to confirm that tested cells have a genotype distinct from that of the mother and are a valid sample from the fetus.[106] Some advocate that identity testing should be performed to detect maternal contamination on all prenatal specimens.[107,108]

Bone Marrow Transplantation (Chimerism)

The main goal of posttransplantation monitoring in allogeneic hematopoietic stem cell transplantation (HSCT) is to assess engraftment and donor chimerism.[109,110] A test for chimerism involves identifying the genetic profiles of the recipient and donor(s) and then evaluating the extent of mixture in the recipient's blood, bone marrow, or other tissue after transplant. Engraftment testing is accomplished by analyzing genomic DNA from recipient and donor(s) pretransplant specimen. Posttransplant, a recipient sample can be analyzed to determine the percent of recipient DNA and the percent of donor(s) DNA. The majority of progenitor cell transplant patients receive only a single graft. However, the use of two cord blood grafts in adult patients requires the analysis of the percent of recipient DNA, donor A DNA, and donor B DNA.

Most laboratories providing clinical engraftment testing currently use commercial STR kits for this testing.[111,112] Analysis of a donor sample and a pretransplant recipient sample allows the laboratory to identify which STR loci are informative (nonidentical genotypes) for the pair. A posttransplant sample can then be studied. If recipient and donor peaks are present, the percent of DNA from each will be reported. If only peaks from the donor are observed, the cells studied are donor in origin. Rarely, only recipient peaks will be seen, consistent with a recipient origin of the cells. These assays can typically detect admixtures of cell populations down to approximately 3% to 5%.

When the donor and the recipient are of different genders, analysis of sex chromosomes in posttransplant hematopoietic cells can be useful in engraftment testing. Alternatively, PCR-based analysis of X or Y chromosome loci can be carried out. The percent donor or recipient cells can be determined by assessing the ratio of cells with XX signals to those with XY signals. Although it may be valuable as a quick and convenient assessment, it should be realized that it is only one of several informative loci that should be used for engraftment studies.

Several clinical questions can be answered with engraftment testing.[113] Is engraftment of donor cells progressing in the weeks following a stem cell transplant? In the setting of a history of successful engraftment, do subsequent studies demonstrate a resurgence of recipient-derived hematopoietic cells indicating a possible relapse or rejection? Has stable chimerism developed following transplantation with the production of hematopoietic cells derived from both the donor and the recipient? A result indicating 85% donor cells and 15% recipient cells might be equally consistent with an engrafting marrow 3 weeks after transplantation, relapse 6 months following transplantation, or stable chimerism. Multiple samples collected at intervals after transplantation and correlation of engraftment testing results with the clinical history are crucial for interpretation of results. Artificial chimerism (from HSCT) and natural chimerism can also contribute to confusing or discrepant results in identity and parentage testing.

Loss of Heterozygosity

Polymorphic sites that exhibit heterozygosity, in which the pair of alleles inherited from each parent differ, will become hemizygous (as opposed to homozygous, in which the same allele is inherited from each parent) when one locus of the chromosomal pair is lost from the genetic pool. This condition is known as LOH,[114] and can occur in a healthy individual. However, a functional tumor suppressor gene can be lost, and the remaining copy can be inactivated by a point mutation, resulting in no functional tumor suppressor gene to protect the body. Copy neutral LOH, also called uniparental disomy (UPD) or gene conversion, is a condition in which both copies of the chromosomal pair come from a single parent and that of the other parent is lost. This can result in an individual acquiring two nonfunctional copies of a tumor suppressor gene—essentially a second hit of the Knudson multiple-hit hypothesis; in other words, the cancer originates from multiple mutational events occurring in a single cell. Acquired UPD is common in both hematologic and solid tumors and constitutes 20% to 80% of the LOH seen in human tumors. LOH is also useful for assessment of malignant degeneration and clonality. UPD can also be seen in parentage testing and is another reason not to use only a single inconsistency for exclusion. LOH testing has been performed using STR kits, but it is now more commonly performed by SNP arrays.[115,116]

REFERENCES

1. Jeffreys AJ, Wilson V, Thein SL. Individual-specific "fingerprints" of human DNA. *Nature* 1985;**316**:76—9.
2. Burch AM, Kelly MRD, Walsh A. Census of publicly funded forensic crime laboratories. NCJ 2012. 238252, http://www.bjs.gov/index.cfm?ty=pbdetail&iid=4412.
3. Milos A, Selmanovic A, Smajlovic L, et al. Success rates of nuclear short tandem repeat typing from different skeletal elements. *Croat Med J* 2007;**48**:486—93.
4. Alonso A, Andolinovic S, Martin P, et al. DNA typing from skeletal remains: evaluation of multiplex and megaplex STR systems on DNA isolated from bone and teeth samples. *Croat Med J* 2001;**42**(3):260—6.
5. Boric I, Ljubkovic J, Sutlovic D. Discovering the 60 years old secret: identification of the World War II mass grave victims from the island of Daksa near Dubrovnik, Croatia. *Croat Med J* 2011;**52**:327—35.
6. Budowle B, Bieber FR, Eisenberg AJ. Forensic aspects of mass disasters: strategic considerations for DNA-based human identification. *Leg Med (Tokyo)* 2005;**7**:230—43.
7. Edwards A, Civitello A, Hammond HA, et al. DNA typing and genetic mapping with trimeric and tetrameric tandem repeats. *Am J Hum Genet* 1991;**49**:746—56.
8. Urquhart A, Kimpton CP, Downes TJ, et al. Variation in short tandem repeat sequences—a survey of twelve microsatellite loci for use as forensic identification markers. *Int J Legal Med* 1994;**107**:13—20.
9. Mizutani M, Yamamoto T, Torii K, et al. Analysis of 168 short tandem repeat loci in the Japanese population, using a screening set for human genetic mapping. *J Hum Genet* 2001;**46**:448—55.
10. Budowle B, Masibay A, Anderson SJ, et al. STR primer concordance study. *Forensic Sci Int* 2001;**124**:47—54.
11. Nadir E, Margalit H, Gallily T, et al. Microsatellite spreading in the human genome: evolutionary mechanisms and structural implications. *Proc Natl Acad Sci USA* 1996;**93**:6470—5.
12. Butler JM. Genetics and genomics of core short tandem repeat loci used in human identity testing. *J Forensic Sci* 2006;**51**:253—65.
13. Saiki R, Gelfand DH, Stoffel S, et al. Primer-directed enzymatic amplification of DNA with a thermostable DNA polymerase. *Science* 1988;**239**:487—91.
14. Kloosterman AD, Kersbergen P. Efficacy and limits of genotyping low copy number (LCN) DNA samples by multiplex PCR of STR loci. *J Soc Biol* 2003;**197**:351—9.
15. Steinlechner M, Grubwieser P, Scheithauer R, et al. STR loci Penta D and Penta E: Austrian Caucasian population data. *Int J Legal Med* 2002;**116**:174—5.
16. Tomsey CS, Kurtz M, Kist F, et al. Comparison of PowerPlex 16, PowerPlex1.1/2.1, and ABI AmpfISTR Profiler Plus/COfiler for forensic use. *Croat Med J* 2001;**42**:239—43.
17. Wang DY, Chang CW, Lagace RE, et al. Developmental Validation of the AmpFℓSTR® Identifiler® Plus PCR Amplification Kit: an established multiplex assay with improved performance. *J Forensic Sci* 2012;**57**:453—65.
18. Budowle B, Moretti TR. Examples of STR population databases for CODIS and for casework. In: *Ninth International Symposium on Human Identification*, 1. Orlando, FL: Promega Corporation; 1998. p. 64—73.
19. Budowle B, Moretti TR, Baumstark AL, et al. Population data on the thirteen CODIS core short tandem repeat loci in African Americans, U.S. Caucasians, Hispanics, Bahamians, Jamaicans, and Trinidadians. *J Forensic Sci* 1999;**44**:1277—86.
20. Perez L, Hau J, Izarra F, et al. Allele frequencies for the 13 CODIS STR loci in Peru. *Forensic Sci Int* 2003;**132**:164—5.
21. Einum DD, Scarpetta MA. Genetic analysis of large data sets of North American Black, Caucasian, and Hispanic populations at 13 CODIS STR loci. *J Forensic Sci* 2004;**49**:1381—5.
22. Lim SES, Tan-Siew WF, Syn CKC, et al. Genetic data for the 13 CODIS STR loci in Singapore Indians. *Forensic Sci Int* 2005;**148**:65—7.

23. Vergara IA, et al. Autosomal STR allele frequencies for the CODIS system from a large random population sample in Chile. *Forensic Sci Int Genet* 2012;**6**:e83–5.
24. Hares DR. Expanding the CODIS core loci in the United States. *Forensic Sci Int Genet* 2012;**6**:c52 4.
25. Hares DR. Selection and implementation of expanded CODIS core loci in the United States. *Forensic Sci Int Genet* 2015;**17**:33–4.
26. Mannucci A, Sullivan KM, Ivanov PL, et al. Forensic application of a rapid and quantitative DNA sex test by amplification of the X-Y homologous gene amelogenin. *Int J Legal Med* 1994;**106**:190–3.
27. Neeser D, Liechti-Gallati S. Sex determination of forensic samples by simultaneous PCR amplification of α-satellite DNA from both the X and Y chromosomes. *J Forensic Sci* 1995;**40**:239–41.
28. Mitchell RJ, Kreskas M, Baxter E, et al. An investigation of sequence deletions of amelogenin (AMELY), a Y-chromosome locus commonly used for gender determination. *Ann Hum Biol* 2006;**33**:227–40.
29. Sullivan KM. Forensic applications of DNA fingerprinting. *Mol Biotechnol* 1994;**1**:13–27.
30. Phillips K, McCallum N, Welch L. A comparison of methods for forensic DNA extraction: Chelex-100® and the QIAGEN DNA Investigator Kit (manual and automated). *Forensic Sci Int Genet* 2012;**6**:282–5.
31. Brevnov MG, Pawar HS, Mundt J, et al. Developmental validation of the PrepFiler Forensic DNA Extraction Kit for extraction of genomic DNA from biological samples. *J Forensic Sci* 2009;**54**:599–607.
32. Turci M, Sardaro MLS, Visioli G, et al. Evaluation of DNA extraction procedures for traceability of various tomato products. *Food Control* 2010;**21**:143–9.
33. Ferrari BC, Power ML, Bergquist PL. Closed-tube DNA extraction using a thermostable proteinase is highly sensitive, capable of single parasite detection. *Biotechnol Lett* 2007;**29**:1831–7.
34. Krnajski Z, Geering S, Steadman S. Performance verification of the Maxwell 16 Instrument and DNA IQ Reference Sample Kit for automated DNA extraction of known reference samples. *Forensic Sci Med Pathol* 2007;**3**:264–9.
35. Butler JM. *Advanced topics in forensic DNA typing: Methodology*. Waltham, MA: Academic Press; 2012. p. 608.
36. Barbisin M, Fang R, Furtado MR, et al. Quantifiler® Duo DNA Quantification Kit: a guiding tool for short tandem repeat genotyping of forensic samples. *J Forensics Res* 2011;**2**:118.
37. Krenke BE, Nassif N, Sprecher CJ, et al. Developmental validation of a real-time PCR assay for the simultaneous quantification of total human and male DNA. *Forensic Sci Int Genet* 2008;**3**:14–21.
38. Andersen JF, Greenhalgh MJ, Butler HR, et al. Further validation of a multiplex STR system for use in routine forensic identity testing. *Forensic Sci Int* 1996;**78**:47–64.
39. Lins AM, Sprecher CJ, Puers C, et al. Multiplex sets for the amplification of polymorphic short tandem repeat loci—silver stain and fluorescence detection. *Biotechniques* 1996;**20**:882–9.
40. Sparkes R, Kimpton C, Watson S, et al. The validation of a 7-locus multiplex STR test for use in forensic casework: (I) mixtures, ageing, degradation and species studies. *Int J Legal Med* 1996;**109**:186–94.
41. Moretti TR, Baumstark AL, Defenbaugh DA, et al. Validation of short tandem repeats (STRs) for forensic usage: performance testing of fluorescent multiplex STR systems and analysis of authentic and simulated forensic samples. *J Forensic Sci* 2001;**46**:647–60.
42. Flores S, Sun J, King J, et al. Internal validation of the GlobalFiler™ Express PCR Amplification Kit for the direct amplification of reference DNA samples on a high-throughput automated workflow. *Forensic Sci Int Genet* 2014;**10**:33–9.
43. Ensenberger MG, Lenz KA, Matthies LK, et al. Developmental validation of the PowerPlex Fusion 6C System. *Forensic Sci Int Genet* 2016;**21**:134–44.
44. Cave CA, Hancock K, Schumm JW. Principles of STR multiplex amplification. *Forensic Sci Int: Genet Supp* 2008;**1**:102–4.
45. Butler JM, Buel E, Crivellente F, et al. Forensic DNA typing by capillary electrophoresis using the ABI Prism 310 and 3100 genetic analyzers for STR analysis. *Electrophoresis* 2004;**25**:1397–412.
46. Gusmao L, Gonzalez-Neira A, Pestoni C, et al. Robustness of the Y STRs DYS19, DYS389 I and II, DYS390 and DYS393: Optimization of a PCR pentaplex. *Forensic Sci Int* 1999;**106**:163–72.
47. Alves C, Gomes V, Prata MJ, et al. Population data for Y-chromosome haplotypes defined by 17 STRs (AmpFlSTR YFiler) in Portugal. *Forensic Sci Int* 2007;**171**:250–5.
48. Iannacone GC, Tito RY, Lopez PW, et al. Y-chromosomal haplotypes for the PowerPlex Y for twelve STRs in a Peruvian population sample. *J Forensic Sci* 2005;**50**:239–42.
49. US Y-STR Database. <https://www.usystrdatabase.org/>.
49a. Schoske R, Vallone PM, Kline MC, et al. High-throughput Y-STR typing of U.S. populations with 27 regions of the Y chromosome using two multiplex PCR assays. *For Sci Int* 2004;**139**:107–21.
50. Olofsson JK, et al. Forensic and population genetic analyses of Danes, Greenlanders and Somalis typed with the Yfiler® Plus PCR amplification kit. *Forensic Sci Int Genet* 2015;**16**:232–6.
51. Just RS, Irwin JA, Parson W. Questioning the prevalence and reliability of human mitochondrial DNA heteroplasmy from massively parallel sequencing data. *Proc Natl Acad Sci U S A* 2014;**111**(43):E4546–7.
52. Chen F, Dang Y, Yan C, et al. Sequence-length variation of mtDNA HVS-I C-stretch in Chinese ethnic groups. *J Zhejiang Univ Sci B* 2009;**10**:711–20.
53. Liu VWS, Yang HJ, Wang Y, et al. High frequency of mitochondrial genome instability in human endometrial carcinomas. *Br J Cancer* 2003;**89**:697–701.
54. Irwin JA, Parson W, Coble MD, et al. mtGenome reference population databases and the future of forensic mtDNA analysis. *Forensic Sci Int Genet* 2011;**5**:222–5.
55. Gill P, Ivanov PL, Kimpton C, et al. Identification of the remains of the Romanov family by DNA analysis. *Nat Genet* 1994;**6**:130–5.
56. Forster L, Thomson J, Kutranov S. Direct comparison of post-28-cycle PCR purification and modified capillary electrophoresis methods with the 34-cycle "low copy number" (LCN) method for analysis of trace forensic DNA samples. *Forensic Sci Int Genet* 2008;**2**:318–28.
57. Whitaker JP, Cotton EA, Gill P. A comparison of the characteristics of profiles produced with the AmpFlSTR® SGM Plus multiplex system for both standard and low copy number (LCN) STR DNA analysis. *Forensic Sci Int* 2001;**123**:215–23.
58. Caragine T, Mikulasovich R, Tamariz J, et al. Validation of testing and interpretation protocols for low template DNA samples using AmpFlSTR® Identifiler®. *Croat Med J* 2009;**50**:250–67.

59. Wickenheiser RA. Trace DNA: a review, discussion of theory, and application of the transfer of trace quantities of DNA through skin contact. *J Forensic Sci* 2002;**47**:442—50.
60. Phipps M, Petricevic S. The tendency of individuals to transfer DNA to handled items. *Forensic Sci Int* 2007;**168**:162—8.
61. Homer N, Szabolcs S, Redman M, et al. Resolving individuals contributing trace amounts of DNA to highly complex mixtures using high-density SNP genotyping microarrays. *PLoS Genet* 2008;**4**:e1000167.
62. Perlin MW, Legler MM, Spencer CE, et al. Validating TrueAllele® DNA mixture interpretation. *J Forensic Sci* 2011;**56**:1430—47.
63. Perlin MW, Kadane JB, Cotton RW. Match likelihood ratio for uncertain genotypes. *Law Probab Risk* 2010;**8**:289—302.
64. Bright J, Evett IW, Taylor D, et al. A series of recommended tests when validating probabilistic DNA profile interpretation software. *Forensic Sci Int Genet* 2015;**14**:125—31.
65. Cooper S, McGovern C, Bright J, et al. Investigating a common approach to DNA profile interpretation using probabilistic software. *Forensic Sci Int Genet* 2015;**16**:121—31.
66. Senge T, Madea B, Junge A, et al. STRs, mini STRs and SNPs—a comparative study for typing degraded DNA. *Leg Med (Tokyo)* 2011;**13**:68—74.
67. Mulero JJ, Chang CW, Lagace RE, et al. Development and validation of the AmpFℓSTR® MiniFiler™ PCR amplification kit: a miniSTR multiplex for the analysis of degraded and/or PCR inhibited DNA. *J Forensic Sci* 2008;**53**:838—52.
68. Butler JM, Budolwe B, Gill P, et al. Report on ISFG SNP panel discussion. *Forensic Sci Int: Genet Supp* 2008;**1**:471—2.
69. Gettings KB, Lai R, Johnson JL, et al. A 50-SNP assay for biogeographic ancestry and phenotype prediction in the U.S. population. *Forensic Sci Int Genet* 2014;**8**:101—8.
70. Butler JM, Coble MD, Vallone PM. STRs vs. SNPs: thoughts on the future of forensic DNA testing. *Forensic Sci Med Pathol* 2007;**3**:200—5.
71. DNA Identification Act of 1994 in U.S.C. 1994. p. 103—322.
72. FBI. Quality Assurance Standards for Forensic DNA Testing Laboratories. <http://www.fbi.gov/about-us/lab/biometric-analysis/codis/qas_testlabs>.
73. NIST. *Organization of Scientific Area Committees (OSAC)*. 2015. http://www.nist.gov/forensics/osac.cfm.
74. NIST. STRBase. <http://www.cstl.nist.gov/strbase/>.
75. Evett IW, Gill PD, Scrange JK, et al. Establishing the robustness of short-tandem-repeat statistics for forensic applications. *Am J Hum Genet* 1996;**58**(2):398—407.
76. *Committee on DNA technology in forensic science. DNA technology in forensic science*. Washington DC: National Academy Press; 1992. p. 185.
77. *Committee on DNA forensic science. An update. The evaluation of forensic DNA evidence*. Washington DC: National Academy Press; 1996. p. 254.
78. Ossmani HE, Talbi J, Bouchrif B, et al. Allele frequencies of 15 autosomal STR loci in the southern Morocco population with phylogenetic structure among worldwide populations. *Leg Med (Tokyo)* 2009;**11**:155—8.
79. Martire KA, Kemp RF, Sayle M, et al. On the interpretation of likelihood ratios in forensic science evidence: presentation formats and the weak evidence effect. *Forensic Sci Int* 2014;**240**:61—8.
80. Vallone PM, Hill CR, Butler JM. Demonstration of rapid multiplex PCR amplification involving 16 genetic loci. *Forensic Sci Int Genet* 2008;**3**:42—5.
81. Giese H, Lam R, Selden R, et al. Fast multiplexed polymerase chain reaction for conventional and microfluidic short tandem repeat analysis. *J Forensic Sci* 2009;**54**:1287—96.
82. Gangano S, Elliott K, Anoruo K, et al. DNA investigative lead development from blood and saliva samples in less than two hours using the RapidHIT™ Human DNA Identification System. *Forensic Sci Int: Genet Supp* 2013;**4**:e43—4.
83. Borsting C, Morling N. Next generation sequencing and its applications in forensic genetics. *Forensic Sci Int Genet* 2015;**18**:78—89.
84. Pitterl F, Schmidt K, Huber G, et al. Increasing the discrimination power of forensic STR testing by employing high-performance mass spectrometry, as illustrated in indigenous South African and Central Asian populations. *Int J Legal Med* 2010;**124**:551—8.
85. Planz JV, Sannes-Lowery KA, Duncan D, et al. Automated analysis of sequence polymorphism in STR alleles by PCR and direct electrospray ionization mass spectrometry. *Forensic Sci Int Genet* 2012;**6**:594—606.
86. Kline MC, Hill CR, Decker AE, et al. STR sequence analysis for characterizing normal, variant, and null alleles. *Forensic Sci Int Genet* 2011;**5**:329—32.
87. Gelardi C, Rockenbauer E, Dalsgaard S, et al. Second generation sequencing of three STRs D3S1358, D12S391 and D21S11 in Danes and a new nomenclature for sequenced STR alleles. *Forensic Sci Int Genet* 2014;**12**:38—41.
88. King JL, LaRue BL, Novsroski NM, et al. High-quality and high-throughput massively parallel sequencing of the human mitochondrial genome using the Illumina MiSeq. *Forensic Sci Int Genet* 2014;**12C**:128—35.
89. Holland MM, McQuillan MR, O'Hanlon KA. Second generation sequencing allows for mtDNA mixture deconvolution and high resolution detection of heteroplasmy. *Croat Med J* 2011;**52**:299—313.
90. Babol-Pokora K, Berent J. SNP-minisequencing as an excellent tool for analysing degraded DNA recovered from archival tissues. *Acta Biochim Pol* 2008;**55**:815—9.
91. Dixon LA, et al. Analysis of artificially degraded DNA using STRs and SNPs—Results of a collaborative European (EDNAP) exercise. *Forensic Sci Int* 2006;**164**:33—44.
92. Gettings KB, Kiesler KM, Vallone PM. Performance of a next generation sequencing SNP assay on degraded DNA. *Forensic Sci Int Genet* 2015;**19**:1—9.
93. Pakstis AJ, Speed WC, Fang R, et al. SNPs for a universal individual identification panel. *Hum Genet* 2010;**127**:315—24.
94. Sanchez JJ, Phillips C, Borsting C, et al. A multiplex assay with 52 single nucleotide polymorphisms for human identification. *Electrophoresis* 2006;**27**:1713—24.
95. Kosoy R, Nassir R, Tian C, et al. Ancestry informative marker sets for determining continental origin and admixture proportions in common populations in America. *Hum Mutat* 2009;**30**:69—78.
96. Walsh S, Liu F, Ballantyne KN, et al. IrisPlex: a sensitive DNA tool for accurate prediction of blue and brown eye colour in the absence of ancestry information. *Forensic Sci Int Genet* 2011;**5**:170—80.
97. Walsh S, Liu F, Wollstein A, et al. The HIrisPlex system for simultaneous prediction of hair and eye colour from DNA. *Forensic Sci Int Genet* 2013;**7**:98—115.
98. van Oorschot RA, Jones MK. DNA fingerprints from fingerprints. *Nature* 1997;**387**(6635):767.

99. The technical working group on biological evidence preservation. Biological evidence preservation: considerations for policy makers. *NISTIR* 2015;**8048**:21.
100. Maryland v. King, 133 S.Ct. 1958, (2013).
100a. Chakraborty R, Stivers D. Paternity exclusion by DNA markers: effects of paternal mutations. *J For Sci* 1996;**41**:671–7.
101. Allen RW, Polesky HF. Chapter 12 parentage and relationship testing. In: Leonard DGB, editor. *Molecular pathology in clinical practice*. 2nd ed. Switzerland: Springer International Publishing; 2016. p. 811–21.
102. Butler JM. Chapter 14 Relationship testing: Kinship statistics. In: *advanced topics in forensic DNA typing*. Amsterdam: Academic Press; 2015. p. 349–401.
103. Evett I, Weir BS. *Interpreting DNA evidence: statistical genetics for forensic scientists*. Sunderland, Mass: Sinauer Assoc; 1998. p. 278.
104. *Standards for relationship testing laboratories*. 12th ed. Bethesda, Maryland: American Association of Blood Banks (AABB); 2016. p. 69.
105. Shibata D. Identification of mismatched fixed specimens with a commercially available kit based on the polymerase chain reaction. *Am J Clin Pathol* 1993;**100**:666–70.
106. Stojilkovic-Mikic T, Mann K, Docherty Z, et al. Maternal cell contamination of prenatal samples assessed by QF-PCR genotyping. *Prenat Diagn* 2005;**25**:79–83.
107. Lamb AN, Rosenfeld JA, Coppinger J, et al. Defining the impact of maternal cell contamination on the interpretation of prenatal microarray analysis. *Genet Med* 2012;**14**:914–21.
108. Schrijver I, Cherny SC, Zehnder JL. Testing for maternal cell contamination in prenatal samples: a comprehensive survey of current diagnostic practices in 35 molecular diagnostic laboratories. *J Mol Diagn* 2007;**9**:394–400.
109. Antin JH, Childs R, Filopovich AH, et al. Establishment of complete and mixed donor chimerism after allogeneic lymphohematopoietic transplantation: recommendations from a workshop at the 2001 tandem meetings. *Biol Blood Marrow Transplant* 2001;**7**:472–85.
110. Scandling JD, Busque S, Dejbakhsh-Jones S, et al. Tolerance and withdrawal of immunosuppressive drugs in patients given kidney and hematopoietic cell transplants. *Am J Transplant* 2012;**12**:1133–45.
111. McCann SR, Crampe M, Molloy K, et al. Hematopoetic chimerism following stem cell transplantation. *Transfus Apher Sci* 2005;**32**:55–61.
112. Jiang Y, Wan L, Qin Y, et al. Donor chimerism of B cells and nature killer cells provides useful information to predict hematologic relapse following allogeneic hematopoetic stem cell transplantation. *PLoS ONE* 2015;**10**:e0133671.
113. Van Deerlin VM, Leonard DG. Bone marrow engraftment analysis after allogeneic bone marrow transplantation. *Clin Lab Med* 2000;**20**:197–225.
114. Archetti M. Recombination and loss of complementation: a more than two-fold cost for parthenogenesis. *J Evol Biol* 2004;**17**:1084–97.
115. Beroukhim R, Lin M, Park Y, et al. Inferring loss-of-heterozygosity from unpaired tumors using high-density oligonucleotide SNP arrays. *PLoS Comput Biol* 2006;**2**:e41.
116. O'Keefe C, McDevitt MA, Maciejewski JP. Copy neutral loss of heterozygosity: a novel chromosomal lesion in myeloid malignancies. *Blood* 2010;**115**:2731–9.

13

Amino Acids, Peptides, and Proteins

Dennis J. Dietzen*

ABSTRACT

Background
Amino acids are the building blocks of proteins, but they also play diverse roles in the provision of energy and the formation of a number of other important biomolecules, including hormones, neurotransmitters, and signaling molecules. The polymers of amino acids, peptides, and proteins orchestrate and control the vast array of human physiologic and biochemical processes. The catalog of amino acids, peptides, and proteins in various biological fluids is a target-rich environment for the detection of pathological states.

Content
This chapter first describes the chemistry, metabolism, transport, and analysis of amino acids. Polymers of amino acids may be relatively short (peptides) or long (proteins). The human genome contains the information to dictate formation of approximately 20,000 polypeptides, but the actual diversity of the human proteome and peptidome is manifold more expansive. Proteome diversity arises from linear amino acid sequence and an array of modifications that include acylation, phosphorylation, glycosylation, and isoprenylation. Systems of short peptides, larger protein monomers, and multimeric protein complexes are the tools that orchestrate and control human physiologic and biochemical processes. Proper synthesis, folding, subcellular targeting, and catabolism of proteins and peptides, therefore, are essential for human health. Analytic exploitation of biologic fluids including blood, urine, and cerebrospinal fluid using chemical, immunologic, and mass spectrometric methods enables informed diagnosis and therapy in a multitude of disease states.

INTRODUCTION

Amino acids, peptides, and proteins are crucial for virtually all biologic processes. Amino acids serve as structural subunits of peptides and proteins but also play diverse roles in metabolism, neurotransmission, and intercellular signaling. Peptides serve as autocrine and endocrine signaling molecules that control appetite, vascular tone, and electrolyte homeostasis, as well as carbohydrate and mineral metabolism. Proteins, longer peptide chains with molecular mass typically greater than approximately 6000 Daltons (Da) serve as (1) intracellular and extracellular structural components, (2) biologic catalysts, (3) mediators of contractility and motility, (4) agents of molecular assembly, (5) ion channels and pumps, (6) molecular transporters, (7) mediators of immunity, and (8) components of intracellular and intercellular signaling networks.

The human genome contains more than 20,000 open reading frames that encode proteins. The actual number of proteins is far greater, however, because of alternative splicing of messenger RNA (mRNA), somatic recombination, mutation, proteolytic processing, and posttranslational modification. The *proteome* represents the complete set of proteins in an organism or compartment of an organism such as the plasma space. Efforts to catalog the proteome include those by the Human Proteome Organization (hupo.org), the National Center for Biotechnology Information (ncbi.nlm.nih.gov), the Swiss Institute of Bioinformatics (expasy.org), and the Healthy Human Individual's Integrated Plasma Proteome Database (bio.informatics.iupui.edu/HIP2). Most databases were designed mainly to assist with peptide and protein identification, but efforts have shifted to characterizing the abundance of specific protein components in healthy and diseased populations, the usual basis for diagnostic applications.

This chapter begins with a discussion of the chemistry, metabolism, and analysis of amino acids. A description of the chemistry and biochemistry of the peptide bond is then followed by a description of several clinically relevant peptide systems and methods for in vitro assessment. The protein narrative begins with an account of protein structure and cellular compartmentalization followed by discussion of co- and posttranslational modifications. Constituents of the proteome in body fluids are also addressed, followed lastly by a description of methods for specific and global assessment of the proteome for clinical purposes.

AMINO ACIDS

Amino acids were likely among the first organic molecules to emerge from the mix of methane, hydrogen, ammonia, and water in earth's primordial atmosphere.[1] Only 20 of the

---

*The author gratefully acknowledges the preceding foundation for this chapter laid by Glen L. Hortin and A. Myron Johnson, as well as generous assistance from Carl H. Smith on the topic of amino acid transport.

hundreds of known amino acids account for the vast majority of residues in human polypeptide chains. Their structure and molecular properties are summarized in Table 13.1. These 20 along with dozens of non–protein-forming amino acids are critical to the form and function of the human body. Disrupted amino acid metabolism is not surprisingly associated with a multitude of pathologic processes.

Basic Biochemistry

Amino acids are organic compounds containing both an amino group ($-NH_2$) and a carboxyl group ($-COOH$) or another acidic group such as a sulfonate group ($-SO_3$). In a majority of biologically relevant amino acids, the amine moiety is primary ($-NH_2$), but some (eg, sarcosine) are secondary ($-NH-$) amines, and others containing tertiary amines (eg, proline) are referred to as imino ($=N-$) acids. With the exception of proline, the amino acids that occur in protein are α-amino acids (below).

The R group represents the unique side chains responsible for the chemical properties of individual amino acids. Not all biologic amino acids are α amino acids. β amino acids such as β-alanine and taurine as well as γ-amino acids such as γ-aminobutyric acid (GABA) also play key biochemical roles (Fig. 13.1).

With the exception of glycine, all α amino acids contain four distinct moieties asymmetrically arranged around the α carbon. As a consequence, amino acids may exist as mirror images (enantiomers) referred to as the D or L configuration.

TABLE 13.1 Structure and Chemical Properties of the 20 Proteogenic Amino Acids

Amino Acid	MW (Da)	Structure (pH 7.0)	pK$_1$	pK$_2$	pK$_3$	pI	HI
Alanine (ALA, A)	89.09		2.4	9.7		6.0	1.8
Arginine (ARG, R)	174.20		2.2	9.0	12.5	10.8	−4.5
Asparagine (ASN, N)	132.12		2.0	8.8		5.4	−3.5
Aspartate (ASP, D)	133.10		2.1	9.8	3.9	2.9	−3.5
Cysteine (CYS, C)	121.16		1.7	10.8	8.3	5.1	2.5
Glycine (GLY, G)	75.07		2.3	9.6		6.0	−0.4
Glutamate (GLU, E)	147.13		2.2	9.7	4.3	3.2	−3.5
Glutamine (GLN, Q)	146.15		2.2	9.1		5.7	−3.5

TABLE 13.1 Structure and Chemical Properties of the 20 Proteogenic Amino Acids—cont'd

Amino Acid	MW (Da)	Structure (pH 7.0)	pK₁	pK₂	pK₃	pI	HI
Histidine (HIS, H)	155.16		1.8	9.2	6.0	7.6	−3.2
Isoleucine (ILE, I)	131.17		2.4	9.7		6.0	4.5
Leucine (LEU, L)	131.17		2.4	9.6		6.0	3.8
Lysine (LYS, K)	146.19		2.2	9.0	10.5	9.7	−3.9
Methionine (MET, M)	149.21		2.3	9.2		5.8	1.9
Phenylalanine (PHE, F)	165.19		1.8	9.1		5.5	2.8
Proline (PRO, P)	115.13		2.1	10.6		6.1	1.6
Serine (SER, S)	105.09		2.2	9.2		5.7	−0.8
Threonine (THR, T)	119.12		2.6	10.4		6.5	−0.7
Tryptophan (TRP, W)	201.22		2.5	9.4		5.9	−0.9
Tyrosine (TYR, Y)	181.19		2.2	9.2	10.5	5.7	−1.3
Valine (VAL, V)	117.17		2.3	9.6		6.0	4.2

HI, Hydropathy index; *MW*, molecular weight; *pk*, acid ionization constant; *pI*, isoelectric point.

With few exceptions, the biologically relevant amino acids exist in the *L* configuration. Small quantities of *D* amino acids occur in physiological fluids but typically do not have specific functions. An exception is *D* serine, which represents 5% to 20% of total serine in cerebrospinal fluid and may serve as a neurotransmitter. Amino acids with the *D* configuration occur in some bacterial products, foods, and pharmaceuticals. *D* amino acid oxidases in liver and kidney convert *D* amino acids to ketoacids, which can be further metabolized. *L* amino acids in proteins undergo slow racemization to a *DL* mixture over many years. Aspartic acid undergoes the most rapid racemization, and this rate can be used to estimate the time of synthesis of proteins with very slow turnover, such as ocular lens proteins or intervertebral collagen in which half-lives may exceed 50 years. Two amino acids, threonine and isoleucine, have a second asymmetric carbon, and their stereoisomers are referred to as *allothreonine* and *alloisoleucine*. The latter compound has utility in the diagnosis of maple syrup urine disease (MSUD; OMIM #248600).

In addition to the 20 well-known protein-forming amino acids, a number of rare amino acids are recovered from protein hydrolysates. For example, 4-hydroxyproline and 5-hydroxylysine are found in collagen lysates, and desmosine and isodesmosine are recovered in elastin hydrolysates. These amino acids are formed by posttranslational mechanisms because no codon is responsible for their incorporation into growing polypeptides. Selenocysteine is a special case of an amino acid synthesized on a specific transfer RNA and incorporated into a few sites in only about 25 proteins that include members of the thyroid hormone deiodinase and glutathione peroxidase families.[2] Some of these rare amino acids are shown in Fig. 13.1.

Acid–base properties of amino acids depend on the amino and carboxyl groups attached to the α carbon and on the basic or acidic groups occurring on some sidechains (R). At a physiologic pH near 7.4, the α-carboxyl group is ionized and carries a negative charge, and the α amino group is protonated and carries a positive charge. Molecules existing simultaneously as cations and anions are referred to as zwitterions (diagrammed below).

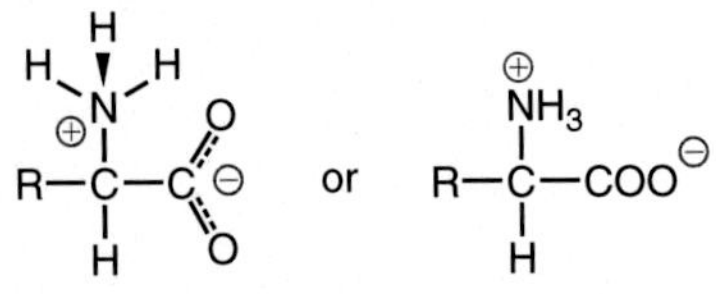

The pH at which ionizable groups exist equally as charged and uncharged forms is referred to as the pK. Amino acids thus have two or more pKs—one for the carboxyl, one for the amino group, and an additional one in the presence of an ionizable side chain. The isoelectric point (pI) is the pH at which an amino acid or other molecule has a net charge of 0. For a typical neutral amino acid such as glycine, the pI of 5.97 is midway between the pK of 2.34 for the carboxylic acid and the pK of 9.60 for the amino group. The pKs of amino acid side chains in proteins vary somewhat from those in free amino acids because of the influence of neighboring amino acids. The buffering capacity of ionizable groups is primarily in a pH range within ±1 unit of the pK for the respective groups. Amino acids and proteins therefore have a limited buffering capacity near physiologic pH. The imidazole side chain of histidine is an exception with a pK near 6.0. Glycine, for example, is used as a buffer near pH 2.5 or 9.5.

The structural diversity of side chains permits formation of proteins with a variety of structure and function. Sidechain diversity is dictated not only by pK but by size and hydrophobicity as well. Amino acids with longer aliphatic or aromatic side chains such as isoleucine, leucine, and phenylalanine have greater hydrophobicity than shorter side chains such as the methyl group found in alanine. Neutral amino acids with polar groups such as hydroxyl or amide groups in their side chains are more hydrophilic. Acidic amino acids have side chains with carboxylic acids, and basic amino acids have side chains with amino, guanidino, or imidazole groups. The thiol side chain (—SH) of cysteine oxidizes easily and may become linked to other molecules via disulfide. In plasma, cysteine occurs as cystine (cysteine homodimer linked via a disulfide) or as a mixed disulfide with albumin or other proteins.

With some exceptions, amino acids are water soluble and stable in plasma. The most soluble amino acids have small side chains with polar or ionizable moieties such as glycine, alanine, arginine, serine, and threonine. Less soluble amino acids such as phenylalanine, tyrosine, leucine, and tryptophan tend to have larger, nonpolar aliphatic or alicyclic side chains. Amino acid solubility is rarely limiting in vivo except in some metabolic disorders. Deposition of tyrosine crystals in the eye and skin is common in tyrosinemia (particularly type II; OMIM #276600). Likewise, cystine may crystallize in the renal parenchyma in patients with cystinuria (OMIM #220100). Structural and chemical details for the 20 protein-forming amino acids are displayed in Table 13.1.

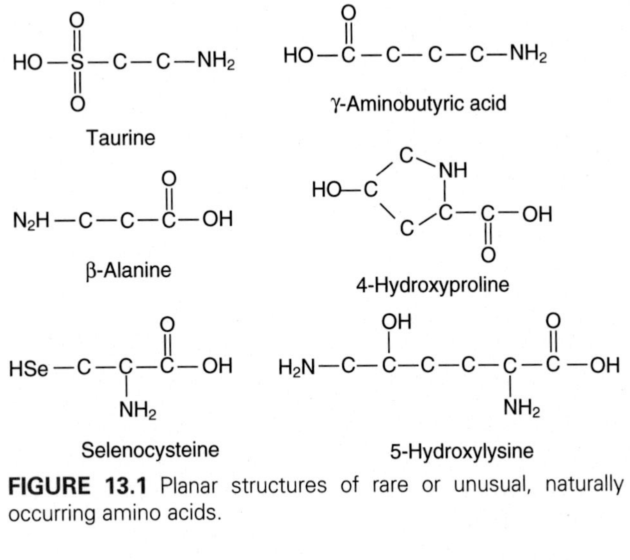

Amino Acid Supply and Transport

Amino acids participate in many metabolic pathways in addition to serving as substrates for protein synthesis. In the

FIGURE 13.1 Planar structures of rare or unusual, naturally occurring amino acids.

healthy state, women require approximately 46 g/d and men approximately 56 g/d of dietary protein (0.8 g/kg body weight), and substantial increases in demand occur during growth and in many disease states.[3] Dietary protein is digested by proteases in the stomach (eg, pepsin) and small intestine (eg, trypsin, chymotrypsin) to yield amino acids. Endogenous protein turnover serves as another source of free amino acids. Eight amino acids used for protein synthesis (isoleucine, leucine, lysine, methionine, phenylalanine, threonine, tryptophan, and valine) are not synthesized by humans and therefore are considered "essential" constituents of the diet. Meat, milk, eggs, and fish contain a full range of essential amino acids. Gelatin is deficient in tryptophan, and some plant sources of protein may be additionally deficient in lysine or methionine. Therefore, diets based on a single source of plant protein may be deficient in some amino acids. When liver function is compromised, cysteine and tyrosine become essential because they are not converted from their usual precursors, methionine and phenylalanine.[4] Arginine may be conditionally essential as well because endogenous rates of synthesis may be insufficient to meet requirements in adults under metabolic stress or in growing children.[5]

Requirements for dietary protein to maintain nitrogen balance increase in infancy and childhood when there are increased demands for growth.[6,7] Daily requirements increase by up to 3.5 to 4 g protein/kg body weight for premature infants, for example.[8] Protein demand is also increased in pregnancy, lactation, and states of protein loss or catabolic states (eg, burn patients). Persistent negative nitrogen balance results in a number of undesirable phenotypic features. A diet severely deficient in protein and consisting primarily of high-starch foods can lead to kwashiorkor, a disorder characterized by decreased serum albumin, immune deficiency, edema, ascites, growth failure, apathy, and many other symptoms.[9] Marasmus results when protein and energy sources such as carbohydrates are deficient, causing wasting of muscles and subcutaneous tissues. Albumin or prealbumin concentrations are used to assess adequacy of the amino acid supply. The shorter biologic half-life of prealbumin compared with albumin (2 vs. 20 days) makes it a valuable marker for acute dietary assessment.[10,11]

Homeostasis of cellular amino acid concentrations is dependent on supply, catabolism, and excretion. Supply and excretion are regulated by a series of transport systems with overlapping substrate specificity, strategic tissue expression, and polarized cellular distribution.[12,13] Amino acids are derived from dietary protein precursors through the action of proteolytic enzymes in the stomach and small intestine that produce shorter oligopeptides and individual amino acids. Enteral absorption of di- and tripeptides is mediated by a single proton-coupled transport system termed PEPT1 (encoded by SLC15A1).[14] Transport of individual amino acids across the intestinal and renal epithelium as well as the blood–brain barrier is far more specialized.

Early biochemical characterization of amino acid transport was technologically limited to studies of the plasma membrane. These broad-specificity transport systems function as co-transporters, exchange transporters, or facilitative transporters. Nomenclature was based on substrate specificity, co-transport requirements, and sensitivity to inhibitors. By convention, capital letters indicate a requirement for Na^+, and lower case descriptors imply the lack of Na^+ dependence. Systems A and ASC are responsible for Na^+-mediated symport of neutral species with small side chains. System L facilitates exchange of amino acids with large, hydrophobic side chains. Cationic amino acid transport is mediated by a system termed y^+. System y^+L facilitates exchange of neutral amino acids with cationic species. System B^o catalyzes Na^+ mediated transport of branched-chain and aromatic amino acids, and system b^o enables exchange of bulky neutral and cationic side chains. Finally, system X^- mediates transport of anionic amino acids. These systems act in a coordinated way to achieve amino acid homeostasis.[15]

Distinct transport systems cooperate to achieve net amino acid transport across epithelial barriers. For example, transcellular transport of cationic amino acids across the renal brush border is achieved by a combination of two exchange systems, b^o and y^+L. On the apical surface, positively charged amino acids are imported in exchange for an uncharged amino acid via system b^o. System b^o uses the transmembrane electrical potential to drive transport. Efflux across the basolateral membrane is also achieved via exchange transport via system y^+L. System y^+L uses the driving force of the transmembrane sodium gradient. Lack of proper polarized expression in appropriate tissue types results in transport disorders such as cystinuria (OMIM #220100) and lysinuric protein intolerance (OMIM #222700).[16]

Intracellular amino acid compartments are maintained by a distinct set of transport systems. These systems are important for metabolizing, sequestering, and recycling various amino acids. Substrate concentrations in the urea cycle are regulated in part via the mitochondrial ornithine-citrulline antiporter. Defects in this transport system leads to HHH (hyperornithinemia, hyperammonemia, homocitrullinuria) syndrome (OMIM #238970). Reclamation of amino acids from lysosomal protein digestion also relies on transport systems. Defects in the CTNS gene, for example, inhibits lysosomal cystine transport and results in the clinical disorder known as cystinosis (OMIM #219750). Finally, neuronal vesicles must concentrate synaptic transmitters to achieve interneuronal communication. Two of these transmitters, glycine and glutamate, are actively transported across vesicle membranes.

The rather coarse biochemical definition of amino acid transport is being redefined as the genetic basis for these systems is clarified.[17] Table 13.2 summarizes selected connections between gene and functional transport systems. Genes of the SLC (solute carrier) family encode integral transmembrane spanning proteins that catalyze amino acid transport. The SLC1 family, for example, encodes transporters primarily responsible for transport of anionic amino acids that are particularly important in brain and neural tissues.[18] Neutral amino acid transporters (eg, System A) are encoded by the SLC38 family.[19] Mitochondrial transport systems are expressed by the SLC25 gene family.[20] The SLC3 and SLC7 gene families encode a wide array of heterodimeric transporters, including the aforementioned b^o and y^+L transport systems.[21] Each of these transporters contains one of two SLC3 genes that encode membrane-targeting subunits that

TABLE 13.2 Genetic Basis of Selected Amino Acid Transport Systems

Gene Family	System	Expression	Substrates	Disease Link
SLC1	X^-, ASC	Brain, gut, kidney, liver	D, E, A, S, T, C, N, Q	
SLC3A1 (rBAT)	y^+, L, y^+L, b^0	Broad	R, K, H, M, L, A, C, L, I, V	Cystinuria
SLC3A2 (4F2hc)				Lysinuric protein intolerance
SLC7 (A1-A13)				
SLC6	B^0	Brain, kidney, gut, liver	F, Y, L, I, V, P, C, A, Q, S, H, G, M	Hartnup disorder Iminoglycinuria
SLC17		Neurons	E	
SLC25	ASP/GLU ORN/CIT	Broad, (mitochondria)	D, E, ornithine, citrulline	Type II citrullinemia HHH syndrome
SLC32		Neurons	G, γ-aminobutyrate	
SLC38	A, N	Broad	Q, A, N, C, H, S, T	
SLC43	L	Liver, kidney, gut, muscle	L, I, V	
CTS		Broad	Cystine	Cystinosis

HHH, Hyperornithinemia, hyperammonemia, homocitrullinuria.

are disulfide linked to one of at least a dozen *SLC7* gene products dictating transport specificity. Metabolic flux of amino acid carbon is critically dependent on the proper expression and regulation of transport as well as enzyme-catalyzed chemical transformation.

Amino Acid Metabolism

Amino acids serve as scaffolds for the synthesis of many hormones, nucleotides, lipids, signaling molecules, and metabolic intermediates that play a role in energy production. As portrayed in Fig. 13.2, the transformation of amino acid carbon to energetic intermediate typically begins with transamination. Excess nitrogen is excreted as urea. Resulting α-ketoacids may enter the Krebs cycle; undergo conversion to ketone bodies, fatty acids, or glucose; or be completely oxidized to CO_2 depending on cellular energy demands. A vast array of enzyme networks has evolved to orchestrate demand for amino acids. Information regarding the substrates, products, kinetics, and inhibitors of these enzymes may be found in multiple databases, including BRENDA (brenda-enzymes.org), ExPASy-enzyme, (enzyme.expasy.org), and ExplorEnz (enzyme-database.org). Pathway databases include KEGG (genome.jp/kegg), GenMAPP (genmapp.org), and BioCyc (biocyc.org).

Glucose, fatty acids, and ketones are primary respiratory substrates in humans. These substrates generate adenosine triphosphate (ATP) via the mitochondrial Krebs cycle. Amino acids play two key roles in energy provision. First,

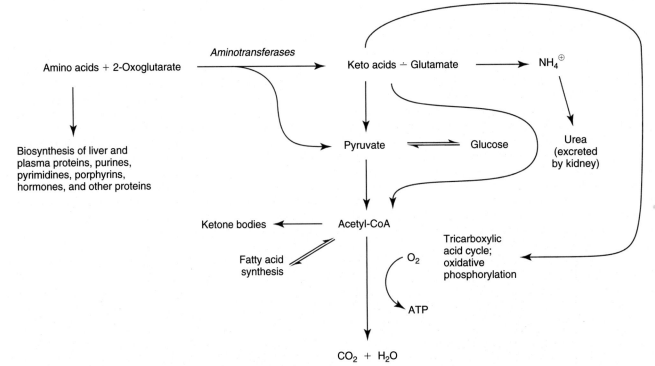

FIGURE 13.2 A generalized scheme of amino acid metabolism in the liver. After transamination, amino acid carbon may be used to in the Krebs cycle directly or transformed into other respiratory fuels such as glucose and ketones. Waste nitrogen is disposed of via the urea cycle. *ATP*, Adenosine triphosphate.

amino acids are converted to Krebs cycle intermediates to maintain the activity of the cycle in a process called anaplerosis. Glutamine and glutamate, for example, are converted to α-ketoglutarate via loss of the epsilon and alpha amino groups. Fumarate may be derived from asparagine and aspartate, and succinate is derived from methionine, threonine, and valine. Second, amino acids may be mobilized to generate fuels for a variety of organ systems. Five amino acids (LEU, ILE, LYS, PHE, and TYR) may be converted to ketones. All of the amino acids except for leucine may be used to produce glucose. In times of high energy demand and limiting fuel sources, therefore, flux of amino acid carbon through proper pathways becomes an important source of respiratory fuel.

Excess tissue nitrogen is disposed of as urea, which contains two moles of nitrogen per mole. Urea production is limited to the liver, so selected amino acids, primarily glutamine and alanine, serve to shuttle excess nitrogen to the liver. Nitrogen in the form of ammonium ion is first converted to carbamoyl phosphate, which is transferred to ornithine to form citrulline. Aspartic acid and citrulline are condensed to form arginosuccinic acid, which, in turn, is cleaved to arginine and fumaric acid. Arginase hydrolyzes arginine to urea and ornithine to allow the cycle to repeat. Urea is usually viewed simply as waste, but it is also the primary contributor to the high osmolality in the renal medulla and enables maximal urinary concentrating ability.

Amino acids are precursors for many hormones and signaling molecules. Tyrosine provides a scaffold for thyroxine, dopamine, and adrenaline synthesis. Tryptophan is a precursor of serotonin and melatonin. The potent vasodilator nitric oxide (NO) is produced from arginine. Glycine, aspartate, glutamine, and serine contribute atoms to purine and pyrimidine synthesis. Glycine and arginine are precursors for creatine synthesis.

Creatine synthesis and many other biochemical processes rely on a series of single-carbon transfer reactions mediated by serine, glycine, histidine, and methionine. Transfer of fully oxidized carbonyl carbon (=C=O) to molecules such as propionyl CoA (to form methylmalonyl CoA) is mediated by biotin. Glycine, serine, and histidine contribute less oxidized carbon units such as methylidine (=CH−), and methylene (−CH$_2$−) groups that enable purine and pyrimidine synthesis via folate derivatives. Folate also mediates transfer of methyl (−CH$_3$) groups to homocysteine to form methionine. The resulting methionine is, in turn, activated to S-adenosylmethionine becoming a methyl donor to a vast array of substrates including DNA, RNA, histones, choline, and catecholamines. The importance of folate and single-carbon metabolism to cell growth and division cannot be overstated. Folate deficiency in a developing embryo can lead to death or severe neurologic birth defects. The use of folate antimetabolites such as methotrexate has been a mainstay in the treatment of cancer for many decades.

Amino Acid Concentrations

Plasma amino acid concentrations collectively span a 4-order-of-magnitude dynamic range from very low micromolar quantities (eg, β-alanine, cystathionine) to near 1 mmol/L (eg, glutamine, glycine). With protein intake of 1 to 2 g/kg, daily variation of approximately 30% in healthy adults has been observed.[22] Concentrations of both essential and nonessential amino acids vary in a coordinated way, suggesting that mechanisms beyond diet and enteral extraction are responsible. Amino acid concentrations tend to peak between 12 and 8 PM with a nadir between midnight and 4 AM.[23,24] After an ingested protein bolus, dietary amino acids rise and tend to return to preprandial levels in 3 to 6 hours. Determination of "fasting" amino acid concentrations, therefore, requires extended periods of dietary abstinence.

Most amino acids in blood undergo glomerular filtration but are efficiently reabsorbed in proximal renal tubules by previously described saturable transport systems. Increased renal excretion of amino acids (aminoaciduria) results from filtration of excessive plasma concentrations, generalized tubular impairment, or heritable defects in amino acid transport systems. Glycine tends to be most abundant in normal urine followed by histidine, glutamine, and serine. Increased concentrations of proteogenic amino acids in plasma tend to precipitate only mildly elevated excretion because of efficient reabsorption. Other amino acids that accumulate in plasma secondary to metabolic errors (eg, argininosuccinate, homocitrulline) demonstrate pronounced excretion because of the absence of specific tubular mechanisms enabling reclamation from the filtrate.

With the exception of glutamine, CSF amino acid concentrations are typically less than 10% of those found in plasma. This high plasma to CSF gradient suggests active net brain to blood transport across the blood–brain barrier.[25,26] Glutamine concentrations in CSF are generally equal to those in plasma, suggesting a bidirectional facilitative transport process. Insofar as CSF concentrations reflect synaptic concentrations, regulation of neurotransmitter amino acid concentrations is critical for normal neural action potential propagation. Glutamate is the most abundant amino acid in the brain and is the primary excitatory transmitter. Glycine and GABA are the predominant inhibitory transmitters. Lumbar puncture to access the CSF amino acid pool must be done with great care to avoid overestimation of central amino acid concentrations secondary to contamination with peripheral blood.

Assessment of amino acid concentrations in blood, urine, and spinal fluid has been historically applied to the detection of inborn errors of metabolism. Aside from the measurement of homocysteine as a marker of vitamin B$_{12}$ and folate status, clinical applications of amino acid measurement are promising but presently limited. Future applications may include the assessment of immune status using tryptophan and its metabolites such as kynurenine.[27] Increased plasma concentrations of α-aminobutyric acid may be useful in detecting liver regeneration.[28] Branched-chain amino acid concentrations may be early indicators of diabetes,[29] while a combination of phenylalanine, glutamate, and alanine has some value to predict the onset of preeclampsia.[30] Finally, quantitation of arginine and its dimethylated derivatives (asymmetric and symmetric dimethylarginine) may have utility in assessing endothelial function.[31,32] These applications require further clinical validation.

Analysis of Amino Acids

For decades, the standard method of amino acid analysis was cation-exchange chromatography with postcolumn spectrophotometric or fluorescent detection of various primary amine derivatives. Derivatizing agents have included dansyl chloride, o-phthalaldehyde, and ninhydrin. The ninhydrin approach developed by Stein and Moore in the 1950s was initially applied to determination of amino acid content of protein hydrolysates and then adapted for profiling of free amino acids in deproteinized body fluids.[33] Other systems commercialized by Beckman (Brea, California) and Hitachi (Tokyo, Japan) were large floor models and required as long as 2 to 3 hours to quantitate 30 to 50 physiologic amino acids in a single patient specimen. These systems have given way to smaller bench-top systems using ninhydrin (Biochrom; Cambridge, United Kingdom) or fluorescent (quinolyl-N-hydroxysuccinimidyl) amine derivatives (Waters; Milford, Massachusetts) that still require 90 to 120 minutes for full sample analysis. In addition to long cycle times, these methods are subject to interference from co-eluting amines, leading to overestimation of some amino acid concentrations. Common co-eluting compounds include methionine with homocitrulline, phenylalanine with aminoglycosides, and histidine with gabapentin.

Mass spectrometry is increasingly being adopted as the method of choice for amino acid profiling. Newborn screening programs quantitate amino acid butyl esters derived from dried blood spots using flow-injection MS protocols. These methods do not use chromatographic separation and so cannot distinguish between isomeric or isobaric amino acids such as leucine, isoleucine, alloisoleucine, and hydroxyproline. Liquid chromatography–tandem mass spectrometry (LC-MS/MS) methods for detection of amino acids in plasma and other body fluids that use liquid chromatography have also been developed.[34-36] Some of these use amine-targeted derivatives, others target the carboxyl group, and some others use no chemical derivatization. Advantages of MS-based techniques include improved analytic specificity, a 3- to 4-order-of-magnitude dynamic range, and rapid (20-minute) throughput. MS methods may also be optimized for profiling multiple molecular species in addition to amino acids. Such approaches promise to improve the scope of metabolic disorders detectable with a single patient specimen in a single analytic run.

PEPTIDES

This section describes the basic biochemistry of peptides. In general, the term *peptide* applies to relatively short polymers of amino acids with molecular weights less than 6,000 Da ($<\sim 50$ amino acid residues). The chemistry of the peptide bond and the physical characteristics of the peptide backbone are discussed in this section along with a number of clinically relevant peptide systems.

Peptide Bond

A peptide bond, also referred to as an amide bond, is formed between the α-nitrogen atom of one amino acid and the carbonyl carbon of a second (diagrammed below).

So-called isopeptide bonds refer to amide bonds between sidechain amines or carbonyl carbons on the side chain rather than α-amine or α-carbonyl. In glutathione, for example, the γ-carboxyl group of glutamic acid is linked to the α-amino group of cysteine. During translation, peptide bonds are formed from the amino (N) to the carboxyl (C) terminus by removal of water (also referred to as dehydration or condensation) and catalyzed by RNA (referred to as a ribozyme) that forms part of the ribosome.[37] Peptides are also synthesized in vitro for therapeutic and experimental purposes. Such chemical peptide synthesis proceeds from C to N terminus using N-protected amino acids and catalyzed by N,N'-dicylohexylcarbodiimide.[38,39] In this scheme, the nucleophilic amine group reacts with a carbodiimide : carbonyl intermediate, resulting in the formation of a new peptide bond and dicyclohexylurea. Dicyclohexylurea is insoluble in most solvents and can be easily removed from the maturing peptide. Cleavage of peptide bonds may be nonspecifically achieved by acid hydrolysis or accomplished specifically by a host of proteolytic enzymes with affinity for bonds between specific amino acid residues. These protease systems are described later in this chapter.

Electron sharing in the amide bond (also known as the ω bond) is delocalized effectively preventing rotation about this bond. This bond is fixed in one plane. Conformational flexibility of the peptide backbone results entirely from rotation about the axes of the two bonds to the α-carbon. Angles of rotation about these bonds are referred to as Ramachandran angles.[40] The nitrogen to α-carbon bond angle is referred to as the Φ angle, and the α-carbon to carbonyl bond is referred to as the Ψ angle (below).

Theoretically, free rotation about these bonds allows angles ranging from -180 to 180 degrees. In reality, steric and energetic factors limit the possible combinations. These bond angles play a key role in dictating the secondary structure of proteins. For example, values of Φ and Ψ in α-helices are approximately −60 and −45 degrees, respectively. Secondary structure is addressed more extensively later in this chapter.

Peptide Heterogeneity and Analysis

Assessment of circulating peptide concentrations has a number of limitations. In the absence of enzymatic activity, peptide measurements have been historically limited to immunologic techniques. Antibodies may recognize linear sequence epitopes or discontinuous, conformational epitopes. These epitopes typically involve 10 to 20 amino acids binding exposed areas of 600 to 1000 Å.[41] Measurement of short peptides (<20–30 amino acids) are therefore limited to single-site, competitive assays that lack the analytic specificity of two-site (sandwich) immunoassays. The molecular specificity issues may be addressed using MS as an alternative. Small peptides may be ionized via electrospray (ESI) or matrix-assisted laser desorption (MALDI) and interfaced to tandem quadrupole mass analyzers.

Absolute analytic specificity is not always ideal when applied to biologic peptide systems. Peptide populations may consist of species with a variable number of amino acid residues with sometimes unknown biologic potency. Hepcidin, for example, is an iron transport regulatory peptide that circulates principally as a 25–amino acid peptide but also as shorter peptides of 22 and 20 amino acids with diminished biologic activity. Likewise, dozens of truncated forms of the mature 32–amino acid B-type natriuretic peptide ranging from 24 to 31 amino acids are detectable in heart failure patients. Some of these truncated forms are present in vivo, and others likely develop in vitro. Thus, narrowly targeted MS assays may exclude active peptide species and run the risk of underestimating bioactive peptide. Cross-reactive immunoassays, on the other hand, may stoichiometrically detect both active and inactive peptide, thus running the risk of overestimating bioactive peptide concentrations. Examples of several important biologic peptide systems and their analytic considerations follow.

Clinically Relevant Peptide Systems
Pro-Opiomelanocortin System

The pro-opiomelanocortin (POMC) gene on chromosome 2 is expressed primarily in the pituitary gland, arcuate nucleus of the hypothalamus, and skin melanocytes. The gene produces a 241–amino acid prohormone that can yield as many as 10 distinct biologically active peptides depending on patterns of cleavage in specific tissue types. The POMC peptides have diverse effects on glucose and electrolyte homeostasis (via adrenocorticotropic hormone [ACTH]), body mass and appetite (via lipotropins and melanocortins), pigmentation (also via melanocortins), and pain (via endorphins).[42-44]

Clinical exploitation of this complex peptide system is currently limited to the impact of ACTH on the adrenal gland and subsequent feedback by cortisol. Measurement of circulating ACTH concentrations may be used to clarify the mechanism of adrenal disease. For example, Cushing's disease may result from autonomous adrenal function or may be fueled by ectopic ACTH production. Likewise, Addison's disease may result from adrenal or pituitary failure. Full-length ACTH consists of 39 amino acid residues and circulates with a half-life ranging from 10 to 15 minutes. The biologic activity of ACTH is contained in residues 1 to18, and the length of the C-terminus mediates circulating half-life. Two-site immunoassays typically target the extreme N and C termini to avoid detection of shorter, inactive circulating species. This approach, however, can lead to cross-reactivity with longer precursor forms of ACTH such as pro-ACTH and the full POMC gene product.[45] These precursors typically circulate at concentrations that are five times greater (5–50 pmol/L) than ACTH (1–10 pmol/L). ACTH assays are poorly standardized.

Natriuretic Peptides

This peptide family consists of atrial natriuretic peptide (ANP), B-type natriuretic peptide (BNP, formerly brain natriuretic peptide), and C-type natriuretic peptide (CNP).[46] ANP and BNP are highly expressed in cardiac tissue relative to CNP, which is expressed at low levels in a broad variety of tissue types. The mature forms of these peptides contain a 17–amino acid loop stabilized by an intramolecular disulfide bond. Each peptide acts via a specific guanylate cyclase-coupled receptor to promote sodium and water excretion, blunt activation of the renin–angiotensin system, and decrease vascular resistance. Circulating concentrations of ANP and BNP (but not CNP) increase rapidly in response to increased cardiac filling pressures that are characteristic of heart failure. Clinical measurement of BNP has become a widely used tool to detect heart failure and monitor its progression.

B-type natriuretic peptide is synthesized as a 108–amino acid precursor (pro-BNP) that is cleaved upon cellular release to the active 32–amino acid peptide and an N-terminal fragment (NT-proBNP), which lacks biologic activity.[47] Two-site immunoassays for both BNP and NT-proBNP are commercially available and widely used. NT-proBNP circulates at higher concentration than BNP by virtue of its longer biologic half-life (~60 minutes vs. ~20 minutes). Recent evidence suggests that the BNP detected immunologically in heart failure is not the bioactive form of BNP. Using MS, almost no mature 32–amino acid peptide was detected in heart failure patients despite the significant presence of immunoreactive BNP.[48,49] Further studies suggest that the immunoreactive BNP in the plasma of heart failure patients may be attributed to higher molecular weight forms such as proBNP that exhibit a fraction of the bioactivity of the mature peptide.[50,51] This new analytic information may prompt a reevaluation of the role of BNP in the pathophysiology, diagnosis, and treatment of heart failure.

Hepcidin

Hepcidin was initially described as an antimicrobial peptide.[52] The role of hepcidin in iron metabolism was noted by Nicolas et al. in 2001.[53] The mature 25–amino acid molecule is derived from a 60–amino acid precursor expressed from 3 exons of the HAMP gene on chromosome 19. The tightly looped structure of hepcidin is stabilized by four intramolecular disulfide bonds. The biologic activity of hepcidin is mediated via its interaction with ferroportin, the transport protein that mediates iron transport from duodenal enterocytes and macrophages. Hepcidin binding promotes the internalization and degradation of ferroportin, thus inhibiting mobilization of iron stores.[54,55] Physiologic states such as chronic inflammation are characterized by microcytic anemia with paradoxically adequate iron stores known as anemia of chronic disease. Increased hepcidin expression is at the pathologic root of this condition. In addition to its diagnostic role in differentiating iron deficiency anemia from the anemia of chronic disease, hepcidin measurement may aid in the treatment of hemochromatosis, transfusion-associated iron overload, and anemia associated with chronic renal failure.[56]

Despite the important role of hepcidin in iron metabolism, its clinical use remains infrequent partly because of difficulties associated with its measurement. Antibodies toward hepcidin have been difficult to develop because of its small, compact size and because it is highly conserved across multiple species. Immunoassays have largely been limited to single-site, competitive formats that cross-react significantly with shorter versions of the molecule (22-, 20-mers).[57] This can lead to overestimation of circulating hepcidin compared with MS techniques[58] when applied to patients with renal failure in whom shorter hepcidin peptides tend to accumulate in the plasma.[59] Improvements in the molecular specificity and harmonization of hepcidin determination will further clarify the role of hepcidin in both normal and pathologic physiology and also promise to enhance its clinical utility.

Angiotensins

Renin is secreted by the afferent arterioles of the kidney in response to decreased flow, pressure, and sodium delivery to the renal juxtaglomerular apparatus.[60] Renin acts on circulating angiotensinogen to initiate the formation of vasoactive peptides that act to reestablish glomerular flow. The N-terminal decapeptide cleaved from the 452–amino acid angiotensinogen molecule by renin is referred to as angiotensin I. Angiotensin-converting enzyme (ACE) cleaves 2 C-terminal residues from angiotensin I to form the octapeptide, angiotensin II, which promotes contraction of vascular smooth muscle and stimulates proximal tubular sodium reabsorption to increase blood pressure. ACE inhibitors (eg, captopril, enalapril, quinapril) are important pharmacologic tools used to treat hypertension.

Although the renin–angiotensin–aldosterone axis is an important therapeutic target, it is a far more infrequent target for diagnostic laboratory studies. Plasma renin activity is assessed to explore the possibility of renovascular hypertension. In this condition, unilateral restriction of renal blood flow results in inappropriate release of renin and severe hypertension. Renin activity in the plasma is normally very low (<10 ng/mL/h) and is typically measured by assessing the production of angiotensin I from endogenous angiotensinogen after a long (>12 hours) incubation period. Angiotensin I generation is most commonly monitored via a competitive immunoassay with the potential to crossreact with shorter peptides. MS approaches that address peptide specificity and stability may mitigate these analytic limitations.[61,62]

Endothelins

Endothelins (ETs) are peptides with 21 amino acids derived from the vascular endothelium (ET-1), intestinal and renal tissue (ET-2), and neural tissue (ET-3).[63] ET-1 is produced from a 203–amino acid precursor (preproendothelin) and smaller, 30– to 40–amino acid "big" ET molecules that are inactive. ET-1 is a potent vasoconstrictor and may mediate pathology associated with diabetic nephropathy and hypertension.[64] Increased circulating concentrations of ET after myocardial infarction suggest a negative survival prognosis.[65] The reliability of these observations using currently available immunoassays and other potential clinical applications for ET measurement remain unclear.

Vasopressin

Vasopressin (arginine vasopressin [AVP]), also known as antidiuretic hormone (ADH), is a nonapeptide stored in and secreted from the posterior pituitary gland. Its primary target organ is the distal convoluted tubule and collecting duct, where it acts to promote water reabsorption. ADH circulates at very low concentrations (<40 pmol/L) and has a very short half-life (15—20 minutes), making routine diagnostic measurement impractical. Diabetes insipidus (DI) may result from faulty secretion (central DI) or from end-organ resistance (nephrogenic DI). Head injury, tumors, and some medications may also induce pathologic secretion of ADH, resulting in fluid overload referred to as the syndrome of inappropriate antidiuretic hormone secretion (SIADH). A synthetic analog referred to as DDAVP (1-desamino, 8-D-arginine vasopressin) is used therapeutically to treat DI and some forms of coagulopathy. DDAVP stimulates release of von Willebrand factor from endothelial cells and extends the half-life of circulating factor VIII, thereby mediating improvements of circulating hemostatic factors associated with various forms of von Willebrand disease and hemophilia A.

Glutathione

Glutathione consists of a glutamate residue linked to cysteine via its γ-carboxyl rather than the α-carboxyl group and followed by a conventional peptide bond between cysteine and glycine.[66] This ubiquitous tripeptide is the most abundant intracellular thiol (1—10 mmol/L) and circulates in the blood at micromolar concentrations. The cellular ratio of reduced glutathione (GSH) to oxidized glutathione (GSSG) ranges from 10 to 100. Intracellular glutathione performs a variety of important functions. It plays an important role in maintaining the proper ratio of oxidized to reduced forms of metabolically important thiols such as coenzyme A. It also provides reducing equivalents that detoxify reactive oxygen species such as peroxides (catalyzed by glutathione peroxidase). Through the activity of glutathione-S-transferase, glutathione also serves to detoxify other xenobiotic compounds via formation of a thioether derivative, which can then be excreted. Amines and peptides are transported across the plasma membrane via the γ-glutamyl moiety of glutathione, a reaction catalyzed by γ-glutamyl-transpeptidase. The tripeptide is then regenerated through the concerted action enzymes in the so-called γ-glutamyl cycle (Fig. 13.3).

Determination of circulating GSH and GSSG is not routinely called for in clinical practice as the site of action is intracellular. Nonetheless, a variety of techniques to measure glutathione have been used.[67] Measurement of total glutathione requires prior reduction of the sample to release all oxidized forms. The simplest techniques employ the colorimetric detection of free thiol using 5,5'-dithio-bis-2-nitrobenzoic acid (DTNB). Reaction of DTNB with thiols results in the formation of 2-thio-5-thiobenzoate, which absorbs with high extinction ($\sim$14,000 L cm^{-1} mol^{-1}) at 410 nm. Other techniques use derivatization and stabilization of GSH followed by high-performance liquid

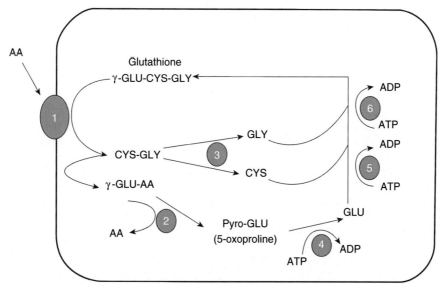

FIGURE 13.3 Transmembrane transport of amino acids (AAs) using the γ-glutamyl cycle. Three-letter AA abbreviations are used. Extracellular AA is transferred to glutathione via activity of membrane-bound γ-glutamyl transpeptidase *(1)*. AA is released in the cytoplasm via the activity of γ-glutamyl cyclotransferase *(2)*, which also results in the formation of pyroglutamate (5-oxoproline). Cysteine and glycine generated via dipeptidase activity *(3)* are recycled with pyroglutamate to reform glutathione via successive activities of 5-oxoprolinase *(4)*, γ-glutamyl-cysteine synthetase *(5)*, and glutathione synthetase *(6)*. *ADP*, Adenosine diphosphate; *ATP*, adenosine triphosphate; *CYS*, cysteine; *GLU*, glutamate; *GLY*, glycine.

chromatography or MS. Inborn errors of glutathione metabolism such as glutathione synthetase deficiency are detected via the accumulation of components of the γ-glutamyl cycle such as pyroglutamic acid (5-oxoproline).B978-0-12-805371-3.16001-7

PROTEINS

The structural diversity of proteins may be described using the following features:

1. Primary structure is the linear sequence of amino acids in a peptide or protein. Posttranslational modifications of amino acids contribute to increased diversity.
2. Secondary structure describes the nature of the peptide backbone dictated by the peptide bond angles described earlier and stabilized by hydrogen bonds. Examples of secondary structure include α-helix, β-sheet, and β-turn. An α-helix has about 3.6 residues per turn and is stabilized by hydrogen bonds between the N—H and C=O group of the fourth following amino acid. A β-sheet involves hydrogen bonds between the peptide bonds of adjacent peptide chains arranged in parallel or antiparallel configurations. Random coils refer to segments of peptide that lack defined secondary structure.
3. Tertiary structure refers to the folding of the polypeptide chain and elements of secondary structure into a compact three-dimensional (3D) shape. Folding is a complex process driven by energy minimization of intramolecular and solvent interactions. Hydrophobic groups tend to fold into the interior with less exposure to solvent, while charged and polar sidechains tend to be located on the surface. The 3D structure is stabilized by intramolecular hydrogen bonds, van der Waals forces, and hydrophobic interactions. Disulfide bonds between cysteine residues also stabilize 3D structure. Denaturation of protein refers to unfolding that occurs with temperature change or in the presence of organic solvents, detergents, or reagents that disrupt hydrogen bonds. Limited denaturation can be reversible, but extensive unfolding and denaturation of proteins often lead to irreversible aggregation and precipitation.
4. Quaternary structure refers to the incorporation of two or more polypeptide chains or subunits into a larger multimeric unit. Examples range from the relatively simple creatine kinase, a heterodimer of M and B subunits, to branched chain α-ketoacid dehydrogenase, which is a heteromeric complex of 12 E1, 24 E2, and 6 E3 subunits.
5. Ligands and prosthetic groups provide additional functional and structural elements, such as metals in metalloenzymes, heme in hemoglobin and cytochromes, and lipids in lipoproteins. Proteins without their associated ligands are often referred to as apoproteins (eg, apotransferrin without iron, apolipoproteins without lipid).

Physical Properties of Proteins

The diverse structural features of proteins result in unique physical properties that can be exploited for analysis. For example, tyrosine and tryptophan residues absorb light at 280 nm, and the abundance of these amino acids determines the extinction coefficient of a peptide or protein. A pure protein, therefore, may be quantitated using A_{280}. Some prosthetic groups such as heme also possess intrinsic absorbance that may be monitored to assess the presence of specific proteins. Automated clinical analyzers assess the presence of hemoglobin at 540—570 nm, for example, to detect hemolyzed plasma or serum specimens. Ionizable groups exert a strong effect on physical properties depending on the pH of the surrounding solution. Differing physical

properties serve as the basis of methods to separate proteins. Some important characteristics include the following:
1. *Differential solubility.* The solubility of proteins is affected by pH, ionic strength, temperature, and the characteristics of the solvent. Changing solvent pH affects the net charge of a protein. Changing ionic strength affects the hydration and solubility of proteins. "Salting-in" and "salting-out" procedures were early methods for separating and characterizing protein. Albumin, for example, stays in solution at high concentrations of ammonium sulfate that precipitate globulins. Addition of organic solvents and polyethylene glycol is also useful for differential precipitation. Fractional precipitation of plasma with ethanol, using protocols developed by Cohn and coworkers,[67a] enables isolation of plasma fractions that are enriched in immunoglobulins, α- and β-globulins, or albumin (fraction V). Polyethylene glycols induce precipitation by steric exclusion and therefore preferentially precipitate large proteins or complexes.
2. *Molecular size.* Separation of small and large molecules is commonly achieved by differential migration through molecular filters. Examples are size exclusion chromatography (also known as gel filtration), ultracentrifugation, and electrophoresis. These techniques may be used under conditions when proteins and peptides are in native globular states or under denaturing conditions. Addition of reducing agents allows separation of disulfide-linked components.
3. *Molecular mass.* Advances in MS allow the determination of masses of peptides and proteins with increasing accuracy. Peptides and proteins can be ionized by MALDI or by ESI.
4. *Electrical charge.* Ion-exchange chromatography, isoelectric focusing, and electrophoresis separate peptides and proteins based on charge.
5. *Surface adsorption.* The affinity of peptides and proteins for a variety of physical surfaces may also be used as the basis for separation. Reverse-phase chromatography, for example, exploits the interaction of hydrophobic molecular moieties with hydrophobic surfaces (C8 or C18 alkyl chains) when the ratio of water to organic solvent is high but not when organic content is increased.
6. *Affinity chromatography.* Specific ligands, antibodies, and other recognition molecules have been used to separate peptides or proteins selectively.

Protein Formation
Folding
Proteins are synthesized by ribosomes reading from the 5′-end of mRNA. Triplet codons in mRNA are matched with complementary sequence in transfer RNA carrying specific amino acids. Protein synthesis begins with an AUG codon encoding methionine, and the polypeptide chain is synthesized from the N-terminus. During translation, the initiator methionine is typically cleaved and the resulting N-terminal residue commonly acetylated. Although 80% to 90% of proteins carry an N-terminal acetyl group, the function of this modification is not entirely clear, but it may play a role in stabilizing the growing peptide chain.

Instructions for folding are largely contained in the primary amino acid sequence of the growing polypeptide chain. The rate of elongation (typically 5–10 amino acids per second in eukaryotes) may have a significant impact on folding. The use of rare codons, secondary structural elements of the mRNA, and polybasic stretches may dictate pauses in translation and enhance formation of secondary structural elements.[68,69] Folding begins as the chain is elongated and still associated with the ribosome, assisted by a family of proteins referred to as chaperones.[70] The function of chaperones was originally ascribed to a group of proteins called "heat shock proteins" that prevented protein denaturation and aggregation in response to heat and other extreme environmental conditions.

Many gene products that share common 3D features have arisen from common ancestral genes. The serpin (serine proteinase inhibitor) superfamily consists of more than 1000 related proteins in different organisms.[71] Humans have 36 serpins, 29 of which are protease inhibitors and 7 of which lack protease inhibitor function. Serpins that act as protease inhibitors in plasma include $α_1$-antitrypsin (AAT), $α_1$-antichymotrypsin, $α_2$-antiplasmin, antithrombin, C1 inhibitor, heparin cofactor II, protein C inhibitor, and plasminogen activator inhibitor-1 (PAI-1). Serpins without known protease inhibitor function are cortisol-binding globulin, thyroxine-binding globulin, angiotensinogen, intracellular proteins, heat shock protein 47, and the tumor suppressor maspin. Serpins illustrate how a common structure motif may be adapted to multiple functions. Other examples of plasma protein families are the albumin and lipocalin families. The albumin family includes albumin, α-fetoprotein, and afamin. The lipocalin family includes several plasma proteins such as $α_1$-acid glycoprotein, retinol-binding protein, apolipoprotein D, $α_1$-microglobulin, prostaglandin D synthase (β-trace), β-lactoglobulin, neutrophil gelatinase-associated lipocalin (NGAL), inter-α-trypsin inhibitor, and C8 γ-chain. Lipocalins generally have a barrel-shaped structure that is well suited to serve as a carrier for small molecules.

Protein folding is an error-prone process, and many molecular chaperones work to refold, prevent aggregation, or degrade misfolded proteins.[72] Several heat shock proteins that increase in response to a variety of stresses are molecular chaperones. Increased accumulation of misfolded proteins induces an adaptive mechanism—the unfolded protein response.[73] This response increases production of chaperones and slows general protein synthesis to allow more time to fold new proteins.

Despite these protective mechanisms, several families of age-related, genetic, and infectious diseases appear connected to disorders of protein folding and protein aggregation. Prion diseases are infectious diseases in which the transmissible protein agent may catalyze misfolding of endogenous proteins. In Alzheimer's disease, deposits of amyloid may contribute to pathogenesis. Polyglutamine diseases result from genetic expansion of repeat units encoding glutamine and are associated with Huntington's disease and other neurodegenerative disorders.[74] These expanded polyglutamine sequences tend to aggregate as β-sheets. TDP-43 proteinopathies include amyotrophic lateral sclerosis (Lou Gehrig's disease), resulting from aggregation of transactive DNA-binding proteins.[75] Several inherited disorders related to mutations in specific proteins probably result from problems in protein folding. In AAT deficiency, hepatic injury results from aggregation and accumulation of misfolded protein.[76,77] The most common cause of cystic fibrosis results from a single amino acid deletion (ΔF508), which results in rapid degradation of the cystic fibrosis transmembrane conductance regulator (CFTR). Accumulation of misfolded proteins has been suggested as a pathogenic mechanism contributing to vascular, cardiac, and β-cell failure in diabetes.[73] Small molecule therapeutics capable of modulating protein folding have shown some promise in mitigating disease caused by abnormal protein aggregation.[78]

Targeting

As originally outlined by Lingappa and Blobel,[79] proteins that are secreted, located in vesicular compartments, or oriented on the external surface of cell membranes usually contain a hydrophobic N-terminal signal peptide about 15 to 30 amino acids in length. Signal peptides interact with signal recognition particles (SRPs) and mediate interaction with the endoplasmic reticulum (ER). Nascent peptide chains are inserted through the membrane of the ER as the protein is synthesized. Signal peptides of most secretory proteins are removed even before synthesis of the entire protein chain is completed. Co-translational membrane retention may be achieved via an uncleaved signal sequence, by one or more hydrophobic transmembrane domains, or by lipid modifications such as N-myristoylation.[80]

Newly synthesized proteins ultimately reside in a number of membranous or soluble compartments, including the nucleus, lysosome, peroxisome, mitochondrion, or plasma membrane. Plasma membrane sorting is further complicated in polarized epithelial cells where proteins may be targeted to basolateral or apical environments. In the so-called "secretory" pathway, proteins are shuttled via membrane-bound vesicles bearing COP II (coat protein II) from ER through the Golgi apparatus.[81,82] Intra-Golgi transport and retrograde Golgi to ER transport is mediated by COP I vesicles. Upon fusion of vesicle with specific membranes, soluble components are extruded, and lipid-associated components take up residence as stable membrane components. Sorting in polarized epithelia is mediated by association of proteins with unique membrane domains. For example, proteins anchored to the membrane via a glycosylphosphatidylinositol (GPI) anchor tend to cluster in cholesterol- and sphingolipid-rich domains called lipid rafts that are selectively sorted to apical surfaces.[83] Proteins destined for mitochondria contain a unique N-terminal targeting sequence that mediates their interaction and import into the proper submitochondrial location (eg, outer membrane, inner membrane, matrix, intermembrane space).[84]

Posttranslational Modifications
Fatty Acylation

The activity and localization of a variety of proteins may be modulated by covalent attachment of fatty acyl chains. Co-translational attachment of myristate via a glycine residue has been previously mentioned as a mechanism for membrane association. The most common acylation of eukaryotic proteins involves thioester linkage of palmitate to membrane proximal cysteine residues.[85] S-palmitoylation reversibly controls localization to membrane microdomains such as lipid rafts and thus regulates interaction of proteins with signaling and other effector molecules. Examples of palmitoylated proteins include caveolin-1, some members of the SRC protein kinase family, nitric oxide synthase (NOS), β-adrenergic receptor, and transferrin receptor. Ghrelin, a potent growth hormone secretagogue, is modified by covalent attachment of an octanoyl moiety at serine 3 of the polypeptide.[86,87] Only octanoylated ghrelin promotes growth hormone release.

Phosphorylation

Reversible phosphorylation may impact as many as one third of all human cellular proteins.[88] O-phosphorylation occurs at serine, threonine, and tyrosine residues. The human genome encodes approximately 1000 kinases, enzymes responsible for phosphorylation, and about 500 phosphatases responsible for removal of covalent phosphate groups. Detailed treatment of reversible phosphorylation is beyond the scope of this chapter but, in general, serine and threonine phosphorylation acutely modifies enzyme activity (eg, glycogen phosphorylase)[89] and subcellular localization (eg, cAMP-dependent protein kinase).[90] Tyrosine phosphorylation, on the other hand, regulates a plethora of signaling pathways (eg, mitogen activated protein kinase, Janus kinase pathways) in part by providing docking site proteins that propagate a transmembrane signal such as those of the SRC kinase family (lyn, lck, fyn).[91] Mitochondria contain members of a primitive kinase family that modulate flux through the pyruvate dehydrogenase and branched chain α-ketoacid dehydrogenase complex via a unique phosphohistidine intermediate.[92-94]

Prenylation

Isoprenoid compounds such as farnesyl pyrophosphate (15 carbons) and geranylgeranyl pyrophosphate (20 carbons) are hydrophobic moieties formed from 3-hydroxy-3-methylglutaryl CoA and mevalonate via HMG CoA reductase. These groups modify more than 300 members of the human proteome via enzymatic attachment to a cysteine residue in a so-called "CaaX" motif where C is cysteine, a is an aliphatic amino acid such as glycine or alanine, and X is typically serine, methionine, glutamine, alanine, or threonine.[95,96] Isoprenylation regulates membrane and molecular association of a number of proteins important for signal transduction (H-Ras, K-Ras), vesicular trafficking (Rab2, Rab3a), cytoskeletal function (RhoA, RhoB), and the integrity of the nuclear membrane (lamin A).[97]

Glycosylphosphatidylinositol Anchor

The GPI anchor is a glycoglycero-phospholipid construct that mediates membrane attachment for a variety of proteins. The anchor is presynthesized in the ER and then transferred to the target protein via a C-terminal hydrophobic signal sequence. After modification, this hydrophobic sequence is removed, leaving a protein that is uniquely membrane anchored via interdigitation of two fatty acyl chains with a single membrane leaflet.[98,99] The purpose of the GPI anchor remains unclear, although it has been proposed that such lipid-anchored proteins diffuse in the lateral plane of biologic membranes more rapidly than transmembrane proteins.[100] GPI-anchored proteins also uniquely associate with cholesterol- and glycosphingolipid-rich plasma membrane domains referred to as lipid rafts and caveolae.[83] Notable examples of GPI-anchored proteins include decay accelerating factor (DAF, CD55), membrane inhibitor of reactive lysis (MIRL,

CD59), alkaline phosphatase, 5'-nucleotidase, and glypican family members. Defects in the PIG-A gene product that mediates GPI synthesis are responsible for paroxysmal nocturnal hemoglobinuria (PNH). PNH is characterized by abnormal complement-mediated lysis of erythrocytes deficient in CD55 and CD59.

γ-Carboxylation

Glutamic acid is a five-carbon α-amino dicarboxylic acid. Vitamin K acts as a cofactor in the enzymatic addition of a second carboxyl group to the γ-carbon.[101] A cluster of γ-carboxylated glutamyl residues is referred to as a "gla" domain. This modification mediates calcium-dependent membrane association and is required for full functional activity. Many proteins of the coagulation cascade contain the gla domain, including thrombin, factors VII, IX, and X, and protein C, S, and Z. Warfarin exerts its anticoagulant effect by abrogating the activities of the vitamin K–dependent factors.

Glycosylation

Secreted proteins and the extracellular domains of transmembrane proteins are commonly modified with carbohydrate. Carbohydrate plays an important role in folding, secretion, and stability of the modified polypeptide. Sugar chains are added in the ER and Golgi apparatus via O-linkage to serine and threonine residues or N-linkage to the amide nitrogen of asparagine residues.[102] O-linked sugar modifications are typically simple and consist of one to four residues. The sugars are transferred to the nascent polypeptide via the energy of the phosphoester bonds of nucleotide sugars such as uridine diphosphate–galactose and guanosine diphosphate–mannose. N-linked sugars are far more complex. A core glycan consisting of 2 N-acetylglucosamine, 9 mannose, and 2 glucose moieties is pre-assembled on a membrane embedded isoprenoid molecule, dolichol, and then transferred en bloc to the newly synthesized peptide chain. The core glycan structure may then be lengthened and trimmed by a host of enzymatic processes. Approximately 200 human gene products are involved in glycosylation, and more than 100 genetic defects in this process have now been documented.[103]

Sulfation

Sulfation of proteins on tyrosine residues was originally described in 1954.[104,105] Nearly 50 secreted and transmembrane proteins carry this irreversible modification that occurs in the Golgi apparatus. Examples of sulfated proteins include coagulation factors V, VIII, and IX, fibrinogen, thyroglobulin, and α-fetoprotein. The sulfation reaction is catalyzed by two widely expressed isoforms of tyrosylprotein phosphotransferase (TPST) and uses 3'-phosphoadenosine-5'-phosphosulfate as the sulfonic acid donor.[106] The consensus peptide sulfation site is poorly understood aside from the fact that acidic residues surrounding the target tyrosine appear to promote this modification. The function of sulfation is likewise not well understood beyond its capacity to enhance the affinity of molecular recognition events. A naturally occurring Tyr → Phe at position 1680 of factor VIII, for example, prevents sulfation and weakens its interaction with von Willebrand factor, causing a mild form of hemophilia.[107]

Hydroxylation

Hydroxylation is the most prevalent posttranslational modification of human proteins. Hydroxylation most commonly occurs at the 4-position of the proline ring. Approximately 30% of proline residues in collagen are modified in this way. Hydroxyproline residues are, therefore, more common in proteins than many other common amino acid residues, including cysteine, histidine, methionine, phenylalanine, tryptophan, and tyrosine.[108] Hydroxylation of proline residues is thought to alter the flexibility ("pucker") of the pyrrolidine ring, thereby stabilizing the triple helical structure of collagen[109] and other connective tissue proteins (eg, elastin).

Proline hydroxylation also plays a unique role in mediating the cellular response to oxygen through hypoxia inducible factor-1 (HIF-1). Under hypoxic conditions, HIF-1α/HIF-1β heterodimers induce expression of genes that mediate angiogenesis, erythropoiesis, vascular tone, citric acid cycle activity, and iron metabolism.[110] During normoxia, proline residues at position 402 or 564 are targets for enzymatic hydroxylation, which mediates ubiquitination by a protein complex containing von Hippel-Lindau (VHL) protein.[111] Ubiquitination of HIF-1α mediates its degradation via the proteasome (see section on protein catabolism) and subsequent blunting of the genetic response to hypoxia.

Nitrosylation

Nitric oxide is a volatile free radical produced from arginine via three human NOS enzyme systems: neuronal (nNOS), endothelial (eNOS), and an inducible form (iNOS). NO exerts its primary biologic effects (eg, vasodilation) via guanylate cyclase–coupled receptors, but it also forms covalent nitrosothiol (S—N=O) derivatives with free cysteine thiol groups.[112,113] Thousands of proteins with a diverse array of functions are reversibly modified in this way.[114] Although no rigorous determinants of nitrosylation sites have been defined, solvent exposed cysteine residues in alpha helices within 6 Å of charged residues seem to be preferred targets.[112] Nitrosylation may occur directly with the possible aid of metal catalysis via transfer of SNO groups from low molecular weight thiols such as glutathione, or through exchange mediated by disulfide reducing enzymes such as thioredoxin.[115,116] Effects of nitrosylation are pleiotropic. Whereas nitrosylation of the N-methyl-D-aspartate receptor in neurons blunts its activity,[117] nitrosylation of matrix metalloprotease-9 (MMP-9) stimulates its activity.[118] The role of protein nitrosylation in health and disease remains to be fully clarified.

Protein Catabolism

The steady-state concentration of any specific intra- or extracellular protein reflects not only its rate of synthesis but also its rate of degradation. The degradative process is much more than a passive, nonspecific mechanism for disposing of unwanted cellular material. It is highly specific and tightly controlled. As supporting evidence, consider that the human genome contains more than 500 proteases. These proteases belong to one of four families based on the catalytic mechanism for hydrolyzing peptide bonds. Representative members of the four protease families are detailed in Table 13.3.

TABLE 13.3 Selected Members of the Four Protease Families

Protease/Family	Gene Loci	Tissue Expression	Subcellular Localization	Substrate, Function, Pathology
Serine				
Corin	4p13	Broad	Plasma membrane	Natriuretic peptides (heart failure)
Trypsin	7q34	Pancreas	Secreted	Promiscuous (cleavage of LYS-X, ARG-X)
Chymotrypsin	16q23	Pancreas	Secreted	Promiscuous (cleavage of TRP-X, PHE-X, TYR-X)
Neutrophil elastase	19p13	Myeloid cells	Cytoplasm, secreted	Promiscuous (cleavage of VAL-X, ALA-X)
Factor IX	Xq27	Liver	Secreted, plasma membrane	Conversion of factor X to Xa (hemophilia B)
Activated protein C	2q14	Liver		Factor V_a, factor VIII_a (coagulopathy)
PSA (kallikrein)	19q13	Prostate	Secreted	Semen liquefaction; marker of prostate mass and cancer
C1s	12p13	Broad	Secreted	Complement C1r, C2, and C4 (angioedema)
Cysteinyl				
Caspase-3	4q34	Broad	Cytosol, nucleus, mitochondria	ASP-X-X-ASP (apoptosis)
Cathepsin C	8p23	Broad	Lysosome	Amyloid precursor (Alzheimer disease)
Aspartyl				
Renin	1q32	Ovary, broad	Secreted	Angiotensinogen (hypertension)
Pepsin	6p21	GI tract, lung	Secreted	Cleavage at adjacent hydrophobic residues (PHE-VAL, ALA-LEU, LEU-TYR)
Presenilin	14q24	Broad	Endoplasmic reticulum/Golgi	Amyloid precursor (Alzheimer disease)
Metallo				
ADAMTS13	9q34	Liver, erythroid precursors, broad	Secreted	von Willebrand factor (thrombotic thrombocytopenic purpura)
MMP-1	11q22	Muscle	Secreted	Collagen (tissue remodeling, embryogenesis, metastasis)
Angiotensin I converting enzyme	17q23	Testes, broad	Secreted	Angiotensin I (hypertension)

Protease Families

Serine Proteases. The serine proteases are the most abundant family in humans and are so named because serine serves as the nucleophilic residue at the active site of the enzyme.[119] Peptide bond hydrolysis is achieved via a conserved "catalytic triad" of spatially adjacent histidine, serine, and aspartate residues. These enzymes enable a vast array of physiologic processes, including protein digestion, complement activation, and blood coagulation. A family of endogenous serine protease inhibitors (serpins) inactivates a broad variety of protease enzymes via formation of a covalent complex with the active site serine. Members of the serpin superfamily with antiprotease activity include AAT, antithrombin, PAI-1, and protein C. Some members of the serpin family, including angiotensinogen and thyroxine-binding globulin, do not possess known activity against specific proteases. Serine protease activity in vitro can be mitigated with a number of inhibitors, including phenylmethanesulfonylfluoride (PMSF), [4-(2-aminoethyl)benzenesulfonyl fluoride] (AEBSF), aprotinin, and leupeptin.

Cysteinyl Proteases. This group of proteases uses a cysteine thiol in nucleophilic attack on the peptide bond.[120] A thioester intermediate with the carbonyl carbon of the peptide bond is formed before hydrolysis and formation of two peptides with new C and N termini. In humans, cysteine proteases mediate apoptosis via a series of enzymes referred to as caspases.[121] Other notable cysteine proteases include some in the cathepsin family and interleukin 1β converting enzyme (ICE). Cystatins are endogenous inhibitors of cysteine protease activities. Some chemical inhibitors of serine proteases such as PMSF and leupeptin also exhibit activity toward cysteine proteases. Unique in vitro inhibitors of cysteine proteases include L-trans-epoxysuccinyl-leucylamide-(4-guanido)-butane (E-64) and N-[N-(N-acetyl-L-leucyl)-L-leucyl]-L-norleucine (ALLN).

Aspartyl Proteases. In contrast to the serine and cysteinyl proteases, aspartyl proteases do not act via a covalent acyl enzyme intermediate. Instead, peptide bond lysis is achieved in a single step through the coordination of a water molecule between two highly conserved aspartate residues. One aspartate residue abstracts a proton from water, which then becomes a nucleophile, attacking the carbonyl carbon of the peptide bond.[122] Human aspartyl protease enzymes include members of the pepsin, cathepsin, and renin families. The target for HIV protease inhibitors (eg, indinavir, ritonavir) is also an aspartyl protease. Pepstatin is a potent hexa-peptide inhibitor of aspartyl protease activity.

Metalloproteases. Members of this protease family, commonly referred to as MMPs, include approximately 25 Zn^{2+} dependent enzymes sub-classified by their reactivity to collagen, gelatin, and other extracellular matrix proteins.[123-125] MMPs contain a conserved motif in which a Zn atom is coordinated by three histidine residues and a glutamate residue. MMPs achieve peptide bond fission through nucleophilic activation of a water molecule bound between Zn and the γ-carboxyl of glutamate.[126] MMPs play a multitude of roles in wound healing and repair, pathogen defense, cancer metastasis, and rheumatoid arthritis. Endogenous control of MMP activity is achieved through an endogenous group of four proteins referred to as tissue inhibitors of matrix metalloproteases (TIMPs). TIMPs act via formation of an equimolar complex with target MMPs. Chemical inhibition of MMP activity may be achieved by metal chelation and hydroxamic acid derivatives.

The Ubiquitin-Proteasome System

Ubiquitin is a highly conserved 76–amino acid polypeptide containing seven lysine residues encoded by four human genes.[127] Intracellular polypeptide chains may be modified with ubiquitin through the concerted action of three ubiquitin ligase enzymes (E1–E3). In the presence of ATP, ubiquitin is first activated via formation of a thioester bond between E1–cysteine residue and the C-terminal carboxyl of ubiquitin. Activated ubiquitin is then trans-esterified to the E2 ligase, which cooperates with E3 to effect the formation of an isopeptide bond between the C-terminal glycine of ubiquitin and a lysine residue on the target protein.[128] The target protein may be modified by a single ubiquitin moiety, or multiple molecules may be added to the original ubiquitin at one or more of the seven ubiquitin lysine residues. Limited ubiquitination of target proteins generally modifies their subcellular location or intermolecular interactions, and polyubiquitination targets the protein for destruction. Proteolysis occurs in the 26S proteasome, consisting of a 20S catalytic and 19S regulatory subunit constructed from greater than 60 polypeptides.[129] After recognition of the protein target by the 19S subunit, the unfolded polypeptide is threaded into the 20S subunit, where peptide bond fission occurs via a threonine-mediated nucleophilic attack.

Proteins in Human Serum and Plasma

The circulating proteome is a complex mixture of thousands of gene products. The most abundant products are proteins secreted directly into the circulation primarily by the liver and immunoglobulins contributed by lymphatic tissue. Classical methods for protein fractionation and purification over several decades led to isolation and characterization of about 100 of the most abundant proteins.[130] The 12 most abundant proteins represent more than 95% of total protein mass. Albumin alone represents more than 50% of the total mass of protein and an even higher proportion of the number of molecules so that albumin is the main contributor to colloid osmotic pressure (oncotic pressure). The distinction between mass and molar concentrations of circulating proteins may be significant in considering oncotic pressure, protease inhibition, and the binding capacity for ions, drugs, or small molecules. Table 13.4 lists the 30 most abundant proteins by mass and molecular abundance. An exhaustive list of the contents of the circulating proteome would exceed 12,000 entries.[131]

The range of protein concentrations measured in clinical assays spans more than 10 orders of magnitude and thus poses a significant analytic challenge. In decreasing order of abundance, the source of circulating proteins include (1) proteins secreted directly into plasma, (2) proteins associated with the cell membrane and shed into the circulation, (3) secretory proteins in exocrine secretions, (4) high-abundance cytoplasmic proteins, (5) low-abundance cytoplasmic proteins, (6) transmembrane proteins, and (7) organellar proteins that must traverse more than one membrane to exit cells. Many of these serve as useful markers of physiology and disease.

Circulating concentrations of proteins depend not only on rates of production and efficiency of entry to the circulation but also on rates of clearance. Proteins and peptides substantially smaller than albumin are cleared from the circulation by glomerular filtration unless they are bound to larger carrier molecules. Peptides and small proteins not bound to carriers are cleared with half-lives of about 2 hours under conditions of normal kidney function and accumulate in kidney failure. Examples of proteins and peptides that increase dramatically in renal failure include $β_2$-microglobulin (BMG), cystatin C, immunoglobulin light chains, parathyroid hormone fragments, complement factor D, atrial natriuretic peptide, and interleukins.[132,133] Other proteins and bioactive peptides such as insulin, intact parathyroid hormone, and growth hormone have much shorter circulating half-lives of only a few minutes, indicating receptor-mediated uptake or degradation by exopeptidases or endopeptidases.[134]

For all circulating proteins, the choice of measurement using plasma or serum is not without consequence. Plasma refers to the fluid portion of blood in the presence of an anticoagulant after the cells are removed by centrifugation. Common anticoagulants include dry heparin or ethylenediaminetetraacetic acid (EDTA) or solutions containing citrate. Small molecular weight additives such as EDTA may introduce osmotic effects on plasma volume, and citrate solutions introduce some specimen dilution. Variation in platelet concentration is noted, depending on the time of centrifugation and the force applied. Preparation of platelet-poor plasma commonly requires spinning specimens twice to ensure platelet removal and remove their contribution to plasma proteins. Serum is the fluid component of blood after blood is allowed to clot. It differs from plasma in several respects. An approximate 4% decrease in total protein content is related mainly to removal of fibrinogen during coagulation. Because of the absence of fibrinogen, serum has a lower viscosity than plasma. Intact clotting factors are consumed during the clotting process and in their place proteolytic fragments and the contents of platelet granules produced by the clotting process may be recovered in serum.[135]

Abundant Components of the Circulating Proteome
Prealbumin (Transthyretin)

Prealbumin (molecular weight, 35 kDa) is composed of four identical noncovalently bound subunits with the capacity to bind and transport 10% of circulating triiodothyronine

TABLE 13.4 High-Abundance Plasma Proteins

	RANKED BY MASS ABUNDANCE (mg/L)		RANKED BY MOLECULAR ABUNDANCE (µmol/L)	
Rank	Protein	Concentration	Protein	Concentration
1	Albumin	35,000–52,000	Albumin	500–800
2	Immunoglobulin G	7000–16,000	Immunoglobulin G	40–120
3	Transferrin	2000–3600	Apolipoprotein A-I	30–70
4	Immunoglobulin A	700–4000	Apolipoprotein A-II	30–60
5	α_2-Macroglobulin	1300–3000	Transferrin	25–45
6	Fibrinogen	2000–4000	α_1-Antitrypsin	18–40
7	α_1-Antitrypsin	900–2000	Haptoglobin	6–40
8	Apolipoprotein A-I	910–1940	α_1-Acid glycoprotein	15–30
9	C3	900–1800	α_2HS-glycoprotein	9–30
10	IgM	400–2300	Immunoglobulin A	5–30
11	Haptoglobin	300–2000	Hemopexin	9–20
12	Apolipoprotein B	600–1550	Apolipoprotein C-III	6–20
13	α_1-Acid glycoprotein	500–1200	Fibrinogen	5–18
14	α_2HS-glycoprotein	400–1300	Gc-globulin	8–14
15	Hemopexin	500–1150	Apolipoprotein C-I	6–12
16	Gc-globulin (vitamin D–BP)	400–700	C3	5–10
17	Factor H	240–740	α_1-Antichymotrypsin	4–9
18	α_1-Antichymotrypsin	300–600	Apolipoprotein D	2–10
19	Inter-α-trypsin inhibitor	200–700	Prealbumin	4–8
20	Apolipoprotein A-II	260–510	β_2-Glycoprotein I	3–6
21	C4b-binding protein	200–530	Apolipoprotein A-IV	3–6
22	Ceruloplasmin	200–500	Apolipoprotein C-II	2–7
23	Factor B	180–460	Serum amyloid A4	3–6
24	Prealbumin	200–400	Inter-α-trypsin inhibitor	3–5
25	Gelsolin	200–400	Antithrombin III	3–5
26	Fibronectin	300	α_1B-glycoprotein	3–5
27	C1 inhibitor	190–370	Gelsolin	3–5
28	C4	100–400	Ceruloplasmin	2–5
29	Plasminogen	150–350	Factor H	2–5
30	Antithrombin III	170–300	Factor B	2–5

Data from in Hortin GL, Sviridov D, Anderson L. High-abundance polypeptides of the human plasma proteome comprising the top 4 logs of polypeptide abundance. *Clin Chem* 2008;54:1608–1616.

(T_3) and thyroxine (T_4). Prealbumin concentrations are often used as an indicator of adequacy of protein nutrition because of its relatively short half-life (~2 days), high proportion of essential amino acids, and small pool size.[6,10,11] Concentrations also fall in cirrhosis of the liver and protein-losing diseases of the gut or kidneys. Prealbumin migrates as a minor component anodal to albumin on routine serum electrophoresis and is not routinely observed by most methods. It is a proportionately greater component of CSF. Prealbumin is most commonly assessed using immunonephelometric or immunoturbidimetric methods.

Albumin

The name *albumin* (L. *albus,* meaning white) originated from the white precipitate formed during the boiling of acidic urine from patients with proteinuria. Normally, albumin is the most abundant plasma protein from the fetal period onward, accounting for about half of the plasma protein mass. It is a major component of most body fluids, including interstitial fluid, CSF, urine, and amniotic fluid. More than half of the total pool of albumin is in the extravascular space.

Albumin has a nonglycosylated polypeptide chain of 585 amino acids and a calculated molecular weight of 66,438 Da. It is synthesized in the liver and has a 3D structure stabilized by 17 intrachain S-S bonds.[136] It is both chemically and biologically stable because it resists denaturation at higher temperatures than most plasma proteins and circulates with a half-life of 15 to 19 days. Albumin has a high abundance of charged amino acids that contribute to high solubility, and it has a net negative charge of about −12 at neutral pH.[137] Albumin therefore contributes about 6 to 10 mmol/L to the anion gap at normal albumin concentrations of 0.5 to 0.8 mmol/L (3.5–5.2 g/dL) and lesser amounts at lower albumin concentrations. At a pH of 8.6 for alkaline electrophoresis, albumin has a net charge of about −25, resulting in high mobility toward the anode. One unpaired cysteine at position 34 occurs partially in reduced form and partially in exchangeable disulfide bonds with other free thiols.

Albumin has two critical biologic functions. First, it serves as the major component of colloid osmotic pressure. Patients with hypoalbuminemia caused by nephrotic syndrome, for example, develop edema. Second, it serves as a transporter for a diverse range of substances, including fatty acids and other lipids, bilirubin, foreign substances such as drugs, thiol-containing amino acids, tryptophan, calcium, and metals. Some of these substances, such as fatty acids and unconjugated bilirubin, have very low solubility in water in the absence of a carrier molecule.

Most clinical laboratories assay albumin in plasma or serum samples by dye-binding methods, which rely on a shift in the absorption spectrum of dyes such as bromcresol green (BCG) or purple (BCP) upon albumin binding. The affinity of these dyes is higher for albumin than for other proteins, providing some specificity for albumin. BCP generally is slightly more specific for albumin and yields lower values than BCG, particularly for patients with kidney failure.[138] Heparin in collection tubes is reported to affect some dye-binding methods.[139] Dye-binding assays also tend to be less accurate when the serum or plasma protein composition is abnormal. The many ligands of albumin do not typically affect dye-binding assays of serum or plasma significantly unless their concentrations are very high.

α_1-Antitrypsin

Schultze et al.[140] described the inhibition of trypsin in the 1950s. However, AAT inhibits a variety of serine proteinases, leading to the term α_1-proteinase inhibitor. AAT is the term commonly used by clinicians and clinical laboratorians, but phenotypes are commonly abbreviated as Pi (protease inhibitor). AAT is synthesized by hepatocytes as a single polypeptide chain with 394 amino acid residues and 3 N-linked oligosaccharides, yielding a total molecular weight of approximately 51 kDa. It belongs to the serpin superfamily, a group of suicide inhibitors of serine proteases, which abortively cleave the inhibitor at the reactive site residue but remain covalently linked to the reactive site. Serpins usually occur in a "stressed" conformation, and cleavage leads to a dramatic conformational shift to a "relaxed" form. AAT and other serpins serve as important models for conformational change in protein function and aggregation.[76]

More than 75 genetic variants of AAT have been noted. Clinical AAT deficiency is inherited in an autosomal codominant fashion with a prevalence of 1 in 3000 to 5000 in the United States. AAT alleles are designated B to Z in order of decreasing electrophoretic mobility.[141] The allele designated M is the wild type. Clinically important variants are the S allele characterized by a Glu → Val substitution at position p264 and the Z allele characterized by a Lys → Glu missense mutation at position p340. Whereas individuals homozygous for the S allele have serum concentrations approximately 60% of normal, those homozygous for the Z allele possess about 15% of normal levels.[77] The lung disease of AAT deficiency is thought to be associated with unchecked elastase activity. Pulmonary disease is characterized by onset of emphysema in the third to fifth decades of life. The onset is particularly early in smokers. Liver manifestations include neonatal cholestasis or hepatitis, cirrhosis, and hepatocellular carcinoma.

Assessment of circulating AAT concentrations is usually performed by immunoturbidimetry or immunonephelometry. Normal concentrations range from 70 to 200 mg/dL (14–50 μmol/L). Concentrations are higher in patients with inflammatory disorders, malignancy, or trauma and in women who are pregnant, on estrogen therapy, or taking oral contraceptives. Neonates also have increased concentrations, possibly secondary to maternal estrogen. Individuals with decreased AAT may be phenotyped using electrophoresis or MS or alternatively, genotyped.

Ceruloplasmin

Ceruloplasmin (Cp) is an α_2-globulin that contains about 95% of serum copper. Each molecule of Cp contains 6 to 8 tightly bound copper atoms. A solution of Cp is blue (L. caeruleus, meaning blue), and increased concentrations of Cp may lend plasma a green tint. Cp is a polypeptide chain with 1046 amino acids and three asparagine-linked oligosaccharides yielding a molecular weight of approximately 132 kDa. Cp is synthesized primarily by hepatic parenchymal cells, with small amounts from macrophages and lymphocytes. The peptide chain is formed first and then copper is added via the activity of an intracellular ATPase (ATP7B). ATP7B is commonly mutated in Wilson disease.[142] Cp synthesized in the absence of copper or the ATPase is degraded intracellularly, but some apoCp reaches the circulation, where it has a shortened half-life of a few hours (normal, 4–5 days). Consequently, serum Cp is low in those with Wilson disease.[143]

Haptoglobin

Haptoglobin (Hp) is an α_2-glycoprotein that binds hemoglobin (G. haptein, meaning to bind). Hp is synthesized as a single peptide chain by hepatocytes and cleaved into α- and β-chains. Variable numbers of α- and β-chains combine and become covalently linked by disulfides to form Hp.[144] Hp scavenges hemoglobin in the vascular space. Hp–hemoglobin complexes are bound by CD163 receptors and rapidly cleared by the reticuloendothelial system, which degrades protein and recycles heme and iron. This process prevents renal clearance of hemoglobin until Hp-binding capacity is exceeded. Because Hp is degraded after complexing with hemoglobin, its concentration drops severely in the event of intravascular hemolysis. The normal plasma half-life of Hp is about 5.5 days. Hemolysis of specimens after blood collection does not decrease Hp. Hp has a capacity to bind only about 1% of the hemoglobin in RBCs at usual hematocrits, so minimal intravascular lysis completely depletes plasma Hp. When Hp capacity is exceeded, free hemoglobin in the circulation increases. Free hemoglobin can oxidize to methemoglobin (Fe^{3+}) followed by dissociation of metheme from globin. Metheme binds to hemopexin (high affinity) or albumin (low affinity), keeping it in solution. Hemopexin–heme complexes are removed by the reticuloendothelial system. Hp depletion is usually a sensitive biochemical indicator of intravascular hemolysis followed by hemopexin depletion and finally by the presence of methemalbuminemia, hemoglobinuria, or both. Hp measurement, typically by immunoturbidimetry or immunonephelometry, is therefore part of the assessment of possible transfusion reactions or other causes of hemolysis.

Transferrin

Transferrin (originally named siderophilin) is the principal plasma transport protein for iron (Fe^{3+}). Transferrin has a molecular weight of 79.6 kDa, including 5.5% carbohydrate. It is a single polypeptide chain, with two N-linked oligosaccharides and two homologous domains, each with an Fe^{3+}-binding site. It is synthesized mainly in the liver and circulates with a half-life of 8-10 days. Transferrin reversibly binds two ferric (Fe^{3+}) ions with high affinity at physiologic pH but lower affinity at decreased pH, which allows release of iron in intracellular compartments. After cellular delivery of iron via receptor-mediated endocytosis, apotransferrin is recycled back into the circulation. Clinical indications for direct measurement of transferrin are few. Indirect assessment of transferrin concentration may be inferred by total iron-binding capacity (TIBC).

Under normal circumstances, transferrin oligosaccharides terminate in four sialic acid residues, but this pattern is altered under both normal and pathologic conditions. Transferrin glycan structure is widely used to detect congenital disorders of glycosylation.[145] These forms are generically referred to as carbohydrate-deficient transferrin (CDT).[146] A desialated version of transferrin, termed tau-transferrin or β2-transferrin, is a substantial component of CSF but not serum. It has been used as an indicator of CSF leakage in fluid from the ear or nasal passages (see later section on CSF proteins).[147,148] Increased CDT in plasma also has been used as an indicator of alcohol abuse.[149] The decreased amount of sialic acid and the reduced negative charge of CDT may be detected by electrophoresis, isoelectric focusing, ion-exchange chromatography, or MS.

β2-Microglobulin

β2-Microglobulin is a small, nonglycosylated 99 residue protein with a molecular weight of 11.8 kDa. It is the noncovalently bound light chain subunit of class I major histocompatibility complex molecules present on the surface of all nucleated cells. BMG is shed into the blood, particularly by B lymphocytes, and some tumor cells. Its small size allows efficient glomerular filtration, resulting in a plasma half-life of approximately 100 minutes. In addition to renal failure, therefore, high plasma concentrations occur in inflammation and neoplasms, especially those associated with B lymphocytes. In patients with chronic lymphocytic leukemia, high BMG concentrations are a negative prognostic marker for decreased survival. Plasma BMG concentrations are used as a staging criterion in multiple myeloma.[150] BMG concentrations are also increased in states of immune activation and have been applied as an indicator of transplant rejection.

C-Reactive Protein

In 1930, Tillet and Francis described a substance in the sera of acutely ill patients that bound cell wall C-polysaccharide of *Streptococcus pneumoniae* and agglutinated the organisms.[151] In 1941, the substance was shown to be a protein and named C-reactive protein (CRP).[152] CRP consists of five identical, nonglycosylated 23 kDa subunits noncovalently associated to form a disk-shaped structure with radial symmetry and total mass of approximately 115 kDa. CRP aids in nonspecific host defense against infectious organisms by activating the classical complement pathway.

C-reactive protein is one of the strongest acute-phase reactants, with plasma concentrations rising up to 1000-fold after myocardial infarction, stress, trauma, infection, inflammation, surgery, or neoplastic proliferation.[153] Concentrations greater than 5 to 10 mg/L suggest the presence of an infection or inflammatory process. Concentrations are generally higher in bacterial than viral infection, although concentrations greater than 100 mg/L (10 mg/dL) may be seen in uncomplicated influenza and infectious mononucleosis. The increase with inflammation occurs within 6 to 12 hours and peaks at about 48 hours and is generally proportional to the extent of tissue damage. Because the increase is nonspecific, however, it cannot be interpreted without other clinical information.

C-reactive protein is normally present in plasma at a concentration below 5 mg/L. High concentrations in inflammatory states are measured with direct immunoturbidimetric or immunonephelometric assays. Epidemiologic studies demonstrate that mildly increased CRP concentrations are associated with risk of cardiovascular disease.[154] Increased concentrations may reflect low-grade, chronic intimal inflammation. The use of CRP for these purposes requires assays with detection limits below 0.3 mg/L that generally are referred to as high-sensitivity CRP assays. High-sensitivity assays require particle-enhanced (also termed *latex-enhanced*) light scattering assays or sandwich-type immunoassay formats.

Complement

The complement system consists of more than 20 proteins, synthesized primarily by the liver. A basic schematic of the complement cascade is presented in Fig. 13.4. The basic functions of the complement cascade are to recruit effector phagocytes for opsonization and clearance of foreign pathogens as well as trigger direct destruction of the foreign organism. Activation of the cascade proceeds by three different stimuli. The classical pathway is activated primarily by IgM or IgG binding to antigens, which activates a complex consisting of C1q, C1r, and C1s. The alternative pathway is triggered by activation of C3, factor B, and factor D on a variety of pathogenic surfaces in the absence of antibodies. The lectin pathway is activated by binding of mannan-binding protein (MBP) to mannose-rich oligosaccharides that are present in the cell walls of many microorganisms.[155] This event triggers activation of MBP-associated serine proteases termed MASP-1 and MASP-2.

During activation, many complement components are enzymatically cleaved into two or more fragments. The larger fragments are designated by a lowercase *b* and the smaller ones by a lowercase *a*. The larger fragments usually contain a binding site for membranes, immune complexes, and protein association, or in some cases, yield new protease activities that activate subsequent component(s) of the cascade. Smaller fragments typically serve as anaphylatoxic or chemotactic peptides. Inactivated fragments are designated by the letter *i* (eg, C3bi). Activated complexes are indicated by a bar over the components (eg, $\overline{c567}$).

Via coordinated proteolysis, all three pathways converge at the C3 convertase. The classical and lectin pathway C3 convertase consists of C4b and C2b, and the alternative pathway C3 convertase consists of C3b and factor Bb. C3 convertase catalyzes formation of C3b and C3a. Surface-bound C3b serves as a docking site for phagocyte receptors. The C3 convertase also triggers the formation of the classical (C4b2a3b) and alternative (C3bBb3b) C5 convertase, which ultimately leads to the formation of a membrane attack complex (MAC) consisting of components C5 to C9. The MAC forms

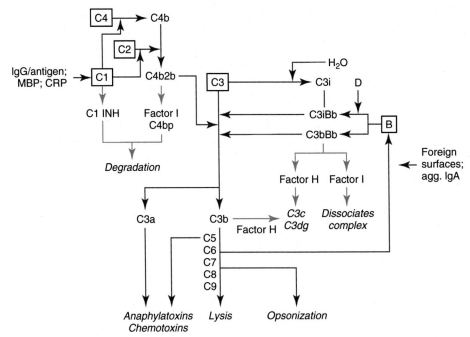

FIGURE 13.4 The complement cascade. Activation via the classical pathway is shown on the *left* and via the alternative pathway on the *right*. Both pathways converge at the level of C3 convertase. Direct activation of C3 by neutrophil and plasma proteases also may occur. *CRP*, C-reactive protein; *IgA*, immunoglobulin A; *MBP*, mannan-binding protein. (Courtesy J.W. Whicher, with modifications.)

an ion channel in the membrane of the foreign pathogen triggering lysis and destruction of the target organism.

The constant slow ongoing activation of complement factors would have devastating circumstances were it not for a host of regulatory proteins designed to limit complement activity. A few of these factors include decay-accelerating factor (DAF or CD55), membrane inhibitor of reactive lysis (CD59), membrane cofactor protein (MCP), C1 esterase inhibitor, and factor H. Deficiency of DAF and CD59 can lead to abnormal lysis of RBCs.[156] Defects in C1 esterase inhibitor are linked to hereditary angioedema.[157] Factor H plays a role in age-related macular degeneration.[158] Proper function of complement regulatory factors prevents destruction of endogenous cells at the same time that foreign cells are destroyed by complement.

Despite its complexities, the function of the complement system in vivo is probed with relatively few tools. Total serum complement is a screening test that demonstrates an intact complement pathway from activation to formation of the MAC. The classic version of this assay tests the capacity of patient sera to lyse sheep RBCs coated with rabbit antisheep antibodies. Progress of the reaction is monitored using hemoglobin release from the RBCs. CH50 refers to the amount of serum required to lyse 50% of the added erythrocytes. Assessment of circulating C3 and C4 is also common. C3 and C4 are typically determined with immunoturbidimetry or immunonephelometry. Measurement of C3 may be complicated by in vitro conversion of C3 to C3c. Because of differences in antibody reactivity toward C3 and C3c, C3 concentrations measured on fresh samples may be lower than those determined after long-term storage. Measurement of C3 and C4 are used to assess activation of the alternative and classical pathways, respectively. Low concentrations are observed in complement deficiency, glomerulonephritis, lupus erythematosus, and sepsis. Clinical disorders of inherited complement deficiency are listed in Table 13.5.

Immunoglobulins

Immunoglobulins (antibodies) are generated against foreign immunogens and initiate clearance of the foreign molecule or organism. Human immunoglobulin molecules consist of one or more basic units built of two identical heavy (H) chains and two identical light (L) chains. Each of the four chains has one variable and one (L chain) or three to four (H chain) constant domains, with the variable region involved in antigen recognition and binding. Extensive diversity in the variable domains is generated by somatic recombination and mutation of the immunoglobulin genes. Individual plasma cells or clonally expanded cells are committed to synthesis of a single variable domain sequence for heavy and light chains. The amino acid sequences of the variable domains at the N-terminal ends of the four chains form two antigen-binding sites with a high degree of variation in binding specificity. The constant domains are the same for every immunoglobulin molecule of a given subclass and carry sites for binding to complement receptors and activation of complement.

The variable domains contain the antigen-binding regions and the constant domains of the heavy chains contain sites for complement activation and receptor binding. Cleavage of immunoglobulins with pepsin or papain can yield antigen-binding fragments (Fab) and constant region fragments (Fc). Variations in the constant domains of heavy chains (Fc region) result in the classes and subclasses into which immunoglobulins are grouped: IgM, IgG (four subclasses), IgA

TABLE 13.5 Inherited Deficiencies of Complement Components

Component	Frequency of Deficiency	Associated Disorders
Ficolins 1-3	Rare?	Recurrent infection
MBP	5%	Infection in infancy; less effect on adults
MASP	Rare?	Recurrent infection (eg, pneumococcal); inflammation
C1q*	Relatively rare	SLE; DLE; GN
C1r, C1s	Rare	SLE; DLE; infection
C2	≥0.0003% (homozygous)	Recurrent infection, vasculitis; SLE, DLE (no antinuclear antibody); half of affected individuals are asymptomatic
C3	Rare	Severe and recurrent bacterial infection, especially with encapsulated, pyogenic bacteria
C4a	13% (heterozygous)	SLE, DLE
C4b	13% (heterozygous)	IgA nephropathy; infection
Combined C4	35% one null; 8%–10% 2 nulls; ≈1% 3 nulls; <0.1% 4 null alleles	Total deficiency: SLE, GN, DLE (many are anti-dsDNA negative but anti-Ro/SSA positive)
C5-C9	Rare	Severe or recurrent infection with Neisseria species
Properdin	Rare	X-linked; neisserial infection
Factor D	Rare	Recurrent infection
Factor H or I	Rare	Hypercatabolism of C3; recurrent bacterial infection; HUS in factor H deficiency
C1 inhibitor	0.002%	Hereditary angioedema (autosomal dominant)
Decay accelerating factor (DAF, CD 55)	Rare	PNH related to decreased DAF and CD59 on cell surfaces

DLE, Discoid lupus erythematosus; *GN*, glomerulonephritis; *HUS*, hemolytic-uremic syndrome; *MBP*, mannan-binding protein; *PNH*, paroxysmal nocturnal hematuria; *SLE*, systemic lupus erythematosus (or SLE-like disease).
*Both quantitative and qualitative (functional) deficiencies reported.
Data from Colten HR, Rosen FS. Complement deficiencies. Annu Rev Immunol 1992;10:809–834; and Unsworth DJ. Complement deficiency and disease. J Clin Pathol 2008;61:1013–1017.

(two subclasses), IgD, and IgE, respectively. Light chains, which are produced independently and in slight excess of heavy chains, are of two types—kappa (κ) and lambda (λ). The heavy-chain genes are located on chromosome 14; whereas κ light chains are encoded by a gene on chromosome 2, the λ-chain gene is on chromosome 22.

Immunoglobulins are synthesized by plasma cells, the progeny of B-lymphocyte stem cells in bone marrow. More mature B lymphocytes, found mainly in lymph nodes and in blood, develop receptor immunoglobulins on their surface membranes. Upon binding a target antigen, these B lymphocytes proliferate and develop into a clone of plasma cells, producing antibody to the target antigen. Somatic mutation of immunoglobulins leads to further diversity of immunoglobulin variable region and antibody maturation, generally leading to antibodies with higher affinity. B lymphocytes at first have IgM surface receptors and secrete IgM as the first or "primary" response to an antigen. Membrane and secreted forms of the antibody arise from differential splicing of the messenger RNA for heavy chains, which adds a transmembrane segment to the membrane-bound form. Heavy chains of the IgM surface receptor molecules undergo class switching to produce immunoglobulins with γ- or α-heavy chains (IgG or IgA), but the variable regions remain unchanged; as the cells change into plasma cells, second exposure to the same antigen causes a larger secondary or anamnestic response of IgG secretion.

Individual Immunoglobulins and Light Chains

Immunoglobulin G. Immunoglobulin G (IgG) accounts for 70% to 75% of the total immunoglobulins in plasma. Only 35% is found in the plasma space, and 65% is extravascular. IgG consists of two γ-heavy and two light chains, linked by disulfides. The molecular weight of IgG is approximately 150 kDa, including one N-linked oligosaccharide on each heavy chain. The oligosaccharide structure may change in inflammatory states and affect interactions with receptors.[159] On agarose gel electrophoresis, IgG migrates broadly in the γ- and slow β-regions as a result of its heterogeneity of charge from sequence variation.

Immunoglobulin G has four subclasses: IgG_1, IgG_2, IgG_3, and IgG_4. The circulating half-life of IgG_1, IgG_2, and IgG_4 is about 22 days. IgG_3 has a half-life of 7 days. IgG_1 and IgG_3 strongly activate complement via the classical pathway, IgG_2 is weakly complement fixing, and IgG_4 does not activate complement. Clustering of multiple IgG molecules is required to activate complement. Both IgG_1 and IgG_3 bind Fc receptors on phagocytic cells, activate killer monocytes, and cross the placenta via receptor-mediated active transport. IgG_1 is the principal IgG to cross the placenta, and neonatal concentrations are similar to maternal concentrations. Neonates have low production of IgG as the result of immaturity of their immune systems, and IgG concentrations fall through infancy as maternally acquired antibody is cleared.

Immunoglobulin M. Immunoglobulin M (IgM) is produced at earlier stages of B-cell development. In the immature immune systems of neonates, IgM is the major immunoglobulin synthesized. In adult serum, it is the third most abundant immunoglobulin, usually accounting for 5% to 10% of total circulating immunoglobulins. IgM as a membrane receptor molecule is monomeric, but most of the serum IgM is a pentamer containing five monomers linked via disulfides to the small J (joining) chain. Plasma cell malignancies may secrete monomeric IgM in addition to, or instead of, pentamers. The high molecular weight of IgM (970 kDa; approximately 10% carbohydrate) prevents its ready passage into extravascular spaces. IgM is not transported across the placenta and therefore is not involved in hemolytic disease of neonates. It activates complement even more efficiently than IgG. Binding of one IgM molecule may be adequate to activate complement. In rare hyper-IgM syndromes, class switching to IgG and IgA is deficient. Affected patients have deficiency of IgG and IgA and increased susceptibility to infection.[160]

Immunoglobulin A. Immunoglobulin A (IgA) has a molecular weight of 160 kDa, including about 10% carbohydrate derived from both N- and O-linked oligosaccharide chains. IgA accounts for about 10% to 15% of serum immunoglobulin and has a half-life of 6 days. In its monomeric form, its structure is similar to that of IgG, but 10% to 15% of IgA in serum is dimeric, particularly IgA_2, which is more resistant to destruction by pathogenic bacteria than IgA_1. On electrophoresis, IgA migrates in the β-γ region, anodal to most IgG. IgA is an important component of mucosal immunity.[161] Secretory IgA is found in tears, sweat, saliva, milk, colostrum, and gastrointestinal (GI) and bronchial secretions. Secretory IgA has a molecular weight of 380 kDa and consists of two molecules of IgA: a secretory component (70 kDa) and a J chain (15.6 kDa). It is synthesized mainly by plasma cells in the mucous membranes of the gut and bronchi and in the ductules of the lactating breast. The secretory component assists with transport of secretory IgA across mucosal epithelium and into secretions. Secretory IgA in colostrum and milk is more abundant than IgG and may aid in protection of neonates from intestinal infection. IgA can activate complement by the alternative pathway, but the exact role of IgA in serum is not clear.

Immunoglobulin D. Immunoglobulin D (IgD) accounts for less than 1% of serum immunoglobulin. It is monomeric, contains about 12% carbohydrate, and has a molecular weight of 184 kDa. Its structure is similar to that of IgG. Similar to IgM, IgD is a surface receptor for antigen on B lymphocytes, but its primary function is unknown.

Immunoglobulin E. Immunoglobulin E (IgE) contains 15% carbohydrate and has a molecular weight of 188 kDa. IgE is so rapidly and firmly bound to specific IgE receptors on mast cells that only trace amounts of it are normally present in serum. IgE binds to mast cells via sites on its Fc region. When the antigen (allergen) cross-links two of the attached IgE molecules, the mast cell is stimulated to release histamine and other vasoactive amines that increase vascular permeability and smooth muscle contraction, mediating type 1 hypersensitivity reactions such as hay fever, asthma, urticaria, and eczema. Rare regulatory disorders with hyperproduction of IgE lead to a primary immunodeficiency disorder, Job's syndrome, with eczema, recurrent infection, and markedly increased IgE.[162] IgE molecules specific for particular allergens are analyzed to identify the specificity of allergies. The total serum concentration of IgE may be increased in individuals with allergic disorders.

Free Immunoglobulin Light Chains. Light chains are usually synthesized in slight excess versus quantities required for intact immunoglobulins. Consequently, small amounts of free light chain, representing only about 0.1% of total immunoglobulin, are present in serum or plasma. Amounts in plasma are kept low by renal clearance. Free κ-light chains (23 kDa) are cleared about two to three times faster than free λ-light chains (a disulfide-linked dimer of 46 kDa), which have a half-life of 4 to 6 hours. Consequently, even though production of κ-light chains is about twice as great as that of λ-light chains, the plasma concentration of free λ-light chains is usually higher, except in renal failure.[163] Free light chains usually are not functional, but immunoassays specific for free light chains are applied to detect plasma cell disorders.[164]

Clinical Utility of Immunoglobulin Measurement. Serum normally contains a diverse, polyclonal mixture of antibodies with varying amino acid sequences, which represent multiple "idiotypes" (ie, the products of many different clones of plasma cells, each producing a specific immunoglobulin molecule). Benign or malignant proliferation of one such clone produces a high concentration of a single monoclonal antibody, which may appear as a sharp, narrow band on protein electrophoresis. Unbalanced production of free light chains might also lead to a second band representing free light chains. If a few clones proliferate, several sharp bands may be seen. Therefore, disease may be associated with a decrease or an increase in normal polyclonal immunoglobulins or an increase in one or more monoclonal immunoglobulins. These disease states are detailed below.

Immunoglobulin Deficiency. Immunodeficiency states may be the result of deficiency of a single factor or combinations affecting multiple systems of immune defense. The diagnosis of major deficiencies in immunoglobulin production is clinically important to avoid infection as their maternally acquired antibodies decline. Severe combined immune deficiency (SCID) is a disorder of B-cell development or activation affecting 1 in 100,000 newborns and resulting in broad-spectrum immunoglobulin deficiency. The more common primary deficiencies[165] involve only one or two immunoglobulin classes (IgA) or subclasses (IgA or IgG subclasses) or ability to generate antibodies against polysaccharide antigens. IgA deficiency occurs in about 1 in 500 whites but much less frequently in Asian populations. IgA deficiency usually is not associated with severe infection, but the risk of clinically mild infection with *Giardia* or other organisms may be increased. IgA deficiency may lead to false-negative assays for celiac disease detection, and some affected individuals are at risk for anaphylaxis if they receive blood products containing IgA.[166] Selective deficiency of IgG subclasses is not rare, but it is unclear whether it is an important risk for infection. Deficiency of IgG2 may be related to poorer responses to polysaccharide antigens and increased risk of infection with encapsulated organisms.[165]

Infants have transient physiologic deficiency of IgG, with a nadir at about 3 months of age (Fig. 13.5). Prolonged or severe physiologic deficiency may be associated with increased infection rates, especially with encapsulated bacteria. Concentrations of maternal IgG, transferred across the placenta, rise rapidly in the fetus during the last half of pregnancy but then drop over a few months after birth. Two groups of neonates are at risk for clinically significant IgG deficiency: premature infants who begin life with less maternal IgG

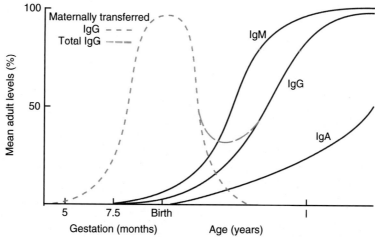

FIGURE 13.5 Serum immunoglobulin concentrations as percent of mean adult concentrations before birth and for the first year of life. *IgA*, Immunoglobulin A; *IgG*, immunoglobulin G; *IgM*, immunoglobulin M.

and infants with delayed initiation of IgG synthesis. Monitoring of IgG concentrations can identify this problem. Rising IgM and normal salivary IgA concentrations at 6 weeks of age suggest a favorable prognosis. Contact of the neonate with environmental antigens normally causes B lymphocytes to begin to multiply and IgM concentrations to start to rise, followed weeks to months later by IgA and IgG.

Polyclonal Hyperimmunoglobulinemia. Polyclonal increases in plasma immunoglobulins are the normal response to infection. IgG predominates in autoimmune responses. IgA is increased in skin, gut, respiratory, and renal infections, and IgM is increased in primary viral infections and bloodstream infection with parasites such as malaria. Chronic bacterial infection may cause increased concentrations of all immunoglobulins. Measurements of total IgE are used in the management of asthma and other allergic conditions, especially in children. Measurements of allergen-specific IgE assist in identifying the stimulus for hypersensitivity responses.

Monoclonal Immunoglobulins (Paraproteins). A single clone of plasma cells produces immunoglobulin molecules with a single defined amino acid sequence. If the clone expands greatly, it may produce a discrete band on electrophoresis, often referred to as an M-spike or M-protein. These monoclonal immunoglobulins, termed *paraproteins*, may be polymers, monomers, individual immunoglobulin chains such as free light chains or heavy chains, or fragments of immunoglobulins. Clinical, epidemiologic, and biochemical characteristics of monoclonal paraprotein diseases are summarized in Table 13.6. About 60% of paraproteins are associated with plasma cell malignancies (multiple myeloma or solitary plasmacytoma), and approximately 15% are caused by overproduction by B lymphocytes, mainly in lymph nodes (lymphomas, chronic lymphocytic leukemia, Waldenström's macroglobulinemia, or heavy-chain disease). Up to 25% of paraproteins are benign and have been termed *monoclonal gammopathy of undetermined significance (MGUS)*. The incidence of MGUS increases with age, with 1% incidence for people 50 to 70 years of age and 3% incidence for people older than 70. The occurrence of MGUS is associated with increased risk of progression to multiple myeloma that should be monitored. Multiple myeloma appears to be preceded consistently by MGUS.[167]

TABLE 13.6 Monoclonal Immunoglobulins (Paraproteins) in Multiple Myeloma

Plasma Paraprotein	Incidence* (%)	Mean Age of Occurrence* (yr)	Incidence of Free Light Chain in Urine (%)	Comments
IgG	50	65	60	Patients more susceptible to infection; paraproteins reach highest concentrations
IgA	25	65	70	Tend to have hypercalcemia and amyloidosis
Free light chain only	20	56	100	Often renal failure; bone lesions; amyloidosis; poor prognosis
IgD	2	57	100	90% λ type; often have extraosseous lesion, amyloidosis, renal failure; poor prognosis
IgM	1	—	100	May or may not have hyperviscosity syndrome
IgE	0.1	—	Most	—
Biclonal	1	—	—	—
None detected	<1	—	0	Usually reduction of normal immunoglobulins; increased plasma cells in bone marrow biopsy

Ig, Immunoglobulin.
*Approximate.

The primary clinical interest in identifying paraproteins is to detect or monitor proliferative disorders of B cells. However, from the laboratory standpoint, paraproteins are also significant as a potentially unpredictable source of interference with many assays. Paraproteins may aggregate or precipitate, causing interference in a variety of photometric reactions and in light-scattering hematology analyzers.

Many patients with paraproteins have nonspecific presentations such as anemia or infection. Identification of paraproteins in serum usually is based on serum protein electrophoresis and immunofixation electrophoresis (IFE) (described in a later section). Urine protein electrophoresis and urine IFE are helpful mainly in identifying patients with free immunoglobulin light chains. Urinary free light chains, as described by Bence Jones in the 1850s, were the first tumor marker. Free light chains are often referred to as *Bence Jones proteins*.[168]

Acute Phase Response

Systemic inflammation in response to infection, tissue injury, or inflammatory disease triggers changes in hepatic production of multiple plasma proteins, as indicated in Box 13.1. This process, mediated by the action of interleukin-6 (IL-6) and other cytokines has been termed the *acute-phase response* (APR).[153] It is a nonspecific reaction to inflammation, analogous to the increase in temperature or leukocyte count. In the APR, synthesis of a few proteins, including albumin, transferrin, and prealbumin, is downregulated. These proteins are termed *negative acute-phase reactants*. Albumin concentrations fall because of acutely decreased synthesis and from redistribution to extracellular fluids. Production of a number of proteins, including AAT, α_1-acid glycoprotein (AAG), Hp, Cp, C4, C3, fibrinogen, and CRP, increases several-fold. Plasma concentrations of individual acute-phase proteins rise at different rates, and all reach maxima within 2 to 5 days after an acute insult. Qualitative changes in proteins, such as altered glycosylation, are also observed secondary to the APR. Changes to glycosylation of immunoglobulins may have an immunomodulatory effect.

Methods for Analyzing Proteins
Determination of Total Protein

Plasma normally contains about 6.5 to 8.5 g/dL protein and serum about 4% less. Determination of total protein in biologic fluids in some respects represents a greater challenge than analysis of a specific protein because variable protein composition of biologic fluids leads to variable carbohydrate composition, charge, and physical characteristics of proteins in the mixture. Many methods of protein analysis respond differentially to different proteins and present problems when applied to specimens of varying protein composition. Most methods other than the biuret method have not been thoroughly examined for interactions with small peptide components as well as intact proteins, and this may become a significant issue in renal failure with increased accumulation of peptides and small proteins. Many methods have been developed to measure the total protein content of biologic fluids. Several are reviewed here.

Kjeldahl Method. This method is rarely used in clinical laboratories but is of historical importance and is sometimes used as a reference method. Protein nitrogen is converted to ammonium ion by heating with sulfuric acid in the presence of a catalyst. Ammonium ion is measured by alkalinization, distillation, and acid titration or by Nessler's reagent. Protein is estimated to contain 16% nitrogen. Errors in protein estimation occur if the amino acid composition is unusual and if nitrogen content differs from 16%. Nonprotein nitrogen, such as from urea and amino acids, also is measured, so a protein precipitation step may be required. The Kjeldahl method was one of the first methods used for reproducible total protein measurement, but it is time consuming and impractical for routine use. This method has been used to assign values to reference materials for the biuret method.

Biuret Method. Under strongly alkaline conditions, Cu^{2+} ions form multivalent complexes with peptide bonds in proteins. Binding shifts the absorption spectrum of Cu^{2+} ions to shorter wavelengths, leading to a color change from blue to violet that has been termed the *biuret reaction*. Absorbance attributable to protein is measured spectrophotometrically at 540 nm. Absorbance changes at 540 nm also result from binding of Cu^{2+} ions by many compounds that can form chelates

---

BOX 13.1 The Acute Phase Response: Changes in Plasma Protein Concentrations

Positive Acute Phase Response
C-reactive protein (extreme)
Serum amyloid A (extreme)
α_1-Acid glycoprotein
α_1-Antitrypsin
α_1-Antichymotrypsin
Antithrombin III
C3, C4, and C9
C1 inhibitor
C4b-binding protein
Ceruloplasmin
Factor B
Ferritin
Fibrinogen
Haptoglobin
Hemopexin
Lipopolysaccharide-binding protein
Mannan-binding protein (lectin)
Plasminogen
Procalcitonin

Negative Acute Phase Response
Albumin
Apolipoprotein A-I
Apolipoprotein B
α_2-HS glycoprotein
Insulin-like growth factor I
Prealbumin
Retinol-binding protein
Thyroxine-binding globulin
Transferrin

Data from Craig WY, Ledue TB, Ritchie RF. Plasma proteins: clinical utility and interpretation. Scarborough, ME: Foundation for Blood Research, 2001; Gabay C, Kushner I. Acute-phase proteins and other systemic responses to inflammation. N Engl J Med 1999;340:448–454; and Vollmer T, Piper C, Kleesiek K, Dreier J. Lipopolysaccharide-binding protein: a new biomarker for infectious endocarditis? Clin Chem 2009;55:295–302

with five- or six-member rings where amino, amide, or hydroxyl groups bind to Cu^{2+} ions. Such compounds include serine, asparagine, ethanolamine, and TRIS buffer, among others.[169] Small compounds lead to a smaller spectral shift than proteins. Historically, the biuret method was not considered to react with free amino acids and dipeptides, but absorbance changes occur with some amino acids, with amino acid amides, and with dipeptides. The biuret action also was considered to react equally with all proteins and peptides longer than two amino acids, but subsequently, peptides containing proline were noted to have reduced reactivity. As long as proteins are not extremely proline rich or do not have a very unusual composition, different proteins probably have similar reactivities as long as the biuret reaction is performed in the typical endpoint manner. Biuret rate assays are also available but should be considered as a separate category from endpoint assays. Cu^{2+} ions complex with small molecules and accessible sites in proteins almost instantaneously while additional absorbance change depends on the rate of unfolding of a protein and exposure of additional binding sites for Cu^{2+} ions under strongly alkaline conditions.

Direct Optical Methods. Absorbance of ultraviolet (UV) light between 200 to 225 nm and 270 to 290 nm has been used to measure protein concentrations and is commonly applied to monitor chromatographic separations of peptides and proteins. Absorbance at 280 nm depends primarily on the tryptophan and tyrosine content of a protein. This technique works best for purified proteins with known absorptivity. For complex mixtures, accuracy and specificity suffer from variable content of tryptophan and tyrosine and from absorbance of low molecular weight compounds such as free amino acids, uric acid, and bilirubin. From 200 to 225 nm, peptide bonds are chiefly responsible for UV absorbance. Absorptivity by proteins at these shorter wavelengths is 10 to 30 times greater than at 280 nm. Many low molecular weight compounds such as urea also have absorbance at wavelengths below 220 nm. Accurate measurement of proteins by this method may require removal of low molecular weight molecules before absorbance measurements are performed. Several other optical methods using infrared or Raman analysis of specimens offer methods for total protein determination based on complex spectral analysis.

Dye-Binding Methods. Dye-binding methods depend on shifts in the absorbance spectra of dyes when they bind to proteins. Variable binding of dyes to different proteins is a limitation. Calibration with a protein mixture is particularly difficult to define consistently. Calibration with a pure protein such as albumin may not simulate binding to the complex mixture of proteins in serum or plasma. Using Coomassie blue, for example, immunoglobulins often give only 60% of the response of an equivalent concentration of albumin or transferrin. This dye binds to polypeptide chains under acidic conditions, resulting in decreased absorbance at 465 nm and increased absorbance at 595 nm. Some dyes offer sensitivity at low protein concentrations. Pyrogallol red, for instance, has become one of the most commonly used dyes for analysis of fluids with lower protein concentrations such as urine and CSF.

Lowry (Folin-Ciocalteu) Method. The detection limit of the Lowry method is about 100 times lower than that of the biuret method.[170] In this technique, specimens are mixed with an alkaline copper solution followed by addition of the Folin-Ciocalteu reagent. Phosphotungstic acid and phosphomolybdic acid are reduced to tungsten blue and molybdenum blue by copper complexed with peptide and by tyrosine and tryptophan residues. Absorbance of products is measured between 650 and 750 nm. Reactivity of proteins varies with the content of tyrosine and tryptophan. Low molecular weight compounds, including tryptophan and tyrosine as well as drugs such as salicylates, chlorpromazine, tetracyclines, and some sulfa drugs, also interfere. Analysis of a fluid such as urine with high concentrations of phenolic compounds requires removal of low molecular weight substances before protein is measured.

Refractometry. Refractometry is a method used to rapidly estimate protein at high concentrations. Accuracy decreases at protein concentrations below 3.5 g/dL, where salts, glucose, and other low molecular weight compounds have a larger proportional effect on refractive index. Refractometry is used more often in clinical laboratories to assess the concentration of solutes in urine specimens than for determining total protein.

Light-Scattering Methods. Many different reagents have been used to aggregate protein for turbidimetric or nephelometric assays, including trichloroacetic acid, sulfosalicylic acid, benzethonium chloride, and benzalkonium salts under alkaline conditions. Precipitation methods for total protein assay depend on forming a suspension of uniform, insoluble protein particles, which scatter incident light. Albumin and globulins often give different reactivities in precipitation methods.

Variables Affecting Measured Protein Concentrations. Calibration of biuret methods commonly uses bovine or human albumin. Protein mixtures with specific albumin-to-globulin ratios often have been recommended for calibration of other methods. Using these calibration schemes, the total protein concentration of plasma obtained from healthy ambulatory adults is typically 6.5 to 8.5 g/dL. Serum usually contains a protein concentration about 0.3 g/dL less because of the content of fibrinogen and other proteins removed during clotting to form serum. Hemoconcentration and relative hyperproteinemia, with increased concentrations of all plasma proteins, occur with inadequate water intake or excessive water loss as in severe vomiting, diarrhea, Addison's disease, or diabetic ketoacidosis. Some hemoconcentration also occurs with standing (reduced intravascular volume) or prolonged tourniquet time during blood collection. Hemodilution and relative hypoproteinemia, with decreased concentrations of all plasma proteins, occur with water intoxication or salt retention syndromes, during massive intravenous infusions, and physiologically when a recumbent position is assumed. A recumbent position decreases total protein concentration by 0.3 to 0.5 g/dL and many individual proteins, including albumin, by up to 10%. This reflects the redistribution of extracellular fluid from the extravascular space to the intravascular space and therefore dilution of a constant amount of plasma protein in a larger volume.

Immunochemical Techniques for Specific Proteins

Nephelometric and turbidimetric methods are performed as equilibrium or rate methods for measuring the amount of light scattering by antigen–antibody complexes. Limits of detection of approximately 10 mg/L are attained with routine nephelometric and turbidimetric methods using antibodies in solution. Binding antibodies to particles of latex or other materials enhances light scattering and can lower limits of detection by 10- to 100-fold. Such assays may be described as latex-enhanced or as particle-enhanced assays. Nephelometric and turbidimetric assays commonly offer within-run coefficients of variation (CVs) of less than 5% except as limits of detection are approached. Turbidimetric methods can be applied on most automated chemistry analyzers capable of performing photometric methods. Nephelometry requires instrumentation capable of measuring light scattering at an angle to the incident light.

Electrophoresis

Electrophoresis is used to separate proteins by charge and thereby to assess protein variants or the concentrations of specific components in serum or other fluids. Electrophoretic techniques commonly performed in clinical laboratories include nondenaturing electrophoresis on cellulose acetate strips or agarose gels, capillary electrophoresis (CE), immunofixation, and "Western blotting." Protein separation in one or two dimensions with polyacrylamide gel is a powerful separation technique frequently used for research.

Serum Protein Electrophoresis. Generally, serum rather than plasma is used for electrophoresis of proteins on agarose gels to avoid the fibrinogen band at the β-γ interface. Fig. 13.6 illustrates examples of serum electrophoretic patterns for normal and pathologic specimens. Analysis usually is performed with low ionic strength buffers (0.05 mol/L) at slightly basic pH (~ 8.6). For agarose gels, the usual sample is 3 to 5 μL. Much smaller volumes are injected for CE. Separation and processing times are typically about 1 hour for agarose gels and a few minutes for CE. A variety of stains are used to visualize proteins in gels, including amido black, Ponceau S, acid violet, and Coomassie blue. Levels of detection and linearity of protein detection vary with different dyes. Only a few of the most abundant proteins are visualized, and intensities of bands with protein stains usually relate to the mass of peptide; oligosaccharides and lipids may reduce rather than contribute to band intensities. Therefore, glycoproteins with a high proportion of carbohydrate (eg, α_1 acid glycoprotein) have lower detection responses than a nonglycosylated protein such as albumin. Quantitative analysis relies on densitometry, which provides relative proportions of different components rather than absolute amounts. Quantitation of individual components relies on calculations derived from total protein concentration. Lipophilic stains such as Sudan black are needed to visualize lipoproteins such as high-density lipoprotein (α_1-lipoprotein), very-low-density lipoprotein (pre-β-region), low-density lipoprotein (LDL; β-lipoprotein), or chylomicrons (origin). CE detects proteins passing through a flow cell by their light absorbance at wavelengths below 220 nm. This offers a more unbiased assessment of protein concentration than staining intensity because absorbance of proteins in the low ultraviolet region is more consistently related to mass. Small molecules at high concentrations, such as metabolites, radiocontrast dyes, or drugs, may also yield absorbance peaks, and this presents problems with analysis of urine specimens.

The major clinical application of serum protein electrophoresis is the detection of monoclonal immunoglobulins (paraproteins) to assist in the diagnosis and monitoring of multiple myeloma and related disorders. Most monoclonal immunoglobulins are observed in the β-region or γ-region. Quantitation of monoclonal components serves as a means of monitoring disease progression and response to therapy. Identification of paraproteins requires distinction from a variety of other sources of additional bands or pseudoparaproteins by means such as IFE as described later. Incompletely clotted specimens contain fibrinogen. Genetic or posttranslational variants of proteins such as transferrin, Hp, and C3 may migrate in different positions than usual. Large increases in CRP may yield a detectable band in the β- or γ-region. Increased lysozyme in monocytic leukemia may produce a band in the post-γ-region. Hemoglobin will yield a band in hemolyzed specimens.

Protein electrophoresis is capable of being informative in many other clinical circumstances. Changes in the α_1-region are typically related to AAT. Decreases are associated with AAT deficiency or protein-losing disorders. Increases are related to inflammation. Changes in the α_2-region usually relate to changes in Hp and α_2 macroglobulin (AMG). Migration of Hp varies with genotype. Hp decreases with in vivo hemolysis and increases with inflammation. AMG and high molecular weight forms of Hp increase in nephrotic syndrome, but most other protein components decrease. Bands in the β-region are related to transferrin, C3, and LDL. Migration of transferrin may change from the β_1-region to the β_2-region when significant carbohydrate deficient forms are present. An increase between β- and γ-bands, so-called bridging of β- and γ-bands, suggests an increase in IgA as seen with cirrhosis, respiratory tract or skin infection, and rheumatoid arthritis. Finally, increases or decreases in the γ-region suggest changes in immunoglobulins. Increases result from chronic infection or paraproteins. Decreases occur with many immunodeficiency states. Multiple myeloma may suppress global production of immunoglobulins other than from the expanded clone. A decrease in the γ-region may suggest the need for additional studies such as IFE to detect paraproteins.

Immunofixation Electrophoresis. Immunofixation electrophoresis is complementary to serum protein electrophoresis and employs antisera targeted to specific proteins rather than nonspecific dyes. Examples of IFE from two patients with monoclonal gammopathies are shown in Fig. 13.7. Specific lanes of the gel are overlaid with antisera against κ and λ light chains or γ, μ, and α-immunoglobulin heavy chains. Antibody dilutions are adjusted to provide approximate equivalence with immunoglobulins separated in the gel so as to form immune complexes precipitated in the gel. Proteins not precipitated are washed from the gel to lower background, and precipitated proteins are stained. Paraproteins characteristically yield sharper precipitin bands than the heterogeneous polyclonal immunoglobulins. IFE provides more sensitive detection of paraproteins because of the lower background and signal amplification from immune complex formation. Additionally, immunofixation helps identify the

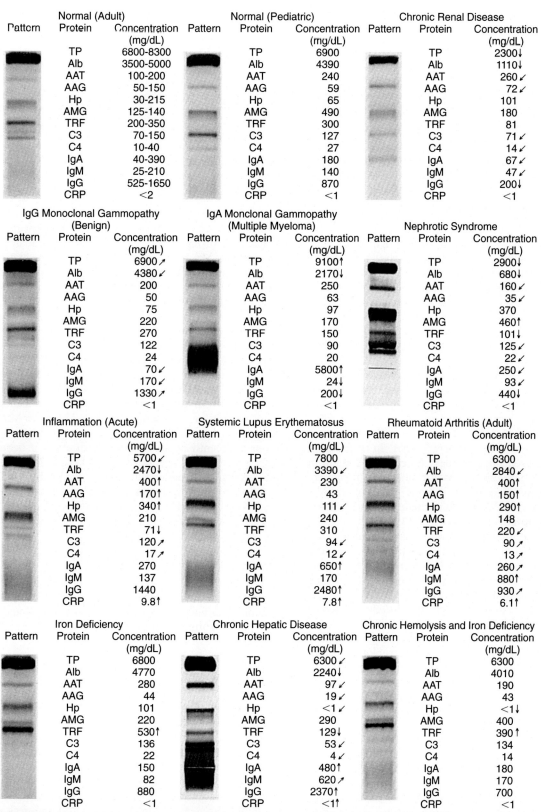

FIGURE 13.6 Electrophoretic patterns typical of normal conditions and of some pathologic conditions (agarose gel). The *upward-* and *downward-pointing arrows* indicate increase and decrease from the reference interval, respectively. *Right-* and *left-slanting arrows* indicate variation from normal to an increase or from normal to a decrease from the reference interval, respectively. *AAG,* Alpha-1 acid glycoprotein; *AAT,* alpha-1 antitrypsin; *Alb,* albumin; *AMG,* alpha-2 macroglobulin; *C3,* complement component 3; *C4,* complement component 4; *CRP,* C-reactive protein; *Hp,* haptoglobin; *Ig,* immunoglobulin; *TP,* total protein concentration; *TRF,* transferrin.

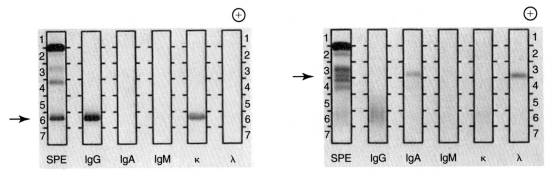

FIGURE 13.7 Immunofixation electrophoresis (IFE). *Left,* Patient specimen with an immunoglobulin G (IgG; κ) monoclonal protein. *Right,* Patient specimen with an immunoglobulin A (IgA; λ) monoclonal protein. The *arrow* indicates the position of monoclonal protein.

immunoglobulin type of the paraprotein. Sometimes more than one clone may be expanded, or free light chain may occur together with intact immunoglobulin. Uncommonly, paraproteins may be of the IgD or IgE class. These paraproteins should be detected by antisera versus light chains but require δ or ε heavy chain–specific antisera to distinguish them from free light chains. High concentrations of paraproteins may interfere with fixation (prozone effect) requiring specimen dilution for optimal studies. For paraproteins present in high concentrations, quantitative analysis of immunoglobulins can confirm the unbalanced production of a specific class of immunoglobulin and may assist in proper dilution of specimens for immunofixation.

Capillary Electrophoresis. Capillary electrophoresis of proteins relies on zone electrophoresis in small-bore (10–100 μm), fused silica capillary tubes 20 to 200 cm in length. Electrokinetic or hydrostatic injection introduces a small amount of protein that is resolved rapidly under high voltage. One of the challenges is to avoid adsorption of proteins to the surface of the capillary. CE is suitable for automation and offers rapid analysis with no need for gel handling or staining. Direct UV detection offers slightly different specificity than protein staining and offers better reproducibility of quantitation than densitometry. Immunofixation cannot be performed with CE. Immunosubtraction with specific antisera is used as an alternative procedure to identify paraproteins.[171]

Western Blotting. For Western blotting, proteins separated on a gel are transferred by diffusion or electroblotting onto a membrane made of nitrocellulose or polyvinylidene fluoride. Proteins bound on the membrane are identified with specific antibodies conjugated to enzymes such as peroxidase or alkaline phosphatase that act on photometric, fluorescent, or chemiluminescent signaling molecules.

Mass Spectrometry. Multiple types of MS instrumentation provide qualitative or quantitative information about proteins. An advantage of MS is the ability to analyze a large number of components in a single analysis, including rapid-sequence analysis of peptides. MS, therefore, has been an enabling technology in *proteomics,* defined as the effort to study the complete set of proteins in an organism or in subcompartments of an organism such as plasma. Ionization of peptides and proteins has been accomplished by electrospray or MALDI sources. Electrospray is better suited to analyzing small peptides than larger intact proteins. After ionization, proteins can be separated by quadrupoles, ion traps, time-of-flight, and other types of mass analyzers. Use of tandem MS with an intermediate fragmentation step between two stages of MS separation offers high sensitivity and specificity for the quantitative analysis of peptides, as it does for most small molecules.

Advantages of using MS for quantitative analysis include the ability to analyze components without developing specific antibodies and the ability to multiplex a large number of measurements. MS can provide information about posttranslational modifications that is difficult to assess by immunoassays and chromatographic or electrophoretic techniques. Examples of clinical applications include identification of genetic variants of prealbumin and CDT. The use of MS is likely to increase for accurate determination of protein concentrations as recently applied for standardization of hemoglobin A_{1c}, insulin, and C-peptide.[172,173] Likewise, MS is also able to distinguish peptides differing in length by one or two amino acids or by a posttranslational modification. For these reasons, MS is likely to find increased use for clinical laboratory analysis of bioactive peptides and other components of the peptidome.

Proteins in Other Body Fluids

Complex mixtures of proteins are present in all biologic fluids; analysis of a variety of other specimens is diagnostically useful, including analyses of urine, CSF, pleural and peritoneal fluids, amniotic fluid, saliva, and feces.

Saliva

Saliva has a very different protein composition compared with plasma.[174,175] Protein composition varies with the site and method of sampling of saliva. In addition to the well-known presence of amylase, proteomic approaches have detected sequence from hundreds of different proteins in saliva. Proteins involved in host defense against pathogens such as immunoglobulins, lysozyme, and lactoferrin are also particularly abundant. Efforts are under way to exploit the salivary proteome to detect and characterize susceptibility to dental caries, periodontal disease, head and neck cancers, diabetes, and cystic fibrosis. Patients with Sjögren syndrome exhibit increased concentrations of $β_2$ MG and other inflammatory proteins compared with normal patients. Interrogation of saliva for the presence of secretory IgA is common in the diagnostic workup of hypogammaglobulinemia.

Cerebrospinal Fluid

Cerebrospinal fluid is the extracellular fluid around the brain and spinal column. CSF usually has total protein concentrations about 100-fold lower than plasma and a different protein composition because most proteins have limited passage across the blood—brain barrier. CSF for testing is most frequently obtained by a spinal tap in the lumbar region. CSF is secreted by the choroid plexus, around the cerebral vessels, and along the walls of the ventricles of the brain. It fills the ventricles and cisternae, bathes the spinal cord, and is reabsorbed into the blood through the arachnoid villi. CSF turnover is rapid, exchanging totally about four times daily. More than 80% of CSF protein content originates from plasma by ultrafiltration and pinocytosis; the remainder is derived from intrathecal synthesis. The lowest protein concentration and the smallest proportion of larger protein molecules are in the ventricular fluid. As the CSF passes down to the lumbar spine (where specimens are usually collected), the protein concentration increases.

Low to intermediate molecular weight plasma proteins such as prealbumin, albumin, and transferrin normally predominate because CSF is mainly an ultrafiltrate of plasma. No protein with a molecular weight greater than that of IgG is present in sufficient concentration to be visible on electrophoresis. The normal plasma/CSF gradient is about 14 for prealbumin, 240 for albumin, 140 for transferrin, 800 for IgG, and more than 1000 for larger proteins such as IgA, AMG, fibrinogen, IgM, and β-lipoprotein.[176]

The electrophoretic pattern of normal CSF after concentration of the fluid has two striking features—a prealbumin band and two transferrin bands—one at β_2 in addition to the usual β_1 position. As mentioned previously, β_2-transferrin has decreased charge because its glyosyl chains lack terminal sialic acid residues. Both β_2-transferrin and another relatively abundant CSF protein, prostaglandin D synthase or β-trace, have been used to determine whether clear fluids from nasal or ear passages represent leakage of CSF, so-called CSF rhinorrhea and otorrhea.[147] The utility of β2-transferrin in this context is compromised in patients with congenital disorders of glycosylation. Prostaglandin D synthase is a protein of about 30 kDa that is relatively enriched in CSF relative to serum or plasma, where it normally is cleared rapidly by glomerular filtration. Apolipoprotein E (ApoE) is also abundant in CSF because it plays a primary role in lipid transport rather than apo AI— and apo B100—containing lipoproteins, which are relatively scarce in CSF.

The blood—brain or blood—CSF barrier limits exchange of many compounds, particularly large compounds such as proteins. Analyses of total protein and specific proteins in CSF are used primarily to detect increased permeability of the blood—CSF barrier, increased intrathecal synthesis, or increased release of proteins from neural and glial tissue. Conditions such as viral meningitis, encephalitis, increased intracranial pressure, trauma, and hemorrhage may all compromise the blood—brain barrier, resulting in increased CSF protein. Protein concentrations associated with these conditions are displayed in Table 13.7. Increased intrathecal synthesis of immunoglobulins, particularly IgG, occurs in demyelinating diseases of the CNS, especially multiple

TABLE 13.7 Cerebrospinal Fluid Total Protein in Various Diseases

Clinical Condition	Appearance and Cells × 10^6/L	Total Protein, mg/dL
Normal	Clear, colorless; 0–5 lymphocytes	15–45*
INCREASED ADMIXTURE OF PROTEINS FROM BLOOD		
Increased capillary permeability		
• Bacterial meningitis	Turbid, opalescent, purulent, usually >500 polymorphs	80–500
• Cryptococcal meningitis	Clear or turbid; 50–150 polymorphs or lymphocytes	25–200
• Leptospiral meningitis	Clear to slight haze; polymorphs early, then 5–100 lymphocytes	50–100
• Viral meningitis	Clear or slight haze, colorless; usually ≤500 lymphocytes	30–100
• Encephalitis	Clear or slight haze, colorless; usually ≤500 lymphocytes	15–100
• Poliomyelitis	Clear, colorless; ≤500 lymphocytes	10–300
• Brain tumor	Usually clear; 0–80 lymphocytes	15–200 (usually normal)
Mechanical obstruction:		
• Spinal cord tumor†	Clear, colorless, or yellow	100–2000
Hemorrhage		
Cerebral hemorrhage	Colorless, yellow, or bloody; blood cells	30–150
Local immunoglobulin production		
• Neurosyphilis	Clear, colorless; 10–100 lymphocytes	50–150
• Multiple sclerosis‡	Clear, colorless; 0–10 lymphocytes	25–50
Both increased capillary permeability and local immunoglobulin production:		
• Tuberculous meningitis	Colorless, fibrin clot, or slightly turbid; 50–500 lymphocytes	50–300 (occasionally ≤1000)
• Brain abscess	Clear or slightly turbid	20–120
After myelography (inflammatory reaction)		Slight increase

*Premature infants: ≤400 mg/dL; children: 30–100 mg/dL; older adults: ≤60 mg/dL.
† Froin syndrome: Lumbar fluid values are much higher than cisternal fluid values.
‡ Similar values may occur in certain other chronic inflammatory conditions of the nervous system.

sclerosis. Methods used to assess potential abnormalities in CSF protein content are detailed next.

Total Protein in Cerebrospinal Fluid. The total protein concentration in CSF is an indicator of blood–CSF permeability. The protein concentration of CSF usually is more than 100-fold lower than for plasma, and methods with greater sensitivity or increased specimen volume are required for measuring total serum or plasma protein. In practice, methods commonly used by clinical laboratories to measure total CSF protein include (1) pyrogallol red, (2) benzethonium chloride, (3) reverse-biuret, and (4) biuret. The reverse-biuret method measures free copper remaining after formation of biuret complexes by reduction and complexation with a chelating dye. Analyzers often use the same method, possibly with some adjustment of specimen volume, for CSF and urine protein. The usual reference interval for CSF total protein is 15 to 45 mg/dL. Total protein concentrations are considerably higher in neonates and in healthy elderly adults. In CSF from premature and full-term neonates, concentrations up to 400 mg/dL are observed. In term newborns, a progressive decline in the reference interval CSF protein is seen over the first few weeks of life, with values approaching adult concentrations after 4 months of age.[177]

Permeability of the Blood–Brain Barrier. A more specific measure of the permeability of the blood–CSF barrier involves determination of the ratio of albumin concentration in CSF versus plasma. This ratio, the *CSF/serum albumin index*, usually is calculated for CSF albumin in milligrams per deciliter and for serum protein in grams per liter, effectively multiplying values by 1000. A CSF–serum albumin index less than 9 indicates an intact blood–CSF barrier. Values of 9 to 14 represent slight impairment, 14 to 30 moderate impairment, and greater than 30 severe impairment. The index helps correct for variation in serum albumin concentration.

Intrathecal Protein Synthesis. Measurement of the CSF–serum immunoglobulin ratio and assessment of oligoclonal immunoglobulin bands on electrophoretic separations of CSF serve as assays for intrathecal immunoglobulin synthesis. At least 90% of cases of multiple sclerosis give positive findings, but increased immunoglobulins and oligoclonal immunoglobulins may be found in other inflammatory diseases of the CNS, such as infection caused by bacteria, viruses, fungi, or parasites; neurosyphilis; subacute sclerosing panencephalitis; and Guillain-Barré syndrome. Multiple sclerosis is less likely if CSF total protein exceeds 100 mg/dL or if the CSF leukocyte count is greater than 50/μL. The CSF albumin concentration in 70% of cases of multiple sclerosis is within the reference interval. Increases in CSF IgG concentration or in the CSF–serum IgG ratio may result from increased permeability of the blood–CSF barrier, increased local production of IgG, or both. To identify intrathecal production specifically, correction for increased permeability is necessary. Corrections use *CSF and serum* albumin and IgG concentrations in one of several ways:

1. Concentrations *in CSF* of IgG and albumin are measured, and the IgG–albumin ratio is calculated. A ratio greater than 0.27 is considered indicative of increased synthesis; in about 70% of cases of multiple sclerosis, the ratio exceeds 0.27.

$$\text{Ratio} = \frac{\text{IgG}_{\text{CSF}}(\text{mg/dL})}{\text{Albumin}_{\text{CSF}}(\text{mg/dL})}$$

2. Concentrations in *CSF and serum* of IgG and albumin are measured, and the CSF immunoglobulin index is calculated.

$$\text{Index} = \frac{\text{IgG}_{\text{CSF}}(\text{mg/dL}) \times \text{Albumin}_{\text{serum}}(\text{g/dL})}{\text{Albumin}_{\text{CSF}}(\text{mg/dL}) \times \text{IgG}_{\text{serum}}(\text{g/dL})}$$

The reference interval for the index is 0.30 to 0.70. Values greater than 0.70 indicate increased IgG synthesis; in more than 80% of cases of multiple sclerosis, this index exceeds 0.70. This CSF immunoglobulin index is now frequently used.

3. The rate of intrathecal IgG synthesis is estimated by Tourtellotte's formula.[178] The rate of synthesis of IgG in milligrams per day is equal to:

$$5\,\text{dL/d}\left[\left\{\text{IgG}_{\text{CSF}} - \frac{\text{IgG}_{\text{serum}}}{369}\right\} - \left\{\left(\text{Albumin}_{\text{CSF}} - \frac{\text{Albumin}_{\text{serum}}}{230}\right)\right.\right.$$
$$\left.\left.\times \frac{0.43(\text{IgG}_{\text{serum}})}{\text{Albumin}_{\text{serum}}}\right\}\right]$$

where protein concentrations are in mg/dL. The 5 dL/d term is daily CSF production. The first bracketed term represents the difference between IgG found in CSF and the IgG expected if the blood–brain barrier is intact. The second bracketed term represents the same for albumin but is corrected by a factor of 0.43, corresponding to the ratio of molecular weights of albumin to IgG, assuming that 1 mole of IgG accompanies every mole of albumin that passes the blood-brain barrier. The constants 369 and 230 originate from the normal serum–CSF ratios for IgG and albumin. The reference interval for the synthesis rate is 0 to 3.3 mg/d. Values exceeding 8 mg/d are found in most cases of multiple sclerosis. This estimator provides no more clinical information than the IgG index; the complex formula merely rearranges the results for serum and CSF IgG and albumin and then factors in several constants.

In addition to albumin and IgG, the concentrations of other CSF proteins may shed light on other disease processes. Myelin basic protein concentrations may be an indicator of myelin turnover in multiple sclerosis.[179] A number of other proteins such as S100B and neuron-specific enolase are potential markers of traumatic or ischemic brain injury.[180] In acute leukemia and lymphoma with central nervous system involvement, the concentration of β2 MG is increased in CSF. Concentrations of tau and β-amyloid isoforms may be useful in the diagnosis and prognosis of Alzheimer disease.[181]

Peritoneal and Pleural Fluids

Pathologic accumulations of fluid in the peritoneal and pleural cavities or elsewhere vary greatly in protein content. These fluids may be ultrafiltrates with low-protein concentrations relative to plasma and scant amounts of large proteins (transudates). Alternatively, fluids may have protein concentrations approaching those of plasma and significant amounts of large proteins such as immunoglobulins and AMG (exudates) in response to local inflammation and increased vascular permeability. Distinction between transudates and exudates assists in diagnosing the cause of fluid accumulation. The major cause of pleural transudates is congestive heart failure. Exudates occur with infection, pleuritis, pulmonary embolism, and cancer. Criteria for identifying exudates in pleural fluid were proposed by Light.[182] According to these criteria ratios of (1) pleural fluid protein to serum protein greater than 0.5, (2) pleural lactate dehydrogenase (LDH) to serum LDH activity greater than 0.6, and (3)c) total pleural fluid LDH greater than 200 IU/L likely result from an exudative process.[183] Pleural fluid may be turbid from large numbers of white blood cells (WBCs), fibrin particles, or chylomicrons. Chylous effusions (containing chylomicrons) result from lymphatic obstruction related to cancer, surgery, trauma, sarcoidosis, or other causes.[184] Lymphatics, particularly the major trunk, the thoracic duct, serve as the major route for chylomicrons from the intestines to the blood circulation. Entry of chylomicrons into the pleural space, therefore, is related in part to dietary fat intake to generate chylomicrons, and fasting lowers the fat content of chylous effusions. Chylomicrons in fluids may be identified by separation of a cream layer upon standing or by triglycerides analysis. For peritoneal fluid, the serum to ascites albumin gradient (ie, the difference between serum and peritoneal fluid albumin) helps distinguish transudates (mainly in portal hypertension from cirrhosis) from infection and other causes of ascites. Usually, the serum to ascites albumin gradient is greater than 1.1 g/dL in portal hypertension and is lower than 1.1 g/dL for other causes that generate exudates.[185]

Fecal Material

The use of fecal material for protein analysis is relatively infrequent but indicated in some specific clinical circumstances. Assessment of protein content in feces is fraught with a number of limitations. Results of fecal protein content are often normalized to fecal weight. Thus, watery stool specimens generally provides decreased estimates of protein concentration. Extraction is likewise dependent on stool consistency and homogeneity, leading to considerable within-subject and between-subject variability.

Protein loss in the GI tract may be assessed using an assay of fecal AAT. The amount of fecal AAT excreted over time is determined as a function of the serum AAT concentration. Correction for serum AAT concentration is necessary because of variation in serum AAT from severe enteric losses or from acute-phase responses. AAT is relatively stable in the lower digestive tract but is digested in the stomach at acid pH. Therefore, suppression of acid secretion is necessary to use AAT excretion as a measure of gastric protein-losing enteropathy.[186] An alternative measure of increased GI loss of protein has been the assessment of stool radioactivity following the injection of radiolabeled albumin.

Inflammatory bowel disease (IBD) includes Crohn's disease and ulcerative colitis. Although GI histology remains the gold standard for diagnosis, a number of fecal protein markers have been employed as tools for screening and response to therapy. In the setting of IBD, it may be more useful to have indicators of inflammation rather than leakage of plasma proteins. Fecal products secreted by WBCs, such as lactoferrin and calprotectin, have been used as a measure of disease activity in IBD.[187] Lactoferrin has diagnostic sensitivity and specificity ranging from 47% to 92% and 60% to 100%, respectively, for the diagnosis of IBD. The sensitivity and specificity of calprotectin is similar and ranges from 61% to 100% and 72% to 100%, respectively.[188]

Exocrine pancreatic disease occurs in the setting of multiple pathologic states, including chronic alcoholism, diabetes, HIV, celiac disease, and cystic fibrosis. Determination of fecal proteins such as elastase that are derived from pancreatic secretions provides an alternative for the diagnosis of pancreatic insufficiency compared with gold standard duodenal sampling after secretin administration.[189,190] For example, fecal elastase concentrations of less than 100 μg/mg stool suggest severe functional deficiency with sensitivity ranging from 72% to 100% and specificity ranging from 29% to 96%.[188]

POINTS TO REMEMBER

- Amino acids are the building blocks of protein but also serve as substrates for energy generation and other important biomolecules.
- Amino acid homeostasis depends on enteral extraction, a host of broad-specificity transport systems, as well as metabolism.
- An array of small peptide systems derived from larger protein precursors dictate control of numerous physiologic processes, including glucose and electrolyte homeostasis (ACTH), fluid retention (natriuretic peptides, vasopressin), iron metabolism (hepcidin), and vascular tone (angiotensins, ETs).
- Protein diversity is derived from linear amino acid sequence and an array of cotranslational and posttranslational modifications, including acylation, prenylation, phosphorylation, and glycosylation.
- Current catalogs of the circulating human proteome contain more than 12,000 components.

REFERENCES

1. Miller SL. A production of amino acids under possible primitive earth conditions. *Science* 1953;**117**(3046):528–9.
2. Berry MJ, Larsen PR. Recognition of UGA as a selenocysteine codon in eukaryotes: a review of recent progress. *Biochem Soc Trans* 1993;**21**(4):827–32.
3. Otten JJHJ, Meyers LD. *Dietary Reference Intakes: the essential guide to nutrient requirements*. Washington, DC: National Academy of Sciences Press; 2006.
4. Vermeulen MA, van de Poll MC, Ligthart-Melis GC, et al. Specific amino acids in the critically ill patient—exogenous glutamine/arginine: a common denominator? *Crit Care Med* 2007;**35**(9 Suppl.):S568–76.
5. Morris Jr SM. Arginine: beyond protein. *Am J Clin Nutr* 2006; **83**(2). 508s–12s.
6. Devoto G, Gallo F, Marchello C, et al. Prealbumin serum concentrations as a useful tool in the assessment of malnutrition in hospitalized patients. *Clin Chem* 2006;**52**(12):2281–5.
7. Kurpad AV. The requirements of protein & amino acid during acute & chronic infections. *Indian J Med Res* 2006;**124**(2): 129–48.
8. Hay Jr WW. Strategies for feeding the preterm infant. *Neonatology* 2008;**94**(4):245–54.
9. Jahoor F, Badaloo A, Reid M, et al. Protein metabolism in severe childhood malnutrition. *Ann Trop Paediatr* 2008;**28**(2):87–101.

10. Beck FK, Rosenthal TC. Prealbumin: a marker for nutritional evaluation. *Am Fam Physician* 2002;**65**(8):1575–8.
11. Bernstein LH, Ingenbleek Y. Transthyretin: its response to malnutrition and stress injury: clinical usefulness and economic implications. *Clin Chem Lab Med* 2002;**40**(12):1344–8.
12. Broer S. Amino acid transport across mammalian intestinal and renal epithelia. *Physiol Rev* 2008;**88**(1):249–86.
13. Palacin M, Estevez R, Bertran J, et al. Molecular biology of mammalian plasma membrane amino acid transporters. *Physiol Rev* 1998;**78**(4):969–1054.
14. Adibi SA. The oligopeptide transporter (Pept-1) in human intestine: biology and function. *Gastroenterology* 1997;**113**(1):332–40.
15. Collarini EJ, Oxender DL. Mechanisms of transport of amino acids across membranes. *Annu Rev Nutr* 1987;**7**:75–90.
16. Broer S, Palacin M. The role of amino acid transporters in inherited and acquired diseases. *Biochem J* 2011;**436**(2):193–211.
17. Palacin M, Nunes V, Font-Llitjos M, et al. The genetics of heteromeric amino acid transporters. *Physiology (Bethesda)* 2005;**20**:112–24.
18. Kanai Y, Clemencon B, Simonin A, et al. The SLC1 high-affinity glutamate and neutral amino acid transporter family. *Mol Aspects Med* 2013;**34**(2–3):108–20.
19. Schioth HB, Roshanbin S, Hagglund MG, et al. Evolutionary origin of amino acid transporter families SLC32, SLC36 and SLC38 and physiological, pathological and therapeutic aspects. *Mol Aspects Med* 2013;**34**(2–3):571–85.
20. Palmieri F. The mitochondrial transporter family SLC25: identification, properties and physiopathology. *Mol Aspects Med* 2013;**34**(2–3):465–84.
21. Fotiadis D, Kanai Y, Palacin M. The SLC3 and SLC7 families of amino acid transporters. *Mol Aspects Med* 2013;**34**(2–3):139–58.
22. Wurtman RJ, Rose CM, Chou C, et al. Daily rhythms in the concentrations of various amino acids in human plasma. *N Engl J Med* 1968;**279**(4):171–5.
23. Hussein MA, Young VR, Murray E, et al. Daily fluctuation of plasma amino acid levels in adult men: effect of dietary tryptophan intake and distribution of meals. *J Nutr* 1971;**101**(1):61–9.
24. Feigin RD, Beisel WR, Wannemacher Jr RW. Rhythmicity of plasma amino acids and relation to dietary intake. *Am J Clin Nutr* 1971;**24**(3):329–41.
25. Hawkins RA, O'Kane RL, Simpson IA, et al. Structure of the blood-brain barrier and its role in the transport of amino acids. *J Nutr* 2006;**136**(1 Suppl.). 218s–26s.
26. Oldendorf WH, Szabo J. Amino acid assignment to one of three blood-brain barrier amino acid carriers. *Am J Physiol* 1976;**230**(1):94–8.
27. Dharnidharka VR, Al Khasawneh E, Gupta S, et al. Verification of association of elevated serum IDO enzyme activity with acute rejection and low CD4-ATP levels with infection. *Transplantation* 2013;**96**(6):567–72.
28. Rudnick DA, Dietzen DJ, Turmelle YP, et al. Serum alpha-NH-butyric acid may predict spontaneous survival in pediatric acute liver failure. *Pediatr Transplant* 2009;**13**(2):223–30.
29. Wang TJ, Larson MG, Vasan RS, et al. Metabolite profiles and the risk of developing diabetes. *Nat Med* 2011;**17**(4):448–53.
30. Odibo AO, Goetzinger KR, Odibo L, et al. First-trimester prediction of preeclampsia using metabolomic biomarkers: a discovery phase study. *Prenat Diagn* 2011;**31**(10):990–4.
31. Mangoni AA. The emerging role of symmetric dimethylarginine in vascular disease. *Adv Clin Chem* 2009;**48**:73–94.
32. Blackwell S. The biochemistry, measurement and current clinical significance of asymmetric dimethylarginine. *Ann Clin Biochem* 2010;**47**(Pt 1):17–28.
33. Moore S, Stein WH. Photometric ninhydrin method for use in the chromatography of amino acids. *J Biol Chem* 1948;**176**(1):367–88.
34. Dietzen DJ, Weindel AL, Carayannopoulos MO, et al. Rapid comprehensive amino acid analysis by liquid chromatography/tandem mass spectrometry: comparison to cation exchange with post-column ninhydrin detection. *Rapid Commun Mass Spectrom* 2008;**22**(22):3481–8.
35. Casetta B, Tagliacozzi D, Shushan B, et al. Development of a method for rapid quantitation of amino acids by liquid chromatography-tandem mass spectrometry (LC-MSMS) in plasma. *Clin Chem Lab Med* 2000;**38**(5):391–401.
36. Nagy K, Takats Z, Pollreisz F, et al. Direct tandem mass spectrometric analysis of amino acids in dried blood spots without chemical derivatization for neonatal screening. *Rapid Commun Mass Spectrom* 2003;**17**(9):983–90.
37. Rodnina MV, Beringer M, Bieling P. Ten remarks on peptide bond formation on the ribosome. *Biochem Soc Trans* 2005;**33**(Pt 3):493–8.
38. Merrifield RB, Stewart JM. Automated peptide synthesis. *Nature* 1965;**207**(996):522–3.
39. Sheehan JC. Forty years of peptide chemistry. *Biopolymers* 1986;**25**(Suppl.):S1–10.
40. Ramakrishnan C, Ramachandran GN. Stereochemical criteria for polypeptide and protein chain conformations. II. Allowed conformations for a pair of peptide units. *Biophys J* 1965;**5**(6):909–33.
41. Kringelum JV, Nielsen M, Padkjaer SB, et al. Structural analysis of B-cell epitopes in antibody:protein complexes. *Mol Immunol* 2013;**53**(1–2):24–34.
42. Dores RM, Baron AJ. Evolution of POMC: origin, phylogeny, posttranslational processing, and the melanocortins. *Ann N Y Acad Sci* 2011;**1220**:34–48.
43. Mountjoy KG. Functions for pro-opiomelanocortin-derived peptides in obesity and diabetes. *Biochem J* 2010;**428**(3):305–24.
44. Cone RD. Anatomy and regulation of the central melanocortin system. *Nat Neurosci* 2005;**8**(5):571–8.
45. Talbot JA, Kane JW, White A. Analytical and clinical aspects of adrenocorticotrophin determination. *Ann Clin Biochem* 2003;**40**(Pt 5):453–71.
46. Pandey KN. Biology of natriuretic peptides and their receptors. *Peptides* 2005;**26**(6):901–32.
47. Hall C. Essential biochemistry and physiology of (NT-pro)BNP. *Eur J Heart Fail* 2004;**6**(3):257–60.
48. Hawkridge AM, Heublein DM, Bergen 3rd HR, et al. Quantitative mass spectral evidence for the absence of circulating brain natriuretic peptide (BNP-32) in severe human heart failure. *Proc Natl Acad Sci USA* 2005;**102**(48):17442–7.
49. Niederkofler EE, Kiernan UA, O'Rear J, et al. Detection of endogenous B-type natriuretic peptide at very low concentrations in patients with heart failure. *Circ Heart Fail* 2008;**1**(4):258–64.
50. Macheret F, Boerrigter G, McKie P, et al. Pro-B-type natriuretic peptide(1-108) circulates in the general community: plasma determinants and detection of left ventricular dysfunction. *J Am Coll Cardiol* 2011;**57**(12):1386–95.
51. Cauliez B, Santos H, Bauer F, et al. Cross-reactivity with endogenous proBNP from heart failure patients for three

52. Park CH, Valore EV, Waring AJ, et al. Hepcidin, a urinary antimicrobial peptide synthesized in the liver. *J Biol Chem* 2001;**276**(11):7806−10.
53. Nicolas G, Bennoun M, Devaux I, et al. Lack of hepcidin gene expression and severe tissue iron overload in upstream stimulatory factor 2 (USF2) knockout mice. *Proc Natl Acad Sci USA* 2001;**98**(15):8780−5.
54. Ganz T. Hepcidin, a key regulator of iron metabolism and mediator of anemia of inflammation. *Blood* 2003;**102**(3):783−8.
55. Ganz T, Nemeth E. Hepcidin and iron homeostasis. *Biochim Biophys Acta* 2012;**1823**(9):1434−43.
56. Kroot JJ, Tjalsma H, Fleming RE, et al. Hepcidin in human iron disorders: diagnostic implications. *Clin Chem* 2011;**57**(12):1650−69.
57. Zipperer E, Post JG, Herkert M, et al. Serum hepcidin measured with an improved ELISA correlates with parameters of iron metabolism in patients with myelodysplastic syndrome. *Ann Hematol* 2013;**92**(12):1617−23.
58. Wolff F, Deleers M, Melot C, et al. Hepcidin-25: measurement by LC-MS/MS in serum and urine, reference ranges and urinary fractional excretion. *Clin Chim Acta* 2013;**423**:99−104.
59. Peters HP, Rumjon A, Bansal SS, et al. Intra-individual variability of serum hepcidin-25 in haemodialysis patients using mass spectrometry and ELISA. *Nephrol Dial Transplant* 2012;**27**(10):3923−9.
60. Fyhrquist F, Saijonmaa O. Renin-angiotensin system revisited. *J Intern Med* 2008;**264**(3):224−36.
61. Hoofnagle AN. Peptide lost and found: internal standards and the mass spectrometric quantification of peptides. *Clin Chem* 2010;**56**(10):1515−7.
62. Bystrom CE, Salameh W, Reitz R, et al. Plasma renin activity by LC-MS/MS: development of a prototypical clinical assay reveals a subpopulation of human plasma samples with substantial peptidase activity. *Clin Chem* 2010;**56**(10):1561−9.
63. Khimji AK, Rockey DC. Endothelin—biology and disease. *Cell Signal* 2010;**22**(11):1615−25.
64. Rubanyi GM. The discovery of endothelin: the power of bioassay and the role of serendipity in the discovery of endothelium-derived vasoactive substances. *Pharmacol Res* 2011;**63**(6):448−54.
65. Haaf P, Zellweger C, Reichlin T, et al. Utility of C-terminal proendothelin in the early diagnosis and risk stratification of patients with suspected acute myocardial infarction. *Can J Cardiol* 2014;**30**(2):195−203.
66. Meister A, Anderson ME. Glutathione. *Annu Rev Biochem* 1983;**52**:711−60.
67. Pastore A, Federici G, Bertini E, et al. Analysis of glutathione: implication in redox and detoxification. *Clin Chim Acta* 2003;**333**(1):19−39.
67a. Cohn EJ, Strong LE. Preparation and properties of serum and plasma proteins; a system for the separation into fractions of the protein and lipoprotein components of biological tissues and fluids. *J Am Chem Soc* 1946;**68**:459−75.
68. Gloge F, Becker AH, Kramer G, et al. Co-translational mechanisms of protein maturation. *Curr Opin Struct Biol* 2014;**24**:24−33.
69. Kramer G, Boehringer D, Ban N, et al. The ribosome as a platform for co-translational processing, folding and targeting of newly synthesized proteins. *Nat Struct Mol Biol* 2009;**16**(6):589−97.
70. Saibil H. Chaperone machines for protein folding, unfolding and disaggregation. *Nat Rev Mol Cell Biol* 2013;**14**(10):630−42.
71. Law RH, Zhang Q, McGowan S, et al. An overview of the serpin superfamily. *Genome Biol* 2006;**7**(5):216.
72. Dobson CM. Protein folding and misfolding. *Nature* 2003;**426**(6968):884−90.
73. Scheuner D, Kaufman RJ. The unfolded protein response: a pathway that links insulin demand with beta-cell failure and diabetes. *Endocr Rev* 2008;**29**(3):317−33.
74. Williams AJ, Paulson HL. Polyglutamine neurodegeneration: protein misfolding revisited. *Trends Neurosci* 2008;**31**(10):521−8.
75. Kwong LK, Uryu K, Trojanowski JQ, et al. TDP-43 proteinopathies: neurodegenerative protein misfolding diseases without amyloidosis. *Neurosignals* 2008;**16**(1):41−51.
76. Carrell RW, Lomas DA. Alpha1-antitrypsin deficiency—a model for conformational diseases. *N Engl J Med* 2002;**346**(1):45−53.
77. Silverman EK, Sandhaus RA. Clinical practice. Alpha1-antitrypsin deficiency. *N Engl J Med* 2009;**360**(26):2749−57.
78. Gomes CM. Protein misfolding in disease and small molecule therapies. *Curr Top Med Chem* 2012;**12**(22):2460−9.
79. Lingappa VR, Blobel G. Early events in the biosynthesis of secretory and membrane proteins: the signal hypothesis. *Recent Prog Horm Res* 1980;**36**:451−75.
80. Gordon JI, Duronio RJ, Rudnick DA, et al. Protein N-myristoylation. *J Biol Chem* 1991;**266**(14):8647−50.
81. van Vliet C, Thomas EC, Merino-Trigo A, et al. Intracellular sorting and transport of proteins. *Prog Biophys Mol Biol* 2003;**83**(1):1−45.
82. Venditti R, Wilson C, De Matteis MA. Exiting the ER: what we know and what we don't. *Trends Cell Biol* 2014;**24**(1):9−18.
83. Brown DA, Rose JK. Sorting of GPI-anchored proteins to glycolipid-enriched membrane subdomains during transport to the apical cell surface. *Cell* 1992;**68**(3):533−44.
84. Schatz G. How mitochondria import proteins from the cytoplasm. *FEBS Lett* 1979;**103**(2):203−11.
85. Aicart-Ramos C, Valero RA, Rodriguez-Crespo I. Protein palmitoylation and subcellular trafficking. *Biochim Biophys Acta* 2011;**1808**(12):2981−94.
86. Delhanty PJ, Neggers SJ, van der Lely AJ. Mechanisms in endocrinology: ghrelin: the differences between acyl- and desacyl ghrelin. *Eur J Endocrinol* 2012;**167**(5):601−8.
87. Sato T, Nakamura Y, Shiimura Y, et al. Structure, regulation and function of ghrelin. *J Biochem* 2012;**151**(2):119−28.
88. Cohen P. The regulation of protein function by multisite phosphorylation—a 25 year update. *Trends Biochem Sci* 2000;**25**(12):596−601.
89. Krebs EG, Fischer EH. Phosphorylase and related enzymes of glycogen metabolism. *Vitam Horm* 1964;**22**:399−410.
90. Walsh DA, Brostrom CO, Brostrom MA, et al. Cyclic AMP-dependent protein kinases from skeletal muscle and liver. *Adv Cyclic Nucleotide Res* 1972;**1**:33−45.
91. Graves JD, Krebs EG. Protein phosphorylation and signal transduction. *Pharmacol Ther* 1999;**82**(2−3):111−21.
92. Harris RA, Popov KM, Zhao Y, et al. A new family of protein kinases—the mitochondrial protein kinases. *Adv Enzyme Regul* 1995;**35**:147−62.
93. Patel MS, Korotchkina LG. Regulation of the pyruvate dehydrogenase complex. *Biochem Soc Trans* 2006;**34**(Pt 2):217−22.
94. Machius M, Chuang JL, Wynn RM, et al. Structure of rat BCKD kinase: nucleotide-induced domain communication in a

94. mitochondrial protein kinase. *Proc Natl Acad Sci USA* 2001;**98**(20):11218−23.
95. Wright LP, Philips MR. Thematic review series: lipid post-translational modifications. CAAX modification and membrane targeting of Ras. *J Lipid Res* 2006;**47**(5):883−91.
96. McTaggart SJ. Isoprenylated proteins. *Cell Mol Life Sci* 2006;**63**(3):255−67.
97. Perez-Sala D. Protein isoprenylation in biology and disease: general overview and perspectives from studies with genetically engineered animals. *Front Biosci* 2007;**12**:4456−72.
98. Yu S, Guo Z, Johnson C, et al. Recent progress in synthetic and biological studies of GPI anchors and GPI-anchored proteins. *Curr Opin Chem Biol* 2013;**17**(6):1006−13.
99. Paulick MG, Bertozzi CR. The glycosylphosphatidylinositol anchor: a complex membrane-anchoring structure for proteins. *Biochemistry* 2008;**47**(27):6991−7000.
100. Lublin DM, Coyne KE. Phospholipid-anchored and trans-membrane versions of either decay-accelerating factor or membrane cofactor protein show equal efficiency in protection from complement-mediated cell damage. *J Exp Med* 1991;**174**(1):35−44.
101. Bandyopadhyay PK. Vitamin K-dependent gamma-glutamylcarboxylation: an ancient posttranslational modification. *Vitam Horm* 2008;**78**:157−84.
102. Fares F. The role of O-linked and N-linked oligosaccharides on the structure-function of glycoprotein hormones: development of agonists and antagonists. *Biochim Biophys Acta* 2006;**1760**(4):560−7.
103. Hennet T, Cabalzar J. Congenital disorders of glycosylation: a concise chart of glycocalyx dysfunction. *Trends Biochem Sci* 2015.
104. Stone MJ, Chuang S, Hou X, et al. Tyrosine sulfation: an increasingly recognised post-translational modification of secreted proteins. *N Biotechnol* 2009;**25**(5):299−317.
105. Seibert C, Sakmar TP. Toward a framework for sulfoproteomics: synthesis and characterization of sulfotyrosine-containing peptides. *Biopolymers* 2008;**90**(3):459−77.
106. Moore KL. The biology and enzymology of protein tyrosine O-sulfation. *J Biol Chem* 2003;**278**(27):24243−6.
107. Leyte A, van Schijndel HB, Niehrs C, et al. Sulfation of Tyr1680 of human blood coagulation factor VIII is essential for the interaction of factor VIII with von Willebrand factor. *J Biol Chem* 1991;**266**(2):740−6.
108. Arsenault PR, Heaton-Johnson KJ, Li LS, et al. Identification of prolyl hydroxylation modifications in mammalian cell proteins. *Proteomics* 2015;**15**(7):1259−67.
109. Berg RA, Prockop DJ. The thermal transition of a non-hydroxylated form of collagen. Evidence for a role for hydroxyproline in stabilizing the triple-helix of collagen. *Biochem Biophys Res Commun* 1973;**52**(1):115−20.
110. Kaelin WG. Proline hydroxylation and gene expression. *Annu Rev Biochem* 2005;**74**:115−28.
111. Jaakkola P, Mole DR, Tian YM, et al. Targeting of HIF-alpha to the von Hippel-Lindau ubiquitylation complex by O2-regulated prolyl hydroxylation. *Science* 2001;**292**(5516):468−72.
112. Gould N, Doulias PT, Tenopoulou M, et al. Regulation of protein function and signaling by reversible cysteine S-nitrosylation. *J Biol Chem* 2013;**288**(37):26473−9.
113. Foster MW, Hess DT, Stamler JS. Protein S-nitrosylation in health and disease: a current perspective. *Trends Mol Med* 2009;**15**(9):391−404.
114. Hess DT, Stamler JS. Regulation by S-nitrosylation of protein post-translational modification. *J Biol Chem* 2012;**287**(7):4411−8.
115. Sengupta R, Holmgren A. Thioredoxin and thioredoxin reductase in relation to reversible S-nitrosylation. *Antioxid Redox Signal* 2013;**18**(3):259−69.
116. Stubauer G, Giuffre A, Sarti P. Mechanism of S-nitrosothiol formation and degradation mediated by copper ions. *J Biol Chem* 1999;**274**(40):28128−33.
117. Lipton SA, Choi YB, Pan ZH, et al. A redox-based mechanism for the neuroprotective and neurodestructive effects of nitric oxide and related nitroso-compounds. *Nature* 1993;**364**(6438):626−32.
118. Gu Z, Kaul M, Yan B, et al. S-nitrosylation of matrix metal-loproteinases: signaling pathway to neuronal cell death. *Science* 2002;**297**(5584):1186−90.
119. Di Cera E. Serine proteases. *IUBMB Life* 2009;**61**(5):510−5.
120. Chapman HA, Riese RJ, Shi GP. Emerging roles for cysteine proteases in human biology. *Annu Rev Physiol* 1997;**59**:63−88.
121. Nicholson DW, Thornberry NA. Caspases: killer proteases. *Trends Biochem Sci* 1997;**22**(8):299−306.
122. Klebe G. Aspartic Protease Inhibitors. In: Klebe G, editor. *Drug design*. Berlin Heidelberg: Springer; 2013. p. 533−64.
123. Verma RP, Hansch C. Matrix metalloproteinases (MMPs): chemical-biological functions and (Q)SARs. *Bioorg Med Chem* 2007;**15**(6):2223−68.
124. Varghese S. Matrix metalloproteinases and their inhibitors in bone: an overview of regulation and functions. *Front Biosci* 2006;**11**:2949−66.
125. Das S, Mandal M, Chakraborti T, et al. Structure and evolutionary aspects of matrix metalloproteinases: a brief overview. *Mol Cell Biochem* 2003;**253**(1−2):31−40.
126. Pelmenschikov V, Siegbahn PE. Catalytic mechanism of matrix metalloproteinases: two-layered ONIOM study. *Inorg Chem* 2002;**41**(22):5659−66.
127. Kleiger G, Mayor T. Perilous journey: a tour of the ubiquitin-proteasome system. *Trends Cell Biol* 2014;**24**(6):352−9.
128. Hochstrasser M. Origin and function of ubiquitin-like proteins. *Nature* 2009;**458**(7237):422−9.
129. Bedford L, Paine S, Sheppard PW, et al. Assembly, structure, and function of the 26S proteasome. *Trends Cell Biol* 2010;**20**(7):391−401.
130. Anderson NL, Anderson NG, Pearson TW, et al. A human proteome detection and quantitation project. *Mol Cell Proteomics* 2009;**8**(5):883−6.
131. Saha S, Harrison SH, Shen C, et al. HIP2: an online database of human plasma proteins from healthy individuals. *BMC Med Genomics* 2008;**1**:12.
132. Schiffer E, Mischak H, Vanholder RC. Exploring the uremic toxins using proteomic technologies. *Contrib Nephrol* 2008;**160**:159−71.
133. Vanholder R, De Smet R, Glorieux G, et al. Review on uremic toxins: classification, concentration, and interindividual variability. *Kidney Int* 2003;**63**(5):1934−43.
134. Hortin GL, Jortani SA, Ritchie Jr JC, et al. Proteomics: a new diagnostic frontier. *Clin Chem* 2006;**52**(7):1218−22.
135. Coppinger JA, Maguire PB. Insights into the platelet releasate. *Curr Pharm Des* 2007;**13**(26):2640−6.
136. Peters T. *All about albumin: serum albumin-biochemistry, genetics, and medical applications*. Washington, DC: AACC Press; 1996.

137. Fogh-Andersen N, Bjerrum PJ, Siggaard-Andersen O. Ionic binding, net charge, and Donnan effect of human serum albumin as a function of pH. *Clin Chem* 1993;**39**(1):48–52.
138. Labriola L, Wallemacq P, Gulbis B, et al. The impact of the assay for measuring albumin on corrected ('adjusted') calcium concentrations. *Nephrol Dial Transplant* 2009;**24**(6):1834–8.
139. Meng QH, Krahn J. Lithium heparinised blood-collection tubes give falsely low albumin results with an automated bromcresol green method in haemodialysis patients. *Clin Chem Lab Med* 2008;**46**(3):396–400.
140. Schultze HE, Heide K, Haupt H. alpha1-Antitrypsin from human serum. *Klin Wochenschr* 1962;**40**:427–9.
141. Bornhorst JA, Calderon FR, Procter M, et al. Genotypes and serum concentrations of human alpha-1-antitrypsin "P" protein variants in a clinical population. *J Clin Pathol* 2007;**60**(10):1124–8.
142. Mak CM, Lam CW, Tam S. Diagnostic accuracy of serum ceruloplasmin in Wilson disease: determination of sensitivity and specificity by ROC curve analysis among ATP7B-genotyped subjects. *Clin Chem* 2008;**54**(8):1356–62.
143. Mak CM, Lam CW. Diagnosis of Wilson's disease: a comprehensive review. *Crit Rev Clin Lab Sci* 2008;**45**(3):263–90.
144. Langlois MR, Delanghe JR. Biological and clinical significance of haptoglobin polymorphism in humans. *Clin Chem* 1996;**42**(10):1589–600.
145. Marklova E, Albahri Z. Screening and diagnosis of congenital disorders of glycosylation. *Clin Chim Acta* 2007;**385**(1–2):6–20.
146. Biffi S, Tamaro G, Bortot B, et al. Carbohydrate-deficient transferrin (CDT) as a biochemical tool for the screening of congenital disorders of glycosylation (CDGs). *Clin Biochem* 2007;**40**(18):1431–4.
147. Bachmann-Harildstad G. Diagnostic values of beta-2 transferrin and beta-trace protein as markers for cerebrospinal fluid fistula. *Rhinology* 2008;**46**(2):82–5.
148. Mantur M, Lukaszewicz-Zajac M, Mroczko B, et al. Cerebrospinal fluid leakage—reliable diagnostic methods. *Clin Chim Acta* 2011;**412**(11–12):837–40.
149. Bortolotti F, De Paoli G, Tagliaro F. Carbohydrate-deficient transferrin (CDT) as a marker of alcohol abuse: a critical review of the literature 2001-2005. *J Chromatogr B Analyt Technol Biomed Life Sci* 2006;**841**(1–2):96–109.
150. Kyle RA, Rajkumar SV. Criteria for diagnosis, staging, risk stratification and response assessment of multiple myeloma. *Leukemia* 2009;**23**(1):3–9.
151. Tillett WSFT. Serological reactions in pneumonia with a non-protein somatic fraction of pneumococcus. *J Exp Med* 1930;**52**:561–71.
152. Abernathy TJAO. The occurrence during acute infections of a protein not normally present in the blood. II. Isolation and properties of the reactive protein. *J Exp Med* 1941;**73**:183–90.
153. Gabay C, Kushner I. Acute-phase proteins and other systemic responses to inflammation. *N Engl J Med* 1999;**340**(6):448–54.
154. Mora S, Musunuru K, Blumenthal RS. The clinical utility of high-sensitivity C-reactive protein in cardiovascular disease and the potential implication of JUPITER on current practice guidelines. *Clin Chem* 2009;**55**(2):219–28.
155. van Asbeck EC, Hoepelman AI, Scharringa J, et al. Mannose binding lectin plays a crucial role in innate immunity against yeast by enhanced complement activation and enhanced uptake of polymorphonuclear cells. *BMC Microbiol* 2008;**8**:229.
156. Parker C, Omine M, Richards S, et al. Diagnosis and management of paroxysmal nocturnal hemoglobinuria. *Blood* 2005;**106**(12):3699–709.
157. Caccia S, Suffritti C, Cicardi M. Pathophysiology of hereditary angioedema. *Pediatr Allergy Immunol Pulmonol* 2014;**27**(4):159–63.
158. Donoso LA, Vrabec T, Kuivaniemi H. The role of complement Factor H in age-related macular degeneration: a review. *Surv Ophthalmol* 2010;**55**(3):227–46.
159. Kaneko Y, Nimmerjahn F, Ravetch JV. Anti-inflammatory activity of immunoglobulin G resulting from Fc sialylation. *Science* 2006;**313**(5787):670–3.
160. Durandy A, Peron S, Fischer A. Hyper-IgM syndromes. *Curr Opin Rheumatol* 2006;**18**(4):369–76.
161. Woof JM, Mestecky J. Mucosal immunoglobulins. *Immunol Rev* 2005;**206**:64–82.
162. Freeman AF, Domingo DL, Holland SM. Hyper IgE (Job's) syndrome: a primary immune deficiency with oral manifestations. *Oral Dis* 2009;**15**(1):2–7.
163. Hutchison CA, Harding S, Hewins P, et al. Quantitative assessment of serum and urinary polyclonal free light chains in patients with chronic kidney disease. *Clin J Am Soc Nephrol* 2008;**3**(6):1684–90.
164. Dispenzieri A, Kyle R, Merlini G, et al. International Myeloma Working Group guidelines for serum-free light chain analysis in multiple myeloma and related disorders. *Leukemia* 2009;**23**(2):215–24.
165. Stiehm ER. The four most common pediatric immunodeficiencies. *J Immunotoxicol* 2008;**5**(2):227–34.
166. Brown R, Nelson M, Aklilu E, et al. An evaluation of the DiaMed assays for immunoglobulin A antibodies (anti-IgA) and IgA deficiency. *Transfusion (Paris)* 2008;**48**(10):2057–9.
167. Landgren O, Kyle RA, Pfeiffer RM, et al. Monoclonal gammopathy of undetermined significance (MGUS) consistently precedes multiple myeloma: a prospective study. *Blood* 2009;**113**(22):5412–7.
168. Bence Jones H. On a new substance occurring in the urine of a patient with Mollities Ossium. *Phil Trans R Soc Lond* 1848;**138**:55–62.
169. Hortin GL, Meilinger B. Cross-reactivity of amino acids and other compounds in the biuret reaction: interference with urinary peptide measurements. *Clin Chem* 2005;**51**(8):1411–9.
170. Lowry OH, Rosebrough NJ, Farr AL, et al. Protein measurement with the Folin phenol reagent. *J Biol Chem* 1951;**193**(1):265–75.
171. McCudden CR, Mathews SP, Hainsworth SA, et al. Performance comparison of capillary and agarose gel electrophoresis for the identification and characterization of monoclonal immunoglobulins. *Am J Clin Pathol* 2008;**129**(3):451–8.
172. Miller WG, Thienpont LM, Van Uytfanghe K, et al. Toward standardization of insulin immunoassays. *Clin Chem* 2009;**55**(5):1011–8.
173. Kaiser P, Akerboom T, Ohlendorf R, et al. Liquid chromatography-isotope dilution-mass spectrometry as a new basis for the reference measurement procedure for hemoglobin A1c determination. *Clin Chem* 2010;**56**(5):750–4.
174. Xie H, Rhodus NL, Griffin RJ, et al. A catalogue of human saliva proteins identified by free flow electrophoresis-based peptide separation and tandem mass spectrometry. *Mol Cell Proteomics* 2005;**4**(11):1826–30.

175. Al Kawas S, Rahim ZH, Ferguson DB. Potential uses of human salivary protein and peptide analysis in the diagnosis of disease. *Arch Oral Biol* 2012;**57**(1):1–9.
176. Felgenhauer K. Protein size and cerebrospinal fluid composition. *Klin Wochenschr* 1974;**52**(24):1158–64.
177. Biou D, Benoist JF, Nguyen-Thi C, et al. Cerebrospinal fluid protein concentrations in children: age-related values in patients without disorders of the central nervous system. *Clin Chem* 2000;**46**(3):399–403.
178. Tourtellotte WW, Staugaitis SM, Walsh MJ, et al. The basis of intra-blood-brain-barrier IgG synthesis. *Ann Neurol* 1985;**17**(1):21–7.
179. Luque FA, Jaffe SL. Cerebrospinal fluid analysis in multiple sclerosis. *Int Rev Neurobiol* 2007;**79**:341–56.
180. Kochanek PM, Berger RP, Bayir H, et al. Biomarkers of primary and evolving damage in traumatic and ischemic brain injury: diagnosis, prognosis, probing mechanisms, and therapeutic decision making. *Curr Opin Crit Care* 2008;**14**(2):135–41.
181. Zetterberg H. Update on amyloid-beta homeostasis markers for sporadic Alzheimer's disease. *Scand J Clin Lab Invest* 2009;**69**(1):18–21.
182. Light RW. Clinical practice. Pleural effusion. *N Engl J Med* 2002;**346**(25):1971–7.
183. Heffner JE. Discriminating between transudates and exudates. *Clin Chest Med* 2006;**27**(2):241–52.
184. Maldonado F, Hawkins FJ, Daniels CE, et al. Pleural fluid characteristics of chylothorax. *Mayo Clin Proc* 2009;**84**(2):129–33.
185. Runyon BA, Montano AA, Akriviadis EA, et al. The serum-ascites albumin gradient is superior to the exudate-transudate concept in the differential diagnosis of ascites. *Ann Intern Med* 1992;**117**(3):215–20.
186. Takeda H, Nishise S, Furukawa M, et al. Fecal clearance of alpha1-antitrypsin with lansoprazole can detect protein-losing gastropathy. *Dig Dis Sci* 1999;**44**(11):2313–8.
187. Sutherland AD, Gearry RB, Frizelle FA. Review of fecal biomarkers in inflammatory bowel disease. *Dis Colon Rectum* 2008;**51**(8):1283–91.
188. Ayling RM. New faecal tests in gastroenterology. *Ann Clin Biochem* 2012;**49**(Pt 1):44–54.
189. Leeds JS, Oppong K, Sanders DS. The role of fecal elastase-1 in detecting exocrine pancreatic disease. *Nat Rev Gastroenterol Hepatol* 2011;**8**(7):405–15.
190. Daftary A, Acton J, Heubi J, et al. Fecal elastase-1: utility in pancreatic function in cystic fibrosis. *J Cyst Fibros* 2006;**5**(2):71–6.

14

Proteomics

Andrew N. Hoofnagle and Cory Bystrom

ABSTRACT

Background

Clinical proteomics has traditionally referred to experiments that attempt to discover novel biomarkers for disease diagnosis, prognosis, or therapeutic management by using tools that measure the abundance of hundreds or thousands of proteins in a single sample. These discovery experiments began with protein electrophoresis, particularly two-dimensional (2D) gel electrophoresis, and have evolved into workflows that rely very heavily on mass spectrometry (MS). Using the workflows developed for discovery proteomics, clinical laboratories have developed quantitative assays for proteins in human samples that solve many of the issues associated with the measurement of proteins by immunoassay. The technology is changing clinical research and is poised to significantly transform protein measurements used in patient care.

Content

This chapter begins with the history of clinical proteomics, with a special emphasis on 2D gel electrophoresis of serum and plasma proteins. It then describes discovery techniques that use MS, including data-dependent acquisition and data-independent acquisition. It finishes with a discussion of targeted quantitative proteomic methods, both bottom-up and top-down, as replacement methodologies for immunoassays and Western blotting. Special attention is paid to peptide selection, denaturation and digestion, peptide and protein enrichment, internal standards, and calibration.

HISTORICAL PERSPECTIVE

The word *proteome* is a combination of the words *protein* and *genome*, first coined by Marc Wilkins in 1994.[1] Wilkins used the term to describe the entire complement of proteins expressed by a genome, cell, tissue, or organism, and *proteomics* refers to the comprehensive identification and quantitative measurement of these proteins. Today, the term encompasses separation science, protein microchemistry, bioinformatics, and mass spectrometry (MS) as the fundamental techniques used in the large-scale study of protein identity, abundance, structure, and function.

Analysis of proteomes, and the human proteome in particular, was a logical extension of the completion of the human genome, which revealed the genetic blueprint from which the proteome is constructed. The promise of proteomics for clinical research was an outgrowth of the understanding that many clinically relevant markers are present in blood and that disease-related aberrations in protein abundance might be quantified by comparative analysis of proteomes.

Early Proteomics

The earliest investigations of the human serum proteome preceded the completion of the human genome by 24 years and used two-dimensional gel electrophoresis (2D gel) to provide a high-resolution snapshot of serum proteins (Fig. 14.1).[2] In this technique a complex pool of proteins was first resolved by isoelectric point in the first dimension and by molecular mass in the second dimension followed by staining for visualization. This spread the proteins in the sample into an array in which each spot is associated with one protein. Applying this technique to serum illuminated its complexity, revealing over 300 spots, which were suggested to arise from 75 to 100 unique proteins. Some of these proteins were identified by comparison with the migration of purified protein standards, comparison with immunoprecipitated proteins, or immunoblotting.

Although powerful, 2D gels did not gain widespread use for comparative studies until the early 1990s. The main limitation was a lack of tools for the rapid identification of the proteins contained in a single spot, which could link the location, intensity, and identity of a gel spot to biology. Before the advent of MS-based tools, identification of a single protein took several weeks of dedicated labor starting with an effort to isolate sufficient protein for subsequent analysis. Once isolated, the extracted protein was proteolyzed with trypsin and two or more peptides were isolated using preparative HPLC separation. These peptides were then analyzed with Edman degradation to obtain short N-terminal amino acid sequences.[3] The sequence data, along with approximate molecular weight, isoelectric point, and any other available data, were used to search available databases and infer protein identity.

Proteome analysis accelerated dramatically through the 1990s largely because of the availability of an array of technologies that transformed the protein identification problem from an arduous task to a scalable, simple procedure that could be completed in a day or two. First, improvements in

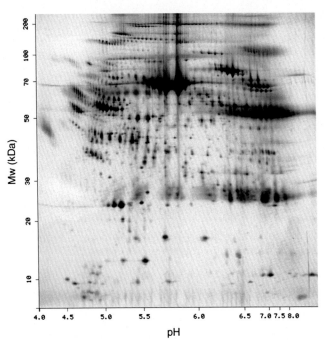

FIGURE 14.1 Two-dimensional gel electrophoresis. Before the advent of mass spectrometric methods to probe the proteome, proteins were resolved and quantified using two-dimensional polyacrylamide gel electrophoresis. Proteins were first separated based on isoelectric point (horizontally) and then based on size (vertically). Hundreds of spots were visible. (Reprinted by permission of the publisher from Hoogland C, Mostaguir K, Sanchez JC, Hochstrasser DF, Appel RD. SWISS-2DPAGE, ten years later. *Proteomics* 2004;4:2352–2356. Copyright 2004 WILEY-VCH Verlag GmbH. http://world-2dpage.expasy.org/swiss-2dpage/viewer&map=PLASMA_HUMAN&ac=all).

MS, which combined cost-effective, semiautomated, high-performance instruments equipped with gentle ionization methods (matrix-assisted laser desorption ionization [MALDI] and electrospray ionization [ESI])[4,5] allowed laboratories to collect easily interpretable mass spectra. Second, protein microchemistry techniques allowed sufficient sample to be extracted from single gel spots to yield interpretable mass spectra. Finally, high-quality genomic databases and novel statistical algorithms could compare mass spectrometric data to database entries to provide protein identifications with unprecedented speed and accuracy.[6,7] Together, these technologies merged into a powerful analytical platform that enabled protein identification and characterization in a manner that matched the throughput and number of protein spots that could be resolved on a 2D gel. A number of public 2D gel databases were published on the World Wide Web, some of which are still available (eg, http://world-2dpage.expasy.org/list/).

Progress continued into the mid-2000s, with advances in 2D gels and associated technologies that allowed the reproducible resolution of hundreds to thousands of protein spots in complex samples. Efforts to apply the techniques in clinical research quickly expanded to more groups and more diverse sample types, including tissue,[8,9] blood,[10,11] tears,[12] and urine.[13] The general approach was to obtain samples representing healthy and disease states, perform preanalytical fractionation, separate the individual protein fractions by 2D electrophoresis, and stain the gels (see Fig. 14.1). Using imaging techniques, comparative analysis of gels identified comigrating spots with different intensity or changes in spot position. Many studies reported statistically significant differences in protein abundance associated with disease, but validation of a primary discovery leading to a biomarker with sufficient promise for clinical studies has not been realized. Frequently, the biomarker proteins were acute-phase proteins that were known to generally associate with many diseases. Alternatively, high-abundance proteins with no clear biological significance to the disease were identified, suggesting experimental artifact. The constraints on proteome depth and breadth and the inability to perform sufficient technical and experimental replicates to overcome variability left many gel-based experiments significantly underpowered to successfully detect biomarkers.

As an analytical tool for proteome analysis, 2D gels suffer from variability, significant demands on labor, and limited dynamic range. Since biological samples had a tremendous range of protein abundances (7 to 12 orders of magnitude), the 1 to 2 orders visible on a 2D gel made a deep analysis of the proteome essentially impossible.[14,15] Unfractionated samples of tissue or blood often have 10 to 20 proteins that account for 80% or more of the total protein content of the sample.[16] As a result, the observable proteome on a 2D gel is severely limited. Loading greater amounts of sample onto 2D gels causes distortions during separation, dramatically reduces reproducibility, and can contaminate large portions of the gel, leading to ubiquitous identification of the abundant proteins, even in unexpected regions on the gel.

To improve the depth of analysis, creative fractionation and depletion strategies were developed to work around the protein abundance problem. These approaches focused on separating protein pools into subfractions based on chemical,[17] structural,[18] or biophysical features.[19,20] Depletion of highly abundant proteins using immunoaffinity[21] and semispecific chemical affinity approaches[22] also became commonplace. These approaches were effective but added cost, complexity, and variability to an already challenging technique. Although this discussion is historical, challenges of proteome dynamic range are still a consideration in modern proteomics workflows.

It was soon recognized that complex mixtures of proteins could be directly ionized and analyzed with MALDI MS to give complex spectra comprising the masses and crude abundance of many proteins.[23] Although much more limited in resolution than 2D gels (as a result of reducing separation to mass alone), it was appealing to suggest that MALDI spectra could substitute for 2D gels and resolve at least some of the long-standing technical issues that compromised gel-based analysis: labor intensiveness and poor throughput. With the ability to apply the MALDI technique to biomarker discovery, efforts could be directed at a smaller set of technical challenges, namely sample preparation.

Biomarker Discovery in the Post–Two-Dimensional Gel Era

With the increased throughput and improved precision that appeared possible with MALDI analysis of complex mixtures, manufacturers developed integrated systems of reagents and instruments for performing discovery experiments. Although mass spectrometers are subject to constraints similar to those with 2D gels regarding the observable dynamic range of protein abundances in a single spectrum, the ability to automate upstream sample preparation or carry out microscale fractionation directly on MALDI targets and commercial availability of "chips" made the technical aspects of biomarker discovery much less daunting.[24,25] Although identification of proteins in this workflow was generally not possible without substantial effort, it was argued by some that the identity of proteins giving rise to individual discriminatory peaks in a spectrum was not important, and instead the pattern was the biomarker that linked phenotype to a snapshot of relative protein abundance. Critics pointed out that in a spectrum of a complex matrix such as serum or plasma, a peak was unlikely to be a single protein and instead would be composed of dozens, if not hundreds, of proteins, making it impossible to create the link between the biology and biochemistry of a disease process and the change in abundance of any specific peak.[26] The dawn of high-throughput proteome analysis led to a shift in thinking about biomarker discovery from hypothesis-driven to hypothesis-generating. In the former, biochemical pathways associated with disease are rigorously interrogated to link observations and biology. In the latter, an unbiased comparison of samples without consideration of biology is used to identify putative differences in protein abundance between sample types. At around the same time, the field also adopted the hopeful concept of multiprotein biomarkers or changes in abundance of multiple proteins, none of which might be individually valuable as a marker that could be combined to provide important clinical information. Unbiased discovery and reduced stringency promised to accelerate biomarker discovery and feed the pipeline of biomarker validation.

Given the apparent technical benefits, and ignoring the probable compromises, the approach was applied to many diseases that had defied biochemical diagnosis or those that would benefit from early detection.[27,28] Initial studies on breast and prostate cancer were promising, and results from one particular study electrified both the clinical and proteomics research communities. The results from that study suggested that ovarian cancer could be unambiguously identified, even at the earliest stages, in almost all patients.[29] Efforts to validate this study led to an acrimonious public debate, and eventually it was accepted that the apparent clinical utility of the test was merely an artifact arising from poor experimental design and fundamental misunderstandings of MS data and processing.[30-33] Biomarker discovery studies using these methods continued for several years after this upheaval, but the approach was largely abandoned by 2007.

A pivotal moment in the early days of proteomics was the development of integrated liquid chromatography–tandem mass spectrometry (LC-MS/MS) systems with automated data acquisition and automated data processing. Pioneering work by Washburn and Yates[34] provided a way to perform proteome analysis that truly overcame many of the shortcomings of 2D gels. The strategy started with a complex protein mixture, which was digested with trypsin. The peptides were then fractionated by multidimensional chromatography, with each fraction being directed to the mass spectrometer. Although it is counterintuitive to make a complex mixture even more complex via proteolysis, the technique was highly automated, scaled easily, and used to generate protein catalogs of thousands of proteins at a fraction of the labor required for a similarly complex 2D gel analysis. Although not originally developed to perform quantitative analysis, subsequent enhancements of this technique provided the innovations that are at the core of most proteomics discovery experiments carried out today.

BIOMARKER PIPELINE

The goal of proteomic biomarker discovery is to identify and then demonstrate clinical utility of a protein or combination of proteins that provide useful diagnostic or prognostic information regarding disease.[35-37] At each step of this process, experimental tools and performance expectations change. At the earliest stages of discovery, dozens of samples are interrogated at the level of thousands of proteins (Fig. 14.2). Discovery workflows from preanalytical sample preparation to complete data analysis are very time consuming, requiring hours to days of data collection for each individual specimen that is being processed. It is not uncommon for discovery experiments to require nonstop data acquisition for a month or longer, which constrains the size of the discovery experiment. After the raw data are collected, they are reduced to give a list of each peak by mass, retention time, intensity, and identity (if available). These lists are then mathematically manipulated to align common features between multiple specimens

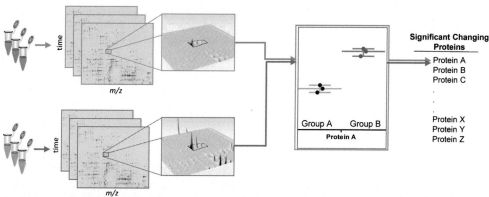

FIGURE 14.2 Proteomics biomarker discovery pipeline. To discover potential novel protein biomarkers, samples can be prepared for liquid chromatography–mass spectrometry (LC-MS) and analyzed as intact proteins. Replicates of pooled samples from two different disease states or unpooled samples from individual patients from two disease states may be compared (eg, healthy vs. control). Signals from the mass spectrometer are integrated and compared *in silico* to identify protein features that are significantly different among the disease states, which can be subsequently identified in later analyses. This type of analysis of intact proteins is limited to small proteins. Alternatively, samples can be prepared by using proteolysis before being analyzed as peptides by LC-tandem MS (LC-*MS/MS*). Using this approach, proteins are identified and quantified to find proteins that are different between pathophysiologic states. (Figure courtesy Dr. Tim Collier, PhD. Used with permission, copyright Tim Collier, 2015.)

and to normalize the overall intensity between each acquired data set. Finally, the feature lists are compared to generate a list of peaks that are reproducibly and significantly different between the two clinical states. Frequently, these discovery experiments yield 10 to 100 candidate proteins suitable for evaluation in preclinical studies. The investigator may choose to use other data to help refine this list by focusing on known or postulated biological relevance.

In the next stage of the pipeline, specific targeted assays for selected candidate proteins are developed with methods that achieve higher throughput and precision, with the goal of verifying the potential utility of the candidate marker proteins in a series of pilot studies that will incorporate samples from hundreds of patients. This will often reduce the number of proteins to 5 to 20 from the original group.

In the final phase of biomarker development, the most neglected phase to date, an assay for one or a panel of markers is developed and rigorously validated to establish analytical performance. The validated assay is then used to analyze hundreds to thousands of samples from prospective and/or retrospective clinical studies to establish the utility (or lack thereof) of the biomarker or panel of biomarkers. On completion of this work, a successful biomarker or biomarker panel assay will be described in a standard operating procedure (SOP) that includes preanalytical requirements, a description of the intended use, and the expected clinical performance of the assay as described by metrics for sensitivity, specificity, positive predictive value, and negative predictive value.

Discovery Experiment

Similar to the first comparative studies with 2D gels, the aim of present-day discovery proteomics experiments is to identify and quantify all of the proteins in complex mixtures to find biomarker proteins that differ in relative abundance between two biological states (eg, diseased and normal). The discovery pipeline begins with the selection of useful samples. Ideally, clinically relevant samples are selected that have been collected, prepared, and stored in an identical fashion along with a dossier of information regarding demographic and clinical history. Preanalytical steps are generally applied to fractionate or otherwise modify the sample to improve depth of analysis to lend the greatest opportunity to measure proteins that may differ by disease state. Subsequently, the samples are then digested from whole proteins into a complex pool of peptides destined for LC-MS or LC-MS/MS experiments to provide data sets that can be compared to identify significant differences between samples.

Requirement for Separations

As mentioned earlier, given the enormous complexity of biological specimens and the additional complexity that is derived from proteolysis of proteins into peptides, separation technologies must be used to allow a mass spectrometer to specifically detect and identify peptides from among a complex mixture of peptides. This is due to the limited duty cycle of mass spectrometers and the need to acquire one tandem mass spectrum at a time (also called a fingerprint spectrum, further described later).

High-performance liquid chromatography (HPLC) using reverse-phase chemistry has been the chief tool for performing peptide and protein separations directly interfaced to mass spectrometers.[38] The compatibility of reverse-phase solvent systems with LC-MS, commercial availability of a wide range of column dimensions and packing materials, and the ease with which high-resolution separations are achieved with modest method development are key advantages compared to alternative separation technologies. Chromatography can be performed under a wide range of flow rates from 0.5 μL/min or less for maximum sensitivity but low throughput, to greater than 1 mL/min for maximal throughput at reduced sensitivity.[39,40] Alternative separation techniques such as capillary electrophoresis have successfully been interfaced with mass spectrometers[41] and offer very-high-resolution separations, but challenges in maintaining robust performance and complexities in developing routine separation methods have prevented wider adoption.

Data Acquisition Strategies

MS experiments in biomarker discovery studies are typically conducted using one of three data collection strategies: (1) full scan only LC-MS, (2) data-dependent LC-MS/MS, or (3) data-independent LC-MS/MS. Each workflow has different strengths and weakness, and laboratories select workflows based on experience and specific assay requirements.

In a full-scan acquisition experiment, a high-resolution mass spectrometer continually collects spectra over a wide mass-to-charge (m/z) range, capturing signals from all ions that are presented to the mass spectrometer as the chromatogram is developed. No tandem mass spectra are acquired, which are needed for the identification of the peptides eluting from the chromatographic column. Instead, very-high-resolution contour maps of mass spectrometric data are collected across three dimensions: time, intensity, and m/z. If a peak appears to be different between two clinically relevant groups in this type of discovery experiment, a second round of analysis is required to identify the peptide of interest.[42]

Data-dependent acquisition (DDA) is an algorithm-guided strategy in which a wide m/z range spectrum or survey scan is collected and used to identify candidate peaks for subsequent MS/MS in real time (Figs. 14.3 and 14.4A). The survey scan is rapidly processed to generate a list of observed m/z for each detected peak ranked in order of intensity. This list is then used to assemble a queue for directed tandem mass spectrometric experiments. Iteratively, the instrument then performs selection and fragmentation of each precursor ion stored in the queue. To prevent the instrument from repeatedly analyzing the same set of high-abundance precursors, specific m/z values can be placed on an exclusion list for a fixed duration, which prevents further analysis until the exclusion criteria expire. Using this experimental strategy, the final data set contains information on the m/z, intensity, retention time of each individual precursor ion, and related fragmentation data.

Finally, data-independent acquisition (DIA) strategies use a full scan precursor survey, followed by the sequential isolation/fragmentation (MS/MS) of all ions in a small window of the precursor scan (eg, 5 to 20 Da) until the entire m/z range of the initial full scan is covered (see Fig. 14.4B). The survey scan and MS/MS scans are repeated throughout the chromatographic run. In contrast to DDA, DIA product ion spectra are composite spectra that include all fragment ions from all precursors isolated in the small window rather than just one precursor, as in DDA. To interpret the data, extensive processing deconvolutes precursor and fragment ion data to yield time, m/z, and intensity for each precursor peptide, as well as fragmentation data inferred from the composite spectra.

For both DDA and DIA experiments, peptide identification relies on the ability of the mass spectrometer to create fragments in the gas phase to generate product ion spectra,

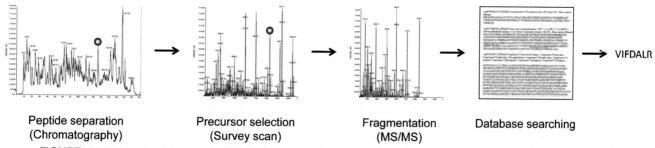

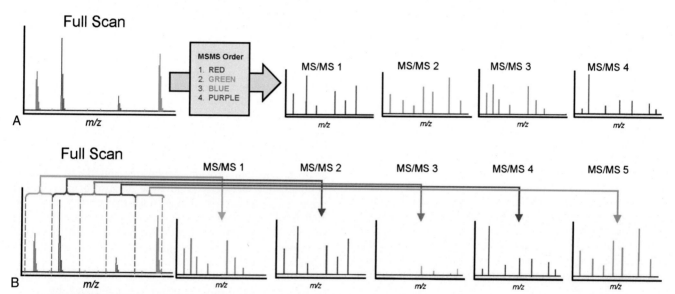

FIGURE 14.3 Data-dependent acquisition. Peptide identification in discovery proteomics experiments often uses software to drive the mass spectrometer in the selection of the precursor peptides to be fragmented. As peptides elute from the chromatographic column, the mass spectrometer first performs a survey scan to assess peptide precursor *m/z* and their abundance. In this theoretical example a survey scan is performed at 105 minutes and the most abundant peaks are selected for subsequent tandem mass spectrometry *(MS/MS)* analysis, including the specific peptide that is fragmented in the third panel. The resulting spectrum is compared against a theoretical database of proteins to identify the peptide from the spectrum using various statistical approaches. *VIFDALR* is the example peptide fragment identified (using single letter amino acid codes).

FIGURE 14.4 Comparison of data-dependent acquisition (DDA) and data-independent acquisition (DIA). During DDA and DIA discovery experiments, the mass spectrometer begins with a high-resolution, high mass accuracy survey scan of the peptides eluting off the chromatographic column. **A,** In DDA, the survey scan is used to build a list of the precursors (using a 0.7 to 2.0 Da window) that will be fragmented/analyzed by tandem mass spectrometry *(MS/MS)* in subsequent steps in the mass spectrometer (typically 2 to 8 fragments are targeted from each precursor survey scan). This cycle of precursor/survey scan and MS/MS steps (typically ≥ 7 Hz) is repeated throughout the chromatographic run. **B,** In contrast, after the precursor scan in DIA, every part of the *m/z* range (typically 400 to 1000 Da wide) is then sampled by stepping through the *m/z* range using windows of 10 to 20 Da, collecting all of the precursors in each window and fragmenting/analyzing them by MS/MS. This cycle of precursor/survey scan and MS/MS steps (typically ≤ 2 Hz) is repeated throughout the chromatographic run. Methods to improve the specificity of the DIA approach include overlapping windows and randomization of the MS/MS windows (not shown). For both methods, there is significant post–data acquisition analysis using software to determine the identity and abundance of peptides in the sample (not shown). (Figure courtesy Dr. Tim Collier, PhD. Used with permission, copyright Tim Collier, 2015.)

also called MS/MS spectra. This fragmentation process occurs by imparting energy into a peptide by accelerating it and colliding it with an inert gas. Once it is at a higher energy state, a peptide will dissociate into two fragments in a thermodynamically probabilistic fashion, most commonly cleaving at specific points along the amide backbone to yield b-ions and y-ions, which respectively include the amino- and carboxyl-terminus of the peptide (Fig. 14.5). The mass differences between peaks in the spectrum, which are associated with fragmentation at each amino acid in the peptide, can then be used to construct the amino acid sequence of the peptide. The resulting product ion spectra are also called *fingerprint* spectra because each spectrum has features that can be associated with a specific peptide without deriving the amino acid sequence by manual inspection. Software algorithms are used to accomplish the process of matching mass spectra to peptide sequences in a database. The databases are generated by *in silico* digestion of all theoretical proteins that are derived from genomic data. In some cases, theoretical databases are supplemented with empirical MS data from a variety of sources, including public repositories representing hundreds of millions of processed mass spectra. Protein and peptide identification software such as Sequest,[43] MASCOT,[7] X! Tandem,[44] Andromeda,[45] and OMSSA[46] have been rigorously evaluated and are widely used by the MS community.

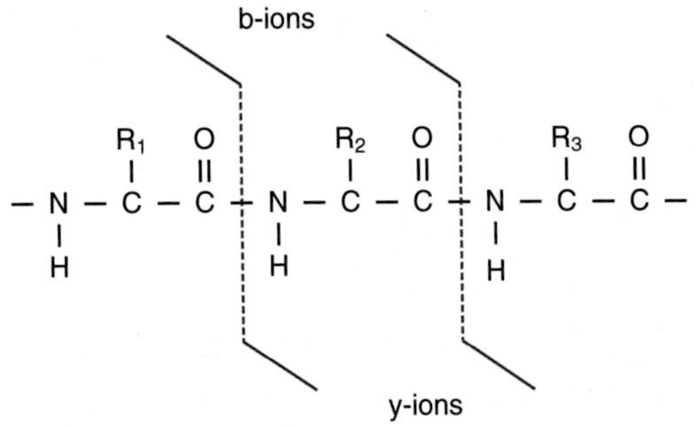

FIGURE 14.5 Fragmentation of peptides in the gas phase. Peptides are fragmented after being excited in the gas phase within the mass spectrometer. The most common ion fragments analyzed in triple-quadrupole mass spectrometers for targeted assays are *b-ions* and *y-ions*, which include the amino- and carboxyl-terminus, respectively. Other ions are formed and may be more predominant in other types of mass analyzers (eg, a-, c-, x-, and z-ions). The characteristic fragmentation patterns of peptides make it possible to search databases for peptide identification from the fingerprint spectra.

In each of these experimental workflows, isotope labels[47,48] and chemical tags[49,50] also have been used to allow the multiplex analysis of many samples simultaneously and/or to provide a reference against which both experimental sample types (eg, disease and healthy) can be compared to improve precision. For example, two samples, one from a diseased patient and one from a healthy patient, are chemically labeled during the preanalytical steps (ie, after proteolysis of the samples). In one case, the chemical label contains isotopes of natural abundance and the other is enriched in heavy isotopes. If the two samples are mixed after labeling, they will contain many identical peptides, but each peptide in the different samples will have a different mass because of the mass difference of the chemical label. Because this mixture of case and control is analyzed by LC-MS/MS, pairs of peaks will appear in the spectra, each representing the same peptide but at slightly different masses, which can be easily resolved by the mass spectrometer. These strategies help increase the number of technical and experimental replicates that can be achieved on any given instrument and also can facilitate the comparison of data acquired over weeks or months where instrument drift can make comparison of data sets more difficult.

Processing of Discovery Proteomics Data

Postprocessing of these data sets is required before final analysis. First, the precursor *m/z*, intensity, and retention time for each data set is arrayed in three dimensions and overlaid using software tools. These software tools can be used to adjust the data to correct for known experimental artifacts, such as drift in retention time. Second, the data are normalized to correct for intensity differences that arise from day-to-day changes in sample preparation efficiency and mass spectrometer performance. Third, the precursor mass and associated fragmentation data are used for database searching to identify *peptides*, which can be assembled to generate protein identifications. With appropriate data processing, relative abundance and identities of thousands of proteins can be achieved in a single discovery experiment. Once these steps are completed, an array of biostatistics tools is used to find proteins where significant abundance increases or decreases are observed between case and control samples.

Although there are commercial software packages that attempt to provide comprehensive solutions to these diverse workflows, many academic and industrial groups build data processing pipelines from custom-built, open-source, and commercial software packages.

Variations and Details

The experimental strategy that is ultimately selected is defined by the availability of specific instrumentation. There are subtle differences in each of the workflows that reflect tradeoffs in data depth versus breadth. For example, the greatest precision is generally observed in full-scan-only experiments, which yield no protein identifications without follow-up MS/MS analyses. The two drawbacks to this approach are (1) the reduced specificity of the method for low-abundance peaks, particularly when multiple peptides elute nearby and it becomes difficult to pick which peak corresponds to the peptide of interest, and (2) subsequent MS/MS analyses that can be helpful in identifying the peaks that differ between two samples or two groups of samples may have a shift in retention time between injections, which can make it difficult to assign MS/MS data to the particular peak of interest. DDA provides peptide and protein identification, but because of greater variability can have less power to discriminate abundance differences. The reason that DDA is more variable is because the automated process of picking peaks to analyze by MS/MS during the analysis is stochastic. More specifically, if more abundant peptides elute at the same time, the peptide of interest may not be selected for MS/MS analysis. DIA can provide the relative abundance of many peptides, but specificity is an issue because of overlapping peaks. In other words, the peptides that coelute with the analyte of interest will be fragmented at the same time and these fragments can make it difficult to tease out the signal of a single peptide from the signals present as a result of other peptides. Overall, significant recent improvements in instrumentation have enabled greater depth of proteome analysis with fewer compromises, but limits remain and the ability to exhaustively interrogate a proteome in a single LC-MS run is not yet achievable.

An alternative to these approaches is to use targeted mass spectrometric experiments to discover protein biomarkers.[51,52] With a list of predefined proteins of interest derived from previous proteomics experiments or biological insight, a tandem mass spectrometric (ie, MS/MS) method using a triple quadrupole or quadrupole–high mass accuracy analyzer hybrid instrument can be developed to detect only specific peptides in the sample and quantify them as chromatographic peak areas. The resulting peak areas are then normalized in some fashion to provide the relative abundance of the peptides of interest (ie, representing a subset of the total proteome). This type of discovery experiment is distinguished from fully quantitative experiments in that it lacks internal standards (ISs) to control for sample preparation and mass spectrometer performance and external calibrators to control for day-to-day digestion variability. It also typically lacks the quality control materials that are the cornerstone of longitudinal monitoring of assay quality in the clinical laboratory. Unlike DDA and DIA experiments, the decision to specifically target individual peptides places a limit on the total number of observations that can be achieved in a run. Although the precision of this approach is often good for research purposes (CV <25%), the breadth of protein coverage is limited to a few hundred peptides in a single run.

> **POINTS TO REMEMBER**
>
> **Discovery Proteomics**
> - It is generally used to identify new biomarkers.
> - Methods are less specific and precise compared to targeted quantitative methods.
> - Gel electrophoresis was one of the first techniques used to examine the proteome.
> - MALDI–time-of-flight (TOF) MS and LC-MS/MS replaced 2D gels.

Targeted Quantitative Experiment

After discovery, targeted quantitative mass spectrometric approaches are used in the development of specific, precise assays. These higher throughput assays, which are designed for improved specificity and precision, provide a mechanism to verify or validate initial proteomic discoveries by running hundreds of patient samples. During this stage, many candidates will fail. In some cases, proteins with marginal statistical significance will not hold up to further testing. In others, proteins will be removed because of poor performance characteristics (ie, stability, degradation, and modification), which would make them unsuitable for further assay development.

Targeted assays are typically performed using triple-quadrupole instruments after a well-developed technique known as stable isotope dilution. In this experiment, calibration materials with known concentrations of protein analyte are processed in parallel with patient samples where each sample and calibrator is spiked with an IS at a constant concentration. The IS is most often chemically and structurally identical to the analyte of interest, with the exception that stable (nonradioactive) heavy isotopes are incorporated. This modification changes only the mass of the peptide and not the chemical characteristics. Samples and calibrators are analyzed by MS to determine the intensity of signals specific for the analyte of interest and the IS, which are used in the determination of the peak area ratio (analyte peak area ÷ IS peak area). Using the calibration materials, a response curve is generated by plotting the peak area ratio versus concentration, which can be fit with a line by standard regression techniques. From this response curve, the concentration of the analyte of interest in an unknown sample can be determined from the peak area ratio.

The selection of MS to perform validation and clinical studies should be evaluated against the possibility of using commercially available immunoassays. Well-validated immunoassays with good sensitivity and specificity could alleviate the need for development and validation of high-quality MS-based assays, which can be time consuming (ie, 6- to 12-month development time for very-high-quality assays for low-abundance proteins is not unexpected). Frequently, though, discovery experiments yield proteins for which no well-validated immunoassay is commercially available, and in these cases the path through assay development can be significantly shorter with MS when compared to the development of an acceptable immunoassay.[35,36] Even when commercial kits are available, there is a growing appreciation of the substantial limitations afforded by immunoassays,[53] including lack of quality control for commercial research–grade enzyme-linked immunosorbent assay (ELISA) kits, nonspecific recognition of nontarget proteins, autoantibody and heterophilic antibody interference, poor concordance between kits for the same analyte, and saturation of sandwich assay reagents leading to falsely low results.[54] A well-designed mass spectrometric assay can avoid these issues, which makes it an attractive alternative to immunoassays.

> **POINTS TO REMEMBER**
>
> **Targeted Proteomics**
> - Mass spectrometry (MS) is sometimes used without internal standards in discovery experiments.
> - May or may not include proteolysis before liquid chromatography (LC).
> - Without significant sample preparation, LC-MS or LC-MS/MS is more specific and precise than MALDI-TOF MS.

Bottom-Up and Top-Down Experiments

From the preceding sections, it may be apparent that proteins can be quantified by MS using two different approaches. In some cases, proteins are of sufficiently low molecular weight to be detected in their intact state, with or without dissociation in the gas phase in the mass spectrometer. This approach to quantification has been termed *top-down* proteomics to contrast with *bottom-up* proteomics, which relies on proteolysis and the quantification of surrogate peptides to determine protein concentration (Fig. 14.6). Both approaches are discussed in the following sections.

BOTTOM-UP TARGETED PROTEOMICS

Workflow Overview

The quantification of proteins is predicated on the stable, reproducible liberation of specific peptides under carefully optimized digestion conditions, which may be different from the conditions used in the discovery experiment. Before digestion, proteins are chemically treated to make them more amenable to digestion. This often includes denaturation, reduction of disulfide bonds, and alkylation of reduced cysteine side chains to prevent reformation of disulfide bonds. Once prepared in this manner, the proteins in the sample can be digested using one or more of a number of different enzymes.[55] Each of these steps is optimized with the goal of maximizing specific rather than global peptide yield.

After digestion, it may be necessary to enrich the peptides of interest away from unwanted potential interferences in the sample to improve the robustness of the measurement. Salts, carbohydrates, and lipids may be removed using bulk chromatographic techniques such as solid-phase extraction. If the peptide of interest is very low in abundance and not detectable over the noise from other peptides, an antibody can be used to enrich the peptide.[56,57] During development of a bottom-up assay, peptides are selected, tested for

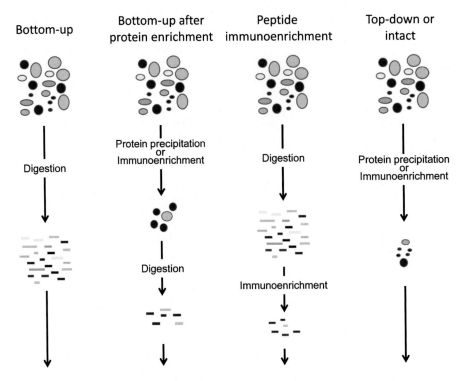

FIGURE 14.6 Common workflows for proteomic assays by mass spectrometry (MS). Bottom-up proteomics assays incorporate a proteolytic digestion step and use peptides detected by the mass spectrometer as surrogates for the protein contained in the sample. Intact or top-down proteomics assays require protein enrichment (either biochemical or immunoaffinity) before analysis of the protein by MS. Internal standard peptides (not shown) can be added before or after digestion (ideally before). Intact protein internal standards (not shown) are ideally added before protein enrichment.

performance, optimized, reevaluated, and reselected if necessary.[58,59] This cycle may be repeated iteratively until a set of suitable peptides has been identified that provides suitable performance. Once peptides are identified, ISs are synthesized and a calibration strategy is devised, both of which are discussed in more detail later. Before full validation, prevalidation experiments provide initial benchmarks that can be used to build confidence that the validation will be successful according to established guidelines (eg, Liquid Chromatography-Mass Spectrometry Methods [Clinical Laboratory Standards Institute, C62-A]). These prevalidation experiments are also described in C62-A and investigate precision, linearity, analytical sensitivity (ie, lower limit of quantification, lower limit of the measuring interval), stability, and matrix effects. A simple set of experiments also has been described in the literature.[60]

Peptide Selection

Development of a bottom-up assay begins with the selection of the peptides that will be surrogates for the protein measurement. Optimal peptides for quantitative protein analysis are freely liberated and unique within the human proteome, have good chromatographic and ionization properties, lack potentially complicating amino acid residues or modifications, present a linear and precise response, are free from interferences, and correlate with other peptides in the protein. Peptides that satisfy the first two criteria have been called *proteotypic* peptides,[44] with the other criteria arising from the basic requirements for a stable analyte suitable for quantitative analysis.

It is difficult to predict the peptides that will be well liberated from the protein during digestion and that have favorable chromatographic and ionization properties. Beyond data that are collected during discovery experiments, numerous algorithms exist to digest proteins *in silico* and predict which peptides will be observed in a typical proteomics experiment.[59,61-65] Several large public databases catalog peptides that have been observed in DDA proteomics experiments.[66] These resources provide spectrometric data and information on frequency of observation. More recently, a database with information regarding the basic assay characteristics (ie, limit of quantification, precision, accuracy) for many peptides under standard conditions has been made freely available.[65] Because of the enormous number of potential peptides that could be used to build an assay, heuristic rules, authentic data, curated MS databases, and protein sequence databases are often combined to prioritize peptides in a largely empirical process. However, a simplified approach using Skyline that evaluates all tryptic peptides in a targeted fashion and relies solely on empirical data is successful in identifying robust peptides.[67]

Any list of candidate peptides can be narrowed based on uniqueness and chemical stability. To evaluate whether a peptide is unique and therefore offers the best potential for specificity, database searching (eg, using the BLAST algorithm) of target peptides against the human genome can identify any other proteins that share sequence identity for the selected peptide.[68] Typically, peptides that are longer than 6 to 8 amino acids have the highest likelihood to be unique in the proteome. Some amino acids or amino acid combinations are not ideal residues for target peptides in bottom-up proteomics assays because of likely preanalytical modifications or heterogeneity that can lead to errors in quantitation. These amino acids are often reactive: methionine can be irreversibly oxidized to the sulfoxide form; glutamine and asparagine can be deamidated. Both of these modifications change the mass of the peptide, rendering them undetectable in an assay designed specifically for the unmodified form. Cysteines are often chemically alkylated during sample preparation, but accurate quantitation of a cysteine-containing peptide assumes stoichiometric conversion to the chemically modified form, which actually can be quite variable. Although the presence of reactive residues does not render them completely unsuitable for use, candidate peptides containing them do require extra attention during development.

A narrowed list of candidate peptides can be further honed empirically. Digestion time-courses, during which aliquots of the digest are removed and analyzed over the course of the digestion, help identify peptides that are liberated quickly from the protein and that are resistant to nonspecific degradation during the digestion.[67,69,70] Although these are ideal characteristics of peptides used in bottom-up assays, the actual rates of cleavage and degradation are strongly dependent on the denaturing conditions and the amount and quality of trypsin used in the digestion. The precision of the amount of peptide liberated (and degraded) is assessed by digesting several samples over several days. Linearity can be assessed using mixing studies of samples with different concentrations of endogenous protein or samples with a relevant protein-free matrix to help identify peptides and transitions that have interference.[58] Spiking purified or recombinant protein into human matrix also helps assess linearity. In addition to being interference free, robust peptides for the quantification of protein concentration often correlate very strongly with other peptides in the protein when measured across samples collected from multiple individuals.[71] Two peptides that correlate strongly suggest that they are liberated from the target protein similarly and when combined in the estimate of protein concentration (eg, averaged), may reduce the variability of the measurement. There are, of course, instances in which there is only one useful peptide that is liberated or when there is a particular posttranslational modification or polymorphism of interest.[67,72] In these cases, the chromatographic and mass spectrometric methods need extra attention to optimize performance, because there are no other peptides against which to compare. In these cases, alternative proteases with different cleavage specificity may be necessary.

Denaturation

Quaternary, tertiary, and secondary protein structural properties can interfere with digestion and can be minimized using chaotropic agents, organic solvents, and detergents (Table 14.1). This "linearization" of globular proteins improves reactivity toward reducing and alkylating agents as well as the efficiency and rate of proteolytic digestion. Some proteases retain activity under these conditions and can dramatically enhance digestion efficiency.

TABLE 14.1 Denaturants Used in Bottom-Up Proteomics Experiments

Reagent or Equipment	Notes
Urea	Must be prepared fresh and used at lower temperatures because of the potential for carbamylation of proteins and peptides.
Trifluoroethanol	Must be diluted to <5% for proteolysis with trypsin.
RapiGest, PPS	Compatible detergents. Hydrolyze under acidic conditions. Products are retained on reverse-phase columns but typically washed away at the end of the gradient.
Deoxycholate	Anionic, weak detergent. Precipitates on the addition of acid facilitating removal before LC-MS.
SDS (RIPA buffer)	Strong anionic detergent with very strong solubilizing activity. Removal before LC-MS is required.
CHAPS	Zwitterionic detergent. Used at <1%. Retained on reverse-phase columns but typically washed away at the end of the gradient.
NP-40	Nonionic detergent. Used at <1%. Removal before LC-MS is required.
Acetonitrile	Denatures proteins but causes precipitation at high concentrations.
Heat	Can be used alone or in combination with an additional denaturant (not urea).

LC-MS, Liquid chromatography–mass spectrometry.

Typical denaturants, such as urea and guanidine, are used in concentrations up to 8 M, although urea predominates because it is less inhibitory to proteases. Trifluoroethanol, acetonitrile, methanol, and ethanol all have denaturing properties and are used as additives up to 80%. Detergents such as 3-[(3-cholamidopropyl)dimethylammonio]-1-propanesulfonate (CHAPS) and deoxycholate also have been used, but they are often avoided because of the potential for strong suppression of ionization. Novel detergents that degrade to noninterfering molecules on the addition of acid are commercially available. In general, the additives used during the digestion steps are generally incompatible with LC-MS/MS analysis and require significant dilution or removal by solid-phase extraction (SPE) or an alternative chromatographic step.

Reduction and Alkylation

In addition to denaturation, the reduction of protein disulfide bonds can help in the linearization process by further dissociating protein complexes and unfolding target proteins. The most common reducing agents used in protein biochemistry laboratories are 2-mercaptoethanol, dithiothreitol (DTT), and tris(2-carboxyethyl)phosphine (TCEP). Although any of these reducing agents are effective, DTT and TCEP have become popular in proteomics assays. The concentration needed to fully reduce disulfide bonds in each sample can be estimated from the protein concentration in the solution but generally ranges from 5 to 50 mM. When urea is not used as the denaturant, heat and agitation can be applied to accelerate the reaction.

After reduction, the free cysteine thiol groups can be covalently blocked by alkylation with iodoacetamide to form a carbamidomethyl cysteine. This easily detected modification blocks the re-formation of disulfide bonds and adds 57.03 Da to the mass of each cysteine (this is the monoisotopic mass; because of the presence of ^{13}C in iodoacetamide, the average mass added is 57.07 Da). The addition of a free amine group to the peptide adds another position to bind a proton and can make ionization of the peptide more favorable. As mentioned earlier, the reaction is variable and the alkylated cysteine is still sensitive to oxidation of sulfur. The usefulness of alkylation after reduction should be evaluated empirically. When choosing the amount of iodoacetamide to add to the reaction, it is important to remember that the number of reactive thiol groups in the sample is affected by the DTT added to the sample. A threefold excess of iodoacetamide (eg, 15 mM iodoacetamide after the addition of 5 mM DTT) provides sufficient alkylating reagent to react with DTT and the cysteines in the sample. Iodoacetamide is selective for cysteine but when present in large excess or at high pH, residues other than cysteine can be alkylated. In some protocols, additional DTT is added after alkylation to quench the remaining iodoacetamide to prevent undesirable side reactions. In other protocols, alkylation is avoided altogether.

Protease Selection and Activity

The most commonly used protease for bottom-up proteomics assays is trypsin, which cleaves C-terminally to arginine and lysine residues (except when proline is adjacent on the carboxyl side). Trypsin has been studied over many years and is readily available in highly purified and well-characterized forms from bovine, porcine, and recombinant sources at very reasonable costs. These preparations are also available in modified and covalently bound forms, which reduce autolysis and minimize nonspecific digestion activity. Trypsin frequently liberates peptides of suitable length for mass spectrometric analysis. The liberated peptides have a carboxyl-terminal lysine or arginine, which imparts a positive charge to the peptide at low pH (in addition to the amino terminus), aiding fragmentation. Finally, trypsin has limited secondary structure specificity, meaning that most lysines or arginines will be cleaved with similar efficiency.

Fidelity of digestion is essential to maximize peptide yield and minimize the generation of undesired peptides that complicate analysis. Preparation of trypsin from natural sources will often contain trace amounts of chymotrypsin, which has very different proteolytic sequence specificity compared to trypsin. This undesired activity can be essentially eliminated by treatment with tosyl phenylalanyl chloromethyl ketone (TPCK). Although very-high-purity preparations of various proteases are available, all show varying degrees of nonspecific activity, which arises during extended incubation and can be monitored by the loss of the desired peptide and appearance of nonspecific digestion products.[70]

The use of multiple proteases can enhance digestion. For sample preparation with 8 M urea, Lys-C (cleaving C-terminally of lysine) retains activity and can be used first for partial digestion. Subsequently, the digest can be diluted to 1 to 4 M urea and trypsin is added to complete the digestion.

There is no requirement for an assay to use trypsin. Other proteases can be used if proper attention is paid to selectivity and efficiency (Table 14.2). Their use can help resolve issues that arise from nonoptimal peptide length or difficulties in observing specific peptides.[55] However, less common proteases may have specific limitations such as poor availability, high cost, and unpredictable behavior.

Digestion Optimization

Although desirable, it is very unusual for proteolysis of complex mixtures to go to completion. Protein structure, protein sequence, protease autolysis, buffer conditions, and product inhibition can inhibit digestion. Given the many variables that contribute to incomplete digestion, optimization of digestion conditions for a specific protein aims to meet analytical performance goals.[73] Robustness testing with variable amounts of denaturants, reducing and alkylating reagents, trypsin, buffers, enzyme substrate ratios, temperature, cationic and anionic additives, as well as variable incubation times, will identify the most sensitive elements of the digestion process. The use of digest progress curves and extensive optimization have yielded an array of protocols that are capable of high-fidelity digestion in minutes to hours.

TABLE 14.2 Proteolytic Enzymes

Enzyme	Specificity	Notes
Trypsin	C side of lysine and arginine	May have contaminating enzymes (eg, chymotrypsin)
LysC	C side of lysine	Can be used in conjunction with trypsin. Active in urea up to 8 M
LysN	N side of lysine	Relatively new
AspN	N side of aspartic acid	Selectivity issues
GluC	C side of glutamic acid	Selectivity issues
ArgC	C side of arginine	Uncommon
Chymotrypsin	Preference at hydrophobic residues	Uncommonly used because of cleavage specificity
Elastase	Preference at hydrophobic residues	Uncommon
Pepsin	Preference at hydrophobic residues	Secondary structure is important; complicated digests with many overlapping peptides; generally not used for quantitative analysis

Internal Standards and Peptide Quality

Stable isotope-labeled ISs are added to the sample in quantitative bottom-up proteomics assays to control for variability in system performance (ie, volume of injection by the autosampler, retention characteristics of the chromatographic column, electrospray stability, and sensitivity of the mass spectrometer), in ion suppression, and, increasingly, to control for digestion variability. Five types of ISs used in proteomics experiments include (1) isotope-labeled peptides with sequences identical to those of the target peptides, (2) isotope-labeled peptides with additional residues at the C- and N-termini of the peptide that get cleaved off during digestion, known as *winged* peptides,[74] (3) intact recombinant isotope-labeled protein,[75-78] (4) isotope-labeled concatenated peptides that liberate a labeled analog of the target peptide on digestion,[79] and (5) mutated peptides with a similar sequence but different mass to the target peptide.

Isotope-labeled peptides are most commonly used in the development of clinical assays because of their availability and acceptable performance when it is not necessary to control for the variability of digestion conditions. Recent experiments suggest that adding labeled peptides before digestion can be beneficial to overall assay performance.[70] Typically, amino acids with carbon-13 (^{13}C) and nitrogen-15 (^{15}N) incorporation (commonly Arg, Lys, Leu, Ile, Phe, Pro, and Val) are used in peptide synthesis. Deuterated amino acids and other amino acids are available in labeled forms but are used less frequently because of cost or subtle changes in chromatographic behavior. Winged peptides are synthesized in a similar fashion by including extra amino acids on each end of the tryptic analyte peptide that correspond to the native sequence of the protein. Recombinant protein expression systems are used to generate authentic protein or other proteins consisting of concatenated peptides labeled at specific amino acids or completely labeled with replacement of all ^{14}N with ^{15}N and/or all ^{12}C with ^{13}C. Synthetic peptides, winged peptides, and labeled proteins have all been used in clinical bottom-up proteomics assays.[72,74,77,80-84] Mutated peptides are generally not considered viable ISs because of shifts in chromatographic properties and the possibility of differential response to interferences.

To the extent that digestion efficiency is being recognized as a key element of assay precision, the utility of winged peptides, concatenated peptides, and fully isotope-labeled recombinant proteins to faithfully reflect the full complexity of digestion in matrix is under investigation. As described earlier, subtle differences in digestion efficiency associated with polypeptide structure are well-known and it is unlikely that any IS will behave identically to a native sample.

Once an IS is selected, adequate quality assurance during production ensures that IS protein and peptide quality is sufficient for reliable measurements. For each batch of ISs, four useful analyses to characterize the ISs before use include (1) amino acid analysis of the actual concentration of the IS, (2) HPLC with ultraviolet (UV) detection for peptides and proteins or sodium dodecyl sulfate polyacrylamide gel electrophoresis (SDS-PAGE) for proteins to give an estimate of purity, (3) MALDI MS or LC-MS/MS analysis to confirm sequence, and (4) correlation of the protein concentration in patient samples using the old versus the new IS (N determined by the precision of the assay). Peptides and proteins often adhere to storage containers, aggregating or precipitating, and becoming oxidized (cysteine, methionine, and tryptophan), deamidated (asparagine and glutamine), or degraded by light (tryptophan and phenylalanine), and these problems are exacerbated at low concentration.[84a] To help minimize these issues, stocks of ISs can be kept in an appropriate solvent at high concentration in amber vials under an inert gas below $-60°C$.

Calibration

There are many possible approaches to calibrating quantitative bottom-up proteomics assays, including (1) a simple calculation in which the ratio of peak area of the endogenous protein-derived peptide to the peak area of the IS peptide is multiplied by the concentration of the added IS, which is added at a known concentration; (2) calibration materials comprising peptides spiked into digested matrix; (3) calibration materials made by spiking purified or recombinant proteins into a relevant matrix; (4) native human samples with given concentrations (pooled or in singlicate); and (5) native human samples diluted into a relevant matrix. Other approaches are certainly conceivable.

Many research assays aimed at identifying novel protein biomarkers for disease use the first and second options for ease and availability of reagents. These options rely on a number of assumptions: the digestion goes to completion, the spiked peptide and IS are stable, and the relationship between analyte in calibrator matrix and experimental sample matrix is stable and linear. These approaches do not include the necessary controls for a validated clinical quantitative assay but can be useful in biomarker verification studies in which relative protein concentration is compared among samples.

Depending on the protein, spiking purified protein into a relevant matrix can be a very useful method for calibration of bottom-up proteomics methods.[77,81,85-88] It is important that the purified or recombinant protein be properly folded and be posttranslationally modified similarly to the endogenous protein in actual patient samples so that the proteins in the calibration materials are digested equivalently. For other proteins, such as those that exist in macromolecular complexes or those that are strongly bound to other molecules, it may be more difficult to generate useful calibrators if the native structure cannot be established.[71] In these cases, value-assigned human samples can be used. Calibrators made from spiked proteins or native samples are useful for improving between-day precision[71,73,89] and between-laboratory precision.[77] Importantly, the proper folding and posttranslational modifications of proteins in native samples and the near identity of the matrix between calibrator and patient samples allows the calibration material to behave as much like patient samples as possible. This helps control for the day-to-day differences in digestion that are the largest source of imprecision.

Preparing calibrators requires assignment of concentration. Comparison to certified or standard reference materials using reference measurement procedures would be ideal for this purpose, but these resources are not widely available for almost all protein targets. When available, certified or standard reference materials can be used to make matrix-matched calibrators. In the absence of reference materials, highly purified or recombinant proteins should be well-characterized by amino acid analysis, Karl-Fischer analysis (if crystalline), SDS-PAGE, HPLC-UV, and LC-MS to assign the concentration of the calibrator based on the mass of protein added to a blank matrix. A final option is to use an immunoassay to assign the concentration, but accuracy can suffer as a result of known issues with lot-to-lot variability and other well-known issues with immunoassay performance.

Protein or Peptide Enrichment

Similar to the issues encountered in discovery proteomics in which highly abundant proteins predominate the observable proteome, the sensitivity and selectivity of mass spectrometric measurements of proteins using bottom-up proteomics is limited by ion suppression and interference from other peptides. Generally speaking, without enrichment before LC-MS/MS, the lowest abundance serum and plasma proteins that can be reliably measured are in the low micromolar range.[71,72,90,91] Biochemical enrichment of peptides and small proteins can enhance sensitivities to the low nanomolar range.[75-77,92,93] Antibodies can be used to enrich proteins before digestion or specific peptides after digestion that can enable limits of quantitation in the low picomolar range.[37,74,80,81,86,94]

The biochemical enrichment of proteins and peptides relies on substantial differences in physical and chemical properties. Small proteins can be separated from the larger proteins in a sample by protein precipitation using organic solvents or polyethylene glycol. Ion-exchange SPE in any form (ie, strong vs. weak, cation vs. anion) can be an effective technique to achieve crude fractionation based on charge, pH, type, and concentration. Reverse-phase SPE enables separation by hydrophobicity. Finally, chemical affinity techniques such as immobilized metal affinity chromatography (IMAC) can selectively enrich acidic peptides, especially those that are phosphorylated. Using nonaffinity approaches, resolution of the separation is modest but can enrich proteins and peptides of interest more than 10-fold. In contrast, affinity techniques using antibodies or other highly selective chemistries can achieve enrichment of 1000- to 10,000-fold. The higher performance of antibody-based approaches is due to selectivity. Although differences in chemical properties are suitable to separate peptides and proteins by class (hydrophobic/hydrophilic, acidic/basic), the ability of antibodies to bind specific amino acid sequences allows for very selective purification even in samples in which target analytes are present in trace amounts. Antibodies, whether traditional immunoglobulins or immunoglobulin fragments,[95] can be conjugated to a solid phase and used in formats common in bioanalytical laboratories. Single-use and multiuse columns in 96-well format or inline flow through columns are available.[88] In particular, paramagnetic particles with conjugated antibody allow for easy automation of the binding, washing, and elution steps used. Both monoclonal and polyclonal antibodies have been used successfully, but success rates are higher with monoclonal antibodies, which can be selected to have higher affinity.

One of the best examples of a clinical assay that uses peptide immunoaffinity enrichment after trypsin digestion for protein quantification is the cancer biomarker thyroglobulin. As one of the best-studied biomarkers in laboratory medicine, particularly with respect to the sample-specific interferences that render the immunoassay measurement meaningless in a significant number of patients, the protein was an early target for MS. More specifically, in 25% to 30% of patients, autoantibodies can mask the epitopes recognized by reagent antibodies, leading to false results. Using the approaches described previously, peptides were selected and antibodies were generated in rabbits for the enrichment of specific peptides in a targeted bottom-up proteomics assay. The resulting workflow[81] and the iterations that came after it[74,80,86] laid the groundwork for other clinical assays using this technology. One of the assays developed for thyroglobulin actually uses rabbit antithyroglobulin antiserum to coat all of the analyte in a serum sample before protein precipitation of large proteins, digestion, and specific peptide enrichment.[74] The assay takes advantage of the fact that many patient samples already have autoantibodies coating the analyte of interest and that the binding of thyroglobulin to rabbit antibodies allows it to be separated from bulk protein more easily.

The coupling of the mass spectrometer to an antibody enrichment step is likely to provide the most sensitive and specific biomolecular detection available. Although antibodies can have significant issues with nonspecific interactions when used in immunoassays relying on secondary detection, the addition of mass and sequence information derived directly from the peptide in the mass spectrometer means that any nonspecific binding of protein or peptide to the antibody will not contribute to the signal of an assay.

Multiplexing

The ability to quantify many analytes in the same assay is as straightforward for peptides derived from proteins as it is for small molecules. Similar to small molecules, if preanalytical preparation techniques can recover peptides similarly, many proteins can be quantitated in a single assay. Development of multiplexed assays is much more complex and requires attention to all of the details mentioned earlier, as well as issues that can arise when attempting to optimize mutually exclusive assay parameters. However, good examples of assays that multiplex protein quantification have been described with and without peptide immunoaffinity enrichment.[71,73,89,94]

Quality control for a multiplexed assay, particularly when results from multiple proteins will be considered together, can be complicated and may involve decision trees or other approaches in addition to standard Westgard rules. For example, a peptide may have an IS peak area that is below the established cutoff, but the ratio of the peptide signal versus another peptide from the same protein may fall at the mean of the established acceptability criteria. These situations should be evaluated for their effects on the accuracy and precision of the assay during method development and evaluation.

Polymorphisms

Genetic polymorphisms can alter the amino acid sequence of proteins of interest. In some cases, changes are relevant to disease by affecting cellular localization, enzyme activity, or binding affinities. As a result, the identification of polymorphisms by MS can complement the quantification of protein concentration in the diagnosis of disease. One example of a bottom-up proteomics assay that quantifies the protein concentration and identifies important polymorphisms is an assay developed for α_1-antitrypsin.[72] In this assay, protein concentration is quantified by the peptide SASLHLPK, the S isoform (containing the E287V mutation) is detected by monitoring for the peptide LQHLVNELTHDIITK, and the Z isoform (containing the E365K mutation) is detected by monitoring for the peptide AVLTIDKK. The ability to quantify protein concentration and identify isoforms in the same assay eliminates the need for a second gel-based assay or a genetic assay for most patients.

When they are not important for the diagnosis of disease, amino acid changes secondary to genetic polymorphisms can interfere with protein quantification. A change in peptide mass as a result of amino acid substitution or the elimination of a cleavage site to produce a new peptide will reduce the amount of peptide detected in the digest. For this reason, databases of common polymorphisms should be considered before selecting a peptide for protein quantification. In addition, designing a quantitative assay with more than one peptide allows one to monitor for the heterozygous loss of a peptide (the ratio of two peptides should be constant from sample to sample and will deviate from the norm when a polymorphism is present in one of the peptides).

Chromatography

For high-throughput targeted clinical proteomics, the use of narrow-bore columns and flow rates approximately 100 to 500 μL/min are common. At these flow rates, injection-to-injection cycle times are minimized and robust operation is achievable. Whereas sensitivity is reduced compared with the nanoflow chromatographic separations and ionization methods common in discovery experiments, new ionization sources and high-efficiency ion transmission instrument designs now offer sensitivities near parity to nanospray techniques.

Summary

To summarize, an optimal workflow for a peptide immunoaffinity assay might begin with the denaturation and digestion of a human sample using an optimized digestion protocol, which includes reduction and alkylation, if necessary. IS is added before digestion (note that an intact protein or winged peptide would allow for normalization of digestion variability and account for the degradation of liberated peptide during the digestion). Matrix-matched calibration material is processed in parallel with patient samples so that the protein concentration can be calculated based on calibration curves rather than the simple ratio of the endogenous peak area to the IS peak area. Matrix-matched quality control materials with endogenous protein are included to evaluate the quality of the entire process. Monoclonal antibodies conjugated to paramagnetic beads are incubated with the digest. A magnet is used to isolate the beads to the side of the vessel, the beads are washed, and then the peptides are eluted with an acid (eg, 5% acetic acid). The eluted endogenous and IS peptides liberated during the digestion step are quantified by LC-MS/MS, and the peak area ratio is compared with the calibration curve to establish the protein concentration in each patient sample.

> **POINTS TO REMEMBER**
>
> **Bottom-Up Quantitative Assays**
> - Internal standards help control for the variability in peptide liberation, peptide degradation, sample handling, and chromatograph and mass spectrometer performance.
> - External calibration material greatly minimizes day-to-day variation.
> - Native human samples can help reduce variability across laboratories.
> - Bottom up quantitative assays are being used clinically in the detection and therapeutic management of disease.

TOP-DOWN PROTEOMICS

Workflow Overview

In proteomics, the nature of the sample is generally categorized based on whether the peptide or protein of interest has been proteolytically cleaved before analysis. For simplicity, polypeptides first might be categorized based arbitrarily on their molecular weight, with peptides being less than 3 kDa, small proteins being 3 to 35 kDa, and large proteins being greater than 35 kDa. The top-down approach to proteomic analysis refers to the introduction of undigested proteins into the mass spectrometer and often uses high-resolution, high mass accuracy instruments with or without dissociation or fragmentation.[96-99] This is in contrast to the bottom-up methods using proteolysis, as described previously. High-throughput, automated analysis of intact proteins is a recent development in the clinical proteomics field, and the high-performance mass spectrometers needed for these analyses represent significant advances over previous technologies, with sufficient resolution, mass accuracy, and sensitivity to be effective. The widespread commercial availability of these instruments has occurred only in the past several years.

The general approach to the analysis of intact proteins is no different than for other molecules. With the availability of a suitable IS, calibration material, separation strategy, and high-performance mass spectrometer, a quantitative analysis can be achieved. Similar to peptides, proteins also can be fragmented in the gas phase using collision-induced dissociation or electron transfer dissociation, which are both types of gas phase fragmentation. These techniques allow a mass spectrometric analysis analogous to selected reaction monitoring, with the detection of precursor-fragment pairs. Although this scheme is accurate in concept, it is very challenging to realize in practice. First, when proteins are ionized, they adopt a distribution of charge states that is roughly correlated with the number of basic amino acids. Whereas a tryptic peptide will typically contain 90% of its signal in a +2 charge state, a protein can spread its signal across 10 or more charge states, which reduces the total signal if only a single charge state is selected for analysis. Second, the fragmentation of very large molecules in the gas phase often yields few fragments with very low sensitivity, frequently making quantitative analysis impractical. Electron capture dissociation may have advantages over collision-induced dissociation (ie, what is used in triple-quadrupole instruments) for fragmentation efficiency, but the ability to perform selection reaction monitoring is still largely experimental for quantitative analysis.[100] For applications in which quantitative analysis is not necessary, such as the determination of clinically relevant amino acid polymorphisms in hemoglobin, top-down analysis allows for unambiguous identification of sequence changes associated with disease.[101-103] The success of this application relies on the analysis of a very-high-abundance protein where quantitative information is generally not needed.

For quantitative analysis, accurate mass, high-resolution full-scan data can be used to quantify the peptide without any fragmentation in the gas phase and has been applied to insulin and insulin-like growth factor-1 (IGF-1).[75,104,105] Although this type of analysis might be better described as intact protein analysis, it is considered with top-down methods because of the similarities in sample preparation. This workflow is appealing because it does not require digestion or the laborious effort to produce, identify, and verify signature peptides. The collection of full-scan data also provides the opportunity for retrospective analysis of other signals in the spectra that were not originally analyzed.[75] Drawbacks include potentially lower sensitivity as a result of the presence of multiple charge states, poor ionization, and difficulties achieving chromatographic separations of the same resolution as those that are routinely achieved for small molecules and peptides.

Sample preparation for the analysis of intact proteins may be similar to that for intact peptides (eg, removal of large proteins by protein precipitation), and many examples of enrichment of small proteins and their analysis by high-resolution mass spectrometers also have been described or are available from reference laboratories.

Assay Specificity

In the absence of the selectivity afforded by the generation of highly specific fragment ions from a selected precursor, additional information from spectra is sought to ensure the integrity of the measurement. In a top-down analysis, high mass accuracy and isotope ratio information can be used to gain confidence in the specificity of measurement.[76] Although accurate mass determination alone cannot allow unequivocal confirmation of the molecular formula of large molecules, it provides a substantial constraint on the number of proteins in a proteome that could give rise to an observed signal. As mass accuracy is a function of resolution, a high-quality spectrum can be interrogated by inspection of very small m/z ranges reducing the possibility that any coeluting molecule will contribute to the signal. In addition, isotope ratios of the intact analyte are characteristic of the molecular formula and can be used to detect the presence of an interfering signal via perturbation of the expected ratio, similar to the use of fragment ion ratios in a selected reaction monitoring experiment.

Calibration

For proper calibration of a top-down protein assay, it is essential to have highly purified protein that is chemically and biologically identical to the human protein of interest. These proteins can be purified from pooled human specimens or obtained by heterologous expression and purification.

At present, commercial sources of proteins purified directly from human matrix are limited. Several organizations, such as the National Institute for Biological Standards and Control, offer reference materials (eg, 83/500 World Health Organization International Standard for Insulin) that are isolated from a human source. Other organizations dedicated to specific clinical diseases are also sources of reference materials. However, the difficulty of isolating large quantities of low-abundance proteins from very complex matrices makes commercial supply of highly purified material a challenging prospect.

Recombinant DNA techniques and related protein expression systems have made commercial availability of proteins more commonplace and are available through extensive searchable catalogs online. For proteins that are not found through commercial sources, the production of recombinant proteins can be outsourced. However, it is essential to carefully characterize these materials for identity, proper folding, posttranslational modifications, and purity before use in assay development. The quality of the quantitative information describing the amount of protein provided is generally insufficient (or nonexistent) to allow direct use of these proteins in clinical assays. Therefore further characterization using well-calibrated and quality-controlled amino acid analysis, Karl-Fischer analysis, spectrophotometry, and/or LC-MS analysis is needed to correctly assign calibrator concentrations.

Compared with bottom-up assays, it is even more important for top-down protein assays to use calibrators with the correct secondary and tertiary structure. Very small changes in α-helix content, β-sheet content, protein folding, and disulfide arrangement often lead to dramatic changes in chromatographic behavior. Thus, if a protein is enriched from human samples and chromatographed under conditions that retain elements of tertiary and secondary structure during analysis, these elements will need to be similarly conserved in proteins used for calibration. For recombinant proteins, correct folding and formation of disulfide bonding is not guaranteed and structural studies of commercially derived material (eg, functional assays, native gel electrophoresis or circular dichroism) may be required before successfully using the materials in a quantitative assay.

Internal Standards

Quantitative analysis of intact proteins by isotope dilution requires the availability of chemically or isotopically labeled proteins with a different mass but intact structure. Using ^{15}N ammonium chloride as the sole nitrogen source in cell culture, it is possible to replace all ^{14}N atoms during protein expression to yield a heavy labeled protein. A small number of proteins are now commercially available (eg, ApoA-I, ubiquitin, and IGF-1 from Cambridge Isotopes or Sigma). For novel targets, it is possible to outsource the production of labeled proteins, though this can be very costly. Although chemical and structural identity is the gold standard for an IS, reasonable compromises for proteins have been successfully accommodated. Homologous proteins with very closely related amino acid sequences have been used; similarly, recombinant proteins with conservative amino acid changes, insertions, or deletions can be used. However, extensive validation to demonstrate identical recovery, response, and chromatographic behavior to the native analyte is required.

Protein Enrichment

Quantitative top-down or intact mass spectrometric measurements of proteins require significant simplification of the matrix before being analyzed by MS. For some peptides and small proteins, this can be achieved simply by precipitating larger proteins using acidic solvents or other chemical treatment before chromatography. Alternatively, proteins can be enriched using immunoaffinity methods, which will necessarily purify any protein isoform that contains the epitope(s). This means that in one assay, all relevant posttranslationally modified protein isoforms could be identified and quantified. Many proteins have been evaluated using this technology including apolipoproteins,[106,107] serum amyloid A,[108] cell signaling molecules,[109] insulin,[110] IGF-1,[111] and parathyroid hormone.[112]

> **POINTS TO REMEMBER**
>
> **Top-Down Quantitative Assays**
> - Current methods rely on high-resolution instrumentation for intact peptides and small proteins.
> - Fragmentation of large proteins in the gas phase is still experimental.
> - It is absolutely imperative that internal standards and calibration materials be properly folded and contain the correct complement of posttranslational modifications.
> - Top-down assays are being used clinically for intact small proteins, including insulin and IGF-1.

PREANALYTICAL AND OTHER TECHNICAL CONSIDERATIONS

Specimen Collection Considerations

Obtaining high-quality specimens for discovery experiments is a crucial component of a successful biomarker pipeline. Although banked serum and plasma samples are the most readily available specimens for discovery studies, the suitability of such specimens has been called into question over a wide range of potential shortcomings, such as the lack of SOPs for collection protocols and collection devices, undocumented clot times for serum, unknown effects of storage time and temperature, and stability to freeze—thaw events.

Recommendations for SOPs for many clinical sample types have been published by the Human Proteome Organization, Early Detection Research Network, and Clinical Proteomic Tumor Analysis Consortium and provide a starting point for investigators evaluating previously banked specimens or embarking on a new collection effort.[113,114]

Urine as a Matrix for Proteomics Assays

For many pathologic conditions, biomarker discovery and clinical testing of urine could be the ideal matrix. As a body fluid, urine is generally much simpler than other specimens. Filtered in the glomerulus and extensively modified throughout its path to the collecting ducts, urine typically contains salts and small molecules diluted in a large volume of water. Depending on the hydration and metabolic status of the individual, urine can be significantly more concentrated and have extremes of pH. In addition, many different pathologic conditions can cause the release of serum proteins into the urine, particularly the conditions that would make urine a useful specimen. The large variability of the constituents in urine make it a very challenging matrix for proteomics experiments. As a general approach, proteins in urine are concentrated before analysis by ultrafiltration or protein precipitation; albumin, and uromodulin (Tamm-Horsfall protein) are the most abundant proteins present. Immunodepletion of urine can be used to remove high-abundance serum proteins, with the associated costs and risk for depletion of the proteins of interest. Besides the measurement of albumin and uromodulin in urine,[115,116,116a] clinical proteomics assays in urine continue to be quite challenging.

Tissues

Clinical proteomic studies of tissues have been limited compared with traditional specimen types.[117] This is partially due to the infrequent use of quantitative diagnostic testing to diagnose disease in whole tissues. Much more commonly, the diagnosis of disease from surgically collected tissues is achieved by histologic and immunohistochemistry examination with the identification of morphologic elements and cellular organization that are associated with pathologic conditions.[118-120] In an effort to couple MS with microscopic analysis by a pathologist, laser capture microdissection has been used to isolate selected groups of cells for proteomic analysis.[121]

One challenge in the application of MS to tissue analysis has been the chemical treatments used to preserve and fix tissue before microscopy. Techniques such as formalin fixation and paraffin embedding (FFPE) dramatically alter proteins through cross-linking and profound dehydration. These processes yield proteins that are highly resistant to tryptic digestion and also heavily modified, which complicates any mass spectrometric analysis.[122] Chemical strategies for analysis of FFPE tissues have been reported that can improve the yield of tryptic peptides and quantitative analysis of selected cancer-related protein markers.

Application of MALDI MS to fresh tissue sections has been explored extensively. In pioneering work at Vanderbilt University, thin sections of frozen tissue collected on slides and overlaid with matrix were used to collect MS images directly from the surface of the tissue.[123] In the absence of any accompanying separation technology, tissue images are often predominated by highly abundant constituents of cells such as membrane lipids. Extensive development of this technology has improved image quality, resolution, and the ability to probe the metabolites and proteins with greater depth. The clinical applications of tissue imaging are promising[124,125] but have yet to make an impact on the diagnosis of disease or to be adopted by the pathology community.

Posttranslational Modifications

Up to this point in the chapter, methods describing the analysis of proteins and peptides have been in the context of ideal situations with individual, simple proteins. However, the number of known biologically relevant protein modifications is substantial and includes highly studied modifications such as phosphorylation and glycosylation. The presence of posttranslational modifications is key in regulating protein function and the aberrant modification of proteins may be central to the disease process, with temporal and abundance changes being relevant.

MS has played a vital role in the identification and analysis of posttranslational modifications, and it is reasonable to envision that selected reaction monitoring, along with chemically synthesized ISs and calibrators, would facilitate the facile development of clinical assays to detect and quantify posttranslationally modified proteins. In principle this is true, but the highly dynamic nature and low stoichiometry of many relevant modifications exacerbates the fundamental issues that challenge any measurement technology. For example, consider a protein present at 1 ng/mL in serum in which 0.1% of the protein is phosphorylated in the normal state and phosphorylated at 0.05% in a pathologic state. In this case, the investigator is faced with the challenge of trying to measure a small change under extremes of dynamic range.

Stability

As for any assay, an understanding of preanalytical stability of the analyte is extremely important. It is well known that many proteins are labile and that they must be quantified in specimens that are carefully drawn and processed.[126,127] Others are much more stable. Indeed, broad surveys of protein abundances suggest that under ideal storage conditions, few proteins experience dramatic changes.[128,129] However, previous comparisons of immunoassays and mass spectrometric assays in regular clinical samples demonstrate a positive bias in mass spectrometric assays,[72] suggesting the possibility that protein fragments are detectable by MS that are not detected by immunoassay. As a result, it is essential to validate that protein measurements by MS are unaffected by common specimen processing and handling procedures.

In addition to protein stability, it is important to consider peptide stability, particularly in bottom-up proteomics methods. After digestion, peptide recovery can suffer because of poor physiochemical properties, which leads to aggregation, peptide oxidation, and adsorption to surfaces of pipettes and vials. Validation studies must therefore include an assessment of peptide stability during analysis by LC-MS, including incubation in the autosampler.

Interferences

The challenges associated with interferences in LC-MS for clinical proteomics are similar to those faced in small molecule analysis—signal suppression, signal enhancement, and presence of isobaric coeluting species (ie, molecules with the same retention time and precursor m/z). Signal suppression is commonly observed when salts, phospholipids, or other nonanalyte compounds coelute with the analyte of interest and inhibit the process of ionization. This results in loss of signal and in severe cases can fully interrupt the ionization process, leading to a complete loss of signal. Signal enhancement is the opposite effect wherein the interference causes an increase in observed signal. While absolute elimination of suppression is a goal of method development, it is seldom achieved and the effects of suppression on assay performance must be carefully evaluated. The use of a coeluting, isotopically labeled IS minimizes the influence of these effects. In cases of suppression and enhancement, the IS signal can be used to correct for a modest degree of perturbation because it will behave identically to the analyte of interest. Coelution of species with the same m/z values also can disrupt measurements by contributing signal where none is expected. Because fragmentation of molecules by collision-induced dissociation follows thermodynamic and kinetic rules specific to their structure, the abundance of various fragment ions remains relatively constant under fixed conditions. This property allows for the use of fragment ion ratios to contribute additional information about molecular identity. Coeluting compounds that interfere can contribute abundances that perturb ion ratios and allow for their presence to be identified.[58] The appearance of these types of interferences is not an uncommon occurrence in clinical proteomics, wherein the massive number of peptides in a complex digest leads to peptides with nearly equal precursor masses and shared fragment ion masses. Only with routine inspection of isotope ratios can these interferences be identified.

Instruments

The choice of instruments for discovery experiments, subsequent biomarker verification and validation experiments, and clinical implementation depends on how they will be used (Table 14.3). For discovery, instruments that collect high-resolution precursor scans and high-speed MS/MS fragmentation spectra in a data-dependent fashion are most desirable and include orbitrap hybrid instruments and high-performance TOF instruments. These instruments provide

TABLE 14.3 Mass Spectrometers

Instruments	Notes
Single quadrupole	Not typically used for peptide analysis in the clinical laboratory
Triple quadrupole	Very common instrument for targeted methods, particularly isotope dilution
Ion trap	Commonly used for discovery experiments, but quantitative performance is limited because tandem-in-time operation
Time-of-flight	Can be used for quantitative analysis with adequate internal standards
Orbitrap	New high-resolution and high mass accuracy instrument that has the same limitations as an ion trap
Hybrid-orbitrap (eg, LTQ Orbitrap, Thermo Fisher Scientific, Waltham Massachusetts)	Commonly used for discovery experiments, lacks the precision needed for quantitative assays
Hybrid quadrupole high-resolution and high mass accuracy analyzer (eg, Q-TOF or Q-Orbitrap [Thermo Fisher])	An emerging competitor to triple quadrupole instruments, can be used for the quantitative analysis of intact proteins or peptides after digestion

data of sufficient quality to achieve high-fidelity peptide identification and relative quantitation.

For biomarker verification and validation, as well as clinical assay implementation, instruments that are able to perform high-speed MS/MS data collection are common. Triple-quadrupole tandem mass spectrometers are ideal for this application, although hybrid quadrupole–high resolution and high mass accuracy instruments are emerging as alternatives.

CONCLUSIONS

Current Clinical Proteomics Assays

Despite more than 20 years of work, the number of new clinical biomarkers derived purely from discovery proteomics is limited. So far, only two lung cancer diagnostics[130,131] and an ovarian cancer test[132] are commercially available. However, many single protein and peptide diagnostic markers measured by MS are offered by a number of reference laboratories. These include thyroglobulin,[74,80-82,84-87] IGF-1,[75,76,133] angiotensin-I (plasma renin activity),[92] insulin,[134] C-peptide, PTHrP, α_1-antitrypsin,[72] ADAMTS13,[135] and amyloidosis[136] and the identification of hemoglobin variants. The amyloidosis assay is particularly important to highlight as the first of its kind to employ a discovery proteomics–type platform in the regular clinical evaluation of surgical pathologic samples. The ADAMTS13 and plasma renin activity assay measure endogenous enzyme activity, and the IGF-1, insulin, C-peptide, and hemoglobin assays are all examples of top-down proteomics methods. This list of assays thus includes almost every imaginable workflow for the quantification of a protein by MS, which speaks to the versatility of MS for the measurement of proteins.

New Directions

In the coming years, MS will become a central method for protein and peptide quantification in the clinical laboratory. Improvements in instrumentation speed and sensitivity will complement the specificity inherent in the technique. It seems very likely that clinical methods will extend well beyond those currently used to measure proteins in blood, urine, and certain tissues to replace many immunoassays and immunohistochemical stains. The ability of immunoenrichment to achieve very low limits of detection will be instrumental in the application of the technique in these new areas. Advances in data processing software and the development of instruments to uniformly apply matrix to tissue samples also will help make MALDI imaging more acceptable to general pathology practices. The reality of protein MS for the care of patients is certainly exciting.

REFERENCES

1. Wasinger VC, Cordwell SJ, Cerpa-Poljak A, et al. Progress with gene-product mapping of the mollicutes: *Mycoplasma genitalium*. *Electrophoresis* 1995;**16**:1090–4.
2. Anderson L, Anderson NG. High resolution two-dimensional electrophoresis of human plasma proteins. *Proc Natl Acad Sci USA* 1977;**74**:5421–5.
3. Aebersold RH, Teplow DB, Hood LE, et al. Electroblotting onto activated glass: high efficiency preparation of proteins from analytical sodium dodecyl sulfate-polyacrylamide gels for direct sequence analysis. *J Biol Chem* 1986;**261**:4229–38.
4. Tanaka K, Waki H, Ido Y, et al. Protein and polymer analyses up to m/z 100,000 by laser ionization time-of-flight mass spectrometry. *Rapid Commun Mass Spectrom* 1988;**2**:151–3.
5. Yamashita M, Fenn JB. Electrospray ion source: another variation on the free-jet theme. *J Phys Chem* 1984;**88**:4451–9.
6. Eng JK, McCormack AL, Yates JR. An approach to correlate tandem mass spectral data of peptides with amino acid sequences in a protein database. *J Am Soc Mass Spectrom* 1994;**5**:976–89.
7. Perkins DN, Pappin DJ, Creasy DM, et al. Probability-based protein identification by searching sequence databases using mass spectrometry data. *Electrophoresis* 1999;**20**:3551–67.
8. Meehan KL, Holland JW, Dawkins HJS. Proteomic analysis of normal and malignant prostate tissue to identify novel proteins lost in cancer. *Prostate* 2002;**50**:54–63.
9. Shen J, Person MD, Zhu J, et al. Protein expression profiles in pancreatic adenocarcinoma compared with normal pancreatic tissue and tissue affected by pancreatitis as detected by two-dimensional gel electrophoresis and mass spectrometry. *Cancer Res* 2004;**64**:9018–26.
10. Li J, Zhang Z, Rosenzweig J, et al. Proteomics and bioinformatics approaches for identification of serum biomarkers to detect breast cancer. *Clin Chem* 2002;**48**:1296–304.
11. Lin Y, Goedegebuure PS, Tan MCB, et al. Proteins associated with disease and clinical course in pancreas cancer: a proteomic analysis of plasma in surgical patients. *J Proteome Res* 2006;**5**:2169–76.
12. Koo B-S, Lee D-Y, Ha H-S, et al. Comparative analysis of the tear protein expression in blepharitis patients using two-dimensional electrophoresis. *J Proteome Res* 2005;**4**:719–24.
13. Thongboonkerd V, McLeish KR, Arthur JM, et al. Proteomic analysis of normal human urinary proteins isolated by acetone precipitation or ultracentrifugation. *Kidney Int* 2002;**62**:1461–9.
14. Corthals GL, Wasinger VC, Hochstrasser DF, et al. The dynamic range of protein expression: a challenge for proteomic research. *Electrophoresis* 2000;**21**:1104–15.
15. Gygi SP, Corthals GL, Zhang Y, et al. Evaluation of two-dimensional gel electrophoresis-based proteome analysis technology. *Proc Natl Acad Sci USA* 2000;**97**:9390–5.
16. Hortin GL, Sviridov D, Anderson NL. High-abundance polypeptides of the human plasma proteome comprising the top 4 logs of polypeptide abundance. *Clin Chem* 2008;**54**:1608–16.
17. Zuo X, Speicher DW. A method for global analysis of complex proteomes using sample prefractionation by solution isoelectrofocusing prior to two-dimensional electrophoresis. *Anal Biochem* 2000;**284**:266–78.
18. Cox B, Emili A. Tissue subcellular fractionation and protein extraction for use in mass-spectrometry-based proteomics. *Nat Protoc* 2006;**1**:1872–8.
19. Chernokalskaya E, Gutierrez S, Pitt AM, et al. Ultrafiltration for proteomic sample preparation. *Electrophoresis* 2004;**25**:2461–8.
20. Greening DW, Simpson RJ. A centrifugal ultrafiltration strategy for isolating the low-molecular weight (<or=25K) component of human plasma proteome. *J Proteomics* 2010;**73**:637–48.

21. Hinerfeld D, Innamorati D, Pirro J, et al. Serum/plasma depletion with chicken immunoglobulin Y antibodies for proteomic analysis from multiple mammalian species. *J Biomol Tech* 2004;**15**:184–90.
22. Björhall K, Miliotis T, Davidsson P. Comparison of different depletion strategies for improved resolution in proteomic analysis of human serum samples. *Proteomics* 2005;**5**:307–17.
23. Hutchens TW, Yip T-T. New desorption strategies for the mass spectrometric analysis of macromolecules. *Rapid Commun Mass Spectrom* 1993;**7**:576–80.
24. Chapman K. The ProteinChip biomarker system from Ciphergen Biosystems: a novel proteomics platform for rapid biomarker discovery and validation. *Biochem Soc Trans* 2002;**30**:82–7.
25. Fung E. Ciphergen ProteinChip technology: a platform for protein profiling and biomarker identification. *Nat Genet* 2001;**27**:54.
26. Diamandis EP. Proteomic patterns in biological fluids: do they represent the future of cancer diagnostics? *Clin Chem* 2003;**49**:1272–5.
27. Fung ET, Yip T-T, Lomas L, et al. Classification of cancer types by measuring variants of host response proteins using SELDI serum assays. *Int J Cancer* 2005;**115**:783–9.
28. Li J, White N, Zhang Z, et al. Detection of prostate cancer using serum proteomics pattern in a histologically confirmed population. *J Urol* 2004;**171**:1782–7.
29. Petricoin EF, Ardekani AM, Hitt BA, et al. Use of proteomic patterns in serum to identify ovarian cancer. *Lancet* 2002;**359**:572–7.
30. Anonymous. Proteomic diagnostics tested. *Nature* 2004;**429**:487.
31. Baggerly KA, Coombes KR, Morris JS. Bias, randomization, and ovarian proteomic data: a reply to "producers and consumers." *Cancer Inform* 2005;**1**:9–14.
32. Baggerly KA, Morris JS, Coombes KR. Reproducibility of SELDI-TOF protein patterns in serum: comparing datasets from different experiments. *Bioinformatics* 2004;**20**:777–85.
33. Sorace JM, Zhan M. A data review and re-assessment of ovarian cancer serum proteomic profiling. *BMC Bioinformatics* 2003;**4**:24.
34. Washburn MP, Wolters D, Yates JR. Large-scale analysis of the yeast proteome by multidimensional protein identification technology. *Nat Biotechnol* 2001;**19**:242–7.
35. Paulovich AG, Whiteaker JR, Hoofnagle AN, et al. The interface between biomarker discovery and clinical validation: the tar pit of the protein biomarker pipeline. *Proteomics Clin Appl* 2008;**2**:1386–402.
36. Rifai N, Gillette MA, Carr SA. Protein biomarker discovery and validation: the long and uncertain path to clinical utility. *Nat Biotechnol* 2006;**24**:971–83.
37. Whiteaker JR, Lin C, Kennedy J, et al. A targeted proteomics-based pipeline for verification of biomarkers in plasma. *Nat Biotechnol* 2011;**29**:625–34.
38. Covey TR, Huang EC, Henion JD. Structural characterization of protein tryptic peptides via liquid chromatography/mass spectrometry and collision-induced dissociation of their doubly charged molecular ions. *Anal Chem* 1991;**63**:1193–200.
39. Chervet JP, Ursem M, Salzmann JP. Instrumental requirements for nanoscale liquid chromatography. *Anal Chem* 1996;**68**:1507–12.
40. Wilm MS, Mann M. Electrospray and Taylor-Cone theory, Dole's beam of macromolecules at last? *Int J Mass Spectrom Ion Process* 1994;**136**:167–80.
41. Loo JA, Udseth HR, Smith RD. Peptide and protein analysis by electrospray ionization-mass spectrometry and capillary electrophoresis-mass spectrometry. *Anal Biochem* 1989;**179**:404–12.
42. Strittmatter EF, Ferguson PL, Tang K, et al. Proteome analyses using accurate mass and elution time peptide tags with capillary LC time-of-flight mass spectrometry. *J Am Soc Mass Spectrom* 2003;**14**:980–91.
43. Yates 3rd JR, Eng JK, McCormack AL, et al. Method to correlate tandem mass spectra of modified peptides to amino acid sequences in the protein database. *Anal Chem* 1995;**67**:1426–36.
44. Craig R, Cortens JP, Beavis RC. The use of proteotypic peptide libraries for protein identification. *Rapid Commun Mass Spectrom* 2005;**19**:1844–50.
45. Cox J, Neuhauser N, Michalski A, et al. Andromeda: a peptide search engine integrated into the MaxQuant environment. *J Proteome Res* 2011;**10**:1794–805.
46. Geer LY, Markey SP, Kowalak JA, et al. Open mass spectrometry search algorithm. *J Proteome Res* 2004;**3**:958–64.
47. Geiger T, Velic A, Macek B, et al. Initial quantitative proteomic map of 28 mouse tissues using the SILAC mouse. *Mol Cell Proteomics* 2013;**12**:1709–22.
48. Jiang H, English AM. Quantitative analysis of the yeast proteome by incorporation of isotopically labeled leucine. *J Proteome Res* 2002;**1**:345–50.
49. Ross PL, Huang YN, Marchese JN, et al. Multiplexed protein quantitation in *Saccharomyces cerevisiae* using amine-reactive isobaric tagging reagents. *Mol Cell Proteomics* 2004;**3**:1154–69.
50. Thompson A, Schäfer J, Kuhn K, et al. Tandem mass tags: a novel quantification strategy for comparative analysis of complex protein mixtures by MS/MS. *Anal Chem* 2003;**75**:1895–904.
51. Anderson L, Hunter CL. Quantitative mass spectrometric multiple reaction monitoring assays for major plasma proteins. *Mol Cell Proteomics* 2006;**5**:573–88.
52. Ang C-S, Nice EC. Targeted in-gel MRM: a hypothesis-driven approach for colorectal cancer biomarker discovery in human feces. *J Proteome Res* 2010;**9**:4346–55.
53. Rifai N, Watson ID, Miller WG. Commercial immunoassays in biomarkers studies: researchers beware! *Clin Chem* 2012;**58**:1387–8.
54. Hoofnagle AN, Wener MH. The fundamental flaws of immunoassays and potential solutions using tandem mass spectrometry. *J Immunol Methods* 2009;**347**:3–11.
55. Tsiatsiani L, Heck AJR. Proteomics beyond trypsin. *FEBS J* 2015;**282**:2612–26.
56. Anderson NL, Anderson NG, Haines LR, et al. Mass spectrometric quantitation of peptides and proteins using stable isotope standards and capture by anti-peptide antibodies (SISCAPA). *J Proteome Res* 2004;**3**:235–44.
57. Becker JO, Hoofnagle AN. Replacing immunoassays with tryptic digestion-peptide immunoaffinity enrichment and LC–MS/MS. *Bioanalysis* 2012;**4**:281–90.
58. Abbatiello SE, Mani DR, Keshishian H, et al. Automated detection of inaccurate and imprecise transitions in peptide quantification by multiple reaction monitoring mass spectrometry. *Clin Chem* 2010;**56**:291–305.
59. Bereman MS, Maclean B, Tomazela DM, et al. The development of selected reaction monitoring methods for targeted proteomics via empirical refinement. *Proteomics* 2012;**12**:1134–41.

60. Grant RP, Hoofnagle AN. From lost in translation to paradise found: enabling protein biomarker method transfer by mass spectrometry. *Clin Chem* 2014;**60**:941–4.
61. de Graaf EL, Altelaar AF, van Breukelen B, et al. Improving SRM assay development: a global comparison between triple quadrupole, ion trap, and higher energy CID peptide fragmentation spectra. *J Proteome Res* 2011;**10**:4334–41.
62. Eyers CE, Lawless C, Wedge DC, et al. CONSeQuence: prediction of reference peptides for absolute quantitative proteomics using consensus machine learning approaches. *Mol Cell Proteomics* 2011;**10**. M110 003384.
63. Marx H, Lemeer S, Schliep JE, et al. A large synthetic peptide and phosphopeptide reference library for mass spectrometry-based proteomics. *Nat Biotechnol* 2013;**31**:557–64.
64. Qeli E, Omasits U, Goetze S, et al. Improved prediction of peptide detectability for targeted proteomics using a rank-based algorithm and organism-specific data. *J Proteomics* 2014;**108**:269–83.
65. Whiteaker JR, Halusa GN, Hoofnagle AN, et al. CPTAC Assay Portal: a repository of targeted proteomic assays. *Nat Methods* 2014;**11**:703–74.
66. Desiere F, Deutsch EW, King NL, et al. The PeptideAtlas project. *Nucleic Acids Res* 2006;**34**:D655–8.
67. Henderson CM, Lutsey PL, Misialek JR, et al. Measurement by a novel LC-MS/MS methodology reveals similar serum concentrations of vitamin D-binding protein in blacks and whites. *Clin Chem* 2016;**62**:179–87.
68. Altschul SF, Gish W, Miller W, et al. Basic local alignment search tool. *J Mol Biol* 1990;**215**:403–10.
69. Proc JL, Kuzyk MA, Hardie DB, et al. A quantitative study of the effects of chaotropic agents, surfactants, and solvents on the digestion efficiency of human plasma proteins by trypsin. *J Proteome Res* 2010;**9**:5422–37.
70. Shuford CM, Sederoff RR, Chiang VL, et al. Peptide production and decay rates affect the quantitative accuracy of protein cleavage isotope dilution mass spectrometry (PC-IDMS). *Mol Cell Proteomics* 2012;**11**:814–23.
71. Agger SA, Marney LC, Hoofnagle AN. Simultaneous quantification of apolipoprotein A-I and apolipoprotein B by liquid-chromatography-multiple-reaction-monitoring mass spectrometry. *Clin Chem* 2010;**56**:1804–13.
72. Chen Y, Snyder MR, Zhu Y, et al. Simultaneous phenotyping and quantification of alpha-1-antitrypsin by liquid chromatography-tandem mass spectrometry. *Clin Chem* 2011;**57**:1161–8.
73. van den Broek I, Romijn FPHTM, Smit NPM, et al. Quantifying protein measurands by peptide measurements: where do errors arise? *J Proteome Res* 2015;**14**:928–42.
74. Kushnir MM, Rockwood AL, Roberts WL, et al. Measurement of thyroglobulin by liquid chromatography-tandem mass spectrometry in serum and plasma in the presence of antithyroglobulin autoantibodies. *Clin Chem* 2013;**59**:982–90.
75. Bystrom C, Sheng S, Zhang K, et al. Clinical utility of insulin-like growth factor 1 and 2; determination by high resolution mass spectrometry. *PLoS ONE* 2012;**7**. e43457.
76. Bystrom CE, Sheng S, Clarke NJ. Narrow mass extraction of time-of-flight data for quantitative analysis of proteins: determination of insulin-like growth factor-1. *Anal Chem* 2011;**83**:9005–10.
77. Cox HD, Lopes F, Woldemariam Ga, et al. Interlaboratory agreement of insulin-like growth factor 1 concentrations measured by mass spectrometry. *Clin Chem* 2014;**60**:541–8.
78. Hoofnagle AN, Becker JO, Oda MN, et al. Multiple-reaction monitoring-mass spectrometric assays can accurately measure the relative protein abundance in complex mixtures. *Clin Chem* 2012;**58**:777–81.
79. Pratt JM, Simpson DM, Doherty MK, et al. Multiplexed absolute quantification for proteomics using concatenated signature peptides encoded by QconCAT genes. *Nat Protoc* 2006;**1**:1029–43.
80. Clarke NJ, Zhang Y, Reitz RE. A novel mass spectrometry-based assay for the accurate measurement of thyroglobulin from patient samples containing antithyroglobulin autoantibodies. *J Investig Med* 2012;**60**:1157–63.
81. Hoofnagle AN, Becker JO, Wener MH, et al. Quantification of thyroglobulin, a low-abundance serum protein, by immunoaffinity peptide enrichment and tandem mass spectrometry. *Clin Chem* 2008;**54**:1796–804.
82. Hoofnagle AN, Roth MY. Improving the measurement of serum thyroglobulin with mass spectrometry. *J Clin Endocrinol Metab* 2013;**98**:1343–52.
83. Kaiser P, Akerboom T, Ohlendorf R, et al. Liquid chromatography-isotope dilution-mass spectrometry as a new basis for the reference measurement procedure for hemoglobin A1c determination. *Clin Chem* 2010;**56**:750–4.
84. Kushnir MM, Rockwood AL, Straseski JA, et al. Comparison of LC-MS/MS to immunoassay for measurement of thyroglobulin in fine-needle aspiration samples. *Clin Chem* 2014;**60**:1452–3.
84a. Hoofnagle AN, Whiteaker JR, Carr SA, et al. Recommendations for the generation, quantification, storage, and handling of peptides used for mass spectrometry-based assays. *Clin Chem* 2016;**62**:48–69.
85. Netzel BC, Grant RP, Hoofnagle AN, et al. First steps toward harmonization of LC-MS/MS thyroglobulin assays. *Clin Chem* 2016;**62**:297–9.
86. Netzel BC, Grebe SK, Algeciras-Schimnich A. Usefulness of a thyroglobulin liquid chromatography-tandem mass spectrometry assay for evaluation of suspected heterophile interference. *Clin Chem* 2014;**60**:1016–8.
87. Netzel BC, Grebe SK, Carranza Leon BG, et al. Thyroglobulin (Tg) testing revisited: Tg assays, TgAb assays, and correlation of results with clinical outcomes. *J Clin Endocrinol Metab* 2015;**100**:E1074–83.
88. Neubert H, Gale J, Muirhead D. Online high-flow peptide immunoaffinity enrichment and nanoflow LC-MS/MS: assay development for total salivary pepsin/pepsinogen. *Clin Chem* 2010;**56**:1413–23.
89. van den Broek I, Nouta J, Razavi M, et al. Quantification of serum apolipoproteins A-I and B-100 in clinical samples using an automated SISCAPA–MALDI-TOF-MS workflow. *Methods* 2015:1–12.
90. Bondar OP, Barnidge DR, Klee EW, et al. LC-MS/MS quantification of Zn-alpha2 glycoprotein: a potential serum biomarker for prostate cancer. *Clin Chem* 2007;**53**:673–8.
91. Keshishian H, Addona T, Burgess M, et al. Quantitative, multiplexed assays for low abundance proteins in plasma by targeted mass spectrometry and stable isotope dilution. *Mol Cell Proteomics* 2007;**6**:2212–29.
92. Bystrom CE, Salameh W, Reitz R, et al. Plasma renin activity by LC-MS/MS: development of a prototypical clinical assay reveals a subpopulation of human plasma samples with substantial peptidase activity. *Clin Chem* 2010;**56**:1561–9.

93. Gerber SA, Rush J, Stemman O, et al. Absolute quantification of proteins and phosphoproteins from cell lysates by tandem MS. *Proc Natl Acad Sci USA* 2003;**100**:6940–5.
94. Kuhn E, Addona T, Keshishian H, et al. Developing multiplexed assays for troponin I and interleukin-33 in plasma by peptide immunoaffinity enrichment and targeted mass spectrometry. *Clin Chem* 2009;**55**:1108–17.
95. Whiteaker JR, Zhao L, Frisch C, et al. High-affinity recombinant antibody fragments (Fabs) can be applied in peptide enrichment immuno-MRM assays. *J Proteome Res* 2014;**13**:2187–96.
96. Kelleher NL. Top-down proteomics. *Anal Chem* 2004;**76**:197A–203A.
97. Kellie JF, Higgs RE, Ryder JW, et al. Quantitative measurement of intact alpha-synuclein proteoforms from post-mortem control and Parkinson's disease brain tissue by intact protein mass spectrometry. *Sci Rep* 2014;**4**:5797.
98. Savaryn JP, Catherman AD, Thomas PM, et al. The emergence of top-down proteomics in clinical research. *Genome Med* 2013;**5**:53.
99. Sze SK, Ge Y, Oh H, et al. Top-down mass spectrometry of a 29-kDa protein for characterization of any posttranslational modification to within one residue. *Proc Natl Acad Sci USA* 2002;**99**:1774–9.
100. Ji QC, Rodila R, Gage EM, et al. A strategy of plasma protein quantitation by selective reaction monitoring of an intact protein. *Anal Chem* 2003;**75**:7008–14.
101. Edwards RL, Creese AJ, Baumert M, et al. Hemoglobin variant analysis via direct surface sampling of dried blood spots coupled with high-resolution mass spectrometry. *Anal Chem* 2011;**83**:2265–70.
102. Rai DK, Griffiths WJ, Landin B, et al. Accurate mass measurement by electrospray ionization quadrupole mass spectrometry: detection of variants differing by <6 Da from normal in human hemoglobin heterozygotes. *Anal Chem* 2003;**75**:1978–82.
103. Shackleton CHL, Falick AM, Green BN, et al. Electrospray mass spectrometry in the clinical diagnosis of variant hemoglobins. *J Chromatogr B Biomed Sci Appl* 1991;**562**:175–90.
104. Darby SM, Miller ML, Allen RO, et al. A mass spectrometric method for quantitation of intact insulin in blood samples. *J Anal Toxicol* 2001;**25**:8–14.
105. Mazur MT, Cardasis HL, Spellman DS, et al. Quantitative analysis of intact apolipoproteins in human HDL by top-down differential mass spectrometry. *Proc Natl Acad Sci USA* 2010;**107**:7728–33.
106. Niederkofler EE, Tubbs KA, Kiernan UA, et al. Novel mass spectrometric immunoassays for the rapid structural characterization of plasma apolipoproteins. *J Lipid Res* 2003;**44**:630–9.
107. Trenchevska O, Schaab MR, Nelson RW, et al. Development of multiplex mass spectrometric immunoassay for detection and quantification of apolipoproteins C-I, C-II, C-III and their proteoforms. *Methods* 2015;**81**:86–92.
108. Kiernan UA, Tubbs KA, Nedelkov D, et al. Detection of novel truncated forms of human serum amyloid A protein in human plasma. *FEBS Lett* 2003;**537**:166–70.
109. Oran PE, Sherma ND, Borges CR, et al. Intrapersonal and populational heterogeneity of the chemokine RANTES. *Clin Chem* 2010;**56**:1432–41.
110. Oran PE, Jarvis JW, Borges CR, et al. Mass spectrometric immunoassay of intact insulin and related variants for population proteomics studies. *Proteomics Clin Appl* 2011;**5**:454–9.
111. Oran PE, Trenchevska O, Nedelkov D, et al. Parallel workflow for high-throughput (>1,000 samples/day) quantitative analysis of human insulin-like growth factor 1 using mass spectrometric immunoassay. *PLoS ONE* 2014;**9**:e92801.
112. Lopez MF, Rezai T, Sarracino Da, et al. Selected reaction monitoring-mass spectrometric immunoassay responsive to parathyroid hormone and related variants. *Clin Chem* 2010;**56**:281–90.
113. Rai AJ, Gelfand Ca, Haywood BC, et al. HUPO Plasma Proteome Project specimen collection and handling: towards the standardization of parameters for plasma proteome samples. *Proteomics* 2005;**5**:3262–77.
114. Tuck MK, Chan DW, Chia D, et al. Standard operating procedures for serum and plasma collection: early detection research network consensus statement standard operating procedure integration working group. *J Proteome Res* 2009;**8**:113–7.
115. Beasley-Green A, Burris NM, Bunk DM, et al. Multiplexed LC-MS/MS assay for urine albumin. *J Proteome Res* 2014;**13**:3930–9.
116. Singh R, Crow FW, Babic N, et al. A liquid chromatography-mass spectrometry method for the quantification of urinary albumin using a novel 15N-isotopically labeled albumin internal standard. *Clin Chem* 2007;**53**:540–2.
116a. Fu Q, Grote E, Zhu J, et al. An empirical approach to signature peptide choice for selected reaction monitoring: quantification of uromodulin in urine. *Clin Chem* 2016;**62**:198–207.
117. Schoenherr RM, Whiteaker JR, Zhao L, et al. Multiplexed quantification of estrogen receptor and HER2/Neu in tissue and cell lysates by peptide immunoaffinity enrichment mass spectrometry. *Proteomics* 2012;**12**:1253–60.
118. Azimzadeh O, Barjaktarovic Z, Aubele M, et al. Formalin-fixed paraffin-embedded (FFPE) proteome analysis using gel-free and gel-based proteomics. *J Proteome Res* 2010;**9**:4710–20.
119. Scicchitano MS, Dalmas DA, Boyce RW, et al. Protein extraction of formalin-fixed, paraffin-embedded tissue enables robust proteomic profiles by mass spectrometry. *J Histochem Cytochem* 2009;**57**:849–60.
120. Sprung RW, Brock JWC, Tanksley JP, et al. Equivalence of protein inventories obtained from formalin-fixed paraffin-embedded and frozen tissue in multidimensional liquid chromatography-tandem mass spectrometry shotgun proteomic analysis. *Mol Cell Proteomics* 2009;**8**:1988–98.
121. Guo T, Wang W, Rudnick PA, et al. Proteome analysis of microdissected formalin-fixed and paraffin-embedded tissue specimens. *J Histochem Cytochem* 2007;**55**:763–72.
122. Nirmalan NJ, Harnden P, Selby PJ, et al. Mining the archival formalin-fixed paraffin-embedded tissue proteome: opportunities and challenges. *Mol Biosyst* 2008;**4**:712–20.
123. Caldwell RL, Caprioli RM. Tissue profiling by mass spectrometry: a review of methodology and applications. *Mol Cell Proteomics* 2005;**4**:394–401.
124. Balluff B, Rauser S, Meding S, et al. MALDI imaging identifies prognostic seven-protein signature of novel tissue markers in intestinal-type gastric cancer. *Am J Pathol* 2011;**179**:2720–9.
125. Rauser S, Marquardt C, Balluff B, et al. Classification of HER2 receptor status in breast cancer tissues by MALDI imaging mass spectrometry. *J Proteome Res* 2010;**9**:1854–63.

126. Yi J, Kim C, Gelfand CA, et al. Inhibition of intrinsic proteolytic activities moderates preanalytical variability and instability of human plasma. *J Proteome Res* 2007;**6**:1768−81.
127. Yi J, Liu Z, Craft D, et al. Intrinsic peptidase activity causes a sequential multi-step reaction (SMSR) in digestion of human plasma peptides. *J Proteome Res* 2008;**7**:5112−8.
128. Pasella S, Baralla A, Canu E, et al. Pre-analytical stability of the plasma proteomes based on the storage temperature. *Proteome Sci* 2013;**11**:10.
129. Zimmerman LJ, Li M, Yarbrough WG, et al. Global stability of plasma proteomes for mass spectrometry-based analyses. *Mol Cell Proteomics* 2012. M111.014340.
130. Gregorc V, Novello S, Lazzari C, et al. Predictive value of a proteomic signature in patients with non-small-cell lung cancer treated with second-line erlotinib or chemotherapy (PROSE): a biomarker-stratified, randomised phase 3 trial. *Lancet Oncol* 2014;**15**:713−21.
131. Vachani A, Pass HI, Rom WN, et al. Validation of a multi-protein plasma classifier to identify benign lung nodules. *J Thorac Oncol* 2015;**10**:629−37.
132. Kim KH, Alvarez RD. Using a multivariate index assay to assess malignancy in a pelvic mass. *Obstet Gynecol* 2012;**119**:365−7.
133. Hines J, Milosevic D, Ketha H, et al. Detection of IGF-1 protein variants by use of LC-MS with high-resolution accurate mass in routine clinical analysis. *Clin Chem* 2015;**62**:990−1.
134. Chen Z, Caulfield MP, McPhaul MJ, et al. Quantitative insulin analysis using liquid chromatography-tandem mass spectrometry in a high-throughput clinical laboratory. *Clin Chem* 2013;**59**:1349−56.
135. Jin M, Cataland S, Bissell M, et al. A rapid test for the diagnosis of thrombotic thrombocytopenic purpura using surface enhanced laser desorption/ionization time-of-flight (SELDI-TOF)-mass spectrometry. *J Thromb Haemost* 2006;**4**:333−8.
136. Vrana JA, Gamez JD, Madden BJ, et al. Classification of amyloidosis by laser microdissection and mass spectrometry-based proteomic analysis in clinical biopsy specimens. *Blood* 2009;**114**:4957−9.

INDEX

Page numbers followed by *f* indicate figures, *t* indicate tables, and *b* indicate boxes.

A

ABCB1. *See* ATP binding cassette transporter B1
Absorbance, 368–369
Accreditation, forensic science laboratories and, 336
ACE. *See* Angiotensin-converting enzyme
Achondroplasia, 133t
Activation-induced cytidine deaminase (AID), 202
Acute myeloid leukemias (AML), 207–209
Adenine, 1, 2f
Adenomatous polyposis coli (APC) gene, 158
ADME (adsorption, distribution, metabolism, elimination), 295–296
AdnaTest, 254
Affinity chromatography, 38–39, 356
Affirm VPIII Microbial Identification test, 101
Agarose, 60
AID. *See* Activation-induced cytidine deaminase
β-Alanine, 348f
Alanine, 346
ALCL. *See* Anaplastic large cell lymphoma
Alkylation, 389–390
Allele-specific PCR (AS-PCR), 77, 77t
Alleles, 2
Alternative splicing, 9
Amelogenin, 330–331
Amide bonds. *See* Peptide bonds
Amino acids. *See also* Peptides; Proteins
  analysis of, 351–352
  basic biochemistry, 346–348, 348f
  concentrations, 351
  metabolism, 350–351, 350f
  overview, 345–346, 346t–347t
  supply and transport, 348–350, 350t
  translation and, 9–11
α-Amino acids, 346, 348f
β-Amino acids, 346
γ-Amino acids, 346
γ-Aminobutyric acid (GABA), 348f
Amitriptyline, 308–309
AML. *See* Acute myeloid leukemias
Amniocentesis, 283
Amplicon melting method, 77, 77t
AMPLICOR HIV-1 DNA PCR test, 88
Amplification methods
  common, 50–51, 50t
  endpoint quantification in, 58–59
  massively parallel sequencing and, 79–80
  probe, 59
  signal
    branched-chain DNA method, 58
    serial invasive amplification method, 58
  target
    loop-mediated, 58
    PCR. *See* Polymerase chain reactions (PCR)
    strand displacement, 58
    transcription-mediated, 57, 57f
    whole-genome and whole-transcriptome, 58
AmpliPrep/cobas TaqMan assays, 88, 92, 95
Anaplastic large cell lymphoma (ALCL), 206
Ancestry informative single-nucleotide repeats (AISNP; AIM), 335
Androgen receptor (AR), 227, 255
Androgen receptor splice variant-7 (AR-V7), 255
Aneuploidy screening, circulating cell-free fetal DNA testing for, 286
Angelman syndrome, 161–163, 162t
Angiotensin-converting enzyme (ACE), 354
Angiotensins, 354
ANP. *See* Atrial natriuretic peptide
Antibiotics, resistance to, 113–114
Anticipation, 133–134
Antidepressants, 308–310, 309t
Antidiuretic hormone (ADH). *See* Vasopressin
α-Antitrypsin disease, 126t
Aortopathies, 137
Apoptosis, 2, 254
Aptima HPV assay, 103
APTIMA Trichomonas vaginalis assay, 101
AR. *See* Androgen receptor
AR-V7. *See* Androgen receptor splice variant-7
Arginine, 346
Array comparative genomic hybridization (aCGH) assays, 69–70, 249, 250f
AS-PCR. *See* Allele-specific PCR
Ashworth, Thomas, 237
ASXL1 mutations, 219
Asymmetric PCR, 55
ATM protein kinase, 7
ATP binding cassette transporter B1 (ABCB1), 298–300
ATR protein kinase, 7
Atrial natriuretic peptide (ANP), 353
Autosomal dominant diseases, 132–139, 132f–133f, 133t
Autosomal recessive disorders, 125–132, 126t

B

B-cell lymphoproliferative disorders, 220
B-lymphoblastic leukemia/lymphoma, 202–204, 203t, 221–222
Bacterial genomes, 21
Balanced cryptic translocation, 202–203
Bart-Pumphrey syndrome, 132
BD MAX EBP test, 112–113
Becker muscular dystrophy (BMD), 140–141
Biochips. *See* Microarrays
BioCyc database, 350
Biofire Film Arrays. *See* FilmArray panels
Biomarker discovery pipeline. *See also* Genomics; Proteomics
  data acquisition strategies, 384–386, 385f
  data processing, 386
  discovery experiments, 384
  overview of, 383–384, 383f
  separations, 384
  targeted quantitative experiments, 387
  variations, 386–387
Biopsy, 236f. *See also* Liquid biopsy
Bisulfite treatment, 50, 50f
Biuret method of total protein determination, 368–369
BK virus, 97–99
Bladder cancer, 258–259
BLAST searches, 388–389
Blood
  amino acid concentrations in, 351
  isolation of nucleic acids from, 40
Bloodstream infections
  direct pathogen identification, 108–110
  positive blood culture identification, 107–108, 108t, 109b
BMD. *See* Becker muscular dystrophy
Boceprevir, 91
Body fluids, detection of cancer mutations in, 197
Bordetella pertussis, 106–107
BRAF gene, 155f–156f, 220
Branched-chain DNA (bDNA) method, 58
BRCA genes, 150–152, 153t–154t
Breast cancer
  circulating tumor cells and, 253–255
  hereditary, 150–152, 153t–154t
BRENDA database, 350
Bridge amplification, 79, 80f
Browser Extensible Data format files, 196
Burkitt lymphoma, 204

C

C-reactive protein (CRP), 363
C-type natriuretic peptide (CNP), 353
C3 convertase, 363–364
CaaX motifs, 357
Calibration
  bottom-up targeted proteomics and, 387–393
  top-down proteomics and, 393–395
CALR mutations, 216–217
Canavan disease, 126t
Cancer. *See also* Solid tumor cells; Tumor markers
  nucleic acid isolation and, 35
Cancer Genome Atlas Project, 14, 193
Cancer stem cells (CSC), 252
CANCER-ID consortium, 267
CAP/CTM CMV test, 97
Carbapenemases, 114
Carbohydrate-deficient transferrin (CDT), 363
Carbon metabolism, 351
χ-Carboxylation, 358
Carcinoma
  head and neck squamous cell, 258
  hepatocellular, 258
Carrier screening
  cystic fibrosis, 127t, 128–129
  cytogenomics, 167–169
  expanded. *See* Expanded carrier screening
  massively parallel sequencing, 164–166, 165t
  regulation of, 170
  reporting of results, 169–170
  whole-exome sequencing, 166–167
Catabolism, 358–360. *See also* Proteases
Catalytic triad, 359
Cathepsin family, 359
CCNA. *See* Cell culture cytotoxicity test neutralization assay
CCR5 entry inhibitors, 89t
CDT. *See* Carbohydrate-deficient transferrin
CEBPA mutations, 218–219
Ceiling principle, 336
Cell culture cytotoxicity test neutralization assay (CCNA), 112
Cell cycle, 7
Cell of origin testing, 224
CellCollector, 239f, 244–246
CellSearch system, 238–241, 240f–241f
CellSieve system, 242–243
CellSpotter Analyzer, 238–239, 240f
Centromeres, 4f, 5–6, 18
Cerebrospinal fluid (CSF)
  circulating tumor DNA and, 261
  proteins in, 373–374, 373t
  total protein in, 374
Ceruloplasmin (Cp), 362
Cervical cancer, 102
Cervista HPV HR assay, 102

403

CFTR. *See* Cystic fibrosis transmembrane conductance regulator
Chains of custody, 338
Chargaff, Erwin, 1
Charge, proteins and, 356
CHARGE syndrome, 133t
Chimpanzee genome, 19–20
Chlamydia infection, 99–101, 100t
Chorionic villus sampling (CVS), 283, 339
Chromatin, 4f, 5–6, 12–13
Chromatin modifiers, 218–219
Chromodomain, 13
Chromosomal instability (CIN) pathway, 152
Chromosome microarray (CMA) technology, 168f
Chromosomes (human), 5–6, 6f, 18
Chronic lymphocytic leukemia, 221
Chronic myelogenous leukemia (CML), 202, 209
Chymotrypsin C (CTRC) gene, 149
CIMP. *See* CpG island methylator phenotype
Circulating cell-free fetal DNA
   analytical aspects
      DNA analysis, 291
      massively parallel maternal plasma DNA sequencing, 291–292
      measuring fetal DNA fraction, 291
      sample collection and processing, 290–291
   biological properties, 284–285, 285f
   diagnostic applications
      fetal 'omics', 290
      fetal chromosomal aneuploidy screening, 286, 287f
      fetal rhesus D status determination, 286–289
      fetal sex assessment for sex-associated disorders, 286
      overview, 285–286
      single-gene diseases, 289–290
   overview, 284, 292
Circulating tumor cells
   analytical methods
      CellSearch system, 238–241, 240f
      detection and molecular characterization, 246–250, 247f
      isolation/enrichment systems, 241–246
      overview, 237–238, 239f
   cancer stem cells and, 252
   challenges, 238
   circulating tumor DNA vs., 259–266, 262t
   clinical significance
      bladder cancer, 258–259
      breast cancer, 253
      colorectal cancer, 256
      cutaneous melanoma, 257
      head and neck cancer, 258
      hepatocellular carcinoma, 258
      lung cancer, 256–257
      ovarian cancer, 258
      pancreatic cancer, 258
      prostate cancer, 254
   as companion blood biomarkers, 267–268
   epithelial and mesenchymal transitions, 251–252, 252f
   historical background, 237, 237f
   overview, 236f–239f
   quality control and logistic issues, 267
   as window to metastasis, 250–252
Circulating tumor DNA (ctDNA). *See also* Liquid biopsy
   circulating tumor cells vs., 237, 267t
   clinical oncology trials, 266
   as companion blood biomarkers, 267–268
   copy number variants, 262, 263t
   future challenges, 266
   gene amplification, 262
   half-life, 265
   integrity, 265, 265t
   isolation of nucleic acids from, 260
   massively parallel sequencing, 261

Circulating tumor DNA (ctDNA) (*Continued*)
   methylated, 262–264, 264t
   microsatellite instability, loss of heterozygosity, 262, 263t
   noncoding, 264
   origin and function, 259
   overview, 259f
   plasma *vs.* serum as source, 261
   quality control and logistic issues, 267
   quantification, 260–261
   single-marker *vs.* multimarker approach assays, 265–266
   tumor-specific mutations, 260f, 261–262, 262t
   types, 259–260, 259f–260f
   viral, 264–265, 264t
Clinical Variant (ClinVar) database, 29
Clonality testing, 224–228
Cloned probes, 67
Clopidogrel, 309–310
Clostridium spp.
   C. difficile, 111–112
Clustering algorithms, 193
CML. *See* Chronic myelogenous leukemia
CMV RGQ MDx test, 97
CNP. *See* C-type natriuretic peptide
Co-amplification at lower denaturation temperature (COLD) PCR, 55–56
Coat proteins, 357
Cobas tests, 103
Codeine, 298–300, 299f, 308
CODIS. *See* Combined DNA index system
Cohesin complex, 218
Cologuard assay, 197
Colony stimulating factors (CSF), 215–216
Colorectal cancer, 152, 153t–154t, 264
Combined DNA index system (CODIS), 330–331, 331t
Companion diagnostics, 296, 297t
Comparative genomic hybridization (CGH) arrays, 69–70
Competitive displacement probes, 71–73
Complement system, 363–364, 365t
Compound mutations, 222
Conformation-sensitive gel electrophoresis (CSGE). *See* Heteroduplex analysis
Connexin genes, 131
Copy number variant (CNV) analysis, 24–25, 69–70, 257–258, 263t, 302
Counting Method, 337
Cp. *See* Ceruloplasmin
CpG island methylator phenotype (CIMP), 152
Crick, Francis, 1, 6
Crigler-Najjar syndrome, 315–316
CRISPR/Cas system, 8
Crizotinib, 257
CRP. *See* C-reactive protein
CSC. *See* Cancer stem cells
CSF. *See* Colony stimulating factors
CSF3R mutations, 214
CTC-Chips, 237–250, 238f
CTCE. *See* Cycling temperature capillary electrophoresis
Cutaneous melanoma, 257–258
CVS. *See* Chorionic villus sampling
Cycling temperature capillary electrophoresis (CTCE), 61–62
Cystatins, 359
Cysteine, 355f
Cysteinyl proteases, 359, 359t
Cystic fibrosis (CF)
   genetics of, 125–129, 127t, 130f
   protein folding and aggregation and, 356–357
Cystic fibrosis transmembrane conductance regulator (CFTR), 126, 127t, 128b, 319
Cytochrome P450 enzymes
   CYP2C19, 309–310, 309t
   CYP2C9, 309

Cytochrome P450 enzymes (*Continued*)
   CYP2D6, 303, 305t, 306–309, 309t
   CYP3A, 306, 309
   nomenclature of, 307f–308f
   pharmacokinetics and, 303–317
Cytogenetics, 211
Cytogenomics, 167–169
Cytokeratins, 238–239, 241–242
Cytomegalovirus (CMV)
   transplant recipients and, 96–99
   viral infections, 96–99
Cytosine, 3, 3f

D

D amino acids, 346–348
D-loop. *See* Displacement loop
DAA. *See* Direct-acting antivirals
DAPC. *See* Dystrophin-associated protein complex
Data-dependent acquisition (DDA), 384, 385f
Data-independent acquisition (DIA), 384, 385f
Database of Genomic Structural Variation-NCBI (dbVAR), 29
Database of Genomic Variation-European Molecular Biology Laboratory (DGV), 29
Database of Single Nucleotide Polymorphisms (dbSNP), 29
Databases
   for gene nomenclature, 28–29
   human genome and, 29, 29b, 31t
   protein, 345
   for tertiary analysis, 31t
dbSNP. *See* Database of Single Nucleotide Polymorphisms
dbVAR. *See* Database of Genomic Structural Variation-NCBI
DDA. *See* Data-dependent acquisition
DDAVP, 354
DDGE. *See* Denaturing gradient gel electrophoresis
Deafness, 131–132
DeaFNess autosomal recessive (B) locus 1 (DFNB1), 131
Deep vein thrombosis (DVT), 147
Demethylation, 12
Denaturation, 389, 389t
Denaturing gradient gel electrophoresis (DDGE), 61–62
Denaturing HPLC (dHPLC), 65
Densitometry, 370
Density gradient separations, 242
Deoxyribose, 2
DEPArray, 243, 243f
Derivatization, amino acid analysis and, 351–352
DFNB1. *See* DeaFNess autosomal recessive (B) locus 1
DGV. *See* Database of Genomic Variation-European Molecular Biology Laboratory
dHPLC. *See* Denaturing HPLC
DIA. *See* Data-independent acquisition
Dideoxy-termination sequencing. *See* Sanger sequencing
Dideoxynucleotides, 62, 62f
Dielectrophoresis, 243
Differential extraction, 331
Diffuse large B-cell lymphoma (DLBCL), 205
Digital PCR, 56–57, 116
Dihydropyrimidine dehydrogenase (DPD), 311
Dinucleotide repeats, 18
Direct-acting antivirals (DAA), 91
Direct-to-consumer (DTC) genetic testing, 170
Discrimination methods
   nucleic acid
      conformation-sensitive scanning techniques, 61–62
      dideoxy-termination sequencing. *See* Sanger sequencing
      electrophoresis. *See* Electrophoresis

Discrimination methods (Continued)
    high performance liquid chromatography. See
        High-performance liquid chromatography
    hybridization assays. See Hybridization assays
    mass spectrometry. See Mass spectrometry
    multiplex ligation-dependent probe amplification
        (MLPA), 63
    oligonucleotide ligation assays, 63, 65f
    PCR product length, 61
    pyrosequencing, 64, 65f
    restriction fragment length polymorphism. See
        Restriction fragment length polymorphism
    single nucleotide extension assays, 64–65
Discriminatory power, 336
Displacement loop (D-loop), 334
Disseminated tumor cells (DTC), 237
DLBCL. See Diffuse large B-cell lymphoma
DMD. See Duchenne muscular dystrophy
DNA (deoxyribonucleic acid). See also Nucleic acids
    circulating cell-free fetal. See Circulating cell-free fetal
        DNA
    circulating tumor. See Circulating tumor DNA
    coding for RNA but not protein, 23
    epigenetics, 11–13, 12f
    gene structure, 8
    history of, 1–2
    modification enzymes, 8
    repair of, 7–8
    replication of, 6–7, 7f
    structure of, 3, 47
    types of, 4–5
DNA arrays/chips. See Microarrays
DNA glycosylases, 8
DNA Identification Act of 1994, 335
DNA methyltransferases, 223–224
DNA polymerases, 6, 50
DNA sequencing. See Sequencing
DNAScan Rapid DNA Analysis System, 337
DNMT mutations, 217, 224
Dormancy, 250–251
Dot blot assays, 68
Double helix structure, 3–6, 3f–4f
Down syndrome
    aneuploidy screening and, 286
    circulating nucleic acids and diagnosis of, 283
    mitigating risks of prenatal diagnosis and, 283–284
DPD. See Dihydropyrimidine dehydrogenase
Drug resistance, 206
DTC. See Disseminated tumor cells
Dual hybridization probes, 71–73, 77t
Duchenne muscular dystrophy (DMD), 139–142, 141f
Duplications, segmental, 18
DVT. See Deep vein thrombosis
Dye-binding methods of protein measurement, 369
Dystrophin-associated protein complex (DAPC), 140

E

EBV. See Epstein-Barr virus
EGFR. See Epidermal growth factor receptor
Ehlers-Danlos type IV syndrome, 137
Elastases, 375
Electrochemistry, 59, 93
Electronic Medical Records and Genomics Network, 14
Electrophoresis
    commonly used techniques, 61t
    methods for protein
        capillary electrophoresis, 372
        immunofixation, 370–372
        mass spectrometry, 372
        serum protein, 370
        Western blotting, 372
    for nucleic acid discrimination, 60–82, 60f
Elongation factors, 10–11

EMPOP. See European DNA Profiling Group mtDNA Population Database
EMT. See Epithelial mesenchymal transition
Emulsion PCR, 57, 79f
Enantiomers, 346–348
Encyclopedia of DNA Elements (ENCODE) project, 6, 14, 25–26
Endocrine treatment, 254
Endonucleases, 8
Endothelial NOS (eNOS), 358
Endothelins, 354
Endoxifen, 306–308
Energy, 13. See also Homeostasis
Engraftment testing, 339
Enhancers, 25–26
eNOS. See Endothelial NOS
Enrichment, 55–56, 108–109
Enteroviruses, 111
EPHX1. See Epoxide hydrolase 1
Epidermal growth factor receptor (EGFR), 241–242, 254
Epifluorescence, 68
Epigenetics, 11–13, 223–224
Epilepsy, 162
EPISPOT assay, 249
Epithelial cell adhesion molecule (EpCAM) antibodies, 154
Epithelial cell adhesion molecule (EpCAM)-based immunologic separation, 238, 239f
Epithelial-mesenchymal transition (EMT), 251–252, 252f
Epitopes, 352–353
Epoxide hydrolase 1 (EPHX1), 300, 301f
Epstein-Barr virus (EBV), 97, 221
ER. See Estrogen receptor
ERCC1 protein, 258
Erythroblasts, 218
Erythrocytes, 242, 357–358
Erythrocytosis, 214
Erythropoiesis, 358
Erythropoietin, 215
Escherichia coli, 21
ESensor system, 105
Estrogen receptor (ER), 262–264, 306–308
Estrogens, 160
Ethanol, 37, 39
Ethidium bromide, 44, 61
Euchromatin DNA, 6
European DNA Profiling Group mtDNA Population Database (EMPOP), 334
Evidence types, 252, 302t
ExAC. See Exome Aggregation Consortium
Examination process, 30–32
Exclusion criteria, 384
Exclusion identity testing, 136
Excretion, 349
Exome Aggregation Consortium (ExAC), 164–165
Exome sequencing, 300–302
Exons, 5, 8
Exonuclease probes. See Hydrolysis probes
Expanded carrier screening, 164
ExPASy-enzyme database, 350
ExplorEnz database, 350
Extensive metabolizers, 306–308
Extracellular matrix, 136–137, 360
Extrinsic coagulation pathway, 220
EZH2 mutations, 219

F

Fabry disease, 139t
FACS. See Fluorescence-activated cell sorters
Factor IX deficiency. See Hemophilia B
Factor V Leiden mutation detection, 164
Factor VIII, 354, 358. See also Hemophilia A

Familial adenomatous polyposis (FAP), 153t–154t, 157–159
Familial hypercholesterolemia, 133t
FAP. See Familial adenomatous polyposis
Farnesyl pyrophosphate, 357
Fastidious bacteria, 107
Fatty acids, 350–351, 361–362
Fatty acylation, 357
Favism, 314–315
Fecal immunochemical tests (FIT), 197
Feces, 115, 197, 372, 375
Ferroportin, 353
Fetal fibronectin (FN), 361t
Fetal fraction, 284, 289
Fetomaternal hemorrhage (FMH) screening, 286
Fetus, 339
Fibrillin-1 (FBN1) gene, 136–137
Fibrin, 147–148
Fibrinogen, 360
Fibronectin, 361t
Fibrosis, 94, 98, 139–140
FilmArray panels, 104–105, 105t, 108
Filtration, 39, 261
Fingerprints, 21
First-order kinetics, 67
FISH. See Fluorescence in situ hybridization
FIT. See Fecal immunochemical tests
Flaviviridae, 90
Flaviviruses, 90
Flies, medically important, 142–143, 219
Flow cytometers, 68
Flow cytometry, 63, 220, 242
FLT3 mutations, 208–209, 216
Fluconazole, 309
Fluorescence in situ hybridization (FISH), 211, 239f, 249, 256
Fluorescence resonance energy transfer (FRET), 59, 73f
Fluorescence-activated cell sorters (FACS), 223
Fluorescent labels, 59
Fluorescent staining, 44
Fluorophores, 194
Fluoropyrimidines, 311, 312t–313t
FMR1 gene, 25, 142, 144, 145f
FN. See Fetal fibronectin
Folding, protein, 356
Folin-Ciocalteu method, 369
Follicular lymphoma, 205
Forensic identification
    analytical process, 331–334
    bone marrow transplantation/chimerism, 339–340
    legal issues, 338
    loss of heterozygosity, 340
    massively parallel sequencing, 337–338
    mitochondrial DNA, 334–335
    overview, 329
    parentage and kinship, 338–339
    quality assurance, 335–336
    rapid DNA technology, 337
    real time PCR kits for, 77t
    sample verification, 339
    short tandem repeats and amelogenin, 330–331, 330f
    single-nucleotide polymorphisms, 338
    statistical interpretation, 336–337
    touch samples and DNA mixtures, 335
    Y-short tandem repeats, 334
Formaldehyde, 191–192
Formalin, 40
Formalin-fixed paraffin-embedded (FFPE) tissue blocks, 40, 191–192
Formulations, 308–309
Förster resonance energy transfer. See Fluorescence resonance energy transfer

Fragile X associated tremor and ataxia syndrome (FXTAS), 144
Fragile X syndrome, 142–146, 143f–145f
Fragmentation, 49–50, 78f, 393
Franklin, Rosalind, 1
Free radicals, 159–160
Frequency, 142
Frequency distribution, 141f, 142, 208f, 222–223, 338–339
FRET. See Fluorescence resonance energy transfer
Friedrich ataxia, 126t
Fume hoods, 37
Fungi
　genomes of, 20–21
　in human disease, 20–21
　taxonomy and nomenclature, 20–21
Fusion inhibitors, 89t
Fusions, genetic, 25
FV gene, 147
FXTAS. See Fragile X associated tremor and ataxia syndrome

G

G6PD. See Glucose-6-phosphate dehydrogenase
GABA. See γ-Aminobutyric acid
Gain of function mutations. See Hot spot mutations
Gastric acid, 111
Gastrinomas, 137–138
Gastroenteritis, 111–113
Gastrointestinal stromal tumors (GIST), 199, 215
Gaucher disease type I, 126t
Gaussian distribution, 228
Gc-globulin, 361t
Gel diffusion (passive), 372
Gel filtration, 37, 356
Gemins, 130
Gene expression profiles, 249–250
GeneOhm assays, 108, 114
Genes, 28–29, 128, 133, 319
Genetic code, 10, 10f
Genetic polymorphisms, 392–393. See also Single nucleotide variants/polymorphisms
GeneXpert system, 106
GenMAPP database, 350
GenMark eSensor RVP assay, 105
Genome-wide association studies (GWAS), 69–70, 147, 302
Genomes (human)
　chromosomes and, 18
　coding for RNA but not protein, 23
　databases, 29–30, 29b, 31t
　fusions, 25
　informatics, 30–32, 30f
　nomenclature
　　gene names, 28–29
　　large structural variants, 27–28
　　small variants, 27
　overview, 17–18
　sequencing approaches, 22f
　variations in
　　copy number variants, 24–25
　　epigenetic alterations, 25
　　fusions, 25
　　overview, 17–18
　　short tandem repeats, 25
　　single nucleotide variants, 23–24
　　specific populations, 18
　　transposable elements, 25, 27f
Genomes (nonhuman)
　bacterial, 21–22
　fungal, 20–21
　primate, 19–20
　rodent, 20
　size range, 21f
　viral, 22–23

Genomics. See also Metagenomics
　fetal, 284
　mass spectrometry and, 64–65
　overview of, 193
　of solid tumors, 193–194
Genotyping, 75, 77
Ghrelins, 357
Giemsa stain, 6, 28
Gilbert syndrome, 315–316
GIST. See Gastrointestinal stromal tumors
GLD. See Glutamate dehydrogenase
Globins, biosynthesis of, 26–27
Glomerulonephritis, 364, 365t
Glomerulus, 395
Glucocorticoids, 139–140
Glucose, 350–351
Glucose-6-phosphate dehydrogenase (G6PD), 312t–313t, 314–315
Glutamate, 350f
Glutamate dehydrogenase (GLD), 112
Glutamine, 10f
Glutathione pathway enzymes, overview of, 304–306, 354–355
Glutathione peroxidase (GSH Px), 348, 354
Glutathione S-transferase (GST), 128, 354
　acute kidney injury and, 317
Glutathione synthetase (GSH-S), 354–355
Glycine, 10f
Glycogen storage disease, 126t
Glycogen synthase kinase 3 β (GSK-3 β), 158
Glycosylation, 358
Glycosylphosphatidylinositol (GPI) anchor, 357
Gout, 315
Gram-negative bacteria, 36, 113–114
Gram-positive bacteria, 35–36
Granulocyte collection, 223
Grapefruit juice, 310–311
Green, Phil, 31b
GSH Px. See Glutathione peroxidase
GSH-S. See Glutathione synthetase
GSK-3 β. See Glycogen synthase kinase 3 β
GST. See Glutathione S-transferase
Guanine, 2f, 3
Guidelines, 59
GWAS. See Genome-wide association studies

H

Hairpin probes, 71–73, 77t
Hairy cell leukemia, 220
Haplotyping, 302–303
Haptoglobin (Hp), 362
HCC. See Hepatocellular carcinoma
HD. See Huntington's disease
HDA. See Helicase-dependent amplification; Heteroduplex analysis
HDM. See Histone demethylases
Head and neck cancer, 258
Healthy Human Individual's Integrated Plasma Proteome Database, 345
Hearing loss, 131–132
Helicase-dependent amplification (HDA), 58
Hematopoietic malignancies
　analytical methods
　　cytogenetics, 211
　　fluorescence in situ hybridization, 211
　　massively parallel sequencing, 228
　　polymerase chain reaction, 211–212, 228
　　single nucleotide polymorphism arrays, 211
　　southern blotting, 227–228
　　clonality testing in, 224–228
　gene mutations in
　　acquired drug resistance, 222
　　cohesin complex, 218
　　detection of single-gene variants, 222–223

Hematopoietic malignancies (Continued)
　epigenetic regulators, 217
　lymphoid diseases, 222–223
　mechanisms, 220
　myeloid diseases, 214
　myeloid transcription factors and chromatin modifiers, 218–219
　overview, 212–223, 213t
　RNA spliceosome, 212–223
　signal transduction pathways, 214–215
　genetic regulatory mechanisms in, 223–224
　massively parallel sequencing for somatic variants in, 223
　recurrent translocations, structural chromosomal abnormalities
　　analytical methods, 211–212
　　lymphoid disorders, 202–206, 203t, 210–211
　　myeloid disorders, 203t, 206–211
　　overview, 201–212
Hematoxylin and Eosin (H&E) staining, 192–193, 192f
Hemochromatosis, hereditary, 126t
Hemophilia A, 139t
Hemophilia B, 139t
Hepatitis viruses, circulating tumor DNA and, 235
Hepatocellular carcinoma (HCC), 258
HER2. See Human epidermal growth factor receptor-2
Hereditary hemochromatosis (HH), 126t
Hereditary hemorrhagic telangiectasia, 133t
Hereditary nonpolyposis colorectal cancer (HNPCC). See Lynch syndrome
Hereditary pancreatitis (HP), 149–150
Heterochromatic regions, 17–18
Heterochromatin DNA, 6
Heteroduplex analysis (HDA), 48b–49b, 61–62
Heteroduplexes, 55–56
Heterogeneous nuclear RNA (hnRNA), 5, 8f, 23t
Heteroplasmy, 19b, 159, 334
HGMD. See Human Gene Mutation Database
HGNC. See HUGO Gene Nomenclature Committee
HGVS. See Human Genome Variation Society
HH. See Hereditary hemochromatosis
High-performance liquid chromatography (HPLC)
　biomarker discovery pipeline and, 384
　for nucleic acid discrimination, 60
Highly quenched probes, 72–73
Histidine, 346t–347t
Histone demethylases (HDM), 12
Histone methyltransferases (HMT), 12
Histones, 12–13
HMT. See Histone methyltransferases
hnRNA (heterogeneous nuclear RNA), 5
Homeostasis
　amino acids and, 349
　electrolyte, 345
Homologous recombination repair pathway, 7
Homoplasmy, 334
Hormones
　amino acids and, 350
Hot spot mutations, 194
Hot start PCR, 54
HotAir, 13
Hp. See Haptoglobin
HP. See Hereditary pancreatitis
HTT gene, 134
HUGO. See Human Genome Organization
HUGO Gene Nomenclature Committee (HGNC), 28–29
Human epidermal growth factor receptor-2 (HER2), 238–239
Human Gene Mutation Database (HGMD), 29–30
Human Genome Organization (HUGO), 27, 29b
Human Genome Project, 5–6, 13–14
Human Genome Variation Society (HGVS), 27, 28b
Human immunodeficiency viruses (HIV)
　HIV-1, 87–90

Human leukocyte antigen (HLA) complex, 295–296, 309, 319–320
Human Microbiome Project, 115
Human papillomaviruses (HPV)
  circulating tumor DNA and, 265
  overview of, 102–103, 103t
  tests for, 100–101, 103t
Human Proteome Organization, 37
Hungerford, David, 209
Huntington's disease (HD), 132–136
Hurler syndrome, 126t
Hybrid Capture 2 (HC2) test, 102
Hybridization assays
  dot blot assays. See Dot blot assays
  kinetics, 67
  line probe assays, 68
  medium-density arrays, 68
  microarrays. See Microarrays
  overview, 65–68
  probes, 67–68
  real-time PCR. See Real-time PCR
  single copy visualization, 71
  in situ hybridization, 70–71
  for solid tumor samples, 71–73
  solid-phase vs. solution phase, 68–71, 68b
  thermodynamics, 66–67, 66f
Hybridization probes, 71–73
Hydrolysis probes, 71–73, 77t
Hydroxylation, 358
5-Hydroxylysine, 348, 348f
4-Hydroxyproline, 348, 348f
Hypercholesterolemia, familial. See Familial hypercholesterolemia

I
IBD. See Inflammatory bowel disease
ICE. See Interleukin I β converting enzyme
Identity testing. See Forensic identification
IDH mutations, 217–218
IFE. See Immunofixation electrophoresis
IISNP. See Individual identification single-nucleotide repeats
IKZF mutations, 221–222
IL-18. See Interleukin 18
IMDx assay, 114
Immunodeficiency, overview of, 366–367
Immunofixation electrophoresis (IFE), 370–372, 372f
Immunofluorescence, 238–239
Immunoglobulins
  clinical utility of measurement
    immunodeficiency, 366–367
    overview, 366–368
    paraproteins, 367–368
    polyclonal hyperimmunoglobulinemia, 367
  free light chains, 366
  IgA, 366
  IgD, 366
  IgE, 366
  IgG, 365
  IgM, 365–366, 367f
Immunohistochemistry, 193
Immunomagnetic separation, 39
Immunoreceptor tyrosine-based activating motifs (ITAM), 221
Imprinting disorders, 161–163
In situ hybridization, 70–71
Inclusion identity testing, 336
Incontinentia pigmenti, 139t
Individual identification single-nucleotide repeats (IISNP), 335
Inducible NOS (iNOS), 358
Inflammatory bowel disease (IBD), 375
Informatics, 30–32, 30f
INO80 proteins, 13
iNOS. See Inducible NOS

INR. See International normalized ratio
Insulators, 8
Integrase inhibitors, 88
Integrity index, 265
Interferences, biomarker discovery pipeline and, 396
Intergenic DNA, 18
Interleukin 18 (IL-18), 128
Interleukin I β converting enzyme (ICE), 359
Intermediate metabolizers, 308–309
Internal standards (IS), 386–387, 391, 394
International Cancer Genome Consortium, 193
International HapMap Project, 13
International normalized ratio (INR), 296, 296b
Introns, 8, 18
Irinotecan, 315–316
Iron metabolism, transferrin and, 362–363
IS. See Internal standards
ISET system, 242–243
Isoleucine, 346t–347t
Isopeptide bonds, 352
ISWI proteins, 13
ITAM. See Immunoreceptor tyrosine-based activating motifs
Ivacaftor, 319

J
Jackpotting, 195
JAK2 mutations, 215
Jeffreys, Alec, 329
"John Doe" warrants, 338

K
Kearns-Sayre syndrome (KSS), 161
KEGG database, 350
Kidney disorders, polycystic kidney disease, 133t
King, Maryland v., 338
Kinship testing, 338–339
KIT mutations, 215
Kjeldahl total protein determination, 368
Klebsiella pneumoniae carbapenemase (KPC), 114
Kornberg, Arthur, 1–2
KPC. See Klebsiella pneumoniae carbapenemase
Krebs cycle, 217–218, 350–351, 350f
KSS. See Kearns-Sayre syndrome

L
L amino acids, 346–348
Lactate dehydrogenase (LDH), cutaneous melanoma and, 257
LAMP. See Loop-mediated amplification
Laser capture microdissection (LCM), 192–193, 192f
LCM. See Laser capture microdissection
Leber hereditary optic neuropathy (LHON), 160
Leiden Open Variation Database (LOVD), 24, 24f
Leigh syndrome, 160
Leukapheresis, 149–150
LHON. See Leber hereditary optic neuropathy
Ligation, sequencing by, 80–81, 81f
Light chains, free, 366
Likelihood ratios, 336
LINE. See Long interspersed nuclear elements
Line probe assays, 68
Lineage informative single-nucleotide repeats (LISNP), 335
Linear after the exponential (LATE) PCR, 55
Lipocalins, 356
Liquid biopsy. See also Circulating tumor cells
  overview, 236–237, 236f
  tumor biopsy vs., 267
Liquid chromatography-tandem mass spectrometry (LC-MS/MS)
  history of proteomics and, 383–384
LISNP. See Lineage informative single-nucleotide repeats
Liver, amino acid metabolism in, 350, 350f

LobSTR, 25
Locus-specific control regions, 8
Loeys-Dietz syndrome, 137
Long interspersed nuclear elements (LINE), 25, 26t, 264
Long QT syndrome, 133t
Long terminal repeats (LTR), 25, 26t
Loop-mediated amplification (LAMP), 57
LOVD. See Leiden Open Variation Database
Low copy number (LCN) DNA, 335
Low template (LT) DNA, 335
LS. See Lynch syndrome
LTR. See Long terminal repeats
Luminex assays, 113, 248–249
Lung cancer, 256–257
Lymphoid disorders, 202–206, 203t, 210–211, 220–222, 224–227
Lymphoplasmacytic lymphoma, 220
Lynch syndrome (LS), 152–157
Lysine, 346t–347t

M
Makorin ring finger protein 3 (MKRN3) gene, 162
MammaPrint tests, 194
Mantle cell lymphoma, 205
MAP. See MYH-associated polyposis
Marfan syndrome (MFS), 136–137
Marginal zone lymphoma, extranodal, 205
Maryland v. King, 338
Massively parallel sequencing
  for bacterial identification, 22, 22f
  circulating cell-free fetal DNA and, 285–292, 287f
  circulating tumor cells and, 243–244, 245f
  circulating tumor DNA analysis and, 261
  clonal sequencing, 78–79
  for genetic disorders, 164–166, 165t
  for hematopoietic malignancies, 223, 228
  identity testing and, 337–338
  overview, 78–82, 78f, 78t
  sequencing by ligation, 80–81, 81f
  sequencing by synthesis, 79–80
  single-molecule sequencing, 81–82
  of solid tumor samples, 194–197
MATCH design, 199
Matrix, nucleic acid isolation and, 40
Matrix-assisted laser desorption ionization time-of-flight mass spectrometry (MALDI-TOF-MS)
  for sequence variant genotyping, 64–65
MCAD. See Medium-chain acyl-coenzyme A dehydrogenase
McKusick, Victor, 29
MDS. See Myelodysplastic syndromes
Medium-chain acyl-coenzyme A dehydrogenase (MCAD), 126t
Medium-density arrays, 68
Melanoma, 257–258
Melanoma antigen family L2 (MAGEL2) genes, 162
MELAS. See Mitochondrial encephalomyopathy, lactic acidosis, and stroke-like episodes
Melting analysis
  high-resolution, 77, 77f
  for homogeneous single nucleotide variant typing, 75–77, 76f, 77t
  overview, 74–75
Melting temperature estimation, 68
MEN. See Multiple endocrine neoplasia
MEN1. See Wermer syndrome
MEN2. See Sipple syndrome
Mendel, Gregor, 1
MERRF. See Myoclonic epilepsy associated with ragged red fibers
Metagenomics, 114–115
Methicillin-resistant Staphylococcus aureus (MRSA), 114
Methyl-CpG binding protein 2 (MECP2) gene, 146

Methylation
  bisulfite treatment for analysis of, 50, 50f
  circulating tumor cells and, 254
  circulating tumor DNA and, 262—264, 264t
  genomic variation and, 25
  hematopoietic malignancies and, 223
  overview of, 12, 12f
5-Methylcytosine, 25
Methylome, fetal, 290
MFS. See Marfan syndrome
Mi prostate score test, 197
Microarrays, 69—70, 69f
Microbiome, human, 114—115, 115t
Microfluidic systems
  for circulating tumor cells, 243—244, 244f
  nucleic acid isolation and, 42
MicroRNA (miRNA), 5, 11, 13, 23, 224
Microsatellite instability (MSI)
  circulating tumor DNA and, 262, 263t
  colorectal cancer and, 152, 156f
Microsatellite loci. See Short tandem repeats
Miescher, Friedrich, 35
Minisequencing, 63, 291
Minor groove binder hydrolysis probes, 77, 77t
miRNA. See MicroRNA
Mismatch repair (MMR), 7, 152
Mitochondrial ATPase 6 (MT-ATPase6) gene, 160
Mitochondrial DNA (mtDNA), 159—161, 334—335
Mitochondrial encephalomyopathy, lactic acidosis, and stroke-like episodes (MELAS), 160—161
Mitochondrial tRNA-encoding gene (MT-TL1) gene, 161
Mitochondrially Encoded tRNA Lysine (MT-TK) gene, 161
Mitogen-activated protein (MAP) kinases, 9
MLPA. See Multiplex ligation-dependent probe amplification
MMR. See Mismatch repair
Molecular beacons. See Hairpin probes
Molecular biology
  basic principles, 2—3, 2f
  central dogma
    DNA modification enzymes, 8
    DNA repair, 7—8
    DNA replication, 6—7, 7f
    gene structure, 8
    RNA transcription and splicing, 8—9, 8f
    RNA translation, 9—11, 10f—11f
  epigenetics
    chromatin conformation regulation, 12—13
    DNA methylation, 12, 12f
    noncoding RNA, 13
  genomics, 13—14
  historical perspective, 1—2
  nucleic acid structure and function
    DNA structure, 3, 3f—4f
    DNA types, 4—5
    human chromosomes, 5—6
    RNA associated with protein production, 5
    RNA molecular composition, 5
    RNA structure, 5
Monoclonal B cell lymphocytosis, 226—227
Monoclonal immunoglobulins. See Paraproteins
Morgan, Thomas, 1
Mosaicism, 288
Mouse genome, 20
mRNA (messenger RNAs), 5
MRSA. See Methicillin-resistant Staphylococcus aureus
MS/MS. See Tandem mass spectrometry
MT-ND5 gene, 161
MTD test. See Mycobacterium tuberculosis Direct test
mtDNA. See Mitochondrial DNA
Muir-Torre syndrome, 157
Multiple endocrine neoplasia (MEN), 137—139
Multiple myeloma, 206, 224

Multiple-displacement amplification, 58
Multiplex ligation-dependent probe amplification (MLPA), 63
Multiplex PCR, 55, 196
Mutation Database Initiative, 27
Mycobacteria spp
  M. tuberculosis, 106
Mycobacterium tuberculosis Direct test (MTD test), 106
MYD88 mutations, 221
Myelodysplastic syndromes (MDS), 209—210
Myeloid and lymphoid neoplasms with eosinophilia, 210—211
Myeloid disorders, 206—211, 207t, 214, 227
MYH-associated polyposis (MAP), 158
Myoclonic epilepsy associated with ragged red fibers (MERRF), 161
Myotonic dystrophy type 1, 133t

N
N-Acetyltransferases (NAT), 297, 297f, 304b, 311—315
Nanopore sequencing, 81—82
Nasopharyngeal cancer, 264—265
NAT. See N-Acetyltransferases
National DNA Index System (NDIS), 330
NDIS. See National DNA Index System
Negative selection, 242
Nested PCR, 55
Neurofibromatosis type 1, 133t
Newborn screening programs, cystic fibrosis, 125—129
Next generation sequencing. See Massively parallel sequencing
Niemann-Pick type C disease, 126t
Nirenberg, Marshall, 1—2
NK cell lymphoproliferative disorders, 221
NNRTIs. See Non-nucleoside reverse transcriptase inhibitors
Non-nucleoside reverse transcriptase inhibitors (NNRTIs), 88
Non-small cell lung cancers (NSCLC), 198, 256
Noncoding DNA, 264
Noncoding RNAs, 13
Nonhistone proteins, 5—6
Nonsyndromic hearing loss and deafness, 131—132
Nowell, Peter, 209
NSCLC. See Non-small cell lung cancers
NTRIs. See Nucleoside reverse transcriptase inhibitors
Nucleases, 8
Nucleic acid isolation
  commonly used procedures, 35
  flexibility for different matrices, 42
  history of, 35
  integrated, 42
  at low concentrations, 41
  measurement of quality and quantity, 42—44
  method selection, 42—44
  microfluidics and, 42
  processing throughput, 40—41
  quality and quantity considerations, 42
  impact of sample matrix on
    blood, 40
    DNA vs. RNA, 40
    human sample complexity, 40
    paraffin embedded tissue with fixatives, 40
  specific applications of, 41—42
  steps of
    concentration, 39
    lysis, 36
    overview, 35—39, 36f
    protein removal, 36—37
    separation methods, 37—39, 38f
    storage, 39

Nucleic acids
  amplification methods. See Amplification methods
  bisulfite treatment for methylation analysis, 50, 50f
  circulating cell-free fetal. See Circulating cell-free fetal DNA
  detection methods, 59
  discrimination methods. See Discrimination methods
  DNA. See DNA
  fragmentation of, 49—50
  history of, 47—48
  RNA. See RNA
Nucleophosmin, 206, 220
Nucleoside reverse transcriptase inhibitors (NTRIs), 88
Nucleosomes, 4f, 5—6, 6f
Nucleotide excision repair, 7

O
Off-label indications, 199
OLA. See Oligonucleotide ligation assays
Oligonucleotide ligation assays (OLA), 63, 65f
Oligonucleotide probes, 67
OncoQuick system, 242
Online Mendelian Inheritance in Man (OMIM) database, 251
Orangutan genome, 20
Origin of replication, 6
Ovarian cancer, 150—152, 153t—154t, 258
Oxidative phosphorylation, 159

P
PABP. See Poly-adenosine-binding protein
Pancreatitis, hereditary, 149—150
Paraproteins
  overview of, 367—368
  serum protein electrophoresis and, 370
Parentage testing, 338—339
Paternity testing, 339
Pediatric Cancer Genome Project, 193
Peptide bonds, 10—11
Peptide nucleic acid FISH (PNA-FISH) probes, 107—108
Peptides
  bonding, 352
  bottom-up targeted proteomics and, 388—389
  clinically relevant systems
    angiotensins, 354
    endothelins, 354
    glutathione, 354—355
    hepcidin, 353—354
    natriuretic peptides, 353
    pro-opiomelanocortin system, 353
    vasopressin, 354
  overview, 345, 352—355
Peritoneal fluid, proteins in, 374—375
PGD. See Preimplantation genetic diagnosis
Pharma GKB. See Pharmacogenetics and Pharmacogenomics Knowledge Base
Pharmacodynamics
  examples of studies, 295—296
  overview of, 300
Pharmacogenetics
  analytical strategies
    genome-wide association studies. See Genome-wide association studies
    haplotyping, 302—303
    reporting, 303
    targeted testing, 300
    whole exome/genome sequencing, 300—302
  clinical interpretation, 303
  drug responses and, 295—296
  examples, 303—320, 304b
  implementation, 300
  logistics, 300

Pharmacogenetics (Continued)
    nucleic acid isolation and, 41
    principles. See Pharmacodynamics; Pharmacokinetics
Pharmacogenetics and Pharmacogenomics Knowledge Base (Pharma GKB), 303
Pharmacogenetics Research Network, 303
Pharmacokinetics
    drug metabolizing enzymes and, 303–316, 304b
    drug transporters and, 316–317
    overview, 299f, 302, 304t
Phenotype informative single-nucleotide repeats (PISNP; PIM), 303
Phenylalanine, 346t–347t
Philadelphia chromosome, 209
Phosphodiester bonds, 2f, 3
Phosphorylation, overview of, 357
Photoacoustic flow cytometry, 244–246, 246f
Phred quality scores, 19b, 31b, 194
PI. See Protease inhibitors
piRNA (piwi interacting RNAs), 13
piwi interacting RNAs. See piRNA
pK, 348
Plasma. See also Peritoneal fluid
    amino acid concentrations in, 351
    circulating tumor DNA from, 252
    proteins in, 360, 361t
Plasmids, 21
Platinum genomes, 18
Pleural fluid, proteins in, 374–375
Polonies, 57
Poly-adenosine-binding protein (PABP), 10
Polyacrylamide, 60–61
Polyclonal hyperimmunoglobulinemia, 367
Polycomb group protein mutations, 219
Polycomb repressive complex (PRC1 and 2) mutations, 219
Polycystic kidney disease, 133t
Polymerase chain reactions (PCR)
    asymmetric, 55
    for circulating tumor cells, 247–249
    contamination, inhibition, and controls, 55
    detection limits, 56
    digital, 56–57
    for hematopoietic malignancies, 211–212, 228
    identity testing and, 332, 332t
    kinetics, 52–53, 53f
    massively parallel sequencing and, 79, 79f
    multiplex and nested, 55
    optimization, 53–54
    overview, 51–57
    primer design, 54–55
    probe generation using, 67
    process details, 51–52, 52f
    product length for nucleic acid discrimination, 61
    real-time. See Real-time PCR
    restriction fragment length polymorphism, 61, 62f
    reverse transcriptase, 51
    selective amplification of sequence variants, 55–56
Polyomavirus-associated nephropathy (PVAN), 97–98
Polyribosomes, 10–11
POMC. See Proopiomelanocortin
Poor metabolizers, 297–298
Posttranscriptional processing, 9
Posttranslational modifications
    biomarker discovery pipeline and, 395–396
    χ-carboxylation, 358
    fatty acylation, 357
    glycosylation, 358
    glycosylphosphatidylinositol anchor, 357–358
    hydroxylation, 358
    nitrosylation, 358
    phosphorylation, 357
    prenylation, 357
    sulfation, 358

Posttransplantation lymphoproliferative disease (PTLD), 97
Prader-Willi syndrome (PWS), 161–163, 162t
Prealbumin. See Transthyretin
Precision Medicine Initiative, 295
Preimplantation genetic diagnosis (PGD), 136
Prenatal diagnosis
    history of, 283
    mitigating risks of invasive, 283–284
    noninvasive fetal DNA analysis, 284
        circulating cell-free fetal nucleic acids in maternal plasma. See Circulating cell-free fetal DNA
        fetal 'omics', 290
    specific applications of
        cystic fibrosis, 125–129
        Huntington disease, 132–136
Primary protein structure, 355
Primate genomes, 19–20
Primer-dimers, 54
Probe Tec TV Q Amplified DNA assay, 101
Probes
    cloned, 67
    indirect detection of hybridized, 59
    melting temperature estimation, 68
    for nucleic acid detection, 59
    oligonucleotide, 67
    PCR-generated, 67
    purity, 68
    for real-time PCR, 71–73, 72f
ProbeTec tests, 100
Prodesse ProGraseto SSCS assay, 112–113
Prodrugs, 297, 298f, 298t
Product Rule, 336
Proficiency testing (external quality assessment)
    forensic science laboratories and, 336
    for HCV detection, 93
Progensa assay, 197
Proline, 346t–347t, 358
Promoters, 8
Proopiomelanocortin (POMC), 353
Prostate cancer, 254
Protease inhibitors (PI), 88
Proteases
    aspartyl, 359–360, 359t
    bottom-up targeted proteomics and, 387–393, 389t
    cysteinyl, 359, 359t
    metallo, 359t, 360. See also Tissue inhibitors of matrix metalloproteases
    serine, 359, 359t
Proteins. See also Total protein determination
    acute phase response, 368, 368b
    analysis methods
        electrophoresis, 370–372
        immunochemical for specific proteins, 369–370
        total protein determination, 368–369
    in body fluids
        cerebrospinal fluid, 373–374, 373t
        fecal material, 375
        peritoneal and pleural cavities, 374–375
        saliva, 372
    catabolism, 358–360
    of circulating proteome
        α1-antitrypsin, 362
        albumin, 361–362
        β2-microglobulin, 363
        C-reactive protein, 363
        ceruloplasmin, 362
        complement system, 363–364, 365t
        haptoglobin, 362
        immunoglobulins, 364–368, 367f
        prealbumin, 360–361
        transferrin, 362–363
    formation, 356–357

Proteins (Continued)
    in human serum and plasma, 360, 361t
    overview, 345, 355–375
    physical properties, 355–356
    posttranslational modifications, 357–358
    removal of during nucleic acid isolation, 36–37
    RNA associated with production of, 5
    ubiquitin-proteasome system, 360
Proteome, 345
Proteomics
    biomarker pipeline
        data acquisition strategies, 384–386, 385f
        data processing, 386
        discovery experiments, 384
        overview of, 383–387, 383f
        separations, 384
        targeted quantitative experiments, 387
        variations, 386–387
    bottom-up targeted
        calibration, 391–392
        chromatography, 393
        denaturation, 389, 389t
        digestion optimization, 390
        internal standards and peptide quality, 391
        multiplexing, 392
        overview of, 387–393
        peptide selection, 388–389
        polymorphisms, 392–393
        protease selection and activity, 390, 390t
        protein or peptide enrichment, 392
        reduction and alkylation, 389–390
        workflow overview, 387–388, 388f
    current clinical assays, 397
    future of, 382
    history of, 381–383
    instruments for, 396–397, 396t
    interferences and, 396
    posttranslational modifications, 395–396
    specimen collection and processing and, 395
    stability and, 396
    tissues and, 395
    top-down
        calibration, 394
        internal standards, 394
        protein enrichment, 395
        specificity, 394
        workflow overview, 393–394
    urine as matrix, 395
PRSS1 gene, 149–150
PTLD. See Posttransplantation lymphoproliferative disease
Purines, 2, 2f
PVAN. See Polyomavirus-associated nephropathy
PWS. See Prader-Willi syndrome
Pyrimidines, 2, 2f
Pyrosequencing, 64, 65f, 79–80

Q
Q-Scores, 30–32, 31b
qPCR. See Real-time quantitative PCR
Quaternary protein structure, 355

R
R groups, 346, 348
Random match probability (RMP), 336
Rapid DNA Index System (RDIS), 337
Rapid DNA technology, 337
RapidHIT System, 337
Ras proteins, 9
Rasburicase, 315
Rat genome, 20
RB1. See Retinoblastoma gene
RDIS. See Rapid DNA Index System

Real-time PCR
  accuracy and precision, 74
  for BKV testing, 98–99
  detection and quantification, 73–74, 75f
  dyes and probes, 71–73, 73f
  for HBV detection, 95
  for hematopoietic malignancies, 211
  for HIV-1, 87–90
  for homogeneous single nucleotide variant typing, 75–77, 76f, 77t
  for HPV, 103
  with melting analysis. See Melting analysis
  overview of, 48b–49b, 73–74, 98, 286
Real-time quantitative PCR (qPCR), 331–332, 332t
Rearrangements, clonal, 225
Recombinase polymerase amplification (RPA), 58
Recombination, 2
Reduced penetrance, 132
Refractoriness, 369
Relative haplotype dosage analysis (RHDO), 290
REMAGUS02 neoadjuvant study, 253
Remodeling enzymes, 13
Renin, 354
Resistance. See also Drug resistance
  antibiotic, 113–114
  gene mutations in, 222
  viral, 89, 92–93
Resistance testing, 89
Restriction endonucleases, 8
Restriction fragment length polymorphism (RFLP), 61, 62f
RET gene, 138–139
Retinoblastoma, 133t
Retinoblastoma gene (RB1), 7
Retrotransposons, 25, 26t
Rett syndrome, 146–147
Reverse transcriptase (RT) -PCR, 51
Reversible terminators, 80
RFLP. See Restriction fragment length polymorphism
RHDO. See Relative haplotype dosage analysis
Rhesus D (RhD) status determination, 286–289
Ribosomes, 5, 10, 10f
Risk of recurrence (ROR) scores, 194
RMP. See Random match probability
RNA (ribonucleic acid). See also Nucleic acids
  associated with protein production, 5
  electrophoresis and, 60
  history of, 1–2
  molecular composition, 5
  noncoding, 13, 22
  oncogene characterization approaches based on, 194, 195f
  structure of, 5
  transcription and splicing of, 8–9, 8f
  translation of, 9–11, 11f
RNA in situ hybridization, 249
RNA polymerases, 2, 6, 8–9
RNA sequencing, 196–197
RNA spliceosome, 218
RNA-guided engineered nuclease system, 8
RNA-induced silencing complexes, 23
Roberts, Richard, 8
Rodent genomes, 20
Rolling circle amplification, 58
Rowley, Janet, 209
RPA. See Recombinase polymerase amplification
RUNX1 mutations, 219

S
Saccharomyces cerevisiae genome, 21
Salting-in/salting-out procedures, 356
Sample collection
  biomarker discovery pipeline and, 346–348
  identity testing and, 331
Sanger sequencing, 22f, 62–63, 62f, 64f

Satellite DNAs, 6
SCID. See Severe combined immunodeficiency
Scientific Working Group on DNA Analysis Methods (SWGDAM), 334
SCLC. See Small cell lung cancers
SCODA. See Synchronous coefficient of drag alteration
ScreenCell filters, 242–243
SDA. See Strand displacement amplification
Second derivative analysis, 73–74
Segmental duplications, 18
Selenocysteine, 348, 348f
Self-probing amplicons, 77, 77t
Self-probing primers, 71–73
Self-sustained sequence replication (3SR). See Transcription-mediated amplification
Semiconductor sequencing, 80
SEPT9 methylation, 264
SeptiFast system, 109
Sequencing. See also Massively parallel sequencing; Sanger sequencing
  of genomes, 22, 22f
Serial invasive amplification, 48b–49b, 50t, 58, 77t
Serine, 346t–347t
Serine protease inhibitor, Kazal type 1 (SPINK1) gene, 149–150
Serine proteases, 359, 359t
SERPINA1 gene, 126t
Serpins, 356, 362
Serum
  circulating tumor DNA from, 260–261
  proteins in, 369
Serum protein analysis, 370
Severe combined immunodeficiency (SCID), 366–367
Sharp, Phillip, 8
Short interspersed nuclear elements (SINE), 25, 26t
Short tandem repeats (STR)
  identity testing and, 330–332, 330f, 331t
  intergenic DNA and, 18
  overview of, 25
Shotgun sequencing, 21
Signal transduction, 9, 214
Silencers, 8
Simeprevir, 91–92
Simvastatin, 317
SINE. See Short interspersed nuclear elements
Single hybridization probes, 76, 77t
Single nucleotide extension assays. See Minisequencing
Single nucleotide variants/polymorphisms
  database of, 29
  for hematopoietic malignancies, 212
  identity testing and, 335, 338
  methods for detecting, 222–223
  overview of, 13, 23–24
  real-time PCR with melting analysis, 75–76, 76f
  research studies, 13
Single-copy visualization, 71
Single-molecule PCR, 57
Single-molecule sequencing, 81–82
Single-stranded conformational polymorphism (SSCP), 61–62
Sipple syndrome (MEN2), 137
siRNA (small interfering RNAs), 13
Skin, cancers of, 257
SLCO1B1. See Solute carrier organic anion transporter 1B1
SMA. See Spinal muscular atrophies
Small cell lung cancers (SCLC), 256
small interfering RNAs. See siRNA
SMN. See Survival motor neuron genes
Snapback primers, 76, 77t
SNRPN-SNURF gene, 162
Sofosbuvir, 91–92
Solenoids, 4f
Solid affinity isolation, 39

Solid tumor cells
  genomic analysis of, 193–194
  sampling and preservation methods, 191–192
  staining and selection of, 192–193, 192f
Solid-phase extraction (SPE), nucleic acid isolation and, 37
Solute carrier organic anion transporter 1B1 (SLCO1B1), 298–300
Spectrophotometry for nucleic acid measurement, 42
Spinal muscular atrophies (SMA), 129–131
Spliceosome mutations, 218
Splicing, 9
SSCP. See Single-stranded conformational polymorphism
Staining of tumor cells, 192–193, 192f
STAT3 mutations, 221
Stem cells, 252
Stevens-Johnson syndrome, 319
Storage, nucleic acid isolation and, 39
Strand displacement amplification (SDA), 58
Subacute necrotizing encephalopathy, 160
SUCCESS trial, 253
Surface binding methods, 39
Survival motor neuron genes (SMN), 129–130
SUZ12 mutations, 219
SWGDAM. See Scientific Working Group on DNA Analysis Methods
SWI/SNF proteins, 13
Swiss Institute of Bioinformatics, 345
SYBR Green I, 44
Synchronous coefficient of drag alteration (SCODA), 37–38
System A, 349–350
System ASC, 349
System B^o, 349
System b^o, 349
System L, 349
System X$^-$, 349
System y$^+$ L, 349
System y$^+$, 349

T
T cell large granular lymphocyte (LGL) lymphoma, 221
T-cell lymphoproliferative disorders, 221
T-lymphoblastic leukemia/lymphoma, 204, 221–222
T2 magnetic resonance (T2MR)-based biosensing, 110
Tacrolimus, pharmacokinetic studies, 297–300
Tamoxifen, 306–308, 307f–308f
Tandem mass spectrometry (MS/MS), 384. See also Liquid chromatography-tandem mass spectrometry
TARGET database, 198
TATA boxes, 9
Taurine, 348f
Taxanes, 255
Tay Sachs disease, 126t
TCA. See Tricyclic antidepressants
Telaprevir, 91
Telomeres, 2, 4f, 6, 18
Temperature-gradient gel electrophoresis (TGGE), 61–62
Ten-eleven translocation (TET) dioxygenases, 12, 28
Termination factors, 10–11
Terstappen, Leon, 240–241
Tertiary protein structure, 355
TGGE. See Temperature-gradient gel electrophoresis
Thiopurine S-methyltransferase (TPMT), 302t, 304t, 315
1000 Genomes Project, 24–25
Threonine, 346t–347t
Threshold analysis, 73–74
Threshold effects, 159
Thrombophilia, 147–149, 148f
Thymine, 2f, 3
TIBC. See Total iron-binding capacity

INDEX

TIMP. *See* Tissue inhibitors of matrix metalloproteases
Tissue factor pathway. *See* Extrinsic coagulation pathway
Tissue inhibitors of matrix metalloproteases (TIMP), 360
TLR. *See* Toll-like receptors
TMA. *See* Transcription-mediated amplification
Toll-like receptors (TLR), 228
Total iron-binding capacity (TIBC), 362–363
Total protein determination
  Biuret method, 368–369
  cerebrospinal fluid and, 374
  direct optical methods, 369
  dye-binding methods, 369
  Kjeldahl method, 368
  light-scattering methods, 369
  Lowry method, 369
  methods for, 368–372
  refractometry, 369
  variables affecting, 369
TPMT. *See* Thiopurine S-methyltransferase
Transcription, 210–211, 212f
Transcription factors, 210–211, 218–219
Transcription-mediated amplification (TMA), 57, 57f
Transcriptome, 18, 290
Transferrin, 362–363
Transition mutations, 7
Translation, 9–11, 10f–11f
Transplantation
  BK virus and, 97–99
  cytomegalovirus and, 96–97
  Epstein-Barr virus and, 97
  identity testing and, 339–340
Transporter systems, 348–350, 350t
Transposable elements, 25, 27f
Transposons, 26t
Transthyretin (TTR; prealbumin), 360–361
Transversion mutations, 21
Tricarboxylic acid cycle. *See* Krebs cycle
Trichomoniasis, 101
Tricyclic antidepressants (TCA), 108–110
Triple test, 284
Trisomy 21, 286, 288
tRNA (transfer RNAs), 5, 10, 11f
Trugene genotyping tests, 93, 95

Trypsin (TRY), bottom-up targeted proteomics and, 387–393, 389t
Tryptophan, 346t–347t
Tumor protein p53 mutations, 219–220, 257
Turbidimetry for protein measurement, 369–370
Tyrosine, 346t–347t
Tyrosine kinase inhibitors, 222, 261–262
Tyrosine kinase receptors, 9
UBE3A gene, 162–163

U

Ubiquitin-proteasome system, 360
Ultraviolet absorbance, 42–44
Ultraviolet absorbance, 369
Uniparental disomy (UPD), 162–163
Unlabeled probe genotyping, 77, 77t
UPD. *See* Uniparental disomy
Uracil DNA glycosylase, 8
Urea, 351, 389t
Uridine diphosphate glucuronosyl transferases, 298–300, 315–316
Urine as matrix for proteomics assays, 395

V

V(D)J recombination, 201–202, 225, 226t, 227f
Valine, 346t–347t
Vallone, Peter, 337
Vancomycin-resistant enterococci (VRE), 113–114
Variant call files (VCF), 30–32
Variants, genomic, 23–26, 55–56
Variants of unknown significance (VUS), 198–199
Vasopressin (antidiuretic hormone), 354
VCF. *See* Variant call files
Venous thromboembolism (VTE), 147–149
Venter, Craig, 5–6, 14
Verigene tests, 105–106, 108, 113
Versant HIV-1 RNA 3.0 assay, 88
Viral blips, 88
Viral genomes, 22
Viral infections
  commercially available load assays, 89t
  hepatitis. *See* Hepatitis viruses
  of respiratory tract, 104–106

Viral load testing
  Epstein-Barr virus, 97
  Hepatitis C, 91
  HIV-1, 88–89, 89t
Virologic failure, 88
Virtual karyotyping, 69–70
Vitamin K epoxide reductase (VKOR), 301f, 317–319
VKOR. *See* Vitamin K epoxide reductase
Vohwinkel syndrome, 132
VRE. *See* Vancomycin-resistant enterococci
VTE. *See* Venous thromboembolism
VUS. *See* Variants of unknown significance

W

Waldenstrom macroglobulinemia, 220
Warfarin, pharmacodynamics, 300, 301f, 318t
Watson, James, 1
Wermer syndrome (MEN1), 137
WES. *See* Whole-exome sequencing
Western blotting, 372
Whole genome sequencing, 195–196
Whole-exome sequencing (WES), 166–167
Winkler, Hans, 17

X

X-chromosome inactivation assays, 227
X-inactive specific transcript (XIST), 13
X-linked disorders, 139–147, 139t
XIST. *See* X-inactive specific transcript
Xpert MRSA/SA BC assay, 108
XTAG Respiratory Viral Panel (RPV) v1 assay, 104

Y

Y-short tandem repeats (Y-STR), 334
YFiler Plus, 334

Z

Zwitterions, 348

Printed in the United States
By Bookmasters